Phylogeny and Evolution of the Angiosperms

Phylogeny and Evolution of the Angiosperms

Revised and Updated Edition

Douglas Soltis, Pamela Soltis,
Peter Endress, Mark Chase, Steven Manchester,
Walter Judd, Lucas Majure, and Evgeny Mavrodiev

The University of Chicago Press
Chicago and London

The University of Chicago Press, Chicago 60637
The University of Chicago Press, Ltd., London
First edition published by Sinauer Associates, Inc, 2005.

Published 2018
Printed in China

26 25 24 23 22 21 20 19 18 1 2 3 4 5

ISBN-13: 978-0-226-38361-3 (cloth)
ISBN-13: 978-0-226-44175-7 (e-book)
DOI: 10.7208/chicago/9780226441757.001.0001

Library of Congress Cataloging-in-Publication Data

Names: Soltis, Douglas E., author.
Title: Phylogeny and evolution of the angiosperms / Douglas Soltis [and seven others].
Description: Revised and updated edition. | Chicago ; London : The University of Chicago Press, 2018. | Includes bibliographical references and index.
Identifiers: LCCN 2016046547 | ISBN 9780226383613 (cloth : alk. paper) | ISBN 9780226441757 (e-book)
Subjects: LCSH: Angiosperms—Phylogeny. | Angiosperms—Evolution.
Classification: LCC QK989 .P49 2017 | DDC 581.3/8—dc23
LC record available at https://lccn.loc.gov/2016046547

♾ This paper meets the requirements of ANSI/NISO Z39.48–1992 (Permanence of Paper).

Contents

1 Relationships of Angiosperms to Other Seed Plants

Seed plants are of fundamental importance both evolutionarily and ecologically. They dominate terrestrial landscapes, and the seed has played a central role in agriculture and human history. There are five extant lineages of seed plants: angiosperms, cycads, conifers, gnetophytes, and *Ginkgo*. These five groups have usually been treated as distinct phyla—Magnoliophyta (or Anthophyta), Cycadophyta, Coniferophyta, Gnetophyta, and Ginkgophyta, respectively. Cantino et al. (2007) used the following "rank-free" names (see Chapter 12): *Angiospermae*, *Cycadophyta*, *Coniferae*, *Gnetophyta*, and *Ginkgo*. Of these, the angiosperms are by far the most diverse, with ~14,000 genera and perhaps as many as 350,000 (The Plant List 2010) to 400,000 (Govaerts 2001) species. The conifers, with approximately 70 genera and nearly 600 species, are the second largest group of living seed plants. Cycads comprise 10 genera and approximately 300 species (Osborne et al. 2012; Fragnière et al. 2015). Gnetales consist of three morphologically disparate genera, *Gnetum*, *Ephedra*, and *Welwitschia* (~90 species total) that are so distinctive that each has been placed in its own family (Gnetaceae, Ephedraceae, and Welwitschiaceae). There is a single living species of *Ginkgo*, *G. biloba*. Each of these extant lineages has a rich fossil history (T. Taylor et al. 2009; Friis et al. 2011); we cover the fossil record of the angiosperms in more detail in Chapter 2 and also in those chapters focused on specific angiosperm clades (Chapters 4–10).

There are also many extinct lineages of seed plants (Crane 1985; Decombeix et al. 2010; E. Taylor and T. Taylor 2009; T. Taylor et al. 2009). Although extant gymnosperms appear to be monophyletic (below and Chapter 2), all gymnosperms (living and extinct) together are not monophyletic. Importantly, several fossil lineages, Caytoniales, Bennettitales, Pentoxylales, and Glossopteridales (glossopterids), have been proposed as putative close relatives of the angiosperms based on phylogenetic analyses (e.g., Crane 1985; Rothwell and Serbet 1994; reviewed in Doyle 2006, 2008, 2012; Friis et al. 2011). These fossil lineages, sometimes referred to as the para-angiophytes, will therefore be covered in more detail later in this chapter. Another fossil lineage, the corystosperms, has been proposed as a possible angiosperm ancestor as part of the "mostly male hypothesis" (Frohlich and Parker 2000), but as reviewed here, corystosperms usually do not appear as close angiosperm relatives in phylogenetic trees.

The seed plants represent an ancient radiation, with the first seeds appearing near the end of the Devonian (~370 million years ago; mya). By the Early to Middle Carboniferous, a diversity of seed plant lineages already existed (e.g., *Cordaites* and walchian conifers; Thomas 1955; Bhatnagar and Moitra 1996; Kenrick and Crane 1997; Davis and Kenrick 2004; T. Taylor et al. 2009). Heterospory, prerequisite to evolution of the seed, evolved in parallel in different major clades, including lycophytes, water ferns (e.g., *Marsilea*), sphenophytes, and aneurophytes, and seed-like structures, with a retained endosporic megagametophyte nearly surrounded by an integument-like covering, occur in some lycophytes (e.g., *Lepidocarpon*). *Lepidocarpon* is not considered a true seed, but is an example of convergence. Importantly, phylogenetic analyses that include the five clades of living seed plants show that they indeed form a clade, indicating that all have inherited seeds from a common ancestor—and that these seeds did not evolve in parallel. Phylogenetic analyses including extinct seed plants also place these groups in the same clade as extant seed plants (see below). Thus, analyses support the hypothesis that fossil and extant seed plants (Spermatophyta) had a single origin.

The first seed-like structures, observed in the Late Devonian to early Carboniferous, apparently ancestral to true seeds, had free integumentary lobes and lacked a micropyle; the pollen-receptive structure, lagenostome, was formed by the nucellus rather than the integument. The fusion of integumentary lobes, except for a micropylar channel, led to the formation of true seeds as in lyginopterid seed ferns.

The fossil record of conifers dates to the Late Carboniferous and that of true cycads to the Early Permian. Available data indicate that by the Permian (~299–251 mya), at least three (cycads, conifers, *Ginkgo*) of the five extant lineages of seed plants had probably diverged (Kenrick and Crane 1997; Donoghue and Doyle 2000). In contrast, the angiosperms are relatively young—their earliest unambiguous fossil evidence is from the Early Cretaceous (~130 mya) although molecular dating methods infer older dates for their origin (see Chapter 2).

Relationships among the lineages of extant seed plants, as well as the relationships of living groups to fossil lineages, have been issues of longstanding interest and debate. A topic of particular intrigue has been the closest relatives of the angiosperms. Angiosperms are responsible either directly or indirectly for the majority of human food and account for a huge proportion of photosynthesis and carbon sequestration. They have diversified to include 350,000–400,000 species in perhaps 130–170 myr and now occur in nearly all habitable terrestrial environments and many aquatic habitats. Understanding how angiosperms accomplished this is of fundamental evolutionary and ecological importance.

At some point, nearly every living and fossil group of gymnosperms has been proposed as a possible ancestor of the angiosperms (e.g., Wieland 1918; Thomas 1934, 1936; Melville 1962, 1969; Stebbins 1974; Meeuse 1975; Long 1977; Doyle 1978, 1998a,b; Retallack and Dilcher 1981; Crane 1985; Cronquist 1988; Crane et al. 1995; reviewed in Doyle 2006, 2008, 2012; Friis et al. 2011). Among extant seed plants, the relationship between angiosperms and Gnetales has received considerable attention.

Ascertaining the closest relatives of the angiosperms is not only of great systematic importance but also critical for assessing character evolution. For example, the outcome of investigations of character evolution among basal angiosperms, including studies focused on the origin and diversification of crucial angiosperm structures (e.g., floral organs, endosperm, vessel elements), may be influenced by those taxa considered their closest relatives. The effect of outgroup choice on the reconstruction of character evolution within angiosperms is readily seen via the widespread use of Gnetales as an outgroup for angiosperms. As reviewed below, for nearly two decades beginning in the 1980s, Gnetales were considered by many to represent the closest living relatives of the angiosperms. The use of Gnetales as an angiosperm outgroup profoundly influenced character-state reconstruction within the flowering plants (see "The Anthophyte Hypothesis" section).

Clarifying relationships among seed plants, both extant and fossil, has been extremely difficult. Factors that have contributed to the difficulties in phylogeny reconstruction of seed plants (living and extinct) include the great age of these groups and the considerable morphological divergence among them, as well as the extinction of many lineages. The tremendous morphological gap among extant and fossil seed plant lineages has complicated and ultimately compromised efforts to reconstruct relationships with morphology because of homoplasy and uncertainty about the homology of structures (e.g., Doyle 1998a, 2006, 2012; Donoghue and Doyle 2000; Soltis et al. 2005b, 2008b; Friis et al. 2011).

Although progress has been made in elucidating relationships among extant seed plants using DNA sequence data, relationships remain problematic. Even with the addition of more taxa and more genes representing all three plant genomes, issues remain. Resolution of relationships among extant seed plants with DNA sequence data has also been difficult because some lineages have relatively short branches (e.g., angiosperms or Pinaceae), whereas other clades (e.g., Gnetales) have long branches. This problem is further compounded by the presence in most analyses of long branches to the sister group of seed plants (monilophytes). In groups such as the angiosperms and conifers, more taxa can be added to break up long branches, but this is not possible across seed plants as a whole given the considerable extinction that has occurred. Another concern given the ancient divergences in seed plants is multiple substitutions per site leading to saturation of base substitutions. Hence, whereas the use of morphological characters has been criticized in seed plant phylogeny (and in a global sense by Scotland et al. 2003), DNA has its own problems and certainly has not been a consistent solution to resolve relationships among extant seed plants (see Burleigh and Mathews 2004, 2007; Mathews 2009; Mathews et al. 2010; Soltis et al. 2005b, 2008b).

As stressed by other investigators, a complete understanding of seed plant phylogeny is not possible without the integration of fossils. Many investigations have attempted this integration (e.g., Crane 1985; Doyle and Donoghue 1986; Doyle 1996, 1998a,b, 2001, 2006, 2008, 2012; Frohlich 1999; Donoghue and Doyle 2000; Hilton and Bateman 2006; Magallón 2010); we cover these analyses in more detail later in this chapter. Despite these efforts, the integration of fossils into studies of seed plant phylogeny remains an area where more research is needed. Seed plant relationships and the closest relatives of the angiosperms have been the focus of many reviews (e.g., Crane 1985; Doyle and Donoghue 1986; Doyle 1996, 1998a, b, 2001, 2006, 2008, 2012; Frohlich 1999; Donoghue and Doyle 2000; Mathews 2009; Friis et al. 2011) and continue to spawn new analyses (e.g., Hilton and Bateman 2006; Doyle 2008, 2012; Magallón 2010; Mathews et al. 2010). We will consider seed plant relationships in general (living and extinct), but a major focus of this chapter is discussing the closest relative(s) of the angiosperms.

PHYLOGENETIC STUDIES: EXTANT TAXA

We first review the considerable effort devoted to reconstructing the phylogeny of living seed plants. Given the immense debate regarding the relationships of Gnetales, we also provide a brief history of the placement of Gnetales relative to the angiosperms. We then focus on cladistic analyses that include fossil as well as extant seed plants.

Molecules and morphology have so far yielded different conclusions about the relationships of Gnetales and angiosperms. Whereas analyses of morphology have consistently placed Gnetales sister to angiosperms (but see review by Rothwell et al. 2009), molecular data support alternative placements (see below). We are strong advocates for the use of morphology in phylogenetic analyses. However, based on the morphological characters so far used, the coding employed, and analyses now available, one could legitimately conclude that to this point seed plants represent an example in which cladistic analyses of morphological characters alone have failed to resolve major relationships in congruence with molecular-informed analyses.

PLACEMENT OF GNETALES

A close relationship of angiosperms and Gnetales was first proposed by Wettstein (1907) and by Arber and Parkin (1908) based on several shared features: vessels, net-veined leaves (present in *Gnetum* as well as angiosperms), and "flower-like" reproductive organs (Fig. 1.1) (see also reviews by Doyle 1996; Frohlich 1999). However, the reasoning that Wettstein (1907) and Arber and Parkin (1908) each used to explain the close relationship of Gnetales and angiosperms differed dramatically. Wettstein (1907) proposed that Gnetales were ancestral to the angiosperms based on the view that the formerly recognized angiosperm group Amentiferae, a group that included wind-pollinated taxa such as Juglandaceae, Betulaceae, and Casuarinaceae, are the most "primitive" living angiosperms. We stress throughout that which extant group exhibits the most "primitive" morphological traits and which is sister to all others are not equivalent, but these statements are often confounded. We can infer ancestral character states via character-state reconstruction using the best estimate of phylogeny, as we have done throughout (Chapter 6). Wettstein maintained that the distinctive inflorescences (termed aments) of Amentiferae, consisting of simple, unisexual flowers, are homologous with the unisexual strobili of Gnetales. Arber and Parkin (1908) also proposed a close relationship of angiosperms and Gnetales, but, in contrast, argued that the reproductive structures of Gnetales are not primitively simple, but reduced, derived from ancestral lineages having more parts.

By the mid-1900s, most authors no longer considered Gnetales and angiosperms closest relatives. Bailey (1944b, 1953) noted that the vessels in the two groups are derived from different kinds of tracheids and hence are not homologous. In addition, Gnetales bear ovules directly on a stem tip, whereas in angiosperms, the ovules are produced within the carpel, the latter structure possibly representing a modified leaf. Views on the earliest angiosperms also changed, with Magnoliaceae and other angiosperms with large, strobiloid flowers considered most ancient, whereas the simple flowers found in Amentiferae were considered secondarily reduced rather than ancestrally simple (e.g., Arber and Parkin 1907; Cronquist 1968; Takhtajan 1969). This new view disrupted the link between Gnetales and angiosperms (via a basal Amentiferae) envisioned by Wettstein.

Issues became more complex when Eames (1952) proposed that the three lineages of Gnetales were not monophyletic. Eames considered *Ephedra* to be related to the fossil group Cordaites and conifers while *Gnetum* and *Welwitschia* were thought to be closer to another extinct lineage of seed plants, Bennettitales. Although morphology and DNA later confirmed the monophyly of Gnetales (below), the work of Eames (1952) shifted interest away from Gnetales as an angiosperm relative. Concomitantly, paleobotanists focused attention on fossils such as *Caytonia* and Glossopteridales as the closest relatives of angiosperms

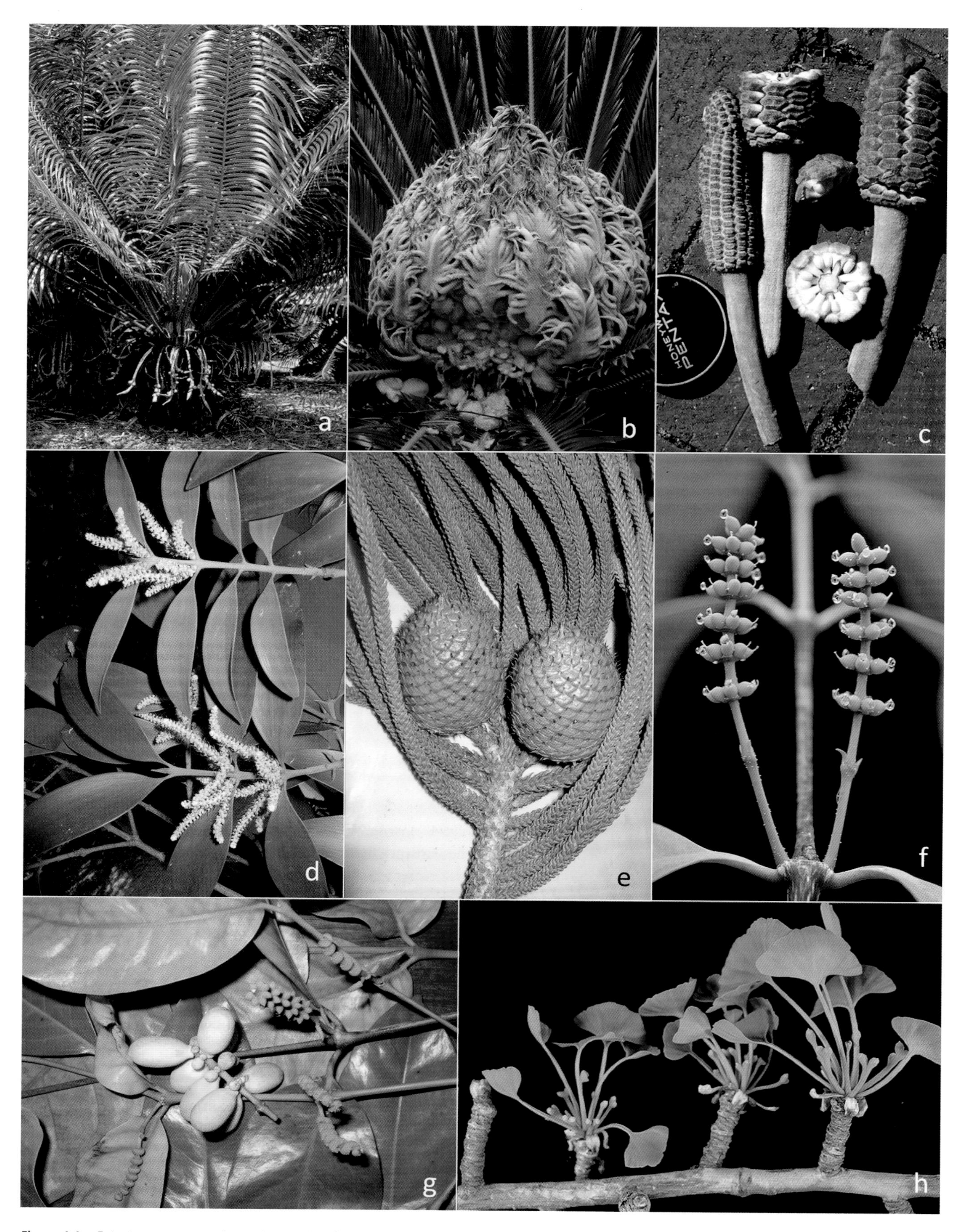

Figure 1.1. Extant gymnosperms: Araucariaceae, Cycadaceae, Ginkgoaceae, Gnetaceae, Podocarpaceae, and Zamiaceae. a. *Cycas circinalis* L. (Cycadaceae), whole plant with megasporophylls. b. *Cycas circinalis*, leaflike megasporophylls and pinnately compound leaves. c. *Zamia furfuracea* L. (Zamiaceae), three ovulate strobili. d. *Nageia nagi* Kuntze (Podocarpaceae), simple microsporangiate strobili and multi-veined leaves. e. *Araucaria subulata* Vieill. (Araucariaceae), female "cones." f. *Gnetum gnemon* L. (Gnetaceae), compound ovulate "cones." g. *Gnetum gnemon*, with multi-veined leaves, young compound ovulate "cones," mature compound ovulate "cones," and young compound microsporangiate "cones" from another plant. h. *Ginkgo biloba* L. (Ginkgoaceae), short shoots with foliage and young ovules.

(Doyle 1996; Frohlich 1999; see below), further diverting attention from Gnetales as possible close relatives of the angiosperms. Gnetales re-emerged, however, as putative close relatives of angiosperms when cladistic approaches were first used to investigate seed plant relationships (below).

THE ANTHOPHYTE HYPOTHESIS

Seed plant relationships were first assessed by cladistic methodology using morphological characters in the 1980s. Several of these early studies included both extant and fossil taxa (e.g., Crane 1985; Doyle and Donoghue 1986). These studies revealed that the three morphologically disparate members of Gnetales (*Ephedra*, *Gnetum*, and *Welwitschia*) are monophyletic (illustrated in Fig. 1.1), a finding now well supported by both morphology and molecules. Only

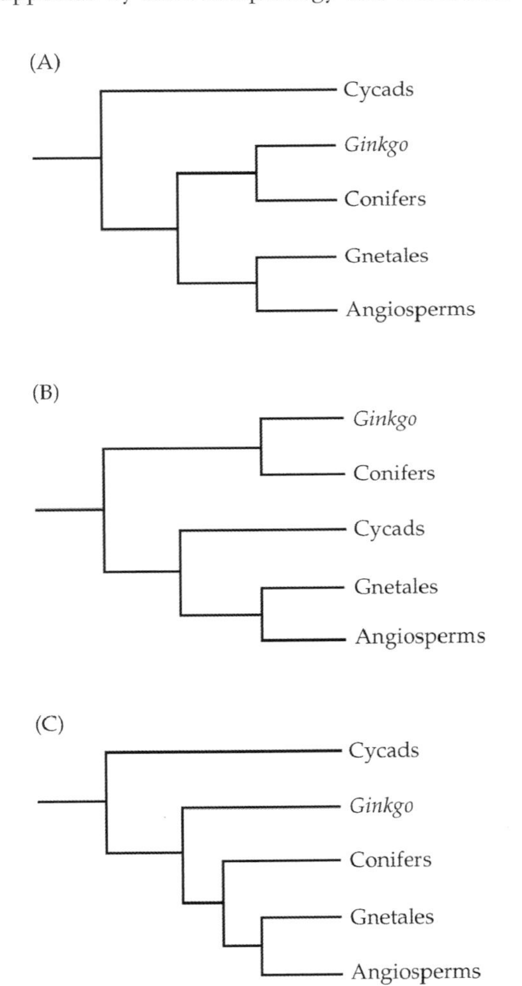

Figure 1.2. Simplified topologies depicting relationships among extant seed plants based on phylogenetic analyses of morphological data. Fossil taxa have been removed from these topologies. (A) Parenti (1980); Crane (1985); Doyle and Donoghue (1986, 1992); Doyle (1996). (B) Doyle and Donoghue (1986, 1992); Doyle (1996). (C) Loconte and Stevenson (1990).

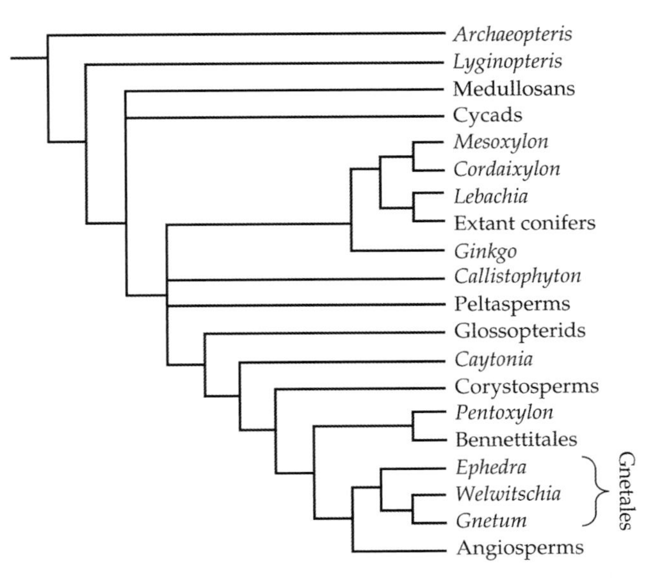

Figure 1.3. One of two shortest trees recovered by Crane (1985) in a cladistic analysis of extant and fossil seed plants (representing the "preferred topology" of Crane 1985). Redrawn from Crane (1985).

Nixon et al. (1994) found Gnetales not to be monophyletic. However, Doyle's (1996) subsequent reanalysis of the data used in Nixon et al. (1994) found a monophyletic Gnetales.

Early phylogenetic studies relying on morphological characters (Parenti 1980; Hill and Crane 1982; Crane 1985; Doyle and Donoghue 1986) recovered Gnetales as the closest living relatives of angiosperms (Fig. 1.2). Perhaps the best known is Crane (1985), which also included fossil seed plants. Crane (1985) recovered Gnetales as the sister group to angiosperms (Fig. 1.3). Subsequent phylogenetic analyses of morphological characters (e.g., Loconte and Stevenson 1990; Doyle and Donoghue 1992; Doyle 1994, 1996; Hilton and Bateman 2006), some of which also included fossils, continued to recover this Gnetales + angiosperm relationship (Fig. 1.2); as summarized by Rothwell et al. (2009; p. 296), "the anthophyte topology of the seed plant tree continues to be supported by morphological analyses of living and extinct taxa."

However, these same early cladistic studies often differed in the relationships suggested among extant seed plants (see Fig. 1.2). In morphological cladistic analyses, extant gymnosperms do not form a clade distinct from the angiosperms, and the positions of some lineages were unstable. Considering extant seed plant lineages, Crane (1985) found that cycads are sister to other extant seed plants and that conifers + *Ginkgo* form a clade that is sister to angiosperms + Gnetales (Fig. 1.2A). In contrast, the shortest trees of Doyle and Donoghue (1986) indicated that conifers + *Ginkgo* are sister to a clade in which cycads are the sister to angiosperms + Gnetales (Fig. 1.2B). Loconte and Stevenson (1990) found cycads followed by *Ginkgo*, then conifers, to be subsequent sisters to Gnetales + angiosperms (Fig. 1.2C).

Crane (1985) conducted two cladistic analyses of extant and fossil seed plants and in one analysis recovered a clade of Bennettitales, *Pentoxylon*, and Gnetales + angiosperms; a second analysis recovered a clade of Glossopteridales, *Caytonia*, corystosperms, Bennettitales + *Pentoxylon*, and Gnetales + angiosperms (Fig. 1.3). Doyle and Donoghue (1986, 1992) similarly found shortest trees in which Gnetales and angiosperms appeared in a clade with the fossil taxa Bennettitales and *Pentoxylon*. However, when fossil lineages were considered, Gnetales were not the immediate sister group to angiosperms (Fig. 1.4). Angiosperms and then *Pentoxylon* and Bennettitales were subsequent sisters to Gnetales (Fig. 1.4). Doyle and Donoghue (1986) named this clade of angiosperms, Gnetales, Bennettitales, and *Pentoxylon* the "anthophytes" in reference to the flower-like reproductive structures shared by all members (see Fig. 1.1 for Gnetales and sections below for fossil groups). Rothwell and Serbet (1994) later recovered the same anthophyte clade. The repeated recognition of this clade resulted in the formulation of the anthophyte hypothesis—that angiosperms are sister to Gnetales within a clade that also included Bennettitales and *Pentoxylon*. As reviewed below, the anthophyte hypothesis subsequently influenced interpretation of character evolution in seed plants and thus had a profound and long impact on studies of angiosperm evolution.

Doyle and Donoghue (1992) and Doyle (1996) recovered Glossopteridales and then *Caytonia* as subsequent sisters to their anthophyte clade (see Doyle 1996). However, Doyle (1996) did not expand the definition of the anthophyte clade to include *Caytonia* or Glossopteridales. This ultimately resulted in confusion given that in most subsequent studies (reviewed below), phylogenetic analyses consistently recovered a revised or modified anthophyte clade that includes Glossopteridales, *Caytonia*, *Pentoxylon*, and Bennettitales as sister groups to the angiosperms, but with Gnetales more distantly related (e.g., Hilton and Bateman 2006; Magallón 2010; Doyle 2008, 2012). For clarity, we refer to this modified or revised anthophyte clade as the "para-angiophytes" (see Doyle 2012), encompassing Glossopteridales, *Caytonia*, *Pentoxylon*, Bennettitales, and angiosperms, but not Gnetales (see below). Hilton and Bateman (2006) refer to this same clade as the "glossophytes." Earlier (pre-cladistic) investigations had also pointed to most of these same fossil groups as close relatives of angiosperms. Arber and Parkin (1907) proposed that Bennettitales and angiosperms shared a common ancestor. Several early workers also suggested *Caytonia* as a close angiosperm relative (Thomas 1925; Harris 1941; Gaussen 1946; see also Stebbins 1974; Doyle 1978). Stebbins (1974) and Retallack and Dilcher (1981) pointed to similarities between Glossopteridales and angiosperms.

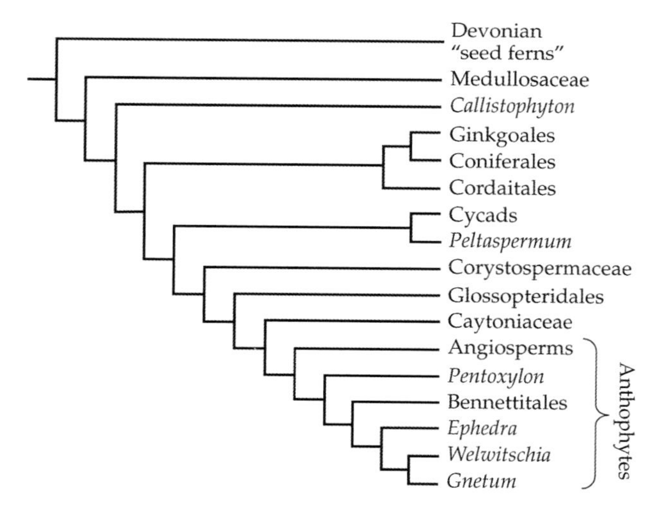

Figure 1.4. Shortest tree recovered by Doyle and Donoghue (1992). Note the composition of the anthophyte clade. Redrawn from Doyle and Donoghue (1992).

Although the anthophyte clade as originally defined by Doyle and Donoghue (1992) remained a focal point of study and controversy for about 15 years, the close relationship inferred between Gnetales and angiosperms was not well supported in any morphological cladistic analysis. Doyle and Donoghue (1986, 1992), for example, found topological differences in trees that were only one or two steps longer than the shortest trees they obtained (i.e., suboptimal trees). In some trees only one step longer than the shortest trees, angiosperms appeared as sister to *Caytonia* and Glossopteridales, rather than with Gnetales, Bennettitales, and *Pentoxylon*. In other one-step-longer trees, the anthophyte clade was retained, but relationships among anthophyte taxa varied (Doyle and Donoghue 1992; reviewed in Doyle 1996). In some studies, Gnetales appeared sister to the angiosperms even when data for fossils were included (e.g., Crane 1985), whereas in others (e.g., Doyle and Donoghue 1986) the sister relationship between Gnetales and angiosperms emerged only when the fossils were removed. Another fossil group that may deserve more attention is Erdtmanithecales, a fossil group putatively closely related to Gnetales or Bennettitales (Friis et al. 2007, 2011), although the group is contentious (Rothwell et al. 2009).

One limitation of early cladistic studies of morphology is that investigators often treated the angiosperms as a single terminal rather than employing multiple representatives. This approach required assumptions about the ancestral states of the angiosperms. Criticism of this approach prompted additional analyses in which several different, putatively basal angiosperm lineages were represented (e.g., Doyle et al. 1994; Nixon et al. 1994; Doyle 1996,

1998a). Although the sister relationship of Gnetales and angiosperms was again recovered in these analyses, strong bootstrap support for this relationship was still lacking, and suboptimal trees again included diverse topologies (the importance of suboptimal trees is noted in Chapter 3).

Despite the lack of internal support and some concerns regarding character homology, the anthophyte hypothesis quickly became widely accepted. Concomitantly, acceptance of the anthophyte hypothesis and a sister relationship between angiosperms and Gnetales had a profound and broad impact. This acceptance stimulated the reinterpretation of character evolution in angiosperms (reviewed in Frohlich 1999; Donoghue and Doyle 2000; Soltis et al. 2005b, 2008b; Doyle 2008, 2012), including the origin of the carpel, the angiosperm leaf (Doyle 1994, 1998a), and double fertilization (Friedman 1990, 1992, 1994). For example, the "double fertilization" process in Gnetales was considered a possible precursor to the double fertilization of angiosperms (Friedman 1990, 1992, 1994) and ultimately a putative synapomorphy for Gnetales + angiosperms (Doyle 1996).

DEMISE OF THE ANTHOPHYTE HYPOTHESIS

Numerous studies (molecular and molecular + morphological) have tried to resolve the relationships among living seed plants—cycads, *Ginkgo*, Gnetales, conifers, and angiosperms—with the caveat that some analyses were much broader in scope, focusing on all land plants or all green plants (e.g., Hamby and Zimmer 1992; Albert et al. 1994; P. Soltis et al. 1999a; Goremykin et al. 1996; Malek et al. 1996; Chaw et al. 1997, 2000; Qiu et al. 1999; Bowe et al. 2000; Nickrent et al. 2000; Magallón and Sanderson 2002; D. Soltis et al. 2002b, 2005b; Burleigh and Mathews 2004; Rydin et al. 2002; Rai et al. 2008; Hajibabaei et al. 2006; Hilton and Bateman 2006; Wu et al. 2007; Zhong et al. 2010; Magallón 2010; Finet et al. 2010; Lee et al. 2011; reviewed in Doyle 2008, 2012). Although considerable progress has been made in resolving relationships (reviewed below), these studies yielded a diversity of results, which highlights the difficulties inherent in resolving relationships among extant seed plants, as well as seed plants in general (reviewed in Burleigh and Mathews 2004; Mathews 2009; Soltis et al. 2005b, 2008b; Doyle 2008, 2012). In this section, we summarize the many molecular phylogenetic analyses of seed plants and discuss the uncertain position of Gnetales, as well as our best current estimate of phylogeny among extant seed plants.

Numerous molecular and morphological phylogenetic studies have provided strong support for the monophyly of Gnetales, despite the pronounced morphological differences among the three genera *Ephedra*, *Gnetum*, and *Welwitschia* (e.g., Hamby and Zimmer 1992; Hasebe et al. 1992; Chase et al. 1993; Albert et al. 1994; Goremykin et al. 1996; Chaw et al. 1997, 2000; Stefanović et al. 1998; Hansen et al. 1999; Winter et al. 1999; Qiu et al. 1999; P. Soltis et al. 1999a,b; Bowe et al. 2000; D. Soltis et al. 2000, 2007c, 2011; Burleigh and Mathews 2004; Magallón 2010; Zhong et al. 2010). The fossil record has brought forth additional extinct genera, indicating a greater diversity of the Gnetales during the Mesozoic (Crane and Upchurch 1987; Friis et al. 2011). However, ascertaining the relationships of Gnetales to other seed plants, as well as determining seed plant relationships in general, has been more problematic. Nonetheless, a close relationship of angiosperms and Gnetales has not been recovered by molecular studies.

As single-gene phylogenetic trees began to appear, it became apparent that they did not support placement of Gnetales as sister to the angiosperms, although the position of Gnetales among other seed plant lineages varied from study to study. Single-gene investigations of plastid (ITS, *rpoC1*), nuclear (18S rDNA), and mitochondrial (*cox1*) sequences indicated a sister-group relationship between Gnetales and conifers (Fig. 1.5). Some analyses of *rbcL* alone and some analyses of 18S and 26S rRNA sequences placed Gnetales as sister to all other seed plants, with angiosperms as sister to the remaining gymnosperms (i.e., a clade of cycads, *Ginkgo*, and conifers; Hamby and Zimmer 1992; Albert et al. 1994; Fig. 1.5A). One parsimony analysis of *rbcL* placed angiosperms as sister to a clade of gymnosperms; within the latter clade, Gnetales were sister to cycads plus (*Ginkgo* + conifers) (Hasebe et al. 1992; Fig. 1.5B). Maximum likelihood analysis of *rbcL* also placed angiosperms as sister to the monophyletic gymnosperms, but relationships among gymnosperms were different than in the parsimony topology (Hasebe et al. 1992; compare Figs. 1.5C and 1.5B). A study using partial 26S rDNA data (Stefanović et al. 1998) did recover a topology with angiosperms sister to Gnetales, but subsequent analyses with more complete 26S rDNA data did not recover this same topology (e.g., Soltis et al. 2011).

Few of these single-gene studies provided strong internal support for relationships. As exceptions, two analyses of 18S rDNA sequences provided some support for a Gnetales + conifers sister-group relationship, with bootstrap percentages of 84 and 64, depending on size of the dataset (Fig. 1.5, E and F; Chaw et al. 1997; P. Soltis et al. 1999b). However, few other studies using single genes provided support > 60% for relationships (Fig. 1.5). Other early DNA sequence analyses provided additional evidence for the monophyly of the living gymnosperms and for a close

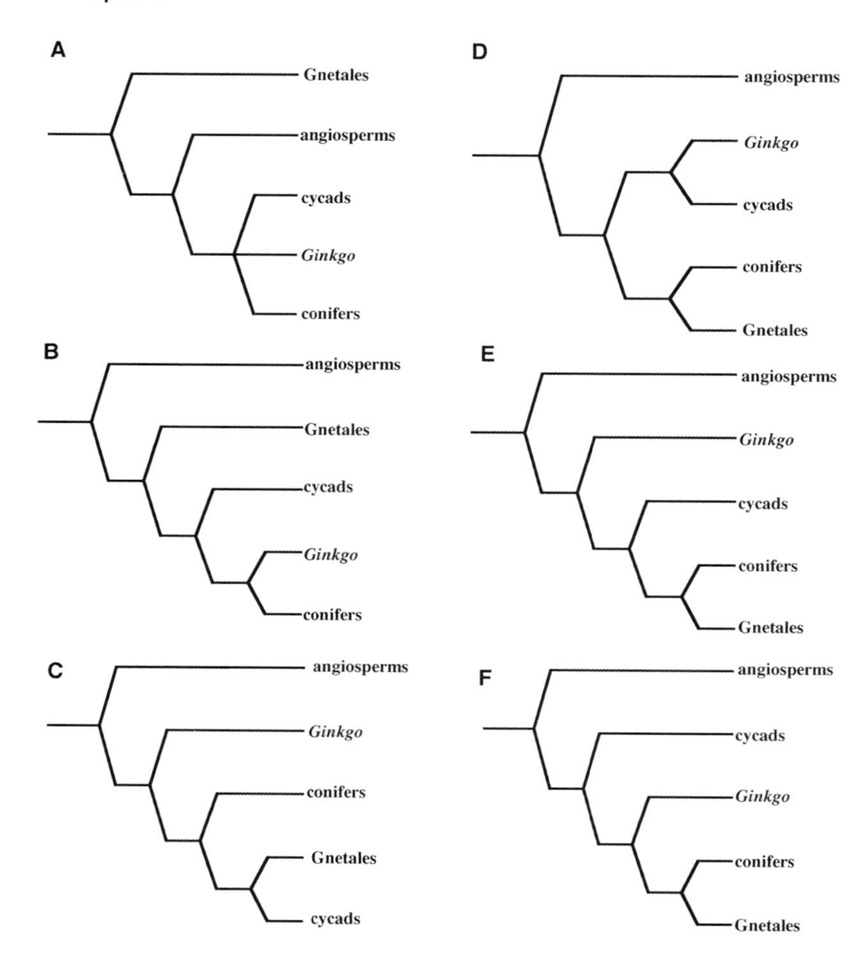

Figure 1.5. Simplified topologies depicting relationships among extant seed plants based on early phylogenetic analysis of gene sequence data using single genes showing diverse topologies. (A) rRNA sequence data, Hamby and Zimmer (1992); *rbcL*, Albert et al. (1994). (B) *rbcL* with parsimony, Hasebe et al. (1992). (C) *rbcL* with maximum likelihood, Hasebe et al. (1992). (D) Plastid ITS (= cpITS), Goremykin et al. (1996). (E) 18S rDNA, Chaw et al. (1997). (F) 18S rDNA, P. Soltis et al. (1999a).

relationship of Gnetales and conifers. However, taxon sampling in early studies was often sparse (e.g., Hansen et al. 1999; Winter et al. 1999; Frohlich and Parker 2000). For example, Hansen et al. (1999) obtained sequence data for a 9.5-kb portion of the plastid genome, but included only *Pinus*, *Gnetum*, and three angiosperms, and used *Marchantia* (a liverwort, distantly related to seed plants) as the outgroup.

Despite limitations of single-gene studies, the anthophyte hypothesis was not supported. However, because sample sizes were often small and internal support low, these analyses were considered equivocal (Doyle 1998a; Donoghue and Doyle 2000). Nonetheless, the results of these single-gene analyses posed a serious challenge to the widespread acceptance of the anthophyte hypothesis.

The lack of support for Gnetales + angiosperms increased with the addition of taxa and the use of multiple genes (see reviews by Doyle 2008, 2012). Analyses in which multiple genes were combined (Fig. 1.6) have repeatedly indicated, with strong support, that Gnetales are not closely related to angiosperms (e.g., Qiu et al. 1999; P. Soltis et al. 1999a; Bowe et al. 2000; Chaw et al. 2000; Graham and Olmstead 2000; Nickrent et al. 2000; D. Soltis et al. 2000, 2002b, 2007c, 2011; Pryer et al. 2001; Magallón and Sanderson 2002; Rydin et al. 2002; Burleigh and Mathews 2004, 2007; Rai et al. 2008; Hajibabaei et al. 2006; Wu et al. 2007; Zhong et al. 2010; S. A. Smith et al. 2010; Magallón 2010; Lee et al. 2011; Ruhfel et al. 2014). These molecular analyses resulted in the rapid demise of the anthophyte hypothesis. However, the placement of Gnetales varied among these analyses. Also, gymnosperm taxon sampling was sparse in studies aimed at the angiosperms (e.g., Qiu et al. 1999; P. Soltis et al. 1999a; D. Soltis et al. 2000, 2007c, 2011; Graham and Olmstead 2000) or vascular plants (Pryer et al. 2001), or all land plants (e.g., Qiu et al. 2007), or even all green plants (e.g., Wickett et al. 2014; Ruhfel et al. 2014).

To summarize, although many different topologies have now been obtained for extant seed plants, several features typically are recovered. Angiosperms are sister to extant gymnosperms; among gymnosperms, cycads or cycads plus *Ginkgo* are sister to the remainder. Placement of Gnetales remains problematic; four hypotheses typically emerge from multi-gene datasets (Fig. 1.6 A-D). Gnetales are 1) sister to all other seed plants, as in some analyses of D. Soltis et al. (2002b), Burleigh and Mathews (2004), and Magallón and Sanderson (2002); 2) sister to all conifers (Gnetifer hypothesis), as in some analyses of Chaw et al. (2000), some analyses of D. Soltis et al. (2002b), Rydin et al. (2002), and S. A. Smith et al. (2010); 3) within conifers, sister to Pinaceae (Gne-pine hypothesis, which was first seen in single-gene trees, such as Chaw et al. 1997; P. Soltis et al. 1999a), as in Qiu et al. (1999), Bowe et al. (2000), some analyses of D. Soltis et al. (2002b), some analyses of Burleigh and Mathews (2004), Chaw et al. (2000), Hajibabaei et al. (2006), Magallón (2010), some analyses of Zhong et al. (2010), and Wickett et al. (2014); and 4) within conifers, sister to cupressophytes or conifers other than Pinaceae (the Gne-cup hypothesis), as in Nickrent et al. (2000), some analyses of Zhong et al. (2010), Doyle (2006), and Ruhfel et al. (2014).

Other, more unusual, relationships based on DNA sequence data have sometimes been recovered for extant gymnosperms, such as cycads + angiosperms (Mathews et al. 2010). The molecular-only analysis of Magallón (2010) recovered a Gne-pine tree. However, a cycad + angiosperm relationship was obtained in a maximum parsimony total evidence analysis (DNA + morphology) of living and fossil seed plants, but only if fossils were subsequently removed from the tree (Magallón 2010) (Fig. 1.7; the fossils Glossopteridales, *Pentoxylon*, Bennettitales, and *Caytonia* are immediate sisters to angiosperms, followed by cycads). Several studies of numerous nuclear genes suggested a placement of Gnetales as sister to all other living gymnosperms, with angiosperms still sister to all living gymnosperms (de la Torre-Barcena et al. 2009; Lee et al. 2011). However, both analyses have important caveats (see below).

Most molecular phylogenetic studies (and molecular + morphology) suggest that Gnetales were derived from within conifers and are sister to Pinaceae (Gne-pine). Analyses of 18S rDNA alone first recovered this topology with BS > 50% (Chaw et al. 1997; P. Soltis et al. 1999b). Multi-gene analyses subsequently recovered this topology (Fig. 1.6), including Qiu et al. (1999, 2007), Bowe et al. (2000), Chaw et al. (2000), Burleigh and Mathews (2004, 2007), Magallón (2010), and Wickett et al. (2014). Bowe et al. (2000) analyzed a four-gene dataset (*rbcL*, 18S rDNA, and mitochondrial *atpA* and *cox1*) and found strong support for Pinaceae + Gnetales. Analysis of combined sequences of mitochondrial small subunit (SSU) rDNA, 18S

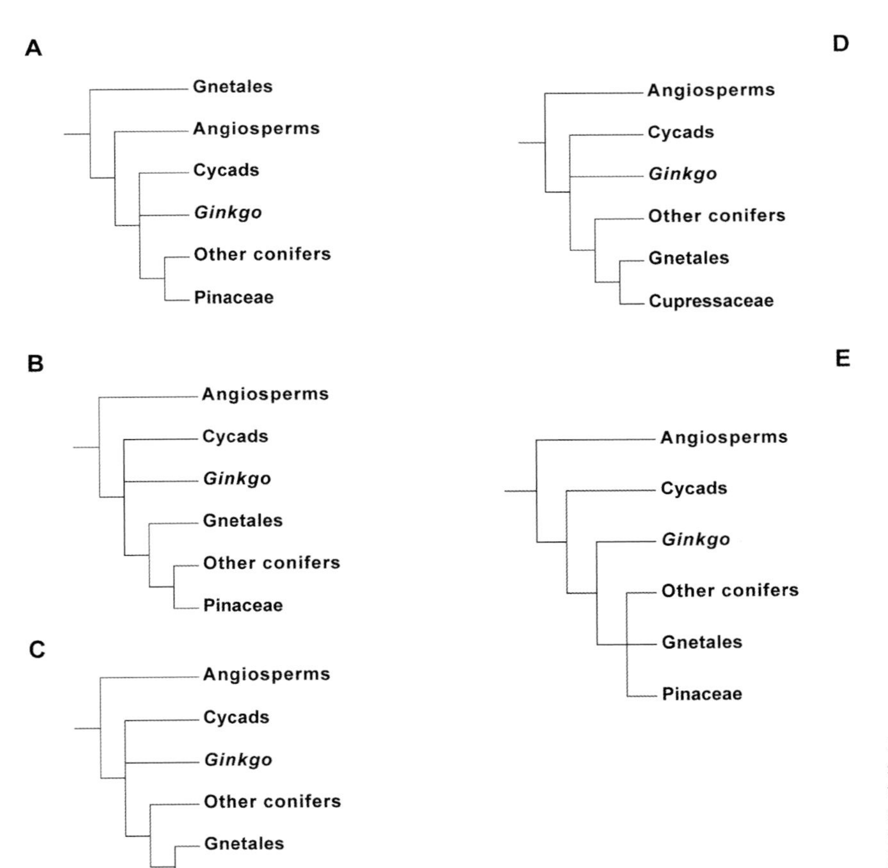

Figure 1.6. Summary of commonly recovered relationships among extant seed plants with an emphasis on the placement of Gnetales. A. Gnetales sister to other seed plants. B. Gnet-ifer topology with Gnetales sister to conifers. C. Gne-pine topology with Gnetales sister to Pinaceae. D. Gne-cup topology with Gnetales sister to Cupressaceae. E. Conservative seed plant summary tree.

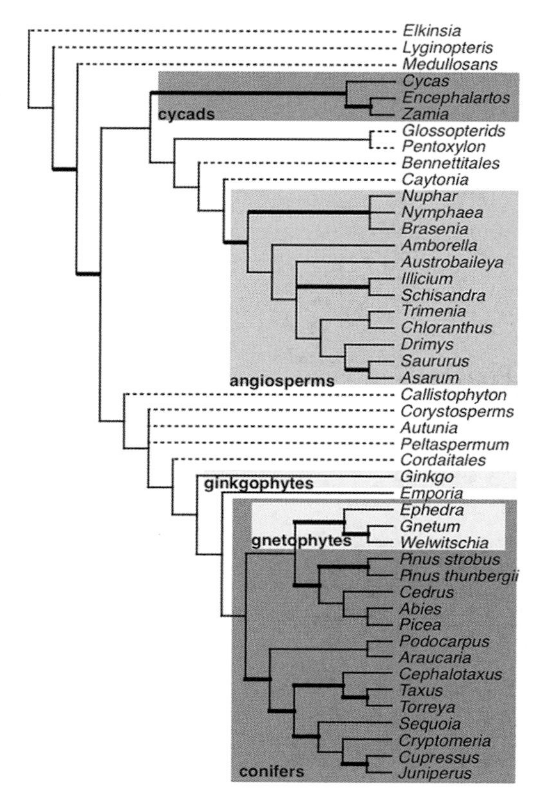

Figure 1.7. Total evidence tree (DNA + morphology) for seed plants based on analyses of Magallón (2010). Modified from Magallón (2010, Fig. 2).

rDNA, and *rbcL* similarly provided strong internal support for Gnetales + Pinaceae (Chaw et al. 2000; Fig. 1.6). An eight-gene analysis (D. Soltis et al. 2002b), involving four plastid genes, three mtDNA genes, and one nuclear gene, also provided strong support for Gnetales + Pinaceae. Qiu et al. (2007) examined seven plastid genes, three mtDNA genes, and nuclear 18S rDNA from 192 land plants and found Gnetales sister to Pinaceae, but with low support (BS = 67%). Burleigh and Mathews (2004) examined a 13-gene dataset representing all three genomes using various partitioning schemes and methods of analysis (see below). ML analyses of the combined datasets and from each partition recovered well-supported Gne-pine trees. Removal of rapidly evolving sites also favored Gne-pine trees.

Although multiple analyses of plastid, mitochondrial, and nuclear genes support the Gne-pine hypothesis, most molecular studies of seed plants included few conifers. With a dataset that included 15 conifers and 13 loci, Burleigh and Mathews (2004) found strong support for Gnetales + Pinaceae. However, Rydin et al. (2002), using 30 conifers and a four-gene dataset, found strong support for Gnetales as sister to conifers (Gnet-ifer hypothesis) (see Rydin and Korall 2009), suggesting that the Gne-pine hypothesis might be an artifact of inadequate taxon sampling. Furthermore, use of the parametric bootstrap in analyses of a dataset involving many base pairs but few taxa revealed that the placement of Gnetales as sister to conifers (rather than as sister to Pinaceae) could not be rejected (D. Soltis et al. 2002b).

A structural mutation in the plastid genome (Raubeson and Jansen 1992) also indicates that a Gnetales-conifer sister-group relationship may be a more parsimonious explanation of the data than a placement of Gnetales within conifers. Most land plants, including Gnetales, have two copies of the ribosomal genes in the plastid genome (the inverted repeat region), but conifers have only a single ribosomal coding region. Placement of Gnetales within conifers (e.g., either Gne-pine or Gne-cup) would necessitate that the ribosomal genes were lost in the conifers and then subsequently regained in Gnetales, which is less parsimonious. However, Wu and Chaw (2014) recently proposed independent losses of the inverted repeat in Pinaceae and cupressophytes.

Some fossil evidence also supports a close relationship of Gnetales and conifers (Wang 2004). However, most morphological and anatomical characters are so different between these groups that morphological data to this point have had little impact in resolving these relationships. Nonetheless, we stress that the Gne-pine hypothesis would necessitate either that the cone evolved twice (once in Pinaceae, and again in all other conifers) or that the cone was lost in Gnetales. The cone is a fairly complex morphological structure (with bracts associated with ovulate scales), and there is no indication that such a structure is present in the very different reproductive axes of Gnetales. Thus, in some ways morphology argues more strongly for the Gnetifer hypothesis.

The Gne-cup topology (Fig. 1.6) was recovered by Nickrent et al. (2000), Zhong et al. (2010), and Ruhfel et al. (2014). The last study employed nearly complete plastid genomes for 360 species of green plants, or *Viridiplantae*, including 311 seed plants. Ruhfel et al. (2014) found strong support for a placement of Gnetales with Cupressaceae (BS = 87%) and BS = 100% for a sister group of this Gne-cup clade with all remaining conifers. As a point of caution, however, Zhong et al. (2010) suggested in another analysis of plastid genomes (but with limited taxon sampling) that support for Gne-cup may be the result of long-branch attraction; by removing rapidly evolving proteins, support for Gne-cup decreased (see also Yang and Rannala 2012). Furthermore, by removing what they considered parallel substitutions between lineages leading to Gnetales and to *Cryptomeria* (the sole cupressophyte in their analyses), a Gne-pine topology was recovered by Zhong et al. (2010).

Trees based on large numbers of nuclear genes (de la Torre-Barcena et al. 2009; Lee et al. 2011; Wickett et al. 2014) merit more discussion. The first two of these studies recovered Gnetales sister to all other living gymnosperms—an unusual placement. However, de la Torre-Barcena et al. (2009) employed only 16 taxa. Using a large number of nuclear genes (22,833 sets of orthologs) from 101 land plant genera, Lee et al. (2011) found strong support for Gnetales sister to remaining all gymnosperms ([cycads + *Ginkgo*] + conifers); this gymnosperm clade is then sister to angiosperms. In another independent analysis of hundreds of nuclear genes, Wickett et al. (2014) found strong support for the monophyly of extant gymnosperms and for the Gne-pine hypothesis. The Lee et al. (2011) dataset is characterized by extensive missing data, whereas in Wickett et al. (2014), more data cells are filled; this difference could result in different topologies.

Resolving seed plant relationships with DNA data is difficult because the signal in datasets may be complex. Sanderson et al. (2000), Magallón and Sanderson (2002), and Rydin et al. (2002) reported conflict between first and second versus third codon positions in plastid genes. Although third codon positions of plastid genes generally have most of the phylogenetic signal (e.g., Källersjö et al. 1998; Olmstead et al. 1998), the third positions may be saturated in some instances (Rydin et al. 2002), depending on taxon sampling. These results may also reflect short branches within the seed plant radiation, as well as high rates of molecular evolution in Gnetales and the outgroups (reviewed in Palmer et al. 2004). Adding to the complexity of the conflict between first plus second versus third positions is the fact that transitions within each codon position conflict with transversions (Chaw et al. 2000; Rydin et al. 2002). Conflicting signal in the datasets could also be explained by differences in trees obtained with rapidly versus slowly evolving sites. Burleigh and Mathews (2004) found that trees in which Gnetales are sister to all other seed plants appear to be the result of signal in the most rapidly evolving sites, whereas when these sites are excluded, Gne-pine trees are obtained.

Given the diversity of studies, in both genes and taxa, as well as the diversity of results, can we make any firm statements regarding the position of Gnetales? In addition, what do most analyses suggest regarding extant seed plant relationships in general? Most analyses now favor some type of a relationship of Gnetales with conifers—a close relationship with angiosperms can be ruled out. Many analyses, including those of numerous nuclear genes (e.g., Wickett et al. 2014), as well as studies involving particularly rigorous examination of the underlying molecular data and signal (e.g., Burleigh and Mathews 2004), appear to favor Gne-pine. However, the largest plastid datasets so far employed favor Gne-cup (e.g., Ruhfel et al. 2014), so different genomes may be telling different stories.

Can we provide any firm summary regarding the overall picture of relationships among extant seed plants (see Fig. 1.6)? It is clear that angiosperms appear sister to a clade of extant gymnosperms. Within living gymnosperms, cycads and *Ginkgo* are then sisters to conifers + Gnetales, but the exact placement of Gnetales to the conifers remains unclear. Similarly, the relationship between cycads and *Ginkgo* also remains uncertain. Some analyses indicate that cycads and *Ginkgo* are successive sisters to other living gymnosperms (e.g., Burleigh and Matthews 2004, 2007; Graham and Iles 2009; Nickrent et al. 2000; Qiu et al. 2007; Ran et al. 2010), often with strong internal support. Other analyses indicate, however, that cycads and *Ginkgo* form a clade that is sister to other living gymnosperms. Internal support for the latter relationship is weak in some studies (e.g., Rydin and Källersjö 2002; Qiu et al. 2006), but in other multigene analyses it is very strong (e.g., Wu et al. 2007; Finet et al. 2010; Zhong et al. 2010; Lee et al. 2011; Ruhfel et al. 2014; Wickett et al. 2014). In still other studies, cycads and *Ginkgo* form a trichotomy with a clade of other living gymnosperms (Rydin and Korall 2009). Accordingly, we have depicted the relationship of cycads and *Ginkgo* as uncertain in a conservative summary tree (Fig. 1.6E).

To summarize, a conservative phylogenetic tree for extant seed plants (Fig. 1.6E) reveals major uncertainties. In fact, this overall summary tree is comparable to what was depicted more than a decade ago (Soltis et al. 2005b). Resolving seed plant relationships and the placement of Gnetales remains problematic, despite intensive study. As a caveat, although the number of base pairs included in seed plant analyses has steadily increased, taxon sampling has remained low in most of the analyses conducted to date, with a few noteworthy exceptions. This sparse sampling should be remedied in future studies. More problematic is the effect of taxa that cannot be sampled for DNA due to their extinction—a real problem in that most seed plant clades are extinct.

Whereas living gymnosperms consistently appear monophyletic, gymnosperms as a whole (fossil and extant) are clearly paraphyletic (Figs. 1.6–1.9; Doyle 2006, 2012). Some extinct gymnosperm lineages attach along the branch to angiosperms, whereas others attach near the base of the seed plant tree. In recent analyses, an "acrogymnosperm" clade (*Acrogymnospermae*; see Cantino et al. 2007) is recovered that includes some fossil gymnosperms as well as extant gymnosperms; a second subclade (sometimes called para-angiophytes) contains angiosperms and other

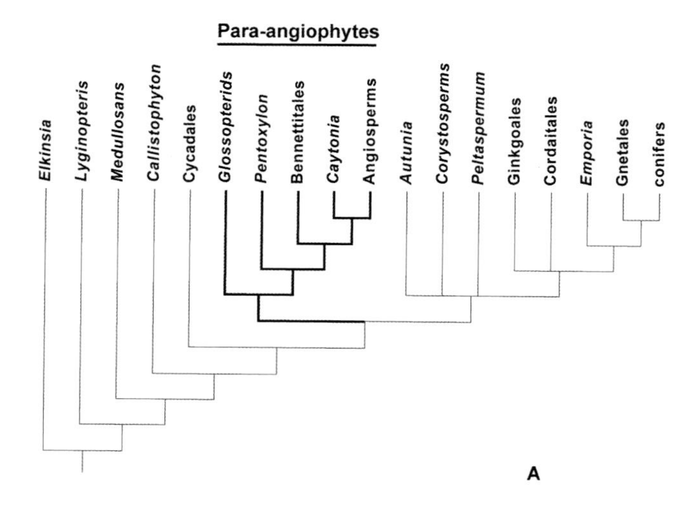

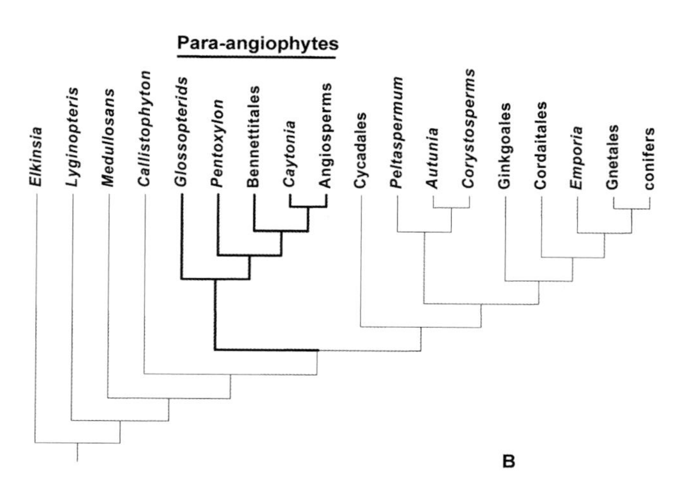

Figure 1.8. Revised views of the phylogeny of seed plants showing putative closest relatives of angiosperms obtained by using both the morphological matrix of Doyle (1996) and molecular data (reviewed in Soltis et al. 2005). A revised anthophyte clade—the para-angiophytes—is depicted. (A) Tree modified from Soltis et al. (2005), in which molecular data for seed plants are combined with the morphological matrix of Doyle (1996). (B). Tree modified from Doyle (2001), in which a molecular constraint was used, placing Gnetales with other extant gymnosperms. Redrawn from Soltis et al. (2008; Fig. 1).

fossil gymnosperms (e.g., Caytoniales) (Figs. 1.8–1.9). However, based on analyses conducted to date, fossil and seed plants as a whole (Spermatophyta) are monophyletic.

MORPHOLOGY REVISITED

Whereas cladistic analyses of morphological data have indicated that angiosperms and Gnetales are closest living relatives (albeit without strong support), most studies of DNA sequences and DNA plus morphology have placed Gnetales with conifers despite their obvious morphological differences. Doyle (1998a) initially attempted to reconcile this conflict, but as more molecular studies argued against a close relationship between Gnetales and angiosperms, such efforts diminished (Donoghue and Doyle 2000).

It is instructive to reconsider those few characters that were thought to unite angiosperms and Gnetales in morphological cladistic analyses (Crane 1985; Doyle and Donoghue 1987, 1992; Loconte and Stevenson 1990; Taylor and Hickey 1992; Nixon et al. 1994; Doyle 1996). Some of these features are actually shared by angiosperms, Gnetales, Bennettitales, *Pentoxylon*, and Caytoniales. As reviewed below, careful scrutiny leads to the conclusion that the homology of many of these shared characters is, in fact, dubious.

Following Crane (1985), two key features unite angiosperms, Gnetales, and Bennettitales and *Pentoxylon*: a distinctively thin megaspore membrane and microsporophylls aggregated in a whorl, or pseudo-whorl, distinct from the helical arrangement in conifer and cycad pollen cones. Doyle and Donoghue evaluated the relationship between Gnetales and angiosperms in several papers (e.g., Doyle 1978, 1994, 1996, 1998a,b; Doyle and Donoghue 1986, 1992); we summarize here the non-DNA characters that support a sister group of Gnetales + angiosperms in Doyle's (1996) analysis. Angiosperms and Gnetales share similar lignin chemistry (i.e., the presence of a Mäule reaction, which is absent from other seed plants; McLean and Evans 1934; Gibbs 1957), double fertilization, microsporangia fused at least basally, an embryo derived from a single uninucleate cell via cellular divisions, a thin megaspore wall (as in Crane 1985 and Loconte and Stevenson 1990), siphonogamy, and a granular exine structure. Doyle (1996) also scored vessels in angiosperms and Gnetales as homologous. Loconte and Stevenson (1990) analyzed only extant taxa and provided three synapomorphies of Gnetales and angiosperms—thin megaspore wall (following Crane 1985), short cambial initials, and lignin syringial groups (equivalent to the Mäule reaction of Doyle 1996).

Some putative synapomorphies for angiosperms and Gnetales from Crane (1985), Doyle (1996), and Loconte and Stevenson (1990) are more complex than initially suggested in these analyses and may in fact not be homologous (see also Donoghue and Doyle 2000). For example, although the angiosperms and Gnetales were coded the same for the presence of a tunica layer in the vegetative shoot apex, the tunica is two cells thick in many angiosperms and only one cell thick in Gnetales. Similarly, although angiosperms and Gnetales were coded as having the same state for the thickness of the megaspore wall, the megaspore wall is thin in Gnetales and absent in angiosperms. The pollen exine character used in some studies is now known to be inappropriate because a granular exine is not ancestral in angiosperms, as once was hypothesized (e.g., Doyle and Endress 2000; Doyle 2001, 2009; see also Chapter 6). Furthermore, the homology of vessels in angiosperms and Gnetales has long been doubted (Bailey 1944b, 1953), and Carlquist (1996) concluded that they are not homologous. Angiosperms and Gnetales should therefore not be scored identically for these features of the tunica, megaspore, and vessel elements. The homology of double fertilization in angiosperms and Gnetales has also been questioned (see Friedman 1994, 1996; Doyle 1996, 2000), but this issue is complex. In most angiosperms, a second sperm nucleus fuses with two nuclei of the megagametophyte (producing triploid endosperm), whereas in Gnetales a second sperm fuses with only one nucleus of the megagametophyte, yielding a diploid nucleus. However, in the basal angiosperm clades Nymphaeales and Austrobaileyales (but not Amborellaceae), a second sperm nucleus fuses with only a single megagametophyte nucleus (Chapter 4). Furthermore, double fertilization events that seem similar to those documented for *Ephedra* (Gnetales) have been reported for conifers, including *Thuja* and *Abies* (Friedman and Floyd 2001). In addition, developmental events in cycads and *Ginkgo* are consistent with double fertilization (reviewed in Friedman and Floyd 2001). Thus, double fertilization may be a synapomorphy for all extant seed plants (Friedman and Floyd 2001), although the formation of endosperm is an exclusively angiosperm feature.

The anthophyte hypothesis continues to have an impact on interpretation of morphology. Friis et al. (2007) proposed a new synapomorphy for some of the traditional anthophytes, namely Gnetales and Bennettitales. Using phase-contrast X-ray microtomography, they found a distinctive seed architecture shared by Gnetales and Bennettitales (as well as Erdtmanithecales, another putatively closely related fossil group). Friis et al. (2007) argued that this seed feature "defines a clade containing these taxa." Nearly all recent topologies indicate a distant relationship of Gnetales and Bennettitales (Figs. 1.8, 1.9) so in our view this seed feature may best be interpreted as homoplasious, having evolved independently in Gnetales and the Bennettitales (Ertmanithecales might, however, share a common origin of the feature with either of them). However, phylogenetic analysis using the seed plant matrix of Hilton and Bateman (2006) supports the grouping of Erdtmanithecales, Bennettitales, and Gnetales (Friis et al. 2011, p. 104).

INTEGRATING FOSSILS

We now revisit the possible close relatives of the angiosperms that are known from the fossil record. The importance of integrating fossils, and thus morphology, into datasets to understand the phylogeny of seed plants has long been noted and continues to be emphasized (e.g., Doyle 2006, 2008, 2012; Doyle and Donoghue 1987; Kenrick and Crane 1997; Donoghue and Doyle 2000; Rydin et al. 2002; D. Soltis et al. 2002b, 2005b; Crane et al. 2004; Magallón 2010; Hilton and Bateman 2006; Mathews et al. 2010). Even if DNA sequence data largely resolved relationships among living seed plant groups (this is not the case; see above), a complete understanding of seed plant relationships and the origins of angiosperm structures such as floral organs still requires the integration of fossil taxa because the extinct taxa can affect the phylogenetic placement of extant taxa. Relationships among lineages of anthophytes have varied among studies depending on whether or not fossils were included. In some cases, Gnetales were sister to the angiosperms even when fossils were included (e.g., Crane 1985); in other analyses, the sister relationship between Gnetales and angiosperms appeared only when fossils were removed from the matrix (e.g., Doyle and Donoghue 1986).

Significantly, most phylogenetic analyses that include fossils (whether morphology or morphology + DNA) reveal the same cast of characters as close, now extinct, relatives of the angiosperms: Glossopteridales, *Pentoxylon*, Bennettitales, and Caytoniales (D. Soltis et al. 2005b, 2008b), referred to as the para-angiophytes (Doyle 2012; Fig. 1.8). Initial studies of morphology alone placed Bennettitales and *Pentoxylon* as subsequent sisters to angiosperms + Gnetales (Crane 1985; Doyle and Donoghue 1986, 1992; Rothwell and Serbet 1994). Doyle (1996) found *Caytonia* sister to *Pentoxylon*, followed by Glossopteridales as sister to Gnetales + angiosperms. However, other studies suggest that Caytoniales are the immediate sister to angiosperms, with Bennettitales sister to this clade. Doyle (2001, 2008) and Soltis et al. (2005b) constructed a molecular scaffold on the basis of molecular phylogenetic analyses, constraining Gnetales to be sister to the conifers, and then analyzed Doyle's (1996) original morphological matrix. Both found a revised anthophyte clade of angiosperms + *Caytonia*, Bennettitales, and *Pentoxylon*, with Glossopteridales sister to the remaining members of this expanded anthophyte clade (Fig. 1.8). Other studies using morphology (Hilton and Bateman 2006), and morphology + DNA (Magallón 2010) (Fig. 1.7), plus studies using a molecular backbone for the living taxa (fixing Gnetales in a Gne-pine position; Doyle 2008, 2012) (Fig. 1.9), also have recovered a para-angiophyte clade with *Pentoxylon* + Glossopteridales sister to Bennettitales, followed by Caytoniales, with the last the immediate sister to angiosperms (Fig. 1.9).

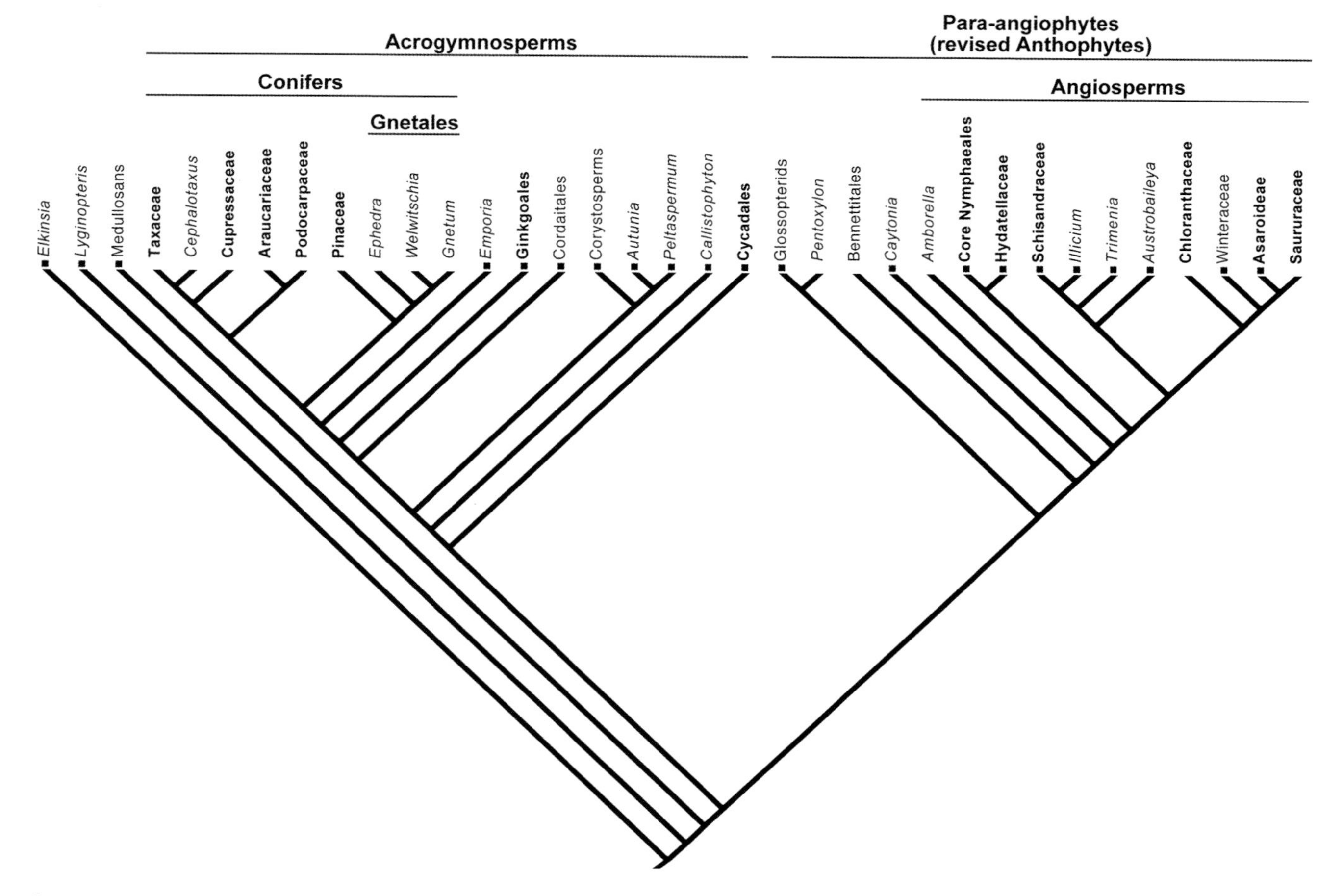

Figure 1.9. Seed plant phylogenetic tree (from analysis of morphological data of Doyle 2008) with the relationships of living gymnosperms constrained to show a Gne-pine backbone. Redrawn from Doyle (2012).

PARA-ANGIOPHYTES: A NEW LOOK AT CHARACTER EVOLUTION

The phylogenetic results reviewed above are crucial for reconstructing morphological evolution of the angiosperms. Topologies that recover the para-angiophyte clade are consistent with the hypothesis that the cupule of Glossopteridales and *Caytonia* is homologous with the outer integument of the angiosperm bitegmic ovule (reviewed in Doyle 2006, 2008, 2012) (Figs. 1.10, 1.11), a hypothesis that had been proposed much earlier (e.g., Gaussen 1946; Stebbins 1974; Doyle 1978). To review, whereas angiosperm ovules generally have two protective layers (integuments), all gymnosperm ovules have a single integument (Fig. 1.10). [Some angiosperms, such as asterids, have only a single integument due either to loss of an integument or to fusion of the two integuments (Endress 2011a; see Chapters 11, 14).] Furthermore, the gymnosperm micropyle is located opposite the stalk bearing the ovule (except in Podocarpaceae), whereas in many angiosperms, the ovule is curved back on itself with the micropyle close to the stalk (i.e., anatropous rather than orthotropous; Fig. 1.10B). Angiosperm ovule types are reviewed in Chapter 4 (see Fig. 4.8).

To address the origin of the outer integument of the angiosperm ovule, Doyle (2012) reconstructed the evolution of the ovule-bearing surface across a phylogenetic tree for fossil and extant seed plants (Fig. 1.9); we summarize his findings here. In *Caytonia* and Glossopteridales, the ovule-bearing surface is the adaxial or upper surface, whereas in those other seed plants that have been considered angiosperm ancestors, such as peltasperms and corystosperms, ovules are borne on the abaxial surface (underside) (Fig. 1.11). Other seed plants bear ovules in an apical or marginal position. Hence, the reconstruction of Doyle (2012) is consistent with the hypothesis that the second or outer integument of angiosperms is homologous with the cupule in *Caytonia* and Glossopteridales—with the cupule in these plants representing a leaf or leaf segment with ovules on the upper surface.

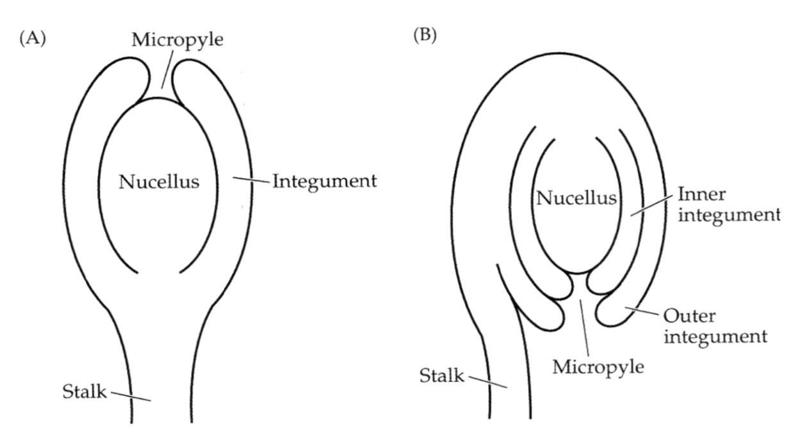

Figure 1.10. Ovules of gymnosperms and angiosperms. Gymnosperms (A) have one integument whereas angiosperms (B) have two—an inner and an outer integument.

The cupule of *Caytonia* is similar to the bitegmic ovule of flowering plants (Stebbins 1974; Doyle 2012). The shape of the *Caytonia* cupule resembles an anatropous ovule in angiosperms (compare Figs. 1.11, 1.12; see below). The *Caytonia* cupule contains multiple ovules, however, so ovule number would have been reduced to one with the origin of the angiosperms (see Stebbins 1974). Doyle (2012) further noted that in Glossopteridales, the cupule is "most easily interpreted" as a fertile leaf "borne on an axillary branch that is adnate to the subtending leaf"; in contrast, cupules in *Caytonia* have been interpreted as leaflets borne along the rachis of a compound leaf (see Stebbins 1974; Doyle 1978; Retallack and Dilcher 1981; Doyle 1978, 2012). Bennettitales are problematic, however, in that the female

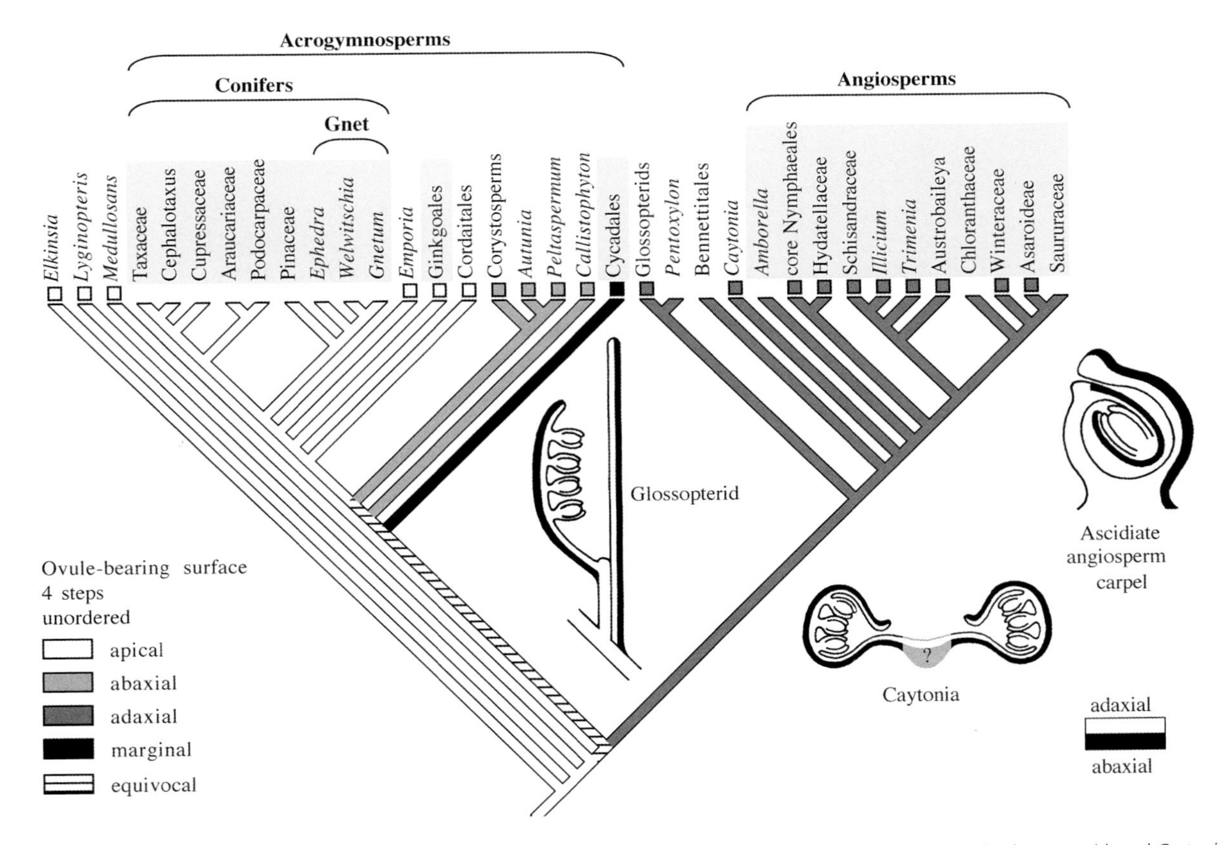

Figure 1.11. Putative reconstruction of ovule evolution using topology shown in Figure 1.9. Diagrams show ovulate structures in glossopeterids and *Caytonia* as well as an ascidiate angiosperm carpel. Abaxial surfaces are shown in black; ovules are borne on the adaxial surface in glossopterids, *Caytonia*, and angiosperms, but not in acrogymnosperms. Based on this reconstruction, the angiosperm bitegmic ovule and carpel may be derived from ovulate structures in glossopterids and *Caytonia*; see text (taken from Doyle 2012).

reproductive structures of this group show no clear relation to structures in these other plants (see Doyle 2012 and section on Bennettitales below).

Doyle (2008, 2012) also used these same phylogenetic results to attempt to infer the homology of the angiosperm stamen. Using the current "best" topology for angiosperms (chapter 3), which has Amborellaceae, Nymphaeales, and Austrobaileyales as basalmost branches, Endress and Doyle (2009) inferred that, in the earliest angiosperms, sporangia were originally located in a lateral or adaxial position. Doyle (2008, 2012) therefore hypothesized that the angiosperm stamen may be "comparable with male structures in Glossopteridales, which had a sporangium-bearing branch on the adaxial side of a leaf, or with those in Bennettitales, which had synangia on the adaxial side of a sporophyll" (Doyle 2012, p. 316) (see also illustrations later in this chapter).

FOSSILS: IN SEARCH OF THE SISTER GROUP OF ANGIOSPERMS

We provide brief coverage of four fossil lineages that may be the closest relatives of angiosperms as inferred from cladistic analyses and/or noteworthy morphological similarities: Caytoniales, Bennettitales, Pentoxylales, and Glossopteridales. Traditionally, the search for fossil groups as possible angiosperm ancestors focused on features of the ovule and structures that may be homologous to the angiosperm carpel, although many other characters, such as pollen morphology, stem anatomy, and leaf venation patterns, are also important in phylogenetic considerations. Several lines of evidence indicate that the outer integument of angiosperms has a different origin from the inner integument. Developmental genetic studies indicate that the inner and outer integuments are under separate genetic control (see discussion of *INO* below and McAbee et al. 2006). Other possible evidence for this distinct origin is the mismatch of the two integuments seen in many basal angiosperms, producing a "zig-zag" micropyle (Davis 1967; Stebbins 1974). This feature is actually widely distributed in the angiosperms, occurring in diverse eudicots, including Bixaceae, various Brassicales, Fabaceae, Malvaceae, Hamamelidaceae, and Dilleniaceae (Endress 2011a). Another distinctive feature is that stomata are occasionally produced on the outer, but not the inner, integument (see Stebbins 1974; Corner 1976).

Developmental studies have shown that in many angiosperms, the outer integument is hood-shaped, indicating an origin from a leaf (Yamada et al. 2001a,b). The gene *INNER NO OUTER* (*INO*) participates in the regulation of dorsoventrality of lateral organs and is expressed in the abaxial side of leaves, as well as in the outer epidermis of the outer integument in *Arabidopsis* (e.g., Bowman 2000). This same expression pattern has been observed in ovules of *Nymphaea*, supporting the hypothesis that the outer integument is homologous with a leaf and that ovules are located on the adaxial surface, away from the zone of expression (Yamada et al. 2003). This differential pattern of *INO* gene expression in inner and outer integuments is consistent with separate origins of the two integuments of angiosperms.

Fossil groups such as Caytoniales, Glossopteridales, *Pentoxylon*, corystosperms, and Bennettitales have been proposed as possible close relatives of angiosperms because the ovule is surrounded by a "cupule," a structure that may be homologous to the outer integument of angiosperms. In Caytoniales (and also corystosperms, which are not close angiosperm relatives in phylogenetic studies), the cupule plus ovule structure is curved, which has been considered a possible antecedent of the anatropous ovule of angiosperms. Corystosperms produced ovules on the abaxial surface of the megasporophyll and therefore probably were not angiosperm ancestors. However, *Caytonia*, *Petriellaea* (a Mesozoic seed fern; Taylor et al. 1994), and *Glossopteris* (Taylor and Taylor 1992) produced ovules on the adaxial surface of a megasporophyll and, on this basis, could be angiosperm relatives (see Fig. 1.11). The seed-enclosing structures of *Petriellaea* and *Glossopteris* seem to have evolved via different structural modifications: in *Petriellaea* by transverse folding of the leaf and in some Glossopteridales by longitudinal enrolling of the leaf margin (Taylor and Taylor 1992). The folding of the megasporophyll in both fossils differs from the presumed origin of the angiosperm carpel via longitudinal folding of a megasporophyll (this assumes plicate carpel development—the situation may be different in ascidiate carpels; see Chapter 4). This leaves *Caytonia* as a possible ancestor or close relative of angiosperms.

CAYTONIALES

Before morphological cladistic analyses placed Gnetales as sister to the angiosperms (Crane 1985; Doyle and Donoghue 1986), Caytoniales had already received considerable attention as a possible angiosperm ancestor (e.g., Thomas 1925; Gaussen 1946; Harris 1951; Stebbins 1974; Doyle 1978, 1996). Although not initially considered an "anthophyte" in early cladistic analyses (Crane 1985; Doyle and Donoghue 1986), Caytoniales emerged as part of a clade (the para-angiophytes; see above and Figure 1.8) that included anthophytes in later analyses (Doyle and Donoghue 1992), appearing in some analyses as sister to the angiosperms (Doyle 1996).

In Caytoniales, male and female reproductive structures do not appear to have been produced together; in fact, neither has been found attached to stems (Fig. 1.12). Caytoniales were considered a possible ancestor of the angiosperms because of their cupules and ovules. The morphology of Caytoniales seemed to explain the origin of both the two integuments and the anatropous ovule characteristic of most angiosperms. In Caytoniales, each cupule contained several ovules, and the cupule enclosed these ovules, leaving only a small opening between the ovule and the stalk to the cupule. Within the cupule, numerous unitegmic ovules were present, each with a micropyle oriented toward the opening (or mouth) of the cupule (Fig. 1.12). It was argued that if the *Caytonia* cupule contained only a single ovule rather than multiple ovules, then the resultant structure would resemble a typical angiosperm ovule in being anatropous and possessing two integuments (Gaussen 1946; reviewed in Stebbins 1974; Doyle 1978; Frohlich and Parker 2000).

One criticism of Caytoniales as a close angiosperm relative or precursor is that no Caytoniales fossils have been reported that possess just one ovule per cupule. The counter argument is that only a simple reduction in ovule number (from several to one) is required. Furthermore, the origin of the angiosperm carpel from Caytoniales cannot easily be explained because the ovules of the latter were located on opposite sides of a narrow stalk, a structure that is difficult to envision as forming a carpel (Fig. 1.12B). The hypothesis put forward was that this Caytoniales stalk became wide and flat and eventually enclosed the ovules, forming the angiosperm carpel (Gaussen 1946; Doyle 1978; Crane 1985). Another difficulty with Caytoniales as a possible angiosperm ancestor is that the microsporophylls in Caytoniales were highly divided, differing greatly in morphology from angiosperm stamens, and bore bisaccate pollen, the only similarity being the occurrence of four microsporangia per structural unit (Fig. 1.12A). Leaves of *Sagenopteris*, the foliage of *Caytonia*, bore leaflets with venation resembling that of *Glossopteris* leaves, lacking the multiple orders of venation typical of angiosperms. Because of the large morphological gap between Caytoniales and angiosperms, the hypothesis that Caytoniales are closely related to angiosperms remains problematic; it is not known whether caytonialean plants produced vessels in their xylem, as do most angiosperms, or lacked them, as do Glossopteridales. However, presence of vessels in an angiosperm relative no longer seems crucial given the absence of vessels in some early-branching angiosperms (Chapter 4). Nevertheless, this group consistently appears close to angiosperms in phylogenetic analyses.

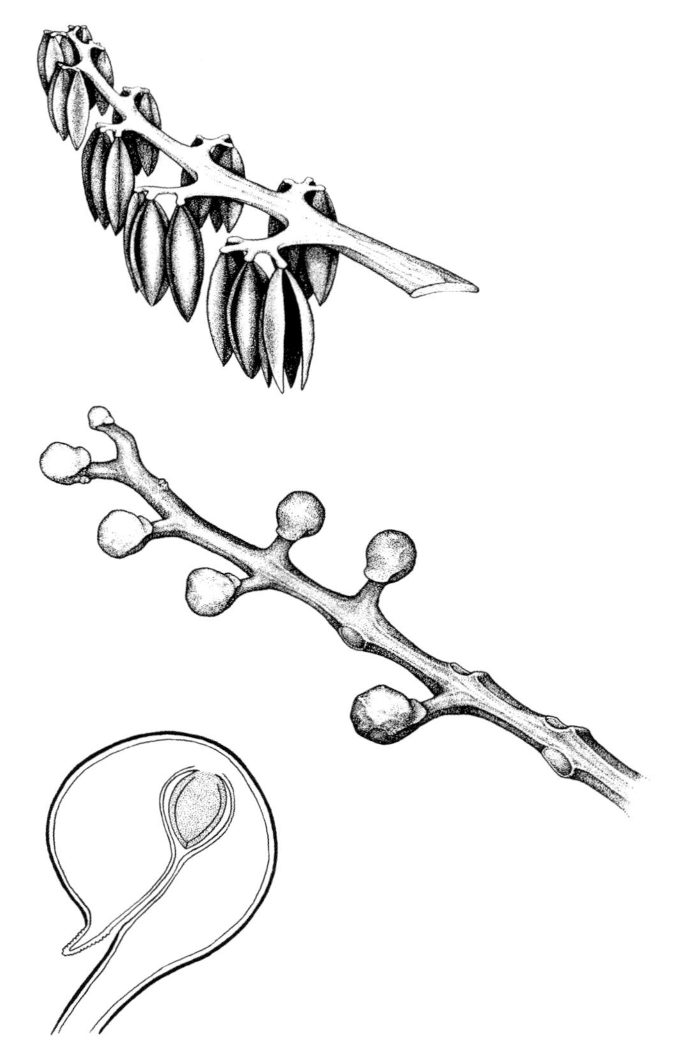

Figure 1.12. Reconstructions of Caytoniales (from Crane 1985). (Top) Male synangia of *Caytonanthus arberi*, based on Harris (1941). (Middle) *Caytonia nathorstii* megasporophylls, based on Harris (1964). (Bottom) *Caytonia* cupule containing seeds, based on Reymanówna (1973).

BENNETTITALES

Bennettitales have long been considered close relatives of angiosperms. Arber and Parkin (1908) proposed a close relationship between angiosperms, Gnetales, and Bennettitales. In Crane (1985), Bennettitales and *Pentoxylon* appeared as sister to Gnetales + angiosperms. The analyses of Doyle and Donoghue (1992) and Rothwell and Serbet (1994), as well as most subsequent analyses, similarly placed angiosperms, *Pentoxylon*, Bennettitales, and Gne-

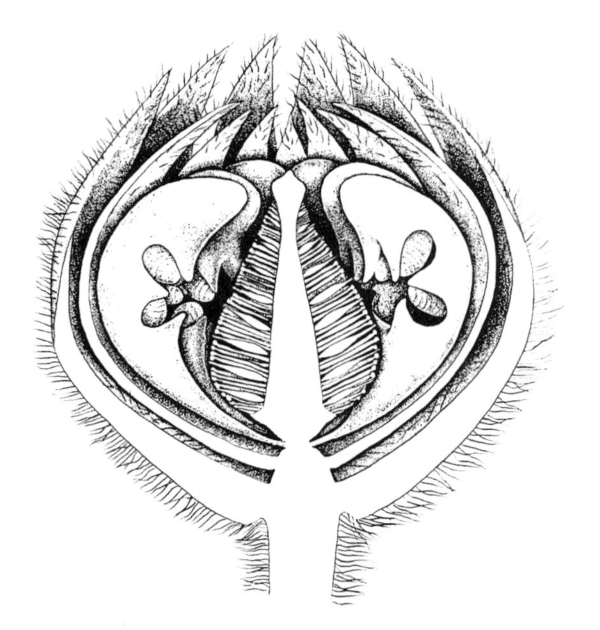

Figure 1.13. Reconstruction of Bennettitales (from Crane 1985). *Williamsoniella coronata*, longitudinal section through "flower," based on Harris (1964).

tales together in one clade. Rothwell et al. (2009) reviewed the unusual reproductive structures of Bennettitales.

Bennettitales possessed strobili that were unisexual in some representatives and bisexual in others. In bisexual (hermaphroditic) forms, the microsporangia and ovules were associated on lateral branches that arose among the leaf bases (Fig. 1.13). This strobiloid reproductive structure was important in the formulation of the "anthostrobilus" theory for the origin of the angiosperm flower (Arber and Parkin 1907). The strobiloid reproductive structure of Bennettitales was considered similar to the strobiloid flowers of Magnoliaceae, and this similarity was the rationale for considering members of "Ranales" (Schisandraceae, Magnoliaceae, Ranunculaceae, and their relatives) as the most "primitive" extant angiosperms (e.g., Delpino 1890; Bessey 1915; reviewed in Endress 1993). Many Bennettitales had flat microsporophylls with adaxial sporangia, containing monosulcate pollen; these structures could be considered suitable precursors to the angiosperm stamen. In other regards, Bennettitales are problematic as close relatives of the angiosperms. Ovules of Bennettitales had an orthotropous orientation rather than potentially anatropous as in Caytoniales, corystosperms, and most angiosperms (see Crane 1985; with notable exceptions including Chloranthaceae, *Ceratophyllum*, and possibly *Amborella*: Endress and Doyle 2009). Discounting Bennettitales on the basis of ovule orientation assumes, however, that the anatropous ovule is ancestral for angiosperms (see Chapters 4 and 6). Furthermore, assessing ovule orientation is problematic when carpels are not present. Finally, Bennettitales did not possess cupules (Rothwell and Stockey 2002; Rothwell et al. 2009). Rothwell et al. (2009) summarized features of Bennettitales—the seeds were produced terminally on sporophylls and were unique in having a nucellus with a solid apex, no pollen chamber, and a single integument; the seeds were not enclosed by a cupule or any other specialized structures. As a result, Bennettitales differ substantially from Gnetales, angiosperms, and *Caytonia* and have sometimes been considered part of a separate lineage, perhaps sharing a common glossopterid ancestor with the angiosperms (see "Glossopterids," below), but not as close relatives of the angiosperms. However, morphological cladistic analyses consistently place Bennettitales close to the angiosperms.

PENTOXYLALES

Before Crane's (1985) phylogenetic analyses, *Pentoxylon* was considered an isolated gymnosperm of uncertain affinity. As background, various detached organs have been inferred to be part of the "*Pentoxylon*" plant: *Pentoxylon* (stem), *Nipaniophyllum* (leaf), *Sahnia* (male flower), and *Carnoconites* (female cone) (Bose et al. 1985). *Pentoxylon* was first associated with anthophytes in Crane (1985) because of its flower-like arrangement of microsporophylls, aggregation of ovules into a head (as in Bennettitales), and similarity of ovules to those of Bennettitales (Fig. 1.14). *Pentoxylon* was also placed in the anthophyte clade of Doyle and Donoghue (1992) and Rothwell and Serbet (1994). Bose et al. (1985) challenged some of the anthophyte-like anatomical features of *Pentoxylon*. However, when Doyle (1996) rescored *Pentoxylon* with the data of Bose et al. (1985) and Rothwell and Serbet (1994), the genus continued to show a close relationship to Gnetales, angiosperms, Bennettitales, and *Caytonia*.

In *Carnoconites*, the ovules were sessile and putatively helically arranged into compact heads. The ovulate heads were borne terminally on short shoots. One problem with Pentoxylales as a close relative of the angiosperms is that there is no clear carpel prototype in *Pentoxylon*. Furthermore, like Bennettitales, the ovules of *Carnoconites* had an orthotropous, rather than anatropous, orientation. Researchers also debate the presence of a cupule in *Carnoconites*, which would ultimately form the outer (second) angiosperm integument. Crane (1985) suggested that the ovules in *Carnoconites* were surrounded by a cupule, whereas others maintained that *Carnoconites* did not possess a cupule (Nixon et al. 1994; Rothwell and Serbet 1994). In addition, the megasporophylls of *Carnoconites* were not leaf-like in

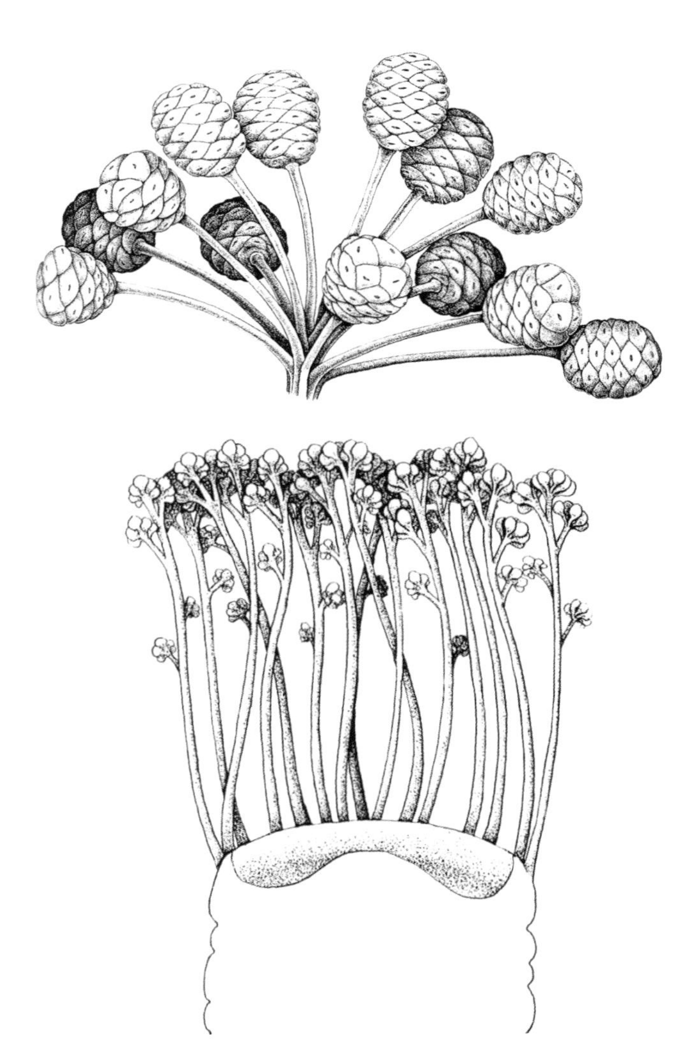

Figure 1.14. Reconstructions of *Pentoxylon* plants (from Crane 1985). (Top) Ovulate heads of *Carnoconites cranwelliae*, based on Harris (1964). (Bottom) Microsporangiate structures of *Sahnia*, based on Vishnu-Mittre (1953).

appearance; rather, each ovulate head bore 10 to 20 spirally arranged (note that Crane 1985, Fig. 19, depicts these as whorled), stalked, unilocular megasporangia (Fig. 1.14; Crane 1985). As a result of these concerns, Pentoxylales, like Bennettitales, have been envisioned as part of a separate lineage, perhaps sharing a common glossopterid ancestor with angiosperms (Fig. 1.14), but not as a close relative or direct ancestor of the angiosperms.

GLOSSOPTERIDALES (GLOSSOPTERIDS)

Plumstead (1956) and Melville (1962) suggested a close relationship of angiosperms and Glossopteridales partly on the basis of reticulate leaf venation, although the venation of glossopterid leaves is much more like that of *Caytonia* leaflets than that of any angiosperms. Retallack and Dilcher (1981) revived interest in a Glossopteridales ancestry of angiosperms, suggesting that the ovule-bearing organs of Glossopteridales have structures that might be homologous with both the outer integument and the carpel of angiosperms (Fig. 1.15). They described the fossil glossopterid *Denkania* using terminology applied to angiosperms, stating that this fossil taxon had bitegmic, orthotropous ovules with the cupule homologous with the outer integument of angiosperms. These "bitegmic" ovules were arranged on a leaf surface but were not enclosed. The leaf surface was considered homologous with the angiosperm carpel (Fig. 1.15). This close relationship is questioned because the initial interpretations of glossopterid fructifications were found to be incorrect (see Doyle and Hickey 1976; Schopf 1976; Taylor and Taylor 1992), and the homologies inferred by Retallack and

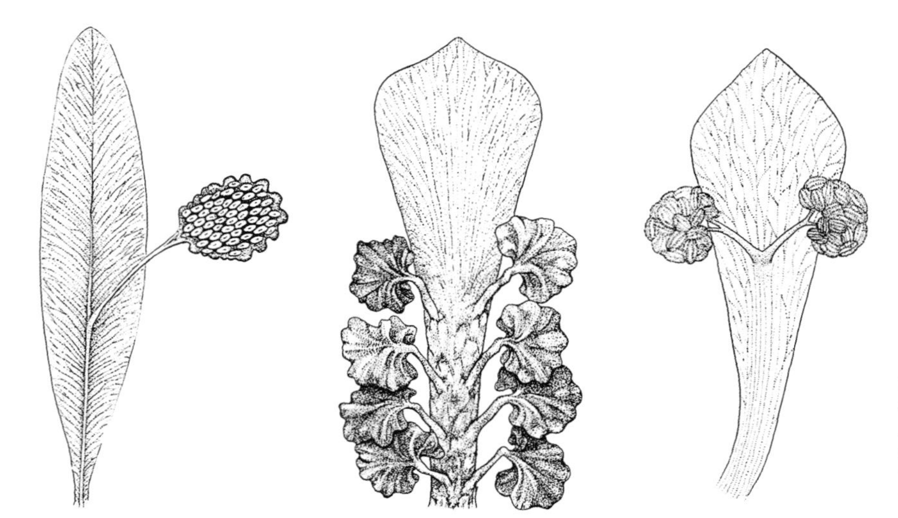

Figure 1.15. Reconstructions of glossopterids (from Crane 1985). (Left) Megasporophyll of *Ottokaria* and associated leaf, redrawn from Pant (1977). (Center) Megasporophyll of *Lidgettonia africana*, based on Thomas (1958). (Right) Microsporophyll of *Eretmonia*, redrawn from Surange and Chandra (1975).

Dilcher have been questioned (Taylor and Taylor 2009). Doyle (1996, 2006) also suggested a close relationship of glossopterids and angiosperms. He proposed not only that Glossopteridales might have been the ancestral group from which Gnetales arose via intermediate fossil forms such as *Piroconites* and *Dechellyia*, but also that the common ancestor of *Caytonia* and the angiosperms might have had "glossopterid-like bract-sporophyll complexes" bearing several cupules per bract (leaf) that would correspond to the "anatropous cupules" of *Caytonia*, which have often been considered homologous to the bitegmic, anatropous ovules of angiosperms (Fig. 1.11). The underlying bract would then only have to be folded lengthwise to produce a carpel or reduced to produce a *Caytonia* sporophyll. Doyle (1996) maintained that this scenario is perhaps more plausible than the origin of a carpel through the expansion and folding of the *Caytonia* rachis (Stebbins 1974; Doyle 1978). In later analyses and reviews (see above and Doyle 2012), Gnetales were removed from this evolutionary scenario, but Glossopteridales remain close to angiosperms and hence the above hypothesis retains some credibility. The bisaccate pollen produced by Glossopteridales is comparable to that in other gymnosperm groups but is not found in any angiosperms. The motile sperm documented in well-preserved *Glossopteris* seeds (Nishida et al. 2004) is a feature shared with cycads and *Ginkgo*, distinguishing them from conifers, Gnetales, and angiosperms.

Cladistic analyses have often indicated a more distant relationship between glossopterids and angiosperms than between Bennettitales, *Pentoxylon*, Caytoniales, and angiosperms (Crane 1985; Rothwell and Serbet 1994; Doyle 1996, 1998a,b; Soltis et al. 2005b). Glossopteridales were not, for example, considered part of the "anthophyte clade" as originally defined (Doyle and Donoghue 1992). However, in Doyle and Donoghue (1992), Caytoniales followed by glossopterids were the sister taxa to the anthophyte clade as originally defined. In a later analysis, Doyle (1996) recovered what he termed a "glossophytes" clade in which glossopterids were sister to angiosperms, Gnetales, Bennettitales, and *Caytonia*. This topology is consistent with the hypothesis that both angiosperms (Retallack and Dilcher 1981) and Gnetales (Schopf 1976) were derived from glossopterids with a loss, in each case, of saccate pollen and motile sperm. However, these hypotheses are now unsupported, with the molecular placement of Gnetales with conifers refuting the anthophyte hypothesis. Nevertheless, it is possible that glossopterids are the immediate sister to a clade that includes angiosperms and their closest relatives and may, therefore, have played a crucial role in angiosperm origins.

MOSTLY MALE

The origin of the flower has long been problematic—as made famous by Darwin in his reference to the origin and early diversification of the angiosperms as "an abominable mystery" (Darwin's letter of July 22, 1879, to Joseph Hooker). Clearly, floral organs are derived from some organ in their gymnosperm precursors, but by what mechanism? Several hypotheses have been proposed and in some

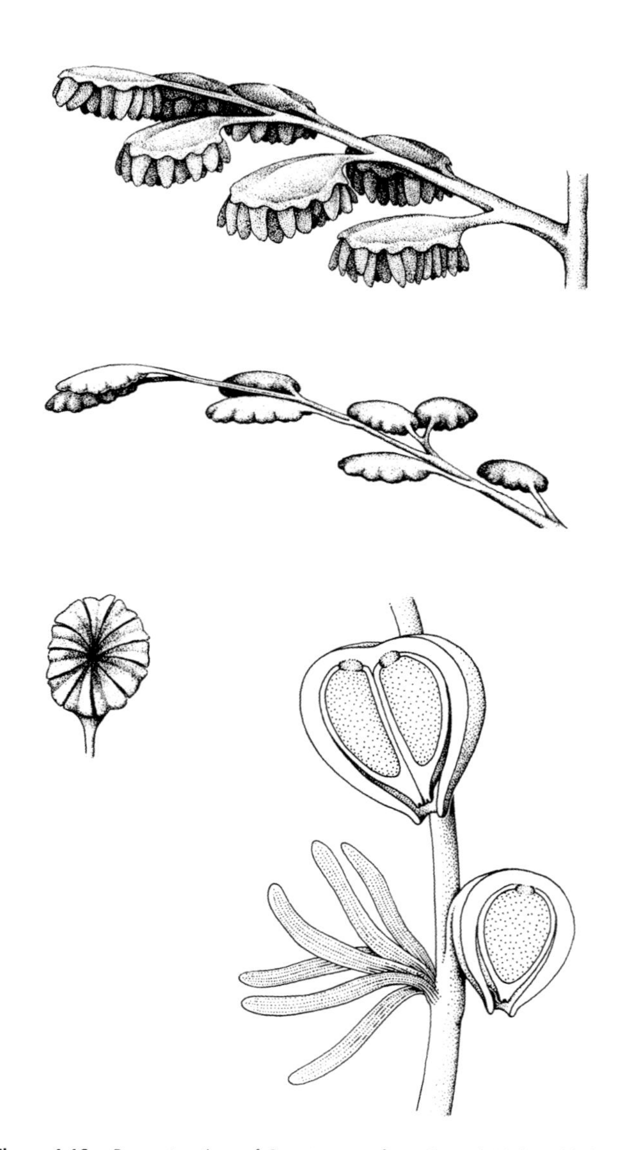

Figure 1.16. Reconstructions of Corystoperms from Crane (1985); Frohlich and Parker (2000). Upper diagram: male structures of *Pteruchus africanus*, redrawn from Townrow (1962). Second from top and left: Synangia of *Pteroma thomasii*, redrawn from Harris (1964). Bottom right: Cupules of *Ktalenia*, from Taylor and Archangelsky (1985).

cases bring fossil evidence from extinct gymnosperm lineages into the proposed evolutionary scenarios. We review leading hypotheses below.

Frohlich and Parker's (2000) mostly male hypothesis for the origin of the flower (Fig. 1.16) was based on their work with the homeotic gene *Floricaula/LEAFY* (*FLO/LFY*). Gymnosperms generally have two copies of the *LFY* gene (although *Gnetum* appears to have only a single copy; see Fig. 1.16A), referred to as the "needle" and "leaf" families. The function of the *LFY* genes of pine has been studied in detail. The "needle" (*NEEDLY*) paralogue is expressed only in early-developing female reproductive structures (hereafter referred to as LFY_f) (Mouradov et al. 1998). The "leaf" paralogue is expressed in male reproductive structures of gymnosperms (hereafter LFY_m). Frohlich and Meyerowitz (1997) speculated that these two gene families show such specialization because gymnosperms have had separate male and female reproductive structures since the Devonian. The *LFY* duplication in gymnosperms may have accompanied or perhaps facilitated the specialization of separate male and female reproductive structures (Fig. 1.16B).

Angiosperms have only one copy of *LFY*, which is most closely related to LFY_m, having apparently lost the female-specifying "needle" paralogue (LFY_f). These data prompted Frohlich and Parker (2000) to propose the mostly male hypothesis of floral origins (Fig. 1.16). Their hypothesis offers one developmental genetic mechanism by which a plant with separate male and female reproductive structures, a feature of all extant gymnosperms, could produce a bisexual structure such as a flower. Frohlich and Parker (2000) suggested that developmental control of floral organization derives more from systems operating in the male reproductive structures of the gymnosperm ancestor of angiosperms than from the female reproductive structures—hence, "mostly male." In the mostly male hypothesis, ovules are considered ectopic in origin on male reproductive structures (i.e., stamens) in early flowers (Fig. 1.16). An ectopic origin of angiosperm ovule position was earlier hypothesized by Meyen (1988).

Researchers have challenged the mostly male hypothesis on the basis of additional studies of *FLO/LFY*. Shindo et al. (2001) demonstrated that, although the *FLO/LFY* gene *GpLFY* from *Gnetum* is in the "leaf" clade (LFY_m) of Frohlich and Parker (2000), *GpLFY* is expressed in female strobili of *Gnetum*. This does not agree with the Frohlich and Parker assumption that only gene members of the "needle" clade (i.e., LFY_f) were expressed in gymnosperm female structures. These observations prompted Shindo et al. (2001) to conclude that the mostly male theory is not plausible.

A scenario somewhat similar to the mostly male hypothesis was proposed by Theissen et al. (2002; Theissen and Melzer 2007), termed "out of male." Following the "out-of-male" hypothesis, the bisexual flower of angiosperms originated from a male gymnosperm cone; reduction of the expression of B-class genes in the upper region of the male cone then resulted in the development of female, rather than male, reproductive units in this upper portion of the structure that became the flower.

Albert et al. (2002) provided another alternative to the mostly male hypothesis. Their model assumes that pleiotropic interactions between LFY_m and LFY_f were critical for stabilizing the retention of these two genes in gymnosperms and suggests that disruption of this delicate balance between the two *LFY* genes occurred in an ancestor of modern angiosperms. This ancestral taxon might have had unisexual flowers together on the same plant, or might have been loosely bisexual. LFY_f would then have been lost through selection for an integrated bisexual reproductive axis.

2 The Age and Diversity of Early Angiosperms

Integration of the Fossil Record and Molecular Dates

INTRODUCTION

Sedimentary rocks commonly contain fossil remains of plants and animals preserved at the time those rocks were deposited. These fossils provide a record of past biota that give insights into the morphology and anatomy of long-extinct organisms as well as representatives of plant and animal groups that are still living today. Fossil plant remains include various parts of the organism, including leaves, wood, flowers, fruits, seeds, and pollen. They may be preserved simply as imprints or molds in the sediment, retaining details of surface morphology and vein patterns, or with organic material remaining, which can provide details of tissues and anatomy. The identification and systematic placement of fossil plants rely on detailed morphological comparisons with other fossil and modern plants.

Because of the tendency of plants to disaggregate, only rarely is a "whole plant" preserved; this limits the number of morphological characters available for study to those preserved in the organs that happen to be recovered from the fossil record. When rare specimens that bear both leaves and reproductive organs are found, the extinct plant can be reconstructed more completely, providing a greater number of characters useful in placing the fossil in relation to extant species. In the absence of physically connected organs, plants may be hypothetically "reconstructed" from isolated fossil organs based on comparative work with extant relatives and co-occurrence data, but often the identification of early records of particular clades relies on isolated organs (Friis et al. 2011).

Among the more commonly preserved isolated organs, pollen is particularly informative because it can be transported long distances (especially those kinds adapted for wind rather than insect, pollination) and can record the presence of plants that might not be represented in the megafossil record. Most pollen grains have a resistant outer coating of sporopollenin that is highly resistant to decay and can therefore be preserved for millions of years. Palynomorphs are recovered by disaggregating and/or dissolving sediments, such as siltstones, claystones, and coals, in acid treatments and/or by floating the organic residues in heavy liquids. Depending on the extent of natural oxidation during fossilization, the residues may require additional maceration steps, comparable to those used in studying modern pollen grains (Traverse 1988). Fossil pollen grains can be studied and compared with modern grains by light and scanning and transmission electron microscopy. Use of both light and electron microscopy provides the most information and is advisable for systematic investigations (Ferguson et al. 2007), but most palynological studies routinely use light microscopy exclusively, as it can provide quicker results of economic value (e.g., recognition of potential oil-bearing strata and dating of sediments by certain indicator grains).

Leaves are commonly preserved as fossils, but vary in the systematic resolution they can provide. Leaves of some families (e.g., Platanaceae and Altingiaceae) are readily recognized and useful in tracing their fossil history, but others, such as those with entire margins and pinnate venation, occur in many unrelated families and are of limited systematic value unless unique features of higher-order venation and/or epidermal characters are preserved. Epidermal characters, such as the arrangement of stomata, the configuration of cells surrounding the stomata, and trichomes, are commonly retained in leaf cuticles, which are often preserved in leaf compression fossils (Dilcher 1974). Paleobotanists specializing in the investigation of extant and fossil angiosperm leaves have adopted a standard terminology useful for the diagnosis and description of leaves and leaflets, emphasizing patterns of venation and the kinds of marginal lobes, teeth, and glands (Ellis et al. 2009). Physiognomic features of the leaves in a fossil flora, such as leaf sizes, the proportions of serrate- vs. entire-margined leaves, and vein density, are useful for interpretations of paleoclimatic variables such as mean annual temperature and precipitation.

When permineralized (naturally infiltrated, e.g., with silica, calcium carbonate, or pyrite) or charcoalified, stems and other woody tissues can be preserved in excellent cellular detail. Stem anatomical characters (e.g., sizes and distribution patterns of vessels, fibers, axial parenchyma and rays in the secondary xylem) permit confident recognition of some families and genera of angiosperms in the fossil record. As with pollen and leaves, there are many instances of convergent patterns of anatomy and morphology among angiosperms. Hence, considerable care and broad comparative work must be conducted to justify assignment to particular clades. Fagales, Sapindales, and Malvales are examples of clades with distinctive wood anatomy readily recognized in the fossil record. Because of the commercial importance of wood identification in application to the timber industry, a standardized terminology has been developed for the distribution and nature of growth rings, vessels, fibers, and axial and ray parenchyma. These characters, typically observed in transverse, tangential, and radial sections by transmitted light microscopy, reveal suites of traits often unique to particular clades.

The age of fossils is determined by dating the rocks in which they are preserved (Fig. 2.1). Relative dating relies on the simple principle of superposition: those fossils at the bottom of a sequence of rock will be the oldest; those occurring higher in the sequence will have been laid down after the lower layers. Geologists assign geologic formations of the Earth to particular stratigraphic stages based on correlation from one place to another. As dating methods and correlation techniques are refined, the consensus on numerical age assignments for the boundaries between successive periods has changed. Currently accepted million-year age boundaries (Gradstein et al. 2012) for successive geologic periods are indicated in Fig 2.1.

The assignment of rocks to a particular period (e.g., Cretaceous) and particular stages within the period (e.g., Maastrichtian) is typically based on correlation to the type areas where those strata were originally recognized, often based on fossils (biostratigraphy). Certain marine fossils, such as ammonites, nannoplankton, and dinoflagellates, are particularly useful in these correlations because they were

Era	Period	Epoch
CENOZOIC (66)	Quaternary (2.6)	Holocene (0.78) Pleistocene (2.6)
	Tertiary (66)	Pliocene (5.3) Miocene (23) Oligocene (34) Eocene (56) Paleocene (66)
MESOZOIC (252)	Cretaceous (145) Jurassic (201) Triassic (252)	
PALEOZOIC (541)	Permian (299) Carboniferous Pennsylvanian (323) Mississippian (359) Devonian (419) Silurian (444) Ordovician (485) Cambrian (541)	
PRECAMBRIAN (4560)		

Period	Epoch/Stage	Age Mya
Cretaceous	Maastrichtian	71.1
	Campanian	83.6
	Santonian	86.3
	Coniacian	89.8
	Turonian	93.9
	Cenomanian	100.5
	Albian	113.0
	Aptian	126.3
	Barremian	130.8
	Hauterivian	133.9
	Valanginian	139.4
	Berriasian	145.0

Figure 2.1. Overview of the geologic time scale, with details for the Cretaceous period. Numbers in parentheses indicate the age in millions of years at which the stage or period began (modified from Harlan et al. 1989; with age assignments from Gradstein et al. 2012).

widespread and evolved quickly, providing a useful "stratigraphic clock." Among terrestrial fossils, mammals are useful in relative dating. Plants, particularly pollen and spores, have a long history of application in biostratigraphy, but to evaluate the evolution and timing of plant radiations, we should use dates obtained via independent criteria (e.g., animals, radiometric dating, and/or paleomagnetic dating).

Radiometric dating is the primary basis for modern numerical age assignments. The familiar carbon-14 method is not useful for the study of fossil angiosperms because the half-life is so short that the decay products become immeasurably small after about 50,000 years. Other radiometric methods, such as argon-argon, uranium-lead, and rubidium-strontium, use elements with longer half-lives and are suitable for dating Cretaceous and Cenozoic rocks. Generally, radiometric techniques apply to igneous rocks and can be used to infer the time that lava cooled and the crystals within it formed. However, the fossils themselves are usually limited to sedimentary rather than igneous rocks. Therefore, the geologic relationships of the dated igneous rocks and the fossil-bearing sedimentary rocks must be assessed. Ideally, the sedimentary rocks in question will be sandwiched between successive layers of dated volcanic rock, providing maximum and minimum ages. When this is not the case, biostratigraphic correlation is used to assess age based on sediments from other regions having better age control.

BIOGEOGRAPHY

The paleobotanical record contains important geographic information on former distribution patterns. As in the case of *Ginkgo*, now found only in central China but widely distributed around the world in the fossil record (Crane 2013), there are numerous examples of angiosperms for which modern distributions are relictual. The hickories (*Carya*), which have no species native to Europe today, were diverse there 20 million years ago, with several species recognizable from fossil nuts in the Tertiary of Germany and adjacent countries (Manchester 1987). *Symplocos* (Ericales) exhibits the same pattern (Mai and Martinetto 2006). Many angiosperm taxa are likewise now endemic to a particular region but were once widespread. Regional extirpations of formerly wide-ranging taxa have resulted in disjunct and endemic distributions. For example, many angiosperm genera now endemic to eastern Asia (e.g., *Cercidiphyllum*, *Cyclocarya*, *Platycarya*, *Rehderodendron*, and *Tetracenton*) have well-documented fossil records in North America and/or Europe (Manchester et al. 2009). The broader geographic ranges documented for many woody taxa in the Northern Hemisphere reflect the presence of former connections between northern continents during the late Cretaceous and early Cenozoic, particularly the North Atlantic Land Bridge linking North America and Europe via Greenland and the Bering Land Bridge linking North America and Asia during the late Cretaceous and early Tertiary (Tiffney 1985; Tiffney and Manchester 2001).

Former land connections in the Southern Hemisphere also facilitated angiosperm dispersal. The tectonic separation of Africa and South America occurred early in angiosperm history, apparently predating most modern families. However, the spread of angiosperms between Australia and South America via Antarctica was apparently possible until the early Oligocene, when Antarctica and South America separated. Thus, *Eucalyptus* (Myrtaceae) and *Gymnostoma* (Casuarinaceae), now native to Australasia, have well-documented fossils in the Eocene of South America (Gandolfo et al. 2011; Zamaloa et al. 2006). Modern patterns of disjunction and endemism of particular clades provide a single window (that of the present day) on plant distributions. In many cases, we can detect earlier patterns through observations of the fossil record.

In some instances fossils help document morphological and biogeographic diversity within extant clades. Modern Platanaceae contain the single living genus *Platanus* with simple unlobed (subg. *Castaneophyllum* Leroy) and palmately lobed (subg. *Platanus*) leaves (Nixon and Poole 2003; Feng et al. 2005). Although simple-leaved Platanaceae have an excellent fossil record extending back to the early Cretaceous with a diversity of palmately lobed forms (see Fig. 8.4), several species with the extinct trait of compound leaves were present in the family during the late Cretaceous and early Tertiary—for example, *Platanites* (Fig. 8.4) and *Dewalquea*.

THE AGE OF THE ANGIOSPERMS: FOSSIL EVIDENCE

Darwin referred to "the origin and early diversification of angiosperms" as "an abominable mystery" in a well-known quotation from a letter to the botanist J. D. Hooker (July 22, 1879). This statement reflects Darwin's frustration with the fossil record of flowering plants as a key to their origin; based on the record at that time, angiosperms appeared to have originated suddenly and then radiated rapidly during the mid-Cretaceous (for a review, see Friedman 2009). This view was due partly to the mis-assignment of many Cretaceous angiosperm leaves to modern genera without critical comparative work and to the lack of reproductive structures for more secure systematic assessment. Although the origin of the angiosperms remains a mystery, considerable progress in understanding this longstanding problem has been made in the past 15–20 years due to combined contributions from paleobotany, molecular systematics, genomics, and developmental genetics (see Frohlich and Chase 2007; Soltis et al. 2008a,b; Chapters 1, 3, 14). Here we review the age of the angiosperms based on both fossil evidence and molecular clock estimates.

In discussing the age of the angiosperms, it is important to review basic terms and concepts. As has been well reviewed (e.g., Doyle and Donoghue 1993, 2012), the *crown* group is defined as the most recent common ancestor of a clade of living taxa (Fig. 2.2). Extinct taxa descended from the most recent common ancestor of living species are still considered part of a crown group. Molecular and morphological phylogenetic analyses of extant taxa are therefore focused on the angiosperm crown group. The earliest diverging angiosperms (e.g., Amborellaceae, Nymphaeales) are covered in Chapters 3 and 4. In this chapter, however, we are interested in pivotal now-extinct stem and crown group angiosperms (Fig. 2.2). Molecular and morphological phylogenetic analyses of extant taxa are therefore focused on the angiosperm crown group, and the placement of some fossils, whether best placed as stem relative or as a member of the crown group, can be controversial due to missing characters that might be required to confirm the position. However, these interpretations are fundamental to the age-calibration of particular nodes in the phylogeny. In Chapter 1, we considered the closest living and extinct relatives of the angiosperms.

The earliest evidence of angiosperms comes from the pollen fossil record (see Friis et al. 2011). The oldest unambiguous angiosperm fossils are pollen grains (~131.8 mya) from the Hauterivian stage of the Early Cretaceous (Hughes 1994) (Fig. 2.1). These fossil pollen grains have a reticulate-columellate wall structure also found in extant angiosperms. Older pollen grains from the Valanginian are less secure in being of angiosperm origin (Brenner 1996). From the Late Barremian and into the Albian, angiosperm pollen diversity increases extensively. The oldest flowers, fruits, and seeds of angiosperms are from the Barremian or Aptian (130–115 mya) (Doyle 1992; Hughes 1994; Brenner 1996; Friis et al. 1999, 2011). Many molecular clock estimates (see below) suggest that angiosperms arose before the Cretaceous, with many dates inferring a late Jurassic, or even Triassic (S. A. Smith et al. 2010), origin. However, no unambiguous fossil evidence supports a pre-Cretaceous origin of flowering plants. Reports of older angiosperm fossils (Sun et al. 1998; Wang 2010) have been considered problematic, due to problems with the age assessment of the strata bearing the fossils or with the interpretation that the fossils represent angiosperms.

Sun et al. (1998) proposed that angiosperms originated in the late Jurassic based on the discovery of *Archaefructus liaoningensis* (Fig. 2.3) (first estimated to be 144 mya). However, the age of *Archaefructus* was revised and given a younger estimate of 125 mya, indicating a Barremian–Aptian age and placing it in the Early Cretaceous (Swisher et al. 1999; Sun et al. 2002). *Archaefructus* is an herbaceous plant, and its phylogenetic placement has been contentious. Plants may have been aquatic as suggested by highly dissected leaves, flowers appear to lack a perianth, and the arrangement of the carpels is uncertain. Because the phylogenetic placement of *Archaefructus* is uncertain, it is unclear whether flowers of *Archaefructus* represent an ancestral state for angiosperms or are modifications associated with a submerged aquatic habitat. The latter possibility must be strongly considered, given the major modifications exhibited by extant submerged aquatics (e.g., Hydrostachyaceae); the relationships of these families could be ascertained only with DNA, and even then with some difficulty (D. Soltis et al. 2000, 2011; Burleigh et al. 2009).

In a phylogenetic analysis involving extant angiosperms, using DNA data plus a small number of morphological characters, Sun et al. (2002) placed *Archaefructus* as sister

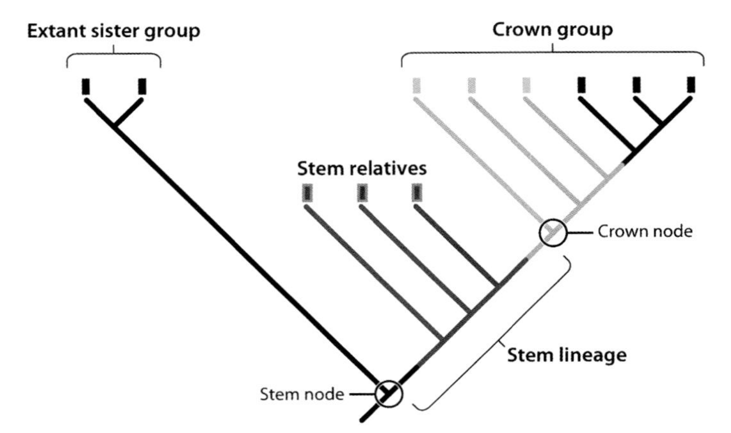

Figure 2.2. A summary of the basic concepts and ideas involved in relating fossil taxa and phylogenetic trees of living taxa. Based on Doyle (2012).

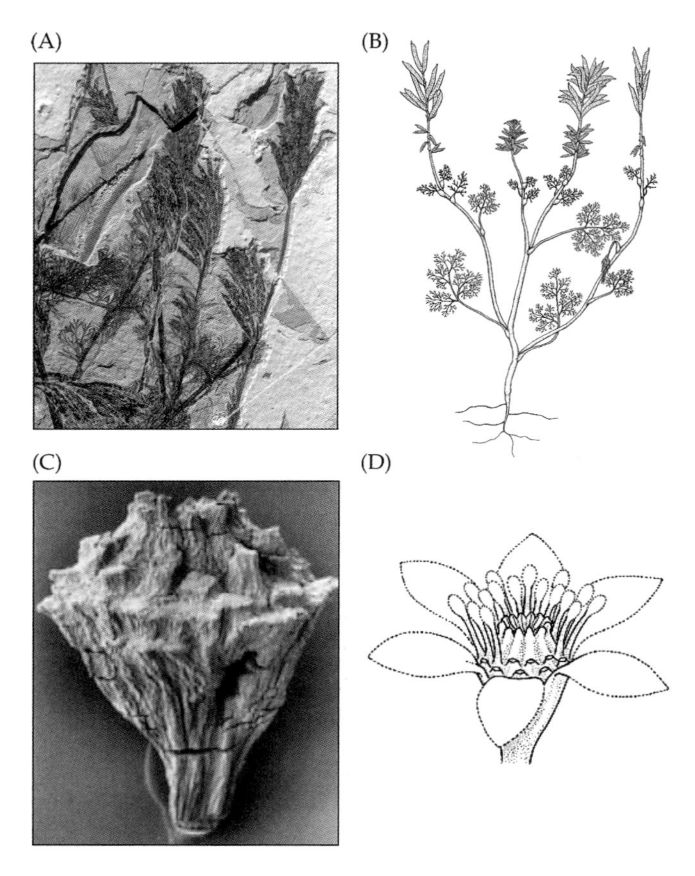

Figure 2.3. Fossils of early angiosperms. A. *Archaefructus sinensis*, photograph of reproductive axis (courtesy of D. Dilcher). B. Reconstruction of *Archaefructus sinensis* (courtesy of D. Dilcher; diagram by K. Simons and D. Dilcher). C. Lateral view of fossil water lily flower (from Friis et al. 2001; reproduced from Soltis et al. 2005b). D. Reconstruction of fossil water lily flower (from Friis et al. 2001; courtesy of P. von Knorring and E. M. Friis; reproduced from Soltis et al. 2005b).

to extant flowering plants, raising the possibility that the first angiosperms were aquatic. However, the original interpretation of the reproductive axes of these fossils has been challenged, and the placement of *Archaefructus* as sister to all living angiosperms hotly debated. The reproductive axis of this plant was originally interpreted as an elongate floral axis. However, it has been reinterpreted to represent an inflorescence axis with simple, unisexual flowers (Friis et al. 2003a). The absence of subtending bracts makes it difficult to confirm this, but features such as paired stamens makes this interpretation very likely. Friis et al. (2003a) concluded that *Archaefructus* was possibly related to Nymphaeales (water lilies), with its unusual floral features reflecting its aquatic habit. More recent analyses using DNA and morphological characters also place *Archaefructus* within Nymphaeales (Endress and Doyle 2009; Doyle 2012).

Nonetheless, despite the problematic placement of *Archaefructus* (Fig. 2.3), it represents one the most complete early angiosperm fossils (Sun et al. 2002). Two other species of *Archaefructus* have now been discovered. In addition to *A. liaoningensis* (Sun et al. 1998), *A. sinsenis* (Sun et al. 2002) and *A. eoflora* (Ji et al. 2004) have also been described from the early Cretaceous (Early Aptian to early Late Aptian).

Importantly, extinct members of several extant basal angiosperm clades (see Chapters 3 and Chapter 4), including Amborellaceae, Nymphaeaceae, Austrobaileyaceae, and Schisandraceae, were present in the Early Cretaceous. Hence, there is congruence between the general framework of angiosperm phylogeny based on analysis of extant species and the early fossil record. There is, in fact, some evidence for more diversity in some of these basal angiosperm groups (e.g., Amborellaceae, Nymphaeaceae, and Austrobaileyales) in the fossil record than is exhibited by these same lineages today (Friis et al. 2000, 2001, 2011). Fossil leaves identified as Illiciaceae (now in Schisandraceae) and Austrobaileyaceae have been reported from the Early Cretaceous (latest Barremian or earliest Aptian; Upchurch 1984; Friis et al. 2011). Stamens similar to those of *Amborella* and *Illicium* have also been recovered from the Early Cretaceous (Friis et al. 2000, 2011). Pollen of Hauterivian age similar to that of *Amborella* has been identified (Doyle 2001). Still another fossil consisting of densely aggregated, spirally arranged carpels shows similarity to Schisandraceae (Friis et al. 2000, 2011).

The discovery of a fossil flower of Nymphaeaceae (Fig. 2.3) from the Early Cretaceous (125–115 mya; Friis et al. 2001) adds to the overall correspondence between recent molecular phylogenetic results (see Chapters 3 and 4) and the paleobotanical record. Early Cretaceous floras also included fossils that are attributable to Chloranthaceae, including flowers that are similar to *Hedyosmum* and stamens similar to those of *Ascarina* (Friis et al. 1986, 1997, 1999; Eklund 1999). The early appearance of Chloranthaceae in the fossil record also agrees with molecular phylogenetic results in that Chloranthaceae diverge just a few nodes after Amborellaceae, Nymphaeaceae, and Austrobaileyales in DNA-based topologies (e.g., Soltis et al. 2008b, 2011).

The monocots have been considered a possible exception to the rough correspondence between early fossils and DNA-based topologies for extant taxa. That is, until recently, the Cretaceous fossil record of monocots had been considered poor, with some of the earliest monocot fossils not appearing until the Late Cretaceous (Gandolfo et al. 1998c, 2000, 2004; Stockey 2006; Friis et al. 2011). However, recent fossil discoveries of monocots from the Early Cretaceous now suggest that monocots are older and were more diverse than earlier appreciated (Friis et al. 2011; Chapter 7). For example, fossils of Araceae (a basal monocot lineage) date to the Early Cretaceous (Hesse and Zetter 2007). Hence, the rough correspondence between fossils and DNA-based topologies for extant taxa extends to the monocots as well. In addition, these Early Cretaceous monocot fossils agree with estimated ages of monocots, including an Early Cretaceous origin, based on molecular dating (K. Bremer 2000; Bell et al. 2010; see Chapter 7).

We stress that although some fossils show similarities to extant basal lineages such as Amborellaceae, Nymphaeaceae, and Schisandraceae, few of these early fossils can be unambiguously assigned to any extant group. Significant extinction also characterizes early angiosperm lineages (Friis et al. 2011). These observations point to the difficulty of using extant basal lineages to infer characteristics of the earliest angiosperms. Although it is tempting to conclude from these fossil data that the earliest angiosperms were morphologically similar to extant Amborellaceae, Nymphaeaceae, and Austrobaileyales, caution must be exercised because many early angiosperm fossils cannot be unambiguously associated closely with any extant taxa. These difficulties are sometimes the result of a poor understanding of the fossil but in many cases arise because the character combinations observed in the fossils differ from those of extant angiosperms. Thus, some early angiosperm fossils exhibit a mosaic of features associated with multiple modern groups. Furthermore, very few fossils have actually been included in phylogenetic analyses with living taxa; hence, the placements of few fossils have been tested rigorously.

Examination of fossil flowers has provided important generalizations about the structural organization and diversity of early flowers (Friis et al. 2000, 2011). Many early fossil flowers and fruits are preserved as charcoalified materials (Friis et al. 2010, 2011). Flowers may be exceptionally well preserved by this mechanism; they are three-dimensional and may still retain all floral organs. Numerous charcoalified flowers have been reported from the Early Cretaceous, and these flowers are small (Fig. 2.4). Even allowing for some shrinkage during fossilization, these early fossil flowers are still small. Larger flowers, such as the Late Aptian *Archaeanthus* (Magnoliaceae) and "Rose Creek flower" (Chapter 10), were preserved in a different mode, as impression specimens rather than charcoalified. The methods for recovery of charcoalified flowers favor small flowers, whereas larger flowers would become fragmentary and difficult to recover by sieving. Thus, divergent perspectives on the size of early angiosperm fossils may result from variation in preservation and recovery of large versus small flowers. Regardless, complete reconstructions of floral organization are possible for fewer than 10% of the angiosperm taxa present in Early Cretaceous floras, which suggests caution is needed and indicates that, for some characters, it is premature to offer many generalizations. Nonetheless, it is clear that many angiosperm reproductive structures from the Early Cretaceous were small. For example, none of the fossil seeds or fruits in the famous Famalicão, Portugal, assemblage exceeds 3.9 mm in length (Eriksson et al. 2000; Friis et al. 2011).

The small size of many Early Cretaceous flowers, and indeed some whole plants (Jud and Hickey 2013), generally agrees with the floral characteristics of the basalmost extant angiosperm lineages, Amborellaceae, Austrobaileyales, and some Nymphaeales (Cabombaceae) in having small or moderate-sized flowers. Furthermore, if larger flowers were present in the Early Cretaceous, pieces would be expected, but these have not been detected. The fossil record therefore argues against the older view that large, magnolia-like flowers represent the ancestral flower (Cronquist 1968; Takhtajan 1969).

The fossil record is more ambiguous on the organization (spiral vs. whorled) of floral parts in Early Cretaceous flowers because this feature can be difficult to reconstruct even in well-preserved fossils (Friis et al. 2000). However, several other characteristics of Early Cretaceous flowers are clear. Both unisexual and bisexual flowers are present in Early Cretaceous floras (Friis et al. 2011). In addition, flowers with few floral parts predominate; flowers with many parts are uncommon. Differentiation into sepals and petals has not been observed in Early Cretaceous flowers. Stamens in Early Cretaceous fossils typically exhibit poor differentiation between filament and anther, and early fossils also appear to lack a well-differentiated style. Both morphological features agree with characteristics seen in the flowers of living basalmost angiosperms (ANA grade: *Amborella*, Nymphaeales, Austrobaileyales) and magnoliids (see Chapter 4). It is often not possible to determine whether fossil carpels were ascidiate or plicate because ontogenetic data are usually needed to make this determination. However, it appears that both types were present in Early Cretaceous floras.

Most early fossil fruits have a single seed, but a few have two to as many as eight seeds. These data suggest that the early carpel possessed a small number of ovules, again in agreement with the general framework of angiosperm phylogeny in that living basal lineages generally have few ovules per carpel (Chapter 4). Friis et al. (2000) inferred that both insect-pollinated and wind-pollinated taxa were present in Early Cretaceous floras and that pollen was probably the major reward for visiting insects. Ethereal oil cells, typical of some extant basal lineages and magnoliids (Chapter 4), have also been documented in early fossil flowers (e.g., *Archaeanthus*).

Crepet (1996) concluded that there was a "dramatic modernization" of the angiosperms by the Turonian (~90 mya), which appears to coincide with a much-improved fossil record of insects (Grimaldi 1999). Our understanding of

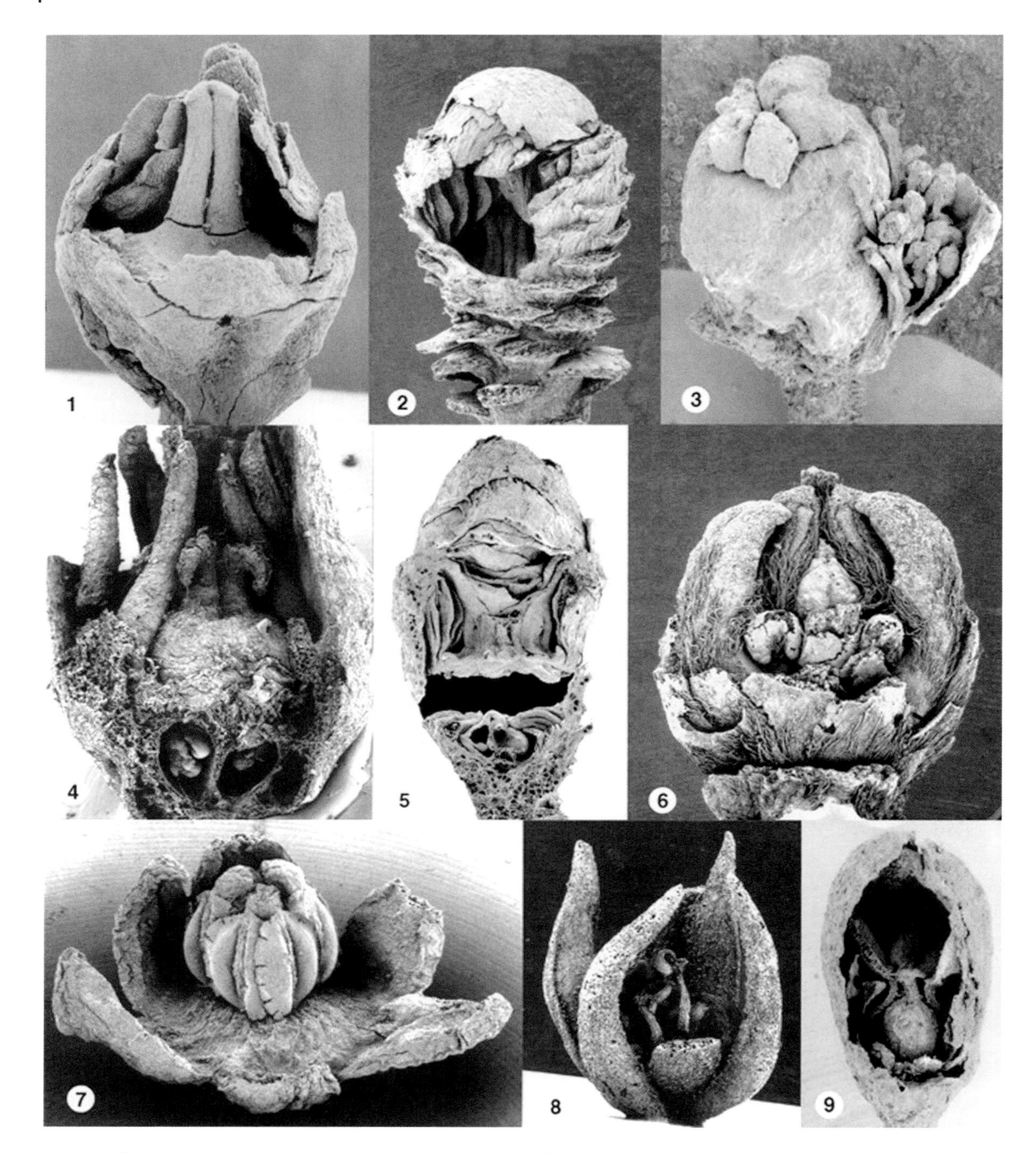

Figure 2.4. Charcoalified fossil flowers from Sayreville, Raritan Formation (Turonian, Upper Cretaceous), New Jersey, USA. 1. *Divisestylus brevistamineus* Hermsen, Gandolfo, Nixon and Crepet (Iteaceae), 80×. 2. *Detrusandra mystagoga* Crepet and Nixon, a magnoliid flower with a unique combination of characters relative to extant magnoliids, 20×. 3. *Paleoclusia chevalieri* Crepet and Nixon, (Clusiaceae), 64×. 4. Pentamerous flower with rosid affinities, 51×. 5. *Microvictoria svitkoana* Gandolfo, Nixon and Crepet (Nymphaeaceae), 33×. 6. *Perseanthus crossmanensis* Herendeen, Nixon and Crepet (Lauraceae), 25×. 7. *Mabelia archaia* Gandolfo, Nixon and Crepet, the oldest monocot flower in the fossil record (Triuridaceae), 40×. 8. Staminate Fagales-like flower, 39×. 9. *Dressiantha bicarpellata* Gandolfo, Nixon and Crepet, attributed to Brassicales, 60×. Photographs and modified legend from Crepet et al. (2004).

the timing of angiosperm diversification versus insect diversification (and the timing of angiosperm–pollinator relationships) has improved (see Misof et al. 2014); the pattern and timing of early angiosperm diversification is consistent with a similar pattern of radiation in anthophilous insects (Crepet 2000). Furthermore, there is "a compelling similarity between the rate of floral innovation per million years and the rate of angiosperm diversification during the Cenomanian–Turonian interval, which coincided with the first occurrences of many derived insect pollinators"

(Crepet 2000). Significantly, a recent phylogenomic study of insects indicates that the major diversification of holometabolous insects, those insects characterized by complete metamorphosis [e.g., Lepidoptera (butterflies and moths), Coleoptera (beetles), Hymenoptera (ants, wasps, and bees), Diptera (true flies)], occurred in the Early Cretaceous (Misof et al. 2014), a timeframe that corresponds to the rise and early diversification of the angiosperms.

DIVERGENCE TIMES AND THE AGE OF THE ANGIOSPERMS

Molecular estimates of the divergence time between angiosperms and gymnosperms have varied dramatically, from 346 to more than 900 mya (see P. Soltis et al. 2002; D. Soltis 2005b; S. A. Smith et al. 2010). Clearly, the latter age estimates, which are much older than the oldest fossils of land plants, are too old. Most estimates for the age of seed plants range from 340 to 400 mya, a period that corresponds to the Devonian and Early Carboniferous. These estimates also agree with the fossil record and the earliest records of gymnosperm seeds. However, the seeds and pollen of these earliest gymnosperms (now extinct) do not correspond to those of crown group seed plants, strengthening the view that the extant seed plant clade is not as old as the oldest seed plants.

Molecular results provide strong evidence for the monophyly of extant gymnosperms and for a sister-group relationship of angiosperms and extant gymnosperms (see Chapter 1). By the Late Carboniferous (~290–320 mya) or Permian, cycads, conifers, and Ginkgoales were represented in the fossil record. The age of some of these older fossils representing extant gymnosperm lineages suggested to some that the angiosperms may be much older than the fossil record now indicates. Some researchers proposed, in fact, that angiosperms must be as old as their gymnosperm sister group (see Axsmith et al. 1998; Doyle 1998b).

However, the fact that extant gymnosperms are monophyletic and may extend to the Carboniferous or Permian does not imply that the angiosperm clade similarly dates to the same time. Several now-extinct lineages of seed plants occur along the branch to the angiosperms, with Caytoniales, Bennettitales, glossopterids, and *Pentoxylon* consistently appearing on the branch to the angiosperms (Chapter 1). Therefore, the argument for a Carboniferous origin of the angiosperm stem lineage is unrealistic. This represents a misinterpretation of DNA-based support for the monophyly of extant gymnosperms as concomitantly providing support for the monophyly of all gymnosperms. Clearly, all gymnosperms (living and extinct) do not constitute a monophyletic group, but living gymnosperms (acrogymnosperms or *Acrogymnospermae* of Cantino et al. 2007) are monophyletic in most analyses (Chapter 1).

Given the numerous diverse angiosperm fossils reported from as early as 115–125 mya, the earliest angiosperms were almost certainly somewhat older than the oldest estimate based on the fossil record of approximately 132 mya. We review below attempts to estimate the age of the angiosperms using molecular data and provide a useful compendium of estimates of ages for angiosperm subclades (orders and families) taken from Bell et al. (2010) and Magallón et al. (2015).

Molecular dating approaches have seen much progress over the past 15 years and also have garnered controversy. Whereas the fossil record may provide underestimates of clade ages, molecular methods tend to overestimate ages (e.g., Rodríguez-Trelles et al. 2002; Bell et al. 2010). As a result, refinements of dating approaches have been developed in an effort to compensate for this bias. As reviewed below, attempts to estimate the timing of angiosperm origins have employed a variety of molecular datasets and estimation procedures.

We also review the many caveats and sources of error and bias that can affect both fossil-based and molecular estimates of divergence times (Sanderson and Doyle 2001; P. Soltis et al. 2002; Sanderson et al. 2004; Bell and Donoghue 2005a; Bell et al. 2005). The fossil record and the interpretation of that record are possible sources of error. In molecular dating studies, fossil dates should represent only minimum age calibrations for a clade because the possibility of the very first member of a lineage to have been fossilized is very slim. Similarly, molecular clock estimates may be affected by several factors that also merit consideration. An incorrect topology will yield erroneous age estimates, and the magnitude of the error in those estimates will of course depend on the extent of the error in the underlying topology (Sanderson and Doyle 2001). Inaccurate fossil calibrations represent another source of error and may also introduce bias in the resulting age estimates. The fossil calibration age itself may be erroneous, or the fossil age may be affixed to the wrong part of the tree. As shown in Figure 2.2, does a fossil calibration represent a minimum age of the crown group or the stem group? Heterogeneous rates of evolution are well known among lineages (e.g., Clegg et al. 1994; Gaut et al. 1996; Li 1997; P. Soltis et al. 2002) and are a major problem in age estimation. Incorrect assessment of gene orthology can also introduce error into inferences of both topology and age (e.g., Sanderson and Doyle 2001; Gaut et al. 1996). Inadequate taxon sampling not only is a problem in phylogeny reconstruction, but also can compound problems of age estimation. Finally, molecular data are often extremely complex (Chapter 1); estimates of divergence times may vary among genes or other data partitions (e.g., among codon positions; Sanderson and Doyle 2001; Bell and Donoghue 2005a,b).

Initial efforts to date the origin of angiosperms as well as divergences within angiosperms relied on a strict molecular clock (i.e., assuming rate constancy among nucleotide sites and among lineages). Those efforts yielded age estimates that not only vary dramatically, but also in some cases are highly inconsistent with the fossil record (see Magallón 2004; Bell et al. 2005; D. Soltis et al. 2008b; Doyle 2012). With a strict molecular clock, the age of the angiosperms has been estimated as 420–350 mya (Ramshaw et al. 1972), 354–300 mya (Martin et al. 1989, 1993; Brandl et al. 1992), and 200 mya (Wolfe et al. 1989; Laroche et al. 1995). All these dates suggest an origin of the angiosperms that predates the angiosperm fossil record (135–125 mya). In some cases the molecular clock estimates are older than the estimated ages of all seed plants (~390–350 mya), all extant seed plants (about 309 mya), and even all vascular plants (~415 mya), based on the fossil record.

Because of the problems with strict molecular clock methods, most recent efforts to date the origin of the angiosperms used improved dating methods that accommodate rate heterogeneity. These approaches have converged on more reasonable estimates between 180—140 mya, predating the dates inferred from the fossil record by only 48 to 8 myr (Sanderson and Doyle 2001; Wikström et al. 2001; P. Soltis et al. 2002; Chaw et al. 2004; Schneider et al. 2004; Sanderson et al. 2004; Bell et al. 2005, 2010; Magallón et al. 2015; but see S. A. Smith et al. 2010; for reviews see Bell et al. 2010; Soltis et al. 2005b, 2008; Magallón 2010; S. A. Smith et al. 2010; Doyle 2012).

Methods that relax the assumption of a strict molecular clock (Table 2.1) include nonparametric rate smoothing (NPRS; Sanderson 1997), local clocks (Yoder and Yang 2000; Yang and Yoder 2003), penalized likelihood (PL; Sanderson 2002), Bayesian "relaxed clock" (BRC) approaches (e.g., Thorne et al. 1998; Huelsenbeck et al. 2000; Aris-Brosou and Yang 2002; Thorne and Kishino 2002), which include the very popular program BEAST (Drummond and Rambaut 2007; Drummond et al. 2012; Rambaut and Drummond 2007; see below and Table 2.1), the Li-Tanimura method (Li and Tanimura 1987), and "PATH," a phylogenetic dating method with confidence intervals that employs mean path lengths (Britton et al. 2002). Smith and O'Meara (2012) have developed treePL, a method for estimating divergence times using PL for large phylogenetic trees—it compares well to other methods, and the approach works well with over 10,000 taxa. Investigators who employed alternatives to a strict molecular clock have generally obtained more reasonable estimates of the age of the angiosperms that are closer to the age of the oldest fossils;

Table 2.1. Commonly used methods of estimating divergence times using DNA sequence data

Method	Reference	Brief Summary
Strict molecular clock		Uses fossil constraints and rates of molecular change to estimate the time of divergence.
Nonparametric rate smoothing (NPRS)	Sanderson 1997	Uses a least squares smoothing approach that penalizes rates that change too quickly from one branch to neighboring branch. Implemented in r8s (loco.biosci.arizona.edu/r8s/).
Local Clocks	Yoder and Yang 2000 Yang and Yoder 2003	Employs maximum likelihood models of local molecular clocks within the overall phylogeny.
Penalized likelihood (PL)	Sanderson 2002	A "semi-parametric" approach that combines maximum likelihood approach with the penalty function noted above with NPRS. One can specify the relative weight of the penalty function and the ML component in which parameters are being fitted. The parametric model has a different substitution rate for each branch. Implemented in r8s (loco.biosci.arizona.edu/r8s/).
Bayesian approaches	Thorne et al. 1998 Huelsenbeck et al. 2000 Aris-Brosou and Yang 2002 Thorne and Kishino 2002 Drummond and Rambaut 2007	Accurate placement of fossil calibrations is a crucial part of dating, but entails uncertainty. Bayesian methods provide a mechanism of accommodating the degree to which fossil evidence approximates the date or time of divergence that they constrain. Commonly used programs include the Bayesian dating method implemented in Multidivtime (Thorne et al. 1998) and the very popular program BEAST (beast.bio.ed.ac.uk/Main_Page; Drummond and Rambaut 2007).
PATH	Britton et al. 2002	The branch lengths of the input tree are interpreted as (mean) numbers of substitutions. The mean path lengths (MPL) method estimates the age of a node with the mean of the distances from this node to all tips descending from it. Under the assumption of a molecular clock, standard-errors of the estimates node ages can be computed (Britton et al. 2002).

the estimates generally range from 250–125 mya (Sanderson and Doyle 2001; Wikström et al. 2001; P. Soltis et al. 2002; Chaw et al. 2004; Schneider et al. 2004; Bell et al. 2005, 2010; but see S. A. Smith et al. 2010, Magallón 2010).

In one of the first rigorous studies to estimate the age of the angiosperms, Sanderson and Doyle (2001) used *rbcL* and 18S rDNA sequences and employed an ML approach; rather than assuming equal rates across sites, they used a gamma distribution of rates to correct for site-to-site variation. They found younger estimates than strict molecular clock approaches; their estimates of the age of crown group angiosperms ranged from 68 to 281 mya, depending on data, tree, and assumptions, with most estimates ranging from 190–140 mya.

Wikström et al. (2001) used NPRS to estimate the age of the angiosperms and divergence times for major angiosperm clades. They used *rbcL*, *atpB*, and 18S rDNA sequences from 560 angiosperms—the dataset and shortest tree from D. Soltis et al. (2000). Wikström et al. (2001), using a single calibration point, fixed the split between Fagales and Cucurbitales at 84 mya based on two fossils, *Protofagacea* (Herendeen et al. 1994) and *Antiquacupula* (Sims et al. 1998) and estimated an age for the angiosperms of 179–158 mya. The estimates provided by Wikström et al. (2001) across angiosperms were of great utility to diverse investigators (e.g., molecular biologists and ecologists) and were widely used for nearly a decade. A comparable comprehensive analysis that employed newer approaches was later provided by Bell et al. (2010; below).

Sanderson (2003) used 27 proteins and PL, as noted above, another method that relaxes the assumption of a strict molecular clock. The major focus was to estimate the origin of land plants (embryophytes); however, the study also estimated the age of crown group angiosperms to be ~200 mya. Aoki et al. (2004), using sequences from MADS-box genes and two other methods, a linearized trees (Takezaki et al. 1995) estimation procedure and local clocks (Yoder and Yang 2000), estimated the age of the most recent common ancestor of all extant angiosperms as 210–140 mya, depending on the tree used and assumptions regarding rates of molecular evolution.

Chaw et al. (2004) used the Li-Tanimura unequal rate method to date the divergence of monocots from other angiosperms, such as core eudicots (see Chapter 3 for overview of angiosperm phylogeny), 61 plastid genes from 12 completely sequenced plastid genomes of land plants, and three calibration points and concluded that monocots diverged from other angiosperms 150–140 mya. Their estimate for this early split among major lineages within angiosperms is in agreement with slightly older estimates for the origin of angiosperms.

Bell et al. (2005) used a BRC method to date the origin of the angiosperms. The BRC approach allows rates to evolve among lineages along the tree and has the advantages of taking into consideration the variation in molecular evolution among genes and allowing the researcher to incorporate fossil evidence as either maximum or minimum age constraints at several nodes on the topology. Bell et al. (2005) tested for the effects of different genes, different methods, different calibrations/constraints, and different prior distributions of parameters on the resulting age estimates. A topology based on analysis of genes from all three plant genomes for 71 taxa was used as a backbone. With BRC, dates for the age of the angiosperms varied across datasets, from 144.5 to 202.6 mya depending on the dataset and the fossil constraint. Bell et al. (2005) compared these results with those obtained from PL using the same dataset and multiple fossil constraints. Estimates of the age of the angiosperms varied among genes and constraint treatments (150.20–275.71 mya) and were generally older than ages inferred using the BRC method. The results of Bell et al. (2005) indicate that very different age estimates can be obtained when different methods (198–139 mya) and different sources of data (275–122 mya) are employed; results also varied with the use of temporal constraints on the trees. Most estimates from this study, however, are between 180–140 mya, suggesting a Middle Jurassic-Early Cretaceous origin of flowering plants. These estimates predate the oldest unequivocal fossil angiosperms by about 48–8 million years.

More recent Bayesian analyses also provide a wide range of dates for the angiosperm crown group. Bell et al. (2010) used BEAST (Drummond and Rambaut 2007; Drummond et al. 2012) to estimate the phylogeny and divergence times of angiosperms simultaneously using 36 calibration points for 567 taxa. They performed two analyses: in one, they estimated rates and ages from the DNA sequences, modeling fossils as exponential priors, and in a second analysis, they set fossil priors to fit a lognormal distribution. Based on the analysis with lognormal priors, the estimated age of the angiosperms was 199–167 mya (Early to Late Jurassic), and the following age estimates for major angiosperm clades were *Mesangiospermae* (140–128 mya), *Eudicotyledoneae* (143–116), *Gunneridae* (124–111 mya), *Rosidae* (117–107 mya), and *Asteridae* (119–101 mya). With the exception of the age of the angiosperms themselves (which is older than the oldest fossils), these age estimates are generally younger than other recent molecular estimates and very close to dates inferred from the fossil record. The chronogram from Bell et al. (2010) and its accompanying table of age estimates have been useful references for researchers (Fig. 2.5).

As a resource, we provide a list of ages for clades recognized as families and orders as well as comparison with

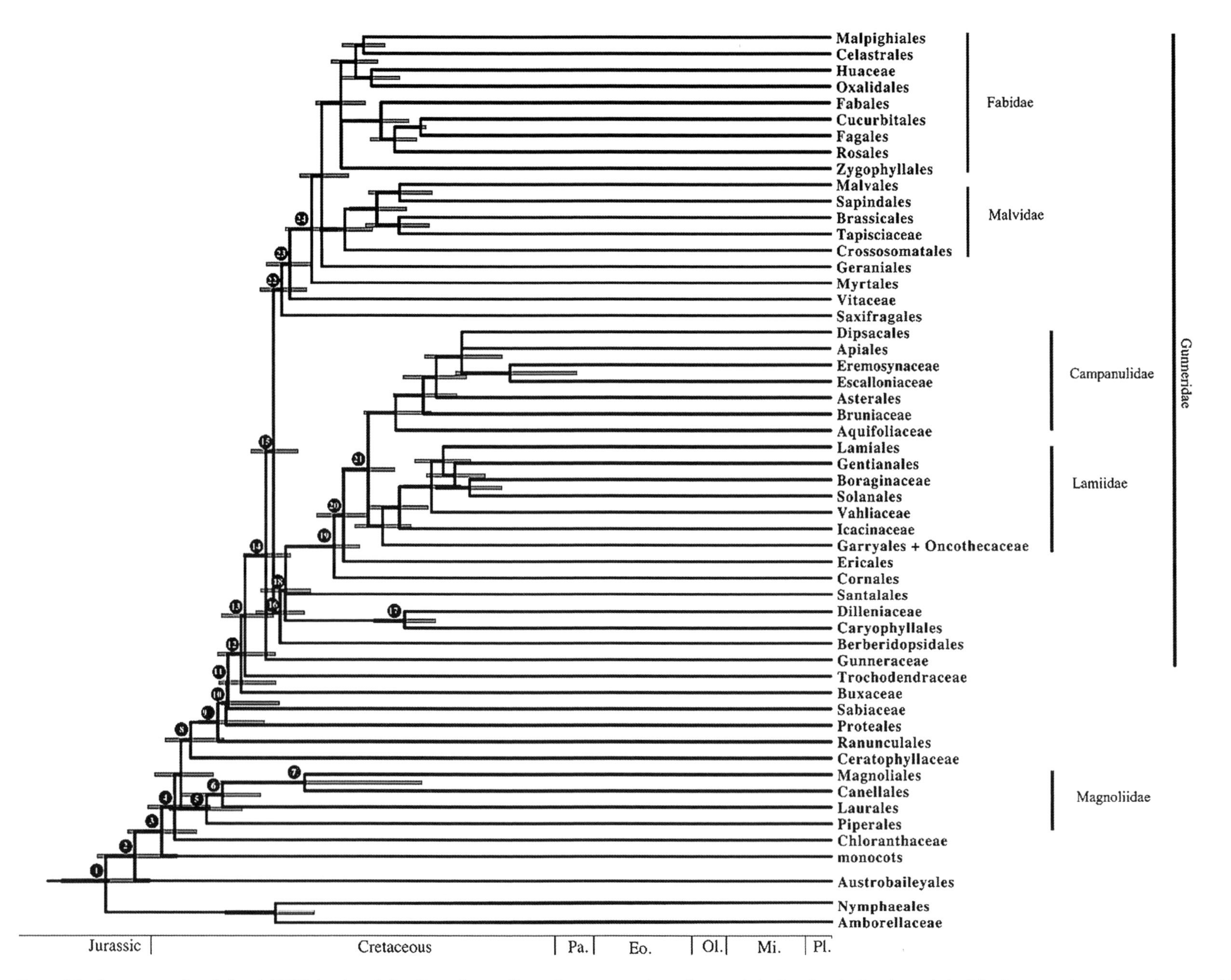

Figure 2.5. Summary tree from Bell et al. (2010) showing relationships and ages with error bars among major angiosperm clades, based on the Bayesian analysis of the three-gene data set for angiosperms of Soltis et al. (2000). Note that some relationships differ from those of more recent analyses (e.g., Soltis et al. 2011 and others), but this figure still provides a reasonable estimate of clade ages. Names for clades recognized as families and orders follow APG III (2009), and names for more inclusive clades follow Cantino et al. (2007). Node numbers of major clades correspond to age estimates (see Table 2.2). Error bars around nodes correspond to 95% highest posterior distributions of divergence times based on 36 fossils using the program BEAST. Outgroup taxa have been pruned. Eo = Eocene, Mi = Miocene, Ol = Oligocene, Pa = Paleocene, Pl = Pliocene. (From Bell et al. 2010.)

recent estimates from Magallón et al. (2015) (Supplementary Table 2.1; we also provide accompanying tree figures from Bell et al. 2010: Appendices. 2.S.1–2.S.12; available on book website at press.uchicago.edu/sites/soltis/).

S. A. Smith et al. (2010) and Magallón (2010) provided much older estimates for the age of the angiosperm crown group than do the above-mentioned studies. Estimates of S. A. Smith et al. (2010) ranged from 228–217 mya (Late Triassic), and those of Magallón (2010) ranged from 275 mya (Early Permian) to 221.5 mya (Late Triassic). Both studies used interesting approaches. Magallón (2010) used fossils to break the angiosperm stem branch (see Fig. 2.2). In phylogenetic trees, the branch to the angiosperms in phylogenetic trees is very long and consequently may hamper divergence time estimation. As reviewed in Magallón (2010), the typical approach to break a long branch is to increase taxonomic sampling, so that the added taxa may break up the long branch into a series of shorter branches. However, the species that can break the long stem branch to angiosperms are now extinct. Therefore, to break the long branch to angiosperms, Magallón (2010) simulated sequences for angiosperm stem relatives. S. A. Smith et al. (2010) did not follow the widely used approach of accepting the age of the oldest known tricolpate pollen (125 mya) as a maximum age for the eudicots; instead they dated crown eudicots as 153 mya (Late Jurassic) using molecular estimates, which may be problematic (see below).

Doyle (2012) reviewed how the estimates of S. A. Smith et al. (2010) and Magallón (2010) compare to the fossil record; we revisit that comparison. S. A. Smith et al. (2010) indicated that angiosperms arose in the Triassic and called into question the longstanding use of the age of the oldest known tricolpate pollen (125 mya) as a maximum age for the eudicot crown clade. S. A. Smith et al. (2010) dated that node as 153 mya (Late Jurassic) and also indicated that the eudicots (*Eudicotyledoneae*) began to diversify in the earliest part of the Cretaceous (the Barremian). In Magallón (2010), the eudicot crown node was given a uniform distribution bounded between 122 and 128 Ma, so the approach of breaking up the angiosperm stem considerably pushed back the estimated age of origin of the angiosperms to the Triassic or even Permian. Doyle (2012) noted that tricolpate pollen, which characterizes the eudicots, is perhaps the best fossil synapomorphy for any angiosperm subclade, as it is conspicuous and has a dense fossil history. If there were already basal eudicots in the Late Jurassic and eudicots in the Barremian, as the molecular dating estimates of S. A. Smith et al. (2010) indicate, "the temporal sequence of monosulcate, tricolpate, and tricolporate pollen and its congruence with the evolutionary series inferred from molecular phylogenies would be a mystery" (Doyle 2012, p. 314). "It is also difficult to imagine why angiosperms would remain hidden for so long if they had already diversified into clades that rose so rapidly in the mid-Cretaceous" (Doyle 2012, p. 314).

THE EARLIEST ANGIOSPERMS: HABIT AND HABITAT

Two different views have been expressed on the general habit and habitat of the earliest angiosperms: woody and terrestrial or herbaceous and aquatic (reviewed in Soltis et al. 2005b, 2008; see Chapter 6). The longstanding hypothesis that the earliest angiosperms were woody is supported by the fact that most extant basal angiosperms are woody (although the second-branching clade, Nymphaeales, are herbaceous), and all gymnosperms are woody, as are the fossil lineages that are considered most closely related to angiosperms (see Chapter 1). *Amborella*, the sister to all other living angiosperms (see Chapter 3), is woody, as are Austrobaileyales, another early-branching extant lineage. However, the strictly aquatic water lily lineage (Nymphaeales) follows *Amborella* as the subsequent sister to all other flowering plants (see Chapter 3). As a result of the placement of Nymphaeales, the habit of the first angiosperms is reconstructed as equivocal (see Chapter 3).

A recent variant on the hypothesis that the first angiosperms were woody/terrestrial employs physiological features. Anatomical and physiological features of the woody *Amborella* and Austrobaileyales are noteworthy. *Amborella* lacks vessel elements, and Austrobaileyales have vessel elements with features intermediate between tracheids and typical vessel elements (Carlquist and Schneider 2001, 2002). These anatomical characters and associated physiological features led to the hypothesis that the earliest angiosperms were understory shrubs, perhaps lacking vessel elements and having lower transpiration and stem water movement than found in most living angiosperms (Feild et al. 2000a,b, 2003a,b, 2004). These plants were found in habitats referred to as "dark and disturbed." The first angiosperms may therefore have been understory shrubs living in partially shaded habitats and depending on disturbance to open up the canopy, providing new sites for colonization.

The alternative hypothesis is that the earliest angiosperms were aquatic. An aquatic origin of angiosperms is supported by the fact that several of the earliest known fossil angiosperms were aquatic. *Archaefructus* represents one of the oldest, most complete angiosperm fossils (Sun et al. 2002; see above). On the basis of morphology, it clearly was aquatic. The phylogenetic placement of *Archaefructus* as either sister to all extant angiosperms (Sun et al.

2002) or within Nymphaeaceae (Friis et al. 2003a), plus the phylogenetic position of extant Nymphaeales (water lilies, below) as an early-diverging branch, lends support to the view that the aquatic habit arose early in angiosperm history and that perhaps the earliest angiosperms were aquatic. The discovery that Hydatellaceae are part of the water lily clade (Saarela et al. 2007) greatly increased the extant morphological diversity encompassed by the Nymphaeales and also raised questions concerning the extent of diversity in extinct members of the clade, as well as other basal angiosperm lineages. Such findings raise the possibility that Nymphaeales were once much more diverse and could have encompassed *Archaefructus* as well as other now-extinct lineages (Doyle and Endress 2010; see Chapters 3 and 4). *Archaefructus* represents well the difficulty in placing fossil lineages that are morphologically distinct with no clear synapomorphies with extant taxa. It is prudent to keep an open mind regarding the placement of *Archaefructus* and other fossils. Nonetheless, the early appearance of Nymphaeales (Friis et al. 2001), together with the discovery that some of the oldest angiosperm fossils are aquatic, reinforced the hypothesis that the earliest angiosperms may have been aquatic—that is, "wet and wild" (D. Dilcher pers. comm.; see Coiffard et al. 2007).

3 Phylogeny of Angiosperms

An Overview

INTRODUCTION

Enormous interest and effort have been invested since the 1990s in reconstructing phylogenetic relationships of the angiosperms. The importance of using a phylogenetic approach to evaluate evolutionary hypotheses has long been recognized, and many comparisons are best made in light of a firm understanding of angiosperm relationships. A robust phylogenetic underpinning for flowering plants is of value not only to systematists but also to scientists in diverse disciplines, including evolutionary biologists, physiologists, ecologists, molecular biologists, and genomicists (see "Implications for the Study of Model Organisms," below).

Here we review efforts to reconstruct angiosperm phylogeny using morphology, RNA (in early studies) and particularly DNA sequence data, as well as combined morphology and DNA datasets. Historically, some of the largest phylogenetic analyses conducted to date for any group of organisms have involved the angiosperms. This was true in the early stages of molecular phylogenetics (e.g., Chase et al. 1993; Källersjö et al. 1998; D. Soltis et al. 2000) and continues today with large supertree and supermatrix analyses (e.g., Davies et al. 2004; S. A. Smith et al. 2009, 2011). As a result, valuable lessons in data analysis have also been learned, and these have general implications that we also briefly review. In addition, we review the best estimate of phylogeny now available for the angiosperms and compare these relationships with traditional, morphology-based classifications from the late 1900s. The molecular estimates of phylogeny provided the stimulus for a revised classification of angiosperms (see Chapter 12). The broad DNA-based topologies also provide evidence for multiple episodes of rapid radiation in the flowering plants and have facilitated estimation of the ages of major diversification events.

HISTORY OF EFFORTS TO RECONSTRUCT ANGIOSPERM PHYLOGENY

MORPHOLOGY AND "MODERN" CLASSIFICATIONS

Many intuitive reconstructions of angiosperm evolution and relationships (e.g., Bessey 1915; Cronquist 1968, 1981; Takhtajan 1969, 1980, 1987, 1997; Stebbins 1974; Thorne 1976) appeared during the past century. These classifications were based on the broad plant knowledge of the investigator and his "gut feelings" as to the big picture of angiosperm relationships based on the characters he emphasized. It was not until the late 1980s that explicit analyses of angiosperm phylogeny were initiated. These pioneering investigations relied solely on evidence from morphology (e.g., Donoghue and Doyle 1989a,b; Loconte and Stevenson 1991) and set the stage for large phylogenetic analyses based on DNA sequences, as well as DNA plus morphology. These studies provided important initial insights, indicating, for example, that dicotyledonous angiosperms (traditionally referred to as dicots) were not monophyletic. Instead, the monocots were nested among clades of what were referred to as "primitive dicots." Although dicots as a whole were not monophyletic, most dicots formed a well-marked clade that was named the eudicots (Doyle

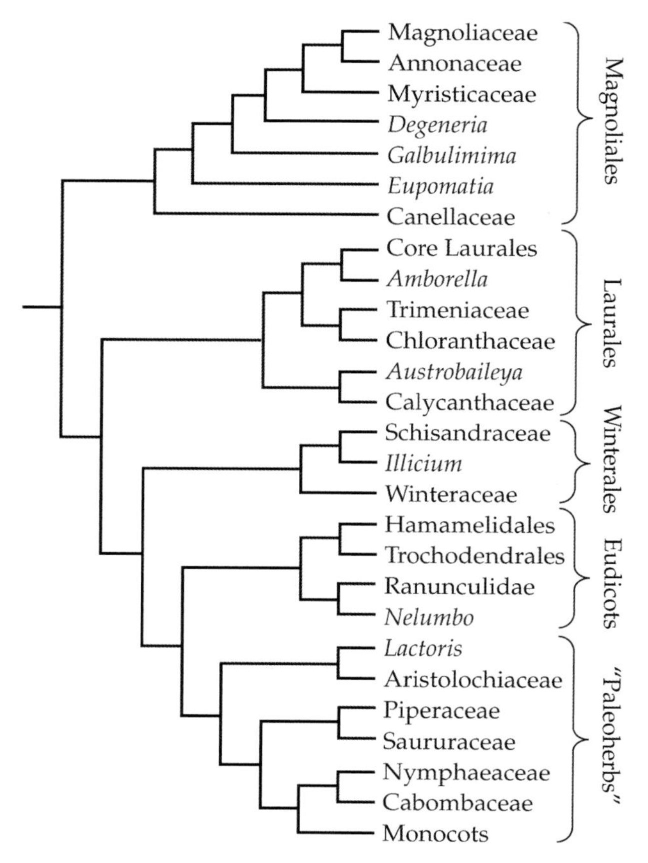

Figure 3.1. Representative most parsimonious angiosperm cladogram of Donoghue and Doyle (1989a) based on morphological data (redrawn from Donoghue and Doyle 1989a). Clade names are those of Donoghue and Doyle (1989a), but the composition does not reflect our current understanding of relationships (see text). The term "paleoherb" has been used to define different groups by different authors and is not of systematic value.

and Hotton 1991) or tricolpates (Donoghue and Doyle 1989a,b; Judd and Olmstead 2004).

Initial phylogenetic studies using morphology provided conflicting results regarding the root of the angiosperm tree, focusing attention on several possible groups. The analysis of Donoghue and Doyle (1989b), which used 62 binary characters for 20 placeholders, placed a Magnoliales clade as sister to all other angiosperms, followed by other woody magnoliids such as Laurales and a winteroid group (see also Doyle and Donoghue 1993) (Fig. 3.1). However, the composition of these clades did not agree with DNA-based circumscriptions that soon followed (see below and Chapter 5). Donoghue and Doyle (1989a) also focused attention on a group they referred to as paleoherbs that included Piperaceae, Saururaceae, Aristolochiaceae, Nymphaeales (sensu APG III 2009; APG IV 2016), and also monocots. But again, DNA results that soon followed did not support such a grouping (although Piperaceae, Saururaceae, and Aristolochiaceae are a clade).

In contrast, the phylogenetic analysis of morphological data conducted by Loconte and Stevenson (1991) placed Calycanthaceae as sister to all other angiosperms (this family was later shown to a member of Laurales; see Chapter 5). Analysis of a larger morphological dataset for basal angiosperms (108 morphological characters for 52 taxa; Doyle and Endress 2000) found trees that were more similar to those obtained with DNA sequence data. Analysis of a large non-DNA dataset, consisting of morphological, anatomical, and chemical data for 151 angiosperms (Nandi et al. 1998), recovered a topology that was similar to trees obtained with molecular datasets. For example, a grade of angiosperms was recovered that included basal lineages as well as the monocots; the eudicots formed a clade. Thus, there appears to be considerable phylogenetic signal in these non-DNA datasets that in many instances agrees with topologies derived from sequence data alone (Chase et al. 2000b; also discussed below in "Single-Gene Studies" and "Combining DNA Datasets").

SINGLE-GENE STUDIES: A HISTORICAL OVERVIEW

Initial efforts to reconstruct angiosperm phylogeny on a broad scale with sequence data actually employed RNA sequences. Hamby and Zimmer (1992) constructed a dataset of partial 18S and 26S rRNA sequences for 60 taxa. Although small by present standards, this analysis was a huge undertaking at that time. Hamby and Zimmer (1992) identified Nymphaeales as sister to all other angiosperms (*Amborella* was not sampled), showed monocots to be nested within a grade of early-branching "dicots," and recovered a eudicot clade.

The phylogenetic analysis of *rbcL* sequence data for 499 species of seed plants (Chase et al. 1993) is a landmark study for several reasons. First, it was a huge dataset for that time period and for many years remained one of the largest analyses conducted for any group of organisms. Chase et al. (1993) ushered in a new era of investigation in which numerous representatives were used to reconstruct phylogeny for large, enigmatic groups. From a historical standpoint, this study was significant in that the prevailing view at that time was that large datasets could not be analyzed phylogenetically (e.g., Graur et al. 1996). Instead, it was recommended that they be broken into smaller datasets or pruned to include only a few exemplars. However, it was evident to many angiosperm systematists that smaller matrices did not produce the same topologies obtained with more exhaustive sampling (reviewed in Chase and Albert 1998; D. Soltis et al. 1998), so they went against the prevailing view of the time and constructed increasingly larger datasets, combining genes for more taxa.

The study by Chase et al. (1993) also stretched the limits of analytical capabilities available at that time (reviewed in Chase and Albert 1998). Initial efforts to analyze the large *rbcL* data matrix actually failed because versions of the phylogenetic program PAUP (Swofford 1991) then available could not handle the number of terminals (500 sequences). Thus, coincident with the compilation of this large matrix was the development of software and search strategies that were appropriate for such large datasets. Of course, rapid developments in analytical capabilities have occurred since that time—but the Chase et al. (1993) analysis served as a stimulus to the community. As reviewed in "Lessons Learned" below, incredible advances in data analysis have continued, and at this point numerous tools and programs are available, permitting the analysis of genome-scale data (e.g., ASTRAL; Mirarab et al. 2014). Nevertheless, this movement was spurred, in part, by the large datasets assembled early on by angiosperm systematists. Finally, Chase et al. (1993) represented a massive collaboration of 42 investigators, most of whom contributed unpublished sequences to the effort because they appreciated the central importance of such an endeavor to angiosperm systematics. This spirit of collaboration among botanists in the early stages of the molecular era has continued and enabled plant systematists to rapidly achieve a tremendous understanding of angiosperm, as well as green plant, phylogeny.

This collaborative approach emerged as a model for other large-scale botanical projects, including the Deep Green, Deep Time, and Deep Gene initiatives, as well as the Grass Phylogeny Working Group (2000, 2012), the Legume Phylogeny Working Group (2013), Ericaceae Working Group (Kron et al. 2002), the Compositae Alliance (Funk et al. 2012), and multiple tree of life efforts (e.g., Angiosperm Tree of Life; Monocot Tree of Life). The impact of Chase et al. (1993) and subsequent multi-authored collaborations on the angiosperms (e.g., D. Soltis et al. 1997a,b, 2000; Savolainen et al. 2000a,b) was also much broader than angiosperm phylogeny, in that these efforts ushered in a cultural change, demonstrating that big phylogeny was also big science and providing a framework for similar efforts in diverse groups of organisms and a model for the National Science Foundation's Assembling the Tree of Life Program.

In the trees of Chase et al. (1993), *Ceratophyllum* was sister to all other angiosperms (Fig. 3.2A). However, this placement was soon shown to be an artifact (see Chapter 4). The remaining noneudicot angiosperms—several clades of traditional dicots (e.g., clades referred to then as Laurales, Magnoliales, paleoherbs) as well as the monocots—formed a clade sister to the eudicots. Within the eudicots, several large clades were recovered, including the asterid, rosid, and caryophyllid clades. Large subclades were also suggested within these major clades (e.g., rosid I, rosid II, asterid I, asterid II). Much of the basic topology from this initial broad analysis of *rbcL* sequences has been further supported and refined by additional analyses (reviewed in "Combining DNA Datasets" and "Lessons Learned" sections).

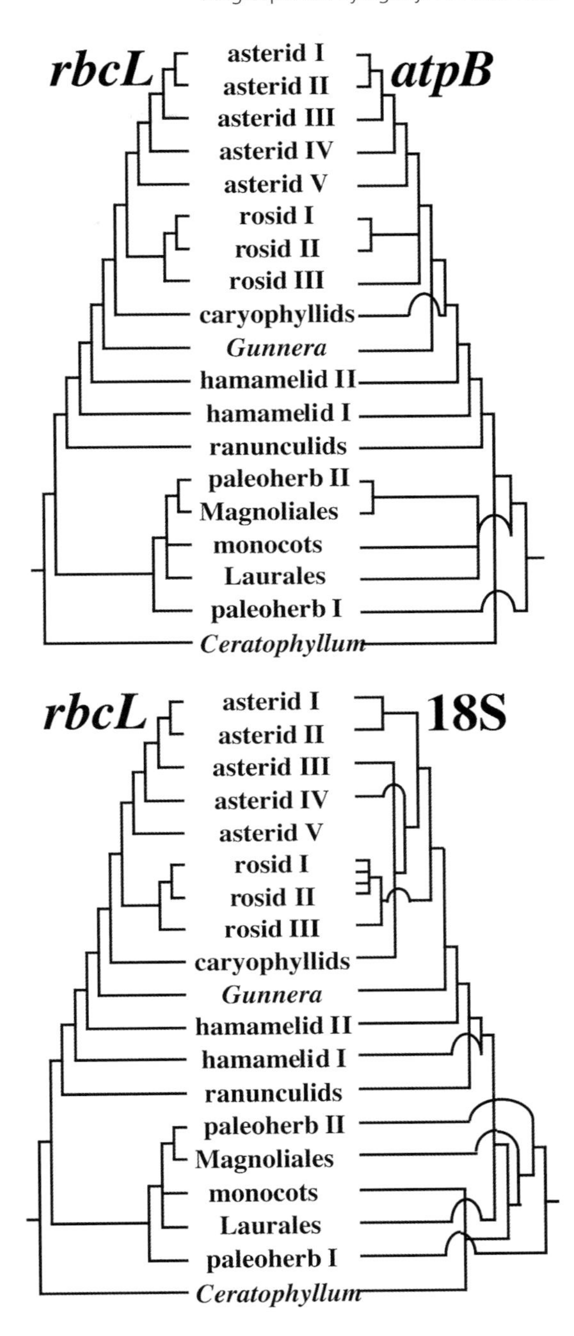

Figure 3.2. Comparison of simplified, general topologies for angiosperms based on three analyses of single-gene datasets: *rbcL* (the 500-terminal search of Chase et al. 1993); 18S rDNA (D. Soltis et al. 1997a); *atpB* (Savolainen et al. 2000a). (Modified from Chase and Albert 1998.) (A) Comparison of *rbcL* and *atpB*. (B) Comparison of *rbcL* and 18S rDNA.

The success of Chase et al. (1993) prompted the question of what topologies would be recovered by other genes, especially nuclear genes. Although partial sequences of 18S rRNA had been used to infer angiosperm relationships (Hamby and Zimmer 1992), doubts persisted about the usefulness of 18S rDNA sequences for phylogeny reconstruction (reviewed in Nickrent and Soltis 1995; D. Soltis et al. 1997a,b). Analysis of 232 entire 18S rDNA sequences (D. Soltis et al. 1997a) revealed that, although 18S rDNA provided lower resolution than *rbcL*, the 18S rDNA and *rbcL* topologies were highly concordant (Fig. 3.2). Differences were minor, and none of these received bootstrap support above 50%. Over a decade later, similar topologies have now been recovered with very large datasets of nearly complete 18S + 26S rDNA sequences (e.g., Soltis et al. 2011; Maia et al. 2014).

Sequences for *atpB*, a second plastid gene similar in rate of evolution to *rbcL* (Ritland and Clegg 1987; Hoot et al. 1995a), revealed relationships highly similar to those recovered with 18S rDNA and *rbcL* alone (Savolainen et al. 2000a) (Fig. 3.2). Use of the partition homogeneity test revealed that the greatest incongruence among the datasets actually involved the two plastid genes (*atpB* and *rbcL*), datasets that many would argue could be combined a priori (see D. Soltis et al. 1998).

Criticisms of Chase et al. (1993) had less to do with the topology itself than with issues of data analysis. Some maintained that the analysis was flawed in that the shortest trees had not been obtained. In fact, Chase et al. (1993) knew that shorter trees could be obtained but did not feel that lengthier searches for shorter trees would significantly alter the outcome (Chase and Albert 1998). Rice et al. (1997) conducted a lengthy reanalysis of the Chase et al. (1993) *rbcL* data matrix. Although they did find shorter trees than those reported in 1993, these did not significantly alter the big picture of angiosperm phylogeny and essentially reinforced the conclusions of Chase et al. (1993). Subsequent studies indicated that lengthy searches of matrices with many taxa and an insufficient number of characters are ineffective (e.g., D. Soltis et al. 1997a,b, 1998; Chase and Cox 1998; Nei et al. 1998; Savolainen et al. 2000a; Takahashi and Nei 2000; see "Lessons Learned," below).

EARLY EFFORTS IN COMBINING DNA DATASETS

Combining DNA datasets is now common, but this process represented a profound change in mindset 20 years ago. Although the single-gene trees (*rbcL*, *atpB*, 18S rDNA) for angiosperms were highly similar (Fig. 3.2), internal support for many major clades and the spine of the tree was lacking (e.g., D. Soltis et al. 1997a,b, 1998; Chase and Albert 1998). Thus, an immediate interest in combining datasets emerged. Because of the cooperation that existed among many angiosperm systematists, the sequences for *rbcL*, *atpB*, and 18S rDNA were often from the same species or genus and in many cases from the same DNA extraction, facilitating direct combination of multiple molecular datasets. This phase of angiosperm phylogenetics started with the combination of data for these few genes and continues today with the assembly of ever-larger datasets involving numerous genes (e.g., the complete plastid genome or hundreds of nuclear genes) for hundreds (and in some cases thousands) of exemplars.

Initial combinations of large DNA datasets were conducted primarily as experiments to assess the effect of additional data on the phylogenetic analysis of large datasets (D. Soltis et al. 1997a, 1998; Chase and Cox 1998) and provided the impetus for two projects that combined DNA datasets on a large scale across the angiosperms. One combined *rbcL* and *atpB* sequences for 357 taxa (Savolainen et al. 2000a); the second combined *rbcL*, *atpB*, and 18S rDNA sequences for 567 taxa (P. Soltis et al. 1999a; D. Soltis et al. 2000). These analyses provided topologies with greater resolution and support than obtained in the single-gene trees.

Combined datasets continued to grow in terms of numbers of genes. A 17-gene angiosperm phylogenetic framework was assembled for 640 species (Soltis et al. 2011), the major features of which are discussed in "The Angiosperm Phylogenetic Framework: An Overview," below. Other studies have compiled supermatrices for thousands of angiosperms (see below), while still others have combined multiple molecular datasets for large clades of interest, such as monocots, rosids, and asterids, and these efforts are discussed in later chapters. Large datasets of complete plastid genomes and large numbers of nuclear genes (see below) have also been compiled, but generally for smaller numbers of taxa than in the 17-gene framework, although a recent analysis of over 1000 complete (or nearly so) plastid genomes and a comparable dataset of nuclear genes (Gitzendanner, Wickett, and Leebens-Mack et al. in prep.) represent a dramatic up-scaling in "big data" and certainly represent the norm for the future.

Molecular and morphological datasets for angiosperms have also been combined (e.g., morphology and partial 18S and 26S rRNA sequences, Doyle et al. 1994; non-DNA characters and *rbcL*, Nandi et al. 1998; morphology and *rbcL*, *atpB*, and 18S rDNA sequences, Doyle and Endress 2000); several analyses have used *rbcL*, *atpB*, 18S rDNA, and morphology and added fossils (e.g., Doyle and Endress 2010; Endress and Doyle 2009).

LESSONS LEARNED

THE ANALYSIS OF LARGE DATASETS

Phylogeny reconstruction of angiosperms prompted the development of methods for analyzing large datasets. Only two decades ago, analyses of datasets of 500 or more terminals were considered impossible (e.g., Patterson et al. 1993; Hillis et al. 1994; Hillis 1995). Simulation studies (Hillis et al. 1994) indicated that in some instances correct phylogeny reconstruction for only four taxa would require more than 10,000 bp of DNA sequence data, implying much greater difficulty with large datasets and stimulating the suggestion that phylogenetic problems be broken into a series of smaller problems, one extreme being a large number of four-taxon questions (Graur et al. 1996). Large datasets also pose problems because of the large number of possible trees that arise with increasing numbers of taxa (Felsenstein 1978a) and a consequent lack of confidence that an analysis has found the optimal tree or trees. For example, for 228 taxa (the number of species analyzed by D. Soltis et al. 1997a), the number of possible trees is 1.2×10^{502}, which is more than the number of atoms in the universe (Hillis 1996). Despite these dire predictions, angiosperm systematists continued to conduct analyses involving hundreds of species, with promising results (reviewed below).

The systematics community as a whole continues to learn valuable lessons from analyses conducted on angiosperms that now involve thousands of taxa. Now, two decades after the first large trees were produced, systematists are routinely building trees with many hundreds to thousands of taxa, including comprehensive trees for all green plants, and even the Tree of Life for all organisms (below).

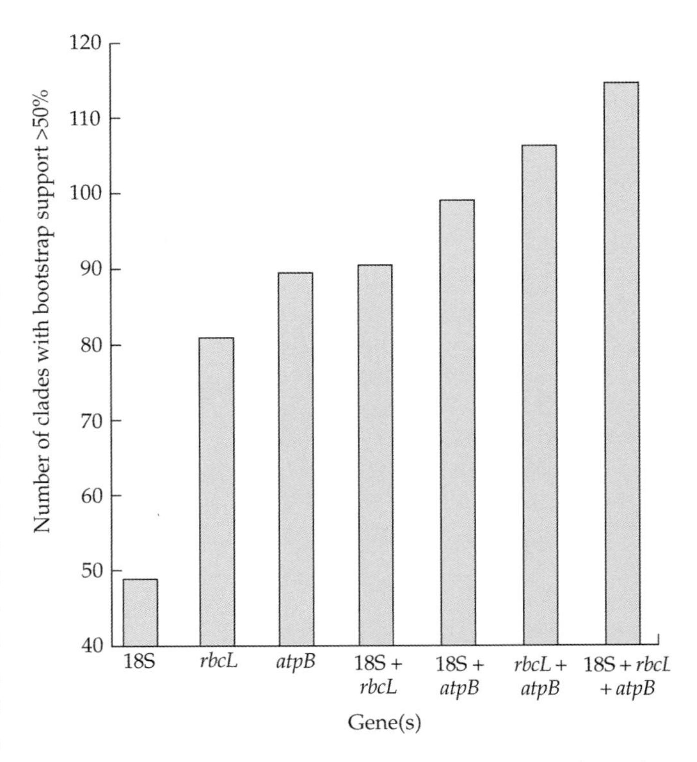

Figure 3.3. More is better—the importance of adding characters. The number of angiosperm clades (y-axis) having bootstrap support above 50% in phylogenetic analysis of separate and combined DNA sequence datasets increases as genes are added. (Modified from D. Soltis et al. 1998.)

MORE IS BETTER (USUALLY)

Initial combinations of large DNA datasets for two and three genes revealed that one solution to the computational problems posed by large datasets was simply to add both taxa and characters (D. Soltis et al. 1997a, 1998; Chase and Cox 1998). This approach resulted in much faster computational run times, which have the advantage that more thorough analyses of tree space can be conducted in a given time. In addition, internal support for clades is much higher in trees from combined datasets than in trees from separate datasets (Fig. 3.3; D. Soltis et al. 1998). In analyses of *rbcL*, *atpB*, and 18S rDNA, some clades received internal support above 50% in the combined trees, but not in trees from the separate datasets. Thus, through combining datasets for two or three genes, strong jackknife or bootstrap support was evident for the first time for most of the major clades of flowering plants. The greater signal present in the combined dataset relative to individual datasets (see Chase and Cox 1998; D. Soltis et al. 1998; Bremer et al. 1999) produced starting trees much closer to the best trees ultimately obtained, thereby speeding up analyses and allowing for more thorough exploration of tree space. Furthermore, the combined signal from multiple genes clarifies historical patterns relative to noise in the dataset. These empirical results for large datasets complement the simulation studies of Hillis (1996) and Graybeal (1998).

Appropriate taxon sampling is crucial for successful phylogeny reconstruction (e.g., Pollock et al. 2002; Zwickl and Hillis 2002; Hillis et al. 2003; D. Soltis et al. 2004; but see Rosenberg and Kumar 2001). However, this appears to be a lesson that has to be learned repeatedly. Early analyses of angiosperm phylogeny with single genes sometimes employed only a few taxa and yielded spurious results. Similarly, early analyses of whole plastid genomes (e.g., Goremykin et al. 2003, 2004) relied on a large number of characters, but for a small number of taxa. The Goremykin et al. analyses yielded trees that were clearly at odds with phylogenetic trees based on more taxa but fewer genes. The Goremykin et al. (2003, 2004) topologies were shown to be attributable to limited taxon sampling (D. Soltis and Soltis 2004b; D. Soltis et al. 2004; Leebens-Mack et al. 2005). More recent efforts using complete plastid genomes and broader taxon sampling (Jansen et al. 2007; Moore et al. 2007, 2010; Ruhfel et al. 2014) have produced nearly identical topologies to those realized with fewer genes but broad sampling. However, complete plastid genome sequencing has yielded greatly improved bootstrap support. Similarly, some large datasets of nuclear genes suffered from poor taxon sampling in some parts of the tree. However, as taxon sampling has improved, topologies derived from nuclear, plastid, as well as mitochondrial genes have largely converged, as reviewed below. But there are important areas of disagreement (e.g., the COM clade; *Mesangiospermae*; see below and Chapter 10), and these are foci for future study. These recent studies of large nuclear vs. plastid datasets also raise cautionary notes regarding the a priori combination of datasets from multiple subcellular genomes.

THE FUTURE OF PARSIMONY ANALYSIS

Most of the phylogenetic results described in the preceding sections were obtained via parsimony analysis, the method of choice in the 1990s to early 2000s for large datasets, for both philosophical and practical reasons. Although maximal likelihood (ML) and Bayesian methods now offer feasible alternatives, the lessons learned through parsimony analyses of angiosperms remain relevant, and we here provide a historical perspective on the development of phylogenetic methods for large datasets. Programs for finding most parsimonious (i.e., shortest) trees, such as Hennig86 (Farris 1988), NONA (Goloboff 1993), and PAUP (Swofford 1993), were developed for use on what would now be considered small datasets. The numerous improvements available in PAUP* 4.0, such as the faster speed at which heuristic parsimony searches are conducted, were a great asset to those interested in analyzing large datasets with parsimony. An important development in the analysis of large datasets via parsimony was the RATCHET, which greatly facilitates the ability to find shorter trees (Nixon 1999; implemented in Winclada, Nixon 2000; and with NONA, Goloboff 1993; www.cladistics.com). When applied to the 567-taxon matrix of three genes for angiosperms, the RATCHET quickly found trees shorter than those found using PAUP* 4.0 (Swofford 1999; see D. Soltis et al. 2000). The useful TNT program has been widely used in parsimony analyses (Goloboff et al. 2008) and is very effective in the analysis of large datasets, including a study of over 73,000 taxa (Goloboff et al. 2009).

A significant problem with the analysis of large datasets had long been the difficulty in assessing internal support for clades. Large datasets are not amenable to standard bootstrapping (Felsenstein 1985) and Bremer support (decay) analysis (Bremer 1988). The large angiosperm datasets were therefore also a stimulus for the adoption of computational approaches, such as "fast" or "quick" search techniques using bootstrapping (Swofford 1999) or jackknifing (Farris et al. 1996). The rationale for quick searches is that they sample trees from many starting points at the expense of extensive swapping on trees from a single starting point. Only those clades above some minimal threshold of internal support (e.g., a bootstrap or jackknife value of 50%) appear in the consensus of trees from these replicate searches. The well-supported clades appear relatively quickly in the analysis of big datasets—continued branch-swapping involving poorly supported branches can never result in strongly supported clades. Simulation studies (Nei et al. 1998; Takahashi and Nei 2000), as well as analyses of the large angiosperm datasets (e.g., D. Soltis et al. 1998, 2000; Savolainen et al. 2000a), provided support for the premise that lengthy parsimony analyses of large datasets may be unnecessary. The parsimony jackknife method (Farris 1996; Farris et al. 1996), for example, was well suited for the quick analysis of large datasets. Mort et al. (2000) found that the fast and standard methods are highly comparable, particularly at higher levels of support (85% and higher). However, as support for clades decreases, percentages from the fast methods tended to be lower than values obtained from the standard bootstrap. For a further comparison of fast and standard methods on empirical and simulated datasets, see Mort et al. (2000), DeBry and Olmstead (2000), and Salamin et al. (2003).

Although parsimony continues to be used in phylogenetic analyses (particularly with morphological data), both Bayesian and ML methods, with their model-based foundations, are now more commonly used than parsimony in molecular phylogenetics. As ML approaches, in particular, improve (see below), an intriguing question arises: is parsimony analysis headed for extinction in phylogenetic analysis? In the following sections, we review Bayesian and ML methods and their contributions to our understanding of angiosperm phylogeny.

BAYESIAN INFERENCE

Bayesian approaches to phylogeny reconstruction were first implemented by Huelsenbeck et al. (2001, 2002) with the popular program MrBayes. This approach affords the opportunity to analyze relatively large datasets and obtain a measure of confidence (posterior probability values, PP), although the MrBayes program has large computer memory requirements. Using MrBayes and an *atpB* dataset for more than 300 angiosperms (from Savolainen et al. 2000a), Huelsenbeck et al. (2001, 2002) obtained a topology with *atpB* alone that was well supported and highly similar to that obtained using three genes (D. Soltis et al. 2000). However, the PP for clades obtained in Bayesian analyses are generally higher than corresponding bootstrap or jackknife percentages (e.g., Huelsenbeck et al. 2001; Miller et al. 2002; Wilcox et al. 2002), perhaps providing misleading estimates of clade support (Suzuki et al. 2002; Erixon et al. 2003).

Soltis et al. (2007c) conducted a Bayesian reanalysis of the 3-gene, 567-taxon dataset for angiosperms (P. Soltis et al. 1999a; D. Soltis et al. 2000), revealing the analytical challenges posed by such large datasets with MrBayes. For example, determining stationarity in Markov chains in analyses of large datasets is difficult. Significantly, Bayesian analyses recovered a topology highly similar to that found previously with parsimony (compare D. Soltis et al. 2000 with Soltis et al. 2007c). However, a few topological differences were found between the Bayesian and shortest parsimony trees obtained for the same dataset, the most noteworthy of which is that a clade (PP = 0.99) of Amborellaceae + Nymphaeales is sister to all other extant angiosperms (PP = 1.0) in the Bayesian tree, whereas Amborellaceae alone are sister to all other extant angiosperms with parsimony. The relationship of Amborellaceae and Nymphaeales is discussed below in more detail. Additionally, the Bayesian analysis indicated that magnoliids and Chloranthaceae (as part of an unresolved trichotomy) are sister to a major clade of *Ceratophyllum* + eudicots. In contrast, magnoliids and Chloranthaceae were sister to the monocots in the original parsimony analyses. Many clades that received moderate to low jackknife support in parsimony analyses received PP values of 1.0. Recovery of high PP values relative to bootstrap values has been a common feature of Bayesian vs. parsimony comparisons.

BEAST (Bayesian Evolutionary Analysis Sampling Trees; Drummond and Rambaut 2007) is an alternative Bayesian analysis package that simultaneously reconstructs phylogeny and estimates divergence times. Application of BEAST to the 567-taxon angiosperm dataset for *rbcL*, *atpB*, and 18S rDNA obtained divergence times for major clades close to those inferred from the fossil record (Bell et al. 2010).

NEW DEVELOPMENTS IN ML INFERENCE

Ten years ago, both parsimony and Bayesian methods were useful means of analyzing very large datasets. In fact, until recently, parsimony and Bayesian approaches were the only means of analyzing large datasets; most datasets were simply too large for ML, which was once limited to investigations of 50 or fewer taxa. However, with recent advances in ML approaches, it is now possible to conduct analyses of thousands of taxa in a timely fashion (e.g., GARLI, Zwickl 2006; RAxML, Stamatakis 2014). A fast bootstrapping option can also be employed, even in large ML analyses (Stamatakis et al. 2008). For example, S. A. Smith et al. (2011) conducted an ML analysis of over 55,000 seed plants. As a result, ML has now emerged as the phylogenetic method of choice over Bayesian approaches, which typically take much longer.

GENE TREE PARSIMONY

The current capability to construct massive alignments of genome-scale data provides opportunities as well as challenges. Phylogenetic analyses using genome-scale datasets must somehow deal with incongruence among gene trees; this can be particularly challenging when many nuclear gene trees are constructed, especially in plants, due to the high frequency of gene duplications and losses (many due to polyploidy) (e.g., D. Soltis et al. 2009; Jiao et al. 2011, 2012). As a result, genes may have complex evolutionary histories. One method for reconstructing species trees from nuclear genes is to identify the species tree that implies the minimum number of events that cause conflict among the underlying gene trees (e.g., Goodman et al. 1979; Maddison 1997). An approach called Gene Tree Parsimony (GTP) uses a collection of gene trees and then attempts to find a species tree that contains all of the taxa present in the gene trees and that also contains the fewest gene duplications or duplications and losses across the tree.

Long computer run times initially limited the use of the GTP approach on large-scale datasets (Burleigh et al. 2011). However, using new software that incorporates recent algorithmic advances, Burleigh et al. (2011) tested the performance of GTP on a vascular plant dataset (focused mainly on seed plant relationships, angiosperms in particular) of 18,896 gene trees and 510,922 protein sequences from 136 vascular plant taxa, with a resulting combined alignment length of >2.9 million characters. Significantly, the inferred relationships from the GTP analysis were highly similar to previous large-scale studies, with the phylogenetic framework essentially the same as in Soltis et al. (2011) and Moore et al. (2007).

COALESCENCE METHODS

For the past decade or more, systematists have been routinely combining multiple molecular datasets via concatenation. Coalescence methods, which stress the distinction between gene trees and species trees, have recently emerged as an intriguing alternative approach to simple concatenation (e.g., Liu et al. 2009, 2010; Mirarab et al. 2014). These methods therefore attempt to take into account and accommodate heterogeneity among individual gene trees.

Coalescence methods have recently been applied to green plant phylogeny, and we will undoubtedly see many more applications in the near future. Zhong et al. (2013) applied the multispecies coalescent to examine the sister group to land plants, stating that concatenation methods that had been conducted to address this problem were basically flawed due to conflicting underlying gene trees (reviewed in Springer and Gatesy 2014), and inferred that Zygnematales are sister to land plants. Springer and Gatesy (2014) criticized several aspects of the Zhong et al. (2013) analysis, including missing sequences, and suggested that coalescent approaches may not be reliable for such deep questions of phylogenetic inference (but see Zhong et al. 2014). In their own coalescent analyses, Springer and Gatesy (2014) recovered Zygnematales + Coleochaetales as sister to land plants. Interestingly, recent large-scale concatenation analyses of complete plastid genome sequences (Ruhfel et al. 2014) and large nuclear gene datasets (Wickett et al. 2014) have both recovered Zygnematales as sister to land plants, suggesting that in this case the two methods have largely converged on the same answer. Whereas Wickett et al. (2014) employed both concatenation and coalescent methods in their examination of land plant relationships and found comparable results, especially for angiosperms, Xi et al. (2014), in their coalescent analysis to address the root of the angiosperms, found *Amborella* + *Nuphar* as sister to the rest of the angiosperms rather than Amborella alone, as supported in the coalescent analyses of Wickett et al. (2014); see Chapter 4 for further discussion of the application of coalescent methods to deep divergences.

MEGATREES AND SUPERTREES

Large phylogenetic trees have seen increasing use in recent years and have become fundamental for addressing broad evolutionary and ecological questions (see below). Two basic methods have been used to build very large trees: supertrees and supermatrices, with pros and cons to both approaches.

Megatrees (or grafted supertrees) should not be confused with supertrees (e.g., Sanderson et al. 1998; Liu et al. 2001; Salamin et al. 2002). With the megatree approach, a backbone topology is used, and other topologies representing subsets of the group under investigation are simply grafted to that backbone. An example of a megatree that is a valuable resource is that for the Asteraceae (Funk et al. 2009a,b).

As an alternative to megatrees, previously reconstructed topologies (source trees) that share some but not necessarily all component taxa can be combined using one of several available algorithms to create a supertree. Although large multigene (supermatrix) analyses are often favored and the best methods of supertree construction are still debated, combinable data are not always available, and megatrees and supertrees thus offer possible solutions.

Salamin et al. (2002) showed that supertrees offer a valuable approach for a large number of taxa. Their supertrees for grasses closely matched the combined "total evidence" phylogenetic tree provided by the Grass Phylogeny Working Group (GPWG 2001, 2012). A supertree for angiosperms (Davies et al. 2004) using 27 source trees represented the first attempt to assemble a comprehensive tree for angiosperms (Davies et al. 2004) and has been a useful tool for comparative studies. Using *rbcL* sequence data and fossil calibrations, Davies et al. (2004) estimated the relative timing of branching events. Although diversification rates are highly constant across time windows, there is evidence for an apparent increase in diversification rates within more recent timeframes. The top 10 major shifts in diversification rates cannot easily be attributed to the action of a few key innovations but instead are likely attributed to a more complex process of diversification, reflecting the interaction of biological traits and the environment. These diversification results are largely in agreement with other, more recent efforts to examine diversification in angiosperms (e.g., Smith et al. 2011; Tank et al. 2015; see below).

Although megatrees and supertrees both have important possible roles as exploratory tools for developing new hypotheses, current efforts seem to be primarily focused on supermatrices. However, the largest supertree yet constructed—the Open Tree of Life—includes all named species of angiosperms and provides a new resource for angiosperm phylogenetics and comparative analyses (htttp://tree.opentreeoflife.org).

SUPERMATRICES AND MEGAPHYLOGENY

One important advantage of a supermatrix approach compared to supertree approaches is that with the former the resulting trees are based on the underlying data and as a result have branch lengths, which are fundamental for many downstream analyses (e.g., diversification rates). We note it is possible to add branch lengths to supertrees (e.g., Torices 2010; Brinkmeyer et al. 2011), but a supermatrix is nonetheless preferred if branch lengths are desired. But how best to construct these supermatrices? The standard supermatrix approach infers trees from large matrices of concatenated alignments of genes with partial taxonomic overlap (de Queiroz and Gatesy 2007). S. A. Smith et al. (2009) describe what they refer to as a "modified supermatrix method" that they have termed megaphylogeny. The megaphylogeny approach uses sequences from databases in conjunction with taxonomic hierarchies—in this way, S. A. Smith et al. (2009) made very large phylogenetic trees with denser matrices than might be present in standard supermatrices using an algorithm that greatly reduced the run time of phylogeny construction. This approach assumes that the sample of species in GenBank is at least roughly proportional to the actual species diversity of different lineages—so this may be a possible weakness. Despite potential problems in tree inference at massive scales, the megaphylogeny approach has been very useful (e.g., Smith and Donoghue 2008; Smith and Beaulieu 2009; Edwards and Smith 2010; Goldberg et al. 2010; see examples below in the Big Trees section).

LARGE AMOUNTS OF MISSING DATA CAN BE TOLERATED

In the early years of molecular phylogenetics, every effort was made to ensure that all cells of a data matrix were complete, with every gene sequence obtained for every taxon investigated (e.g., D. Soltis et al. 2000). But as datasets grew to include more and more taxa and genes, and as supermatrices were constructed from data taken from GenBank (see Supermatrices and Megaphylogeny section), this approach became impractical. Whether missing data are problematic in phylogenetic inference is therefore an important issue. Importantly, both simulation and empirical studies generally agree that surprisingly large amounts of missing data have little (if any) negative effect (reviewed in Wiens and Morrill 2011), with implications for inclusion of both incompletely known fossils and samples for which molecular and/or morphological characters are missing. As examples, in Saxifragales a supermatrix was constructed with 950 species and a length of 48,465 characters. The supermatrix contains 2,379,996 nucleotides and has 94.83% missing data. The resulting topology was in close agreement with studies using far fewer taxa but many more nucleotides per species (Soltis et al. 2013). Similarly, in a supermatrix analysis of legumes having 228 terminals, 95% of the data were missing (McMahon and Sanderson 2006), yet a reasonable topology was obtained. Similar results were obtained for all green plants, with a matrix with ~84% missing data (Driskell et al. 2004).

MOLECULAR EVOLUTION

Large DNA datasets, for plastid, mtDNA, and nuclear genes, have provided a wealth of data for studies of molecular evolution, a large topic for which we provide only a short overview with appropriate references (see Palmer 1987; D. Soltis and Soltis 1998 for general overview; see also Bousquet et al. 1992; Albert et al. 1994; Hoot et al. 1995a; Kellogg and Juliano 1997; Laroche et al. 1997; Chase and Albert 1998; Chase and Cox 1998; Olmstead et al. 1998; P. Soltis and Soltis 1998; Savolainen et al. 2000a,b; Cho et al. 2004; Lancaster 2010).

Rates of molecular evolution vary considerably between closely related lineages and between genes and genomes. For example, mitochondrial substitution rates are extraordinarily elevated and variable in the flowering plant genus *Plantago* (Plantaginaceae), with the fastest rates ~4000 times faster than those reported in most other seed plants and an order of magnitude faster than the fastest-evolving animal mitochondrial genome (Cho et al. 2004). Enormous variation in rates of mitochondrial evolution has also been detected in *Pelargonium* (Geraniaceae) (Parkinson et al. 2005), in which fast rates were detected in some lineages but reversed in others. Further, evolutionary rates of mtDNA were decoupled from rates of plastid and nuclear gene evolution, both of which were essentially "normal" (comparable to most plants).

Fundamental differences in molecular evolution, perhaps related to functional constraints, may exist between the plastid genome of angiosperms and the mitochondrial genome of some animals, such as vertebrates (Savolainen et al. 2002). The large datasets of plastid gene sequences for angiosperms have made it possible to conduct rigorous comparisons of plant and animal organellar gene evolution. Whereas some have suggested that functional requirements, including chemical properties, charge, and hydrophobicity, in the mtDNA genome of animals have seriously compromised estimation of relationships (Naylor and Brown 1997), such is not the case with the plastid genome (Savolainen et al. 2002). Thus, much of the success in phylogeny reconstruction at deep levels in flowering plants may be due to properties of the plastid genome itself.

Large-scale analyses of angiosperm datasets have also prompted a reassessment of the utility of third codon positions for phylogeny reconstruction. Rapidly evolving nucleotide sites have often been considered less informative than those that evolve more slowly, particularly in the analysis of deep branches and especially in studies using animal mtDNA. However, several studies have shown that third codon positions of plastid genes possess the most phylogenetic information in angiosperms (Källersjö et al. 1998; Olmstead et al. 1998); in the animal mtDNA genes, in contrast, third positions performed most poorly (Savolainen et al. 2002). Källersjö et al. (1998, 1999) demonstrated the utility of third positions on a broad scale in their analysis of 2538 *rbcL* sequences across photosynthetic life. Although rapidly evolving and prone to homoplasy, third positions contained most of the phylogenetically informative sites and performed well. With broad sampling, homoplasy is dispersed across the tree, allowing the historical signal carried in third positions to emerge.

Some life-history strategies appear to be associated with changes in rate of molecular evolution. Long-lived organisms (perennials) have slower rates than do annuals (see section "Big Trees" below; Smith and Donoghue 2008). Rates of molecular evolution in parasitic plants may be faster than those in non-parasitic lineages (e.g., Nickrent and Starr 1994; de Pamphilis et al. 1997; Young and de Pamphilis 2005). In addition, carnivorous lineages with morphologically complex traps have higher relative rates of gene substitutions than do those with simple sticky "fly-paper" traps (Ellison and Gotelli 2009). Müller et al. (2004) reported that *Genlisea* and *Utricularia*, both with specialized traps, have relative rates of nucleotide substitutions in some plastid genes that are 63% higher than in the related *Pinguicula*, with sticky traps (see also Jobson and Albert 2002).

Unusual Features of the mtDNA Genome

Put simply, the plant mitochondrial genome is a very "different beast" from the plastid genome, and its structure and dynamics are very different from those of the other two plant genomes (Palmer 1985, 1990, 1992) and from the mtDNA of animals and fungi. The noteworthy features of the plant mitochondrial genome have been well summarized elsewhere, but two key points of caution should be considered when using mtDNA genes in reconstructing phylogeny: 1) many non-mitochondrial DNA sequences, both plastid and nuclear, may be present in plant mtDNA; and 2) the plant mitochondrial genome often houses many foreign copies of mitochondrial genes that were acquired via horizontal gene transfer (HGT) from other species. Examples of these issues are noted below.

Amborella, the extant sister group to all other living angiosperms (below), has a very large, complex mitochondrial genome (Rice et al. 2013), with many foreign copies of mitochondrial genes that were acquired via HGT from a diverse array of donors, including other angiosperms as well as mosses (Bergthorsson et al. 2003, 2004; Rice et al. 2014). Other studies indicate that HGT events involving mitochondrial genes have been widespread, at least in flowering plants (Fig. 3.4) (Bergthorsson et al. 2003; Davis and

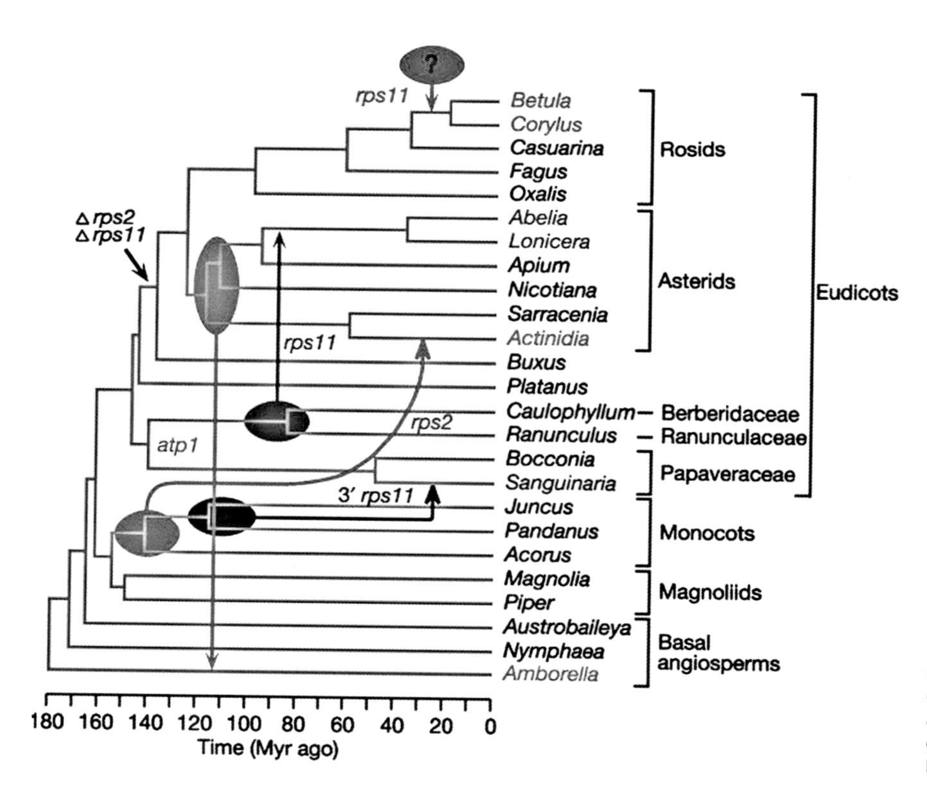

Figure 3.4. Approximate timing and donor-recipient relationships of five HGT "events" in angiosperm mitochondrial DNA. Shadowed ovals indicate rough identity of donor groups. The exact placement of arrowheads on recipient lineages is arbitrary. If correct, the older ages of donors relative to recipients for the *rps2* and 3′ *rps11* transfers imply the existence of the transferred gene in an intermediate, unidentified vectoring agent or host plant for millions of years, but these discrepancies could easily be due to imprecision in the gene trees and/or in molecular-clock-based estimates of divergence times. (Figure and legend [modified] from Bergthorsson et al. 2003.)

Wurdack 2004). In addition, Won and Renner (2003) documented HGT of mitochondrial genes from flowering plants to the gymnosperm *Gnetum*. Mitochondrial genomes frequently contain integrated plastid DNA, which accounts for an additional 25,281 bp (4.44%) of the mitochondrial DNA of maize (Clifton et al. 2004). The topic of HGT is covered in more detail in Chapter 15.

The plant mitochondrial genome is also dynamic at the gene level with gene loss common in angiosperms. Adams et al. (2000) also showed repeated loss and transfer of the mitochondrial gene *rps10* to the nucleus and subsequently (Adams et al. 2002) demonstrated that, whereas basal lineages of angiosperms have the same set of mitochondrial genes as found in their algal ancestors, more derived lineages of flowering plants exhibit losses of many mitochondrial genes, again apparently the result of transfer of genes to the nucleus.

THE ANGIOSPERM PHYLOGENETIC FRAMEWORK: AN OVERVIEW

We provide a brief overview of angiosperm relationships that relies primarily on the 640-taxon, 17-gene topology based on genes from all three plant genomes (i.e., 11 plastid genes, 4 mitochondrial genes, 2 nuclear genes) (Soltis et al. 2011; see Fig. 3.5). Because of its more extensive taxon sampling, the Soltis et al. (2011) study remains the most useful published analysis of angiosperm phylogeny, but we also highlight analyses employing entire or nearly entire plastid genome sequences (Jansen et al. 2007; Moore et al. 2007, 2010; Ruhfel et al. 2014), including a recent noteworthy analysis of over 1000 nearly complete plastid genomes for green plants (Gitzendanner et al. in prep.), mitochondrial gene datasets for angiosperms (Qiu et al. 2010), and recent nuclear phylogenomic studies (Lee et al. 2011; Zeng et al. 2014, see Fig. 3.6; Wickett et al. 2014).

MOLECULAR PHYLOGENIES COMPARED TO TRADITIONAL CLASSIFICATIONS

Here we also compare the current best estimate of phylogeny with the major features of the most prominent morphology-based classification systems from the late 1900s—Dahlgren (1980), Cronquist (1981), Takhtajan (1987, 1997), and Thorne (1992a,b, 2001, 2007) (see also Dahlgren et al. 1985; Dahlgren and Bremer 1985; Dahlgren 1989) (Figs. 3.7, 3.8). In many ways, these four systems were highly similar; each recognized a basic split between monocotyledons and dicotyledons (recognized as separate classes; Magnoliopsida and Liliopsida of Cronquist 1981), except for Thorne (2007), which was influenced by molecular analyses. We now know, however, that dicots are nonmonophyletic (see below) so that group is no longer recognized. The monocots and dicots were, in turn, divided into a series of subclasses or superorders. Relationships among these subclasses were often depicted in terms of phylogenetic shrubs or "bubble" diagrams that represented each author's intuition of evolutionary relationships (see also Stebbins 1974). The six subclasses of dicots recognized by Cronquist (1981) and their interrelationship are depicted in Figure 3.7A; ordinal interrelationships were expressed in a similar fashion (Fig. 3.7A and B). Numerical analyses of the content of orders and the overall structure of these four traditional classifications have indicated that they are not significantly different (Cuerrier et al. 1997), despite some noteworthy variation among them. Dahlgren et al. (1985) and Dahlgren and Bremer (1985), for example, represent the only angiosperm classifications to emphasize chemical characters, as well as to incorporate cladistic principles (Fig. 3.8).

There are both striking similarities and differences between the long-used morphology-based classifications (e.g., Cronquist 1981; Takhtajan 1987; Thorne 1992a,b) and DNA-based phylogenetic hypotheses. At this point, these older classifications have been superseded by a series of classifications based on molecular phylogenies. The reclassification of angiosperms is discussed in more detail below in "A Reclassification Based on Sequence Data." A useful source of phylogenetic trees and other information for angiosperms is provided by Stevens (2001 onwards; www.mobot.org/MOBOT/ research/APweb/).

Throughout this discussion, we generally refrain from using phrases such as "basal group" or "early-diverging group" because a node marks the divergence of two lineages, neither of which is basal to the other or diverges earlier than the other. It is more appropriate to refer to clades as sister to other clades or groups of clades. However, some lineages clearly diverged before others, such as the origins of Nymphaeaceae and Austrobaileyales before the origins of any of the eudicot lineages. Thus, nodes or branches can be "early-diverging" relative to other clades, even though the extant groups, which are terminals in the tree, are not. As reviewed below (and Chapters 4 and 7), a graded series of three clades (Amborellaceae, Nymphaeales, Austrobaileyales; the ANA grade) is sister to all other angiosperms (*Mesangiospermae*) and for convenience we will refer to this grade as the "basal angiosperms" or "basalmost angiosperms." Similarly, within the eudicot clade, a graded series of clades appears as successive sisters to the core eudicots. For convenience, we will sometimes refer to this grade as the "early-diverging" (or "basal") eudicots.

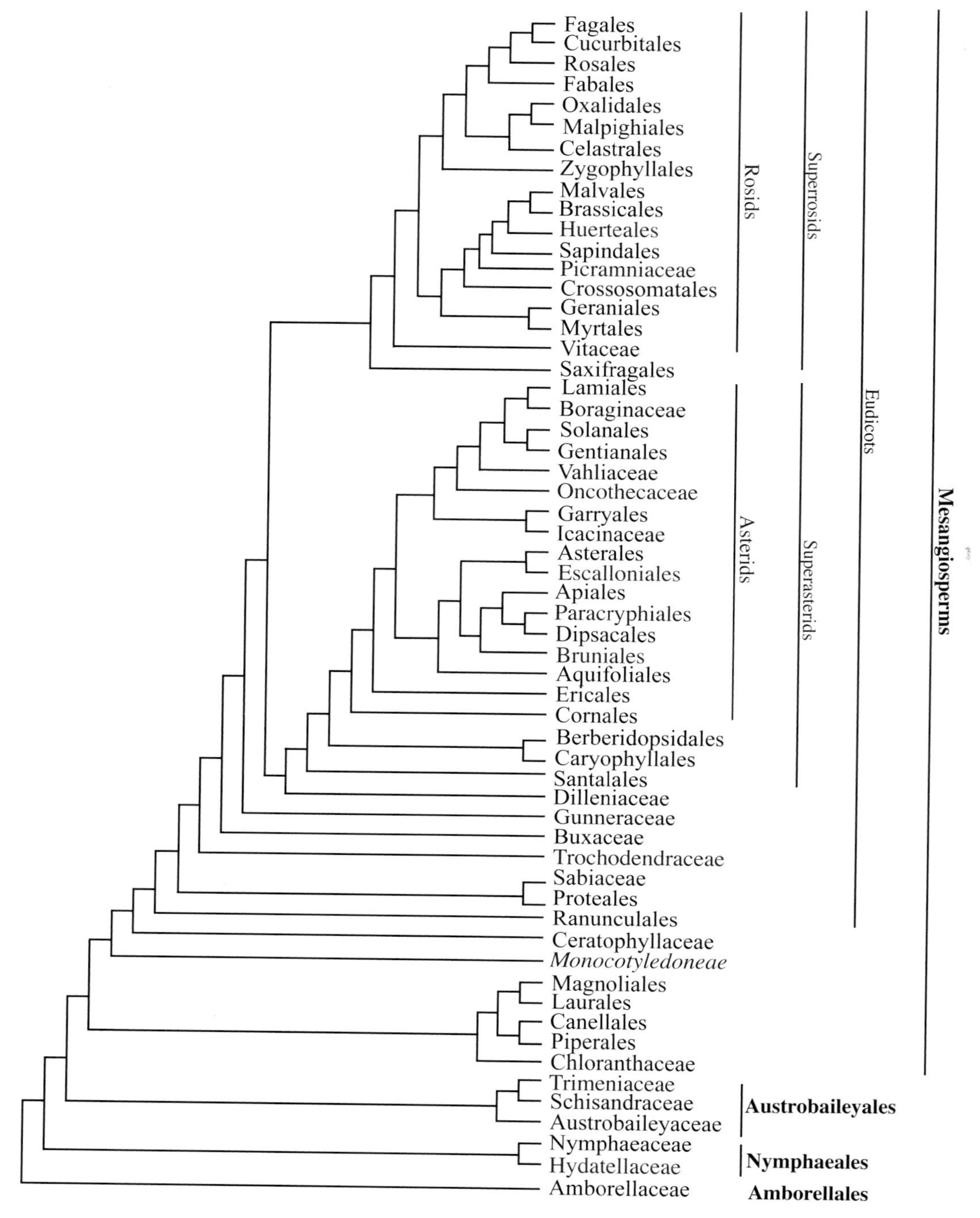

Figure 3.5. Summary of the maximum likelihood majority-rule consensus from the 17-gene analysis with mtDNA data removed for *Polysoma*. Names of the orders and families follow APG III (2009); other names follow Cantino et al. (2007). (Modified from Soltis et al. 2011.)

GENERAL COMMENTS

The overall framework of angiosperm phylogeny is now largely well-supported, with only a few problematic issues remaining at deep levels (below). Recent analyses (e.g., Moore et al. 2007; Jansen et al. 2007; Jian et al. 2008; Soltis et al. 2011) demonstrate that with enough sequence data (20,000 or more base pairs), most problematic, deep-level problems in the flowering plants could be resolved. Especially helpful have been analyses involving complete sequencing of the plastid genome (e.g., Jansen et al. 2007; Moore et al. 2007, 2010; Ruhfel et al. 2014), as well as

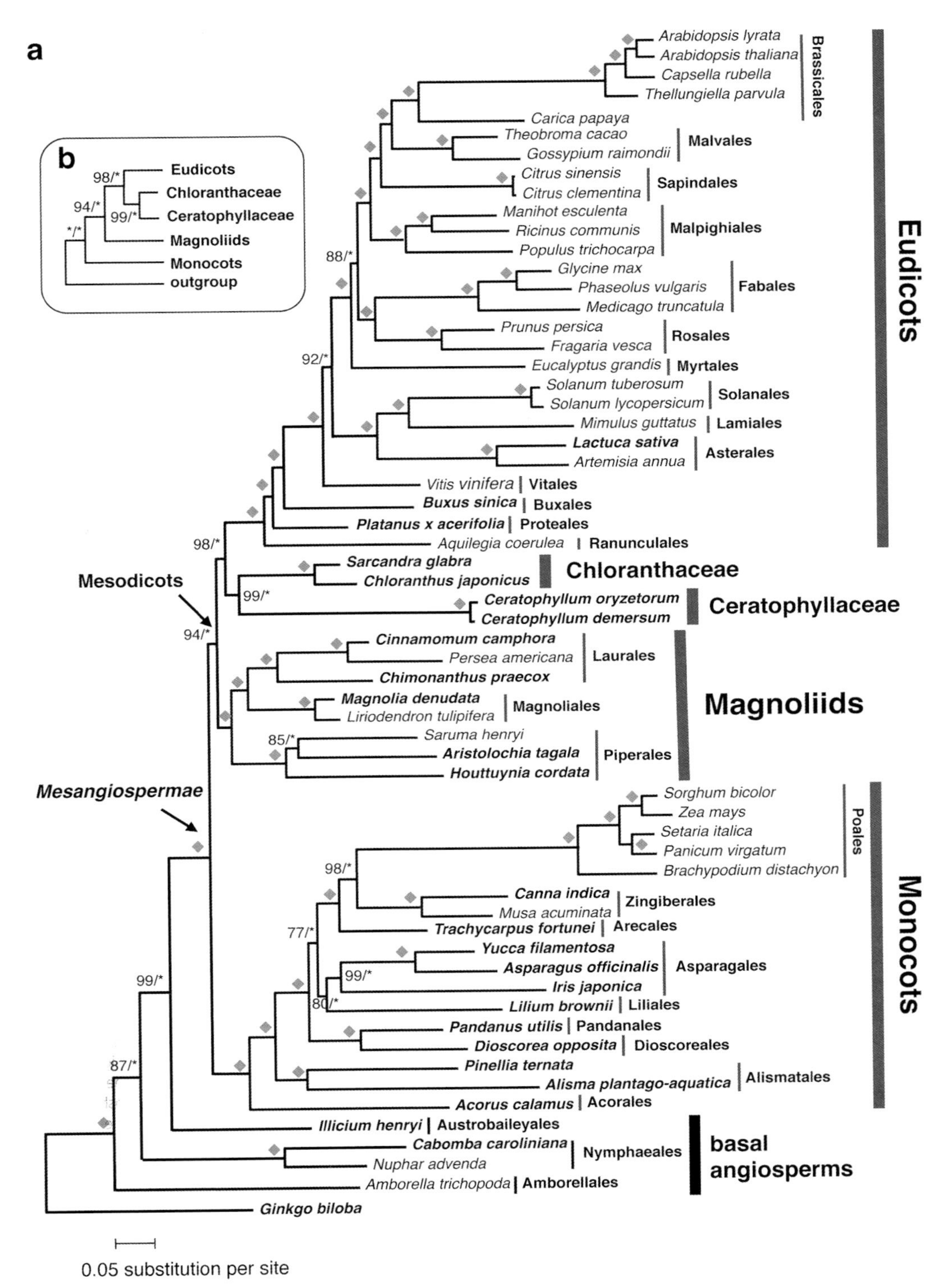

Figure 3.6. (a) Phylogenetic tree based on 61 taxa using 59 genes (from Zeng et al. 2014). Numbers associated with nodes are ML bootstrap value (BS; on left) and the posterior probability (PP, on right). Asterisks indicate either BS of 100% or PP of 1.0. Diamonds indicate both BS of 100% and PP of 1.0. (b) A summary tree showing the relationship of the five lineages of *Mesangiospermae*, with associated support values (BS/PP).

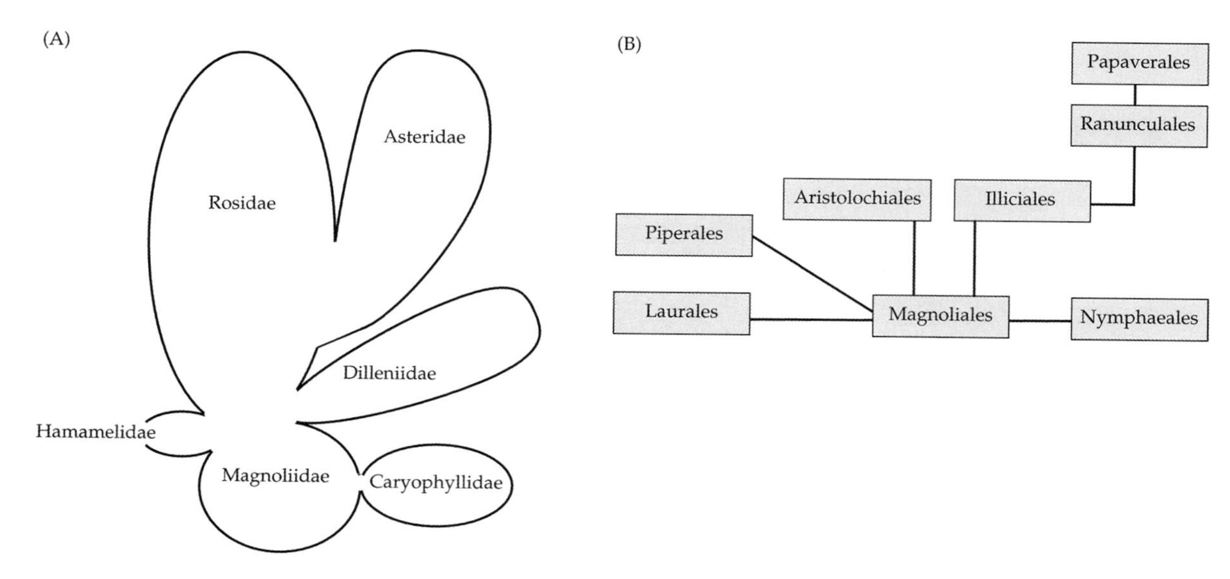

Figure 3.7. Relationships among major groups of angiosperms as proposed by Cronquist (1981). (A) Bubble diagram illustrated putative evolutionary relationships among subclasses of "dicots." The size of the balloons is proportional to the number of species in each group (modified from Cronquist 1981). (B) Proposed relationships among orders of Cronquist's Magnoliidae. (Modified from Cronquist 1981.)

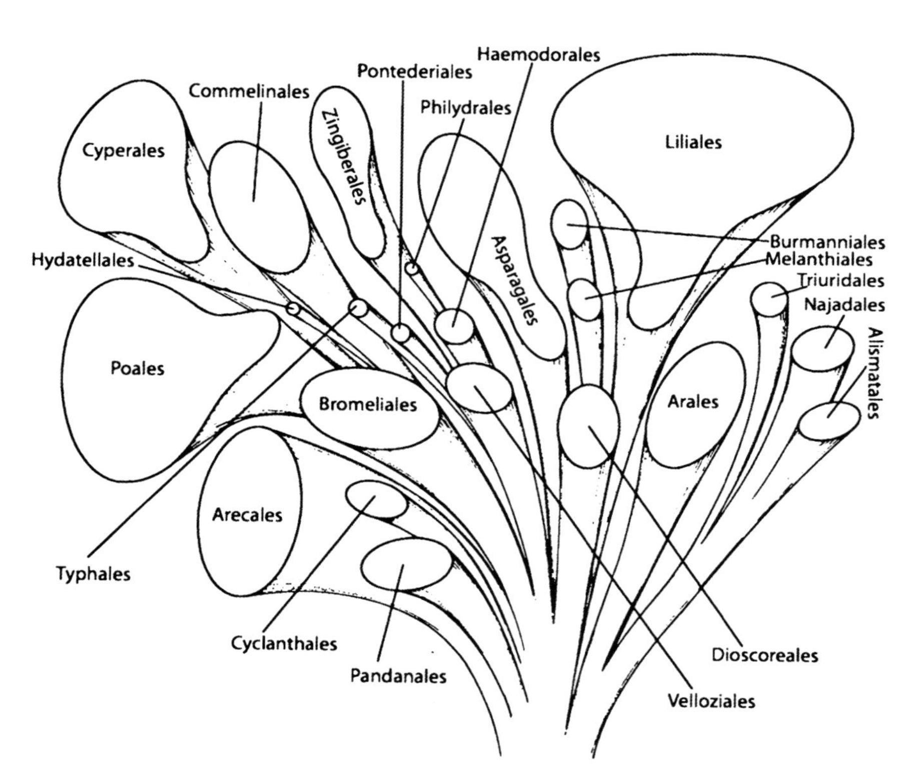

Figure 3.8. A diagram illustrating the circumscription and putative phylogenetic relationships among orders of monocotyledons as circumscribed by Dahlgren et al. (1985; Fig. 26); names and groupings are superimposed on the three-dimensional diagram of phylogenetic relationships (Dahlgren et al. 1985; Fig. 25). The diagram is envisioned as the transection of a phylogenetic shrub. The transected branches are positioned according to degree of similarity and have a size roughly approximating the number of species. We have combined figures from Dahlgren et al. (1985; Figs. 25 and 26) to make this novel figure depicting their view of relationships.

the compilation of large datasets of nuclear genes (e.g., Lee et al. 2011; Wickett et al. 2014) using various next-generation sequencing approaches. Smaller datasets of highly conserved low-copy genes have also been informative (e.g., N. Zhang et al. 2012; Zeng et al. 2014). This enormous progress on the backbone does not imply that angiosperm phylogenetic knowledge is now complete. Instead, as noted below, unresolved problems with the backbone remain. In some cases, differences exist between nuclear- and plastid-based trees, and it remains unclear whether this is actual conflict between these different genomic compartments. Furthermore, our understanding of relationships within many major subclades (recognized as families and orders) is often very limited, and at the generic and species levels, the angiosperms remain poorly understood.

As noted above, a conspicuous feature of DNA-based topologies is the absence of a major split between monocots and dicots (Figs. 3.5, 3.6; see also Figs. 3.1 and 3.2). Instead, essentially all trees indicate that, whereas the monocots (~22% of all angiosperm species) are monophyletic, the dicots are not. The monocots are derived from within a basal grade of taxa that are all traditional dicots corresponding largely to Magnoliidae of Cronquist (1981) (Fig. 3.7). The remaining traditionally recognized dicots, now known as the eudicots, which include about 75% of angiosperm species (Drinnan et al. 1994), form a well-supported clade (Figs. 3.5, 3.6).

Most combinations of plastid, nuclear (18S + 26S rDNA), and mtDNA genes, as well as large datasets of only plastid genes, recover a grade of Amborellaceae, Nymphaeales, and Austrobaileyales, followed by a magnoliid clade (*Magnoliidae* sensu Cantino et al. 2007) plus Chloranthales as subsequent sisters to two large clades: 1) Ceratophyllaceae + eudicots and 2) monocots. However, several recent trees based on low-copy nuclear genes suggest alternative topologies. In N. Zhang et al. (2012), Amborellaceae, Nymphaeales, and Austrobaileyales are subsequent sisters to a large clade comprising two major subclades: 1) eudicots and 2) monocots, magnoliids, Ceratophyllaceae, and Chloranthaceae. In Lee et al. (2011), the ANA grade is also followed by two large clades: 1) eudicots + magnoliids and 2) monocots (Fig. 3.5). Both of these nuclear gene analyses suffer from limited taxon sampling in crucial areas of the tree (e.g., no Chloranthaceae or Ceratophyllaceae in Lee et al. 2011). Analyses of a large nuclear dataset (1KP data; 1000 plant transcriptomes; www.onekp.com), but with limited taxon sampling, also reveals instability at this point in some of these same relationships (Wickett et al. 2014), with some analyses placing magnoliids + Chloranthaceae as sister to eudicots and others placing magnoliids followed by Chloranthaceae + eudicots.

In a much larger sampling of angiosperm taxa (but many fewer genes) than analyzed by Lee et al. (2011), Zeng et al. (2014) used 59 carefully selected nuclear genes and analyzed 61 taxa (60 angiosperms plus *Ginkgo* as an outgroup). With this sampling (Fig. 3.6), the nuclear framework is very similar to that obtained with datasets largely based on plastid genes. However, the Zeng et al. (2014) tree places magnoliids sister to ([Ceratophyllaceae + Chloranthaceae] + eudicots), a clade they refer to as mesodicots. Clearly, deeper taxon sampling is needed for the picture of relationships based on nuclear genes to stabilize.

Most of the non-eudicot angiosperms are characterized by uniaperturate pollen, although there is considerable variation in pollen morphology among these taxa (see Sampson 2000). In contrast, the eudicots are characterized by triaperturate or triaperturate-derived pollen (Doyle and Hotton 1991), which some authors have suggested might have been a crucial factor for eudicot success (e.g., Furness and Rudall 2004). Other morphological features are shared by a major subset of the eudicots, called the core eudicots or *Gunneridae* sensu Cantino et al. (2007; see below) (reviewed in Nandi et al. 1998; Savolainen et al. 2000a; and Chapters 7 and 8); these features include ellagic acid (apparently replaced in the euasterids by iridoids and other compounds), non-laminar/marginal placentation, and differentiation of the calyx and corolla (Albert et al. 1998; Nandi et al. 1998).

In the 17-gene summary tree (Fig. 3.5), *Ceratophyllum* is sister to the large clade of eudicots; this clade of *Ceratophyllum* + eudicots is sister to the monocots. Those angiosperms that are not part of the eudicot and monocot clades form a grade, not a clade, and for convenience in coverage we simply refer to them here as the non-eudicot and non-monocot angiosperms—that is, everything other than the monocots and eudicots. This same basic topology has been recovered in analyses of complete plastid genome sequences (Leebens-Mack et al. 2005; Jansen et al. 2007; Moore et al. 2007, 2010). Several noteworthy non-eudicot, non-monocot groups are only briefly highlighted below as they are discussed in more detail in subsequent chapters: the basalmost angiosperm lineages, the magnoliid clade, and Chloranthales.

BASALMOST ANGIOSPERMS

The basalmost angiosperms (Amborellaceae, Nymphaeales, Austrobaileyales) typically form a grade, even in most early analyses (Mathews and Donoghue 1999; Parkinson et al. 1999; Renner 1999; Qiu et al. 1999, 2000; P. Soltis et al. 1999a; Barkman et al. 2000; Doyle and Endress 2000; Graham and Olmstead 2000; D. Soltis et al. 2000; Zanis et al. 2002, 2003; Leebens-Mack et al. 2005; Jansen et al. 2007; Moore et al. 2007; Burleigh et al. 2009; Lee et al. 2011; Soltis et al. 2011; Drew et al. 2014; Wickett et al. 2014). Generally, Amborellaceae are sister to all other angiosperms (see Chapter 4), followed in succession by Nymphaeales (the water lilies), and then Austrobaileyales (Austrobaileyaceae, Trimeniaceae, and Schisandraceae, which now include Illiciaceae; APG IV 2016 and earlier). The grade itself was initially referred to as the "ANITA" grade (Qiu et al. 1999), named for Amborellaceae, Nymphaeales, Illiciaceae, Trimeniaceae, and Austrobaileyaceae, but has also been referred to as the ANA grade, for Amborellaceae, Nymphaeales, and Austrobaileyales (see Soltis et al. 2005b)—the latter is more appropriate, given the revised classification of Austrobaileyales. The term "basalmost angiosperms" has been more widely used for this grade in recent years, and we will follow that here (see also Chapter 4).

Early analyses involving single genes, as well as initial analyses of combined datasets (Qiu et al. 1999; Barkman et al. 2000; D. Soltis et al. 2000; Zanis et al. 2002, 2003), usually recovered these basal nodes with internal support above 50%. As datasets increased in size (more taxa and particularly more data), strength for these basalmost branches increased. One debate has involved the relationship of Amborellaceae and Nymphaeales. In some early analyses, they appeared as sister taxa—that is, together they are sister to all other angiosperms; this topology has also been seen in analyses with mtDNA data. We review this debate in more detail in Chapter 4, but stress that nearly all recent analyses, including large datasets of nuclear genes, support Amborellaceae alone as sister to all other angiosperms (reviewed in Drew et al. 2014; see also Wickett et al. 2014; Zeng et al. 2014).

MAGNOLIIDS/CHLORANTHALES

There is generally strong support for a magnoliid clade (*Magnoliidae* sensu Cantino et al. 2007) of Magnoliales, Laurales, Canellales (Winterales of some recent publications), and Piperales. Within this clade, Canellales and Piperales are well supported as sister taxa, as are Laurales and Magnoliales. Also well supported is Chloranthales, a clade consisting of four genera in one family (Chloranthaceae; see Chapter 5).

In early analyses based on one or a few genes, the placements of the magnoliids and Chloranthales were unclear because the spine of the angiosperm tree was uncertain (e.g., Chase et al. 1993; P. Soltis et al. 1999a; D. Soltis et al. 2000). But with the addition of more genes, the relationships of these two lineages became better resolved and supported (e.g., Zanis et al. 2002; Jansen et al. 2007; Moore et al. 2007, 2010; Soltis et al. 2007c, 2011; Burleigh et al. 2009). In most trees based on many plastid genes, Chloranthaceae are sister to magnoliids, and this clade is then sister to the remaining angiosperms; with 17 genes and complete plastid genome sequences, these relationships receive strong internal support (e.g., Zanis et al. 2002; Jansen et al. 2007; Moore et al. 2007, 2010; Soltis et al. 2011).

However, recent nuclear gene trees suggest alternative relationships. For example, Zeng et al. (2014) place Ceratophyllaceae sister to Chloranthaceae with strong support, with magnoliids and Ceratophyllaceae + Chloranthaceae sister to the eudicots (Fig. 3.7). That is, magnoliids and Chloranthaceae are not sisters in this analysis. It will be crucial to examine these relationships with nuclear gene datasets with larger taxon sampling.

MONOCOTS (*MONOCOTYLEDONEAE*)

Phylogenetic analyses of monocots (Chase et al. 2000c, 2006; D. Soltis et al. 2000; Chase 2004; Givnish et al. 2006, 2010; Graham et al. 2006; Janssen and Bremer 2004; Soltis et al. 2011) have indicated some relationships and circumscriptions that differ substantially from most morphology-based classification systems. The classification of Dahlgren et al. (1985) most closely resembles the topologies obtained; those of other prominent authors sometimes disagree substantially with molecular results (Cronquist 1981; Takhtajan 1987, 2009; Thorne 1992a,b). One conspicuous difference between DNA topologies and morphology-based classifications involves the enigmatic genus *Acorus*. Traditionally, *Acorus* was placed in Araceae. However, *Acorus* appears with strong support as sister to all other monocots, a result that was apparent with just a few genes (see Chapter 7). Following *Acorus* (Acoraceae), a clade of Alismatales is sister to all other monocots. The Alismatales corresponds largely to Aranae and Alismatanae of Dahlgren et al. (1985). Following Alismatales, there are several well-supported clades of monocots (commelinids, Asparagales, Liliales, Dioscoreales, Pandanales), the interrelationships of which were unresolved even in earlier multi-gene analysis (e.g., D. Soltis et al. 2000). But the addition of more genes has resolved most deep-level monocot relationships, nearly all with high bootstrap support (e.g., Chase et al. 2006; Givnish et al. 2010; Soltis et al. 2011; see Chapter 7).

Although Lilianae of Dahlgren et al. (1985) are paraphyletic in phylogenetic analyses, they include groups that correspond well to the clades retrieved in molecular phylogenetic analyses (Dioscoreales, Liliales, and Asparagales). The phylogenetic placements of only a few families and genera in recent phylogenies do not agree with the circumscriptions of these orders by Dahlgren et al. (1985). For example, Iridaceae and Orchidaceae were placed in Liliales by Dahlgren et al. (1985), but appear in Asparagales in DNA-based trees. Although Dahlgren et al. also recognized the presence of a commelinid group, some of their ordinal circumscriptions differ from those inferred from molecular analyses. Within the commelinid clade of molecular-based trees are several well-supported subclades: Arecales, Zingiberales, Poales, and Commelinales. Arecales and Zingiberales correspond well to the Arecanae and Zingiberanae of Dahlgren et al. (1985), whereas their Bromelianae and Commelinanae are both polyphyletic. Chapter 7 covers the monocots in more detail.

EUDICOTS (*EUDICOTYLEDONEAE*)

The monophyly of eudicots, recognized first based on morphology, was apparent in molecular analyses of single genes, but typically without strong support (e.g., Chase et al. 1993; Chase and Albert 1998; D. Soltis et al. 1997b; Savolainen et al. 2000a,b). With only three genes, the monophyly of the eudicot clade was recovered with a high degree of confidence (100%; Fig. 3.6; P. Soltis et al. 1999a; D. Soltis et al. 2000; see also D. Soltis et al. 1998 and Hoot et al. 1999). Large datasets of many genes have now consistently recovered the eudicot clade with strong support (e.g., Jansen et al. 2007; Moore et al. 2010; Lee et al. 2011; Soltis et al. 2011).

Within the eudicots, there is a basal grade of ancient (based on the fossil record) lineages (Fig. 3.6): Ranunculales, Proteales, Sabiaceae (now included in Proteales; Chapter 12), Trochodendraceae (including *Tetracentron*), and Buxaceae (including *Didymeles*) (see Chapter 8). With the exception of Ranunculales and Proteales, most of these clades consist of relatively few genera and species. Ranunculales are well supported and sister to the remaining eudicots. The relationships among the remaining basal eudicots have been much more difficult to resolve and were unclear with three genes (Chapter 8), but large datasets of numerous genes have resolved most of these problematic relationships. Proteales (including Sabiaceae) follow Ranunculales as sisters to the remaining eudicots. In some analyses Proteales sensu APG III (Proteaceae, Platanaceae, Nelumbonaceae) and then Sabiaceae appear as subsequent sisters to all other angiosperms. However, recent analyses involving 17 genes (Soltis et al. 2011) and complete plastid genome sequences (Moore et al. 2010; Sun et al. 2013) recovered a weakly supported Proteales (sensu APG III) + Sabiaceae clade that is sister to all remaining eudicots, which are strongly supported as monophyletic (see Chapter 8) (Fig. 3.6). Bootstrap support for Sabiaceae + Proteales (sensu APG III) is > 90 % in a recent analysis of complete plastid genome sequences (Sun et al. 2016). This prompted inclusion of Sabiaceae in an expanded Proteales (APG IV 2016; see Chapter 12). Trochodendraceae and Buxaceae then typically appear in recent large datasets as successive sister groups to a well-supported clade of core eudicots (Chapters 8, 9).

Within the core eudicots (*Gunneridae* sensu Cantino et al. 2007), Gunnerales (Myrothamnaceae and Gunneraceae) are sister to the remainder of the clade (*Pentapetalae* sensu Cantino et al. 2007) (Fig. 3.6; see Chapter 9). With three, four, or five genes, relationships within the core eudicots were unclear, but with larger datasets (17 genes and complete plastid genomes; Soltis et al. 2011; Moore et al. 2010), resolution and support were obtained for major clades. The core eudicot clade represents a rapid radiation (see below). Complete plastid genome sequence data indicate that soon after its origin, *Pentapetalae* diverged into two major clades: (i) a "superrosid" clade (*Superrosidae* sensu Soltis et al. 2011) consisting of rosids (*Rosidae* sensu Cantino et al. 2007) and Saxifragales; and (ii) a "superasterid" clade (*Superasteridae* sensu Soltis et al. 2011) comprising Berberidopsidales, Santalales, Caryophyllales, and asterids (*Asteridae* sensu Cantino et al. 2007). The exact position of Dilleniaceae remains unclear (see remaining deep-level problems, below). Molecular dating analyses suggest that the major lineages of both superrosids and superasterids arose in as few as 5 million years—clearly a rapid radiation (Moore et al. 2010).

The rosid, asterid, and Caryophyllales clades identified with DNA sequence data correspond roughly to Rosidae, Asteridae, and Caryophyllidae, respectively, of traditional morphology-based classifications of Cronquist, Takhtajan, and Thorne (see above). However, the DNA-based versions are all significantly larger than their counterparts in these classifications. For example, in addition to Rosidae of morphology-based classifications, the rosid clade also contains many families of Dilleniidae and Hamamelidae (neither group is monophyletic nor currently recognized; see below). A few families typically placed in Rosidae in morphology-based classifications actually occur outside of the rosid clade, such as some members of Saxifragales (e.g., Crassulaceae, Haloragaceae, and Saxifragaceae), as well as the small clades of Gunnerales and Berberidopsidales. Santalales, placed in Rosidae by Cronquist and others (Fig. 3.7), form a distinct clade as part of the superasterids in DNA analyses. The asterid clade is also larger than Asteridae of morphology-based classifications due to the inclusion of several groups of Dilleniidae, such as Ericales, and also some Rosidae, such as Cornales and Apiales (Fig. 3.6; see Chapter 12).

Caryophyllales are also an expanded version of Caryophyllidae of traditional classifications. In addition to core families of the old Caryophyllidae (e.g., Caryophyllaceae, Aizoaceae, Nyctaginaceae, Cactaceae, and close relatives, plus Plumbaginaceae and Polygonaceae), the DNA-based Caryophyllales clade also includes the carnivorous plant families Nepenthaceae and Droseraceae, as well as several other former families of Dilleniidae such as Tamaricaceae and Frankeniaceae. Subclasses Hamamelidae and Dilleniidae of modern classifications are both grossly polyphyletic. Families of Hamamelidae are scattered across the topology, with some occurring as early-diverging eudicots (e.g., Platanaceae and Trochodendraceae; see Chapter 8), another (Eucommiaceae) as part of the asterid clade (see Chapter 11), and several families as part of Saxifragales (e.g., Cercidiphyllaceae, Hamamelidaceae, and Daphniphyllaceae; see Chapter 10). However, most families assigned previously to Hamamelidae are found in the rosid clade (e.g., Betulaceae, Casuarinaceae, Fagaceae, Juglandaceae, Ulmaceae, and Urticaceae; see Chapter 10), but even these families of Hamamelidae do not form a monophyletic group. Betulaceae, Casuarinaceae, Fagaceae, and Juglandaceae are in Fagales, whereas Ulmaceae and Urticaceae are in Rosales. Families of Dilleniidae also appear in several different places in the eudicots, occurring in the asterid, rosid, and Caryophyllales clades. Paeoniaceae, treated as a dilleniid by Cronquist, are placed in Saxifragales.

REMAINING DEEP-LEVEL PROBLEMS

Although the framework of angiosperm phylogeny has largely been resolved with strong support, several problematic areas remain. We briefly mention prominent examples here; others are discussed in more detail in the specific chapters dealing with those taxa. Despite some contention on the placement of *Amborella*, large datasets with appropriate taxon sampling consistently place it alone as sister to all other angiosperms (Chapter 4). One vexing problem is the position of Dilleniaceae, a family variously placed with the superrosids or superasterids. Complete plastid genome sequencing did not resolve its placement (Moore et al. 2010). Maximum-likelihood analyses support Dilleniaceae as sister to superrosids, but topology tests did not reject alternative positions of Dilleniaceae as sister to *Asteridae* or all remaining *Pentapetalae* (see Moore et al. 2010; Soltis et al. 2011). Relationships within several major subclades have also been difficult to resolve, even with numerous taxa and genes. Prominent examples include Malpighiales and Lamiales, each of which is covered in more detail in Chapters 10 and 11, respectively.

As molecular datasets continue to grow, it is also becoming apparent that different genomes may tell different stories at deep levels, just as they do near the tips of the tree (e.g., Rieseberg and Soltis 1991). The COM clade (Celastrales, Oxalidales, and Malpighiales) is an excellent example of this problem. Plastid data place the COM clade with the fabid subclade of rosids, whereas large nuclear datasets, mtDNA, and morphology favor a placement with the malvid subclade of rosids (Sun et al. 2015; see Chapter 10). Likewise, conflict between nuclear and plastid datasets is evident for *Mesangiospermae* (see Zeng et al. 2014), as noted above.

CIRCUMSCRIPTIONS OF FAMILIES AND ORDERS BASED ON MOLECULAR DATA

DNA sequence data provide strong support for the monophyly of most flowering plant families recognized in traditional morphology-based classifications. However, some long-recognized families are clearly not monophyletic based on broad molecular phylogenetic analyses. More focused molecular and morphological analyses, using higher taxon density, often provide additional crucial information regarding familial composition. In a few cases, well-known families have been shown to be highly polyphyletic. Both Scrophulariaceae and Liliaceae as long recognized are polyphyletic, consisting of several distantly related lineages. In other instances, phylogenetic analyses have indicated broader circumscriptions of traditional families (e.g., Apocynaceae, Lamiaceae, and Malvaceae; Judd et al. 1994; Judd and Manchester 1997; Bayer et al. 1999). Asclepiadaceae are nested within Apocynaceae (Judd et al. 1994). Lamiaceae are monophyletic only with the inclusion of some genera of Verbenaceae, such as *Callicarpa* and *Clerodendrum* (Wagstaff and Olmstead 1997). The distinction of Malvaceae from Bombacaceae, Tiliaceae, and Sterculiaceae is not supported by phylogenetic analyses (e.g., Judd and Manchester 1997; Bayer et al. 1999), which indicate instead that a single broadly defined family, Malvaceae, should be recognized. Other angiosperm families as traditionally recognized are also now known to be non-monophyletic; these include Sapindaceae, Flacourtiaceae, Euphorbiaceae, Chenopodiaceae, Ericaceae, and Saxifragaceae (discussed in Chase et al. 2000b; D. Soltis et al. 2000; Savolainen et al. 2000a,b; see also reclassification of angiosperms below by the Angiosperm Phylogeny Group, APG 1998 and APG II 2003; APG III 2009; APG IV 2016; see Chapter 12).

Whereas many families of angiosperms have received strong support in DNA-based analyses, most traditionally recognized orders have not. Well-supported clades of families often show little agreement with orders recognized in morphology-based classification systems (e.g., Dahlgren 1980; Cronquist 1981; Takhtajan 1987, 1997; Thorne 1992a,b, 2001). For example, Saxifragales as delineated by DNA sequence analyses are a highly eclectic assemblage of taxa previously placed in three subclasses (sensu Cronquist 1981). The Ericales clade recovered in phylogenetic analyses contains families that were placed in several orders of subclasses Dilleniidae and Rosidae. Rosales of molecular analyses bear only slight similarity to Rosales of morphology-based classifications. In addition to Rosaceae, the Rosales clade also contains several families of Cronquist's (1981) Hamamelidae (e.g., Ulmaceae, Moraceae), as well as additional families of Rosidae (sensu Cronquist 1981) previously placed in other orders (e.g., Elaeagnaceae were once placed in Proteales). However, some noteworthy instances in which ordinal circumscriptions in the traditional classifications largely agree with those derived from DNA-based topologies include Myrtales, Sapindales, and Zingiberales.

A RECLASSIFICATION BASED ON SEQUENCE DATA

The high degree of resolution and internal support obtained for the angiosperms through molecular investigations indicated the need for a revised higher-level classification. As molecular phylogenetic studies continued at a rapid pace, different names were used by different laboratories for the same groups of angiosperms. The rapidly growing need for a standardized system of names that corresponded to the DNA-based clades prompted the Angiosperm Phylogeny Group (APG), a group of angiosperm systematists, to provide a revised higher-level (family and above) system of classification for flowering plants (APG 1998). The enormous progress that angiosperm systematists have continued to make since the APG (1998) classification has, as anticipated, prompted updating of that classification (APG II 2003; APG III 2009; APG IV 2016); a recent update is discussed in Chapter 12. A classification based on phylogenetic nomenclature treats major groups of angiosperms, as well as other tracheophytes (Cantino et al. 2007) (see Chapter 12).

The angiosperms are the first major group of organisms to have been classified solely on the basis of DNA sequence data. As with most large molecular phylogenetic analyses, this classification was constructed collaboratively by many systematists, some of whom were molecular systematists and others who were classically trained. This collaborative approach is a major departure from both typical classifications in which one or a few "experts" provided systematic treatments of large groups such as the angiosperms. In contrast, higher-level classification of the angiosperms now involves large collaborative networks of systematists, and this trend (at least at higher levels) will continue. This approach may, in fact, serve as a useful model for investigators of other large problematic groups of organisms. For example, a revised classification of the ferns has been developed in a manner similar to the APG format used for angiosperms (Smith et al. 2006).

IMPLICATIONS FOR THE STUDY OF MODEL ORGANISMS

Examples of the importance of a phylogenetic underpinning to diverse areas of research abound. One obvious example is the value of placing model organisms in the appropriate phylogenetic context to obtain a better understanding of both patterns and processes of evolution. For example, most model plants represent highly nested angiosperm lineages when viewed in the context of the angiosperm phylogenetic tree (Fig. 3.9). For example, *Arabidopsis* (Brassicaceae) is a derived lineage within a derived family (Brassicaceae) of Brassicales, and *Fragaria* (strawberry) represents a derived lineage within Rosales; both models are deeply embedded within the rosid clade. Similarly, *Oryza* and *Zea* (rice and corn) are derived members of Poaceae, a family deeply embedded within the monocot clade.

The importance of phylogenetic inference can be seen at multiple scales with model systems. The fact that tomato, long known as *Lycopersicon esculentum*, is actually embedded within a well-marked subclade within *Solanum* (and therefore is now appropriately referred to as *Solanum lycopersicum*) is a powerful statement (Spooner et al. 1993). This phylogenetic result is important to geneticists, molecular biologists, and plant breeders in that it points to a few close relatives of *Solanum lycopersicum* (out of a genus of ~1500 species) as focal points for comparative research.

A phylogenetic perspective also provides unique opportunities for comparative genomics (e.g., D. Soltis and Soltis 2000, 2003; Walbot 2000; Kellogg 2001; A. Hall et al. 2002; Mitchell-Olds and Clauss 2002; Pryer et al. 2002; Doyle and Luckow 2003; Wojciechowski et al. 2004; Albert et al. 2005; Lavin et al. 2005; Bowman et al. 2007). Considering *Arabidopsis* (Brassicaceae), Brassicaceae are part of a well-supported clade, Brassicales (also referred to as the glucosinolate clade), that also contains Akaniaceae, Bataceae, Caricaceae, Limnanthaceae, Tropaeolaceae, and several other families. This knowledge of phylogenetic relationships should be used in efforts to extend the knowledge garnered from detailed genomic and developmental analyses of *Arabidopsis* to close relatives in Brassicaceae, as well as other Brassicales (cf. A. Hall et al. 2002; Mitchell-Olds and Clauss 2002; Bailey et al. 2006; Schranz et al. 2006).

Excellent opportunities for comparative genomics are also afforded by other families that are home to model organisms, such as Plantaginaceae (*Antirrhinum*; Reeves and Olmstead 1998), Fabaceae (legumes; Doyle and Luckow 2003; Cronk et al. 2006; Cannon 2008), and Poaceae (*Zea, Oryza, Sorghum*, others; Grass Phylogeny Working Group 2001, 2012), and Rosaceae (Shulaev et al. 2008). Importantly, the most fruitful phylogenomic comparisons will be

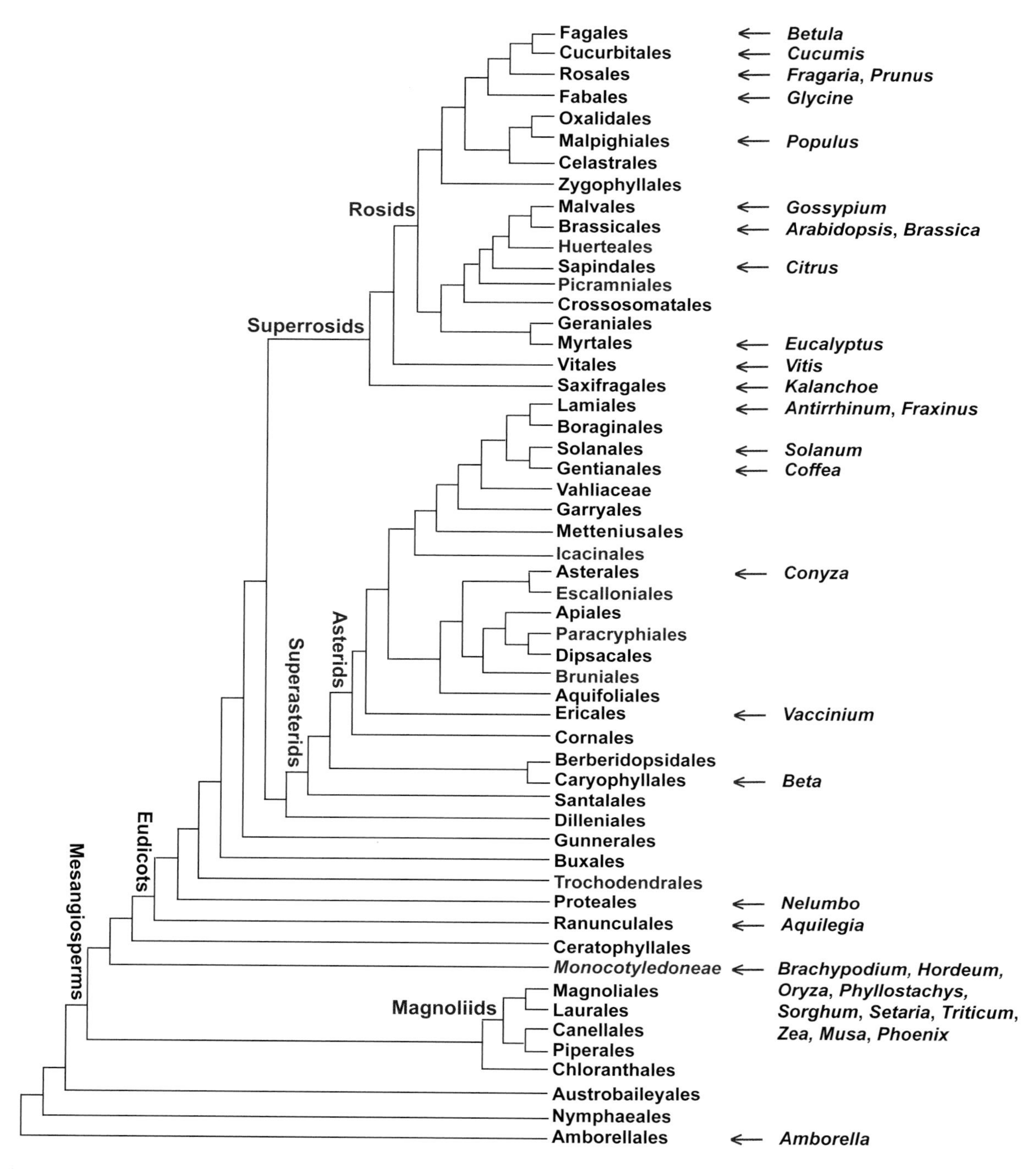

Figure 3.9. Summary tree of angiosperms (Soltis et al. 2011) updated reflecting recent findings; tree shows the placements of some model organisms as well as the placements of some other angiosperms with completely sequenced nuclear genomes; many other angiosperm genomes have now been sequenced (see https://www.ncbi.nlm.nih.gov/genome/browse/).

those made in light of the best tree, and our phylogenetic understanding of groups containing model organisms has improved dramatically over the past decade.

The snapdragon, *Antirrhinum majus*, represents one of the best model systems for the study of floral developmental genetics (e.g., Whibley et al. 2006); researchers can extrapolate from what is known about the genetics of floral development in *Antirrhinum* to other asterids (Donoghue et al. 1998; Reeves and Olmstead 1998). Although *Antirrhinum* was long placed in Scrophulariaceae, the family is now known to be polyphyletic, comprising perhaps four distinct lineages (Olmstead and Reeves 1995; Young et al. 1999; Olmstead et al. 2001; Tank et al. 2006; Soltis et al. 2011; Chapter 11). One clade (now Orobanchaceae; APG II 2003; APG III 2009; APG IV 2016) contains the parasitic members of the family (*Pedicularis*, *Castilleja* and Orobanchaceae s.s.), and a second clade (now Scrophulariaceae) contains *Scrophularia* and *Verbascum* (and now includes groups such as *Buddleja* and *Myoporum*, previously placed in Buddlejaceae and Myoporaceae, respectively). A third clade (now Plantaginaceae) contains many well-known "scrophs," such as the model system *Antirrhinum*, as well as *Digitalis*, *Veronica*, and genera usually placed in their own families (*Plantago*, *Callitriche*, and *Hippuris*). *Mimulus*, another model system and another former "scroph," is best placed in Phrymaceae, although bootstrap support for this relationship is low (Beardsley et al. 2001; Beardsley and Olmstead 2002; Tank et al. 2006; Schäferhoff et al. 2010; Chapter 11).

Both broad and focused phylogenetic analyses also have implications for the study of close relatives of crops that have become the focus of considerable genetic research, such as cotton (*Gossypium*), now known to be part of an expanded Malvaceae (see Chapter 10). The closest relatives of Poaceae, the grass family, have been identified (Flagellariaceae, Joinvilleaceae, and Ecdeiocoleaceae), and phylogenetic relationships within Poaceae are now generally well resolved (Grass Phylogeny Working Group 2001, 2012; see Chapter 7). Numerous legumes are of economic importance, and genome structure and evolution have been studied intensively in several species (e.g., Cannon et al. 2006, 2010). The closest relatives of Fabaceae have long been debated; however, phylogenetic analyses point to a clade of Quillajaceae, Polygalaceae, and Surianaceae as sister to Fabaceae (Soltis et al. 2011; see Chapter 10). Additional examples of phylogenetic studies encompassing model organisms include Solanaceae (Schlueter et al. 2004) and Asteraceae (Soltis and Soltis 2003; Kane et al. 2011).

A phylogenetic framework also is crucial for the investigation of genome evolution across the angiosperms (e.g., Soltis and Soltis 2013b; see Chapters 15 and 16). The origin and evolution of the angiosperms is one of the greatest terrestrial radiations. Flowering plants are responsible for most of our food and also account for much of the land-based photosynthesis and carbon sequestration. Angiosperms have also diversified to occupy nearly every terrestrial environment, and many aquatic ones. Understanding how angiosperms have accomplished this over such a short evolutionary timeframe may clarify many of the key processes underlying the assembly of many plant/animal associations and entire ecosystems. As noted, the angiosperm nuclear genomes that were initially sequenced not only are highly derived lineages, but also represent just a few branches within the angiosperm Tree of Life and as a result provide limited insights into features of the most recent common ancestor of all angiosperms. Key angiosperm innovations, including the origin of the flower and fruit, pollination systems and double fertilization, vessel elements, diverse biochemical pathways, and many of the specific genes that regulate key growth and developmental processes, appeared first among the basal angiosperm lineages. A thorough understanding of processes shaping genes and genomic features, and of the many similarities and differences between model monocot and eudicot plants, requires a perspective based on phylogeny. Such perspectives can be obtained only through analysis of an appropriately broad sampling of genomes, including lineages branching from the most basal node on the angiosperm tree (see Soltis et al. 2008a,b). This background was the basis for the sequencing of the nuclear genome of *Amborella* (*Amborella* Genome Project 2013), which offers the unique opportunity to "root" the analyses of diverse angiosperm features, including gene families and genome structure, as well as physiology and morphology (Soltis et al. 2008a) (Fig. 3.10). In the same way, the duck-billed platypus was chosen for nuclear genome sequencing in mammals (Warren et al. 2008) because it represents the same crucial phylogenetic position as sister to all other mammals.

SEQUENCE VARIATION: GENERAL PATTERNS, RAPID RADIATIONS, AND DIVERSIFICATION

The oldest angiosperm fossils date from the early Cretaceous, 130–136 mya, although molecular estimates for the age of the clade are typically somewhat older (see Chapter 2), followed by a rise to ecological dominance in many habitats before the end of the Cretaceous. The origin and early diversification of the angiosperms (Darwin's "abominable mystery") represent one of several key radiations in angiosperm evolution. Through contributions from paleobotany, phylogenetics, classical developmental biology, and

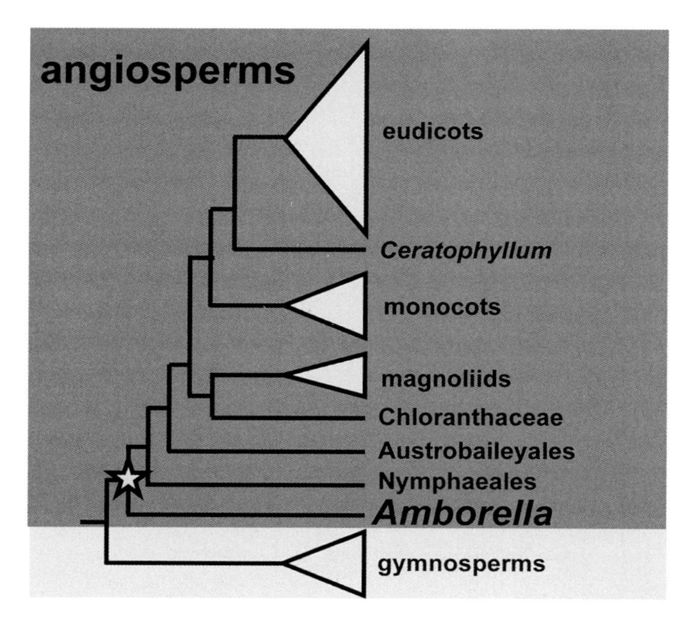

Figure 3.10. Simplified phylogenetic tree for flowering plants showing position of *Amborella* as sister to all other angiosperms. Modified from Soltis et al. (2008). Flowers of *Amborella* are shown in Fig. 4.7.

modern developmental genetics (evo–devo), tremendous progress has recently been made in resolving both early and more recent radiations. As reviewed below (see also Soltis et al. 2005b, 2008b), it is clear that there was not one radiation of the angiosperms, but layer upon layer of radiation.

A striking aspect of angiosperm phylogenies—both early and recent (e.g., Chase et al. 1993; D. Soltis et al. 1997a,b, 2000, 2011; P. Soltis et al. 1999a; Savolainen et al. 2000a,b)—is the highly uneven pattern of branch lengths (reviewed in Chase et al. 2000b; D. Soltis et al. 2000; Soltis et al. 2008b, 2011). Groups that are highly divergent for plastid genes are typically highly divergent for nuclear genes, although this is not always the case, as some lineages have experienced an acceleration in one genome but not another (see below). Nonetheless, this pattern provides the impression that both morphological and molecular evolution sometimes experience parallel alterations of evolutionary rates.

In contrast to clades exhibiting pronounced bursts of molecular evolution, other taxa display a marked slowdown in rates. Such taxa have been termed "molecular living fossils," and examples have been provided in vascular plants (e.g., tree ferns; P. Soltis et al. 2002), basal angiosperms including Winteraceae (Suh et al. 1993), and the early-diverging eudicots *Platanus* and *Nelumbo* (Sanderson and Doyle 2001). Long-lived perennials may have shorter branches than do closely related annuals (Smith and Donoghue 2008; see below). In addition to rate accelerations and decelerations, there is evidence that in some taxa, change at some loci is nearly clock-like or "quasi-ultrametric" (first noted by Albert et al. 1994). Hence, there is a mixture of long and short branches when one takes a global view of the angiosperm tree and focuses on the major groups. The combination of a long branch to a given clade and a starburst of short branches therein, with a repeated nesting of this general pattern, provides evidence for several episodes of rapid radiation in the flowering plants.

One major angiosperm radiation occurred soon after the divergence of the basalmost lineages. The root of the angiosperms is now clear, with the initial lineages well supported: Amborellaceae, Nymphaeales, and Austrobaileyales (Fig. 3.6). In contrast, relationships among the lineages of *Mesangiospermae* (monocots, Chloranthaceae, magnoliids, Ceratophyllaceae, eudicots) have been much more difficult to tease apart. The *Mesangiospermae* appear to represent the "big bang" of angiosperm evolution, having diversified rapidly, yielding the major branches of angiosperms and 97% of all angiosperm species within less than 5 million years (Moore et al. 2007; Soltis et al. 2008b). Putting this in perspective, 5 million years represents a timeframe comparable to the rapid radiation of the Hawaiian silversword alliance (Asteraceae–Madiinae), which putatively arose from a North American ancestor 5 mya (Baldwin and Sanderson 1998; Barrier et al. 1999). The fact that the first branches of the angiosperm topology (Amborellaceae, Nymphaeales, Austrobaileyales) are well supported and not species-rich and are followed by several species-rich clades indicates that the initial explosive radiation of the angiosperms did not coincide directly with the origin of flowering plants, but likely occurred later (cf. Mathews and Donoghue 1999; P. Soltis et al. 1999a, 2000).

Subsequent radiations within the clades of *Mesangiospermae* include, for example, the monocots, which also exemplify a mixture of long and short branches and appear to represent another rapid radiation (see Chapter 7). This same basic pattern is repeated in the eudicots, indicating another major deep-level radiation of flowering plants (Fig. 3.6). The grade of early-diverging eudicots, with some relationships still unresolved, is followed by the well-supported core eudicots, which include most (~70%) of the angiosperms. The core eudicot clade consists of several well-supported major lineages, including the rosids, asterids, Santalales, Caryophyllales, and Saxifragales. Although the monophyly of each subclade of core eudicots is strongly supported, relationships among these lineages were poorly resolved and weakly supported until recently (Moore et al. 2007, 2010; Lee et al. 2011; Soltis et al. 2011). This hypothesized radiation is in general agreement with the fossil record: the oldest eudicot fossils represent basal lineages (Platanaceae and Buxaceae). The fossil record also suggests an uneven distribution of species diversity across the major clades of eudicots (Davies et al. 2004), with the most species-rich groups known only from relatively young fossils, suggesting that a large proportion of eudicot diversity is the result of recent radiations (Magallón et al. 1999).

Simple inspection of phylogenetic trees reveals that this trend is common within major eudicot clades (P. Soltis et al. 1999a, 2000, 2005b; D. Soltis et al. 2000, 2011). For example, the Saxifragales and rosid clades also diversified over narrow time spans of several million years (Jian et al. 2008; H. Wang et al. 2009). The rapid diversification within the rosids (originated 110 [±6] to 93 [± 6] Ma, followed by rapid diversification into the malvid and fabid subclades around 108 [±6] to 91[±6] Ma and 107 [±6] to 83 [± 7] Ma, respectively) (H. Wang et al. 2009) may be of particular importance in that the inferred bursts of diversification correspond in timing with the rapid rise of angiosperm-dominated forests, as suggested by the fossil record (Crane 1987; Upchurch and Wolfe 1993; see Chapter 10). The rosid diversification also corresponds to the diversification of several other lineages that apparently evolved in parallel with the diversification of angiosperm forests (Chapter 10). Woody species in clades other than the rosids (e.g., Cornales, Bremer et al. 2004; J. Xiang pers. comm.; Saxifragales (e.g., Altingiaceae), Jian et al. 2008) seem to have diversified during the same window in time, probably indicative of the rapid rise of angiosperm-dominated forests on a worldwide basis.

The mixture of long and short branches noted for both basal angiosperms and eudicots continues for those clades typically recognized at the level of orders and families (Chase et al. 2000b; D. Soltis et al. 2000). For example, Malpighiales remain "one of the most recalcitrant clades in the angiosperm tree of life" (Wurdack and Davis 2009; p. 1551). It was early noted that the branch to Malpighiales is long, and within this clade there are also well-supported lineages (most recognized as families), but relationships among these lineages and within this clade have been extremely difficult to resolve. In early studies, the topology for the clade (when considering clades with support > 50%) was essentially a huge polytomy (D. Soltis et al. 2000). The same pattern can also be seen in Lamiales (Tank et al. 2006; Schäferhoff et al. 2010), Ericales (Schönenberger et al. 2005; Soltis et al. 2011), and Saxifragales (Jian et al. 2008) and is also observed within some large clades corresponding to families: Asteraceae (Jansen and Kim 1996; Panero and Funk 2002; Funk et al. 2009a,b), Orchidaceae (Cameron et al. 1999), Crassulaceae (Mort et al. 2000), Rhamnaceae (Richardson et al. 2000), and Fabaceae (Lavin et al. 2005).

Diversification in the angiosperms has been revisited in a rigorous statistical framework with bigger and bigger trees. Smith et al. (2011) used a large phylogenetic tree of over 50,000 terminals to examine shifts in diversification, and the results complement those noted above. Many major named clades are associated with shifts or bursts in diversification, but these rate shifts are not directly associated with major named clades (with a few exceptions) and seem to occur somewhat later (Fig. 3.11). As Smith et al. (2011)

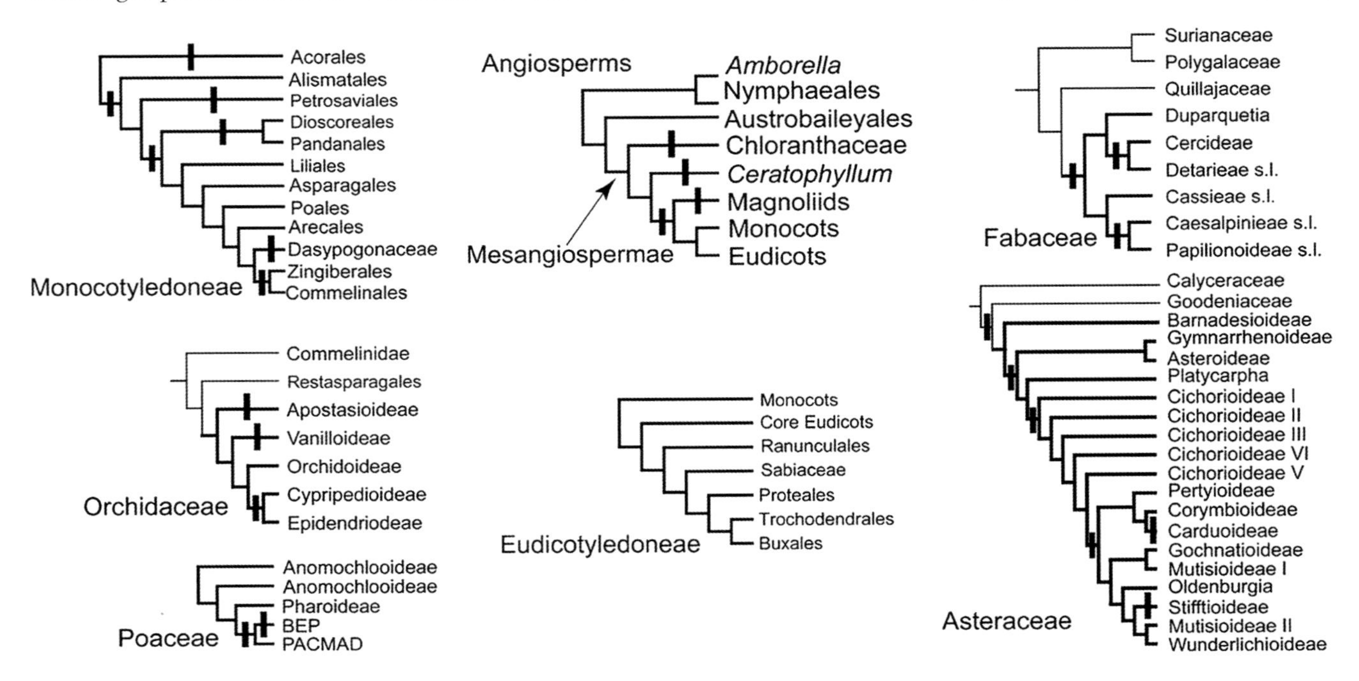

Figure 3.11. Rate shifts in diversification. The trees represent the phylogenetic relationships among the eight clades that were analyzed using large-scale phylogenetic approaches (Smith et al. 2011). Results of analyses of a backbone or summary tree approach are not shown. Black bars denote the locations of rate shifts in diversification. Note that shifts are not associated with the origin of flowering plants or the origin of several highly diverse, named groups nested within them. Instead, significant rate shifts are usually nested somewhat within each named group. (From Smith et al. 2011.)

summarize, "this may support a generality about angiosperm evolution: major shifts in diversification may not be directly associated with major named clades, but rather with clades that are nested not far within these groups" (p. 404). A similar hypothesis has been posed for bursts of diversification following polyploidy—there often seems to be a lag between the genome doubling event and the actual increase in species diversification (see Chapter 16). Alternatively, there has been an increase in extinction rates in the early-diverging lineages within these major clades. Using the Soltis et al. (2011) framework and then building a larger tree, Tank et al. (2015) found evidence for repeated nested shifts of diversification in angiosperms, in agreement with inferences of nested radiations based on visual inspection. The general phylogenetic pattern that occurs repeatedly in the flowering plants is suggestive of episodic radiations (resulting in short branches) that produce lineages that exist for long periods (creating the long branches), followed by another phyletic burst, and so on. A key innovation in a nested subclade may be most likely to spur on another radiation, and so on. Based on Tank et al. (2015), polyploidy may be one such key innovation.

But how do these nested rapid radiations in angiosperms compare to what is observed in other major clades? Do other major clades of terrestrial life show evidence of a similar "big bang" of major lineages (*Mesangiospermae*)? If so, when did they occur relative to those in angiosperms? Also, do these occur in a similar short time frame of a few million years? It appears that some lineages have tracked the angiosperm diversification events—or diversified "in the shadow of the angiosperms" (see Schneider et al. 2004). In fact, the timing of the inferred rapid radiation of the large rosid clade and the rise of angiosperm-dominated forests [108 to 91 million years ago (mya)] corresponds to the concomitant diversification of other clades that inhabit these forests, including amphibians, ants, placental mammals, and ferns (reviewed in H. Wang et al. 2009; see also Chapter 10).

Mammals may have similarly had bursts and slowdowns in diversification. Fossil evidence has suggested an explosive Tertiary radiation of placental mammals (Archibald and Deutschuman 2001; Bininda-Emonds et al. 2007; Bloch et al. 2007; Wible et al. 2007; Meredith et al. 2011). Several broad DNA-based phylogenetic analyses of mammals have been conducted, one using a supertree approach (Bininda-Emonds et al. 2007) and a more recent study using a supermatrix approach (Meredith et al. 2011). These two analyses have some similar conclusions, as well as striking differences. The Meredith et al. (2011) analysis agrees with the "long-fuse model" for the diversification of mammals, in which interordinal diversification occurred during the Cretaceous, with subsequent diversification within orders occurring in the Cenozoic (last 65 myr). In this scenario, many mammal lineages appear to have diversified following the major diversifications reviewed above for the angiosperms. In addition, within some subclades of placental mammals, the data also suggest that rapid radiations occurred in a similar, short window (~5 myr) to several major radiations in angiosperms (e.g., Klaus and Miyamoto 1991; Allard et al. 1992; Meredith et al. 2011).

Phylogenetic analyses of birds also indicate rapid radiation, but with some debate on the timing of these events. Although birds originated perhaps 160 mya, based on the fossil record, the clade showed a strong increase in diversification starting only 50 million years ago and continuing to about 5 million years ago (Hackett et al. 2008; Jetz et al. 2012; Jarvis et al. 2014; Zhang et al. 2014). This overall increase in diversification is more recent than most major angiosperm radiations, but as in angiosperms, radiations in birds are also the result of a number of significant rate increases within multiple lineages (e.g., within songbirds, waterfowl, gulls, and woodpeckers). Mayr (2013) suggested that, while there is debate, crown group passerine birds (songbirds) may not have diverged until the Cenozoic. The fossil record provides evidence for an early Tertiary (~65 mya) explosion of modern bird orders that may have occurred over a timeframe of just 5–10 myr (Feduccia 2003). This "big bang for birds" is more recent than that of angiosperms, but is comparable in terms of duration (a few million years) to that for the major deep radiations estimated for angiosperms (e.g., the diversification of *Mesangiospermae* and rosids).

In contrast to reports for mammals and birds, molecular and fossil data have generally been in close agreement for both the origin and early diversification of flowering plants (Chapter 2), although molecular dates for the origin of the angiosperms are generally older than the oldest fossils. Furthermore, the speed (often ~5 myr) and magnitude (ultimately ~ 350,000–400,000 extant species) of the explosive major radiations of flowering plants may be unique.

The data for angiosperms may be similar to many other groups of eukaryotes, but seem to contrast with the red algae (Freshwater et al. 1994), in which the pattern of change in *rbcL* across all branches gives the appearance of slow and clock-like lineage production without the periodic "starbursts" present in the angiosperm tree—or visible in trees for other organisms. However, phylogenetic analyses based on 18S rDNA sequence data suggest more of a mix of long branches subtending clades, with subsequent short branches within those clades (Ragan et al. 1994), although 18S rDNA sequences are highly divergent in red algae; Ragan et al. (1994, p. 7278) reported that "Rhodophyta are more divergent among themselves than are fungi, green algae and green plants considered together."

BIG TREES

Only a decade ago, phylogenetic trees of a few hundred taxa were considered large. However, with current next-generation sequencing (NGS) technologies and analytical capabilities, phylogenetic trees based on hundreds of genes and thousands of species are readily feasible. The workhorse of plant systematics has long been the plastid genome, and following the advent of 454 (Moore et al. 2006) and Illumina (Cronn et al. 2008; Parks et al. 2009) technologies, sequencing complete plastid genomes has become commonplace. With various means of enriching for plastid genes, sequencing entire genomes at a low price is now possible, even from samples with limited DNA or from herbarium specimens (Stull et al. 2013). The number of taxa in complete plastid genome sets has grown dramatically as a result of the application of NGS approaches. For example, only a few years ago, Moore et al. (2010) analyzed complete plastid genomes for 86 seed plants; recently, Ruhfel et al. (2014) examined 360 complete plastid genomes for green plants, and a dataset of over 1000 nearly complete plastomes is being assembled and analyzed as of this writing as a result of 1kp data (see below).

As the pace of sequencing continues to increase via technological advances, investigators are not only adding more and more genes to clarify relationships in problematic parts of the tree, but also obtaining data for numerous taxa in a short period of time, accelerating efforts to build trees for thousands of species. A series of recent studies has demonstrated the value of such enormous trees. Trees with thousands of terminals have now been constructed with diverse foci, not just resolving relationships per se. Big trees have been used to conduct analyses of molecular evolution in relationship to angiosperm life history (Smith and Donoghue 2008), patterns of biodiversity and conservation (For-

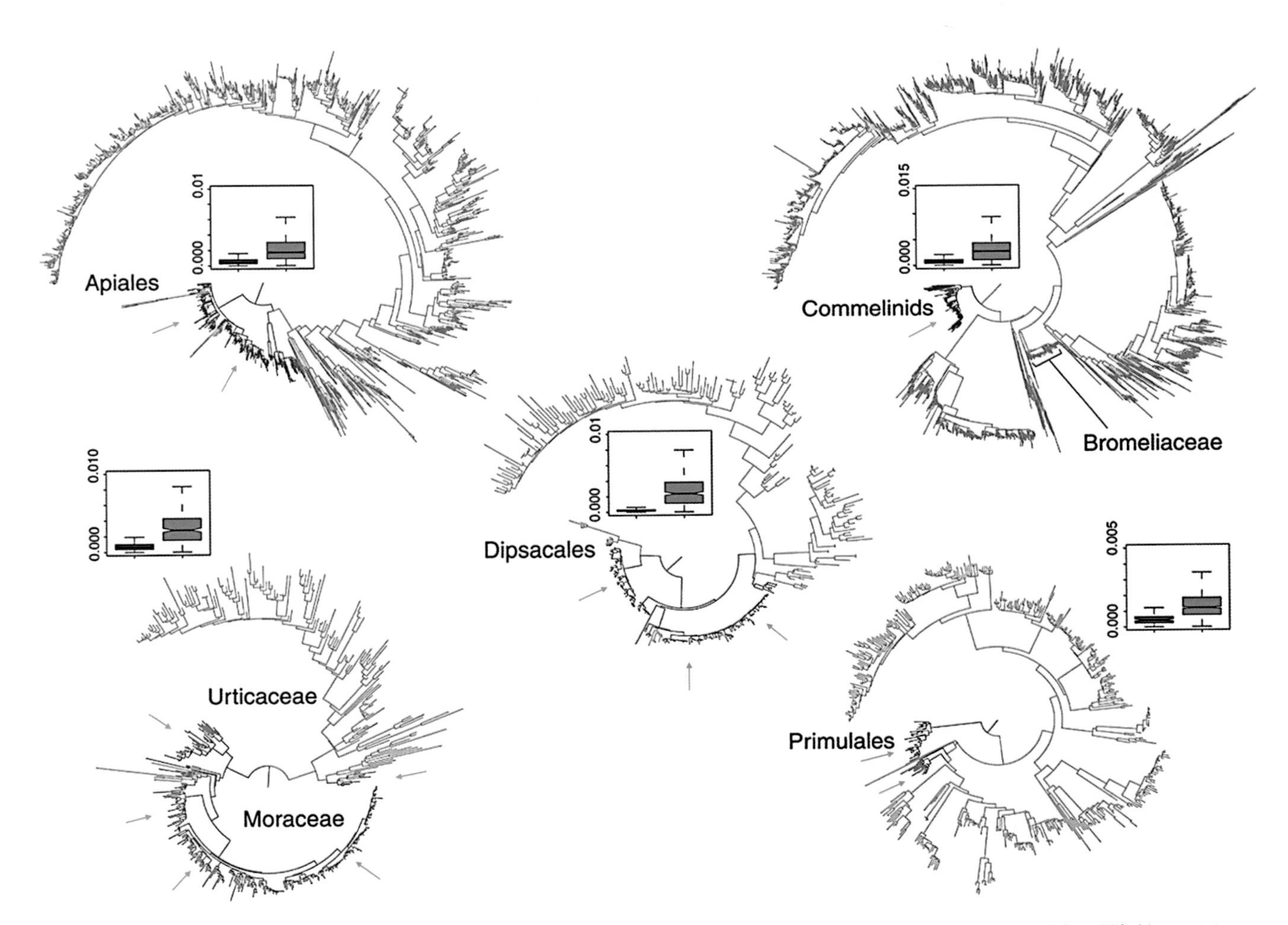

Figure 3.12. Phylogenies of five angiosperm clades with branch lengths proportional to substitutions per site. Branch colors represent inferred life-history states (dark and arrows for trees/shrubs; gray for herbs). Box plots show substitutions per site per million years for the inferred life-history categories; the center line represents the median, hinges mark the first and third quartiles, whiskers extend to the lowest and highest non-outlier. Outliers (not shown) have values >1.5 times beyond the first or third quartiles. (Modified from Smith and Donoghue 2008.)

est et al. 2007), responses to climate change (Willis et al. 2008; Thuiller et al. 2011), rates of diversification (Smith et al. 2011), and analyses of character evolution on a very large scale (Edwards and Smith 2010; Soltis et al. 2013). We review a few examples here briefly to illustrate the utility of such trees.

Using large trees, Smith and Donoghue (2008) further substantiated the association between life history and rates of molecular evolution (see Gaut et al. 1992). Large phylogenies for five major angiosperm clades, each with woody and herbaceous members (see Fig. 3.12), clearly demonstrated that rates of molecular evolution are consistently low in trees and shrubs (with relatively long generation times), compared to the rates found in related herbaceous plants from the same major clades (generally with shorter generation times).

An exciting recent use of big trees is in the study of phylogenetic diversity (PD) and conservation. PD is a biodiversity index that considers the length of the evolutionary branches on a tree that connect a given set of species. Forest

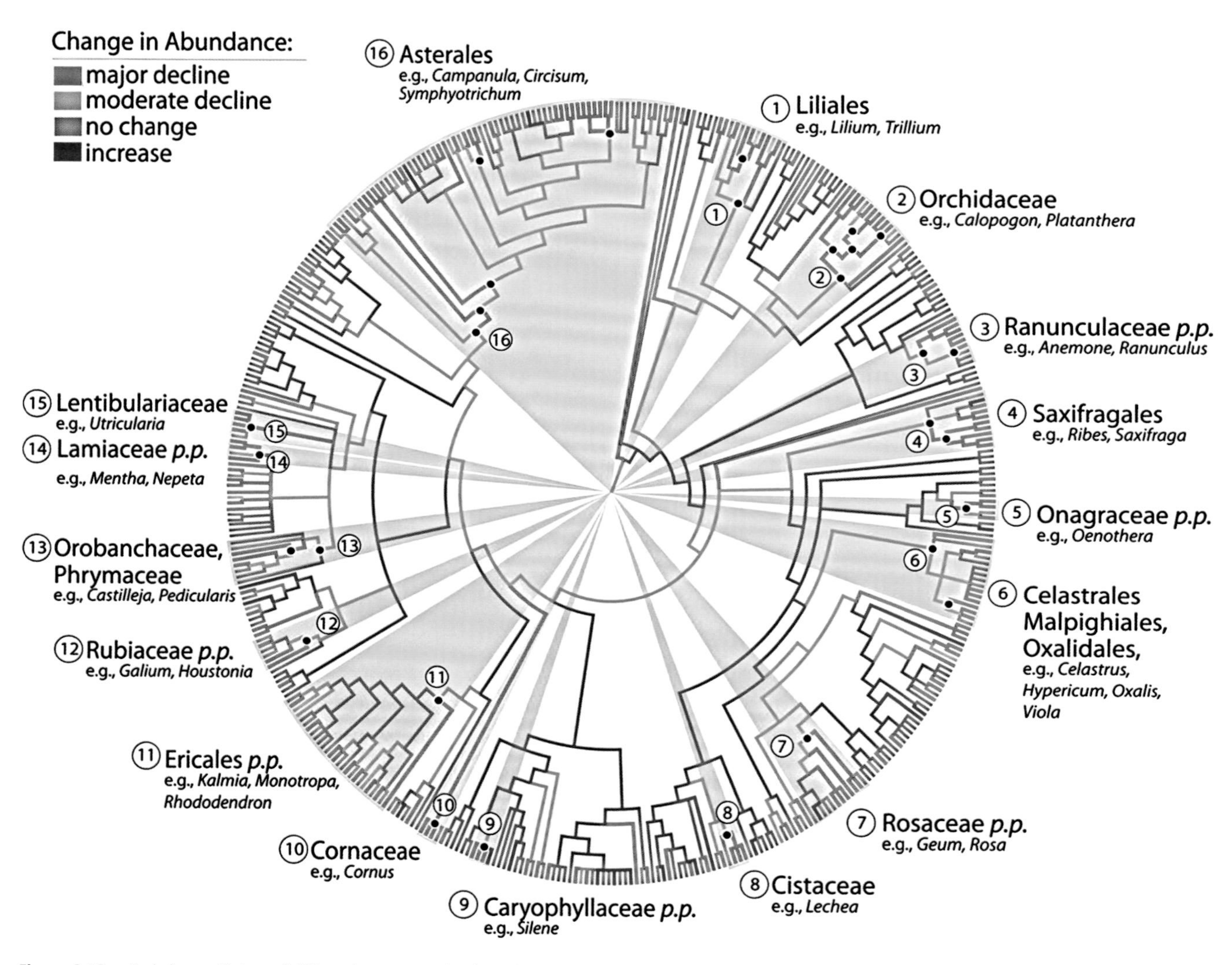

Figure 3.13. A phylogenetic tree of 429 angiosperm species from the Concord flora showing changes in plant abundance from 1900 to 2007. Change in abundance ranges from −5 to +4 and was calculated as the difference in abundance for each taxon in 1900 and 2007 based on 7 abundance categories (0 to 6). Branch color indicates parsimony character-state reconstruction of change in abundance: red (major decline, −5 to −3), pink (moderate decline, −2), gray (little to no change, −1 to +1), and blue (increase, +2 to +4). Average decline in abundance was calculated for all internal nodes as the mean change in abundance of descendant nodes weighted with branch length information ascertained from divergence time estimates. An average decline of 2.5 or greater corresponds to a decline in abundance of 50% or greater, based on conservative scoring using 6 abundance categories (0 to 5). Clades exhibiting these major declines are indicated with black dots. Each of the most inclusive clades exhibiting these declines is indicated in pink and referenced numerically to their clade name. Subclades in major decline that are nested within more widely recognized clades are labeled with the more familiar name followed by pro parte (p.p.). These clades include diverse New England wildflower species, including members of the following clades: Ranunculaceae, various Asterales, Rubiaceae, Lentibulariaceae, Cornaceae, Liliales, Orobanchaceae, Lamiaceae, Orchidaceae, Rosaceae, Saxifragales), and Ericales. (Figure taken from Willis et al. 2008; legend modified from Willis et al. 2008.)

et al. (2007) demonstrated the application of PD inferred from big trees in conservation biology in a study of the flora of the Cape region of South Africa, a well-known biodiversity hotspot. The western part of this region, with winter rainfall, has approximately twice the number of plant species compared to the eastern region (with year-round rainfall). Forest et al. (2007) found a marked east-west regional difference in the distribution of PD that roughly corresponds to climatic zones, with PD being higher in the eastern region than in the western area. Hence, species richness was decoupled from PD. That is, the region with higher species richness had lower PD, with major implications for conservation efforts and clearly revealing the value of constructing large trees for a geographic area rather than a specific clade.

Willis et al. (2008) (Fig. 3.13) demonstrated the utility of big phylogenetic trees for a small geographic area for addressing climate change in their investigation of flowering plants in Thoreau's woods in Massachusetts, United States. Due to the efforts of Henry David Thoreau, data on the

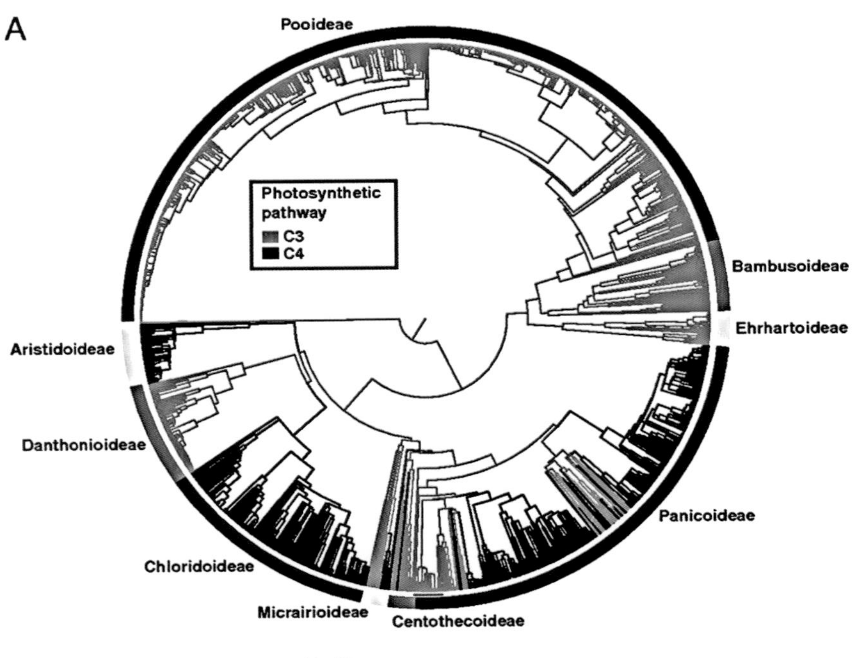

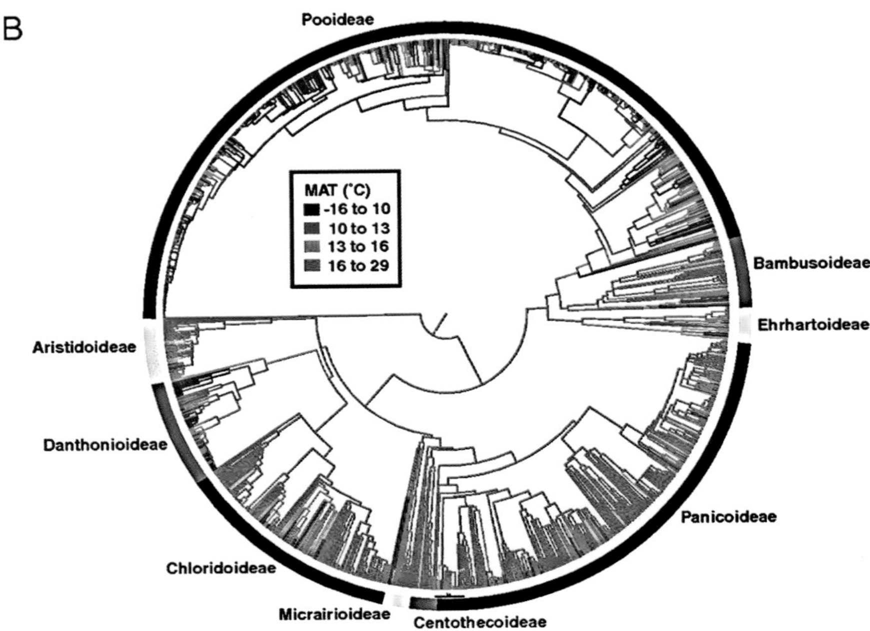

Figure 3.14. The evolution of photosynthetic pathway and temperature niche in grasses. (A) Green lines indicate C_3 photosynthesis; black lines indicate C_4 photosynthesis. Maximum likelihood methods reconstructed 20 origins of C_4 photosynthesis and one reversal to C_3 photosynthesis. (B) Maximum likelihood reconstructions of mean annual temperature (MAT), using the mean values of species generated from 1,146,612 georeferenced herbarium specimens (taken from Edwards and Smith 2010).

flora for that area span the past 150 years with valuable information on changes in species abundance and flowering time. Willis et al. (2008) analyzed these data in a phylogenetic context and noted that change in abundance over that time period is strongly correlated with flowering-time response. Species that do not respond to temperature have greatly decreased in abundance; these include members of Ranunculaceae (anemones and buttercups), Cornaceae (dogwoods), Lamiaceae (mints), Saxifragaceae, and some Liliales. Because flowering-time response traits are shared among closely related species, this study suggests that "climate change has affected and will likely continue to shape the phylogenetically biased pattern of species loss in Thoreau's woods" (Willis et al. 2008, p. 17029). The use of large phylogenetic trees to examine patterns of climate change in a phylogenetic as well as an ecological context continues to be an exciting area of research (e.g., Edwards et al. 2007; Thuiller et al. 2011; Beaulieu et al. 2012).

Edwards and Smith (2010) used a large phylogenetic tree in a study of the evolution of C_3/C_4 grasses relative to habitat features (Fig. 3.14). As background, C_4 grasses dominate tropical and subtropical grasslands and savannas, whereas C_3 grasses typify cooler temperate grassland regions. The evolution of the C_4 pathway has long been thought to have enabled C_4 grasses to persist in warmer environments. Edwards and Smith (2010) tested this hypothesis using a 1,230-taxon tree of grasses with accompanying climate data and found that grasses are ancestrally a warm-adapted clade and that C_4 evolution was not correlated with shifts between temperate and tropical biomes. This result can be seen by comparing the ancestral-state reconstructions of C_4 and mean annual temperature (Fig. 3.14); 18 of 20 inferred C_4 origins were correlated with marked reductions, rather than increases, in mean annual precipitation.

These examples illustrate well the rapid changes that are occurring in plant systematics. Because phylogenetic trees have become foundational in studies of ecology and climate change, increasing efforts are focusing on building large phylogenetic trees for all of the plants of large geographical regions, a marked departure from the traditional systematics goal of generating a tree for a specific clade.

These advances set the stage for rapid progress in filling in major gaps in the green plant branch of the tree of life, as well as the entire tree of all life. The Open Tree of Life (http://opentreeoflife.org/) initiative seeks to provide an initial first-pass tree of life, as well as the infrastructure and computational tools to facilitate updating and annotating the tree. In addition, efforts will be made to engage the systematics community on a broad scale, given that the long-term success of the project hinges on community buy-in and involvement, which require and lead to the deposition of data and alignments. At this point, the botanical community, as well as systematists in general, have done a very poor job of submitting DNA alignments and trees to public repositories such as Dryad—this should not be confused with submitting the gene sequence data to GenBank, which is often mandatory. However, only about 17% of trees and alignments generated by systematists are actually retrievable; that is, over 80% of our underlying datasets and trees are essentially lost (Drew et al. 2013).

In summary, this is an exciting time in systematics. The systematics community is now in position to take a true "moon shot," that is, using cutting-edge tools to assemble and analyze enormous datasets of all the named ~500,000 species or more of green plants. In fact, the first draft of the Tree of Life (http://tree.opentreeoflife.org) represents an important milestone—a starting point for biodiversity science comparable to the first draft of the human genome sequence. This enterprise will not only facilitate a better understanding of the diversification of green plants, but will also permit the building of a cyberinfrastructure toolkit to navigate and employ big trees.

4 The ANA Grade

INTRODUCTION

A longstanding goal has been to determine which extant taxa are sister to all other taxa in the angiosperm tree of life and concomitantly to use that information to infer ancestral morphological, chemical, pollination (and now genomic) attributes of the earliest angiosperms. Before modern phylogenetics, prominent authors (e.g., Takhtajan, and Cronquist) frequently wrote in terms of the "most primitive living angiosperms." We avoid such terminology today because, as we note throughout, taxa attached to basal nodes may exhibit an array of both ancestral ("primitive") as well as derived ("advanced") traits.

Because studies using combined datasets of multiple genes have indicated that the basalmost angiosperms form a grade rather than a clade (reviewed below), no formal name should be applied to these taxa. For convenience, we refer to these early-branching lineages (Amborellaceae, Nymphaeales, and Austrobaileyales) as the "basalmost" or "basal angiosperms," or simply as the **ANA** grade (**A**mborellaceae, **N**ymphaeales, **A**ustrobaileyales) (Soltis et al. 2005b; or ANITA; Qiu et al. 1999, in which Austrobaileyales was replaced by its component families, Illiciaceae [now Schisandraceae], Trimeniaceae, and Austrobaileyaceae).

Much of the interest in determining the basalmost branches of angiosperms has involved attempts to use that information on branching order to reconstruct the hypothetical ancestral angiosperm. We do not mean to imply, however, that these (or any) extant species are at the base of the angiosperm tree (i.e., that others were derived from them). There has been enormous interest however in examining key angiosperm attributes, such as the origin and diversification of the flower, in light of these basal lineages. Studies of floral diversity in basal angiosperms have, in turn, influenced a broad range of disciplines, including plant systematics, ecology, pollination biology, and developmental biology.

Most morphology-based classifications of the twentieth century (e.g., Cronquist 1981, 1988; Takhtajan 1987, 1997; Thorne 1992a,b, 2001) placed most of the taxa now recognized as basal (based on phylogenetic analyses of molecular data), in what was termed subclass Magnoliidae (not to be confused with the clade now recognized as *Magnoliidae*, Cantino et al. 2007; Chapter 3). This Magnoliidae of traditional classifications also contained Ranunculales, now recognized to be basal eudicots (Chapter 8). Endress (1986b) proposed three potential groups of basalmost angiosperms: (1) Degeneriaceae, Himantandraceae, Eupomatiaceae, Austrobaileyaceae; (2) Chloranthaceae, Trimeniaceae, Amborellaceae; (3) Winteraceae.

The basalmost angiosperm lineages are now clear, based on numerous molecular studies, with support from molecular + morphological analyses: the ANA grade. Our overview is historical, beginning with morphological cladistic analyses, followed by early use of DNA data, then larger and larger DNA datasets, and DNA + morphology.

EARLY ANALYSES OF MORPHOLOGY

In early phylogenetic analyses of morphology, angiosperms were scored as a single entity rather than as several exemplars (Chapters 1, 3). Doyle and Donoghue (1986, 1992) scored angiosperms as a composite of Magnoliales and

Winteraceae, both of which were considered "primitive" at that time. Donoghue and Doyle (1989b) considered a composite Magnoliales clade (scored on the basis of several families, including Degeneriaceae, Magnoliaceae, Myristicaceae, Eupomatiaceae, and Annonaceae). To correct this shortcoming, Doyle et al. (1994) and Doyle (1996) included several angiosperms considered "primitive" (e.g., Nymphaeales, Piperaceae, Saururaceae, Aristolochiaceae, Winteraceae, Magnoliales, Calycanthaceae, Chloranthaceae, and Laurales). Although these were improvements, the small number of families involved in early analyses illustrates one limitation of these morphological cladistic studies—not all potentially basal angiosperm families were sampled. In contrast, in many molecular phylogenetic studies (see below), angiosperms were broadly sampled, and as a result, representatives of all or most basal families were included; for the larger families, several genera were often included. Another limitation of these early morphological analyses is that many families that we now consider critical, such as Amborellaceae and Schisandraceae (including Illiciaceae; APG III 2009 and APG IV 2016; covered later in this chapter), either were not included or were included in orders with taxa to which they are now known not to be closely related.

With these caveats, early studies using morphology recovered an array of putatively basalmost extant angiosperms. Donoghue and Doyle (1989b) placed a composite Magnoliales clade as sister to all other angiosperms (Fig. 3.1). This result was consistent with views at that time regarding the most primitive angiosperms based on evolutionary taxonomy, represented as "bubble diagrams" (Fig. 3.7) (e.g., Cronquist 1968, 1981; Takhtajan 1969, 1991). Magnoliales of Donoghue and Doyle (1989b) were followed by two other clades of woody magnoliids: a composite Laurales that also included Chloranthaceae and a winteroid group (Illiciaceae/Schisandraceae, Winteraceae). However, the taxonomic makeup of these "composite" clades does not agree with modern DNA-based circumscriptions (Chapters 3 and 5). The remaining major clades of angiosperms identified in Donoghue and Doyle (1989a,b) were "dicots" with triaperturate pollen (now eudicots; labeled as such in Figure 3.1; Doyle and Hotton 1991) and a "paleoherb" group (now known to be polyphyletic) consisting largely of herbaceous magnoliids (Aristolochiaceae, Lactoridaceae), Nymphaeaceae, and monocots. Donoghue and Doyle (1989a) stressed that their results were not robust; in trees one step longer, paleoherbs were sister to all other angiosperms.

Other early phylogenetic analyses of morphology (Loconte and Stevenson 1991; Loconte 1996) placed Calycanthaceae as sister to all remaining angiosperms, followed by a grade of Magnoliales, Laurales, and Illiciaceae + Schisandraceae. Most of these lineages are now in the DNA-based *Magnoliidae* (Cantino et al. 2007); only Illiciaceae + Schisandraceae represent a basalmost lineage. Taylor and Hickey (1990, 1992, 1996a,b) recovered Chloranthaceae, an herbaceous family, as sister to all other extant angiosperms. As a result, attention was focused on herbaceous taxa as possible basalmost angiosperms. Not only are herbaceous taxa found early in the fossil record (Taylor 1990; Taylor and Hickey 1992), but these morphological analyses (e.g., Taylor and Hickey 1992) also placed Chloranthaceae, followed by Piperaceae (another herbaceous lineage), as sister to all other angiosperms (Hickey and Taylor 1996).

Doyle et al. (1994) analyzed morphological data (and also combined rRNA sequence data plus morphology; see below). Analyses of morphological data alone again indicated that herbaceous taxa were early-diverging, with monocots + Nymphaeaceae sister to all other flowering plants, followed by Piperaceae + Saururaceae and then Aristolochiaceae as subsequent sisters to the remaining angiosperms. However, in trees just one step longer, a very different topology was recovered, with Magnoliales sister to all other taxa, followed by Laurales, then Calycanthaceae (all woody magnoliids; Chapter 5).

Nixon et al. (1994) also provided evidence that herbaceous lineages might be basalmost. Their cladistic analyses of morphological data included extant and fossil seed plants. When fossils were included, a large polytomy was obtained. When fossils were excluded, *Chloranthus* (Chloranthaceae) appeared as sister to the remaining angiosperms, followed by the herbaceous lineage, *Ceratophyllum*.

Nandi et al. (1998) conducted a large morphological cladistic analysis of 162 living angiosperms—and combined these data with *rbcL* sequences (below). This study remains the most comprehensive morphological cladistic analysis for angiosperms. Their topology shows strong similarities to DNA-based trees. With morphology alone, Nandi et al. (1998) recovered Ceratophyllaceae, followed by Chloranthaceae, Amborellaceae, Winterales, and Austrobaileyales + Illiciales as subsequent sisters to all other living angiosperms. Thus, there is strong signal in morphology alone for the basal (or nearly so) placement of Amborellaceae and Austrobaileyales.

EARLY DNA ANALYSES

In contrast to morphology-based analyses, DNA-based studies were much more comprehensive in coverage, often including all families placed in Cronquist's Magnoliidae and multiple exemplars of families. Individual analyses of the plastid genes *rbcL* and *atpB* and nuclear 18S rDNA (e.g., Hamby and Zimmer 1992; Chase et al. 1993; Qiu et al. 1993; Soltis et al. 1997b; Savolainen et al. 2000a) all agreed in recognizing the same suite of basal angiosperms as either a clade (*rbcL*) or a grade (*atpB*; 18S rDNA).

The *rbcL* studies of Chase et al. (1993) and Qiu et al. (1993) placed *Ceratophyllum* (Chapter 8) as sister to all other extant flowering plants and indicated that all other angiosperms sometimes regarded as "basal" or "primitive" formed a clade sister to the eudicots. *Amborella* and water lilies (Nymphaeaceae) were part of a clade referred to as "paleoherb 1" (although *Amborella* is woody; Figs. 4.7 and 6.1). 18S rDNA showed that basal angiosperms formed a grade rather than a clade and was the first gene to place *Amborella* near the base of the angiosperm phylogeny (Soltis et al. 1997b). That study revealed a clade of Austrobaileyales (Chapter 3), followed by *Amborella* as successive sister taxa to all other angiosperms. However, internal support was low in 18S and *rbcL* trees. *AtpB* sequences (Savolainen et al. 2000a) provided results similar to 18S rDNA, indicating that basal angiosperms formed a grade, with Amborellaceae, followed by Nymphaeaceae, and then Austrobaileyales as successive sisters to all other angiosperms. With *atpB*, *Ceratophyllum* plus the monocot *Acorus* were sister to all other monocots.

Using only 1000 bp of *matK* sequence data, Hilu et al. (2003) reconstructed a phylogenetic tree for angiosperms similar to that obtained using *rbcL*, *atpB*, and 18S rDNA (P. Soltis et al. 1999a,b; D. Soltis et al. 2000). *matK* placed Amborellaceae, followed by Nymphaeaceae and Austrobaileyales, as successive sisters to all other angiosperms with moderate bootstrap support. Similarly, phylogenetic analysis of *trnL-F*, a plastid intron and spacer region, yielded *Amborella*, followed by Nymphaeaceae and then Austrobaileyales, as subsequent sisters to all other angiosperms (Borsch et al. 2003).

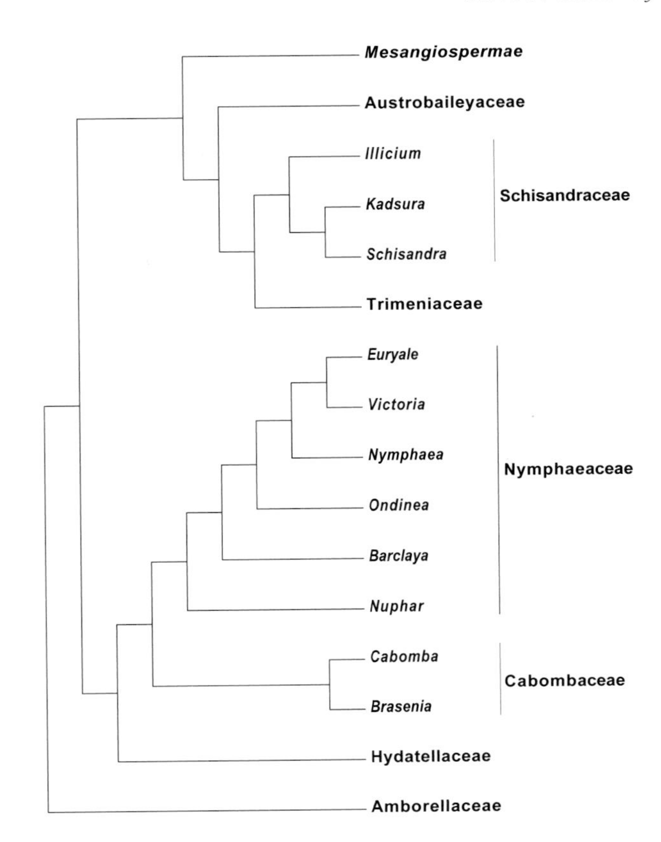

Figure 4.1. Summary phylogeny for basalmost angiosperms (the ANA grade). Basic topology is a composite of Soltis et al. (2011), Drew et al. (2014), and Yoo et al. (2005).

COMBINED DNA DATASETS

Single-gene studies were rapidly superseded by combined datasets; the sequence of nodes at the base of extant angiosperms is now known with high confidence (Fig. 4.1; see Chapter 3): Amborellaceae, Nymphaeales, Austrobaileyales. This is remarkable in that some investigators had suggested that elucidating the root of extant angiosperms might not be possible because of rapid radiation and extinction. Several groups of investigators independently and simultaneously arrived at the same conclusion regarding these nodes using different datasets and approaches (Mathews and Donoghue 1999; Parkinson et al. 1999; Qiu et al. 1999; P. Soltis et al. 1999a; D. Soltis et al. 2000; Graham and Olmstead 2000; Graham et al. 2000). Mathews and Donoghue (1999, 2000) identified these same basal nodes by analyzing duplicated phytochrome genes using a reciprocal rooting strategy. Simultaneous analysis of sequences from a gene pair that results from a duplication event in the lineage leading to extant angiosperms should yield two identical or similar gene trees; the angiosperm topology could therefore be rooted without outgroups on the branch connecting the duplicated genes. In addition to the many studies reviewed above, the placement of *Amborella* alone as sister to all other flowering plants is supported by analyses of structural features of B-class genes (Kim et al. 2004a).

The high level of sequence divergence between angiosperms and their gymnosperm outgroups also raised the question of whether the rooting with *Amborella* sister to other extant angiosperms was affected by long-branch attraction. Qiu et al. (2001) addressed this question using several types of artificial (random and nonrandom) sequences, as well as sequences of a lycopod and a bryophyte as outgroups. They concluded that the *Amborella* rooting was based on historical signal rather than an artifact of long branches.

Datasets have become increasingly larger in terms of

the numbers of genes sequenced and taxa included. In the span of just a few years, the number of genes combined and brought to bear on the question of basal angiosperm relationships (often as part of broader surveys—Chapter 3) quickly increased from two (e.g., Mathews and Donoghue 1999; Savolainen et al. 2000a), three (P. Soltis et al. 1999a; D. Soltis et al. 2000), 11 (Zanis et al. 2002), 17 (Graham and Olmstead 2000; Graham et al. 2000; Soltis et al. 2011; Fig. 3.6), to surveys of complete plastid genomes (e.g., Jansen et al. 2007; Moore et al. 2007, 2010; Drew et al. 2014; Ruhfel et al. 2014) to datasets of numerous nuclear genes (Lee et al. 2011; N. Zhang et al. 2012; Wickett et al. 2014; Zeng et al. 2014).

There are important differences among these studies in the genes used, as well as in taxon sampling and methods of phylogenetic analysis. Genes representing the nuclear, mitochondrial, and plastid genomes were sometimes combined, whereas in other cases, the genomic compartments were kept separate. In some cases, few taxa were used, whereas other studies sampled all families of basal angiosperms and many representatives of larger families. Diverse phylogenetic approaches have also been used. In addition to parsimony (used in early studies), investigators have applied maximum likelihood (e.g., Parkinson et al. 1999; Barkman et al. 2000; Qiu et al. 2000; P. Soltis et al. 2000; Stefanović et al. 2004; Zanis et al. 2002, 2003; Leebens-Mack et al. 2005; Lee et al. 2011; Soltis et al. 2011; N. Zhang et al. 2012; Ruhfel et al. 2014; Wickett et al. 2014; Zeng et al. 2014) and Bayesian methods (Zanis et al. 2002; Soltis et al. 2007c; Wickett et al. 2014).

Nearly all large-scale molecular studies (Qiu et al. 1993, 1999, 2005; Parkinson et al. 1999; Savolainen et al. 2000a; P. Soltis et al. 1999; D. Soltis et al. 2000, 2007c, 2011; Zanis et al. 2002; Borsch et al. 2003; Hilu et al. 2003; Leebens-Mack et al. 2005; Jansen et al. 2007; Moore et al. 2007, 2011; Lee et al. 2011; N. Zhang et al. 2012) from the past 20 years, including recent studies involving dense sampling of genes and taxa (e.g., Ruhfel et al. 2014; Wickett et al. 2014; Zeng et al. 2014) have reached the same conclusion, that *Amborella* alone is sister to all other angiosperms. This result is bolstered by analyses of complete plastid genomes (Ruhfel et al. 2014) and large datasets of nuclear genes (Wickett et al. 2014; Zeng et al. 2014).

However, internal support for this placement of *Amborella* varies considerably among studies. With three genes, the support for the monophyly of all angiosperms except *Amborella* received 65% jackknife support (P. Soltis et al. 2000); with 17 genes, BS support was 83% (Soltis et al. 2011). With complete plastid genomes and large datasets of nuclear genes, support typically is > 95% BS (e.g., Moore et al. 2007; Drew et al. 2014; Ruhfel et al. 2014; Wickett et al. 2014; Zeng et al. 2014).

Several analyses support *Amborella* + Nymphaeales as sister to all other living angiosperms (e.g., Barkman et al. 2000; P. Soltis et al. 2000; Goremykin et al. 2009, 2013; Finet et al. 2010; Qiu et al. 2010; D. Soltis et al. 2007c; Wu et al. 2007). Some of these results may reflect the choice of genes or genomes used. Plastid analyses, especially complete plastid genomes (Leebens-Mack et al. 2005; Cai et al. 2006; Jansen et al. 2007; Moore et al. 2007; Ruhfel et al. 2014), have generally provided strong support for *Amborella* as sister to all other angiosperms, followed by Nymphaeales and then Austrobaileyales (Savolainen et al. 2000a; Borsch et al. 2003; Hilu et al. 2003; Moore et al. 2011; Soltis et al. 2011).

The nuclear genome also supports *Amborella* alone as sister to all other living angiosperms. Primarily due to the difficulty in confidently assessing orthology across a diverse clade such as seed plants, nuclear gene sequence data, with the exception of 18S and 26S rDNA, have not been widely used in large-scale phylogenetic studies of angiosperms until quite recently. Early analyses of 18S rDNA (see above) were the first to show the ANA grade to be subsequent sister taxa to all other living angiosperms (Soltis et al. 1997b). Subsequent analyses of 18S and 18S + 26S rDNA continued to point to the ANA grade as basalmost lineages, but support for relationships was low (Qiu et al. 1999; Barkman et al. 2000; Zanis et al. 2002; Qiu et al. 2005; Soltis et al. 2011), which is not surprising given the low signal in these regions (P. Soltis and Soltis 1998). Nonetheless, a recent large analysis of numerous complete 18S + 26S rDNA sequences found *Amborella* alone (but with bootstrap support of only 56%) as sister to all other angiosperms (Maia et al. 2014). Two phytochrome genes were also employed and also suggested *Amborella* sister to other angiosperms (Mathews and Donoghue 1999).

Recent phylogenetic studies have used low-copy nuclear genes to investigate angiosperm relationships, as well as the root of the angiosperms (Finet et al. 2010; Lee et al. 2011; Morton 2011; N. Zhang et al. 2012; Wickett et al. 2014; Leebens-Mack et al. in prep.). Most of these studies have supported *Amborella* alone as sister to all other angiosperms with varying degrees of internal support (Lee et al. 2011, >65% Maximum Likelihood (ML) BS; N. Zhang et al. 2012, ML BS = 83%, Posterior Probability (PP) in Bayesian analysis = 0.99; Wickett et al. 2014, ML BS >90%). In contrast, Finet et al. (2010) recovered *Amborella* + Nymphaeales as sister to other angiosperms but without BS support above 50% (ML BS = 49%); however, they also employed low taxon density.

The mitochondrial genome provides the best support for the *Amborella* + Nymphaeales relationship, but that evidence is not strong. Typically mtDNA sequence data have been used in combination with sequences from other compartments, and the combined data have typically recovered *Amborella* alone as sister to all other angiosperms, probably due to the strength of the plastid signal (Qiu et al. 1999, 2005; Zanis et al. 2002; Soltis et al. 2011). Qiu et al. (2010) included sequence data from four mtDNA genes and broad taxonomic sampling and in an unpartitioned analysis found that *Amborella* plus Nymphaeales form a clade (ML BS = 79%) that is sister to all other angiosperms. However, the mtDNA data may be more complex than initially thought. An analysis of third codon positions from the mtDNA data of Qiu et al. (2010) actually supports *Amborella* alone as sister to other angiosperms (ML BS = 66%; Drew et al. 2014). The *Amborella* mitochondrial genome is also an excellent example of horizontal gene transfer (Chapter 15).

Method of phylogenetic analysis may also play a role in which topology is recovered at the base of the angiosperms. For example, parsimony analysis of the three-gene dataset (P. Soltis et al. 2000) recovered *Amborella* alone as sister to other angiosperms, whereas a Bayesian analysis of the same dataset recovered *Amborella* + Nymphaeales (Soltis et al. 2007c). Analysis of a large five-gene dataset with parsimony provided strong support for *Amborella* alone as sister (BS = 83%), whereas ML analysis of the same dataset recovered a weakly supported (BS = 65%) clade of *Amborella* + Nymphaeales (Burleigh et al. 2009).

Reducing highly variable sites ("noise"; below) in datasets has also recovered *Amborella* + Nymphaeales. One application of coalescent methods recovered *Amborella* + Nymphaeales (Xi et al. 2014), but taxon sampling was sparse and other issues problematic (Simmons and Gatesy 2015). In contrast, a coalescence-based analysis with greater taxon sampling yielded *Amborella* as sister to all remaining extant angiosperms (Wickett et al. 2014).

NOISE-REDUCED DNA DATASETS

As reviewed, studies using broad taxon sampling have typically recovered *Amborella* as sister to other living angiosperms, whereas noise-reduced datasets (highly variable sites are removed) often yield *Amborella* + Nymphaeales. Barkman et al. (2000) employed the noise-reducing program RASA (Relative Apparent Synapomorphy Analysis; Lyons-Weiler et al. 1996) and recovered *Amborella* + Nymphaeales as sister to other angiosperms. However, the results of Barkman et al. (2000) varied depending on the method of analysis (e.g., parsimony vs. likelihood).

More recently, Goremykin et al. (2013) investigated the effects of iteratively reducing "noise" in plastid data through removal of highly variable sites (per Goremykin et al. 2009) and concluded that *Amborella* + Nymphaeales formed a clade sister to all remaining angiosperms (with 94% bootstrap support in one analysis). However, Graham and Iles (2009) investigated noise-reduction effects on the angiosperm rooting; their results suggesting the premise of this approach is problematic. Drew et al. (2014) carefully reexamined the analyses and findings of Goremykin et al. (2013). Significantly, analysis of the discarded characters from the noise-reduced alignments of Goremykin et al. (2012) produced a well-resolved tree with all major angiosperm subgroups. The tree built on "noise" correctly recovered the accepted topology for angiosperms (e.g., Stevens 2001 onwards; APG III 2009, APG IV 2016; Soltis et al. 2011), confirming that these variable sites were, in general, not noise. Notably, *Amborella* appeared as sister to all other angiosperms. Analysis of larger plastid datasets with greater taxon sampling than used by Goremykin et al. (2012) provides further support for *Amborella* alone as sister to all other angiosperms (Drew et al. 2014; Ruhfel et al. 2014).

MORPHOLOGY + DNA DATA

Nandi et al. (1998) conducted a morphological cladistic analysis of 162 extant angiosperms (Chapter 3)—and also combined these data with *rbcL* sequences. Although these analyses provided a framework tree for all major groups of angiosperms (Chapter 3), we review here only the results for the basalmost lineages. The non-molecular analyses revealed some relationships similar to those recovered with DNA sequences, such as an early-branching clade of Austrobaileyaceae and Schisandraceae (including *Illicium*), whereas other relationships are clearly spurious, such as the placement of the eudicot Eupteleaceae as an early-branching angiosperm and the recovery of a clade dominated by magnoliids, but that also contains eudicots (e.g., Berberidopsidaceae, Nelumbonaceae). The combined morphology + *rbcL* tree placed Ceratophyllaceae, followed by a clade of Winterales, Austrobaileyales, Amborellaceae, and Nymphaeaceae, as sisters to all other extant angiosperms. Hence, this combined tree shows many similarities to more widely accepted DNA-based trees.

Doyle et al. (1994) combined a morphological dataset with a dataset of partial 18S and 26S rRNA sequences from Hamby and Zimmer (1992). However, the Hamby and Zimmer dataset did not contain sequences for several important taxa (e.g., *Amborella*). Doyle et al. (1994) deleted some of the taxa for which sequence data were available, but morphological data were lacking, to produce a molecular + morphological dataset that matched in taxon composition. As a result, some critical taxa were deleted from the molecular matrix (e.g., *Illicium*; see Hamby and Zimmer 1992). Analysis of the combined dataset yielded a topology in which Nymphaeales appeared as sister to all other angiosperms (Doyle et al. 1994), followed by monocots, a clade of Piperaceae + Saururaceae, and then Aristolochiaceae as subsequent sisters to the rest of the angiosperms (the latter three families are covered in Chapter 5). Thus, the Doyle et al. (1994) analysis again placed herbaceous taxa as sisters to all other extant angiosperms.

Doyle and Endress (2000) conducted an extensive cladistic analysis of angiosperms (including basal angiosperms) using morphological and molecular data. Compared to the earlier study of Donoghue and Doyle (1989b), Doyle and Endress (2000) increased the number of taxa from 27 to 52 and the number of morphological characters from 54 to 108. The shortest trees based on morphology revealed what they termed as "shifts toward molecular results," due in some cases directly to the addition of new characters, such as carpel form. With morphology alone, *Amborella*, followed by Chloranthaceae, and then Trimeniaceae (Austrobaileyales) were subsequent sisters to all other angiosperms. However, in some instances the non-DNA characters resulted in important differences in the placement of taxa compared to DNA-based trees. A significant conclusion of Doyle and Endress (2000) and earlier analyses of DNA + morphology is that morphology may be more consistent with molecular data than was previously thought (Nandi et al. 1998; Chase et al. 2000b).

Doyle and Endress (2000) also combined morphology with sequences for three genes (18S rDNA, *rbcL*, *atpB*), resulting in shortest trees with *Amborella*, Nymphaeales, and Austrobaileyales as subsequent sisters to other angiosperms—identical to the DNA-based topology; the remainder of their tree also matched results from DNA-based trees (Fig. 4.2). That this matrix with ~100 morphological characters and ~4700 nucleotides should produce a topology similar to that based on DNA sequences alone is not surprising.

Doyle and Endress continued to refine the morphological matrix (Endress and Doyle 2009; Doyle and Endress 2010, 2014; Doyle 2012), for example, adding Hydatellaceae and *Ceratophyllum*. However, they used the Doyle and Endress (2000) combined morphology + DNA tree as a constraint and not surprisingly recovered *Amborella*, Nymphaeales, and Austrobaileyales as subsequent sisters to other angiosperms. They noted that there were continuing problems with the overall topology when a DNA constraint was not used.

An important use of the Endress and Doyle (2009) combined matrix and tree of basal angiosperm relationships was the placement of fossil taxa (Endress and Doyle 2009; Doyle and Endress 2010, 2014; Doyle 2012). For example, Doyle and Endress (2010) attempted to place the herbaceous fossil *Archaefructus* (Chapter 3). Importantly, in their analyses *Archaefructus* appears either as sister to Hydatellaceae (Nymphaeales) (Fig. 4.3) or as sister to *Ceratophyllum*—not as sister to all other angiosperms.

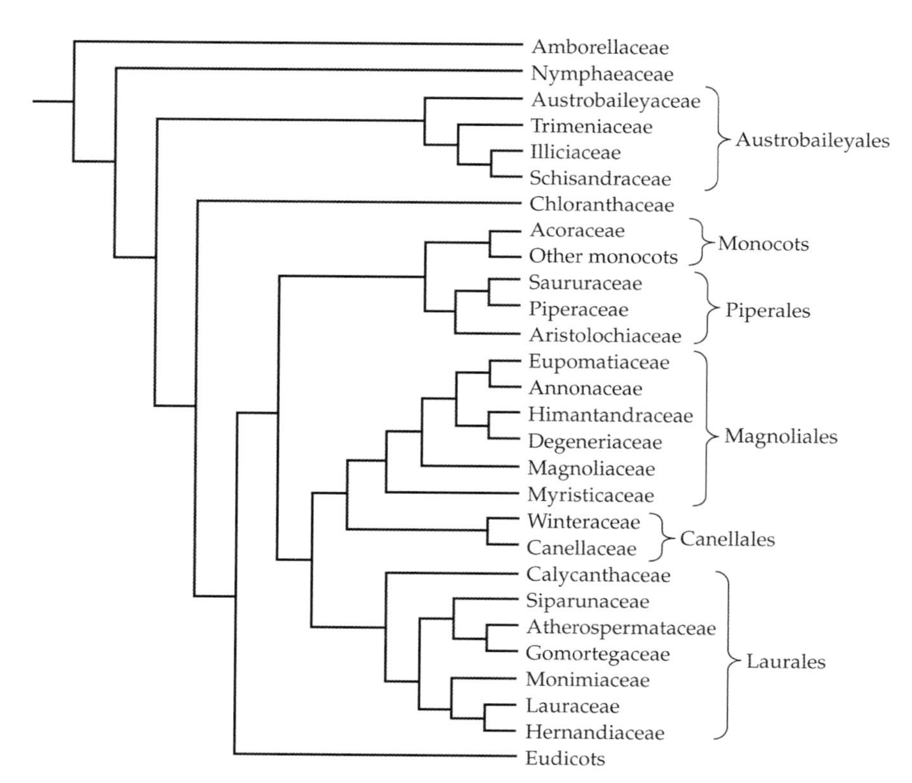

Figure 4.2. Single most parsimonious tree from analysis of a combined data set of morphology plus DNA (*rbcL*, 18S rDNA, *atpB*) (modified from Doyle and Endress 2000; from Soltis et al. 2005).

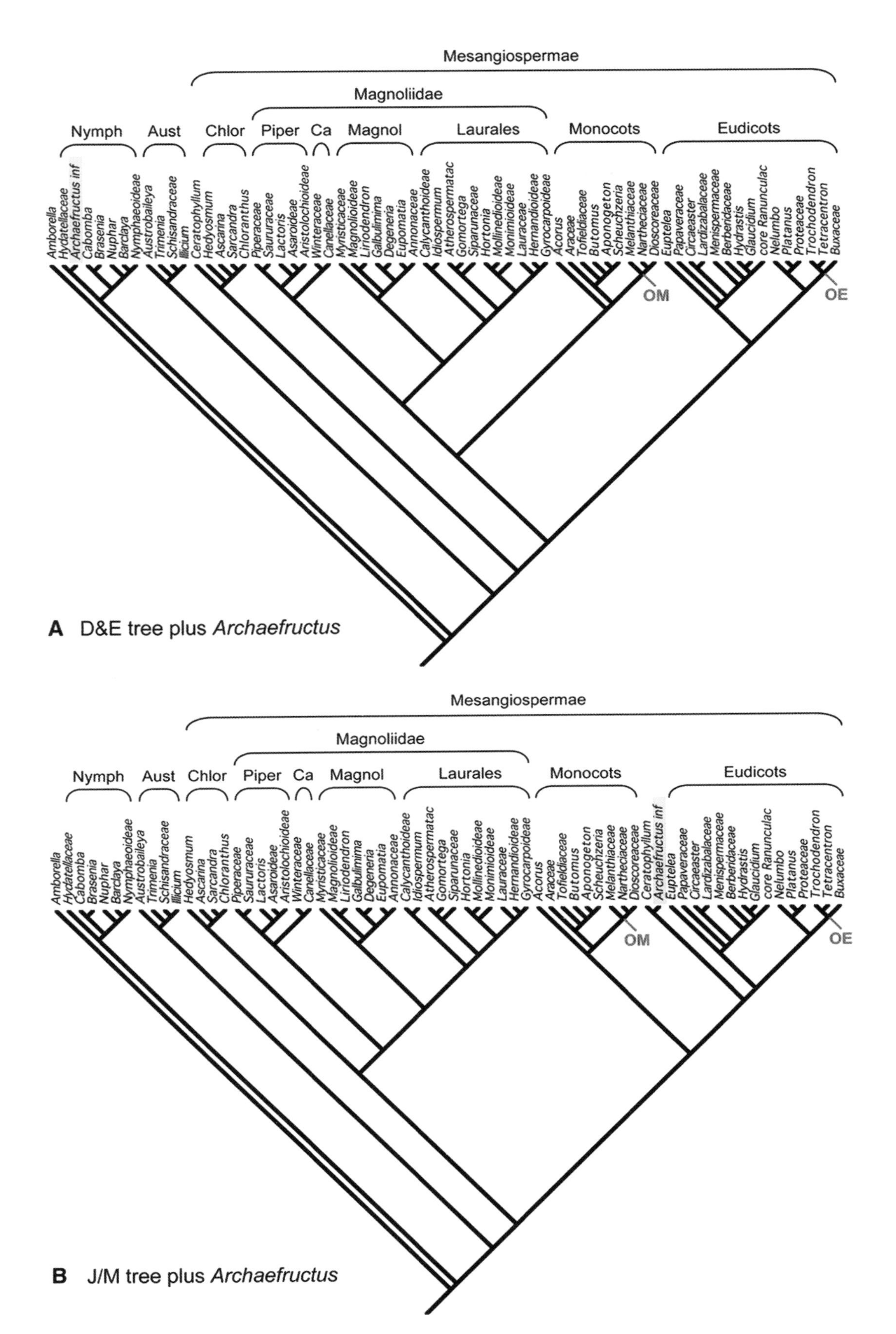

Figure 4.3. Representative most parsimonious trees after the addition of morphological data for the fossil *Archaefructus* using a backbone constraint tree (from Doyle and Endress 2000). A. One analysis placed this fossil in Nymphaeales. B. Other analyses they conducted placed *Archaefructus* with *Ceratophyllum*. (Fig. 2 from Endress and Doyle 2009.)

THE IMPORTANCE OF KNOWING THE ANGIOSPERM ROOT

Angiosperms represent one of the great terrestrial radiations with an enormous global economic and ecological impact. How did angiosperms accomplish this over a relatively short span of evolutionary time—from an origin in the late Jurassic/Cretaceous to present? Understanding the root of the angiosperms is critical to clarifying not only the diversification of angiosperms, but also the evolution of diverse morphological and chemical features, and aspects of angiosperm genome evolution. In addition, this knowledge will help elucidate many key processes whereby plant/animal associations and entire ecosystems assembled.

It was clear in early DNA investigations that understanding the structure and evolution of flowering plant genomes (as well as seed plant genomes in general) would contribute to agriculture and forestry. However, until recently, the angiosperm genomes that have been sequenced represent phylogenetically derived clades of the angiosperm tree (Fig. 3.9). As a result, these genomes provide few, if any, insights into the characteristics of the "ancestral angiosperm" (Soltis et al. 2008a; *Amborella* Genome Project 2013). For example, a complete understanding of any gene or genomic feature requires a phylogenetic perspective based on evolutionary relationships.

The analyses (summarized here) showing *Amborella* as sister to other flowering plants, focused attention on *Amborella*, as well as on Nymphaeaceae and Austrobaileyales, as sisters to all other angiosperms. The realization that *Amborella* is the sister to all other living angiosperms offers the unique ability to "root" analyses of all angiosperm features, from gene families to genome structure to morphology, chemistry, pollination, and more. For example, floral morphology, physiological ecology, and wood anatomy of *Amborella* have been carefully investigated multiple times (Endress and Igersheim 2000a,b; Feild et al. 2000a; Carlquist and Schneider 2001; Posluszny and Tomlinson 2003; Soltis et al. 2008a). Floral morphology and endosperm development have been compared in *Amborella* and Nymphaeales (Floyd and Friedman 2000; Endress 2001a; Williams and Friedman 2002; Buzgo et al. 2004b; Friedman 2006, 2008), and both lineages have also become the focus of analyses of floral organ identity genes (e.g., D. Soltis et al. 2002b; Kim et al. 2004a; Stellari et al. 2004; Chanderbali et al. 2010; *Amborella* Genome Project 2013).

Here we use this newly gained phylogenetic perspective to discuss the evolution of key morphological and genomic features of angiosperms.

MORPHOLOGICAL FEATURES OF THE BASALMOST ANGIOSPERMS

As it became clear that *Amborella*, Nymphaeales, and Austrobaileyales represent the basalmost angiosperm lineages, researchers began to examine morphological and other non-DNA features to corroborate these placements (e.g., Doyle and Endress 2000; Feild et al. 2000a; Soltis et al. 2005b). Importantly, *Amborella* possesses vesselless wood—the plants have only tracheids for water conduction (Bailey and Swamy 1948; Doyle and Endress 2000), and in that way are similar to most gymnosperms (Figs. 4.4). This feature therefore supports *Amborella* as sister to other angiosperms. More detailed investigations (Herendeen and Miller 2000; Carlquist and Schneider 2001, 2002) revealed that wood anatomy is complex and not a simple binary character, but rather as a suite of characters (Table 4.1). Carlquist (2012) more recently reviewed wood anatomy, inferring what early angiosperm wood may have been like and stressing that cell types other than vessel elements may make important contributions to conductive efficiency.

Nonetheless, wood anatomy is consistent with the position of *Amborella* branching near the base of the angiosperm tree, although not necessarily as the sole basal lineage (Carlquist and Schneider 2001, 2002) (Fig. 4.4). The aquatic habit of Nymphaeales makes comparisons of wood anatomy among basal angiosperms more complicated in that these plants lack vessels in stem tissue, although vessels are found in their roots. Are vesselless stems the ancestral state for Nymphaeales, or could this represent a secondary loss due to the aquatic habit? This is unclear. Members of Austrobaileyales have water-conducting cells that are intermediate between tracheids and vessels. Hence, in stem tissue, members of the ANA grade have either no vessels, or cells that are intermediate between those of tracheids and vessels, arguing that the earliest angiosperms stems may have lacked vessels.

Several other morphological features also support a basal branching position for *Amborella* (as well as Nymphaeales and Austrobaileyales). Briefly, there are two extremes of carpel form (Doyle and Endress 2000), plicate and ascidiate. Ascidiate carpels are formed via invagination to form a pitcher-like, hollow structure with ovules inside. The plicate carpel resembles a folded leaf and is formed by one-sided elongation of the terminal part of the ascidiate carpel. Ascidiate carpels characterize *Amborella*, most Nymphaeaceae, Austrobaileyaceae, most Schisandraceae, and Trimeniaceae (Endress and Igersheim 2000b) (Fig. 4.5). Intermediate carpel forms are present in *Barclaya* (Nymphaeaceae) and also in *Illicium* (Schisandraceae). Following Amborellaceae, Nymphaeaceae, and Austrobaileyales,

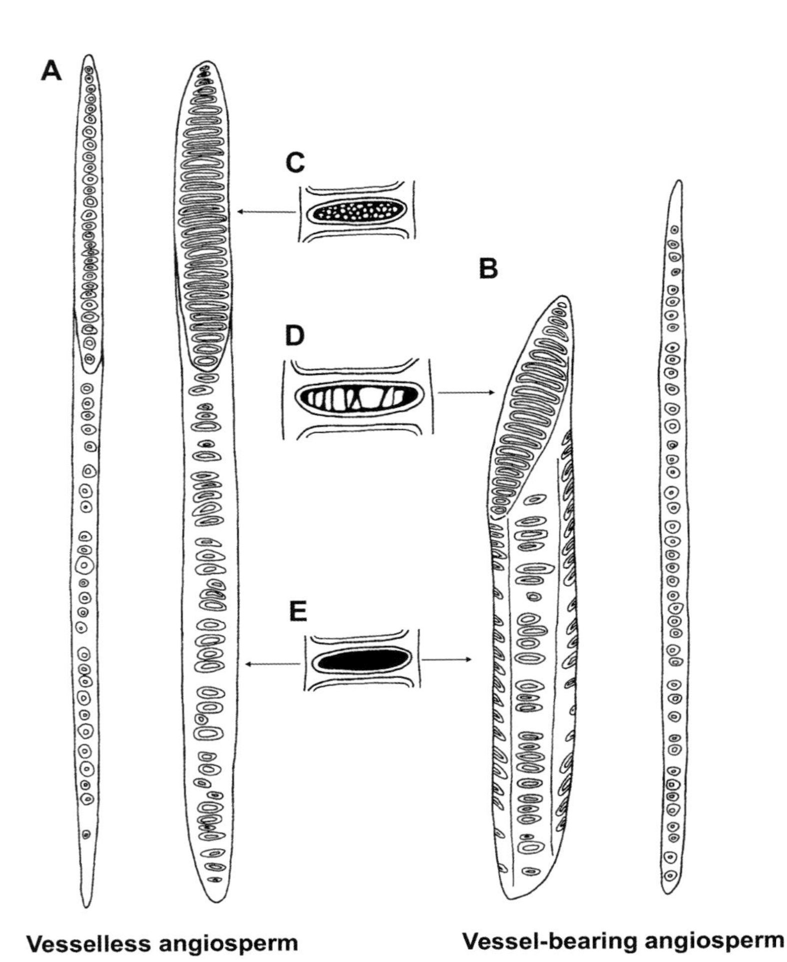

Figure 4.4. Changes proposed to have occurred during evolution of vessel elements from tracheids in vesselless angiosperms (modified from Carlquist and Schneider 2002). (A) Vesselless angiosperm; tracheids similar to those present in *Amborella*. (B) Vessel-bearing-angiosperm. The vessel (on the left) and the tracheid (on the right) are similar to those found in *Illicium* in which the vessels exhibit "tracheid" features (see text). The six changes between tracheids and vessels are illustrated: (1) Pit membranes of end walls of tracheids typically are nonporose (not shown), but in some taxa the pit membranes become porose (C). As vessels originate, varying degrees of pit membrane remnants are visible in perforations (D). In contrast, pit membranes of the lateral walls of vessels remain intact and nonporose, just as they are on lateral walls of tracheids (E). (2) Conductive area of the end wall of the wider tracheid in A is larger than the conductive area of pits of the tracheid lateral walls because the pits are larger on the end walls. This is interpreted as an incipient form of vessel-like structure in *Amborella*. As vessels originate, the differences in size of perforations become more pronounced. (3) There are proposed changes in the morphology of perforations of end walls compared to lateral walls when vessels originate. One difference is the narrower borders on the perforations of the vessel compared to borders of lateral pit walls (shown); other differences in morphology of lateral wall pits also occur (see Table 4.1). (4) Vessel elements become wider than the tracheids that they accompany. The vessel element in B is wider than the associated tracheid. (5) Vessel elements are shorter than the tracheids that they accompany. Whereas tracheids are all the same length in wood that lacks vessels (A), vessel-bearing wood has vessel elements shorter than the tracheary elements they accompany (B). (6). Tracheids in vesselless wood (A) are longer than both the vessel elements and the tracheids of vessel-bearing wood (B). That is, the origin of vessels results in shorter vessel elements and also shorter tracheid elements. (Legend modified from Carlquist and Schneider 2002.) (Fig. 3.10 from Soltis et al. 2005).

Table 4.1. Characters that distinguish vessels from tracheids (modified from Carlquist and Schneider 2002)

1. Pit membranes of end walls
   - lacking porosities (tracheid)
   - porose or web-like (intermediate)
   - entirely lacking (true vessel)
2. Pits of end walls
   - with the same conductive area per unit wall area as pits of lateral walls
   - with more conductive area per unit wall area than the pits of lateral walls
3. Pit morphology of end walls
   - identical to pitting of lateral walls
   - different from that of end walls
4. Tracheary element diameter
   - uniform
   - bimodal, with wider elements possessing one or more other features of vessel elements

most remaining angiosperms (part of a large clade termed *Mesangiospermae*; see Chapter 3) have plicate carpels. This inference also agrees with the prevalence of putative ascidiate carpels in fossils from the Early Cretaceous (Crane et al. 1995). In addition, Amborellaceae, as well as most Nymphaeaceae, and Austrobaileyales have carpels that are not sealed by postgenital fusion but by secretion—that is, the carpel is essentially glued shut (Fig. 4.6); in contrast, most *Mesangiospermae* have carpels that are postgenitally fused (Fig. 4.6). *Amborella* and *Austrobaileya* (see below) have distinctly laminar (leaf-like) stamens as do some Nymphaeaceae (Fig. 4.7; see also Fig. 4.10, below).

We review some of the general features of ANA grade members below. We also have conducted extensive character-state reconstructions across the angiosperms in an effort to infer the ancestral states for a suite of morphological and other features. These are discussed in a separate chapter (Chapter 6).

AMBORELLACEAE

We describe here some of the general features of *Amborella trichopoda* (the sole species of *Amborella*). Because of the enormous interest in this taxon we dedicate more space to our coverage of *Amborella trichopoda* than other species.

Amborella is an understory shrub to small tree found

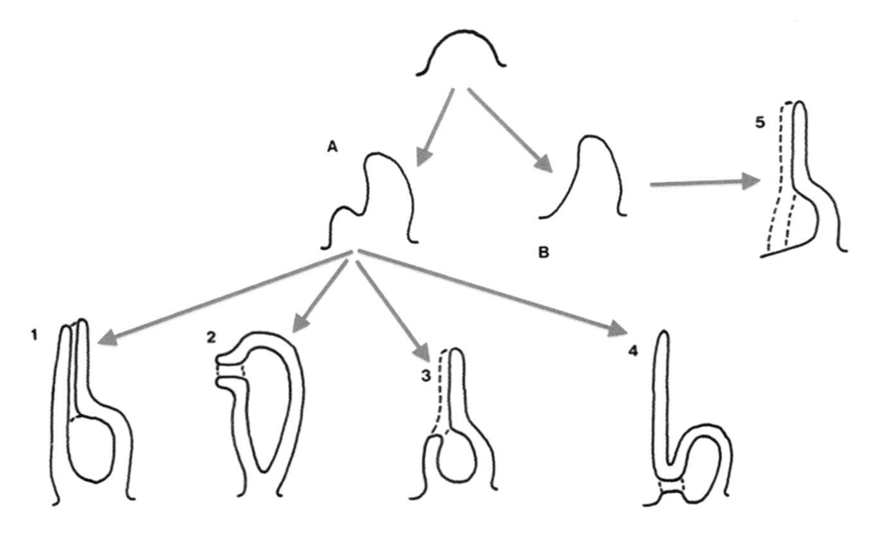

Figure 4.5. Diversity of carpel development seen in median longitudinal sections. A. Ascidiate (peltate). B. Completely plicate (non-peltate). 1. Completely ascidiate. 2. Strongly ascidiate with dorsal bulge. 3. Half ascidiate. 4. Weakly peltate with dorsal bulge. 5. Epeltate. Examples: Austrobaileyaceae (1), Winteraceae (2,3), Monimiaceae (3,4), Annonaceae (3,5). Stippled: area without any organ part exactly in the median plane (in the plicate zone of carpels). (Modified from Endress1994a.)

only in New Caledonia, where fewer than 20 populations are known. The alternate, simple, pinnately veined leaves of *Amborella* possess chloranthoid teeth, the function of which is unknown, but from evidence in ferns (Sperry 1983) may function as release valves for root pressure (Feild et al. 2003a, 2004). Chloranthoid teeth are present in several basalmost angiosperm lineages, including not only *Amborella* (Bailey and Swamy 1948) but also *Kadsura* and *Schisandra* (Schisandraceae; Hickey and Wolfe 1975) and species of *Trimenia* (Trimeniaceae; Rodenburg 1971; Endress 1987c). Doyle and Endress (2000) suggested that chloranthoid leaf teeth are ancestral in the angiosperms although reconstructions remain equivocal (Soltis et al. 2005b; Chapter 6). As noted earlier, *Amborella* possesses vesselless wood (see above; Table 4.1).

The inflorescences of *Amborella* are produced in the axil of foliage leaves and are monotelic and considered botryoids or poorly ramified panicles (Endress and Igersheim 2000). *Amborella* is dioecious but the carpellate flowers have sterile stamens (= staminodes or staminodia) (Fig. 4.7), suggesting that the ancestral condition in angiosperms is bisexual (perfect) flowers (Endress and Igersheim 2000b; Buzgo et al. 2004b). The flowers are small (3–5 mm in diameter) and appear to have many features of the small flowers that appear early in the fossil record (see chapter 2 and Friis et al. 2011). The flowers possess an undifferentiated, spirally arranged perianth, subtended by spirally arranged bracts (Fig. 4.7). The staminate flowers have 9–11 tepals and possess 12–21 laminar stamens. The inner stamens are gradually smaller than the outer ones.

Williams (2009) found that the in vivo pollen tube growth rate of *Amborella* is similar to that of many other early-diverging angiosperms, but much faster than that of any gymnosperm. However, the rate of pollen tube growth in *Amborella* is among the slowest in angiosperms, prompting Williams (2009) to propose that the earliest angiosperms had very slow pollen tube growth rates.

The carpellate flowers possess 7–8 tepals. One to several staminodia are present, followed by 4–6 carpels, each with a single ovule. Each carpel develops into a small drupe (resulting, potentially, in an aggregate of drupes). Ascidiate carpels characterize *Amborella* (as well as most Nymphaeaceae, Austrobaileyaceae, most Schisandraceae, and Trimeniaceae; Endress and Igersheim 2000b). As with most ascidiate carpels in the ANA grade, the carpels of *Amborella* are sealed by secretion rather than by postgenital fusion (Fig. 4.6; Endress and Igersheim 2000b). The ovules in *Amborella* have been variously described as orthotropous, anatropous, or intermediate (Tobe et al. 2000), but are here considered orthotropous (see Endress and Igersheim 2000b; Endress 2011a; Fig. 4.8).

The embryo sac (female gametophyte) of *Amborella* is distinctive (Tobe et al. 2000; Friedman 2006; Rudall 2006). Friedman (2006) reported a seven-celled and eight-nucleate stage in young *Amborella* flowers with three cells at each pole of the embryo sac as well as a large central cell with two nuclei—comparable to what is observed in most angiosperms. However, this is not the final stage in development of the *Amborella* embryo sac. A mitotic cell division at the micropylar end of the embryo sac produces a ninth nucleus and an eighth cell (Friedman 2006). The end result is that *Amborella* possesses an embryo sac with eight cells and nine nuclei. This contrasts with (but is most similar to) the seven-celled and eight-nucleate embryo sac typical of most angiosperms. Interestingly, however, Nymphaeales and Austrobaileyales are both characterized by a four-nucleate embryo sac (see below; reviewed in Friedman and Ryerson 2009). There is also strong evidence that

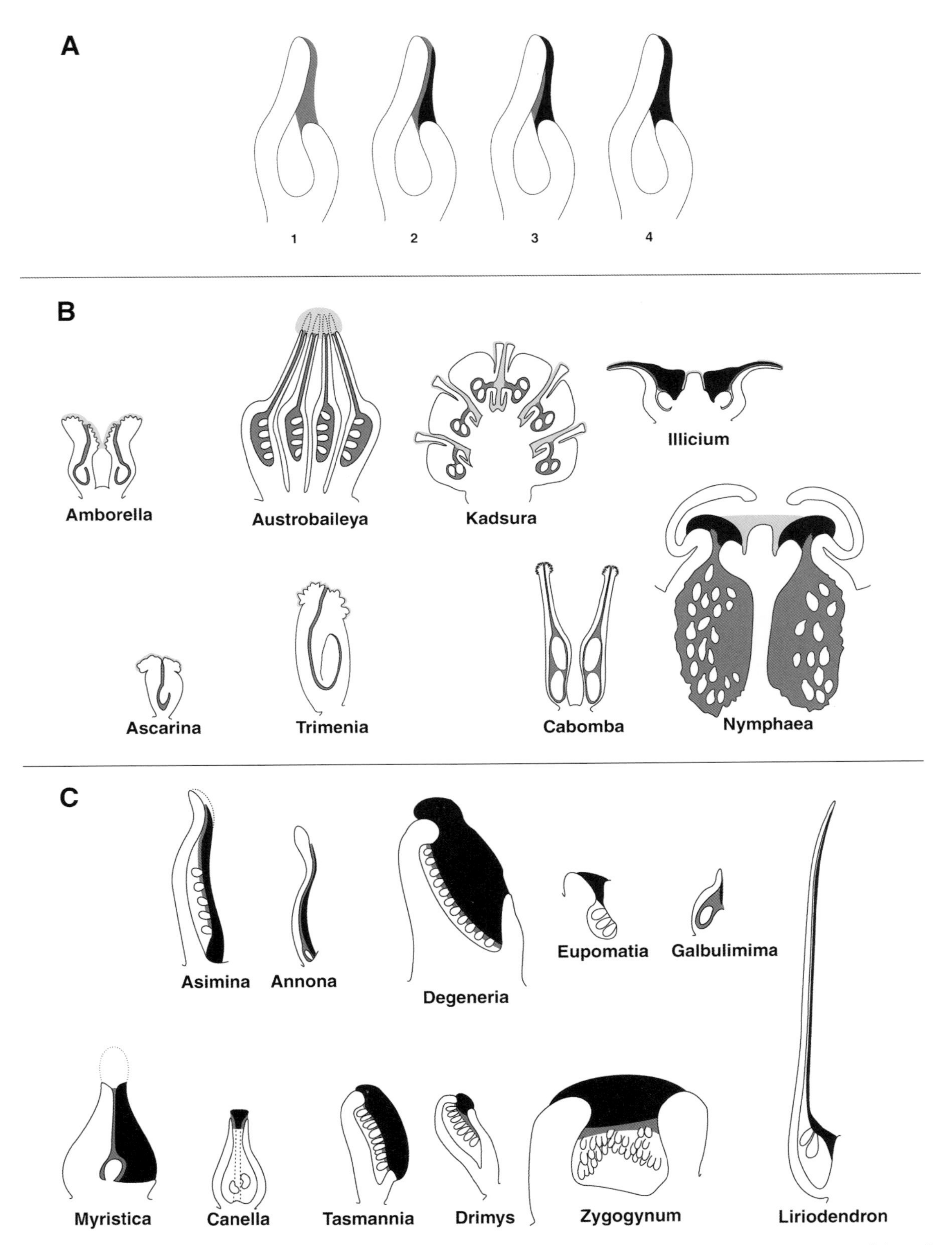

Figure 4.6. Carpel sealing. Carpel form and sealing in basal angiosperms. (From Igersheim and Endress 2000a.) (A) Types of angiospermy. Carpels in median longitudinal section. Gray areas: ventral slit closed by secretion; dark areas: ventral slit closed by postgenital fusion. Type 1: angiospermy by secretion; Type 2: angiospermy by a continuous secretory canal and partial postgenital fusion at the periphery; Type 3; angiospermy by a partial secretory canal and complete postgenital fusion at the periphery; Type 4: angiospermy by complete postgenital fusion. (B) Illustrations of carpel types in Amborellaceae, Nymphaeaceae, Austrobaileyales, and Chloranthaceae (see Chapter 5). Median longitudinal sections of gynoecia showing carpel and gynoecial structure. Gray areas designate secretion inside of carpels; light gray areas: secretion at the outer surface of carpels; dark areas: postgenital fusion; not drawn to scale. (C) Illustrations of carpel types in magnoliids (see Chapter 5). Median longitudinal sections of carpels, ventral side at right (the figure of *Canella* represents an entire gynoecium because the carpels are completely united). Gray areas: secretion inside of carpels; dark areas: postgenital fusion; secretion at the outer carpel surface is not indicated. In microtome sections of *Eupomatia* and *Liriodendron*, secretion is not visible inside the carpels, although it is likely to be present (from Endress and Igersheim 2000b) (Fig. 3.16 from Soltis et al. 2005).

Figure 4.7. Amborellaceae (*Amborella trichopoda* Baill.): a. Shoot and axillary infructescences with immature fruits; b. Staminate inflorescences, leaves with chloranthoid teeth; c. Staminate flower; d. Side view of carpellate flower; e. Flower bud showing bracts; f. Top view of carpellate flower; g. Fruits (drupes).

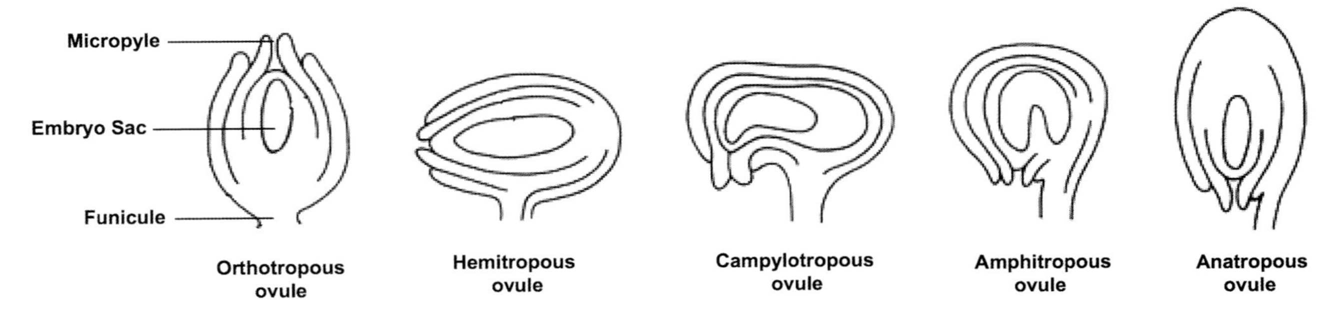

Figure 4.8. Diagrams produced for this volume reviewing ovule types in angiosperms. See text.

double fertilization occurs in *Amborella*, resulting in triploid endosperm, whereas diploid endosperm is present in Nymphaeales and Austrobaileyales (reviewed in Williams 2009). Taken together, these studies reveal considerable developmental lability in the embryology of basal angiosperms (Friedman 2006; Rudall 2006; Williams 2009).

There has been considerable interest in the pollination biology of basal angiosperms (e.g., Thien et al. 2000, 2003, 2009). Thien et al. (2003, 2008) showed that flowers of *Amborella* are both wind- and insect-pollinated. *Amborella* flowers appear to offer only pollen as the resource for visiting insects—neither floral odor nor heat (another insect attractor) was detected (Endress and Igersheim 2000a; Thien et al. 2003). However, Collett (1999) noted that scent was produced at night, with visits by numerous moths in greenhouse-grown plants (reviewed in Thien et al. 2008). However, no insects were observed at night on the flowers of *A. trichopoda* in nature (Thien et al. 2003)—all the insects observed as visitors (Coleoptera, Homoptera, Hemiptera, Microlepidoptera, Hymenoptera, and Diptera) were captured during the day (Thien et al. 2008).

THE *AMBORELLA* NUCLEAR GENOME

The recently sequenced *Amborella* genome has provided crucial new insights into the earliest angiosperm genomes while serving as an evolutionary reference for comparisons with other angiosperms (*Amborella* Genome Project 2013). To briefly summarize here, the complete genome of *Amborella* facilitated comparative genomic analyses that confirmed that an ancient genome duplication predated all living angiosperms (see Jiao et al. 2011). However, unlike all other sequenced angiosperm genomes, the *Amborella* genome shows no evidence of more recent, lineage-specific genome duplications. This feature of the *Amborella* genome makes *Amborella* well suited as a reference genome for the interpretation of genomic changes after polyploidy in other angiosperms. Via comparison to the *Amborella* genome, an ancient hexaploidy event (i.e., two polyploidizations in close succession) was confirmed in the eudicots. Furthermore, synteny is highly conserved in *Amborella* and the genomes of other angiosperms (see Chapter 15). This conserved synteny between the genomes of *Amborella* and other angiosperms has enabled reconstruction of ancestral gene order in eudicots. Annotation of the *Amborella* genome and comparative analysis across sequenced plant genomes has enabled reconstruction of the ancestral angiosperm gene set of at least 14,000 genes. Analysis of the *Amborella* genome revealed that over 1200 gene lineages were gained in the ancestral angiosperm relative to other seed plants, with many of the genes gained including those important in flowering. The *Amborella* genome has retained ancient transposons, but, significantly, the genome lacks evidence for any recent transposon insertions (Chapter 15).

"Resequencing" of the *Amborella* genome from 12 plants representing most known natural populations has also been informative. These 12 individuals harbor levels of genetic diversity similar to those reported for species of *Populus*, which are also outcrossing perennials. Genetic variation among *Amborella* populations is significantly structured into three geographic clusters of populations on New Caledonia (Fig. 4.9), corresponding roughly to populations in (1) the northern and eastern part of the range, (2) the southern and western part of the range, and (3) a single disjunct location at the southern end of the distribution. These results generally support those obtained through analysis of 10 microsatellite loci (Poncet et al. 2012, 2013). Conservation efforts in New Caledonia should focus on conserving and managing the genetic diversity found in the geographic clusters of this relictual angiosperm species.

The mitochondrial genome of *Amborella* is unusually large and contains genes of several other diverse lineages of land plants, including mosses and other angiosperms (Bergthorsson et al. 2004).

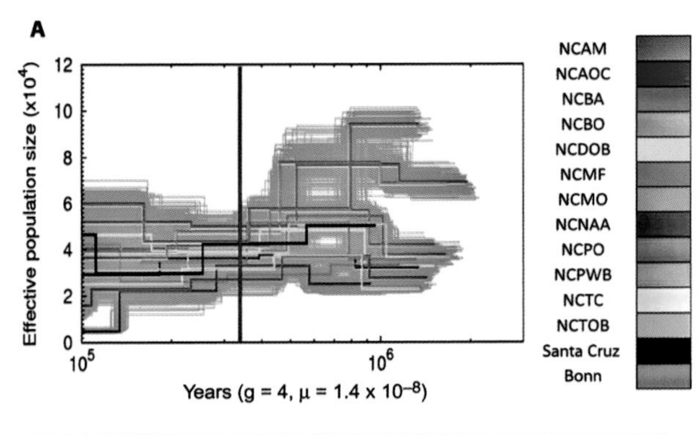

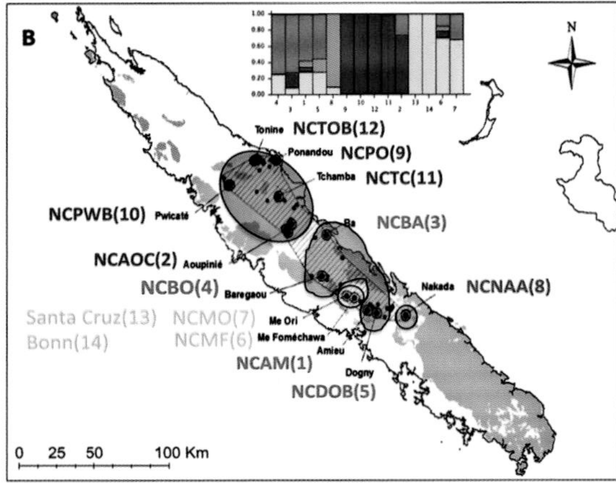

Figure 4.9. Population genomic diversity in *Amborella*. A. Density plot of nucleotide diversity (π), Watterson estimator (θ_w), and Tajima's *D* statistics for the 50 largest scaffolds. Red lines indicate nucleotide diversity (π); the two red bounding lines represent maximum and minimum values, max = 0.08, min = 0. Blue lines represent Watterson estimator (θ_w); the two blue bounding lines represent maximum and minimum values, max = 0.006, min = 0. Green lines indicate Tajima's *D* statistics, the two green bounding lines represent maximum and minimum values, max = 3.24, min = –1.67; the black line represents a value of Tajima's *D* of 0. Each scaffold unit represents 100,000 bp. Inset shows close-up of scaffold 1. B. Results of Structure analysis, showing three significant genetic clusters of individuals from natural populations (blue, red, green areas) and additional cluster (yellow) of cultivated individuals (Fig. 6 from the *Amborella* Genome Project 2013).

NYMPHAEALES (INCLUDING HYDATELLACEAE)

Nymphaeales (water lilies; Figs. 4.1, 4.10) have long attracted interest as a putatively ancient lineage of angiosperms, given various flower features considered "primitive" by earlier generations of researchers. Some investigators once suggested that Nymphaeaceae might be closely allied with monocots; some even suggested that monocots might be derived from Nymphaeaceae or a now-extinct Nymphaeaceae-like lineage of ancient angiosperms (Hallier 1905; Eber 1934; Arber 1920; Cronquist 1968; Takhtajan 1969, 1991; Burger 1977). Interest in Nymphaeaceae increased after the earliest cladistic analyses based on morphology (e.g., Donoghue and Doyle 1989a,b) and molecules (Hamby and Zimmer 1988) indicated that these plants might be sister to all other angiosperms. As reviewed above, a series of phylogenetic studies using both non-DNA and extensive DNA characters have only strengthened support for Nymphaeaceae as one of the basalmost lineages of extant angiosperms. The clade also appears early in the fossil record. Friis et al. (2001, 2009) provided the earliest unequivocal fossil evidence of Nymphaeaceae, extending the history of the group into the Early Cretaceous (125–115 mya) and into the oldest fossil assemblages that contain stamens and carpels.

Traditional treatments of the water lilies have varied considerably (reviewed in Les et al. 1991, 1999). Cronquist (1981, 1988) recognized several families (e.g., Nymphaeaceae, Cabombaceae, and Barclayaceae), while also considering Ceratophyllaceae closely related to these families; he also included the eudicot family Nelumbonaceae within Nymphaeales. Takhtajan (1997) recognized six families of water lilies as part of his subclass Nymphaeidae.

As now recognized (APG III 2009; APG IV 2016), Nymphaeales include Nymphaeaceae, Cabombaceae, and the small aquatic family Hydatellaceae, which was previously placed in Poales. As reviewed elsewhere, Hydatellaceae were traditionally considered to be monocots, based on a morphological similarity to the monocot family Centrolepidaceae. However, sequence data and a combined morphological and DNA analysis revealed that *Trithuria* is sister to Nymphaeaceae (Saarela et al. 2007). Previous analyses that had placed the family in Poales appear to be the result of sequencing of a polymerase chain reaction (PCR)–recombinant. Hydatellaceae include only the genus *Trithuria* (~12 species), which has been recently redefined to include the genus *Hydatella* (Sokoloff et al. 2008) (see Fig. 4.10).

Plants of Hydatellaceae are small, simple in overall form, and aquatic or semi-aquatic annuals (Figs. 4.10, 4.11); they occur in Australasia and India. The linear leaves are clustered at the base of a short stem. Species are monoecious or dioecious and pollination has been hypothesized to be by wind, water, or selfing, but analyses of the reproductive ecology of *T. submersa* reveal selfing as well as outcrossing by wind, but not by water or insects (Taylor et al. 2010). Although plants of *Trithuria* are highly modified in morphology, they nonetheless exhibit embryological features of Nymphaeales including a very small biparental endosperm and embryo (Friedman et al. 2012). Despite some debate in terms of the number of cotyledons (Tillich et al. 2007), careful investigation reveals that the embryo initiates two cotyledons (Friedman et al. 2012; Sokoloff et al. 2013a).

The floral/inflorescence morphology of Hydatellaceae is unusual (Fig. 4.11). The reproductive units of the family consist of perianth-like bracts that enclose several carpels

Figure 4.10. Representatives of Nymphaeales (Nymphaeaceae, Cabombaceae, Hydatellaceae): a. *Brasenia schreberi* J.F.Gmel. (Cabombaceae), flower. b. *Trithuria submersa* Hook.f. (Hydatellaceae), habit. c. *Victoria amazonica* (Poepp.) Klotzsch (Nymphaeaceae), flower and leaves. d. *Nymphaea odorata* Aiton (Nymphaeaceae), flower and leaves. e. *Nuphar advena* (Aiton) W.T.Aiton. (Nymphaeaceae), flower, with one inner tepal pulled back to reveal staminodia, stamens, and syncarpous gynoecium. f. *Nymphaea* sp. (hybrid) (Nymphaeaceae), flower.

Figure 4.11. SEM of anthetic and postanthetic reproductive units of *Trithuria submersa*. (A) Anthetic reproductive unit with two prophylls (pr) on its peduncle; a, b. outer-whorl involucral bracts; stars indicate inner-whorl involucral bracts; white arrowhead indicates filament of first-formed stamen (anther abscised); black arrowhead indicates second-formed stamen (anther not yet dehisced). (B) Top view of typical postanthetic reproductive unit with four involucral bracts, two stamens (arrowheads), and several fruits, some of them already dehisced (Fig. 4 of Rudall et al. 2007). (C) Unusual reproductive unit with three rather than four bracts. Arrowheads point to the filaments of the two stamens. (D) Unusual reproductive unit with three stamens. Arrowheads point to the stamen filaments.

and/or stamens. Hamann (1975) and Rudall et al. (2007, 2009) suggest that one interpretation of these structures is that each reproductive unit represents an aggregation of reduced, unisexual, perianthless flowers—thus, each unit may represent a pseudanthium (an inflorescence that appears to be a single flower). The floral/inflorescence morphology of Hydatellaceae is therefore very different from other Nymphaeales and indicates that the clade may have been once far more diverse than today (Fig. 4.3). In this regard, it is noteworthy that some analyses have potentially placed the fossil genus *Archaefructus* within Nymphaeales, perhaps as sister to Hydatellaceae (Endress and Doyle 2009; Doyle and Endress 2010; see Chapter 3). Chromosomal studies of *Trithuria submersa* reveal an unusual meiotic behavior (Kynast et al. 2014); the authors speculate that this could provide new insights into the ancestral microsporogenesis of angiosperms, but acknowledge that this could simply be a derived feature.

Many phylogenetic analyses (focused studies as well as broad analyses of angiosperms) have found strong support for a split of *Brasenia* and *Cabomba* (Cabombaceae) from Nymphaeaceae (in the strict sense) (e.g., Les et al. 1999; D. Soltis et al. 2000, 2011; Yoo et al. 2005; Löhne et al. 2007). We follow the suggested treatment of APG III (2009) and APG IV (2016) and use the family name Nymphaeaceae to designate a narrowly defined family and use Cabombaceae for *Cabomba* and *Brasenia* (see Les et al. 1999). Nymphaeaceae in the strict sense consist of six genera: *Barclaya*, *Euryale*, *Nuphar*, *Nymphaea*, *Ondinea*, and *Victoria* (Fig. 4.1). The two families are recognized as two subfamilies (Nymphaeoideae and Cabomboideae) by others; see Les et al. 1999; Judd et al. 2008. A number of features readily distinguish the two families (or subfamilies). In some molecular phylogenetic analyses, *Victoria*, *Euryale*, and *Ondinea* are derived from within *Nymphaea* s.l. (Löhne et al. 2007, 2008, 2009; Borsch et al. 2008).

There has been debate as to when the major split between Cabombaceae and Nymphaeaceae occurred. Löhne et al. (2008) proposed a divergence time of ~56.4 (with a range of 75–38) mya. In contrast, Yoo et al. (2005) determined a somewhat younger timeframe for this split, estimating that extant Nymphaeales diversified into two major clades corresponding to Cabombaceae and Nymphaeaceae during the Eocene (44.6 ± 7.9 mya); extant genera of Nymphaeaceae date to 41.1 ± 7.7 mya, and extant Cabombaceae diversified during the Miocene (19.9 ± 5.6 mya). Importantly, both of these estimates (Yoo et al. 2005; Löhne et al. 2008) suggest that whereas the stem lineage of Nymphaeales is old based on fossil evidence (125–115 mya), extant Nymphaeales may have diversified relatively recently (but see Iles et al. 2014). Similar results have been reported for other "old" basal lineages, such as Chloranthaceae (see Chapter 5) and *Illicium* (Schisandraceae).

The flowers of Cabombaceae have few parts. *Cabomba* flowers usually have two whorls of three petaloid tepals, 3 or 6 whorled stamens, and 3 to 18 whorled carpels (Fig. 4.10). Flowers of Nymphaeaceae are diverse, with a large range in the number and appearance of floral parts. The perianth is composed of 4 to many tepals (typically petal-like) that may intergrade with the staminodia and stamens (e.g., *Nymphaea*) (Fig. 4.12); in fact, many of these petal-like structures are not true tepals but instead are petaloid staminodia. In *Nuphar*, the outer perianth organs are large and green and yellow (petal-like in appearance) whereas the inner perianth organs are petal-like and small (Fig. 4.10e) (Hiepko 1965b; Endress 2008). Stamens are numerous, the innermost sometimes represented by staminodia (Fig. 4.10e). Stamens vary from laminar to having a well-differentiated filament and anther. The carpels are many (ovaries semi-inferior to inferior) and connate, in contrast to the separate carpels of Cabombaceae.

The floral phyllotaxis in Nymphaeaceae has been debated. Ronse De Craene et al. (2003) considered *Brasenia* and *Cabomba* to have whorled phyllotaxis and the phyllotaxis of other genera (*Nymphaea*, *Victoria*, *Nuphar*) to be polymorphic, with early organs appearing in a whorled arrangement and later ones in a spiral. Zanis et al. (2003), in contrast, scored *Nymphaea*, *Victoria*, and *Nuphar* as spiral. In what represents the best assessment of these flowers, Endress (2001a) indicated that the phyllotaxis of *Nymphaea*, *Victoria*, and *Nuphar* is whorled and then may become irregular in development, but not spiral.

On the basis of longstanding views of angiosperm floral evolution, flowers such as *Nymphaea* and *Victoria* with numerous floral parts were considered "primitive" (see Bessey 1915; Cronquist 1968, 1981, 1988; Takhtajan 1969, 1987, 1991, 1997; Stebbins 1974). However, using the well-resolved tree available for Nymphaeaceae and estimates for stamen number from Les et al. (1999), our simple character reconstructions (Fig. 4.13) indicate instead that numerous floral parts are derived, rather than ancestral, in Nymphaeaceae. This result contrasts with the traditional view (see Stebbins 1974) that secondary increases in number could occur in some floral organs, such as stamens, but that such increases were likely rare. The greatest number of petals, stamens, and carpels occurs in *Nymphaea* and *Victoria*, which represent derived genera within the family (we illustrate only stamen number; Fig. 4.13). These character reconstructions support Schneider (1979), who earlier suggested that numerous floral parts within the family were of secondary origin.

The number of floral parts present in genera of Nymphaeaceae may be associated with pollination (Gottsberger 1977, 1988; Les et al. 1999; Lippok et al. 2000). Large flowers with numerous parts, such as *Nymphaea* and *Victoria*, may represent an evolutionary response to herbivory by beetles, which are major pollinators. Genera with lower numbers of floral organs have different pollinators or pollination systems. The smaller number of floral parts in *Euryale*, another derived genus (Fig. 4.1), appears to represent a decrease associated with cleistogamy and self-pollination. Also, the increase in number of stamens in *Brasenia* versus *Cabomba* may relate to the shift from insect to wind pollination (see Williamson and Schneider 1993a; Les et al. 1999).

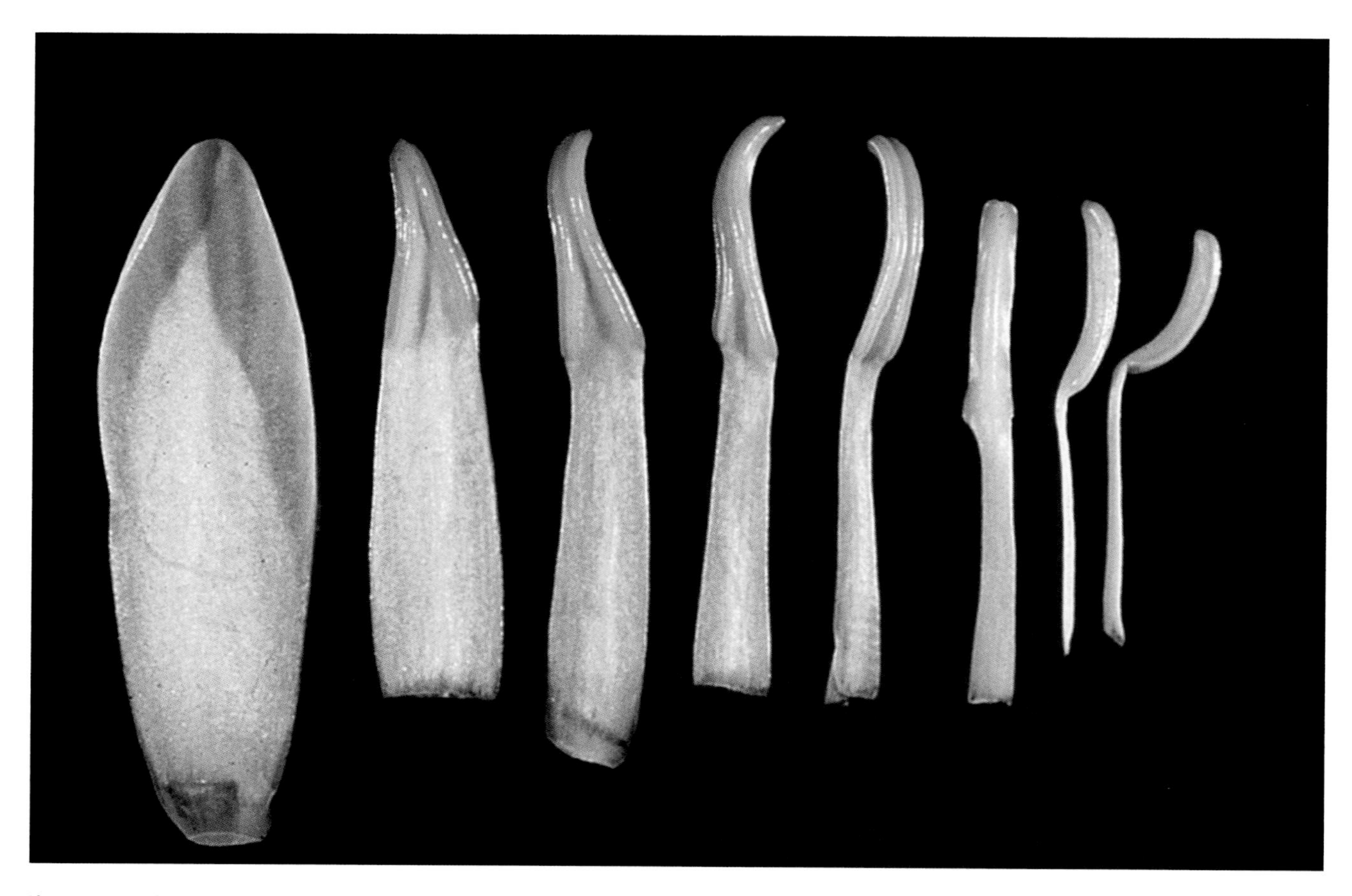

Figure 4.12. Photographs of floral part variation in *Nymphaea tuberosa* (Nymphaeaceae). Dissected floral parts showing variation from white petals, petaloid staminodia, to functional stamens.

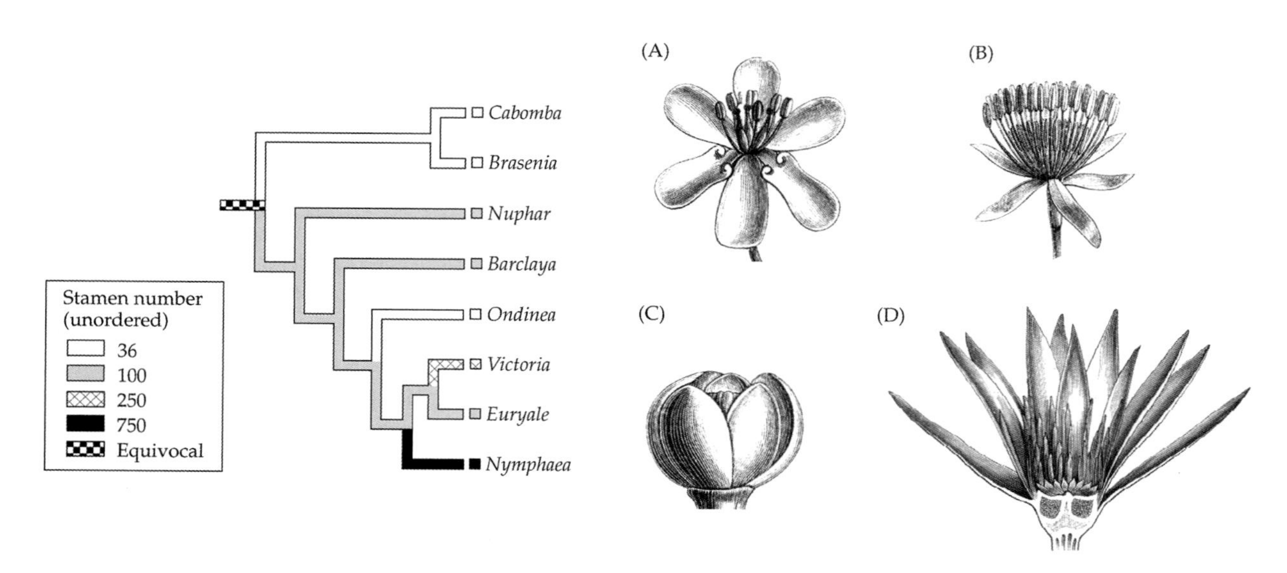

Figure 4.13. Stamen diversification in Nymphaeaceae. (Left) Simple MacClade reconstruction of stamen evolution in Nymphaeaceae. Numbers are maximum number of stamens. Topology is the shortest tree of Les et al. (1999); stamen character states were compiled from data in Les et al. (1999). (Right) Line drawings of Nymphaeaceae from Engler and Prantl (1898). (A) *Cabomba aquatica*. (B) *Brasenia purpurea*. (C) *Nuphar pumilum*. (D) *Nymphaea coerulea* (Fig. 3.5 from Soltis et al. 2005).

(A) Typical double fertilization

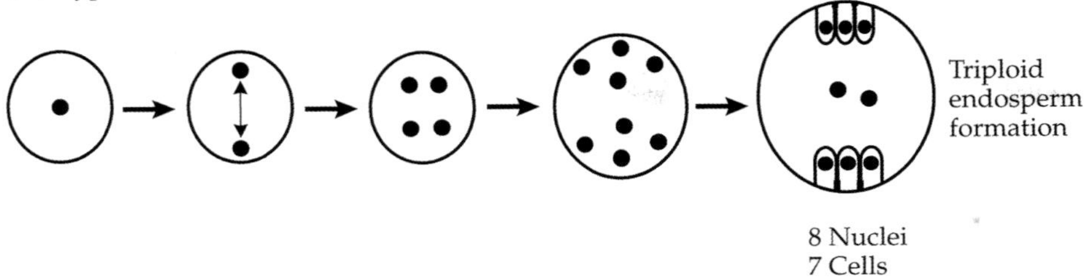

(B) Double fertilization in Nymphaeaceae

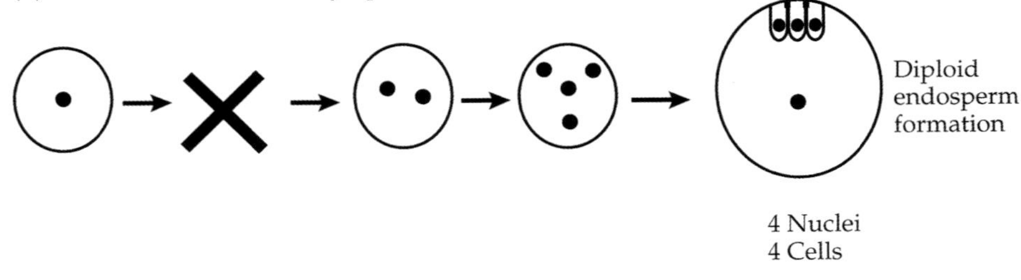

Figure 4.14. Embryo sac (female gametophyte) formation in *Nuphar* compared with other angiosperms. (A) Seven-celled (eight-nucleate) female gametophyte and triploid formation typical of most angiosperms. (B) Four-celled (four-nucleate) female gametophyte and diploid endosperm formation (as reported in *Nuphar*) (see text and Williams and Friedman 2002; Figure 3.18 from Soltis et al. 2005).

Embryo sac development in Nymphaeales is noteworthy (Fig. 4.14). Whereas most angiosperms have an eight-nucleate and seven-cell embryo sac (including *Amborella*), Nymphaeales (including Hydatellaceae) and some Austrobaileyales have a 4-nucleate and 4-celled embryo sac (Winter et al. 1991; Tobe et al. 2007; see also Friedman 2006, 2008; Rudall et al. 2008). In addition, Nymphaeales produce diploid endosperm, which is also characteristic of Austrobaileyales (see Tobe et al. 2007; Williams and Friedman 2002), but not Amborellaceae. Although it is tempting to speculate that early angiosperms therefore had diploid endosperm, the reconstructions are equivocal (because of triploid endosperm in *Amborella*; see Williams and Friedman 2002). In Nymphaeales, the endosperm is poorly developed, with perisperm (a nutritive tissue derived from nucellus) playing a major nutritive role (Friedman 2008).

Nymphaeales have also been of interest in terms of floral organ identity and developmental genetics. Yoo et al. (2010) investigated the expression patterns of floral organ identity genes in *Cabomba*, *Nuphar*, and *Nymphaea*. They focused on gene expression associated with (1) perianth differentiation in Nymphaeales, (2) the transition of petaloid staminodia to stamens, and (3) patterns of expression of organ identity genes in Nymphaeaceae and Cabombaceae. In *Cabomba*, the expression patterns of B-class gene homologues fit the "sliding boundaries" model, with B-class gene expression in sepals and petals (see Chapter 14). In contrast, *Nuphar* and *Nymphaea* exhibit broad B-class gene expression that extends across all floral organs (i.e., "fading borders" model). This broad pattern of MADS-box gene expression suggests that the innermost petals of *Nymphaea* originated from petaloid staminodia (Yoo et al. 2010).

AUSTROBAILEYALES

Bootstrap support for Austrobaileyales is consistently very high (95–100%)—even in early analyses of combined DNA datasets (e.g., D. Soltis et al. 2000; Zanis et al. 2002, 2003). The clade comprises Schisandraceae (including *Illicium*, which is often placed in Illiciaceae) as well as two small families from Australasia, Austrobaileyaceae and Trimeniaceae. However, Trimeniaceae and Austrobaileyaceae were not considered closely related to Schisandraceae in any morphology-based classifications (e.g., Cronquist 1981, 1988; Takhtajan 1997). Trimeniaceae were usually placed in Laurales, separate from Schisandraceae (Cronquist 1981; Takhtajan 1997). A cladistic analysis of non-DNA characters (Doyle and Endress 2000) did not recover an Austrobaileyales clade. However, the cladistic analysis of non-DNA characters by Nandi et al. (1998) did reconstruct a clade of Austrobaileyaceae and Schisandraceae. However, non-DNA synapomorphies uniting Trimeniaceae with other members of this clade are still unclear, except for caducousness of tepals (see Endress 2008). Although Illiciaceae and Schisandraceae s.s. have been placed in Illiciales, the oldest ordinal name for the clade comprising Austrobaileyaceae, Trimeniaceae, and Schisandraceae is Austrobaileyales.

Within Austrobaileyales, relationships are strongly supported, with Austrobaileyaceae (with a single genus, *Aus-*

Figure 4.15. Austrobaileyales (Austrobaileyaceae, Trimeniaceae, Schisandraceae): a. *Austrobaileya scandens* C. T. White (Austrobaileyaceae), flower. b. *Trimenia moorei* (Oliv.) Philipson (Trimeniaceae), inflorescence. c. *Illicium floridanum* J. Ellis (Schisandraceae), flower. d. *Illicium parviflorum* Michx. (Schisandraceae), flower in bud. e. *Kadsura japonica* (L.) Dunal. (Schisandraceae), staminate flower. f. *Schisandra rubriflora* (Franch.) Rehder.&E.H.Wilson., habit and flowers.

trobaileya) sister to the rest of the clade (e.g., Qiu et al. 1999; Renner 1999; P. Soltis et al. 1999a; Savolainen et al. 2000a,b; D. Soltis et al. 2000). Trimeniaceae (comprising the single genus, *Trimenia*) are then sister to a well-supported Schisandraceae (*Schisandra* and *Kadsura*), which as noted above also includes the formerly recognized Illiciaceae (which itself comprises only *Illicium*). Even in morphology-based classifications that recognized Illiciaceae and Schisandraceae, the two families exhibit many morphological similarities and were considered very closely related and placed in Illiciales (Cronquist 1981, 1988; Takhtajan 1997). Both Illiciaceae and Schisandraceae have pollen similar to the triaperturate pollen of eudicots—a feature not found elsewhere among basal angiosperms (Cronquist 1981; Sampson 2000). However, this is a convergence; the configuration of apertures in the tetrad of Schisandraceae differs from eudicots. Phylogenetic analyses confirmed that Illiciaceae and Schisandraceae s.s. are sister taxa, and because of their morphological similarity, they have been considered a single family, Schisandraceae (see APG II 2003; APG III 2009; APG IV 2016).

As in Nymphaeales, the embryo sac in Austrobaileyales is four-celled and four-nucleate, as shown for *Schisandra* (Swamy 1964), *Illicium* (Williams and Friedman 2004), *Austrobaileya* (Tobe et al. 2007), and *Trimenia* (Friedman and Bachelier 2013) (Fig. 4.14). Although it is tempting to speculate that the four-celled and four-nucleate condition is ancestral for angiosperms, *Amborella* possesses an eight-celled, nine-nucleate embryo sac that is, in some ways, more typical of most angiosperms—that is, it possesses antipodals and two polar nuclei. As a result, ancestral-state reconstructions for the angiosperms are equivocal (Chapter 6).

The endosperm in those Austrobaileyales analyzed is diploid (Williams and Friedman 2004), as in Nymphaeales (see above), and is also well developed (with the exception of *Trimenia*), and no perisperm is produced. In this regard, Austrobaileyales are more like *Amborella*; Nymphaeales, in contrast, produce little endosperm and do produce perisperm (see above; Floyd and Friedman 2001; Friedman 2008).

Pollination has been examined in several Austrobaileyales (Endress 2010b). Diptera (flies and especially gall-midges) and Coleoptera (beetles) have been reported as pollinators (Endress 1980b; Yuan et al. 2007); flowers produce odor with pollen as the floral reward. In *Trimenia*, wind, flies, and bees have been reported, again with odor production and pollen as the reward. Schisandraceae primarily are pollinated by flies, as well as beetles, with both pollen and nectar feeding reported (Thien et al. 2009). Flowers of Schisandraceae are also noteworthy in that floral heat is used as a mechanism to attract pollinators (reviewed in Thien et al. 2009). *Austrobaileya* is fly pollinated (Endress 1980b).

Austrobaileya comprises one species of woody vines that produces ethereal oils. The flowers are large with a spirally arranged perianth of numerous parts (the outer more sepal-like and the inner more petal-like). Stamens are also numerous, spirally arranged, laminar, petaloid, and not differentiated into filament and anther; carpels are also numerous and spirally arranged (Fig. 4.15), developing into berries. Based on these floral features, the family was often considered "primitive" in traditional treatments (e.g., Cronquist 1981; Takhtajan 1987).

Trimeniaceae comprise shrubs and vines with ethereal oils. The flowers are small, hypogynous, and either perfect or unisexual. Tepals are numerous, spirally arranged, with a gradual transition of bracts subtending them, and early caducous (Fig. 4.15). The stamens are also numerous and spirally arranged, and the typically single carpel develops into a drupe.

Schisandraceae consist of vines (*Schisandra* and *Kadsura*) and trees and shrubs (*Illicium*) that produce ethereal oils. Flowers are bisexual or unisexual, actinomorphic with numerous, spirally arranged tepals, the outer more sepal-like (Fig. 4.15). The gradual transition from outer bracts to outer tepals and then inner tepals in some members is similar to the bracts and tepals of *Amborella* (Fig. 4.15). The stamens are numerous, spirally arranged, poorly differentiated into filament and anther (laminar-like), and distinct (in *Illicium*) to variably connate. The carpels are 7 to numerous, distinct, and in a single series, which is not a whorl in the morphological sense because phyllotaxis is spiral (the carpels having a distinctive appearance in fruit in *Illicium*), to spirally arranged (in *Schisandra* and *Kadsura*); the ovary is superior. Fruit is an aggregate of follicles (with explosive seed release) or berries, borne on a short to elongate receptacle.

5 Magnoliids and Chloranthales

INTRODUCTION

The magnoliids (*Magnoliidae* of Cantino et al. 2007) consist of Magnoliales, Laurales, Piperales, and Canellales; some analyses indicate they are sister to Chloranthales. Magnoliids + Chloranthales occupy a pivotal position—they may be sister to the eudicots + Ceratophyllaceae plus monocots (Chapter 3). In early phylogenetic analyses (e.g., D. Soltis et al. 2000), a more broadly defined magnoliid clade (then called the eumagnoliids) included not only Magnoliales, Laurales, Piperales, and Canellales, but also Chloranthaceae and monocots. However, analyses of additional genes have indicated that this broader circumscription is inappropriate (e.g., Qiu et al. 1999, 2000; Zanis et al. 2002, 2003; Soltis et al. 2007c, 2011; Li et al. 2011; Chapter 3).

In early studies, the clade of Magnoliales, Laurales, Piperales, and Canellales did not receive jackknife support above 50% in three-gene analyses (P. Soltis et al. 1999a; D. Soltis et al. 2000), but with the addition of more genes, bootstrap support for this clade increased to 67% (Qiu et al. 1999; 5 genes) and to 100% in a compartmentalized analysis of six genes (Zanis et al. 2003). With a 104-taxon dataset of 5 to as many as 11 genes (up to 15,000 bp of sequence data per taxon), support for the magnoliid clade was 78% (Zanis et al. 2002). With larger datasets (ranging from 17 genes to phylogenomic datasets of nuclear and plastid genes), maximal support has been obtained for this clade (Soltis et al. 2011; Moore et al. 2010; Li et al. 2011; Ruhfel et al. 2014; Wickett et al. 2014). Thus, the term "magnoliids" (*Magnoliidae* of Cantino et al. 2007) is currently restricted to denote the clade of Magnoliales, Laurales, Canellales, and Piperales (Fig. 5.1).

After the monocots, the magnoliids contain the largest number of non-eudicot genera and species. Some of the families of magnoliids are large, such as Annonaceae (120 genera, 2,000 species), Lauraceae (32 genera, 2,500 species), Piperaceae (5 genera, 2,000 species), and Myristicaceae (16 genera, 380 species). The well-known Magnoliaceae comprise over 200 species.

There are several possible non-DNA synapomorphies for the magnoliids (Nandi et al. 1998; Judd et al. 2002, 2008), although more survey work is needed to establish the full extent of the occurrence of these characters. Likely synapomorphies of the magnoliids include the phenylpropane compound asarone, the lignans galbacin and veraguensin, and the neolignan licarin (Hegnauer 1962–1994). In addition, sieve-tube plastids of the P-type are present in most of these families (Behnke 1988a), and all but Piperales share stratified phloem with wedge-shaped phloem rays (Nandi et al. 1998).

The magnoliids comprise two well-supported sister groups, Magnoliales + Laurales and Canellales + Piperales. Although some cladistic analyses of morphological characters (e.g., Doyle and Endress 2000) recovered the same Magnoliales, Laurales, Canellales, and Piperales clades, that study resulted in different relationships among these four clades. Doyle and Endress (2000) found that Magnoliales + Canellales were sister to Laurales, and Piperales were more distantly related, falling in a large polytomy with monocots, Nymphaeaceae, and several clades of eudicots; none of these non-DNA relationships received bootstrap support above 50%. However, later analyses that incorporated DNA data along with morphology (Endress and Doyle 2009; Doyle and Endress 2010) recovered the topol-

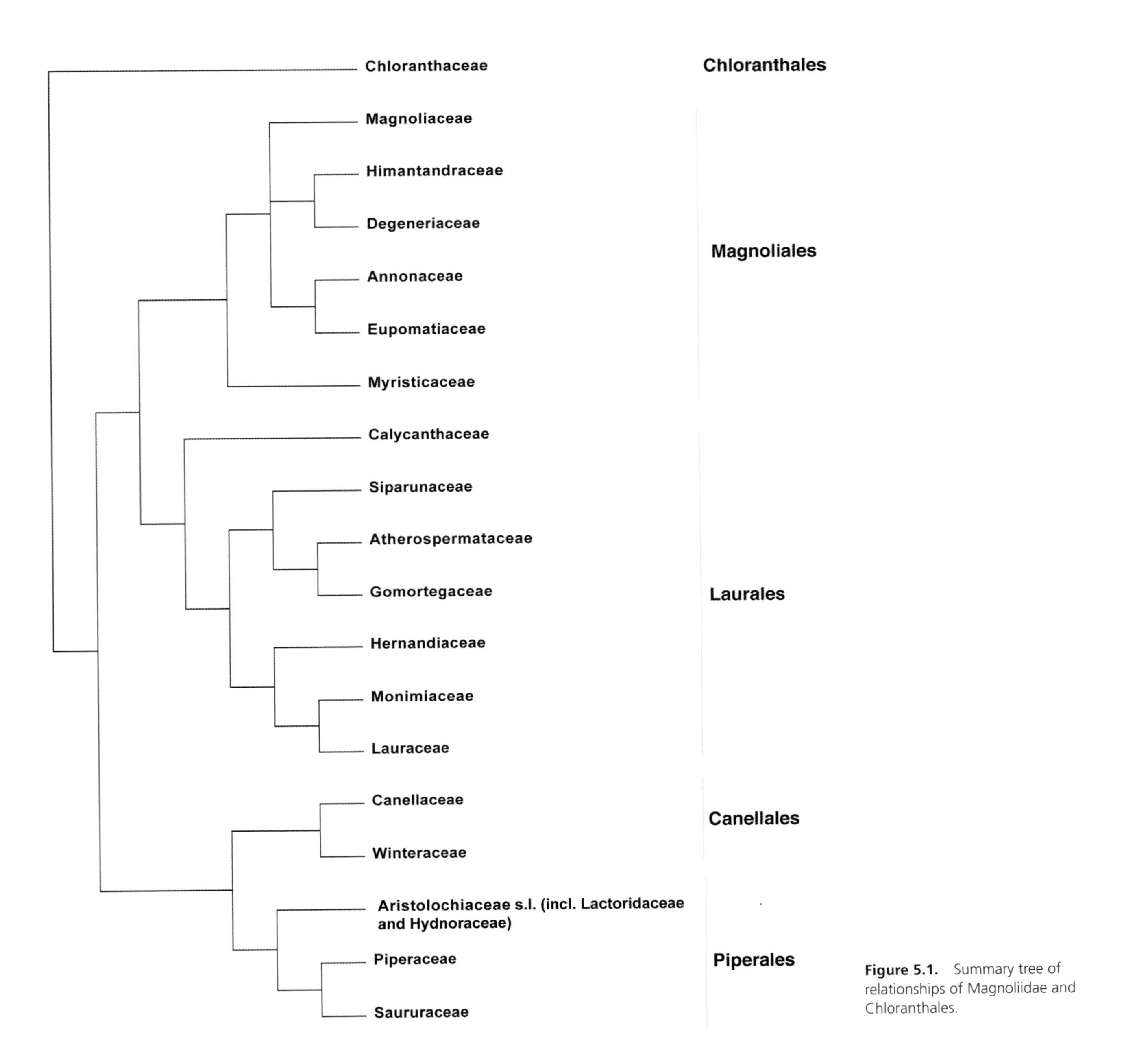

Figure 5.1. Summary tree of relationships of Magnoliidae and Chloranthales.

ogy depicted here ([Magnoliales + Laurales], [Canellales + Piperales]).

These sister-group relationships within the magnoliids ([Magnoliales + Laurales] and [Canellales + Piperales]) were largely unexpected from the perspective of traditional morphology-based classifications. Winteraceae were considered a close relative of Magnoliaceae (e.g., Cronquist 1968, 1981, 1988), and families of Piperales (Piperaceae, Aristolochiaceae, Lactoridaceae, Saururaceae) were not considered closely related to either Winteraceae or Canellaceae in these classifications (e.g., Cronquist 1981, 1988; Takhtajan 1987, 1997).

Non-DNA synapomorphies for the two subclades of magnoliids, Magnoliales + Laurales and Canellales + Piperales, respectively, are still unclear, although the presence of aporphine alkaloids may be synapomorphic for the latter ordinal pair. Relationships within each of the four orders, Magnoliales, Laurales, Piperales, and Canellales, are discussed below. Representatives of the families of these four orders are depicted in Figures 5.2 and 5.3.

MAGNOLIALES

Magnoliales comprise Myristicaceae, Degeneriaceae, Himantandraceae, Magnoliaceae, Eupomatiaceae, and Annonaceae (Figs. 5.1, 5.2), and the same Magnoliales clade was indicated in a phylogenetic analysis of non-DNA characters (Doyle and Endress 2000). Many of these families have generally been considered closely related (e.g., Cronquist 1981, 1988; Takhtajan 1987, 1997). Cronquist's (1981) Magnoliales represented a diverse assemblage, consisting of these six families, plus Winteraceae, Canellaceae, and Lactoridaceae. Winteraceae and Canellaceae are now placed in Canellales (see below), and Lactoridaceae are in Piperales; the remaining families constitute Magnoliales as currently recognized (APG 1998; APG II 2003; APG III 2009; APG IV 2016). Takhtajan placed only Degeneriaceae, Himantandraceae, and Magnoliaceae in his concept of Magnoliales. On the basis of morphological attributes, some families of Magnoliales (e.g., Magnoliaceae, Degeneriaceae, and Himantandraceae) had been considered to be "among the most archaic living angiosperms" (e.g., Cronquist 1968, 1981, 1988; Takhtajan 1969, 1991, 1997). However, recent topologies have revealed that these families are nested higher within the angiosperm tree (Fig. 5.1).

Members of Magnoliales are united by a few non-DNA putative synapomorphies—that is, stratified phloem, 2-ranked leaves, and ruminate endosperm. Other noteworthy characters include reduced fiber pit borders, an adaxial plate of vascular tissue in the petiole, palisade parenchyma, asterosclereids in the leaf mesophyll, and continuous tectum in the pollen (Doyle and Endress 2000).

DNA-based topologies with extensive sampling placed Myristicaceae as sister to the remaining members of Magnoliales (Sauquet et al. 2003; Müller et al. 2006). This position of Myristicaceae is also moderately supported by several non-DNA characters (Doyle and Endress 2000). Other DNA-based analyses have recovered alternative placements of Myristicaceae, although generally with lower support (e.g., Hilu et al. 2003; Qiu et al. 2010; Soltis et al. 2011).

Both sister groups (Annonaceae + Eupomatiaceae) and (Degeneriaceae + Himantandraceae) are well supported (D. Soltis et al. 2000; P. Soltis et al. 2000; Doyle and Endress 2000; see also Sauquet et al. 2003; Müller et al. 2006; Massoni et al. 2014), but the phylogenetic placement of Magnoliaceae remains somewhat unclear (Massoni et al. 2014). Alternative moderately supported relationships of Magnoliaceae within Magnoliales have been proposed in various analyses, all with good taxon sampling (e.g., Hilu et al. 2003; Qiu et al. 2010; Soltis et al. 2011).

Magnoliaceae, Annonaceae, and Myristicaceae have also been the focus of more detailed phylogenetic analyses (e.g., Doyle and Le Thomas 1997; J. A. Doyle et al. 2000, 2004; Kim et al. 2001; Sauquet et al. 2003; Mols et al. 2004; Richardson et al. 2004). See also Kim and Suh (2013) for a recent update for Magnoliaceae, and Chatrou et al. (2012) and Pirie and Doyle (2012) for Annonaceae.

The fossil record of Magnoliales includes members of Magnoliaceae, Myristicaceae, and Annonaceae (Supplementary Table 5.1, at press.uchicago.edu/sites/soltis/). The *Liriophyllum-Archaeanthus* plant (Dilcher and Crane 1984) from the late Albian of Kansas was reconstructed from several different isolated organs that were found together at the same localities and are dotted with resin dots believed to represent the hardened contents of ethereal oil cells. The *Liriophyllum* leaves have a prominent distal notch similar to the leaves of extant *Liriodendron*, but are distinguished in the primary venation pattern. *Archaeanthus* flowers show an arrangement of floral parts similar to Magnoliales, but the carpels bore multiple rather than single ovules. Seeds and multifollicles clearly corresponding to extant *Magnolia* are known by the middle Eocene (Manchester 1994b). The fossil record of Annonaceae extends back to the late Cretaceous based on a flower from the Coniacian of Japan (Takahashi et al. 2008), and a seed cast showing the distinctive pattern of ruminate endosperm is known from the Maastrichtian of Nigeria (Chesters 1955). Seeds preserved in this way, usually placed in the highly inclusive genus *Anonaspermum*, are common in the Paleogene (Friis et al. 2011).

LAURALES

Laurales (Figs. 5.1, 5.3) consist of Calycanthaceae (including Idiospermaceae), Monimiaceae, Gomortegaceae, Atherospermataceae, Lauraceae, Siparunaceae, and Hernandiaceae. Circumscription of Laurales has varied greatly among recent morphology-based classifications (e.g., Cronquist 1981, 1988; Takhtajan 1987, 1997; reviewed in Renner 2001). Cronquist (1981, 1988) placed Amborellaceae and Trimeniaceae in his Laurales together with Monimiaceae, Gomortegaceae, Calycanthaceae, Idiospermaceae, Lauraceae, and Hernandiaceae. *Amborella* and *Trimenia* were placed in Monimiaceae in earlier works (e.g., Perkins 1925). Others have placed Chloranthaceae in Laurales (e.g., Thorne 1974; Takhtajan 1987, 1997).

The best putative synapomorphy for the seven families of Laurales is a perigynous flower—that is, with a concave to cup-shaped or urn-shaped receptacle (hypanthium) in

Figure 5.2. *Magnoliidae*: Magnoliales (Annonaceae, Magnoliaceae, Myristicaceae): a. *Liriodendron tulipifera* L. (Magnoliaceae), terminal flower. b. *Asimina parviflora* Dunal (Annonaceae), seed in cross-section, showing ruminate endosperm. c. *Magnolia virginiana* L. (Magnoliaceae), fruit, an aggregate of follicles, and seeds with colorful sarcotesta. d. *Myristica fragrans* Houtt. (Myristicaceae), fruit dehiscing to reveal seed and red aril. e. *Annona squamosa* L. fruit, an aggregate of berries, partially connate. f. *Myristica insipida* (Myristicaceae) R. Br., shoot and inflorescences. g. *Cananga odorata* (Lam.) Hook.f. & Thomson (Annonaceae), flowers, note 2-ranked leaves and already-open immature flowers. h. *Monoon longifolium* (Sonn.) B.Xue & R.M.K.Saunders (Annonaceae), inflorescence and flowers. i. *Magnolia champaca* (L.) Baill. ex Pierre, flowers, note the elongate receptacle.

Figure 5.3. *Magnoliidae*: Canellales (Canellaceae, Winteraceae), Piperales (Aristolochiaceae), and Laurales (Lauraceae, Hernandiaceae): a. *Tasmannia insipida* R.Br. ex DC. (Winteraceae), inflorescence and flowers. b. *Canella winterana* (L.) Gaertn. (Canellaceae), inflorescence and flowers. c. *Drimys winteri* J.R.Forst. & G.Forst. (Winteraceae), shoot and inflorescence. d. *Aristolochia maxima* Cham. (Aristolochiaceae), capsule and leaves, note the flattened seeds. e. *Persea americana* Mill. (Lauraceae), flower, note the staminal flaps and staminodial nectar glands. f. *Hernandia nymphaeifolia* (Hernandiaceae), inflorescence with staminate and carpellate flowers.

which the gynoecium is usually deeply embedded (Endress and Igersheim 1997; Renner 1999). This receptacle sometimes becomes woody in fruit (e.g., Calycanthaceae). Other putative synapomorphies are the unilacunar nodes and opposite and decussate leaves. Several other non-DNA characters that unite the clade are the presence of inner staminodia, ascendant ovules, tracheidal endotesta (Doyle and Endress 2000), and a carpel that is intermediate between ascidiate and plicate. Additionally, most members of Laurales (i.e., all but Calycanthaceae) are united by the additional apomorphies of anthers opening by 1, 2, or 4 flaps, and associated with a pair of glands (of staminodial origin, producing nectar or odor); inaperturate, often spinulose pollen, with a more-or-less reduced exine; and carpels with only a single median ovule. Other morphological characters (e.g., uniaperturate pollen) that were considered to typify Laurales as previously circumscribed not only vary within Laurales as considered here but are also found in other basal angiosperm lineages.

Phylogenetic analyses of Laurales were provided by Renner (1999) and Chanderbali et al. (2001), who used sequences from multiple DNA regions. Within Laurales, the deepest split is between Calycanthaceae and the remaining six families, which in turn form two clades (Fig. 5.1): (1) Siparunaceae, which are sister to Atherospermataceae + Gomortegaceae; and (2) Hernandiaceae sister to Monimiaceae + Lauraceae. Resolution within the Monimiaceae, Lauraceae, Hernandiaceae clade has been difficult using DNA. Based on morphology there is actually support for Hernandiaceae + Lauraceae. For example, these two families share reduction to a single carpel whereas Monimiaceae have retained the ancestral condition of numerous separate carpels. We have depicted the accepted molecular topology in Figure 5.1.

Monimiaceae as traditionally recognized are clearly polyphyletic. Phylogenetic studies indicate that Atherospermataceae and Siparunaceae (then considered subfamilies within Monimiaceae, but see Schodde 1970) must be recognized as distinct families. Several morphological character-state changes are congruent with the molecular tree (Renner 1999). Calycanthaceae have disulculate tectate-columellate pollen, whereas their sister clade has inaperturate, thin-exined pollen (with the exception of Atherospermataceae, which have columellate, but meridianosulcate or disulcate pollen). Calycanthaceae also have two ovules, whereas the remaining Laurales have a solitary ovule in each carpel. Monimiaceae, Lauraceae, and Hernandiaceae have apical ovules, whereas the ancestor of the clade of Siparunaceae and Atherospermataceae + Gomortegaceae is inferred to have had basal ovules, a condition lost in *Gomortega*, the only lauralean genus with a syncarpous ovary. Lauraceae and Hernandiaceae share flowers with a single carpel. Depending on the correct placement of Calycanthaceae-like fossil flowers, tetrasporangiate anthers with valvate dehiscence (with the valves laterally hinged) may be ancestral in Laurales and lost in most modern Calycanthaceae (but present in *Sinocalycanthus*, Staedler et al. 2007) and most Monimiaceae (but present in *Peumus*; Endress and Hufford 1989).

Recent updates on the phylogeny of Laurales include Soltis et al. (2011) and Massoni et al. (2014). Recent phylogenetic trees for the families of the order have been published by Li et al. (2004) (Calycanthaceae) and Rohwer and Rudolph (2005) (Hernandiaceae). See also Renner (2004, 2010), Doyle and Endress (2010), and Massoni et al. (2014).

Fossils of Laurales include numerous examples of Lauraceae plus a few of Calycanthaceae and Hernandiaceae. Lauraceae have a particularly rich record recognized by their flowers, leaves, fruits, and wood, but their pollen is rarely preserved (Supplementary Table 5.1).

PIPERALES

Piperales consist of Aristolochiaceae, Lactoridaceae, Piperaceae, Saururaceae, and Hydnoraceae (Figs. 5.1, 5.3). Molecular analyses have provided strong support for the monophyly of Piperales (e.g., Qiu et al. 1999; P. Soltis et al. 1999a; Barkman et al. 2000; D. Soltis et al. 2000, 2011; Zanis et al. 2002, 2003). Molecular analyses also place Hydnoraceae, a family of parasitic plants (see Chapter 11), within Piperales, apparently close to Aristolochiaceae (Nickrent and Duff 1996; Nickrent et al. 1998, 2002). Phylogenetic analyses of morphological characters likewise recognized a clade of Piperaceae, Saururaceae, Aristolochiaceae, and Lactoridaceae (Doyle and Endress 2000); Hydnoraceae were not included in that analysis.

The circumscription of Piperales has varied. Saururaceae and Piperaceae have consistently been considered closely related (Cronquist 1981, 1988; Takhtajan 1987, 1997). However, the relationships of Aristolochiaceae, Lactoridaceae, and Hydnoraceae have been problematic. Cronquist (1981, 1988) placed Lactoridaceae in Magnoliales close to Magnoliaceae (see also Lammers et al. 1986) and put Aristolochiaceae in their own order; he also added Chloranthaceae to Piperales. Some have suggested a close relationship of Lactoridaceae and Piperaceae (Weberling 1970; Carlquist 1990) and of Aristolochiaceae to Annonaceae (e.g., Dahlgren 1980; Thorne 1992a,b; Takhtajan 1997). Whereas Takhtajan (1987, 1997) considered Hydnoraceae to be closely related to Aristolochiaceae, others (e.g., Cronquist 1981; Heywood 1998) placed Hydnoraceae in their Rosidae.

The monophyly of Piperales is supported by leaves that are often cordate, with palmate to pinnate venation and widely spaced vascular bundles, in one, two, or several concentric rings, to ± scattered, or at least wood with broad rays. Other possible synapomorphies include sympodial growth, distichous phyllotaxis, and possibly a single prophyll. Piperaceae and Saururaceae form a well-supported clade (Fig. 5.1). *Lactoris* is closely related to Aristolochiaceae. In early DNA analyses, *Lactoris* appeared within Aristolochiaceae, as sister to *Aristolochia* + *Thottea* (Qiu et al. 1999; Zanis et al. 2003) or *Aristolochia* alone (D. Soltis et al. 2000; *Thottea* was not included), but support for a placement of *Lactoris* within Aristolochiaceae was weak in those studies, even with five genes.

The close relationship of Lactoridaceae to Aristolochiaceae is often considered a "surprise" result of molecular phylogenetics, given what appear to be major gross morphological differences between the two families. However, morphological similarities between Aristolochiaceae and *Lactoris* include strongly extrorse anthers with broad connective, almost sessile anthers, and stamens basally fused with the gynoecium (Endress 1994c). González and Rudall (2001) examined branching pattern, inflorescence morphology, and stipule development in *Lactoris* and other Piperales. They determined that whereas sympodial growth and a sheathing leaf base are present in all Piperales, the presence of stipules is confined to *Lactoris*, Saururaceae, and some Piperaceae. These characters are consistent with the placement of *Lactoris* within Piperales. In morphology-based cladistic analyses, *Lactoris* appears with Piperaceae + Saururaceae in some most-parsimonious trees and with Aristolochiaceae in others (Doyle and Endress 2000).

Given the uncertain position of *Lactoris* in both molecular and morphological trees, APG (1998) and APG II (2003) recommended that Lactoridaceae be retained until more convincing evidence of its placement is obtained. More recently, however, Jaramillo et al. (2004), Wanke et al. (2007), and Massoni et al. (2014) increased the sampling of both genes and taxa and provided improved resolution of phylogenetic relationships within *Magnoliidae* in general and Piperales in particular. Their analyses place *Lactoris* as sister to Aristolochioideae and also show Hydnoraceae nested within Aristolochiaceae. Based on these results, both Lactoridaceae and Hydnoraceae are no longer treated as distinct families but are included in Aristolochiaceae (see APG IV 2016 and Chapter 12).

The most recent phylogenetic updates of Aristolochiaceae are those of Kelly and González (2003) and Neinhuis et al. (2005); see also Jaramillo et al. (2004), Wanke et al. (2007), and Massoni et al. (2014) for recent phylogenetic analyses of Piperales.

CANELLALES

A sister-group relationship of Canellaceae + Winteraceae received bootstrap or jackknife support of 99 or 100% in all multigene analyses (e.g., Qiu et al. 1999; P. Soltis et al. 1999a; D. Soltis et al. 2000; Zanis et al. 2002, 2003) (Figs. 5.1, 5.3); Doyle and Endress's (2000) morphological analysis also found this sister group (bootstrap support <50%). However, morphology-based classifications generally considered Winteraceae closely related to Magnoliaceae (e.g., Cronquist 1981, 1988; Heywood 1993). Canellaceae, in contrast, were often placed near Myristicaceae (e.g., Wilson 1966; Cronquist 1981, 1988). Takhtajan (1997) placed Canellaceae and Winteraceae in separate orders, noting similarities of Canellaceae to both Winteraceae and Illiciaceae. Canellaceae + Winteraceae was initially referred to as Winterales (e.g., Qiu et al. 1999; Doyle and Endress 2000; D. Soltis et al. 2000; Zanis et al. 2002). However, the name "Canellales" is older than "Winterales," and the former name has been used in APG III (2009) and APG IV (2016; see Chapter 12).

The phylogenetic position of Winteraceae within the magnoliid clade, well removed from the base of the angiosperms, is noteworthy in that some classifications have considered the family to be perhaps the "most archaic" living family of angiosperms (Cronquist 1981; see also Endress 1986b). Winteraceae have also been of interest because of their vesselless xylem and (in some taxa) plicate carpels. The phylogenetic position of Winteraceae indicates, however, that these features are apparently secondarily derived (Young 1981; Doyle 2000).

Putative synapomorphies of Canellales include branched sclereids, a nearly glabrous surface, a well-differentiated perianth of calyx and corolla (unusual in non-eudicots), well-differentiated pollen tube transmitting tissue, an outer integument with only two to four cell layers, and seeds with a palisade exotesta (Doyle and Endress 2000). Winteraceae and Canellaceae share a similar irregular, "first-rank" leaf venation (Hickey and Wolfe 1975; see also Doyle 2000) and similar stelar and nodal anatomies (Keating 2000). Vascularization of the seeds in Winteraceae also indicates a close relationship to Canellaceae (Deroin 2000). Phylogenetic studies using morphology (Vink 1988, 1993) indicated that *Takhtajania*, a genus until recently thought to be extinct in nature, was sister to all other Winteraceae. With the rediscovery of *Takhtajania* (Schatz 2000), material became available for molecular phylogenetic and other analyses, and the placement of *Takhtajania* as sister to other Winteraceae was confirmed by DNA sequence data (Karol et al. 2000). An issue of the *Annals of the Missouri*

Botanical Garden (2000;87[3]:297–432) was dedicated to an overview of *Takhtajania*, including discussion of its rediscovery and relationships. The issue contained papers discussing the morphology, anatomy, palynology, cytology, and ecology of both Canellaceae and Winteraceae (Carlquist 2000a; Doust 2000; Doyle 2000; Endress et al. 2000b; Feild et al. 2000b; Karol et al. 2000; Sampson 2000; Tobe and Sampson 2000).

Recent phylogenetic analyses of Canellales include Marquínez et al. (2009), Salazar and Nixon (2008), Massoni et al. (2014), and Müller et al. (2015). Müller et al. (2015) suggest that both vicariance and long-distance dispersal from the Old to New World have played important roles in the evolutionary history of this clade. Although fossil occurrences of Canellales are rare, the group can be traced to the Early Cretaceous (Late Barremian—Early Aptian) based on pollen tetrads distinctive for Winteraceae from the Gabon and Israel (Doyle et al. 1990).

CHLORANTHALES

Chloranthales comprise Chloranthaceae, with four genera: *Ascarina*, *Chloranthus*, *Hedyosmum*, and *Sarcandra*. The position of Chloranthaceae (Fig. 5.4) has long been problematic. Takhtajan (1987, 1997) placed the family in its own order and suggested a close relationship with Trimeniaceae and Piperales. Others have also suggested a close relationship of Chloranthaceae and Trimeniaceae (Endress 1986b, 1987c; Thorne 2001). Cronquist (1981) placed Chloranthaceae in his Piperales. Taylor and Hickey (1992, 1996b), Doyle et al. (1994), and Hickey and Taylor (1996) focused attention on Chloranthaceae as the possible sister group to all other extant angiosperms.

Even with recent multigene analyses, the placement of Chloranthaceae remains uncertain (e.g., Moore et al. 2011; Soltis et al. 2011; N. Zhang et al. 2012; Ruhfel et al. 2014;

Figure 5.4. Chloranthaceae: a. *Chloranthus japonicus* Siebold, inflorescence. b. *Sarcandra glabra* (Thunb.) Nakai, plants in fruit. c. *Hedyosmum nutans* Sw., branch with young infructescence (fruits immature) on the right side and male inflorescences on the left side (one anthetic and two preanthetic).

Wickett et al. 2014). Large plastid datasets and datasets dominated by plastid genes favor a sister-group relationship of Chloranthaceae with magnoliids with strong bootstrap support (Soltis et al. 2011; Ruhfel et al. 2014). However, a recent analysis of conserved nuclear genes places Chloranthaceae sister to Ceratophyllaceae (Chapter 3); this clade is then sister to eudicots (N. Zeng et al. 2014) (Fig. 3.6). More large datasets of nuclear genes having broad taxon coverage are needed to evaluate this placement further. Phylogenetic analyses of non-DNA characters indicated that Chloranthaceae occupy an isolated position as an early-diverging lineage (e.g., Nixon et al. 1994; Nandi et al. 1998; Doyle and Endress 2000). Phylogenetic analysis of non-DNA plus DNA characters (Doyle and Endress 2000) placed Chloranthaceae "above" Austrobaileyales as sister to all remaining angiosperms.

Burger (1977) proposed that Chloranthaceae may be closely involved in the origin of the monocots. Based on topologies recovered in recent years (see Fig. 5.1), however, this scenario would seem unlikely. In addition, several morphological features are shared by Chloranthaceae and Amborellaceae, Nymphaeaceae, and Austrobaileyales, including ascidiate carpels, unilacunar nodal anatomy, and chloranthoid leaf teeth, many of which likely represent symplesiomorphies in these groups.

Some of the earliest known fossil angiosperm material is similar to modern Chloranthaceae, including mesofossils as well as dispersed pollen (e.g., Doyle 1969; Walker and Walker 1984; Friis et al. 1986, 1994b, 2011; Crane et al. 1995; Brenner 1996), in agreement with phylogenetic inferences that point to the antiquity of the lineage.

DNA analyses have clarified relationships among the four extant genera of Chloranthaceae (Zhang and Renner 2003; Eklund et al. 2004). Kong et al. (2002a) studied relationships and character evolution in *Chloranthus*. Others have examined floral evolution and integrated fossils of Chloranthaceae into datasets and topologies of extant taxa (Doyle et al. 2003; Eklund et al. 2004). Available data indicate that whereas Chloranthaceae represent an old lineage, the extant taxa represent a much more recent diversification, a theme that has been observed in other "old" lineages, including cycads (see Chapter 1) and Nymphaeaceae (Yoo et al. 2005; see Chapter 3).

Stamen morphology is highly variable within Chloranthaceae. *Sarcandra* and *Ascarina* have "normal" stamens with two pairs of microsporangia (Fig. 5.5), but most species of *Chloranthus* have stamens that are three-lobed with four thecae (Fig. 5.5). It has been controversial whether the stamens of *Chloranthus* represent a single stamen with four pairs of sporangia or three independent stamens that have become fused at the base (reviewed in Kong et al. 2002a; Doyle et al. 2003). On the basis of fossil evidence, most researchers have favored the latter hypothesis (e.g., Crane et al. 1989; Herendeen et al. 1993; Eklund et al. 1997), but Endress (1987c) considered the structure to represent a single tripartite stamen. Phylogenetic studies of extant *Chloranthus* (Kong et al. 2002a), as well as analysis of floral organogenesis of extant *Chloranthus sessilifolius*

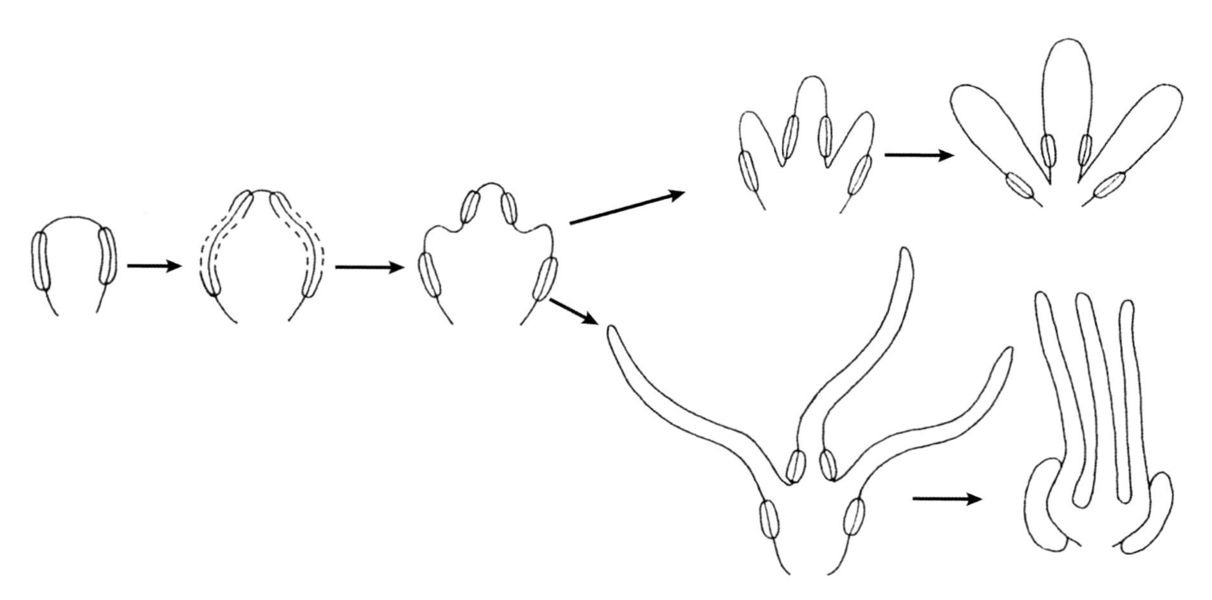

Figure 5.5. Hypothesis of Endress (1987c) for the origin of the unusual tripartite androecium present in *Chloranthus* (Chloranthaceae). Molecular and developmental data support this hypothesis (see text).

(Kong et al. 2002b), a species with a tripartite androecium, have provided support for Endress's (1987c) hypothesis (Fig. 5.5).

Q. Zhang et al. (2011) and Doyle and Endress (2014) provide recent phylogenetic analyses of Chloranthaceae. Doyle and Endress (2014) examined Cretaceous fossils of Chloranthaceae and confirmed the early diversity of the family in a phylogenetic analysis of living and fossil taxa. Furthermore, their examination of the fossil record elucidated floral evolution indicating that a shift to unisexual flowers preceded the loss of the perianth in this clade (note that some extant Chloranthaceae still have a perianth—the carpellate flowers of *Hedyosmum*).

6 Character Evolution

The Ancestral Angiosperm and General Trends

INTRODUCTION

The well-supported framework of angiosperm relationships now available affords the opportunity to reconstruct ancestral character states for extant angiosperms and to investigate character evolution throughout the angiosperms as a whole. In this chapter, having just reviewed current hypotheses of phylogeny for the ANA grade and magnoliids, we focus on character evolution across these lineages to elucidate the morphological and anatomical features of early angiosperms. Our reconstructions include monocots and eudicots, but patterns of evolutionary change in these clades will be described in later chapters that focus on these groups.

Attempts to examine character evolution across the angiosperms for key features (e.g., merism = merosity, symmetry, and other floral characters; genome size) appeared soon after the first multi-gene analyses solidified the angiosperm backbone tree (e.g., Ronse De Craene et al. 2003; Soltis et al. 2005b; Doyle 2009). Since those reconstructions were published, the framework tree has been further resolved (see Chapter 3); we base our primary reconstructions on these recent trees (e.g., Moore et al. 2010; Soltis et al. 2011) with the caveat that some recent analyses have recovered slightly different topologies (Chapter 3).

There are several important caveats in these ancestral state reconstructions. The sister group of the angiosperms is unknown (other than "all extant gymnosperms" in molecular-based trees; when fossils are considered this is even more of a problem), which compromises efforts to infer ancestral angiosperm states. Furthermore, many characters of interest are confined to the angiosperms and cannot be scored in the outgroup. For example, extant gymnosperms lack a perianth. If the outgroup of the angiosperms is coded as having an indeterminate number of perianth parts, then an indeterminate number is also ancestral for the angiosperms. Alternatively, if the outgroup for the angiosperms is considered to lack a perianth, then it is equally parsimonious for the ancestral state of the angiosperms to be either trimerous or indeterminate in perianth merism (see Zanis et al. 2003; Soltis et al. 2005b).

Alternative methods of character-state reconstruction that do not rely on outgroup coding, such as maximum likelihood (ML) (e.g., Pagel 1998, 1999), may be more useful for inferring patterns of character evolution in the earliest angiosperm lineages. In contrast to our earlier reconstructions and those of others (e.g., Doyle and Endress 2000; Endress and Doyle 2009), which relied solely on maximum parsimony (MP), we have employed maximum likelihood (ML) methods here. In the reconstructions for this book, we used Mesquite (Maddison and Maddison 2011, 2015), for both MP analyses with all most parsimonious states and ML optimizations. In figures depicting reconstructions, MP = maximum parsimony reconstruction; ML = maximum likelihood reconstruction. For maximum likelihood reconstructions, the Mk model was employed.

There are other limitations to reconstructing patterns of character evolution. Extant basal angiosperms (e.g., *Amborella* and Nymphaeales) are almost certainly not the earliest angiosperm lineages and may not even represent or resemble the products of the earliest diversification events. The likelihood of extinction of many basal angiosperm lineages (e.g., Friis et al. 2000, 2011), with some intercalated among the extant crown, also compromises the accuracy of any character-state optimizations, as does the paucity of well-placed fossils of early angiosperms. We are reconstructing the character states of the common ancestor of all extant angiosperms, not necessarily those of the first angiosperms. We therefore often state that our reconstructions suggest that "early angiosperms" possessed a certain feature based on our reconstructions.

With these limitations, we focus primarily on characters that are among those often discussed as critical for understanding the morphology and anatomy of the earliest angiosperms and their subsequent diversification. These include habit, the presence of vessel elements, and diverse aspects of floral morphology, including perianth and stamen phyllotaxis, perianth merism, perianth differentiation, stamen form, carpel form and sealing, and endosperm characters. We have updated the character matrix from Soltis et al. (2005b) when new data were available. As noted, the tree topology has been updated to correspond to Soltis et al. (2011) (Fig. 3.6; 6.1). In addition, *Trithuria* (Hydatellaceae) has been added.

Many authors have reconstructed the evolution of diverse floral characters or reviewed patterns of character evolution (e.g., Albert et al. 1998; Doyle and Endress 2000, 2010; Ronse De Craene et al. 2003; Zanis et al. 2003; Ronse De Craene 2007, 2008, 2013; Doyle 2008, 2012; Endress and Doyle 2009; Soltis et al. 2005b). However, the topologies used in earlier studies varied, and many of the trees used deviated significantly from what the community now accepts as the framework of angiosperm phylogeny (e.g., Stevens 2001 onwards; http://www.mobot.org/MOBOT/research/APweb/; APG III 2009; APG IV 2016; Soltis et al. 2011). Albert et al. (1998) used a topology with *Ceratophyllum* sister to all other angiosperms, reflecting the understanding of relationships at that time. Doyle and Endress (2000) standardly used a combined morphology plus three-gene dataset (D. Soltis et al. 2000). However, critical taxa (e.g., *Ceratophyllum*) were omitted from Doyle and Endress (2000). Importantly, the topology they used differs substantially from our current understanding of relationships. Albert et al. (1998) and Doyle and Endress (2000) used families as terminals, either exclusively, or in large part, rather than individual genera or species, making it difficult to assess patterns of variation for characters that are polymorphic within a family. In subsequent analyses (e.g., Endress and Doyle 2009), crucial taxa (e.g., *Trithuria* and *Ceratophyllum*) were added, families were replaced by exemplar species and genera, and the topology used was updated to reflect the currently accepted backbone tree. However, as in most studies, only MP was used for ancestral-state reconstruction. Nonetheless, Endress and Doyle (2009) remains an excellent overview of floral character evolution, and we refer to those results here.

Character-state reconstructions have benefited greatly from the recent clarifications of angiosperm relationships (see Chapter 3). Phylogenetic studies focused attention on Amborellaceae, Nymphaeales, and Austrobaileyales (the ANA grade) as subsequent sisters to the remaining extant angiosperms. As a result, a wealth of new data was obtained for these taxa, including important information on floral development (e.g., Endress and Igersheim 2000a,b; Endress 2001a; Posluszny and Tomlinson 2003; Buzgo et al. 2004a; Endress and Doyle 2007). To reflect these new data, authors have used different codings for several characters (e.g., merism, phyllotaxis) from those used earlier (e.g., Albert et al. 1998; P. Soltis et al. 2000; Ronse De Craene et al. 2003; Zanis et al. 2003).

For our optimizations, we used either a 64-taxon or 53-taxon tree and two alternative topologies. One topology follows recent summary trees for angiosperms (e.g., Moore et al. 2010; Soltis et al. 2011) (Fig. 3.6), in which after the ANA grade, Chloranthaceae + magnoliids are sister to all other angiosperms (i.e., monocots plus [eudicots + *Ceratophyllum*]). This is the topology we present here (for both 64-taxon and 53-taxon analyses). However, because the sister group of Chloranthaceae + magnoliids has only moderate support (see Chapter 3), a second topology was used in which after the ANA grade, Chloranthaceae and then magnoliids are sister to all other angiosperms (reconstructions not shown). In addition, we conducted other analyses, with *Amborella* + Nymphaeales as sister to all other angiosperms, although data strongly support *Amborella* alone as sister to all other angiosperms (Chapter 3) (reconstructions not shown). Generally (as discussed below), these changes in topology had little impact on the major themes of character evolution discussed here. In another set of analyses, we added the fossil *Archaefructus* as sister to all other angiosperms, although some analyses have suggested that this is better placed in Nymphaeales (see Chapters 3 and 4).

We typically used genera as terminals. For the 53-taxon analysis, families were sometimes used as terminals because few data were available for some characters. A list of generic terminals, as well as the familial and ordinal designations for each genus (sensu APG III 2009; APG IV 2016), is provided in Table 6.1.

We used a reduced version of the underlying matrix assembled by Ronse De Craene et al. (2003) for the study of floral evolution, with the addition of taxa (e.g., *Trithuria*). Additional data are from Endress (2001a), Rudall et al. (2007), Endress and Doyle (2009), and the primary sources cited therein. We present only some of our reconstructions here, but describe the robustness of the analyses to differences in topology or optimization (MP or ML).

Table 6.1. Taxa employed in character state reconstructions, providing family and ordinal designations (following APG IV 2016)

Order (if assigned)	Family	Genus	Order (if assigned)	Family	Genus
Acorales	Acoraceae	*Acorus*	Laurales	Lauraceae	*Cinnamomum*
Alismatales	Alismataceae	*Sagittaria*	Laurales	Lauraceae	*Cryptocarya*
Amborellales	Amborellaceae	*Amborella*	Laurales	Lauraceae	*Laurus*
Magnoliales	Annonaceae	*Annona*	Magnoliales	Magnoliaceae	*Magnolia*
Magnoliales	Annonaceae	*Asimina*	Magnoliales	Magnoliaceae	*Liriodendron*
Magnoliales	Annonaceae	*Cananga*	Ranunculales	Menispermaceae	*Tinospora*
Magnoliales	Annonaceae	*Monodora*	Laurales	Monimiaceae	*Hortonia*
Magnoliales	Annonaceae	*Polyalthia*	Laurales	Monimiaceae	*Kibara*
Alismatales	Aponogetonaceae	*Aponogeton*	Laurales	Monimiaceae	*Tambourissa*
Alismatales	Araceae	*Spathiphyllum*	Magnoliales	Myristicaceae	*Knema*
Piperales	Aristolochiaceae	*Aristolochia*	Magnoliales	Myristicaceae	*Myristica*
Piperales	Aristolochiaceae	*Asarum*	Proteales	Nelumbonaceae	*Nelumbo*
Piperales	Aristolochiaceae	*Saruma*	Nymphaeales	Nymphaeaceae	*Brasenia*
Piperales	Aristolochiaceae	*Thottea*	Nymphaeales	Nymphaeaceae	*Cabomba*
Laurales	Atherospermataceae	*Daphnandra*	Nymphaeales	Nymphaeaceae	*Nuphar*
Laurales	Atherospermataceae	*Doryphora*	Nymphaeales	Nymphaeaceae	*Nymphaea*
Asparagales	Asphodelaceae	*Bulbine*	Nymphaeales	Nymphaeaceae	*Victoria*
Austrobaileyales	Austrobaileyaceae	*Austrobaileya*	Piperales	Piperaceae	*Piper*
Alismatales	Butomaceae	*Butomus*	Proteales	Platanaceae	*Platanus*
Buxales	Buxaceae	*Buxus*	Poales	Poaceae	*Oryza*
Buxales	Buxaceae	*Didymeles*	Ranunculales	Ranunculaceae	*Glaucidium*
Austrobaileyales	Calycanthaceae	*Calycanthus*	Ranunculales	Ranunculaceae	*Ranunculus*
Austrobaileyales	Calycanthaceae	*Chimonanthus*	Proteales	Sabiaceae	*Sabia*
Austrobaileyales	Calycanthaceae	*Idiospermum*	Piperales	Saururaceae	*Anemopsis*
Canellales	Canellaceae	*Canella*	Piperales	Saururaceae	*Houttuynia*
Canellales	Canellaceae	*Cinnamodendron*	Piperales	Saururaceae	*Saururus*
Ceratophyllales	Ceratophyllaceae	*Ceratophyllum*	Austrobaileyales	Schisandraceae	*Illicium*
Chloranthales	Chloranthaceae	*Ascarina*	Austrobaileyales	Schisandraceae	*Kadsura*
Chloranthales	Chloranthaceae	*Chloranthus*	Austrobaileyales	Schisandraceae	*Schisandra*
Chloranthales	Chloranthaceae	*Hedyosmum*	Dioscoreales	Dioscoreaceae	*Tacca*
Magnoliales	Degeneriaceae	*Degeneria*	Alismatales	Tofieldiaceae	*Pleea*
Magnoliales	Eupomatiaceae	*Eupomatia*	Austrobaileyales	Trimeniaceae	*Trimenia*
Ranunculales	Eupteleaceae	*Euptelea*	Trochodendrales	Trochodendraceae	*Tetracentron*
Ranunculales	Papaveraceae	*Dicentra*	Trochodendrales	Trochodendraceae	*Trochodendron*
Laurales	Gomortegaceae	*Gomortega*	Canellales	Winteraceae	*Bubbia*
Laurales	Hernandiaceae	*Gyrocarpus*	Canellales	Winteraceae	*Drimys*
Laurales	Hernandiaceae	*Hernandia*	Canellales	Winteraceae	*Pseudowintera*
Laurales	Hernandiaceae	*Sparattanthelium*	Canellales	Winteraceae	*Tasmannia*
Magnoliales	Himantandraceae	*Galbulimima*	Canellales	Winteraceae	*Takhtajania*
Nymphaeales	Hydatellaceae	*Trithuria*	Canellales	Winteraceae	*Zygogynum*
Piperales	Lactoridaceae	*Lactoris*			

HABIT

The hypothesized habit (woody vs. herbaceous) of the earliest angiosperms has long been debated. Cronquist (1968, 1988), Takhtajan (1969), Stebbins (1974), Thorne (1976), and Loconte (1996) maintained that the ancestral angiosperms were likely woody. The logic for this hypothesis was sound—all gymnosperms are woody, so the first angiosperms were likely woody as well. Furthermore, the woody habit is prevalent in the extant angiosperms these workers considered to be archaic (e.g., Magnoliales as traditionally defined). In contrast, as reviewed earlier (see Chapters 2, 3, and 4), some phylogenetic studies and fossil data focused attention on a few herbaceous families (e.g., Chloranthaceae, Piperaceae, Nymphaeaceae) or the aquatic fossil *Archaefructus*, as possible sister groups to all other extant angiosperms; these results indicated the possibility of a herbaceous origin of angiosperms (e.g., Taylor and Hickey 1992, 1996a,b; Doyle et al. 1994; Sun et al. 2002; Goremykin et al. 2013; reviewed in Soltis et al. 2008b). Once phylogenetic trees became available for basal (and other) angiosperms, efforts were made to reconstruct the ancestral habit for angiosperms. As reviewed below, these efforts typically yield an unsatisfying conclusion—the ancestral habit is difficult to infer unequivocally.

Doyle and Endress (2000) reconstructed the evolution of habit using a summary tree then available and recognizing two states: (1) tree or shrub (woody) or (2) rhizomatous, scandent, or acaulescent (herbaceous). In their reconstructions, the ancestral habit for angiosperms is equivocal, with the herbaceous habit established as ancestral for all angiosperms except *Amborella*; the woody habit was derived independently in magnoliids and eudicots. Doyle and Endress (2000) scored *Chloranthus* and *Ascarina* (Chloranthaceae) as herbaceous and *Hedyosmum* as polymorphic because the young stages of *Hedyosmum* have been considered rhizomatous and the mature plants woody (Blanc 1986; see also discussion in Endress and Doyle 2015).

As in Soltis et al. (2005b), we found that the coding of Chloranthaceae for habit has a large effect on character-state reconstruction. If *Hedyosmum* is scored as polymorphic for woody and herbaceous forms, or if all three representatives of Chloranthaceae (*Chloranthus*, *Ascarina*, *Hedyosmum*) are scored as polymorphic, then the ancestral habit is equivocal with both MP and ML optimization (Fig. 6.1). However, if *Hedyosmum* is scored as woody, then the woody habit is reconstructed as ancestral for extant angiosperms with MP and the woody condition is favored as ancestral with ML (Fig. 6.2). This is also the case if a topology is used in which Nymphaeales are treated as sister to *Amborella* (not shown). The sensitivity of our results to single changes in the data matrix is further illustrated by adding the fossil *Archaefructus* as sister to all extant angiosperms. The ancestral state for angiosperms is then reconstructed as herbaceous with both MP and ML (not shown). Thus, it is possible to obtain either a woody or herbaceous ancestral state for angiosperms. However, if gymnosperm outgroups are added (all are woody) the ancestral state for angiosperms is woody with MP and ML.

CHLORANTHOID LEAF TEETH

Chloranthoid teeth (Fig. 6.3; Chapter 4) are present in *Amborella* (Bailey and Swamy 1948) and most Austrobaileyales (*Kadsura*, *Schisandra*, and *Illicium* [Schisandraceae; Hickey and Wolfe 1975]; *Trimenia* [Trimeniaceae; Rodenburg 1971; Endress 1987c]). Both *Trimenia* and *Illicium* were scored as polymorphic for presence/absence of chloranthoid teeth. Chloranthoid teeth are also present in Chloranthaceae, but are not found in *Austrobaileya* or Nymphaeales.

Among basal eudicots, a similar tooth type is found in Trochodendraceae, Buxaceae, and some Ranunculales (Berberidaceae, Papaveraceae, Ranunculaceae, and Lardizabalaceae). Doyle and Endress (2000) and Soltis et al. (2005b) scored the leaf teeth of these eudicots as homologous to the chloranthoid type; we followed that approach here (but see below).

Doyle and Endress (2000) suggested that chloranthoid leaf teeth are ancestral in the angiosperms. However, our analyses indicate that the ancestral condition of extant angiosperms is equivocal with both MP and ML (although ML favors the absence of chloranthoid teeth) (Fig. 6.4). This is the case even though *Amborella* has chloranthoid teeth as do some Austrobaileyales (see above). The ancestor of Austrobaileyales is also reconstructed as ambiguous, regardless of whether *Illicium* and *Trimenia* are scored as polymorphic for this feature or as possessing chloranthoid teeth. After *Amborella*, the ancestral state for all other angiosperms is the absence of chloranthoid leaf teeth (with both MP and ML). If a topology is used in which *Amborella* and Nymphaeales are sisters, then the ancestral state for extant angiosperms is the absence of chloranthoid teeth. Regardless of topology, Chloranthaceae represent a separate origin of this tooth type. As noted, if chloranthoid teeth function to release root pressure (Feild et al. 2003a, 2004), they would not be needed in aquatics and may have been lost.

The reconstructions indicate multiple origins of chloranthoid teeth (Fig. 6.4), although the exact number is unclear. In addition to their presence in *Amborella*, Austrobaileyales, and Chloranthaceae, they appear to have arisen

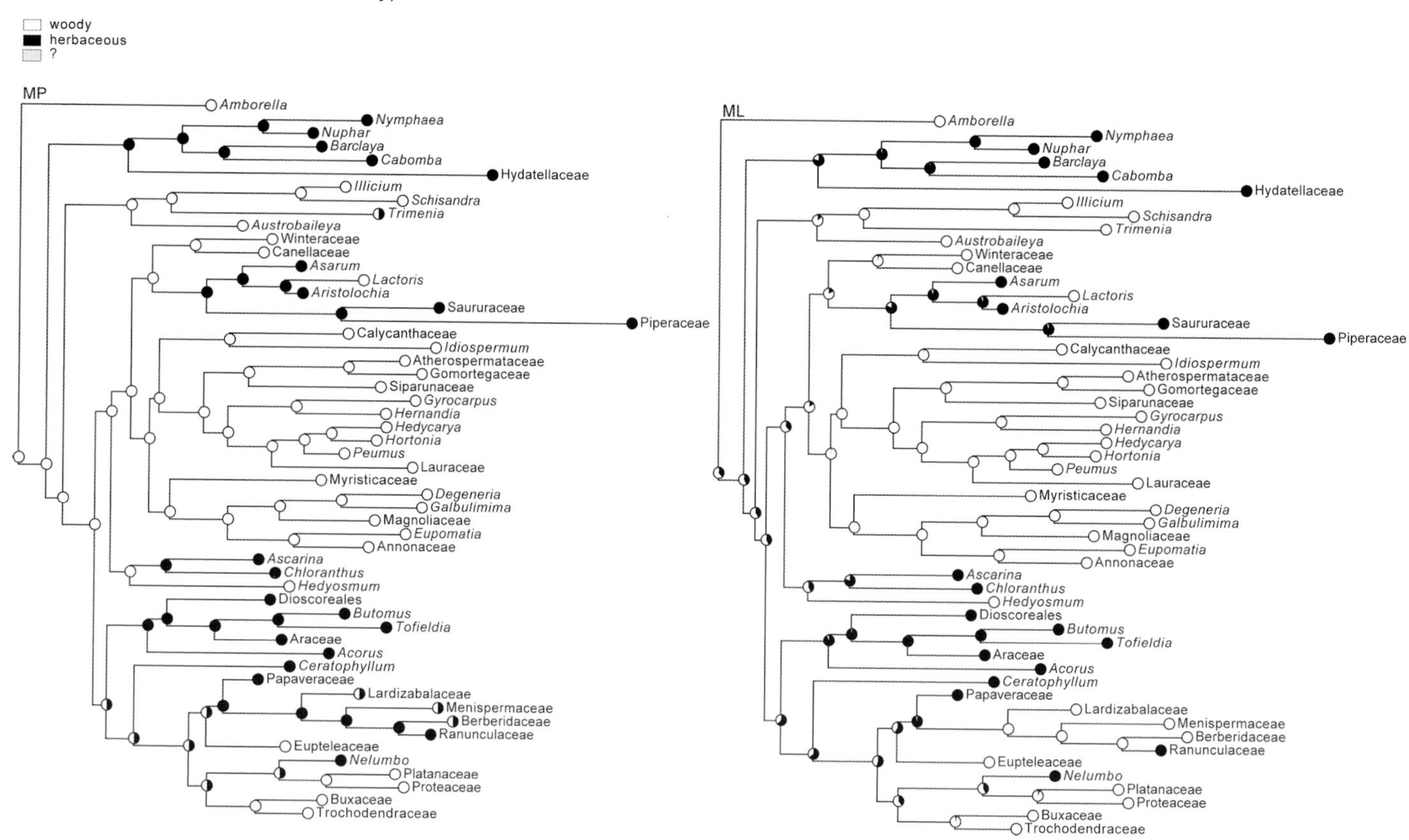

Figure 6.1. MP and ML reconstruction of the evolution of habit (woody vs. herbaceous) in the ANA grade, magnoliids, monocots, and basal eudicots. The topology follows Soltis et al. (2011). Note that *Lactoris* is placed within Aristolochiaceae in these trees, but may be sister to that family. Data are from Doyle and Endress (2000), Endress and Doyle (2009), and Rudall et al. (2007) with modifications (see text). In this reconstruction, *Hedyosmum* is scored as woody. In all figures MP = maximum parsimony reconstruction; ML = maximum likelihood reconstruction. For maximum likelihood reconstructions, the Mk model was employed here and in all other ML reconstructions.

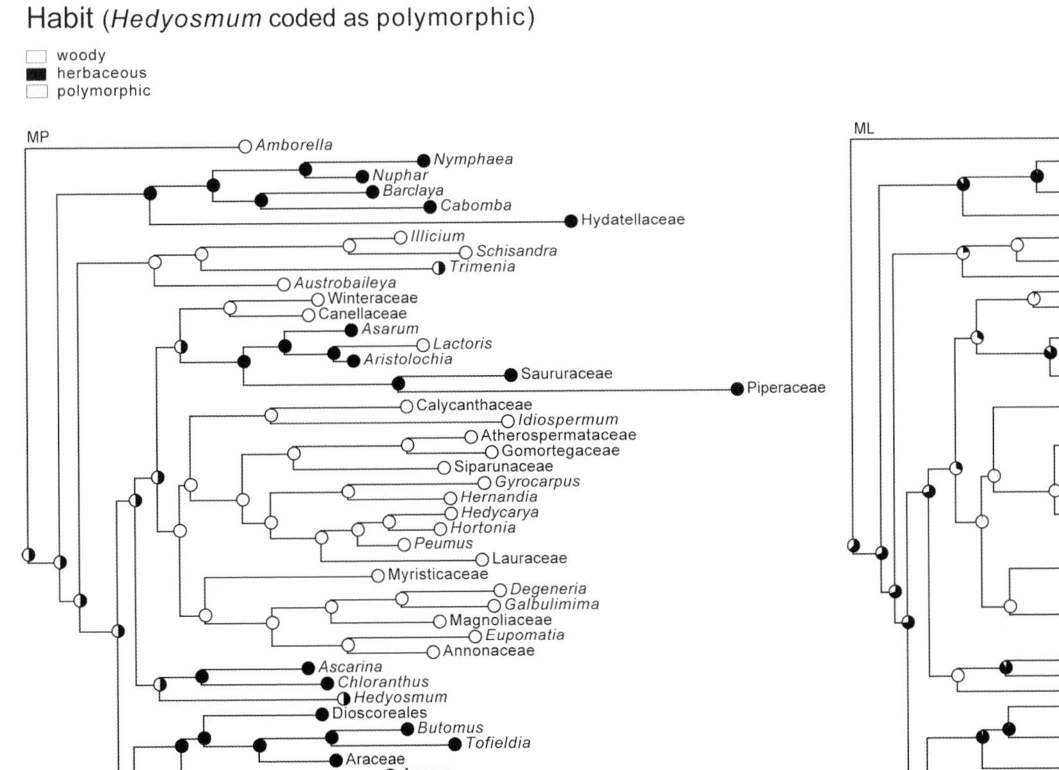

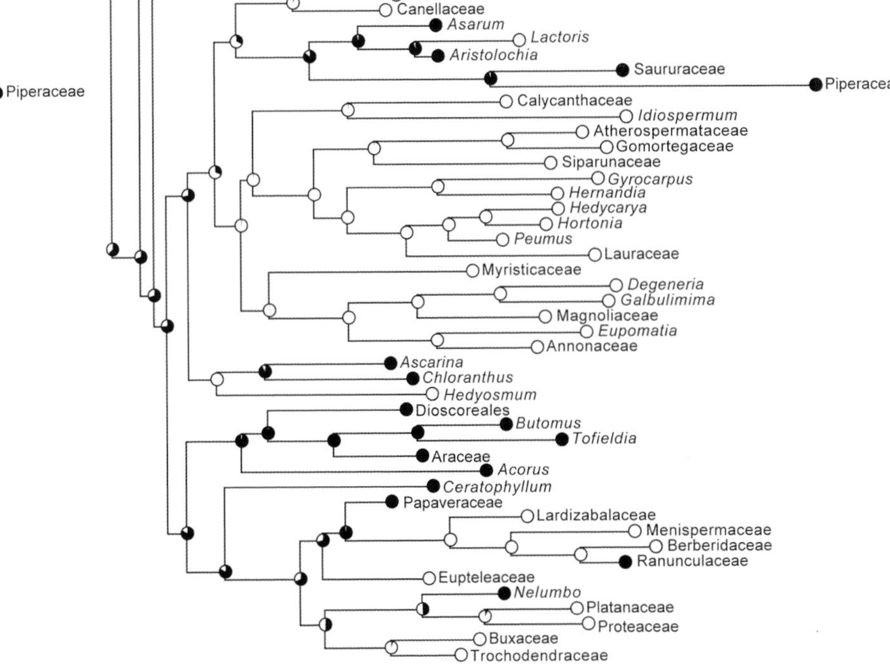

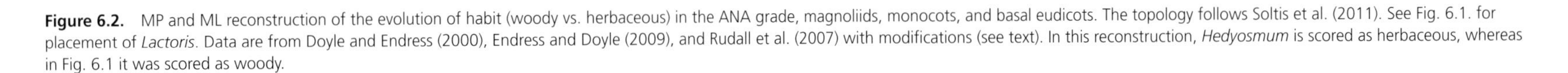

Figure 6.2. MP and ML reconstruction of the evolution of habit (woody vs. herbaceous) in the ANA grade, magnoliids, monocots, and basal eudicots. The topology follows Soltis et al. (2011). See Fig. 6.1. for placement of *Lactoris*. Data are from Doyle and Endress (2000), Endress and Doyle (2009), and Rudall et al. (2007) with modifications (see text). In this reconstruction, *Hedyosmum* is scored as herbaceous, whereas in Fig. 6.1 it was scored as woody.

Figure 6.3. Chloranthoid teeth. *Amborella*, leaf photograph, showing leaf margin with chloranthoid teeth. Insert: drawing showing one chloranthoid gland (from Soltis et al. 2005).

multiple times in the eudicots: in the common ancestor of Papaveraceae, the ancestor of Berberidaceae and Ranunculaceae, and the ancestors of Trochodendraceae and Buxaceae. Given the apparent independent derivation of chloranthoid teeth in the eudicots, this character does not appear to be homologous to the chloranthoid teeth in Amborellaceae, Austrobaileyales, and Chloranthaceae (and they may not be homologous in these latter groups either). Studies are needed to assess the homology of chloranthoid teeth in basal angiosperms and eudicots.

ETHEREAL OILS

Ethereal oils (Chapter 4) are prevalent in all Austrobaileyales and all magnoliids and are present in *Acorus*, the sister to all other monocots (see Chapter 7). However, ethereal oil cells are not present in *Amborella* (Bailey and Swamy 1948), Nymphaeales, or Ceratophyllaceae. Thus, even though early workers suggested that early angiosperms would have produced these oils (e.g., Cronquist 1988), presumably as a defensive compound, the ancestral condition for angiosperms is reconstructed as the absence of ethereal oils with both MP and ML (Fig. 6.5).

The ancestral states for mesangiosperms and mesangiosperms + Austrobaileyales are equivocal. The absence of these oils in Nymphaeales and Ceratophyllaceae could easily be due to losses associated with the aquatic habit. Oil cells are reconstructed as ancestral for the ancestor of Chloranthaceae + magnoliids with both MP and ML (Fig. 6.5), and this feature represents a possible synapomorphy for this clade.

Oil cells also appear to have been present early in the fossil record; they are abundant, for example, on carpels, petals, and leaves of *Archaeanthus* (Dilcher and Crane 1984), as expected for a member of Magnoliaceae. The ancestral state for the monocots is unclear because of the presence of these compounds in Acoraceae, while the compounds are absent in other early-branching monocots (but occur in Zingiberales; see Chapter 7). Regardless of optimization or topology, the ancestral state for eudicots is the absence of ethereal oils.

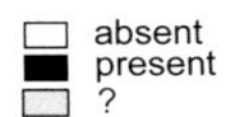

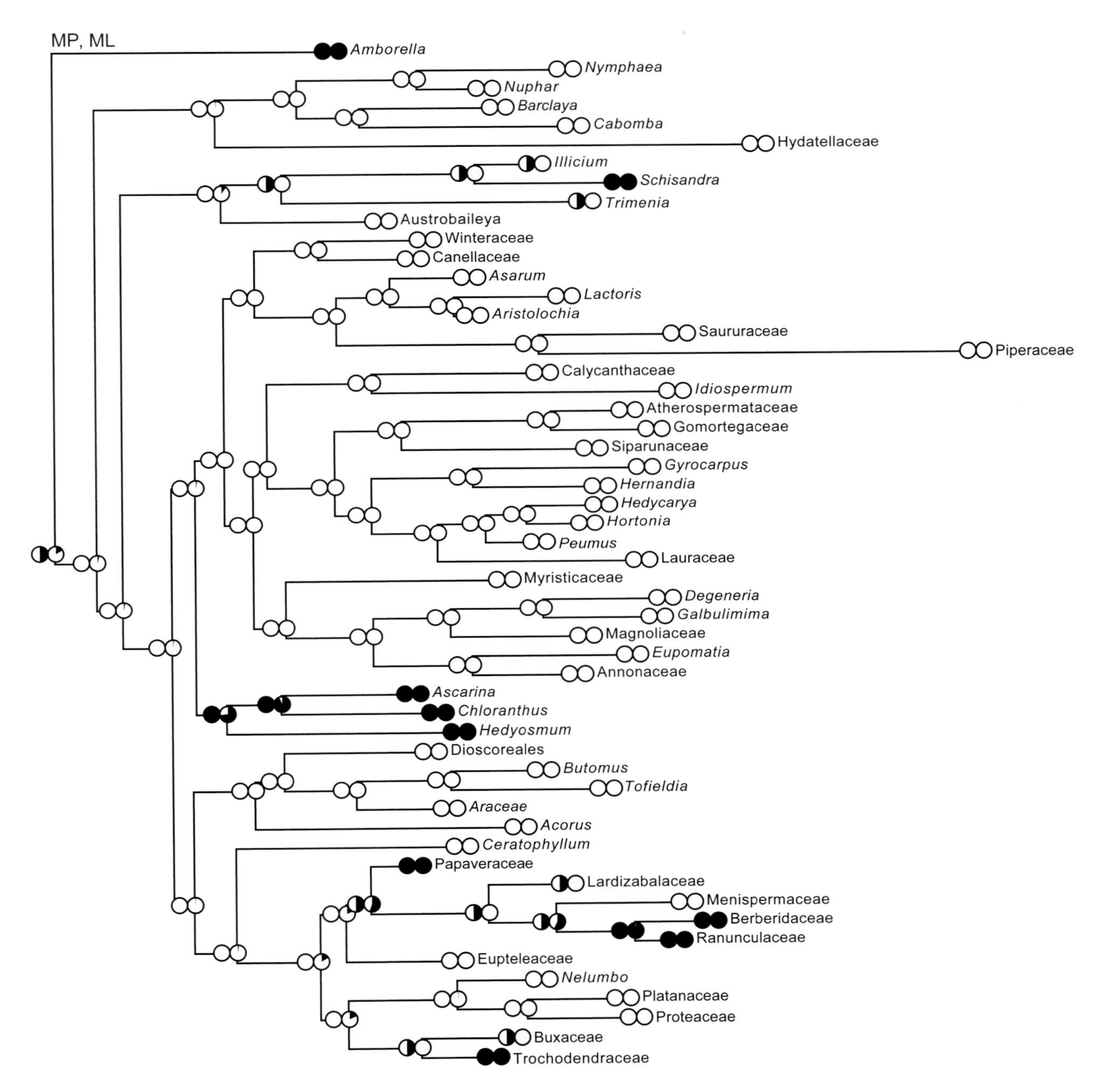

Figure 6.4. MP and ML reconstruction of the evolution of chloranthoid leaf teeth (present vs. absent) in the ANA grade, magnoliids, monocots, and basal eudicots. The topology follows Soltis et al. (2011). See Fig. 6.1. for placement of *Lactoris*. Data are from Doyle and Endress (2000), Endress and Doyle (2009), and Rudall et al. (2007) with modifications (see text). In this and most subsequent reconstructions in the book we place both the MP and ML reconstructions on the same tree (the MP circle to the left and the ML circle to the right).

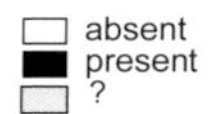

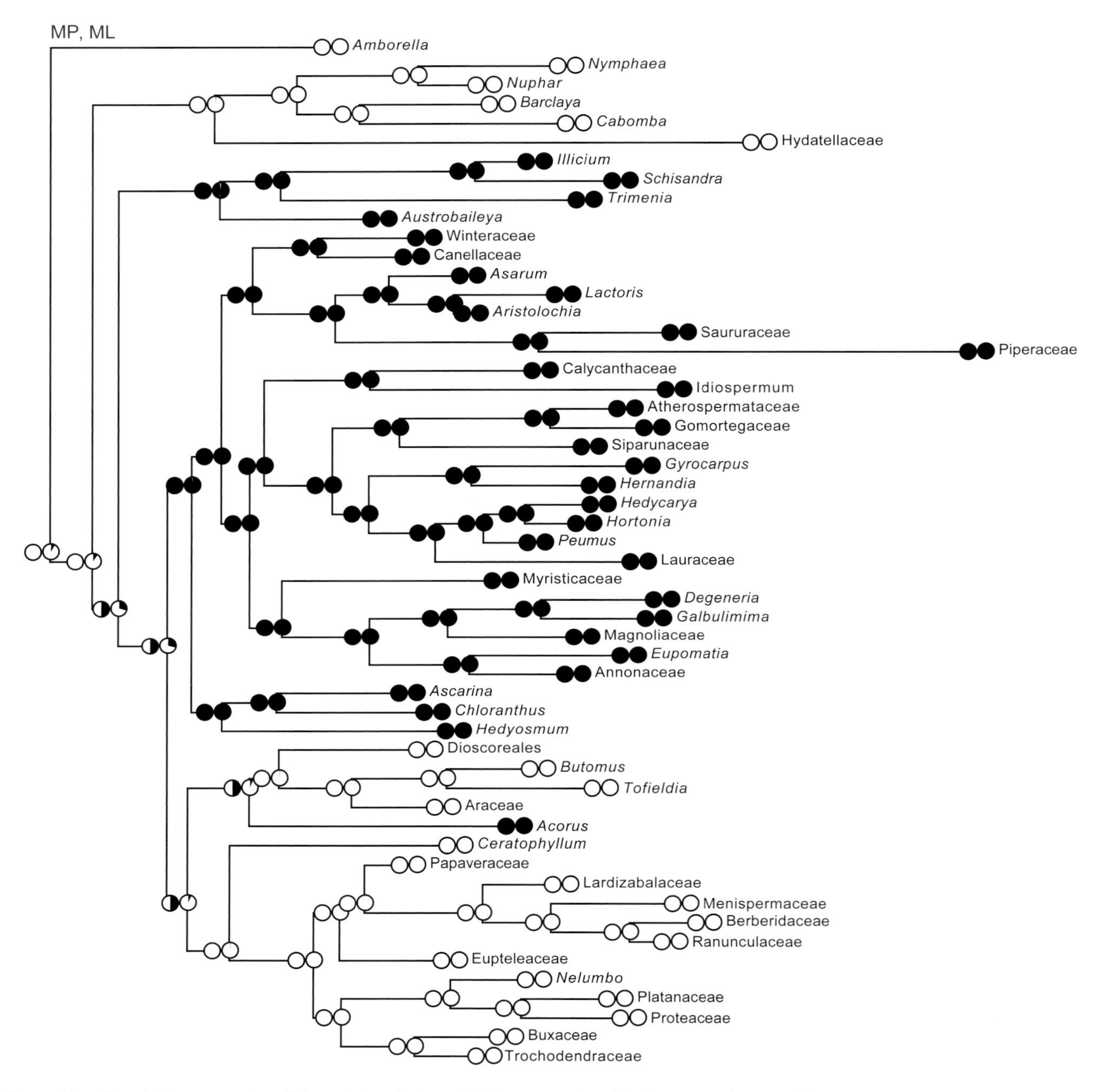

Figure 6.5. MP and ML reconstruction of the evolution of ethereal oils (present vs. absent) in the ANA grade, magnoliids, monocots, and basal eudicots. The topology follows Soltis et al. (2011). See Fig. 6.1. for placement of *Lactoris*. Data are from Doyle and Endress (2000), Endress and Doyle (2009), and Rudall et al. (2007) with modifications (see text).

NODAL ANATOMY

The nodal anatomy of angiosperms has long been of research interest (e.g., Canright 1955; Bailey 1956; Benzing 1967; Cronquist 1968), although the adaptive significance of changes in nodal anatomy is uncertain. Branch systems in the stele in ferns, gymnosperms, and many angiosperms usually have two traces from a single gap. The most common nodal types are one-trace unilacunar, two-trace unilacunar, trilacunar, and multilacunar (Fig. 6.6). Bailey (1956) proposed that unilacunar nodes with two traces are ancestral in the angiosperms. Cronquist (1968, 1988) suggested that either unilacunar nodes with two traces or perhaps trilacunar nodes were ancestral. In their reconstructions of nodal anatomy, Doyle and Endress (2000) provided support for the former hypothesis—unilacunar nodes with two traces are ancestral.

Analyses conducted for this book generally agree with those of Doyle and Endress (2000), but because of the different topologies used, there are also differences. Furthermore, Doyle and Endress (2000) scored Amborellaceae as having unilacunar nodes with two traces, following Bailey (1956). However, Carlquist and Schneider (2001) indicated that *Amborella* has unilacunar nodes with one trace—*Amborella* may have both states. We therefore used multiple scorings of *Amborella*. In one optimization, we scored *Amborella* as polymorphic, having unilacunar nodes with both one trace and two traces. We also compared alternative reconstructions in which *Amborella* was coded as having unilacunar nodes with one trace vs. unilacunar nodes with two traces.

Because of the uncertainty surrounding the distinction between unilacunar nodes with one versus two traces, we conducted two reconstructions. In the first, we followed Doyle and Endress (2000) and recognized four states: unilacunar with one trace, unilacunar with two traces, trilacunar, and multilacunar. In the second, we combined one- and two-trace types into a single category, unilacunar (see also Dickison and Endress 1983), and considered three basic states: unilacunar, trilacunar, and multilacunar.

Combining unilacunar one-trace and two-trace nodes into a single character state (unilacunar) gives a much clearer picture of nodal anatomy evolution, unambiguously reconstructing the ancestral state of the angiosperms through the ANA grade as unilacunar (Fig. 6.7). The ancestral state of the mesangiosperms is ambiguous; the ancestral state for Chloranthaceae is unilacunar. There are multiple origins of trilacunar nodal anatomy: ancestor of Canellales + Piperales, some Magnoliales (independently in Myristicaceae and Himantandraceae), Proteales, and Ranunculales, with other scattered origins (Fig. 6.7). These results contrast with Doyle and Endress (2000), who used a topology that did not place Canellales (their Winterales) with Piperales or Laurales with Magnoliales. Multilacunar nodal anatomy also evolved multiple times, including within the monocots and the ancestor of Piperaceae/Saururaceae.

When two types of unilacunar nodes are considered, the ancestral state for angiosperms is less clear, as is the ancestral condition for most major subclades (not shown). Unilacunar nodes with two traces are reconstructed as ancestral when MP is used (but not ML), but only if *Amborella* is coded as polymorphic, possessing both one and two traces, or as having two traces. However, if *Amborella* is coded as having only nodes that are unilacunar with one trace, the ancestral state for the angiosperms becomes ambiguous (not shown).

If *Amborella* is coded as polymorphic, possessing both one and two traces, or as having two traces, the ancestral state of all angiosperms except *Amborella* as well as all angiosperms except *Amborella* and Nymphaeales is also reconstructed with MP (but not ML) as unilacunar with two traces. Unilacunar nodes with two traces are present in some Nymphaeaceae (other members of the family have trilacunar nodes) and in *Trimenia* (Austrobaileyales; Marsden and Bailey 1955); other Austrobaileyales (*Schisandra* and *Illicium*) have unilacunar nodes with one trace; *Austrobaileya* is intermediate (see above) (Dickison and Endress 1983).

Both sets of reconstructions indicate that nodal anatomy may provide synapomorphies for some clades. Unilacunar nodes unite Laurales when unilacunar one-trace and two-trace nodes (Fig. 6.7) are combined. When one- and two-trace nodes are scored separately, the ancestor of Laurales is reconstructed as having unilacunar nodes with two

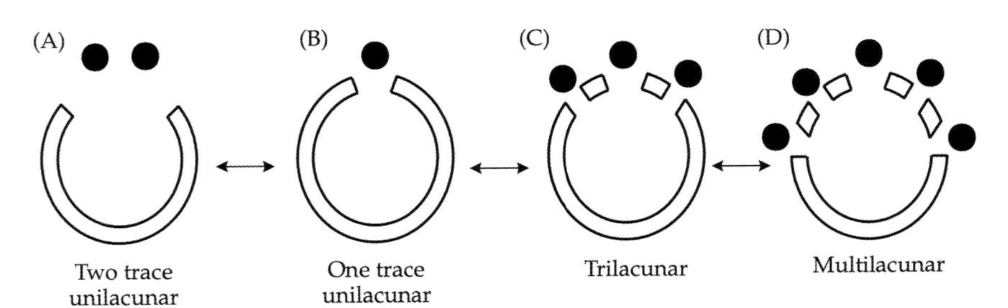

Figure 6.6. Major types of nodal anatomy in angiosperms and the traditional view of nodal anatomy evolution are depicted using arrows.

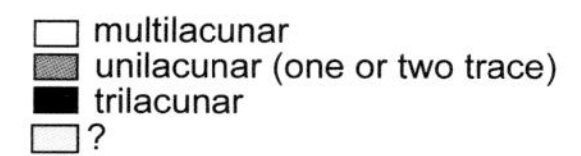

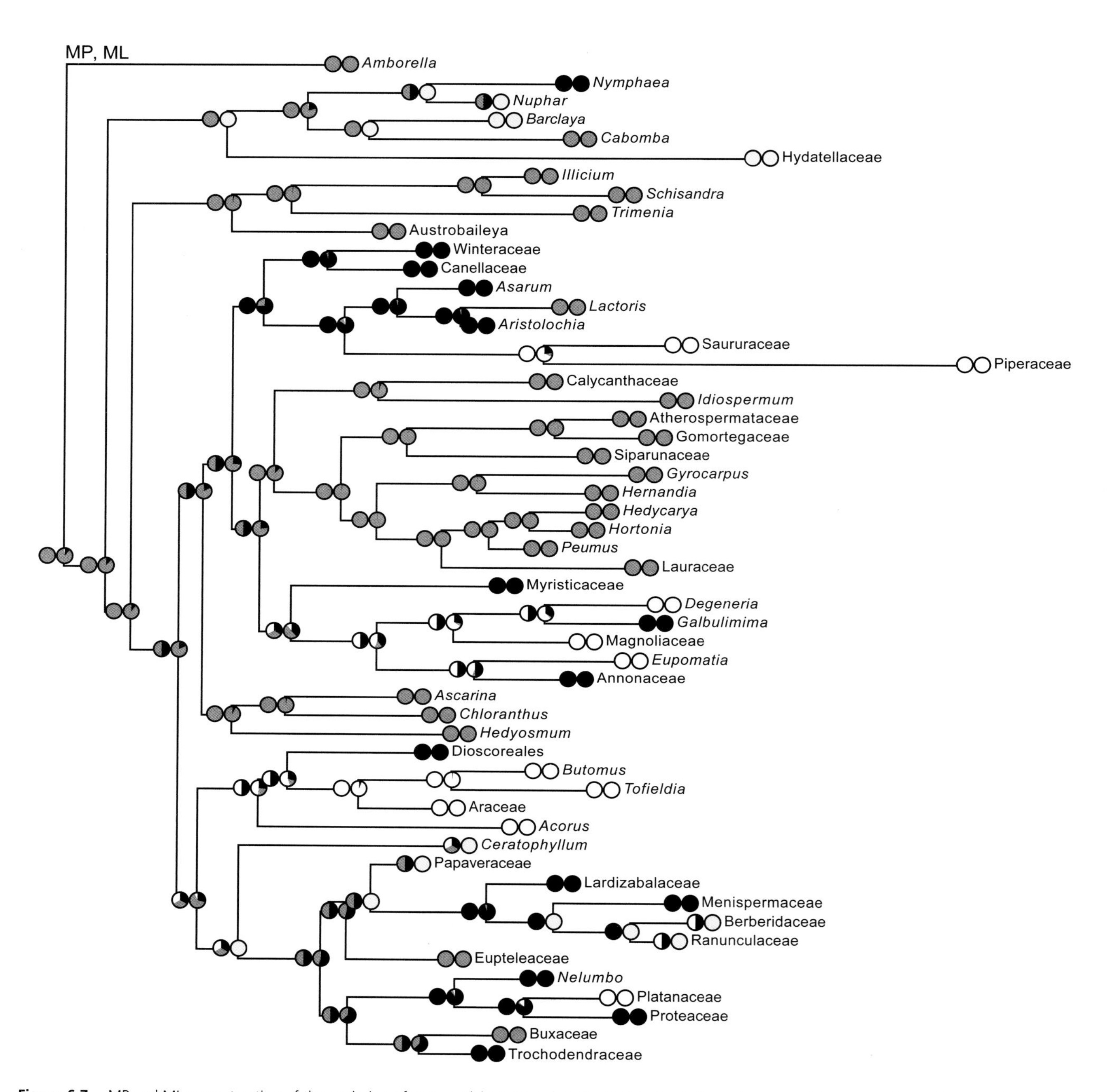

Figure 6.7. MP and ML reconstruction of the evolution of stem nodal anatomy (see Fig. 6.6) in the ANA grade, magnoliids, monocots, and basal eudicots. The topology follows Soltis et al. (2011). See Fig. 6.1. for placement of *Lactoris*. Data are from Doyle and Endress (2000), Endress and Doyle (2009), and Rudall et al. (2007) with modifications (see text).

traces; unilacunar nodes with one trace were derived in the ancestor of Hernandiaceae.

VESSEL ELEMENTS

We covered vessels in Chapter 4 (Fig. 4.4) and here provide a broader phylogenetic perspective. Several families lack vessels; Cronquist (1968, 1988), and Takhtajan (1969) suggested that some families lacking vessels (e.g., Winteraceae) might be primitively vesselless. However, a morphological cladistic analysis (Young 1981) showed that it was more parsimonious to have vessels present in the common ancestor of the angiosperms and then lost several times in diverse taxa, including *Amborella*, Winteraceae, and *Sarcandra* (Chloranthaceae). Donoghue and Doyle (1989a) provided additional support for Young's hypothesis of many losses of vessels.

The discovery that *Amborella*, a vesselless angiosperm (see Chapters 3 and 4), is sister to all other extant angiosperms has renewed interest about whether the angiosperms were ancestrally vesselless. Adding to the intrigue is the discovery in some basal angiosperms of tracheary elements that are intermediate between true vessels and tracheids (see Schneider et al. 1995; Schneider and Carlquist 1996). A perforation plate (Fig. 4.4) is a feature of a vessel; it is considered to be one, or a series, of modified "pits" or perforations in the end wall of a water-conducting cell. Pit membranes are lacking in the perforations of a true vessel, but some taxa, such as Nymphaeaceae, have porose pit membranes in the end walls. Adding to the complexity is a combination of porose and nonporose pit membranes in *Bubbia* of Winteraceae (Carlquist 1983) and *Amborella* (Feild et al. 2000b; Carlquist and Schneider 2001). Due to this complexity, Doyle and Endress (2000) recognized an "intermediate" condition between true vessels and tracheids. They found that the vesselless condition is ancestral in angiosperms, with both *Amborella* and Nymphaeaceae lacking vessels in stems; vessels were also lost in Winteraceae. Vessels were previously thought to have been lost in Trochodendraceae, but recent studies reveal that vessels are present in this lineage (Chapter 8) (Li et al. 2011).

Carlquist and Schneider (2002) proposed that definitions of vessels are overly simplistic. When all sources of data are considered, these authors found that changes in not one, but at least four and perhaps as many as six, characters can occur in the evolution of a vessel from a tracheid (see Table 4.1 and Fig 4.4). For four of these characters, delimiting the differences in character states is straightforward. Two other differences between tracheids and vessels are relative and therefore cannot be easily coded. In vessel-bearing woods that also contain tracheids, vessel elements are shorter than the tracheids in any given sample, whereas in vesselless wood, lengths of the water-conducting elements (tracheids only) in any given sample follow a unimodal distribution. In addition, tracheids of vesselless woods are much longer than both the tracheids and vessel elements of vessel-bearing woods.

Hence, "vessel" should not be treated as a single character with two states, presence or absence (some other characters may be similarly complex). Furthermore, adding a third state, intermediate, does not capture the complexity of the transition from tracheid to vessel. Unfortunately, adequate data are not available to map any of the characters that Carlquist and Schneider (2002) recognized as distinguishing "vessels" from "tracheids." Porosity, for example, can be measured only by SEM examination, and the number of taxa examined for this feature is still small.

Ecophysiological insights into the function of water-conducting cells in basal angiosperms also bear on the evolution of vessel elements. It is possible, perhaps probable, that the early angiosperms lacked vessels, because *Amborella* and Nymphaeaceae lack vessels and other early-diverging lineages such as Schisandraceae and Austrobaileyaceae have what are regarded as "primitive" vessel types with tracheid-like features (Carlquist and Schneider 2002). In *Austrobaileya*, the vessels have gradually tapered end walls and numerous scalariform perforation plates and are generally small in diameter (Bailey and Swamy 1948; Carlquist 2001a). *Illicium* and *Kadsura* have vessel elements with long, oblique perforation plates that partially retain pit membranes and hence are tracheid-like (Carlquist 1983; Carlquist and Schneider 2002).

In summary, although it is difficult to define a "vessel" and it should not be treated as a single character with several states, basic analyses suggest that simple vessels are present in the common ancestor of Austrobaileyales + mesangiosperms and that "typical" vessels originated in the common ancestor of mesangiosperms. This is one of several key features associated with the origin of mesangiosperms (Chapter 3).

PERIANTH PHYLLOTAXIS

To explore the evolution of perianth phyllotaxis (Endress and Doyle 2007; Staedler and Endress 2009), we used both multistate and binary character codings. With binary coding, we scored the perianth as spiral or whorled. With multistate coding, phyllotaxis was scored as absent (no perianth), a single whorl, two whorls, multiple whorls (more than two), and spiral. We also used a second multistate coding in which we reduced the number of forms of whorled phyllotaxis (i.e., single whorl, two whorls, multiple whorls) to a single character state—whorled. In this second multistate coding, phyllotaxis was scored as absent, spiral, or whorled.

Amborella has spiral phyllotaxis, as do members of Austrobaileyales (Chapter 4; Endress and Igersheim 2000a). In some clades, phyllotaxis is complex and does not fit cleanly into any of the states specified for either multistate or binary coding. For example, in Nymphaeaceae, phyllotaxis has often been considered to be spiral, but it now appears to be primarily whorled or in some cases irregular (Endress 2001a). In Winteraceae, phyllotaxis is primarily whorled but occasionally spiral (Doust 2000). In *Drimys winteri*, flowers within one tree vary between spiral and whorled phyllotaxis (Doust 2001). In Hydatellaceae, inflorescence/floral morphology has been hard to interpret (Rudall et al. 2007), but as each flower has only a single organ (either a stamen or a carpel), phyllotaxis is not present. For other taxa, phyllotaxis has not been carefully examined, and critical studies are still needed. In addition, the distinction between spiral and whorled is not always clear. Ontogenetic studies have revealed that in some cases floral organs that appear to be whorled in mature flowers actually result from spiral initiation of primordia and the rhythmic occurrence of a long plastochron (the time interval between the initiation of two consecutive organs) after several short plastochrons (e.g., Tucker 1960; Huber 1980; Leins and Erbar 1985; Endress 1994a). Thus, both spiral and whorled phyllotaxis might ultimately have organs developing in a spiral sequence (Endress 1987a,b). *Illicium* has spiral phyllotaxis of all floral organs; in mature flowers the carpels form a single series and thus simulate a whorl, but they still have spiral phyllotaxis (Fig. 4.17).

We scored *Ceratophyllum* as lacking a perianth (Endress 1994b, 2001a), although the whorl of scales that precedes the reproductive organs has been variously interpreted as perianth or bracts (Aboy 1936; Les 1988, 1993; Les et al. 1991; Endress 1994b, 2001a; Albert et al. 1998). Most Chloranthaceae lack a perianth; however, the pistillate flowers of *Hedyosmum* have three structures variously interpreted as a perianth or as staminodes (Endress 1987c). We scored *Hedyosmum* as having an undifferentiated perianth. We interpreted Hydatellaceae flowers to be in compact aggregations (pseudanthia) that comprise unisexual flowers with no perianth, with each male flower consisting of a single stamen and each female flower a single carpel (Hamann 1975; Rudall et al. 2007). As noted, the interpretation of perianth phyllotaxis in Nymphaeaceae has been debated. Ronse De Craene et al. (2003) considered *Brasenia* and *Cabomba* to have whorled phyllotaxis and scored *Nymphaea*, *Victoria*, and *Nuphar* as polymorphic, with early organs appearing in a whorled arrangement. Zanis et al. (2003), in contrast, scored *Nymphaea*, *Victoria*, and *Nuphar* as spiral. We followed Endress (2001a), who suggested that the phyllotaxis of *Nymphaea*, *Victoria*, and *Nuphar* is whorled and then becomes irregular, but not spiral.

Spiral phyllotaxis is present in the ANA grade, including *Amborella*, *Trimenia*, *Austrobaileya*, and all Schisandraceae (see Chapter 4). However, because of the whorled phyllotaxis of Nymphaeaceae and Cabombaceae, the ancestral reconstruction for perianth phyllotaxis for the angiosperms is not straightforward. Reconstruction of phyllotaxis with more than one whorled state (e.g., single whorl, two whorls, multiple whorls) often results in ambiguous character-state optimizations, and evolutionary inference is difficult. Most of our discussion below focuses on the results of reconstructions that involve three states—spiral, whorled, and absent (Fig. 6.8); for more detail on multistate reconstructions, see Albert et al. (1998), Ronse De Craene et al. (2003), Zanis et al. (2003), and Endress and Doyle (2009, 2015).

In our analyses, the ancestral phyllotaxis for the angiosperms is generally reconstructed as equivocal. For binary coding (not shown), the phyllotaxis of extant angiosperms was reconstructed as equivocal with both MP and ML (see also Ronse De Craene et al. 2003; Soltis et al. 2005b; Endress and Doyle 2007, 2009; Doyle and Endress 2011). This is also the case using three states (spiral, whorled, and absent) with both ML and MP. Reconstructions with multiple whorls coded (one, two, or three) again suggests an ambiguous ancestral state with MP, although ML favors a spiral perianth in the ancestor of extant angiosperms (Fig. 6.8).

Using parsimony, Endress and Doyle (2009) also reconstructed the ancestral state as ambiguous. In P. Soltis et al. (2000) the ancestral condition for angiosperms was reconstructed as spiral because different codings for Nymphaeaceae were used. If *Victoria*, *Nymphaea*, and *Nuphar* were scored as having a spiral perianth, then a spiral perianth is reconstructed as ancestral throughout the ANA grade (e.g., P. Soltis et al. 2000; Zanis et al. 2003). However, if the perianth phyllotaxis of these genera is considered to be primarily whorled (see Ronse De Craene and Smets 1993; Endress 2001a,b), which developmental data indicate is the case, then a whorled perianth is present throughout the family, and the ancestral phyllotaxis for angiosperms is reconstructed as equivocal (Fig. 6.8). The same ambiguous reconstruction for angiosperms is obtained if the perianth phyllotaxis of Nymphaeaceae is considered polymorphic rather than whorled. In summary, the ancestral state for phyllotaxis in the angiosperms is best considered equivocal (see also Endress and Doyle 2007).

Following the ANA grade, whorled perianth phyllotaxis is reconstructed as ancestral for the mesangiosperms

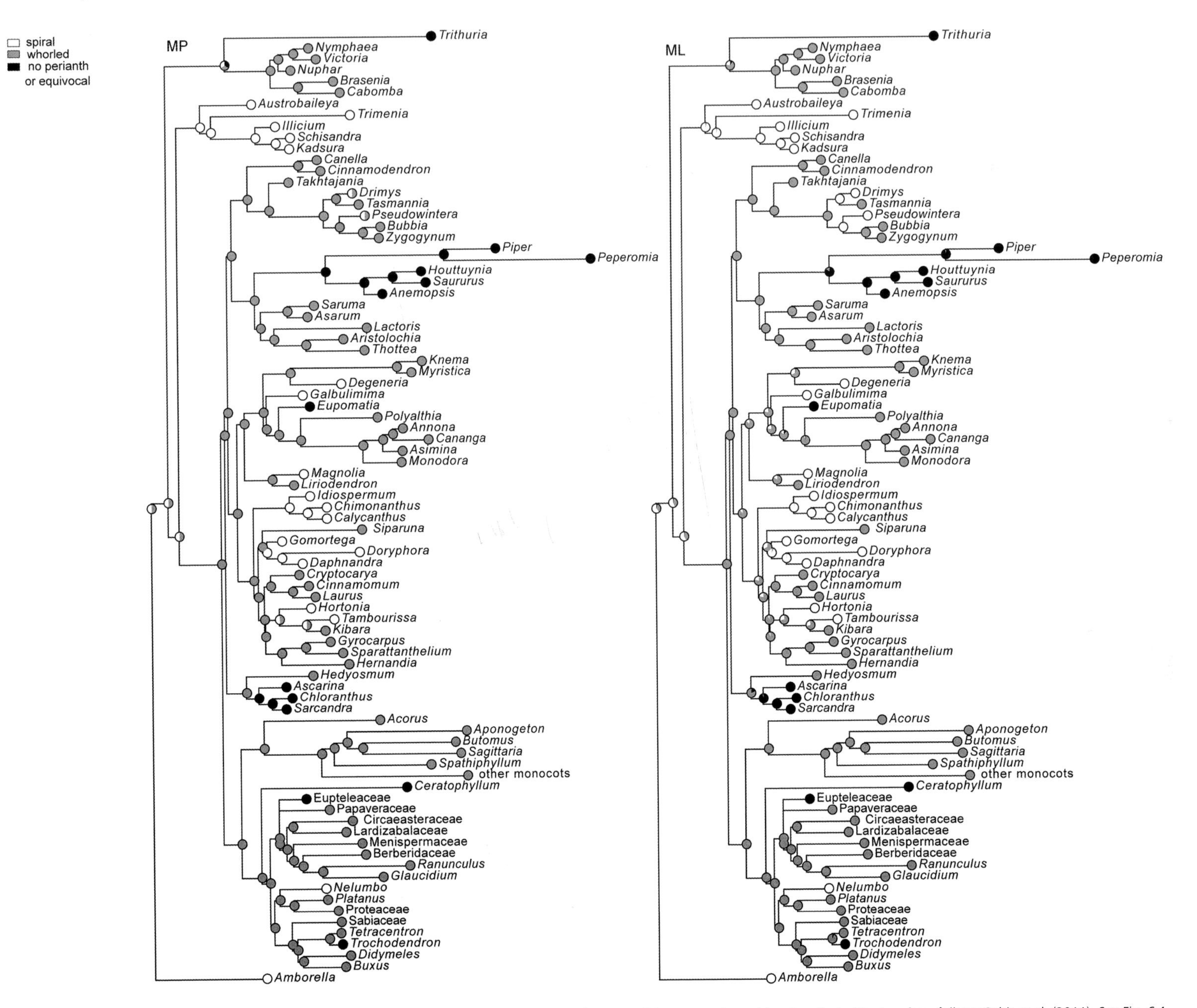

Figure 6.8. MP and ML reconstruction of the evolution of perianth phyllotaxis in the ANA grade, magnoliids, monocots, and basal eudicots. The topology follows Soltis et al. (2011). See Fig. 6.1. for placement of *Lactoris*. Data are from Doyle and Endress (2000), Endress and Doyle (2009), and Rudall et al. (2007) with modifications (see text).

(see also Endress and Doyle 2009). This is the case with ML when binary coding is used and also with simple coding (spiral, whorled, and absent) with both MP and ML (Fig. 6.8). Following the origin of a whorled perianth in mesangiosperms, there are multiple shifts to a spiral perianth in various magnoliids, such as Calycanthaceae (Staedler et al. 2007), some Atherospermataceae (e.g., *Daphnandra*; not all members of the Atherospermataceae have a spiral perianth—*Dryadodaphne* has whorled phyllotaxis), Gomortegaceae, some Monimiaceae (e.g., *Hortonia*; Endress 1980a,b), Degeneriaceae (*Degeneria*), and Magnoliaceae (some species of *Magnolia*; Fig. 6.8).

Another possible transformation from whorled to spiral phyllotaxis may have occurred in *Drimys* and *Pseudowintera* (Winteraceae), which have a labile phyllotaxis involving spirals and multiple whorls (Doust 2000, 2001; Endress et al. 2000b). Thus, perianth phyllotaxis is highly labile in the ANA grade and magnoliids, as has long been suggested (Endress 1987a,b, 1994a,c; Albert et al. 1998; Zanis et al. 2003; Soltis et al. 2005b; Endress and Doyle 2009). In fact, studies of floral development indicate that some basal angiosperms are not fully committed to either spiral or whorled phyllotaxis (Buzgo et al. 2004).

A perianth has also been lost frequently in non-eudicot and non-monocot angiosperms (Fig. 6.8). Reconstruction of phyllotaxis with more than one whorled state (e.g., single whorl, two whorls, multiple whorls) is noteworthy in showing that a perianth of a single whorl has likely evolved multiple times: in the ancestors of Aristolochiaceae, Myristicaceae, and Hernandiaceae, respectively, and in several basal eudicot lineages (e.g., Platanaceae + Proteaceae and Circaeasteraceae) (the eudicots are covered in more detail in Chapters 8 and 9).

PERIANTH MERISM (MEROSITY)

We coded perianth merism as dimerous, trimerous, tetramerous, pentamerous, or indeterminate, the latter referring to a variable and typically large number of perianth parts. Data for merism were taken from Endress (2001a), Ronse De Craene et al. (2003), Zanis et al. (2003), and Rudall et al. (2007). The flowers of many basal angiosperms have multiple parts, whereas those of other clades are clearly trimerous in their basic floral plan. However, the coding of merism is sometimes problematic. In some Nymphaeaceae (e.g., *Victoria* and *Nymphaea*), many perianth parts are present, but the basic plan for these genera is trimerous or tetramerous (Endress 2001a). Other families are complex. In some Winteraceae, the outermost floral organs are in dimerous whorls in some genera, whereas others possess tetramerous, and even pentamerous, whorls (Endress et al. 2000b). In Magnoliaceae, the perianth of some species of *Magnolia* is indeterminate, whereas that of *Liriodendron* and other Magnolias is in three trimerous whorls; Magnoliaceae may represent a transition from whorled to spiral phyllotaxis (Tucker 1960; Erbar and Leins 1981, 1983). More data are also needed. Only one species of Siparunaceae has been investigated (*Siparuna nicaraguensis*, now called *S. thecaphora*; Endress 1980a; Staedler and Endress 2009). Staminate flowers have an undifferentiated perianth of four (or five) tepals and could be interpreted as dimerous. However, pistillate flowers have five or six tepals, and merism is unclear. We did not include Siparunaceae in Fig. 6.9; however, Zanis et al. (2003) attempted reconstructions with Siparunaceae scored as either dimerous or uncertain.

Amborella exhibits indeterminate merism (see Chapter 4). Other early-branching lineages also have indeterminate merism, with Austrobaileyales uniformly indeterminate. *Trithuria* (Hydatellaceae; Nymphaeales) has but a single stamen or carpel per flower (Hamann 1975; Rudall et al. 2007) and lacks a perianth. Within Nymphaeaceae and Cabombaceae, *Cabomba*, *Brasenia*, and *Nuphar* are trimerous; other genera (e.g., *Victoria*, *Nymphaea*) also appear trimerous or tetramerous (see Chapter 4; Fig. 4.11; see Endress 2001a). Hence, trimery appears to be ancestral for Nymphaeaceae + Cabombaceae. Importantly, these results also illustrate that trimery arose early in angiosperm evolution, and multiple times (see below).

In our reconstructions, binary coding (indefinite vs. whorled or fixed) results in a reconstruction of indefinite merism for the ancestral state of angiosperms with both MP and ML. In addition, ML reconstructs the ancestor of mesangiosperms as having a fixed perianth merism, although this is equivocal with MP (not shown).

With multistate coding, the ancestral state of the angiosperms is equivocal with both MP and ML. Above the ANA grade, both MP and ML give highly similar results with the ancestral character state for the mesangiosperms being a trimerous perianth (Fig. 6.9; see also Ronse De Craene et al. 2003; Zanis et al. 2003). This is the case regardless of whether *Amborella* and Nymphaeales are treated as a clade or as successive sisters to all other angiosperms and also regardless of whether Chloranthales follow ANA as sister to all other angiosperms or if Chloranthales + magnoliids are sister to other remaining angiosperms. Thus, although the trimerous condition is typically associated with monocots, our results further demonstrate that trimery has played a major role in the early evolution and diversification of the flower (see also Kubitzki 1987; Ronse De Craene and Smets 1994; Soltis et al. 2005b, 2008b; Endress and Doyle 2009). Endress and Doyle (2007, 2009) and Doyle and Endress (2011) reconstructed the ancestral state of the angiosperms as trimerous, but only if taxa with spiral phyllotaxis were

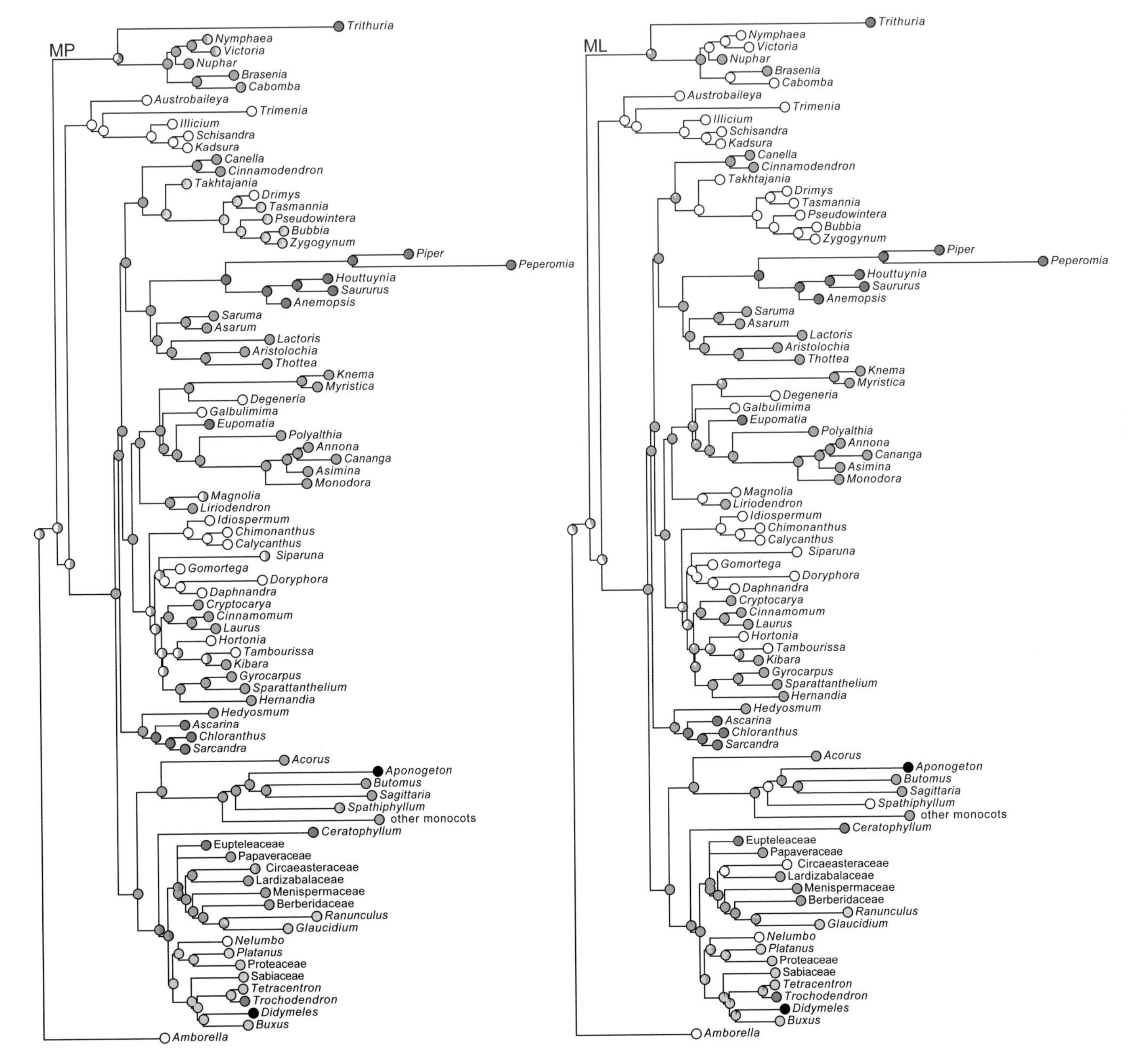

Figure 6.9. MP and ML reconstruction of the evolution of perianth merism in the ANA grade, magnoliids, monocots, and basal eudicots. The topology follows Soltis et al. (2011). See Fig. 6.1. for placement of *Lactoris*. Data are from Doyle and Endress (2000), Endress and Doyle (2009), and Rudall et al. (2007) with modifications (see text).

scored as "unknown." Our reconstructions also indicate that trimery arose independently in Nymphaeales and the ancestor of mesangiosperms (Fig. 6.9).

Following the origin of a trimerous perianth, there was a return to an indeterminate perianth in several magnoliid lineages, including Calycanthaceae, the clade of Atherospermataceae (represented by *Daphnandra*) and Gomortegaceae (*Gomortega*), Himantandraceae (*Galbulimima*), some Monimiaceae (e.g., *Hortonia*), and some Magnoliaceae (*Magnolia*). A perianth has also been lost several times (e.g., Eupomatiaceae, Piperaceae, Ceratophyllaceae, and most Chloranthaceae; Fig. 6.9).

Our reconstructions indicate that perianth merism is labile in non-eudicot and non-monocot angiosperms (see also Endress 1987a,b, 1994a,c; Albert et al. 1998; Zanis et al. 2003; Soltis et al. 2005b). This labile condition then continues through the basal nodes of the eudicots, where trimery as well as dimery predominate (Fig. 6.9). In fact, the evolution of dimery immediately precedes the appearance of pentamery in core eudicots (see Chapter 8). This lability in basal eudicots contrasts with the condition of core eudicots and monocots (this is discussed further in Chapters 7 and 8).

PERIANTH DIFFERENTIATION

A differentiated or bipartite perianth has an outer whorl of sepals clearly differentiated from the inner whorl(s) of petals. In contrast, an undifferentiated perianth lacks clear distinction between the outer and inner whorls, or the perianth may consist of undifferentiated spirally arranged parts; these undifferentiated perianth organs have traditionally been referred to as tepals. The term "tepal" was coined by de Candolle (1827) to describe a calyx and corolla that are not differentiated; thus, the entire perianth may be petaloid. Takhtajan (1969, 1987, 1997), in contrast, used the term tepal in a phylogenetic sense such that all monocots have tepals, as derived from magnoliid tepals or bracts (Endress 1994c). Takhtajan's definition of tepal limits the application of this term to specific groups of angiosperms and requires different terms for an undifferentiated perianth in other groups. We use tepal as defined by de Candolle—an undifferentiated calyx and corolla.

Distinguishing sepals from petals is not always straightforward (Endress 1994c; Albert et al. 1998). Whereas sepals and petals are readily distinguished in most eudicots, this is not the case in basal angiosperms (Endress 2008), many of which have numerous spirally arranged, undifferentiated perianth parts, a condition long considered ancestral (e.g., Bessey 1915; Cronquist 1968, 1988; Takhtajan 1969). Perianth differentiation has also been of interest with regard to the origin of the sepal and petal (e.g., Eames 1931; Hiepko 1965b; Kosuge 1994; Albert et al. 1998; Kramer and Irish 1999, 2000; Zanis et al. 2003; Kim et al. 2004a; P. Soltis et al. 2009; Irish 2009). It has been suggested that petals evolved first and that sepals evolved later (e.g., Albert et al. 1998) and that petals have evolved multiple times from different floral organs in different groups (e.g., Eames 1961; Takhtajan 1969; Kosuge 1994; Albert et al. 1998; Zanis et al. 2003; Ronse De Craene and Brockington 2013). The petals present in some lineages, such as Magnoliales and Laurales, are proposed to have evolved from an undifferentiated perianth, whereas in other groups (Ranunculales and other eudicots) petals are similar to stamens in morphology and ontogeny and may, in fact, be derived from stamens. Takhtajan (1969, 1987) suggested two types of petal origins, one from stamens (e.g., Ranunculales) and one from bracts (e.g., Magnoliales). Hiepko (1965b) and Endress (2008) recommend use of "tepals" for basal angiosperms (ANA grade and magnoliids), and "sepals" and "petals" for eudicots; monocots are more difficult, however.

Support for an independent origin of petals in eudicots has come from morphological studies showing that "petals" of various angiosperms exhibit major differences (Endress 1994c; Kramer and Irish 2000; Brockington et al. 2010, 2012; Ronse De Craene and Brockington 2013) and can be grouped into two basic classes. In one group are petals that resemble stamens. The characters shared by these petals and stamens include the following: the petals are developmentally delayed relative to the stamens; young petals are similar in appearance to stamen primordia at inception; petals are supplied by a single vascular trace; and, in some cases, petals possess nectaries (Endress 1994c). These petals have sometimes been termed "andropetals." The second type of petaloid organ is found in undifferentiated perianths (standardly termed tepals; Cronquist 1981) and is more leaf-like in general characteristics—that is, "bracteopetals." These structures mature much earlier than the stamens, often have three vascular traces, and are generally more leaf-like in appearance than other petals (Smith 1928; Tucker 1960; Takhtajan 1969).

According to Albert et al. (1998), two or more whorls of perianth parts must be present for an unambiguous interpretation of sepals and petals. If only a single perianth whorl is present, it may be difficult to score as "sepals" or "petals" (Endress 1994a,c, and Chapter 14). Is a single whorl an undifferentiated perianth of neither sepals nor petals? Or, is the single whorl either sepals or petals with the other perianth whorl absent? A single-whorled perianth has traditionally been referred to as composed of "sepals" (e.g., Cronquist 1968). Although the number of vascular traces (one vs. three) has sometimes been considered an important distinguishing feature, studies in rosids showed that this distinction between one- and three-traced organs may not be of deep phylogenetic significance (e.g., Endress et al. 2013), but this needs further study.

Families of basal angiosperms with a single whorled perianth include nearly all Aristolochiaceae (except *Saruma*), all Myristicaceae, and Chloranthaceae (*Hedyosmum*). Sometimes the nature of a single-whorled perianth can be determined through comparison with the perianths of closely related taxa.

In Aristolochiaceae, most taxa have a single-whorled perianth that has been considered a calyx (Cronquist 1968, 1981; Takhtajan 1991; Tucker and Douglas 1996). This determination is supported by observations for one genus (*Saruma*), which has two perianth whorls differentiated into sepals and petals. Furthermore, in some species of *Asarum*, petals begin to develop, but the only traces are small threadlike structures (Leins and Erbar 1985). Topologies place *Asarum* and *Saruma* as sister to the rest of the family (see Chapter 5); hence, the ancestral condition for the family is a differentiated perianth, with the single-whorled perianth derived through the loss of petals.

We scored the single whorls of Myristicaceae and *Hedyosmum* (Chloranthaceae) as representing an undifferentiated perianth, although others have considered these to be sepals for the reasons discussed above (e.g., Cronquist 1981; Zanis et al. 2003). The perianths of some taxa are difficult to characterize. In Himantandraceae (comprising only *Galbulimima*), Cronquist (1981) described the perianth as consisting of both sepals and petals. Others have considered the outer structures bracts, rather than sepals, meaning that the remaining petaloid structures represent an undifferentiated perianth or staminodes (e.g., Endress 1977). We have tentatively interpreted these structures as an undifferentiated perianth. Adding to the complexity is the recent observation that even in the "undifferentiated" perianth of *Amborella*, there is some morphological differentiation between inner and outer members of the perianth (Endress and Igersheim 2000a; Endress 2001a; Buzgo et al. 2004b).

We conducted two sets of analyses. In one, perianth was coded as either undifferentiated or sepals + petals. In a second, the perianth was coded as either undifferentiated, sepals + petals, or absent; only this latter reconstruction is shown here.

In all of our MP and ML reconstructions (Fig. 6.10) the ancestral state for the angiosperms is an undifferentiated perianth. This is the case regardless of the number of states employed (above), whether *Amborella* and Nymphaeaceae are treated as sister taxa or as successive sister groups to all remaining angiosperms, or the placement of Chloranthales.

Amborella and Austrobaileyales have an undifferentiated perianth. *Trithuria* (Hydatellaceae) lacks a perianth. In this reconstruction, the ancestral state for Nymphaeales, as well as Nymphaeaceae + Cabombaceae, is determined to be ambiguous in most analyses. Although the ancestor of Nymphaeaceae + Cabombaceae is reconstructed as having a differentiated perianth with ML with binary states, a differentiated perianth for the clade is strongly favored with ML when multiple states are employed. Cabombaceae (e.g., *Cabomba*, *Brasenia*) and *Nuphar* have sometimes been considered to have an undifferentiated perianth, whereas more derived Nymphaeaceae (*Victoria*, *Nymphaea*) have been considered to have more of a differentiated perianth. However, as noted in Chapter 4, some authors consider all Nymphaeales to have a differentiated perianth.

Following the ANA grade, an undifferentiated perianth is ancestral for the remaining angiosperms (mesangiosperms) in all reconstructions (Fig. 6.10); a differentiated perianth subsequently evolved many times. Separate origins of a differentiated perianth occurred in the monocots, some Magnoliaceae, Annonaceae, Canellales, and some Aristolochiaceae with additional origins in early-diverging eudicots and core eudicots (see Chapter 8, 9). The data indicate that "sepals" and "petals" in these lineages are not homologous, a point long stressed (see Ronse De Craene and Brockington 2013).

Our coding of a single-whorled, trimerous perianth as undifferentiated in *Hedyosmum* (Chloranthaceae) and Myristicaceae is a conservative approach. Many basal angiosperms such as Canellaceae, Magnoliaceae, and Nymphaeaceae have a trimerous outer whorl and an inner whorl or whorls that are variable in merism (Zanis et al. 2003). Thus, the trimerous outer whorl in basal angiosperms may be homologous to the trimerous perianth found in taxa having a single-whorled perianth. Assuming that the single-whorled perianth is in all cases sepaloid, the ancestral state above Austrobaileyales changes to a differentiated perianth (not shown).

STAMEN PHYLLOTAXIS

Reconstructions of the evolution of stamen phyllotaxis (not shown) mirror the differentiation of perianth phyllotaxis discussed above (see also Ronse De Craene et al. 2003). We reconstructed stamen phyllotaxis as both a binary character (spiral [including chaotic] or whorled) and a multistate character (spiral, polycyclic, tetracyclic, tricyclic, dicyclic, monocyclic, and chaotic). However, Endress and Doyle (2007) found the phyllotaxis of perianth and androecium somewhat decoupled in Magnoliales.

With binary coding, the interpretation is straightforward compared to multistate coding. The ancestral state for angiosperms is ambiguous with MP and ML using binary states (Fig. 6.11); with multiple states, the ancestor

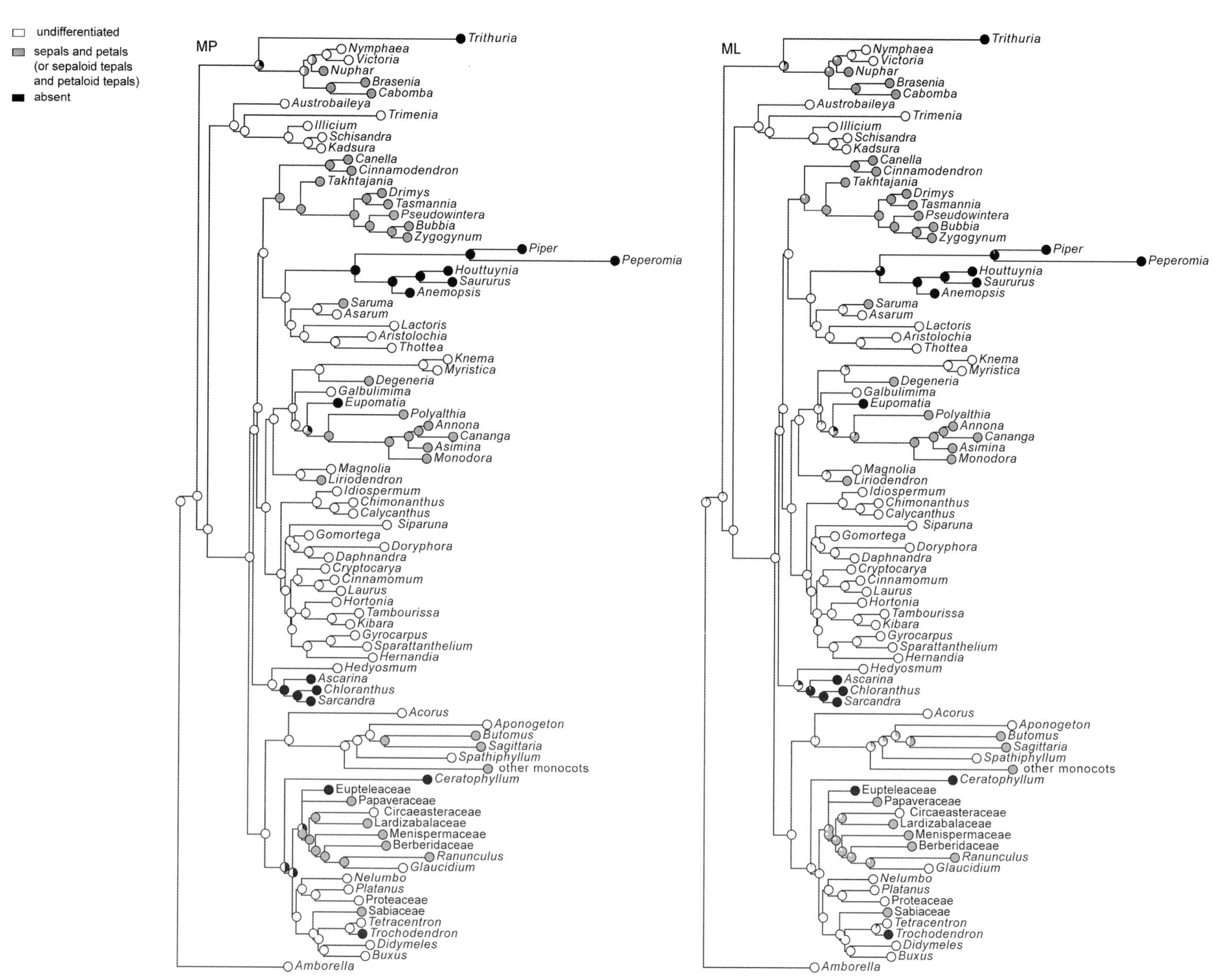

Figure 6.10. MP and ML reconstruction of the evolution of perianth differentiation in the ANA grade, magnoliids, monocots, and basal eudicots. The topology follows Soltis et al. (2011). See Fig. 6.1. for placement of *Lactoris*. Data are from Doyle and Endress (2000), Endress and Doyle (2009), and Rudall et al. (2007) with modifications (see text).

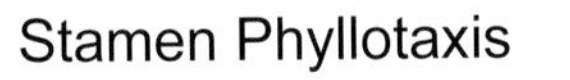

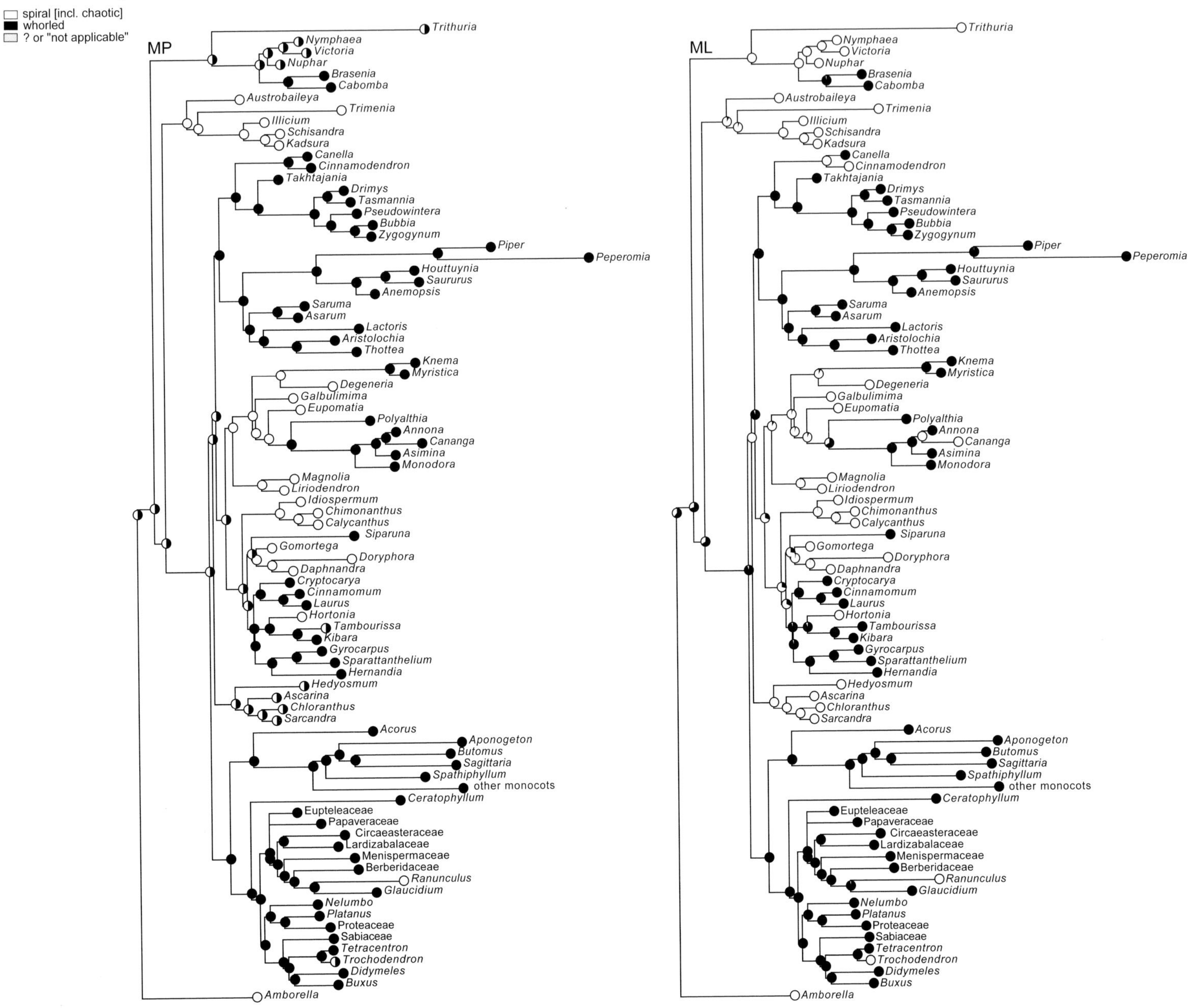

Figure 6.11. MP and ML reconstruction of the evolution of stamen phyllotaxis in the ANA grade, magnoliids, monocots, and basal eudicots. The topology follows Soltis et al. (2011). See Fig. 6.1. for placement of *Lactoris*. Data are from Doyle and Endress (2000), Endress and Doyle (2009), and Rudall et al. (2007) with modifications (see text).

is spiral with MP; spiral is favored with ML. With binary states, the ancestor of the mesangiosperms is unambiguously whorled with both MP and ML (Fig. 6.11); in the multistate analyses, the mesangiosperm ancestral state is ambiguous (not shown), probably because of the number and complexity of the states (a problem seen with perianth phyllotaxis). The multistate analysis is constructive in showing that polycycly has arisen multiple times (e.g., Winteraceae, Annonaceae, within Nymphaeaceae), as has monocycly (Hernandiaceae, Chloranthaceae, as well as some basal eudicots—see Chapter 8).

STAMEN MERISM

As with perianth merism, stamen merism was also coded as binary (indefinite in number and fixed in number) and multistate (indefinite, dimerous, trimerous, and pentamerous). With binary states, MP reconstructions are ambiguous for the ancestral state of angiosperms, but ML favors indefinite (not shown). Multistate reconstructions (both MP and ML) indicate that the ancestor of angiosperms had an indefinite number of stamens (not shown).

Reconstruction of ancestral stamen merism for mesangiosperms varies greatly depending on the states assigned and analysis and therefore remains unclear. What is clear from multistate analyses is that the trimerous condition has evolved multiple times: Cabombaceae, Piperales + Aristolochiales, Lauraceae, monocots, and Ranunculales (see Chapter 8). These results parallel those (above) for perianth merism and again illustrate that trimery has played a major role in the early evolution and diversification of the flower. Multistate coding also illustrates the lability of stamen merism, a result also stressed above with perianth merism.

STAMEN FORM

The stamens of many non-eudicot and non-monocot angiosperms are distinctive because of their large size, extensive quantity of sterile tissue surrounding the pollen sacs, and the lack of differentiation between filament and anther (Endress and Hufford 1989). The leaf-like, or laminar, stamens present in many of the non-eudicot/non-monocot taxa have long been of interest (Fig. 6.12). Authors have long proposed that the earliest angiosperms possessed laminar stamens (e.g., Canright 1952; Eames 1961; Cronquist 1968; Takhtajan 1969) (see also Chapter 4). From this putatively ancestral type of stamen, a gradual modification in size and shape was proposed, resulting in the typical stamen present in most angiosperms, a stamen composed of a well-defined anther and filament (Fig. 6.12).

Hufford (1996a) reconstructed stamen evolution across basal angiosperms using topologies from Donoghue and Doyle (1989a,b) with either Magnoliales or Nymphaeaceae and monocots as sister to other angiosperms; two basic states were used, laminar and non-laminar. With the first topology, Hufford reconstructed laminar stamens as ancestral; with the second topology, as not ancestral.

We did not reconstruct the evolution of stamen form in basal angiosperms because of the numerous intermediates between laminar stamens and those with well-defined anther and filament regions (Fig. 6.12; Endress and Hufford 1989; Endress 1994c). Appropriate assessments of characters and character states require additional developmental studies. Until those data are available, character-state reconstructions have to be viewed with caution. We review the diversity of stamen form in basal angiosperms.

Among basal angiosperms, *Amborella*, *Austrobaileya*, and some Nymphaeaceae have laminar stamens (Chapter 4; Fig. 6.12). Stamen form in Nymphaeaceae is diverse, with laminar stamens in *Nuphar* and *Victoria* (Endress and Hufford 1989; Hufford 1996a) and stamens with well-defined anther and filament regions present in *Cabomba* and *Brasenia* (Cronquist 1988); laminar stamens and stamens intermediate between laminar and and non-laminar occur in *Nymphaea* (Chapter 4; Hufford 1996a). In Trimeniaceae, the stamens have well-defined anther and filament regions. The stamens of Schisandraceae (including *Illicium*) are "relatively undifferentiated," with the filament often broader than the "anther" region (Endress and Hufford 1989). Hence, within Amborellaceae, Nymphaeales, and Austrobaileyales, laminar, linear, and intermediate stamens are found.

Extreme examples of laminar stamens are present in Degeneriaceae, Himantandraceae, and Eupomatiaceae (all Magnoliales). In Winteraceae, the stamens are only slightly differentiated into anther and filament (Fig. 6.12). Many magnoliid families have stamens with well-differentiated filament and anther: Monimiaceae, Myristicaceae, Lauraceae, Piperaceae, Saururaceae, Aristolochiaceae, and Lactoridaceae (Fig. 6.12). The stamens of *Ceratophyllum* lack a filament region, but basal monocots possess stamens with well-differentiated anther and filament regions (Fig. 6.12).

Laminar stamens have originated multiple times, with independent origins in at least the ANA grade, Winteraceae, and Magnoliales (Fig. 6.12; see also, e.g., Endress and Hufford 1989; Hufford 1996a; Endress and Doyle 2009); therefore, not all laminar stamens are homologous (Endress and Hufford 1989; Hufford 1996a). For example, the laminar stamens of *Galbulimima* and *Degeneria* dif-

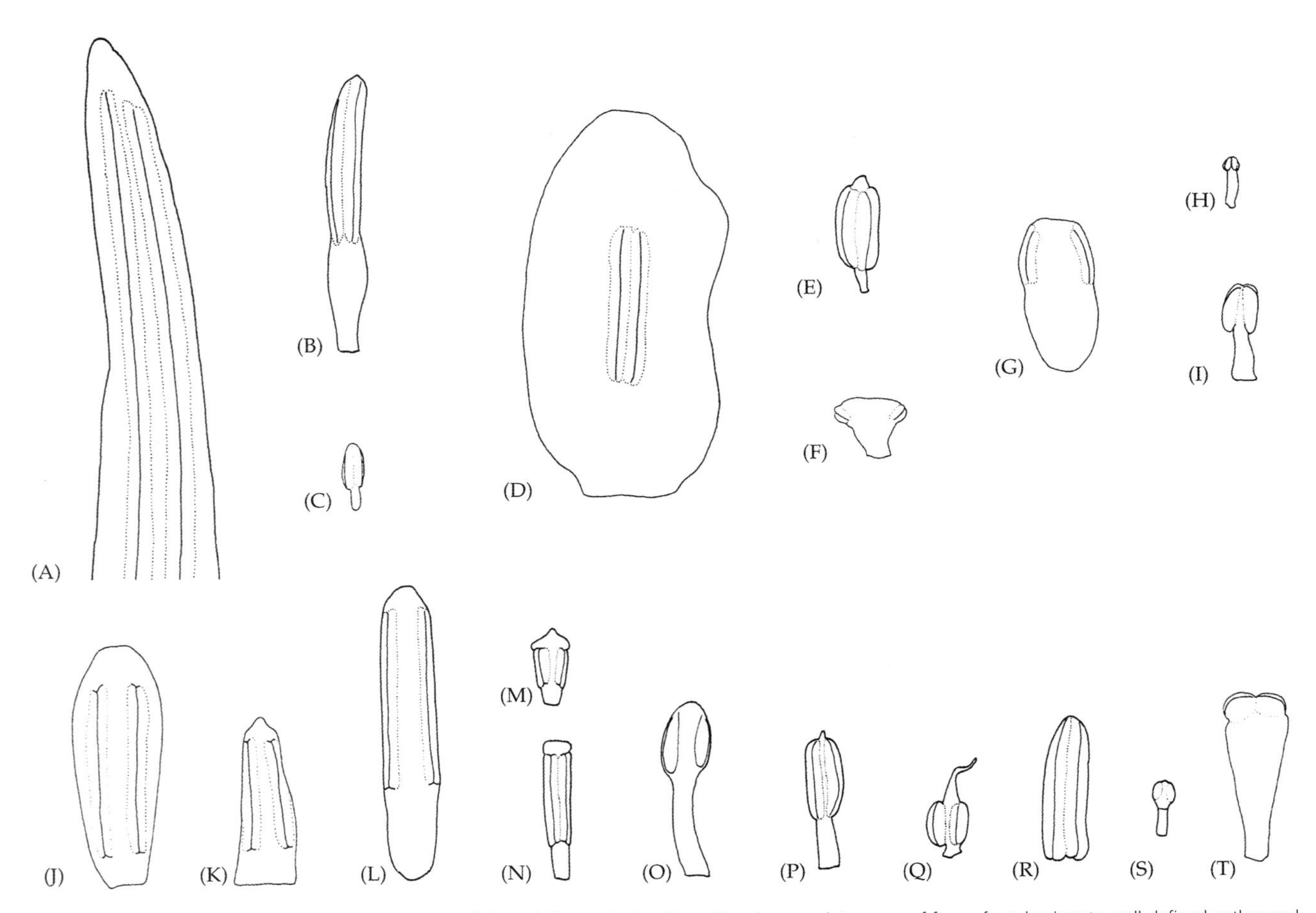

Figure 6.12. Stamen form in basal angiosperms, magnoliids, and Chloranthales, illustrating the complete range of forms from laminar to well-defined anther and filament, as well as intermediate morphologies. Stamens are drawn to comparable scale and are arranged in a generally phylogenetic manner. A–C, Nymphaeales; D–F, Austrobaileyales; G, Chloranthaceae; H–I, monocots; J–N, Magnoliales; O–P, Laurales; Q–S, Piperales; T, Canellales. (From Endress 1994a, 1996a.) (A) *Nymphaea* (Nymphaeaceae). (B) *Nymphaea* (Nymphaeaceae). (C) *Cabomba* (Nymphaeaceae). (D) *Austrobaileya* (Austrobaileyaceae). (E) *Trimenia* (Trimeniaceae). (F) *Kadsura* (Schisandraceae). (G) *Sarcandra* (Chloranthaceae). (H) *Acorus* (Acoraceae). (I) *Sagittaria* (Alismataceae). (J) *Degeneria* (Degeneriaceae). (K) *Eupomatia* (Eupomatiaceae). (L) *Magnolia* (Magnoliaceae). (M) *Artabotrys* (Annonaceae). (N) *Annona* (Annonaceae). (O) *Laurus* (Lauraceae). (P) *Chimonanthus* (Calycanthaceae). (Q) *Asarum* (Aristolochiaceae). (R) *Aristolochia* (Aristolochiaceae). (S) *Piper* (Piperaceae). (T) *Zygogynum* (Winteraceae). Reproduced from Soltis et al. (2005; Fig. 3.15).

fer substantially from those in *Austrobaileya* (Fig. 6.12) in radial thickness, sporangial dehiscence, location of pollen sacs, and the manner in which the sporangia are embedded within the microsporophyll (Endress 1994c).

The prevalence of laminar and intermediate stamen morphologies in the basalmost angiosperms agrees with the fossil record. Angiosperm stamens in Early Cretaceous fossil floras typically exhibit little differentiation between anther and filament (e.g., Friis et al. 2000, 2011). It is tempting to speculate that laminar stamens are ancestral in the angiosperms. However, there is room for caution, given the diversity of forms present within just the ANA grade. Furthermore, *Cabomba* and *Brasenia*, the sister group to other Nymphaeaceae, both have stamens with well-differentiated anther and filament regions, and it appears that laminar stamens are derived within Nymphaeaceae in association with beetle pollination (Gottsberger 1988).

POLLEN

Early angiosperms likely had uniaperturate (one area of weakness) pollen, which is typical of most basal angiosperms, magnoliids, and Chloranthales and is also present in early angiosperm fossils (Friis et al. 2011). Triaperturate pollen is a synapomorphy for the eudicots. Schisandraceae also have triaperturate pollen, but of a unique type distinct from the pollen typical of eudicots (Huynh 1976).

There has been considerable speculation as to the fine-scale features of the pollen (pollen infratectum) in the ancestral angiosperms. Cronquist (1988) considered pollen with granular exine to be ancestral in the angiosperms because of its prevalence in his Magnoliales. Furthermore, granular pollen was one of the putative synapomorphies of Gnetales and angiosperms (Crane 1985; Doyle and Donoghue 1986;

Doyle 1996) and a feature that placed Magnoliales as sister to other angiosperms in earlier investigations (Donoghue and Doyle 1989a,b). However, Doyle (2005, 2009) found granular pollen to be derived.

Doyle and Endress (2000) examined the evolution of infratectal pollen wall structure. In addition to the granular and columellar states recognized in an earlier study (Donoghue and Doyle 1989a), they recognized an intermediate condition characterized by irregular radial (columellar) elements mixed with apparent granules. It is not entirely clear whether this "intermediate" condition is an appropriate independent state; additional developmental studies are needed to evaluate this state more critically.

Our reconstructions generally agree with Doyle and Endress (2000) although our ML reconstructions provide a new perspective. The intermediate condition is found in *Amborella* and some Nymphaeaceae, indicating that this could be the ancestral state for angiosperms and is reconstructed as such with MP. However, with ML the ancestral state for angiosperms is unclear (Fig. 6.13). Significantly, above the node to Nymphaeales, the ancestral state for all remaining angiosperms is columellar with both MP and ML. Most of the transitions to granular and intermediate forms occur within Laurales + Magnoliales; Piperales + Canellales share columellar pollen. These results were not evident, however, in Doyle and Endress (2000) because they used a tree in which neither Laurales + Magnoliales nor Canellales + Piperales appeared as sister groups.

Our reconstructions, as in Doyle (2005, 2009), indicate that a granular infratectum is not ancestral in the angiosperms but likely arose independently within some Nymphaeaceae (e.g., *Nuphar* and *Nymphaea*), Laurales (e.g., Lauraceae and Hernandiaceae), and Magnoliales (Annonaceae, Eupomatiaceae, Degeneriaceae, Himantandraceae).

CARPEL FORM

We discuss carpel form briefly in Chapter 4 (basal angiosperms). Here we focus on our reconstructions of this feature. There are two extremes (Doyle and Endress 2000), plicate and ascidiate (Fig. 6.14; see Fig. 4.5). The plicate carpel form can be compared to a leaf folded down the middle with ovules on the folded (plicate) zone or area (Fig. 4.5). The plicate condition was considered ancestral for angiosperms by Bailey and Swamy (1951) and Eames (1961). In the ascidiate form, in contrast, the carpel grows like a tube; this form was considered ancestral by Leinfellner (1969) and van Heel (1981, 1983). Doyle and Endress (2000) noted that intermediate carpel conditions also exist and are found in several basal angiosperms, including *Barclaya* (Nymphaeaceae), *Illicium* (Schisandraceae, sensu APG IV 2016), and *Acorus* (Acoraceae). We follow Doyle and Endress (2000) and Endress and Doyle (2009) in recognizing an intermediate state, but it is not clear whether this is an appropriate, independent state; additional developmental studies are encouraged.

Reconstructions indicate that ascidiate carpels are ancestral in the angiosperms (Fig. 6.14; Doyle and Endress 2000); they characterize *Amborella*, most Nymphaeaceae, Austrobaileyaceae, most Schisandraceae, and Trimeniaceae (Figure 3.16; Endress and Igersheim 2000b). Intermediate carpel forms are present in *Barclaya* (Nymphaeaceae) and also in *Illicium* (Schisandraceae). Above ANA, *Ceratophyllum* and Chloranthaceae also have ascidiate carpels. The first derivation of plicate carpels occurs in the monocots. Our reconstruction also agrees with the prevalence of putative ascidiate carpels in fossils from the Early Cretaceous (Crane et al. 1995; Friis et al. 2011).

The ancestral state for mesangiosperms is uncertain (Fig. 6.14). The plicate form is reconstructed as ancestral for the magnoliids and for eudicots; these derivations appear to be independent origins. Several reversals to the ascidiate form are also apparent in magnoliids (Monimiaceae), monocots (Araceae), and early-diverging eudicots (Berberidaceae and *Nelumbo*).

An intermediate carpel type characterizes much of Laurales and represents a possible morphological synapomorphy of that clade. Intermediate carpel form has also evolved independently in *Acorus*, Myristicaceae, and the eudicot *Euptelea* (Ranunculales). It would be of interest to characterize these intermediate types in more detail.

CARPEL SEALING

Carpel sealing was initially discussed in chapter 4—here we provide additional detail, including ancestral-state reconstruction. In most taxa having ascidiate carpels, the carpels are sealed by secretion rather than by postgenital fusion (see Fig. 4.5; Endress and Igersheim 2000b). Carpel form and carpel sealing are therefore closely associated. Endress and Igersheim (2000b) recognized four types of carpel sealing: (1) a complete lack of postgenital fusion but occlusion of the inner space by secretion; (2) a combination of postgenital fusion at the periphery and no fusion but occlusion by secretion in the center with a complete unfused canal up to the stigma; (3) a combination of postgenital fusion encompassing the entire periphery and an unfused secretory canal that is not complete but ends below the stigma; and (4) complete postgenital fusion of the inner space between the ovary and the stigma. *Amborella* exhibits the first of these (Fig. 4.5), as do most Austrobaileyales (however,

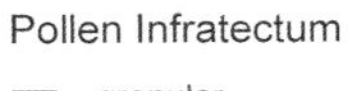

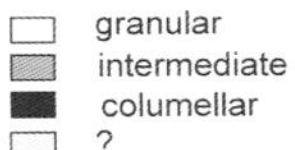

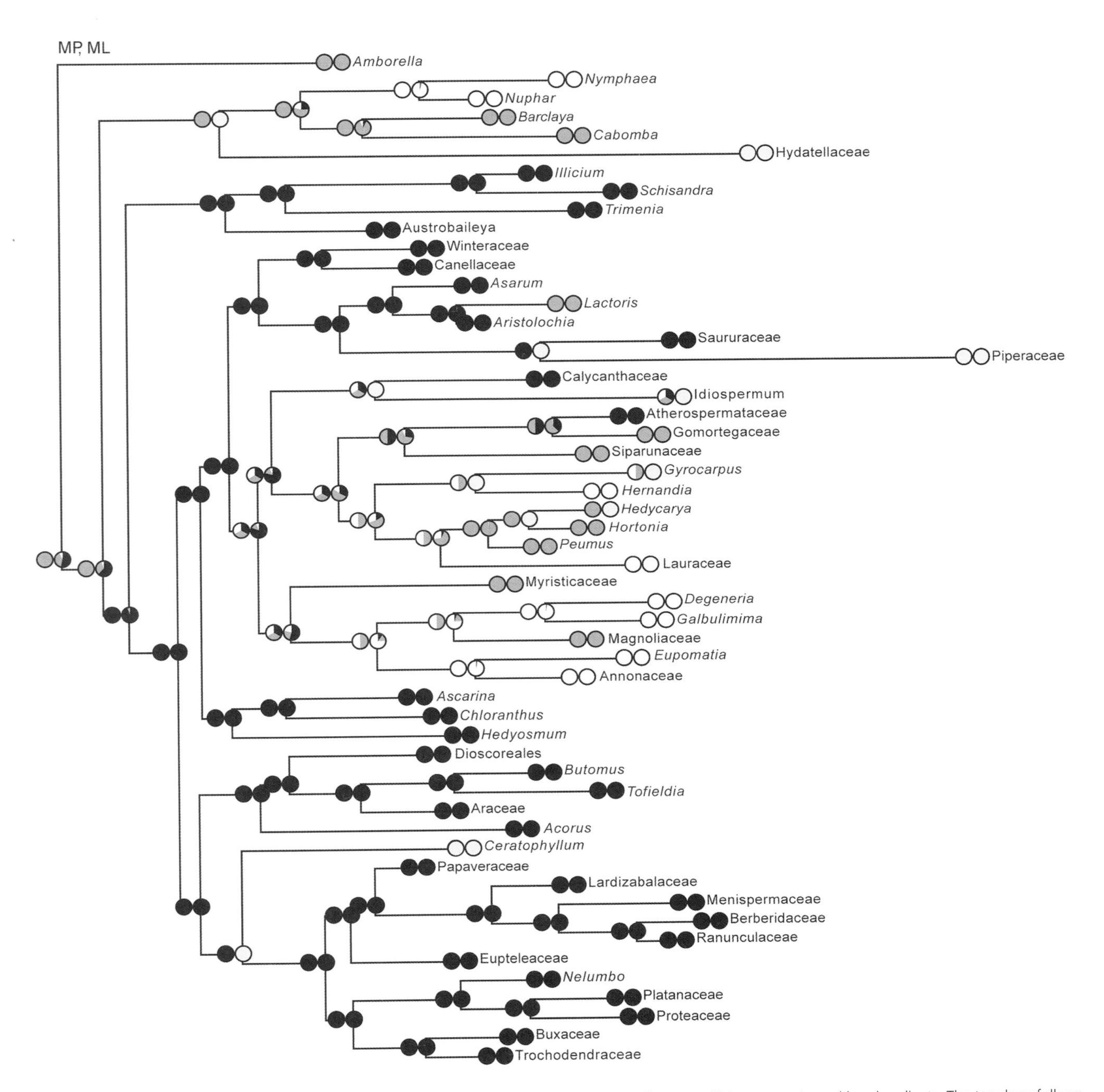

Figure 6.13. MP and ML reconstruction of the evolution of pollen infratectum in the ANA grade, magnoliids, monocots, and basal eudicots. The topology follows Soltis et al. (2011). See Fig. 6.1. for placement of *Lactoris*. Data are from Doyle and Endress (2000), Endress and Doyle (2009), and Rudall et al. (2007) with modifications (see text).

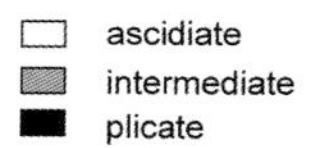

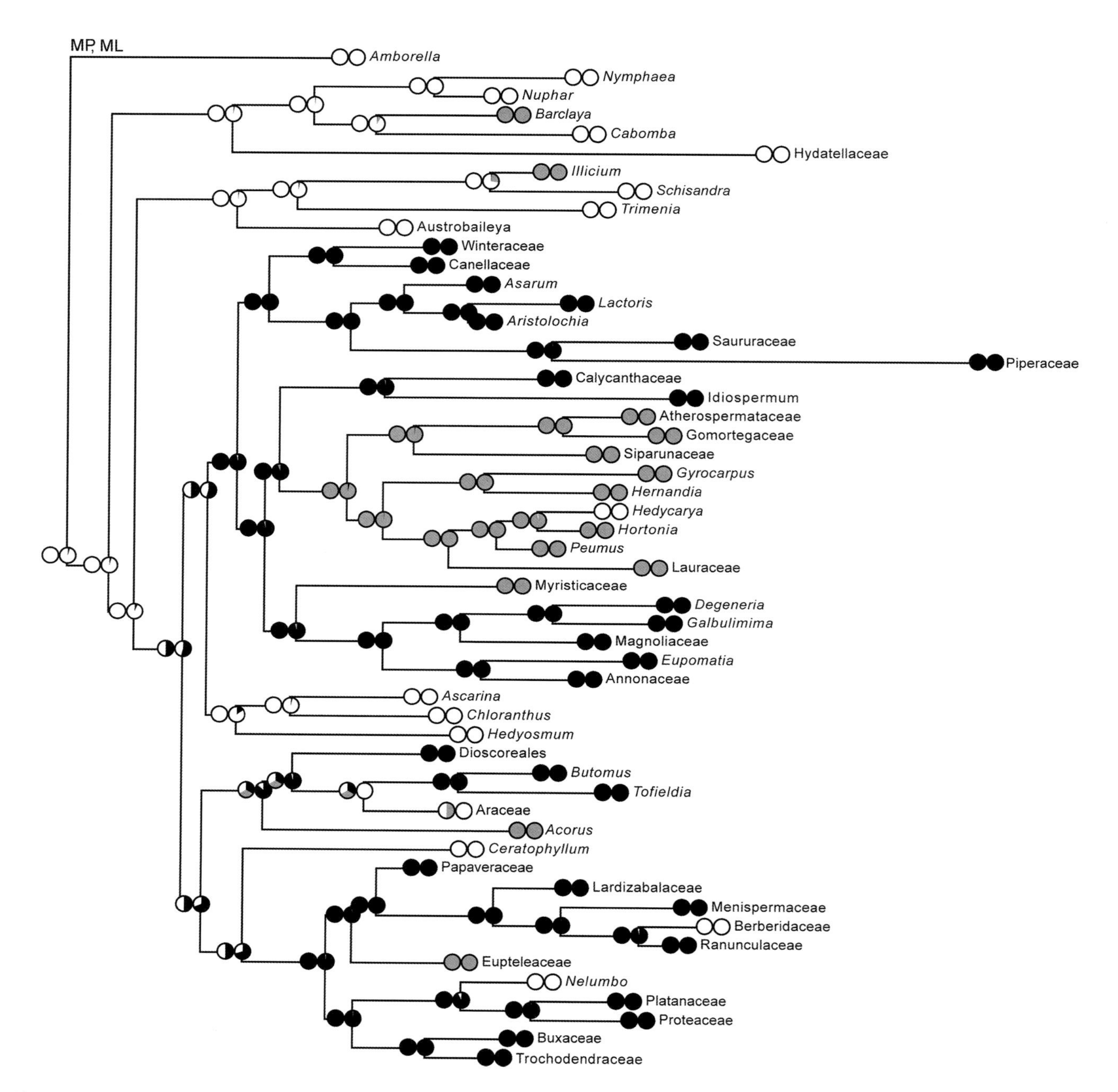

Figure 6.14. MP and ML reconstruction of the evolution of carpel form in the ANA grade, magnoliids, monocots, and basal eudicots. The topology follows Soltis et al. (2011). See Fig. 6.1. for placement of *Lactoris*. Data are from Doyle and Endress (2000), Endress and Doyle (2009), and Rudall et al. (2007) with modifications (see text).

Illicium has type 2) and *Cabomba* (Nymphaeaceae). Other Nymphaeaceae (*Barclaya* and *Nymphaea*) have type 3. Most eudicots have type 4.

We followed Endress and Doyle (2009) and scored postgenital sealing as follows: none, partial, complete (combining 2 and 3 above). Our MP reconstructions (Fig. 6.15) are comparable to those of Endress and Doyle (2009) and indicate that carpels sealed by secretion (no postgenital fusion) represent the ancestral condition for angiosperms. ML reconstructions are less clear, however, with some probability of partial fusion or even complete fusion. This is likely due to the presence of some derived members of Nymphaeaceae having complete postgenital fusion. Nonetheless, the fact that Amborellaceae, some Nymphaeaceae, and most Austrobaileyales have carpels that are not postgenitally sealed but are fused by secretion has important evolutionary implications, suggesting that this is ancestral.

The ancestral state for mesangiosperms is ambiguous—complete postgenital fusion may have evolved multiple times (Fig. 6.15). These data imply that angiospermy evolved first in early flowering plants and that postgenital fusion of the carpel evolved later (Endress and Igersheim 2000b).

APOCARPY VERSUS SYNCARPY

It has long been suggested that the hypothetical ancestral angiosperms had numerous free carpels (apocarpy; Cronquist 1968; Takhtajan 1969; Stebbins 1974). Our reconstructions unambiguously reconstruct the ancestral state of the angiosperms, as well as that of mesangiosperms and early-diverging eudicots, as apocarpous (Fig. 6.16). We followed Doyle and Endress (2000; see Endress and Doyle 2009), who recognized three states: apocarpous, in which the carpels are free (distinct), parasyncarpous (a gynoecium in which the carpels are united and a unilocular ovary is present, i.e., no fusion in the center, no septa or only partial septa), and eusyncarpous (a gynoecium with a plurilocular ovary in which the septa meet in the center and are fused).

Amborella, *Cabomba*, and *Brasenia* (Nymphaeales) have free carpels, although other Nymphaeales have united carpels. Austrobaileyales, Laurales, Magnoliales, and most Ranunculales are characterized by apocarpy. The reconstructions indicate that syncarpy is derived in Nymphaeales; the monocots may be ancestrally syncarpous (*Acorus* and Araceae are both syncarpous), with a reversal to apocarpy in *Tofieldia*, Alismataceae, and *Butomus*. Parasyncarpy may also be ancestral for Piperaceae and Canellales as well as Papaveraceae (cf. Doyle and Endress 2000). Eusyncarpy has arisen separately within the monocots and eudicots.

OVULE NUMBER

Doyle and Endress (2000) (see also Endress and Doyle 2009) reconstructed the evolution of ovule number and scored character states as one, mostly two, and more than two ovules. With these states and their topology, the ancestral state for the angiosperms is equivocal. This equivocality appears to be due more to variability in this character and the character-state coding used than to the topology.

Many non-eudicot angiosperms have a small number of ovules per carpel. *Amborella* and Trimeniaceae have a single ovule per carpel; Schisandraceae have one or two. Chloranthaceae and *Ceratophyllum* also have one ovule per carpel. However, the basal angiosperm *Austrobaileya* has 8 to 15 ovules per carpel; ovule number per carpel in *Acorus* (sister to other monocots) is 2 to 6; the number in Nymphaeales ranges from 1 to 300, but in *Cabomba* and *Brasenia* the number is low, typically 1 to 3, and in Hydatellaceae it is 1. Scoring these critical taxa as "more than two" ovules per carpel hinders attempts to reconstruct the number of ovules per carpel in early angiosperms by obscuring diversity; this single large category (more than two) includes a low to moderate number of ovules per carpel as well as a very high number of ovules per carpel. The fact that some basal angiosperms have a low number of ovules (5–20) is lost by lumping all taxa together that have more than two ovules per carpel. In addition, some families have substantial variation in ovule number (e.g., Nymphaeaceae, families of Canellales and Piperales, and the basal eudicot family Ranunculaceae).

In our reconstruction, we added additional states: 1, mostly 2 (but including 3 or 4), 5 to 20, and more than 20 ovules per carpel. We coded taxa of Nymphaeaceae following information in Endress (2001a) and coded the polymorphism evident in some families, which was not captured by Doyle and Endress (2000).

In our reconstruction (Fig. 6.17), a single ovule is ancestral not only for the angiosperms but also for mesangiosperms. The ancestral state for mesangiosperms was ambiguous in Endress and Doyle (2009). In our reconstructions the ancestral state for eudicots may be two (this is favored by ML, but not MP), but is best considered equivocal.

OVULE CURVATURE

Most angiosperms have anatropous ovules; however, some, including several early-branching lineages, have orthotropous ovules. Gymnosperms also have orthotropous ovules (Fig. 1.10). Other types of ovules (see Chapter 4; Fig. 4.8)

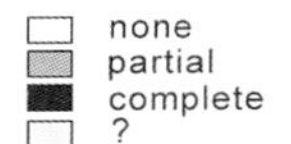

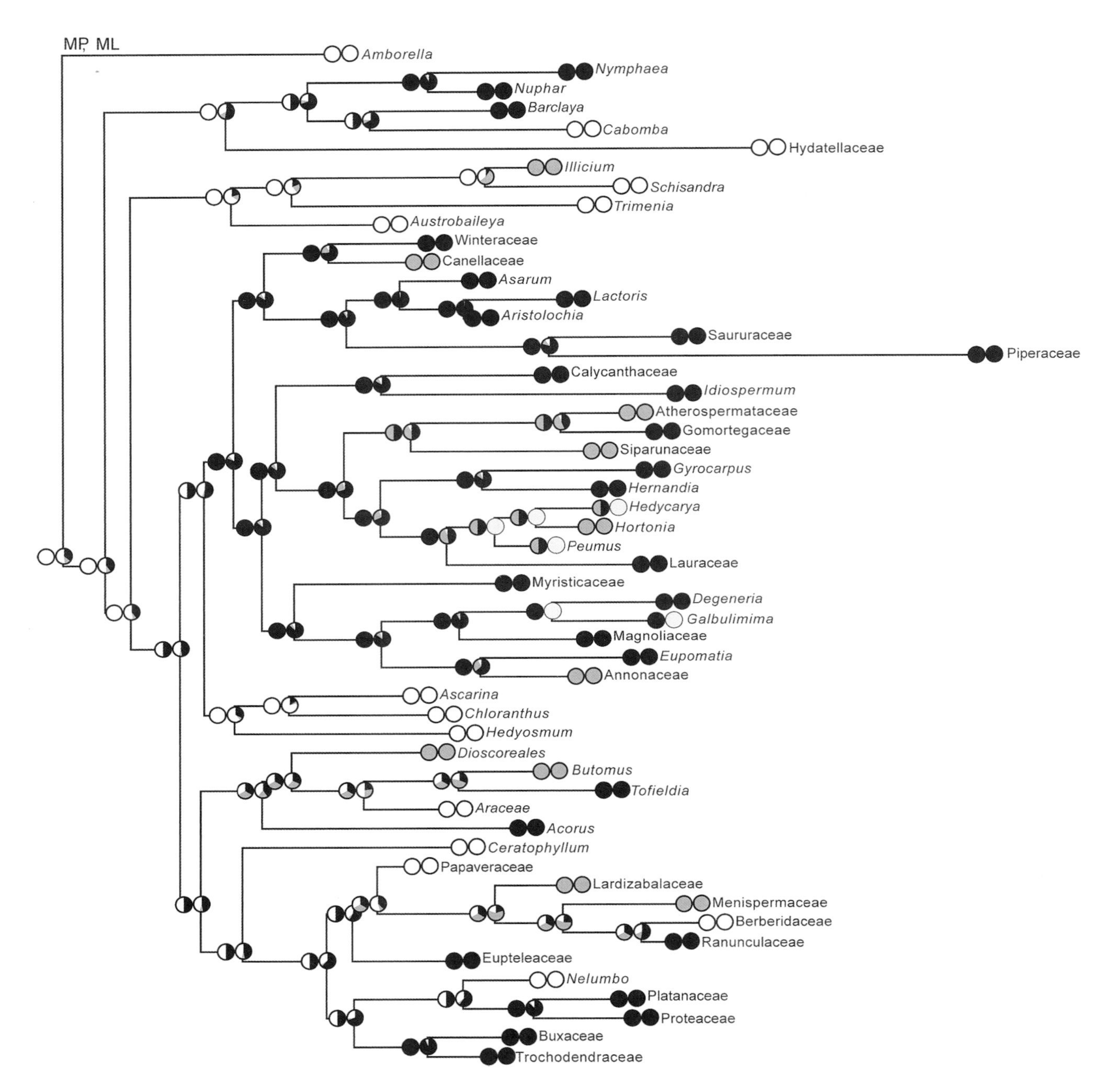

Figure 6.15. MP and ML reconstruction of the evolution of carpel sealing in the ANA grade, magnoliids, monocots, and basal eudicots. The topology follows Soltis et al. (2011). See Fig. 6.1. for placement of *Lactoris*. Data are from Doyle and Endress (2000), Endress and Doyle (2009), and Rudall et al. (2007) with modifications (see text).

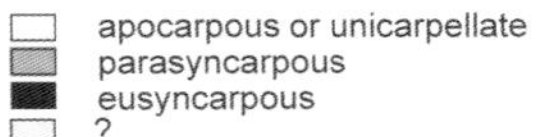

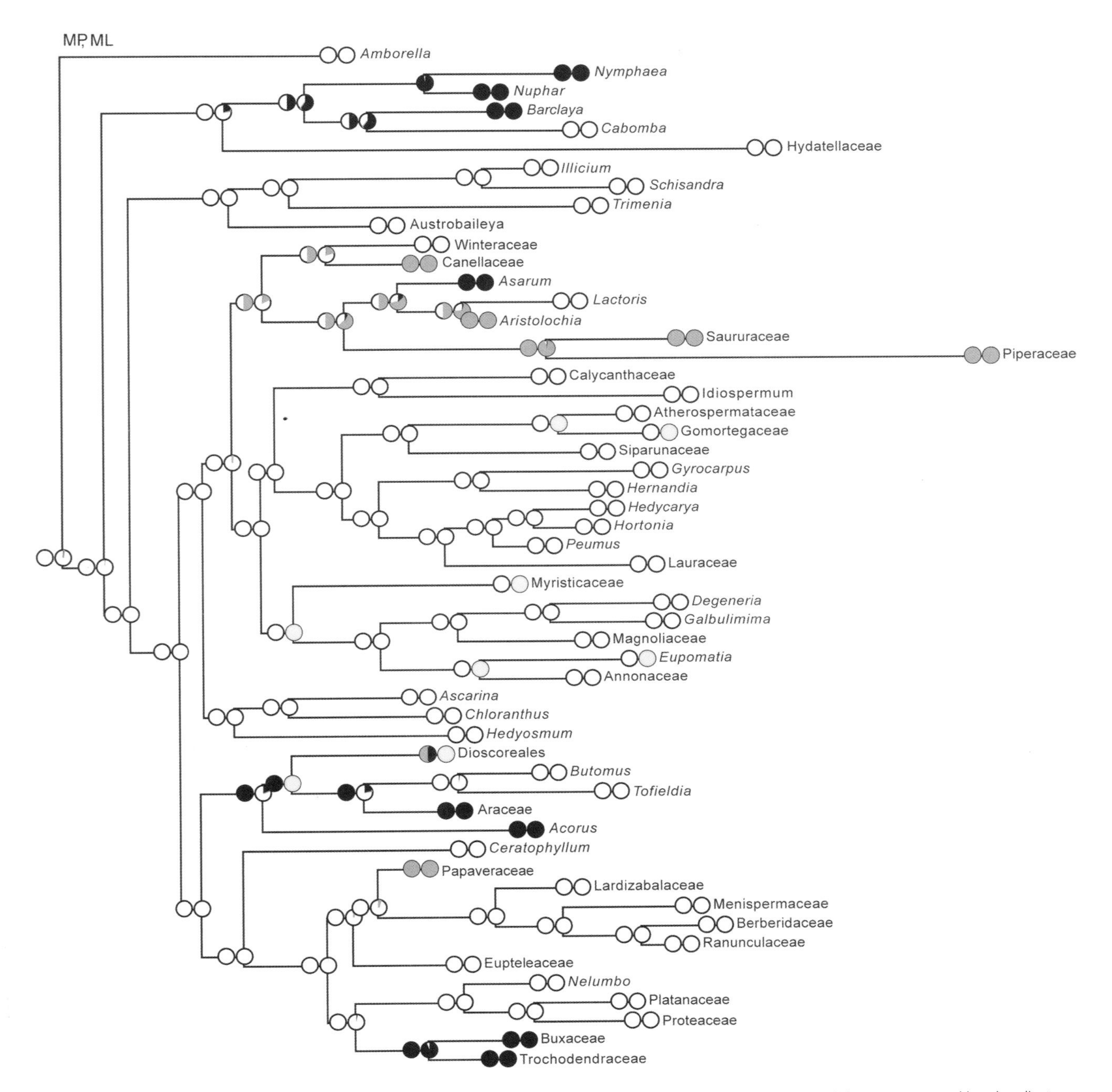

Figure 6.16. MP and ML reconstruction of the evolution of carpel fusion (apocarpy vs. syncarpy) in the ANA grade, magnoliids, monocots, and basal eudicots. The topology follows Soltis et al. (2011). See Fig. 6.1. for placement of *Lactoris*. Data are from Doyle and Endress (2000), Endress and Doyle (2009), and Rudall et al. (2007) with modifications (see text).

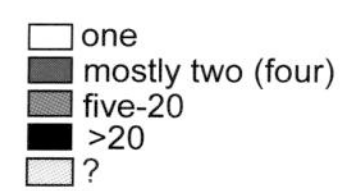

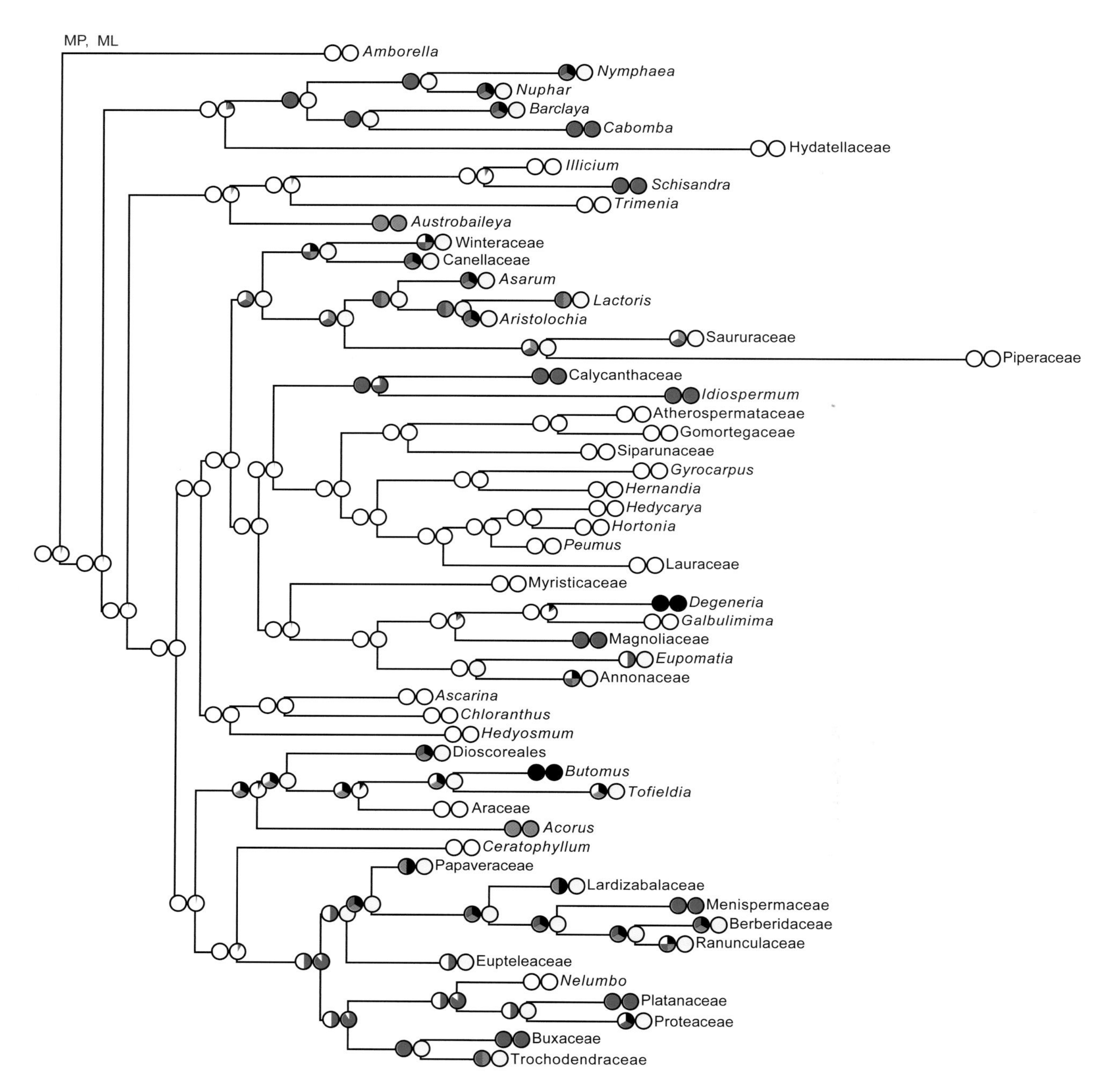

Figure 6.17. MP and ML reconstruction of the evolution of ovule number in the ANA grade, magnoliids, monocots, and basal eudicots. MP is the circle to the left and ML the circle to the right. The topology follows Soltis et al. (2011). See Fig. 6.1. for placement of *Lactoris*. Data are from Doyle and Endress (2000), Endress and Doyle (2009), and Rudall et al. (2007) with modifications (see text).

are also found in various angiosperm lineages (e.g., campylotropous, Caryophyllales). Because of the prevalence of anatropous ovules across all of the angiosperms, this state typically has been considered the ancestral condition (Cronquist 1968; Takhtajan 1969; Stebbins 1974). However, *Amborella* and several other non-eudicots, including Chloranthaceae, some Nymphaeaceae (i.e., *Barclaya*), *Ceratophyllum*, and *Acorus* (sister to all other monocots), possess ovules considered orthotropous (discussion in Endress 2011).

We followed Doyle and Endress (2000) and Endress and Doyle (2009) in scoring ovules as anatropous or orthotropous (including hemitropous in this category). If *Amborella* is scored as orthotropous (Endress and Igersheim 2000b; Yamada et al. 2001b), the ancestral condition for the angiosperms with both MP and ML is equivocal, although ML favors anatropous (Fig. 6.18). Following *Amborella*, the anatropous ovule is consistently reconstructed as ancestral for the remaining angiosperm nodes (mesangiosperms, magnoliids, monocots, eudicots) (Fig. 6.18). Doyle and Endress (2009) obtained similar results.

Developmental data have indicated that the interpretation of ovule evolution is more complex than our reconstructions indicate. In many non-eudicots, the outer integument is not cup-shaped, but rather markedly asymmetric and hood-shaped. Yamada et al. (2001b) found that the outer integument of both *Amborella* and *Chloranthus* is asymmetrical and concluded that the developmental pattern of the outer integument and the ovule developmental pattern seen in both *Amborella* and *Chloranthus* are not equivalent to those of the orthotropous ovules of eudicots. Developmental data indicate that the orthotropous ovules found in ascidiate carpels (as in *Amborella* and *Chloranthus*, see above) are derived (see Endress 1986b, 1994c, 2011a; Yamada et al. 2001b). Further support for the antiquity of ascidiate carpels with anatropous ovules is provided by the fossil record, including the fossil *Couperites* (Chloranthaceae) with an anatropous ovule (Pedersen et al. 1994) and fruits with anatropous ovules from the Barremian of Portugal (one of the oldest strata bearing angiosperm fossils; Friis et al. 1999, 2000, 2011).

ENDOSPERM

Until a decade ago any review of angiosperm evolution would have indicated that the ancestral state for angiosperms was double fertilization, with the second fertilization event forming triploid endosperm (the latter formed via fusion of a sperm and two polar nuclei from the female gametophyte). However, recent papers demonstrated that diploid endosperm is formed in *Nuphar* (Nymphaeaceae) as well as Austrobaileyales (Williams and Friedman 2002; Friedman 2006, 2009; Tobe et al. 2007; Rudall et al. 2008; Friedman and Ryerson 2009; Williams 2009; Bachelier and Friedman 2011).

Rather than possessing a female gametophyte of eight nuclei and seven cells, which is typical of angiosperms, ANA lineages other than *Amborella* have four nuclei in four cells (see Figs. 4.15 and 6.19; e.g., Yoshida 1962, for *Schisandra*; Winter and Shamrov 1991, for *Nuphar*). The second fertilization involves a sperm nucleus and just one nucleus of the female gametophyte, resulting in diploid rather than triploid endosperm. Previous studies provide indirect developmental data consistent with a four-celled female gametophyte in representatives of the ANA grade: *Nymphaea*, *Cabomba*, *Schisandra*, and *Illicium* (see Williams and Friedman 2002). Other analyses indicated that a four-celled female gametophyte is present in *Austrobaileya*, *Illicium*, and *Trimenia* (Tobe et al. 2000, 2007; Friedman and Ryerson 2009; Williams 2009; Bachelier and Friedman 2011). Many of these taxa also produce diploid endosperm (in some taxa analyzed embryologically such as *Trimenia*, the nature of the endosperm [diploid vs. triploid] is unclear). However, *Amborella* has a female gametophyte of nine nuclei and eight cells and triploid endosperm (Friedman 2006; Williams 2009).

Because some basal angiosperms have four-celled female gametophytes, and others seven-celled, MP reconstructions indicate that the ancestral state for the female gametophyte (four-celled vs, seven-celled) and endosperm (diploid vs, triploid) in the angiosperms is equivocal (see Williams and Friedman 2002; Soltis et al. 2005b). In our analyses, MP reconstructions similarly indicate that the ancestral state of angiosperms is equivocal. With ML reconstructions, the results vary. When relative branch lengths are employed (based on the Soltis et al. 2011 topology; Fig. 6.19), a four-celled female gametophyte and diploid endosperm are slightly favored as ancestral; in contrast, when equal branch lengths are used, the ancestral state is more likely to be eight nuclei and seven cells and triploid endosperm. The results are best summarized as indicating that the ancestral condition for angiosperms is equivocal.

Following the ANA grade, however, the ancestral state for the remaining angiosperms is clear: a female gametophyte of eight nuclei and seven cells and triploid endosperm. These states are characteristic of lineages of mesangiosperms, including Chloranthaceae and *Magnoliidae*.

In addition to the discovery of diploid endosperm, subsequent patterns of early endosperm development are more complex than described in the traditionally used categories of cellular, free nuclear, and helobial (Floyd and Friedman 2000). Floyd and Friedman (2000) suggested that basal angiosperms have retained plesiomorphic features. Most

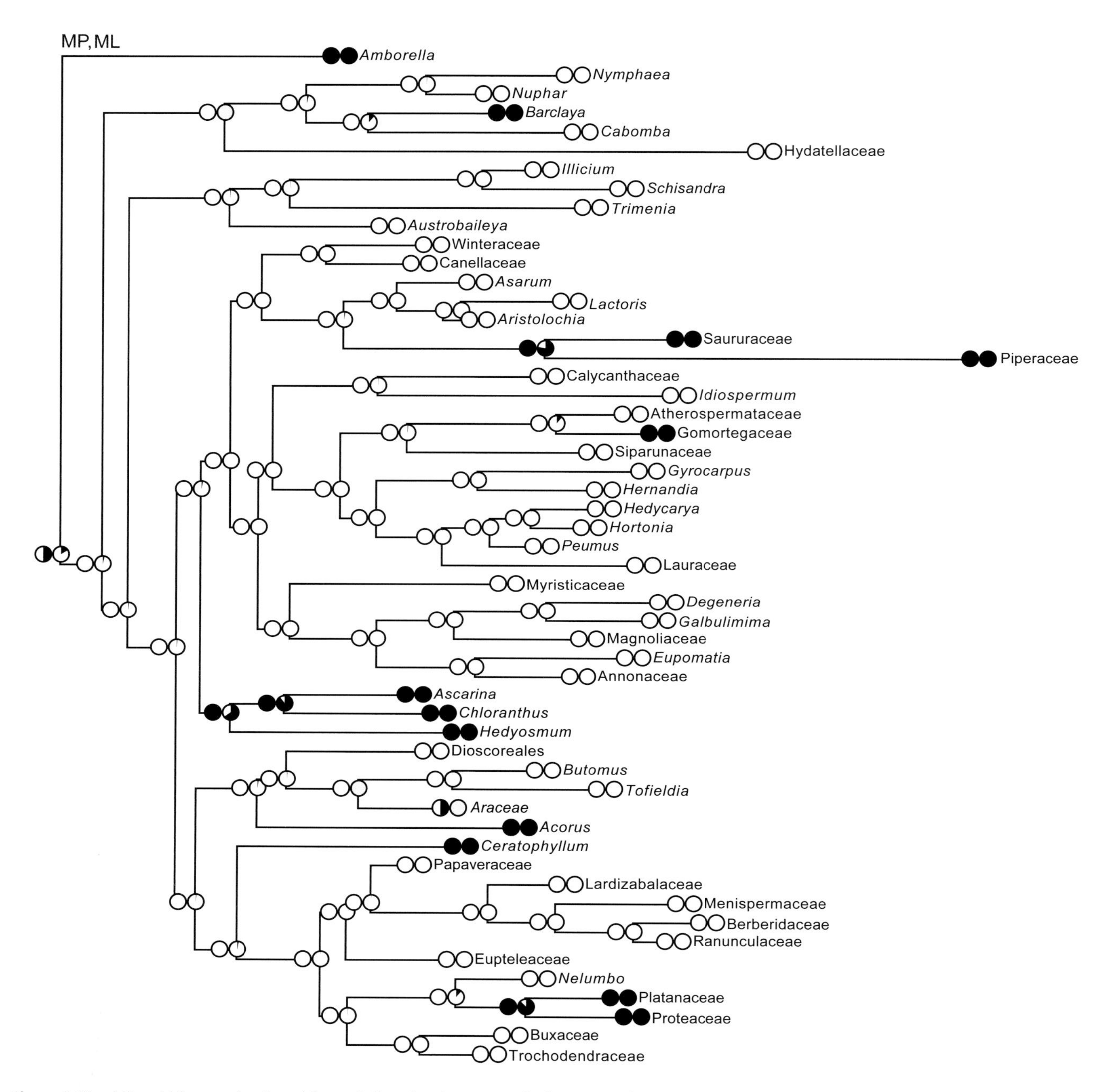

Figure 6.18. MP and ML reconstruction of the evolution of ovule curvature in the ANA grade, magnoliids, monocots, and basal eudicots. The topology follows Soltis et al. (2011). See Fig. 6.1. for placement of *Lactoris.* Data are from Doyle and Endress (2000), Endress and Doyle (2009), and Rudall et al. (2007) with modifications (see text).

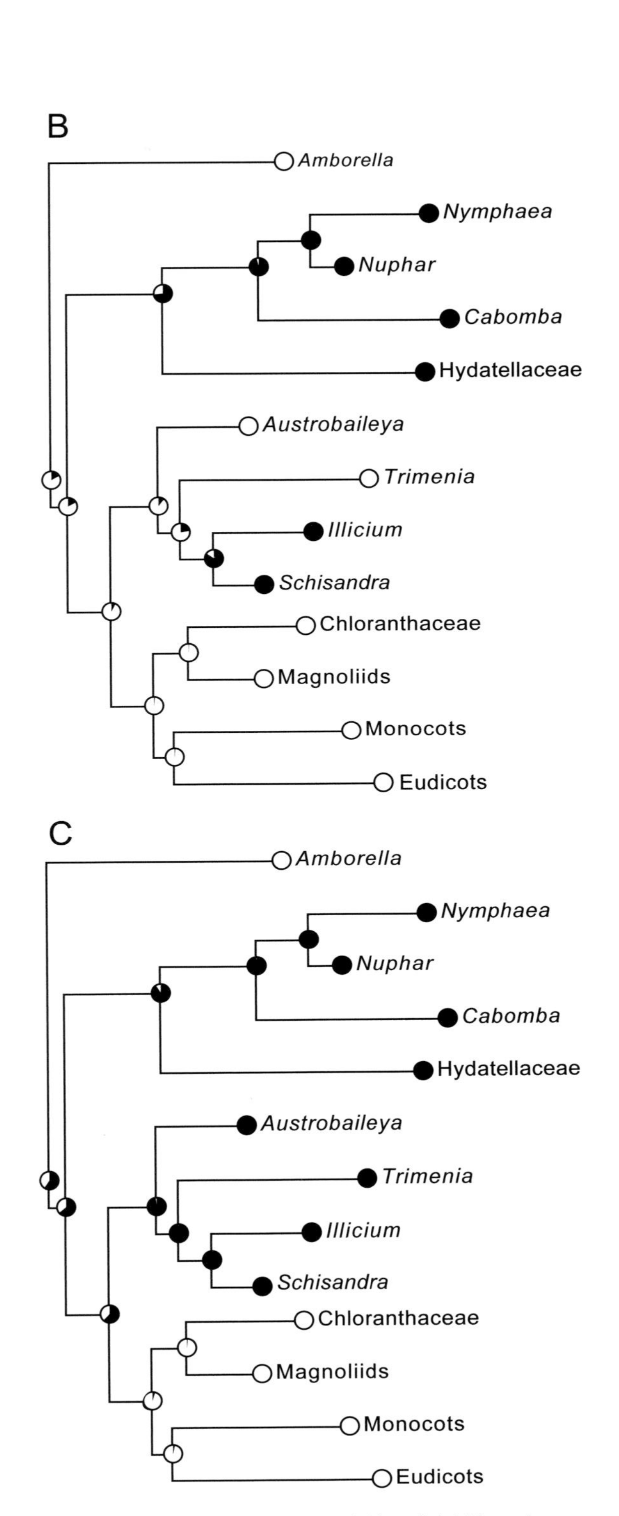

Figure 6.19. ML reconstruction of the diversification of female gametophyte (seven-celled vs. four-celled) and endosperm (triploid vs. diploid) in angiosperms. (A) Reconstruction based on existing data. The number of cells in the female gametophyte of several critical basal lineages is unknown (e.g., Chloranthaceae, Trimeniaceae, Austrobaileyaceae). (B) Reconstruction with the assumption that Chloranthaceae, Trimeniaceae, and Austrobaileyaceae have seven-celled female gametophytes and triploid endosperm. (C) Reconstruction with the assumption that Chloranthaceae have seven-celled female gametophytes and triploid endosperm and that Trimeniaceae and Austrobaileyaceae have four-celled female gametophytes and diploid endosperm.

basal angiosperms exhibit extensive endosperm development (but not Nymphaeales), as well as differential developmental fates of the chalazal and micropylar regions of the endosperm (bipolar endosperm development). In most basal angiosperms, endosperm development in both domains is cellular (cell walls are formed), an ontogeny referred to as "bipolar cellular." Furthermore, endosperm development appears to be highly conserved throughout the radiation of basal angiosperms, with a few exceptions (e.g., *Cabomba*, *Saururus*, and Lauraceae). The diversification of endosperm development coincided with the radiations of monocots and eudicots (Floyd and Friedman 2000).

RECONSTRUCTING THE EARLIEST ANGIOSPERMS

HISTORICAL VIEWS

There have been frequent attempts to reconstruct the floral morphology, habit, ecology, and other characteristics of the earliest angiosperms (e.g., Arber and Parkin 1907; Axelrod 1952; Cronquist 1968, 1981, 1988; Takhtajan 1969; Stebbins 1974; Doyle and Hickey 1976; Doyle 1978; Retallack and Dilcher 1981; Taylor and Hickey 1996b; Soltis et al. 2005b, 2008b; Endress and Doyle 2007, 2009). Cronquist (1968, 1988), Takhtajan (1969), and Stebbins (1974) attempted to reconstruct a hypothetical ancestral angiosperm based on their views of extant "primitive" angiosperms. They considered extant members of their Magnoliales to be the most archaic living angiosperms, with each species retaining some ancestral features. They suggested that the earliest angiosperms had simple, entire leaves, were evergreen, woody plants, but "not necessarily trees," of moist tropical habitats with an equable climate. Cronquist maintained that the nodes were unilacunar with two leaf traces or perhaps trilacunar with three traces. The wood lacked vessels, but possessed long, slender tracheids, with long, tapering ends, and numerous scalariform bordered pits. The flowers were large, bisexual, and strobiloid, with numerous spirally arranged parts (e.g., *Magnolia*-like). The perianth was not differentiated into sepals and petals, but consisted of spirally arranged tepals. The stamens were large and laminar with no differentiation into filament and anther. The pollen had a single long sulcus; the exine had a raised-reticulate surface ornamentation and internally was either solid or, more probably, tectate-granular. No nectar or nectaries were present. The flowers contained several to many carpels, each of which was folded along the midrib (plicate) and unsealed, with the margins loosely appressed rather than "anatomically joined," effectively closed by "a tangle of hairs." The embryo was small; double fertilization occurred, and the endosperm was triploid. The endosperm was probably copious and of the nuclear type (i.e., it had a free-nuclear stage early in ontogeny).

Stebbins (1974) had a different view of the ancestral angiosperm, reflecting his belief that novelty evolved in marginal habitats. Nonetheless, he also maintained that these first angiosperms were similar in many respects to modern Magnoliales. He suggested that the first angiosperms were shrubs that grew in semi-arid or seasonally dry tropical or subtropical habitats, with leaves strongly reduced in size in association with this dry habitat. A seasonal climate favored compression of the reproductive cycle into a relatively short period. The distinctive characteristics of angiosperms arose in response to seasonal drought (Axelrod 1970, 1972; see also Hickey and Doyle 1977), including closure of the carpel, reduction in the size of gametophytes, evolution of double fertilization, and a shift from wind to insect pollination.

Retallack and Dilcher (1981) maintained that a "xeromorphic bottleneck" (Hickey and Doyle 1977) played a major role in angiosperm origins; early evolution was induced by near-marine environments rather than by arid environments. Building on their "herbaceous origin" hypothesis, Taylor and Hickey (1996b) suggested that angiosperms evolved in a fluvial regime, in sites characterized by high disturbance and moderate amounts of alluviation. These areas would have had high nutrient levels and frequent loss of plant cover due to periodic disturbance.

Whereas Cronquist, Takhtajan, and Stebbins focused on Magnoliales and Taylor and Hickey (1996b) focused on Piperales for inferring ancestral characteristics of the angiosperms, molecular phylogenetic studies conducted over the past 20 years convincingly point to the ANA grade as the basalmost extant angiosperms. Endress (1986b) was perhaps the only investigator to suggest, from morphology, that some of these groups were among the earliest-diverging extant angiosperms. He indicated that Amborellaceae, Austrobaileyaceae, Trimeniaceae, and Winteraceae (which is not part of the ANA grade, but a magnoliid) were potential candidates as some of the earliest angiosperm families.

A SUMMARY HYPOTHETICAL RECONSTRUCTION OF EARLY ANGIOSPERMS

Based on the overview provided above and information in Chapters 3 and 4, what can we say about the ancestor of living angiosperms? We attempt an overall summary. With improved understanding of angiosperm phylogeny, it is now possible to make more informed reconstructions of the morphology and ecology of the ancestors of extant angiosperms, as well as subsequent evolutionary patterns. Of course, extant basal angiosperms may not be representative of the earliest, now extinct, lineages and may not even represent the earliest diversification of the angiosperms. Furthermore, some Early Cretaceous fossils cannot be clearly assigned to any extant families (Friis et al. 2000); *Archaefructus* is one example (Sun et al. 2002; Friis et al. 2003a, 2011).

Although we have made great advances in our understanding of many aspects of angiosperm phylogeny, morphology, floral genetics, and paleobiology, in many respects the origin of the angiosperms remains unclear. We and others have gone to great lengths to reconstruct various features of early angiosperms, yet it is not at all clear that these reconstructions are representative of the earliest angiosperms. Because gymnosperms lack many characters of interest in early angiosperms, our reconstructions are ambiguous regarding the ancestral state for angiosperms. Furthermore, reconstructions based on extant taxa do not take into account the diversity of fossils. *Archaefructus* further reveals the enormous diversity in morphology and form present in early angiosperms.

With these important caveats, we offer a reconstruction of early angiosperms that is based on the features of extant species with input from the fossil record. Perhaps the best we can hope to do is to reconstruct early angiosperm motifs.

Whether the first angiosperms were forest shrubs (dark-and-disturbed hypothesis) or aquatic herbs (wet-and-wild hypothesis) remains unclear (Feild et al. 2004; Coiffard et al. 2007; Soltis et al. 2008b). The phylogenetic position of Nymphaeales (water lilies) indicates that the aquatic habit arose early (Endress and Doyle 2009). As a result of the near-basal placement of Nymphaeales, the habit of the first flowering plants is reconstructed as equivocal in many analyses (Soltis et al. 2005b and here). However, it takes only minor changes in character coding to yield a woody ancestral angiosperm. In addition, there may be more reasons to favor a woody ancestral state—that is, because all gymnosperms are woody. Prominent authors favored a terrestrial origin for this reason (Cronquist 1968; Takhtajan 1969; Stebbins 1974), and thought that the first angiosperms were shrubs or small trees, which agrees with early-diverging extant taxa such as *Amborella* and members of Austrobaileyales. As noted, reconstructions with gymnosperm outgroups added result in a woody ancestral state for angiosperms.

However, others proposed that angiosperms were ancestrally herbaceous (e.g., Taylor and Hickey 1992, 1996a,b; Doyle et al. 1994). Nymphaeales, which follow *Amborella* as sister to all remaining angiosperms, are aquatic herbs. The early angiosperm *Archaefructus* was also herbaceous and aquatic. The aquatic habit clearly appeared very early in the diversification of flowering plants. Some of our reconstructions suggest that angiosperms were ancestrally woody; only if Chloranthaceae are coded as herbaceous is the ancestral state for angiosperms ambiguous. *Amborella* and Austrobaileyales are small trees, shrubs, or woody vines; all are shade tolerant and exhibit morphological and physiological traits that enhance the capture of light in a forest understory (Feild et al. 2000a, 2003a,b). Based on characteristics of Amborellaceae and Austrobaileyales, Feild et al. (2004) proposed that angiosperms arose in "dark and disturbed" understory habitats.

Character-state reconstructions indicate that early angiosperms also possessed unilacunar (one- or two-trace) nodal anatomy, although the evolutionary significance of this anatomical characteristic is unknown. It remains unclear whether the ancestor of extant angiosperms possessed leaves with chloranthoid teeth, although ML analyses favor that reconstruction. The early angiosperms did not produce ethereal oils—those appeared slightly later in angiosperm evolution.

The earliest angiosperms may have lacked vessels in the xylem. Vessels are considered more efficient in conducting water than tracheids and provide water to leaves at an increased rate (Carlquist 1975; Takhtajan 1969). The xylem features of *Amborella*, *Austrobaileya* (Austrobaileyaceae), *Illicium*, and *Kadsura* (Schisandraceae) suggest that early angiosperms either lacked vessels or instead possessed "primitive" vessel elements with tracheid features. However, if many early angiosperms lived in wet understory conditions as hypothesized, a low water-transport capacity of the stems would not have been problematic with xylem of this nature (Feild et al. 2000a, 2003a,b, 2004). For example, *Amborella* lacks vessels, and thus the water-conducting capacity of the stems is low, but the plants grow in areas of low evaporative demand and high rainfall. Vessels are present in *Austrobaileya*, but have tracheid-like features such as gradually tapered end walls, frequent scalariform perforation plates, and small diameters (Bailey and Swamy 1948; Carlquist 1996; Carlquist and Schneider 2002). Although the hydraulic capacity of the xylem in *Austrobaileya* is higher than that of *Amborella*, the values are still lower than in other climbing angiosperms (Tyree and Ewers 1991, 1996). The stem xylem water-transport capacities of *Illicium* and *Kadsura* are also much lower than values for most vessel-bearing shrubs and deciduous angiosperm trees and lianas, respectively (Tyree and Ewers 1991, 1996; Sperry et al. 1994; see Feild et al. 2000a, 2003a,b, 2004). The low water-conducting values in *Illicium* and *Kadsura* are probably the result of the unspecialized vessels, which are short, with long, oblique scalariform perforation plates that partially retain pit membranes (Carlquist 1982).

Early terrestrial angiosperms may have been successful because they exploited wet, understory forest environments more effectively than ferns and other seed plants, such as Bennettitales, Caytoniales, cycads, and Gnetales, present in understory environments at that time. Furthermore, these early angiosperms may have been successful because they had more developmental flexibility than these other lineages. Production of different types of leaves in varying light environments may have enabled early angiosperms to compete more effectively across a wider range of light regimes than other groups with more limited morphological plasticity (Feild et al. 2004).

In contrast to *Amborella* and Austrobaileyales, Nymphaeales possess ecological and physiological traits associated with sunny to shady aquatic habitats, but some authors have proposed that their ecology and morphology may not be representative of the earliest angiosperms (e.g., Schneider and Williamson 1993). Nymphaeaceae are often believed to represent a separate and distinct evolutionary experiment that does not represent the ecology associated with the origin of the angiosperms (Feild et al. 2004). However, as we have stressed, *Amborella* and Austrobaileyales may also not be representative of the earliest angiosperms. Fossils such as *Archaefructus* (Sun et al. 2002) and those attributable to Nymphaeaceae (Friis et al. 2001, 2009) indicate that the aquatic habit appeared early in angiosperm evolution.

Several lines of evidence indicate that flowers of the hypothetical early angiosperms were likely small. The floral features of *Amborella*, Nymphaeales, and Austrobaileyales suggest that the flowers of early angiosperms were not the large strobiloid structures envisioned by Cronquist (1968, 1981) and Takhtajan (1969), but more likely were small to moderate in size. Further support for this view comes from the fossil record. The angiosperm reproductive structures (flowers, fruits, seeds) from the Early Cretaceous are also small (Friis et al. 2000, 2011; see Chapter 3).

Our reconstructions of ancestral angiosperm floral phyllotaxis were equivocal—spiral vs. whorled. *Amborella* and Austrobaileyales have spiral phyllotaxis, but Nymphaeales have whorled. However, different flowers on the same plant of *Drimys winteri* may have spiral or whorled phyllotaxis (Doust 2001). Ontogenetic studies have revealed that, in some cases, floral organs that appear in a series in mature flowers actually result from spiral phyllotaxis; both spiral and whorled phyllotaxis may result from spiral organ development. Phyllotaxis is equivocal in some early angiosperm fossil flowers (Friis et al. 2000, 2009, 2011).

The perianth of early angiosperms was likely undifferentiated (a perianth of tepals), a view supported by our reconstructions and early fossils (Friis et al. 2011). Based on the flowers of *Amborella* and Schisandraceae, bracts preceding the flower may have been scarcely differentiated from the tepals. Alternatively, early angiosperms did not possess a perianth—this was added later.

Cronquist (1968), Takhtajan (1969), and Stebbins (1974) envisioned early angiosperms with numerous (indeterminate) perianth parts, but in most of our reconstructions the ancestral perianth merism is equivocal. Although *Amborella* and Austrobaileyales have an indeterminate number of perianth parts, this is not generally the case in non-eudicots. Importantly, our reconstructions point to an early and significant role of trimery in angiosperms. The ancestral condition for Nymphaeales is a trimerous perianth, as is the ancestral state for mesangiosperms.

The multiple carpels in early angiosperms, as well as other floral organs, would have been free (not fused).

Stamens likely were undifferentiated into anther and filament regions—that is, they were leaf-like (laminar) and similar to the inner tepals. The gradual transition from bracts to tepals to stamens and ultimately carpels is currently reflected in the similarity and overlap in gene expression patterns among these organs in living basal angiosperms—in contrast, in eudicots and monocots the patterns of gene expression are highly canalized and differ between floral organs (Chanderbali et al. 2010; see Chapter 14).

Pollen was likely uniaperturate in early angiosperms and exhibited a reticulate or "intermediate" morphology (Doyle 2005, 2009) characterized by irregular radial (columellar) elements mixed with granules.

The early carpel was ascidiate and sealed by secretion (Endress and Igersheim 2000b). The actual number of ovules per carpel was low, perhaps one. The ancestral ovule curvature is unclear (orthotropous vs. anatropous).

The ancestral state for the female gametophyte (four-celled vs. seven-celled) and endosperm (diploid vs. triploid) remains equivocal; triploid and diploid endosperm are equally parsimonious alternatives (Williams and Friedman 2002). Endosperm formation was likely "bipolar cellular" (Floyd and Friedman 2000); seeds were likely few in number and small, with copious endosperm produced.

Major changes accompanied the evolution of the mesangiosperms. Well-developed vessel elements were apparent. Perianth phyllotaxis was whorled, and our reconstructions favor a trimerous ancestral state for mesangiosperms. Carpels became more or less plicate in several groups. Patterns of morphological and anatomical evolution within monocots and eudicots are covered in more detail in later chapters.

Ancestral-state reconstructions are needed over trees that are more densely sampled than provided here. In addition, these reconstructions may need to be repeated as the nuclear-based phylogeny of angiosperms becomes clear. Initial studies of numerous nuclear genes suggest possible differences in the placement of major clades (e.g., Chloranthales) compared to the topologies based on plastid genes (see Chapter 3). Furthermore, more studies of anatomy and morphology are needed so that we have a clearer understanding of these features as well as data for more than just a few exemplars per clade. In addition, the underlying developmental genetics of many of these characters remains unknown and ripe for investigation.

7 Monocots

INTRODUCTION

Several studies have examined phylogenetic placement of the monocotyledons (often shortened to monocots) relative to Ceratophyllaceae, eudicots, magnoliids, and Chloranthaceae and found well-supported results (see also Chapter 3), but there still is no consensus about the exact placement of monocots relative to these other clades. Soltis et al. (2011) and recent analyses of plastid genomes (Gitzendanner pers. comm.) placed the monocots sister to Ceratophyllaceae/eudicots. However, recent phylogenomic analyses of nuclear genes place monocots after the ANA grade as sister to all other angiosperms (Zeng et al. 2014; Leebens-Mack pers. comm.; see Chapter 3). Given that there are at least incongruent placements of monocots, more research is clearly needed before we can say with confidence where the monocots fit into the bigger picture of angiosperm relationships. It is possible to rule out some of the previous hypotheses (i.e., to Nymphaeales, see below, or Piperales, Burger 1977).

Regardless of which of the above-noted hypotheses is correct, the monocots certainly represent one of the oldest lineages of angiosperms. The origin of the monocot clade must have occurred no later than the origin of the eudicot clade (see Chapters 3 and 8). Hence, given that eudicot pollen dates to at least 125 million years ago (mya), the monocot lineage must also be that old, and molecular-based age estimates have supported this inference. Using varied datasets and approaches, estimates range from 134 mya (Bremer 2000, 2002), 127–141 mya (Sanderson 1997; Wikström et al. 2001; Sanderson et al. 2004), to 123–138 mya (Bell et al. 2010). Zeng et al. (2014) estimated the crown group age of the monocots to be in the Jurassic at 150 mya, and Hertweck et al. (2015), using the dataset of Chase et al. (2006) plus the low-copy nuclear gene PHYC, found the crown node to be 131 mya. These studies reach relatively consistent results, given the different datasets and methods employed.

There are about 52,000 species of monocots (Mabberley 1993), representing 22% of all angiosperms (assuming only ~240,000 species of angiosperms, although many estimates are much higher). Half of the monocots can be found in the two largest families, Orchidaceae and Poaceae, which include 34% and 17%, respectively, of all monocots. These two families are also among the largest families of angiosperms, and Poaceae are dominant members of many plant communities. Although monocots are largely herbaceous, some, particularly agaves, palms, pandans, and bamboos, can reach great heights, lengths, and masses. Monocots provide most of the world's staple foods (e.g., grains and starchy root crops), abundant building materials, and a great number of medicines.

The single cotyledon is one of several putative synapomorphies for the monocots. Phylogenetic studies of nonmolecular data (Donoghue and Doyle 1989a,b; Loconte and Stevenson 1991; Doyle and Donoghue 1992) have identified many putative synapomorphies for the monocots (Table 7.1), including a single cotyledon, parallel-veined leaves, sieve-cell plastids with several cuneate protein crystals, scattered vascular bundles in the stem, and an adventitious root system.

Recognition of the monocots as a distinct group dates from Ray (1703) and was based largely on their having a single cotyledon instead of the two cotyledons typical of

Table 7.1. Putative synapomorphies for the monocots from Donoghue and Doyle (1989b), Loconte and Stevenson (1991), and Doyle and Donoghue (1992)

1. Presence of calcium oxalate raphides
2. Absence of vessels in the leaves
3. Monocotyledonous anther wall formation*
4. Successive microsporogenesis
5. Syncarpous gynoecium
6. Parietal placentation
7. Monocotyledonous seedling
8. Persistent radicle
9. Haustorial cotyledon tip
10. An open cotyledon sheath
11. Presence of steroidal saponins*
12. Fly pollination*
13. Diffuse vascular bundles (lack of secondary growth)

*Indicates characters that are synapomorphies for topologies in which *Acorus* (Acoraceae) is not sister to the rest of the monocots. *Acorus*, for example, lacks steroidal saponins and has dicotyledonous anther formation, so if *Acorus* is sister to the rest of the monocots, then these are not synapomorphies for the monocots as a whole.

the dicotyledons or "dicots." (As reviewed in Chapter 3, the dicot grouping is now known to be a non-monophyletic group that is not a relevant point of comparison.) Monocot seedlings display a great diversity of form (Tillich 1995), however, and not all possess an obvious single cotyledon. In the grasses, for example, the single cotyledon has become an absorptive organ (for the food stored in the endosperm) within the seed.

Another distinctive monocot trait is their vascular system, characterized by vascular bundles that are scattered throughout the medulla and cortex and closed (i.e., they do not contain an active cambium). In contrast, most basal angiosperms (e.g., members of the magnoliid clade, Amborellaceae, Austrobaileyales) and eudicots possess vascular bundles in an open ring. Tomlinson (1995) argued that the vascular system of monocots is unique among the seed plants and so different that there is "no homology of organization between the primary vascular system of monocotyledons and dicotyledons." However, scattered vascular bundles occur outside of the monocots (e.g., Nymphaeaceae and some Piperaceae). Tomlinson's statement emphasizes the distinctiveness of the monocot habit and vegetative organization and implies a fundamental developmental reorganization. However, the floral features of monocots are similar to many other basal angiosperms, particularly members of the magnoliid clade such as Aristolochiaceae and Annonaceae. It is easy to see how one gets from a gymnosperm vascular system to that of a typical (non-monocot) angiosperm but not from this type to that of a monocot. This also indicates the need to use genomic tools to better understand how the standard angiosperm habit became so reorganized that understanding of its homology is unclear.

The fact that the vascular bundles of monocots are scattered rather than arranged in a ring as in the woody basal angiosperms and eudicots means that orderly, bifacial growth of the vascular cambium is not possible. That is, secondary growth does not occur in monocots as it does in eudicots and magnoliids. As a result, many arborescent monocots achieve their stature by primary gigantism. Palm "trees" (Arecaceae) and grass trees (*Kingia*, members of Dasypogonaceae) are overgrown herbs with primary "woody" stems; once laid down, the stems cannot increase in diameter through secondary accretion. Stems of palms and others such as the pandans (Pandanaceae) do become stiffer with age by means of increased cell-wall thickness and lignin deposits, but this is an entirely different process from that occurring in plants with a bifacial cambium.

However, primary gigantism is not the only method by which monocots can achieve tree status. Other arborescent monocots, which are almost all members of Asparagales (see below), such as *Agave*, *Cordyline*, *Dracaena* (all Asparagaceae), and *Aloe* (Asphodelaceae), achieve their large stature by means of an "anomalous" secondary growth produced by an etagen or tiered cambium. This growth is unidirectional and therefore still not like that of other angiosperms. Outside of Asparagales, an etagen cambium is reported only in *Dioscorea* (Dioscoreales). The inability to produce a well-organized bifacial cambium has limited the evolution of growth form in the monocots, but they nevertheless exhibit considerable habit diversity. How this radical reorganization of the monocot stem occurred and from what sort of ancestral state it evolved are major unanswered problems in angiosperm evolution.

Another widely cited monocot character is their particular form of sieve cell plastids (Behnke 1969), which are triangular and have cuneate proteinaceous inclusions. Similar sieve cell plastids are found in Aristolochiaceae (Piperales; Dahlgren et al. 1985). We now infer that this similarity between monocots and Aristolochiaceae represents convergence, not shared ancestry, because DNA phylogenetic studies have demonstrated a strongly supported relationship of Aristolochiaceae to other Piperales within the magnoliid clade to the exclusion of monocots (Chapter 3).

Trimerous flowers have long been considered a uniting characteristic of the monocots, but it is not an exclusive feature because many other basal angiosperms, including Nymphaeaceae and magnoliids, also exhibit trimery. Character-state reconstructions indicate that trimery arose early in the angiosperms; it may be ancestral for angiosperms branching above *Amborella* (see Chapter 6). Trimery appears therefore to be a symplesiomorphic feature in monocots and other angiosperms and is not therefore a "monocot character." Given the number of monocots, most with trimerous floral organization, and the relatively small number of trimerous non-monocots, this character still works most of the time for identification, but it should always be used in combination with the other traits discussed here.

An often-overlooked synapomorphy for monocots is their sympodial growth. There are other sympodial angiosperms, but monocots are nearly exclusively so. Even branched, arborescent genera, such as *Aloe* (Asphodelaceae), are sympodial; new sympodia arise near the apex of the previous one and "displace" the terminal inflorescence into a lateral position, but these plants are nonetheless sympodial. Branching in arborescent monocots is achieved by production of more than one terminal sympodium, but such branching is limited by the demands it makes on the vascular system of older sympodia, which cannot expand to meet increased requirements (see Tomlinson 1995).

PHYLOGENETIC ANALYSES USING MORPHOLOGY

Analysis of morphological characters to estimate relationships within monocots has been limited to those of Stevenson and Loconte (1995; 101 characters) and Chase et al. (1995b; 103 characters, compiled collectively by the participants of the 1993 Monocot Symposium at the Royal Botanic Gardens, Kew). These analyses reached similar conclusions about higher-level relationships in the monocotyledons, based on similar information. Both analyses used the same outgroups (Nymphaeales, Piperales, Lactoridales, Aristolochiales, sensu Takhtajan), but arranged them in different ways. Stevenson and Loconte (1995) specified *Lactoris* (as Lactoridales at that time) as the ultimate outgroup; Chase et al. (1995b) used a polytomy. As a result of this difference in rooting, Aristolochiales alone appeared as the sister group of the monocots in Stevenson and Loconte (1995). Stevenson and Loconte (1995) then claimed that their analysis demonstrated that Aristolochiales were the sister group of the monocots, but this conclusion was not a result of their analysis but a product of assumptions made before the analysis about how to arrange the outgroups. Chase et al.'s (1995b) results could have been arranged in the same way.

The alismatids have commonly been considered as the most "primitive" monocots (Hallier 1905; Arber 1925; Hutchinson 1934; Cronquist 1968, 1981; Takhtajan 1969, 1991; Stebbins 1974; Thorne 1976). Net-veined groups, such as Dioscoreales (Dahlgren et al. 1985) and Melanthiales (Thorne 1992a,b), have also been considered most "primitive" among the monocots. As we stress throughout this book, it is pointless to refer to living taxa as "primitive" because all are mosaics of primitive and derived characters. The issues of which extant group exhibits the most "primitive" morphological traits and which is the sister of all others are not equivalent, although these statements have often been confounded. We can infer ancestral character states via character-state reconstruction using the best estimate of phylogeny, as we have done throughout.

Characters ancestral for monocots could be present in some derived groups via secondary derivation. Concomitantly, basal taxa can exhibit many morphological autapomorphies (derived traits; see Chapters 3, 4, and 6). Although *Acorus* has now been shown to be sister to all other monocots (e.g., Chase et al. 1993, 1995a,b, 2000a, 2006; Duvall et al. 1993a,b; Davis et al. 1998; Fuse and Tamura 2000; Tamura et al. 2004; Givnish et al. 2005; Soltis et al. 2011; Zeng et al. 2014), this does not imply that *Acorus* is "the most primitive monocot" in morphological terms. In fact, character-state reconstructions indicate that *Acorus* in general is highly derived in most morphological characters, but many of these are autapomorphies.

A major effect of using a few basal angiosperms such as Aristolochiaceae as outgroups in strictly morphological analyses is a rooting among the monocots with net-veined leaves. The net-veined monocots were lumped in taxonomic treatments from the mid-1980s to mid-1990s (e.g., Cronquist 1981), and were largely referable to Dioscoreales sensu Dahlgren et al. (1985). The net-veined monocots include members of Dioscoreaceae, Smilacaceae, Stemonaceae, and Melanthiaceae (especially tribe Paridae, often treated as family Trilliaceae), as well as several smaller families. Any other rooting would have been difficult to imagine because most of the other major groups of monocots, including *Acorus* and the alismatids, have no counterpart in any other group of angiosperms; their habits are unique to monocots, a feature that is likely to polarize characters in such a manner as to make these groups occupy a derived place in cladograms based solely on morphology. Thus, it should have been no surprise to see *Acorus* and the alismatids occupy relatively derived positions in the trees produced by Chase et al. (1995b) and Stevenson and Loconte (1995).

One advantage of using characters totally divorced from morphology in phylogenetic studies is that this approach can lead to new perspectives on old problems. Such is the case for the hypotheses that the alismatids are "primitive" monocots and that monocots were ancestrally aquatic (Henslow 1893; Hallier 1905; Arber 1925; Hutchinson

1934; Cronquist 1968, 1981; Takhtajan 1969, 1991; Stebbins 1974; Thorne 1976). The position occupied by alismatids in the DNA analyses (see below) is consistent with them exhibiting a preponderance of traits that are ancestral among monocots and with the hypothesis that monocots had an aquatic origin, but neither of these ideas is one that could be reached by outgroup comparison. Ideas of these earlier authors were influenced by the morphological similarities of alismatid taxa such as *Sagittaria* and *Butomus*, which have magnoliid-type flowers with follicular fruits.

Plesiomorphic characters in both Chase et al. (1995b) and Stevenson and Loconte (1995) included stem vessels present, leaf petiole of the Dioscoreales type, primary venation palmate, secondary venation reticulate, and a eustele. These are all traits associated with plants occurring in forest understory, and their occurrence is due to convergence (Givnish et al. 2006). Although not used in these analyses, a vining habit could be added to this list; Dioscoreales sensu Dahlgren et al. (1985) are mostly understory vines. Reproductive traits were also highly influenced in these analyses by the choice of outgroups (i.e., Nymphaeales, Piperales, Lactoridales, Aristolochiales sensu Takhtajan), and the following traits were plesiomorphic: differentiated perianth, stamen connectives protruding, lack of septal nectaries, parietal placentation, and septicidal capsules (i.e., a syncarpous fruit). These are not characters that would be considered ancestral within monocots and are either absent in, or not applicable to, the alismatids. Stevenson and Loconte (1995) posited Stemonaceae and Trilliaceae (now a synonym of Melanthiaceae) as the most early-diverging monocots. Taccaceae (now considered a synonym of Dioscoreaceae; APG IV 2016 and Chapter 12) exhibited more ancestral states in Chase et al. (1995b).

Both morphological analyses then followed with either a grade (Chase et al. 1995b) or clade (Stevenson and Loconte 1995) of the lilioid monocots (treated as Asparagales and Liliales below, and including Zingiberales in the trees of Stevenson and Loconte 1995) sister to a clade of alismatids and commelinids. Characters for the lilioid clade were not specified, but the short branch shown in Stevenson and Loconte (1995) indicates that there were probably only two synapomorphies. In the clade sister to the lilioids of Stevenson and Loconte (1995) was a grade of taxa associated with either Velloziaceae or Bromeliaceae followed by clades composed of alismatids, including Araceae, and commelinids (see below), except for Zingiberales, but including *Acorus* as sister to *Typha* and *Sparganium* (Typhaceae). In the Chase et al. (1995b) results, the lilioid grade was followed by a clade of Bromeliaceae/Velloziaceae, Zingiberales, and alismatids and commelinids; Araceae were again associated with alismatids and *Acorus* with Sparganiaceae and Typhaceae. These associations of Alismatales (including Araceae) and *Acorus* with Typhaceae and commelinids were mostly due to concerted convergence on traits adaptive in aquatic or semi-aquatic habitats (several Commelinales, in particular, exhibit similar aquatic adaptations; e.g., Philydraceae and Pontederiaceae).

Neither the Chase et al. (1995b) nor Stevenson and Loconte (1995) morphological analysis surmounted the issues of how to address the unique characters of the monocots, particularly those of *Acorus* and Alismatales, and the problems this posed for determining the root by outgroup comparison. The many earlier authors who supported the hypothesis of the "primitiveness" of the alismatids did not use an outgroup approach to arrive at this conclusion. Tomlinson (1995) likewise had problems with the use of an extant "dicot" as a model starting point for addressing the issue of monocot vegetative architecture.

MOLECULAR ANALYSES

Due to intensive molecular study and frequent collaboration, the monocots emerged early on as the best-characterized major clade of angiosperms (Chase 2004; Chase et al. 2006; Graham et al. 2006). The earliest well-sampled DNA analyses of monocots were the *rbcL* analyses of Duvall et al. (1993a,b) and Chase et al. (1993). These analyses concluded that *Acorus* was sister to the rest of the monocots, but several critical taxa were still missing (e.g., Triuridaceae) or under- represented (e.g., only three alismatids were included). Duvall et al. (1993a,b) and Chase et al. (1993) did include at least some representatives of each of the three main contenders for "the most primitive monocot": alismatids, *Veratrum* (Melanthiaceae), and Dioscoreales. Bharathan and Zimmer (1995) found that *Acorus* did not fall with the other monocots in their study of partial 18S rDNA sequences, but they did not sample many monocots (only 24). A large-scale sampling of 18S rDNA sequences for monocots did not emerge until the studies of D. Soltis et al. (2000) and Chase et al. (2000a); the latter used 159 species of monocots.

In contrast to the morphological analyses of angiosperm relationships (Donoghue and Doyle 1989a,b; Loconte and Stevenson 1991; Doyle and Donoghue 1992), the *rbcL* analyses produced radically different ideas about the relationships of monocots. Morphological analyses alternated between two positions for the monocots (e.g., Donoghue and Doyle 1989a,b): sister to either Nymphaeaceae or Aristolochiaceae. These studies coded monocots as a single terminal (as was generally done in morphological studies at that time to reduce the number of terminal taxa; see Chapter 3) and so presumed ideas of what constituted ancestral monocot traits undoubtedly played a major role in where the monocots were positioned. For the reasons stated above, these studies have a degree of circularity that is difficult to avoid, but certainly such problems are exacerbated by the use of single, synthetic terminals coded with putatively ancestral monocot traits. Use of *rbcL* avoided such circularity and arrived at novel conclusions, not only for the position of monocots, but also for relationships within the monocots. In Chase et al. (1993), the monocots were positioned with some relationship to the magnoliids: Laurales, Magnoliales, and Piperales (at that time termed the "paleoherb I clade"), but none of these relationships received bootstrap support above 50% (Chase and Albert 1998).

Since the Chase et al. (1993) analysis, several studies of single, as well as combined, gene datasets have been analyzed (D. Soltis et al. 1998, 2000, 2011; Qiu et al. 1999; P. Soltis et al. 1999a; Savolainen et al. 2000a; Zanis et al. 2002, 2003; Hilu et al. 2003; Duvall et al. 2006; Zeng et al. 2014), and only the last two of these studies plus Soltis et al. (2011) provided clear evidence for monocot relationships (see also Chapter 3). What is clear is that an exclusive relationship to either Nymphaeaceae or Aristolochiaceae is excluded.

RELATIONSHIPS AND CHARACTER EVOLUTION WITHIN MONOCOTS

Monocot phylogenetics have made immense strides over the past 20 years due primarily to the foci provided by the international monocot symposia held in 1993, 1998, 2003, 2008, and 2013 (at the Royal Botanic Gardens, Kew, Rudall et al. 1995; the Royal Botanic Garden, Sydney, Wilson and Morrison 2000; Rancho Santa Ana Botanical Garden, Columbus et al. 2006; Natural History Museum of Denmark, Seberg et al. 2010; and the New York Botanical Garden, respectively). These meetings have focused attention both on what was known and, more importantly, on which groups needed additional research. As a result, we now know more about monocots than about any other major group of angiosperms, a remarkable achievement given the paucity of information available in 1985 (Dahlgren et al. 1985).

The overview presented here (Fig. 7.1) is based in large part on the three-gene analyses of Chase et al. (2000a) and D. Soltis et al. (2000), the seven-gene analysis of Chase et al. (2006), a large plastid DNA dataset of Graham et al. (2006), and a three-genome analysis of Hertweck et al. (2015, which used the seven genes of Chase et al. 2006, plus low-copy nuclear gene *PHYC*). The two most recent broad analyses of angiosperms, Soltis et al. (2011) and Zeng et al. (2014), have included enough monocots to confirm the general patterns observed in the monocot-focused earlier studies. Givnish et al. (2010) used 81 plastid genes for 83 taxa, focusing on the commelinid order Poales. Kim et al. (2013) focused on Liliales and Seberg et al. (2012) and S. Chen et al. (2013) on Asparagales. Chase et al. (2006) included plastid *rbcL*, *atpB*, *matK*, and *ndhF*, mitochondrial *atpA*, 18S rDNA, and partial 26S rDNA. The patterns obtained in Chase et al. (2006) are like those in the other analyses, but the sampling was the most complete, except for the narrower work focusing on Poales (Givnish et al. 2010), Alismatales (Iles et al. 2013; Les and Tippery 2013), Liliales (Kim et al. 2013), and Asparagales (Seberg et al. 2012; S. Chen et al. 2013). Davis et al. (2013) assembled a large matrix of taxa but had reasonably complete sampling for only *rbcL* and *matK* and a set of 60 plastid genes for about one-third of the taxa. Their interests were in examining contrasting patterns of support among these datasets rather than examining relationships of monocots per se. However, their overall results do not alter our ideas about relationships, and so this paper will not be discussed further. Below, we mention newer results only when they offer different relationships or stronger support than the earlier well-sampled (taxonomically) studies.

All but two DNA analyses of monocots have placed *Acorus* alone as sister to all other monocots (Fig. 7.1). The first exception was the18S rDNA analysis of Bharathan and Zimmer (1995), in which *Acorus* was placed outside of the monocots, a result that must be considered spurious. Combining the 18S rDNA sequence data with sequences from *rbcL* and *atpB* (Chase et al. 2000a; D. Soltis et al. 2000) resulted in strong support for the monophyly of monocots, as well as strong support for the monophyly of all monocots excluding *Acorus*. A re-analysis of *rbcL* and *atpA* (Davis et al. 2004) recovered an alismatid clade that contained *Acorus*. This deviating result is perplexing because neither *rbcL* (Chase et al. 1993, 1999; Duvall et al. 1993a,b) nor *atpA* (Davis et al. 1998) analyzed alone produced such a position for *Acorus*. In contrast, studies that have used more genes have consistently found *Acorus* sister to the remaining monocots (e.g., Qiu et al. 1999, 2000; Zanis et al. 2002, 2003; Soltis et al. 2011; Iles et al. 2013; Zeng et al. 2014).

The placement of *Acorus* as sister to all other monocots (Fig. 7.1; see also Fuse and Tamura 2000) strongly contradicts the morphological analyses of Chase et al. (1995b) and Stevenson and Loconte (1995), but is consistent with the combined morphology/*rbcL* analysis of Chase et al. (1995b). Chase et al. (1995b) noted that the ingroup networks of the morphological and molecular trees were highly similar and that they differed most significantly in the point at which they were rooted by the outgroups. The potentially misleading effects of outgroup rooting from morphology-based data are discussed above. Chase et al.

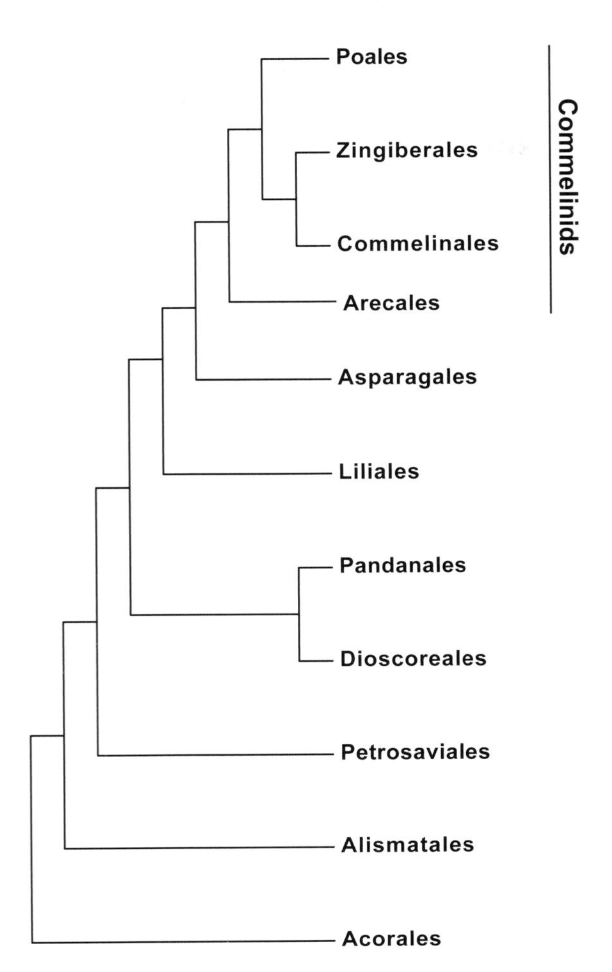

Figure 7.1. Summary tree for relationships among the orders of monocots. *The position of Dasypogonales as shown here is based on the results of Hertweck et al. (2015), which is contradicted by those of Givnish et al. (2010), but in the latter study the position of Dasypogonales was weakly supported.

(1995b) also pointed out that the aquatic alismatids and the largely aquatic (helophytic) aroids and *Acorus* were nested in the morphological trees among other aquatic or semiaquatic taxa such as Pontederiaceae, Philydraceae, and Typhaceae. Discounting these two differences, all other patterns of relationship were similar.

Following *Acorus*, monophyly of the remaining monocots is again strongly supported. Alismatales are sister to the remaining monocots (Fig. 7.1), and the remaining monocots form a well-supported clade. Within this remaining large clade are several component subclades: commelinids (including Arecales [now including Dasypogonaceae; see Chapter 12], Commelinales, Poales, and Zingiberales), Dioscoreales, Petrosaviales, Pandanales, Liliales, and Asparagales. All these groups now receive maximum support (in MP, ML, and Bayesian studies), and in general their interrelationships also receive strong support (Fig. 7.1), with the exception of Dasypogonaceae and Arecales, which even with the 81 plastid genes (Givnish et al. 2010) received only 57% and 56% bootstrap support as sister to Commelinales/Zingiberales and Poales, respectively. The branching order above Alismatales is Dioscoreales/Pandanales, Liliales, and Asparagales/commelinids. The large low-copy nuclear dataset of Zeng et al. (2014) placed Liliales and Asparagales as sister taxa, but although support was high, taxon sampling was low. Soltis et al. (2011) recorded only moderate bootstrap support (86%) for the position of Liliales, but Asparagales as sister to the commelinids received bootstrap support of 94%. Within the commelinids, relationships of Arecales (including Dasypogonaceae, which was placed in Dasypogonales) relative to Commelinales/Zingiberales and Poales must still be considered unresolved. Only the eight-gene matrix of Hertweck et al. (2015) resolved the base of the commelinids with a relevant level of support; they placed Dasypogonaceae as sister to Arecaceae with 81% bootstrap support and this pair of families sister to the rest of the commelinids with 92% bootstrap support.

ACORALES

Acorales consist of the single family Acoraceae and one genus *Acorus*. Prior to the molecular era, *Acorus* was typically considered an aberrant member of the aroid family, Araceae (Grayum 1987). On the basis of its highly divergent seedling morphology, Tillich (1985) stated that *Acorus* should be excluded from Araceae (Acoraceae was proposed by Martynov in 1820; see APG II 2003). The treatment followed here, placing *Acorus* in its own family and monofamilial order, Acorales, has now been widely accepted (see Chapter 12). *Acorus* has many atypical morphological features for a basal monocot, but it also has some that might be expected: dicotyledonous anther wall formation and ethereal oils, both of which are unique to *Acorus* among monocots at the basal nodes. Its habit with unifacial, ensiform leaves, inflorescence structure (a spadix-like structure of flowers with an undifferentiated perianth), syncarpous ovary, and berry-like fruits are not what would be expected for a "primitive" monocot based on traditional views (Fig. 7.2a).

ALISMATALES

Alismatales include Alismataceae, Aponogetonaceae, Araceae, Butomaceae, Cymodoceaceae, Hydrocharitaceae, Juncaginaceae, Maundiaceae, Posidoniaceae, Potamogetonaceae, Ruppiaceae, Scheuchzeriaceae, Tofieldiaceae, and Zosteraceae (several of these are illustrated in Fig. 7.2, and their relationships are depicted in Fig. 7.3). The monophyly of the order is well supported by molecular analyses (e.g., Chase et al. 1993, 1995a, 2000c; Duvall et al. 1993a; D. Soltis et al. 2000; Iles et al. 2013; Les and Tippery 2013).

Figure 7.2. Acorales (Acoraceae), Petrosaviales (Petrosaviaceae), Alismatales (Alismataceae, Araceae, Butomaceae, Cymodoceaceae, Hydrocharitaceae, Juncaginaceae, Potamogetonaceae, Ruppiaceae). a. *Acorus calamus* L. (Acoraceae), spadix-like inflorescence. b. *Japonolirion osense* Nakai (Petrosaviaceae), inflorescence. c. *Sagittaria lancifolia* L. (Alismataceae), staminate and carpellate flowers, and milky latex (inserted); note flowers with calyx and corolla, and numerous, distinct stamens and carpels. d. *Hydrocleys nymphoides* (Willd.) Buchenau (Alismataceae), perfect flower. e. *Butomus umbellatus* L. (Butomaceae), flowers and umbellate inflorescence; note that perianth parts of both the outer and inner whorls are petaloid but different in shape. f. *Triglochin maritima* L. (Juncaginaceae), spicate inflorescence; note inconspicuous flowers with penicellate stigmas. g. *Ruppia megacarpa* R. Mason (Ruppiaceae), stipitate fruits. h. *Potamogeton nodosus* Poir (Potamogetonaceae), inflorescence of sessile, four-merous flowers. i. *Potamogeton americanus* Roem. & Schult. (Potamogetonaceae), aquatic habit. j. *Limnobium spongia* Steud. (Hydrocharitaceae), carpellate flower; note inferior ovary and divided styles. k. *Syringodium filiforme* Kütz. (Cymodoceaceae), aquatic habit. l. *Anthurium superbum* Madison (Araceae), inflorescence a spathe and spadix.

This circumscription of Alismatales includes those families that most authors have previously placed in the order (e.g., Cronquist 1981; Thorne 1992a,b; Takhtajan 1997) plus Araceae and Tofieldiaceae. At least some authors on morphological grounds referred Araceae to their own order with a possible close relationship to Arecales (e.g., Cronquist 1981); hence, the results of all molecular studies that placed Araceae with Alismatales were unanticipated (Fig. 7.1). The placement of Tofieldiaceae in this clade was also unexpected; these genera had been considered a subfamily or tribe of Melanthiaceae.

Families of Alismatales share scales or glandular hairs at the nodes within the sheathing leaf bases, and the embryos are uniquely green. Araceae (Fig. 7.2l) and Tofieldiaceae (3 genera, 27 species; North and South America, as well as Eurasia) are otherwise different from the remaining families in their inflorescence, floral structure, and habit (they are not aquatic and lack lacunae in their stems). Flowers of Tofieldiaceae are nondescript and typical of monocots, except that some species have up to 12 stamens and a septicidal capsule. A "calyculus" (a calyx-like structure composed of bracts or bractlets) underneath the flowers is present in most species.

Araceae are by far the largest family of the order (106 genera, 4,025 species; cosmopolitan); they are particularly diverse in the wet tropics, but some temperate genera (e.g., *Arisaema*) are also species-rich. Typically, Araceae are herbs with a distinct petiole and expanded lamina; their most distinctive feature is their inflorescence, which has an often petaloid spathe and spadix of sessile flowers with an expanded sterile terminal portion that emits odors (Fig. 7.2l). Many genera have separate male and female zones on the spadix. The duckweeds, five genera formerly placed in Lemnaceae (Les et al. 1997b; Rothwell et al. 2003), are embedded in Araceae and included there; these are the smallest of all angiosperms. Even in Soltis et al. (2011) and Iles et al. (2013), the position of Tofieldiaceae relative to Araceae and the core alismatids continued to be unresolved.

The remaining Alismatales have often been termed "core Alismatales" and are well supported by DNA sequence data as well as non-DNA characters (Fig. 7.3). All these families have seeds that lack endosperm and root hair cells that are shorter than other epidermal cells (Judd et al. 2002). Members of this large subclade occur in wetland or aquatic habitats, and Judd et al. (2002, 2008) referred to these families as the aquatic clade; Les and Tippery (2013) referred to them as Alismatidae and the subclades within as orders. The aquatic clade is divided into two subclades. The first subclade, the petaloid clade, is smaller and comprises Alismataceae (Fig. 7.2c,d), Butomaceae (Fig. 7.2e), and Hydrocharitaceae (Fig. 7.2j); these have a scapose inflorescence, a differentiated calyx and corolla (in most), and exotestal seeds. Alismataceae (12 genera; 88 species, including Limnocharitaceae; Soros and Les 2002; Iles et al. 2013; Les and Tippery 2013) are emergent aquatics and are linked by having latex, leaves with a pseudopetiole with a midvein and cross veins, spinose-pantoporate pollen, and a follicular fruit. Hydrocharitaceae are the largest family of this subclade (18 genera; 116 species) and are submersed aquatics, some of which are marine (e.g., *Halophila*, *Enhalus*, and *Thalassia*), with petiolate, usually undifferentiated leaves, an inflorescence with two bracts underneath, and an inferior ovary (the rest of this subclade has superior ovaries). Pollination biology of this family is particularly diverse (Tanaka et al. 2013). Butomaceae (monospecific) are monopodial with two-ranked, three-angled leaves and an umbel-like inflorescence. As in Araceae and Alismataceae, members of Butomaceae have a differentiated perianth and a follicular fruit.

The second subclade, the tepaloid clade, consists of Aponogetonaceae, Cymodoceaceae (Fig. 7.2k), Juncaginaceae (Fig. 7.2f), Maundiaceae, Posidoniaceae, Potamogetonaceae (Fig. 7.2h,i), Ruppiaceae (Fig. 7.2g), Scheuchzeriaceae, and Zosteraceae (Fig. 7.3). Scheuchzeriaceae are highly

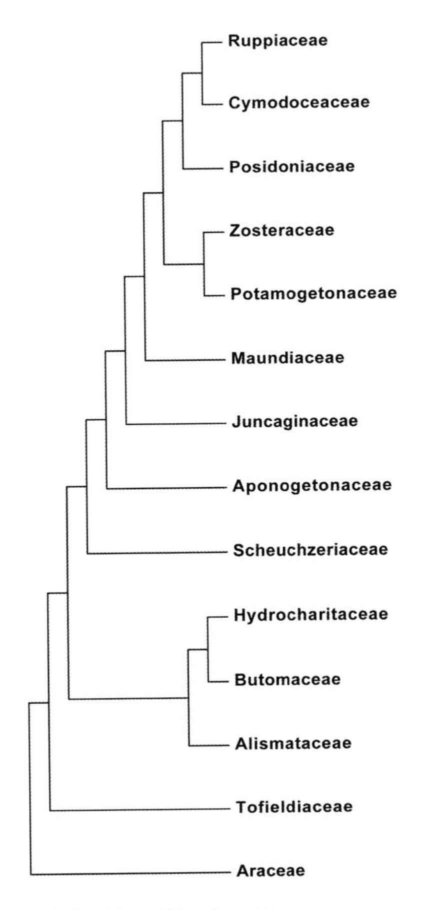

Figure 7.3. Relationships within Alismatales

apomorphic and not clearly linked by morphological characters to the rest of this subclade. The family is monospecific and has two-ranked, auriculate leaves, and a racemose inflorescence with several bracts; as in the other subclade, the fruit is a follicle. The rest of the subclade shares more or less linear leaves and (except for Aponogetonaceae) united carpels and an unusual condition in which the perianth appears to be an "outgrowth" of the stamens. Aponogetonaceae are monogeneric (43 species) and typically have petiolate leaves, parallel venation and cross veins, and small spicate flowers with tepals. Juncaginaceae (3 genera; 14 species) and Maundiaceae (1 species) are emergent aquatic to terrestrial plants (but still of wet sites) and have imperfect flowers often lacking a perianth or with the perianth internal to the stamens. Maundiaceae were only recently recognized as a distinct family (von Mering and Kadereit 2010; Iles et al. 2013); formerly included in Juncaginaceae, they fall as sister to the clade of Cymodoceaceae, Posidoniaceae, Potamogetonaceae, Ruppiaceae, and Zosteraceae, and we have thus recognized them here. The rest of the families are all rhizomatous, water-pollinated aquatics, with ascidiate carpels. Posidoniaceae (monogeneric; 5 species), Ruppiaceae (perhaps monospecific), and Cymodoceaceae (5 genera; 16 species) are largely marine with two-ranked leaves and filiform pollen. Ruppiaceae (Fig. 7.2g) might better be included in Posidoniaceae; these families are highly similar in most respects. Zosteraceae (2 genera; 14 species) and Potamogetonaceae (7 genera; 122 species; Fig. 7.2h,i) share a leaf with an apical pore (i.e., their sheaths are closed), although the former is marine and the latter in freshwater. Zosteraceae are seagrasses with linear leaves, leaf-opposed branches, and spadix-like inflorescences enclosed in a spathe (see analysis by Les et al. 2002). Potamogetonaceae have petiolate leaves with a midvein and a spicate inflorescence densely packed with flowers in which there is a tepal opposing each stamen. Pollination is commonly by wind, on the water surface, or in the water.

Some have suggested that the "flowers" of Alismatales, except for Araceae and Tofieldiaceae, are modified inflorescences (Buzgo 2001; Rudall 2003), which could explain the wide variety of floral types in these families. For a small clade, floral diversity is exceptionally high in Alismatales, and their pollination mechanisms are among the most bizarre in the angiosperms (Cox and Humphries 1993).

Alismatales contain several families adapted to marine habitats. This feature has apparently evolved independently several times in the order, once in Hydrocharitaceae, and perhaps several additional times in the seagrass families Zosteraceae, Cymodoceaceae, Ruppiaceae, and Posidoniaceae (Les et al. 1997a). Marine angiosperms are confined to this order.

PETROSAVIALES

Relationships of Petrosaviaceae (*Petrosavia*, 3 species, and *Japonolirion*, one species, Fig. 7.2b) were uncertain until Chase et al. (2006). They are strongly supported as the sister group of all monocots except Acorales and Alismatales (Fig. 7.1; Chase et al. 2006; Graham et al. 2006), which necessitates their elevation to an order. Until DNA placed the two genera together (Chase et al. 2000a), *Petrosavia* and *Japonolirion* had not been considered closely related. Dahlgren et al. (1985) placed *Petrosavia* in Melanthiales, but did not list *Japonolirion* at all. Takhtajan (1997) placed Petrosaviaceae in their own order in Triurididae and Japonoliriaceae in Melanthiales of Liliidae. Cameron et al. (2003) reviewed the morphology of *Petrosavia* and *Japonolirion* and found several similarities, including a lack of completely fused carpels. *Japonolirion* is photosynthetic, whereas *Petrosavia* is achlorophyllous and mycotrophic. Both are rhizomatous herbs with spirally arranged leaves or bracts on racemes; microsporogenesis is simultaneous.

PANDANALES

In Chase et al. (2006), Dioscoreales + Pandanales was found to be sister to all remaining monocots (Figs. 7.1, 7.4). A close relationship of Pandanales to Dioscoreales was not observed until Chase et al. (2006), in which this pair of orders received 92% bootstrap support; it has been well supported in all subsequent analyses. However, non-DNA characters linking these two orders are unknown.

Pandanales comprise Cyclanthaceae (12 genera, 225 species; tropical America; Figs. 7.4, 7.5j), Pandanaceae (3 genera, 805 species; West Africa to the Pacific; Fig. 7.5h), Stemonaceae (3 genera, 25 species; tropical East Asia to Australia with a disjunct genus in eastern North America; Fig. 7.5i), Triuridaceae (8 genera, 48 species; pantropical), and Velloziaceae (9 genera, 240 species; nearly pantropical; Fig. 7.5g). Within Pandanales, Cyclanthaceae and Pandanaceae form a well-supported clade to which Stemonaceae are sister. The closest relatives of Triuridaceae and Velloziaceae within Pandanales remain unclear (see below).

A close relationship of Cyclanthaceae (Fig. 7.5j) and Pandanaceae (Fig. 7.5h) had long been hypothesized (e.g., Dahlgren et al. 1985); the two families are sisters in the morphological analyses of Chase et al. (1995b) and Stevenson and Loconte (1995). Cyclanthaceae and Pandanaceae share compound bundles in their stems, imperfect flowers, an indehiscent syncarpous fruit, and a nonphotosynthetic cotyledon. These two families have several instances of

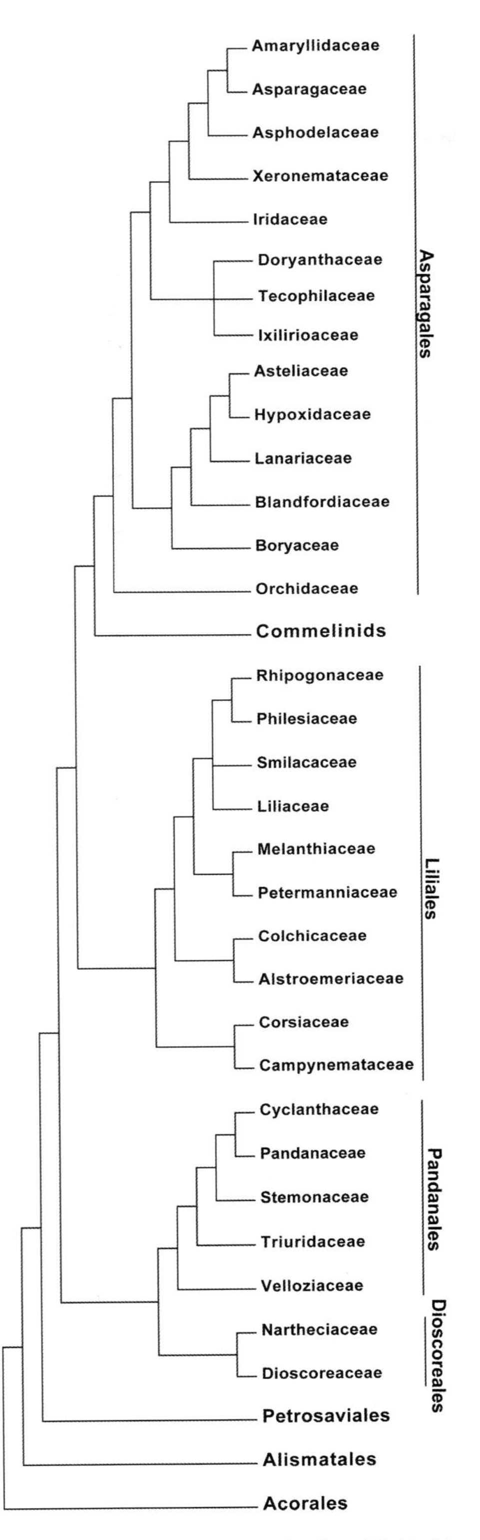

Figure 7.4. Summary tree for the orders and families of lilioids, Dioscoreales, Pandanales, Liliales, and Asparagales.

gigantism and can be either long vines (Cyclanthaceae and Pandanaceae) or large tree-like herbs (Pandanaceae).

Stemonaceae, a small family of mostly vines, have previously been associated with Dioscoreaceae (e.g., Cronquist 1981; Dahlgren et al. 1985); their association with Cyclanthaceae and Pandanaceae in molecular analyses was unexpected. In Chase et al. (2000a) and D. Soltis et al. (2000), Pandanales received very strong (99% bootstrap or jackknife) support. Cyclanthaceae and Pandanaceae share tetramerous (or at least not trimerous) flowers and parietal placentation; they differ in having broad leaves with a petiole and net venation and a dehiscent fruit. Velloziaceae had been frequently placed with Bromeliaceae, as in Dahlgren et al. (1985). In D. Soltis et al. (2000), Velloziaceae are sister to Stemonaceae, a relationship that received low (75% jackknife) support, but in Chase et al. (2006), the former is well supported (100% bootstrap) as sister to all other Pandanales if Triuridaceae (represented only by 18S rDNA sequence data; see next paragraph) were excluded from the analysis. Graham et al. (2006) also had Velloziaceae well supported as sister to the rest of Pandanales. Characters supporting the position of Velloziaceae in Pandanales are unknown; all Pandanales have small embryos and a nucellar cap. Velloziaceae can become woody, but they achieve this as in Pandanaceae, by being giant herbs with toughened stems. Flowers of Velloziaceae (Fig. 7.5g) are the most plesiomorphic in Pandanales (Sajo et al. 2013), which explains why they were difficult to place, given that many of their characteristics are also plesiomorphic across monocots. Some (e.g., Behnke et al. 2000) have considered *Acanthochlamys* (one species, from China) too different from the rest of Velloziaceae to be included in the family (and treated it as Acanthochlamydaceae), but its position as sister to Velloziaceae s.s. is well supported, so it seems better to keep it as a divergent genus in the family (Mello-Silva et al. 2011).

The position of Triuridaceae is mostly based on 18S rDNA sequence data (Chase et al. 2000a, 2006), and this relationship was also unsuspected; most authors considered Triuridaceae to be related to Alismatales because they shared several characters with those families (and were placed there in the morphological analyses of Chase et al. 1995b, but not in Stevenson and Loconte 1995). The position of the anthers inside the carpel whorl in *Lacandonia* and the complexity and diversity of flowers in Triuridaceae have led some authors to speculate that, like several other families of Pandanales (e.g., Cyclanthaceae and Pandanaceae), the "flowers" of Triuridaceae are really inflorescences (Rudall 2003). In Mennes et al. (2013) and Hertweck et al. (2015), *Sciaphila* (Triuridaceae) is sister to the rest of Pandanales except for Velloziaceae (Fig. 7.4).

Figure 7.5. Dioscoreales (Dioscoreaceae, Nartheciaceae) and Pandanales (Cyclanthaceae, Pandanaceae, Stemonaceae, Triuridaceae, Velloziaceae). a. *Burmannia coelestis* D. Don (Dioscoreaceae), habit and flower; note small leaves, flowers with inferior and winged ovary. b. and c. *Dioscorea bulbifera* L. (Dioscoreaceae), habit, a twining vine with tubers; note petiolate leaves with palmate venation, the major veins converging and connected by a network of higher-order veins (b), carpellate flowers, note inferior, winged ovaries (c). d. *Tacca integrifolia* Ker Gawl. (Dioscoreaceae), habit and umbellate inflorescence with broad bracts and filamentous bracts. e. *Thismia* sp. (Dioscoreaceae), mycotrophic habit, flowers with apex of inner perianth whorl long and slender, connate basally. f. *Aletris lutea* Small (Nartheciaceae), racemose inflorescence of small flowers with connate tepals. g. *Vellozia bahiana* L. B. Sm. & Ayensu (Velloziaceae), large flowers with inferior ovary. h. *Pandanus tectorius* Parkinson (Pandanaceae), carpellate plant with drupaceous, multiple fruit; note also woody habit, spirally, three-ranked, spiny, M-folded leaves. i. *Stemona tuberosa* Lour. (Stemonaceae), petiolate leaves with palmate venation and flowers with four tepals and carpels/styles. j. *Carludovica palmata* Ruiz & Pav. (Cyclanthaceae), spadix-like infructescence (of berries).

DIOSCOREALES

Dioscoreales have a long history of recognition, but not in the circumscription used in APG (APG 1998; APG II 2003; APG III 2009; APG IV 2016). In APG III (2009) and APG IV (2016), the order included only Burmanniaceae, Dioscoreaceae, and Nartheciaceae. Some may want to include Burmanniaceae in Dioscoreaceae because Merckx et al. (2006) demonstrated a more complex set of relationships than had previously been shown by Caddick et al. (2002a). Most previous circumscriptions (e.g., Dahlgren et al. 1985) included in Dioscoreales nearly all lilioid monocots with net-veined leaves, and thus they comprised a set of families now known to be unrelated, such as Smilacaceae, Stemonaceae, and Trilliaceae. Nartheciaceae were not included in Dioscoreales by Chase et al. (1993, 1995a,b), but other analyses have consistently placed them as sister to Dioscoreaceae (Chase et al. 2000c, 2006; Caddick et al. 2002a), although not with bootstrap support above 70% until the seven-gene analysis of Chase et al. (2006) put them there with 99% bootstrap support. Nartheciaceae are a small family (four or five genera; Fig. 7.5f) that deviates in morphology from the general patterns found in the other family of the order, Dioscoreaceae, but they do share with Dioscoreaceae a perianth persistent on the fruit, as well as micromorphological characters (e.g., a tanniferous exotegmen and a short embryo). Like Tofieldiaceae (see above), Nartheciaceae are relatively poorly studied.

Dioscoreaceae in the broad sense (17 genera; 926 species; Fig. 7.5c,d) could include Burmanniaceae (13 genera; 126 species). Dioscoreaceae s.s. and Burmanniaceae share reflexed stamens, inferior, winged ovaries (although some genera of each family lack this feature), and simultaneous microsporogenesis (the last trait rare in monocots); the groups are otherwise dissimilar, but this would not be unexpected given their different life-history strategies. Burmanniaceae are rhizomatous mycoparasites, but some of them are green and have small leaves (Fig. 7.5a), whereas Dioscoreaceae s.s. are free-living and often vines (Fig. 7.5c), some with large underground storage organs. On the basis of characters that are convergent due to the mycoparasitic syndrome, many authors had suggested that Burmanniaceae were related to the orchids (e.g., Cronquist 1981), but this is now completely discredited.

Many authors have separated out several additional families from Dioscoreaceae (e.g., Taccaceae, Trichopodaceae, Thismiaceae), but they do have shared characters (Caddick et al. 2002a,b). Taccaceae (Fig. 7.5d) were included in Dioscoreaceae (Caddick et al. 2002a,b; APG II 2003; APG III 2009). Likewise, several authors (e.g., Dahlgren et al. 1985) separated Thismiaceae from Burmanniaceae, but because they fell together in Caddick et al. (2002a,b) and share most of their characters (*Thismia*, Fig. 7.5e, and its relatives, such as *Afrothismia*, *Triscyphus*, and *Geomitra*, being only somewhat more reduced vegetatively than many Burmanniaceae s.s.), APG (APG II 2003; APG III 2009) placed these two families together in Burmanniaceae. Merckx et al. (2006) showed that *Thismia* and relatives fell with *Tacca*, the pair then sister to *Dioscorea*, and to this clade *Burmannia* and its relatives were sister. Given that recognition of Thismiaceae would then require resurrection of Taccaceae, it seems more practical to some investigators to lump them all into Dioscoreaceae, based on their readily observable shared characteristics (reflexed stamens and inferior, winged ovaries, although these features may not be consistently present). Based on the lack of resolution of this issue, APG IV (2016) has retained Burmanniaceae, Dioscoreaceae, and Nartheciaceae in Dioscoreales and noted that more data are needed before changes can be intelligently made (see Chapter 12).

In addition, Merckx et al. (2006) did not include the other genera of Dioscoreaceae, which raises the issue of having to resurrect them as families (e.g., Trichopodaceae), leading to many narrowly defined families in this order. In Hertweck et al. (2015), *Thismia/Burmannia* are sister to *Tacca* + *Trichopus* and *Dioscorea*. These differing results could be due to the different taxon sampling or the genes/regions used, but given the problems with circumscribing families when the phylogenetic trees are different, one option is to recognize Dioscoreaceae s.l. and let specialists sort out these problematic relationships. However, others have already recognized the separate families (but not so far Trichopodaceae), and clearly more research is needed. In Chapter 12 we provide the option of recognizing Burmanniaceae, Dioscoreaceae s.s., Nartheciaceae, and Taccaceae. The evidence that they are related is clear, but in exactly which way is disputed. There are other families with mycoparastic as well as photosynthetic genera, and a transition from forest-adapted, broad-leaved herbs such as *Tacca*, *Avetra*, and some species of *Dioscorea* to dark forest-dwelling mycoparasitic herbs is similar to that observed in orchids.

LILIALES

Liliales sensu APG appear to be sister to Asparagales plus the large commelinid clade (Figs. 7.1, 7.4), although as mentioned above, this is not as strongly supported as other parts of the monocot tree; Zeng et al. (2014) with minimal sampling of taxa showed Liliales and Asparagales to be sister clades. Liliales comprise 10 families and about 1,300 species. The concept of Liliales as distinct from other lilioid taxa such as Asparagales and Dioscoreales originated with Huber (1969, 1977), whose ideas were then adopted by Dahlgren et al. (1985). Nearly all of these taxa were considered to be closely related, perhaps even members of a single family, Liliaceae. Cronquist (1981) lumped these and other families such as Velloziaceae and Pontederiaceae into his subclass Liliidae, but if they were not arborescent or did not have net-veined leaves, then he dumped them in Liliaceae. The broad concepts of Liliidae and Liliaceae were formulated because the patterns of characters did not indicate clear-cut groupings, and those authors who did segregate some genera into other families never did so consistently. For example, Cronquist (1981) segregated *Aloe* and some of its close relatives into Aloaceae, largely because *Aloe* had some arborescent species and the other genera were highly succulent (e.g., *Haworthia* and its relatives), but retained the closely related *Bulbine* in Liliaceae because it was strictly herbaceous. However, *Kniphofia* is herbaceous and not succulent but was retained in Aloaceae because its flowers and inflorescences were "aloeoid." Dahlgren et al. (1985) kept both *Bulbine* and *Kniphofia* in Asphodelaceae, whereas he placed the others in Aloaceae because of their shared strongly bimodal and nearly identical karyotypes (both families were in Asparagales due to their phytomelanous seeds; see below).

Chase et al. (1993) and Duvall et al. (1993a) identified a Liliales clade containing Alstroemeriaceae, Colchicaceae, Liliaceae, Melanthiaceae, and Smilacaceae. Huber (1977) and Dahlgren et al. (1985) also recognized Liliales, but they included Iridaceae and Orchidaceae, whereas in the DNA studies these two families appeared as part of Asparagales (see below). Huber (1977) and Dahlgren et al. (1985) excluded Melanthiaceae and Campynemataceae (as Melanthiales), which the DNA studies have consistently indicated to be members of Liliales. Chase et al. (1995a) added Luzuriagaceae, Philesiaceae, Smilacaceae, and Rhipogonaceae to the order, but they also placed here genera that previously had been referred to Calochortaceae (now in Liliaceae), Trilliaceae (now in Melanthiaceae), and Uvulariaceae (genera now in Colchicaceae or Liliaceae). Recently, Kim et al. (2013) produced a better-supported analysis of this order based on four plastid DNA regions, but relationships and family circumscriptions remain as they were in APG III (2009) and APG IV (2016).

Liliales sensu APG (APG 1998; APG II 2003; APG III 2009; APG IV 2016) are mostly geophytes bearing elliptical leaves with fine reticulate venation; their flowers have various forms of tepalar nectaries, extrorsely dehiscent anthers, and, often, spotted tepals. The morphological analyses of Chase et al. (1995b) and Stevenson and Loconte (1995) variously placed Alstroemeriaceae, Colchicaceae, Liliaceae, and Melanthiaceae in a clade, but the other families of Liliales sensu APG fell elsewhere; Iridaceae were also included in Liliales in morphological studies. Relationships of the families within this clade are discussed in Rudall et al. (2000), Vinnersten and Bremer (2001), and Kim et al. (2013). Alstroemeriaceae (5 genera, 170 species; tropical and temperate Central and South America and Australia/New Zealand; Fig. 7.6k) include Luzuriagaceae (formerly 2 genera; 5 species); they share vegetative features such as being vines with the leaves twisted such that the developmentally upper surface is lowermost at maturity, although the ovary is superior in *Luzuriaga* and *Drymophila*. Sister to these is Colchicaceae (18 genera, 224 species; Fig. 7.6a,c), which are native principally to the Old World, the only exception being *Uvularia* (north temperate; Fig. 7.6b). Colchicaceae have been studied in more detail in Chacón et al. (2014). Some genera of Colchicaceae exhibit the twisted leaves of Alstroemeriaceae. Colchicine alkaloids (which are used to inhibit spindle formation and cause nondisjunction at meiosis, leading to polyploid offspring) are found in all members of the family. *Petermannia* was included in Colchicaceae in APG (APG 1998; APG II 2003), but it is now known that the DNA used came from a plant of *Tripladenia cunninghamii* (Colchicaceae) that had been misidentified as *Petermannia* (Kim et al. 2013). Real material of *Petermannia* is sister to Melanthiaceae (Kim et al. 2013), so Petermanniaceae (Fig. 7.4) has been reinstated (APG III 2009; APG IV 2016).

Melanthiaceae (Fig. 7.6f,g; 16 genera; 170 species) were studied by Zomlefer et al. (2001), Pellicer et al. (2013), and Kim et al. (2016); the family now includes the members of former Trilliaceae (APG II 2003; APG IV 2016; see Chapter 12). All genera have a north temperate distribution; a single genus, *Schoenocaulon*, occurs as far south as Peru, but it may have been taken there by humans (they are medicinal plants). The family contains potent alkaloids that provide some of their common names (e.g., fly poison, death camas). *Xerophyllum* (two species in North America) is sister to the genera of the former Trilliaceae (Chase et al. 1995a,b; Rudall et al. 2000; Pellicer et al. 2013; Kim et al. in press). Species of *Xerophyllum* are xerophytically adapted (narrow, grass-like leaves and tufted rosette habit), which contrasts sharply with *Trillium* and its relatives, which are adapted to forest understories and have net-veined leaves. *Paris japonica* holds the record for the largest plant genome (Pellicer et al. 2010), and other members of the former Trilliaceae are similarly large, although not *Xerophyllum*, their sister genus. Some authors (Thorne 1992a,b) considered *Veratrum* (Melanthiaceae) to be one of the most "primitive" monocots because of its plicate, net-veined leaves and mostly unfused carpels, but it is deeply embedded within Liliales and not at all close to the root of the monocots, meaning that it is unlikely to have retained these as primary ancestral features.

Figure 7.6. Liliales (Alstroemeriaceae, Colchicaceae, Liliaceae, Melanthiaceae, Philesiaceae, Smilacaceae). a. *Gloriosa superba* L. (Colchicaceae s. s.), habit, a vine with leaf tendrils, from a corm, and flower. b. *Uvularia grandiflora* Sm. (Colchicaceae s. l., former Uvulariaceae), habit, herb from a rhizome, with immature capsule. c. *Tulipa* sp. (Liliaceae s. s.), center of flower with six stamens. d. *Lilium catesbaei* Walter (Liliaceae s. s.), flower; note spotted tepals and perigonal nectaries. e. *Tricyrtis hirta* Hook (Liliaceae s. l., former Uvulariaceae), flowers with apically divided styles. f. *Schoenocaulon dubium* (Michx.) Small (Melanthiaceae s. s.), small flowers; note anthers opening by a single slit, resulting in a peltate appearance. g. *Trillium maculatum* Raf. (Melanthiaceae s. l., or often recognized in Trilliaceae), herb with three-whorled leaves and flower with sepals and petals. h. *Lapageria rosea* Ruiz & Pav. (Philesiaceae), pendulous flowers. i. *Smilax smallii* Morong (Smilacaceae), vine with umbellate inflorescences. j. *Smilax auriculata* Walter (Smilacaceae), vine with few-seeded berries. k. *Bomarea dulcis* (Hook.) Beauverd (Alstroemeriaceae), flowers; note inferior ovary and resupinate leaves.

Liliaceae sensu APG (see Chapter 10) are composed of many fewer genera than in most previous circumscriptions (e.g., Cronquist 1981). However, the APG concept is broader than that of others who would limit the family to just the core genera related to *Lilium* (Fig. 7.6d; Tamura 1998) by excluding *Calochortus*, *Prosartes*, *Tricyrtis* (Fig. 7.6e), and others, and placing these in Calochortaceae and/or Tricyrtidaceae. Liliaceae are exclusively north temperate and composed of geophytes with spotted flowers, extrorsely dehiscing anthers, and a superior ovary. Related to Liliaceae are Smilacaceae (Fig. 7.6i,j; monogeneric, 315 species; nearly cosmopolitan), Philesiaceae (2 monospecific genera; southern South America; Fig. 7.6h), and Rhipogonaceae (monogeneric, 6 species; Australasia), but non-DNA characters that reflect this pattern are unknown. Smilacaceae, Philesiaceae, and Rhipogonaceae all have unique spiny pollen (Rudall et al. 2000), but in most analyses they have not formed a clade. In Kim et al. (2013), they do form a clade but with poor support; *Rhipogonum* is recovered as sister to *Philesia*/*Lapageria*, so it could be combined with them, but *Smilax* has generally been found to be sister to Liliaceae (but never with good support). If the Kim et al. (2013) topology is supported by subsequent studies with greater amounts of data, then these three could be combined, given their shared characters.

Melanthiaceae, Campynemataceae (2 genera, 4 species; Australasia), and Corsiaceae (3 genera, 30 species; China, South America, and Australasia) have had an unclear pattern of relationships to the other members of Liliales (Fig. 7.4), but in Kim et al. (2013), without including Corsiaceae, the result is Campynemataceae sister to the rest, followed by Colchicaceae, Pertermanniaceae/Melanthiaceae, and the rest of Liliales. Mennes et al. (2015) demonstrated that Corsiaceae (represented by *Corsia* and *Arachnitis*) are monophyletic and sister to Campynemataceae using nuclear, mitochondrial, and plastid sequence data.

Campynemataceae were previously considered to be related to Melanthiaceae (Dahlgren et al. 1985) because of their mostly unfused carpels, whereas Corsiaceae were considered to be related to Burmanniaceae because of their shared mycoparasitic life history. However, both suites of characters are unreliable; unfused (free) carpels are potentially a symplesiomorphy, whereas the syndrome of traits associated with mycoheterotrophy is convergent even between eudicots and monocots with this life history. Campynemataceae and Corsiaceae generally fit the pattern of characters observed among the families of Liliales, and Hertweck et al. (2015) found *Arachnitis* to be sister to Campynemataceae (Fig. 7.4).

ASPARAGALES

Asparagales had originally been suggested by Huber (1977) and later adopted by Dahlgren et al. (1985). Asparagales are sister to the commelinid clade (Figs. 7.1, 7.4) in most analyses, except for Zeng et al. (2014), which has sparse taxonomic sampling, and in which they are sister to Liliales. Asparagales comprise 14 families: Amaryllidaceae (including Agapanthaceae and Alliaceae; 60 genera, 1,605 species; nearly cosmopolitan), Asparagaceae (including Agavaceae, Aphyllanthaceae, Hyacinthaceae, Laxmanniaceae, Ruscaceae, and Themidaceae; 120 genera, 2,640 species; nearly cosmopolitan, but rare in the wet tropics), Asteliaceae (4 genera, 36 species; mostly Australasia), Blandfordiaceae (monogeneric with 4 species; southeastern Australia), Boryaceae (2 genera, 12 species; Australia), Doryanthaceae (monogeneric with 2 species; southeastern Australia), as well as the closely related Ixioliriaceae and Tecophilaeaceae (9 genera, 24 species; Central Asia, Africa, Madagascar, Chile, and California), Hypoxidaceae (7 genera, 200 species; nearly cosmopolitan but not in Europe), Iridaceae (67 genera, perhaps more, 1,800 species; cosmopolitan), Lanariaceae (monospecific; South Africa), Orchidaceae (736 genera, 27,135 species; cosmopolitan, but particularly diverse in the cool wet tropics), Asphodelaceae (including Xanthorrhoeaceae and Hemerocallidaceae; 35 genera, 900 species; mostly Old World, particularly Australia, mostly temperate), and Xeronemataceae (monogeneric with 2 species; an island off New Zealand and New Caledonia). Another option preferred by some is to maintain many of the families subsumed above—see Chapter 12 (compare Fig. 7.4 with Supplementary Fig. 7.1 at press.uchicago.edu/sites/soltis/).

The preeminent characters uniting Asparagales are their phytomelanous (a dark, non-cellular material) seed coat and collapsed/obliterated outer epidermis. Many Asparagales with superior ovaries have septal nectaries, and geophytes are common, although in this case their leaves are mostly linear without reticulate fine-scale venation. Anomalous secondary growth with a tiered (etagen) meristem is nearly confined to the genera of this clade (outside this clade, it is known only in *Dioscorea*). Some species of *Agave*, *Aloe*, *Cordyline*, *Dracaena*, *Nolina*, and *Yucca* become massive, short trees (rarely more than five meters tall), generally, but not always, with limited branching (e.g., *Aloe*, *Dracaena*, and *Yucca*). Other genera (e.g., *Aphyllanthes* and *Lomandra*) have this same type of secondary growth confined to their underground stems. In the morphological analyses of Chase et al. (1995b) and Stevenson and Loconte (1995), at least some of these genera of Asparagales formed clades, although members of Liliales (Chase et al. 1995b) and Zingiberales and Pandanales (Stevenson and Loconte 1995) also fell among them. Seed anatomy would appear to be the best set of characters for the order, the members of which are otherwise heterogeneous, particularly considering that Orchidaceae are members of this clade (see below).

Orchidaceae (Fig. 7.7g) are one of the two largest families of angiosperms (the other is Asteraceae), and their infrafamilial relationships are clearly resolved and well supported (e.g., Dressler 1983, 1993; Chase 1986, 1988; Chase and Hills 1992; Chase and Palmer 1992; Cameron et al. 1999; Kores et al. 2000; Whitten et al. 2000; Salazar et al. 2003; Cameron 2004, 2006; Górniak et al. 2010; Chase et al. 2015). Some (e.g., Dahlgren et al. 1985) recognized three families of orchids on the basis of anther morphology: Apostasiaceae (2 genera; 2 or 3 anthers only partially fused to the gynoecium), Cypripediaceae (5 genera; 2 anthers fused to the gynoecium), and Orchidaceae (the rest, all with a single anther fused to the gynoecium). All molecular evidence (summarized in Chase et al. 2003) shows that this three-family view (Dahlgren et al. 1985) is non-monophyletic. Analyses indicate instead a broadly defined Orchidaceae; within this family, five subfamilies are now recognized with the following relationships: Apostasioideae (Vanilloideae (Cypripedioideae (Orchidoideae, Epidendroideae))). Thus, reduction to one anther occurred at least twice in the evolution of the orchids. A new phylogenetic classification of Orchidaceae was published by Chase et al. (2003, with an update in 2015).

Orchidaceae are sister to the rest of Asparagales (Fig. 7.4); Chase et al. 2006; Graham et al. 2006; Pires et al. 2006; S. Chen et al. 2013; Hertweck et al. 2015), and it is clear that no single family of Asparagales is closely related to orchids. In spite of being a member of the same clade as the rest of Asparagales, orchids lack most of the non-DNA synapomorphies of the order. Seeds of Orchidaceae lack phytomelan due to their dust-like nature (like most groups of mycoparasites), and their nectaries are only rarely septal. Orchids do have simultaneous microsporogenesis and inferior ovaries, characters that are typical of the clades at the basal nodes of Asparagales (Fig. 7.4; the "lower Asparagales" of Rudall 1997), but it is not clear whether these characters can be considered strict synapomorphies of the entire order. The only apparent synapomorphy for Orchidaceae is their protocorm (the structure produced by the growth of their undifferentiated embryos before roots and shoots develop), although bilaterally symmetrical flowers and the reduction to three or fewer stamens are also likely synapomorphic. The hallmark of orchids is the fusion of androecium and gynoecium, but this feature is present only to varying degrees in members of subfamily Apostasioideae. Dust seeds distributed by the wind occur in most subfamilies of Orchidaceae, but there are crustose seeds in some members of subfamilies Cypripedioideae and Vanilloideae (probably an association with seed dispersal, which in these cases is due to their fleshy fruit fermenting in situ and releasing fragrant compounds attractive to birds and mammals, i.e., vanillin). The effect of particular characters that have been considered to be important (key characters) for Orchidaceae has recently been studied (Freudenstein and Chase (2015). They concluded that a shift to epiphytism in the largest subfamily by far, Epidendroideae, is the character most strongly associated with species richness, followed by expansion into the New World and then a number of anther characters involved with pollinator specificity. All these characters show significant association with speciation in Epidendroideae, suggesting that no single character accounts for the success of this group and instead a number of key features contributed to diversification, often in parallel.

Asteliaceae (Fig. 7.8d), Blandfordiaceae, Boryaceae (Fig. 7.8h), Hypoxidaceae (Fig. 7.7d), and Lanariaceae form a well-supported clade (Fig. 7.4) in molecular phylogenetic analyses. Blandfordiaceae are sister to a well-supported clade of Asteliaceae, Hypoxidaceae, and Lanariaceae. Morphology provides some support for these relationships. Asteliaceae and Hypoxidaceae are rosette-forming, covered with branched, multicellular hairs, and have root canals filled with mucilage; the former is also true for *Lanaria* (Lanariaceae), but *Blandfordia* (Blandfordiaceae) does not share these traits. A potential synapomorphy for these four families is ovule structure: these families all have a chalazal constriction and a nucellar cap. In S. Chen et al. (2013), this clade, including *Blandfordia*, received maximal support, so one could merge the four families in Hypoxidaceae (which is the only conserved name). *Lanaria* (Lanariaceae) has in the past been considered to be a member of Haemodoraceae (Hutchinson 1967) or Tecophilaeaceae (Dahlgren et al. 1985).

Boryaceae (two genera, *Borya* and *Alania*, 12 species) are sister to the rest of the hypoxid clade (above) with good support in Pires et al. (2006) and S. Chen et al. (2013). *Borya* (Fig. 7.8h) is typically a "resurrection plant" found on rocky slopes; during the dry season, they fall to a fraction of their normal water content and turn rusty-orange, but they quickly become green and active again once it rains. Boryaceae are known to be mycorrhizal, but this is of the standard vesicular-arbuscular (VA) type and not like that of the orchids (their own unique type, endomycorrhizal). Boryaceae were previously thought to be members of Anthericaceae (Dahlgren et al. 1985), a family shown to be grossly polyphyletic (Chase et al. 1996).

The next node in Asparagales (Fig. 7.4; all families other than Orchidaceae and the clade of Asteliaceae, Blandfordiaceae, Boryaceae, Hypoxidaceae, and Lanariaceae)

Figure 7.7. Asparagales (Asparagaceae, Hypoxidaceae, Iridaceae, Orchidaceae, and Asphodelaceae). a. *Asparagus aethiopicus* L. (Asparagaceae; Asparagoideae), flowers, phylloclades, and reduced leaves (spines and reduced brown structures). b. *Dianella tasmanica* Hook.f. (Asphodelaceae: Hemerocallidoideae), flower with fleshy bodies on anthers. c. *Lachenalia aloides* (L.f.) Engl. (Asparagaceae: Scilloideae), racemose inflorescence. d. *Hypoxis juncea* Sm. (Hypoxidaceae), flower. e. *Cordyline terminalis* Kunth (Asparagaceae: Lomandroidae), berry fruits and woody habit. f. *Iris lutescens* Lam. (Iridaceae) and *Ixia dubia*. (Iridaceae, inserted). Flower of *I. lutescens* with petaloid tepals, differentiated into a claw and a broadened blade. The inner tepals are erect and arch over together to form a "dome"; the "falls" or outer tepals are spreading-deflexed, bearded with a band of hairs. Style deeply divided, with petaloid style branches that together with "falls" form a labiate "flower." Crest is curved over the stamen, which stands in front of it and is appressed to its lower side. Open actinomorphic flower of *Ixia dubia* (inserted) with three stamens. g. *Cyrtopodium aureum* L. C. Menezes (Orchidaceae). Typical orchid zygomorphic flower with the outer whorl of three tepals (or sepals) and the inner whorl of three tepals (or petals). The enlarged upper medial petal of the resupinate flower forms a lip (or labellum). h. *Dichelostemma ida-maia* Greene (Asparagaceae: Brodieaoideae), umbellate inflorescence, but with a small bract subtending each flower; umbel convergent with Amaryllidaceae but the latter with two large bracts subtending the whole umbel. i. *Xanthorrhoea resinosa* Pers. (Asphodelaceae: Xanthorrhoeoideae), portion of inflorescence with many small flowers.

Figure 7.8. Asparagales (Asparagaceae, Asteliaceae, Boryaceae, Doryanthaceae, Tecophilaeaceae, Xeronemataceae). a. *Beaucarnea recurvata* Lem. (Asparagaceae: Nolinoideae), woody habit with anomalous secondary growth, and (insert) dry, indehiscent, three-parted fruits (a dry "berry"). b. *Convallaria majalis* L. (Asparagaceae: Nolinoideae), herbaceous habit, berries and (insert) flowers. c. *Dracaena draco* L. (Asparagaceae: Nolinoideae), berries (non-phytomelanous seeds, not observable), and leaf bases with resinous red sap. d. *Astelia neocaledonica* Schltr. (Asteliaceae), habit and inflorescence. e. *Xeronema callistemon* W. R. B. Oliv. (Xeronemataceae), inflorescence. f. *Doryanthes palmeri* W. Hill ex Benth. (Doryanthaceae), flowers with inferior ovaries, but the top of the ovary is seen here, making it appear as though there might be a superior ovary. g. *Odontostomum hartwegii* Torr. (Tecophilaeaceae), flowers with inferior ovary on a raceme. h. *Borya sphaerocephala* R. Br. (Boryaceae), habit with involucrate inflorescence.

is maximally supported (Chase et al. 2006; S. Chen et al. 2013), but there are no known non-DNA synapomorphies for this clade. Within this large subclade, some analyses have found that Ixioliriaceae are sister to the rest (Chase et al. 2006; Graham et al. 2006; Pires et al. 2006). Ixioliriaceae (Fig. 7.8f) share corms, a leafy inflorescence, and, often, a nearly capitate inflorescence. However, relationships of Doryanthaceae to Ixioliriaceae and Tecophilaeaceae remain unclear. Although they form a clade in Hertwick et al. (2015), support is weak; their relationship to Iridaceae is also unclear from both morphological and molecular analyses. After Ixioliriaceae, Graham et al. (2006) found the next diverging clade is Doryanthaceae, followed by Iridaceae, with 92% and 97% BS support, respectively. However, Chase et al. (2006) and S. Chen et al. (2013) recovered other relationships, but none with high support. All these families are excluded from the next node in Asparagales (Fay et al. 2000b; Chase et al. 2006; Graham et al. 2006; Pires et al. 2006; S. Chen et al. 2013). The two species of *Doryanthes* are enormous, rosette-forming herbs (Fig. 7.8f); they are a conspicuous floristic element around Sydney, Australia, and their flowering is difficult to miss. Iridaceae (Fig. 7.6f) are one of the largest and best-studied families of Asparagales (e.g., Goldblatt 1990, 1991; Goldblatt et al. 1998, 2002; Reeves et al. 2001). Iridaceae consist of three subfamilies, Crocoideae (including Nivenioideae), Iridoideae, and Isophysidoideae, with the last sister to the other two. Iridaceae are distinctive among Asparagales because of their unique inflorescence structure (a rhipidium) and their combination of inferior ovaries and three stamens (Fig. 7.6f); unifacial leaves are also common in the family, whereas bifacial leaves are the norm in other Asparagales (an obvious exception is *Xeronema*, Xeronemataceae).

Character Evolution in Asparagales

It was long known that *Allium* (Amaryllidaceae subf. Allioideae) lacked the standard plant telomeric repeats, termed the *Arabidopsis*-type, that cap the ends of chromosomes. Then, it was discovered (Adams et al. 2001) that *Aloe* (Asphodelaceae) also lacked this standard repeat type. Comparing the position of these two genera on the plastid DNA tree of Fay et al. (2000b) led to the hypothesis that the intervening genera of Asparagales should also lack these standard sequences. Representatives of all Asparagales families were then examined by a combination of fluorescent in situ hybridization (FISH) and slot blot Southern hybridization, and it was found that all but a single genus, *Ornithogalum* (Asparagaceae subf. Scilloideae), of those predicted to lack the *Arabidopsis*-type repeats, did indeed lack them. Sykorova et al. (2003) discovered that *Ornithogalum* and several other related genera had replaced the standard plant repeat with a human-type repeat and, with more sensitive probing, that nearly all genera of Asparagales have at least some copies of the human-type repeat as well as a variety of others located at their chromosome tips. Thus, Asparagales represent a dynamic situation in which a variety of telomeric cap motifs are maintained in low copy, in spite of the presence of the dominant *Arabidopsis*-type repeat. However, in some cases, one of the low-copy types is amplified and replaces the standard type. The presence and composition of the various repeat types provide a great deal of phylogenetic structure within Asparagales.

Character Evolution in Iridaceae

Asparagales offer many examples of the elucidation of character evolution in a phylogenetic context. Orchidaceae and pollination are discussed in Chapter 14; Iridaceae have been well studied (Goldblatt and coworkers). Goldblatt et al. (2002) dated the radiation of *Moraea* (Iridaceae; nearly 200 species in Africa and Eurasia, but particularly diverse in the Cape area of South Africa), and evaluated patterns of pollination and chromosomal evolution. *Moraea* and its sister genus *Ferraria* split from each other about 25 mya, and the early radiation of *Moraea* took place against a background of increasing aridification and the spread of desert, shrublands, and fynbos (a diverse, shrubby community found in South Africa). An *Iris*-type flower (Fig. 7.7f, one with three separate pollination units) is ancestral in *Moraea*. This type of flower is pollinated by long-tongued bees foraging for nectar; pollen deposition is passive. In multiple, unrelated lineages, this flower type has been replaced by open, *Homeria*-type flowers (Fig. 7.7f insert), in which pollen is actively sought by short-tongued bees; pollen is taken to their nests and fed to the developing larval bees. In addition, some lineages have become adapted to nectar-seeking flies or scarab beetles, the latter using the flowers as sites for mate selection and mating.

Chromosome number change in *Moraea* occurred late in the evolution of the genus; most nodes near the base of the genus were $n = 8$. Some clades experienced increases and others decreases around the same time, so that lineage diversification over a short period fixed these differences in separate clades, after which little further change took place. Species-rich clades, mostly containing taxa endemic to the Cape, also contain species with ranges outside southern Africa. Some species managed to reach Europe and western Asia, but this occurred late in the evolution of *Moraea*. Species-level analyses, such as that of Goldblatt et al. (2002), demonstrate how a much-improved understanding of evolutionary patterns and processes, as well as molecular clock dating, can be established, which in turn can provide insights into geographic patterns of species diversity.

The next node of Asparagales, a clade of Xeronemataceae, Asphodelaceae (including Xanthorrhoeaceae and Hemerocallidaceae), Amaryllidaceae (including Agapanthaceae and Alliaceae), and Asparagaceae (including Agavaceae, Anthericaceae, Anemarrhenaceae, Aphyllanthaceae, Behniaceae, Herreriaceae, Hyacinthaceae, Laxmanniaceae, Ruscaceae, and Themidaceae) is also well supported (e.g., Fay et al. 2000b; Chase et al. 2006; Graham et al. 2006; S. Chen et al. 2013). If an inferior ovary is a synapomorphy of Asparagales (Fig. 7.9; Chase et al. 1995b), then this node would mark the transition to a superior ovary (although there are inferior ovaries in Amaryllidoideae of Amaryllidaceae and *Yucca* in Agavoideae of Asparagaceae, but these are obviously embedded clades). *Xeronema* (Xeronemataceae; Fig. 7.8e) was previously considered a close relative of *Phormium* (usually Phormiaceae; Dahlgren et al. 1985; in APG, *Phormium* is in Hemerocallidoideae of Asphodelaceae), but although there are superficial similarities (e.g., unifacial leaves), they are not identical in detail. Chase et al. (2000a) described the family Xeronemataceae, and it falls within the general patterns of variation within Asparagales.

The clades at this node all have infralocular septal nectaries, which Rudall (2000) interpreted to be the result of these ovaries being secondarily superior (see Chapter 10).

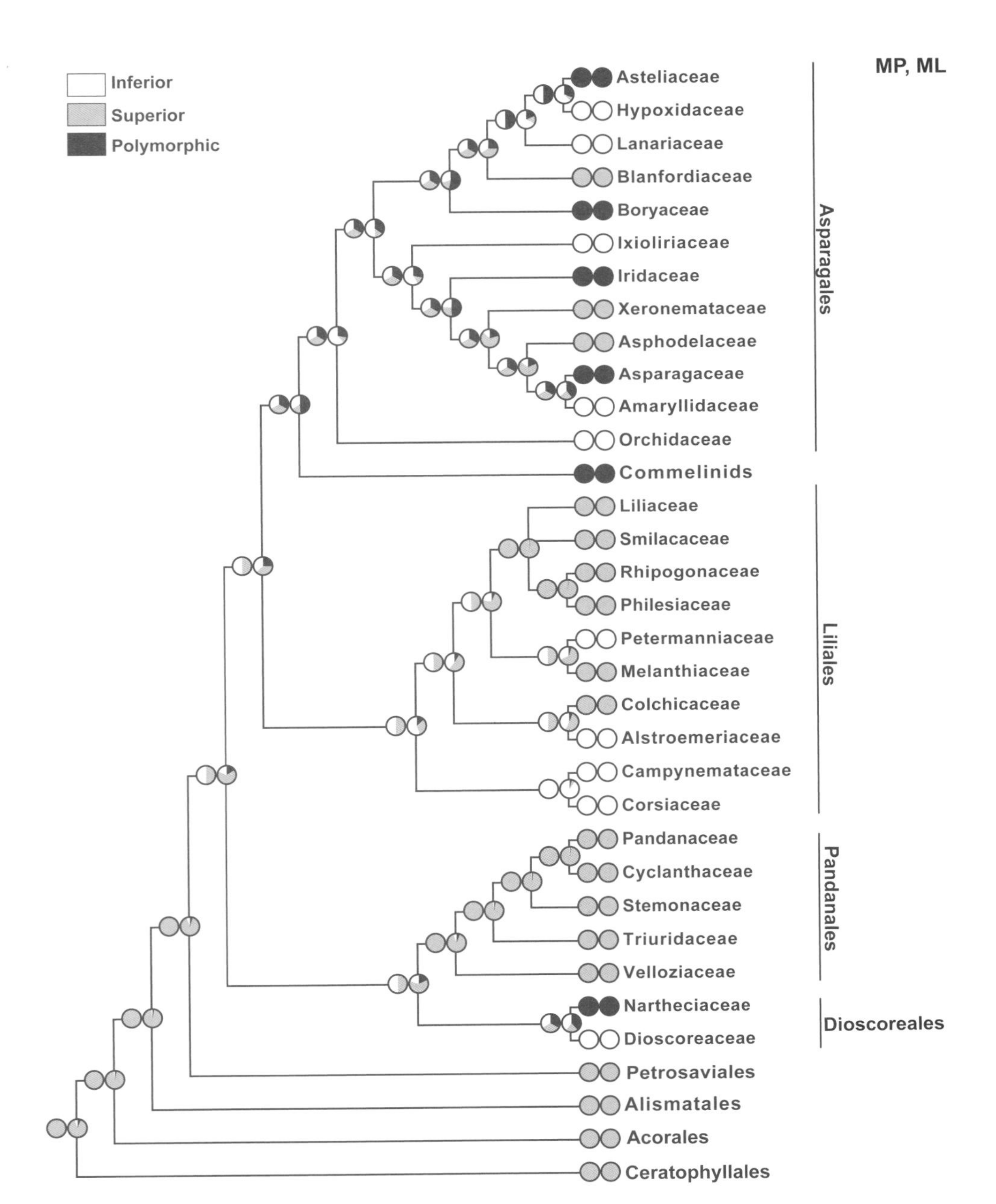

Figure 7.9. One possible reconstruction of ovary position evolution in monocots.

Since nearly all early-diverging nodes in Asparagales are characterized by inferior ovaries, this seems a reasonable interpretation and demonstrates the reversibility of this trait that has often been emphasized in monocot systematics (Fig. 7.9; Cronquist 1981; Dahlgren et al. 1985).

Asphodelaceae (Figs. 7.7b,i, 7.10h,j) are characterized by anthraquinones. The value of this chemical character is seen in Asphodeloideae, a subfamily now considered part of a broadly defined Asphodelaceae (and including both Hemerocallidaceae and Xanthorrhoeaceae), but historically placed with Anthericaceae (a polyphyletic concept, most genera now in Asparagaceae). Because of the presence of anthraquinones, Asphodelaceae s.s. were kept separate from Anthericaceae by Dahlgren et al. (1985; in addition, Asphodelaceae have simultaneous microsporogenesis, whereas Anthericaceae have successive). Within Asphodelaceae, secondary growth occurs in *Aloe*, *Phormium*, and *Xanthorrhoea* (Fig. 7.7i); some species of *Aloe* (Fig. 7.10h,j) can attain enormous size. Included in Xanthorrhoeaceae s.s. in the past have been several other arborescent taxa such as *Kingia* and *Cordyline* (Fig. 7.7e), but the former does not have secondary growth and is now placed in Dasypogonaceae (now placed in Arecales; see below) and the latter is only distantly related (Lomandroideae of Asparagaceae, which = Laxmanniaceae).

If recognized, Hemerocallidaceae (included in Xanthorrhoeaceae sensu APG III 2009 and Asphodelaceae in APG IV 2016—Xanthorrhoeaceae s.l. is a synonym of Asphodelaceae s.l.) have often included only *Hemerocallis* (Dahlgren et al. 1985), but even in the APG (1998) circumscription, Hemerocallidaceae included a larger number of genera, such as *Dianella* (Fig. 7.7b), *Johnsonia*, and *Phormium*. All these genera share trichotomosulcate pollen, a rare condition in monocots, and early DNA studies (Chase et al. 1995a) were critical in determining whether this was an important character for these genera (Rudall et al. 1997). Asphodelaceae have many genera with distinctive bimodal karyotypes (Chase et al. 2000a; Pires et al. 2006), and this trait has been argued as the basis for separating some genera as Aloaceae from the rest (Dahlgren et al. 1985). Molecular studies have played an important role in identifying the members of this clade and changing its circumscription. Bimodal karyotypes of different fundamental number have been recorded from several of the families and subfamilies of Asparagales (Pires et al. 2006), and these have been studied in more detail using next-generation transcriptomic methods and found (McKain et al. 2012) to possibly be the result of a paleopolyploid event. These bimodal karyotypes may be the result of hybridization between taxa with two radically differently sized chromosomes and subsequent polyploidy (polyploidy is discussed in Chapter 16). The mystery is why the differently sized chromosomes have been so faithfully maintained in these taxa following this event (Chase et al. 2000a).

The remaining Asparagales (broadly defined Amaryllidaceae and Asparagaceae, sensu APG IV 2016; see Chapter 12) are characterized by successive microsporogenesis (the "higher asparagoids"; Rudall et al. 1997), but steroidal saponins are also a potential synapomorphy. These two families can be distinguished by umbelloid inflorescences with generally two large enclosing bracts in Amaryllidaceae versus racemes in Asparagaceae (although subfamily Brodiaeoideae have umbelloid inflorescences, they have in this case a bract associated with each flower, which is taken to indicate that they are condensed racemes). Papers dealing with phylogenetic relationships within these expanded Amaryllidaceae, Asparagaceae, and Asphodelaceae include Meerow et al. (1999), Chase et al. (2000c), Yamashita and Tamura (2000), Reeves et al. (2001), Pires and Sytsma (2002), Kim et al. (2010), and S. Chen et al. (2013).

COMMELINID MONOCOTS

The existence of a commelinid clade (Figs. 7.1, 7.11) had long been suspected (Dahlgren et al. 1985) because several characters occur exclusively in these taxa: cell walls with ultraviolet-fluorescent ferulic (Fig. 7.12) and coumaric acids, silicon dioxide bodies in leaves, and epicuticular waxes of the *Strelitzia* type. Many authors confused the monophyly of this group relative to these characters because of the gross morphological similarity of Pandanaceae/Cyclanthaceae (Pandanales) to the clearly commelinid Arecaceae (Arecales). These three families share an arborescent or vining herbaceous habit and inflorescences with a spathe, so it is not surprising that many people saw them as forming a natural group. Some authors (e.g., Cronquist 1981) also linked the aroids to Pandanaceae/Cyclanthaceae because of their shared inflorescence morphology (presence of a subtending spathe), but this character occurs in unrelated monocots (e.g., some alismatid families, aroids, cyclanths, pandans, palms). On the basis of their commelinid characters, Dasypogonaceae also were placed here (Rudall and Chase 1996), whereas earlier authors (Dahlgren et al. 1985) had associated them with other groups, such as Xanthorrhoeaceae-now Asphodelaceae (Asparagales) because of similarities in gross habit, although Dasypogonaceae do not exhibit secondary growth.

The orders of commelinids (Fig. 7.11) have typically been associated in various ways by previous authors. Cronquist (1981) associated his Commelinidae (which approximates Poales sensu APG, but without Bromeliaceae, which he placed in Zingiberidae) with Zingiberidae and then Liidae. According to Cronquist (1981), Arecidae included

Figure 7.10. Asparagales. Amaryllidaceae, Asparagaceae, Asphodelaceae. a. *Agapanthus africanus* Hoffmanns. (Amaryllidaceae: Agapanthoideae), umbellate inflorescence of blue flowers with superior ovaries. b. *Agave macrantha* Zucc. (Asparagaceae: Agavoideae), habit, a rosette of succulent, fibrous, and spinose leaves. c. *Hesperoaloe campanulata* G. D. Starr (Asparagaceae: Agavoideae), flower with superior ovary. d. *Agave lechuguilla* Torr. (Asparagaceae: Agavoideae), capsules with black (phytomelanous) seeds. e. *Allium tricoccum* Aiton (Amaryllidaceae: Allioideae), scape, umbel, and three-celled glossy black capsules, each with one seed. f. *Allium canadense* L. (Amaryllidaceae: Allioideae), flower with superior ovary. g. *Crinum americanum* L. (Amaryllidaceae: Amaryllidoideae), habit with linear leaves, scape, umbel with a pair of large subtending bracts, and typical flower with an inferior ovary. h. *Aloe saponaria* (Aiton) Haw. (Asphodelaceae: Asphodeloideae) habit (rosette of succulent, non-fibrous, spinose leaves). i. *Crinum* hybrid 'Ellen Bosanquet' (Amaryllidaceae: Amaryllidoideae), flower with inferior ovary. j. *Aloe dyeri* Schönland (Asphodelaceae: Asphodeloideae), flowers with fused tepals and superior ovary.

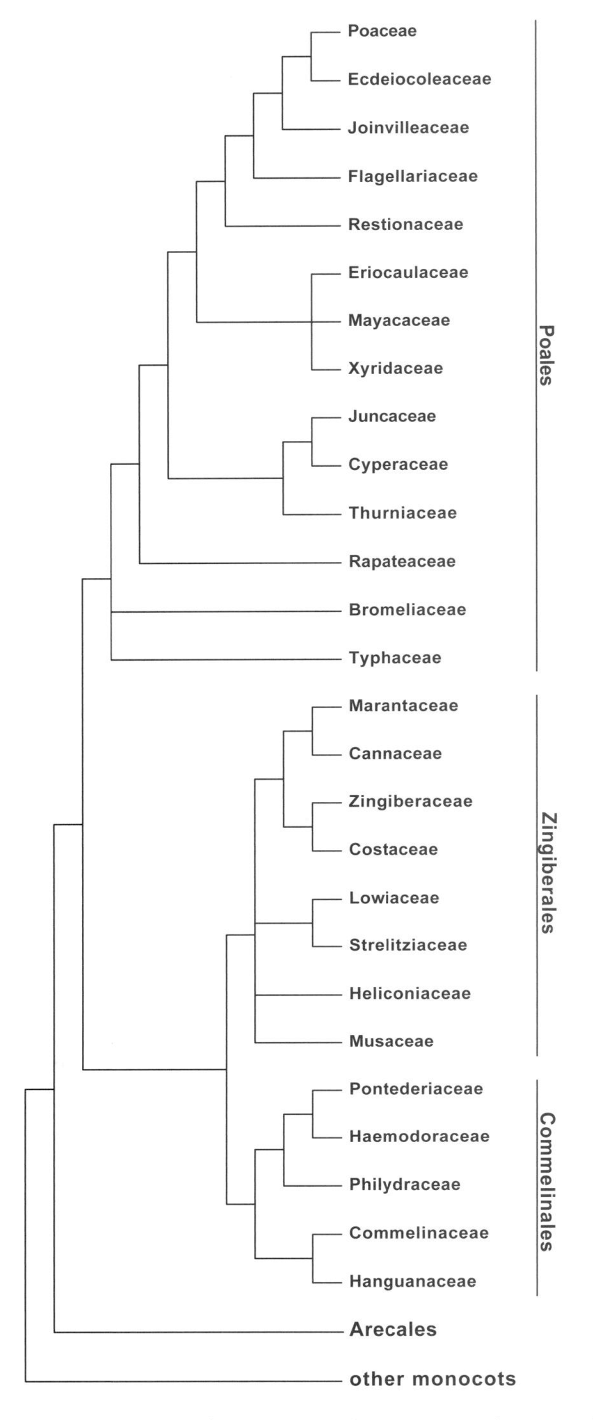

Figure 7.11. Summary tree for the commelinid monocots: Arecales, Commelinales, Poales, and Zingiberales.

Pandanales (Pandanaceae/Cyclanthaceae) and Arales (Araceae) and were not related to Commelinidae/Zingiberidae. The other families of the commelinids that Cronquist (1981) did not associate with Commelinidae/Zingiberidae

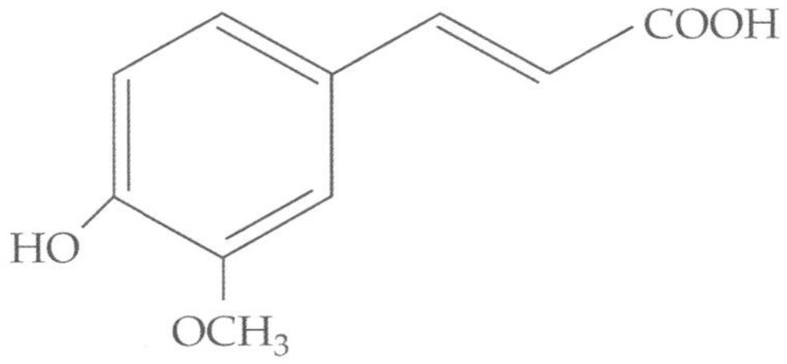

Figure 7.12. Chemical structure of ferulic acid, a putative synapomorphy for the commelinid clade of monocots: Arecales, Commelinales, Dasypogonales, Poales, and Zingiberales.

were Dasypogonaceae (see above) and Hanguanaceae, Philydraceae, Pontederiaceae, and Haemodoraceae, the last three of which he considered to be members of Liliidae because they have large tepals and look like "lilies."

Arecales

Arecales as considered in APG III (2009) and earlier consist only of Arecaceae (or Palmae; 190 genera, 2,000 species), a family well supported as monophyletic (Chase et al. 1995a,b; Asmussen and Chase 2001). Recent DNA evidence, however, provides strong support for Dasypogonaceae (below) as sister to Arecales (Barrett et al. 2015), and the former have been treated as part of Arecales in APG IV (2016; see Chapter 12).

Arecaceae (Fig. 7.13a) are easily recognized as trees, shrubs, or lianas with unbranched or rarely branched trunks with the stem apex consisting of a large apical meristem. It remains unclear how Arecales (including Dasypogonaceae), Commelinales/Zingiberales, and Poales are interrelated, but the best-supported estimate (Hertweck et al. 2015) places Arecales and Dasypogonaceae (former Dasypogonales) as sisters, and this pair sister to Poales and Commelinales/Zingiberales. Givnish et al. (2010) used 81 plastid genes but did not obtain >60% bootstrap support for any of these interrelationships; their results also differ from Hertweck et al. (2015).

Flowers are trimerous in Arecaceae with three sepals, usually three petals, and three or six to numerous stamens; carpels are typically three, but sometimes there are as many as ten. The pseudomonomerous condition occasionally occurs (Fig. 7.13d). The family is also well known from the fossil record (Iles et al. 2015), with fossils from Europe and North America dating to the Late Cretaceous (90 mya). These fossils have provided important calibration points in many studies attempting to date the crown node of the monocots (e.g., Zeng et al. 2014; Hertweck et al. 2015) (Suppl. Table 7.1).

Relationships within Arecaceae have also been investigated (e.g., Asmussen et al. 2000; Asmussen and Chase

Figure 7.13. Arecales (Arecaceae), Dasypogonales (Dasypogonaceae), and Commelinales (Commelinaceae, Haemodoraceae, Philydraceae). a. *Roystonea borinquena* O. F. Cook (Arecaceae: Arecoideae), habit and pinnate leaves. b. *Veitchia arecina* Becc. (Arecaceae: Arecoideae), a floral triad. c. *Sabal palmetto* (Walter) Lodd. ex Schult. & Schult.f. (Arecaceae: Coryphoideae), habit and costapalmate, induplicate leaves. d. *Serenoa repens* (W. Bartram) Small (Arecaceae: Coryphoideae), perfect flowers with three sepals, three petals, six stamens, and a tricarpellate gynoecium. e. *Metroxylon vitiense* (H. Wendl.) Hook.f. (Arecaceae: Calamoideae), distinctive scales on the germinating fruit. f. *Nypa fruticans* Wurmb (Arecaceae: Nypoideae), unusual inflorescence with catkin-like clusters of staminate flowers, and also a single cluster of carpellate flowers, forming a dense head. g. *Pseudophoenix sargentii* H. Wendl. (Arecaceae: Ceroxyloideae), flower buds on pseudopedicels, the three-lobed drupes, and pinnate leaves (reduplicate; in the background). h. *Kingia australis* R.Br. (Dasypogonaceae), habit with capitate, scapose inflorescences. i. *Callisia graminea* (Small) G.C.Tucker (Commelinaceae), flowers with three sepals, three petals, and moniliform hairs on the filaments. j. *Tradescantia fluminensis* Vell. (Commelinaceae), flowers, boat-shaped bracts, and sheathing leaves. k. *Philydrum lanuginosum* Banks ex Gaertn. (Philydraceae), zygomorphic flowers, each with a single stamen. l. & m. *Lachnanthes caroliniana* (Lam.) Dandy (Haemodoraceae), habit and hairy inflorescence (l) and rhizomes and roots (with orange-red to purple arylphenalenones) (m).

2001; Hahn 2002), but some other published studies have used non-coding plastid DNA regions that cannot be aligned well against the outgroup taxa, so a robust evaluation of where the root should be placed within the family has been lacking. *Nypa* (Fig. 7.13f) had been arranged as sister to the rest of Arecaceae because the genus had been used that way by Uhl et al. (1995) on the basis of plastid DNA restriction site studies. Plastid DNA has been shown to evolve slowly in the palms (Wilson et al. 1990), such that most researchers gave up using conserved regions, such as *rbcL*, early in the history of the use of this gene in angiosperm systematics. Asmussen and Chase (2001) found that Calamoideae (Fig. 7.13e) were sister to the rest of Arecaceae, but without bootstrap support above 50%. Following Calamoideae, *Nypa* (the sole member of Nypoideae) was then sister to the remaining taxa (71% bootstrap), which include two large clades—Coryphoideae (with induplicately plicate, and usually palmate or costapalmate leaves, and flowers not in triads; Fig. 7.13c,d) and the Arecoideae (with reduplicately plicate and pinnate leaves, and flowers usually in triads or lines; Fig. 7.13a,b), along with a small one, Ceroxyloideae (Fig. 7.13g). In a combined analysis of three DNA regions, resolution and bootstrap support for large portions of the rest of the topology were poor, so little can be said with confidence regarding other relationships within Arecaceae.

Baker and Couvreur (2013a) used ancestral area reconstruction and a dated phylogenetic tree to hypothesize that palm dispersals began in the Late Cretaceous, in agreement with the first palm fossils and rain forests at that time in South America and Africa. Contrary to what had been assumed, palms in Africa did not have higher extinction rates than other regions; the paucity of taxa in that continent is due to much higher rates of in situ diversification in Asia, the Americas, and the Pacific and Indian Oceans. These conclusions may have implications for other organisms exhibiting a similar paucity of taxa in Africa. Studies that attempt to integrate phylogenetic, biogeographic, macroecological, and diversification studies across many tropical rain forest lineages present us with the possibility of understanding global evolution and ecology of the tropical rain forest biome (Couveur and Baker 2013).

Dasypogonaceae (4 genera, 16 species) are a family of short trees (less than five meters), stout shrubs, and herbs, restricted to Australia (mostly Western Australia). They were previously linked to other, similar-looking groups, such as Xanthorrhoeaceae s.s. (now in an expanded Asphodeleaceae) and Lomandraceae (the latter Laxmanniaceae in APG 1998; Rudall and Chase 1996; and Asparagaceae s. l. in APG III 2009 and APG IV 2016; see Chapter 12), families now known to be in Asparagales (see above). The genus *Kingia* (Fig. 7.13h) is a small tree, but this growth form is achieved through just primary growth, whereas the other three genera (*Baxteria*, *Dasypogon*, and *Calectasia*) are rhizomatous herbs or tough shrub-like herbs. Floral characters of the family are like those of the lilioid monocots, with which they have always been placed previously, but Dasypogonaceae were found to have the diagnostic characters of the commelinids, which were reviewed above (Rudall and Chase 1996). Given that large datasets (Givnish et al. 2010; Barrett et al. 2013; Hertweck et al. 2015) were unable to find a well-supported placement of Dasypogonaceae (and recognized Dasypogonales) but recent data (Barrett et al. 2015) suggest the family is sister to Arecaceae, APG IV (2016) expanded Arecales to include Dasypogonaceae (Chapter 12).

Commelinales

Commelinales consists of five families (Figs. 7.11, 7.13i-m): Commelinaceae (38 genera, 640 species; worldwide except for Europe), Haemodoraceae (14 genera, 116 species; pantropical to warm temperate regions), Hanguanaceae (monogeneric, 6 species; Southeast Asia and northern Australia), Philydraceae (4 genera, 5 species; Australia to Southeast Asia), and Pontederiaceae (9 genera, 33 species; nearly cosmopolitan). Monophyly of Commelinales has been well supported in DNA studies (e.g., Kellogg and Linder 1995; Chase et al. 2000a, 2006; D. Soltis et al. 2000; Graham et al. 2006; Hertweck et al. 2015). However, non-DNA synapomorphies for Commelinales are cryptic and thus far confined to chemistry (phenylphenalenones) and seed characters (e.g., abundant endosperm formed helobially). The families also share tannin cells in the perianth and sclereids in the placentae (Judd et al. 2002, 2008).

This association of families of Commelinales was another unpredicted phylogenetic result of early molecular studies. Dahlgren et al. (1985) allied not only Haemodoraceae, Philydraceae, and Pontederiaceae but also Bromeliaceae, Typhaceae, and Velloziaceae (families of Poales and, the last, Pandanales). Within their concept of Bromeliiflorae, Dahlgren et al. placed each of the last three of these families in a monofamilial order, signifying the degree to which they thought these families were isolated from each other. Cronquist (1981) and others considered the tepalar perianth of Haemodoraceae, Philydraceae, and Pontederiaceae to indicate a relationship to Liliidae. Dahlgren et al. (1985) placed Commelinaceae in Commeliniflorae together with Eriocaulaceae, Mayacaceae, Rapateaceae, and Xyridaceae, which are members of Poales (sensu APG). Hanguanaceae were considered part of Asparagales by Dahlgren et al. (1985) and Cronquist (1981) thought them to be related to *Lomandra* (Lomandroideae of Asparagaceae sensu APG III 2009; APG IV 2016).

Commelinaceae and Hanguanaceae are sisters (BS = 74% in Chase et al. 2006; BS = 100% in Saarela et al. 2008), but synapomorphies are thus far unknown. Commelinaceae (Fig. 7.13i,j) are generally soft and fleshy herbs, some with a degree of succulence, and their inflorescences have cymose units; flowers have a differentiated perianth (Fig. 7.13i) and are typically fugacious (short lived); their hairy anthers are well known. *Hanguana* is an often massive, coarse herb with petiolate leaves and many parallel secondary veins; its inflorescence is a panicle with sessile, nondescript flowers; its most distinctive trait is that the fruits are berries with a single, bowl-shaped seed. In many characters, *Hanguana* is a better match for Zingiberales (Rudall et al. 1999) than Commelinales, but molecular results are clear about its position in Commelinales.

Haemodoraceae, Philydraceae, and Pontederiaceae share a few characters, such as styloids and tanniniferous tepals that are typically persistent in fruit. These three families are relatively unusual in the commelinids due to their tepalar perianth. Philydraceae (Fig. 7.13k) are sister to the other two and characterized by distichous phyllotaxis (except in *Philydrella*), pilose inflorescences, and monosymmetric flowers with a single stamen. The outer perianth whorl is petaloid, but much smaller than the inner whorl. In Haemodoraceae and Pontederiaceae, the outer whorl is nearly the same as the inner. Haemodoraceae (Fig. 7.13l,m) grow typically in dry sites and have two-ranked, unifacial leaves and cymes; most genera have mostly inferior ovaries. Roots of Haemodoraceae have an orange to red color (Fig. 7.13m) that is the basis for the family name. Pontederiaceae are emergent aquatics or wetland plants with sheathing leaf bases and a distinct petiole and blade; flowers exhibit tristyly, enantiostyly, and monosymmetry.

Zingiberales

Zingiberales are a well-defined clade of eight families (Figs. 7.11, 7.14): Cannaceae (monogeneric with 19 species), Costaceae (4 genera, 110 species), Heliconiaceae (monogeneric with 100 to 200 species), Lowiaceae (monogeneric with 15 species), Marantaceae (31 genera, 550 species), Musaceae (3 genera, 35 species), Strelitziaceae (3 genera, 7 species), and Zingiberaceae (50 genera, 1,300 species).

A close relationship among these eight families has long been recognized from their morphology (Tomlinson 1962; Dahlgren and Rasmussen 1983; Kress 1990; Stevenson and Loconte 1995). Molecular phylogenetic studies involving single genes recovered a Zingiberales clade (Chase et al. 1993, 1995a; Duvall et al. 1993a; D. Soltis et al. 1997b), and combined molecular datasets provided strong support for this clade (Chase et al. 2000c; D. Soltis et al. 2000). These families are well known for their androecial modifications, such that what appears to be the perianth is often petaloid stamens (Fig. 7.14h,j). The flowers are thus difficult to interpret and are infamous among taxonomy students for not being what they appear to be.

Non-DNA synapomorphies for Zingiberales include presence of silica cells in the bundle sheath; leaves well differentiated into a petiole and blade (Fig. 7.14b,e; with a reversal in *Costus*, Fig. 7.14i); leaf blade with pinnate venation, often tearing between the secondary veins (Fig. 7.14b,e); leaf blade in bud rolled into a tube (Fig. 7.14i); petiole with enlarged air canals; flowers often with bilateral symmetry (Fig. 7.14d,f,j); ovary inferior (Fig. 7.14a,d,h); and arillate seeds with endosperm (Fig. 7.14c; Judd et al. 2002, 2008). These plants are mostly herbs, although some achieve large size, such as *Musa*, *Ravenala*, and *Strelitzia* (Fig. 7.14e), which are monstrous herbs.

Relationships within Zingiberales have been partially resolved in molecular studies (Chase et al. 2000c; D. Soltis et al. 2000; Kress et al. 2001; Barrett et al. 2013). Even with a complete plastid gene analysis, Barrett et al. (2013) could not sort out the basal nodes. A tetrachotomy is present of Lowiaceae sister to Strelitziaceae, Musaceae, Heliconiaceae, and a clade with Cannaceae sister to Marantaceae and Costaceae sister to Zingiberaceae (Fig. 7.11). The clade of Cannaceae, Marantaceae, Costaceae, and Zingiberaceae is supported by several putative synapomorphies, including androecium of a single functional stamen (Fig. 7.14j), showy staminodes, seeds with more perisperm than endosperm, and absence of raphides in vegetative tissue (Tomlinson 1962, 1969; Kress et al. 2001; Judd et al. 2002, 2008).

Musaceae (Fig. 7.14b-d) are often large herbs with spirally arranged leaves in which the secondary veins are at right angles to the midvein; their inflorescences have large, deciduous bracts that subtend fascicles of flowers with five fused tepals and six stamens. Their stamen number is higher than other Zingiberales and responsible for the family being viewed as "primitive" within the order. Heliconiaceae (Fig. 7.14a) are similar to Musaceae, except that the former have distichously arranged leaves and different fruit types. Like Musaceae, Heliconiaceae have large bracts subtending fascicles of flowers, but in Heliconiaceae the bracts persist and are highly colored. In both families, floral parts are in fives with the androecium basally fused to the tepals.

Strelitziaceae (Fig. 7.14e-g) and Lowiaceae are sisters and share a petiole with air canals, a floral column that is formed from the sterile apex of the ovary, and a capsular fruit. Strelitziaceae have two-ranked, long-petiolate leaves, inflorescences that are fibrous and boat-shaped, and flowers in which the lateral petals are connate. Lowiaceae have shorter-petiolate leaves with prominent veins and flowers with the median sepal uppermost and the median petal lowermost, forming a lip.

Figure 7.14. Selected members of Zingiberales. a. *Heliconia lingulata* Ruiz & Pav. (Heliconiaceae), flowers subtended by colorful, 2-ranked bracts. b. *Musa ornata* Roxb. (Musaceae), showing colorful, spirally arranged bracts, pinnately veined leaves. c. *M. balbisiana* Colla (Musaceae), fruit (berry), cross-section showing seeds. d. *M. balbisiana* (Musaceae), flowers with single tepal of inner whorl holding nectar. e. *Strelitzia nicolai* Regel & Körn. (Strelitziaceae), tree with 2-ranked leaves. f. *Strelitzia reginae* Banks (Strelitziaceae), zygomorphic flowers with calyx (orange) and corolla (blue). g. *Ravenala madagascariensis* Adans. (Strelitziaceae), dehisced capsules showing iridescent, blue arils. h. *Canna × indica* L. (Cannaceae), flowers with no plane of symmetry, sepals, petals, petaloid staminodia, a broad and flat style, and inferior, papillose ovary. i. *Costus lucanusianus* J. Braun & K. Schum. (Costaceae), spiromonostichous phyllotaxy. j. *Costus malortieanus* H. Wendl. (Costaceae), flowers, note showy, fused staminodia and single stamen grasping the style.

Cannaceae and Marantaceae share flowers that lack any plane of symmetry (asymmetrical flowers; Fig. 7.14h); the single fertile stamen is modified with half of the structure expanded and staminodial. Cannaceae differ from the rest in their flattened style and muricate ovary (Fig. 7.14h). Marantaceae differ in having clearly petiolate leaves with a pulvinus (which facilitates their closing at night) and flowers in mirrored pairs. Several morphological characters unite Zingiberaceae and Costaceae: connate sepals, fused staminodes, reduction of two of the three stigmas, and a ligule at the apex of the leaf sheath. In addition, in Costaceae and Zingiberaceae the single functional stamen grasps the style. Costaceae differ from Zingiberaceae in having spiromonostichous leaves (Fig. 7.14i), whereas Zingiberaceae have more or less distichously arranged leaves and often a branched inflorescence. Costaceae have been studied in detail by Specht (2006), Marantaceae by Prince and Kress (2006a,b), Zingiberaceae by Harris et al. (2006), and Lowiaceae by Johansen (2005).

Poales

Poales (sensu APG) comprise 14 families (Figs. 7.11, 7.15): Bromeliaceae (57 genera, 1,400 species; Neotropical, but 1 species in West Africa), Cyperaceae (98 genera, 4,350 species; cosmopolitan), Ecdeiocoleaceae (2 genera, 2 species; Western Australia), Eriocaulaceae (10 genera, 1,160 species; pantropical to temperate), Flagellariaceae (monogeneric, 4 species; Paleotropics), Joinvilleaceae (monogeneric, 2 species; Malay Peninsula to the Pacific), Juncaceae (7 genera, 430 species; cosmopolitan), Mayacaceae (monogeneric, about 4 species; Neotropics, but 1 species in southeastern North America and 1 in West Africa), Poaceae (650 genera, 9,700 species; cosmopolitan), Rapateaceae (16 genera, 94 species; Neotropical with 1 genus in West Africa), Restionaceae (64 genera, 660 species; East Asia, Indomalaysia to Australia and New Zealand, but 2 genera in South America; including Anarthriaceae and Centrolepidaceae), Thurniaceae (bigeneric, 4 species; South Africa and eastern South America), Typhaceae (2 genera, 65 species—*Typha* could be split into two genera based on some analyses; nearly cosmopolitan; including Sparganiaceae), and Xyridaceae (5 genera, 260 species; pantropical with a few species in the temperate zone). Hydatellaceae (monogeneric, 12 species; India to New Zealand) were considered to be members of the order, but they are not even monocots and belong instead to Nymphaeales (Saarela et al. 2007; Chapters 3 and 4).

A broadly defined Poales is well supported by DNA analyses (e.g., Chase et al. 2000c; D. Soltis et al. 2000, 2011; Givnish et al. 2010; Hertweck et al. 2015). Several non-DNA characters support this broad circumscription of Poales, including silica bodies in the epidermis and strongly branched styles. Poales are also united by the loss of raphide crystals and sepalar nectaries. Some authors have objected to the inclusion of so many families in Poales (Judd et al. 1999, but not Judd et al. 2002, 2008). Chase et al. (2006) and Givnish et al. (2010) found support for two subclades that Judd et al. (1999) recognized as orders: Juncales with Cyperaceae, Juncaceae, and Thurniaceae and Poales with Flagellariaceae, Joinvilleaceae, Ecdeiocoleaceae, Poaceae, and Restionaceae. However, the presence in Judd et al. (1999) of several monofamilial orders (e.g., Bromeliales, Typhales) argues in favor of the broader circumscription. The argument that the order is too large and diverse does not hold up against the comparable levels of diversity and numbers in Asparagales and even the much smaller Commelinales (sensu APG). Poales s.l. also compare well with other recognized monocot orders if one considers naming clades of equal ages (Bremer 2002).

Within Poales sensu APG (Fig. 7.11), there are two well-supported and competing topologies for the relationship for Bromeliaceae and Typhaceae. One (Givnish et al. 2010) is that Bromeliaceae are sister to the rest, followed by Typhaceae (including *Sparganium*, which has sometimes been placed in its own family, Sparganiaceae), and the other (Soltis et al. 2011) is that these two are sister families; both topologies are well supported in their respective analyses. Here we illustrate this conflict by showing the base of Poales as unresolved (Fig. 7.11). Following the divergence of Bromeliaceae and Typhaceae (in either of the two above incongruent topologies), Rapateaceae are sister to the remaining Poales. Bromeliaceae have long been viewed as an isolated family of unclear relationships (Dahlgren et al. 1985), with possible links to the lilioids (through Velloziaceae), Commelinales (through Haemodoraceae and Pontederiaceae), and Poales (through Eriocaulaceae and Xyridaceae). It is clear that Bromeliaceae are isolated morphologically and could share characters with these groups (those shared with the lilioid orders, Asparagales and Liliales, and Commelinales most likely being symplesiomorphies).

Typhaceae (Fig. 7.15a) include two small genera, both of which are rhizomatous, floating or emergent, marsh-loving herbs with two-ranked leaves and small, chaffy flowers arranged in complex inflorescences with the female flowers below the zone of male flowers. Bromeliaceae (Fig. 7.15b) are mostly rosette-forming, usually epiphytic herbs with lepidote hairs and a bracteate inflorescence. That Bromeliaceae may be related exclusively to Typhaceae is reflected in that both lack the mitochondrial gene *sdh4* (Adams et al. 2002) and possess helobial endosperm formation and an amoeboid tapetum; they differ principally in that Typhaceae are wind-pollinated as opposed to the primarily entomophilous Bromeliaceae.

Figure 7.15. Selected members of Poales. a. *Typha angustifolia* L. (left) and *T. latifolia* L. (right) (Typhaceae), densely packed fruits in an elongated spike, associated with hair-like perianth parts. b. *Vriesea sintenisii* (Baker) L. B. Sm. & Pittendr. (Bromeliaceae), strap-shaped leaves, tank-like epiphytic habit. c. *Eriocaulon decangulare* L. (Eriocaulaceae), leaves in basal rosette and capitate inflorescences on long scapes. d. *E. decangulare* (Eriocaulaceae), close-up of capitate inflorescence. e. *Xyris platylepis* Chapm. (Xyridaceae), a head (cone-like) with only a single open flower, with three showy petals (connate, but only the three lobes are evident), three stamens, three staminodia. f. *Mayaca fluviatilis* Aubl. (Mayacaceae), flower of a clubmoss-like aquatic. g. *Juncus dichotomus* Elliott (Juncaceae), capsular fruit, each with six persistent tepals. h. *Carex raynoldsii* Dewey (Cyperaceae), showing perigynia enclosing nutlets and divided stigma. i. *Carex hamata* Sw. (Cyperaceae), showing sterile and apically uncinate rachillae protruding from perigynia. j. *Anarthria scabra* R. Br. (Restionaceae), carpellate inflorescence. k. *Anarthria scabra* (Restionaceae), staminate inflorescence. l. *Chondropetalum tectorum* (L.f.) Pillans (Restionaceae), the leaf blades are reduced, represented primarily by the sheaths. m. *Centrolepis aristata* (R. Br.) Roem. & Schult. (Restionaceae), reduced flowers. n. *Flagellaria indica* L. (Flagellariaceae), leaves 2-ranked, and apically coiled, an adaptation for climbing. o. *Ecdeiocolea monostachya* F. Muell. (Ecdeiocoleaceae), inflorescence. p. *Joinvillea gaudichaudiana* Brongn. & Gris (Joinvilleaceae), plant in fruit (note drupes). q. *Coix lacryma-jobi* L. (Poaceae), showing involucres (modified leaf sheaths) encasing sessile carpellate spikelet, and the pedicellate staminate spikelets protruding. r. *Lolium perenne* L. (Poaceae), spikelets with two bracts per flower, sagittate anthers, and feathery stigmas. s. *Uniola paniculata* L. (Poaceae), spikelets, sagittate anthers.

Bromeliaceae have been the focus of several recent studies that have included large numbers of (usually) plastid DNA regions. Givnish et al. (2014) assembled the largest dataset so far, 9341 aligned nucleotide positions (from eight loci) and 90 species of Bromeliaceae to look at correlations between species numbers and ten hypotheses that have sought to explain species diversity in this mostly epiphytic family. They found that all of these anticipated patterns were significantly correlated with increases in net diversification. These were tied to life in fertile, moist, large mountain ranges, with additional ones connected to epiphytism, avian pollination, and possession of tanks. They detected six large adaptive radiations, and the highest rates of net diversification were observed in the tank clade (subfamily Bromelioideae) in the mountain ranges of coastal Brazil (Mata Atlantica) and the core tillandsioids (subfamily Tillansioideae), which occur principally in the Andes and mountains of Central America. Carnivory has been documented in *Brocchinia* and may have evolved more than once (Chapter 13).

Rapateaceae are rosette-forming herbs that can reach large sizes; their flowers are arranged in scapose inflorescences and are large with distinct perianth whorls and six poricidally dehiscing anthers. Many authors have compared them to Xyridaceae (Cronquist 1981; Dahlgren et al. 1985), but their position as sister to a large clade containing Cyperaceae, Poaceae, Xyridaceae, and related families is clear, indicating that the similarities are parallelisms. The peculiar distribution of Rapateaceae in isolated areas of South America and West Tropical Africa compares well with that of Bromeliaceae and especially Xyridaceae.

Following these early-branching Poales, a large clade of Juncaceae, Cyperaceae, Thurniaceae, Eriocaulaceae, Xyridaceae, Flagellariaceae, Restionaceae, Ecdeiocoleaceae, Joinvilleaceae, and Poaceae is supported by DNA data (Fig. 7.11). Non-DNA characters uniting this large clade are unclear, however. Within this large clade, several clear subclades are apparent.

Xyridaceae and Eriocaulaceae are strongly supported (Chase et al. 2006, but not in Graham et al. 2006 or Givnish et al. 2010). A relationship of Xyridaceae to Eriocaulaceae has been widely accepted (Dahlgren et al. 1985). The two families (Fig. 7.15c-e) share a similar habit (rosette-forming herbs), strictly basal leaves with paracytic stomata, capitate inflorescences with dimerous flowers in which the anthers are adnate to the corolla, a tepalar perianth (calyx and corolla indistinguishable), ovules with thin-walled megasporangia, and spinulate/echinate pollen. However, the relationship of this pair to Mayacaceae is unclear. Bremer (2002) had difficulties obtaining sequences from Mayacaceae, and Chase et al. (2006) and Graham et al. (2006) had them placed outside this clade but still in Poales. Givnish et al. (2010) placed Mayacaceae in a novel but well-supported position as sister to Eriocaulaceae, to which Xyridaceae were a weakly supported sister taxon. Clearly, the relationships of Mayacaceae are still to be resolved. Mayacaceae (Fig. 7.15f) are morphologically unusual, looking much like a clubmoss, except that the plants are aquatic. Members of Mayacaceae have spirally arranged leaves with apical teeth and axillary flowers with clearly distinct calyx and corolla and three stamens. Here we show these three families as a trichotomy (Fig. 7.11).

Cyperaceae, Juncaceae, and Thurniaceae form a well-supported clade in multi-gene analyses (Figs. 7.11, 7.15g-i). *Prionium* had been placed in Juncaceae, but molecular results have indicated that it should be a member of Thurniaceae. Thurniaceae are sister to Juncaceae and Cyperaceae (Plunkett et al. 1995; Munro and Linder 1997; Chase et al. 2006; Graham et al. 2006; Givnish et al. 2010). Non-DNA features also unite the three families: solid stems with three-ranked leaves and pollen in tetrads. The families share an unusual feature of chromosomes with diffuse centromeres (Plunkett et al. 1995; Simpson 1995; Munro and Linder 1998). Juncaceae were found in two studies not to be monophyletic. *Oxychloe* (Juncaceae) was embedded in or sister to Cyperaceae (Plunkett et al. 1995; Muasya et al. 1998). The former study may have used a mixed leaf collection of *Oxychloe* and a sedge and almost certainly sequenced the sedge; the latter was based on a contaminated sample. True *Oxychloe* has now been sequenced, and it goes with other genera of Juncaceae, so both families are monophyletic (Jones et al. 2007) and should be retained.

Core Poales consist of Restionaceae (see below) as sister to Flagellariaceae, Joinvilleaceae, Ecdeiocoleaceae, and Poaceae (Fig. 7.11). The monophyly of core Poales is well supported by DNA analyses (e.g., Chase et al. 2000c, 2006; Graham et al. 2006; D. Soltis et al. 2000, 2011; Givnish et al. 2010). However, a close relationship among these families had long been recognized (e.g., Dahlgren and Rasmussen 1983; Dahlgren et al. 1985).

Most families of core Poales are herbaceous plants native to the Southern Hemisphere and are small in terms of number of genera and species. The two largest families are Restionaceae and Poaceae. Non-DNA synapomorphies of core Poales include two-ranked leaves with an open sheath around the stem, stomata with dumbbell-shaped guard cells, a single apical orthotropous ovule per carpel, monoporate pollen with a rim around the pore, a complex (pinnately branched) stigma, and nuclear endosperm development (Endress 1995a; Kellogg and Linder 1995; Soreng and Davis 1998).

Restionaceae (Fig. 7.15j-m) as treated here include Anarthriaceae (Fig. 7.15j) and Centrolepidaceae, which share dioecy, peg cells in their chlorenchyma, and dorsifixed anthers. Linder et al. (2000) suggested that these should be considered members of one family, Restionaceae because they have the distinctive culm anatomy of that family. Centrolepidaceae, which are mostly small plants, could well be just paedomorphic Restionaceae (Linder et al. 2000). Relationships among the sometimes-recognized families Anarthriaceae, Centrolepidaceae, and Restionaceae s.s. have been problematic and varied depending on which methods and loci were used (Briggs et al. 2010, 2014). Given this uncertainty, one option is to recognize the broader circumscription of Restionaceae, which do have some obvious synapomorphies (Chapter 12).

Restionaceae (including Anarthriaceae and Centrolepidaceae) are sister to the remainder of core Poales, which includes Flagellariaceae (Fig. 7.15n), Joinvilleaceae (Fig. 7.15p), and Ecdeiocoleaceae (Fig. 7.15o) as successive sister groups to Poaceae (Fig. 7.11; Chase et al. 2006; Givnish et al. 2010). Poaceae (Gramineae; Fig. 7.15q-s) are one of the largest and most important angiosperm families. A clear picture of the phylogenetic relationships has emerged for Poaceae from the combination and analysis of molecular datasets by a consortium of researchers, the Grass Phylogeny Working Group (GPWG 2000, 2012; www.virtualherbarium.org/grass/gpwg/; see also Chapter 13).

The origin of Poaceae can be constrained in part by the appearance of grass pollen, which is distinctive, in the fossil record, with the earliest accepted records from the Paleocene of South America and Africa (60–55 mya; Thomasson 1987). The oldest known grass fossils are provided by silicified cuticles with diagnostic grass phytoliths in dinosaur coprolites from the latest Cretaceous (Late Maastrichtian, ca. 66 mya) of India (Prasad et al. 2005, 2011) (Suppl. Table 7.1).

CHARACTER EVOLUTION IN POACEAE

Because Poaceae have enormous ecological and economic significance, it is important to consider their origin and evolution; Kellogg (2000) and Bouchenak-Khelladi et al. (2014) provided excellent overviews. Using the GPWG phylogeny and by comparing grasses with their closest relatives, Kellogg (2000) reconstructed the types of changes that likely occurred in the early evolution of Poaceae. A major change occurred in the timing of embryo development. Most monocotyledons have undifferentiated embryos; seed maturation begins after the embryo has formed a shoot apical meristem, but the differentiation of cotyledon, leaves, root meristem, and vasculature largely occurs after the seed is shed from the parent plant. In the grasses, however, embryo development is accelerated relative to seed maturation (Kellogg 2000).

The immediate ancestors of the grasses had ovaries of three united carpels; each carpel possessed one locule with one ovule (Kellogg and Linder 1995). In the grasses, only one locule and one ovule form. As the ovule develops, the outer integument fuses with the inner ovary wall to form the distinctive fruit of the grasses, known as the grain or caryopsis. The structure is unique among flowering plants.

Perhaps the most striking characteristic of grasses today is their floral and inflorescence structure. Grass flowers are generally arranged in spikelets; each spikelet consists of one or more flowers (termed florets in Poaceae) and associated bracts (Fig. 7.15r,s). Phylogenetic trees reveal that the spikelet must have originated in several steps (GPWG 2000, 2012). The earliest grasses had three stigmas, a relic of the three fused carpels they inherited from their ancestors; this number was reduced to two after the divergence of *Pharus*. The earliest species also had, like their non-grass ancestors, six stamens. It is unclear precisely when the shift from six to three stamens occurred, but it must have been after the divergence of the *Guaduella*/*Puelia* group. Poaceae also provide a model for the evolution of C_4 photosynthesis (covered in Chapter 13; Fig. 13.13).

FOSSIL HISTORY

Some putative monocot fossils have been found in strata as old as the eudicots (Herendeen et al. 1995 and Iles et al. 2015; Supplementary Table 7.1 at press.uchicago.edu/sites/soltis/), and what we know of the angiosperm tree (Chapter 3) indicates that monocots should be among the oldest lineages of angiosperms. Iles et al. (2015) reviewed monocot fossils that are well determined in terms of ages and taxonomic identity; they recommend the use of 34 fossils from 19 families and eight orders as constraints in molecular dating analyses. Conran et al. (2015) also documented a number of well-identified fossils from Australia and New Zealand that are suitable for calibration points. We provide here a critically reviewed set of monocot fossils, following Friis et al. (2011), suitable to use as calibration points (Supplementary Table 7.1).

Wikström et al. (2001) produced dates ranging from 158 to 141 mya for the stem node of the monocots and from 141 to 127 mya for the crown group (the split of *Acorus* and the rest of the monocots). Bremer (2000) put the crown group node of the monocots at 134 mya, which indicates that the stem lineage of the monocots goes back to ~200 mya. Bell et al. (2010) found estimates from 130–146 mya (Chapter 2; Supplementary Table 7.1). Zeng et al. (2014) put the crown group node at 150 mya. These dates agree well with the estimates of Magallón (2010;

140 mya for the monocot crown node) and Iles et al. (2015; 156–164 mya).

Fossils of the palms were once thought to be the oldest reliable monocots; they first appeared around 90 mya. However, Iles et al. (2015) cited the 113–125 mya pollen of *Liliacidites* (Doyle and Hickey 1976) as a valid constraint for the stem node of monocots. Recently, the fossil record of Araceae has been extended well into the early Cretaceous. Distinctive inaperturate striate fossil pollen and infructescence fragments attributable to Araceae have been recognized from the early Cretaceous (Barremian-Aptian) of Portugal (ca. 125 mya; Friis et al. 2004, 2010, 2011). The identify of fossil flowers initially attributed to Triuridaceae (Pandanales) from the Upper Cretaceous of New Jersey, ca. 90 mya (Gandolfo et al. 2002), has been questioned due to differences in pollen morphology and the lack of information on pistillate flowers (Rudall 2003); Friis et al. (2011) have noted that some of the distinctive features of these fossils also occur in magnoliids. Older dates for many individual families have been reported, such as 100–110 mya for a fossil proposed to be a bambusoid grass (Poinar 2004), but it had none of the clearly distinctive features of a grass (S. Y. Smith et al. 2010); the earliest well-documented grass occurrence, mentioned above, is that from the latest Cretaceous (Late Maastrichtian, ca. 66 mya) of India (Prasad et al. 2005, 2011). Our earliest records for many important monocot clades are Cenozoic (Supplemenary Table 7.1), indicating the need for intensified investigation of Cretaceous strata that may reveal older representatives.

8 Eudicots (+ Ceratophyllaceae)

Introduction and Early-Diverging Lineages

INTRODUCTION

Eudicots (*Eudicotyledoneae* sensu Cantino et al. 2007) represent ~75% of all angiosperm species (Drinnan et al. 1994). Analyses of three or more genes provided strong support for their monophyly (see Chapter 3), within which multiple datasets have identified a basal grade of eudicots, followed by a clade of core eudicots, as well as several major subclades of core eudicots (Fig. 8.1).

Eudicots are one of the few major angiosperm clades to have a clear unambiguous synapomorphy, triaperturate pollen (Doyle and Hotton 1991) (Fig. 8.2). The lines of weakness (apertures) represent areas for pollen tube growth. Possession of three apertures is hypothesized to be advantageous and may represent a key factor in the success of eudicots (Furness and Rudall 2004). Although triaperturate pollen is a synapomorphy for eudicots, not all eudicots have three-grooved pollen grains. Some other types (e.g., inaperturate, polyporate, and polycolporate) also occur due to subsequent evolution of pollen structure.

Because triaperturate pollen is so distinctive, the estimated age of crown eudicots is perhaps the firmest date in the paleobotanical record. Fossilized pollen documents the presence of the eudicots at the Early Cretaceous (Barremian-Aptian boundary, 125 mya; Crane et al. 1995; Magallón et al. 1999). The fossil record indicates an uneven distribution of species diversity across the major clades of eudicots, with the most species-rich groups known only from the relatively recent fossil record, suggesting that most eudicot species diversity is the result of relatively recent radiations (e.g., 50–70 mya; Magallón et al. 1999; Bell et al. 2005, 2010).

We focus here on the basal branches of the eudicot clade. Several lineages consistently emerge as a grade, sister to the core eudicots (*Gunneridae*): Ranunculales, Proteales (including Sabiaceae; APG IV 2016), Trochodendrales, and Buxales (depicted in Figs. 8.3–8.5). Gunnerales and Dilleniaceae are introduced in Chapter 9; major clades of eudicots (superrosids, superasterids) are discussed in Chapters 10 and 11 (Judd and Olmstead 2004; Stevens 2004; Endress 2010a).

The early-diverging eudicots (Figs. 8.1, 8.3–8.5) deserve special attention because of their critical phylogenetic position and morphological diversity. The term "early-diverging eudicots" has been used to refer to this basal grade (e.g., D. Soltis et al. 2000, 2005b), and we use this term here. However, others may prefer "basal eudicots" for consistency with use of the terminology "basal angiosperm grade" in Chapter 4. We do not advocate applying names to grades and do so here only as a convenient way to consider these lineages for discussion.

THE ENIGMATIC CERATOPHYLLACEAE

Ceratophyllaceae consist of the single genus *Ceratophyllum* (~30 species of aquatics), which is highly distinctive morphologically (Fig. 8.5). Traditionally, the genus was associated with water lilies (Nymphaeaceae and segregate families; e.g., Cronquist 1981, 1988; Takhtajan 1997). Initial analyses of *rbcL* sequences indicated the family might be sister to all other angiosperms (e.g., Les et al. 1991; Chase et al. 1993; Qiu et al. 1993). This position received added attention because fossils with distinctive horned fruits simi-

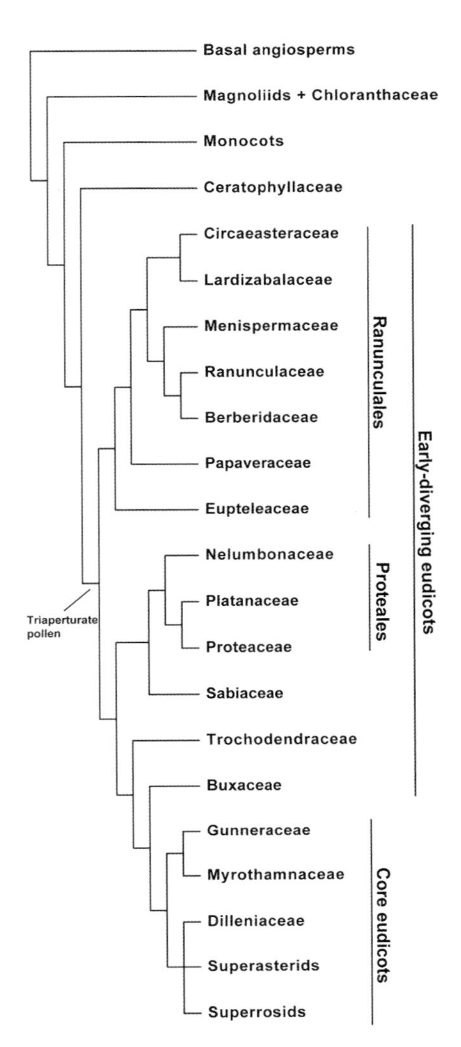

Figure 8.1. Phylogenetic summary tree depicting relationships among early-diverging eudicots, based on analyses of 17 genes (Soltis et al. 2011) and complete plastid genomes (based on Moore et al. 2010).

lar to those of *Ceratophyllum* are among early fossil angiosperms (e.g., Dilcher 1989; Dilcher and Wang 2009) (Supplementary Table 8.1 at press.uchicago.edu/sites/soltis/). The earliest unequivocal *Ceratophyllum* fruits are from the Late Cretaceous (Late Campanian) Horseshoe Canyon Formation of Alberta (Aulenback 2009); it is particularly common in Eocene lake deposits of the Green River Formation in southwestern Wyoming (Fig. 8.5g,h).

However, subsequent analyses of other genes revealed long-branch issues (Felsenstein 1978b) with the initial placement of *Ceratophyllum* and suggested alternative placements. A sister-group relationship between monocots and Ceratophyllaceae was tentatively suggested by several early studies using nuclear, mitochondrial, and plastid gene sequences (Soltis et al. 1997b; Qiu et al. 1999; Savolainen et al. 2000a). P. Soltis et al. (2000) obtained low support (BS = 71%) for this sister-group relationship. Using 17 plastid genes, Graham and Olmstead (2000) also found low BS support (54%) for monocots + Ceratophyllaceae, as did Zanis et al. (2002) using 11 genes (BS = 57%). The strongest support (>90%) for Ceratophyllaceae + monocots was provided by Zanis et al. (2003). Most studies employed parsimony, but some maximum likelihood analyses also indicated that monocots and *Ceratophyllum* are sister taxa (Graham and Olmstead 2000; Qiu et al. 2000; P. Soltis et al. 2000; Zanis et al. 2002, 2003), as did a Bayesian analysis (Zanis et al. 2002). Because taxon sampling was often sparse and internal support for this relationship was generally low, the result was considered tenuous.

In contrast, *matK* alone (using broad sampling) placed Ceratophyllaceae sister to eudicots (parsimony jackknife 67–71%; Hilu et al. 2003). Other multigene studies with broader taxon sampling also placed Ceratophyllaceae sister to eudicots. In D. Soltis et al. (2000), Ceratophyllaceae were sister to the eudicots with low jackknife support (53%, parsimony; 0.91 Bayesian, Soltis et al. 2007c). With five genes, ML BS support for Ceratophyllaceae + eudicots was 65% (Burleigh et al. 2009); with 17 genes, a placement of Ceratophyllaceae with eudicots remained weakly supported (BS = 68%; Soltis et al. 2011). With complete (or nearly so) plastid genome sequences, a clade of Ceratophyllaceae + eudicots was again weakly supported (BS = 64%; Moore et al. 2007). Moore et al. (2007) analyzed the underlying signal in complete plastid genomes and found that the placement of Ceratophyllaceae varied with parsimony depending on codon position and also whether rapidly or slowly evolving genes were used. ML, in contrast, consistently placed Ceratophyllaceae sister to eudicots. Recent analyses of four mitochondrial genes placed Ceratophyllaceae sister to Chloranthaceae with weak support (ML BS = 63%; Qiu et al. 2010; Soltis et al. 2011). Morphological analyses of Endress and Doyle (2009) suggested that *Ceratophyllum* may be related to Chloranthaceae.

A large dataset of complete nuclear 18S + 26S rDNA sequences variously placed Ceratophyllaceae sister to monocots + eudicots or within monocots (Maia et al. 2014). Lee et al. (2011) and Wickett et al. (2014) did not include Ceratophyllaceae in their analysis of numerous nuclear genes. The analysis of many nuclear genes by Zeng et al. (2014) placed *Ceratophyllum* sister to Chloranthaceae; this clade is then sister to the monocots. A recent analysis of numerous nuclear genes for over 1000 species of *Viridiplantae* placed *Ceratophyllum* sister to monocots (Leebens-Mack et al. pers. comm.), whereas datasets of nearly complete plastid genomes for thousands of taxa place *Ceratophyllum* sister to eudicots (Gitzendanner et al. unpubl.).

In summary, although some recent molecular analyses place Ceratophyllaceae sister to eudicots, this placement

Figure 8.2. SEM micrograph of triaperturate pollen. *Rumex* (Chenopodiaceae = Amaranthaceae, APG III 2009; APG IV 2016).

depicted in APG III (2009) and APG IV (2016), internal support for relationships is not strong, and that inference is largely based on plastid data. Hence, the placement of Ceratophyllaceae remains somewhat uncertain. Although the prevailing view is to place Ceratophyllaceae with eudicots (followed here), we urge caution. Furthermore, non-DNA synapomorphies for the placement of Ceratophyllaceae with eudicots are unknown; as submerged water plants, Ceratophyllaceae have lost many structural features (e.g., roots).

EARLY-DIVERGING EUDICOTS

The early-diverging eudicots are Ranunculales, Proteales (incl. Sabiaceae), Buxales (including Didymelaceae; APG III 2009; APG IV 2016), and Trochodendrales (including Tetracentraceae). (See Figs. 8.3–8.5 for photos of extant and fossil early-diverging eudicots.) Most of these lineages are small; only Ranunculales, Buxaceae, and Proteaceae contain more than a few genera and species.

These lineages generally have a rich fossil history (Supplementary Table 8.1). The oldest fossils of early-diverging eudicots are the extinct *Leefructus* (ca. 122.6 to 125.8 mya; Sun et al. 2011) and well-preserved apocarpous fruits called *Ranunculaecarpus samylina* from the early Middle Albian of Siberia. Floral remains of both Platanaceae (Proteales) and Buxaceae have been reported from the Albian (108 mya; Magallón et al. 1999), and fossils of Proteaceae are known from the mid-Cretaceous (97 mya). Fossils of Ranunculales and Sabiaceae are considerably younger than those of other early-diverging eudicots; both have been reported from the Maastrichtian (69.5 mya; see Fig. 2.1 for time scale; reviewed in Magallón et al. 1999).

Some early-diverging eudicots have figured prominently in older discussions of the evolution of the angiosperms (e.g., Bessey 1915; Cronquist 1968, 1981, 1988; Takhtajan 1969, 1991). Ranunculales were typically placed in Magnoliidae in morphology-based classifications because they were thought to possess ancestral floral features (e.g., Stebbins 1974; Cronquist 1981, 1988; Thorne 1992a,b, 2001; Heywood 1993, 1998). Takhtajan (1997) recognized a distinct subclass, Ranunculidae, closely related to his Magnoliidae.

Ranunculaceae and relatives (e.g., Menispermaceae, Berberidaceae, and Papaveraceae) were central members of

Figure 8.3. Basal eudicots. a. *Grevillea robusta* A. Cunn. ex R. Br. (Proteaceae), inflorescence. b. *Nelumbo nucifera* Gaertn. (Nelumbonaceae), flower. c. *Platanus occidentalis* L. (Platanaceae), leaves and pistillate inflorescence. d. *Euptelea pleiosperma* Hook.f. & Thomson. (Eupteleaceae), fruits. e. *Ficaria verna* Huds. (Ranunculaceae), flower. f, g. *Argemone albiflora* Hornem. (Papaveraceae). f. Capsule and immature seeds, showing parietal placentation. g. Flowers; note wrinkled petals and numerous stamens. h. *Cocculus carolinus* s. (Menispermaceae), endocarps. i. *Fumaria officinalis* L. (Papaveraceae), flowers and fruits. j. *Cocculus carolinus* DC. (Menispermaceae), leaves and fruits.

Figure 8.4. Fossil (a-g; i-l) and living (h) Platanaceae. a. "*Platanus*" *primaeva* Lesq. from Cenomanian of Kansas (note expanded petiole base that probably enveloped axillary bud as in most modern *Platanus* spp.), USNM coll. b. *Macginitiea wyomingensis* from Eocene of Utah. UCMP coll. c. *Platanites raynoldsii* from Paleocene of Saskatchewan, one of several extinct platanaceous species with compound leaves; note small lateral leaflets. Univ. Sask coll. 3-66. d. Fruiting axis of *Macginitiea wyomingensis*, UCMP coll. e. *Macginicarpa* fruiting head showing protruding groups of 5-styled florets (florets delimited and confined by surrounding perianth seen in i, j), unlike extant *Platanus* spp., Eocene of Oregon. UF 5148. f. A single *Macginicarpa* fruitlet, showing persistent style and absence of dispersal hairs. g. *Platanus* fruitlet, Eocene of Oregon, showing basal tuft of dispersal hairs, UF 5156. h. Extant *Platanus* fruitlet for comparison. i., j. Transverse sections of permineralized *Macginicarpa* fruiting head showing florets of 5 fruitlets each, delimited by well-developed perianth; note absence of dispersal hairs around the achene, Eocene of Oregon. k. l, Corresponding staminate head, with 5 stamens per floret, and with pollen preserved in anthers, Eocene of Oregon.

Figure 8.5. Basal eudicots and *Ceratophyllum* (Ceratophyllaceae). a. *Buxus microphylla* Siebold & Zucc. (Buxaceae), fruits and leaves. b. *Meliosma impressa* Krug & Urb. (Sabiaceae), inflorescence. c. *Trochodendron aralioides* Siebold & Zucc. (Trochodendraceae), branch and inflorescence; d. *Berberis julianae* C. K.Schneid. (Berberidaceae), flowers and young fruits. e. *Ceratophyllum demersum* L. (Ceratophyllaceae), stems and linearly segmented leaves. f. *Ceratophyllum echinatum* A. Gray. (Ceratophyllaceae), fruits. g. *Ceratophyllum* sp. (Ceratophyllaceae), vegetative branches, early Eocene of Wyoming. h. *Ceratophyllum* sp. (Ceratophyllaceae), fruit from Early Eocene of Wyoming.

Bessey's (1915) Ranalean complex, which contained those angiosperms he considered most "primitive." Ranunculales include taxa with numerous stamens and carpels, which are often spirally arranged, consistent with interpretations of "primitive" angiosperm traits. Cronquist (1968, 1981, 1988), for example, considered members of Ranunculales to represent the herbaceous equivalent of his woody Magnoliales. Cronquist (1968, 1981) and Takhtajan (1987, 1991) suggested that Ranunculales originated from Magnoliales via Illiciales or some close relative of the latter. There has been renewed interest in Ranunculales (see below) in light of their phylogenetic placement as sister to all other eudicots. *Trochodendron* and *Tetracentron* have also attracted special attention because they possess several "magnoliid" features (e.g., the production of ethereal oil cells, chloranthoid leaf teeth, and valvate anther dehiscence) long considered "primitive" or plesiomorphic. These features were once thought to have been retained from magnoliid ancestors (Cronquist 1981; Endress 1986a; Crane 1989).

Because of their pivotal phylogenetic position, early-diverging eudicots now figure prominently in investigation of floral developmental genetics and angiosperm genome evolution. Aspects of floral structure, such as merism, are labile in these early-branching eudicots; this lability is comparable to that observed in basal angiosperms and magnoliids (see below and Chapters 4, 6, and 14). Similarly, flower symmetry is of interest in understanding angiosperm diversification. Phylogenetic analysis of a particular (*CYC/TB1*) clade of TCP transcription factors important in specifying dorsal identity in bilateral flowers reveals the occurrence of gene duplications just prior to the origin of core eudicots (Howarth and Donoghue 2006) (Chapter 14). In addition, the evolution of basal eudicots seems to have preceded two closely coinciding ancient whole-genome duplications (a so-called "paleohexaploidy" event) that characterize core eudicots (Jiao et al. 2012). The exact phylogenetic placement of this key event remains uncertain (Chapter 16).

Molecular as well as morphological (e.g., Hufford 1992) phylogenetic studies have consistently identified the same families as successive sisters to all other eudicots (see Chapter 3). Ranunculales have consistently appeared as sister to all other eudicots in DNA analyses. Molecular studies consistently recover Proteales (incl. Sabiaceae), Buxaceae, and Trochodendraceae as the remaining lineages of early-diverging eudicots. These results were apparent even with early analyses of one or a few genes (e.g., Chase et al. 1993; D. Soltis et al. 1997a,b, 2000, 2003c; Hoot et al. 1999; Savolainen et al. 2000a,b; Hilu et al. 2003; Kim et al. 2004b; Worberg et al. 2007) and have been reinforced with larger datasets (e.g., P. Soltis et al. 1999a; D. Soltis et al. 2000, 2007c, 2011; Burleigh et al. 2009; Moore et al. 2010; Qiu et al. 2010; Maia et al. 2014; Ruhfel et al. 2014; Wickett et al. 2014). Datasets from all three genomes also support this result, including recent analysis of numerous nuclear and plastid genes with good taxon sampling (Leebens-Mack et al. pers. comm.; Gitzendanner et al. unpubl.).

Placement of Ranunculales as sister to all other eudicots receives maximal bootstrap support in analyses of combined genes (e.g., Soltis et al. 2011; Moore et al. 2010). However, following Ranunculales, relationships among Proteales, Buxaceae, Trochodendraceae, and the core eudicots have been more difficult to resolve. The best current estimates of phylogeny are those based on 17 genes and complete plastid genomes (Moore et al. 2010; Soltis et al. 2011); both recovered Proteales + Sabiaceae following Ranunculales as sister to all other eudicots (which receive BS = 100%). In Moore et al. (2010), Proteales + Sabiaceae received BS = 70–80%; in Soltis et al. (2011), BS = 59%. Ruhfel et al. (2014) also recovered Proteales + Sabiaceae (BS = 63%). Bootstrap support for Proteales + Sabiaceae was therefore weak to moderate in several studies; however, support for this clade was >90% in a recent analysis of complete plastid genome sequences (Sun et al. 2016), suggesting that Proteales should be expanded to include Sabiaceae, as adopted by APG IV (2016; see Chapter 12). In analyses with fewer genes, Proteales and Sabiaceae sometimes formed a grade (Proteales followed by Sabiaceae), as in P. Soltis et al. (1999a), D. Soltis et al. (2000), and Burleigh et al. (2009). Conflicting results were obtained in these earlier studies based on one or a few genes (e.g., Chase et al. 1993; D. Soltis et al. 1997a,b, 2000, 2007c; Savolainen et al. 2000a,b; Kim et al. 2004b).

Following Proteales (sensu APG IV 2016), there is strong support (BS = 98–100%) in multigene datasets for Trochodendraceae (which includes *Tetracentron*) and then Buxaceae as subsequent sisters to all other eudicots (e.g., Moore et al. 2010; Soltis et al. 2011; Ruhfel et al. 2014). This relationship was not clear in early analyses with fewer genes (e.g., Hoot et al. 1999; P. Soltis et al. 1999a; D. Soltis et al. 2000; Kim et al. 2004b). Conflicting results (albeit without strong support) were obtained in analyses that used individual genes (e.g., Chase et al. 1993; Soltis et al. 1997a,b; Savolainen et al. 2000a,b). Following Buxaceae and Trochodendraceae, there has long been strong support (BS = 100%) for the monophyly of the core eudicots (*Gunneridae* sensu Cantino et al. 2007)—this was apparent even in early studies of one or a few genes (e.g., Hoot et al. 1999; P. Soltis et al. 1999a; Savolainen et al. 2000a; Soltis et al. 2000, 2003c; Hilu et al. 2003).

Early-diverging eudicots share putatively plesiomorphic features with basal angiosperms (Chapters 4 and 6). Ellagic acid and gallic acid (Fig. 8.6) are absent from the early-diverging eudicots, as they are from most basal an-

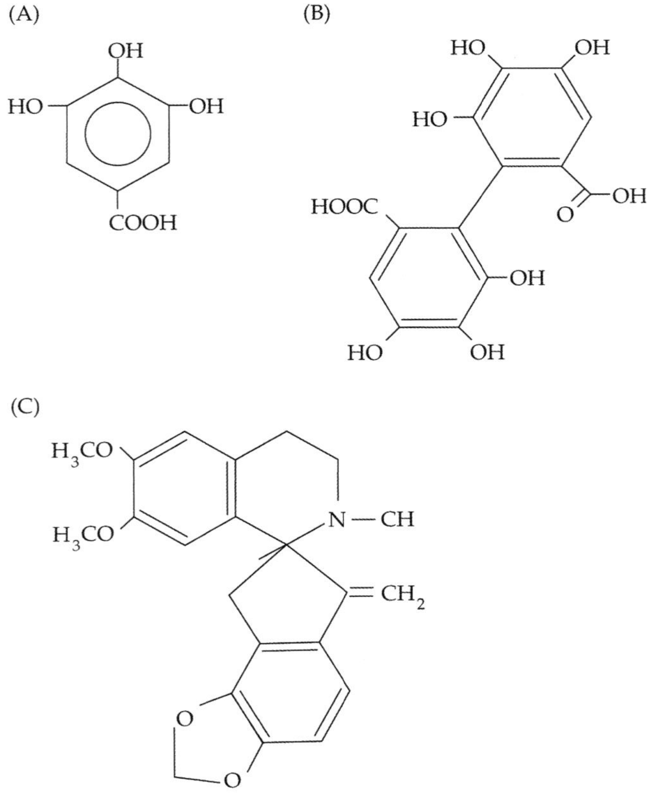

(A)

(B)

(C)

Figure 8.6. Chemical structure of (A) gallic acid, (B) ellagic acid, and (C) ochotensimine, a representative isoquinoline alkaloid (from Soltis et al. 2005).

giosperms, but these acids appear to be prevalent throughout core eudicots (see Fig. 8.7) (e.g., Chase et al. 1993; Savolainen et al. 2000a; D. Soltis et al. 2000). Using data from Nandi et al. (1998), we reexamined the phylogenetic distribution of these compounds. Ellagic acid is absent from basal angiosperms, with the exception of Nymphaeaceae, and from all early-diverging eudicots, but is common throughout core eudicots. With MP, ellagic acid is reconstructed as a putative synapomorphy for core eudicots, although this is equivocal with ML (Fig. 8.7A). The subsequent diversification of ellagic acid production is complex and beyond the scope of our analyses, but several putative losses and secondary gains have clearly occurred in core eudicots (Fig. 8.7A). In fact, the absence of ellagic acid is reconstructed as ancestral for the large asterid clade. Our reconstructions indicate that the evolution of ellagic acid production in some large clades (e.g., Malpighiales and Ericales) is particularly dynamic.

Production of gallic acid displays a similarly complex pattern of evolutionary diversification (Fig. 8.7B). We also reexamined this character—with MP and ML, gallic acid is also reconstructed as a putative synapomorphy for core eudicots (*Gunneridae*) (Fig. 8.7B).

RANUNCULALES

Ranunculales are strongly supported as sister to all other eudicots. Ranunculales consist of seven families representing ~1.6% of eudicot diversity (Fig. 8.1)—Berberidaceae, Circaeasteraceae (including *Kingdonia*, placed in Kingdoniaceae by Takhtajan 1997), Eupteleaceae, Lardizabalaceae (including *Sargentodoxa*), Menispermaceae, Papaveraceae, and Ranunculaceae. Papaveraceae are broadly defined to include Fumariaceae, *Hypecoum* (treated as a distinct family, Hypecoaceae, in Takhtajan 1997), and the monotypic *Pteridophyllum* (placed in its own family, Pteridophyllaceae, in Takhtajan 1997). Ranunculaceae are also broadly defined to include *Hydrastis* and *Glaucidium*, monotypic genera sometimes placed in their own families, Hydrastidaceae and Glaucidiaceae (Takhtajan 1997). Ranunculales have been the focus of several phylogenetic investigations, both morphological and molecular (e.g., Drinnan et al. 1994; Hoot and Crane 1995; Loconte et al. 1995).

With the exception of Eupteleaceae, most families now placed in Ranunculales were considered closely related in morphology-based classifications (e.g., Cronquist 1981; Takhtajan 1987, 1997). Families of Ranunculales share the presence of benzyl isoquinoline alkaloids of the berberine and morphine type (Fig. 8.6; Jensen 1995), primarily herbaceous habit, hypogynous flowers, unusually large and homogeneous S-type sieve element plastids (Behnke 1995), epicuticular wax tubules (also found in some families outside of Ranunculales, such as Nelumbonaceae; Barthlott and Theissen 1995), and seeds with small embryos and copious endosperm (see Hoot et al. 1999). Eupteleaceae, which are sister to the rest of the order, have scalariform perforation plates, whereas the rest of Ranunculales possess simple perforation plates. The epicuticular wax in tubules is also not in Eupteleaceae, but otherwise would be synapomorphic for the order. As a result, no obvious synapomorphies are indicated for Ranunculales by floral morphology or other characters (Endress 1995b), although there are synapomorphies for Ranunculales minus Eupteleaceae.

As noted, Eupteleaceae represent an unexpected addition to Ranunculales. Because its anemophilous (i.e., wind-pollinated) or partly entomophilous (insect-pollinated) flowers lack a perianth (Fig. 8.3), *Euptelea* (the single genus in the family) had traditionally been placed in the old Hamamelidae, close to Trochodendraceae, Cercidiphyllaceae, and Platanaceae (Cronquist 1981, 1988; Endress 1986a, 1993; Takhtajan 1987, 1997). However, molecular studies clearly place *Euptelea* in Ranunculales (e.g., Chase et al. 1993; Qiu et al. 1993; Hoot et al. 1999; D. Soltis et al. 2000, 2011), with Trochodendraceae, Platanaceae, and Cercidiphyllaceae representing more distantly related

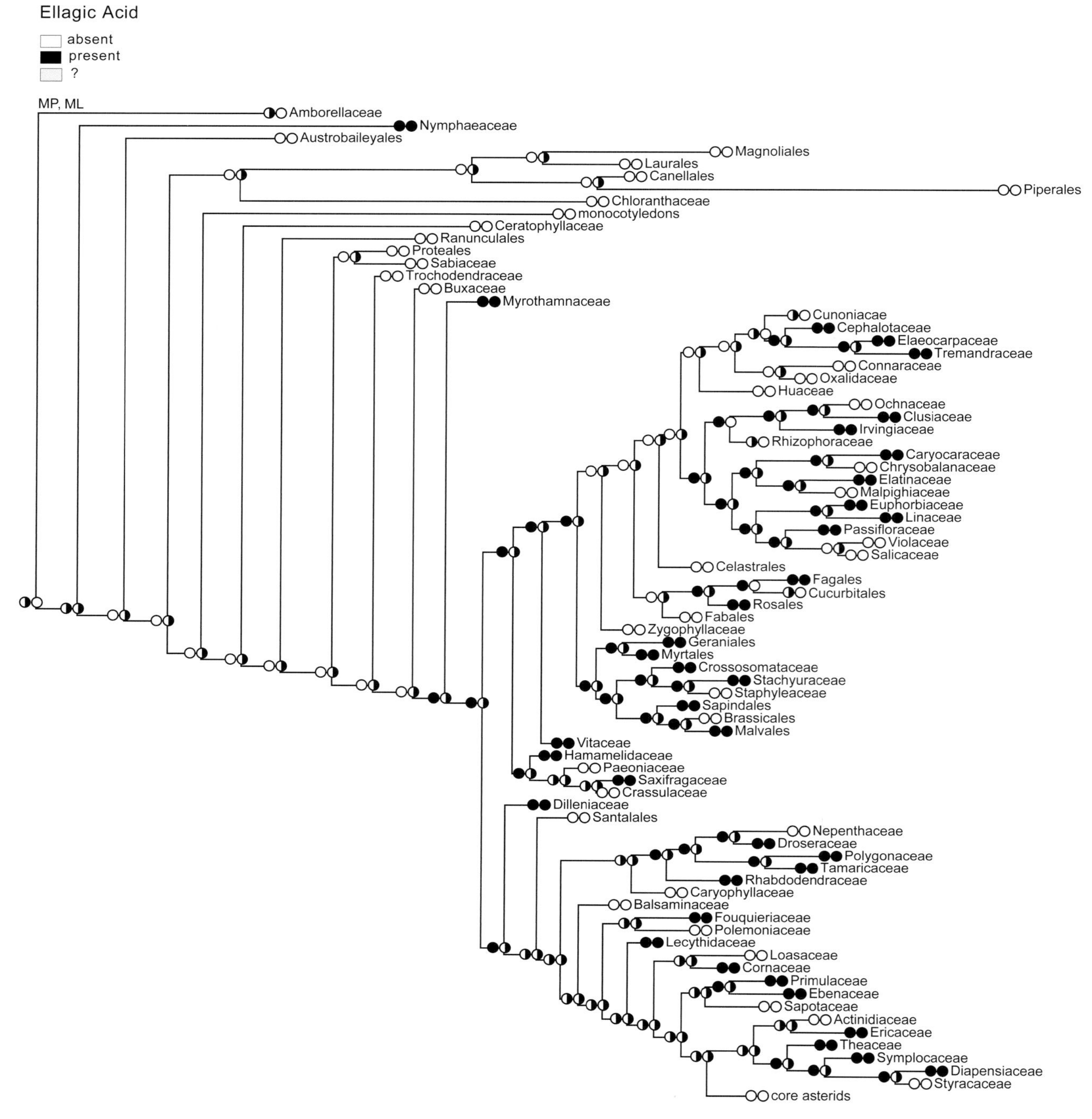

Figure 8.7. Reconstruction of the diversification of ellagic acid (A) and gallic acid (B) in the angiosperms with a focus on early-diverging eudicots. Data taken from Nandi et al. (1998), but using the general topology of Soltis et al. (2011). Some terminals used in Nandi et al. (1998) have been renamed to reflect the APG III (2009) and APG IV (2015) nomenclature and circumscription; other terminals have been excluded from reconstruction.

lineages (see below, Trochodendraceae; Proteales; and Chapter 6).

Molecular studies have also confirmed that *Sargentodoxa* (Lardizabalaceae) belongs to Ranunculales as sister to all other Lardizabalaceae (Hoot et al. 1995b; D. Soltis et al. 1997b, 2000). *Sargentodoxa* had typically been placed in its own family close to Lardizabalaceae in some morphology-based classifications (Cronquist 1981, 1988; Takhtajan 1987, 1997).

The age of Ranunculales is controversial. Using molecular approaches, Anderson et al. (2005) dated the crown group of Ranunculales as 114–121 mya. Magallón and Castillo (2009) obtained a similar molecular estimate for the crown group (~113.2 mya) as did Bell et al. (2010):

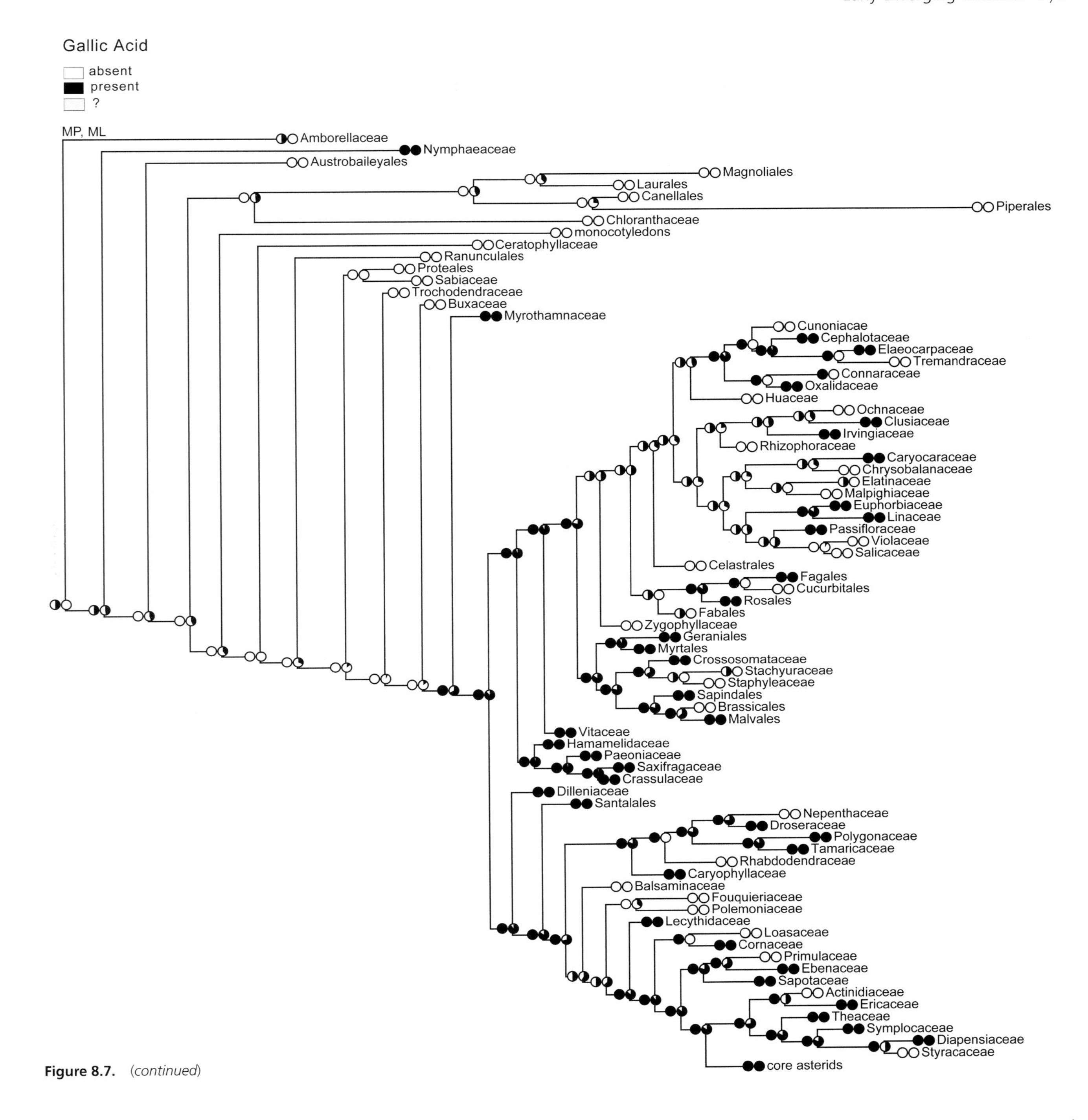

Figure 8.7. *(continued)*

100 (85–115) mya and 108 (94–122) mya, depending on the method used. However, the well-preserved fossil, *Leefructus* (Sun et al. 2011) is dated at 122.6 mya and was placed in stem Ranunculaceae, although the fossil may have affinities elsewhere in Ranunculales. Importantly, this fossil suggests that the age of Ranunculales may be older than previously suggested.

Given the pivotal phylogenetic placement of Ranunculales and their floral characters (see below), there has been considerable interest in the floral developmental genetics of Ranunculales, as well as basal eudicots in general. Within Ranunculales, attention has focused on Ranunculaceae (e.g., Kramer and Irish 1999; Kramer et al. 2003; Kramer and Zimmer 2006; Rasmussen et al. 2009; Galimba et al. 2012; see also discussion of *Aquilegia* and *Thalictrum*, below).

Most relationships within Ranunculales are well supported in trees inferred from multiple genes (e.g., Hoot et al. 1999; D. Soltis et al. 2000, 2011; Kim et al. 2004b; Wang et al. 2009). Eupteleaceae are sister to all remaining Ranunculales (shown in Fig. 8.3). After Eupteleaceae, Papaveraceae (comprising three subclades) are sister to the remaining members of the clade (shown in Fig. 8.3). Following Papaveraceae and Eupteleaceae, the remaining Ranunculales form a well-supported clade of Berberidaceae, Circaeasteraceae, Lardizabalaceae, Menispermaceae, and Ranunculaceae (Fig. 8.1). Nectaries on staminodes occur at least partially in all five families and may represent a synapomorphy of this clade (Endress 2010a). Within this clade, Lardizabalaceae + Circaeasteraceae are sister to a clade in which Menispermaceae are sister to Ranunculaceae + Berberidaceae (Fig. 8.1).

Phylogenetic analyses have also been conducted for the larger families of Ranunculales. Papaveraceae (44 genera and ~760 species), as well as several genera within the family (e.g., *Papaver* and *Meconopsis*—neither is monophyletic as traditionally circumscribed), have been the subject of phylogenetic analyses using morphology and DNA characters (e.g., Kadereit and Sytsma 1992; Kadereit 1993; Kadereit et al. 1994, 1995, 1997, 2011; Loconte et al. 1995; Hoot et al. 1997; Carolan et al. 2006; Kadereit and Erbar 2011). Papaveraceae comprise three subclades: Pteridophylloideae are likely sister to Papaveroideae + Fumarioideae.

Several phylogenetic studies have focused on Ranunculaceae (Hoot 1991, 1995; Johansson and Jansen 1993; Tamura 1993; Hoot and Crane 1995; Jensen et al. 1995; Johansson 1995; Wang et al. 2005). Analyses support the placement of the morphologically similar *Hydrastis* and *Glaucidium* as a clade sister to the remainder of the family. Other studies have focused on large genera, such as *Anemone* and relatives (e.g., Hoot et al. 1994; Schuettpelz et al. 2002; Meyer et al. 2010), *Clematis* (Miikeda et al. 2006), *Delphinium* s.l., *Aconitum*, and relatives (Delphinieae) (Jabbour and Renner 2011, 2012a), and *Ranunculus* and relatives (e.g., Hörandl et al. 2005; Paun et al. 2005; Hoot et al. 2008; Gehrke and Linder 2009; Emadzade et al. 2010, 2011; Hoffmann et al. 2010; Hörandl and Emadzade 2011, 2012). *Aconitum* appears to be derived from within *Delphinium*, with implications for floral evolution (Jabbour and Renner 2012b). The largely arctic and temperate genus *Ranunculus* has dispersed into the subtropics and tropics multiple times followed by subsequent radiation (Emadzade et al. 2010, 2011; Hörandl and Emadzade 2011).

Phylogenetic relationships and rapid diversification have been analyzed within *Aquilegia* (columbine; Ranunculaceae, Hodges and Arnold 1995; Whittall et al. 2006). The genus has become a model for investigation of adaptive radiation and floral developmental genetics (e.g., Kramer et al. 2007; Kramer and Hodges 2010; Hodges and Derieg 2009; Kramer 2009). Hodges and Arnold (1995) demonstrated, for example, that the evolution of nectar spurs in *Aquilegia* represents a key innovation that was a stimulus for diversification in the genus (Fig. 8.8). The genome of *A. formosa* has been sequenced; given the phylogenetic position of Ranunculales as sister to all other eudicots, this genome sequence is a valuable resource. *Thalictrum* is a model for investigation of dioecy (Di Stilio et al. 2005; Soza et al. 2012). Because Ranunculaceae are large (50 genera, 1,800 species), a comprehensive phylogenetic analysis of the family is still needed.

Berberidaceae (14 genera and ~700 species) have also been the focus of phylogenetic study (Loconte and Estes 1989; Y. Kim and Jansen 1995, 1998) (shown in Fig. 8.3). The fossil record of this family is limited but includes compound leaves corresponding to *Berberis* in the late Eocene to Miocene of western North America (MacGinitie 1953) and the Oligocene to Miocene of central and eastern Europe (Kvaček et al. 2011; Güner and Denk 2012). Three subclades are supported, but the relationships among them are unclear (Kim and Jansen 1996, 1998; Kim et al. 2004b; W. Wang et al. 2007a, 2009). Phylogenetic relationships in Lardizabalaceae have been analyzed, indicating that *Sargentodoxa* is sister to the rest of the family (Kofuji et al. 1994; Hoot et al. 1995a,b).

Menispermaceae (70 genera and ~440 species) have been the focus of several recent analyses that provided evidence for two subclades (Tinosporoideae and Menispermoideae) and several additional subclades within these (Ortiz et al. 2007; W. Wang et al. 2007b, 2012; Hoot et al. 2009; Wefferling et al. 2013). Fruit characters appear to be good synapomorphies for the major clades and subclades (Wefferling et al. 2013) (Fig. 8.3). W. Wang et al. (2012) hypothesized that the diversification of Menispermaceae coincided with the emergence of modern tropical rainforests near the Cretaceous-Paleogene boundary. This family is represented by a diversity of extant and extinct genera by the Eocene in western North America and Europe (Jacques 2009), but has not been traced with confidence to the Cretaceous. Currently the oldest accepted records are from the Paleocene of Colombia, Wyoming, and North Dakota (Crane et al. 1990; Herrera et al. 2011).

PROTEALES

Proteales comprise four families (Proteaceae, Platanaceae, Nelumbonaceae, and the recently included Sabiaceae) (Figs. 8.1, 8.3, 8.4, 8.5) that traditionally were considered distantly related based on morphology. Although the composition of Proteales has to be one of the major surprises

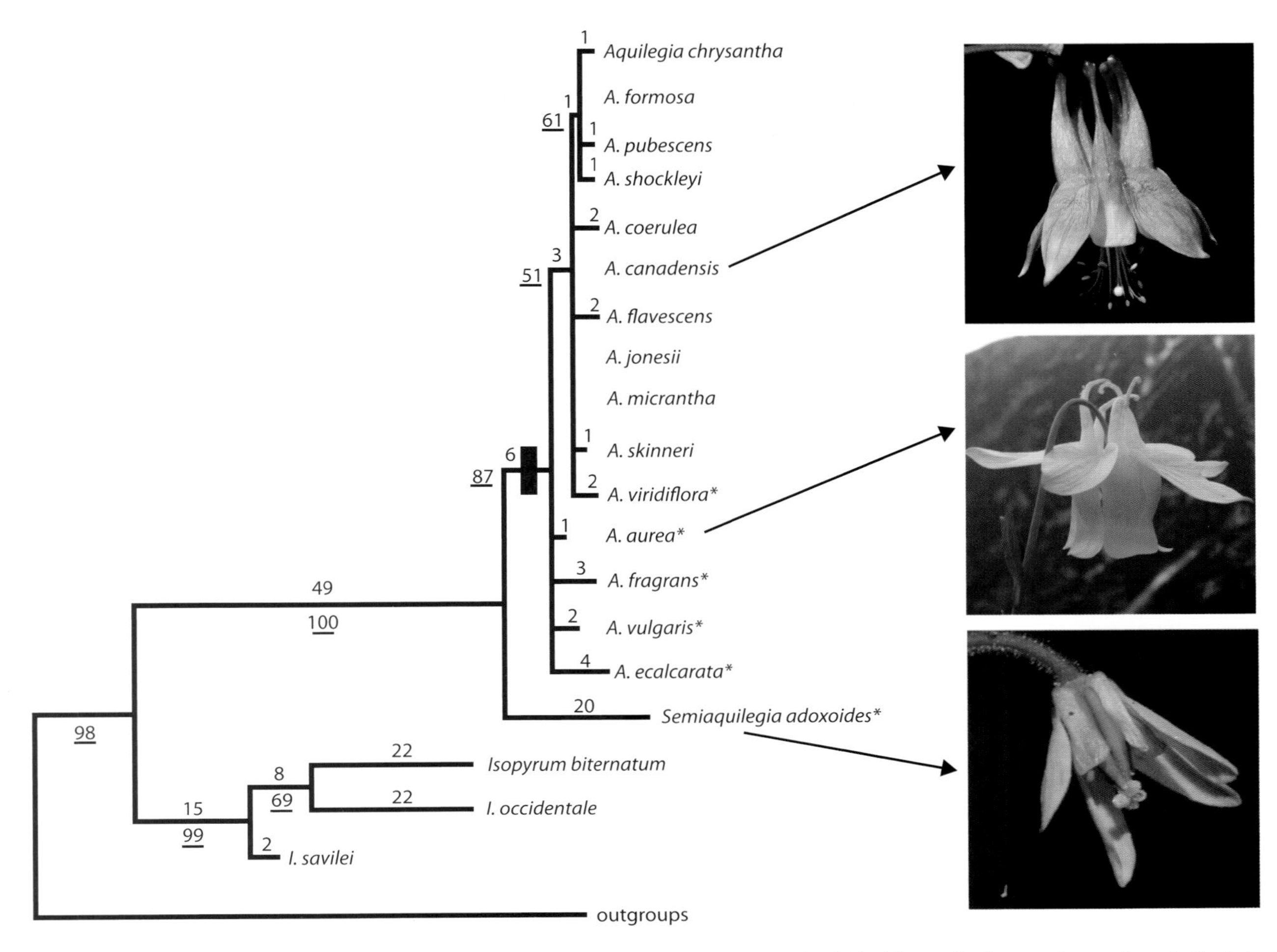

Figure 8.8. Reconstruction of the origin of nectar spurs in *Aquilegia* (Ranunculaceae). Photographs of individual flowers for three species are shown to illustrate shape diversity (not to scale). Modified from Hodges and Arnold (1995).

of molecular phylogenetics, given the diverse habits and morphologies of its members, analyses of numerous DNA datasets provide strong support for the Proteaceae + Platanaceae + Nelumbonaceae clade (e.g., Hoot et al. 1999; Savolainen et al. 2000a; P. Soltis et al. 1999a; D. Soltis et al. 2000, 2011; Burleigh et al. 2009; see Chapter 3). Nelumbonaceae are sister to the strongly supported (BS = 100% in Soltis et al. 2011) clade of Platanaceae and Proteaceae. Bell et al. (2010) estimated 110–116 (101–131) mya for the age of crown group Proteales. Magallón and Castillo (2009) had a comparable crown group estimate (116.4 and 117.4 mya) and ca. 122.8 and 123.6 mya for the stem group.

Although the Proteaceae + Platanaceae + Nelumbonaceae clade (i.e., Proteales s.s.) is an eclectic assemblage, there are possible non-DNA synapomorphies: rod- or tube-shaped epicuticular waxes, seeds with scanty or no endosperm, one or two ovules per carpel, and alternate vessel pitting (Nandi et al. 1998; Savolainen et al. 2000a). Rod- or tube-shaped epicuticular waxes are also widespread in Ranunculales—so this feature may be a plesiomorphy at this level (if lost in Eupteleaceae), but it could be synapomorphic if the waxes evolved independently in these two clades. This feature is therefore somewhat equivocal and needs further study.

Proteaceae and Platanaceae also share carpel closure by complete postgenital fusion and orthotropous ovules (Endress and Igersheim 1999); Nelumbonaceae have anatropous ovules. Proteaceae and Platanaceae share epidermal leaf characters, including compound trichome bases consisting of an annular surface scar associated with more than one underlying epidermal cell (Carpenter et al. 2005).

Platanaceae are among the most common fossils in mid-Cretaceous floras (see Fig. 8.4), and there was greater morphological diversity of foliage types than observed in extant Platanaceae. In addition to simple, trilobed, to palmately lobed leaves similar to those of extant *Platanus* (Fig. 8.4a,b), the late Cretaceous and early Cenozoic diversity included genera with compound leaves (e.g., Fig. 8.4c). Associated inflorescences were heads similar to those of extant *Platanus* (e.g., Fig. 8.4d,e), but in most Cretaceous and early Cenozoic representatives, the individual fruits were glabrous, lacking the dispersal hairs found in all extant species of *Platanus* (cf. Fig. 8.4g,h).

Proteaceae may be a Southern Hemisphere variant of Platanaceae that diverged early in angiosperm evolution (Drinnan et al. 1994); this close relationship could explain the platanoid aspect of many leaf taxa in very Late Cretaceous floras from the Southern Hemisphere (attributed by Drinnan et al. 1994 to Kirk Johnson, pers. comm.) and highlights the need for more studies of morphology (Drinnan et al. 1994).

Given the pivotal phylogenetic position of Proteales s.s., there is considerable interest in the floral developmental biology and genome structure of its members. Most Cretaceous fossils attributed to Platanaceae, and some of those from the Paleocene and Eocene, have flowers with relatively conspicuous perianth parts. Although most are pentamerous, at least one representative was tetramerous (Magallón-Puebla et al. 1997). Extant *Platanus* species are more variable within individuals, mostly ranging from 5 to 9 carpels per floret and 3–5 stamens each (Boothroyd 1930; von Balthazar and Schönenberger 2009). The fixed tetramerous organization (4 stamens per staminate flower and 8 carpels per carpellate floret) seen in *Quadriplatanus* from the Late Cretaceous (Coniacean-Santonian) of Georgia, USA (Magallón et al. 1997) could provide another link with Proteaceae, although, as reviewed in "Merism," below, developmental studies indicate that Proteaceae are dimerous rather than tetramerous (Douglas and Tucker 1996a,b; von Balthazar and Schönenberger 2009). The merism of *Nelumbo* has sometimes been considered indeterminate, but Hayes et al. (2000) noted that *Nelumbo* has only two sepals (see Yoo et al. 2010), suggesting that Proteales may therefore have ancestrally dimerous flowers (see below).

Yoo et al. (2010) showed that despite the phylogenetic distance between *Nymphaea* and *Nelumbo*, floral gene expression patterns are nearly identical; *Nelumbo*, like *Nymphaea*, exhibits broad B-class gene expression that extends across all floral organs (i.e., "fading borders" model). The complete nuclear genome of *Nelumbo nucifera* was sequenced (Ming et al. 2013). Importantly, the "paleohexaploid" event that has been observed in the genomes of diverse core eudicot species is not present in *Nelumbo* and must have occurred after the divergence of Proteales from other extant eudicots (Chapter 16).

Xue et al. (2012) estimated the divergence time between Nelumbonaceae and Platanaceae to be 109 mya, in agreement with the fossil record. The earliest records of *Nelumbites* (an extinct *Nelumbo* relative) are from the Early Cretaceous (Hickey and Doyle 1977; Upchurch et al. 1994). Grimm and Denk (2008) found evidence in *Platanus* for recent as well as ancient hybridization.

The affinities of Proteaceae (80 genera and ~1600 species) were long considered problematic given their distinctive floral morphology (see Fig. 8.3). Cronquist (1981) placed Proteaceae in his Rosidae close to Elaeagnaceae. Proteaceae have a rich fossil history (Hill et al. 1995; Leng et al. 2005; Weston 2007; Carpenter 2012) and are well known for modern and fossil lineages on all southern continents. Traditionally, the modern distribution of Proteaceae was explained by vicariance resulting from the break-up of Gondwana. Barker et al. (2007), based on molecular phylogenetic/dating analyses, concluded that some of the disjunctions in Proteaceae result from Gondwanan vicariance but that transoceanic dispersal also played an important role. Molecular analyses also suggest rapid radiations within Proteaceae at multiple levels, including Grevilleoideae (Hoot and Douglas 1998), *Banksia* (Crisp and Cook 2007), *Hakea* (Mast et al. 2012), and *Protea* (Barraclough and Reeves 2005; Schnitzler et al. 2011).

Sabiaceae are a small tropical and subtropical family of three genera (*Meliosma*, *Ophiocaryon*, and *Sabia*) (see Fig. 8.5). *Meliosma* and *Sabia* have been included in broad phylogenetic studies (e.g., Soltis et al. 2011). Takhtajan (1997) recognized two separate families, Meliosmaceae and Sabiaceae, within his Sabiales. The monophyly of Sabiaceae is strongly supported in DNA phylogenetic studies (e.g., Savolainen et al. 2000b; Soltis et al. 2000, 2011; Burleigh et al. 2009).

The inferred relationships of Sabiaceae varied greatly in morphology-based classifications. Heimsch (1942) and Heywood (1993, 1998) placed the family in Sapindales (subclass Rosidae), whereas Takhtajan (1997) placed the family in its own order within Rosidae. Cronquist (1981, 1988), however, considered Sabiaceae a member of Ranunculales.

Sabiaceae often appear sister to Proteales s.s. (Fig. 8.1), but in some early analyses, Sabiaceae followed Proteales as sister to all other eudicots (e.g., Savolainen et al. 2000a; D. Soltis et al. 2000; see above). However, in other analyses, Sabiaceae and Proteales s.s. form a moderately supported clade (e.g., Chase et al. 1993; Savolainen et al. 2000b; Kim et al. 2004b). Recent multigene analyses (Soltis et al. 2011; Moore et al. 2010; Sun et al. 2016) suggest that Sabiaceae are sister to Proteales s.s. with BS = 59%, 80%, and >90%, respectively, and thus the family recently (APG IV 2016) has been included within Proteales. It is therefore noteworthy that Sabiaceae and Proteaceae both possess wedge-shaped phloem rays (Metcalfe and Chalk 1950; see Nandi et al. 1998) and also share a nectar disk (Haber 1966; Beusekom 1971; Nandi et al. 1998), a rare feature among early eudicot lineages. These features are not apparent in Nelumbonaceae and Platanaceae, so it is unclear if these characters represent potential synapomorphies of Proteales (including Sabiaceae) that were lost in some Proteales or if they simply evolved in parallel in Proteaceae and Sabiaceae. However, a nectar disk would not be expected in a wind-pollinated group such as Platanaceae.

Estimates for the age of stem group Sabiaceae range from 118 to 122 mya (Anderson et al. 2005) and ~123.2 mya (Magallón and Castillo 2009). Bell et al. (2010) provided mean estimates for crown group Sabiaceae of 87–89 mya (57–126); Anderson et al. (2005) suggest 91–119 mya.

Meliosma endocarps from Europe and North America have been reported from the Paleocene and Eocene, with rare occurrences in the latest Cretaceous of Europe (Knobloch and Mai 1986) and Colorado (Crane et al. 1990). Other fossils identified as Sabiaceae include *Sabia* and the extinct *Insitiocarpus* from the latest Cretaceous (Maastrichtian) of Europe (Knobloch and Mai 1986; reviewed in Friis et al. 2011).

TROCHODENDRALES

Trochodendron and *Tetracentron* (historically placed in Trochodendraceae and Tetracentraceae) have long been considered to occupy a pivotal position in angiosperm evolution (Fig. 8.5). Both genera are monotypic and today confined to Southeast Asia and possess features long considered ancestral, such as ethereal oil cells and chloranthoid leaf teeth. However, although the idioblasts in *Tetracentron* look like oil cells, they have not been studied in detail (Endress 1986a) and may not be homologous with those of certain basal angiosperms; our reconstructions (Chapter 6) support this hypothesis.

Cronquist (1981, 1988) considered the two genera to be basal in his Hamamelidae and connecting links between Magnoliidae and the rest of his Hamamelidae. The two genera have long been considered to lack vessels. However, vessel elements with scalariform to scalariform-reticulate perforation plates have now been reported from both *Tetracentron* and *Trochodendron* (Hacke et al. 2007; Ren et al. 2007a; Li et al. 2011) (see Chapter 6).

Phylogenetic analyses based on non-DNA characters provided support for *Trochodendron* and *Tetracentron* as sister taxa (Crane 1989; Hufford and Crane 1989); Crane (1989) placed the families together in one family, Trochodendraceae (see also Endress 1986a, based on floral and pollen structure). These same studies also indicated that Trochodendraceae are sister to all other eudicots (Crane 1989; Hufford and Crane 1989). Molecular data strongly supported a sister-group relationship between *Trochodendron* and *Tetracentron* (e.g., Chase et al. 1993; Drinnan et al. 1994; D. Soltis et al. 1997b, 2000, 2011; Hoot et al. 1999; Savolainen et al. 2000a,b), and the single family Trochodendraceae has been recognized (APG 1998; APG II 2003; APG III 2009; APG IV 2016). Recent molecular studies placed Trochodendraceae as part of the early-diverging eudicot grade, sister to Buxales + the core eudicots. Trochodendraceae have a rich fossil history that includes both extant genera as early as the Eocene and an extinct genus, *Nordenskioldia*, from the Late Cretaceous (Campanian) to the Miocene of the Northern Hemisphere (e.g., Crane et al. 1991; Manchester and Chen 2006; Pigg et al. 2007; Grimsson et al. 2008; Wang et al. 2009).

BUXALES

Recent analyses place Buxales (~80 species, comprising Buxaceae [including *Haptanthus*] and Didymelaceae; see Chapter 12) as the sole sister to the core eudicots (Figs. 8.1, 8.5). The enigmatic *Haptanthus* (recently rediscovered) has been placed within Buxaceae as sister to *Buxus* (Doust and Stevens 2005; Shipunov and Shipunova 2011).

Molecular phylogenetic studies have consistently indicated a well-supported sister-group relationship between Buxaceae as traditionally recognized and Didymelaceae (Hoot et al. 1999; Savolainen et al. 2000a; Soltis et al. 2000, 2011; von Balthazar et al. 2000; Hilu et al. 2003; Worberg et al. 2007). As a result, it has been proposed that the two families be combined into a single family, Buxaceae (APG II 2003; APG III 2009) that also includes Haptanthaceae (APG IV 2016; see Chapter 12).

Buxaceae in the narrow sense have been variously placed. Cronquist (1981, 1988) and Heywood (1993, 1998) placed the family in Euphorbiales, whereas Takhtajan (1987, 1997) placed Buxaceae with the hamamelids close to Myrothamnaceae, the latter family now known to be sister to Gunneraceae (core eudicots). Flowers of Buxaceae exhibit considerable variability in organ number and differentiation (von Balthazar and Endress 2002a,b). Flowers are unisexual with a perianth of small, inconspicuous tepals. Most staminate Buxaceae are dimerous with two stamen whorls, although pistillate flowers are either trimerous or dimerous. The weakly differentiated perianth is spiral in pistillate flowers, whereas two whorls are present in staminate flowers (von Balthazar and Endress 2002a,b).

Didymeles, with two species from Madagascar, was long considered taxonomically isolated and placed in a separate family, Didymelaceae. The plants are dioecious, and the staminate flowers lack a perianth; pistillate flowers sometimes have one to four minute scales, although it is unclear whether these scales represent perianth parts (von Balthazar et al. 2003). Cronquist (1981) placed Didymelaceae in its own order in Hamamelidae. Stebbins (1974) also placed it in Hamamelidae, near Hamamelidaceae. Thorne (1976) considered it part of Euphorbiales, whereas Takhtajan (1997) placed it in its own order, together with Buxales and Simmondsiales, in his superorder Buxanae.

Buxaceae and Didymelaceae are united by distinctive steroid alkaloids, similar wood and leaf anatomy (Sutton 1989; Takhtajan 1997), and a simple bract-like perianth or lack of a perianth (Nandi et al. 1998; Hoot et al. 1999; von Balthazar and Endress 2002a,b; von Balthazar et al. 2003). Both families also have encyclocytic stomata, a rare feature in the angiosperms (see Metcalfe and Chalk 1988/1989) that also unites Aextoxicaceae and Berberidopsidaceae (see Chapters 6, 11).

Bell et al. (2010) estimated the age of crown group Buxaceae as 98 (97–100) mya and 99 (97–103) mya; Magallón and Castillo (2009) suggested ~111.5 mya. Buxaceae are well represented in the fossil record; fossils from the Cretaceous include *Lusicarpus*, *Silucarpus*, *Valecarpus*, and *Spanomera* (Drinnan et al. 1991; Crepet et al. 2004; Pedersen et al. 2007; reviewed in Friis et al. 2011).

CHARACTER EVOLUTION

Families now recognized as early-diverging eudicots were long considered to represent some of the most "primitive" angiosperms (e.g., Cronquist 1968, 1981, 1988; Takhtajan 1969; Stebbins 1974; Dahlgren 1980; Thorne 1992a,b). Trends in floral evolution observed in these plants were considered important in linking floral characters in basal angiosperms (Magnoliidae of those authors) with those of the remaining subclasses of "dicots" (Cronquist 1981, 1988).

As in basal angiosperms and magnoliids, floral organization and development in basal eudicots are considered "open" and highly labile (e.g., Endress 1987a,b, 1990a,b, 1994a,c; see Chapter 4). In contrast, in most core eudicots, numbers of floral parts are low (i.e., 4 or 5) and fixed, with floral organs arranged in whorls, indicating that the basic floral bauplan became canalized during the early diversification of the eudicots (Zanis et al. 2003; Soltis et al. 2005a,b; Endress and Doyle 2009). The early-diverging eudicots have also been considered labile in their basic floral development and more similar to basal angiosperms than to core eudicots in their arrangement and number of floral parts (Endress 1987a,b, 1994a,c). Dimerous, trimerous, tetramerous, and pentamerous floral organizations are all found in the early-diverging eudicots, as are both spiral and whorled phyllotaxis (see Merism, below; see also Chapters 4 and 12).

In contrast to intensive efforts to infer angiosperm phylogeny, surprisingly little recent effort has been made to use DNA-based topologies to examine floral evolution in an explicitly phylogenetic context. This is especially true of early-diverging lineages of angiosperms, which might be expected to yield critical insights into the significance of the early radiation and success of flowering plants (but see Albert et al. 1998; Doyle and Endress 2000; Zanis et al. 2003). Several authors addressed character evolution in the context of phylogeny (e.g., Drinnan et al. 1994; Hoot et al. 1999) but did not conduct explicit character mapping. Drinnan et al. (1994) suggested that (1) there was plasticity in floral form in the early-diverging eudicots, with frequent transitions between dimerous and trimerous forms; (2) a cyclic (whorled) floral plan not only is common in the early-diverging eudicots but also represents the basic formula for the eudicots as a whole; and (3) the helical (spiral) arrangement observed in some early-diverging eudicots (e.g., some Ranunculaceae) is secondary.

Early mapping of floral traits (Albert et al. 1998; Zanis et al. 2003) provided some support for the hypothesized floral lability of early-diverging eudicots, but these studies focused primarily on basal angiosperms. Other studies of character evolution provided additional insights into the early diversification of the eudicots (e.g., Drinnan et al. 1994; Hoot et al. 1999; Soltis et al. 2005b; Endress and Doyle 2009; Doyle and Endress 2011). For our analyses, we mapped floral characters onto a phylogenetic tree for eudicots. We focused on perianth, androecial, and gynoecial phyllotaxis and merism, and perianth differentiation in an effort to provide more explicit hypotheses regarding floral evolution in these plants. We constructed phylogenetic trees using the 17-gene topology (Soltis et al. 2011) as the framework, with additional representatives of Ranunculaceae, Papaveraceae, Lardizabalaceae, and other Ranunculales included to reflect greater diversity within these subclades.

The genera used as placeholders, as well as the family and order (if placed to order) designations (sensu APG III 2009 and APG IV 2016), are provided in Table 8.1. We obtained floral data from Drinnan et al. (1994), Zanis et al. (2003), Ronse De Craene et al. (2003), and primary references therein and reconstructed evolutionary history under both maximum parsimony (MP) and maximum likelihood

Table 8.1. Taxa used in character-state reconstructions (arranged by family in alphabetical order), with family and ordinal designations

Order (if assigned)	Family	Genus
Amborellales	Amborellaceae	*Amborella*
Magnoliales	Annonaceae	*Asimina*
Piperales	Aristolochiaceae	*Asarum*
Austrobaileyales	Austrobaileyaceae	*Austrobaileya*
Buxales	Buxaceae	*Buxus*
Buxales	Buxaceae	*Didymeles*
Buxales	Buxaceae	*Pachysandra*
Laurales	Calycanthaceae	*Calycanthus*
Canellales	Canellaceae	*Cinnamodendron*
Chloranthales	Ceratophyllaceae	*Ceratophyllum*
Choranthales	Chloranthaceae	*Chloranthus*
Choranthales	Chloranthaceae	*Hedyosmum*
Ranunculales	Eupteleaceae	*Euptelea*
Gunnerales	Gunneraceae	*Gunnera*
Laurales	Lauraceae	*Cinnamomum*
Magnoliales	Magnoliaceae	*Magnolia*
Ranunculales	Menispermaceae	*Menispermum*
Ranunculales	Menispermaceae	*Tinospora*
Magnoliales	Myristicaceae	*Myristica*
Gunnerales	Myrothamnaceae	*Myrothamnus*
Proteales	Nelumbonaceae	*Nelumbo*
Nymphaeales	Nymphaeaceae	*Cabomba*
Nymphaeales	Nymphaeaceae	*Nymphaea*
Ranunculales	Papaveraceae	*Dicentra*
Ranunculales	Papaveraceae	*Eschscholzia*
Ranunculales	Papaveraceae	*Hypecoum*
Ranunculales	Papaveraceae	*Platystemon*
Ranunculales	Papaveraceae	*Pteridophyllum*
Ranunculales	Papaveraceae	*Sanguinaria*
Piperales	Piperaceae	*Piper*
Proteales	Proteaceae	*Placospermum*
Proteales	Proteaceae	*Roupala*
Proteales	Platanaceae	*Platanus*
Ranunculales	Ranunculaceae	*Caltha*
Ranunculales	Ranunculaceae	*Coptis*
Ranunculales	Ranunculaceae	*Glaucidium*
Ranunculales	Ranunculaceae	*Ranunculus*
Ranunculales	Ranunculaceae	*Xanthorhiza*
Proteales	Sabiaceae	*Meliosma*
Proteales	Sabiaceae	*Sabia*
Austrobaileyales	Schisandraceae	*Illicium*
Trochodendrales	Trochodendraceae	*Tetracentron*
Trochodendrales	Trochodendraceae	*Trochodendron*
Canellales	Winteraceae	*Takhtajania*

(ML) criteria, as implemented in Mesquite (Maddison and Maddison 2011).

PERIANTH PHYLLOTAXIS

We considered this feature across all angiosperms in Chapter 4. Most earliest-branching lineages (Amborellaceae and Austrobaileyales, but not Nymphaeales; see Chapter 4) exhibit a spiral perianth (although the ancestral state for all angiosperms is equivocal). The whorled perianth is reconstructed as ancestral for mesangiosperms (Chapter 4) with multiple changes to a spiral perianth in magnoliids.

A whorled perianth is also reconstructed as ancestral for the eudicots (Fig. 8.9), which agrees with Endress and Doyle (2009) but who analyzed far fewer basal eudicots. Our results indicate a reversal to a spiral perianth in *Nelumbo* (Williamson and Schneider 1993b; Hayes et al. 2000) and in some Ranunculales, including *Kingdonia* and several Ranunculaceae (e.g., *Caltha* and *Coptis*; Endress 1995b; Fig. 8.9). Several early-diverging eudicots have lost a perianth, including *Trochodendron*, *Euptelea*, and *Didymeles*. However, in *Trochodendron* very young flowers have, in addition to two lateral prophylls, minute rudiments of organs outside of the androecium in apparently irregular arrangement; these organs may represent perianth remnants (Endress 1986a).

We also conducted multistate codings of perianth phyllotaxis (e.g., one whorl = monocycly, two whorls = dicycly, three whorls = tricycly, multiple whorls = polycycly), and these reconstructions (not shown) similarly show a dynamic picture of perianth evolution in early-diverging eudicots comparable to that seen in basal angiosperms and magnoliids (see Chapter 4). For example, polycycly characterizes much of Papaveraceae, as well as *Hydrastis* + *Glaucidium*. Tetracycly has arisen multiple times (e.g., in Lardizabalaceae, in the ancestor of Menispermaceae, Berberidaceae, and Ranunculaceae), and dicycly has arisen multiple times as well (Buxaceae and *Tetracentron*). ML reconstructions indicate with a high probability that a dicyclic perianth is ancestral in the common ancestor of Proteales and all other angiosperms, and this is also reconstructed as ancestral with MP as well as ML in the ancestor of Buxales and all other angiosperms (not shown). This result agrees with Endress and Doyle (2009).

ANDROECIAL AND GYNOECIAL PHYLLOTAXIS

As with perianth phyllotaxis (see above), we coded androecial and gynoecial phyllotaxis as both binary and multi-

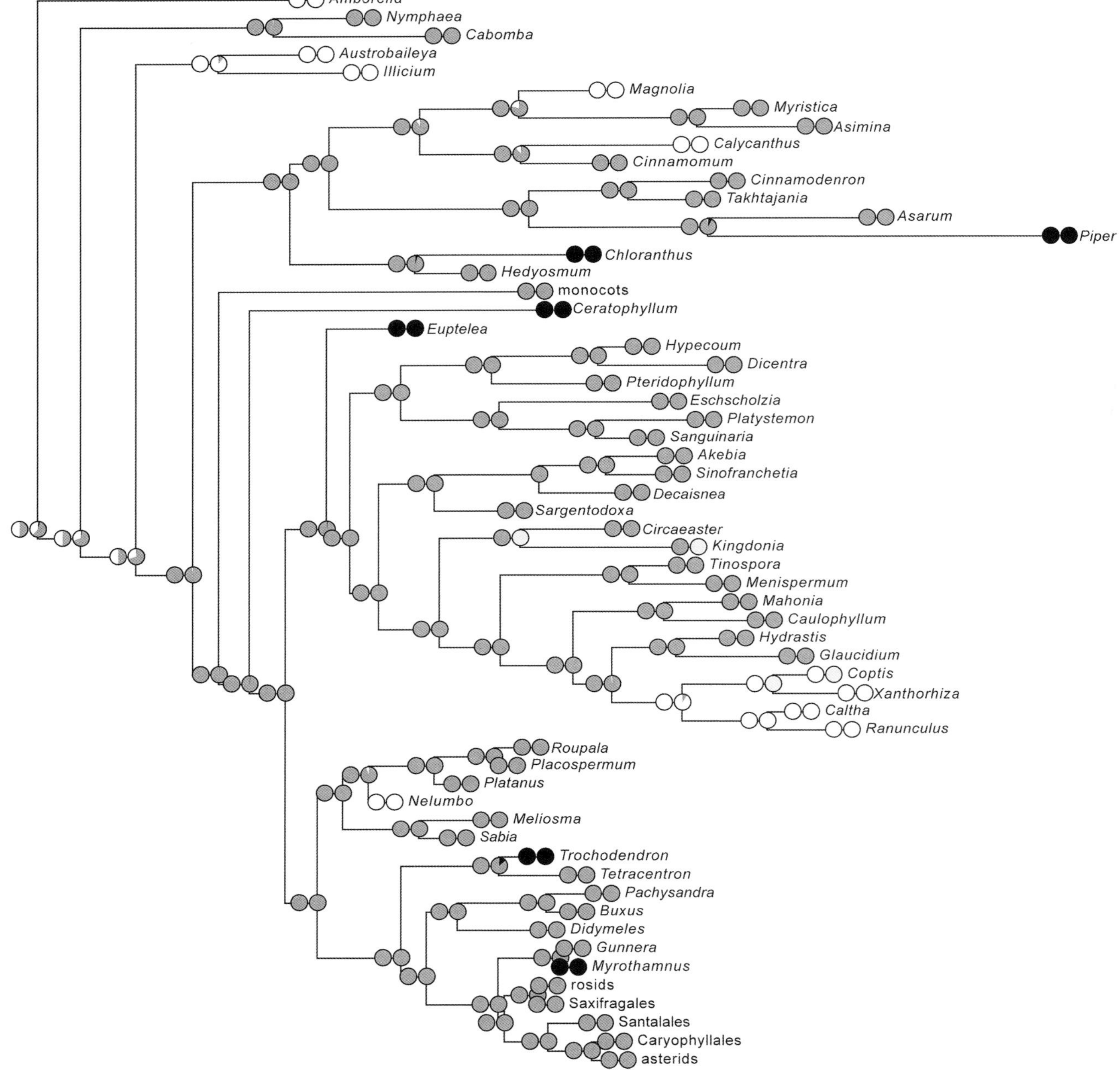

Figure 8.9. Reconstruction of the diversification of perianth phyllotaxis with a focus on early-diverging eudicots.

state characters. Stamen phyllotaxis can be complex and difficult to interpret (see also Chapters 4 and 14). In Ranunculaceae and *Trochodendron*, stamen phyllotaxis may be both spiral and whorled within the same genus or even species (see Chapter 14; Endress 1987a,b, 1990b, 1995b; Ren et al. 2010). We will focus therefore on binary coding.

A whorled androecium (Fig. 8.10) and whorled gynoecium are ancestral for the mesangiosperms (Chapter 4), as well as the eudicots (see Endress and Doyle 2009). Reconstruction of gynoecial evolution (not shown) mirrors that provided for androecial diversification. Almost all Ranunculales have a whorled androecium and gynoecium, and in binary character-state reconstructions, the ancestral state throughout Ranunculales is also whorled. Our mapping studies indicate (Fig. 8.10) that spirally arranged stamens and carpels have been derived independently within the early-diverging eudicots: Ranunculaceae, *Kingdonia* (Circaeasteraceae), Nelumbonaceae, and Trochodendraceae. Our codings for *Kingdonia* follow Ren et al. (2004; see also Kosuge and Tamara 1989; Drinnan et al. 1994). The Menispermaceae included here have whorled phyllotaxis, but some members (e.g., *Hypserpa*) have spiral androecial phyllotaxis (Endress 1995b) and would thus represent another independent origin of spiral phyllotaxis.

Flowers in Ranunculaceae exhibit great variety in form (Chapter 14). Numerous spirally arranged parts were traditionally considered to be ancestral for the family (Tamura 1965; see also Cronquist 1981, 1988), but our reconstructions based on a few exemplars suggest a whorled ancestor with later derivation of a spiral phyllotaxis. *Hydrastis* and *Glaucidium* are sisters to the remainder of Ranunculaceae, and both have whorled phyllotaxis (see Doyle and Endress 2011). Flowers of *Glaucidium* have numerous stamens that are sometimes misinterpreted as spirally arranged, although early development was not studied. Hence, whorled phyllotaxis could be ancestral for the family (see Drinnan et al. 1994). There may be multiple derivations of spiral androecial and gynoecial phyllotaxis just within Ranunculaceae, but more detailed phylogenetic analyses of the family are required (Endress 1987a,b, 1995b).

Ranunculales were long considered to represent an early angiosperm lineage with features of Magnoliales, such as numerous stamens and carpels (Cronquist 1968, 1988; Takhtajan 1969, 1991, 1997; see Chapter 4). Coding androecial merism as a multistate character (not shown) allows us to evaluate whether numerous stamens in these taxa are ancestral or derived. Following the definitions of monocycly and polycycly provided by Ronse De Craene and Smets (1998a,b), our reconstructions indicate that polycyclic androecia are derived, not ancestral, in the early-diverging eudicots. The ancestral androecium is reconstructed as ambiguous for all eudicots, given the extensive variation in Ranunculales (see below).

Within Ranunculales, a polycyclic androecium has been derived several times, at least once each within Ranunculaceae and Papaveraceae. The derivation of a polycyclic androecium from a dicyclic ancestor within Papaveraceae is more clearly seen by adding representatives according to the topology of Hoot et al. (1997; tree not shown). The polymerous androecia and gynoecia of Papaveraceae reflect secondary multiplication (Drinnan et al. 1994; Kadereit et al. 1995; Hoot et al. 1997). Within Ranunculaceae, a polycyclic androecium, rather than spiral phyllotaxis, may be the ancestral state, although more representatives of the family are needed to evaluate this hypothesis. Furthermore, stamen phyllotaxis is so labile that it may be extremely difficult to tease apart "spiral" versus "whorled" within Ranunculaceae. For example, in *Actaea alba*, stamen phyllotaxis is normally whorled but sometimes is spiral; both spiral and whorled stamen phyllotaxis are also present within *Anemone* (Schöffel 1932; Ren et al. 2010).

MERISM (MEROSITY)

Our reconstructions indicate that merism is highly labile in early-diverging eudicots (Fig. 8.11), a finding similar to that reconstructed for basal and magnoliid angiosperms (see Chapter 4). As noted in Chapter 4, a trimerous perianth may be ancestral for mesangiosperms (Fig. 8.11). Trimery has also played a prominent role in Ranunculales and is reconstructed as ancestral for all members of the clade following Papaveraceae. A dimerous perianth has also been important in floral evolution within the early-diverging eudicots (Fig. 8.11; Soltis et al. 2003a, 2005a,b). The extent of the role of dimery depends in large part on the coding of Proteaceae (below) (Fig. 8.11) and the influence of fossils. A dimerous perianth is also prevalent in Ranunculales and is reconstructed as ancestral with MP (but not ML) for Papaveraceae, with an apparent transition to trimery in *Platystemon*. Reconstruction of the ancestral state of Ranunculaceae is equivocal with MP and ML, but the ancestor of *Hydrastis* + *Glaucidium* was dimerous. Both *Hydrastis* and *Glaucidium* are dimerous genera that are sister to the remainder of the family, which in our analyses is a clade (*Coptis*, *Xanthorhiza*, *Caltha*, and *Ranunculus*) of predominantly pentamerous taxa. Bear in mind, however, that many other genera of the family (after *Glaucidium* and *Hydrastis*) are not pentamerous. Nonetheless, this same result is obtained in other mapping studies (not shown) that use additional genera of Ranunculaceae.

The perianth of Proteaceae has been interpreted as

Stamen Phyllotaxis

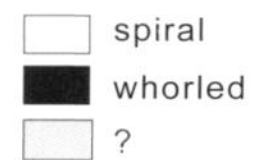

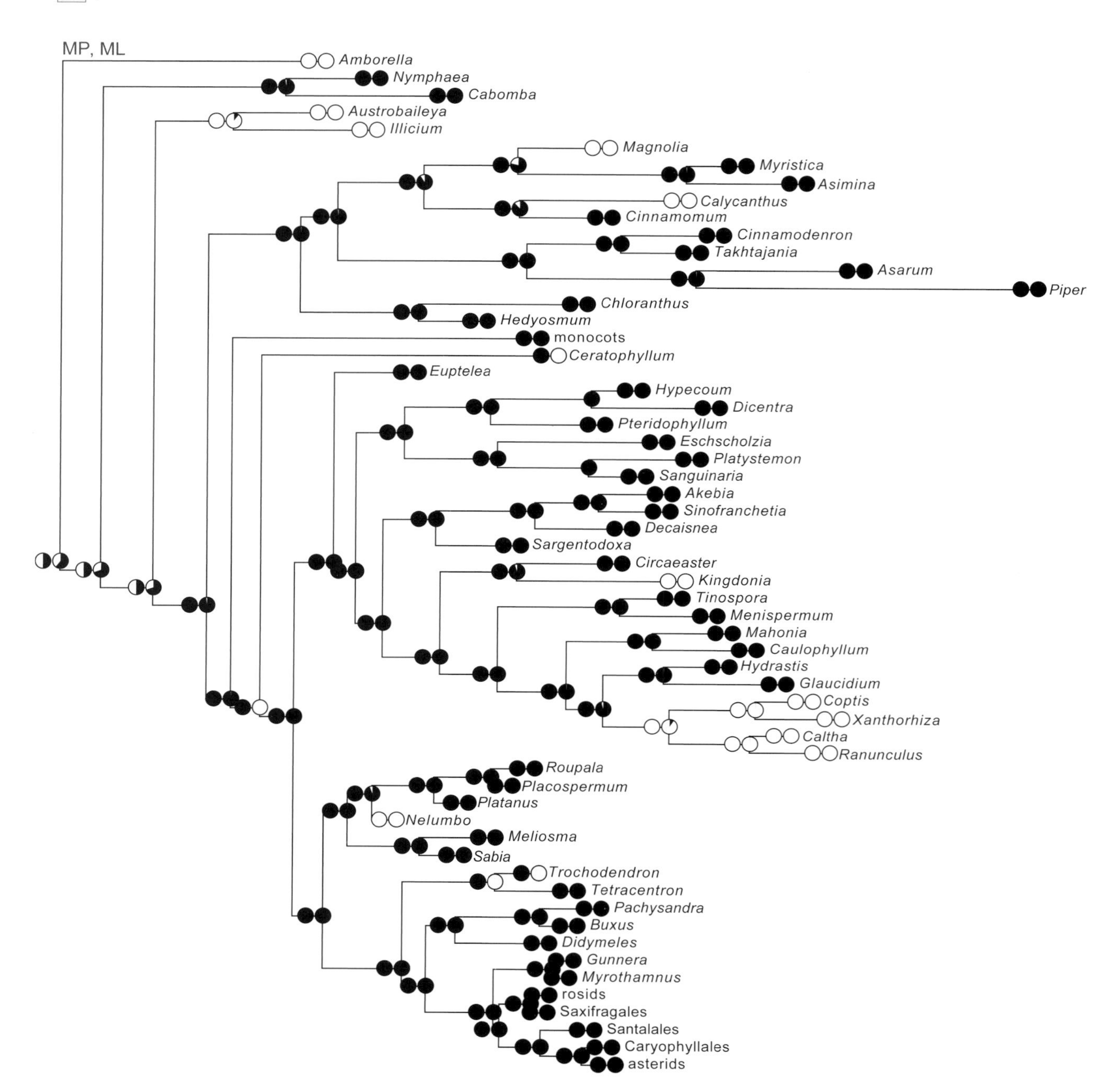

Figure 8.10. Reconstruction of the diversification of stamen phyllotaxis (multistate coding) with a focus on early-diverging eudicots.

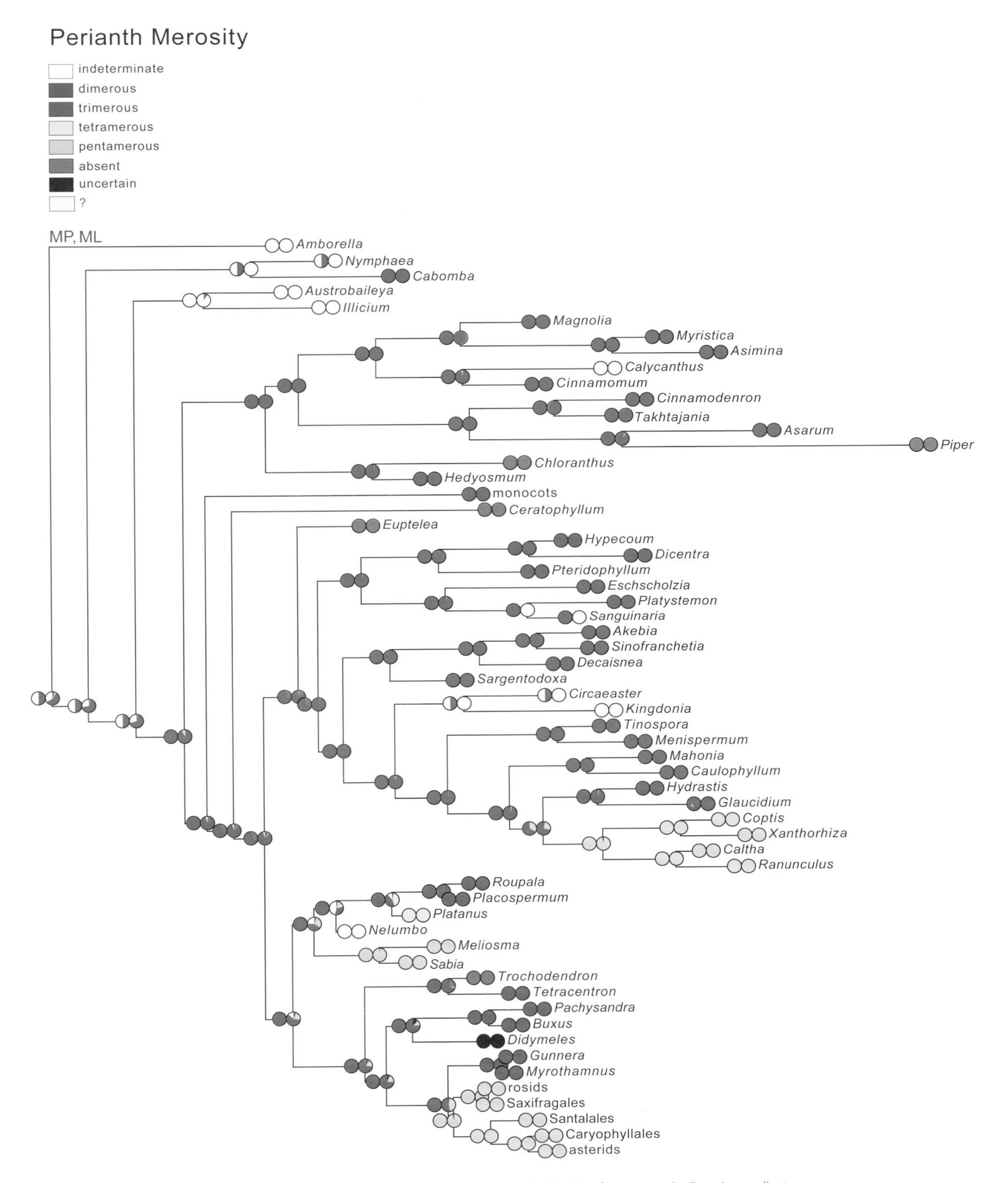

Figure 8.11. Reconstruction of the diversification of perianth merism (multistate coding) with a focus on early-diverging eudicots.

tetramerous by some (Drinnan et al. 1994), but phyllotactically the four tepals "could represent two dimerous whorls of successive primordia" (Douglas and Tucker 1996a,b; see Endress 2010a). The patterns of perianth phyllotaxis in the family appear similar to those observed in Buxaceae and Papaveraceae, which have dimerous (opposite/decussate) arrangements of organs (Douglas and Tucker 1996a,b); we therefore scored Proteaceae as dimerous (Endress and Doyle 2009). However, we also conducted reconstructions coding Proteaceae as tetramerous (not shown). If Proteaceae are tetramerous, then they represent one of the few instances in which this arrangement has evolved in the early-diverging eudicots (in addition to some Ranunculaceae). Platanaceae, sister to Proteaceae, may also be tetramerous (von Balthazar and Schönenberger 2009). Early fossils of Platanaceae are tetramerous, whereas other reports have indicated pentamery (Manchester 1986; Friis et al. 1988; Crane et al. 1993; Magallón-Puebla et al. 1997); we used both conditions in various reconstructions.

If Proteaceae are scored as tetramerous, then the ancestral condition with MP and ML for all eudicots after Ranunculales (i.e., Proteales + all other eudicots) is equivocal. Interpreting the merism of Proteaceae as dimerous (Fig. 8.12B; Douglas and Tucker 1996a,b) has a major impact; the ancestral state of the branch to all eudicots except Ranunculales is then reconstructed as dimerous using MP, as is the ancestor of all subsequent nodes preceding the core eudicots, although this interpretation is uncertain at each node with ML (Fig. 8.11). However, these reconstructions are hampered by equivocal interpretations of "perianth" in some lineages, such as Buxaceae, Trochodendraceae, and Myrothamnaceae. It is unclear whether the apparent perianth in these groups is actually a perianth or an assemblage of bracts, and the relationship between bracts and tepals is uncertain (see von Balthazar and Endress 2002a for discussion of Buxaceae). *Tetracentron* possesses a dimerous perianth (Fig. 8.12C), and a perianth appears to be absent from its sister group, *Trochodendron*. However, rudiments of what may be a perianth are present in young flowers of *Trochodendron*, and these structures are in a dimerous condition (Fig. 8.12D), although it is still unclear if these structures represent perianth parts (Endress 1986a). A dimerous perianth may also be ancestral for Buxaceae (Fig. 8.11). Reconstruction of the ancestral state for Buxaceae is confounded by the uncertain perianth condition in *Didymeles* (see above) (von Balthazar et al. 2003) and female flowers in some Buxaceae, which are not dimerous (von Balthazar and Endress 2002a,b). We scored the merism of *Didymeles* as uncertain; in our reconstruction the ancestral condition for the entire Buxaceae is dimerous with MP; this is also favored by ML.

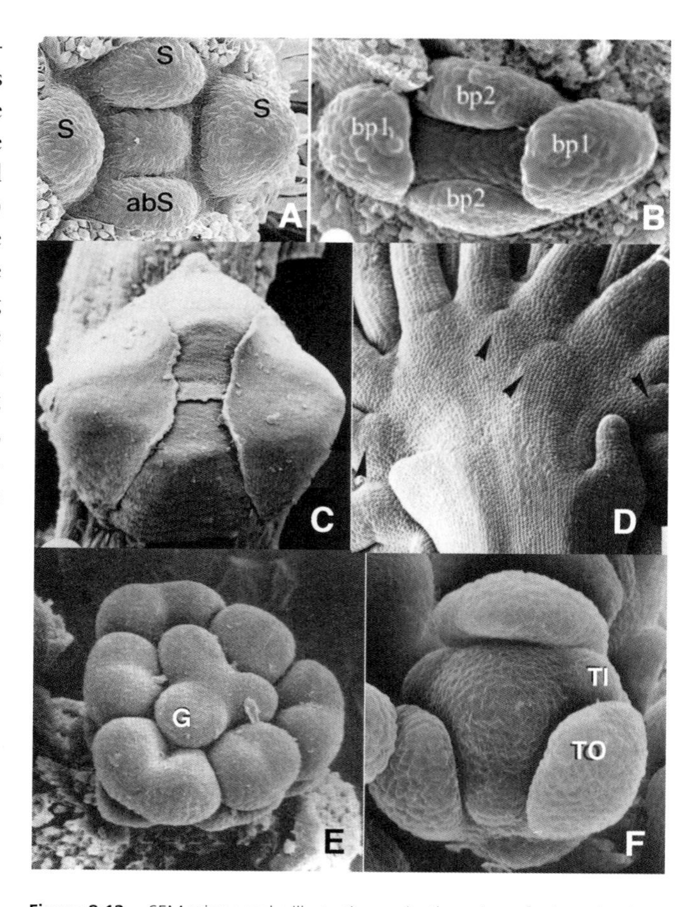

Figure 8.12. SEM micrographs illustrating perianth merism of selected early-diverging eudicots. A. *Persoonia myrtilloides* (Proteaceae), dimerous; S = stamen; abS = abaxial stamen (from Douglas and Tucker 1996b). B. *Buxus balearica* (Buxaceae), dimerous; bp = bract-like phyllome (from von Balthazar and Endress 2002b). C. *Tetracentron sinense* (Trochodendraceae), dimerous (from Endress 1986b). D. *Trochodendron aralioides*, lateral flower from abaxial side showing two prophylls and rudimentary tepals (arrows; from Endress 1986b). E. *Cocculus laurifolius* (Menispermaceae), trimerous; G = gynoecium (from Ronse De Craene and Smets 1993). F. *Decaisnea fargesii* (Lardizabalaceae), trimerous; TI = inner tepal; TO = outer tepal (from Ronse De Craene and Smets 1993).

Regardless of the coding of Proteaceae, our reconstructions indicate that a dimerous perianth may be the immediate precursor to the pentamerous condition characteristic of core eudicots following Gunnerales (see Soltis et al. 2003a, 2005b). Significantly, a dimerous perianth is also found in Gunneraceae and perhaps Myrothamnaceae (Gunnerales), a clade sister to all other core eudicots. The canalization of merism that yielded the pentamerous perianth typical of core eudicots occurred just after the node to Gunnerales (the branch to *Pentapetalae*).

Our reconstructions indicate that a trimerous perianth is likely ancestral for Ranunculales (Fig. 8.11). ML reconstruction provides the strongest support for this ancestral state. The likelihood of this ancestral state increases after the node to Eupteleaceae. Furthermore, the ancestral state of a trimerous perianth is very strong in the ancestor of the clade containing Lardizabalaceae, Circaeasteraceae, Menispermaceae, Berberidaceae, and Ranunculaceae. Trimery characterizes all of these families (Fig. 8.12E,F) except Ranunculaceae and Circaeasteraceae. Ranunculaceae are highly variable in merism, and trimerous, pentamerous, and dimerous flowers are common and occur in different genera (Salisbury 1919; Schöffel 1932; Endress 1987b, 1995b; Ronse De Craene and Smets 1993, 1995; Drinnan et al. 1994). Following the evolution of trimery in this clade, a transition to a dimerous or pentamerous perianth subsequently occurred in the ancestor of Ranunculaceae. Some Ranunculaceae (e.g., *Anemone* and *Clematis*) are trimerous, but this trimery appears to be derived rather than ancestral within the family (see also Endress 1995b). An additional modification of the perianth occurred in *Kingdonia* (Circaeasteraceae), which has a variable number of tepals and indeterminate merism. Endress and Doyle (2009) also reconstructed trimery as ancestral for much of Ranunculales. Ronse De Craene et al. (2015) recently proposed that Sabiaceae may also be ancestrally trimerous (modern flowers appear pentamerous).

Our reconstructions support earlier suggestions that plasticity in floral form was prevalent in the early-diverging eudicots (e.g., Drinnan et al. 1994; Soltis et al. 2005b; Doyle and Endress 2011). Drinnan et al. (1994) suggested that dimerous (opposite/decussate) and trimerous (ternate) arrangements are widespread in early eudicots and "likely primitive" and that the transition from one to the other may have occurred multiple times. Results bolstering this claim were apparent in early reconstructions (but using only MP; Soltis et al. 2005b; Endress and Doyle 2009).

The pentamerous perianth is sometimes considered a synapomorphy for core eudicots—although it has actually arisen one node after the origin of the clade (after the branch to Gunnerales) and hence is best considered a synapomorphy for *Pentapetalae*. Furthermore, pentamery has arisen at least four times within the eudicots: once in Sabiaceae, once in the ancestor of *Pentapetalae*, at least twice within Ranunculaceae (S. B. Hoot pers. comm.), and possibly in Platanaceae. It is noteworthy that the origin of pentamery in both the core eudicots and Ranunculaceae appears to have involved dimerous ancestors (Fig. 8.11); the origin of pentamery in Sabiaceae is unclear. These results provide some support for Kubitzki's (1987) hypothesis that the origin of the pentamerous condition from the trimerous condition was unlikely. Ronse De Craene and Smets (1994) advocated that tetramery is derived from pentamery. Although the interpretation of merism (tetramerous vs. dimerous) in Proteaceae and Platanaceae has been debated, our results provide some support for this suggestion in that tetramerous families of Proteales potentially have as their sister Sabiaceae, which are pentamerous.

These examples of character reconstructions again illustrate the importance of detailed ontogenetic data in the study of angiosperm evolution and diversification (see also Chapter 4 and the section on Saxifragales in Chapter 9). Such data are critical for differentiating among dimerous, trimerous, and pentamerous flowers that may have arisen in different ways (Drinnan et al. 1994). For example, whereas Buxaceae may superficially appear tetramerous, developmental studies demonstrate that the flowers consist of two dimerous whorls of tepals with the position of the stamens (opposite the tepals) arising via the subsequent production of two decussate pairs of stamens (Drinnan et al. 1994; von Balthazar and Endress 2002a). Proteaceae, in contrast, are more difficult to interpret. Drinnan et al. (1994) suggested that Proteaceae may be truly tetramerous and only superficially similar to Buxaceae. However, rigorous developmental studies indicate that flowers of members of Proteaceae may also consist of two dimerous whorls of tepals (Douglas and Tucker 1996a,b); we have followed this dimerous interpretation (Fig. 8.11 and 8.12A). Similarly, trimery may arise in several ways (Ronse De Craene and Smets 1994).

Developmental data also provide insights into the polymerous androecia and gynoecia of some Papaveraceae and Ranunculaceae, suggesting secondary multiplication (e.g., Endress 1987a,b; Ronse De Craene and Smets 1993). High petal numbers also appear to be secondarily derived in these families. In *Sanguinaria* (Papaveraceae), for example, polypetaly appears to be a specialized condition that has arisen through the modification of stamen primordia (Lehman and Sattler 1993). The basic developmental plan for this genus is one of dimery, and we coded it as such (Fig. 8.11).

PERIANTH DIFFERENTIATION

Within Ranunculales, our reconstructions indicate that the evolution of the perianth is highly dynamic. It is unclear whether the ancestor of Ranunculales possessed a differentiated or undifferentiated perianth—the reconstruction is ambiguous with MP, although ML favors an undifferentiated perianth in the ancestral eudicot (Fig. 8.13). Earlier reconstructions suggested that ancestral eudicots may have lacked differentiated perianth whorls (e.g., Albert et al. 1998; Zanis et al. 2003; Soltis et al. 2005b; Endress and Doyle 2009). We still have much to learn about the ancestral eudicot and eudicot perianth evolution.

Within Ranunculales, the ancestral state for Papaveraceae is a differentiated perianth, which is also the case for the ancestor of the clade of Menispermaceae + (Berberidaceae + Ranunculaceae). There are clear transitions between the two states within Ranunculales, with Papaveraceae,

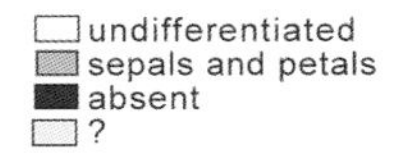

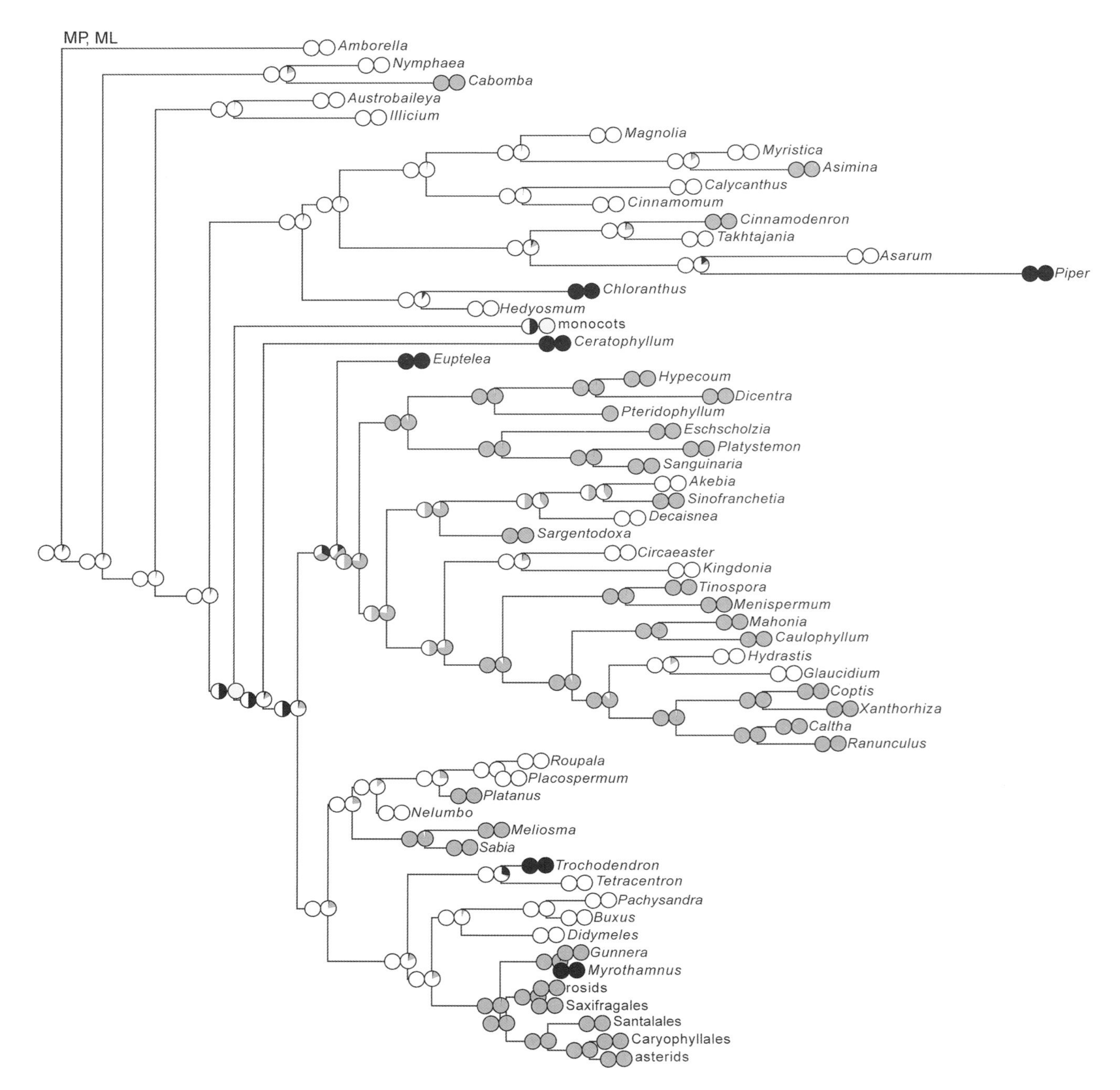

Figure 8.13. Reconstruction of the diversification of perianth differentiation with a focus on early-diverging eudicots.

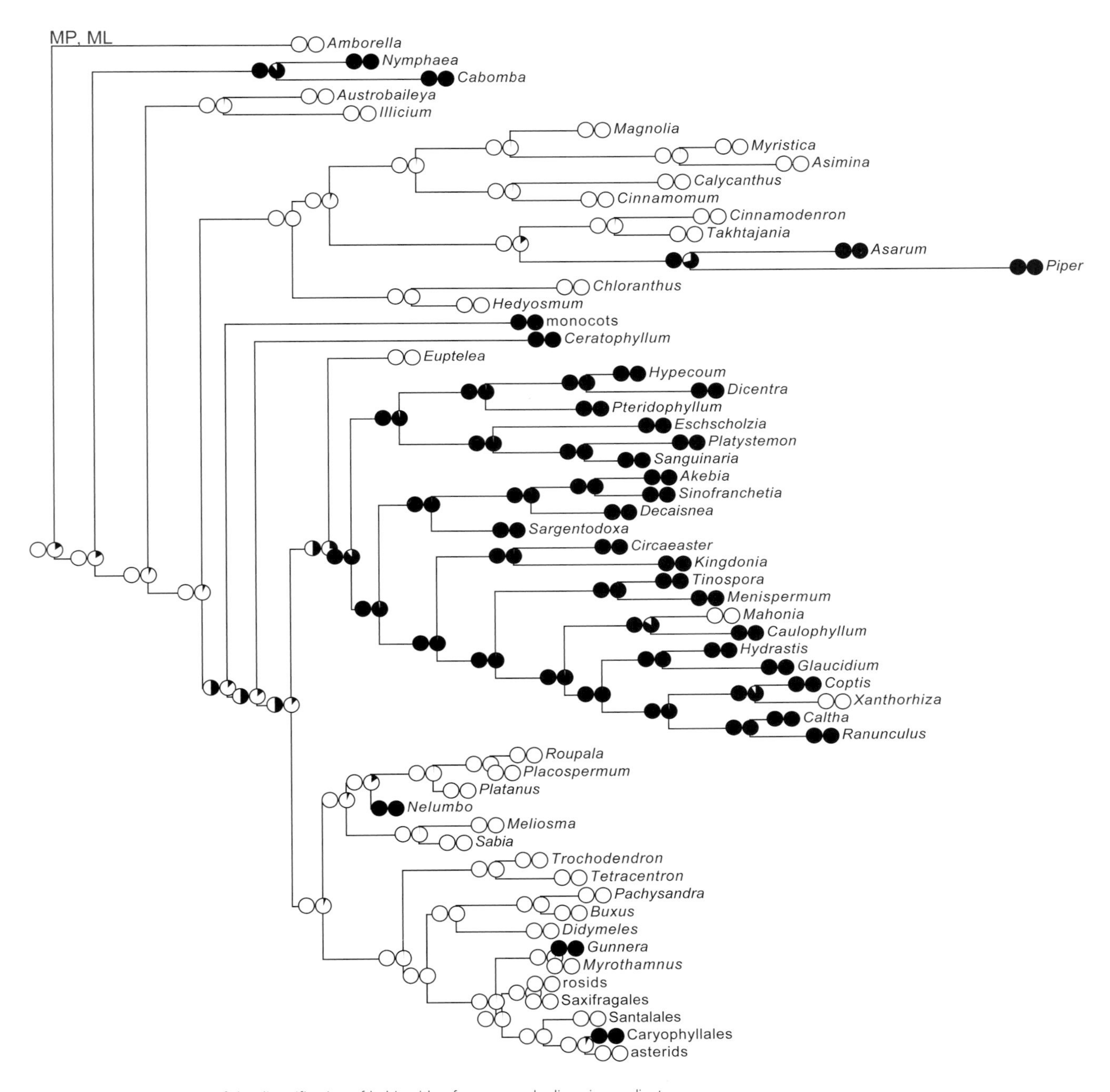

Figure 8.14. Reconstruction of the diversification of habit with a focus on early-diverging eudicots.

Menispermaceae, and some Ranunculaceae having a differentiated perianth, and some Lardizabalaceae and Ranunculaceae having an undifferentiated perianth (Fig. 8.13). Multiple transitions between a differentiated and undifferentiated perianth apparently occurred even within Lardizabalaceae. It is unclear whether the ancestor of Lardizabalaceae possessed a differentiated or undifferentiated perianth, although ML favors a differentiated perianth. This family may be of particular interest for more detailed analyses of perianth diversification/evolution.

The common ancestor of Proteales, Sabiaceae, Trochodendraceae, Buxaceae, and the core eudicots lacked a differentiated perianth with both MP and ML (Fig. 8.13). Our reconstructions indicate that a differentiated perianth evolved independently in Sabiaceae, Platanaceae, and the core eudicots.

Our reconstructions again illustrate that within the early-diverging eudicots, the evolution of a differentiated perianth was highly dynamic (see Soltis et al. 2005b; Endress and Doyle 2009), comparable to that seen in basal angiosperms and magnoliids (Chapter 4). These results are not surprising in light of the fact that petals of angiosperms almost certainly reflect different developmental origins and are likely not homologous in all cases (e.g., Brockington et al. 2012). In Ranunculaceae, petals are extremely similar to stamens in morphology and ontogeny (Hiepko 1965b; Erbar et al. 1999; Endress 2010a), and petals may be derived from stamens in these cases. In other instances, petals may have evolved from undifferentiated tepals or from sepals (see Chapters 4, 6; Endress 1994a,c; Takhtajan 1997; Albert et al. 1998; Kramer and Irish 2000; Ronse De Craene et al. 2003; Soltis et al. 2005a,b; Endress and Doyle 2009; P. Soltis et al. 2009; Brockington et al. 2010, 2012; Ronse De Craene and Brockington 2013).

HABIT

Some reconstructions of habit, when coded as woody versus herbaceous, indicate that the woody habit may be ancestral for eudicots (e.g., Kim et al. 2004b; Soltis et al. 2005b). However, this is highly dependent on topology and method of optimization. Kim et al. (2004b) and Soltis et al. (2005b) used a topology with magnoliids as sister to the eudicots. With the revised topology now employed for angiosperms in which monocots are sister to Ceratophyllaceae + eudicots, the ancestral state for habit in the eudicots is equivocal with both MP and ML (Fig. 8.14; see also Chapter 6). But these reconstructions are based on extant lineages. Importantly, early eudicot megafossils are herbaceous rather than woody and suggest that the herbaceous habit may be ancestral for eudicots and typified their early radiation (Jud 2015).

Ranunculales in the traditional sense were considered to be derived from Magnoliales and "primitively herbaceous" (Cronquist 1968, 1988; Takhtajan 1987, 1991). However, the ancestral state for habit for Ranunculales is equivocal. The woody Eupteleaceae are sister to the remainder of Ranunculales, and in some earlier reconstructions, the ancestral state for the clade was reconstructed as woody (Kim et al. 2004b). After Eupteleaceae, the ancestor of the remainder of Ranunculales is reconstructed as herbaceous (Fig. 8.14).

Following Ranunculales, the ancestral state for the remaining eudicots may be woody. When Sabiaceae are sister to Proteales (the preferred topology—Fig. 8.1), the ancestor of all eudicots after Ranunculales is woody. However, when a topology is used in which Proteales are followed by Sabiaceae as sister to all other eudicots, the ancestral state to all eudicots other than Ranunculales is equivocal with MP, but ML reconstructions continue to favor the woody habit as ancestral—the ambiguity is due to the placement of the herbaceous Nelumbonaceae as sister to the remaining Proteales (Fig. 8.14). Regardless, following Proteales or Proteales + Sabiaceae, the ancestral state of all remaining eudicots is woody, and the ancestral state of the core eudicots is also woody.

9 Core Eudicots

Introduction, Gunnerales, and Dilleniales

INTRODUCTION

Within eudicots (*Eudicotyledoneae*), Ranunculales, Proteales, Trochodendraceae, and Buxaceae are successive sisters to a strongly supported core eudicot clade (*Gunneridae*, see Cantino et al. 2007; Figs. 8.1, 9.1). Diagnosis of the core eudicot clade is not based on any morphological or other non-molecular character, but on the strong internal support obtained for this clade in analyses of DNA datasets. In fact, strong support for *Gunneridae* was apparent in early multigene analyses of just two or three genes (e.g., Hoot et al. 1999; P. Soltis et al. 1999a; Savolainen et al. 2000a; D. Soltis et al. 2000); maximal support has been recovered repeatedly for core eudicots in later multigene analyses employing larger numbers of genes. Within *Gunneridae*, relationships are now largely clarified: Gunnerales are sister to the remainder, a large clade referred to as *Pentapetalae* (Cantino et al. 2007).

Several key events seem to correspond fairly closely to the origin of the core eudicots, including synthesis of ellagic and gallic acids, and perhaps duplication of several floral organ identity genes (Chapter 12), but the precise placement of these events is unclear. Evolution of a highly synorganized flower does not correspond with the origin of core eudicots, but with *Pentapetalae* (see below). Clearly, more research is needed to assess non-DNA synapomorphies for this major clade.

Pentapetalae comprise all core eudicots except Gunnerales (Fig. 9.1) and include ≈70% of angiosperm species. Clear synapomorphies for this clade are a differentiated perianth with calyx and corolla in two whorls and usually five petals (hence the name of the clade). A corolla of four petals in members of *Pentapetalae* is clearly derived, and this modification has occurred multiple times independently from an ancestral state of five petals.

Until relatively recently, with the use of numerous genes, circumscription and relationships among the major lineages of *Pentapetalae* remained at least partially unresolved. Phylogenetic analyses of 83 protein-coding and rRNA genes from the plastid genome for 86 species of seed plants indicated that soon after its origin, *Pentapetalae* diverged into several clades (Moore et al. 2010): (i) a "superrosid" clade consisting of *Rosidae* sensu Cantino et al. (2007), Vitales, and Saxifragales; (ii) a "superasterid" clade consisting of Berberidopsidales, Santalales, Caryophyllales, and *Asteridae* sensu Cantino et al. (2007); and (iii) Dilleniaceae, the placement of which remains somewhat uncertain (see below; Fig. 9.1).

Within the superrosid clade (*Superrosidae*, Soltis et al. 2011), several different topologies have been recovered. With 17 genes (Soltis et al. 2011), Saxifragales are sister to Vitales + *Rosidae*, but other studies have recovered Vitales sister to Saxifragales (Moore et al. 2010; Ruhfel et al. 2014). Both the 17-gene and complete plastid genome studies have only moderate bootstrap support for these alternative placements, so the precise branching pattern of Saxifragales, Vitales, and *Rosidae* remains unclear.

Within the superasterids (*Superasteridae*; Soltis et al. 2011), Santalales are followed by a clade of Berberidopsidales + Caryophyllales as subsequent sisters to *Asteridae* (Soltis et al. 2011). However, the sister-group relationship of Berberidopsidales + Caryophyllales was weakly supported (BS = 75%). With more genes, but fewer samples, Moore et al. (2010) recovered Santalales, Berberidopsida-

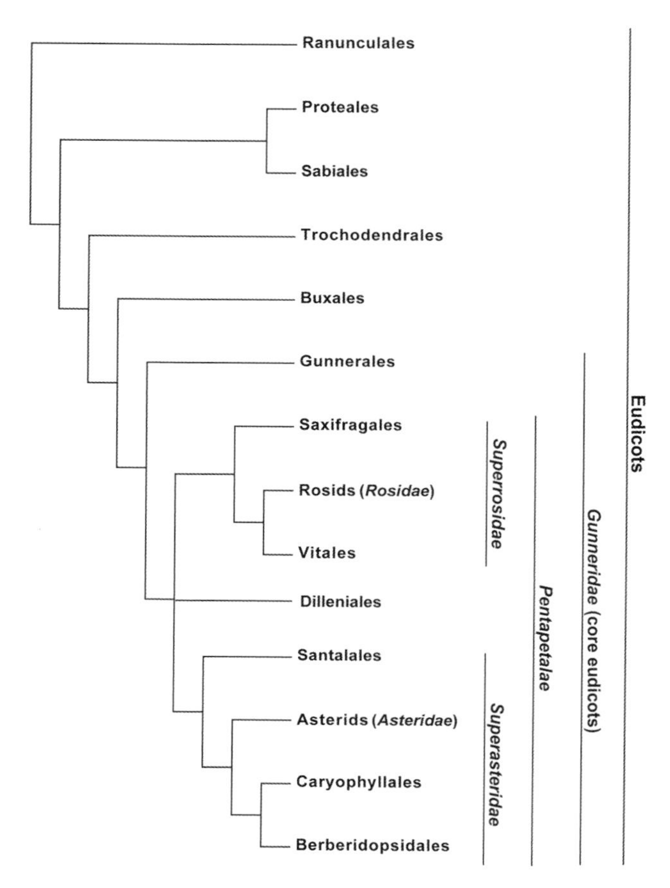

Figure 9.1. Summary tree of eudicot relationships showing unresolved position of Dilleniaceae. Relationships of the basal grade (Proteales and Sabiales, as well as Trochodendraceae and Buxaceae) remain somewhat uncertain, with variation among studies, but we here present the topology of Soltis et al. (2011).

les, and then Caryophyllales as subsequent sisters to *Asteridae*. A larger plastid phylogenomic study with many more angiosperms recovered this same topology with high bootstrap support (Ruhfel et al. 2014).

RAPID RADIATIONS

Darwin referred to the origin and early diversification of the angiosperms as an "abominable mystery," in reference to a rapid radiation that took place early in angiosperm history. However, it is now clear that angiosperm evolution has involved repeated cycles of radiations and stasis (P. Soltis et al. 2004; Tank et al. 2015). Extant angiosperm diversity is therefore the result of multiple radiations within several major clades. Independent radiations within clades of core eudicots occurred in a narrow window of 10 million years or less. For example, molecular clock analyses (Bell et al. 2010) suggested that *Pentapetalae* arose 113–123 mya, with derivative major clades originating within a few million years: *Superasteridae* (104–117 mya) and *Superrosidae* (97–105 mya). Similarly, major lineages within both *Superrosidae* and *Superasteridae* arose in windows of perhaps 4–10 million years (see Chapters 10, 11).

GUNNERALES

Gunnera (Gunneraceae) and *Myrothamnus* (Myrothamnaceae) form a clade, a result that first received support (albeit weak) in analyses of *rbcL* and *rbcL* + *atpB* sequences (Savolainen et al. 2000a,b). Stronger support was achieved as additional genes were added to analyses: 75% jackknife with three genes (D. Soltis et al. 2000) and 85% bootstrap with four genes and >200 eudicot species (D. Soltis et al. 2003c). Analysis of a large *matK* dataset also provided strong (100%) bootstrap support for Gunneraceae + Myrothamnaceae (Hilu et al. 2003), and more recent studies continue to provide strong support for this clade (e.g., Soltis et al. 2011). This Gunneraceae + Myrothamnaceae pair was also strongly excluded from all other eudicot clades, providing support for its isolated status (e.g., D. Soltis et al. 2000, 2011). As a result, it seemed most appropriate to treat these two families as an order, Gunnerales (APG II 2003; APG III 2009; APG IV 2016; see Chapter 12).

Although molecular data indicate a close relationship between Gunneraceae and Myrothamnaceae, the two families exhibit different morphologies and were not, until recently, considered closely related, and have been retained as distinct families by APG. In fact, all recent classifications considered the two families to be distantly related (e.g., Dahlgren 1980; Cronquist 1981; Thorne 1992a,b, 2001; Takhtajan 1997). Cronquist (1981) and Takhtajan (1997) both placed Myrothamnaceae in Hamamelidae and Gunneraceae in Rosidae, although their respective placements of Gunneraceae within the rosids differed. Phylogenetic analysis of non-DNA characters (Nandi et al. 1998) did not indicate a close relationship between the two families, but instead placed Myrothamnaceae with early-diverging eudicots (i.e., Buxaceae, Proteaceae, and Platanaceae) and Gunneraceae as part of *Asteridae*.

The two species of *Myrothamnus* are small xerophytic shrubs (Kubitzki 1993). The flowers of *Myrothamnus* are unisexual and apetalous, with up to four scales interpreted variously as sepals or tepals or bracts (Jäger-Zürn 1966; Cronquist 1981; Drinnan et al. 1994; Takhtajan 1997). Detailed study suggests that these structures may be bracts (Jäger-Zürn 1966; von Balthazar and Endress 2002a). Staminate flowers have three to eight stamens, and pistillate flowers have a gynoecium of three or four united carpels. *Gunnera* (40 species) consists of herbs with short, upright

Figure 9.2. Dilleniaceae and Gunnerales: a. *Dillenia indica* L. (Dilleniaceae), flower. b. *Hibbertia scandens* (Willd.) Dryand. (Dilleniaceae), flower. c. and d. *Gunnera manicata* Linden. (Gunneraceae), habit showing leaves and emerging inflorescence (c) and infructescence (d).

stems (illustrated in Fig. 9.2). In *Gunnera*, the flowers are often unisexual. Typically, the basal flowers of the inflorescence are pistillate, the upper flowers staminate, and the middle flowers bisexual, although all flowers of an inflorescence may be either unisexual or bisexual. Floral parts are in twos, with two sepals, two petals, one or two stamens, and two carpels. *Gunnera* is also of interest in that plants host nitrogen-fixing cyanobacterial symbionts in vegetative parts. *Gunnera* therefore represents an independent evolution of symbiotic nitrogen fixation in the angiosperms (see Chapter 10). Relationships within *Gunnera* have been resolved using gene sequence data (Wanntorp et al. 2002; Wanntorp 2006).

The position of Gunnerales as sister to all other eudicots has important implications for floral evolution. As reviewed in Chapter 5, a dimerous or trimerous perianth is frequently encountered in early-diverging eudicots. In contrast, the pentamerous condition predominates in core eudicots. However, a dimerous perianth is also present in *Gunnera*, and it appears that *Myrothamnus* actually does not possess a perianth (von Balthazar and Endress 2002a). All reconstructions indicate that the canalization of merism that yielded the pentamerous perianth typical of core eudicots occurred following the divergence of Gunnerales from the remaining core eudicots (D. Soltis et al. 2003c; P. Soltis et al. 2004; Chapter 6).

DILLENIALES

Dilleniales (consisting only of Dilleniaceae) are well defined, and the clade can be diagnosed by the leaves with pinnate venation, the secondary veins running straight to the teeth, the showy flowers with petals crumpled in bud (and thus ± wrinkled at maturity), the absence of a nectary, a gynoecium of several separate carpels, and arillate seeds. Several members are illustrated in Figure 9.2. However, relationships of Dilleniaceae to other eudicots have long been problematic, and the phylogenetic placement of the family remains one of the last major mysteries in the angiosperms. Cronquist (1981) considered Dilleniaceae, together with Paeoniaceae, to occupy a basal position within subclass Dilleniidae, whereas Takhtajan (1997) viewed Dilleniaceae alone as the "most archaic family in Dilleniidae." These treatments were based on some noteworthy features of the family, such as often five or more spirally imbricate sepals and petals, often numerous stamens, and 1–20 distinct carpels. These characteristics reminded authors such as Cronquist and Takhtajan of features of Magnoliidae, and hence they considered Dilleniaceae as a "morphological connection" between Magnoliidae and Dilleniidae (see Endress 1997a).

Early DNA studies were unclear about the placement of Dilleniaceae, although with three genes, placement with Caryophyllales was recovered, albeit without strong bootstrap support (D. Soltis et al. 2000). With the addition of more genes, the placement of Dilleniaceae within the core eudicots remained somewhat unclear. Use of 17 genes and over 600 taxa suggested a placement of Dilleniaceae as sister to all other superasterids with high bootstrap support (97%). However, because of the varying position of Dilleniaceae in several studies employing many genes, its position remains unclear. Maximum-likelihood analyses support the position of Dilleniaceae as sister to *Superrosidae*, but topology tests did not reject alternative positions of Dilleniaceae as sister to *Superasteridae* or as sister to all remaining *Pentapetalae*. Hence, we show the placement of Dilleniaceae as unresolved (Fig. 9.1). Relationships within Dilleniaceae have been examined by Horn (2009).

10 Superrosids

INTRODUCTION

The use of many genes revealed the presence of a maximally supported clade of Saxifragales, core rosids (eurosids), and Vitales. Eurosids plus Vitales have been referred to as rosids (APG III 2009; APG IV 2016) or *Rosidae* (sensu Moore et al. 2010; Judd et al. 2013b; Soltis et al. 2013) or superorder Rosanae (Chase and Reveal 2009). *Rosidae* plus Saxifragales have been named *Superrosidae* (Moore et al. 2010; Soltis et al. 2011).

Within *Superrosidae*, interrelationships among Saxifragales, Vitales, and remaining rosids (eurosids) have varied across. Our summary (Fig. 10.1) shows Saxifragales sister to eurosids + Vitales. Strongest support for Saxifragales + (Vitales + eurosids) comes from the analysis of 17 genes (BS support for the placement of Saxifragales = 85%; Vitales + eurosids has BS = 100%). However, with complete plastid genome sequences, *Vitis* is sister to Saxifragales (BS = 82%) with this clade sister to remaining rosids (eurosids) (Moore et al. 2010). Ruhfel et al. (2014) found the same topology with nearly complete plastid genome sequences with increased sampling of angiosperms (BS support for Saxifragales + *Vitis* = 75%). However, in both analyses of plastid genomes, taxon sampling of Saxifragales and Vitales was low, so this topology could reflect poor taxon density.

In studies of numerous nuclear genes, neither Lee et al. (2011) nor Wickett et al. (2014) included representatives of Saxifragales. Hence, support for *Superrosidae* is strong (BS = 100% in Moore et al. 2010 and Soltis et al. 2011) and a topology of Saxifragales + (Vitales + eurosids) is favored; but the position of Vitales is unclear. Because of this uncertainty, Judd et al. (2013b) indicated that *Rosidae* may or may not include Vitales (Fig. 10.1). Also uncertain is the placement of Dilleniaceae. Complete plastid genome sequences place Dilleniaceae sister to *Superrosidae* (BS = 95%, Ruhfel et al. 2014), but Soltis et al. (2011) placed Dilleniaceae as sister to *Superasteridae* (BS = 97%). Because of the uncertain position of the family it is covered in Chapter 9.

SAXIFRAGALES

INTRODUCTION

The composition of Saxifragales is one of the major surprises of DNA phylogenetic analyses because this group of families had never before been associated with one another (Chase et al. 1993; Morgan and Soltis 1993; D. Soltis and Soltis 1997; D. Soltis et al. 1997b, 2000; Qiu et al. 1998; Hoot et al. 1999; P. Soltis et al. 1999a; Savolainen et al. 2000a,b). Saxifragales include Altingiaceae, Cercidiphyllaceae, Crassulaceae, Daphniphyllaceae, Grossulariaceae, Haloragaceae, Hamamelidaceae, Iteaceae, Paeoniaceae, Peridiscaceae, Pterostemonaceae (the last included in Iteaceae; APG II 2003; APG III 2009; APG IV 2016), Saxifragaceae, with recent data confirming an earlier suggestion that Cynmoriaceae are also part of the clade (Fig. 10.2; APG IV 2016).

This unexpected assemblage (representatives are illustrated in Fig. 10.3) consists of taxa placed in three subclasses in previous morphology-based classifications. The DNA circumscription of Saxifragales therefore departs markedly from morphology-based classifications (e.g., Cronquist 1981; reviewed in Morgan and Soltis 1993; Takhtajan

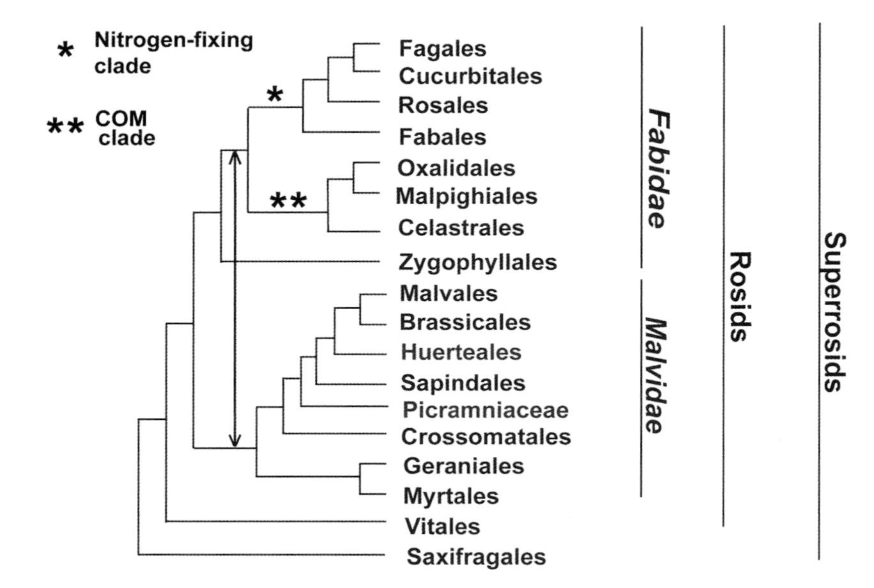

Figure 10.1. A summary tree showing the composition and relationships among major clades of *Superrosidae*. Position of COM clade reflects plastid sequence data. Arrow indicates possible ancient reticulation event (see text).

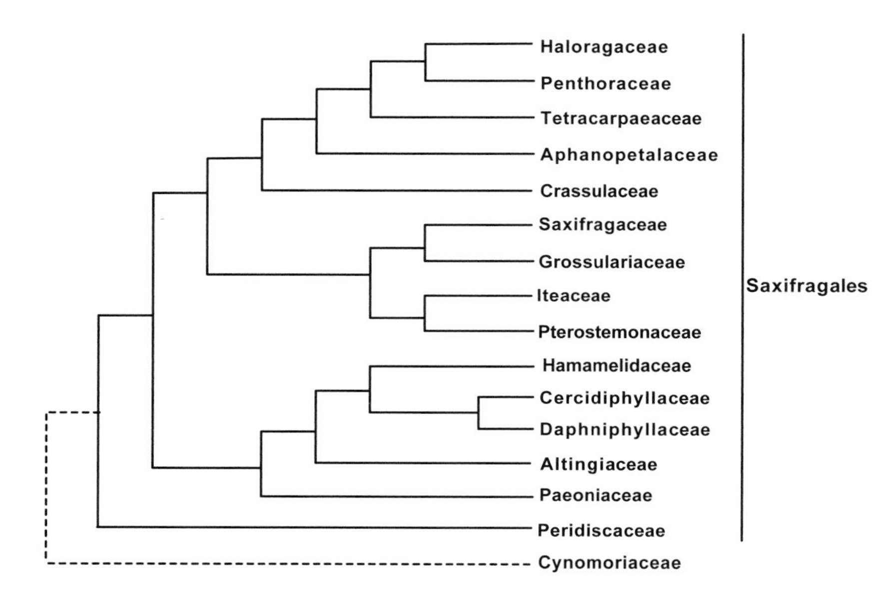

Figure 10.2. A summary tree showing the composition and relationships among families of Saxifragales.

1997; Qiu et al. 1998; Fishbein et al. 2001; Jian et al. 2008; Soltis et al. 2013). Altingiaceae, Hamamelidaceae, Cercidiphyllaceae, and Daphniphyllaceae were previously placed in subclass Hamamelidae; Saxifragaceae, Iteaceae, Pterostemonaceae, Peridiscaceae, Grossulariaceae, Crassulaceae, and Haloragaceae s.s., Tetracarpaeaceae, Penthoraceae, and Aphanopetalaceae were treated as members of subclass Rosidae. Paeoniaceae were variously placed in subclasses Magnoliidae or Dilleniidae (reviewed in Cronquist 1981; Jian et al. 2008; Endress and Matthews 2012).

Analysis of one and two genes first revealed the Saxifragales clade (Chase et al. 1993; Morgan and Soltis 1993; D. Soltis et al. 1997b, 1998; D. Soltis and Soltis 1997; Chase and Albert 1998; Savolainen et al. 2000a,b). Saxifragales received strong support with three genes (Hoot et al. 1999; D. Soltis et al. 2000).

Molecular age estimates (Jian et al. 2008) indicate that Saxifragales originated ~112 (+ 9.7) to 120 (+ 10.2) mya and diversified rapidly. A similar window (100 to 120 mya) is common for the diversification of other major clades of angiosperms, including some rosids (e.g., Malpighiales; Davis et al. 2005b), several major asterid lineages, including Ericales, Cornales, Campanulids (Bremer et al. 2004; Sytsma et al. 2006), and Dipsacales (Bell and Donoghue

Figure 10.3. Saxifragales (Altingiaceae, Cercidiphyllaceae, Crassulaceae, Grossulariaceae, Iteaceae, Hamamelidaceae, Paeoniaceae) and Vitales: a. *Cercidiphyllum japonicum* Siebold & Zucc. ex J.J.Hoffm. & J.H.Schult. (Cercidiphyllaceae), shoot. b. *Liquidambar styraciflua* L. (Altingiaceae), inflorescence. c. *Ribes magellanicum* Poir. (Grossulariaceae), inflorescence showing details of flowers. d. *Saxifraga stolonifera* Curtis (Saxifragaceae), flower. e. *Sedum spurium* M.Bieb. (Crassulaceae), inflorescence showing details of flowers. f. *Fothergilla gardenii* L. (Hamamelidaceae), habit and inflorescence. g. *Itea virginica* L. (Iteaceae), flower. h. *Paeonia lactiflora* Pall. (Paeoniaceae), flower. i. *Leea coccinea* Bojer (Vitaceae), inflorescence with some flowers at anthesis.

2005a; Beaulieu et al. 2013); and most major clades of monocots (Bremer 2000; see Bell et al. 2010). Molecular estimates for Saxifragales are somewhat older than the fossil record estimates in that fossils attributed to Saxifragales are present in Turonian-Campanian strata (89.5 mya) (Hermsen et al. 2006; Friis et al. 2011); the molecular estimate is comparable in age to the oldest fossils of core eudicots (Magallón et al. 1999).

Some Saxifragales were considered closely related in some morphology-based classifications. Saxifragaceae, Grossulariaceae, Iteaceae, Pterostemonaceae, Penthoraceae, and Tetracarpaeaceae were considered part of a much more broadly defined Saxifragaceae s.l. (Engler 1930) that also included families now placed among the asterids. A close relationship of these core families of Saxifragaceae s.l. to Crassulaceae had also been proposed (e.g., Cronquist 1981; Takhtajan 1987, 1997). Hamamelidaceae, Altingiaceae, Cercidiphyllaceae, and Daphniphyllaceae were also considered closely related in these classifications (e.g., the "lower hamamelids" sensu Walker and Doyle 1975). However, Saxifragales also include Haloragaceae, Paeoniaceae, and Peridiscaceae, families that have never been placed with any other members of Saxifragales in previous classifications. Haloragaceae have been placed in or near the rosid order Myrtales (Cronquist 1981), whereas Paeoniaceae were variously considered closely related to Magnoliaceae (e.g., Worsdell 1908; Sawada 1971), Ranunculaceae (e.g., Takhtajan 1997), or Dilleniaceae (e.g., Corner 1946; Melchior 1964; Keefe and Moseley 1978; Cronquist 1981).

Treatment of the three genera of Peridiscaceae (*Peridiscus*, *Soyauxia*, and *Whittonia* from tropical South America and West Africa) is complex (Davis and Chase 2004; Soltis et al. 2007b). Members have been placed in various rosid families (e.g., Passifloraceae, Flacourtiaceae, Medusandraceae). Inclusion of *Aphanopetalum* (previously placed in the rosid family Cunoniaceae) in Saxifragales initially seemed surprising but had been suggested based on anatomical data (Dickison et al. 1994). The inability of classical systematics to reveal close relationships among members of Saxifragales reflects the great morphological diversity encompassed by this small clade. Although including far fewer species than asterids, rosids, and Caryophyllales, Saxifragales exhibit similar levels of diversity in vegetative and reproductive morphology. Saxifragales include trees, shrubs, lianas, annual and perennial herbs, succulents, and aquatics. Flowers vary considerably in arrangement, merism, degree of perianth fusion, stamen and carpel number, ovary position, and degree of syncarpy (Cronquist 1981; Takhtajan 1997; Soltis et al. 2013).

Synapomorphies for Saxifragales are only now becoming clear (Carlsward et al. 2011). One putative synapomorphy is a basifixed anther (Fig. 10.4), sometimes with the filament attached at a basal pit (Endress and Stumpf 1991; Carlsward et al. 2011) although basifixed anthers also occur in many other eudicots; reversals from a basifixed anther to the dorsifixed position have occurred in the clade. A basifixed anther is a synapomorphy of Saxifragales, but not the pit itself, which occurs only in several subgroups (*Aphanopetalum*, *Saxifragaceae*, and *Crassulaceae*). In addition, all but *Peridiscus* (Peridiscaceae) are united by latrorse anther dehiscence and follicular fruits, the latter a feature that correlates with the ovaries being at least distally unfused. The woody clade (Altingiaceae, Cercidiphyllaceae, Daphniphyllaceae, and Hamamelidaceae) is diagnosed by adaxially flattened petioles, polytelic inflorescences, sessile flowers, and the loss of floral nectaries. The woody members are also similar in wood anatomy: solitary vessels, scalariform perforations, opposite to scalariform intervessel pits, nonseptate fibers with distinctly bordered pits, and apotracheal parenchyma (Baas et al. 2000). Core Saxifragales (all members of the clade except Peridiscaceae, Paeoniaceae, and the woody clade) are united by the presence of a hypanthium (except for Crassulaceae).

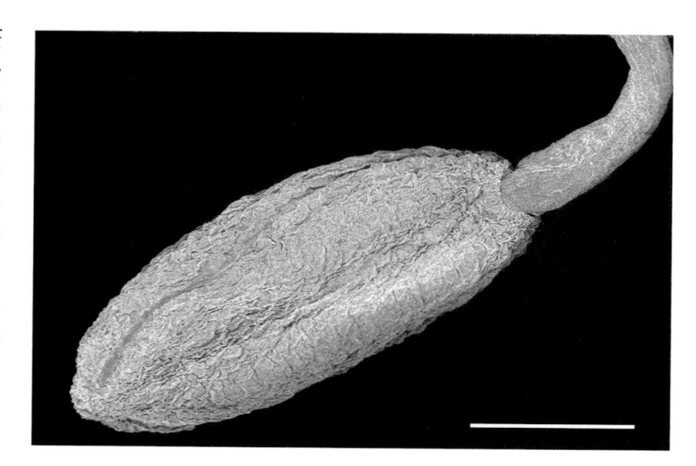

Figure 10.4. Basifixed anthers are a possible synapomorphy for Saxifragales, an unusual feature that many members possess. *Tetracarpaea tasmanica*, scale = 300 µm; see Carlsward et al. (2011); photograph courtesy of B. Carlsward.

RELATIONSHIPS WITHIN SAXIFRAGALES

Relationships within Saxifragales were difficult to resolve due to a rapid radiation (Fishbein et al. 2001; Fishbein and Soltis 2004; Jian et al. 2008). Using numerous genes Jian et al. (2008) and Soltis et al. (2011) resolved relationships (Fig. 10.2). Peridiscaceae are sister to all remaining Saxifragales; Paeoniaceae are sister to a well-supported woody clade (Cercidiphyllaceae, Daphniphyllaceae, Altingiaceae, and Hamamelidaceae). This Paeoniaceae + woody clade is, in turn, sister to a well-supported core Saxifragales. Core Saxifragales comprise two subclades: Crassulaceae + Haloragaceae s.l. (i.e., Aphanopetalaceae, Tetracarpaeaceae, Penthoraceae, and Haloragaceae s.s.—families maintained as distinct in APG III 2009 and APG IV 2016) and the Saxifragaceae alliance. Within the Saxifragaceae alliance, Saxifragaceae + Grossulariaceae are sister to Iteaceae + Pterostemonaceae.

Strong support was found for Grossulariaceae (see Senters and Soltis 2002 and Weigend et al. 2002 for phylogenetic analyses of *Ribes*) as sister to Saxifragaceae (Soltis et al. 2001a), which, in turn, is sister to a clade of Iteaceae + Pterostemonaceae. Crassulaceae (Mort et al. 2001) are well-supported as sister to a clade of Tetracarpaeaceae, Penthoraceae, Haloragaceae s.s., and Aphanopetalaceae. *Tetracarpaea*, *Penthorum*, and *Aphanopetalum* are all small genera that could be included within an expanded Haloragaceae (APG II 2003), but were retained as distinct families in APG III (2009) and APG IV (2016).

Monophyly of Paeoniaceae, consisting solely of *Paeonia*, is strongly supported (e.g., Sang et al. 1997; Fishbein et al. 2001; Jian et al. 2008). In recent analyses the family is sister to the woody clade, but support for this placement remains low (Jian et al. 2008). The relationships of Paeoniaceae have long been problematic and the family is morphologically distinct compared to other Saxifragales. The family is polystemonous (i.e., having many stamens), a feature also found in some woody clade members: Hamamelidaceae and Cercidiphyllaceae have 1 to 24 stamens, although in Cercidiphyllaceae delimitation of flowers is difficult (*Fothergilla*, Hamamelidaceae, has up to 24 stamens; in *Cercidiphyllum* there are no more than 7; Endress 1989a). Paeoniaceae also possess an apocarpous gynoecium of more than two carpels, a feature shared only with Crassulaceae and *Tetracarpaea* of the Crassulaceae + Haloragaceae clade.

Sequence data provided strong support for Altingiaceae (*Liquidambar* s.l. = *Altingia* + *Liquidambar*; *Liquidambar* as long defined is not monophyletic; Ickert-Bond and Wen 2013) and also supported removal of Altingiaceae from Hamamelidaceae (Hoot et al. 1999; D. Soltis et al. 2000; Fishbein et al. 2001), as proposed in studies of morphology, anatomy, cytology, and chemistry (Doweld 1998). Sequence data also supported the inclusion in Hamamelidaceae of *Exbucklandia* and *Rhodoleia*, two genera sometimes excluded from Hamamelidaceae (Doweld 1998). Within Hamamelidaceae support has emerged for four clades corresponding to the four subfamilies, Exbucklandioideae, Mytilarioideae, Disanthoideae, and Hamamelidoideae (e.g., Li et al. 1999a,b; Li and Bogle 2001; Li 2008).

The relationships of Cercidiphyllaceae and Daphniphyllaceae were long contentious (reviewed in Cronquist 1981; Takhtajan 1997; Fishbein et al. 2001). *Cercidiphyllum* was considered closely related to *Myrothamnus* (now in Gunnerales) (Hufford and Crane 1989; Hufford 1992) or to the early-diverging eudicot family Trochodendraceae (see Chapters 8 and 9). It was historically viewed as a derivative of the magnoliids, either distinct from hamamelids (Swamy and Bailey 1949) or transitional between magnoliids and hamamelids (Cronquist 1981). *Daphniphyllum* was variously considered closely related to Euphorbiaceae (Scholz 1964), *Cercidiphyllum* (Croizat 1941), and Hamamelidaceae (Bhatnagar and Garg 1977; Cronquist 1981; Zavada and Dilcher 1986; Sutton 1989; Takhtajan 1997). Gene sequence data indicate that Cercidiphyllaceae and Daphniphyllaceae together are sister to Hamamelidaceae (Fig. 10.2).

Cercidiphyllum can be traced back about 34 million years (Early Oligocene) based on its distinctive leaves and fruits (Fig. 10.5 e–g). Leaves and infructescences of extinct, likely stem-group relatives, of *Cercidiphyllum*, are common in the Late Cretaceous and Paleocene of the Northern Hemisphere (Crane 1984; Crane and Stockey 1986) known as *Trochodendroides* (leaves, Fig. 10.5h), and *Nyssidium* (infructescences, Fig. 10.5i). Inflorescences of Hamamelidaceae are known from well-preserved flowers of *Allonia* and *Androdecidua* from the Late Cretaceous (Late Santonian) Allon site in Georgia, USA (Magallón et al. 1996, 2001), and the family is well represented in the Cenozoic based on fruits and seeds (fig. 10.5d). Altingiaceae are represented by *Liquidambar* leaves in the Eocene (MacGinitie 1941) and younger occurrences of both leaves and fruits (Pigg et al. 2004). Well-preserved fruiting heads of the extinct Altingiaceous genus, *Steinhaurera*, are common in the Eocene of Europe (Mai 1968; Collinson et al. 2012) (Fig. 10.5c).

Placement of the parasite *Cynomorium* (Cynomoriaceae) has been problematic. Some analysis placed *Cynomorium coccineum* (Cynomoriaceae) in Saxifragales (Nickrent et al. 2005), albeit with low support, whereas others (Jian et al. 2008) placed Cynomoriaceae in Santalales, in agreement with some classifications (e.g., Cronquist 1981); sequences from the plastid inverted repeat placed the genus in Rosales close to Rosaceae (Zhang et al. 2009; Moore et al. 2011). Recent studies now place the genus in Saxifragales with strong support, although the closeset relatives of *Cynomorium* within Saxifragales are unclear (Renner pers. comm.; APG IV 2016). We show the family unplaced in Saxifragales (Fig. 10.2). Obviously, more study of Cynomoriaceae is required.

CHARACTER EVOLUTION IN SAXIFRAGALES

Saxifragales are an important clade in which to examine floral evolution not only because they appear to be sister

Figure 10.5. Fossil representatives of rosids including Saxifragales, Vitales, and Fagales. a, b. An early eudicot, possible rosid flower with showy perianth and large stamens situated opposite to the petals, Late Cretaceous Cenomanian of Nebraska, USA. c. Altingiaceae: Infructescence heads of *Steinhauera subglobosa* from the Middle Eocene of Germany. d. Hamamelidaceae. *Hamawilsonia boglei* fruits from the Paleocene of North Dakota. e-g. Leaves and fruits of *Cercidiphyllum crenatum* from the early Oligocene of Oregon. h. Leaf of extinct cercidiphyllaceous genus, *Trochodendroides*. i Racemose infructescence of *Nyssidium*, believed to be extinct representative of Cercidiphyllaceae. j, k. Oldest known Vitaceae: *Indovitis chitaleyae* from late Maastrichtian of India. l, m. *Leea olssonii* from early Oligocene of Peru, external view, and transverse section. n. Cross section of extant *Leea herbacea* [A: Maxwell 89–1051, Chang Mai. Thailand] seed for comparison. o. Fruiting twig of *Ulmus okanaganensis* from the early Middle Eocene of McAbee, British Columbia. p. Extinct Betulaceae *Palaeocarpinus dakotensis* fruits from Paleocene of North Dakota. q. Nut of *Palaeocarpinus dakotensis* sectioned transversely near the apex. r-v. Juglandaceae. r. *Cyclocarya brownii* from Paleocene of North Dakota. s. *Cruciptera simsonii* from middle Eocene of Oregon. t, u. Oldest walnut: *Juglans clarnensis* in lateral view and transverse section, early Middle Eocene of Oregon. v. Transversely sectioned nut of *Cruciptera simsonii*.

to either the rosids alone or the rosids + Vitaceae, but in addition, although possessing only a few species, Saxifragales exhibit levels of diversity in floral morphology comparable to the major eudicot clades. Thus, understanding floral diversification in Saxifragales may elucidate general evolutionary processes that operated early in eudicot diversification. We spend more time on character evolution in this clade than many others.

The evolution of several floral characters has been examined in Saxifragales using summary trees as well as a large tree with 909 species (Soltis et al. 2013): perianth merism, stamen number, anther dehiscence, and several gynoecial features (carpel union, carpel number, and ovary position). Character reconstructions provided support for some longheld ideas about evolutionary trends in angiosperms, but in other cases suggested opposing trends or labile evolution.

Life history—Beginning from a perennial ancestor, there have been multiple shifts to the annual habit in Saxifragales (Soltis et al. 2013). Diversification analyses indicate that transitions from perenniality to the annual habit are rare, but that annual lineages diversify at a greater rate, mostly due to the higher propensity of extinction of perennial lineages. This results in few, but species-rich annual groups among extant taxa, in some cases with a seemingly long history of annuality (e.g., *Crassula*). The reverse transition (from annual to perennial) is more frequent, but leaves a weaker signal because perennial lineages tend to go extinct at a higher rate than annuals. Therefore, transitions to annuality tend to lead to the radiation of annual lineages that only rarely produce perennial descendants. In *Aichryson* (Crassulaceae), from an ancestor that was perennial there has been a shift to the annual habit and then only in a few cases a return to the perennial habit.

Both woodiness and perenniality are associated with lower net diversification rates, although through different mechanisms. Woody lineages appear to have lower speciation rates than herbaceous taxa, which results in lower net diversification in spite of also having lower extinction rates compared to herbaceous taxa. Conversely, perenniality results in higher speciation rates, but this habit is also associated with higher rates of extinction. Annual species appear to have a significantly lower extinction rate, which leads to higher net diversification (Soltis et al. 2013).

Perianth—A longstanding view of angiosperm floral evolution is a trend toward reduction in the number of perianth parts (Bessey 1915; Stebbins 1974; Cronquist 1981, 1988; Takhtajan 1987, 1997). In agreement with these proposals, analyses of Saxifragales revealed a reduction in number of perianth parts. In addition, the complete loss of perianth appears irreversible in this clade (perianth loss, not shown, has occurred independently in Altingiaceae, Cercidiphyllaceae, and some Hamamelidaceae; for the latter see Endress 1978).

From an ancestral state of pentamery, most transitions within Saxifragales (trees not shown) are putative reductions in merism (e.g., to tetramerous in some Crassulaceae and Haloragaceae) or a loss of stability in merism (Paeoniaceae, Hamamelidaceae, staminate *Cercidiphyllum*) (Soltis et al. 2013). Three transitions in petal number are equally common in Saxifragales (all other transition types rare or absent): 5 to 0; 5 to 4; 5 to 6–10 (Fig. 10.6). An increase in petal number above the typical number of five is seen in Paeoniaceae and also in multiple cases in Crassulaceae. In fact, the same Crassulaceae showing an increase in stamen number exhibit an increase in petal number (see below; Soltis et al. 2013). A transition from five to four petals occurred multiple times, with major shifts to four exemplified by Haloragaceae s.s. and the large genus *Kalanchoe* (Crassulaceae). These shifts are also accompanied by a change in carpel number to four, revealing again that some traits appear coupled. The complete loss of petals also is uniform in some clades (e.g., *Cercidiphyllum* + *Dapniphyllum*, *Chrysosplenium*, and *Liquidambar*).

Androecium—Another longstanding view is that angiosperms were ancestrally polystemonous (numerous stamens) with a trend toward reduction in the number of stamens (Bessey 1915; Sprague 1925; Cronquist 1988; Leins and Erbar 1991, 1994; Takhtajan 1991; Ronse De Craene and Smets 1993). Others proposed that both increases and decreases in stamen number have occurred (Stebbins 1974; Takhtajan 1991; Ronse De Craene and Smets 1992), although both Stebbins (1974) and Cronquist (1988) suggested that reductions have been much more common than increases.

In Saxifragales, however, polystemony in *Paeonia* is clearly derived from an ancestral state of diplostemony (10 stamens). Two additional independent origins of polystemony from diplostemony have occurred within Hamamelidaceae, in which *Fothergilla* and *Matudaea* both exhibit polystemony (reconstruction not shown), but with both centrifugal and centripetal development (Endress 1976). Endress assumed that polystemony evolved separately in these two genera, a hypothesis supported by molecular studies because they are not sister taxa (Li et al. 1999a).

Anther dehiscence—Despite the wide occurrence of valvate anther dehiscence among basal angiosperms and early-diverging eudicots, as well as the presence of this pattern in the earliest well-preserved fossil stamens (Crane and Blackmore 1989; Crane et al. 1995), this condition is not necessarily a plesiomorphic feature for the angiosperms. Possibly, only the predisposition for easy development of valvate dehiscence was present in the earliest angiosperms, which may have possessed anthers that dehisced via simple

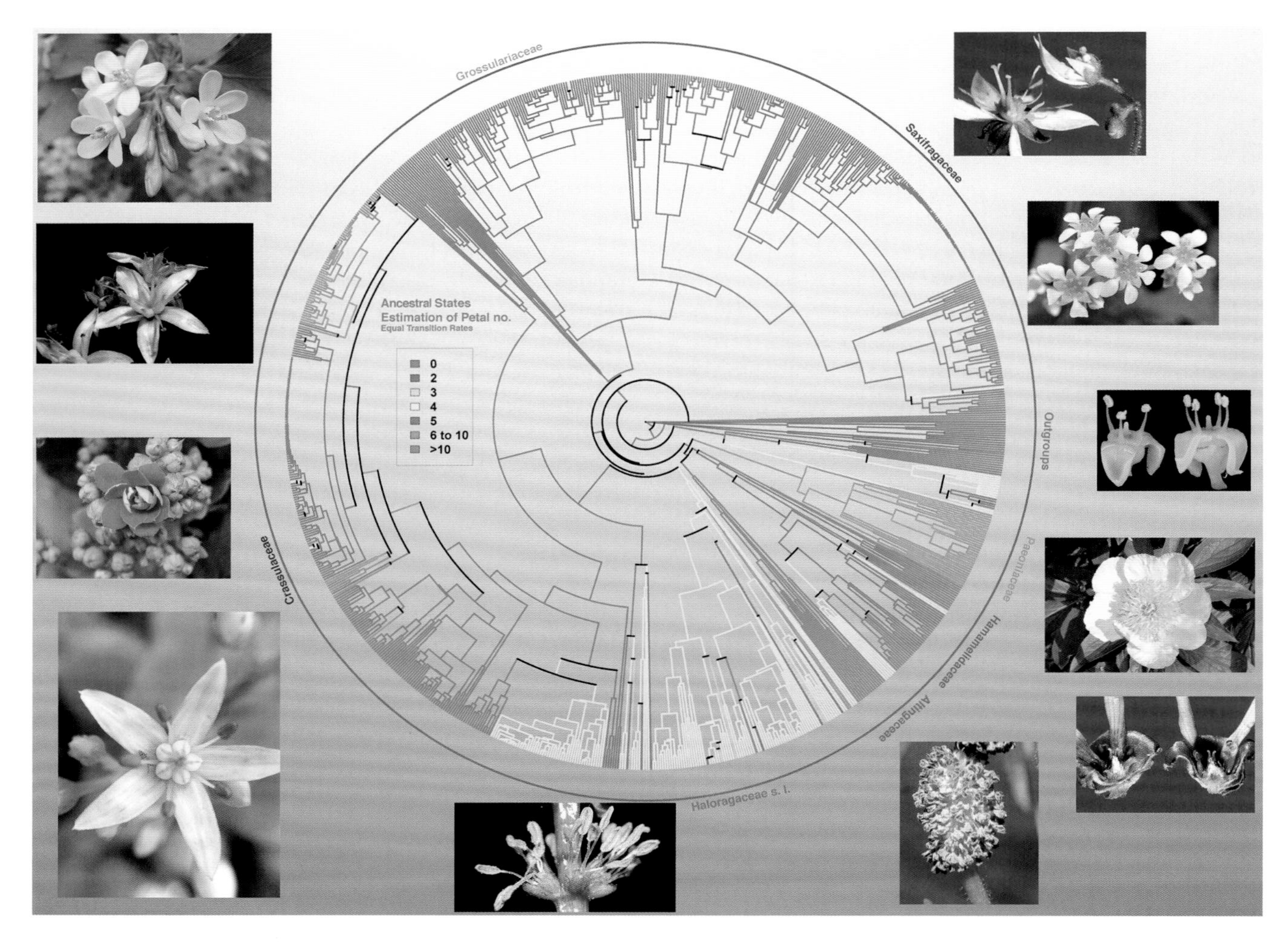

Figure 10.6. Ancestral state reconstruction (ML) of petal number showing patterns of variation across a large tree for Saxifragales (see Soltis et al. 2013).

longitudinal slits. This predisposition would have been lost in more derived angiosperms (Endress 1986a,b, 1989a; Endress and Hufford 1989; Hufford and Endress 1989).

Our reconstructions place longitudinal anther dehiscence as ancestral for Saxifragales with at least one shift to valvate dehiscence occurring in members of the woody clade. The number of times the trait of valvate dehiscence arose in the woody clade is not clear, but it may have arisen in the ancestor of the clade. Valvate anther dehiscence in Saxifragales does not appear to be homologous with that found in basal angiosperms or early-diverging eudicots (e.g., Trochodendraceae, Platanaceae). The presence of longitudinal dehiscence in *Disanthus* (Hamamelidaceae) is noteworthy in that, because of its derived phylogenetic position within Hamamelidaceae, *Disanthus* clearly represents a reversal within the family from valvate back to longitudinal dehiscence. Thus, anther dehiscence within Saxifragales may be more labile than previously envisioned.

Carpel number and union—Reductions in number of carpels and increasing syncarpy have been widely accepted as major trends in the angiosperms, with reversals considered unlikely (e.g., Bessey 1915; Sprague 1925; Grant 1950; Cronquist 1968, 1981, 1988; Takhtajan 1969, 1987, 1991; Stebbins 1974). It has been proposed that syncarpy offers a selective advantage over apocarpy due to increased pollination efficiency (Carr and Carr 1961; Stebbins 1974; Endress 1982; Takhtajan 1991) and centralized selection of male gametophytes (Endress 1982); it is also a precondition for fruit diversification (Endress 1982; Takhtajan 1991). Intermediate stages have also been proposed for the evolution of syncarpy. Cronquist (1988), for example, suggested that partial syncarpy, in which the apical portion of the gynoecium remains free (as in many Saxifragales), is an intermediate between apocarpy and complete syncarpy.

Character mapping provided evidence against the long-standing view that there has been a reduction in carpel number (not shown) in Saxifragaceae and other Saxifragales. Saxifragales were inferred to be ancestrally bicarpellate with at least two (perhaps more—results not shown) independent increases in carpel number.

Analyses of carpel union in Saxifragales also argued against the widely accepted trend of increased syncarpy. The ancestral condition for Saxifragales is carpels that are free apically; the extensive variation in the degree of carpel union in Saxifragales is shown (Fig. 10.7). Several independent derivations of free carpels (apocarpy) have occurred in Saxifragales: in Paeoniaceae, *Tetracarpaea*, and some Crassulaceae. Other putative cases of apocarpy derived from syncarpous ancestors include Apocynaceae, Malvaceae, Rutaceae, and Simaroubaceae (Endress et al. 1983; Fallen 1986; Sennblad and Bremer 1996). Other instances of secondary apocarpy or perhaps secondary partial apocarpy were later shown in Crossosomatales (Matthews and Endress 2005b) and some genera of Sapindales (Bachelier and Endress 2008, 2009). There are several independent transitions to completely fused carpels (*Aphanopetalum*, *Pterosteomon*, *Ribes*) from carpels partially fused (free apically), but no direct transitions from apocarpy to complete syncarpy were reconstructed within Saxifragales, although this transformation was the traditionally accepted pattern of evolution.

The ancestral carpel number in Saxifragales is reconstructed as two (Soltis et al. 2013), which also characterizes Vitaceae (outgroup), as well as the woody clade, and is maintained in major subclades such as Saxifragaceae and Grossulariaceae. Several prominent increases in carpel number occur, and these often characterize major clades and can serve as potential synapomorphies in some cases. There are variable numbers of carpels in Paeoniaceae and four carpels in the clade of Haloragaceae s.s., Apha-

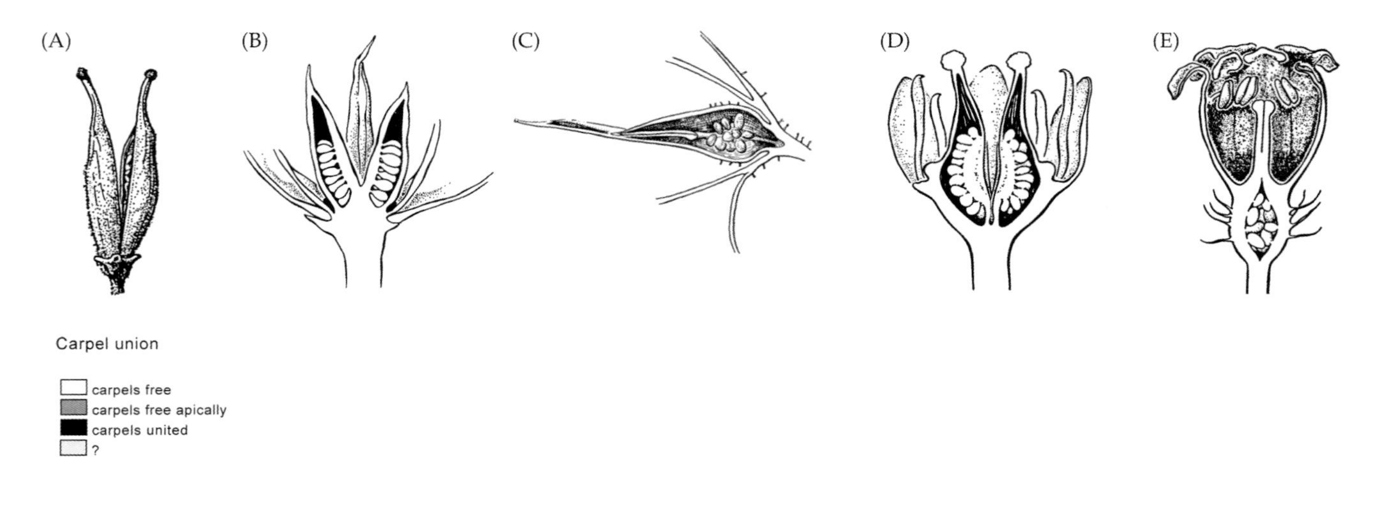

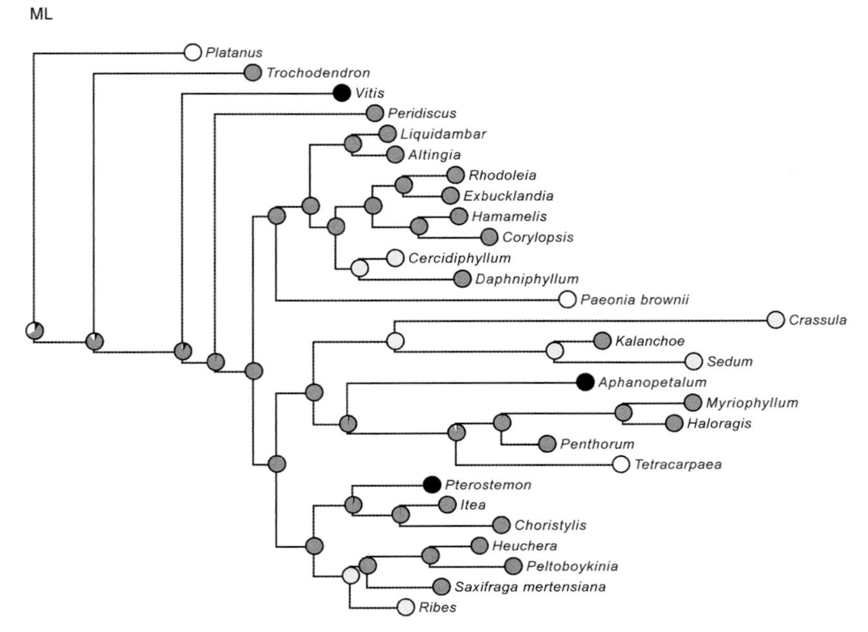

Figure 10.7. Carpel union and analyses of its evolution in Saxifragales. A-E illustrate variation in carpel union and ovary position in Saxifragales. (A) *Itea virginica* (Iteaceae) in fruit, carpels united only at base; ovary with only a short inferior portion (from Spongberg 1972). (B) *Sedum pusillum* (Crassulaceae) in flower, carpels nearly free (slight union at base); ovary superior in appearance (from Wood 1974). (C) *Tiarella cordifolia* (Saxifragaceae) in flower, carpels fused for about one-third their length; ovary with only a short inferior region (perhaps best termed pseudosuperior; see section on ovary position) (from Spongberg 1972; Wood 1974). (D) *Micranthes virginiensis* (Saxifragaceae) in flower, carpels fused only at base; ovary perhaps one-half inferior (from Spongberg 1972; Wood 1974). (E) *Ribes cynosbati* (Grossulariaceae) in flower, carpels completely united; ovary completely inferior (from Spongberg 1972). A reconstruction of the evolution of carpel union (F) shows the extensive variation in the degree of carpel union in Saxifragales.

nopetalaceae, Tetracarpaeaceae, Penthoraceae, with a further increase to five carpels in Penthoraceae. There is also an increase to four carpels in Peridiscaceae and five in all Crassulaceae, in which there is a decrease to four and constituting a synapomorphy for the large genus *Kalanchoe*. A further amplification in carpel number occurs in other Crassulaceae. For more than five carpels in general, see also Endress (2014).

Ovary position—The position of the ovary (gynoecium) (Fig. 10.8) has a major impact on floral architecture, concomitantly influencing various interactions with animals, such as predation, pollination, and seed dispersal (e.g., Grant 1950; Stebbins 1974; Thompson 1994; Thompson et al. 2013). In general, ovary positions have been treated as either superior or inferior according to the point of attachment of the perianth and androecium relative to the ovary in an anthetic (i.e., open or mature) flower.

A superior ovary is one that is situated above the point of attachment of the perianth and androecium to form a hypogynous flower; an inferior ovary has the perianth and androecium attached above the ovary to form an epigynous flower (Fig. 10.8). Flowers in which the outer floral appendages are basally fused to form a floral cup, or hypanthium, that surrounds the ovary are often called perigynous.

Ovary position has long been considered a key descriptive and stable feature (Grant 1950; Stebbins 1974) that is sometimes used to distinguish related families. A common view is that ovary position has evolved in a unidirectional manner from superior to greater inferiority, generally via congenital fusion of the outer floral appendages to the ovary wall (e.g., Langdon 1939; Douglas 1944; Gauthier 1950; Eames 1961); reversals were considered rare or impossible (Bessey 1915; Grant 1950; Cronquist 1968, 1988; Takhtajan 1969, 1991; Stebbins 1974). The putative selective advantage of epigyny is protection of ovules (Grant 1950; Stebbins 1974; Takhtajan 1991). Epigyny also brings the perianth whorl into additional contact with the androecium, permitting more interactions between these whorls in synorganized flowers. Reversals to superior ovaries were thought to be constrained by the complex alterations of developmental pathways presumed to be associated with the formation of inferior ovaries (Grant 1950; Stebbins 1974).

Families and orders that display extensive variation in ovary position have long been of evolutionary interest (e.g., Klopfer 1973; Stebbins 1974; Cronquist 1988; Gustafsson and Albert 1999). Ovary position is highly variable in the small family Saxifragaceae, ranging from inferior to what has been termed superior with a complete range of intermediate positions (Figs. 10.7, 10.8). In fact, this entire range of ovary positions is present within single genera (e.g., *Lithophragma*, *Saxifraga* s.s., *Micranthes*, and *Chrysosplenium*; Hitchcock et al. 1961; Taylor 1965; Elvander 1984; Webb and Gornall 1989; Soltis et al. 1993; Soltis and Hufford 2002). Even at this level, unidirectional ovary evolution from superior to increasing inferiority was invoked to explain the remarkable range of ovary positions (e.g., Taylor 1965; Stebbins 1974).

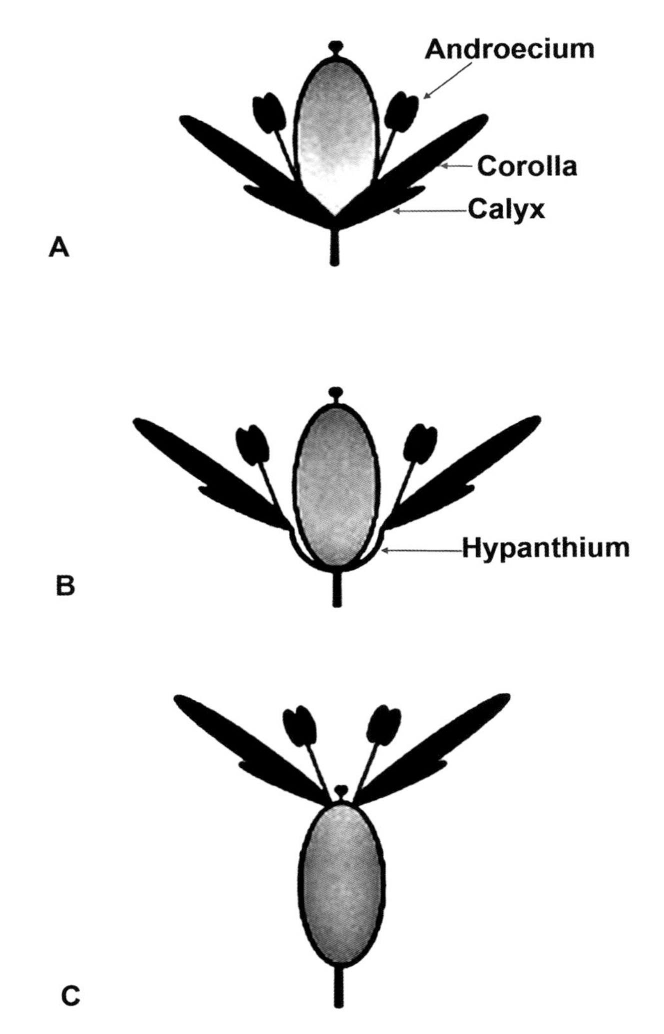

Figure 10.8. Ovary positions in angiosperms. (A) In a flower with a superior ovary the perianth and androecium are inserted beneath the ovary; the flower is hypogynous. (B) Alternatively, these outer appendages can be united to form a hypanthium; the flower is perigynous. (C) A flower with an inferior ovary has the floral appendages inserted at the apex of the ovary; the flower is epigynous.

Reconstructions reveal that many families of Saxifragales are stable in ovary position (superior for many clades). However, both Saxifragaceae and Hamamelidaceae are labile with frequent transitions among the three positions scored (inferior, subinferior [also called semi-inferior], superior) (Fig. 10.9). Attesting to this dynamic pattern, nearly all ovary position transitions occur at an equal frequency (that is, from inferior to subinferior and the reverse; from superior to subinferior and the reverse; and inferior directly to superior) (Soltis et al. 2013), with only direct transitions from superior to inferior half as frequent as the above. Ovary position evolution is thus particularly dynamic in Saxifragaceae, even within genera (Kuzoff et al. 1998, 1999, 2001; Mort and Soltis 1999; Soltis et al. 2001b). For example, ovary position evolution in *Lithophragma* has gone toward greater superiority in some lineages and greater inferiority in others, contrary to the longstanding view of a unidirectional trend in ovary position evolution (Kuzoff et al. 2001; reviewed in Soltis et al. 2003a). The *Heuchera* group had a subinferior ancestor with multiple transitions to a superior and inferior ovary. In contrast, *Micranthes* had a superior ovary as the ancestral state, with transitions to both subinferior and completely inferior ovaries. Similar evolutionary shifts from inferior to superior ovaries have occurred in other groups (Endress 2011b,c).

Gynoecial diversification and development—Studies of Saxifragales also revealed a dynamic picture of floral diver-

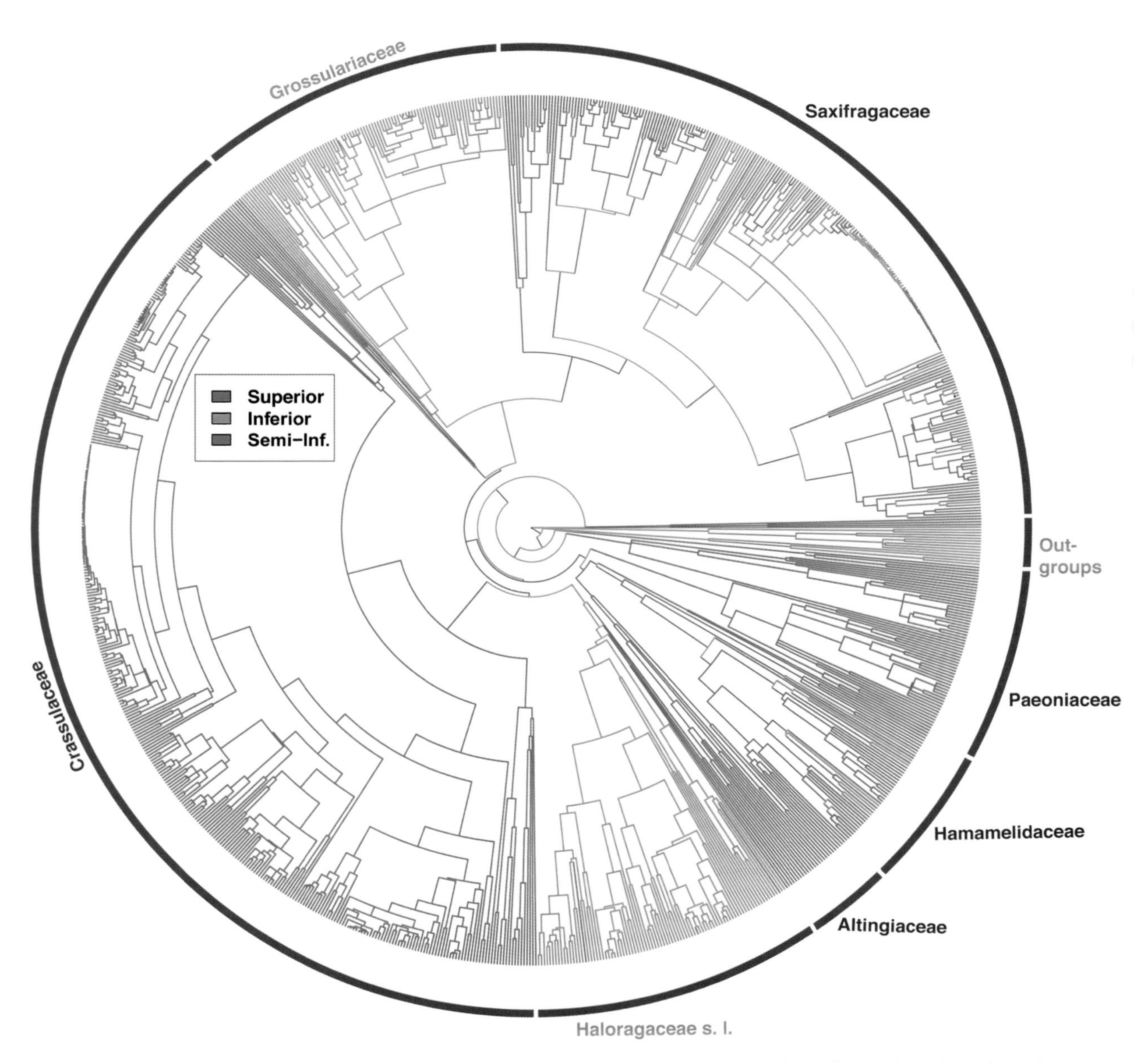

Figure 10.9. Ancestral state reconstruction of ovary position showing patterns of variation across a large tree for Saxifragales (see Soltis et al. 2013).

sification (Kuzoff et al. 2001; Soltis et al. 2003a). Typically, ovary positions have been described and decisions concerning structural homology made exclusively on the basis of anthetic (i.e., mature) floral structure. As reviewed, superior and inferior ovaries typically are distinguished by the point of attachment of other floral organs relative to the ovary in anthetic flowers (Fig. 10.8). Early floral development serves as an additional line of evidence to distinguish between flowers that are hypogynous versus epigynous (Kuzoff et al. 2001; Soltis and Hufford 2002). Boke (1963, 1964) and Kaplan (1967) demonstrated that hypogynous and epigynous flowers actually differ from the time of organogenesis, but the significance of their research has often been overlooked. There are two basic ground plans of floral development, hypogynous and appendicular epigynous (reviewed in Fig. 10.10).

Studies of floral development in Saxifragaceae have indicated that species with putatively superior ovaries actually have an appendicular epigynous ground plan (Figs. 10.10, 10.11; e.g., Kuzoff et al. 2001; Soltis and Hufford 2002; Soltis et al. 2003a). Thus, these ovaries are not truly superior but represent "superior mimics" or "pseudosuperior" ovaries. Pseudosuperior ovaries should not be considered homologous to truly superior ovaries derived from an hypogynous ground plan. Importantly, in Saxifragaceae the ovary position at anthesis is a consequence of the amount of vertical extension of the inferior versus superior region of the ovary. Differences in ovary position are a direct result of allometric shifts in the growth proportions of the superior versus inferior regions (Kuzoff et al. 2001), as observed for *Lithophragma* (Fig. 10.11). The relative ease with which allometric shifts can occur in the course

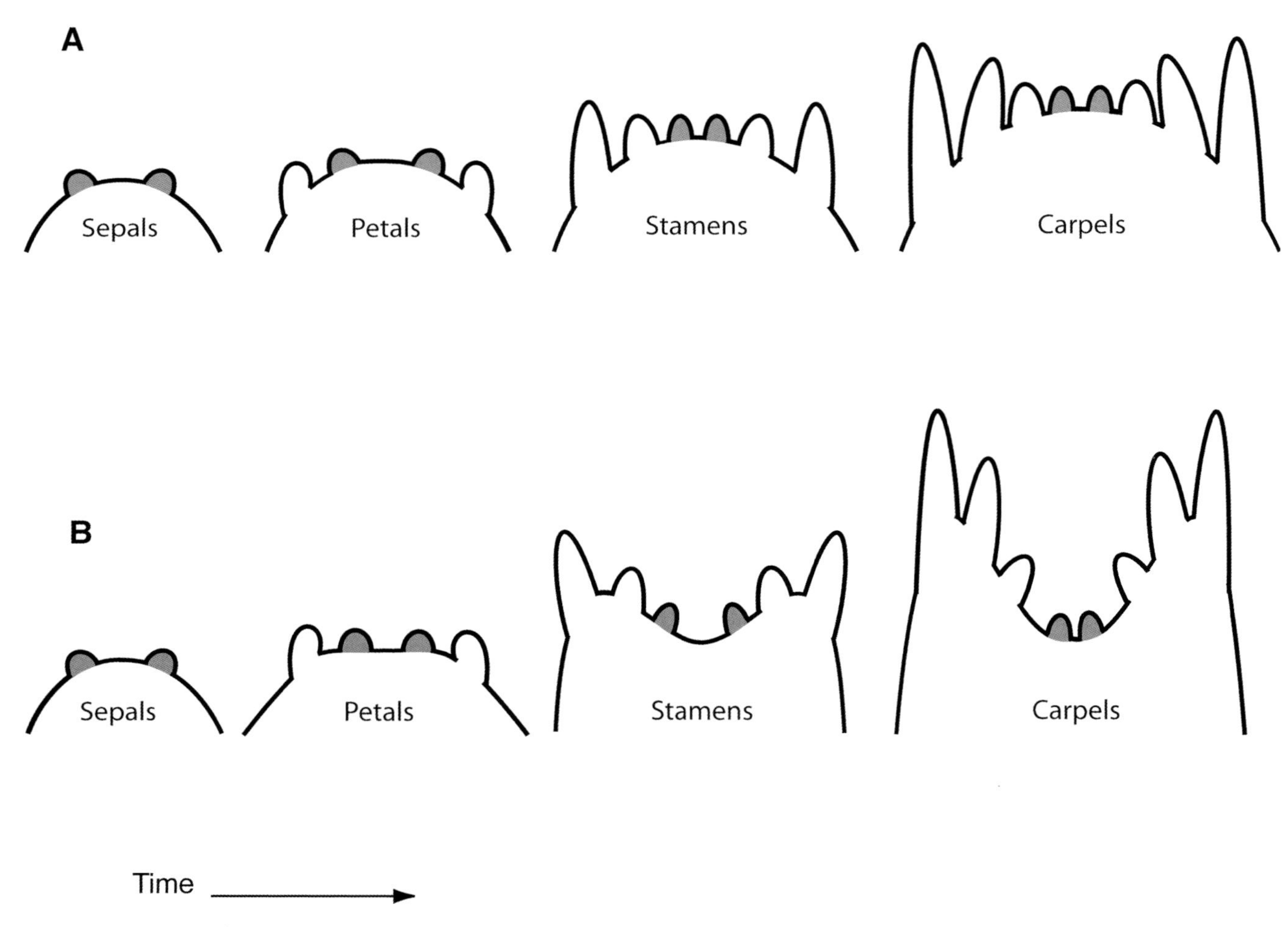

Figure 10.10. Patterns of ovary development in angiosperms. The early floral conformation of most hypogynous flowers represents what is termed a hypogynous ground plan (A). Hypogynous flowers generally exhibit a convex floral apex throughout organogenesis (e.g., Endress 1994c). This ground plan is common in the angiosperms and likely accounts for many flowers with ovaries described as "superior" based on the examination of anthetic flowers. An hypogynous ground plan characterizes many basal angiosperm lineages, such as Piperaceae and Alismataceae, as well as some early-branching eudicots, such as members of Ranunculaceae and Papaveraceae (Sattler 1973). Most inferior ovaries are the result of an appendicular epigynous pattern of floral development (B). Appendicular epigyny also begins with a convex floral apex, but during or just after perianth initiation the floral apex becomes concave (Boke 1964, 1966; Kaplan 1967; Leins 1972; Magin 1977; reviewed in D. Soltis et al. 2003c).

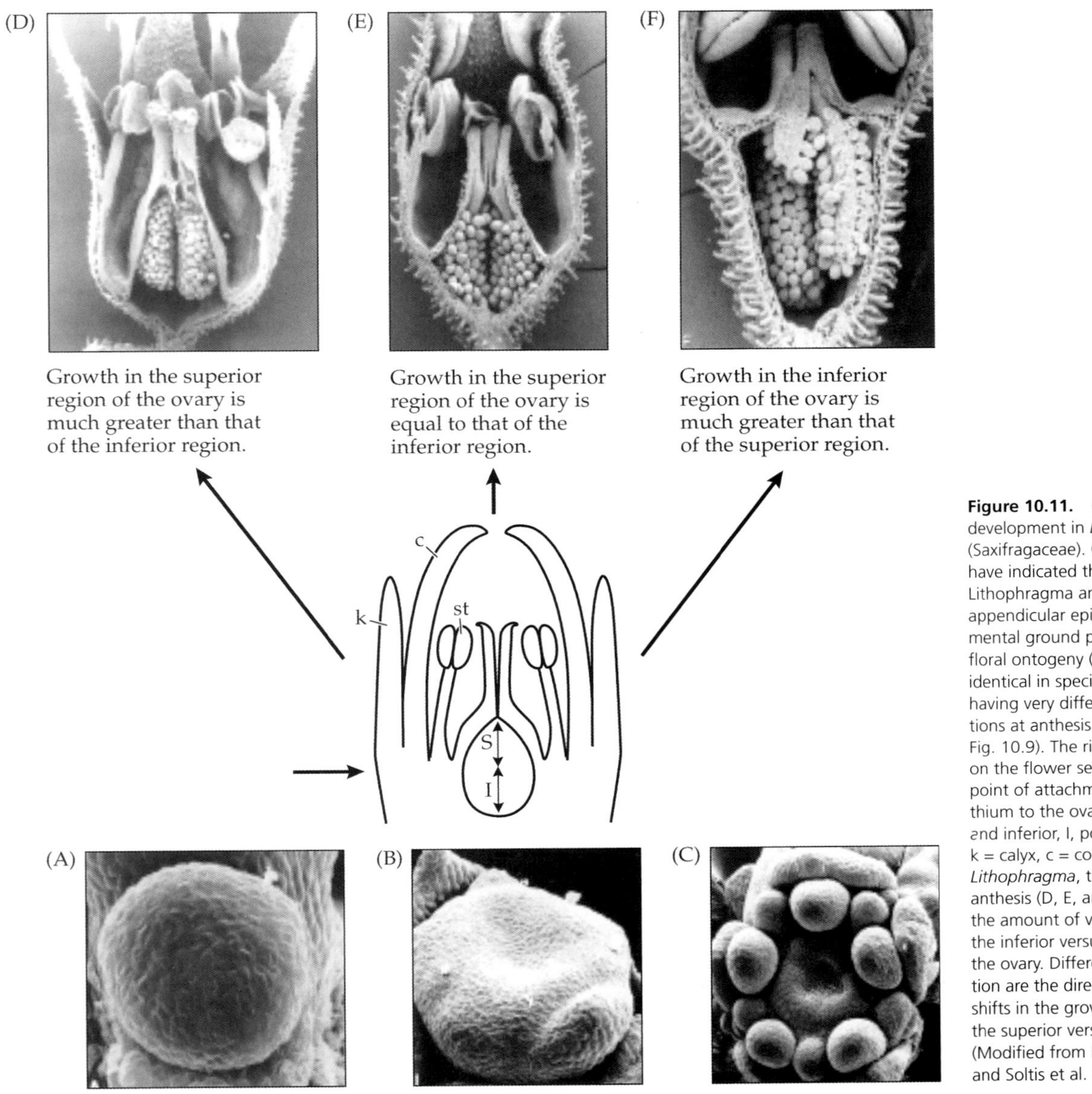

Figure 10.11. Gynoecial development in *Lithophragma* (Saxifragaceae). Ontogenetic studies have indicated that all species of Lithophragma and relatives have an appendicular epigynous developmental ground plan (see text). Early floral ontogeny (A, B, and C) is identical in species of *Lithophragma* having very different ovary positions at anthesis (D, E, and F) (see Fig. 10.9). The right-hand arrow on the flower section indicates the point of attachment of the hypanthium to the ovary (note superior, S, and inferior, I, portions of the ovary); k = calyx, c = corolla, st = stamen. In *Lithophragma*, the ovary position at anthesis (D, E, and F) is the result of the amount of vertical extension of the inferior versus superior region of the ovary. Differences in ovary position are the direct result of allometric shifts in the growth proportions of the superior versus inferior regions. (Modified from Kuzoff et al. 2001 and Soltis et al. 2005.)

of evolution explains the wide range of ovary positions in Saxifragaceae, as well as the numerous reversals that have occurred. Thompson et al. (2013) have shown that insect pollinator interactions with *Lithophragma* can drive ovary position to be more superior or more inferior. These data also indicate that gynoecial evolution may be more rapid than generally maintained, an important result in light of studies in Saxifragaceae that suggest ovary position affects pollinator preferences (e.g., Thompson and Pellmyr 1992; Thompson 1994; Segraves and Thompson 1999).

Data for Saxifragales have broader implications for floral evolution. The same processes described for gynoecia in Saxifragales are likely operating throughout the rosids (Soltis and Hufford 2002). Truly inferior ovaries derived from an epigynous ground plan may actually be the plesiomorphic state for many rosids. Data for the rosid family Vochysiaceae suggested that floral development is the same as in Saxifragaceae (Litt 1999; Litt and Stevenson 2003). Richardson et al. (2000) similarly proposed that the Saxifragaceae model of ovary position diversification also occurs in Rhamnaceae. Similar allometric shifts in development may explain the transition to a superficially superior ovary in *Gaertnera* (Rubiaceae; Igersheim et al. 1994). Developmental complexities are present in Rosaceae, however, in which species of *Physocarpus* and *Oemlera* have flowers that exhibit a concave floral apex before gynoecial initiation, yet have clearly superior ovaries at maturity (Evans and Dickinson 1999a,b). In contrast to the lability noted in Saxifragales and these other examples, ovary position is strongly fixed in many groups. More studies of gynoecial development and development genetics are clearly needed.

ROSIDS

OVERVIEW

Initial DNA analyses (Chase et al. 1993; Savolainen et al. 2000a; 18S rDNA, Soltis et al. 1997b) recovered a rosid clade that included more families than belonged to rosid groups (e.g., Rosidae, Rosanae) in morphology-based classifications (Cronquist 1981, 1988; Dahlgren 1983; Takhtajan 1987, 1997; Thorne 1992a,b, 2001). Analysis of non-DNA characters by Hufford (1992) and Nandi et al. (1998) also recovered a large rosid clade, broader than traditionally recognized. However, the taxon sampling of Hufford (1992) was small and there were odd placements (e.g., the clade also included Caryophyllales). None of these non-DNA analyses provided internal support (e.g., bootstrap or jackknife) above 50% for this expanded rosid clade.

As DNA datasets were combined, strong support ultimately emerged for the monophyly of a rosid clade, excluding Vitaceae (= eurosids). As reviewed, the interrelationships among Saxifragales, Vitales, and remaining rosids (eurosids) have varied across analyses. Here, we follow the 17-gene results (see APG III 2009 and APG IV 2016) and consider Vitales + eurosids to constitute the rosid clade.

The rosid clade as delineated via DNA is enormous, encompassing approximately 140 families and 70,000 species—over one-fourth of all angiosperm species and roughly 39% of eudicot species diversity (Magallón et al. 1999). The clade includes most lineages of extant temperate and tropical forest trees. The oldest fossils representing the clade are impression and compression specimens of the informally named "Rose Creek flower," from the late Albian of the Dakota Formation in Nebraska (ca. 102 mya) (Fig. 10.5a,b; Basinger and Dilcher 1984). A younger, well-preserved charcoalified flower known as *Paleoclusia* from the late Santonian to Turonian (about 84–89.5 mya; Crepet and Nixon 1998; Magallón et al. 1999) appears, as indicated in initial morphological cladistic analyses, to be related to Clusiaceae (Crepet and Nixon 1998) (see Supplementary Table 10.1 for selected rosid fossils; at press.uchicago.edu/sites/soltis/) and may be either a stem of the family or perhaps derived from within the crown group based on combined morphological and molecular analyses (Ruhfel et al. 2013).

In addition to most of those families previously placed in the traditional Rosidae, many families of former Dilleniidae and Hamamelidae (see Chapter 3) are also placed in the rosid clade by DNA data. Casuarinaceae, Fagaceae, Juglandaceae, Moraceae, Myricaceae, Ulmaceae, and Urticaceae are all members of the former Hamamelidae now placed in the rosid clade. Similarly, Brassicaceae, Caricaceae, Datiscaceae, Dipterocarpaceae, Malvaceae, Ochnaceae, Passifloraceae, and Violaceae are some of the families of Dilleniidae now part of the rosid clade. Even a few families formerly placed in Magnoliidae (e.g., Coriariaceae, Corynocarpaceae) are now considered rosids. Conversely, a few families of Rosidae of morphology-based classifications have been shown to be asterids (e.g., Apiaceae, Araliaceae, Cornaceae, Hydrangeaceae, and Pittosporaceae; see Chapter 11). Although great progress has been made in defining clades and resolving relationships within the rosids, some relationships remain poorly understood (see Endress et al. 2013).

Non-DNA synapomorphies are still unclear for the rosid clade, as well as for most subclades (orders) within the rosids. Although several morphological and anatomical features are shared by many rosids, none of these traits represent a clear synapomorphy. Noteworthy shared non-DNA features of rosids include nuclear endosperm development, reticulate pollen exine, generally simple perforations of vessel end-walls, alternate intervessel pitting, mucilaginous leaf epidermis, ellagic acid (see Chapter 8), flowers with separate petals, and often two (diplostemony) or more whorls of stamens (Hufford 1992; Nandi et al. 1998).

One factor contributing to the complexity of the rosids is the sheer size of the group. However, aspects other than size contribute to the poor understanding of rosid relationships. The asterid clade, which is of comparable size, is much better understood in terms of relationships, possible non-DNA synapomorphies, and character evolution (see Chapter 11). Similarly, enormous advances have been made in elucidating monocot relationships (see Chapter 7). Another reason the rosid clade has been so problematic is that, as noted, it contains many morphologically disparate families that were long considered so distinctive they were placed in four different subclasses. These morphologically diverse families are now interdigitated within the rosid clade. Hence in many cases, families now considered closely related in the rosid clade were not considered closely related in the past; as a result, the uniting morphological, chemical, and other non-DNA characters have been little studied, making them difficult to ascertain. The floral structure of a number of newly recognized suprafamilial clades in molecular phylogenetic studies have been studied and thus also morphologically characterized by reproductive structures since the first edition of this book: Celastrales (Matthews and Endress 2005a), some subclades of Malpighiales (Matthews and Endress 2008, 2011, 2013; Matthews et al. 2012;

Endress et al. 2013), and Sapindales (Bachelier and Endress 2007, 2008, 2009; Bachelier et al. 2011).

Further complicating the problem, many non-DNA characters of interest have not been surveyed throughout the rosids. There are so many gaps in taxonomic coverage of non-DNA characters that the potential utility of many traits cannot be meaningfully assessed until more basic survey work is conducted. In contrast, although the asterid clade is larger than the Asteridae of recent classifications, most of the additions (e.g., Cornales, Ericales) have been placed at the basalmost nodes. The core asterids remain highly similar to the Asteridae recognized long before DNA-based phylogenetic analyses.

Early phylogenetic studies revealed two large subclades within the rosids, but neither obtained bootstrap or jackknife support above 50% (e.g., Chase et al. 1993; Savolainen et al. 2000a,b). These large clades were initially termed rosid I and II (see Chase et al. 1993) or eurosid I and II (Savolainen et al. 2000a,b; D. Soltis et al. 2000). Analyses of combined molecular datasets provided the first internal support for these two large subclades of rosids (Fig. 10.1). Multi-gene studies of 3 to 5 genes provided increased resolution and support of rosids and a large subclade of fabids or *Fabidae*, as defined below (e.g., P. Soltis et al. 1999a; D. Soltis et al. 2000; Burleigh et al. 2009), but the exact circumscription of these two large subclades as well as relationships within the subclades remained unclear following these large analyses.

Recent analyses with broad taxonomic sampling and large numbers of plastid genes (Wang et al. 2009; Moore et al. 2010; Soltis et al. 2011) have provided a greatly improved understanding of rosid phylogeny, which form the two large clades noted above, each with maximal bootstrap support: (i) eurosids I (*Fabidae*) include Cucurbitales, Fabales, Fagales, Rosales (the nitrogen-fixing clade), Zygophyllales, and Celastrales, Malpighiales, and Oxalidales (the COM clade of Endress and Matthews 2006), these last three based just on plastid DNA; and (ii) eurosids II (*Malvidae*) include Huerteales, Brassicales, Malvales, Sapindales, Geraniales, Myrtales, Crossosomatales, and Picramniales. As reviewed below, nuclear and mitochondrial DNA data favor a placement of the COM clade with the malvids.

RAPID DIVERSIFICATION

The rosid clade (70,000 species) not only is enormous, but also includes most lineages of extant temperate and tropical forest trees. Sequence data indicate that the rosid clade diversified rapidly into the major lineages (fabids and malvids) over a period of <15 million years, and perhaps in as little as 4 to 5 million years (Fig. 10.1). The timing of the inferred rapid radiation of rosids (108 to 91 million years ago [mya] and 107–83 mya for Fabidae and Malvidae, respectively) corresponds with the rapid rise of angiosperm-dominated forests and the concomitant diversification of other clades that inhabit these forests. Ants (Moreau et al. 2006), beetles, and hemipterans (Farrell 1998; Wilf et al. 2000), amphibians (Roelants et al. 2007), ferns (Schneider et al. 2004), and some mammals (e.g., primates, Bininda-Emonds et al. 2007)—appear to have closely tracked the rise of rosid-dominated forests (H. Wang et al. 2009) and diversified "in the shadow of the angiosperms" (Schneider et al. 2004—original quote as applied to ferns).

VITALES

In morphology-based classifications, Vitaceae have been placed within Rosidae, typically within Rhamnales (Cronquist 1981). However, as reviewed, the DNA placement of Vitaceae, which include *Leea* (alternately recognized in a separate family, Leeaceae; see Ingrouille et al. 2002), is with rosids and Saxifragales, although the position of Vitales differs among DNA phylogenetic analyses. In earlier phylogenetic studies (e.g., Savolainen et al. 2000a,b; D. Soltis et al. 2000; Hilu et al. 2003), as well as some recent ones (e.g., Worberg et al. 2009), the circumscription of "eurosids" and "rosids" tended to exclude Vitaceae. However, there has been a trend towards inclusion of Vitaceae in rosids (e.g., APG II 2003; APG III 2009; APG IV 2016; Wang et al. 2009; Soltis et al. 2011) (Fig. 10.1). As reviewed, more analyses are needed to evaluate this placement further.

Inclusion of *Leea* (Leeaceae) (illustrated in Fig. 10.3i) in Vitaceae has a long history. Bentham and Hooker (1862–1883) recognized only three genera in Vitaceae: *Leea*, *Vitis*, and *Pterisanthes*. Thorne (1992a,b) also placed *Leea* in Vitaceae as a distinct subfamily. Most modern classifications, however, have followed Planchon (1887) in placing *Leea* in its own family, Leeaceae, and recognizing 10 genera in Vitaceae. If treated as a distinct family in classifications, Leeaceae, the author(s) also recognized that it had a close relationship to Vitaceae (e.g., Cronquist 1981; Takhtajan 1997; Judd et al. 2002, 2008). DNA data early indicated that *Leea* (Leeoideae) is sister to the remainder of the family (Vitoideae) (Ingrouille et al. 2002). Subsequent analyses have also revealed a number of subclades within Vitoideae (Wen et al. 2007; Chen et al. 2011; Ren et al. 2011; Trias-Blasi et al. 2012; Liu et al. 2013). For example, Ren et al. (2011) recovered subclades in the Vitoideae subclade that correspond with floral morphology (4-merous flowers vs. 5-merous flowers).

Putative synapomorphies of Vitaceae (including *Leea*) include raphide sacs, specialized stalked, gland-headed hairs, diminutive calyx, valvate corolla, stamens opposite the petals, and berries with distinctive seeds. The seeds have a cordlike raphe on the adaxial surface, extending from the hilum to the seed apex and onto the convex abaxial side, where it joins a round to linear depressed to somewhat elevated chalazal knot. The seeds also have a pair of deep grooves flanking the raphe. The endosperm is thus three-lobed (Judd et al. 2008).

The fossil record of Vitaceae is anomalously young relative to the deep phylogenetic split from other rosids that are represented in the fossil record back to the Cenomanian (Supplementary Table 10.1). Despite a rich paleobotanical record of distinctive vitaceous seeds in the Cenozoic, the seeds are missing from the Cretaceous charcoalified fossil assemblages of Europe and North America that have yielded other early Rosid representatives. The oldest known record for the Vitaceae is fruits with intact seeds from the uppermost Cretaceous of India (Late Maastrichtian, ca. 66 mya; Fig. 10.5j,k). India, at that time, was a tectonically isolated island rafting northward from the Southern Hemisphere. This suggests that the grape family originated in the Southern Hemisphere. The paucity of well-studied Cretaceous floras in the Southern Hemisphere may explain the missing early record of this family (Manchester et al. 2013). The complicated biogeographic history of this family is underscored by the presence of *Leea* (Fig. 10.3i), which is now confined to the Asian-Malesian tropics, in the Oligocene of South America (Fig. 10.5l, m; Manchester et al. 2012).

FABIDS AND MALVIDS

Due to their rapid radiation, resolving relationships within *Rosidae* has been difficult (Hilu et al. 2003; Soltis et al. 2005b, 2007c, 2011; Jansen et al. 2007; Zhu et al. 2007; Wang et al. 2009; Moore et al. 2010, 2011; Qiu et al. 2010; Lee et al. 2011), but multi-gene studies recover two major subclades—*Fabidae* (eurosid I, fabids) and *Malvidae* (eurosid II, malvids) (Jansen et al. 2007; Soltis et al. 2007c, 2011; Burleigh et al. 2009; Wang et al. 2009; Moore et al. 2010, 2011; Qiu et al. 2010; Zhang et al. 2012). However, Celastrales, Oxalidales, and Malpighiales (the COM clade) have sometimes been placed in the malvid clade with nuclear or mtDNA gene sequences, a placement also supported by floral structure (Endress and Matthews 2006). We address this incongruence in more detail below.

Fabids

We follow the circumscription of fabids given in APG III (2009) and APG IV (2016) (Fig. 10.12), but stress that recent evidence indicates that the COM clade may be of ancient reticulate origin (see below)—if this is indeed the case the COM clade could be treated as part of both the malvids and the fabids.

ZYGOPHYLLALES

The morphologically distinct *Krameria* (Krameriaceae) and Zygophyllaceae s.s. have formed a strongly supported (100%) clade in DNA analyses (e.g., Savolainen et al. 2000a; D. Soltis et al. 2000, 2007c, 2011; H. Wang et al. 2009; Fig 10.12; representatives shown in Fig. 10.13h,i). However, a close relationship between these two families had not been suggested in any previous non-DNA based classifications. *Krameria* had been variously placed in Fabaceae (Caesalpinioideae; reviewed in Cronquist 1981), Polygalales (Cronquist 1981), or Vochysiales (Heywood 1993; Takhtajan 1997). Zygophyllaceae, a heterogeneous assemblage of 25 genera, have been variously placed in five different orders, but have most frequently been classified in Sapindales or Geraniales (e.g., Cronquist 1981; Takhtajan 1997).

Synapomorphies for Zygophyllales remain unclear. *Krameria* shares few features with Zygophyllaceae (Sheahan and Chase 2000). Hence, it seems justified to maintain Zygophyllaceae and Krameriaceae as distinct families (as in APG III 2009 and APG IV 2016) rather than combining them into a single broadly defined Zygophyllaceae (as in APG II 2003). Relationships within Zygophyllaceae are discussed in detail by Sheahan and Chase (1996, 2000). Zygophyllaceae and Krameriaceae also share few non-DNA traits with any other rosid lineage. Anthraquinones are present in Zygophyllaceae, and these are also common in the nitrogen-fixing clade (Nandi et al. 1998), but rare elsewhere, so this could be a synapomorphy (Sheahan and Chase 2000). However, there is no evidence from DNA analyses to support a close relationship with the nitrogen-fixing clade (Soltis et al. 2011). In fact, with large datasets Zygophyllaes appear with strong support as sister to rest of the fabid clade (e.g., Soltis et al. 2011; Fig. 10.12).

NITROGEN-FIXING CLADE

DNA analyses strongly support a clade of Rosales, Fabales, Cucurbitales, and Fagales (APG 1998; APG II 2003; APG III 2009; APG IV 2016; Figs. 10.1, 10.12). Fabales, followed by Rosales, are subsequent sisters to Fagales + Cucurbitales. As reviewed below, these orders share a genetic predisposition for nitrogen fixation via root nodules; this condition represents a possible synapomorphy for this group of four orders. Only 10 of the approximately 400 families (APG III 2009; APG IV 2016) of angiosperms form symbiotic associations with nitrogen-fixing bacteria involving root

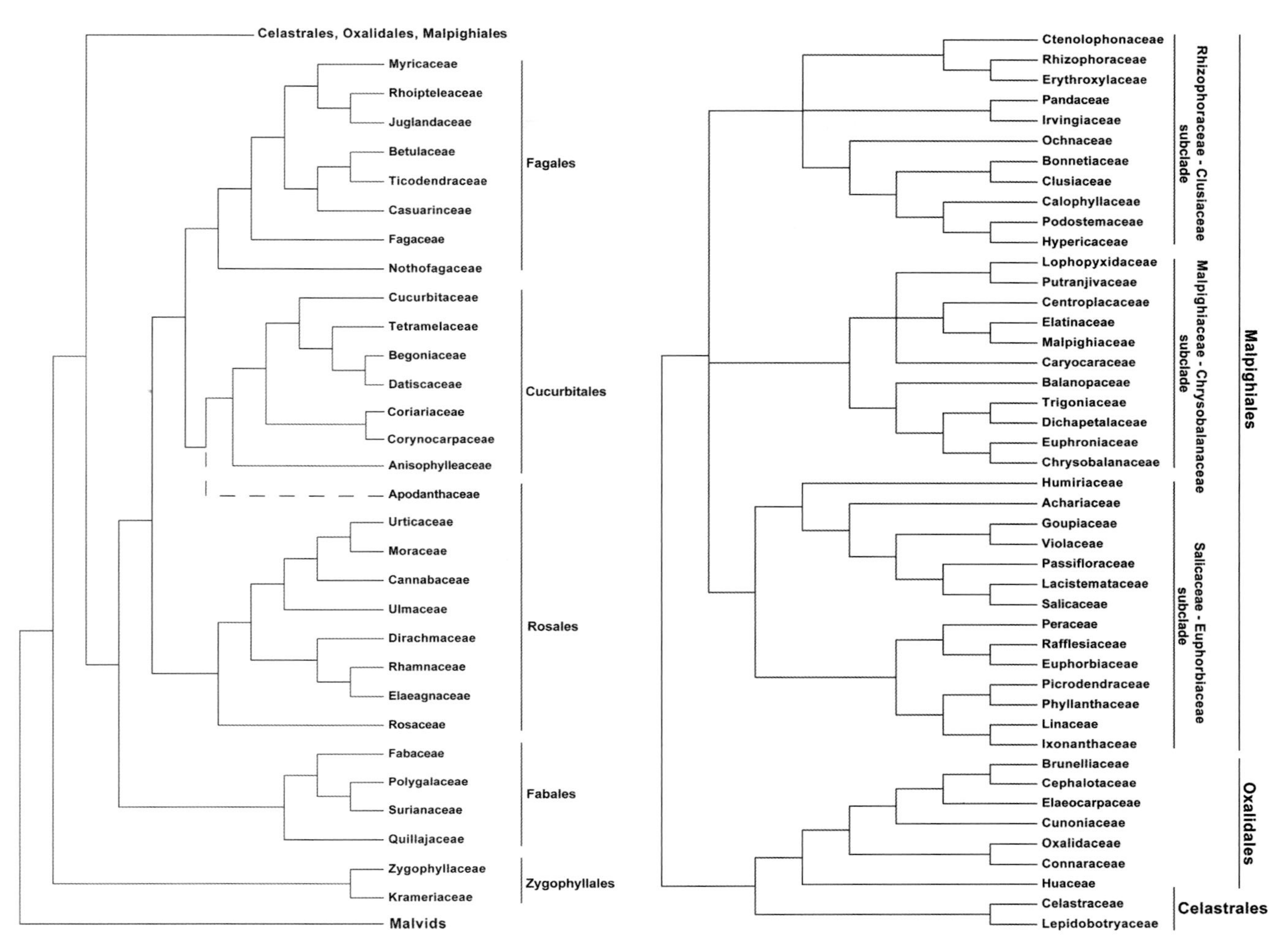

Figure 10.12. Summary tree for rosids with an emphasis on relationships within part of the fabid (*Fabidae*). A. Nitrogen-fixing clade and relationships among Cucurbitales, Fagales, Fabales, Rosales. B. Relationships among members of the COM clade of fabids.

nodules (Table 10.1): Betulaceae, Casuarinaceae, Cannabaceae (including the nitrogen-fixing genus *Parasponia* of the former Celtidaceae), Coriariaceae, Datiscaceae, Elaeagnaceae, Fabaceae, Myricaceae, Rhamnaceae, and Rosaceae (Akkermans and van Dijk 1981; Torrey and Berg 1988). In traditional morphology-based classifications (e.g., Dahlgren 1980; Cronquist 1981, 1988; Takhtajan 1987, 1997; Thorne 1992a,b, 2001), these families were considered distantly related. Cronquist (1981, 1988), for example, distributed these families among four of his six subclasses of dicots. Thus, the presence of these 10 families in one well-defined clade is a remarkable result. The monophyly of those angiosperms forming such symbiotic associations was initially indicated by *rbcL* sequences alone (Soltis et al. 1995). The clade containing these 10 families has been referred to since that time as the nitrogen-fixing clade. Maximal BS support has ultimately emerged for this clade (see Soltis et al. 2011; Wang et al. 2009), reinforcing the hypothesis of a single origin of the predisposition of nitrogen-fixing symbiosis (Soltis et al. 1995; see Werner et al. 2014).

Some of the nitrogen-fixing families (Betulaceae, Casuarinaceae, Fabaceae, Elaeagnaceae, Myricaceae, and Ulmaceae) also share the ability to produce nodular hemoglobin (Landsmann et al. 1986), also known as leghemoglobin, which has clear similarities to hemoglobin (and is also red in color). However, many of the families in the nitrogen-fixing clade do not form symbiotic associations with nitrogen-fixing bacteria; furthermore, many of the genera and species within the 10 nitrogen-fixing families also lack this symbiosis (Table 10.1). The process of nitrogen-fixation in members of the nitrogen-fixing clade has been the subject of multiple reviews (e.g., Sprent 2005; Kneip et al. 2007; Santi et al. 2013) and also many subsequent analyses. Examining the bacteria, Clawson et al. (2004) detected three

Figure 10.13. Rosids. Celastrales (Celastraceae, Parnassiaceae), Oxalidales (Cephalotaceae, Cunoniaceae, Elaeocarpaceae, Oxalidaceae), and Zygophyllales (Krameriaceae and Zygophyllaceae): a. *Parnassia grandifolia* DC. (Parnassiaceae), flower with conspicuous staminodia. b. *Euonymus americanus* L. (Celastraceae), fruits and arillate seeds. c. *Oxalis debilis* Kunth (Oxalidaceae), flower, showing tristyly (mid-length styles, with long and short stamens). d. *Averrhoa carambola* L. (Oxalidaceae), fruits. e. *Cephalotus follicularis* Labill. (Cephalotaceae), traps. f. *Elaeocarpus hygrophyllus* Kurz (Elaeocarpaceae), flower. g. *Weinmannia pinnata* L. (Cunoniaceae), habit and inflorescence. h. *Krameria lanceolata* Torr. (Krameriaceae), flower, with petaloid calyx and two, oil-producing lower petals. i. *Bulnesia arborea* Engl. (Zygophyllaceae), flower with clawed petals.

Table 10.1. Angiosperm families with nodular nitrogen-fixing symbioses and the frequency of this association in each family

Prokaryote	Angiosperm Family	Root nodules / total no. genera
Rhizobiaceae	Fabaceae	530/730
	Cannabaceae (including Celtidaceae)	1/11
	Betulaceae	1/6
Frankia	Casuarinaceae	4/4
	Elaeagnaceae	3/3
	Myricaceae	2/3
	Rhamnaceae	7/55
	Rosaceae	5/100
	Datiscaceae	1/1
	Coriariaceae	1/1

major *Frankia* clades and determined that these diverged early during the emergence of eudicots in the Cretaceous.

The morphologically diverse mix of families present in the nitrogen-fixing clade illustrates well the problem of diagnosing DNA-based clades within the rosids with non-DNA characters. Families of Fagales (Betulaceae, Casuarinaceae, Fagaceae, Juglandaceae, and Myricaceae) represent former higher Hamamelidae. Other former hamamelids are found in Rosales, including Cannabaceae, Ulmaceae, Moraceae, and Urticaceae; Rosales also include members of the formerly recognized subclass Rosidae, such as Elaeagnaceae, Fabaceae, Rhamnaceae, and Rosaceae. Many Cucurbitales were placed in Dilleniidae (e.g., Begoniaceae, Cucurbitaceae, and Datiscaceae), with one family in the order, Coriariaceae, placed in Magnoliidae as part of Ranunculales.

Within the nitrogen-fixing clade, the monophyly of each of the four subclades (Rosales, Cucurbitales, Fagales, and Fabales) is also well supported (e.g., H. Wang et al. 2009; Soltis et al. 2011); Fabales and then Rosales are sister to Fagales + Cucurbitales (Fig. 10.12).

CHARACTER EVOLUTION

NITROGEN-FIXING SYMBIOSIS

Root nodules are induced and inhabited by one of two groups of distantly related bacteria. Species of Rhizobiaceae (gram-negative motile rods) nodulate the legumes and *Parasponia* (Cannabaceae; formerly of Celtidaceae; Trinick and Galbraith 1980). In the legumes, the symbiont is *Rhizobium*, *Bradyrhizobium*, or *Azorhizobium*; only *Bradyrhizobium* is symbiotic with *Parasponia*. Actinomycetes of the genus *Frankia* (gram-positive, non–endospore-forming, mycelial bacteria) nodulate hosts in the remaining families (Table 10.1); these plants are referred to as actinorhizal (Akkermans and van Dijk 1981). Gunneraceae, a family that hosts nitrogen-fixing cyanobacterial symbionts in shoots (rather than in root nodules), were not found to be a member of the nitrogen-fixing clade, but instead appeared (with Myrothamnaceae) as sister to all other core eudicots. Gunneraceae therefore represent an independent evolution of nitrogen-fixing symbiosis in angiosperms.

Putative gains and losses of nitrogen-fixing symbioses within the nitrogen-fixing clade have been discussed by Soltis et al. (1995), Swensen (1996), Swensen and Benson (2008), Doyle (2011), and Werner et al. (2014). The hypothesized predisposition that enabled the evolution of nodulation appeared approximately 100 million years ago (mya) and was retained in the four major lineages (Rosales, Cucurbitales, Fagales, Fabales) that then radiated rapidly after the origin of the nitrogen-fixing clade. The most detailed reconstruction of the evolution of nitrogen-fixing symbiosis is the recent study of Werner et al. (2014), who employed over 3400 species and assembled the largest database of nodulating plant species. Their analyses confirmed a single origin of the predisposition of N_2 fixation (Soltis et al. 1995) in what they term a "cryptic evolutionary innovation driving symbiotic N_2-fixation evolution" (p. 1, Soltis et al. 1995). This innovation was followed by ~8 subsequent actual origins of N_2 fixation with ~10 losses.

To illustrate the complexity of nitrogen-fixation symbiosis across the entire clade, we reconstructed the pattern of gain and loss of symbiosis on a simple summary phylogenetic tree that represents the best estimate of phylogeny (Fig. 10.14). Our goals are to illustrate in a simple fashion the transitions proposed by Werner et al. (2014) based on their extensive taxon sampling and to highlight the complexities of the evoluation of this feature in the nitrogen-fixing clade.

Because the bacterial symbionts involved (Rhizobiaceae vs. *Frankia*) are different, those taxa involving Rhizobiaceae, Fabaceae, and *Parasponia* (Cannabacae) may represent separate origins from those using the genus *Frankia*. Furthermore, Fabaceae and *Parasponia* are distantly related, implying a minimum of two separate origins of nitrogen-fixing symbiosis with Rhizobiaceae.

There is also evidence for multiple origins of nodulation involving Rhizobiaceae within just Fabaceae (J. J. Doyle 1994; Sprent 1994; Doyle et al. 1997; Doyle and Luckow

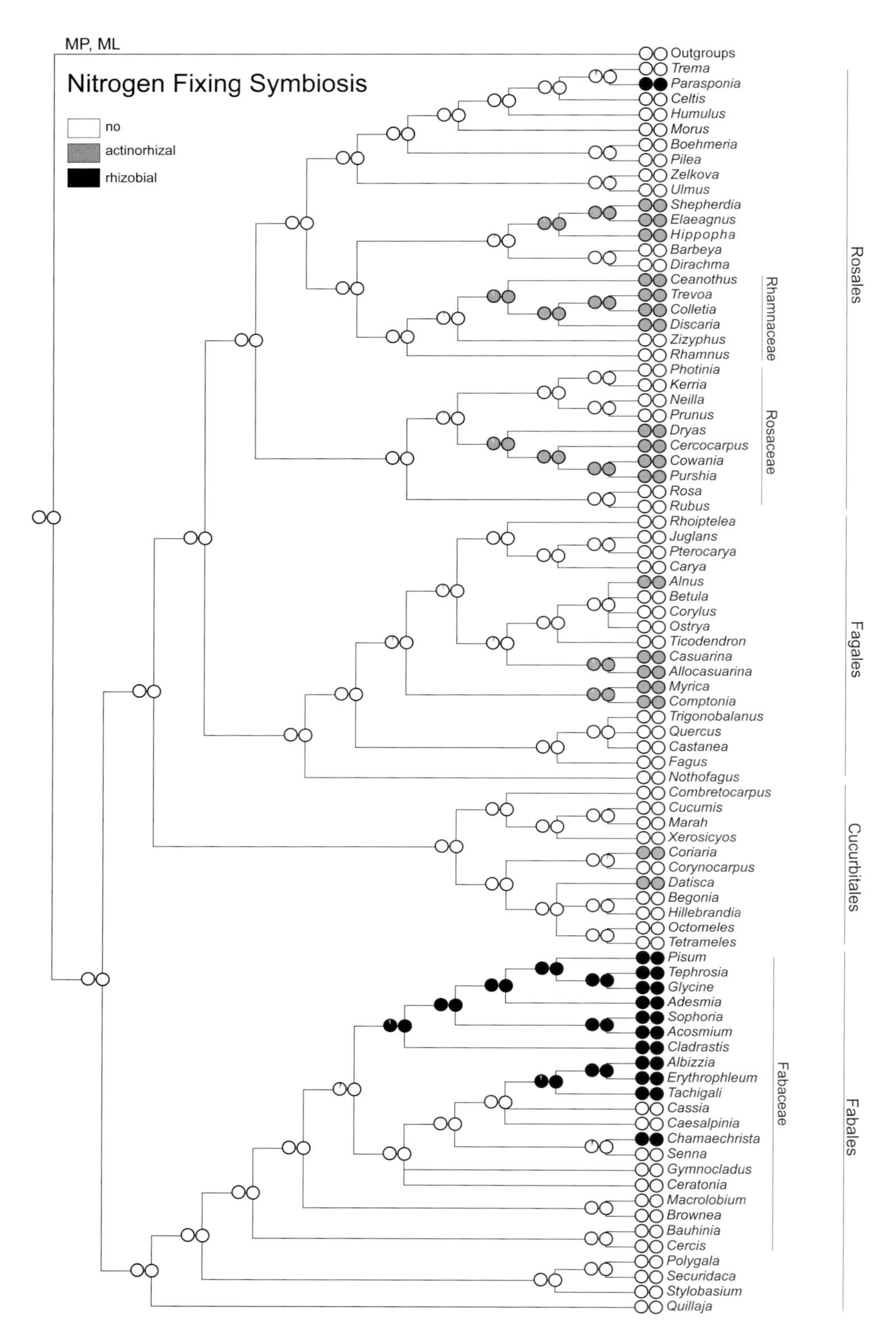

Figure 10.14. Ancestral state reconstructions of the evolution of nitrogen fixation via root nodules in the angiosperms.

2003; Sprent et al. 2013; Werner et al. 2014) (Fig. 10.14). Nodulation of legumes appears to have evolved 55–60 mya (Doyle 2011; Sprent et al. 2013). Most members of subfamilies Mimosoideae (92% of those examined) and Papilionoideae (97% of the taxa investigated) nodulate. Those Papilionoideae that do not nodulate mostly represent early-branching lineages within the subfamily. Caesalpinioideae are paraphyletic to the remainder of the family, and few genera nodulate (approximately 5%), making it clear that nodulation arose only after the major lineages of Fabaceae had diverged (Doyle and Luckow 2003). Caesalpinioids confirmed as nodulating are either sister to mimosoids or belong to the more distantly related genus *Chamaecrista*. There may be at least three origins of nitrogen-fixing symbiosis just in the legumes: (1) the papilionoids; (2) the clade consisting of some caesalpinioids (e.g., the *Sclerolobium* and *Dimorphandra* groups of the highly polyphyletic tribe Caesalpinieae) and Mimosoideae; and (3) *Chamaecrista* (Doyle et al. 1997; Doyle and Luckow 2003) (Fig. 10.14).

Our simple analysis indicates several possible gains of actinorhizal symbiosis (Fig. 10.17). Evolution of this symbiosis apparently occurred in: (1) Elaeagnaceae; (2) some Rhamnaceae; (3) some Rosaceae; (4) some Betulaceae; (5) Casuarinaceae; (6) Myricaceae; (7) *Coriaria*; and (8) *Datisca*. However, our studies actually indicate that the ancestral state for the subclade (within Fagales) of Betulaceae, Casuarinaceae, and Myricaceae is ambiguous. If actinorhizal symbiosis is ancestral for this subclade, then two losses occurred, once in *Ticodendron* and again in Betulaceae. Alternatively, Casuarinaceae (*Casuarina*, *Gymnostoma*, and *Allocasuarina*), *Alnus*, and Myricaceae (*Myrica* and *Comptonia*) could represent separate gains of actinorhizal symbiosis.

Swensen (1996) argued that comparative morphological data provided additional evidence for multiple origins of this symbiotic relationship within the actinorhizal lineages. For example, only host plants in the Fagales clade (Betulaceae, Casuarinaceae, and Myricaceae) are infected via root hairs (all others are infected by intercellular penetration) and possess nodules with a specialized oxygen diffusion–limiting cell layer; these same taxa also have higher levels of hemoglobin than hosts in other clades (reviewed in Swensen 1996), which could be interpreted as evidence for a separate origin of actinorhizal nitrogen-fixing symbiosis. *Alnus* also differs from other members of this clade in several respects involving nodulation, making the number of transitions a complex subject.

Coriaria (Coriariaceae) and *Datisca* (Datiscaceae; both Cucurbitales) share several anatomical nodule features. The infected cells are multinucleate, and the arrangement of vesicles within the cells is unique (reviewed in Swensen 1996). Thus, although our reconstructions and those of others indicate two independent gains, one each in *Coriaria* and *Datisca*, an alternative explanation that may better explain the shared morphology is one gain, followed by two losses (one in *Corynocarpus* and another in the clade of *Begonia*, *Hillebrandia*, *Octomeles*, and *Tetrameles*).

WIND POLLINATION AND FRUIT TYPE

Wind pollination (anemophily) occurs in ~18% of flowering plant families and has evolved many times (Linder 1998; Ackerman 2000; Culley et al. 2002), most likely in response to pollinator limitation and environmental variation. Wind pollination could be facilitated by movement of individuals into a geographic area with a distinct dry season in which conditions are unfavorable for insect pollination (Whitehead 1969, 1983; Weller et al. 1998; Culley et al. 2002). A phylogenetic analysis focused on the nitrogen-fixing clade indicated that wind pollination is more likely to evolve in those groups that have small, simple flowers and dry pollen (Culley et al. 2002). Fagaceae illustrate well this evolutionary tendency. Fagaceae are characterized by small flowers (see below), and ancestral members were apparently insect-pollinated. Molecular analyses indicate multiple derivations of wind pollination within this single family (see Fagales, below).

Similarly, whereas a cupule (derived from a modified inflorescence branch system) associated with one to several nuts evolved only once, and is in fact a synapomorphy of the entire Fagaceae, analyses by Oh and Manos (2008) indicate that the scaly cupule that encloses just a single fruit (acorn) evolved multiple times in the family (see below).

FABALES

Fabales comprise Fabaceae, Polygalaceae, Quillajaceae, and Surianaceae (Fig. 10.12). The DNA-based composition of Fabales is significant in that the closest relatives of Fabaceae have long been debated and were extremely important to identify given the great economic importance of Fabaceae (see J. J. Doyle 1994; Doyle and Luckow 2003). Non-DNA synapomorphies for Fabales are not clear, but may include clawed petals and more or less distinct carpels. Fabales represent another unexpected result of molecular phylogenetic analyses—the four constituent families (Fabaceae, Surianaceae, Polygalaceae, and Quillajaceae) have not been placed together in any morphology-based classifications.

Cronquist (1981, 1988) and Takhtajan (1997) both considered Fabaceae to be close to Connaraceae (see also Dickison 1981), the latter sister to Oxalidaceae (Oxalidales; discussed below). Polygalaceae have also been variously placed in morphology-based classifications because of their unusual, bilaterally symmetric flowers (Fig. 10.15).

Figure 10.15. Rosales-Fabales (Cannabaceae, Elaeagnaceae, Fabaceae, Moraceae, Polygalaceae, Rhamnaceae, Rosaceae, Surianaceae, Ulmaceae, and Urticaceae): a. *Prunus domestica* L. (Rosaceae), flower; note hypanthium and numerous stamens. b. *Rosa bracteata* J.C.Wendl. (Rosaceae), flower; note numerous stamens. c. *Sorbus americana* Marshall (Rosaceae), fruits (pomes). d. *Colubrina arborescens* Brandegee (Rhamnaceae), flowers with nectiferous hypanthium and five stamens opposite the cup-shaped petals. e. *Artocarpus heterophyllus* Lam. (Moraceae), multiple and accessory fruit. f. *Humulus lupulus* L. (Cannabaceae), infructescence. g. *Ulmus parvifolia* Jacq. (Ulmaceae), branch with oblique leaves and fruits (samaras). h. *Pilea formonensis* Urb. & Ekman (Urticaceae), herbaceous habit and inflorescences. i. *Cecropia peltata* L. (Urticaceae), carpellate inflorescences. i. *Elaeagnus umbellata* Thunb. (Elaeagnaceae), habit, flowers with showy sepals and hypanthium. k. *Suriana maritima* L. (Surianaceae), habit, flower. l. *Polygala myrtifolia* L. (Polygalaceae), bilateral flower, note the 2 petaloid sepals and large keel-like, fringed petal. m. *Crotalaria pallida* Klotzsch ex Aiton (Fabaceae), papilionaceous flower. n. *Leucaena leucocephala* (Lam.) de Wit, fruits (legumes), bipinnately compound leaves and young inflorescences. o. *Calliandra haematocephala* Hassk., inflorescence, note the showy androecia.

Fabaceae were often associated with Malpighiaceae and Krameriaceae, two other families exhibiting bilateral floral symmetry (Cronquist 1981). Takhtajan (1997) considered Polygalaceae to be closely related to Xanthophyllaceae, Vochysiaceae, and Emblingiaceae, but only the first of these families is now known to be closely related to Polygalaceae, in which it has been included (APG II 2003; APG III 2009; APG IV 2016; see Chapter 12).

Quillaja had typically been placed in Rosaceae with *Kageneckia*, sometimes as a separate subfamily (Takhtajan 1987, 1997; Thorne 1992a,b) on the basis of several shared features including follicular fruits with numerous pleurotropic ovules that mature into winged seeds. Other features, however, distinguish *Quillaja* from Rosaceae, such as its base chromosome number of $x = 14$, which is not found in Rosaceae. Furthermore, cytological studies indicated that $x = 14$ in *Quillaja* does not represent a tetraploid derivative of $x = 7$, a common number in Rosaceae (Goldblatt 1976). Bate-Smith (1965) noted chemical characters of *Quillaja* that are discordant with Rosaceae, including the presence of saponins and trihydroxy-substituted flavonoids.

The placement within Fabales of Surianaceae, a small, enigmatic family of five genera from Australia and Mexico, was also unexpected. The family had often been placed with Simaroubaceae and Rutaceae in Sapindales (Takhtajan 1997); Cronquist (1981) placed it in Rosales.

Soltis et al. (2011) recovered Quillajaceae + Polygalaceae as sister to Surianaceae + Fabaceae, but BS support was low. Qiu et al. (2010) found weak support for Quillajaceae sister to Fabaceae + (Surianaceae + Polygalaceae). Use of a 12-gene dataset recovered Quillajaceae, Surianaceae, and Polygalaceae as subsequent sisters to Fabaceae (H. Wang et al. 2009) (Fig. 10.12), but again with weak support. Bello et al. (2012) found weak support for Fabaceae and Polygalaceae as successive sisters to Surianaceae + Quillajaceae. Hence, relationships within Fabales still require attention (representatives are shown in Fig. 10.15).

Fabaceae have been the focus of many phylogenetic analyses (see Doyle et al. 2000; Bruneau et al. 2008; Lavin et al. 2004, 2005; Wojciechowski et al. 2004; Cardoso et al. 2012b, 2013a,b; Legume Phylogeny Working Group 2013a,b). Both Mimosoideae and Papilionoideae are monophyletic; Caesalpinioideae are paraphyletic to these (see Doyle et al. 1997; Doyle and Luckow 2003; Legume Phylogeny Working Group 2013a,b). Progress has been made in clarifying relationships across Fabaceae and also within subclades (see Stevens 2012; Angiosperm Phyogeny Website). Nonetheless, considering that Fabaceae are the third largest angiosperm family (~19,500 species), relationships within the family may not be as clearly understood as are those of the two largest angiosperm families, Orchidaceae and Asteraceae.

There are no confirmed Cretaceous records of Fabales, but the legumes are confirmed from pods and leaflets in the Paleocene of Colombia and the UK (Wing et al. 2009) and were abundant and widespread by the Eocene (Lavin et al. 2005).

Several complete nuclear genomes have been sequenced in Fabaceae (e.g., soybean, *Glycine max*; barrel medic, *Medicago truncatula*, and birdsfoot trefoil, *Lotus japonicas*; pigeonpea, *Cajanus cajan*; chickpea, *Cicer arietinum*; common bean, *Phaseolus vulgaris*; Cannon et al. 2009; Varshney et al. 2012, 2013), with more on the way, providing excellent genomic resources for systematic/evolutionary analyses.

ROSALES

Rosales are strongly supported and consist of (Fig. 10.12) Barbeyaceae, Cannabaceae (including Celtidaceae), Dirachmaceae, Elaeagnaceae, Moraceae, Rhamnaceae, Rosaceae, Ulmaceae, and Urticaceae (including Cecropiaceae). Rosales are one of several examples of an order consisting of families of the former subclasses Hamamelidae (e.g., Ulmaceae, Urticaceae, Moraceae) and Rosidae (e.g., Rosaceae, Rhamnaceae, and Elaeagnaceae; some members are shown in Fig. 10.15). Judd et al. (2002, 2008) suggested that a reduction (or lack) of endosperm and a variously shaped hypanthium (a character lost in some wind-pollinated members, however) may be synapomorphies; other potential synapomorphies include valvate sepals and ± clawed petals. Craspedodromous leaf venation and anthraquinones are also potential uniting features. As with other rosid clades, critical analyses are needed to evaluate these and other potential synapomorphies.

Within Rosales relationships are now generally clear (Fig. 10.12) (Savolainen et al. 2000a,b; D. Soltis et al. 2000, 2011; Sytsma et al. 2002; H. Wang et al. 2009; S. Zhang et al. 2011). Rosaceae are sister to the remainder of the order, which comprises several subgroups: 1) Barbeyaceae sister to Rhamnaceae + Elaeagnaceae + Dirachmaceae (see Thulin et al. 1998); and 2) Ulmaceae and Cannabaceae as subsequent sisters to Moraceae + Urticaceae (e.g., Savolainen et al. 2000a,b; D. Soltis et al. 2000, 2011; H. Wang et al. 2009).

Celtidaceae have been subsumed within Cannabaceae (APG III 2009; APG IV 2016), a result supported by ultrastructure and chromosome number (Sytsma et al. 2002). Additional DNA evidence for the recognition of Cannabaceae and Ulmaceae as distinct families was provided by

Wiegrefe et al. (1998); morphological data supporting this distinction were presented by Grudzinskaja (1967). The monophyly of Rhamnaceae and relationships within the family were discussed by Richardson et al. (2000).

Studies have clarified relationships within Rosaceae, drastically revising previous views (e.g., Morgan et al. 1994; Lee and Wen 2001; Potter et al. 2002, 2007). Four subfamilies were traditionally recognized with fruit type the major character used in distinguishing these. Amygdaloideae (Prunoideae) possessed drupes (e.g., *Prunus*), and Maloideae pomes (e.g., *Malus*), whereas Rosoideae have achenes or druplets (e.g., *Fragaria* and *Rubus*) and Spiraeoideae (e.g., *Spiraea*) follicles or capsules.

DNA results prompted a dramatic redefining of subfamilies. Campbell et al. (2007) recognized three subfamilies. Rosoideae are more narrowly defined (and characterized by numerous free carpels); a new subfamily, Dryadoideae, is recognized (which contains all four genera that have nitrogen-fixing symbiosis); and Amygdaloideae (Spiraeoideae of some recent publications) are expanded to include both Maloideae and Amygdaloideae (see Campbell et al. 2007). The clades recovered in molecular analyses do not agree with either fruit type or chromosome number (see below). We also stress the features with which the subclades (subfamilies) detected within Rosaceae do agree. Amygdaloideae are diagnosed by the presence of sorbitol in significant amounts (vs. absent or only in trace amounts in the other two subfamilies), and also many have a base chromosome number of 17 vs. 9 in Dryadoideae and 7 in Rosoideae (see below).

Quillaja, a genus placed in Rosaceae in most morphology-based classifications, is a member of Fabales (discussed above). Chrysobalanaceae (now in Malpighiales) had often been included in Rosaceae in the past. They are superficially similar to Rosaceae, in having a hypanthium, alternate stipulate leaves, 5-merous flowers with showy petals and numerous stamens, but the style is gynobasic, the gynoecium is pseudomonomerous, and they lack cyanogenic compounds.

Multiple Rosaceae crops, including strawberry, *Fragaria vesca* (Shulaev et al. 2011), peach, *Prunus persica* (International Peach Genome Initiative 2013) as well as other species of *Prunus*, and apple, *Malus domestica* (Velasco et al. 2010), have been the focus of complete nuclear genome sequencing, providing excellent genomic resources for systematic/evolutionary analyses. Using these resources, researchers are using gene capture methods to isolate and sequence numerous nuclear genes (Cronn et al. 2012).

Ulmaceae are recognizable from dispersed pollen and leaves in the Paleocene, and crown group representatives (synapomorphies include the following: fruit a pseudomonomerous samara or nutlet, with a flat seed; the leaf having pinnate venation with the secondary veins extending into marginal teeth and terminating medially or submedially to the tooth apecies; and pollen is 4- to 6-porate) allow recognition of *Ulmus* by the early Eocene (see Fig. 10.5 for fossil *Ulmus*). Rosaceae are confirmed by anatomically preserved fruits of *Prunus* in the Eocene (Cevallos-Ferriz and Stockey 1991; Manchester 1994b), by foliage of several modern genera in the Eocene (DeVore and Pigg 2007), and by fruits and foliage of *Rosa* in the early Oligocene (Meyer and Manchester 1997). Cannabaceae are confirmed in the Paleocene by endocarps and leaves of *Celtis* (Manchester et al. 2002). Rhamnaceae are recognizable by their distinctive flowers with petal-opposed stamens by the Eocene (Millan and Crepet 2014) and by fruits of *Paliurus* (Burge and Manchester 2008) and *Berchemia* (Collinson et al. 2012) (Supplementary Table 10.1).

CHARACTER EVOLUTION IN ROSALES

Using the new assessments of relationships for Rosaceae, new insights have been obtained into the evolution of fruit types and chromosome number (Fig. 10.16). Rosaceae represent one of several examples in which base chromosome number agrees well with clades recovered in phylogenetic analyses, indicating that base chromosome number may be a good predictor of relationships in many groups. Other examples include Crassulaceae (Mort et al. 2001), Rutaceae (Chase et al. 1999), and Onagraceae (Sytsma and Smith 1988; Levin et al. 2003).

Our reconstructions indicate that $x = 7$ is ancestral for Rosaceae (Fig. 10.16a). Rosoideae consist primarily of taxa with $x = 7$ and some with changes to $x = 8$ (e.g., *Alchemilla*). The remaining Rosaceae (Dryadoideae + Amygdaloideae) have $x = 9$ as the ancestral state. Within the large Amygdaloideae there have been multiple derivations of $x = 8$ (*Prunus*; a clade of *Prinsepia*, *Oemleria*, and *Exochorda*). One subclade within Amygdaloideae evolved high base chromosome numbers ($x = 15, 17$), the former Maloideae (e.g., *Pyrus*, *Malus*, and their close relatives), discussed below, now treated as subtribe Malinae (some have called this subtribe Pyrinae; Potter et al. 2007).

Our reconstructions of fruit type are similar to those of Potter et al. (2007), although we use fewer character states (Fig. 10.16b). The ancestral fruit type for Rosaceae is reconstructed as equivocal. However, Rosoideae as now defined and Purshiodeae both have achenes (or aggregates of achenes) and the achene appears to be ancestral for these clades. It is unclear if this represents a single or two separate origins. The achene also appears to have evolved independently in some Amygdaloideae (e.g., *Adenostoma*,

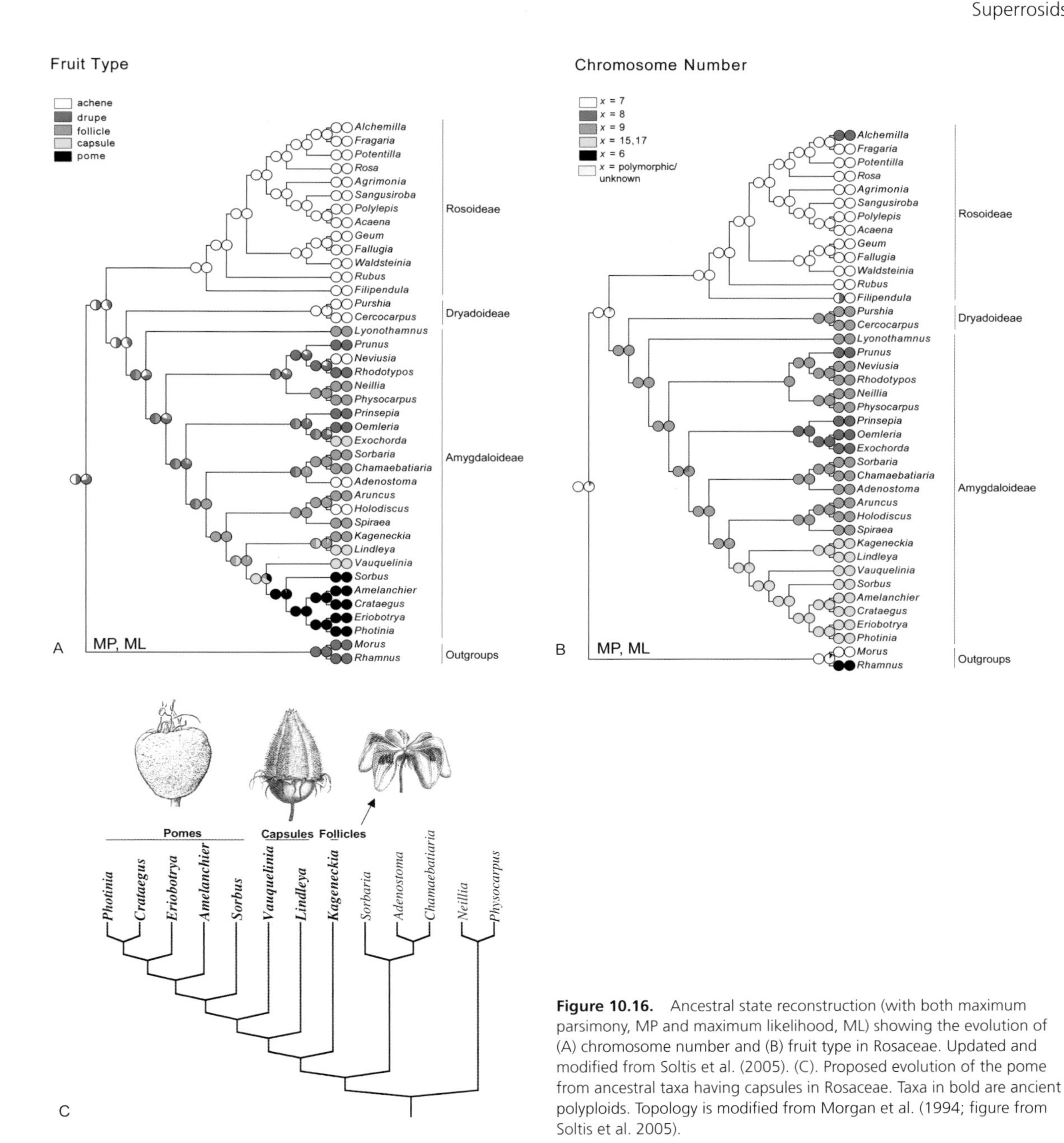

Figure 10.16. Ancestral state reconstruction (with both maximum parsimony, MP and maximum likelihood, ML) showing the evolution of (A) chromosome number and (B) fruit type in Rosaceae. Updated and modified from Soltis et al. (2005). (C). Proposed evolution of the pome from ancestral taxa having capsules in Rosaceae. Taxa in bold are ancient polyploids. Topology is modified from Morgan et al. (1994; figure from Soltis et al. 2005).

Holodiscus; see Potter et al. 2007). The ancestral fruit type for Amygdaloideae as now defined is also uncertain; the drupe and capsule have each evolved multiple times within this large clade (Fig. 10.16b).

There is a single origin of the pome fruit (defining the former Maloideae, current Malinae) and this is a noteworthy story in its own right. Members of the former subfamily Maloideae had traditonally been defined by fruit type (a fleshy fruit, the pome; e.g., apple or pear). Sequence analyses indicated a well-defined core maloid group, all in Malineae (e.g., represented here by *Amelanchier*, *Crataegus*, *Eriobotrya*, *Photinia*, and *Sorbus*) and also placed several traditionally problematic genera (*Vauquelinia*, *Lindleya*, and *Kageneckia*) with capsules and follicles as sister to these core maloids—together these comprise a clade, the tribe Maleae. Interestingly, these three genera possess high chromosome numbers (x = 15, 17) that agree with those in the traditionally recognized Maloideae (x = 17); these high numbers are the result of an ancient polyploid event that occurred in the ancestor of the maloid clade (e.g., Stebbins 1976b; Evans and Campbell 2002). These results also indicated that the distinctive pome fruit was likely derived from dry follicular or capsular fruits (Fig. 10.16c). More clues to the origin of the apple are provided by Harris et al. (2002).

CUCURBITALES

Cucurbitales consist of (APG III 2009) Anisophylleaceae, Begoniaceae, Coriariaceae, Corynocarpaceae, Cucurbitaceae, Datiscaceae, and Tetramelaceae (Fig. 10.12). To this APG IV (2016) has added the holoparasitic Apodanthaceae (see below). Attesting to the morphological complexity of the rosids, these families represent three subclasses in some traditional classifications. Cucurbitaceae, Begoniaceae, Tetramelaceae, and Datiscaceae were placed in Dilleniidae; Coriariaceae and Corynocarpaceae were variously placed within Rosidae or Magnoliidae; Apodanthaceae (as part of Rafflesiaceae) and Anisophylleaceae were often placed in Rosidae (Cronquist 1981, 1988; Takhtajan 1987, 1997; some members are shown in Fig. 10.17).

A close relationship of Cucurbitaceae, Begoniaceae, Tetramelaceae, and Datiscaceae had long been recognized (e.g., Cronquist 1981; Takhtajan 1997). However, Takhtajan (1997) considered Cucurbitaceae to be more closely related to Passifloraceae and Achariaceae (now in Malpighiales) than to Begoniaceae, Tetramelaceae, or Datiscaceae. Corynocarpaceae, Coriariaceae, and Anisophylleaceae are all small, problematic families for which affinities were long considered uncertain, and a diverse array of placements had been proposed (e.g., Cronquist 1981, 1988; Takhtajan 1987, 1997; Thorne 1992a,b, 2001).

An enigmatic parasitic family that appears in Cucurbitales in recent DNA studies is Apodanthaceae (APG IV 2016). DNA analyses now support this placement although the position within the order remains unclear (Nickrent et al. 2004; Barkman et al. 2007; Filipowicz and Renner 2010; Schaefer and Renner 2011). We have tentatively placed the family sister to the rest of Cucurbitales (Schaefer and Renner 2011) (Fig. 10.12).

Non-DNA synapomorphies for Cucurbitales are unclear, although possibly include wood with broad multiseriate rays or stems with separate vascular bundles, and leaves with more or less palmate venation. Cucurbitaceae, Begoniaceae, Tetramelaceae, and Datiscaceae share some distinctive features, including an inferior ovary, imperfect flowers, strongly protruding placentae, a distinctive cucurbitoid tooth, and the presence of curcurbitacins (Nandi et al. 1998; Judd et al. 2008). Comparisons of floral structure (Matthews and Endress 2004) indicate a close relationship between Coriariaceae and Corynocarpaceae, as well as between Tetramelaceae and Datiscaceae; these same comparisons also support a relationship of Tetramelaceae and Datiscaceae to Begoniaceae.

Although the clade is well-supported, most interfamilial relationships within Cucurbitales are not (see also Swensen et al. 1994, 1998; Soltis et al. 2011; H. Wang et al. 2009; Filipowicz and Renner 2010; Schaefer and Renner 2011). After Apodanthaceae, Anisophylleaceae followed by Coriariaceae + Corynocarpaceae are sisters to the rest of the order; the remainder form a clade in which Cucurbitaceae, then Tetramelaceae are sister to a well-supported Datiscaceae + Begoniaceae (Fig. 10.12).

The well-supported sister-group relationship of Coriariaceae and Corynocarpaceae is noteworthy in that the relationships of both families have long been uncertain. Corynocarpaceae, a family of one genus and perhaps seven species, have been placed in diverse orders, including Ranunculales, Sapindales, and Celastrales (Cronquist 1981, 1988; Takhtajan 1997). Coriariaceae are also monogeneric, comprising five species; the placement of the family has been extremely controversial (Sharma 1968; Takhtajan 1997). Whereas some authors proposed relationships with Sapindales (Takhtajan 1997), Cronquist (1981) placed the family in Ranunculales. Coriariaceae and Corynocarpaceae share several features: fibers with simple pits, vasicentric axial parenchyma, small flowers with ± separate carpels, scanty or no endosperm, and bitter chemical compounds (identified as sesquiterpenoids in *Coriaria*).

Schwarzbach and Ricklefs (2000) first showed that Anisophylleaceae, an enigmatic tropical family of four genera, were part of Cucurbitales. The family appears as sister to the rest of Cucurbitales (after Apodanthaceae) in analyses of multiple genes (Filipowicz and Renner 2010; Soltis et al. 2011; Schaefer and Renner 2011). Anisophylleaceae have often been considered either closely related to, or part of, Rhizophoraceae (Schwarzbach and Ricklefs 2000; see also "Malpighiales," below). However, molecular data indicated that Anisophylleaceae are distantly related to Rhizophoraceae. Matthews et al. (2001) and Schönenberger et al. (2001a) showed striking similarities in floral structure of Anisophylleaceae with the distantly related Cunoniaceae (Oxalidales) and a Cretaceous fossil exhibiting features of both families.

Schaefer et al. (2009) proposed an Asian origin of Cucurbitaceae followed by numerous subsequent, oversea dispersal events. Recent studies have clarified relationships

Figure 10.17. Cucurbitales-Fagales (Begoniaceae, Betulaceae, Casuarinaceae, Cucurbitaceae, Fagaceae, Juglandaceae, Myricaceae, and Nothofagaceae). a. *Quercus laevis* Walter. (Fagaceae), habit, staminate catkins and fruits (acorns, i.e., a nut with a cupule) shown as insert. b. *Ostrya virginiana* K.Koch (Betulaceae), staminate and carpellate catkins. c. *Alnus incana* (L.) Moench subsp. *rugosa* (Du Roi) R.T. Clausen (Betulaceae), leaf showing craspedodromous venation and compound serrations. d. *Gymnostoma nobile* (Whitmore) L.A.S.Johnson (Casuarinaceae), infructescence with pair of bony bracteoles enclosing each fruit. e. *Juglans nigra* L. (Juglandaceae), staminate catkin. f. *Carya glabra* (Mill.) Sweet (Juglandaceae), fruits (nuts surrounded by bract-derived husk). g. *Myrica cerifera* L. (Myricaceae), fruits (drupes with wax globules). h. *Nothofagus cunninghamii* (Hook. f.) Heenan & Smissen (Nothofagaceae), vegetative branch, leaves. i. *Cucurbita pepo* L. (squash, pumpkin; Cucurbitaceae), sympetalous, carpellate flower, and fruit. j. *Begonia* sp. (Begoniaceae), oblique-based leaves and flowers, note perianth of tepals and winged, inferior ovaries.

at multiple levels within the family and provided insights into character evolution and evolutionary processes—for example, long-distance dispersal has played a major role at multiple levels within the family (e.g., Renner et al. 2002, 2007; Kocyan et al. 2007; Schaefer and Renner 2008, 2010, 2011a,b; Schaefer et al. 2008a,b, 2009; Duchen and Renner 2010). The family is also of interest in that the mitochondrial genomes of some members are large and variable (*Cucumis*; 379,000–2,900,000 bp; Alverson et al. 2010).

FAGALES

Fagales consist of seven woody families formerly considered part of subclass Hamamelidae (Fig. 10.12): Betulaceae, Casuarinaceae, Fagaceae, Juglandaceae (Rhoipteleaceae now included within Juglandaceae as in APG III [2009] and IV [2016]), Myricaceae, Nothofagaceae, and Ticodendraceae (*Ticodendron*, Ticodendraceae were described by Gómez-Laurito and Gómez 1989). Rhoipteleaceae were bracketed in APG II (2003) and included in Juglandaceae in APG III (2009), but in recent treatments they have been retained as distinct (APG IV 2016; several families are illustrated in Fig. 10.17). Molecular as well as nonmolecular analyses have demonstrated, however, that Fagales are rosids and not closely related to other former Hamamelidae such as Platanaceae, Cercidiphyllaceae, and Daphniphyllaceae (see also Hufford 1992; Nandi et al. 1998).

Members of Fagales share ectomycorrhizal roots, an indumentum of gland-headed and/or stellate hairs, an inferior ovary (except Rhoipteleaceae), one or two ovules per locule, unisexual flowers with tepals reduced or absent, nectaries lacking, delayed ovule development and a pollen tube that enters the immature ovule via the chalaza (chalazogamy), and indehiscent fruits with only one seed (by abortion of additional ovules; Hufford 1992; Loconte 1996; Manos and Steele 1997; Judd et al. 1999, 2008; Sogo and Tobe 2008; Endress 2011a).

Relationships witin Fagales are now well resolved (Fig. 10.12). Nothofagaceae, followed by Fagaceae, are successive sisters to the remainder of the clade. The rest of Fagales consists of two subclades: (1) Myricaceae + (Rhoipteleaceae + Juglandaceae); and (2) Casuarinaceae + (Ticodendraceae + Betulaceae). The distribution of the three types of pollen in the order (colporate, porate, stephanoporate) is in general agreement with these four clades of families (Manos and Steele 1997).

The position of Myricaceae has varied; in some analyses the family has appeared with Casuarinaceae, Ticodendraceae, and Betulaceae (Manos and Steele 1997; Li et al. 2004; Herbert et al. 2006; H. Wang et al. 2009; Soltis et al. 2011; Sauquet et al. 2012). However, Myricaceae share morphological features with Rhoipteleaceae and Juglandaceae, including aromatic glandular hairs and a gynoecium with a single, orthotropous ovule. The sister-group relationship of Ticodendraceae and Betulaceae (Fig. 10.12; see also Conti et al. 1996) is also supported by similarities in floral morphology (Tobe 1991), wood and bark anatomy (Carlquist 1991), sieve element plastids (Behnke 1991), pollen (Feuer 1991), and leaf architecture (Hickey and Taylor 1991).

Molecular data also provided strong support for Juglandaceae + Rhoipteleaceae (Fig. 10.12; Manos and Steele 1997; Chen et al. 1998; Li et al. 2004). Rhoipteleaceae are an enigmatic monogeneric family that could be included in Juglandaceae (APG III 2009), with several features that are unusual in Fagales (e.g., perfect flowers, a superior ovary). However, Juglandaceae (including Rhoipteleaceae) are distinctive in Fagales in possessing compound leaves; a close relationship is also suggested by stem anatomy, floral morphology (Manning 1938, 1940; Withner 1941), and chromosomal data (Oginuma et al. 1995). Rhoipteleaceae are included in Juglandaceae in APG IV (2016).

Fagales are well represented in the fossil record (e.g., Jones 1986; Crepet and Nixon 1989; Herendeen et al. 1995) (Supplementary Table 10.1), and the position of Nothofagaceae as sister to the remainder of the order is in agreement with the relatively early appearance of this family in the fossil record (see Manos and Steele 1997 and Knapp et al. 2005 for phylogenetic relationships and biogeographic patterns within *Nothofagus*). Extinct fagalean taxa likely ancestral to Juglandaceae, Betulaceae, and Fagaceae are represented by several genera of charcoalified flowers and fruits from the late Cretaceous of Europe and North America (Sims et al. 1998; Schönenberger et al. 2001b; Friis et al. 2003b) (see Fig. 10.5 for fossils of Betulaceae and Juglandaceae). Crown group representatives of Betulaceae, Fagaceae, Myricaceae, Casuarinaceae, and Juglandaceae are well represented in the fossil record by the late Paleocene and early Eocene. These include species representing extinct genera (e.g., *Polyptera* and *Cruciptera* of the Juglandaceae; *Cranea* and *Palaeocarpinus* of Betulaceae) along with those corresponding to modern genera (e.g., *Juglans*, *Cyclocarya*, *Alnus*, *Betula*, *Gymnostoma*, *Comptonia*, *Castanea*, *Fagus*, *Quercus*).

The Hardwood Genomics Project (hardwoodgenomics.org) provides useful genomic resources for a number of Fagaceae (e.g., *Castanea*, *Fagus*, *Quercus*).

CHARACTER EVOLUTION IN FAGALES

Analysis of fruit evolution in Juglandaceae indicated a dynamic picture of diversification (Manos and Stone 2001). Although wind dispersal appears to be ancestral within

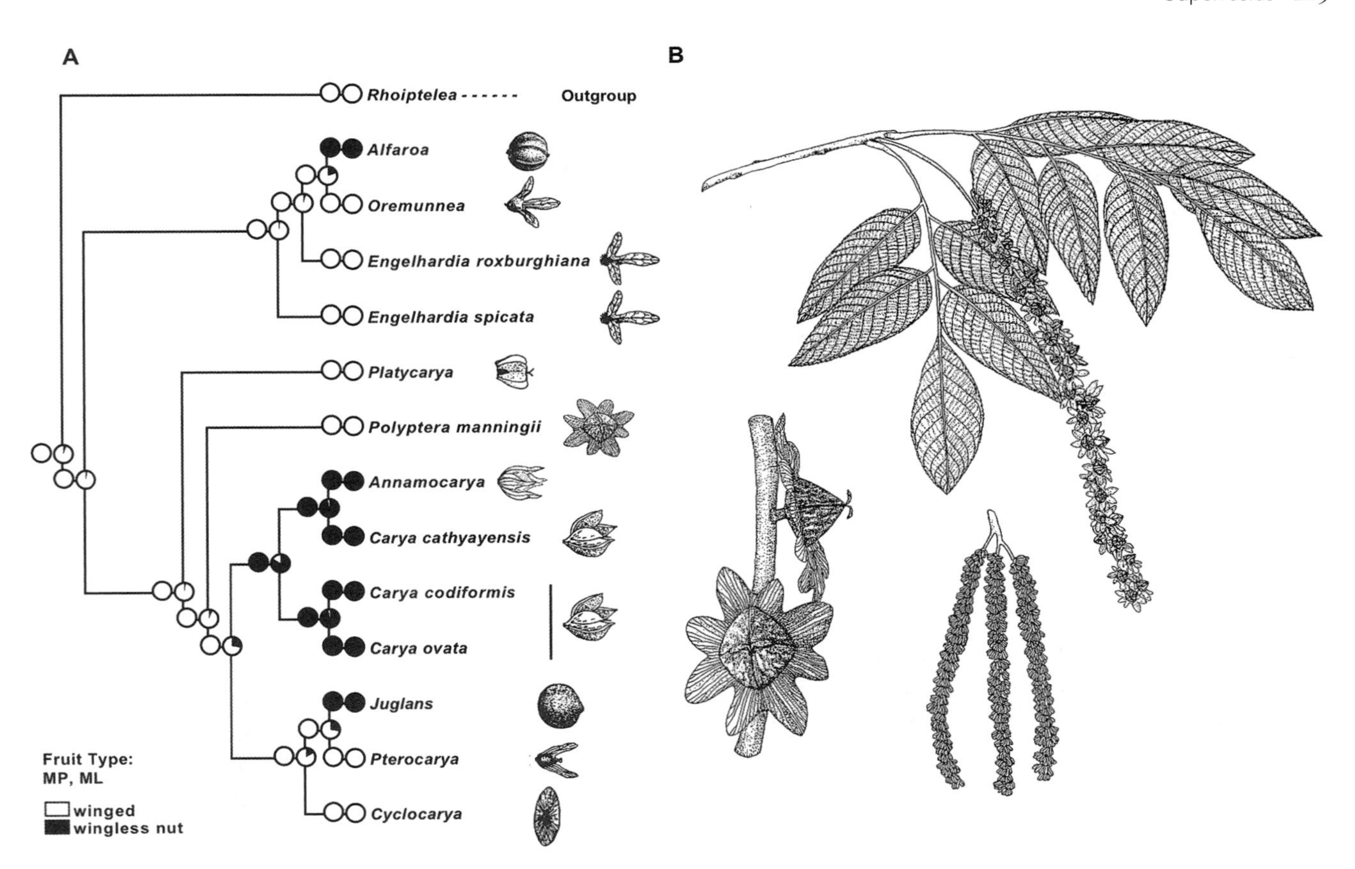

Figure 10.18. Ancestral state reconstruction showing evolution of fruit types in Juglandaceae using living and fossil members of the family; the placement of the fossil *Polyptera manningii* is based on a total evidence analysis (Manos et al. 2007). (A) Fruit type. *Carya cordiformis* and *Carya ovata* are characterized by the same general fruit type. Fruit illustrations are from Manos and Stone (2001). (B) *Polyptera manningii*, a fossil member of Juglandaceae (from Manchester and Dilcher 1997). (Top) Twig with infructescence. (Bottom right) Staminate catkins. Updated from Soltis et al. (2005).

the family, four independent origins of wings from a combination of floral and accessory structures are suggested (Fig. 10.18): (1) *Engelhardia* and *Oreomunnea* have wings formed from a tri-lobed bract; (2) *Platycarya* has two small wings that result from the fusion of bracteoles and lateral sepals; (3) *Platycarya* has wings derived from bracteoles; and (4) *Cyclocarya* has a distinctive circular wing that is the result of fusion of bracts and bracteoles. Our reconstruction indicated that animal dispersal evolved independently three times, resulting in nuts formed uniquely by the fusion of different organs: (1) fusion of four sepals in *Alfaroa*; (2) fusion of bract, bracteoles, and sepals in *Juglans*; and (3) fusion of the bract and bracteoles in *Carya* and *Annamocarya*. In general, wind-dispersed seeds have epigeal germination, and those that are animal-dispersed are hypogeal. *Oreomunnea* and *Cyclocarya* are exceptions in their respective clades in having wind-dispersed seeds with hypogeal germination.

Juglandaceae have been well-investigated using DNA data (Manos and Stone 2001; Sauquet et al. 2012) and have an excellent fossil record, and these fossils have been carefully investigated (e.g., Manchester and Dilcher 1997). Hence, the family affords the opportunity to integrate fossils into the excellent phylogenetic framework for extant taxa. Manos et al. (2007) found that despite large and variable amounts of missing data for the fossils, most methods provided reasonable placement of both fossils and simulated "artificial fossils" in the phylogenetic tree previously inferred only from extant taxa. Their results show that the amount of missing data in any given taxon is not by itself an operational guideline for excluding fossils from analysis. In fact, three fossil taxa (*Cruciptera simsonii*, *Paleoplatycarya wingii*, and *Platycarya americana*) were placed within crown clades containing living taxa for which relationships previously had been suggested based on morphology, whereas *Polyptera manningii*, a mosaic taxon of equivocal affinities, was placed firmly as sister to two modern crown clades (Fig. 10.18).

Fagaceae (~1000 species), the largest family within Fagales, have also been the focus of phylogenetic analyses

(Manos and Steele 1997; Manos and Stanford 2001; Manos et al. 2001; Oh and Manos 2008). Assuming that ancestral Fagaceae were insect-pollinated, as indicated by the fossil record (Herendeen et al. 1995; Sims et al. 1998), then the DNA-based reconstructions support at least three origins of wind pollination in the family. There was also at least one reversion to insect pollination (Juglandaceae: *Platycarya*; Endress 1986c; Fukuhara and Tokumaru 2014).

COM CLADE

The COM clade (Celastrales, Oxalidales, and Malpighiales) (Figs. 10.1, 10.2; representatives are shown in Figs. 10.13, 10.20, and 10.21) was recovered using three genes (D. Soltis et al. 2000) but with low support. The clade ultimately received strong support in large multi-gene studies (H. Wang et al. 2009; Soltis et al. 2011). Morphological characters also united the three orders and indicated a placement closer to malvids than fabids (Endress and Matthews 2006; Endress et al. 2013).

The COM clade contains approximately one-third of all *Rosidae*, 870 genera and ~19,000 species (APG III 2009; APG IV 2016); their placement is one of the most problematic deep-level issues remaining in the angiosperms. Plastid genes indicated a placement in *Fabidae* whereas mitochondrial and nuclear genes suggested placement of the COM clade with *Malvidae* (Zhu et al. 2007; Duarte et al. 2010; Finet et al. 2010; Qiu et al. 2010; Burleigh et al. 2011; Lee et al. 2011; Morton 2011; Shulaev et al. 2011; Zhang et al. 2012). Morphologically, several floral characters also linked the COM clade with *Malvidae*. COM species and *Malvidae* share a type of ovule with a thicker inner integument than the outer one, which is otherwise nearly absent in *Fabidae* and rare in eudicots. Contorted petals and a tendency towards polystemony and polycarpy also suggest a placement of COM members with *Malvidae* (Endress and Matthews 2006; Endress et al. 2013).

To unravel the evolutionary history of the COM clade Sun et al. (2014) assembled taxonomically comparable matrices representing each plant genome; they used extensive tests and analyses (e.g., RY coding, and rapidly evolving sites removed) to explore the underlying causes of incongruence. They found that the two conflicting topologies inferred from the three genomes are accurate; the disagreement is not due to systematic biases or sampling errors, but represents some form of biological incongruence. Their analyses also reveal that two conflicting phylogenetic signals exist in both single-copy and multi-copy nuclear genes and the pattern of gene duplications in multi-copy gene trees also strongly supports placement of the COM clade with *Malvidae*.

The analyses of Sun et al. (2015) suggested a complex evolutionary history in which either ancient lineage sorting, or more likely reticulation/introgressive hybridization and subsequent chloroplast capture occurred during the early and rapid radiation of *Rosidae*. This ancient event resulted in conflict between plastid and mitochondrial gene trees, as well as a mixture of underlying signal in the nuclear genome representing both *Malvidae* and *Fabidae*. Sun et al. (2015) provided one hypothetical ancient hybridization scenario for the origin of COM from the ancestral *Fabidae* and *Malvidae* lineages (Fig. 10.19). Although lineage sorting and reticulation have been long recognized as important (and complex) evolutionary forces at shallow phylogenetic levels, these results reveal their impact at deep levels in the angiosperms. These results also suggest that perhaps the COM clade be placed with both the *Fabidae* and *Malvidae* in classifications—although we continue to follow the APG III (2009) and APG IV (2016) treatment here.

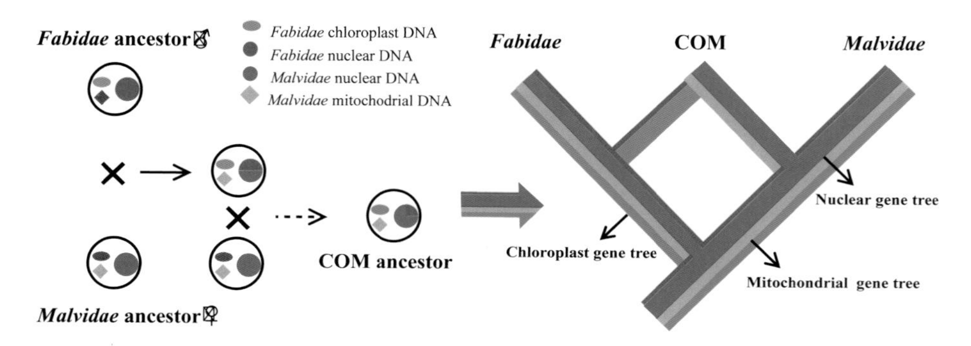

Figure 10.19. Hypothetical reticulation scenario for the origin of COM from the ancestral *Fabidae* and *Malvidae* lineages. Large circles reflect the plant lineages; the small circles represent their nuclear DNA types (the red circle represents the *Fabidae* nuclear DNA and the blue one the *Malvidae* nuclear DNA), the ovals represent the chloroplast (the green ovals represent the *Fabidae* plastid, and the gray oval represents the chloroplast from the *Malvidae* ancestor), and the diamonds represent the mitochondria (the gray diamond represents the *Fabidae* mitochondrion and the orange diamond the *Malvidae* mitochondrion); dashed arrows represent multiple generations of backcrossing. During hybridization, the mitochondrion is maternally inherited from the *Malvidae* ancestor, and the chloroplast is paternally inherited from the *Fabidae* ancestor. After subsequent F_1 backcrosses to the *Malvidae*, the resulting generations contain chloroplasts from *Fabidae*, mitochondria from *Malvidae*, and a majority of the nuclear genes from *Malvidae*, with a smaller number from *Fabidae* (roughly 25%). The reticulate phylogeny at the bottom illustrates this hypothetical introgressive hybridization scenario and shows the phylogenetic incongruence among the three genomes with respect to COM.

Figure 10.20. Malpighiales (Calophyllaceae, Chrysobalanaceae, Clusiaceae, Erythroxylaceae, Euphorbiaceae, Euphorbiaceae, Hypericaceae, Linaceae, Malpighiaceae, Ochnaceae, Passifloraceae, Phyllanthaceae, Podostemaceae, Putranjivaceae, Rafflesiaceae, Rhizophoraceae, Salicaceae, Turneraceae, Violaceae): a. *Mammea americana* L. (Calophyllaceae), fruit and branch with flowers. b. *Clusia rosea* Jacq. (Clusiaceae), flower and fruits (youung, in cross-section, dehiscent), note colored sap. c. *Erythroxylum coca* Lam. (Erythroxylaceae), flower with appendaged petals. d. *Rafflesia arnoldii* R.Br. (Rafflesiaceae), flower (these may be a meter across). e. *Cnidoscolus chayamansa* McVaugh (Euphorbiaceae), flowers, note divided styles. f. *Euphorbia cyathophora* Murray (Euphorbiaceae), cyathia. g. *Geobalanus oblongifolius* (Michx.) Small (Chrysobalanaceae), flower, dissected to show hypanthium and gynobasic style. h. *Hypericum tetrapetalum* Lam. (Hypericaceae), flower, showing asymmetric petals, numerous stamens, and slender style branches with minute stigmas. i. *Malpighia emarginata* DC. (Malpighiaceae), flowers with clawed petals and exposed oil-glands on the sepals. j. *Ochna thomasiana* Engl. & Gilg ex Gilg (Ochnaceae), fruits with drupaceous mericarps on enlarged receptacle. k. *Passiflora alata* Curtis (Passifloraceae), flower with showy corona. l. *Phyllanthus juglandifolius* Willd. (Phyllanthaceae), 3-lobed fruits borne on plagiotrophic branch. m. *Linum kingii* S.Watson (Linaceae), flower with only 5 stamens. n. *Drypetes lateriflora* (Sw.) Krug & Urb. (Putranjivaceae), fruits. o. *Flacourtia indica* (Burm.f.) Merr. (Salicaceae), fruits and carpellate flowers p. *Salix caroliniana* Michx. (Salicaceae), male catkin. q. *Rhizophora mangle* L. (Rhizophoraceae), flower with thick, valvate sepals and hairy petals. r. *Turnera subulata* Sm. (Turneraceae, or Passifloraceae s.l.), flowers. s. *Viola sororia* Willd. (Violaceae), zygomorphic flowers, note nectar spur. t. *Podostemum* sp. (Podostemaceae), habit, an aquatic of rapid-flowing streams.

Figure 10.21. Fossil rosid representatives including Malphighiales, Sapindales, and Malvales. a. Extinct Salicaceae: *Pseudosalix handleyi*, bearing paniculate inflorescence and flowers with perianth, middle Eocene of Colorado. b. *Populus tidwellii* middle Eocene of Utah. c. Euphorbiaceae: *Euphorbiotheca* from middle Eocene of Messel, Germany. d. Oldest known fruit of *Acer*, from the Maastrichtian of North Dakota DMNH coll. e. Transverse section of phyllanthaceous fruit with two seeds per locule, Late Cretaceous (Maastrichtian) of India. f. Sapindaceae: *Dipteronia brownii* from middle Eocene of British Columbia. g. Sapindaceae: *Aesculus hickeyi* from Paleocene of North Dakota. h. Tapisciaceae: *Tapiscia occidentalis*, middle Eocene of Oregon. i. Anacardiaceae: *Anacardium germanicum*, middle Eocene of Messel, Germany. j. *Florissantia* flower from the Eocene of Republic, Washington. k, l. Pollen from the stamen of a *Florissantia* flower from the Oligocene of Oregon.

MALPIGHIALES

Malpighiales are one of the largest angiosperm orders, comprising ~16,000 species (~7.8% of eudicots); they are extremely diverse morphologically. Molecular phylogenetic analyses provided strong support for a Malpighiales clade of 29 families not recognized in any morphology-based classification (Fig. 10.12): Achariaceae, Balanopaceae, Bonnetiaceae, Calophyllaceae, Caryocaraceae, Centroplacaceae, Chrysobalanaceae, Clusiaceae, Ctenolophonaceae, Dichapetalaceae, Elatinaceae, Euphorbiaceae, Euphroniaceae, Humiriaceae, Hypericaceae, Irvingiaceae, Ixonanthaceae, Lacistemataceae, Linaceae (including Hugoniaceae), Lophopyxidaceae, Malpighiaceae, Ochnaceae (including Medusagynaceae, Quiinaceae), Pandaceae, Passifloraceae (expanded to include Malesherbiaceae, Medusandraceae, and Turneraceae; see below), Picrodendraceae, Podostemaceae (including Tristichaceae), Peraceae (recognized here but whether the family should be segregated depends on the position of Rafflesiaceae, see below), Putranjivaceae, Rafflesiaceae, Rhizophoraceae, Erythroxylaceae, and an expanded Salicaceae (nomenclaturally, Salicaceae includes the type *Flacourtia*, so technically Flacourtiaceae is now a synonym of Salicaceae) and Scyphostegiaceae, Trigoniaceae, and Violaceae (see Chapter 12; several families are illustrated in Fig. 10.20 with several fossils shown in Fig. 10.21).

Because the order is so heterogeneous, non-DNA characters uniting the taxa remain poorly understood, although when the leaves have teeth, each tooth usually has a distally expanded vein, with the tooth-apex thus congested (Judd et al. 2008). Many families of the order were previously placed in Violales (subclass Dilleniidae of Cronquist 1981): Flacourtiaceae, Lacistemataceae, Passifloraceae, Malesherbiaceae and Turneraceae (both now included in Passifloraceae; APG III 2009), and Violaceae. Ochnaceae represent another family formerly placed in Dilleniidae in recent classifications (Dilleniales). Malpighiales also include families previously attributed to a diverse array of orders in the traditionally recognized Rosidae, such as Linales (Humiriaceae, Linaceae, and Erythroxylaceae), Polygalales (Malpighiaceae, Trigoniaceae, and Euphroniaceae), Euphorbiales (Euphorbiaceae), and Rosales (Chrysobalanaceae).

The fossil record of Malphigiales includes well-preserved examples of Salicaceae, Euphorbiaceae, Humiriaceae, and Malphighiaceae. Most of these are first seen in the Eocene, but the extinct flower, *Paleoclusia*, mentioned earlier, is known from the Turonian, and distinctive polycolpate pollen of Ctenolophonaceae occurs in the Campanian of Africa (van der Ham 1989). Eocene representatives of Salicaceae include the modern genus *Populus* (Fig. 10.21b), and the extinct genus, *Pseudosalix* (Fig. 10.21a). *Pseudosalix* had leaves resembling *Salix*, and capsular fruits with comose seeds like those of *Populus* and *Salix*, but was distinguished from the modern genera by having branched rather than racemose inflorescence and well-developed perianth (Boucher et al. 2003; Manchester et al. 2006b). The distinctive explosively dehiscent trilocular fruits of Euphorbiaceae are first known from the Early and Middle Eocene (Fig. 10.21c; Collinson et al. 2012), but may have actually originated earlier indicated by the occurrence of fruits conforming to Phyllanthaceae in the late Cretaceous (Maastrichtian of India, Fig. 10.21d). Malpighiaceae are first known based on flowers from the middle Eocene (Taylor and Crepet 1987), and distinctive fruits and wood of Humiriaceae are traced to the late Eocene (Herrera et al. 2014).

Monophyly of Malpighiales first received strong support (99%) in early DNA analyses (D. Soltis et al. 2000), but the topology at that time within the clade was essentially a polytomy. Ten years ago Malpighiales were one of the clades of angiosperms most in need of extensive study, but the use of broad sampling and numerous genes has greatly resolved relationships and provided novel evolutionary insights (e.g., C. Davis et al. 2002, 2004, 2005b, 2007; Wurdack and Davis 2009; Davis and Anderson 2010; Xi et al. 2012b). Malpighiales represent an excellent example of a rapid radiation (Wikström et al. 2001; Davis et al. 2005b; H. Wang et al. 2009; Bell et al. 2010). Davis et al. (2005b) suggested 114–101 mya for the explosive radiation of this clade of ~16,000 species. Xi et al. (2012b) discussed the rapid rise of the clade and provided evidence for multiple bursts in diversification, as well as several slowdowns.

Malpighiales also illustrate the value of broad phylogenetic analyses in placing enigmatic taxa. These include Balanopaceae, Bonnetiaceae, Ctenolophonaceae, and Caryocaraceae, Goupiaceae, Irvingiaceae, Lophopyxidaceae, Pandaceae, Podostemaceae, Rafflesiaceae (Xi et al. 2012b). To give one example, the relationship of *Medusagyne oppositifolia* (formerly the sole species of Medusagynaceae) was unclear, but molecular studies not only placed the species firmly within Malpighiales but also indicated a position within Ochnaceae (Melville 1971; Fay et al. 1997b; Schneider et al. 2014). In fact, Ochnaceae, Medusagynaceae, and Quiinaceae form a distinctive clade (Nandi et al. 1998; Savolainen et al. 2000a; Xi et al. 2012b) in which leaves have well-defined secondary and tertiary venation (for floral features, see, Matthews et al. 2012). It seems appropriate to combine all of these into a single family, Ochnaceae (APG II 2003; APG III 2009; APG IV 2016; Schneider et al. 2014; see Chapter 12).

There are three major clades in Malpighiales (see Xi et al. 2012b): the Rhizophoraceae-Clusiaceae, Malpighiaceae-Chrysobalanaceae, and Salicaceae-Euphorbiaceae clades. We briefly cover each below.

Rhizophoraceae-Clusiaceae—This subclade consists of Ctenolophonaceae, Erythroxylaceae, Rhizophoraceae, Irvingiaceae, Pandaceae, Ochnaceae (expanded to include Medusagynaceae and Quiinaceae, Fay et al. 1997b; Nandi et al. 1998; Savolainen et al. 2000a; Chase et al. 2002; Korotkova et al. 2009; Xi et al. 2012b); Bonnetiaceae, Clusiaceae, Calophyllaceae, Hypericaceae, and Podostemaceae form one large clade.

Ctenolophonaceae, Erythroxylaceae, Rhizophoraceae form one subclade (Wurdack and Davis 2009), well supported by floral morphology (Matthews and Endress 2011). A close relationship between Erythroxylaceae and Rhizophoraceae has long been noted (Schwarzbach and Ricklefs 2000). Phylogenetic analysis of non-DNA characters first recovered Rhizophoraceae + Erythroxylaceae s.s. (Hufford 1992), and their close relationship is supported by embryological features, the presence of the alkaloid hygroline, and a unique sieve-tube plastid type (Behnke 1988b; Nandi et al. 1998; reviewed in Schwarzbach and Ricklefs 2000). The many features shared by these families was recognized by Dahlgren (1988), but he did not consider them closely related. The mangrove habit associated with Rhizophoraceae is restricted to only a few members (Cronquist 1988). Relationships and character evolution within Rhizophoraceae have been analyzed (Schwarzbach and Ricklefs 2000). In addition, the poorly known African genus *Aneulophus* (Erythroxylaceae) is morphologically similar to some Rhizophoraceae.

A well-supported clusioid clade was recovered, which comprises Bonnetiaceae, Calophyllaceae, Clusiaceae s.s., Hypericaceae, as well as the aquatic family Podostemaceae. Species of this subclade are important components of tropical forests (Ruhfel et al. 2011). Bonnetiaceae, a small enigmatic family segregated from Theaceae (Chapter 11), was found to be sister to Clusiaceae s.s. (Savolainen et al. 2000b; Ruhfel et al. 2011, 2013). Bonnetiaceae + Clusiaceae are sister to Calophyllaceae + (Hypericaceae + Podostemonaceae). Ochnaceae are then sister to all of these families (Fig. 10.12). There are non-DNA synapomorphies for some of these relationships. Members of the entire Bonnetiaceae through Podostemaceae clade share distinctive xanthones; tenuinucellate ovules are known from Clusiaceae and Podostemaceae (e.g., Contreras et al. 1993; Jäger-Zürn 1997; Endress et al. 2013). Relationships within Clusiaceae s.s. have been investigated by Gustafsson et al. (2002), Sweeney (2008), and Ruhfel et al. (2011, 2013). DNA studies provided support for an expanded Ochnaceae that includes Quiinaceae and Medusagynaceae (Fay et al. 1997a,b). Based on morphology, there is no clear reason for separating these families (Fay et al. 1997a,b; Nandi et al. 1998; Schneider et al. 2014).

Also well-supported is Calophyllaceae + (Hypericaceae, Podostemaceae). The placement of Podostemaceae with these families is one of several examples of enigmatic aquatics being successfully placed with DNA data. Podostemaceae are highly modified submerged aquatics variously associated with Piperaceae, Nepenthaceae, Polygonaceae, Caryophyllaceae, Scrophulariaceae, Rosaceae, Crassulaceae, or Saxifragaceae in previous classifications (see Ueda et al. 1997; D. Soltis et al. 1999; Gustafsson et al. 2002; Xi et al. 2012b). Molecular phylogenetic analyses placed Podostemaceae in Malpighiales with strong support as sister to Hypericaceae with Calophyllaceae their immediate sister (see Ruhfel et al. 2011). Podostemaceae have been well investigated phylogenetically with three subclades recognized (see Koi et al. 2012).

A close relationship between Irvingiaceae and Pandaceae is strongly supported, but relationships of these two families to other Malpighiales is unclear.

Malpighiaceae-Chrysobalanaceae—This subclade comprises Lophopyxidaceae, Putranjivaceae, Caryocaraceae, Centroplacaceae, Elatinaceae, Malpighiaceae, Balanopaceae, Trigoniaceae, Dichapetalaceae, and Chrysobalanaceae. There are few well-supported relationships within this subclade, although relationships among Centroplacaceae + (Elatinaceae + Malpighiaceae) and Balanopaceae, Chrysobalanaceae, Trigoniaceae, Dichapetalaceae, and Euphroniaceae receive good support (Fig. 10.12). This illustrates well a problem encountered throughout Malpighiales—it contains families not considered closely related in any traditional classifications. According to Cronquist (1981) the monogeneric Balanopaceae belonged in Hamamelidae, with the remaining families placed in different orders of his Rosidae (Dichapetalaceae in Sapindales; Chrysobalanaceae in Rosales; Trigoniaceae and Euphroniaceae in Polygalales). Trigoniaceae, Dichapetalaceae, Chrysobalanaceae, and Euphroniaceae, as well as several other Malpighiales, all have incompletely tenuinucellate ovules, a feature rare outside of asterids (Endress 2003b, 2011a; Matthews and Endress 2008; APG III 2009; APG IV 2016; Chapter 12). However, *Balanops*, which is sister to this clade, has crassinucellate ovules (Merino Sutter and Endress 2003).

Salicaceae-Euphorbiaceae—This third large subclade consists of Humiriaceae, Achariaceae, Goupiaceae, Violaceae, Passifloraceae, Lacistemataceae, Salicaceae, Peraceae, Euphorbiaceae, Rafflesiaceae, Phyllanthaceae, Picodendraceae, Linaceae, and Ixonanthaceae. Several subclades within this clade are well supported: 1) Humiriaceae, Achariaceae, Goupiaceae, Violaceae, Passifloraceae, Lacistemataceae, Salicaceae; 2) Peraceae, Rafflesiaceae, Euphorbiaceae, Phyllanthaceae, Picodendraceae, Linaceae, and Ixonanthaceae. Peraceae are sister to Rafflesiaceae + Euphorbiaceae. Within the second subclade, two small subclades are recovered, but other relationships remain less clear: Phyllanthaceae + Picrodendraceae and Linaceae + Ixonanthaceae; in the second there is a tendency to form secondary septa, in addition to the normal septa, in the ovary (Endress et al. 2013).

One Salicaceae-Euphorbiaceae subclade receiving support (subclade 1 above) includes families traditionally considered closely related based on sharing parietal placentation: Salicaceae, Achariaceae and Violaceae, Flacourtiaceae, Passifloraceae, Malesherbiaceae and Turneraceae (part of the Dilleniidae of Cronquist 1981). Of these, Passifloraceae s.s. are sister to Malesherbiaceae + Turneraceae. A close relationship among these three families has long been recognized (e.g., Cronquist 1981; Takhtajan 1997); they produce structurally related cyanogenic glycosides with a cyclopentenoid ring system (see also Achariaceae) and similar hydrolytic enzymes (Spencer and Seigler 1985a,b; Takhtajan 1997). Turneraceae and Passifloraceae possess foliar glands; both have biparental or paternal transmission of plastids (Malesherbiaceae have not been examined); in most angiosperms plastids are maternally inherited (e.g., Shore et al. 1994). Malesherbiaceae, some Turneraceae, and Passifloraceae also possess an extrastaminal corona, and possess a hypanthium-like structure that does not bear the stamens and a corona, lost in most Turneraceae.

In Chase et al. (2002), Turneraceae plus Malesherbiaceae were sister to Passifloraceae. Historically (e.g., Cronquist 1981), Passifloraceae were considered to differ from Turneraceae on being vining versus herbs, shrubs, or trees, but this is highly artificial; few characters differentiate these families as narrowly circumscribed. Furthermore, unpublished data (M. W. Chase and L. W. Chatrou) had suggested that Passifloraceae were paraphyletic and included both Turneraceae and Malesherbiaceae, which impacted APG classifications (APG II 2003; APG III 2009). Recent studies indicate that Passsifloraceae s.l are monophyletic but also support Turneraceae as sister to Passifloraceae s.s. (Korotkova et al. 2009; Tokuoka 2012) with monogeneric Malesherbiaceae then sister to the rest. Easily recognized characters (above) could be considered synapomorphies for Passifloraceae s.l.

Phylogenetic analyses indicated that major changes needed to be made in the delimitation of Euphorbiacae. Studies have revealed a narrowly defined Euphorbiaceae and recovered several other well-separated clades of former euphorbs (e.g., *Drypetes* and *Putranjiva*; genera of Pandaceae; Phyllanthaceae) that have closest relatives elsewhere in Malpighiales. In APG III (2009) and IV (2016), Euphorbiaceae in the traditional sense were divided into four families: Euphorbiaceae s.s., Phyllanthaceae, Picrodendraceae, and Putranjivaceae. Furthermore, DNA studies indicated that groups within Euphorbiaceae s.s also needed reconsideration (Wurdack and Chase 2002; Wurdack et al. 2005; Tokuoka 2007). Euphorbiaceae s.s. consist of the uniovulate Euphorbioideae, Crotonoideae, and Acalyphoideae. Phyllanthaceae include the biovulate Phyllanthoideae, whereas Picrodendraceae include the biovulate Oldfieldioideae. Furthermore, parasitic Rafflesiaceae need to be accomodated (below). Peraceae are recognized to deal with Rafflesiaceae, but whether they should be segregated depends on the position of Rafflesiaceae—more work is needed.

Multiple complete nuclear genome sequences are available, useful resources in evolutionary studies: rubber, *Hevea brasilensis*, manihot, *Manihot esculenta*, and castor bean, *Ricinus communis*.

Drypetes and *Putranjiva* (Putranjivaceae) represent a distinct clade of former euphorbs (Rodman et al. 1998; D. Soltis et al. 2000). Evidence is available from seed-coat morphology (structure of the exotegmen; seeds exotegmic vs. endotestal) to support some of these groups (Corner 1976; Tokuoka and Tobe 2001), but additional investigations are needed. Genera of Pandaceae have also been placed within Euphorbiaceae (Thorne 1992a,b) or close to the family in Euphorbiales (Cronquist 1981; Takhtajan 1997). However, sequence data reveal that Pandaceae, a family of three genera, may be sister to Lophopyxidaceae (a family of one genus placed in or near Celastraceae by Cronquist 1981), or Lophopyxidaceae are sister to Putranjivaceae (Xi et al. 2012b) (for floral structure, see Matthews and Endress 2013). Euphorbiaceae, in the narrow sense, have been investigated phylogenetically (Wurdack et al. 2004, 2005). Phyllanthaceae comprise two major clades (Wurdack et al. 2004; Kathriarachchi et al. 2005; Samuel et al. 2005; Hoffmann et al. 2006).

Linaceae now include Hugoniaceae; a close relationship of the two was long suggested (Takhtajan 1997) and their separation was largely geographical. Whereas Linaceae s.s. are widely distributed, especially in temperate areas and the subtropics, Hugoniaceae are largely found in the tropics of the Southern Hemisphere.

The Flacourtiaceae problem—Flacourtiaceae, as previously defined (see Gilg 1925; Sleumer 1954, 1980; Lemke 1988; Judd 1997), were shown to be polyphyletic in DNA studies (Chase et al. 2002). The former, broadly defined family is composed of two major subclades that are more closely related to other families in Malpighiales than to each other. Thus, Chase et al. (2002) recognized two families: a broadly defined Salicaceae (which include Salicaceae, some Flacourtiaceae, the tribe Abatieae of Passifloraceae and the monogeneric Scyphostegiaceae; Bernhard 1999) and the expanded family Achariaceae (which include other former Flacourtiaceae, such as *Kiggelaria*, as well as *Acharia* and *Guthriea* of Achariaceae). The name Achariaceae was used because members of this poorly known family with three highly modified genera appeared as sister to *Kiggelaria* plus related genera (D. Soltis et al. 2000; Chase et al. 2002), and so the name of the family becomes the older conserved name, Achariaceae (not Kiggelariaceae as in several papers).

Salicaceae are now broadly defined (APG III 2009; APG IV 2016) to include taxa (but some split Salicaceae into Salicaceae and Samydaceae), formerly placed in Flacourtiaceae, that possess salicoid teeth (Nandi et al. 1998), have centrifugal stamen initiation (Bernhard and Endress 1999), lack cyanogenic glycosides of the gynocardin type, usually have small globose anthers, and possess generally small flowers in which the sepals and petals, if both are present, are equal in number or, as in many Salicaceae s.l. (e.g., *Azara*, *Idesia*), petals are frequently absent. The situation is reversed in Achariaceae; most lack salicoid teeth, have centripetal or simultaneous stamen initiation (Bernhard and Endress 1999), produce cyclopentenoid cyanogenic glycosides, usually have elongate anthers, and have large flowers typically with sepals and petals not equal in number but with petals always present. In addition, members of the lepidopteran genus *Cupha* feed only on species of the expanded Salicaceae (Nandi et al. 1998). Chase et al. (2002) provided an assessment of subclades within the family. Although the expansion of Salicaceae to include the largely tropical tribes of Flacourtiaceae appears to challenge the concept of the family as one of cool, north-temperate regions, the diversity of *Populus* (Salicaceae s.s.) actually reaches its peak in warm temperate to subtropical latitudes, and temperate members occur in China and Chile. Furthermore, those species of *Populus* most resembling former Flacourtiaceae in morphology are tropical (Eckenwalder 1996; Chase et al. 2002). Also, whereas *Salix* is most diverse in the Northern Hemisphere, its distribution includes Africa, Malaysia, and Central and South America (Argus 1997). The large genus *Salix* has also been the focus of phylogenetic study (Azuma et al. 2000; Hardig et al. 2010).

Although Lacistemataceae had been placed in Flacourtiaceae by some (e.g., Sleumer 1980), Cronquist (1981) and Takhtajan (1997) recognized Lacistemataceae as a distinct family consisting of *Lacistema* and *Lozania*, two small woody genera from Central and South America. In initial DNA analyses, the placement of Lacistemataceae within Malpighiales was unclear (D. Soltis et al. 2000), although Chase et al. (2002) suggested a close relationship to the expanded Salicaceae; this relationship was later confirmed (Soltis et al. 2011; Xi et al. 2012b).

The extent of the taxonomic confusion surrounding the former broadly defined Flacourtiaceae is further illustrated by the placements of several other genera formerly placed in the family by some authors (e.g., *Aphloia*, *Asteropeia*, *Berberidopsis*, *Gerrardinia*, *Muntingia*, and *Plagiopteron*, as well as genera of Peridiscaceae; for the last see Saxifragales, above). Molecular analyses indicated that these and other genera must be excluded not only from Flacourtiaceae but also from Malpighiales. For example, *Berberidopsis* (Berberidopsidaceae) is sister to *Aextoxicon* (Aextoxicaceae) (see "Berberidopsidales" in Chapter 11). *Aphloia* (Aphloiaceae), formerly placed in Flacourtiaceae, is in the rosid order Crossosomatales in its own family Aphloiaceae (discussed below). Another former Flacourtiaceae, *Plagiopteron* (sometimes recognized as a distinct family, Plagiopteridaceae; Airy Shaw 1964; Takhtajan 1997), is embedded in a broadly defined Celastraceae (discussed below). *Lethedon* was referred to Flacourtiaceae because it possesses cyanogenic glycosides (Spencer and Seigler 1985b), but the genus is well supported as part of Thymelaeaceae in Malvales (Savolainen et al. 2000b; van der Bank et al. 2002; see below). Alzateaceae were included by Hutchinson (1967) in Flacourtiaceae, but the family is a member of Myrtales (Graham 1984; Conti et al. 1996). *Gerrardinia* has been shown to belong to Huerteales and is now considered a monogeneric family (Woberg et al. 2009; Christenhusz et al. 2010). Both *Asteropeia* and *Dioncophyllum* have been placed in Flacourtiaceae (e.g., Hutchinson 1967). In contrast, Schmid (1964) proposed that *Dioncophyllum* was related to Droseraceae; both *Asteropeia* and *Dioncophyllum* (with Droseraceae) appeared as members of Caryophyllales in DNA analyses (see Chapter 11). *Muntingia* (included in Flacourtiaceae only by Cronquist 1981), together with *Dicraspidia*, were placed in Malvales as a distinct family, Muntingiaceae (Bayer et al. 1998) (see Malvids, below).

CHARACTER EVOLUTION IN MALPIGHIALES

Euphorbiaceae, with diverse growth forms and habit, have been the focus of multiple studies of character evolution. *Euphorbia* (with ~2000 species it is one of the largest genera of angiosperms) has been of particular interest. Horn et al. (2012) revealed that the xeric habit has evolved at least 14 times in this large genus.

Phyllanthaceae provide an excellent example of cospeciation involving plant host and pollinators. Phylogenetic analyses of species of the genus *Glochidion* (Phyllanthaceae) and their coevolving pollinating floral parasitic moths (genus *Epicephala*) have been conducted (Kawakita et al. 2004). Comparisons of the phylogenetic trees of host and pollinator reveal that they match closely (Fig. 10.22). *Glochidion* species and species of *Epicephala* show an overall pattern of co-speciation, but there are also occasional shifts of the moths onto more distantly related plants (Kawakita et al. 2004; Thompson 2013).

Phylogenetic studies have also provided insights into biogeography, as well as fruit and pollen diversification within Malpighiaceae (Cameron et al. 2001; Davis et al. 2001, 2002). Davis et al. (2002, 2005b) showed that the

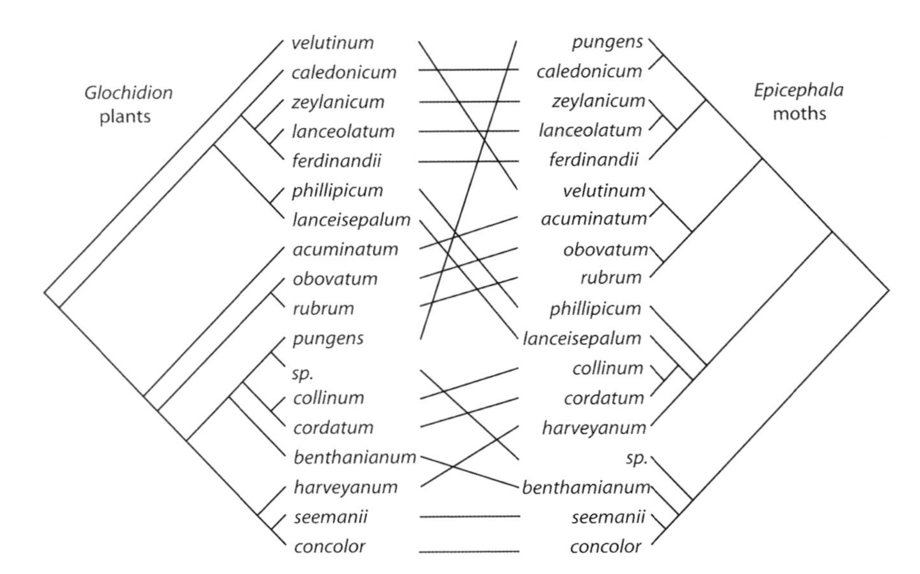

Figure 10.22. Co-speciation in plants of the genus *Glochidion* (Phyllanthaceae) and their coevolving pollinating floral parasitic moths in the genus *Epicephala*. When perfect co-speciation occurs, each speciation event in the plants or the moths would be matched by a speciation event in the other lineage. In this system, *Glochidion* species and *Epicephala* show an overall pattern of co-speciation; but there are also some occasional shifts of the moths onto more distantly related plants. (Modified from Thompson 2012, p. 339, based on Kawakita et al. 2004.)

family was once distributed to the north and then migrated south into the tropics, supporting a boreal tropical hypothesis as applying to Malpighiaceae. The rapid radiation of Malpighiales also has broader implications, supporting a mid-Cretaceous origin of modern tropical rain forests.

Species of Rafflesiaceae possess the largest flowers (up to 1 meter in diameter). Traditionally, the relationships of the family were highly debated. Placement of the parasitic Rafflesiaceae with Euphorbiaceae represents a breakthrough in our understanding of the evolution of parasitic plants (Fig. 10.23). Most Euphorbiaceae have small flowers, so the results suggest that the large flowers of Rafflesiaceae evolved from ancestors with small flowers—a dramatic example of size evolution (Davis et al. 2007; Nikolov et al. 2013, 2014). In fact, the large flowers of *Rafflesia* may be 200× the size of a flower of some Euphorbiaceae (Figs. 10.23 and 10.20d,e; flowers not to same scale).

Davis and Wurdack (2004) documented movement of mitochondrial genes into Rafflesiaceae; these apparently were acquired via horizontal gene transfer (HGT) from their obligate hosts in the grape family, *Tetrastigma*. The results suggested that in some instances HGT events are facilitated by the intimate association between the parasite and the host (see Chapter 15 for discussion of HGT).

OXALIDALES

Phylogenetic analyses (e.g., Savolainen et al. 2000a,b; D. Soltis et al. 2000) indicated that Oxalidales consist of seven families (Fig. 10.12) that had not all previously been considered closely related: Huaceae, Brunelliaceae, the carnivorous Cephalotaceae, as well as Connaraceae, Cunoniaceae (including Eucryphiaceae and Davidsoniaceae), Elaeocarpaceae (including Tremandraceae), and Oxalidaceae (several are illustrated in Fig. 10.13). Brunelliaceae, a monogeneric family from tropical America, were included in Cunoniaceae in APG (1998; see also Hufford and Dickison 1992), but their position in recent analyses as sister to the subclade of Elaeocarpaceae–Cunoniaceae–Cephalotaceae, rather than as part of Cunoniaceae, justifies continued familial status (APG II 2003; APG III 2009; APG IV 2016).

Oxalidales represent another rosid clade composed of families previously placed in several different orders in morphology-based classifications. Most families were placed in the traditionally recognized Rosidae, with Cephalotaceae, Brunelliaceae, and Cunoniaceae all considered part of Rosales (sensu Cronquist 1981); Oxalidaceae were placed in Geraniales; Connaraceae were considered part of Sapindales, and Tremandraceae were placed in Polygalales. Elaeocarpaceae were placed in Malvales of subclass Dilleniidae. The affinities of Huaceae have long been obscure. The two genera (*Hua* and *Afrostyrax*) have been placed in or near Styracaceae (Heywood 1993; a family in Ericales), placed in Violales (Cronquist 1981), or considered closely related to Sterculiaceae in Malvales (Takhtajan 1997).

Thus, this small clade again illustrates a pattern seen throughout the rosids—clades now recognized as orders (sensu APG 1998; APG II 2003; APG III 2009; APG IV 2016) are composed of taxa previously considered distantly related in traditional classifications. As with other orders of rosids, non-DNA synapomorphies for Oxalidales

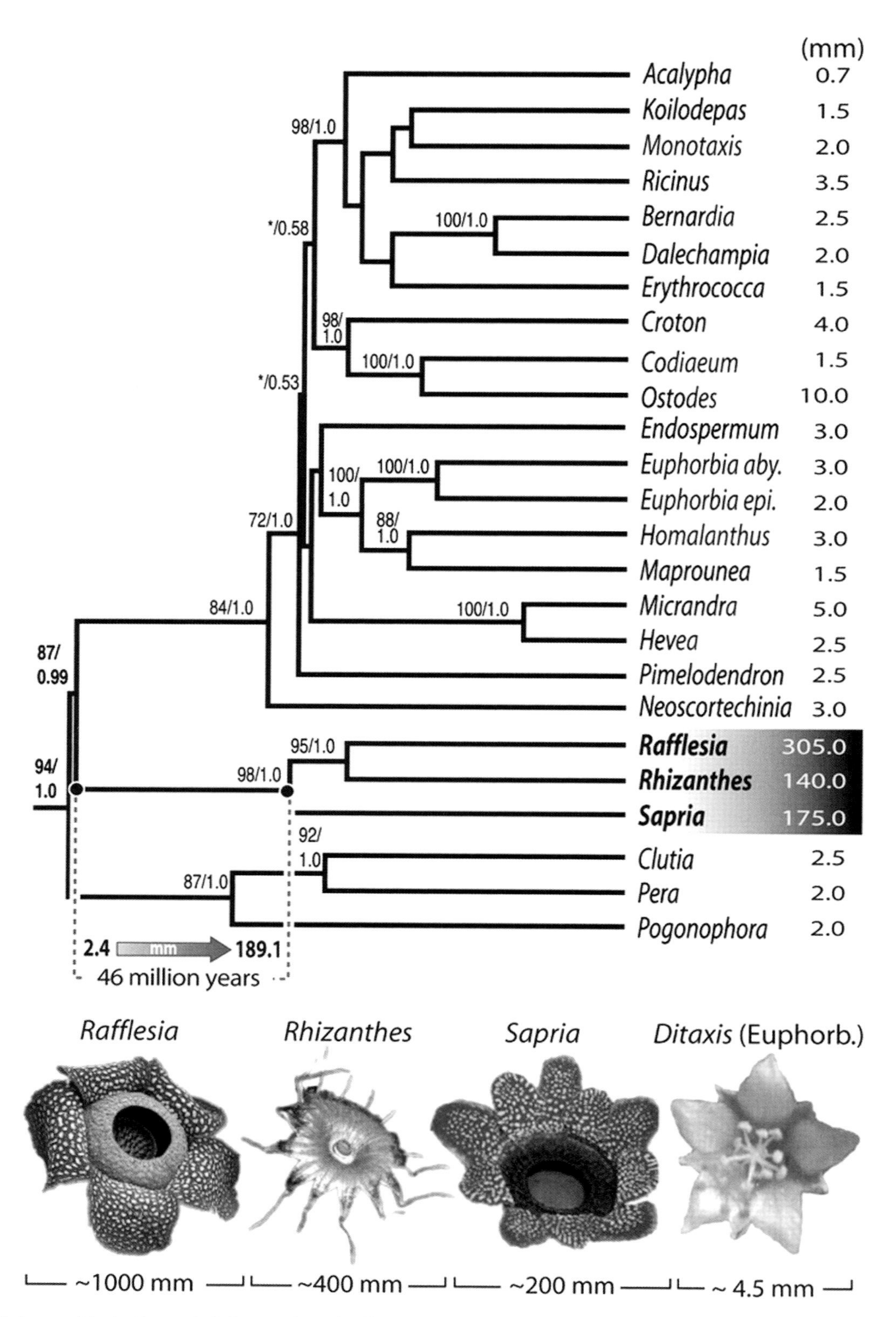

Figure 10.23. Phylogeny of Euphorbiaceae, including members of Rafflesiaceae (in bold). This tree is based on a temporally calibrated ML tree. ML bootstrap and Bayesian Posterior Probability values, respectively, are provided above nodes. Support values ≤ 50%/0.50 are designated with asterisks. Flower size diameters (in mm) are provided (right), and ancestral flower size estimates are indicated at the stem and crown nodes of Rafflesiaceae. Reconstructions indicate a 79-fold increase in floral diameter for stem lineage Rafflesiaceae (with a 95% confidence interval of 74- to 83-fold). For additional ancestral size estimates see SOM text. Color images with scale bars illustrate the approximate sizes of flowers representative of the three genera of Rafflesiaceae (*Rafflesia arnoldii*, *Rhizanthes lowii*, and *Sapria himalayana*), plus a representative of Euphorbiaceae (*Ditaxis neomexicana*), the latter being similar in size to the inferred ancestral flowers at the stem node of Rafflesiaceae. (Modified from Davis et al. 2007.)

are unclear, but see Matthews and Endress (2002) for potential synapomorphies in floral structure (below).

In early DNA analyses the problematic Huaceae appeared as sister to the strongly supported Celastrales clade, but with weak support (e.g., Savolainen et al. 2000a; D. Soltis et al. 2000). With large multi-gene datasets, Huaceae are sister to the remainder of Oxalidales with strong support (BS = 82% in Soltis et al. 2011; BS = 100% in H. Wang et al. 2009). Thus, Oxalidales comprise Huaceae as sister to the remaining families, which comprise two subclades: Connaraceae + Oxalidaceae and Cunoniaceae as sister to (Brunelliaceae + Cephalotaceae) + Elaeocarpaceae (Fig. 10.12).

Phylogenetic studies indicated that Cunoniaceae be broadly defined to include both *Bauera* and *Davidsonia* (but not *Brunellia*, which had sometimes also been included), genera sometimes placed in Baueraceae and Davidsoniaceae, respectively (see also Hufford and Dickison 1992). However, *Aphanopetalum*, previously placed in Cunoniaceae, has been shown to belong in Saxifragales (Fishbein et al. 2001; see above).

The close relationship between Elaeocarpaceae and Tremandraceae (the former now including the latter) was unexpected. Elaeocarpaceae had been placed in Malvales in recent classifications, but they lack many of the non-DNA synapomorphies for the latter group (see "Malvales," below). Tremandraceae had been placed in Polygalales (subclass Rosidae) in recent classifications and therefore were not considered a close relative of Elaeocarpaceae (Cronquist 1981), but they share most of their characters with some Elaeocarpaceae, especially their capsular fruit, and differ only in having a definite number of stamens rather than indefinite numbers as in Elaeocarpaceae.

The clade of Oxalidaceae + Connaraceae is morphologically well separated from the remainder of the order, possessing tristyly, postgenital union of petals into a basal tube, congenitally united stamens, hemianatropous to almost orthotropous ovules, and a special type of sieve tube plastids (Matthews and Endress 2002). Elaeocarpaceae (including the genera of former Tremandraceae) are also well supported, characterized by buzz-pollination and a syndrome of structural features functionally connected with it. In addition, flowers have involute petals that are longer than the sepals in advanced buds, three vascular traces, and stamens partly wrapped by adjacent petals; ovules have a chalazal appendage and a thick inner integument.

Cephalotaceae, an Australian carnivorous family with pitcher-trap leaves comprising one species (*C. follicularis*; see Fig. 10.13e), are morphologically isolated. The family also exemplifies one of several independent origins of carnivory and the pitcher-type trap (see Chapter 13). *Cephalotus* (Cephalotaceae) and Cunoniaceae had been placed in Rosales in recent classifications (e.g., Cronquist 1981) and share several floral features.

CELASTRALES

Celastrales were recognized in Judd et al. (1999). Additional phylogenetic analyses supported recognition of Lepidobotryaceae, Parnassiaceae (including *Lepuropetalon*), and an expanded Celastraceae (discussed below) as an order, Celastrales (APG II 2003; APG III 2009; APG IV 2016; Fig. 10.12; several families are illustrated in Fig. 10.13). This circumscription of Celastrales differs substantially from those in morphology-based classifications (e.g., Cronquist and Takhtajan). The delimitation of the order based on morphology has been highly variable, with several families, including Rhamnaceae, Salvadoraceae, Vitaceae, Aquifoliaceae, and Icacinaceae, placed in Celastrales in various classifications (Hutchinson 1973; Cronquist 1981, 1988; Takhtajan 1987, 1997; Heywood 1993; see Savolainen et al. 1994). However, molecular analyses demonstrated that most families previously placed in Celastrales should be assigned to other orders (see Savolainen et al. 1994, 1997; Simmons et al. 2001).

As with many rosid clades now recognized as orders based on DNA data, nonmolecular synapomorphies for Celastrales are unclear (for floral features, see Matthews and Endress 2005a). Celastrales comprise an expanded Celastraceae sister to Lepidobotryaceae. Lepidobotryaceae consist of two disjunct genera, *Rutiliocarpon* and *Lepidobotrys*, in Central America and tropical Africa, respectively. Cronquist (1981) placed these genera in Oxalidaceae, whereas Takhtajan (1997) considered Lepidobotryaceae and Oxalidaceae closely related. However, Oxalidaceae are placed in a distinct clade (Oxalidales, see above).

Based on molecular analyses, Celastraceae were greatly expanded to include the small families Brexiaceae, Hippocrateaceae, Stackhousiaceae, Plagiopteridaceae, Parnassiaceae, and Canotiaceae (Savolainen et al. 1994, 1997, 2000b; Simmons et al. 2001; APG III 2009; APG IV 2016). Parnassiaceae are morphologically distinctive (e.g., they do not have fiaments inserted at the outer border of a nectar disk).

Four morphological synapomorphies, all of which show reversals, support a broadly defined Celastraceae (Simmons et al. 2001): stamen and staminode number each equal petal number; filaments are inserted at the outer border of, or within, the conspicuous nectar disk; styles are connate; and two to four ovules per locule are present. Other possible synapomorphies include the presence of crystals in leaf epidermis (Baas et al. 1979) and an integumentary tapetum (Johri et al. 1992; Tang 1994; Nandi et al. 1998;

Matthews and Endress 2005a). Members of a broadly circumscribed Celastraceae also share completely syncarpous gynoecia with commissural stigmatic lobes, only weakly crassinucellar or incompletely tenuinucellar ovules, partly fringed sepals and petals, protandry in bisexual flowers combined with herkogamy by the movement of stamens and anther abscission from filaments, and stamens fused with the ovary (Matthews and Endress 2005a).

Questions about the distinctiveness of Celastraceae and Hippocrateaceae have existed since the two families were first described, so it is not surprising that the latter were found to be nested within Celastraceae. Similarly, although *Brexia* was variously placed (e.g., Saxifragaceae s.l., Escalloniaceae, Grossulariaceae, and Brexiaceae, reviewed in Morgan and Soltis 1993), a close relationship between *Brexia* and Celastraceae had also been proposed based on morphology (Perrier de la Bâthie 1933; see Takhtajan 1991, 1997). *Plagiopteron* had been placed in Tiliaceae and Flacourtiaceae, and sometimes recognized as a distinct family, Plagiopteridaceae (Airy Shaw 1965; Hutchinson 1967). However, *Plagiopteron* also fits well in a broadly defined Celastraceae based on leaf and wood anatomy (Baas et al. 1979) and embryology (Tang 1994). The molecular placement of Stackhousiaceae within Celastraceae also was not surprising based on morphological similarities. *Canotia* had been variously placed in Rutaceae, Koeberliniaceae, Canotiaceae, and Celastraceae, but both molecular and embryological data (Tobe and Raven 1993) indicated a close relationship of *Canotia* to *Acanthothamnus* within Celastraceae (Simmons et al. 2001). Parnassiaceae are not recognized in APG IV (2016) or here (see Chapter 12), and this clade, which may be sister to remaining members of Celastraceae, comprises *Parnassia* and the monotypic *Lepuropetalon*, the smallest terrestrial angiosperm (often placed in its own family; see Spongberg 1972; Gastony and Soltis 1977). Cronquist (1981) placed Parnassiaceae with Saxifragaceae, but Takhtajan (1991, 1997) included the family in Celastrales.

CHARACTER EVOLUTION IN CELASTRALES

Simmons et al. (2001) showed the complex nature of fruit and aril diversification within Celastraceae, with multiple origins of most fruit and aril forms. There have been many transitions from dehiscent to indehiscent fruit types. Nuts, drupes, samaras, and berries have arisen many times independently, most often from capsules; similarly, arils (fleshy covering surrounding seed) apparently arose once or twice in the family and have been lost in five or six lineages (Simmons et al. 2001). Other studies have similarly indicated the lability of fruit evolution within families (e.g., Manos and Stone 2001—see "Fagales," above; Richardson et al. 2000; Davis et al. 2001).

Malvids

Three genes provided strong support for a core clade (eurosid II; now malvids or *Malvidae*) that included the well-supported groups Brassicales, Malvales, Sapindales, and Huerteales (e.g., D. Soltis et al. 2000). It was initially unclear, however, if other orders may also be part of eurosid II. In the three-gene analysis, for example, Myrtales appeared as sister in all shortest trees to the eurosid I clade, but Geraniales and Crossosomatales were related to the well-supported eurosid II clade. With larger datasets a well-supported malvid clade ultimately emerged consisting of Geraniales, Crossosomatales, Sapindales, Huerteales, Malvales, Picramniaceae, and Brassicales (Fig. 10.24). Myrtales + Geraniales are sister to a large clade of the remaining members. Within that clade Crossosomatales and then Picramniaceae are subsequent sisters to the remaining members of the clade: Sapindales and a clade of Huertelaes + (Brassicales + Malvales) (H. Wang et al. 2009; Worberg et al. 2009; Soltis et al. 2011). (Members are shown in Figs. 10.25, 10.26.)

MYRTALES

Myrtales consist of 13 families (Fig. 10.24): Alzateaceae, Combretaceae, Crypteroniaceae, Heteropyxidaceae, Lythraceae (including Sonneratiaceae, Punicaceae, and Trapaceae), Melastomataceae (including Memecylaceae—which should be named Olisbeoideae; or these two can be treated as sister families), Myrtaceae, Oliniaceae, Onagraceae, Penaeaceae, Psiloxylaceae, Vochysiaceae, and Rhynchocalycaceae (several are illustrated in Fig. 10.25). DNA data have provided a well-resolved phylogenetic tree for Myrtales (Fig. 10.24; Conti et al. 1993, 1996, 1997, 2002; Sytsma et al. 2004; Savolainen et al. 2000b; D. Soltis et al. 2000, 2011; Clausing and Renner 2001).

Comparable age estimates have been obtained for Myrtales: H. Wang et al. (2009) 89–85 mya; Bell et al. (2010) 99–89 mya; Thornhill et al. (2012) 98–86 mya.

Myrtales have been well investigated. An entire issue of the *Annals of the Missouri Botanical Garden* (1984; vol. 71, no. 3) was dedicated to the order, summarizing morphology, chemistry, anatomy, palynology, and embryology (e.g., Behnke 1984; Cronquist 1984; Dahlgren and Thorne 1984; Johnson and Briggs 1984; Patel et al. 1984; Tobe and Raven 1984; van Vliet and Baas 1984). Nonetheless, disagreement persisted regarding the composition of, and relationships within, the order (Dahlgren and Thorne

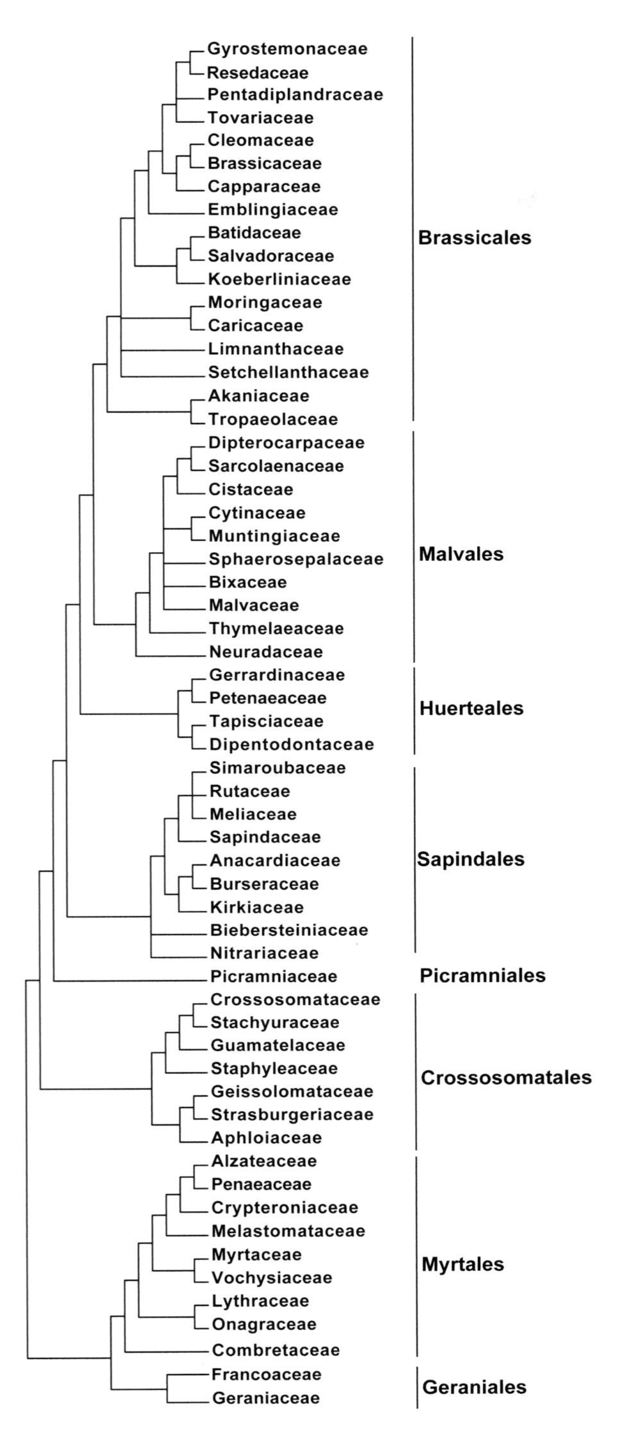

Figure 10.24. Summary tree for malvids (*Malvidae*).

1984; Conti et al. 1996). Treatments agreed in placing 14 families in Myrtales: Myrtaceae, Heteropyxidaceae, Psiloxylaceae, Lythraceae, Punicaceae, Sonneratiaceae, Combretaceae, Melastomataceae, Memecylaceae, Onagraceae, Trapaceae, Crypteroniaceae, Alzateaceae, and Rhynchocalycaceae. Most also included the small South African families Oliniaceae and Penaeaceae. However, additional families were sometimes placed in Myrtales: Thymelaeaceae, Lecythidaceae, Rhizophoraceae, Haloragaceae, and Gunneraceae (e.g., Cronquist 1981, 1988; Heywood 1993).

DNA analyses clarified the composition of Myrtales, as well as relationships within the order (e.g., Conti et al. 1993, 1996, 1997, 1999, 2002; Sytsma et al. 2004; Savolainen et al. 2000b; D. Soltis et al. 2000, 2011; Clausing and Renner 2001; Schönenberger and Conti 2003; Fig. 10.24: APG III 2009). The DNA-based circumscription of Myrtales agreed closely with that of Dahlgren and Thorne (1984), except that molecular studies indicated that Vochysiaceae also belong in Myrtales, a noteworthy result in that this family had never been associated with the order, but was placed in Polygalales (e.g., Cronquist 1981, 1988; Heywood 1993) or close to Polygalales (Takhtajan 1997). The unique floral morphology of Vochysiaceae (strongly asymmetric flowers, one fertile stamen, and the number of petals reduced to three, one, or none; Cronquist 1981) had made it difficult to place the family using morphology. Other families sometimes placed in Myrtales (e.g., Thymelaeaceae, Lecythidaceae, Rhizophoraceae, Haloragaceae, and Gunneraceae) were clearly excluded from the order (Conti et al. 1993; Savolainen et al. 2000a,b; D. Soltis et al. 2000).

Myrtales are characterized by two distinctive features of wood anatomy, presence of bicollateral vascular bundles in the primary stem and vestures in bordered pits of secondary xylem. Significantly, this combination is found in Vochysiaceae. The occurrence of both features outside of Myrtales (e.g., Apocynaceae, Thymelaeaceae) is rare. Myrtales also possess methylated ellagic acids and intraxylary phloem (Nandi et al. 1998). Other synapomorphies include stamens incurved in bud and a single style (carpels completely connate), hypanthium (often with inner surface nectariferous), and opposite leaves (Judd et al. 2008). Still other features are present in most Myrtales (Tobe and Raven 1983, 1984): glandular anther tapetum, an inner integument with two layers, micropyle formed by both integuments, ephemeral antipodal cells, and exalbuminous seeds.

Combretaceae are sister to the remaining members of Myrtales (Fig. 10.23), followed by a strongly supported sister group of Lythraceae (broadly defined to include *Trapa* and *Punica*) + Onagraceae. The remaining families form two major subclades. In one of these subclades, an expanded Myrtaceae (including Heteropyxidaceae and Psiloxylaceae) are sister to Vochysiaceae. The second subclade consists of Melastomataceae (including Memecylaceae) as sister to a strongly supported clade consisting of Crypteroniaceae, Alzateaceae, and Penaeaceae (the latter

Figure 10.25. Geraniales-Picramniales-Crossosomatales-Myrtales-Sapindales (Anacardiaceae, Burseraceae, Combretaceae, Geraniaceae, Lythraceae, Melastomataceae, Meliaceae, Melianthaceae, Myrtaceae, Onagraceae, Picramniaceae, Rutaceae, Sapindaceae, Simaroubaceae, and Staphyleaceae): a. *Geranium maculatum* L. (Geraniaceae), fruit; note distinctive ovary with separating, fertile, lower portion and slender, sterile, upper portion of each carpel. b. *Melianthus comosus* J.Vahl (Melianthaceae), zygomorphic flower. c. *Picramnia pentandra* Sw. (Picramniaceae), fruits (2-carpellate berries). d. *Staphylea trifolia* L. (Staphyleaceae), compound leaf and fruits (inflated capsules). e. *Miconia angustilamina* (Judd & Skean) Judd & Ionta (Melastomataceae), leaves with acrodromous venation and flower with zygomorphic androecium. f. *Chamerion angustifolium* (L.) Holub (Onagraceae), flower. g. *Decodon verticillatus* (L.) Elliott (Lythraceae), flowers showing heterostyly and wrinkled petals. h. *Acca sellowiana* (O.Berg) Burret (Myrtaceae), flower in longitudinal section showing short hypanthium, numerous stamens, inferior ovary, and single style. i. *Combretum indicum* (L.) DeFilipps (Combretaceae), habit and flowers; note elongate hypanthia. j. *Ailanthus altissima* (Mill.) Swingle (Simaroubaceae), pinnate leaf and fruits (i.e., each flower producing a cluster of samaras). k. *Bursera simaruba* (L.) Sarg. (Burseraceae), aromatic, smooth, peeling bark. l. *Rhus typhina* L. (Anacardiaceae), infructescence. m. *Chukrasia tabularis* Juss. (Meliaceae), flower with monadelphous androecium. n. *Citrus × aurantium* L. (Rutaceae), flower with intrastaminal nectar disk. o. *Harpullia pendula* Planch. ex F.Muell. (Sapindaceae), flowers, note appendaged petals.

now includes Rhynchocalycaceae and Oliniaceae) (Conti et al. 1996; Savolainen et al. 2000a,b; Clausing and Renner 2001; Sytsma et al. 2004).

Johnson and Briggs (1984) provided a cladistics analysis of interfamilial relationships within Myrtales based on morphological characters, using the circumscription proposed by Dahlgren and Thorne (1984). Some of their results compare favorably with the DNA-based topology; they recovered a clade of Myrtaceae, Heteropyxidaceae, and Psiloxylaceae (they did not include Vochysiaceae in their non-DNA analysis) as sister to Penaeaceae and Oliniaceae (Schönenberger and Conti 2003). Analyses have also focused on phylogenetic relationships within families of Myrtales. Onagraceae, for example, are one of the best-studied angiosperm families (e.g., Raven 1979; Sytsma et al. 1990; Conti et al. 1993; Levin et al. 2003; Wagner et al. 2007). Other families of the order also have been the focus of phylogenetic analyses, including Myrtaceae (Briggs and Johnson 1979; Wilson et al. 2001, 2005; Wilson 2011), Vochysiaceae (Litt 1999; Litt and Stevenson 2003), and Melastomataceae (Clausing et al. 2000; Clausing and Renner 2001; Renner and Meyer 2001; Renner et al. 2001; Michelangeli et al. 2004; Goldenberg et al. 2008, 2010; see also www.flmnh.ufl.edu/natsci/herbarium/melastomes/default.htm), and Lythraceae (Graham et al. 1993) and Penaeaceae (Schönenberger and Conti 2003).

The rich fossil record of Lythraceae (including Trapaceae) has been carefully reviewed (Graham 2013), including numerous records from the Tertiary, as well as credible Campanian pollen records of *Lythrum* and *Peplis* from Wyoming (Grímsson et al. 2011). Myrtaceae are represented by well-preserved guava-like fruits of *Paleomyrtinaea* from the Paleocene of North Dakota and *Eucalyptus* fruits and foliage from Eocene of Argentina (Pigg et al. 1993; Gandolfo et al. 2011). Pollen of Onagraceae is readily recognizable with occurrences of their distinctive pollen back to the Campanian (reviewed by Martin 2003) and *Fuschia* in the Oligocene of Australia (Berry et al. 1990).

CHARACTER EVOLUTION IN MYRTALES

Fruit characters have been evolutionarily labile in Myrtales; this lability is seen throughout the angiosperms (see above, Fagales, Celastrales). In Myrtaceae, the plesiomorphic state is a dehiscent fruit; indehiscent fruits have evolved in four lineages, with three origins of fleshy fruits (Wilson et al. 2001), but Biffin et al. (2010) showed that fleshy fruits have evolved independently twice in Myrtaceae; both shifts in fruit type has been accompanied by exceptional diversification rate shifts. Biffin et al. (2010) suggested that the evolution of fleshy fruits is a key innovation for clades of Myrtaceae inhabiting rainforest habitats. In Melastomataceae, berries evolved from capsules at least four times (Clausing and Renner 2001). Sytsma et al. (2004), in a broader analysis of Myrtales, also found that fleshy fruits had evolved multiple times.

Renner et al. (2001) examined the historical biogeography of Melastomataceae in light of phylogenetic hypotheses. The family has a pantropical distribution and, in contrast to earlier hypotheses, the results of Renner et al. indicated that the current distribution of Melastomataceae is the result of long-distance dispersal during the Neogene, rather than fragmentation of Gondwana. Similarly, Sytsma et al. (2004) showed that long-distance dispersal was important in other Myrtales. They found that the African lineage of Vochysiaceae is nested within a South American clade and probably arose via long-distance dispersal during the Oligocene. Long-distance dispersal has been shown to be important in establishing the broad distributions of other families of flowering plants, including another rosid family, Simaroubaceae (Clayton et al. 2009; see above).

Goldenberg et al. (2008) examined the complex relationships between anther morphology and pollinators in the genus *Miconia* and relatives (Melastomataceae) and provided evidence for multiple independent derivations of several floral traits, including parallel derivation of stamen features (e.g., elongate to short anthers, minute to broader anther pores); some of these changes may be due to pollinator shifts, especially from buzz pollination to standard, nonvibrational pollination. The loss of a nectary is a synapomorphy of Melastomataceae, but nectar glands have been re-acquired multiple times (Varassin et al. 2008).

GERANIALES

Geraniales comprise Geraniaceae (including Hypseocharitaceae), as sister to a broadly defined Francoaceae (including Bersamaceae, Greyiaceae, Ledocarpaceae, Melianthaceae, Rhyncothecaceae, and Francoaceae s.s.) (Fig. 10.24; APG IV 2016; see Chapter 12). In APG III (2009) Melianthaceae included Francoaceae and Greyiaceae. The old APG delimitation is based on results from studies that often differed in taxonomic sampling (e.g., Savolainen et al. 2000a; D. Soltis et al. 2000, 2011). The topology depicted here (Fig. 10.24) and the broad circumscription of Francoaceae follows APG IV (2016) and is based on the topology of Palazzesi et al. (2012) and Sytsma et al. (2014). The alternative would be three families: Melanthiaceae, Francoaceae, and Ledocarpaceae. Several representatives of Geraniales are illustrated in Fig. 10.25; Francoaceae s. l. comprise relatively few species while Geraniaceae contain ~850 species.

This circumscription of Geraniales represents another

unexpected result of DNA studies (Price and Palmer 1993). Geraniaceae, Hypseocharitaceae (now part of Geraniaceae), and Vivianiaceae were considered closely related in some treatments (e.g., Takhtajan 1997); Cronquist (1981) also placed them together in his Geraniaceae. They share vessels with simple perforations, fibers with simple pores, extrastaminal nectar glands, two ovules per locule, and five connate, more or less lobed carpels (usually). However, Geraniaceae have often been associated with Oxalidaceae, a family now part of Oxalidales. Furthermore, the remaining members of the molecular-based Geraniales clade (i.e., Melianthaceae, Francoaceae, and Greyiaceae) have not been closely associated with Geraniaceae in morphology-based classifications. For example, Francoaceae and Greyiaceae represent former members of Saxifragaceae s.l. (Morgan and Soltis 1993), whereas Melianthaceae were placed in Sapindales (Cronquist 1981; Takhtajan 1997).

The more or less lobed carpels and extrastaminal nectaries are likely synapomorphies for Geraniales. Morphological evidence indicates that some of these families (Melianthaceae, Greyiaceae, and Francoaceae) are closely related (Danilova 1996; Ronse De Craene and Smets 1999; Savolainen et al. 2000b). As a result, in APG II (2003), Greyiaceae were merged with Melianthaceae, together with Francoaceae (comprising *Francoa* and *Tetilla*). However, *Francoa* and *Tetilla* are both from Chile and the only members of Melianthaceae s.l. outside eastern and southern Africa; both are both herbaceous perennials and otherwise dissimilar to the other woody taxa of this group. The relationships of the small family Ledocarpaceae (consisting of *Balbisia* and *Wendtia*) were long obscure. *Wendtia* (*Balbisia* has not been included in most analyses) has been sister to *Viviania* in several studies and is now included in Vivianiaceae. *Wendtia* was not included in the analyses of D. Soltis et al. (2000, 2011). Ledocarpaceae are now included in Vivianiaceae.

Pelargonium is noted for unusual plastid genome evolution (e.g., inverted repeat expansion) (Guisinger et al. 2008, 2011) as well as accelerated rates of sequence evolution for both mitochondrial and plastid genes (Cho et al. 2004; Parkinson et al. 2005; Bakker et al. 2006; Guisinger et al. 2008). In fact, there have been multiple major increases as well as decreases in substitution rates in mtDNA genes in the family, particularly *Pelargonium* (Parkinson et al. 2005).

The plastid genomes of Geraniaceae have undergone extensive genomic changes compared to other plastid genomes. The plastome genomes of *Erodium texanum*, *Geranium palmatum*, and *Monsonia speciosa*, and *Pelargonium hortorum* have all been characterized (Palmer et al. 1987; Guisinger et al. 2011). Geraniaceae plastomes are highly variable in size, gene content and order, repetitive DNA, and codon usage. For example, the plastid genome sequence of *Monsonia speciosa* is very small (128,787 bp; only nonphotosynthetic plants and lineages that have lost one copy of the inverted repeat are smaller); *Pelargonium hortorum*, in contrast, has the largest known plastid genome (217,942 bp) (Guisinger et al. 2011).

Analysis show highly accelerated plastid nucleotide substitution rates, both synonymous and nonsynonymous, throughout Geraniaceae (Guisinger et al. 2008). To explain these unusual substitution patterns in the highly rearranged Geraniaceae plastid genomes, Guisinger et al. (2008) suggested that aberrant DNA repair together with altered gene expression are possible factors.

CROSSOSOMATALES

Phylogenetic analyses indicated the presence of a previously unrecognized assemblage of closely related taxa first designated Crossosomatales in APG II (2003). Circumscription of this order was conservative in APG II (2003), consisting of only Crossosomataceae, Stachyuraceae, and Staphyleaceae (Figs. 10.24, 10.25). Early analyses of *rbcL* sequences alone indicated the presence of this Crossosomatales clade (Savolainen et al. 2000b), as did analysis of three genes (D. Soltis et al. 2000). Early molecular studies indicated, albeit with weak support, that several other families were related to Crossosomatales: Ixerbaceae, Aphloiaceae, Geissolomataceae, and Strasburgeriaceae (Savolainen et al. 2000b; D. Soltis et al. 2000, 2007c; Sosa and Chase 2003). With the sampling of additional genes and taxa, strengthened support for a close relationship of these other families to core Crossosomatales emerged, and the order was expanded (APG III 2009; APG IV 2016). In addition, *Guamatela* was found to be related and placed in its own family (Oh and Potter 2006). The order comprises two subclades: 1) Staphyleaceae + Guamatelaceae subsequent sisters to Stachyuraceae + Crossosomataceae; 2) Aphloiaceae and Geissolomataceae as subsequent sisters to Strasburgeriaceae + Ixerbaceae (Sosa and Chase 2003; Oh and Potter 2006; Oh 2010; Soltis et al. 2011).

Crossosomatales represent another unexpected assemblage revealed by DNA analyses. Crossosomataceae and Stachyuraceae were placed by Heywood (1993) in Dilleniidae, Dilleniales and Theales, respectively. However, other authors placed Crossosomataceae in or near Rosales (Cronquist 1981; Takhtajan 1997). Staphyleaceae have been placed in Sapindales (Rosidae of most authors; Cronquist 1981; Heywood 1993, 1998; Takhtajan 1997). Geissolomataceae and Stachyuraceae have been placed together in Celastrales (Heywood 1993), but in other treatments have been variously placed. *Ixerba* has been placed in its own family, whereas *Brexia* and *Roussea* were in Brexiaceae, or the three genera have each been placed in separate families within Brexiales (Takhtajan 1997). However, based on DNA data these three genera are only distantly related (Koontz and Soltis 1999; Soltis et al. 2011). The monotypic *Aphloia* (Aphloiaceae) has been treated as a member of Flacourtiaceae, but the problems with Flacourtiaceae were reviewed above (see Malpighiales, above). Takhtajan (1997) considered the monotypic *Strasburgeria* to represent "a New Caledonian relict" with "primitive characters"; he placed Strasburgeriaceae close to Ochnaceae in his Ochnales, whereas Cronquist (1981) considered *Strasburgeria* to be a member of Ochnaceae. *Guamatela* was included in Rosaceae (see Oh and Potter 2006).

Crossosomataceae, Stachyuraceae, and Staphyleaceae appear to share a seed type in which the cell walls of the many-layered testa are all, or mostly, lignified, and this may be synapomorphic for the order. It is unclear whether this feature occurs also in Guamatelaceae. This same seed morphology also occurs in *Geissoloma*, *Ixerba*, *Strasburgeria*, and *Aphloia*, strengthening support for a close relationship of these taxa to Crossosomatales and the broad circumscription of the order. This further demonstrates that seed anatomy may be a valuable source of systematic information; seed characters often appear highly congruent with phylogenetic relationships inferred from analyses of molecular data (see Nandi et al. 1998). The presence of a hypanthium with a nectiferous inner surface may also be synapomorphic.

Despite diverse taxonomic placements, most Crossosomatales are similar in floral morphology, possessing four or five sepals and petals, four or five stamens (numerous stamens in Aphloiaceae and some Crossosomataceae), four or five carpels, and a superior ovary, but these characters are, of course, widespread in the eudicots. A comparative study of floral characters in all seven families considered here (Crossosomataceae, Stachyuraceae, Staphyleaceae; Ixerbaceae, Aphloiaceae, Geissolomataceae, and Strasburgeriaceae) revealed potential synapomorphies for subclades recovered with molecular data within this extended Crossosomatales (Matthews and Endress 2005b). Floral structure supports a close relationship among the core Crossosomatales (Crossosomataceae, Stachyuraceae, and Staphyleaceae). In addition, floral data support a particularly close relationship of Ixerbaceae and Strasburgeriaceae, as well as a close relationship of Aphloiaceae and Geissolomataceae to these two families (Matthews and Endress 2005b). Crossosomatales are in need of additional investigation.

PICRAMNIALES

Alvaradoa and *Picramnia* (Picramniaceae) are two poorly understood genera from Central and tropical America, placed by Cronquist (1981) and Takhtajan (1997) in Simaroubaceae. *Alvaradoa* has also been placed in its own family. The two genera appeared as sisters with strong support in early analyses (Savolainen et al. 2000a,b), but were not included in D. Soltis et al. (2000). Recent analyses (H. Wang et al. 2010; Soltis et al. 2011) have placed the Picramniaceae with strong support as sister to a clade of Sapindales, Huerteales, Brassicales, and Malvales. Hence, the family is not closely related to Simaroubaceae as traditionally maintained. In addition, the recently described genus *Nothotalisia* has been included in the family (Thomas 2011) (shown in Fig 10.25).

SAPINDALES

DNA investigations recovered a well-supported Sapindales comprising (Fig. 10.24) Anacardiaceae, Biebersteiniaceae, Burseraceae, Kirkiaceae, Meliaceae, Nitrariaceae (including Peganaceae and Tetradiclidaceae), Rutaceae, Simaroubaceae, and Sapindaceae (broadly defined to include Aceraceae and Hippocastanaceae; APG 1998; D. Soltis et al. 2000, 2011; APG II 2003; APG III 2009; APG IV 2016; several families are illustrated in Fig. 10.25 with fossils of *Acer* and *Aesculus* shown in Fig. 10.5). This composition corresponds closely to Cronquist's (1968, 1981, 1988) broad view of the order, although he included families in Sapindales now considered only distantly related to the group (e.g., Akaniaceae, including Bretschneideraceae, of Brassicales). In contrast, other authors divided the families of Sapindales between Rutales and Sapindales s.s. (e.g., Scholz 1964; Takhtajan 1987, 1997; Dahlgren 1989; Thorne 1992a,b).

Although Sapindales were well supported in early DNA analyses (Gadek et al. 1996; Savolainen et al. 2000a; D. Soltis et al. 2000), non-DNA synapomorphies for the clade are less clear. Several characters are widespread in Sapindales: woodiness, exstipulate and pinnately compound leaves (sometimes becoming palmately compound, trifoliate, or unifoliate), generally small, tetra- or pentamerous flowers with a distinct nectar disc and imbricate perianth parts, and a tendency to apocarpy (but with postgenitally connected carpel tips) (Endress et al. 1983; Bachelier and Endress 2009; Ottra et al. 2013). The pinnately compound leaves and distinct, usually intrastaminal nectar disk (or ring of glands) are likely synapomorphic (Judd et al. 2008). Early studies (Gadek et al. 1996; Savolainen et al. 2000b) indicated several major subclades in Sapindales and these have generally been supported (Fig. 10.24).

Biebersteiniaceae, followed by Nitrariaceae, appear as successive sisters to the remainder of Sapindales. *Biebersteinia* comprises about five species previously placed in Geraniaceae (Heywood 1993). Takhtajan (1997), however, placed the genus in its own family and order close to Geraniales. Biebersteiniaceae appeared as sister to the remaining members of Sapindales in Savolainen et al. (2000b). DNA analyses also provided support for a broadly defined Nitrariaceae that also included Peganaceae and Tetradiclidaceae (APG II 2003; APG III 2009; APG IV 2016). In molecular analyses, Nitrariaceae (consisting of the single genus *Nitraria*) emerged as sister to Peganaceae (a family first recognized by Takhtajan 1969; see also Dahlgren 1980), which also included *Tetradiclis* (Tetradiclidaceae). Both *Nitraria* and *Tetradiclis* had previously been included in Zygophyllaceae (Sheahan and Chase 1996).

Meliaceae, Simaroubaceae, and Rutaceae form a well-supported subclade. Rutaceae include *Cneorum* (Cneoraceae) and *Ptaeroxylon* (Ptaeroxylaceae) (Gadek et al. 1996; Chase et al. 1999; Savolainen et al. 2000a,b). Although some analyses recovered Meliaceae sister to Rutaceae + Simaroubaceae (Soltis et al. 2011), others find Rutaceae sister to Meliaceae + Simaroubaceae (H. Wang et al. 2009). This clade of three families is united by biosynthetically related triterpenoid derivatives, limonoids and quassinoids (Hegnauer 1962–1994). Cronquist (1988) also considered the families closely related because of the absence of resin ducts in the bark, rays, and veins. Relationships, biogeography, and evolution within Rutaceae have been examined in several studies (e.g., Chase et al. 1999; Groppo et al. 2008; Appelhans et al. 2011, 2012); relationships within Meliaceae were analyzed by Muellner et al. (2003, 2006).

Generic relationships within Simaroubaceae have been largely resolved (Clayton et al. 2007); those analyses (see character evolution below) also provided insights into long-distance dispersal events.

Molecular data also indicated a strongly supported clade of Kirkiaceae (consisting of one genus) sister to Anacardiaceae and Burseraceae (D. Soltis et al. 2000, 2011; H. Wang et al. 2009). This close relationship is not evident based on non-DNA characters (but see Bachelier and Endress 2008). Takhtajan (1987, 1997) recognized Kirkiaceae as a distinct family of two genera, *Kirkia* and *Pleiokirkia* (the latter has not been included in phylogenetic analyses), from Africa and Madagascar, respectively. Molecular evidence confirmed that the enigmatic Kirkiaceae represent a distinct family within Sapindales (Gadek et al. 1996; Savolainen et al. 2000b). Cronquist (1983) placed Kirkiaceae in Simaroubaceae and Takhtajan (1997) placed the family near Simaroubaceae; *Kirkia* lacks the distinctive secondary compounds found throughout Sapindales (see Mulholland et al. 2003). Anacardiaceae and Burseraceae do display noteworthy non-DNA similarities, including the presence of vertical intercellular secretory canals in the primary and secondary phloem (Gadek et al. 1996); they also possess resin canals in the leaves and are the only two families of Sapindales from which biflavones are reported (Wannan et al. 1985). Anacardiaceae as defined here and by APG (1998), APG II (2003), APG III (2009), and APG IV (2016) also include Podoaceae, Julianiaceae, and Blepharocaryaceae; these families were typically included within Anacardiaceae in classifications (for floral structure, see Bachelier and Endress 2007, 2009).

DNA data provided strong support for a broadly defined Sapindaceae that include Hippocastanaceae and Aceraceae, a definition similar to Thorne (1992a,b). Cronquist (1981) and Takhtajan (1987, 1997) considered Aceraceae and Hippocastanaceae to be closely related to Sapindaceae. Judd et al. (1994, 1999, 2002, 2008) similarly provided morphological support for a broadly defined Sapindaceae. Non-DNA characters uniting Aceraceae and Hippocastanaceae with Sapindaceae include the production of triterpenoid saponins in secretory cells, cyclopropane amino acids, usually eight stamens, pubescent or papillose filaments, ovules that lack a funiculus and are broadly attached to a protruding portion of the placenta (the obturator), the presence of a nectar disc, and an embryo with the radicle separated from the rest of the embryo by a deep fold or pocket in the seed coat. Hypoglycin, an unusual amino acid, may be another synapomorphy for the clade (Gadek et al. 1996). Aceraceae and Hippocastanaceae are temperate specializations within Sapindaceae, a family with temperate and tropical members. Phylogenetic relationships within a broadly defined Sapindaceae have also been investigated, and *Xanthoceras* is sister to the rest of the family and is diagnosed by its large petals, nectar disk with five horn-like appendages, and seven or eight ovules per locule (Savolainen et al. 2000a; Harrington et al. 2005; Harrington and Gadek 2009, 2010).

Picramnia and *Alvaradoa* were usually included within Simaroubaceae, but phylogenetic studies indicated a placement of these two genera well removed from Sapindales (Savolainen et al. 2000b; Soltis et al. 2011). These taxa are now known to constitute a distinct lineage (Picramniaceae; Picramniales; see above). Savolainen et al. (2000b) initially reported that *Lissocarpa* (Lissocarpaceae) falls within Rutaceae. However, the sequence was obtained from degraded herbarium DNA; further study indicated that *Lissocarpa* is sister to *Diospyros* (Ebenaceae), as expected based on morphology (Berry et al. 2001). APG II (2003), APG III (2009), and APG IV (2016) included *Lissocarpa* in Ebenaceae.

Sapindales may have began to diversify in the latest Cretaceous as suggested by the occurrence of samaras that conform to *Acer* from North Dakota (Fig. 10.21e), but the record is better documented for representatives of most families in the early Cenozoic. Crown group Sapindalean families are commonly represented in the Early Eocene, including members of Anacardiaceae (e.g., *Rhus*, *Anacar-*

dium, *Choerospondias*), Burseraceae, Sapindaceae (e.g., *Aesculus*, *Acer*, *Dipteronia*, *Koelreuteria*), Meliaceae, Simaroubaceae (e.g., *Ailanthus*, *Chaneya*) (summarized in Friis et al. 2011) (see Fig. 10.21e,f for fossils of Sapindaceae).

CHARACTER EVOLUTION IN SAPINDALES

Several studies have provided new insights into the biogeography of families of Sapindales. Simaroubaceae are pantropical with taxa distributed in each of the Neotropics, tropical Africa and tropical Asia and Australia. A number of families have a similar distribution (e.g., Melastomataceae, Malpighiaceae, Meliaceae, Burseraceae, Moraceae), but detailed studies of such families to explore hypotheses of intercontinental migration are few. Clayton et al. (2009) reconstructed a North American origin for Simaroubaceae, with migration via Beringia by ancestral taxa. In contrast to traditional views, long-distance dispersal events were found to be common in the family, particularly in the Late Oligocene and later. Notable dispersals are inferred to have occurred across the Atlantic Ocean in both directions, as well as between Africa and Asia, and around the Indian Ocean basin and Pacific islands (Clayton et al. 2009).

HUERTEALES

Huerteales comprise four little known families: Gerrardinaceae + Petenaeaceae are sister to Tapisciaceae + Dipentodontaceae. D. Soltis et al. (2000) first placed *Tapiscia* (Tapisciaceae) in eurosid II (malvids) with strong support, but *Tapiscia* was the only representative included of what would ultimately be named Huerteales. In Soltis et al. (2011) *Dipentodon* (Dipentodontaceae) appeared as sister to *Tapiscia*. The growing support for a clade of three families (Dipentodontaceae, Tapisciaceae, Gerrardinaceae) in multigene analyses led to the recognition of Huerteales (APG III 2009; APG IV 2016). More recently, *Petenaea* was found to be related to *Gerrardina* (Christenhusz et al. 2010). Worberg et al. (2009) and Christenhusz et al. (2010) recovered, albeit with weak support, the topology shown here (Fig. 10.24) of Gerrardinaceae + Petenaeaceae as sister to Tapisciaceae + Dipentodontaceae.

These are all small, little-known genera, sometimes previously recognized as families that had not been considered closely related. Dipentodontaceae comprises two genera and 16 species. Tapisciaceae consists of *Tapiscia* and *Huertea* and five species. Gerrardinaceae comprises one genus and two species. Peteneaceae consists of one species, *Petenaea cordata*. *Perrottetia* (Dipentodontaceae) had been placed in Celaastraceae; *Dipentodon* was placed in Santalales (Cronquist 1981); *Gerrardina* was previously placed in the problematic Flacourtiaceae (see discussion in Malpighiales). *Petenaea cordata* was first described in Eleocarpaceae and later placed in Tiliaceae. *Tapiscia* and *Huertea* were traditionally placed in or near Staphyleaceae. Cronquist (1981) considered them to represent a distinct subfamily within Staphyleaceae, following Pax (1893). Spongberg (1971) suggested the separation of these two genera into their own family, Tapisciaceae. Takhtajan (1980, 1997) similarly recognized Tapisciaceae, considering it closely allied to Staphyleaceae. Thus, the placement of Tapisciaceae and Huerteaceae with Brassicales, Sapindales, and Malvales, rather than near Staphyleaceae in Crossosomatales, is another unexpected result of phylogenetic analyses, as would the close relationship suggested by DNA results for all four families of Huerteales.

Athough *Tapiscia* is limited to two species in China today, it was present in the Eocene of Oregon, England, and Germany, as determined by its distinctive seeds (Fig. 10.21h; Manchester 1988; Collinson et al. 2012).

BRASSICALES

Molecular studies revealed a strongly supported Brassicales of 17 families (Fig. 10.24): Akaniaceae (including Bretschneideraceae), Batidaceae (= Bataceae), Brassicaceae, Capparaceae, Caricaceae, Cleomaceae (APG II 2003 included both Capparaceae and Cleomaceae in Brassicaceae; this was changed in APG III 2009 and APG IV 2016) Emblingiaceae, Gyrostemonaceae, Koeberliniaceae, Limnanthaceae, Moringaceae, Pentadiplandraceae, Resedaceae, Salvadoraceae, Setchellanthaceae, Tovariaceae, and Tropaeolaceae (some of these families are illustrated in Fig. 10.26). Borthwickiaceae (one species, *B. trifoliata*, from China) were recently described and added to Brassicales (H. Su et al. 2012), although APG IV (2016) include in Resedaceae (see below).

Brassicales include all families that produce mustard-oil glucosides (also called glucosinolates), except for Putranjivaceae of Malpighiales (Rodman 1991a,b; Gadek et al. 1992; Rodman et al. 1993, 1998). Thus, glucosinolate production represents a non-DNA synapomorphy for the Brassicales clade (as is the indeterminate inflorescence) and is a feature that evolved only twice in the angiosperms (see below). Morphology-based classifications placed members of Brassicales in many different orders (e.g., Cronquist 1981, 1988; Takhtajan 1987, 1997; Thorne 1992a,b, 2001). However, Dahlgren (1975, 1977) challenged traditional approaches to the classification of these taxa and placed in his Capparales nearly all plant families producing glucosinolates. He later backed off from this "radical" classification (Dahlgren 1980, 1983) and, like other authors,

Figure 10.26. Brassicales-Malvales (Bataceae, Bixaceae, Brassicaceae, Capparidaceae, Caricaceae, Cistaceae, Cleomaceae, Dipterocarpaceae, Malvaceae, Moringaceae, Thymelaeaceae, and Tropaeolaceae): a. *Batis maritima* L. (Bataceae), succulent habit. b. *Berteroa incana* DC. (Brassicaceae), fruits (siliques and seeds; note replum and false septum) and flowers. c. *Brassica oleracea* L. (Brassicaceae), cruciform flowers and various cultivars. d. *Capparis flexuosa* L. (Capparaceae), 4-petaled flower; note elongate gynophore. e. and f. *Carica papaya* L. (Caricaceae), carpellate flower (e) and berries with mature seeds; note parietal placentation. g. *Moringa oleifera* Lam. (Moringaceae), compound leaves, elongate capsules and winged seeds. h. *Tropaeolum majus* L. (Tropaeolaceae), flower. i. *Cleome domingensis* Iltis (Cleomaceae), 4-petaled flowers; note elongate gynophores. j. *Phaleria octandra* (L.) Baill. (Thymelaeaceae), flowers; note showy sepals and hypanthia. k. *Shorea faguetioides* F.Heim (Dipterocarpaceae), leaf and fruit showing expanded, winged calyx. l. *Bixa orellana* L. (Bixaceae), flower. m. *Crocanthemum carolinianum* Spach (Cistaceae), flower. n. *Lavatera assurgentiflora* Kellogg (Malvaceae), flower with monadelphous androecium. o. *Abutilon theophrasti* Medik. (Malvaceae), schizocarpic fruit.

emphasized the striking morphological differences among these families.

The concept of a broadly defined "glucosinolate" clade (sensu Dahlgren 1975) reemerged after Rodman's (1991a,b) phenetic and cladistic analyses of morphological and phytochemical characters. A similar glucosinolate-producing clade (now termed Brassicales rather than Capparales) was recovered in early DNA analyses (Gadek et al. 1992; Chase et al. 1993; Rodman et al. 1993, 1994; Soltis et al. 1997a,b). Strong support ultimately emerged for this clade and for relationships within it (Rodman et al. 1998; D. Soltis et al. 2000). *Drypetes* (Putranjivaceae), another glucosinolate-producing taxon, is not part of Brassicales, but instead is part of the well-supported Malpighiales.

The framework detected within Brassicales by Rodman et al. (1998) has been largely supported with several modifications and the inclusion of additional taxa (e.g., *Emblingia*, Ronse De Craene and Haston 2006): (1) Tropaeolaceae + Akaniaceae (including Bretschneideraceae) are sister to the remainder of the family; (2) Moringaceae + Caricaceae are then sister to the rest of the order; (3) following a grade of Setchellanthaceae and then Limnanthaceae, a clade of (Koeberliniaceae + (Batidaceae + Salvadoraceae)), followed by Emblingiaceae emerged as subsequent sisters to the remainder of the order. Relationships among the remaining taxa are problematic. Resedaceae + Gyrostemonaceae form a clade, as do Capparidaceae + (Cleomaceae + Brassicaceae); the placement of Pentadiplandraceae and Tovariaceae is uncertain (Fig. 10.24). Adding to the complexity is the recent suggestion of a new family, Borthwickiaceae, related to Resedaceae (see below).

The taxonomic history of the single species of *Emblingia* (*E. calceoliflora*), endemic to Australia, has been complex. It was originally placed in Capparaceae based on the presence of an androgynophore, a structure also seen in Brassicaceae, Capparacaeae, Cleomaceae, and Resedaceae. However, pollen morphology and floral features suggested a relationship with either Polygalaceae or Sapindaceae, respectively. Leaf and stem anatomy supported a relationship to the asterid family Goodeniaceae (Erdtman et al. 1969; reviewed in Chandler and Bayer 2000). Based on *rbcL* sequences Chandler and Bayer (2000) suggested that *Emblingia* (Emblingiaceae) belonged in Brassicales near Resedaceae. The *rbcL* sequence obtained for *Emblingia* by Savolainen et al. (2000b), using different material, resulted in a placement within Gentianales (asterid clade)—this sequence is problematic. Ronse De Craene and Haston (2006), using DNA + morphology, also placed *Emblingia* within Brassicales. Glucosinolates have now been confirmed in *Emblingia* (Mithen et al. 2010); the genus has other Brassicales features including curved embryos, scanty endosperm, and some floral whorls that are tetramerous, whereas others are pentamerous (it has two petals and four stamens, but five sepals) (Takhtajan 1997; Chandler and Bayer 2000).

H. Su et al. (2012) found that *Borthwickia* as well as *Stixis*, *Tirania*, and *Forchhammeria*, genera typically placed in Cleomeaceae, did not appear in that family in their analyses. *Borthwickia* appeared as sister to Resedaceae, *Stixis*, *Tirania*, and *Forchhammeria*. H. Su et al. (2012) placed *Borthwickia* in a new family (Borthwickiaceae), but one could also argue that it, as well as *Stixis*, *Tirania*, and *Forchhammeria*, could all be included in an expanded Resedaceae, as done in APG IV (2016; see Chapter 12).

Olson (2002a,b) analyzed relationships among the 13 species of *Moringa*, the only genus in Moringaceae. Brassicaceae were expanded to include Capparaceae in the initial APG (1998) and APG II (2003) classifications because Capparaceae, long considered the closest relative of Brassicaceae, appeared paraphyletic relative to Brassicaceae (Judd et al. 1994). A broadly defined Brassicaceae including Capparaceae are also indicated by non-DNA characters such as a long gynophore, vacuolar or utricular cisternae of the endoplasmic reticulum, a similar method of glucosinolate production involving synthesis from methionine via long chain-extensions, and the presence of sinapine (Rodman 1991a,b; Judd et al. 1994, 1999). However, J. Hall et al. (2002) found evidence for three well-supported, primary clades within Brassicaceae s.l.: (1) Capparaceae subfamily Capparoideae; (2) Capparaceae subfamily Cleomoideae; and (3) Brassicaceae s.s. These authors suggested that three families be recognized corresponding to each of these clades: Capparaceae, Cleomaceae, and Brassicaceae. Capparaceae typically have fleshy fruits whereas the fruits of Cleomaceae and Brassicaceae are typically dry. Cleomaceae possess bracteate inflorescences as a synapomorphy; Brassicaceae have a false-septum and six stamens (4 long and 2 short) (J. Hall et al. 2002) (six stamens also are present in many Cleomaceae). Capparaceae are sister to Cleomaceae + Brassicaceae (see Fay and Christenhusz 2010; H. Su et al. 2012).

It now appears that Cleomaceae consist of just *Cleome*, which some authors have divided into two to several genera (Hall 2008; Inda et al. 2008; Riser et al. 2013). Enormous progress has been made in understanding relationships in the large family Brassicaceae (e.g., Koch et al. 2001, 2003; Al-Shehbaz et al. 2006; Bailey et al. 2006; Beilstein et al. 2006, 2008; Franzke et al. 2009; Zhao et al. 2010; Al-Shehbaz 2012).

Numerous complete nuclear genome sequences are available for Brassicaceae (e.g., *Arabidopsis thaliana*, *A. lyrata*, *Brassica rapa*, *Capsella rubella*, *Eutrema oparvulum*, *Sisymbrium irio*) given the economic importance of the family, as well as the importance of some members as genetic models (e.g., species of *Arabidopsis* and *Bras-*

sica; e.g., Wang et al. 2006; Gaeta et al. 2009; Ng et al. 2012). The importance of understanding the phylogenetic relationships of these organisms has been repeatedly emphasized (see Chapter 3; see also D. Soltis and Soltis 2000, 2003; Franzke et al. 2009). Recent studies have attempted to date the origin of *Arabidopsis thaliana* (Beilstein et al. 2010) and provide other age estimates in Brassicaceae (Franzke et al. 2009). Phylogenetic inferences for Brassicaceae and relatives allow the *Arabidopsis* model of floral development and genomics to be extended to encompass more diversity in Brassicales as well as other angiosperms (e.g., Mitchell-Olds and Clauss 2002; Schranz et al. 2007; see chapters in D. Soltis et al. 2006). Within Brassicaceae, phylogenetic studies have provided a general framework of generic relationships and elucidated the closest relatives of *Arabidopsis* (e.g., Koch et al. 1999, 2001, 2003; Koch 2003; O'Kane and Al-Shehbaz 2003; Franzke et al. 2009).

The fossil record of Brassicales is poor, but winged fruits of *Thlaspi* from the Oligocene of Montana have been useful for inferring the age of *Arabidopsis* (Beilstein et al. 2010). The well-preserved bicarpellate flower, *Dressiantha*, from the Late Cretaceous (Turonian) of New Jersey, USA provides a suite of features compatible with a placement as sister group to all Brassicales except Gyrostemonaceae (Gandolfo et al. 1998a), but this placement has been questioned in view of a subsequent analysis placing it within Sapindales (Ronse De Craene and Haston 2006).

Table 10.2. Characters and character states in Brassicales[a]

Family	Floral Merosity	Embryo
Tropaeolaceae	5	Straight
Akaniaceae		
Bretschneidera	5	Curved
Akania	5	Straight
Caricaceae	5	Straight
Moringaceae	5	Straight
Limnanthaceae		
Limnanthes	5	Straight
Floerkea	3	Straight
Setchellanthaceae	4	Curved
Koeberliniaceae	4(5)	Curved
Batidaceae	4	Slightly curved
Salvadoraceae	4(5)	Straight
Pentadiplandraceae	5	Curved
Tovariaceae	(8)4	Curved
Emblingiaceae	5[b]	Curved
Gyrostemonaceae	Uncertain	Curved
Resedaceae	4-8[c]	Curved
Brassicaceae		
Cleome	4	Curved
Capparis	4	Curved
Brassica	4	Curved
Arabidopsis	4	Curved

[a]Character states generally follow Cronquist (1981) and Takhtajan (1997).
[b]Reports for the number of sepals and petals vary. Some reports indicate that there are five sepals and two petals (reviewed in Chandler and Bayer 2000).
[c]Variously considered 4- or 6-merous.

CHARACTER EVOLUTION IN BRASSICALES

Core Brassicales (e.g., Koeberliniaceae, Batidaceae, Salvadoraceae, Pentadiplandraceae, Tovariaceae, Emblingiaceae, Gyrostemonaceae, Resedaceae, and Brassicaceae) share several features typically associated with Brassicaceae, such as tetramerous flowers (Table 10.2), seeds with curved or folded embryos and campylotropous ovules, endosperm lacking or nearly absent, vessels with vestured pits, and protein-rich, vascular or utricular cisternae of the endoplasmic reticulum (see Rodman 1991a,b; Judd et al. 1999, 2008). There are important exceptions, however. Pentadiplandraceae and Emblingiaceae have perianth parts in five (the latter having two petals and five sepals); Resedaceae and Tovariaceae have six- to eight-merous flowers; Salvadoraceae have a straight rather than curved embryo; both Gyrostemonaceae and Pentadiplandraceae also have non-vestured pits. Other families illustrate the difficulty of some character codings; in Batidaceae, the embryo is reported to be "slightly curved," not fitting neatly into the categories of "straight" or "curved."

To examine character evolution in Brassicales, we updated our dataset (Soltis et al. 2005b) and conducted new character-state reconstructions using Mesquite (Maddison and Maddison 2011) and the characters and character states of Rodman (1991b), with modifications (sensu Cronquist 1981; Takhtajan 1997) for some characters, such as sepal and petal merism and embryo curvature used in Soltis et al. (2005b). The topology we used is based on Soltis et al. (2011) and Stevens (2014).

We found that some features associated with Brassicaceae, such as embryo morphology, actually evolved in the common ancestor of Brassicaceae and its close relatives. A curved embryo apparently evolved twice in Brassicales (Fig. 10.27A), once in *Bretschneidera* (Akaniaceae) and again in the common ancestor of Setchellanthaceae, Koeberliniaceae, Batidaceae, Salvadoraceae, Pentadiplandraceae, Tovariaceae, Emblingiaceae, Gyrostemonaceae, Resedaceae, Capparidaceae, Cleomeaceae, and Brassicaceae; this feature was lost in Salvadoraceae.

Perianth evolution in Brassicales is similar to the pattern of embryo diversification. A pentamerous perianth appears

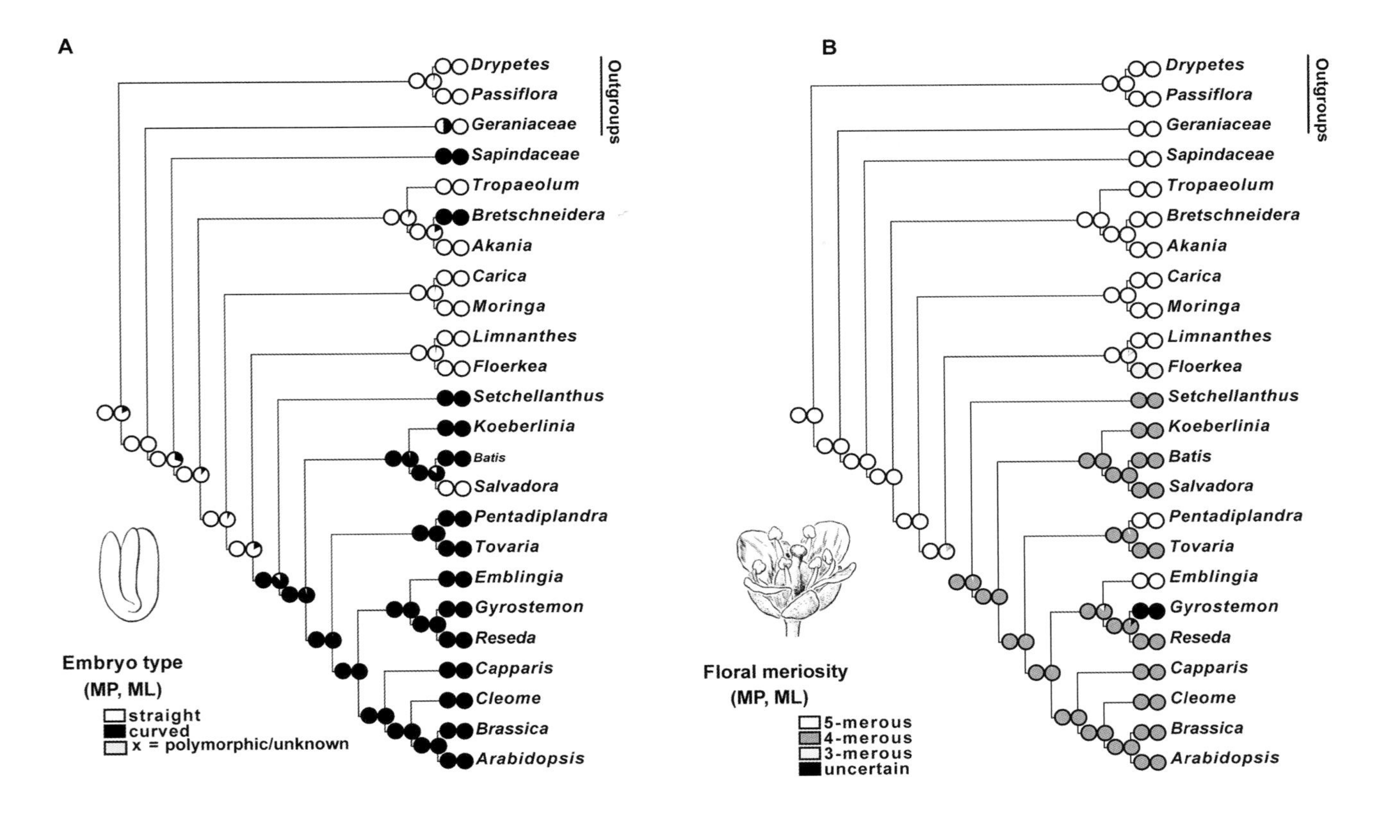

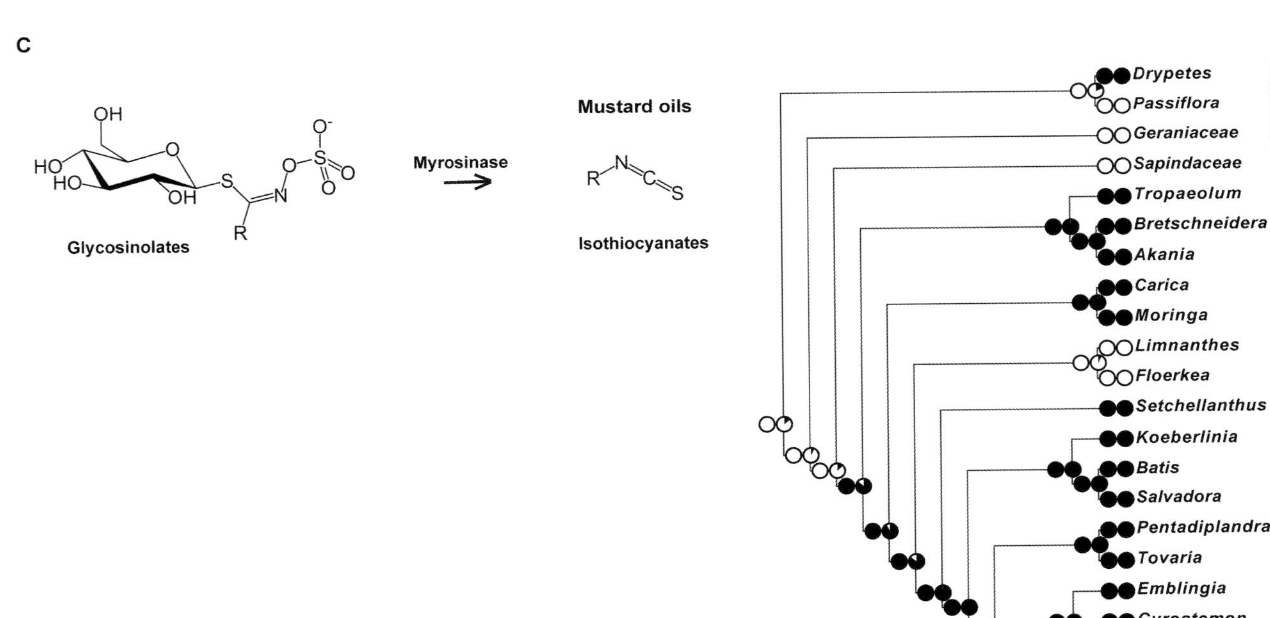

Figure 10.27. Character evolution in Brassicales with both maximum parsimony, MP, and maximum likelihood, ML. The topology was generated for this book. (A) Embryo type (curved versus straight); a curved embryo from *Capsella bursa-pastoris* (Brassicaceae) is illustrated (from Wood 1974). (B) Floral merosity; a flower of *Capsella bursa-pastoris* (Brassicaceae) is illustrated (from Wood 1974). (C) Glucosinolate production.

ancestral in Brassicales with a tetramerous perianth evolving in the common ancestor of Setchellanthaceae, Koeberliniaceae, Batidaceae, Salvadoraceae, and other taxa above these nodes (Fig. 10.27B). A reversal to a pentamerous perianth occurred in Pentadiplandraceae and Emblingiaceae. However, the coding of perianth merism is not always straightforward. *Emblingia* has five sepals and two petals. Similarly, although Resedaceae were scored here as tetramerous, the number of sepals and petals each range from four to eight; the family has been considered hexamerous by some (e.g., Cronquist 1981). This part of the topology is further complicated by the recent placements of *Borthwickia*, *Stixis*, *Tirania*, and *Forchhammeria* (H. Su et al. 2012), which we did not consider.

Mustard-oil glucosides (Ettlinger and Kjaer 1968) are oxime-derived sulfur-containing compounds for which breakdown products include the familiar pungent principles of mustard, radish, and capers. The compounds are usually accompanied in the plant by a hydrolytic enzyme, myrosinase (thioglucoside glucohydrolase, E.C. 3.2.3.1), which may be compartmentalized in special myrosin cells (reviewed in Rodman et al. 1998). This "mustard oil bomb" may deter herbivores and pathogens (Lüthy and Matile 1984). These compounds are costly to produce, perhaps increasing photosynthesis requirements by 15% (Bekaert et al. 2012).

The phytochemical system of glucosinolates with myrosinase enzyme, the latter in special myrosin cells, evolved twice, once in the ancestor of Brassicales and again, independently, in *Drypetes* (Putranjivaceae; Malpighiales; Fig. 10.27c). *Drypetes* is a large pantropical genus (~150 species); the few species screened for mustard-oil glucosides have yielded positive results (Ettlinger and Kjaer 1968). Additional species of *Drypetes*, as well as the related genus *Putranjiva*, should be analyzed for glucosinolates. Ultrastructural studies of *Drypetes* revealed protein-accumulating phloem cells similar to developing myrosin cells, although tests for ascorbate-activated myrosinase enzymes proved negative (reviewed in Rodman et al. 1998). However, no studies have been reported on the enzymology of mustard-oil glucoside biosynthesis in *Drypetes*; in contrast, several members of the Brassicales have been analyzed in detail (Du et al. 1995). The available data indicate that glucosinolate biosynthesis and myrosin cell formation in *Drypetes* are remarkably similar to those in Brassicales.

Within Brassicales, 14 families are known to produce glucosinolates; only *Koeberlinia* does not produce these compounds (Mithen et al. 2010). The deeply embedded position of *Koeberlinia* within Brassicales indicates that it has lost critical gene functions or otherwise repressed the capacity to synthesize glucosinolates (Fig. 10.27C). Significantly, *Koeberlinia* possesses myrosin cells (Gibson 1979), which is consistent with this hypothesis of glucosinolate loss or repression (Rodman et al. 1993, 1998). Also, *Koeberlinia* is thorny, with small, deciduous leaves, so physical means of protection may have been associated with the loss of glucosinolate production.

Two evolutionary origins of glucosinolate production poses an intriguing question. What biochemical pathway was modified for glucosinolate production and is it the same pathway in both *Drypetes* and Brassicales? Kjaer (1973) and Harborne (1982) stressed the similarities in biosynthesis between mustard-oil glucosides and cyanogenic glycosides. The latter are cyanide-releasing compounds widespread in the angiosperms. Kjaer (1973) suggested that glucosinolate biosynthesis evolved via the recruitment and modification of cyanogen biosynthesis. Both glucosinolates and cyanogenic glucosides are groups of natural plant products derived from amino acids and have oximes as intermediates. It is noteworthy that Brassicales are in a clade with both Malvales and Sapindales, lineages with cyanogenic taxa. Similarly, *Drypetes* is part of Malpighiales, which also contain cyanogenic taxa (e.g., *Euphorbia*, *Passiflora*, and Violaceae).

Hansen et al. (2001a,b) showed that cytochromes P450 (belonging to the CYP79 family) are involved in oxime formation in the biosynthetic pathway of both cyanogenic glucosides and glucosinolates. This is consistent with the hypothesis that glucosinolate production evolved from the cyanogenic glucoside pathway and indicates that the oxime-metabolizing enzyme is a possible branch point between the cyanogenic glucoside and glucosinolate pathway (Hansen et al. 2001a,b). Progress has been made in elucidating the biosynthetic pathway of glucosinolates (reviewed in Mithen et al. 2010; see Halkier and Gershenzon 2006). Mithen et al. (2010, p. 2074) reported on the possible evolutionary origin of glucosinolate biosynthesis in Brassicales from cyanogenic glycoside biosynthesis, noting that glucosinolate biosynthesis "is associated with diversion of the aldoxime away from cyanogenic glycosides biosynthesis through the activity of a mutated CYP83-like enzyme. This is followed by subsequent detoxification of the reactive product through sulphation and glycosylation to produce glucosinolates."

MALVALES

Malvales consist of (Fig. 10.24) Bixaceae (including Cochlospermaceae and Diegodendraceae), Cistaceae, the parasitic Cytinaceae, Dipterocarpaceae, Malvaceae (including Bombacaceae, Tiliaceae, and Sterculiaceae), Muntingiaceae, Neuradaceae, Sarcolaenaceae, Sphaerosepalaceae, and Thymelaeaceae (several families are illustrated in Fig. 10.26). Thymelaeaceae were expanded (APG II 2003) to include the enigmatic *Tepuianthus*, a genus of five species from Venezuela and Colombia that had been placed in Celastraceae, but transferred to its own family (Cronquist 1981). Takhtajan (1997) considered Tepuianthaceae closely related to Simaroubaceae. However, DNA studies revealed that *Tepuianthus* was the well-supported sister to other Thymelaeaceae (Savolainen et al. 2000b; Wurdack and Horn 2001). Several phylogenetic studies have focused on Malvales, and the topology shown here is in agreement with the results of Alverson et al. (1998), Bayer et al. (1998, 1999), Dayanandan et al. (1999), and van der Bank et al. (2002).

Cronquist (1981) restricted Malvales to Bombacaceae, Tiliaceae, Malvaceae, Sterculiaceae, and Elaeocarpaceae. However, others included additional families such as Bixaceae, Cistaceae, Cochlospermaceae, Diegodendraceae, Dipterocarpaceae, Dirachmaceae, Huaceae, Peridiscaceae, Plagiopteridaceae, Sarcolaenaceae, Sphaerosepalaceae, and Thymelaeaceae (Dahlgren 1983; Thorne 1992a,b; Huber 1993; Takhtajan 1997).

A suite of morphological and other nonmolecular characters unites the Malvales clade, although in most cases, only one or a few representatives of a family have been examined for these characters, so it is not known if exceptions to the occurrence of these features exist. With this caveat, potential synapomorphies for Malvales include the presence of cyclopropenoid fatty acids in the seeds, complex chalazal anatomy, mucilage (slime) canals, vestured pits, dilated phloem rays, and simple perforations in the secondary xylem (Metcalfe and Chalk 1950; Nandi et al. 1998). Mucilage cells or mucilage cavities (Metcalfe and Chalk 1950; Cronquist 1981) appears to be another synapomorphy. All taxa for which the character is known (all but Sarcolaenaceae and Sphaerosepalaceae) are also characterized by centrifugal polyandry, or rarely, as in Thymelaeaceae, lateral polyandry (e.g., Cronquist 1981, 1988; Ronse De Craene and Smets 1992; Nandi 1998b; Nandi et al. 1998; von Balthazar et al. 2004). Malvales have the exotegmen differentiated as a palisade layer (Corner 1976; Nandi 1998a); this seed feature occurs only rarely outside of Malvales (e.g., in Trochodendraceae, Huaceae, and some Euphorbiaceae; Nandi et al. 1998). Most Malvales (except Thymelaeaceae) possess a stratified phloem (Metcalfe and Chalk 1950), another feature that is uncommon outside of the order (Nandi et al. 1998). Other characters common in Malvales, but also found outside the clade, include stellate hairs, peltate scales, strong phloem fibers, valvate sepals, and an epicalyx (see Alverson et al. 1998; Nandi et al. 1998).

Of those families previously placed in Malvales but now excluded by phylogenetic evidence, Elaeocarpaceae (now placed in Oxalidales) are the most notable because they were frequently associated with families now considered to be the single family Malvaceae (see Cronquist 1981). However, Elaeocarpaceae lack most of the putative nonmolecular synapomorphies of Malvales as now circumscribed. For example, Elaeocarpaceae lack fatty acids, stellate hairs, peltate trichomes, and mucilage cavities; they have unstratified phloem, opposite lateral pitting of the vessels, and a fibrous exotegmen (Takhtajan 1997). Soltis et al. (2011) provided strong support for some subclades but some families of Malvales were not sampled (Muntingiaceae, Cytinaceae, Sphaerosepalaceae). Stevens (2013) represents much of the "backbone" as a polytomy, which may be the best, most conservative assessment.

Neuradaceae typically emerge as sister to the remainder of the clade followed by Thymelaeaceae (Alverson et al. 1998; D. Soltis et al. 2000, 2011; H. Wang et al. 2009). These placements had maximal support in H. Wang et al. (2009), but overall sampling of Malvales was modest. Neuradaceae were often included in or considered a close relative of Rosaceae in previous treatments (Cronquist 1981). However, the family shares two potential synapomorphies with the expanded Malvales, an exotegmic seed coat with a palisade layer and cyclopropenoid fatty acids (Huber 1993). Neuradaceae also possess valvate sepals, lysigenous mucilage canals, stellate hairs, and an epicalyx.

Following Neuradaceae and Thymelaeaceae, relationships become problematic in Malvales—it remains one of the more poorly understood rosid subclades. Several early analyses supported a clade of Bixaceae, Diegodendraceae, and Cochlospermaceae (Alverson et al. 1998; Bayer et al. 1998; Fay et al. 1998a); the last two families have now been included in Bixaceae (see APG II 2003, APG III 2009; APG IV 2016).

Muntingiaceae may be sister to the parasitic Cytinaceae (Nickrent 2007). Cytinaceae comprise two genera, *Cytinus* and *Bdallophytum*, that were formerly placed in the family Rafflesiaceae (now Malpighiales), but molecular data indicated a placement in Malvales. Neotropical Muntingiaceae are poorly known. *Muntingia* is noteworthy in that the genus had historically been placed in Elaeocarpaceae, Tiliaceae, or Flacourtiaceae (see Cronquist 1981; Takhtajan 1997), but molecular results clearly indicate that it, and the related genus *Dicraspidia*, form a distinct lineage of Malvales (Bayer et al. 1998; Savolainen et al. 2000b). Bayer et al. (1998) recognized these two genera as a new family, Muntingiaceae (stating that it may also include *Neotessmannia*).

Although some early studies suggested that Dipterocarpaceae are not monophyletic (Dayanandan et al. 1999; Savolainen et al. 2000b), Cistaceae and then Sarcolaenaceae are subsequent (and well-supported) sisters to a monophyletic Dipterocarpaceae (Soltis et al. 2011), although inclu-

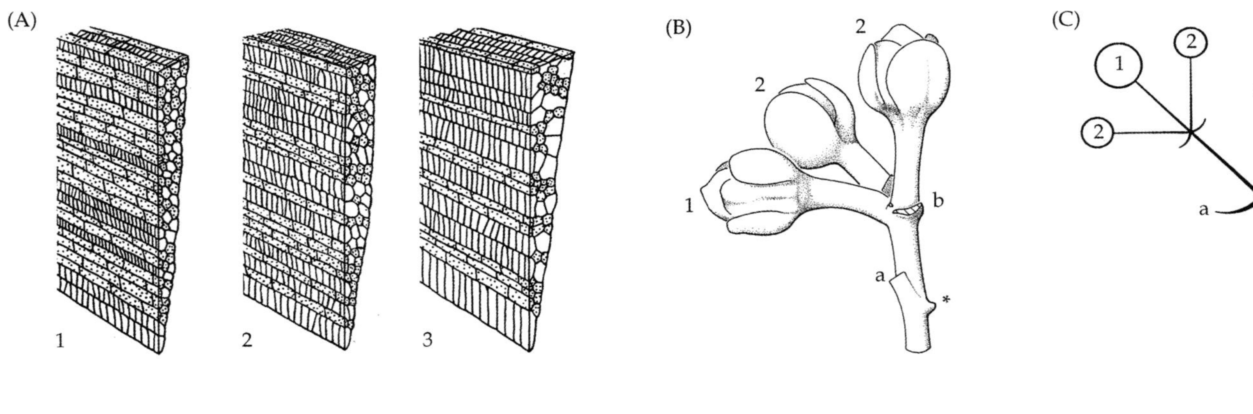

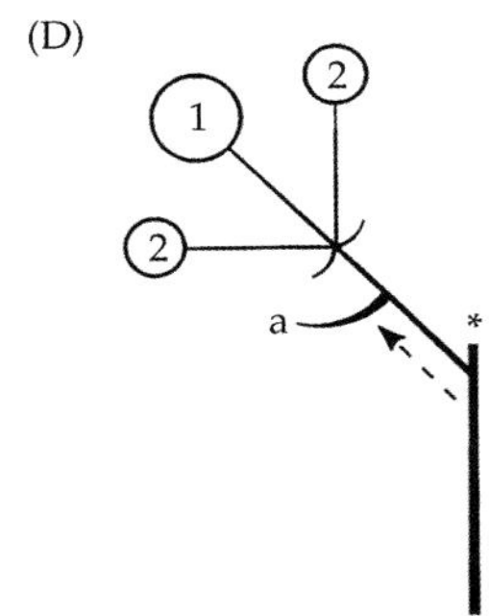

Figure 10.28. Diagnostic morphological features of Malvaceae. (A) The distinctive upright "tile" cells in wood rays. Cutaway diagrams of multiseriate rays showing different types of tile cells. Procumbent cells stippled, tile cells not stippled. 1. Tile cells of the *Durio* type, approximately the same height as the procumbent cells. 2. Tile cells of the intermediate type, slightly higher than the procumbent cells. 3. Tile cells of the *Pterospermum* type, two to several times higher than the procumbent cells (from Manchester and Miller 1978). B–D. Bicolor unit (from Bayer 1999). (B) Inflorescence of *Monotes kerstingii* axillary to a displaced subtending bract (a = scar). Asterisk designates what is interpreted as a rudiment of the apical bud; b = the scar of the subtending bract of a second-order lateral flower. C, D. Possible derivation of the bicolor unit from a lateral cymose inflorescence. (C) The lateral cymose inflorescence is subtended by a bract (a). (D) The bract (a) is shifted to a more distal position (arrow). With the suppression of the apex of the main axis (indicated by asterisk), the bicolor unit appears to occupy a terminal position. Reproduced from Soltis et al. (2005).

sion of *Monotes* and *Pakaraimaea* (representatives of the other two subfamilies of Dipterocarpaceae) may change this result.

A broadly defined Malvaceae received strong support in molecular analyses (Alverson et al. 1998, 1999; Bayer et al. 1999; Savolainen et al. 2000a,b; D. Soltis et al. 2000, 2011). In addition to the traditionally recognized Malvaceae, Malvaceae now include former members of Bombacaceae, Sterculiaceae, and Tiliaceae (Judd and Manchester 1997; Bayer et al. 1999). A close relationship among these four families has been recognized since Linnaeus, although relationships among them have long been problematic. Analyses of DNA sequences and non-DNA characters demonstrated the polyphyly of both Sterculiaceae and Tiliaceae, as well as the paraphyly of Bombacaceae, and indicated that these four families should be merged into one family: Malvaceae (Judd and Manchester 1997; Bayer et al. 1999; see also Judd et al. 1999, 2002, 2008). *Theobroma* of Sterculiaceae received strong support as sister to *Grewia* of Tiliaceae, and *Durio* and its close relatives in Bombacaceae were found to be more closely related to some members of Sterculiaceae than to other Bombacaceae. Several genera treated as Bombacaceae and other genera treated as Sterculiaceae emerged as sister to traditional Malvaceae. These interdigitating relationships among genera assigned to different families in traditional classifications (all recognizing four families) indicated that a broadly defined Malvaceae is required (Judd and Manchester 1997; Alverson et al. 1999; Bayer et al. 1999).

Likely synapomorphies for a broad Malvaceae include distinctive nectaries composed of tightly packed, multicellular hairs, normally found on the adaxial surface of the sepals (Judd and Manchester 1997; Vogel 2000) and the distinctive upright "tile" cells in wood rays (Manchester and Miller 1978; Fig. 10.28). In addition, the inflorescences of all members of a broadly defined Malvaceae are composed of special modules called "bicolor units" (Fig. 10.28B–D; Bayer and Kubitzki 1996). In contrast, there are no clear synapomorphies for Malvaceae s.s.

Malvaceae are readily identified in the late Cretaceous and younger fossil record. Distinctive bombacoid pollen and tilioid/sterculioid wood can be traced to the Maastrichtian of the late Cretaceous (Wolfe 1975; Wheeler et al. 1994). Malvaceous flowers, fruits, seeds and pollen are common from Cenozoic deposits (e.g., the extinct flower known as *Florissantia* from the Eocene of western North America and eastern Asia (Fig. 10.21j–l; Manchester 1992, 1999). *Craigia*, a genus closely related to *Tilia*, that survives only in southern China today, is known from Eocene and younger deposits of North America, and Asia, and from Oligocene and younger deposits in Europe (Kvaèek et al. 2005). *Tilia* itself is recognizable on the basis of distinctive

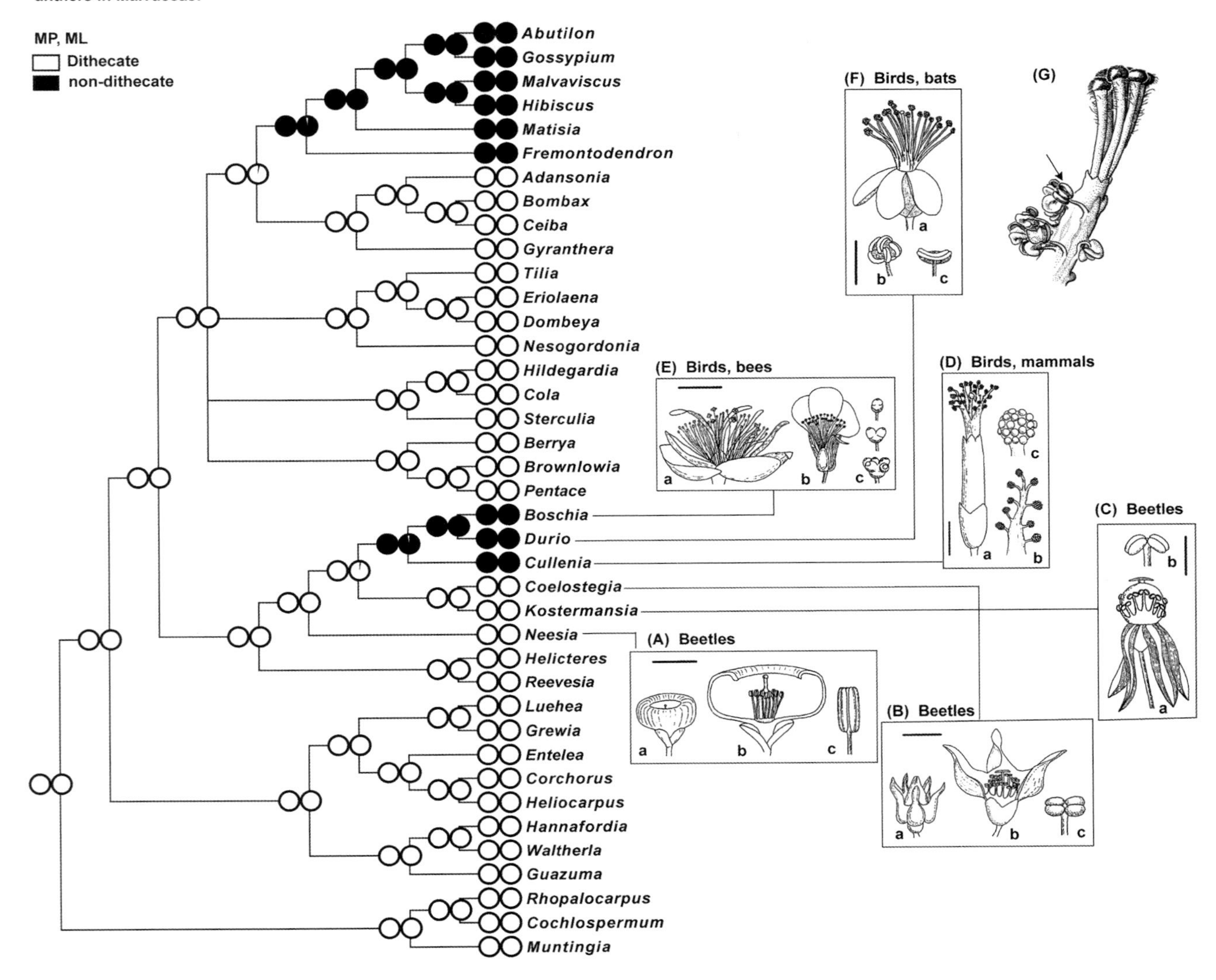

Figure 10.29. Evolution of non-dithecate (monothecate and non-thecal) anthers in Malvaceae. The topology shown is modified from the strict consensus of shortest trees of Alverson et al. (1998). To make the topology easier to visualize, some genera have been removed without altering the general pattern of relationships. Additional genera were added to tribe Durioneae, which contains *Neesia*, *Coelostegia*, *Kostermansia*, *Cullenia*, *Boschia*, and *Durio*; the relationships depicted among members of Durioneae (genera are indicated in bold) follow Nyffeler and Baum (2000). The phylogeny indicates that non-dithecate anthers evolved twice: once in a clade of several genera of Durioneae (see E, F, D) and again in a clade that largely corresponds to Malvaceae in the traditional or narrow sense (G). Floral structure and pollinators are indicated for members of Durioneae. Within this tribe, the molecular phylogeny suggests that beetle pollination was ancestral in the tribe, with subsequent evolution of bat and bird pollination. Furthermore, the transition to bat and bird pollination was accompanied by extensive changes in the androecium. The early-branching genera *Neesia, Coelostegia*, and *Kostermansia* all have dithecate, tetrasporangiate anthers. However, in *Cullenia*, *Boschia*, and *Durio*, the anthers are highly modified (D, E, F). Flowers and stamens of Durioneae (A–F; modified from Nyffeler and Baum 2000). (A) *Neesia glabra*: (a) flower at anthesis; (b) longitudinal section of the flower at anthesis; (c) dithecate stamen. (B) *Coelostegia griffithii*: (a) flower before anthesis; (b) flower at anthesis with part of calyx removed; (c) dithecate stamen. (C) *Kostermansia malayana*: (a) flower at anthesis; (b) dithecate stamen. (D) *Cullenia exarillata*: (a) flower at anthesis; (b) terminal segment of the floral tube with nine clusters of pollen chambers; (c) detail of cluster of locules. (E) *Boschia griffithii*: (a) flower at anthesis; (c) anthers, showing three different forms of poricidal locules; *Boschia grandiflora*: (b) flower at anthesis. (F) *Durio zibethinus*: (a) flower at anthesis; (b) polylocular anther. (G) Stamens characteristic of the monothecate stamen clade, which largely corresponds to narrowly defined Malvaceae. *Kosteletzkya virginica*: arrow designates monothecate anther (from Wood 1974). Reproduced from Soltis et al. (2005).

fruiting bracts as well as leaves from late Eocene and younger rocks (Manchester 1994a).

CHARACTER EVOLUTION IN MALVALES

Malvaceae have experienced many gains and losses of diverse characters, including staminal tubes, palmately compound leaves, spiny pollen, winged seeds or fruits, and non-dithecate anthers. Considering the last feature, it has long been assumed that monothecate anthers (sometimes non-thecate or "half-anthers") were a derived character uniting Malvaceae and Bombacaceae (Cronquist 1981, 1988; Judd et al. 1994; Judd and Manchester 1997). However, DNA-based trees indicated instead that monothecate anthers have been derived at least twice independently (Fig. 10.29): once in core Malvaceae ("Malvoideae" of Alverson et al. 1999) and again in members of tribe Durioneae of former Bombacaceae ("Helicteroideae" of Alverson et al. 1999). Furthermore, in Helicteroideae, modifications in the androecium may be related to pollinator syndromes. Helicteroideae contain *Neesia*, *Coelostegia*, *Kostermansia*, *Cullenia*, *Boschia*, and *Durio* (the durians). Molecular results indicate that beetle pollination is ancestral in the tribe, with subsequent evolution of bat and bird pollination. Furthermore, the transition to bat and bird pollination was accompanied by extensive changes in the androecium. The genera attached to the basalmost nodes, *Neesia*, *Coelostegia*, and *Kostermansia*, all have dithecate, tetrasporangiate anthers, the typical condition in angiosperms, and their flowers are beetle-pollinated. However, in *Cullenia*, *Boschia*, and *Durio*, the anthers are non-dithecate, and the flowers are pollinated primarily by birds and bats (Alverson et al. 1999; Nyffeler and Baum 2000).

A molecular study provided similar insights regarding floral evolution and pollinators in the famous baobabs (*Adansonia*) (Baum et al. 1998). In *Adansonia*, character-state mapping using a molecular tree indicated that hawkmoth pollination is ancestral in the genus and that there were two independent switches to pollination by fruit bats (Baum et al. 1998). Flowers in the two clades with bat pollination differ greatly in morphology.

PARASITISM IN ROSIDS

Recent phylogenetic analyses have now revealed the placement of many parasitic plants that initially were problematic. The reasons for these difficulties are several, but generally involve the fact that these plants have degenerate plastid genomes. As a result, plastid genes, which have been so widely used to infer angiosperm phylogeny, are absent or have accelerated evolutionary rates (e.g., Nickrent et al. 1998; www.parasiticplants.siu.edu/). However, these parasitic taxa can be placed using nuclear genes, or mitochondrial DNA sequences. Sequence data have indicated that several groups of parasitic plants are nested within the asterids (e.g., Mitrastemonaceae, Orobanchaceae, *Cuscuta*; see Chapter 11). As reviewed here, several other lineages of parasitic plants that have been particularly difficult to place using morphological data now appear to be nested within the rosids. Barkman et al. (2004) found evidence for the placement of *Rafflesia* and *Rhizanthes* (Rafflesiaceae) with Malpighiales, most likely close to or derived from within Euphorbiaceae (Davis et al. 2007). These genera of Rafflesiaceae lack leaves, stems, and roots and rely entirely on the host plants for nutrients (see Chapter 13); the genus *Rafflesia* also contains the world's largest flowers (*R. arnoldii*; see Figs. 10.20d; 10.23).

Barkman et al. (2004) showed that *Mitrastema*, another genus sometimes placed in Rafflesiaceae, is distantly related to *Rafflesia* and *Rhizanthes*, exhibiting a close relationship to Ericales in the asterid clade (see Chapter 11). Additional analyses suggested that Cytinaceae are embedded within Malvales (Barkman et al. 2007) as sister to Muntingiaceae (Nickrent 2007). Apodanthaceae appear to be part of Cucurbitales (see Filipowicz and Renner 2010) despite earlier suggestions that it might be a member of Malvales (e.g., Nickrent 2002; Blarer et al. 2004). One caveat of these molecular analyses is that branch lengths to the parasitic taxa are often long; hence, spurious attraction could be occurring. Recent molecular analyses now suggest a placement of Cynomoriaceae within Saxifragales, although the closest relatives of the family are unclear. Many of these parasitic lineages are discussed in more detail in Chapter 13.

11 Superasterids

INTRODUCTION

Superasteridae (i.e., superasterids) have only recently been recognized based on multi-gene phylogenetic and phylogenomic analyses (Moore et al. 2010, 2011; Soltis et al. 2011; Ruhfel et al. 2014; Fig. 11.1) and were also described formally using PhyloCode definitions (Soltis et al. 2011). This clade comprises Santalales, Caryophyllales, Berberidopsidales, *Asteridae* (lamiids and campanulids), and possibly Dilleniaceae (see Chapters 9, 10). Higher-level relationships of Santalales and Caryophyllales were unresolved in previous phylogenetic analyses (see Chapter 3). However, these clades were shown to be well-supported subsequent sisters to Berberidopsidales + asterids in Moore et al. (2008, 2011) using large amounts of plastid gene sequence data. Soltis et al. (2011), using plastid, nuclear, and mitochondrial data, reconstructed Santalales as sister to a (Berberidopsidales + Caryophyllales) + *Asteridae* clade (Fig. 11.1); Ruhfel et al. (2014) found the same topology in their analysis of 360 plastid genomes. As reviewed in Chapters 3 and 9, Dilleniaceae have been variously recovered as sister to *Superasteridae* (Moore et al. 2011; Soltis et al. 2011), as well as sister to *Superrosidae* (Moore et al. 2010; Ruhfel et al. 2014) and sister to a clade composed of *Superrosidae* and *Superasteridae* (Moore et al. 2011). The placement of Dilleniaceae as sister to *Superasteridae* is the best supported (Soltis et al. 2011) but is still unstable (Ruhfel et al. 2014), which led Soltis et al. (2011) to only provisionally place Dilleniaceae within *Superasteridae*. Here Dilleniaceae are treated separately in Chapter 9.

SANTALALES

The composition of Santalales has differed among recent morphology-based classifications (Cronquist 1981; Thorne 1992a,b, 2001; Takhtajan 1997). The order has included Santalaceae, Eremolepidaceae, Loranthaceae, Opiliaceae, Olacaceae, Misodendraceae, and Viscaceae, as well as Balanophoraceae, Medusandraceae, and in some cases Dipentodontaceae (see Cronquist 1981). The monophyly of Santalales was well supported even in initial broad analyses of *rbcL* and 18S rDNA alone (Chase et al. 1993; Nickrent and Soltis 1995; Soltis et al. 1997b; Chase and Albert 1998) with 100% support in analyses of combined datasets (e.g., Nickrent et al. 1998; P. Soltis et al. 1999a; Savolainen et al. 2000a; D. Soltis et al. 2000). Molecular evidence also indicates that families sometimes included in Santalales (such as Medusandraceae and Dipentodontaceae) should be excluded from the order (Nickrent et al. 1998; APG II 2003). The holoparasitic Balanophoraceae were shown to be sister to Santalales (Nickrent et al. 2005) and nested within Santalales by Su and Hu (2012), albeit with limited taxon sampling. Balanophoraceae were subsequently included in Santalales by APG III (2009); however, Nickrent et al. (2010) did not include Balanophoraceae in their recircumscription of the order. Recent research provides new insights, with one study suggesting that Balanophoraceae s.l. comprise two distinct clades, which are sister groups (Su et al. 2015). However, a more recent analysis indicates that Balanophoraceae are monophyletic and perhaps nested within Santalaceae s.l. (Byng unpubl. data).

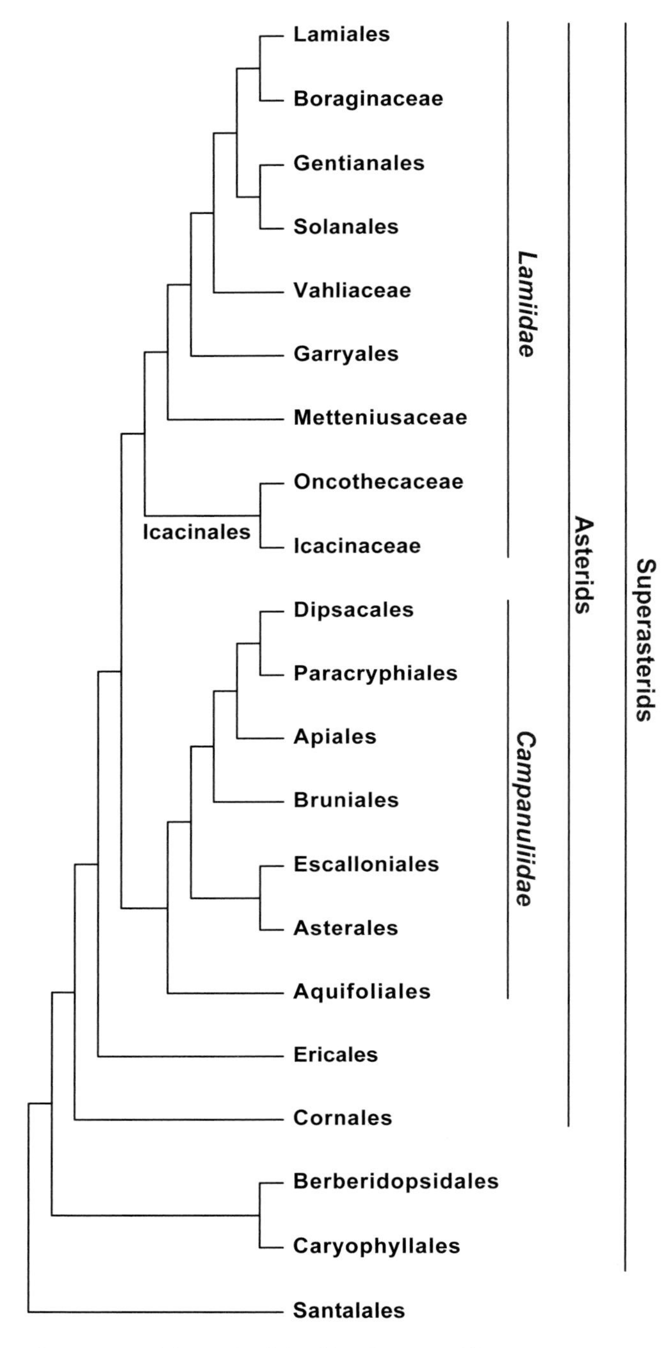

Figure 11.1. Phylogenetic relationships of Superasterids.

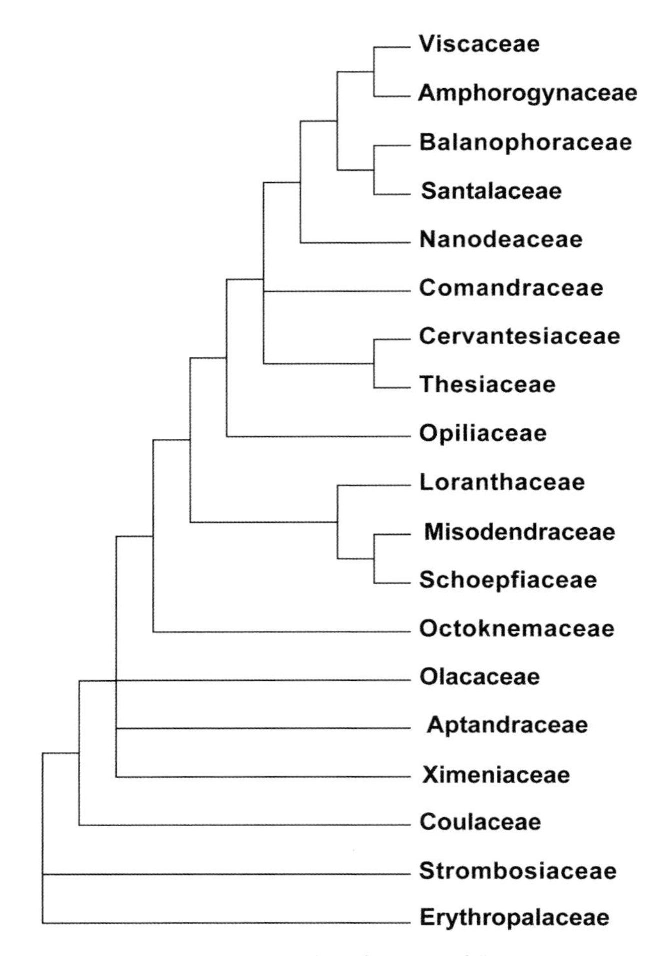

Figure 11.2. Phylogenetic relationships of Santalales, following Nickrent et al. 2010.

A major contribution of early molecular phylogenetic studies was to circumscribe Santalales as a small order that consists of Santalaceae s.l. (including Eremolepidaceae and Viscaceae), Loranthaceae, Opiliaceae, "Olacaceae," and Misodendraceae (Nickrent et al. 1998; Nickrent and Malécot 2001; APG II 2003). The APG II (2003) and APG III (2009) classifications included Viscaceae within Santalaceae to avoid recognition of paraphyletic groups. However, Nickrent et al. (2010) proposed a revised classification of Santalales, wherein several other smaller families were recognized to maintain monophyletic groups, as well as to recognize traditional and easily diagnosable families. Thus, in Nickrent et al. (2010), Santalaceae s.s. is recognized, as well as Viscaceae and Amphorogynaceae, which maintains the monophyly of Santalaceae s.s. This work also separated the traditional (and paraphyletic) Olacaceae into several families (Fig. 11.2). However, most of the backbone of the Nickrent et al. (2010) tree is poorly supported. As a result, APG IV (2016) did not make changes to the families recognized from what is given in APG III (2009). Santalales were inferred to have diverged 108–101 mya (Anderson et al. 2005).

Santalales are fairly easily characterized, possessing several putative non-DNA synapomorphies: plants lacking root hairs, flowers with sepals reduced to a cup-like rim, open aestivation, and petals valvate, the stamens often opposite the petals, ovules pendulous, fruit a drupe (or modified and berry-like), and seeds with small embryo. Most Santalales are hemiparasites (Nickrent et al. 2010; defined in Nickrent et al. 1998 as plants that photosynthesize during part of their life cycle and obtain mainly water and dissolved minerals from their hosts), although certain members are holoparasites (i.e., Balanophoraceae, certain members of Viscaceae; see Ashworth 2000); holoparasites are defined as nonphotosynthetic (some holoparasites are discussed in the chapters on superrosids and parallel evolution; see Chapters 10 and 13).

Ascertaining the relationships of many parasitic plants has long been problematic, because evolutionary modifications associated with the parasitic habit, such as the loss of leaves, perianth parts, integuments, and chlorophyll, may result in extreme morphologies that may be subject to convergence (see Nickrent et al. 1998). The relationships of many holoparasites remain particularly problematic, because the plants are nonphotosynthetic, and the plastid genomes of these plants are highly modified with many genes absent or with accelerated rates of molecular evolution (Nickrent et al. 1998; Su and Hu 2012; see Chapters 10 and 13). Likewise, horizontal gene transfer (HGT) from the hosts has been shown to occur in certain parasitic taxa, especially of mitochondrial but also of nuclear genes (Davis and Wurdack 2004; Mower et al. 2004; Yoshida et al. 2010) (see Chapters 3 and 16), which could lead to erroneously inferred relationships in phylogenetic analyses. Nickrent et al. (2005) showed that aerial hemiparasitism has evolved at least five times in Santalales (see Chapter 13). Thus, a highly specialized character is highly homoplasious in the clade. See also Harbaugh and Baldwin (2007) and Der and Nickrent (2008) for additional phylogenetic analyses in Santalales. See Fig. 11.3 for various members of the Santalales.

The fossil record of this order is known mainly from pollen of Loranthaceae; loranthaceous pollen has distinctive triangular oblate grains that are typically concave sided and syncolpate, with distinctive variation in ornamentation and aperture configuration among genera (e.g., Feuer and Kuijt 1980, 1985). Investigations of dispersed fossil grains demonstrate that the family was widespread by the middle Eocene, including the early Eocene of Australia (Macphail et al. 2012) and middle Eocene of the Northern Hemisphere including West Greenland and central Europe (Zetter et al. 2014; Manchester et al. 2015).

BERBERIDOPSIDALES

Berberidopsidales consist of two small families, Berberidopsidaceae and Aextoxicaceae, considered distantly related in non-DNA classifications, but strongly supported as sister taxa in molecular phylogenetic investigations (Savolainen et al. 2000a; D. Soltis et al. 2000, 2003c; Moore et al. 2011; Soltis et al. 2011). Berberidopsidales was not recognized in APG II (2003) pending additional investigation but was recognized by APG III (2009) and APG IV (2016). Nonmolecular synapomorphies are few, but include cyclocytic stomata and characteristically short filaments.

The precise placement of Berberidopsidales within *Superasteridae* is unclear. Moore et al. (2011) recovered Berberidopsidales as sister to the asterid clade; Moore et al. (2010) and Ruhfel et al. (2014) place the order as sister to Caryophyllales + asterids. Soltis et al. (2011) recovered the order as forming a clade with Caryophyllales that was then sister to the asterids (Fig. 11.1). The stem age for Berberidopsidales has been dated to 113.1–112.5 mya (Magallón and Castillo 2009).

Ronse De Craene (2004) suggested that the flowers of *Berberidopsis* (with 12 tepals, 8 stamens, and 3 carpels) fit the mold for a typical, core eudicot perianth in that the more typical biseriate perianth is thought to have been derived from spirally arranged tepals that subsequently differentiated into two whorls of perianth parts (Ronse De Craene 2007; Ronse De Craene and Stuppy 2010). *Aextoxicon*, however, has well-differentiated sepals and petals and is more representative of a biseriate perianth (Ronse De Craene and Brockington 2013). Members of Berberidopsidales are shown in Fig. 11.4.

CARYOPHYLLALES

Since the 1980s, the name "Caryophyllales" has been applied to an increasingly broad, and correspondingly diverse, set of families. Before that, the concept of Caryophyllales generally corresponded to that of Centrospermae (Harms 1934), a group of families long recognized as closely related (e.g., Braun 1864; Eichler 1875–1878). Morphological and embryological characters unite the core families of Centrospermae, but Centrospermae gained attention as one of the earliest groups for which circumscription was modified based on chemical characters. The discovery that all but two families of Centrospermae—Caryophyllaceae and Molluginaceae—produce betalain pigments instead of the anthocyanin pigments produced in other angiosperms supported a close relationship among the core group of families. Furthermore, Cactaceae and Didiereaceae, not previously considered closely related to Centrospermae, were also discovered to produce betalains. Revised non-DNA-based classifications (e.g., Dahlgren 1975, 1980; Takhtajan 1980; Cronquist 1981; Thorne 1983, 1992a,b), applying the name Caryophyllales and incorporating chemical, morphological, embryological, and anatomical characters, included 12 families.

Recent phylogenetic analyses have further clarified relationships (Applequist and Wallace 2001; Cuénoud et al.

Figure 11.3. Members of Santalales (Amphorogynaceae, Aptandraceae, Balanophoraceae, Cervantesiaceae, Comandraceae, Coulaceae, Erythropalaceae, Loranthaceae, Misodendraceae, Nanodeaceae, Oktonemaceae, Olacaceae, Opiliaceae, Santalaceae, Schoepfiaceae, Strombosiaceae, Thesiaceae, Viscaceae, Ximeniaceae): a. *Nestronia umbellula* Raf. (Santalaceae), showing inflorescence and large photosynthetic leaves, b. *Phoradendron leucarpum* (Raf.) Reveal & M.C. Johnst. (Viscaceae), inflorescence with reduced flowers, showing valvate perianth, c. *Agonandra macrocarpa* L.O. Williams (Opiliaceae), branch with catkin-like inflorescences, d. *Misodendron angulatum* Phil. (Misodendraceae), on *Nothofagus*, e. *Tristerix penduliflorus* Kuijt (Loranthaceae), flowers with showy corollas, f. *Corynaea crassa* Hook. f. (Balanophoraceae), the inflorescence is the only part of the plant produced aboveground in this holoparasite; note lack of chlorophyll. g. *Ximenia americana* L. (Ximeniaceae), branch with flowers, h. *Aptandra zenkeri* Engl. (Aptandraceae), branch with fruits with accrescent calyx.

Figure 11.4. Members of Berberidopsidales (Aexitoxicaceae, Berberidopsidaceae) and Caryophyllales (Achatocarpaceae, Aizoaceae, Amaranthaceae, Anacampserotaceae, Ancistrocladaceae, Asteropeiaceae, Barbeuiaceae, Basellaceae, Nepenthaceae, Cactaceae, Caryophyllaceae, Didiereaceae, Dioncophyllaceae, Droseraceae, Drosophyllaceae, Frankeniaceae, Gisekiaceae, Halophytaceae, Hypertelidaceae, Limeaceae, Lophiocarpaceae, Macarthuriaceae, Microteaceae, Molluginaceae, Montiacae, Petiveriaceae, Physenaceae, Plumbaginaceae, Polygonaceae, Portulacaceae, Nyctaginaceae, Phytolaccaceae, Rhabdodendraceae, Sarcobataceae, Simmondsiaceae, Stegnospermataceae, Talinaceae, Tamaricaceae): a. *Aexitoxicon punctatum* Ruiz & Pav. (Aexitoxicaceae), branch with drupes; note peltate scales, b. *Berberidopsis corallina* Hook. f. (Berberidopsidaceae), branch with flowers; note spiraled tepals, c. *Mirabilis jalapa* L. (Nyctaginaceae), flowers with tepals petaloid, connate, and whorl of calyx-like bracts, d. *Mesambryanthemum cordifolium* L. f. (Aizoaceae), leaves with epidermis with bladderlike cells, and flower with numerous petaloid staminodia, e. *Simmondsia chinensis* C.K. Schneid. (Simmondsiaceae), branch with opposite leaves, axillary and solitary carpellate flower, showing accrescent calyx, f. *Stegnosperma cubense* A. Rich. (Stegnospermataceae), racemes with fruits (capsules) and flowers; note the calyx-like tepals, and staminodial "petals," g. *Gomphrena serrata* L. (Amaranthaceae), flowers with papery tepals, each flower with two bracteoles, h. *Silene scouleri* Hook. (Caryophyllaceae), deeply lobed "petals" of staminal origin.

2002; Brockington et al. 2009; Schäferhoff et al. 2009; Moore et al. 2010, 2011; Soltis et al. 2011; Ruhfel et al. 2014; Yang et al. 2015; Fig. 11.5) although this clade is not well characterized phenotypically (putative synapomorphies are anthers with outer secondary parietal cell layer developed directly into the endothecium, pollen exine of often fine spinules, and gynoecia with the style branches well developed). The order is composed of two major clades, the core Caryophyllales and non-core Caryophyllales (Polygonineae; sensu Judd et al. 2008; Moore et al. 2011), with the non-core Caryophyllales composed of Nepenthaceae, Droseraceae, Drosophyllaceae, Dioncophyllaceae, Polygonaceae, Plumbaginaceae, Frankeniaceae, Ancistrocladaceae, and Tamaricaceae (Fig. 11.6), and the core Caryophyllales composed of Macarthuriaceae, Microteaceae (see Schäferhoff et al. 2009), Caryophyllaceae, Achatocarpaceae, Amaranthaceae, Stegnospermataceae, Limeaceae, Lophiocarpaceae, Kewaceae (see Christenhusz et al. 2014), Barbeuiaceae, Aizoaceae, Phytolaccaceae, Petiveriaceae, Gisekiaceae, Sarcobataceae, Nyctaginaceae, Molluginaceae, Montiaceae, Halophytaceae, Didiereaceae, Basellaceae, Talinaceae, Anacampserotaceae, Portulacaceae, and Cactaceae (Figs. 11.4 and 11.6–11.7).

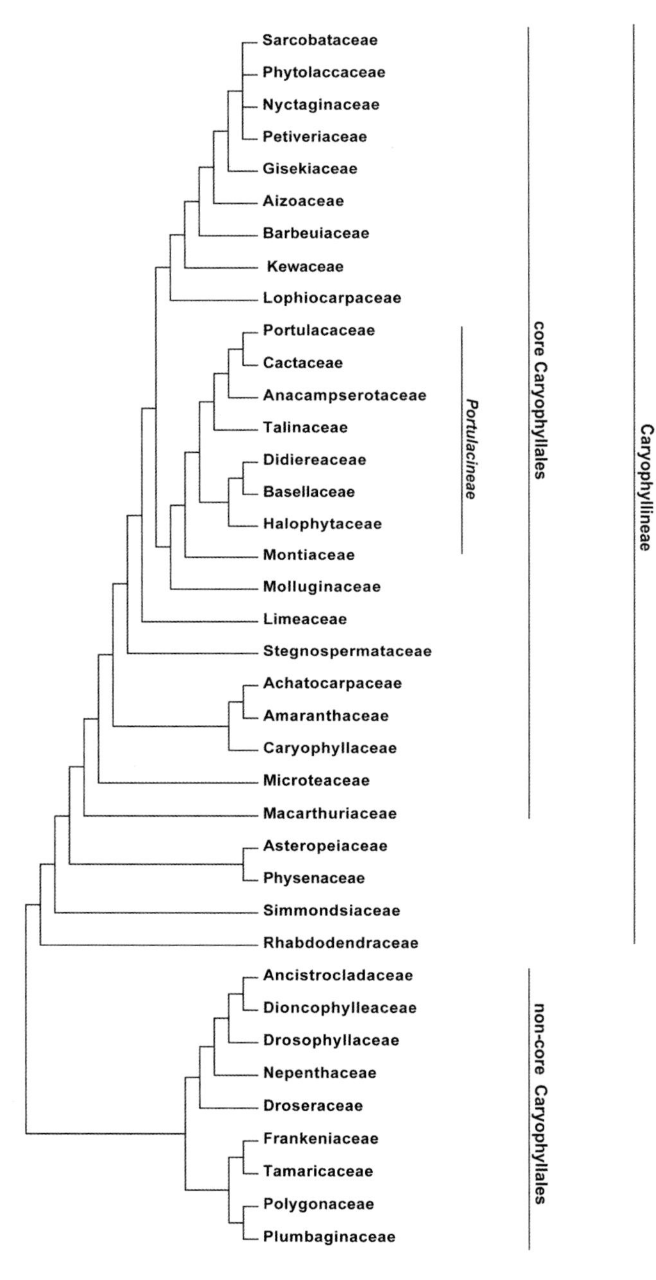

Figure 11.5. Phylogenetic relationships among members of Caryophyllales.

Rhabdodendraceae, Simmondsiaceae (Fig. 11.4), Asteropeiaceae, and Physenaceae are successive sisters to the core Caryophyllales (Fig. 11.5), and the entire clade including core Caryophyllales is often referred to as Caryophyllineae. Successive cambia, plastids of sieve elements with a peripheral ring of proteinaceous filaments, linear stigmas, often extending most of the length of the style or style-branches, ovules campylotropous, nectar produced by glandular adaxial staminal bases, and seeds with a curved embryo and starchy perisperm are synapomorphies for core Caryophyllales, and acetogenic naphthaquinones are probably a synapomorphy of Polygonineae. Many core Caryophyllales are also herbaceous and most have betalains (see also Brockington et al. 2011, 2015).

Applequist and Wallace (2001) showed that Portulacaceae as previously circumscribed were highly polyphyletic and that to recognize the enigmatic Cactaceae and Didiereaceae would mean splitting the portulacaceous cohort, as they called it, into a number of segregate families. Subsequently, Portulacaceae have been split into numerous families (i.e., Anacampserotaceae, Montiaceae, Portulacaceae s.s., Talinaceae), and the portulacaceous cohort was recognized as suborder Portulacineae by Nyffeler and Eggli (2010).

Cuénoud et al. (2002) and Brockington et al. (2009) showed that the traditional circumscription of Molluginaceae and Phytolaccaceae also represented non-monophyletic groups, wherein *Limeum* of Molluginaceae s.l. was not closely related to *Mollugo*, and *Rivina* of Phytolaccaceae was putatively more closely related to Nyctaginaceae and is therefore more appropriately treated within Petiveriaceae. Yang et al. (2015) have further clarified relationships within the phytolaccoid clade (i.e., Sarcobataceae, Phytolaccaceae, Petiveriaceae, and Nyctaginaceae), resolving Phytolaccaceae as sister to Sarcobataceae, and corroborating the sister relationship of Petiveriaceae and Nyctaginaceae. Caryophyllales have been dated to 115–63 mya (Wikström et al. 2001; Anderson et al. 2005;

Figure 11.6. Caryophyllales (cont'd.): a. *Nepenthes truncata* Macfari. (Nepenthaceae), apical extension of the midrib forming insect traps, b. *Drosera capillaris* Poir. (Droseraceae), gland-headed hairs for trapping insects, c. *Drosera brevifolia* Pursh (Droseraceae), flower, d. *Plumbago capensis* Thunb. (Plumbaginaceae), sympetalous corolla and gland-headed hairs on calyx, e. *Tamarix chinensis* Lour. (Tamaricaceae), branch with flowers and scale-like leaves, f. *Coccoloba uvifera* L. (Polygonaceae) node with ocrea (sheathing stipule) characteristic of the family, g. *Antigonon leptopus* Hook. & Arn. (Polygonaceae), flower, h. *Phytolacca heterotepala* H. Walter (Phytolaccaceae), portion of raceme of flowers with divided styles, elongate stigmas and pentamerous perianth, i. *Rivina humilis* L. (Petiveriaceae) raceme of berries, each developed from a single carpel.

Magallón and Castillo 2009; Bell et al. 2010; Moore et al. 2010; Naumann et al. 2013).

Numerous families and intrafamilial groups in Caryophyllales have been the subject of recent phylogenetic investigations. New insights into Cactaceae have been obtained with work done on *Pereskia* s.l. (Edwards et al. 2005), as well as Cactoideae (Butterworth and Wallace 2004; Arias et al. 2005; Harpke and Peterson 2006; Calvente et al. 2011; Albesiano and Terrazas 2012; Schlumpberger and Renner 2012; Frank et al. 2013; Sánchez et al. 2014), Opuntioideae (Griffith 2004; Griffith and Porter 2009; Helsen et al. 2009a,b; Majure et al. 2012b, 2013a,b; Ritz et al. 2013; Majure and Puente 2014; Bárcenas 2015), and Cactaceae in general (Nyffeler 2007; Arakaki et al. 2011; Bárcenas et al. 2011; Hernández-Hernández et al. 2011, 2014). Cytogenetic and population genetic work in the group also has been greatly enhanced (Arakaki et al. 2007; Negrón-Ortiz 2007; Segura et al. 2007; Baker et al. 2009; Helsen et al. 2009a,b; las Peñas et al. 2009, 2014; Majure and Ribbens 2012; Majure et al. 2012a,c), as has anatomical work (Mauseth 2004, 2006); barcoding cacti has been attempted (Yesson et al. 2011). Sanderson et al. (2015) showed marked reduction in the chloroplast genome of saguaro, *Carnegiea gigantea* (Engelm.) Britton & Rose, and Ibarra-Laclette et al. (2015) recently sequenced the transcriptome of peyote, *Lophophora williamsii* (Salm-Dyck) J.M. Coult., which allowed them to discover putative genes involved in alkaloid production. Arakaki et al. (2011) proposed a date of ~35 mya for the origin of the cacti with major diversification events in the family taking place around 5–10 mya; the age of cacti recovered by Arakaki et al. (2011) is in line with previous dates for their origin suggested by Hershkovitz and Zimmer (1997). The closest relatives of cacti had long been debated (Hershkovitz and Zimmer 1997; Applequist and Wallace 2001; Nyffeler 2007; Griffith 2008). Most recent analyses support Portulacaceae s.s. as sister to the family (Brockington et al. 2009; Soltis et al. 2011).

Other groups within "Portulacaceae s.l." have been studied phylogenetically (Montieae, Montiaceae, O'Quinn and Hufford 2005; *Hectorella*, Montiaceae, Applequist et al. 2006; *Portulaca*, Portulacaceae s.s., Ocampo and Columbus 2012; Didiereaceae, Bruyns et al. 2014) further increasing our understanding of evolutionary relationships in Portulacineae. Likewise, other studies have focused on Amaranthaceae (Applequist and Pratt 2005; Müller and Borsch 2005; Ogundipe and Chase 2009), Caryophyllaceae (Fior et al. 2006; Greenberg and Donoghue 2011), Polygonaceae (Kim and Donoghue 2008; Sanchez and Kron 2008; Galasso et al. 2009; Sanchez et al. 2009; Burke et al. 2010), Droseraceae (Rivadavia et al. 2003), Nepenthaceae (*Nepenthes*, Meimberg and Heubl 2006), Aizoaceae (Hassan et al. 2005; Klak et al. 2007; Thiede et al. 2007), and Nyctaginaceae (Douglas and Manos 2007). Sloan et al. (2014) analyzed the enigmatic occurrence of increased rates of plastid genome evolution in *Silene* (Caryophyllaceae), and Yashina et al. (2012) reportedly revived 30,000-year-old seeds of *Silene stenophylla* Ledeb. from Pleistocene permafrost. Yang et al. (2015) have recently reconstructed the phylogeny of Caryophyllales using transcriptome data. Selected members of Caryophyllales are illustrated in Figs. 11.4 and 11.6–11.7.

CHARACTER EVOLUTION IN CARYOPHYLLALES

Chemical Evolution: Anthocyanins versus Betalains

Anthocyanin and betalain production are mutually exclusive chemical pathways. In all core Caryophyllales examined, except Caryophyllaceae and Molluginaceae s.l. (i.e., including Kewaceae), anthocyanin pigments have been replaced by betalains, which have taken on the functions of anthocyanins in both reproductive and vegetative tissues. From phylogenetic analyses, it is clear that betalain pigments probably evolved once and were lost three times in the group, although many taxa still have yet to be screened for pigment type (Brockington et al. 2011, 2015). Several hypotheses have been posited for the evolution of betalain pigments in Caryophyllales, but no clear evidence has strongly supported those ideas. Both anthocyanins and betalains are apparently regulated in a similar, if not identical, manner (e.g., Stafford 1994; Clement and Mabry 1996; Shimada et al. 2007), but the production of anthocyanin pigments appears to be modified during transcription in betalain producers (Shimada et al. 2004, 2005). The biochemical and evolutionary mechanisms responsible for the complete replacement of anthocyanins by betalains in this clade remain unresolved. Anthocyanin production occurs via the flavonoid biosynthetic pathway, but although anthocyanins are absent from betalain-producing plants, the precursors of anthocyanins (proanthocyanidins) are present, especially in seedlings, which often contain both dihydroflavonol-4-reductase (DFS) that converts dihydroflavonols into leucocyanidins and anthocyanidin synthase (ANS) that converts leucocyanidins to anthocyanidins (Shimada et al. 2005), although they both are downregulated (reviewed in Brockington et al. 2011). Transcriptome data suggest that duplications of *DODA* and *CYP76AD1* (those genes that in tandem are responsible for producing red betanidin pigments) in the core Caryophyllales resulted in the neofunctionalization of those duplicates to produce betalains. This is inferred to have happened only once, while the

Figure 11.7. Caryophyllales (cont'd.): a. *Pereskia grandifolia* Haw. (Cactaceae), branch with conspicuous photosynthetic leaves and inflorescence; note flowers with numerous petaloid tepals and stamens, b. *Hylocereus undatus* (Haw.) Britton & Rose (Cactaceae), flower; note hypanthium, with modified leaves (greenish, sepal-like structures), numerous petaloid tepals, and numerous stamens, c. *Opuntia chlorotica* Engelm. & J.M. Bigelow (Cactaceae), showing modified stem tissue (note the areoles) within which the ovary is embedded to form the red fruit, as well as the vegetative stem (= long shoot) producing spirally arranged areoles (= short shoots) with spines and glochids, d. *Didierea trollii* Capuron & Rauh (Didiereaceae), habit, e. *Alluaudia procera* Drake (Didiereaceae), long-shoot (new growth) and short-shoots producing leaves (on older growth), f. *Portulaca grandiflora* Hook. (Portulacaceae), flower with petaloid tepals, conspicuously divided stigma, g. *Portulaca amilis* Speg. (Portulacaceae), conspicuous trichomes produced from leaf axils, h. *Basella alba* L. (Basellaceae), vine demonstrating betalain pigments, i. *Claytonia virginica* L. (Montiaceae), flowers, j. *Talinum fruticosum* (L.) Juss., flowers with petaloid tepals, k. *T. fruticosum* (Talinaceae) showing pair of sepals (derived from bracteoles).

downregulation of those same genes within the betalain-producing clade led to a reversal to anthocyanin production on three separate occasions, in Caryophyllaceae, Kewaceae, and Molluginaceae s.s. (Brockington et al. 2015).

Using maximum parsimony, Brockington et al. (2011) showed that betalains may have evolved once (under ACCTRAN optimization) in the most recent common ancestor of the Microteaceae + Cactaceae clade or twice (under DELTRAN optimization; once in Amaranthaceae and once in the Stegnospermataceae + Cactaceae clade). Using stochastic character mapping, they revealed mostly the same pattern. Considering the scenario where betalains evolved twice in Caryophyllales, there would have been two reversals back to the production of anthocyanin pigments versus three reversals under the scenario where betalains evolved only once. However, Brockington et al. (2011) stressed that we still know little about pigment expression within core Caryophyllales. Only dubious reports for pigment type are available for a number of taxa, and pigment data are completely unknown for others, rendering character reconstruction of pigment type in the group preliminary at best.

Here we provide the results from character mapping of pigment type using both maximum parsimony (MP) and maximum likelihood (ML) analyses (Fig. 11.8). For MP under DELTRAN optimization, betalain pigments were acquired independently, once in Amaranthaceae, and again in the most recent common ancestor of the rest of the betalain-producing members of core Caryophyllales with two reversals to anthocyanin pigments. Under ACCTRAN optimization, the acquisition of betalain pigments is equivocal but may have evolved only once with the most recent common ancestor of *Microtea* + Cactaceae, which would suggest three reversals to anthocyanin pigmentation, as do most recent analyses of transcriptome data (Brockington et al. 2015). Using ML, the origin of betalain pigments also is equivocal, although it is most likely that the common ancestor of the *Stegnosperma* + Cactaceae clade produced betalain pigments with two subsequent reversals in *Hypertelis* and Molluginaceae (Fig. 11.8).

Pollen

No single pollen type characterizes all members of Caryophyllales. However, all core Caryophyllales examined possess pollen with a spinulose and tubuliferous/punctate ektexine and a spinulose aperture membrane or operculum (Nowicke 1975). Within core Caryophyllales, most families exhibit multiple pollen morphologies, indicating that pollen structure is labile (Fig. 11.9). Three basic types of pollen have been observed in core Caryophyllales: tricolpate, pantoporate, and pantocolpate, all with a spinulose and tubuliferous/punctate ektexine (Nowicke 1975). Tricolpate pollen of similar morphology (type I) is found in Aizoaceae and some Cactaceae, Caryophyllaceae, Molluginaceae, Nyctaginaceae, Phytolaccaceae, and Portulacaceae. Pantoporate pollen (type II) occurs in Amaranthaceae (including Chenopodiaceae) and some genera of Caryophyllaceae, Nyctaginaceae, Phytolaccaceae, and Portulacaceae. The pantocolpate pollen (type III) of core Caryophyllales typically has 12 to 15 colpi and is found in Basellaceae and some genera of Cactaceae, Molluginaceae, Nyctaginaceae, Phytolaccaceae, and Portulacaceae.

Two narrowly distributed pollen types have been reported for core Caryophyllaceae (Nowicke 1975): tricolpate with a reticulate ektexine (type IV) in Nyctaginaceae and pantoporate with a reticulate ektexine (type V) in Caryophyllaceae. A divergent form of pantoporate grains with reticulate ektexine is found in some Amaranthaceae and Cactaceae. For example, having pantoporate pollen grains with reticulate ektexine is a synapomorphy for *Opuntia* s.s., the prickly pear cacti, although the reticulate surface has been lost in certain hummingbird-pollinated species of the *Nopalea* clade (Majure and Puente 2014). Certain aspects of structural characteristics of the pollen (e.g., partially fused columellae) unite some members of Basellaceae, Didiereaceae, Portulacaceae, and Cactaceae (Nowicke 1994, 1996; see also Erdtman 1966) in Cronquist and Thorne's (1994) suborder Portulacineae. However, at least some of the exine characteristics that unite these groups are also found elsewhere in core Caryophyllales (Nowicke 1975, 1996). Despite their small size (fewer than 20 species in 4 genera), Basellaceae are highly polymorphic in pollen morphology (Nowicke 1994, 1996). All pollen of Didiereaceae is 5–7-zonocolpate, a pollen type that is apparently unique in core Caryophyllales, with a spinulose and annular perforate or punctate tectum (Nowicke 1996). Cactaceae and Portulacaceae are polymorphic, with pollen types I and III and types II and III, respectively (e.g., Nowicke 1975, 1994).

The pantoporate pollen type with exine ornamentation of scattered spinulae, typical of Amaranthaceae, is well represented in the fossil record of dispersed pollen. Usually given the fossil-genus name *Chenopodipollis,* these fossils provide our only evidence for Amaranthaceae in the late Cretaceous (Maastrichtian; Nichols 2002) and Paleocene (Zetter et al. 2011). Numerous younger occurrences have been recorded (Muller 1981), but the assignment to individual extant genera is not possible, because they overlap in morphology. *Caryophylloflora paleogenica*, an inflorescence containing periporate pollen in situ, was documented from the Eocene of northeastern Tasmania (Jordan and Macphail 2003). In general, however, the megafossil record of Caryophyllales remains poorly documented.

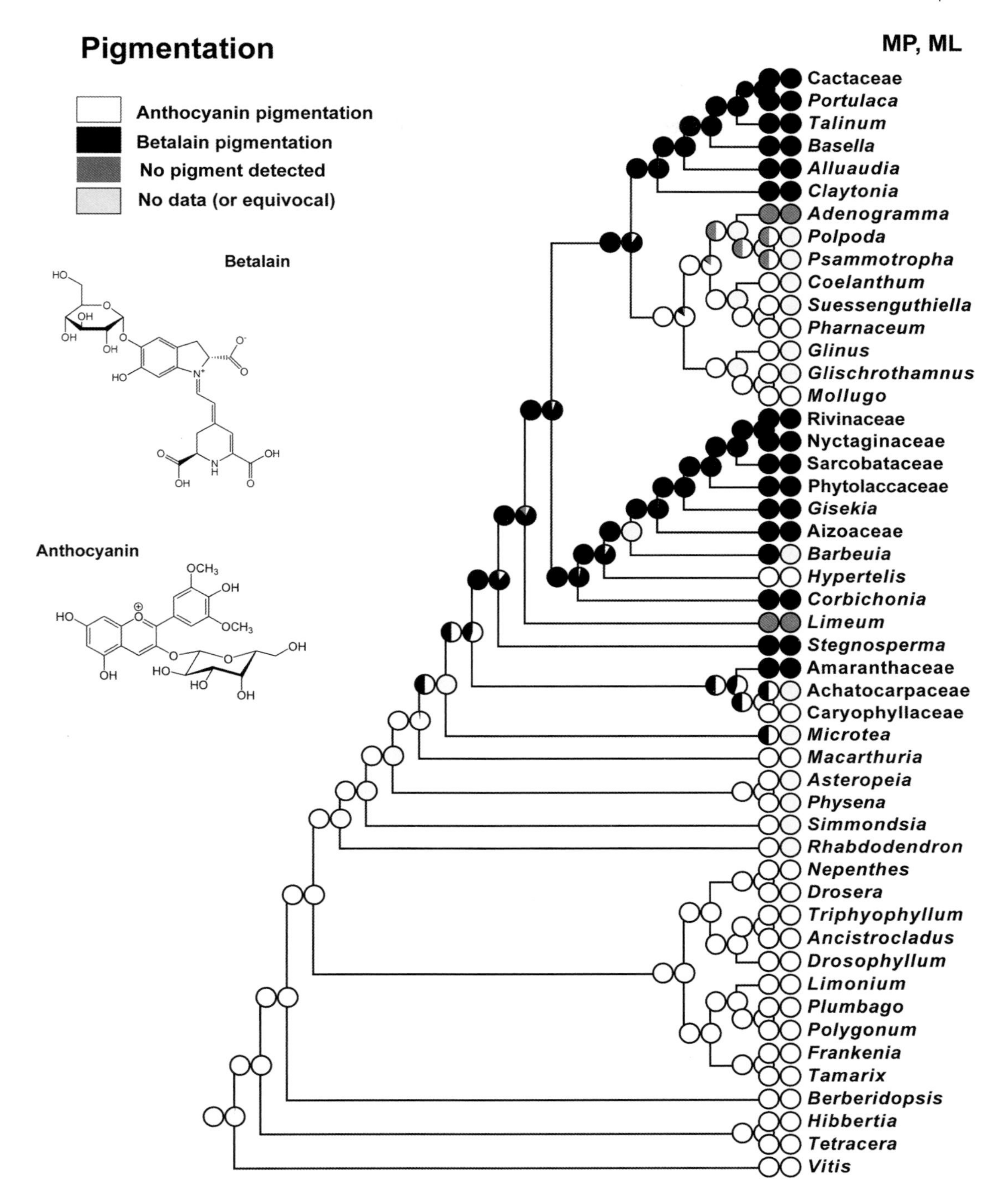

Figure 11.8. Reconstruction of anthocyanin vs. betalain production in Caryophyllales using both MP and ML. Diagrams of both betalain and anthocyanin molecules are given to the left. Under MP, the evolution of betalain pigments is equivocal. It is clear that betalain pigments were present in the most recent common ancestors of Amaranthaceae, as well as the Stegnospermataceae + Cactaceae clade, although with reversals in Molluginaceae and Hypertelidaceae.

Pollen of Achatocarpaceae (*Achatocarpus* and *Phaulothamnus*) differs from that of all core Caryophyllales in being hexaporate with a scabrate ektexine (Bortenschlager et al. 1972; Nowicke 1975; Nowicke and Skvarla 1977, 1979; Skvarla and Nowicke 1976, 1982). Pollen of *Simmondsia* is triporate with an irregularly scabrate tectum (Nowicke and Skvarla 1984), distinct from that of all core Caryophyllales and from Buxaceae, Euphorbiaceae, and Pandaceae with which it had previously been placed (e.g., Hutchinson 1959; Scholz 1964; Takhtajan 1969; Cronquist 1981). A comprehensive comparative analysis of pollen in noncore Caryophyllales has not yet been conducted.

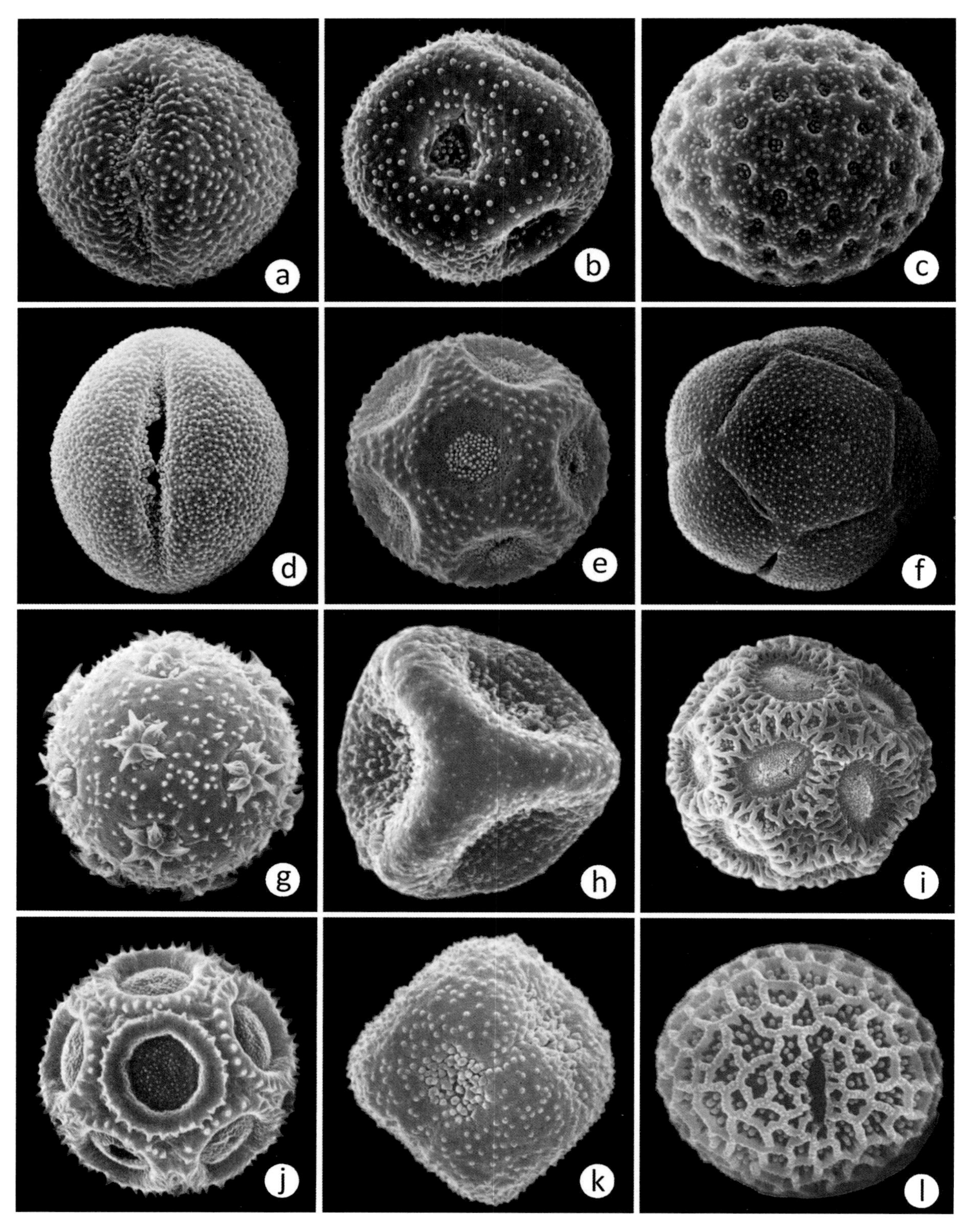

Figure 11.9. SEM photomicrographs of pollen from selected members of Caryophyllales (modified from Nowicke and Skvarla 1977). a. *Mesembryanthemum variable* Haw. (Aizoaceae), b. *Anredera scandens* Moquin (Basellaceae), c. *Chenopodium ambrosioides* L. (Amaranthaceae), d. *Limeum viscosum* Fenzl (Limeaceae), e. *Gymnocarpos fruticosum* Persoon (Caryophyllaceae), f. *Montia parvifolia* (Moc. ex DC.) Greene (Montiaceae), g. *Psilotrichum amplum* (Amaranthaceae), h. *Herniaria glabra* L. (Caryophyllaceae), i. *Opuntia lindheimeri* Engelm. (Cactaceae), j. *Siphonychia americana* (Nutt.) Torr. & A. Gray (Caryophyllaceae), k. *Cardionema ramosissima* (Caryophyllaceae), l. *Abronia angustifolia* Greene (Nyctaginaceae).

Given the many gaps in pollen data across Caryophyllales, reconstruction of pollen evolution is premature.

Wood Anatomy

Some features of stem anatomy appear to be synapomorphies for clades within Caryophyllales. Successive cambia are widespread in core Caryophyllales (Metcalfe and Chalk 1950; Horak 1981; Carlquist 1999a, 2010) and have also been reported from *Simmondsia*, *Rhabdodendron*, *Frankeniaceae* (Carlquist 2001b, 2010), and at least one genus each of Plumbaginaceae (Carlquist and Boggs 1996) and Polygonaceae (see Carlquist 2001b). Successive cambia are absent in Achatocarpaceae, Cactaceae, Didiereaceae, Microteaceae, Portulacaceae, and some Petiveriaceae (*Hilleria*, *Ledenbergia*, *Schindleria*, *Trichostigma*; Carlquist 2000b, 2010 and refs. therein).

Carlquist (2001b) suggested that the genetic mechanisms for the production of successive cambia may have arisen at the base of the core Caryophyllales and were then lost in those families that lack successive cambia. However, it may be that initiation of subsequent cambial layers is delayed, as in *Stegnosperma* (Horak 1981), rather than absent. Successive cambia have not been reported in the woody members of the carnivorous clade, but their presence in Polygonaceae and Plumbaginaceae indicates that the origin of this feature may predate the entire Caryophyllales, with multiple losses of the trait. However, this is not the most parsimonious interpretation of the evolution of the character in Caryophyllales. It appears most likely that the lack of successive cambia is plesiomorphic and that successive cambia has arisen independently throughout Caryophyllales.

Several wood features could be synapomorphies for the entire Caryophyllales and may have been lost in all or most of the core Caryophyllales (depending on the placement of *Rhabdodendron*). Alternatively, these wood features could represent cases of parallel evolution or possibly synapomorphies of Polygonineae. The vestured pits in vessels and tracheids of *Rhabdodendron* also occur in Polygonaceae (Ter Welle 1976; Carlquist 2001b). Rhabdodendraceae, Polygonaceae, Plumbaginaceae, Ancistrocladaceae, and Dioncophyllaceae all have silica bodies (Gottwald and Parameswaran 1968; Ter Welle 1976; Carlquist 1988, 2001b; Carlquist and Boggs 1996). Non-bordered perforation plates are also present in Rhabdodendraceae and are common in much of Caryophyllales (Carlquist 1999b, 2000b, 2001b). Many wood characters, such as simple perforation plates, small pits on lateral vessel walls, libriform fibers, paratracheal axial parenchyma, storying, silica bodies, and dark-staining amorphous deposits, also indicate a close relationship between Polygonaceae and Plumbaginaceae (Carlquist and Boggs 1996).

The carnivorous clade of Droseraceae, Nepenthaceae, Drosophyllaceae, Dioncophyllaceae, and Ancistrocladaceae is also apparently supported by wood anatomical features. Although secondary growth is limited in members of this clade, vessel elements with simple perforation plates, fibriform vessel elements, tracheids with large, fully bordered pits, diffuse (and variously grouped) axial parenchyma, and paedomorphic rays (sensu Carlquist 1988) one to two cells wide are shared by those members of this clade that have been analyzed to date (Gottwald and Parameswaran 1968; Carlquist 1981; Carlquist and Wilson 1995). Some features, such as simple perforation plates, are also common to other members of Caryophyllales and may not be synapomorphies for the carnivorous clade.

Despite the many investigations of various aspects of wood anatomy for members of Caryophyllales, gaps remain for most characters. Even a character such as successive cambia, the presence of which has been investigated in many taxa (see also Carlquist 2007), has not been scored in several other groups, leaving potentially significant gaps in our knowledge. Furthermore, scoring of successive cambia as "present or absent" may introduce additional problems, because evidence of successive cambia may arise at different stages of development and in different tissues in different taxa. Thus, further analysis of the evolution of wood characters requires additional comparative data before optimization of the observed variation across a phylogenetic tree should be attempted.

Extreme Environments

Many members of Caryophyllales are adapted structurally or physiologically to extreme environments such as deserts, high-alkaline soils, high-saline substrates, and nutrient-poor soils. They have conquered these habitats through a variety of adaptations such as unusual photosynthetic pathways (crassulacean acid metabolism, CAM, and C_4 as opposed to C_3 photosynthesis), unusual morphologies (e.g., succulence, spine production), secretion of excessive salt by special glands, and unusual methods of nutrient uptake (e.g., carnivory). Given the distributions of these adaptations across Caryophyllales, it appears that most of these adaptations have arisen many times (see Chapter 13). The fact that many Caryophyllales are adapted to extreme environments may account for the paucity of megafossils of the clade.

Photosynthetic Pathways

C_4 photosynthesis has apparently evolved independently in several lineages of Caryophyllales (see Chapter 13). However, the predisposition to evolve this suite of traits may have arisen in an ancestor common to those lineages that eventually developed this adaptation to high-light, high-temperature environments. The lability of photosynthetic pathways is clear in members of *Salsola* and relatives (formerly in tribe Salsoleae of Chenopodiaceae, but now part of Amaranthaceae). This group exhibits in microcosm the patterns of photosynthetic variation present in the family as a whole. A highly resolved phylogenetic tree of *Salsola* largely agrees with the photosynthetic type and anatomy of leaves and cotyledons and provides strong support for the origin and evolution of two main lineages of plants in "tribe Salsoleae"—the NAD-ME and NADP-ME types of C_4 photosynthesis, respectively (Pyankov et al. 2001). Reconstruction of photosynthetic characters on the ITS phylogenetic tree demonstrates a single origin of C_4 photosynthesis, with subsequent divergence into the NAD-ME and NADP-ME lineages and two reversions to C_3 photosynthesis. Sage et al. (2007) showed that C_4 photosynthesis has evolved at least five times in Amaranthaceae s.s. Ocampo et al. (2013) showed that the ancestral photosynthetic pathway was C_4 in *Portulaca*, and three different types of C_4 Kranz anatomy were revealed in their study. The *Cryptopetala* clade of *Portulaca* showed a shift to C_3-C_4 intermediate photosynthetic types. Likewise, Christin et al. (2011) showed that C_3-C_4 intermediate types evolved at least twice, if not three times, in Molluginaceae s.l. and that C_4 photosynthesis was derived twice in that group.

The topology for core Caryophyllales indicates that CAM has also arisen multiple times within this clade, including in Cactaceae, Didiereaceae, Anacampserotaceae, and Aizoaceae. The similar selection pressures exerted by the arid habitats occupied by members of these clades have resulted in a spectacular convergence in morphology and photosynthetic pathway. Winter et al. (2011) showed that *Opuntia elatior* Mill. (Cactaceae, Opuntioideae) is facultatively CAM, wherein seedlings, before the production of the typical flattened stem segments (cladodes) of the group, undergo C_3 photosynthesis. Those authors correlated the initiation of CAM in *O. elatior* with plant maturity and drought stress; however, more mature plants of the same species under amply watered conditions also displayed C_3 photosynthesis (Winter et al. 2011). This study contrasts with other studies that have shown cacti to be strictly CAM (Gerwick and Williams 1978; Osmond et al. 1979, 2008; Nobel 1988). Winter et al. (2011) suggested that constantly monitoring CO_2 uptake, as well as titratable acidity, from germination through the production of more mature stem segments may reveal that photosynthetic pathways are more labile (i.e., faculatively CAM; see also Winter et al. 2015) than generally presumed in Cactaceae. Hernández-González and Villareal (2007) analyzed members of the subfamily Cactoideae, *Neobuxbaumia*, *Pachycereus*, *Stenocereus*, *Escontria*, *Myrtillocactus*, and *Ferocactus*, which all exhibited CAM photosynthesis, even during the seedling stage. Likewise, most studies monitoring photosynthetic rates in Cactaceae also have focused on the larger subfamily Cactoideae. Thus, there likely are different photosynthetic strategies in the clade, with some taxa more flexible in their photosynthetic pathways than others. Guralnick et al. (2008) also showed that *Grahamia* (Anacampserotaceae) was facultatively CAM, which was induced through water stress. Ocampo and Columbus (2010) reported facultatively CAM species in Anacampserotaceae, Cactaceae (Opuntioideae), and Didiereaceae.

Carnivory

Multiple mechanisms of carnivory have evolved in Caryophyllales (see Chapter 13). Insects are ensnared by pitfall traps in Nepenthaceae, flypaper traps in *Drosera*, *Drosophyllum*, and *Triphyophyllum*, and snap-traps in *Dionaea* and *Aldrovanda*. Although previous classifications considered only *Drosera* and *Dionaea* to be closely related to each other (e.g., Cronquist 1981) and none of these groups to be close to Caryophyllales, all phylogenetic analyses have indicated both their relationship to each other and to Caryophyllidae sensu Cronquist (1981) (e.g., Albert et al. 1992; Chase et al. 1993; Williams et al. 1994; D. Soltis et al. 1997b, 2000; Meimberg et al. 2001; Cuénoud et al. 2002; Brockington et al. 2009; Moore et al. 2010; Soltis et al. 2011). Although relationships among these carnivorous genera have varied with taxon sampling and gene(s) analyzed (e.g., Albert et al. 1992; Williams et al. 1994; Fay et al. 1997a; Lledó et al. 1998; Meimberg et al. 2001; Cameron et al. 2002; Cuénoud et al. 2002; Brockington et al. 2009; Moore et al. 2011; Soltis et al. 2011), their phylogenetic relationships are clear: Droseraceae form a clade with Nepenthaceae, which is sister to Drosophyllaceae + (Ancistrocladaceae + Dioncophyllaceae) (Brockington et al. 2009; Soltis et al. 2011). This topology implies that "carnivory" was gained a single time in Caryophyllales and lost in Ancistrocladaceae and some Dioncophyllaceae (see Meimberg et al. 2001; see Chapter 13, Fig. 13.6).

Carnivory was achieved through several different mechanisms (pitchers, flypaper traps, snap-traps), involving extreme modifications of leaves and glands (see Williams 1976; Juniper et al. 1989; Albert et al. 1992; Chapter 13, Fig. 13.6); each method may have evolved only once in this clade (Cameron et al. 2002 for snap-traps, and inferences from trees of Meimberg et al. 2001 and Cameron et al. 2002 for pitchers and flypaper traps; see Chapter 13), although Renner and Specht (2011) using stochastic character mapping provided strong evidence that flypaper traps—that is, stalked and vascularized, with both xylem and phloem, gland-headed hairs, evolved three times in the clade. The absence of carnivory in Ancistrocladaceae and some Dioncophyllaceae results from the loss of the flypaper trap (i.e., the absence of stalked, vascularized, gland-headed hairs; Renner and Specht 2011).

Floral Development

The perianth of Caryophyllales is highly variable, often monocyclic and sepaloid or sepal-derived, but in other cases dicyclic with both sepaloid and petaloid whorls. However, the nature of the petals in Caryophyllales has long been debated: are they true petals, or staminodes, or appendages of stamens (Ronse De Craene and Smets 1998a,b)? Inferences of perianth evolution require analyses of the androecium as well; stamen number in Caryophyllales ranges from one to more than 4000 (Ronse De Craene 2013). Analyses of floral development in Caryophyllaceae and other families may indicate that the apparent "diplostemonous" condition of Caryophyllales (i.e., five sepals, five petals, ten stamens, five-carpellate gynoecium) differs developmentally from diplostemony in other core eudicots (Ronse De Craene and Smets 1993, 1995; Vanvinckenroye and Smets 1996; Ronse De Craene et al. 1997, 1998). The five "petals" were interpreted as being derived from a hexamerous whorl of three stamen pairs, followed by loss of a stamen. A similar process was suggested to account for the outer (antesepalous) stamens, which likely correspond to one whorl of three stamens and a second whorl of two; loss of one stamen from a hexamerous whorl (or two whorls of three stamens) resulted in five inner (antepetalous) stamens. The result of these modifications is an apparently diplostemonous flower; however, this final form would not be homologous to that of other core eudicots, because it was achieved independently via a different mechanism (as interpreted by Ronse De Craene and Smets 1993).

Developmental genetic work has confirmed that some core Caryophyllales exhibit a perianth that is unique (among eudicots) wherein petaloid organs (i.e., staminodial petaloids) are derived from stamens (Brockington et al. 2013), while still other members of Caryophyllales clearly exhibit petaloid structures derived from sepals (Brockington et al. 2012) (see Chapter 14). The origin of petaloid structures in Caryophyllales is highly variable, and even within Aizoaceae, certain members exhibit sepal-derived petaloid structures (Sesuvioideae and Aizooideae), while others are derived from stamens (i.e., the clade composed of Mesembryanthemoideae + Ruschoideae; Brockington et al. 2012; Ronse De Craene and Brockington 2013).

ASTERIDS

Recognition of a group of angiosperms that corresponds in large part to Asteridae (asterids) traces to de Jussieu's (1789) Monopetalae, those angiosperms with fused petals. Subsequently referred to as Sympetalae (Reichenbach 1827–1829), this group was divided into Pentacyclicae and Tetracyclicae by Warming (1879). Tetracyclicae, with a single series of stamens, formed the core of modern treatments of subclass Asteridae (e.g., Takhtajan 1969, 1980, 1997; Cronquist 1981). Asteridae sensu Takhtajan and Cronquist have been the focus of phylogenetic analysis for more than 25 years, with data from morphology (e.g., Hufford 1992), restriction site analysis of the plastid genome (e.g., Jansen and Palmer 1987, 1988; Downie and Palmer 1992), gene sequences (e.g., Olmstead et al. 1992, 1993, 2000; D. Soltis et al. 2000; Albach et al. 2001c; Bremer et al. 2002; Jansen et al. 2007; Moore et al. 2010, 2011; Soltis et al. 2011; Ruhfel et al. 2014), as well as molecular data plus morphology (Bremer et al. 2001), used to improve understanding of relationships within Asteridae and between Asteridae and other groups. A significant result of these analyses is that Asteridae sensu Takhtajan and Cronquist are not monophyletic; instead, they form the core of a larger asterid clade that also includes members of Cronquist's (1981) Hamamelidae (Eucommiales), Dilleniidae (most Theales, Lecythidales, Ericales, Ebenales, Diapensiales, Primulales, Sarraceniaceae, Fouquieriaceae, Loasaceae), and Rosidae (Hydrangeaceae, Cornales, Pittosporaceae, Apiales, Byblidaceae, Columelliaceae, Alseuosmiaceae, Aquifoliaceae, Icacinaceae, Balsaminaceae, *Escallonia* and *Montinia* from Grossulariaceae, and *Eremosyne* and *Vahlia* from Saxifragaceae; Olmstead et al. 1992, 1993; Chase et al. 1993).

The asterids sensu APG (1998), APG II (2003), APG III (2009), and APG IV (2016) include approximately one-third of all angiosperm species, with almost 80,000 species in nearly 4,700 genera and approximately 111 families (Thorne 1992a,b; Albach et al. 2001c; APG IV 2016). In contrast to the large rosid clade, for which no unifying non-DNA characters are known (see Chapter 10), several features unite all, or most, asterids. The asterid clade includes nearly all species of angiosperms that produce iridoids (Jensen 1992) and tropane alkaloids (Romeike 1978) and most angiosperms that produce caffeic acid (Mølgaard 1985; Grayer et al. 1999). Dahlgren (1975) noted that the occurrence of iridoids is strongly correlated with sympetalous flowers and embryological characters such as unitegmic-tenuinucellate ovules and cellular endosperm development. However, although several non-DNA characters are prevalent throughout the asterids, circumscription of the clade based on morphological data alone has been difficult because of extensive parallelisms (Hufford 1992) and losses of characters (Olmstead et al. 1992; Albach et al. 2001b). Patterns of character evolution are discussed in more detail later in this chapter.

The circumscription of and relationships within asterids sensu APG have been clarified through both broad analyses of the angiosperms (e.g., Chase et al. 1993; D. Soltis et al. 1997a,b, 2000; P. Soltis et al. 1999a; Savolainen et al. 2000a,b; Moore et al. 2010, 2011; Soltis et al. 2011; Ruhfel et al. 2014) and more focused analyses of specific clades. The most comprehensive recent molecular analyses of the asterids themselves (Albach et al. 2001c; Bremer et al. 2002) have built on the earlier work of Olmstead et al. (1992, 1993, 2000).

Most asterids fall into one of four major clades—Cornales, Ericales, lamiids (euasterid I or *Garryidae*; Cantino et al. 2007), and campanulids (euasterid II or *Campanulidae*; Cantino et al. 2007) (Fig. 11.10). With combined datasets of two (Savolainen et al. 2000a), three (D. Soltis et al. 2000), four (Albach et al. 2001c), six (Bremer et al. 2002), and 17 genes (Soltis et al. 2011), internal support is generally high (bootstrap values >95%) for these clades, as well as for the relationships among them. In earlier analyses, lamiids received only weak support (56% jackknife) with three genes (D. Soltis et al. 2000) but very strong support (99% jackknife) with six DNA regions (Bremer et al. 2002). More recent analyses also found strong support for that clade (Jansen et al. 2007; Moore et al. 2011; Soltis et al. 2011).

We discuss phylogenetic relationships of Asteridae following D. Soltis et al. (2011). Cornales, followed by Ericales, are sister to lamiids + campanulids (*Gentianidae* sensu Cantino et al. 2007) (Fig. 11.10). The composition and internal phylogenetic structure of these clades are discussed in more detail below. Fossil elements known for this group are given in Supplementary Table 11.1 (at press.uchicago.edu/sites/soltis/).

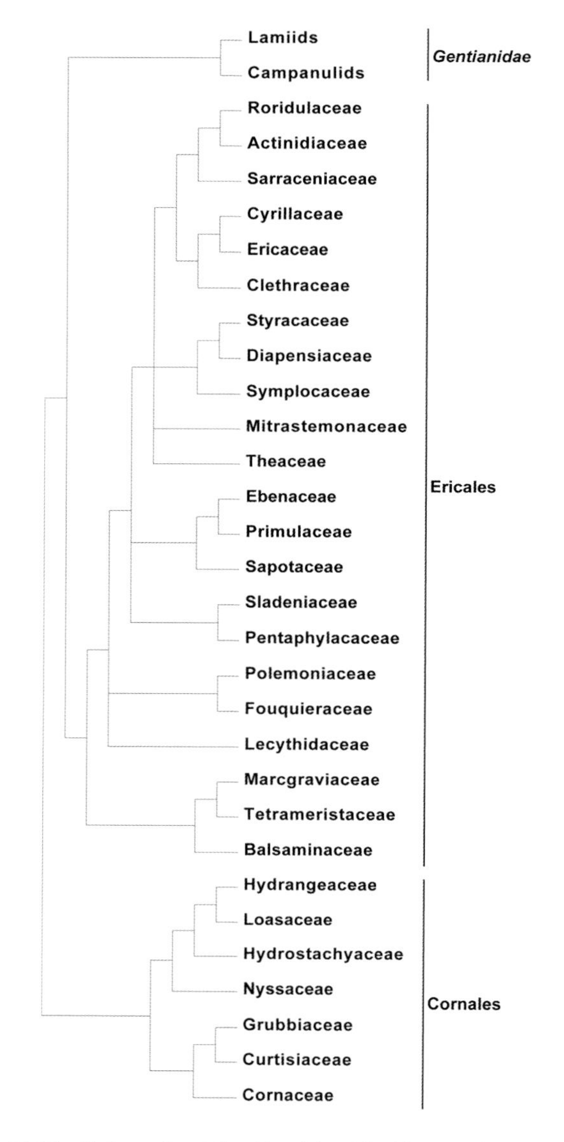

Figure 11.10. Phylogenetic relationships of Asterids with an emphasis on families within Ericales and Cornales.

CORNALES

Cornales consist of six (sensu APG III 2009), seven (sensu Stevens 2001 onwards; APG II 2003; APG IV 2016; Fig. 11.10), or ten families (Xiang et al. 2011) (see Chapter 12): Cornaceae (including *Alangium*, sometimes recognized as Alangiaceae; Xiang et al. 2011), Nyssaceae (including Mastixiaceae and Davidiaceae; Xiang et al. 2011), Loasaceae, Hydrangeaceae, Grubbiaceae, Curtisiaceae, and Hydrostachyaceae (several families are illustrated in Figs. 11.11, 11.12). This circumscription expands Cronquist's (1981) Cornales with the addition of Hydrangeaceae (Cronquist's Rosidae-Rosales), Hydrostachyaceae (Asteridae-Callitrichales), Loasaceae (Dilleniidae-Violales), and *Grubbia* (Dilleniidae-Ericales).

Molecular data have also indicated that three families should be excluded from Cornales sensu Cronquist: Garryaceae, Aucubaceae (now included in Garryaceae), and Helwingiaceae. These three taxa have affinities elsewhere in the asterids (see "Garryales" and "Aquifoliales," below). The DNA-based circumscription of Cornales resembles more closely the Cornales of Takhtajan (1997), although he placed Hydrostachyaceae and Loasaceae in Lamiidae and Grubbiaceae in Bruniales of his Dilleniidae. Thorne's (1992a,b) Cornineae resembled Cronquist's Cornales, but also included Eucommiaceae and Icacinaceae, both of

Figure 11.11. Members of Cornales (Cornaceae, Curtisiaceae, Grubbiaceae, Hydrangeaceae, Hydrostachyaceae, Loasaceae, Nyssaceae). a. *Cornus alternifolia* L.f., (Cornaceae), branch with drupes, b. *Cornus foemina* Mill. (Cornaceae), 4-merous flowers with distinct petals, c. *Philadelphus inodorus* L. (Hydrangeaceae), flowers, d. *Mentzelia floridana* Nutt ex. Torr. & A. Gray (Loasaceae), note uncinate trichomes on the ovary. e. *Davidia involucrata* Baill. (Nyssaceae), drupes; note ridged pit. f. *Nyssa ogeche* Marshall (Nyssaceae), drupes.

which are now known to have relationships elsewhere in the asterids, Garryales and Icacinales (i.e., basal lamiids). Non-DNA synapomorphies include the sepals obsolete or represented by small teeth, pollen grains sometimes with H-shaped thin-regions (endoapertures), an inferior or half-inferior ovary, nectary a more or less epigynous disk, and drupaceous fruits.

Cladistic analyses using restriction sites (Downie and Palmer 1992), morphological and chemical data (Hufford 1992), *rbcL* or *matK* sequences (Olmstead et al. 1992,

1993; Chase et al. 1993; Xiang et al. 1993), and combined DNA datasets involving multiple genes (e.g., Albach et al. 2001a,c; Xiang et al. 2002; Xiang et al. 2011) have revealed a similar Cornales clade. Cornales include 51 genera and approximately 590 species (Stevens 2001 onwards).

Cornales comprised two subclades based on plastid data: Curtisiaceae + Grubbiaceae and Cornaceae (including Alangiaceae); and Nyssaceae (including Mastixiaceae and Davidiaceae), Hydrostachyaceae, Loasaceae, and Hydrangeaceae (Xiang et al. 2011). Curtisiaceae and Grubbiaceae formed a clade sister to Cornaceae (including Alangiaceae), and the clade composed of Mastixiaceae + (Nyssaceae + Davidiaceae) was sister to a clade composed of Hydrostachyaceae + (Loaceae + Nyssaceae). However, based on nuclear ribosomal data (Xiang et al. 2011), Grubbiaceae + Curtisiaceae were recovered as sister to Loasaceae, Nyssaceae and Davidiaceae formed a clade with unresolved sister relationships, Mastixiaceae formed a clade also with unresolved sister relationships, Hydrostachyaceae formed a clade with Alangiaceae that was sister to Cornaceae s.s., and Hydrangeaceae formed a clade sister to the rest of Cornales. Combined analyses of both plastid and nuclear data resolved the same clades as in the plastid analysis, although sister relationships of Nyssaceae (including Davidiaceae and Mastixiaceae) were unresolved.

Cornales are well represented in the fossil record. Among the oldest examples is the fruit *Hironoia* from the Late Cretaceous (Coniacian) of Japan (Takahashi et al. 2002). The fruit was thick-walled, and composed mainly of fibers, with three or four locules each containing one pendulous seed and opening by a single dorsal valve; the fruit also showed preserved persistent, epigynous tepals, and a single, narrow style. Absence of an axial vascular strand is a distinctive feature that it shares with most extant Cornaceae and Nyssaceae. The presence of fibers, rather than isodiametric sclereids, is a feature distinguishing this from *Cornus* and *Alangium* and indicating similarity with Nyssaceae (Takahashi et al. 2002).

Cornus and *Alangium* are recognizable by the early Cenozoic. *Cornus* is confirmed in the Paleocene on the basis of leaves with compass-needle trichomes (Fig. 11.12d,e), and well-preserved fruits (Fig. 11.12g). *Alangium* is recognizable from fruits from the early Eocene London Clay (Reid and Chandler 1933) and middle Eocene of Oregon (Manchester 1994b), as well as by fossil-dispersed pollen (Morley 1982; Martin et al. 1996).

The fossil record of Nyssaceae includes many occurrences representing both extinct and extant genera (Supplementary Table 11.1). Fruits conforming morphologically and anatomically to extant *Nyssa* have been identified from the Campanian of Alberta, Canada. They are trilocular with single-seeded locules that are c-shaped in cross-section, having dorsal, apically positioned germination valves (Fig. 11.12j; Manchester et al. 2015). *Davidia* also is first known from the late Campanian of Alberta, Canada, based on permineralized fruits (Fig. 11.12h; Manchester et al. 2015). These fossil fruits conform to the modern genus in having multiple, radially arranged, single-seeded locules, fibrous endocarp construction, and elongate dorsal germination valves. They closely resemble the modern genus morphologically and anatomically, but differ from the modern species of *Davidia* by their smaller size and fewer locules (5–6 vs 6–9). By the Paleocene, *Davidia* is recognizable on the basis of well-preserved fruits, and the infructescence axis, bearing scars of showy bracts, and cordate, serrate leaves fitting well to the modern genus (Manchester 2002). Extinct members of Nyssaceae include the Paleocene plants *Amersinia* in North America and Asia (Manchester et al. 1999) and *Browniea* in North America (Manchester and Hickey 2007). These plants bore heads of trilocular fruits similar to those of extant *Camptotheca* and had dorsal germination valves conforming with Nyssaceae, but they are distinguished by differences in fruit wall and septum, thickness, perianth development, and the associated foliage and pollen types.

Although Mastixioideae contain only two Old World tropical genera today, *Mastixia* and *Diplopanax*, this subfamily was diverse and widespread in the Northern Hemisphere based on fossil fruits (Mai 1993; Stockey et al. 1998) (Supplementary Table 11.1). The earliest examples are from the Late Cretaceous (Maastrichtian) of Germany (Knobloch and Mai 1986). These fossils show full-length dorsal germination valves, one of the diagnostic features of Mastixioideae and Cornaceae, distinguishing them from the Nyssaceae, which have germination valves confined to the apical half of the fruit. Cenozoic mastixioid fruits include the two modern genera (e.g., Fig. 11.12), as well as several extinct genera with novel combinations of fruit characters (Mai 1993).

Cornales appear to have originated in the mid-Cretaceous, according to divergence time estimates based on numerous fossil calibrations (stem age = ~115 mya, crown age = ~96 mya; Xiang et al. 2011). Others have recovered similar ages for the clade, between 123–89 mya (Wikström et al. 2001; Bremer et al. 2004; Anderson et al. 2005; Janssens et al. 2009; Lemaire et al. 2011a). Possible non-DNA synapomorphies for Cornales include an ovary that is at least partially inferior, an epigynous nectar disk, small sepals, and drupaceous fruits (Fig. 11.11); however, there are deviations from these features, particularly in the highly modified aquatic *Hydrostachys* (see Albach et al. 2001a; Hufford et al. 2001). See Fig. 11.12 for representatives of fossil members of Cornales.

The southern African genus *Curtisia* was included in

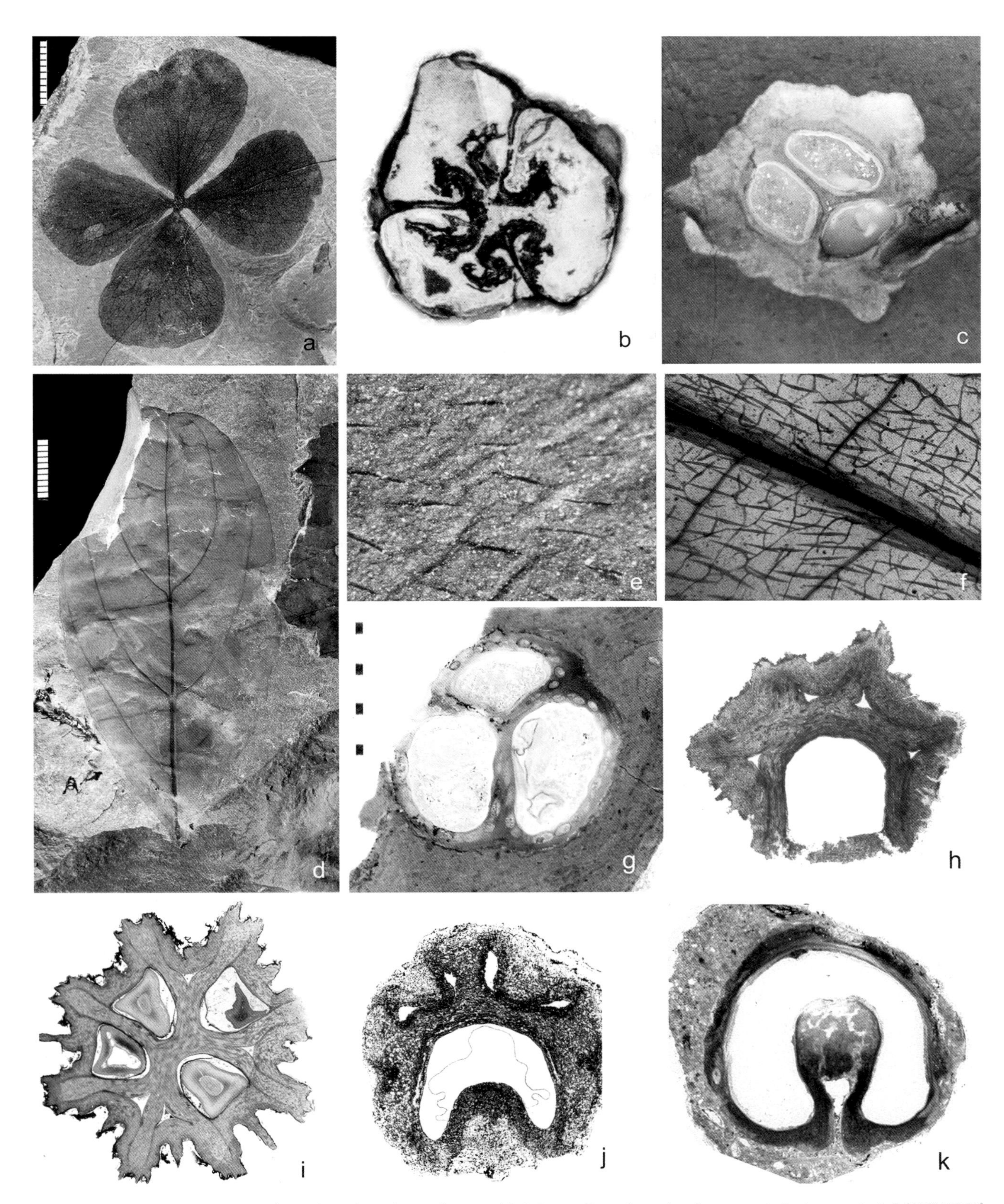

Figure 11.12. Paleobotanical examples of Cornales. a, b. *Hydrangea* flower and fruit Eocene Clarno Formation, Oregon, a. sterile showy perianth [uf 230-19187], b. transverse section of fruit showing placentation, c. Extinct Nyssaceae: *Amersinia* fruit in transverse section [uf 22474], d. *Cornus swingii* leaf from Paleocene of Montana [uf 18126–13266], e. Closeup of leaf impression from D, showing trichome impressions, f. Extant *Cornus controversa* leaf showing double armed trichomes for comparison with e, g. *Cornus piggae* fruit in transverse section showing characteristic lacunose wall and three locules, Paleocene of North Dakota [pp34489], h. *Davidia* fruit in transverse section Late Cretaceous (late Campanian) of Drumheller, Alberta [Tyrrell Museum 1988.232,6505], i. Modern fruit of *Davidia involucrata* in transverse section, j. *Nyssa* fruit in transverse section Late Cretaceous (late Campanian) of Drumheller, Alberta [Tyrrell Museum], k. *Mastixia* sp. from Eocene of Oregon [uf 6303].

Cornaceae by Cronquist (1981) and APG I (1998), but based on more recent analyses, Curtisiaceae were reinstated as distinct in APG II (2003), and that treatment has been maintained. Fossil fruits of *Curtisia* from the early Eocene of England indicate that its modern distribution is relictual (Manchester et al. 2007b). The relationships of *Grubbia*, a genus of three species from southern Africa, have long been problematic using morphology. It has variously been referred to Santalales, Ericales, and Rosales (reviewed in Cronquist 1981; see also Carlquist 1978; Xiang et al. 2002).

Cornaceae (including *Alangium*) form a strongly supported subclade. *Cornus* and *Alangium* resemble each other in floral morphology and embryology but differ in biochemistry and some morphological features (Eyde 1988). Nyssaceae sensu lato, including *Camptotheca, Davidia*, *Mastixia*, and *Diplopanax*, also form a well-supported clade. Some analyses have suggested that Nyssaceae and Cornaceae are sisters; however, bootstrap support for this sister group relationship was less than 50% (Xiang et al. 2002), and more recent analyses do not support that placement (Xiang et al. 2011). Although it had been proposed that Nyssaceae be included in Cornaceae (APG II 2003), that circumscription was shown to be premature using either plastid or nuclear data (Xiang et al. 2011), and Nyssaceae have now been reinstated as a distinct family (APG IV 2016; Chapter 12). However, it should be noted that Cornaceae and Nyssaceae share a number of features such as dorsal germination valves in their fruits, a base chromosome number of 11, and trans-septal vascular bundles to the ovules. Cornaceae and most Nyssaceae also share the presence of ellagitannins (Bate-Smith et al. 1975).

Loasaceae and Hydrangeaceae appear as well-supported sisters in molecular analyses (e.g., Hempel et al. 1995; D. Soltis et al. 1995, 2000; Savolainen et al. 2000a,b; Hufford et al. 2001; Xiang et al. 2002; Xiang et al. 2011), a relationship not recognized before Hufford's (1992) analysis of morphological and chemical data. Both families have rare iridoids (deutzioside and derivatives, C_{10}-decarboxylated iridoids; see "Patterns of Character Evolution," below; El-Naggar and Beal 1980) and share tuberculate trichomes with basal cell pedestals (Hufford 1992). *Fendlera* (Hydrangeaceae) and Loasaceae share similar vessel elements (Quibell 1972). The relationship between Hydrangeaceae and Loasaceae is also supported by floral morphology (Roehls et al. 1997; Albach et al. 2001c). Relationships within Hydrangeaceae and Loasaceae have been the subject of several morphological and molecular analyses (D. Soltis et al. 1995; Hufford et al. 2001; Moody et al. 2001; Xiang et al. 2011).

Hydrangeaceae can be traced to the Late Cretaceous based on charcoalified flowers of *Tylerianthus crossmanensis* Gandolfo from the Turonian of New Jersey (Gandolfo et al. 1998). The extant genus *Hydrangea* is readily recognized by the Eocene based on compressed, showy, sterile flowers and fruits (Fig. 11.12a; Mustoe 2002). Permineralized Eocene *Hydrangea* fruits from the middle Eocene of Oregon show intruded parietal placentation (Fig. 11.12b; Manchester 1994b).

The enigmatic submerged aquatic *Hydrostachys* (ca. 22 species; Xiang et al. 2002; Fan and Xiang 2003), the only member of Hydrostachyaceae, is well supported as a member of Cornales and was recovered as sister to Hydrangeaceae-Loasaceae using plastid data. However, the position of *Hydrostachys* within Cornales remains uncertain, as nuclear and plastid data suggest different relationships (Xiang et al. 2011). The placement of *Hydrostachys* may be difficult to resolve given the long branch to this taxon in most analyses.

ERICALES

Ericales consist of 22 families (Fig. 11.10): Actinidiaceae, Balsaminaceae, Clethraceae, Cyrillaceae, Diapensiaceae, Ebenaceae (including Lissocarpaceae), Ericaceae, Fouquieriaceae, Lecythidaceae (including Scytopetalaceae), Marcgraviaceae, Pentaphylacaceae (including Ternstroemiaceae), Sladeniaceae, Polemoniaceae, Primulaceae (including Maesaceae, Myrsinaceae, Theophrastaceae), Roridulaceae, Sapotaceae, Sarraceniaceae, Styracaceae (including Halesiaceae), Symplocaceae, Tetrameristaceae (including Pellicieraceae), Theaceae and possibly the parasitic Mitrastemonaceae (several members of Ericales are illustrated in Figs. 11.13 and 11.14 and fossils in 11.17) (Chase et al. 1993; Kron and Chase 1993; Olmstead et al. 1993; Johnson and Soltis 1995; Johnson et al. 1996, 1999; Kron 1996, 1997; Morton et al. 1996; D. Soltis et al. 1997b, 2000; Anderberg et al. 1998, 2002; Savolainen et al. 2000a,b; Bremer et al. 2002; Schönenberger et al. 2005; Soltis et al. 2011). However, the exact placement of the genus *Mitrastema* (Mitrastemonaceae) within Ericales remains unclear (Barkman et al. 2004).

Ericales consist mainly of taxa previously placed in Dilleniidae (sensu Cronquist 1981); the clade includes many former members of the dilleniid orders Theales, Lecythidales, Ericales, Diapensiales, Ebenales, and Primulales, and families Fouquieriaceae and Sarraceniaceae, as well as Polemoniaceae (Asteridae) and Balsaminaceae (Rosidae). The monophyly of the clade is strongly supported in multigene analyses (D. Soltis et al. 2000; Albach et al. 2001c; Bremer et al. 2002; Schönenberger et al. 2005; Soltis et al. 2011); weak support is also apparent with *rbcL* alone (e.g., Källersjö et al. 1998; Savolainen et al. 2000b).

Figure 11.13. Members of Ericales (Actinidiaceae, Balsaminaceae, Clethraceae, Cyrillaceae, Diapensiaceae, Ebenaceae, Ericaceae, Fouquieriaceae, Lecythidaceae, Marcgraviaceae, Mitrastemonaceae, Pentaphylacaceae, Polemoniaceae, Primulaceae, Roridulaceae, Sapotaceae, Sarraceniaceae, Sladeniaceae, Styracaceae, Symplocaceae, Tetrameristaceae, Theaceae): a. *Impatiens pallida* Nutt. (Balsaminaceae), herb with zygomorphic flowers; note abaxial sepal with nectar-spur, lateral petals connate in pairs, b. *Marcgravia rubra* Alain (Marcgraviaceae), inflorescence with flowers and nectaries (derived from floral bracts), c. *Gustavia superba* Berg. (Lecythidaceae), flower with numerous stamens, d. *Phlox drummondii* Hook. (Polemoniaceae), flowers with sympetalous corollas, the tube well developed and with a spreading limb, e. *Fouquieria splendens* Engelm. (Fouquieriaceae), flowers modified for bird pollination; note distinct sepals and punctate stigmas, f. *Ternstroemia gymnanthera* (Wight & Arn.) Bedd. (Pentaphylacaceae), fleshy capsules and seeds with colorful sarcotesta, g. *Manilkara zapota* (L.) P. Royen (Sapotaceae), branch with spirally arranged leaves, sympetalous flowers and berry; note latex, brownish hairs, h. *Diospyros digyna* Jacq. (Ebenaceae), branch with distichous leaves, berries with accrescent calyx; note dark-colored naphthaquinones, i. *Diospyros ebenum* J. Koenig (Ebenaceae), carpellate flowers with urceolate corollas, j. *Bonellia macrocarpa* (Cav.) B. Stahl & Kallersjo (Primulaceae), branch with flowers and fruits; fruit showing free-central placentation and flowers with petaloid staminodia and stamens opposite the corolla lobes.

The age of Ericales has been estimated at 126–80 mya (Wikström et al. 2001; Anderson et al. 2005; Sytsma et al. 2006; Soltis et al. 2008b; Janssens et al. 2009; Magallón and Castillo 2009; Bell et al. 2010; Lemaire et al. 2011a). Several extinct genera, based on well-preserved fossil flowers—for example, *Palaeoenkianthus* (Nixon and Crepet 1993), and *Raritaniflora* (Crepet et al. 2013) from New Jersey, extend the Ericales back to the Late Cretaceous (Turonian; ~92 mya) (Martínez-Millán 2010; Friis et al. 2011; see also Schönenberger et al. 2005) (Supplementary Table 11.1).

Ericales differ from most other asterids in the common occurrence of ellagic acid, a compound frequently found outside the clade (see Chapter 8, Fig. 8.7a) and stamens usually twice the number of petals (to numerous; vs. stamen number equaling or fewer than the number of petals in the euasterids). A possible synapomorphy for the clade is the presence of protruding, diffuse placentae (Nandi et al. 1998). Another uniting character is the theoid tooth type in which a single vein enters the tooth and ends in an opaque deciduous cap or gland (Hickey and Wolfe 1975; Judd et al. 2002, 2008).

In Schönenberger et al. (2005) and D. Soltis et al. (2011), a well-supported clade (100% BS in Soltis et al. 2011) composed of Marcgraviaceae + (Tetrameristaceae + Balsaminaceae) is sister to the rest of Ericales (BS = 95%), although interrelationships within the latter are mostly poorly supported or unresolved. Well-supported subclades in the rest of Ericales are Lecythidaceae, Fouquieriaceae + Polemoniaceae, Sapotaceae + (Ebenaceae + Primulaceae s.l.) (well-supported in Schönenberger et al. 2005, but not in Soltis et al. 2011), Pentaphylacaceae, Theaceae, Symplocaceae, Styracaceae, Diapensiaceae, and Sarraceniaceae + (Roridulaceae +Actinidiaceae), which is sister to a clade composed of Clethraceae + (Cyrillaceae + Ericaceae). In D. Soltis et al. (2011) Lecythidaceae form the second diverging clade that is poorly supported (BS = 67%) as sister to the rest, although this is unresolved in Schönenberger et al. (2005). Fouquieriaceae and Polemoniaceae form a well-supported clade that is sister to the remaining members, although this also is very poorly supported (BS = 53%) and unresolved in Schönenberger et al. (2005). The only other resolved relationship is the clade ((Sarraceniaceae + (Roridulaceae +Actinidiaceae) + Clethraceae + (Cyrillaceae + Ericaceae)). Descriptions of these subclades within Ericales follow.

With three, four, six, 11, or 17 genes combined, the backbone of the tree for Ericales is mostly unresolved (Fig. 11.10), with few interfamilial relationships receiving strong support. However, as noted, the clade of Tetrameristaceae (including Pellicieraceae), Marcgraviaceae, and Balsaminaceae is strongly supported (Morton et al. 1996; Savolainen et al. 2000b; D. Soltis et al. 2000; Albach et al. 2001c; Anderberg et al. 2002; Bremer et al. 2002) and occupies a pivotal position as sister to the rest of Ericales. Marcgraviaceae are sister to Balsaminaceae + Tetrameristaceae (including Pellicieraceae). A close relationship among Tetrameristaceae (including Pellicieraceae), Marcgraviaceae, and Balsaminaceae (Fig. 11.10) had never been proposed. These families differ in habit, varying from herbaceous (Balsaminaceae) to woody climbers and shrubs (Marcgraviaceae). However, they share specialized nectaries or glands on leaves, sepals, or petals (Morton et al. 1996), mostly simple perforation plates in vessels, and hypogynous flowers (Albach et al. 2001c). Von Balthazar and Schönenberger (2013) found additional shared features: broad and dorsiventrally flattened filaments, thread-like structures lining the stomia of dehisced anthers, secretory inner primary morphological surfaces of the gynoecium, ovules intermediate between uni- and bitegmic, incompletely tenuinucellar ovules, fruits with persistent style and stigma, and seeds lacking endosperm. Oxalate druses are absent; members of the clade instead form oxalate raphides (Nandi et al. 1998).

Balsaminaceae and Tetrameristaceae share a caducous calyx, postgenital coherence of filaments and ovary, latrorse anther dehiscence, commissural carpel lobes and ovules with a thickened funiculus and a constricted chalazal region (von Balthazar and Schönenberger 2013). Marcgraviaceae and Tetrameristaceae also exhibit several wood features, such as vasicentric axial parenchyma and high and wide multiseriate rays with heterocellular body ray cells in Marcgraviaceae, and apotracheal and paratracheal axial parenchyma in Tetrameristaceae. Both Marcgraviaceae and Tetrameristaceae have raphides (Lens et al. 2007).

Fossil Ericaceae are recognized by the distinctive permanent tetrads known from the Paleocene (Fig. 11.17c) and younger strata and by dispersed winged seeds from the Paleocene of England (Collinson and Crane 1978) and late Eocene of California (Wang and Tiffney 2001).

Pollen of Sapotaceae is sufficiently distinctive to be recognized as dispersed fossil grains. Examples are known from the Paleocene up to the Miocene worldwide (e.g., Muller 1981; Manchester et al. 2015). Ebenaceae pollen have a distinctive micro-ornamentation of organized microrugulate sculpture. Such pollen has been identified as dispersed grains from the late Eocene of Florissant, Colorado (Manchester 2015), and from Eocene flowers named *Austrodiospyros cryptostoma* with in situ pollen from late Eocene of Anglesea, Victoria, Australia (Basinger and Christophel 1985).

Polemoniaceae and Fouquieriaceae form another small clade within Ericales (Fig. 11.10) (Anderberg et al. 2002;

Figure 11.14. Ericales, cont'd. a. *Gordonia lasianthus* (L.) J. Ellis (Theaceae), flower, b. *Halesia carolina* L. (Styracaceae), branch with flowers, c. *Symplocos hartwegii* A. DC. (Symplocaceae), flower; note only slightly connate petals, numerous stamens with globose anthers, d. *Saurauia* sp. (Actinidiaceae), flower with anthers late-inverting, more or less porose, e. *Roridula gorgonias* Planch. (Roridulaceae), gland-headed trichomes on leaf blades, f. *Clethra mexicana* DC. (Clethraceae), flowers, g. *Sarracenia minor* Walter (Sarraceniaceae), pitcher-like leaves forming insect traps and greatly expanded style. h. *Cyrilla racemiflora* L. (Cyrillaceae), plant in bloom, i. *Eubotrys racemosus* Nutt. (Ericaceae), pendulous flowers with sympetalous, urceolate corollas; note inverted anthers opening by seemingly terminal pores, and with awns, j. *Sarcodes sanguinea* Torr. (Ericaceae), a mycoparasite.

Bremer et al. 2002; Schönenberger et al. 2005; Soltis et al. 2011). A sister-group relationship of Polemoniaceae and Fouquieriaceae is consistent with the morphological similarity observed between *Fouquieria* (Fouquieriaceae) and *Acanthogilia,* one of the basalmost nodes within Polemoniaceae (Nash 1903; Henrickson 1967; Hufford 1992; Johnson et al. 1996), and biochemistry (Scogin 1977, 1978). Schönenberger et al. (2010) found more similarities between the two families: cochlear and quincuncial corolla aestivation, connective protrusions, ventrifixed anthers, and nectariferous tissue at the base of the ovary. The fossil record has yielded little information for this clade, with the exception of a single relatively complete fossil *Gilisenium hueberi,* based on a single specimen showing a plant with the roots, leaves, and fruits assignable to Polemoniaceae from the middle Eocene of Utah (Lott et al. 1998).

The relationships of Fouquieriaceae and Polemoniaceae to the primuloid clade (Primulaceae s.l.; discussed below) remain unresolved. Polemoniaceae are sister to the primuloid clade in both D. Soltis et al. (2000) and Albach et al. (2001c); this relationship appears in the strict consensus tree of the former study and received weak (BS = 61%) support in the latter analysis. However, in both studies, Fouquieriaceae appear elsewhere in Ericales. Non-DNA characters also indicate a close relationship of Polemoniaceae to the primuloid group. Primuloids and Polemoniaceae share nuclear endosperm and simple perforation plates in vessels (see Albach et al. 2001c). Neither character is restricted to these two groups within Ericales, but they occur in combination elsewhere only in Sapotaceae.

Another character shared by at least some members of the primuloid clade and Polemoniaceae is the occurrence of separate sepals in *Cobaea, Polemonium,* and some genera of Primulaceae s.l. (Theophrastaceae and Myrsinaceae), generally a rare condition among sympetalous plants (Stebbins 1974). Additionally, Primulaceae and Polemoniaceae both initiate stamens earlier than petal primordia (Nishino 1978, 1983), but this needs more study. Polemoniaceae and Primulaceae share many biochemical characters, such as the presence of cucurbitacins, and plants in both families excrete methylated 6- or 8-oxygenated flavonols from glandular hairs (Hegnauer 1962–1994).

The polyphyly of Theaceae was noted earlier with the placement of *Asteropeia* (now Asteropeiaceae) within Caryophyllales (this Chapter) and the placement of subfamily Bonnetioideae (now Bonnetiaceae) within Malpighiales (Chapter 10; Savolainen et al. 2000b). The two remaining former subfamilies of Theaceae (Theoideae, Ternstroemioideae) have not formed a clade in any molecular analyses (e.g., Morton et al. 1996; Savolainen et al. 2000a,b; D. Soltis et al. 2000; Albach et al. 2001c; Bremer et al. 2002; Schönenberger et al. 2005; Soltis et al. 2011), supporting the recognition of two separate families (APG 1998), Theaceae and Ternstroemiaceae (the latter included in Pentaphylacaceae; APG II 2003; APG III 2009; APG IV 2016).

Theaceae and Pentaphylacaceae differ in embryological characters (curved rather than straight embryo in the latter; Tsou 1995) and floral characters, including fruit and stamen morphology. The phylogenetic positions of the two families within Ericales are not clear, but they repeatedly appear as part of different clades. Savolainen et al. (2000b) first showed that the monogeneric Sladeniaceae should be added to Ericales, as a member of Ternstroemiaceae (now Pentaphylacaceae); however, APG III (2009) and Stevens (2012) recognized *Sladenia* in its own family, separate from Pentaphylacaceae (see also APG IV 2016). The DNA-based position of *Sladenia* in Ericales is in general agreement with the treatments of Dahlgren (1980) and Cronquist (1981), who placed Sladeniaceae in Theales/Theaceae, the component members of which are found in Ericales as circumscribed here. In a more recent study, Anderberg et al. (2002) found *Sladenia* and *Ficalhoa* (Theaceae) to be a weakly supported (51%) sister group to a well-supported (96%) clade of Pentaphylacaceae + Ternstroemiaceae. Bremer et al. (2002) found stronger support (72%) for a clade of *Sladenia* as sister to Pentaphylacaceae + Ternstroemiaceae, but they did not include *Ficalhoa* in their analysis. In Bremer et al. (2001), Ternstroemiaceae appear with Sladeniaceae and Pentaphylacaceae in a weakly supported (72% jackknife) clade now referred to as a broadly defined Pentaphylacaceae (sensu APG II 2003). In D. Soltis et al. (2000), Albach et al. (2001c), and Bremer et al. (2002), the narrowly defined Theaceae are sister to the clade of Symplocaceae (Diapensiaceae + Styracaceae), but this relationship received support above 50% (52% jackknife) only in Bremer et al. (2001) and has not been supported in subsequent multigene analyses (Schönenberger et al. 2005; Soltis et al. 2011) where the relationship of Theaceae and other clades was unresolved.

Diospyros (Ebenaceae) and *Lissocarpa* (Lissocarpaceae) are strongly supported sisters (D. Soltis et al. 2000; Berry et al. 2001; Anderberg et al. 2002; Schönenberger et al. 2005) or sister to a *Diospyros* + (*Euclea* + *Royena*) clade in Duangjai et al. (2006, 2009). Savolainen et al. (2000b) suggested that *Lissocarpa* might be a member of Sapindales, but this placement was based on an *rbcL* sequence obtained from degraded DNA. The placement of Lissocarpaceae as a close relative of Ebenaceae agrees with morphological data; the two families share similar floral morphology and wood anatomy, differing chiefly in that *Lissocarpa* has an inferior ovary (Cronquist 1981; Berry et al. 2001; Anderberg et al. 2002). On the basis of molecular and morphological evidence, Lissocarpaceae have been included within Ebenaceae (APG II 2003; APG III 2009; APG IV 2016). Recent studies have illuminated considerable phylogenetic structure within Ebenaceae, especially in the widespread genus *Diospyros* (Duangjai et al. 2009).

Molecular analyses revealed a well-supported primuloid clade of Primulaceae, Myrsinaceae, and Theophrastaceae, and Maesaceae, a family consisting only of *Maesa* (formerly of Myrsinaceae; Anderberg et al. 2000, 2002; Källersjö et al. 2000), all four of which have been recircumscribed as part of Primulaceae s.l. (APG III 2009; APG IV 2016). This primuloid clade corresponds to Cronquist's (1981) Primulales and is strongly supported by both molecular (D. Soltis et al. 2000; Albach et al. 2001c) and morphological (Cronquist 1981; Anderberg and Ståhl 1995) data. This clade is united by free central placentation and stamens equal in number and opposite the corolla lobes (Fig. 11.13j); free central placentation is an unusual feature, occurring also in Santalales and Lentibulariaceae (as reviewed in this Chapter, some Caryophyllales have secondarily free central placentation). Maesaceae are sister to the remainder of the primuloid clade, followed by Theophrastaceae as sister to Myrsinaceae + Primulaceae (Anderberg et al. 2002; Bremer et al. 2002; Schönenberger et al. 2005; Soltis et al. 2011), all of which now constitute a broadly defined Primulaceae (APG III 2009; APG IV 2016).

Diapensiaceae and Styracaceae (including Halesiaceae) form a clade with low to moderate (D. Soltis et al. 2000; Albach et al. 2001c; Bremer et al. 2002; Soltis et al. 2011) or high support (Schönenberger et al. 2005). Strongest support is provided by the six-gene analysis of Bremer et al. (2002) and the 11-locus analysis of Schönenberger et al. (2005). In contrast to previous analyses based on *rbcL* (Morton et al. 1996) that showed *Halesia* and other genera of Styracaceae to be only distantly related to *Styrax*, multigene analyses of Bremer et al. (2002) and Schönenberger et al. (2005) found Styracaceae (including *Halesia*) to form a well-supported clade with 100% bootstrap support and 1.0 posterior probability, respectively (see also Kron 1996). Characters that unite Styracaceae and Diapensiaceae include calcium oxalate crystals, unitegmic ovules (but not in all Styracaceae), unilacunar nodes, cellular endosperm, and binucleate pollen (Albach et al. 2001c; see also Morton et al. 1996). The two families also share the *Chenopodium* variant of polygonad type embryo formation (Yamazaki 1974). These characters (except polygonad embryo formation) are also present in Symplocaceae. Symplocaceae appear as sister to Styracaceae + Diapensiaceae in the six-gene analysis (Bremer et al. 2002) and the 11-locus analysis of Schönenberger et al. (2005). The fossil record of this clade includes reports of Styracaceae, such as *Rehderodendron* fruits from the Eocene to Miocene of Europe (Mai 1970; Vaudois-Miéja 1983; Manchester et al. 2009), seeds of *Styrax* from the European Miocene (Kirchheimer 1957), and of Symplocaceae with an excellent fossil record of endocarps, reviewed by Mai and Martinetto (2006). The oldest known occurrences of *Symplocos* fruits are from the early Eocene of London Clay (Reid and Chandler 1933; Fritsch et al. 2014).

Both molecular and morphological studies have provided compelling evidence for a broadly defined Ericaceae that includes Pyrolaceae, Monotropaceae, Epacridaceae, and Empetraceae (Anderberg 1993; Judd and Kron 1993; Kron and Chase 1993; Kron 1996; Kron et al. 2002; Kron and Luteyn 2005; Schönenberger et al. 2005; Braukmann and Stefanović 2012). Ericaceae are part of a larger clade (sometimes referred to as the core Ericales; see Judd et al. 2002, 2008) that also includes Actinidiaceae, Sarraceniaceae, Roridulaceae, Cyrillaceae, and Clethraceae (Fig. 11.10; Albach et al. 2000c; Savolainen et al. 2000a,b; D. Soltis et al. 2000; Bremer et al. 2002; Schönenberger et al. 2005; Soltis et al. 2011).

Within the core Ericales, Sarraceniaceae, Actinidiaceae, and Roridulaceae form one subclade (Bremer et al. 2002; Schönenberger et al. 2005), with the second subclade consisting of Clethraceae as sister to a clade of Cyrillaceae and a broadly defined Ericaceae. Core Ericales is characterized by inverted anthers, a usually hollow style that emerges from an apical depression in the ovary, and endosperm with haustoria at both ends (Anderberg 1992, 1993; Judd and Kron 1993; Kron and Chase 1993; Kron 1996); this group also contains all of the iridoid-producing genera of Ericales except *Fouquieria* and *Symplocos* (Albach et al. 2001c).

Sarraceniaceae and Roridulaceae are carnivorous (see Chapter 13). Sarraceniaceae produce water-filled, pitcher-like leaves trapping insects or other small animals that are then digested; Roridulaceae trap insects with glandular hairs on the leaves; the insects are then consumed by species of the insect hemipteran genus *Pameridea* (Miridae). The waste products of the consumed insects are then taken up by the plants in these nutrient-poor environments (Givnish 2014; Sadowski et al. 2014). Eocene fossils of Roridulaceae have recently been found in Baltic amber (Sadowski et al. 2014), which coincide with molecular dates for the Actinidiaceae + Roridulaceae (Ellison et al. 2012), and suggest that the plant family was much more widespread in the past and currently exhibits a relictual distribution (Sadowski et al. 2014). *Actinidia* has also been recognized back to the Eocene based on seeds (Fig. 11.17a,b; Manchester 1994), and Actinidiaceae have been traced to the late Cretaceous (late Santonian) based on flowers and fruits of *Parasaurauia allonensis* (Keller et al. 1996) and *Glandulocalyx* (Schönenberger et al. 2012). Sarraceniaceae, a New World family, also apparently were much more widespread in the Eocene of the Americas but now occupy vastly reduced distributions (Ellison et al. 2012).

Numerous families within Ericales are aluminum accumulators (Jansen et al. 2004), although there does not seem to be any phylogenetic signal to this character. Schönenberger et al. (2005) found that sympetaly likely evolved either two or three times in the clade. The two-whorled androecium of certain clades evolved either once or twice, and the single-integumented ovary evolved four times. Cellular endosperm formation is plesiomorphic, and nuclear endosperm formation derived in the clade; however, a reversal to cellular endosperm formation happened 1–3 times. The study of morphological evolution in Ericales is still premature, however, until more robust phylogenetic relationships among major clades can be established (Schönenberger et al. 2005).

Several families of Ericales have been the focus of more detailed phylogenetic studies, including Ericaceae s.l. (Kron and Chase 1993; Kron 1996, 1997; Kron et al. 2002; Kron and Luteyn 2005; Braukmann and Stefanović 2012), Primulaceae and related families (Anderberg et al. 1998, 2002; Mast et al. 2001; Trift et al. 2002; Martins et al. 2003), Polemoniaceae (Johnson and Soltis 1995; Johnson et al. 1996, 1999; Johnson et al. 2008), Diapensiaceae (Rönblom and Anderberg 2002), Actinidiaceae (Li et al. 2002), Styracaceae (Fritsch 2001), and Ebenaceae and related families (Morton et al. 1996; Duangjai et al. 2006, 2009; Turner et al. 2013). Also, numerous members of Ericales show very interesting patterns of neo- and paleopolyploidy and have been studied in some detail (e.g., *Actinidia*, Actinidiaceae; Chat et al. 2004; Shi et al. 2010).

EUASTERIDS (LAMIIDS AND CAMPANULIDS)

Euasterid families (Fig. 11.10) have flowers with epipetalous stamens that equal the corolla lobes in number and a gynoecium of usually two fused carpels. Molecular analyses have provided strong support for this clade and also indicated the presence of two subclades of euasterids—lamiids (*Garryidae*; euasterid I) and campanulids (*Campanulidae*; euasterid II; Olmstead et al. 1993; D. Soltis et al. 2000; Albach et al. 2001c; Bremer et al. 2002; Jansen et al. 2007; Moore et al. 2010, 2011; Soltis et al. 2011; Ruhfel et al. 2014). Euasterid I and II were used in the APG II (2003) classification. Several recent publications (Moore et al. 2011; Soltis et al. 2011; Ruhfel et al. 2014) have referred to these groups as *Lamiidae* and *Campanulidae* (sensu Cantino et al. 2007). We will use the names lamiids (asterids I) and campanulids (asterids II) (per Bremer et al. 2002), as done in APG III (2009) and APG IV (2016) and in Chapter 12.

An analysis by Bremer et al. (2001) of 142 genera using 143 morphological, anatomical, embryological, palynological, chemical, and RFLP (restriction fragment length polymorphisms) characters recovered some groups (mostly families and small groups of families) supported by DNA sequence data and recognized by APG (1998), but the analysis identified only one order, Dipsacales, of the eight APG orders of euasterids. Furthermore, lamiids and campanulids were not distinguished until *rbcL* and *ndhF* sequences were added to the analysis (Bremer et al. 2001). Thus, although morphological and other nonmolecular data contain phylogenetic signal, it is not sufficiently strong to discern all of the groups identified by molecular data.

Lamiids are generally characterized by opposite leaves, entire leaf margins, hypogynous flowers, "early sympetaly" with a ring-shaped corolla primordium, fusion of stamen filaments with the corolla tube, and capsular fruits (Bremer et al. 2001). Taxa appearing in campanulids typically have alternate leaves, serrate-dentate leaf margins, epigynous flowers, free stamen filaments, indehiscent fruits, and "late sympetaly" with distinct petal primordia (Bremer et al. 2001). Although these morphological features are useful as general descriptors, it is unclear which are synapomorphies and which are symplesiomorphies, and both reversals and parallelisms have generated many exceptions to these general patterns.

Lamiids

The lamiid clade (asterid I sensu Chase et al. 1993; lamiids sensu Bremer et al. 2002; euasterid I sensu APG II 2003) received only weak (56% bootstrap) support with three genes (D. Soltis et al. 2000), but support has increased greatly (86%) in four gene (Albach et al. 2001c), six gene (100% jackknife support; Bremer et al. 2002), 10 gene (96%; Refulio-Rodriguez and Olmstead 2014) and 17 gene analyses (100% bootstrap support; Soltis et al. 2011). Within the lamiid clade, Refulio-Rodriguez and Olmstead (2014) found that Icacinaceae s.s. (including only *Icacina* and *Cassinopsis*) formed a clade sister to a Metteniusaceae-Oncothecaceae clade, which was sister to Garryales plus the rest of the lamiids. However, Icacinaceae s.l. are known to be non-monophyletic (Kårehed 2001; Bremer et al. 2002; Lens et al. 2008; Byng et al. 2014), so additional taxon sampling in future studies will be needed to resolve

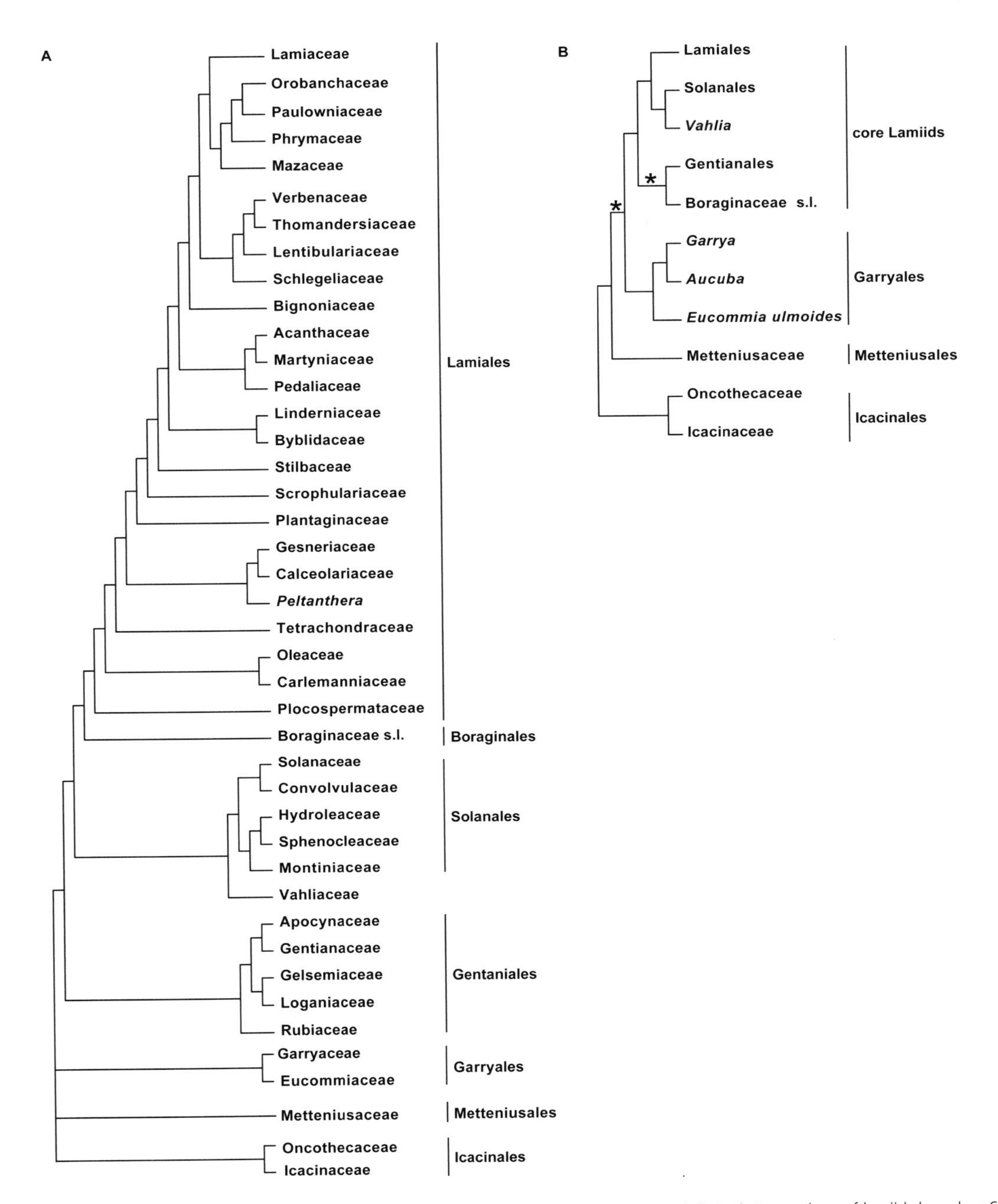

Figure 11.15. Phylogenetic relationships among lamiids. A. Topology of lamiids modified from Soltis et al. (2011), B. topology of lamiids based on Stull et al. (2015). Asterisks indicate topological differences between Stull et al. (2015) and Soltis et al. (2011).

the placement of all members traditionally placed in that family. For example, some traditional members of Icacinaceae are now placed in Stemonuraceae of Aquifoliales (campanulids, see below) and several other members of traditional Icacinaceae have been shown to be actually part of Metteniusaceae (Stull et al. 2015; Byng et al. 2014). Most recent analyses based on plastome data find that Icacinaceae + Oncothecaceae form a well-supported clade sister to Metteniusaceae and Garryales + the rest of the lamiids (Stull et al. 2015; Fig. 11.15).

The remaining lamiids are strongly supported as monophyletic and consist of Gentianales, Solanales, Boraginales,

Lamiales, as well as *Vahlia* (Vahliaceae) (Fig. 11.15; representatives are illustrated in Fig. 11.16–11.19; Soltis et al. 2011, Refulio-Rodriguez and Olmstead 2014). Although previous studies provided poor resolution for relationships among these major clades, especially considering the problematic placement of Boraginaceae, both the broad angiosperm phylogeny of D. Soltis et al. (2011) and the lamiid-focused phylogeny of Refulio-Rodriguez and Olmstead (2014) recovered Boraginaceae as sister to Lamiales. Boraginaceae + Lamiales is sister to Solanales + Vahliaceae in Refulio-Rodriguez and Olmstead (2014) and sister to a Solanales + Gentianales in Soltis et al. (2011). Refulio-Rodriguez and Olmstead (2014) recover Gentianales as sister to the Solanales, Vahliaceae, Boraginaceae + Lamiales clade. Vahliaceae were recovered as sister to Solanales in Refulio-Rodriguez and Olmstead (2014) and sister to a Solanales, Gentianales, Boraginaceae, Lamiales clade in Soltis et al. (2011).

Although APG recognizes a broadly circumscribed Boraginaceae s.l., including Hydrophyllaceae and the parasitic Lennoaceae, others split the group into numerous (ca. eight to eleven) families (Gottschling and Hilger 2001; Gottschling et al. 2001, 2005; Hilger and Diane 2003; Weigend and Hilger 2010; Cohen 2013; Weigend et al. 2013; Refulio-Rodriguez and Olmstead 2014; Luebert et al. 2016). However, topological problems (i.e., incongruence) in resultant phylogenies based on either plastid or nuclear loci (e.g., Hydrophyllaceae s.l. are monophyletic using nuclear ribosomal loci but non-monophyletic using plastid data), make the split into numerous families premature (Refulio-Rodriguez and Olmstead 2014).

ICACINALES AND METTENIUSALES

The relationships of Oncothecaceae, Metteniusaceae, and Icacinaceae to each other and other lamiid families have long been problematic, although they, with Garryales (below), consistently appear as the basalmost lineages of the clade (D. Soltis et al. 2000; Bremer et al. 2002; Gonzáles et al. 2007; Soltis et al. 2011; Refulio-Rodriguez and Olmstead 2014). In some analyses, Icacinaceae appeared (with weak support) as sister to Garryales (D. Soltis et al. 2000; González et al. 2007; Soltis et al. 2011). In the strict consensus of shortest trees obtained in B. Bremer et al. (2002), Oncothecaceae and Icacinaceae form a clade that is sister to the remainder of lamiids, although neither relationship received bootstrap support above 50%. Metteniusaceae are poorly known and only recently have been shown to belong here (González et al. 2007). González et al. (2007) found Oncothecaceae and Metteniusaceae to be the subsequent sisters to all lamiids excluding a clade of Garryales + Icacinaceae, which they found to be sister to the rest of lamiids, including Oncothecaceae and Metteniusaceae. Refulio-Rodriguez and Olmstead (2014) recovered a clade composed of Icacinaceae + (Oncothecaceae + Metteniusaceae), which was supported (96 % bootstrap) as sister to the rest of the lamiids (including Garryales). Most recent analyses, which included numerous members of traditional Icacinaceae s.l., recovered Icacinaceae s.s. + Oncothecaceae as sister to a clade of Metteniusaceae (Garryales + core lamiids) (Stull et al. 2015). These are recognized as Icacinales (Icacinaceae + Oncothecaceae) and Metteniusales (including only Metteniusaceae) in APG IV (2016) and Chapter 12.

Icacinaceae as previously circumscribed are polyphyletic (see also Aquifoliales below, Stemonuraceae). *Icacina* and related genera such as *Cassinopsis* and *Pyrenacantha* (Icacinaceae s.s.) form a well-supported clade and are part of lamiids (Fig. 11.15); several studies pointed to a close relationship of these genera to Garryales, but without strong bootstrap support (D. Soltis et al. 2000; Kårehed 2002; González et al. 2007; Soltis et al. 2011; Byng et al. 2014). Several other genera previously attributed to Icacinaceae are part of either Cardiopteridaceae or a new family, Stemonuraceae, both placed in Aquifoliales of the campanulid clade (Kårehed 2002; APG III 2009; Soltis et al. 2011; APG IV 2016; see Aquifoliales below). Also, as mentioned above, numerous genera attributed to Icacinaceae s.l. belong in Metteniusaceae (Stull et al. 2015; see nomenclature section below). Members of Icacinaceae have been the subjects of systematic and phylogenetic work (Angulo et al. 2013; Byng et al. 2014).

Among the genera retained in Icacinaceae sensu stricto, fruits often have very distinctive endocarp morphology with reticulate external ribbing and/or tubercles that protrude into the locule (Figs. 11.16d, 11.17i). These features facilitate the recognition of fossil representatives including both extinct and extant genera common in the Eocene (Fig. 11.17 h,j–l) and extending back at least to the mid-Paleocene (Stull et al. 2012) (Supplementary Table 11.1).

Oncothecaceae consist of two species (evergreen trees or shrubs) native to New Caledonia. Cronquist (1981) placed the family in his Theales (Dilleniidae), noting that it was poorly studied and distinctive in morphology. Takhtajan (1997) considered Metteniusaceae closely related to Icacinales. Metteniusaceae and Oncothecaceae share pentalacunar nodes and five-carpellate gynoecia (González and Rudall 2010). Also, *Emmotum* (Icacinaceae s.s.) is five carpellate (Endress and Rapini 2014). Additional data will be required to determine how these characters evolved in these basal lamiids. However, considering recent topologies (Fig. 11.15), it is likely that those morphological characters

Figure 11.16. Members of Lamiids (Boraginaceae, Garryales, Gentianales, Icacinaceae, Lamiales, Metteniusaceae, Oncothecaceae, Solanales, Vahliaceae): a. *Eucommia ulmoides* Oliv. (Eucommiaceae), leaf and samaras; note elastic latex (gutta) threads, b. *Garrya elliptica* Douglas ex Lindl. (Garryaceae), branch with opposite leaves and catkin-like staminate inflorescences, c. *Garrya fadyenii* Hook. (Garryaceae), infructescence with berries, d. *Pyrenacantha malvifolia* Engl. (Icacinaceae), branch with fruit and fruit longitudinal section showing sculptured pit, e. *Calotropis procera* (Aiton) W.T. Aiton (Apocynaceae), flowers; note stylar head and five staminal outgrowths, f. *Nerium oleander* L. (Apocynaceae), pair of follicles with wind-dispersed, comose seeds, g. *Gelsemium sempervirens* (L.) J. St.-Hill (Gelsemiaceae), flowers, h. *Sabatia grandiflora* (A. Gray) Small (Gentianaceae), flower, i. *Spigelia loganioides* A. DC. (Loganiaceae), plant with flowers and fruits, j. *Strychnos spinosa* Lam. (Loganiaceae), thick-rinded berries, k. *Cubanola domingensis* (Britton) Aiello (Rubiaceae), flowers; note inferior ovary, l. *Portlandia proctorii* (Aiello) Delprete (Rubiaceae), flowers, m. *Rondeletia odorata* Jacq. (Rubiaceae), interpetiolar stipule showing colleters.

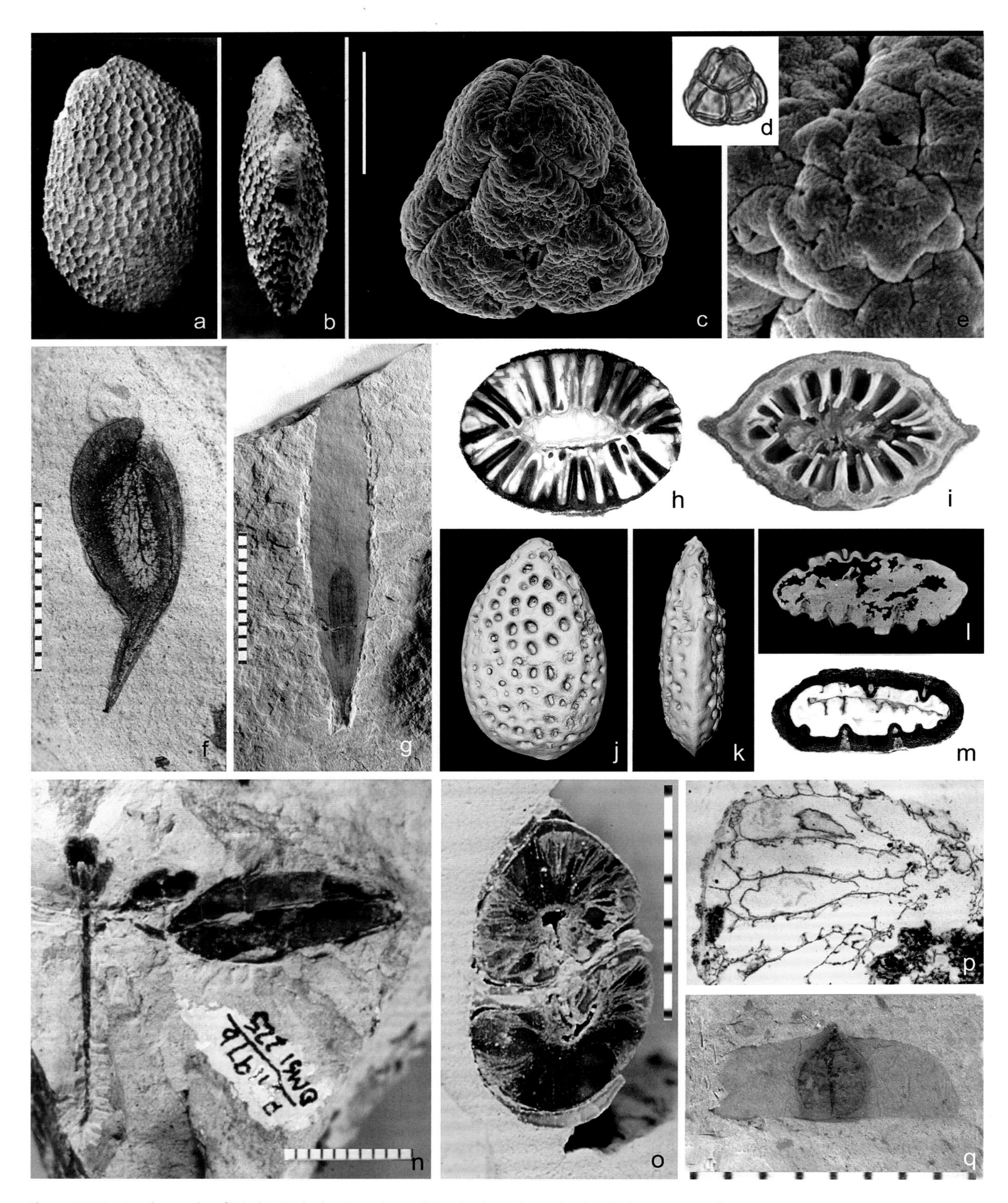

Figure 11.17. Fossil examples of Ericales, Icacinales, Garryales, and Gentianales. a, b. Seeds of *Actinidia oregonensis* from the middle Eocene Clarno Formation, Oregon [uf6292], c. Pollen tetrad of Ericaceae, from early Paleocene of Agatdalen of Iceland [courtesy F. Grimsson], d. Same tetrad by light microscopy, e. Detail of exine ornamentation from d, f. *Eucommia* fruit from Oligocene of Mexico, g. *Fraxinus* fruit from middle Eocene of Oregon, h. *Pyrenacantha occicentalis*, Eocene of Oregon [ucmp 10708], i. Extant *Pyrenacantha grandiflora* [Hardy 5118], S. Africa, j-l. *Palaeophytocrene foveolata* [v 22635] from Early Eocene London Clay flora. NHM specimen images by Margaret Collinson, j, k. surface renderings of locule cast showing holes representing endocarp inner protrusions, l. Same specimen, virtual transverse section showing intrusions into locule, m. *Palaeophytocrene pseudopersica* physical transverse section, Eocene Oregon, n-p. Rubiaceae: *Emmenopterys dilcheri* from the middle Eocene of Oregon, q. Winged seed of Bignoniaceae. Paleocene of Japan, courtesy J. Horiuchi.

exhibited by both Oncothecaceae and Metteniusaceae are either plesiomorphies or homoplasious.

GARRYALES

Garryales consist of Eucommiaceae and Garryaceae (including *Aucuba*) (Figs. 11.15 and 11.16, 11.17). In APG (1998), the order also included Oncothecaceae, but that placement was premature; it did not receive support above 50% in D. Soltis et al. (2000) and has been unsupported in subsequent analyses (González et al. 2007; Soltis et al. 2011; Refulio-Rodriguez and Olmstead 2014; Stull et al. 2015). Eucommiaceae and Garryaceae were not considered closely related in morphology-based classifications, but this clade was recovered in analyses of *rbcL* alone (Olmstead et al. 1993; Xiang et al. 1993; Savolainen et al. 2000b), as well as in combined analyses (e.g., D. Soltis et al. 2000; Albach et al. 2001c; Bremer et al. 2002; González et al. 2007; Soltis et al. 2011; Refulio-Rodriguez and Olmstead 2014; Stull et al. 2015). Most recent analyses suggest a sister relationship of Garryales to the core lamiids (Stull et al. 2015).

Possible morphological synapomorphies for Garryales are unisexual flowers (with dioecious plants), inferior ovaries, the gynoecium with a single short style and apical placentation, as well as the presence of gutta and valvate perianth parts (Stevens 2001 onwards), although there are parallelisms in Aquifoliales and Apiales of the campanulids (Bremer et al. 2001). Eucommiaceae, which have a wide fossil distribution in the Northern Hemisphere (e.g., Fig. 11.17f; Manchester et al. 2009), today consist of a single species from China that was placed in Hamamelidae in morphology-based classifications because of the simple nature of the flowers (Cronquist 1981). A close relationship of the family to Garryaceae is another example of an unexpected result of molecular systematic studies. Garryaceae are now broadly defined to include *Aucuba,* as well as *Garrya* (APG II 2003; APG III 2009; APG IV 2016; see Soltis et al. 2011; see also Chapter 12). Although the sister-group relationship between *Garrya* and *Aucuba* received strong support in molecular phylogenetic studies (e.g., D. Soltis et al. 2000; Savolainen et al. 2000a,b; Albach et al. 2001c; Bremer et al. 2002; Refulio-Rodriguez and Olmstead 2014), the two were not considered closely related in morphology-based classifications (e.g., Cronquist 1981). However, a close relationship between *Garrya* and *Aucuba* was first proposed by Baillon (1866–1895), and this relationship is supported by several non-DNA characters, including palynological data (Ferguson 1977), the presence of petroselinic acid in seed oils (Breuer et al. 1987), two-seeded fruits lacking a septum, and a similar gynoecial vasculature (Eyde 1964). In addition, both *Garrya* and *Aucuba* are dioecious, evergreen shrubs or small trees with tetramerous flowers and opposite leaves. Non-DNA characters also unite Eucommiaceae with *Garrya* and *Aucuba.* All three are dioecious and woody, and contain similar iridoid compounds; aucubin has been detected in all three genera (Jensen et al. 1975), and eucommioside has been found in *Eucommia* and *Aucuba* (Boros and Stermitz 1991). Garryales have been dated to between 96–31 mya (Wikström et al. 2001; Bell et al. 2010; Lemaire et al. 2011a; Naumann et al. 2013). Extant *Eucommia* and *Garrya* are illustrated in Fig. 11.16a–c.

GENTIANALES

Gentianales constitute a well-supported clade of Apocynaceae (including Asclepiadaceae), Gelsemiaceae, Gentianaceae, Loganiaceae, and Rubiaceae (Fig. 11.15; several are illustrated in Fig. 11.16). Gentianales comprise about 1,100 genera and nearly 20,000 species. Although *Emblingia* (Emblingiaceae: Brassicales) was placed in Gentianales by Savolainen et al. (2000b), this appears to be the result of a DNA mix-up (see Chapter 10). Dialypetalanthaceae, consisting of one species in eastern Brazil, should be included in Rubiaceae (Fay et al. 2000a; Savolainen et al. 2000b). A close relationship among these families of Gentianales had long been proposed in classifications based solely on morphology. Cronquist (1981), for example, had a similar Gentianales but also recognized Retziaceae (= Stilbaceae, now included in Lamiales); like most authors at that time, Cronquist also included Asclepiadaceae as a family distinct from Apocynaceae.

Synapomorphies for Gentianales include vestured pits, stems with a nodal line, opposite and decussate leaves, and stipules with thick glandular hairs (i.e., colleters, Fig. 11.16m; these were lost in most Gentianaceae; Backlund et al. 2000), corollas convolute in bud (Bremer and Struwe 1992), and possibly some types of complex indole alkaloids (Bremer et al. 2001; Judd et al. 2002, 2008). Rubiaceae are sister to the remainder of the clade using DNA sequence data (Downie and Palmer 1992; Chase et al. 1993; Olmstead et al. 1993; Bremer et al. 1994; Backlund et al. 2000; Savolainen et al. 2000a,b; D. Soltis et al. 2000; Albach et al. 2001c; Bremer et al. 2002; Soltis et al. 2011; Refulio-Rodriguez et al. 2014). However, a morphological cladistic analysis (Struwe et al. 1994) placed Rubiaceae sister to Gelsemiaceae and Loganiaceae s.s. as sister to the remaining Gentianales. Phylogenetic relationships within Rubiaceae have been analyzed using morphology (e.g., Bremer and Struwe 1992), DNA characters (e.g., Manen et al. 1994; Bremer et al. 1995, 1999; Andersson and Rova 1999; Andersson and Antonelli 2005; Maurin et al. 2007;

Bremer and Eriksson 2009; Groeninckx et al. 2009; Lemaire et al. 2011b), and combined morphology and DNA datasets (Andreasen and Bremer 2000).

The fossil record of Rubiaceae includes dispersed pollen and occasional megafossil occurrences (Graham 2008). A relatively early, secure megafossil record of Rubiaceae is that of permineralized fruiting capsules with intact winged seeds attributed to the Condamineeae genus *Emmenopterys* from the middle Eocene of Oregon (Fig. 11.17n–p; Manchester 1994).

Following Rubiaceae, a clade of the remaining families (Apocynaceae, Gelsemiaceae, Gentianaceae, Loganiaceae) received moderate to strong support in analyses based on three or four genes (D. Soltis et al. 2000; Albach et al. 2001c) and strong support in Bremer et al. (2002), Soltis et al. (2011), and Refulio-Rodriguez and Olmstead (2014). This clade of four families also shares a feature of wood anatomy; all Gentianales except Rubiaceae are characterized by intraxylary phloem (Carlquist 1992). Relationships among these four families, however, are not completely resolved. There is weak bootstrap support (<65%) for Loganiaceae as sister to Gentianaceae and for Apocynaceae as sister to Gelsemiaceae in D. Soltis et al. (2000). In contrast, in Bremer et al. (2002), Apocynaceae are sister to Gentianaceae (95% jackknife support), and Gelsemiaceae are sister to Loganiaceae (85% support), relationships also found by Refulio-Rodriguez and Olmstead (2014), although not well-supported. Soltis et al. (2011) recovered Loganiaceae and Gelsemiaceae as subsequent sisters to a moderately (86%) well-supported Gentianaceae + Apocynaceae clade.

Adding to the uncertainty in relationships is the polyphyly or paraphyly of Loganiaceae as previously defined (e.g., Chase et al. 1993; Olmstead et al. 1993; Bremer et al. 1994; Struwe et al. 1994; Backlund et al. 2000). *Fagraea* and *Potalia* (traditionally placed in Loganiaceae) should be placed in Gentianaceae; *Gelsemium* and *Mostuea* are well separated from other Loganiaceae and have been treated as a separate family, Gelsemiaceae (reviewed in Judd et al. 2002, 2008; APG II 2003; APG III 2009; APG IV 2016).

Phylogenetic analyses of both molecular and non-DNA characters indicated that Asclepiadaceae and Apocynaceae should be combined to form a single family, Apocynaceae (Judd et al. 1994; Endress and Bruyns 2000). Apocynaceae in this broad sense are united by several non-DNA characters, including tissues with laticifers and usually milky sap, carpels postgenitally united by styles and/or stigmas with ovaries usually distinct, and the apical portion of the style expanded and modified to form a secretory head (Fig. 11.16; Judd et al. 1994). Relationships within Apocynaceae have been the subject of numerous phylogenetic analyses (Judd et al. 1994; Endress et al. 1996; Sennblad and Bremer 1996; Civeyrel et al. 1998; Endress and Stevens 2001; Potgieter and Albert 2001; Fishbein et al. 2011), and Straub et al. (2014) recently resolved the backbone phylogeny of the family using complete plastome sequence data. Studies have enhanced our knowledge of the ecological and herbivore defense strategies often employed by members of *Asclepias*, formerly Asclepiadaceae, as well as morphological evolution in the group (Fishbein 2001; Agrawal and Fishbein 2006, 2008; Agrawal et al. 2008, 2009a–c). The family exhibits a stepwise series of morphological modifications to the androecium and gynoecium that appear to be associated with increased specialization and efficiency of pollination (Schick 1980, 1982; Judd et al. 1994, 2002, 2008; M. Endress 2001). Crown group ages of Gentianales range from 104.7–57 mya (Wikström et al. 2001; Bremer et al. 2004; Bremer and Eriksson 2009; Janssens et al. 2009; Bell et al. 2010; Lemaire et al. 2011a).

VAHLIACEAE

Vahliaceae contain five species in the genus *Vahlia* (= *Bistella*) from tropical and southern Africa to northwestern India. *Vahlia* usually has been considered close to Saxifragaceae because of their free sepals and petals, alternatively included there (Cronquist 1981) or treated as a separate related family (e.g., Dahlgren 1980; Thorne 1992a,b; Takhtajan 1997). However, the DNA placement of *Vahlia* in the asterids is supported by the presence of iridoids (Al-Shammary 1991) and tenuinucellate ovules. Nonetheless, *Vahlia* differs from most lamiids in having bitegmic rather than unitegmic ovules; bitegmic ovules also occur in *Emmotum* (Icacinaceae s.s.; Endress and Rapini 2014).

The position of Vahliaceae in the lamiid clade remains uncertain. Based on *rbcL* sequence data, Morgan and Soltis (1993) first suggested that *Vahlia* occupies an isolated position as sister to Lamiales, a position it also occupies in other analyses of *rbcL* alone (e.g., Savolainen et al. 2000a) and in some multigene analyses (D. Soltis et al. 2000; Albach et al. 2001c). However, even in three- and four-gene analyses, this placement did not receive internal support above 50%. In Bremer et al. (2002), *Vahlia* is weakly supported (69%) as sister to Boraginaceae. Refulio-Rodriguez and Olmstead (2014) recovered Vahliaceae as sister to Solanales (with BS < 50%), while Soltis et al. (2011) recovered Vahliaceae as sister to a Solanales, Gentianales, Boraginaceae, Lamiales clade (i.e., core *Lamiidae*) with BS = 100%. We illustrate the former placement of Vahliaceae (Fig. 11.15). Vahliaceae remain unplaced within the core lamiids (APG IV 2016).

SOLANALES

Solanales consist of (APG 1998; APG II 2003; APG III 2009; APG IV 2016; Fig. 11.15) Convolvulaceae, Hydroleaceae, Montiniaceae, Solanaceae (including Duckeodendraceae, Goetzeaceae, and Nolanaceae sensu Cronquist 1981), and Sphenocleaceae (several are illustrated in Fig. 11.18). Solanales include about 165 genera and 4080 species. This molecular-based circumscription of Solanales differs from concepts of the order in morphology-based classifications.

Cronquist (1981) also included Polemoniaceae, Hydrophyllaceae, and Menyanthaceae in his Solanales, but he did not include Montiniaceae. Montiniaceae comprise three genera (*Montinia*, *Kaliphora*, and *Grevea*), and the relationships of the family have long been problematic. Some included the family in a broadly defined Saxifragaceae (reviewed in Morgan and Soltis 1993). Montiniaceae are characterized by unisexual flowers, an unusual feature in asterids, but found also in Garryales and Aquifoliales.

Molecular studies also clearly placed *Hydrolea* in Solanales. The split of Hydrophyllaceae, with *Hydrolea* assigned to Solanales and *Hydrophyllum* and *Phacelia* to Boraginaceae, was first proposed by de Jussieu (1789). However, this inference was not accepted by Gray (1875), and modern morphology-based classifications included *Hydrolea* in Hydrophyllaceae (e.g., Cronquist 1981). Characters separating *Hydrolea* from other Hydrophyllaceae include a bilocular capsule and axile placentation in the former.

Few non-DNA characters have been identified as possible synapomorphies for Solanales. The families of Solanales have radially symmetric flowers (likely plesiomorphic) and a plicate, sympetalous corolla (a likely synapomorphy; Fig. 11.18), and the number of stamens equals the number of petals; the pollen sacs exhibit placentoids, another likely, albeit non-unique, synapomorphy (i.e., pollen sac placentoids also occur in Lamiales). Pollen sac placentoids could be plesiomorphic in Solanales, if Solanales are sister to the Boraginales + Lamiales clade, as found in Refulio-Rodriguez and Olmstead (2014; see below) and would then be synapomorphic for Solanales + (Lamiales + Boraginales). Solanales lack iridoids (discussed below; Albach et al. 2001b; Bremer et al. 2001; Judd et al. 2002, 2008). Alternate leaves may also be a synapomorphy; however, a reversal to opposite leaves has occurred in *Grevea* (Montiniaceae). Solanaceae and Convolvulaceae are united by similar alkaloids and wood with inter- and intraxylary phloem.

The position of Solanales within lamiids is still unresolved. Soltis et al. (2011) recovered a clade of Solanales and Gentianales, which was sister to a clade composed of Boraginaceae and Lamiales, while in Refulio-Rodriguez and Olmstead (2014) Solanales formed a clade with Vahliaceae, which was sister to the Boraginales + Lamiales clade. Neither topology was well supported, however.

Within Solanales are two major clades (Fig. 11.15): Montiniaceae as sister to Sphenocleaceae + Hydroleaceae: Solanaceae and Convolvulaceae (D. Soltis et al. 2000; Bremer et al. 2002; Soltis et al. 2011; Refulio-Rodriguez and Olmstead 2014). Solanaceae and Convolvulaceae are chemically distinct from other members of lamiids in the replacement of iridoids by tropane alkaloids (Jensen et al. 1975; Romeike 1978). Circumscription of the major lineages within Convolvulaceae is provided by Stefanović et al. (2002), and a molecular phylogenetic framework and provisional reclassification are available for Solanaceae (Olmstead et al. 1999; Olmstead and Bohs 2006). Solanaceae have been shown to include Duckeodendraceae (Fay et al. 1998b; APG II 2003). Solanales have been dated to 93–62 mya (Wikström et al. 2001; Bremer et al. 2004; Magallón and Castillo 2009; Bell et al. 2010; Lemaire et al. 2011a).

Numerous analyses have been conducted within Solanaceae (e.g., Spooner et al. 1993; Olmstead and Sweere 1994; Olmstead et al. 1999; Bohs 2005, 2007; Martins and Barkman 2005; Whitson and Manos 2005; Levin et al. 2006; Smith and Baum 2006; Wees and Bohs 2007; Olmstead et al. 2008; Tu et al. 2008; Knapp 2010; Sarkinen et al. 2013). Tomato (formerly *Lycopersicon*) is clearly embedded within the now megadiverse genus *Solanum* (e.g., Spooner et al. 1993; Bohs and Olmstead 1997) that contains between 1250–1700 species. Numerous studies of specific clades within *Solanum* have also been conducted (e.g., Bohs 2001; Knapp 2002; Spooner et al. 2004; Bennett 2008; Peralta et al. 2008; Knapp et al. 2013). *Nicotiana* has been used as a system for the study of polyploidy (Clarkson et al. 2005; Leitch et al. 2008) and chromosomal evolution (Wu and Tanksley 2010). The non-monophyletic genus *Iochroma* has been the subject of numerous investigations of pollinator shifts with concomitant floral morphological diversity (S. Smith et al. 2008a,b; Fenster et al. 2009; Smith 2010; Smith and Rausher 2011; see also Chapter 14, Fig. 14.1). For a discussion on the biogeography of Solanaceae, see Olmstead (2013).

BORAGINALES

The position of Boraginales (i.e., Boraginaceae s.l. including Hydrophyllaceae and several other families as mentioned above; see Gottschling and Hilger 2001; Gottschling et al. 2001, 2005; Hilger and Diane 2003; Weigend and Hilger 2010; Cohen 2013; Weigend et al. 2013; Refulio-Rodriguez and Olmstead 2014; Weigend et al. 2014; Luebert et al. 2016 for alternatives regarding familial circumscription in this group) as sister to Lamiales is well supported (Soltis et al. 2011; Refulio-Rodriguez and Olmstead 2014; Fig. 11.15). Boraginaceae have usually been placed in or close to Solanales in morphology-based classifications (Dahlgren 1980; Thorne 1992a,b, 2001; Takhtajan

Figure 11.18. Lamiids cont'd. (Solanales and Boraginales). a. *Hydrolea corymbosa* J. Macbr. ex Elliott (Hydroleaceae), flowers with two styles and stamens glandular-swollen at base, b. *Ipomoea pes-caprae* (L.) R. Br. (Convolvulaceae), sympetalous corolla with plications, c. *Iochroma cyanea* M.L. Green (Solanaceae), tubular sympetalous flowers; also note connate sepals, d. *Physalis walteri* Nutt. (Solanaceae), sympetalous flower with plications, terete stem with alternate, exstipulate leaves, e. *Solanum tuberosum* L. (Solanaceae), poricidal anthers, f. *Wigandia urens* (Ruiz & Pav.) Kunth (Boraginaceae), flower, g. *Tournefortia staminea* Griseb. (Boraginaceae), scorpioid, cymose inflorescence.

1997) and in some molecular-based trees (e.g., Chase et al. 1993; Olmstead et al. 2000), but previous molecular analyses indicated that Boraginaceae are an isolated family with no clear or close relatives. With *ndhF* sequences, Boraginaceae were recovered as sister to Solanales, but without support above 50% (Olmstead et al. 2000). In other studies, Boraginaceae were sister to Lamiales (*rbcL*; Olmstead et al. 1992) or to Vahliaceae + Lamiales (multigene analyses; D. Soltis et al. 2000; Albach et al. 2001c). However, this close relationship of Boraginaceae to Lamiales did not receive support above 50%. In Bremer et al. (2002), Boraginaceae + Vahliaceae are sister to Solanales + Lamiales, but again without support above 50%. Most recent phylogenetic analyses further support the placement of Boraginaceae s.l. (Boraginales sensu Refulio-Rodriguez and Olmstead 2014) as sister to Lamiales (Soltis et al. 2011; Refulio-Rodriguez and Olmstead 2014). Development of the endosperm may provide synapomorphies for Boraginaceae and Lamiales; Boraginaceae and all Lamiales (except Oleaceae) share terminal endosperm haustoria. Estimated ages for Boraginaceae s.l. range from 88.4–39 mya (Wikström et al. 2001; Bell et al. 2010; Naumann et al. 2013; Nazaire et al. 2014).

Boraginaceae are currently broadly circumscribed to include Hydrophyllaceae (excluding *Hydrolea*, which is in Solanales; APG II 2003; APG III 2009; APG IV 2016), although several authors have advocated for breaking Boraginaceae into numerous families (see Stevens 2001 onwards; Refulio-Rodriguez and Olmstead 2014). Cronquist (1981, 1988) did not consider Boraginaceae and Hydrophyllaceae to be closely related. He placed Hydrophyllaceae in Solanales and Boraginaceae in Lamiales on the basis of similarities between tropical, woody Boraginaceae and tropical, woody Verbenaceae. A morphological character uniting Boraginaceae as now broadly circumscribed (APG II 2003; APG III 2009; APG IV 2016) is a distinctive inflorescence, usually composed of helicoid or scorpioid cymes, which straighten as the flowers mature (see also Judd et al. 2002, 2008; Fig. 11.18).

Boraginaceae offer the opportunity to examine facets of mating system evolution. Although there are obvious genetic benefits to outcrossing, many (20%) angiosperms have evolved predominant selfing (autogamy; Barrett 2002). Multiple origins of small-flowered predominantly autogamous species from outcrossing species is a common feature of herbaceous flowering plant families (Barrett 2002). An excellent example is provided by *Amsinckia* (Boraginaceae).

Phylogenetic analysis and character-state mapping indicated that predominant selfing has evolved from outcrossing at least four times in the genus (Schoen et al. 1997). The outcrossing species are distylous with large flowers; the selfing species are homostylous and have small flowers. In addition, the repeated occurrence of short branches separating selfers from related outcrossers indicates that selfing may be of recent origin (Schoen et al. 1997).

LAMIALES

Lamiales, with almost 18,000 species in about 1,100 genera, are among the most taxonomically difficult clades of asterids, as well as all angiosperms. Lamiales include the former orders Lamiales, Scrophulariales, Callitrichales, and Plantaginales of Cronquist (1981). Lamiales are strongly supported, comprising 25 families in APG III (2009); with a slight modification in APG IV (2016) to 24 families—Orobanchaceae was expanded to include Lindenbergiaceae and Rehmanniaceae (Fig. 11.15). *Sanango* was moved to Gesneriaceae; however, *Peltanthera* is still unplaced to family. The 24 families in APG IV (2016) are Acanthaceae (including Avicenniaceae), Bignoniaceae, Byblidaceae, Calceolariaceae, Carlemanniaceae, Gesneriaceae, Lamiaceae, Lentibulariaceae, Linderniaceae, Martyniaceae, Mazaceae, Oleaceae, Orobanchaceae (including Nesogenaceae), Paulowniaceae, Pedaliaceae, Phrymaceae, Plantaginaceae, Plocospermataceae, Schlegeliaceae, Scrophulariaceae (including Buddlejaceae and Myoporaceae), Stilbaceae, Tetrachondraceae, Thomandersiaceae, and Verbenaceae (several are illustrated in Fig. 11.19).

Most Lamiales share characters that are extremely rare outside of this clade: verbascoside; anthraquinones derived from the shikimic acid pathway; C11-decarboxylated iridoids (Jensen 1991, 1992); oligosaccharides; 6-oxygenated flavones; protein inclusions in nuclei of mesophyll cells; often diacytic stomates; hairs with a stalk of one or more uniseriate cells and a head of two or more vertical cells; young anthers with pollen sac placentoids (also found in the sister clade Boraginaceae s.l.); and fewer than five fertile stamens (Rahn 1996). Much of the taxonomic complexity of Lamiales involves compelling DNA evidence for the polyphyly of Scrophulariaceae in the traditional sense (see paragraphs below; reviewed in Olmstead et al. 2001; and Tank et al. 2006).

The polyphyly of Scrophulariaceae and other recent phylogenetic results resulted in several newly named families in Lamiales. Families added (see APG II 2003; APG III 2009; Refulio-Rodriguez and Olmstead 2014) include Calceolariaceae, Carlemanniaceae, Linderniaceae, Mazaceae, Plocospermataceae, Rehmanniaceae, Tetrachondraceae, and Thomandersiaceae. However, in APG IV (2016), Rehmanniaceae have been included in Orobanchaceae (Fig. 11.15). Other families have been redefined. For example, phylogenetic studies dictate that Buddlejaceae and Myoporaceae be included in Scrophulariaceae (Olmstead

Figure 11.19. Lamiids cont'd. (Lamiales). a. *Cartrema americana* (L.) G.L. Nesom (Oleaceae), branch with flowers with four connate petals and two stamens, b. *Polypremum procumbens* L. (Tetrachondraceae), herb with tetramerous flowers, c. *Calceolaria chelidonioides* Kunth (Calceolariaceae), herb with zygomorphic, 4-merous flowers with saccate lower lip, d. *Columnea schiedeana* Schldtl. (Gesneriaceae), zygomorphic flower with anthers sticking together, e. *Nuttallanthus canadensis* (L.) D.A. Sutton (Plantaginaceae), zygomorphic flowers with two-lobed upper lip, and three-lobed lower lip, f. *Plantago lanceolata* L. (Plantaginaceae), reduced, wind-pollinated flowers, g. *Leucophyllum frutescens* (Berland.) I.M. Johnst. (Scrophulariaceae), branch with flowers, h. *Verbascum virgatum* Stokes (Scrophulariaceae), flowers; note anthers opening by a single slit, i. *Torenia fournieri* Linden ex Fourn. (Linderniaceae), flowers with sensitive stigmas, j. *Uncarina grandidieri* (Baill.) Stapf (Pedaliaceae), flower, k. *Thunbergia grandiflora* Wall. (Acanthaceae), flowers, l. *Ruellia tweedieana* Griseb. & Fernald (Acanthaceae), capsule with retinacula, m. *Kigelia africana* (Lam.) Benth. (Bignoniaceae), zygomorphic flower; note androecium of two long and two short stamens, and foliaceous, sensitive stigma, n. *Spathodea campanulata* P. Beauv. (Bignoniaceae), capsule releasing flattened, winged, wind-dispersed seeds (lacking endosperm), o. *Schlegelia parasitica* Griseb. (Schlegeliaceae), shrub with opposite, simple leaves, flowers, p. *Utricularia cornuta* Michx. (Lentibulariaceae), flowers with nectar spurs, q. *Pinguicula primuliflora* C.E. Wood & R.K. Godfrey (Lentibulariaceae), leaf with gland-headed trichomes trapping insects, r. *Lamium amplexicaule* L. (Lamiaceae), herb with square stem, opposite, reduced, cymose inflorescences, forming pseudoverticillate clusters, the flowers zygomorphic, bilabiate, s. *Callicarpa americana* L. (Lamiaceae), infructescences cymose, drupes, t. *Lantana montividensis* (Spreng.) Brig. (Verbenaceae), indeterminate inflorescence of zygomorphic flowers, u. *Agalinis fasciculata* Raf. (Orobanchaceae), flowers, a hemiparasite, v. *Erythranthe guttata* (DC.) G.L. Nesom (Phrymaceae), zygomorphic flowers.

et al. 1992, 1999, 2001; Oxelman et al. 1999). *Avicennia* (Avicenniaceae) appears in Acanthaceae and has been included in that family (APG II 2003; APG III 2009; APG IV 2016). Orobanchaceae have been expanded to contain parasitic Scrophulariaceae (plus the non-parasitic *Lindenbergia* formerly of Scrophulariaceae and sometimes treated as Lindenbergiaceae; see Chapter 13 for evolution of parasitism). Orobanchaceae also include core genera of the former Cyclocheilaceae. Bremer et al. (2002) found *Cyclocheilon* to be embedded within Orobanchaceae with strong support; *Asepalum* is also nested within Orobanchaceae (Tank et al. 2006). Some also include Rehmanniaceae in Orobanchaceae (see APG IV 2016; Stevens 2001 onwards; Rehmannieae). Holoparasitism has evolved three times in Orobanchaceae (McNeal et al. 2013). Tank and Olmstead (2008, 2009) and Tank et al. (2009) reconstructed phylogenetic relationships within the large subtribe Castillejinae of Orobanchaceae, including *Castilleja* and relatives.

Confusion has been introduced in the naming of clades in this disintegrated Scrophulariaceae. The name Veronicaceae has been resurrected for one clade of the former, broadly defined (and polyphyletic) Scrophulariaceae (Olmstead and Reeves 1995; Olmstead et al. 1999, 2001), but the name with priority is Plantaginaceae (APG II 2003; APG III 2009; APG IV 2016), because it has been conserved over Veronicaceae. The newly recognized Calceolariaceae (slipper flowers) are distinct from Scrophulariaceae s.s. in having four calyx lobes, and the bilateral corolla has also been shown by developmental studies (Mayr and Weber 2006) to be based upon four petals (they also exhibit oil-producing hairs) and include *Calceolaria* (250 species from South America), *Jovellana* (six species from New Zealand and Chile), and *Porodittia* (= *Stemotria*) from Peru (Olmstead et al. 1999, 2001).

The relationships of some genera formerly placed in Scrophulariaceae (e.g., *Mimulus*) remain uncertain. It appears, however, that *Mimulus* should be placed in Phrymaceae (Beardsley et al. 2001; Beardsley and Olmstead 2002; Beardsley et al. 2004; Barker et al. 2012; Refulio-Rodriguez and Olmstead 2014) rather than Plantaginaceae. Sensitive, foliaceous style branches in *Mimulus* versus non-sensitive, papillose stigmatic lobes in Plantaginaceae separate the two morphologically.

Carlemanniaceae consist of five species and two genera (*Carlemannia* and *Silvianthus*) from Southeast Asia to Sumatra; they were formerly included in Caprifoliaceae/Dipsacales by Cronquist (1981), but their connivent anthers are consistent with their placement in Lamiales. Martyniaceae consist of four genera from tropical and subtropical America; they were included in Pedaliaceae in APG (1998), although multigene analyses (Olmstead et al. 1999, 2001; Albach et al. 2001c; Bremer et al. 2002) indicated that their familial status should be maintained. Recent analyses of Lamiales by Refulio-Rodriguez and Olmstead (2014) suggest that Martyniaceae form a clade with Acanthaceae, which is sister to Pedaliaceae, but this topology is not well supported (see Fig. 11.15). Both Acanthaceae and Pedaliaceae contain amyloid, an oligosaccharide not found in other Lamiales (Hegnauer 1962–1994), and share pollen morphology (Erdtman 1952), so a clade containing Acanthaceae and Pedaliaceae is supported by chemical and morphological characters as well.

Plocospermataceae are a monogeneric family comprising three species from Central America; they were included in Loganiaceae by Cronquist (1981) and not assigned to order in APG (1998). Tetrachondraceae are a new family comprising *Tetrachondra* from Patagonia and New Zealand and *Polypremum procumbens* from warm temperate and tropical America (Oxelman et al. 1999; Bremer et al. 2002).

Within Lamiales, relationships among families were relatively well-supported in most recent analyses (Refulio-Rodriguez and Olmstead 2014) (Fig. 11.15). Plocospermataceae appear to be sister to the rest of Lamiales (Bremer et al. 2002; Soltis et al. 2011; Refulio-Rodriguez and Olmstead 2014) followed by a well-supported clade formed by Carlemanniaceae and Oleaceae (Refulio-Rodriguez and Olmstead 2014). Lamiales are inferred to have diverged ca. 97 mya (Bremer et al. 2004).

In contrast to molecular results placing Oleaceae in Lamiales, Oleaceae were often considered closely related to Gentianales in morphology-based classifications (e.g., Dahlgren et al. 1980). Cronquist (1981) placed the family in his Scrophulariales. Oleaceae are distinct from the remaining Lamiales in their embryogeny, which is of the polygonad type, in contrast to the onagrad type found in the remainder of the order. Phylogenetic relationships within Oleaceae have recently been studied (e.g., Wallander and Albert 2000; Heuertz et al. 2006; Wallander 2008; Besnard et al. 2009; Guo et al. 2011; Kim and Kim 2011). Fossil records of the family include fruits of *Fraxinus* from the Eocene (Fig. 11.17g), and dispersed pollen assignable to the family from the late Cretaceous (Campanian) (Manchester et al. 2015).

Following Oleaceae, Tetrachondraceae, both with actinomorphic and tetramerous flowers, are sister to the remaining Lamiales (Fig. 11.15), a placement receiving strong support in Bremer et al. (2002), as well as other studies (Soltis et al. 2011; Refulio-Rodriguez and Olmstead 2014). Following Tetrachondraceae, the remaining members of Lamiales form a well-supported clade (Soltis et al. 2011; Refulio-Rodriguez and Olmstead 2014) that is also supported by two non-DNA characters: (1) flowers having bilateral symmetry (a two plus three pattern); and (2) four stamens in a two-long and two-short configuration (e.g., Judd et al. 2002, 2008). Within this latter clade, Calceolariaceae + Gesneriaceae and Plantaginaceae, are subsequent sisters to the rest of Lamiales (Soltis et al. 2011; Refulio-Rodriguez and Olmstead 2014). Soltis et al. (2011) recovered a poorly supported clade of Plantaginaceae + Scrophulariaceae and Stilbaceae; however, Refulio-Rodriguez and Olmstead (2014) recovered a well-supported topology where Scrophulariaceae and Stilbaceae were subsequent sisters to the rest of Lamiales, after Plantaginaceae (Fig. 11.15).

The South American genus *Peltanthera*, unplaced to family, is strongly supported as sister to the Gesneriaceae + Calceolariaceae (Soltis et al. 2011; Refulio-Rodriguez and Olmstead 2014; Fig. 11.15); APG IV (2016) has included *Sanango* in Gesneriaceae. Gesneriaceae are distinct from most other Lamiales in their parietal placentation (although this also has evolved in Orobanchaceae) and lack of iridoids. Relationships within Gesneriaceae have been the subject of several phylogenetic and systematic investigations (e.g., Smith and Carroll 1997; Smith and Atkinson 1998; Smith 2000a,b; Zimmer et al. 2002; Mayer et al. 2003; Clark 2005, 2009; Clark et al. 2006, 2011, 2012; Perret et al. 2013; Smith and Clark 2013).

Plantaginaceae are morphologically diverse; they include the wind-pollinated *Plantago*, as well as Callitrichaceae and Hippuridaceae, two groups of aquatic herbs. *Aragoa*, a genus of Andean trees, is sister to *Plantago*, and although insect-pollinated, *Aragoa* shares a similar floral morphology with *Plantago* (Bello et al. 2002). *Antirrhinum* (snapdragon), a model organism for the study of floral developmental genetics, is also a member of Plantaginaceae (see "Floral Symmetry," below).

Within the remaining Lamiales, relationships remain poorly understood other than a large clade formed by Lamiaceae as sister to a well-supported clade formed by Orobanchaceae, Rehmanniaceae, Paulowniaceae, Phrymaceae, and Mazaceae (Refulio-Rodriguez and Olmstead 2014; Fig. 11.15). Verbenaceae form a well-supported clade sister to Thomandersiaceae, which may be sister to a clade composed of Lentibulariaceae + Schlegeliaceae (Refulio-Rodriguez and Olmstead 2014), although the last clade was not well supported. Bignoniaceae also form a well-supported clade that is likely sister to a clade formed by the aforementioned families (Fig. 11.15); see Olmstead (2013) for a discussion on the biogeographic history of the group. Refulio-Rodriguez and Olmstead (2014) also found high support for a clade composed of Bignoniaceae and the rest of Lamiales, as well as Acanthaceae, Pedaliaceae, and Martyniaceae, a clade that was moderately supported by Olmstead et al. (2001). Among these families, Bignoniaceae have the most informative fossil record, based on biwinged seeds extending back to the Paleocene (Fig. 11.17q). Scrophulariaceae s.s., including *Buddleja* (formerly Buddlejaceae, a genus with actinomorphic flowers with four petals) and *Myoporum* (formerly Myoporaceae, a clade with pellucid-dotted leaves), form a well-supported clade (Olmstead and Reeves 1995; Oxelman et al. 1999; Albach et al. 2001c; Olmstead et al. 2001; Bremer et al. 2002; Soltis et al. 2011; Refulio-Rodriguez and Olmstead 2014), and they are characterized by the usually confluent pollen sacs. A close relationship between Scrophulariaceae and Buddlejaceae was first noted by Dahlgren (1983). The genus *Androya*, previously placed in Loganiaceae, is also part of the *Myoporum* clade of Scrophulariaceae s.s. (Bremer et al. 2002). Members of this redefined Scrophulariaceae are characterized by acylated rhamnosyl iridoids (Jensen et al. 1998).

Phylogenetic relationships within Acanthaceae have been examined by several investigators (e.g., Scotland et al. 1995; McDade et al. 2000). Schwarzbach and McDade (2002) provided molecular evidence for the close relationship of the mangrove genus *Avicennia* (formerly Avicenniaceae) to subfamily Thunbergioideae of Acanthaceae. In the analysis by Bremer et al. (2002), Acanthaceae and Pedaliaceae are part of a clade with Stilbaceae and Schlegeliaceae. However, with the greater taxon sampling of Olmstead et al. (2001), Soltis et al. (2011) and Refulio-Rodriguez and Olmstead (2014), Stilbaceae and Schlegeliaceae appeared in different positions. Schlegeliaceae, a small family of shrubs or vines with simple leaves, nectar glands on the outside of the calyx, and berry fruits, that sometimes is placed in Bignoniaceae (e.g., Cronquist 1981), may be sister to Lentibulariaceae but do not appear closely related to Bignoniaceae.

Lamiaceae include those taxa traditionally placed in Lamiaceae, as well as many taxa previously placed in Verbenaceae (e.g., *Callicarpa*, *Clerodendrum*, *Tectona*, *Vitex*; Olmstead and Reeves 1995; Wagstaff and Olmstead 1997; Wagstaff et al. 1998; Cantino 1992a). Previously, Verbenaceae were circumscribed much more broadly (e.g., Cronquist 1981), but phylogenetic analyses revealed this broadly defined family to be paraphyletic. The family now includes only the former subfamily Verbenoideae (with the exclusion of tribe Monochileae) (Cantino 1992a,b; Chadwell et al. 1992; Judd et al. 1994; Wagstaff and Olmstead 1997). To make Lamiaceae monophyletic, roughly two-thirds of the genera of the former, broadly defined Verbenaceae were transferred to Lamiaceae. In most studies, the now narrowly defined Verbenaceae and broadly defined Lamiaceae clades appear distantly related within Lamiales (Olmstead and Reeves 1995; Wagstaff and Olmstead 1997; Wagstaff et al. 1998; Savolainen et al. 2000a,b; D. Soltis et al. 2000; Albach et al. 2001c; Soltis et al. 2011; Refulio-Rodriguez and Olmstead 2014).

Verbenaceae in the narrow sense differ from Lamiaceae

(as redefined) in accumulation of 4-carboxy iridoids in the former versus C_4-decarboxylated iridoids in the latter (von Poser et al. 1997). Verbenaceae can also be distinguished from Lamiaceae by the presence in the former of indeterminate racemes, spikes, or heads (vs. inflorescences with an indeterminate main axis and cymosely branched lateral axes), simple style with conspicuous two-lobed stigma (vs. apically forked style with inconspicuous stigmatic region), marginally attached (vs. laterally attached) ovules, pollen exine thickened near apertures (vs. not thickened), and nonglandular hairs exclusively unicellular (vs. multicellular, uniseriate) (Judd et al. 2002, 2008). See Olmstead (2013) for a discussion on the biogeographic history of Verbenaceae and Marx et al. (2010) for a detailed phylogeny of the clade.

The relationships of the carnivorous families, Byblidaceae, Martyniaceae, and Lentibulariaceae, remain unclear (see Chapter 13). In Bremer et al. (2002) they formed a clade with weak support, but Byblidaceae and Lentibulariaceae were not included in Olmstead et al. (2001). Some molecular phylogenetic analyses indicated a close relationship between Byblidaceae and Martyniaceae, although support for this sister group was not apparent in analyses of three genes (D. Soltis et al. 2000). *Byblis* was generally considered related to Pittosporaceae (e.g., Dahlgren 1980; Cronquist 1981; Dahlgren et al. 1981), whereas Martyniaceae were placed near Pedaliaceae and Bignoniaceae, and sometimes even included in Pedaliaceae (Cronquist 1981). More recently, Refulio-Rodriguez and Olmstead (2014) found weak support for a clade of *Byblis* (Byblidaceae) and Linderniaceae (a clade with bilobed, but sensitive stigma and ruminant endosperm, traditionally included within Scrophulariaceae s.l.), while Lentibulariaceae were weakly supported as sister to Schlegeliaceae and Martyniaceae were sister to Acanthaceae. Thus, it appears that the evolution of carnivory in these different groups has evolved several times, although these results are tentative and need further clarification.

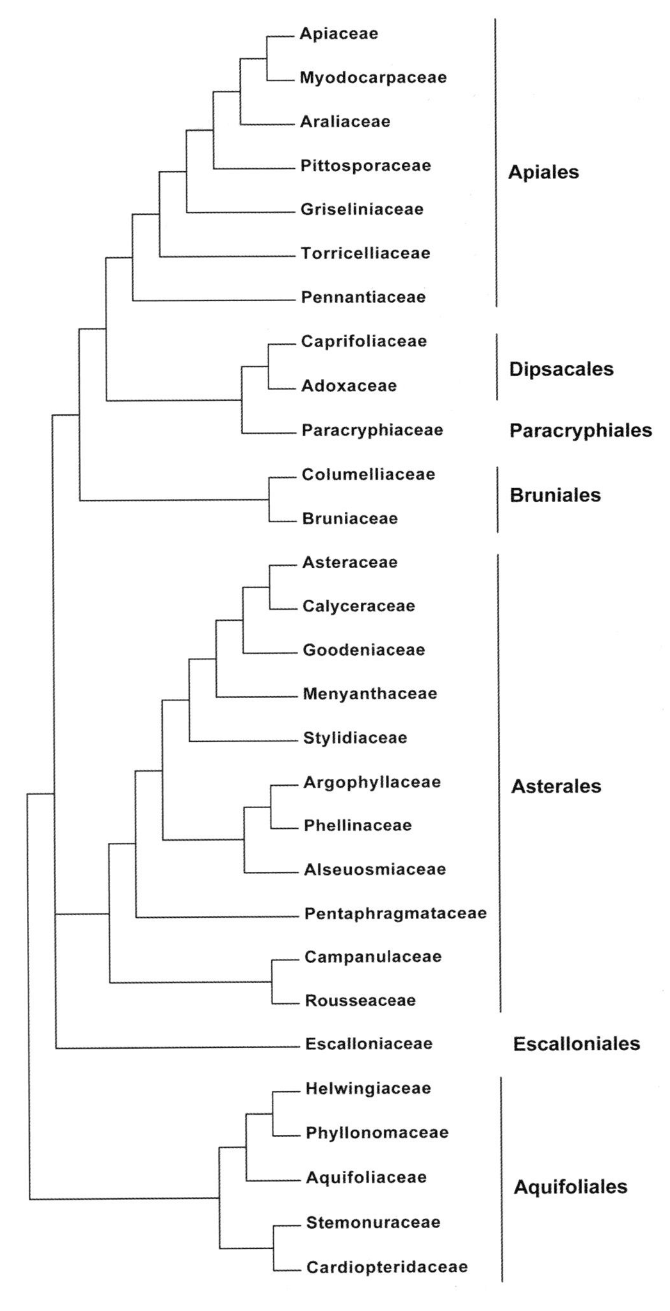

Figure 11.20. Phylogenetic relationships of Campanulids.

Campanulids

The campanulid clade (asterid II sensu Chase et al. 1993; campanulids of B. Bremer et al. 2002; euasterids II sensu APG II 2003) comprises seven major subclades (Fig. 11.20): Asterales, Escalloniales, Aquifoliales, Dipsacales, Paracryphiales, Apiales, and Bruniales (several of these families are illustrated in Figs. 11.21–11.23). Although the compositions of Dipsacales, Asterales (Asteranae sensu Thorne), and Apiales correspond well to the orders described by Cronquist (1981) and Thorne (1992a,b), the campanulid clade does not agree with any group of orders suggested in morphology-based classifications.

The campanulid clade combines members of traditional Asteridae with several groups of Rosidae (e.g., Aquifoliales, Apiales, Escalloniaceae, Icacinaceae; sensu Cronquist 1981, 1988). Monophyly of the campanulid clade was indicated by analyses of single genes, such as *rbcL* (Chase et al. 1993; Olmstead et al. 1993; Cosner et al. 1994; Backlund and Bremer 1997), *atpB* (Savolainen et al. 2000a), and 18S rDNA (Soltis et al. 1997b), and subsequent multigene analyses have provided further support for the clade (Jansen et al. 2007; Moore et al. 2010, 2011; Tank and Donoghue 2010; Soltis et al. 2011; Ruhfel et al. 2014). The clade is also supported by several non-DNA characters, including floral ontogeny (Erbar and Leins 1996, 2011; Roehls and Smets 1996). In addition, campanulids mostly have alternate leaves (with exceptions mainly in Dipsacales) and includes many taxa characterized by either secoiridoids or the apparent loss of iridoid production (the latter in Campanulaceae, Aquifoliaceae, and Apiales except *Griselinia* and *Melanophylla*; Albach et al. 2001b). Carbocyclic iridoids have been found in only four genera of campanulids: three genera of Stemonuraceae (*Cantleya* and *Lasianthera*, formerly Icacinaceae, Kaplan et al. 1991), *Stylidium* (Stylidiaceae), and *Escallonia* (Plouvier and Favre-Bonvin 1971). Iridoids and polyacetylenes seem to be complementary in distribution among campanulid members; they compete for acetyl-CoA and mevalonic acid as biosynthetic precursors (Stuhlfauth et al. 1985; reviewed in Albach et al. 2001b).

Aquifoliales are strongly supported as sister to the other campanulids (e.g., D. Soltis et al. 2000; Bremer et al. 2002; Moore et al. 2010, 2011; Tank and Donoghue 2010; Soltis et al. 2011; Ruhfel et al. 2014). The rest of the orders form a well-supported clade (BS = 100%, Soltis et al. 2011), but subclade relationships are still unclear. Soltis et al. (2011) suggest that the Asterales and Escalloniales may be sister to a clade formed by Dipsacales, Paracryphiales, Apiales, and Bruniales, but those relationships are not well supported. Tank and Donoghue (2010) found a similar topology, although Asterales did not form a clade with Escalloniales, but rather was sister to a clade composed of Escalloniales, Bruniales, Apiales, Paracryphiales, and Dipsacales. That topology also was not well supported (BS = 61%). Tank and Donoghue (2010) also found that Apiales + (Dipsacales + Paracryphiales) formed a well-supported clade (BS = 98%), but this was not well supported in Soltis et al. (2011).

AQUIFOLIALES

Aquifoliales consist of (APG II 2003; APG III 2009; Soltis et al. 2011; APG IV 2016; Fig. 11.20) Aquifoliaceae, Helwingiaceae, Phyllonomaceae, Cardiopteridaceae, and Stemonuraceae (several of these are illustrated in Figs. 11.21 and 11.23). The placement of these families together in the asterid clade was not anticipated from their treatments in morphology-based classifications. Cronquist (1981) placed Aquifoliaceae and Cardiopteridaceae in Celastrales, Helwingiaceae within Cornaceae, and Phyllonomaceae within Grossulariaceae. Morphological synapomorphies for Aquifoliales are stipules (which are rare in asterids), unisexual flowers, and fleshy, drupaceous fruits, which evolved in parallel in other orders (Bremer et al. 2001). They also have apical or apical-axile placentation and 1–2 ovules per locule. Kårehed (2001) indicated that Icacinaceae are polyphyletic. Some are part of lamiids (Icacinaceae s.s. or likely within Garryales; see Stevens 2001 onwards), with the remaining members part of Aquifoliales. Some taxa constitute a new family, Stemonuraceae (sensu Kårehed 2001), with other genera part of an enlarged Cardiopteridaceae.

Stemonuraceae and Cardiopteridaceae form a clade that is sister to the remaining families. The second subclade of Aquifoliales consists of Aquifoliaceae as sister to Helwingiaceae + Phyllonomaceae. These familial relationships are strongly supported (bootstrap above 97%) in Soltis et al. (2011). Helwingiaceae and Aquifoliaceae were weakly supported sisters in D. Soltis et al. (2000). However, other analyses have indicated with strong support that Aquifoliaceae are sister to Phyllonomaceae + Helwingiaceae (Savolainen et al. 2000b; Albach et al. 2001c; Soltis et al. 2011). The association of *Helwingia* and *Phyllonoma* is interesting, given that they share the rare character of flowers borne directly on the leaf blades (Fig. 11.21b; Dickinson and Sattler 1974; Morgan and Soltis 1993; Tobe 2013), as well as inferior ovaries and epigynous disc nectaries (Tobe 2013). Phyllonomaceae also exhibit numerous autapomorphies (Tobe 2013). Stemonuraceae and Cardiopteridaceae exhibit two synapomorphies: 1) two carpels in median position and 2) a pseudomonomerous gynoecium (Tobe 2013). Aquifoliales were inferred to have originated 113–85 mya (Bremer et al. 2004; Bell et al. 2010). *Ilex* (Aquifoliaceae) has been the focus of phylogenetic study (Cuénoud et al. 2000; Manen et al. 2010). This large genus provides the best insight into the fossil history of this group, due to the easily recognized, tricolporate pollen, which is ornamented with distinctive clavae (Fig. 11.23a–c). This pollen type is common in Eocene and younger deposits through the world. There are also reports from the Cretaceous, but these remain to be documented with SEM (Martin 1977).

ASTERALES

Asterales are strongly supported and consist of 11 families (Fig. 11.20): Alseuosmiaceae, Argophyllaceae, Asteraceae, Calyceraceae, Campanulaceae (including Lobeliaceae), Goodeniaceae, Menyanthaceae, Pentaphragmataceae, Phellinaceae, Rousseaceae (including *Carpodetus*), and Stylidiaceae (including Donatiaceae; several of these families are illustrated in Fig. 11.21; see Backlund and Bremer 1997; Gustafsson and Bremer 1997; Kårehed et al. 1999; Soltis et al. 2011). The option of retaining Lobeliaceae as a family distinct from Campanulaceae has been noted (see also Lundberg and Bremer 2003; APG II 2003) but remains included within Campanulaceae in most recent treatments (Stevens 2001 onwards; APG III 2009; APG IV 2016), although see Haberle et al. (2009). Donatiaceae and Stylidiaceae

Figure 11.21. Representatives of Campanulids (Apiales, Aquifoliales, Asterales, Bruniales, Dipsacales, Escalloniales, Paracryphiales). a. *Ilex glabra* A. Gray (Aquifoliaceae), carpellate flower; note sessile, expanded-flattened stigma, b. *Helwingia chinensis* Batalin (Helwingiaceae), flowers borne directly from the leaf adaxial midrib, c. *Wahlenbergia marginata* (Thund.) A. DC. (Campanulaceae), actinomorphic flower, pollen-receptive phase; note expanded stigmas, d. *Lobelia martagon* Hitchc. (Campanulaceae), zygomorphic flower, showing secondary pollen presentation; note connate anthers, with protruding style and stigma, so flower in pollen-receptive phase, e. *Stylidium torticarpum* Lowrie & Kenneally (Stylidiaceae), zygomorphic flowers with two stamens adnate to style, f. *Nymphoides cristata* (Roxb.) Kuntze (Menyanthaceae), flowers with corolla lobes with marginal wings, g. *Scaevola plumieri* Vahl (Goodeniaceae), zygomorphic flower, with a dorsal split, style with pollen-collecting cup, and the corolla lobes with marginal wings, h. *Calycera crassifolia* Hicken (Calyceraceae), heads of small flowers, i. *Vernonia gigantea* (Walter) Trel. (Asteraceae), heads of disk flowers, j. *Chrysogonum virginianum* L. (Asteraceae), heads of disk and ray flowers, i.e., radiate heads, k. *Helenium pinnatifidum* (Schwein. ex Nutt.) Rydb. (Asteraceae), head of disk and ray flowers, the latter with three lobes, l. *Hieracium aurantiacum* L. (Asteraceae), heads of ligulate flowers, each with five lobes.

are sister taxa (Lundberg and Bremer 2003; Soltis et al. 2011); in APG III (2009) the former is placed in synonymy under the latter. Asterales in this circumscription include 1,743 genera and 26,870 species, dominated by Asteraceae. This circumscription differs from that of Lammers (1992) in the addition of Stylidiaceae s.l. and the inclusion of *Alseuosmia* (Alseuosmiaceae) and *Corokia* (Argophyllaceae), which Lammers did not consider. Lammers (1992) did not include Stylidiaceae because of the presence of carbocyclic iridoids in this family (it is absent from the other families of this order). However, the production of carbocyclic iridoids also occurs in Cornales and early-diverging members of campanulids (Stemonuraceae, Aquifoliales), as well as lamiids (*Apodytes* of Icacinaceae s.l.; Kaplan et al. 1991); the presence of carbocyclic iridoids in Stylidiaceae likely represents a separate origin from these lineages.

Possible non-molecular synapomorphies for Asterales are a base chromosome number of $x = 9$, the production of the oligosaccharide inulin as a storage compound (reported from Asteraceae, Calyceraceae, Goodeniaceae, Campanulaceae, and Stylidiaceae), valvate corolla aestivation (with a reversal to imbricate aestivation in Stylidiaceae s.l. and parallel derivation of valvate aestivation in other euasterids), an inferior or partly inferior ovary, and perhaps also secondary pollen presentation (Bremer et al. 2001), sometimes referred to as "plunger pollination" (Judd et al. 2002, 2008), although the last could have developed in parallel within the order. As this is a pollen presentation mechanism rather than a means of pollination per se, "plunger pollen presentation" may be more appropriate (for the diversity of this mechanism, see Erbar and Leins 1995b). Campanulaceae and Stylidiaceae share the character of a filamentous suspensor of the proembryo (Tobe and Morin 1996), although most recent and well-sampled analyses do not support a sister relationship of these two taxa, as was found previously (Albach et al. 2001c; Bremer et al. 2002). Petal venation (Gustafsson 1995) and loss of micropylar endosperm haustoria (Cosner et al. 1994) are synapomorphies for the well-supported clade of Asteraceae, Calyceraceae, Goodeniaceae, and Menyanthaceae. Other characters shared by these four families are multinucleate tapetal cells and the production of secoiridoids (Lammers 1992), although in Asteraceae the production of secoiridoids is mostly replaced by the production of sesquiterpene lactones (but see Wang and Yu 1997).

There have been several comprehensive phylogenetic analyses of Asterales (e.g., Cosner et al. 1994; Gustafsson et al. 1996; Backlund and Bremer 1997; Gustafsson and Bremer 1997; Kårehed et al. 1999; Olmstead et al. 2000; Albach et al. 2001c; Bremer et al. 2002; Lundberg and Bremer 2003; Winkworth et al. 2008a; Tank and Donoghue 2010; Soltis et al. 2011). Despite differences in taxon sampling, these studies reported many congruent relationships, but certain interrelationships among families of Asterales remain poorly understood. For example, the basal relationships are not well supported, even in large multigene analyses (Albach et al. 2001c; Bremer et al. 2002; Soltis et al. 2011).

Within Asterales, Rousseaceae are strongly supported as sister to Campanulaceae (Tank and Donoghue 2010; Soltis et al. 2011). *Roussea*, which contains a single species, *Roussea simplex*, was recognized as part of the Asterales clade by Soltis et al. (1997a,b) and Koontz and Soltis (1999). The genus had often been placed with *Brexia* and *Ixerba* in Saxifragaceae s.l., Grossulariaceae, or Brexiaceae, all Rosidae. *Carpodetus* comprises 10 species and was formerly included in Escalloniaceae. *Roussea*, *Carpodetus*, *Abrophyllum*, and *Cuttsia* have been included in an expanded Rousseaceae (see Stevens 2001 onwards; Soltis et al. 2011).

The sister group to Campanulaceae + Rousseaceae is unresolved; Soltis et al. (2011) supported a sister relationship with all of Asterales except Pentaphragmataceae (which was sister to all Asterales in their maximum likelihood tree), whereas the topology of Winkworth et al. (2008a,b) and Tank and Donoghue (2010) placed the clade as sister to the rest of Asterales + Pentaphragmataceae, and the latter clade is supported by the putative synapomorphy of petal wings. Soltis et al. (2011) found that a clade composed of Phellinaceae, Argophyllaceae, and Alseuosmiaceae was sister to Stylidiaceae, Menyanthaceae, Goodeniaceae, Calyceraceae and Asteraceae. However, Tank and Donoghue (2010) found that the clade of Phellinaceae, Argophyllaceae, and Alseuosmiaceae was sister to Stylidiaceae, which was sister to a clade composed of Menyanthaceae, Goodeniaceae, Calyceraceae, and Asteraceae (as in previous analyses by Savolainen et al. 2000b; Albach et al. 2001b; Bremer et al. 2002).

Donatia + *Stylidium* (i.e., Stylidiaceae s.l.) has been recovered in several analyses (D. Soltis et al. 2000, 2011; Winkworth et al. 2008a,b; Tank and Donoghue 2010), which led to the inclusion of both genera in Stylidiaceae (APG III 2009), but the placement of that clade is problematic (as discussed above). There is strong support in some multigene analyses for a clade of Argophyllaceae, Alseuosmiaceae, and Phellinaceae (Bremer et al. 2002; Winkworth et al. 2008a,b; Tank and Donoghue 2010; Soltis et al. 2011), with Phellinaceae as sister to an Alseuosmiaceae + Argophyllaceae clade (Tank and Donoghue 2010; Soltis et al. 2011) or Alseuosmiaceae sister to a Phellinaceae + Argophyllaceae clade (Fig. 11.20), a putative clade that is supported by the possible synapomorphies of a base chromosome number of eight and glandular-toothed leaves; additionally, these taxa do not show plunger pollen presentation.

Menyanthaceae, Goodeniaceae, Calyceraceae, and Asteraceae form a well-supported clade (Savolainen et al. 2000a,b; D. Soltis et al. 2000; Albach et al. 2001c; Bremer et al. 2002; Lundberg and Bremer 2003; Winkworth et al. 2008a,b; Tank and Donoghue 2010; Soltis et al. 2011). This clade is supported by the herbaceous habit, stems with separate vascular bundles, details of corolla venation, and the antiraphal vascular bundle proceeding to the micropyle; at least the last may be synapomorphic (Stevens 2001 onwards). Menyanthaceae are well supported as sister to the remaining three families (Fig. 11.20)—that is, Goodeniaceae + Calyceraceae + Asteraceae, which are supported by the presence of acetylenes, fused or connivent anthers, pollen exine with bifurcating columellae; additionally they all show secondary pollen presentation (plunger pollen presentation).

The sister group of Asteraceae is either Calyceraceae (Gustafsson and Bremer 1995; Albach et al. 2001c; Bremer et al. 2002; Lundberg and Bremer 2003; Winkworth et al. 2008a,b; Tank and Donoghue 2010; Soltis et al. 2011) or Calyceraceae + Goodeniaceae (Cosner et al. 1994; Gustafsson et al. 1996; Savolainen et al. 2000a; D. Soltis et al. 2000; Albach et al. 2001c). The conflict among these analyses is of interest. Bremer et al. (2002) obtained strong support (88%), as did Tank and Donoghue (2010; ≥80%) and Soltis et al. (2011; 100%) for Calyceraceae alone as sister to Asteraceae; morphology plus DNA sequences also supported this relationship (Lundberg and Bremer 2003). However, D. Soltis et al. (2000) found strong support (88%) for Calyceraceae + Goodeniaceae. Characters supporting the sister-group relationship of Calyceraceae and Asteraceae are pollen morphology (Skvarla et al. 1977), petal venation (i.e., connate, commissural veins), and morphology of the receptacles and receptacle bracts (Hansen 1992), and a persistent calyx that is modified for dispersal of the fruit (Stevens 2001 onwards). However, biochemical evidence seems to support a relationship between Goodeniaceae and Calyceraceae: both produce bis-secoiridoids of the sylvestroside type (Jensen et al. 1979; Capasso et al. 1996; Albach et al. 2001b). In light of the most recent results for this group, perhaps the biochemical properties shared by Goodeniaceae and Calyceraceae represent symplesiomorphies that have been lost in Asteraceae. Ages of Asterales have been estimated from 101–77 mya (Wikström et al. 2001; Bremer et al. 2004; Janssens et al. 2009; Magallón and Castillo 2009; Beaulieu et al. 2013). See Fig. 11.21 for selected members of Asterales.

The large family Asteraceae (approximately 23,000 species) has been the focus of numerous phylogenetic analyses (K. Bremer 1987, 1994, 1996; Jansen et al. 1991, 1992; Kim et al. 1992; K. Kim and Jansen 1995; Watson et al. 2002; Ford et al. 2006; Keeley et al. 2007; Panero and Funk 2008; Funk et al. 2009a,b, 2014; Gruenstaeudl et al. 2009; Mavrodiev et al. 2012). Barnadesioideae, a small South American group comprising mostly trees and shrubs, are sister to the remainder of the family (Jansen and Palmer 1987; Panero and Funk 2008). Funk et al. (2009a,b) covers the phylogenetics, systematics, and biogeography of Asteraceae, including coverage of all major clades.

The fossil record of Asteraceae includes dispersed pollen and a few megafossil examples (Supplementary Table 11.1). The earliest convincing records are pollen grains documented by both scanning electron and light microscopy attributed to *Tubulifloridites lilliei* from the Late Cretaceous (Campanian/Maastrichtian) of Antarctica, suggesting a Southern Hemisphere origin (Barreda et al. 2015). The oldest known macro remains are the impression of a fossil capitulum, *Raiguenrayun cura*, with multiseriate–imbricate involucral bracts and pappus-like hairs from the Eocene of Argentina (Barreda et al. 2012). Dispersed pollen reports provide additional convincing records for the family in various geographic regions, and in some cases can be assigned to specific clades based on distinctive ornamentation types that have been distinguished as *Ambrosia* type, Liguliflorae type, Lactuceae-Veronieae type, and *Mutisia* type (Graham 1996). Early pollen records include examples of Asteraceae pollen from the Eocene Princeton Chert of British Columbia (Manchester et al. 2015). Pollen grains characteristic of tribe Mutiseae have been recognized from the late Paleocene-Eocene of South Africa, treated under the name, *Tubulifloridites antipodica* (Zavada and de Villiers 2000), and from the early Oligocene of northwestern Tasmania, named *Mutisiapollis patersonii* (Macphail and Hill 1994). Pollen from these and other Southern Hemisphere sites allow for recognition of the Barnadesioideae, Nassauvieae, and Mutisieae (Barreda et al. 2010).

Campanulaceae also have been studied phylogenetically by numerous researchers (Cosner et al. 2004; Antonelli 2008, 2009; Roquet et al. 2008, 2009; Cellinese et al. 2009; Givnish et al. 2009; Haberle et al. 2009; Prebble et al. 2011, 2012a,b; Wendling et al. 2011; Oleson et al. 2012; Zhou et al. 2013; Crowl et al. 2014; Lagomarsino et al. 2014). Four primary clades are recovered in most analyses, the lobelioid, platycodonoid, wahlenbergioid, and campanuloid clades (see Haberle et al. 2009). It is clear that the large genera *Lobelia* and *Campanula* are highly polyphyletic and have given rise to highly specialized taxa with a large variety of growth forms, floral modifications for bird pollination, and fleshy fruit for animal dispersal instead of the typical capsular fruit, which are more common in the group (Antonelli 2008; Givnish et al. 2009; Haberle et al. 2009).

Of considerable interest is the radiation of Hawaiian lobeliads (including the Hawaiian silverswords), which have been the focus of several recent studies (Givnish et al. 2004,

2009; Montgomery and Givnish 2008). This group consists of roughly 126 species and had been placed in six genera. Growth forms in these species range from rosette species to trees up to 18 m tall, and fruits range from dry capsules to fleshy berries; most of these taxa are bird-pollinated and occur in the understory of broad-leaved tropical forests (Givnish et al. 2009). Historically this group was considered to have arisen via 4–5 original colonizations to the islands, considering their disparate morphological characters, life history traits, and ecological niches; however, phylogenetic analyses revealed that the group likely originated from one dispersal event to the islands and subsequently formed one of the largest island radiations of plants known (Givnish et al. 2009). Givnish et al. (2009) also demonstrated that ecological preference, fruit type, inflorescence position, habit, and pollination mode were labile characters, having evolved multiple times within the lobelioids in general. Givnish et al. (2004) and Montgomery and Givnish (2008) further showed that the Hawaiian lobeliads—notably *Cyanea* and *Clermontia*—demonstrated functionally different physiological adaptations for light capture and photosynthesis, allowing them to occur in different ecological niches with specific light regimes, suggesting a possible mechanism for such a large adaptive radiation of the group in the Hawaiian Islands.

ESCALLONIALES

Bremer et al. (2002) first suggested that Polyosmaceae, as well as Tribelaceae, were members of a clade with Eremosynaceae + Escalloniaceae. In most recent analyses, Escalloniaceae, Polyosmaceae, Tribelaceae, and Eremosynaceae form a well-supported clade (Fig. 11.20) that is sister to Asterales (Soltis et al. 2011) or to a clade formed by Dipsacales, Apiales, and Bruniales (Tank and Donoghue 2010). A sister relationship of Eremosynaceae (*Eremosyne*) and Escalloniaceae had been supported in most multigene analyses (e.g., D. Soltis et al. 2000; Albach et al. 2001c), although with the inclusion of more taxa in analyses, it appears that *Valdivia* and *Forgesia* could be more closely related to Escalloniaceae s.s. (Soltis et al. 2011; Sede et al. 2013). The close relationship of Escalloniaceae and Eremosynaceae was somewhat unexpected, although both had been placed in Saxifragaceae s.l. by Engler (1930; see Morgan and Soltis 1993).

Treatments of these four families have been diverse. Cronquist (1981), for example, included Escalloniaceae in his Grossulariaceae and Eremosynaceae within Saxifragaceae (both Rosales, Rosidae). Takhtajan (1997), in contrast, placed Escalloniaceae in Cornidae, while similarly considering *Eremosyne* (as Eremosynaceae) close to Saxifragaceae in Saxifragales (Rosidae). Polyosmaceae consisting of one genus, *Polyosma*, and ca. 60 species, has been included in Escalloniaceae by some authors (Takhtajan 1987). Cronquist (1981) placed *Polyosma* in Grossulariaceae with other putative woody relatives of Saxifragaceae, including Montiniaceae, Tribelaceae, and Phyllonomaceae, families now known to be part of the asterid clade. Cronquist (1981) placed Tribelaceae in Grossulariaceae with other putative woody relatives of Saxifragaceae. Takhtajan (1997) placed the family close to Escalloniaceae (Hydrangeales, Cornidae). Eremosynaceae, Polyosmaceae, and Tribelaceae are now recognized under a broader Escalloniaceae considering the most recent phylogenetic findings (Fig. 11.20).

Escalloniaceae as previously circumscribed are polyphyletic; those traditional members have diverse relationships within the campanulids. Other genera formerly placed in the family with relationships elsewhere in campanulids include *Columellia* and *Desfontainia* (included within Columelliaceae; Bruniales), *Phelline* (Asterales), and *Sphenostemon* and *Quintinia* (Paracryphiales) (Backlund and Bremer 1997; APG III 2009; Tank and Donoghue 2010; Soltis et al. 2011). Trichomes with radially arranged apical cells in the glandular head are shared by *Escallonia* and *Eremosyne* (Al-Shammary and Gornall 1994). Synapomorphies for Escalloniaceae s.l. are opposite leaves, tetramerous flowers, valvate corollas (see Fig. 11.22), triporate pollen, starchy endosperm, and one-seeded drupes (Stevens 2001 onwards). Several recent phylogenetic analyses have focused on Escalloniaceae s.s. (i.e., *Escallonia*; Sede et al. 2013; Zapata 2013).

DIPSACALES

Dipsacales consist of Adoxaceae (including *Sambucus* and *Viburnum*, genera formerly placed in Caprifoliaceae) and a broadly circumscribed Caprifoliaceae (Fig. 11.20; Judd et al. 2002, 2008; APG II 2003; APG III 2009; APG IV 2016; several representatives are illustrated in Figs. 11.22, 11.23). The order comprises about 46 genera and around 1090 species. Rather than recognizing a broadly defined Caprifoliaceae, some systematists prefer subdividing Caprifoliaceae into several families: Caprifoliaceae s.s., Diervillaceae, Dipsacaceae, Linnaeaceae, Morinaceae, and Valerianaceae, a treatment that was also provided as an option in the APG II (2003) classification (see Backlund and Pyck 1998; Bremer et al. 2002). Savolainen et al. (2000a,b) suggested that Desfontainiaceae (now included in Columelliaceae), Polyosmaceae, Paracryphiaceae, and Sphenostemonaceae might also be part of Dipsacales, but bootstrap support for this placement was lacking; other multigene analyses placed these families outside of Dipsacales (Soltis et al. 2011). Bremer et al. (2002) found weak support (55%) for a sister-group relationship of Paracryphiaceae

Figure 11.22. Campanulids cont'd. (Escalloniales, Bruniales, Apiales, Dipsacales). a-b. *Escallonia rubra* (Ruiz & Pav.) Pers. (Escalloniaceae), flowers, c. *Brunia* sp. (Bruniaceae), note ericoid habit, d. *Pittosporum pentandrum* (Blanco) Merr. (Pittosporaceae), flowers, sympetalous but with well-developed lobes, and also evident calyx, e. *Pittosporum tenuifolium* Gaertn. (Pittosporaceae), capsular fruits, seeds with viscid, resinous pulp, f. *Aralia nudicaulis* L. (Araliaceae), umbels of drupes, g. *Myodocarpus fraxinifolius* Brongn. & Gris (Myodocarpaceae), woody habit, with pinnately compound leaves and paniculate-umbellate inflorescence, h. *Myodocarpus fraxinifolius* (Myodocarpaceae), samaroid schizocarps; note inferior ovary, well-developed stylopodium, and two recurved style branches, i. *Heracleum lanatum* Michx. (Apiceae), immature schizocarps; note inferior ovary, well-developed stylopodium, and two spreading style branches, j. *Daucus carota* L., (Apiaceae), umbellate inflorescence, flowers with separate petals. k. *Viburnum obovatum* Walter (Adoxaceae), actinomorphic flowers, each with a short style, l. *Diervilla lonicera* Mill. (Caprifoliaceae), flowers slightly zygomorphic, style elongate, stigma expanded and capitate, m. *Valeriana scandens* Loefl. (Caprifoliaceae), opposite-leaved vine, n. *Dipsacus fullonum* L. (Caprifoliaceae), involucrate floral head.

and Dipsacales (Fig. 11.20), and this placement has been further supported by more recent analyses (Tank and Donoghue 2010; Soltis et al. 2011).

Dipsacales (Fig. 11.20) correspond closely to the Dipsacales of Cronquist (1981) and Thorne (1992a,b) and Dipsacanae sensu Takhtajan (1997). Dahlgren's (1980) treatment of Dipsacales differed in that he also included Calyceraceae in the order, a family firmly placed in Asterales (see above). Takhtajan (1997) placed several genera of Dipsacales (i.e., *Sambucus*, *Viburnum*, *Triplostegia*, and *Morina*) in monogeneric families. Morphological synapomorphies for Dipsacales include pericyclic cork and a gynoecium of three or more carpels (with some reversals to two; Bremer et al. 2001), along with opposite and decussate leaves and seeds with a vascularized testa.

The relationships of Dipsacales remain unclear. Analyses based on three (D. Soltis et al. 2000), four (Albach et al. 2001c), and 12 (Tank and Donoghue 2010) loci indicated that Dipsacales and Apiales form a clade with Bruniaceae and Escalloniaceae, although this clade did not receive support above 61% and relationships among these groups were either not highly supported or unclear. Soltis et al. (2011) placed Escalloniaceae as sister to Asterales instead of the former placement with Apiales and Dipsacales, and again, relationships among Dipsacales, Apiales, Bruniales, and Paracryphiales were not well supported. A possible close relationship between Apiales and Dipsacales is supported by the shedding of trinucleate pollen (Albach et al. 2001c).

Dipsacales have been the subject of a series of molecular phylogenetic investigations (e.g., Donoghue et al. 1992, 2001; Backlund and Bremer 1997; Bell and Donoghue 2005a; Clement and Donoghue 2007; Smith and Donoghue 2008; Theis et al. 2008; Winkworth et al. 2008a,b; Carlson et al. 2009), as well as evolutionary developmental studies (Howarth and Donoghue 2005, 2009). Within Dipsacales, Adoxaceae are sister to the remainder (D. Soltis et al. 2000; Albach et al. 2001c; Donoghue et al. 2001; Bremer et al. 2002; Winkworth et al. 2008a,b; Tank and Donoghue 2010; Soltis et al. 2011). Caprifoliaceae s.l. are easily separated from Adoxaceae by bilateral floral symmetry in the former versus radial in the latter; flowers with an elongate versus short style, capitate versus lobed stigma, spiny versus reticulate pollen exine, and differences in the structure of the nectary (Judd et al. 2002, 2008; for the nectary, see also Wagenitz and Laing 1984). A series of morphological and molecular phylogenetic analyses has focused on Caprifoliaceae s.l., some as part of more comprehensive analyses (e.g., Donoghue et al. 1992, 2001; Downie and Palmer 1992; Judd et al. 1994; Backlund and Bremer 1998; Backlund and Pyck 1998; Källersjö et al. 1998; Savolainen et al. 2000b; D. Soltis et al. 2000; Albach et al. 2001c; Bremer et al. 2001; Bremer et al. 2002; Winkworth et al. 2008a,b; Theis et al. 2008; Jacobs et al. 2009). The multigene analysis of Bremer et al. (2002) strongly supported relationships among the component families of Caprifoliaceae s.l. found in a cladistic analysis of non-DNA characters (Judd et al. 1994). Donoghue et al. (2001) provided a comprehensive overview of the order.

Within Caprifoliaceae s.l., Diervillaceae are sister to the rest of the clade. Caprifoliaceae s.s. are sister to the Linnaeaceae, Morinaceae, Dipsacaceae + Valerianaceae clade (Winkworth et al. 2008a,b; Jacobs et al. 2009; Soltis et al. 2011). Caprifoliaceae s.s. are united by 4–5 carpels and an indeterminate inflorescence (Judd et al. 1994, 2002). The clade of Diervillaceae, Linnaeaceae, Valerianaceae, and Dipsacaceae (Backlund and Pyck 1998; Albach et al. 2001c; Bremer et al. 2002) exhibits an achene fruit (rather than a berry or drupe), a base chromosome number of $x = 8$, and abortion of two or three carpels leaving a single-ovuled carpel occupying half of the ovary (Judd et al. 1994). They also produce bis-secoiridoids, such as sylvestroside and related compounds (Jensen et al. 1979; Plouvier 1992; Capasso et al. 1996). Jacobs et al. (2009) showed the fleshy fruits of the *Lonicera* clade to be derived within the order and suggested that the ancestral state may have been achene fruits.

Removal of *Abelia* and tribe *Linnaea* from Caprifoliaceae s.s. was proposed based on morphology (Donoghue 1983), including a putative reduction in the number of seeds, stamens, and fertile locules in the ovary in the former. Morphological and molecular evidence ultimately resulted in the description of the family Linnaeaceae (Backlund and Pyck 1998). However, Linnaeaceae now are recognized within Caprifoliaceae s.l. Strong support for a sister-group relationship between the Dipsacaceae and Valerianaceae clades has been provided in several molecular analyses (e.g., Albach et al. 2001c; Bremer et al. 2002; Winkworth et al. 2008a,b; Jacobs et al. 2009; Soltis et al. 2011). The two clades are united by reduced or absence of endosperm and a modified calyx.

The fossil record of Dipsacales is incomplete, but there are notable examples representing both Adoxaceae and Caprifoliaceae s.l. Pollen of *Viburnum* can be distinctive when studied by both SEM and LM. *Viburnum* pollen has been documented from the middle Eocene Princeton Chert of British Columbia (Manchester et al. 2015), showing the high reticulum with numerous free-standing columellae in the luminae, typical of the *Solenotinus* and *Tinus* clades and in *Viburnum clemensiae* (e.g., Donoghue 1985, pers. comm.). Caprifoliaceae are represented by winged fruits of Linnaeoideae (Caprifoliaceae) from the late Eocene including representing the extant genus *Dipelta* in the late Eocene of Southern England (Reid and Chandler 1926) and Mississippi (Fig. 11.23f; Manchester et al. 2009) and the extinct genus *Diplodipelta* (Manchester and Donoghue 1995)

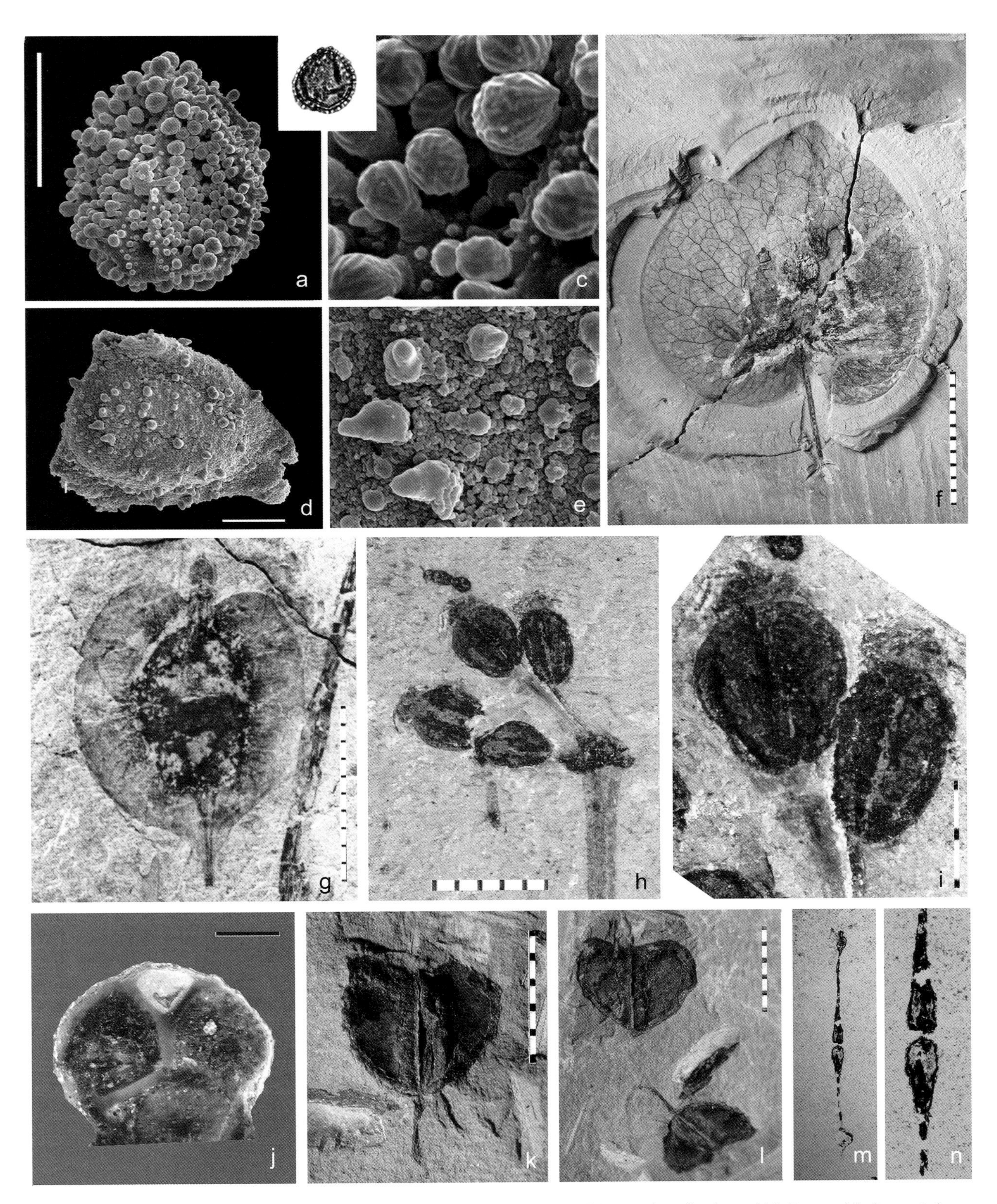

Figure 11.23. Paleobotanical examples of Aquifolialeas, Dipsacales, Apiales. a-c. Pollen of Aquifoliaceae, *Ilex* pollen from middle Eocene of Profen, central Germany by SEM and light microscopy, courtesy Fridgeir Grimsson, d-e. Caprifoliaceae: *Diervilla* pollen from the Eocene of Greenland, f. *Dipelta* sp. Eocene of Mississippi [uf 15737–49026]. g. Apiaceae: winged mericarp *Carpites ulmiformis* from late Cretaceous (Maastrichtian) of Montana, h-i. Araliaceous umbel from early Middle Eocene of Utah, j. Fruit of *Torricellia bonesii* from the middle Eocene of Oregon, transverse section showing one fertile seed-containing locule and two sterile locules [uf 9577], k, l. *Araliaecarpum kolymensis* schizocarpic fruits from Cretaceous of Russia, m, n. Transverse sections of the fruit in l.

from the Eocene, and Miocene of western North America. Several pollen types present among extant Dipsacales are distinctive (Donoghue 1985) and readily recognizable in the paleobotanical record. Examples of the distinctive echinate pollen of Linnaeoideae are known from the Eocene of Greenland (Fig. 11.23d,e), Germany, and North America (Manchester et al. 2015).

PARACRYPHIALES

Paracryphiales consist of the woody species, *Paracryphia alticola* (Paracryphiaceae s.s.), native to New Caledonia, as well as *Sphenostemon* (previously Sphenostemonaceae) and *Quintinia* (previously Quintiniaceae), all now placed within one family, Paracryphiaceae (Stevens 2001 onwards; see Chapter 12). Several phylogenetic analyses have supported this clade as a member of the campanulids (Savolainen et al. 2000b; Bremer et al. 2002; Tank and Donoghue 2010; Soltis et al. 2011) likely as sister to Dipsacales (Tank and Donoghue 2010; Soltis et al. 2011), although support for this placement has been weak; this placement should be considered tentative (Fig. 11.20). *Quintinia* was resolved as sister to *Paracryphia* + *Sphenostemon* (Fig. 11.20).

The relationships of *Paracryphia* had long been debated. Cronquist (1981) placed the genus in Theales (Dilleniidae). Takhtajan (1997) and Thorne (1992a,b 2000) also considered the species a member of Theales (see also Schmid 1978). *Paracryphia* exhibits an unusual combination of characters (Dickison and Baas 1977). In vegetative anatomy, the species is unspecialized, whereas in some aspects of floral morphology, it appears highly specialized (Takhtajan 1997). Likewise, *Sphenostemon*, comprising seven species, is another enigmatic genus. Cronquist (1981) placed the genus in his Aquifoliaceae, as did Takhtajan (1997; his Icacinales + Aquifoliales). *Quintinia* was often included within Hydrangeaceae or Saxifragaceae (Stevens 2001 onwards). Stevens (2001 onwards) includes tetramerous flowers and septicidal capsules as synapomorphies for Paracryphiales, although *Quintinia* often deviates from this morphology. More study of morphological characters would greatly enhance our knowledge of this group.

APIALES

Apiales, with approximately 494 genera and 5489 species, consist of 7 families following APG III and APG IV (2009, 2016; Fig. 11.20): Apiaceae, Araliaceae, Griseliniaceae, Myodocarpaceae, Pennantiaceae, Pittosporaceae, and Torricelliaceae (several are illustrated in Fig. 11.22; see also Plunkett et al. 1997a,b, 2001, 2004a; Plunkett 2001; Plunkett and Lowry 2001; Lowry et al. 2001; Chandler and Plunkett 2004; Winkworth et al. 2008a,b; Tank and Donoghue 2010; Soltis et al. 2011). In recent classifications (e.g., Cronquist 1981), Apiales have been included in Rosidae; their placement within asterids, as evident in all molecular phylogenetic studies, was not anticipated. In morphology-based classifications, the order was often limited to Apiaceae and Araliaceae. Griseliniaceae and Torricelliaceae (including Aralidiaceae and Melanophyllaceae) were often included in Cornaceae (Rosidae); Pittosporaceae were also considered part of Rosidae, and *Pennantia* was placed in Icacinaceae. *Mackinlaya* (Apiaceae) and *Myodocarpus* (Myodocarpaceae) were both formerly placed in Araliaceae.

Apiales consist of taxa with mostly choripetalous flowers, whereas most euasterids are characterized by sympetaly. However, Erbar and Leins (1996) showed that members of Apiales exhibit early sympetaly. Thus, choripetaly (or at least sympetalous flowers with well-developed lobes) is a synapomorphy for Apiales (assuming that sympetaly is a synapomorphy for all asterids), together with sheathing petioles (Bremer et al. 2001). Additional putative synapomorphies may include the presence of polyacetylenes (but some taxa have not been sampled), small flowers with sepals that are reduced in size or even obsolete, a reduction to only two ovules in each locule (but often even more reduced), and a drupe with 1–5 pits (but variously modified in some groups).

The fossil record of Apiales is meager. Pollen of Araliaceae, characterized by thick exine, elongate endoapertures, extine thickening adjacent to apertures, and microreticulate to reticulate or rugulate (perforate, fossulate) ornamentation visible by SEM, has been recognized from the Late Cretaceous (early Campanian) and Eocene of western North America, and the Paleocene and Eocene of Greenland and Eocene of British Columbia (Manchester et al. 2015). Pollen assignable to *Aralia*, distinguished by its thickened exine in polar areas in combination with other general features of Araliaceae, is recognized from the Eocene Princeton Chert of British Colombia (Manchester et al. 2015). Late Cretaceous fruits of Apiaceae include *Carpites ulmiformis* Dorf from the Maastrichtian of Montana and Wyoming based on winged fruits with a persistent epigynous calyx (Fig. 11.23g; Manchester and O'Leary 2010). Schizocarpic apiaceous fruits preserving stylopodium, persistent epigynous perianth and recurved style arms, have been recovered from the Eocene Green River Formation of Utah (Fig. 11.23h,i). Fossil fruits named *Paleopanax* represent the schizocarpic fruit type and persistent epigynous perianth characteristic of Araliaceae s.l. from the Eocene of Oregon (Manchester 1994b). Distinctive trilocular fruits with a pair of sterile locules surrounding a central fertile locule, diagnostic of the extant genus *Torricellia*, are known from the middle Eocene of Oregon (Fig. 11.23j; Manchester 1999) and Middle Eocene to Miocene of Germany (Meller 2006; Collinson et al. 2012).

The oldest potential Apiales fossil, and indeed the oldest potential superasterid, is the fruit called *Araliaecarpum* Samylina from the early Cretaceous (Albian) Buor-Kemiusskaja locality near the Zyrianka River in eastern Siberia (Fig. 11.23k–l; Samylina 1960) (Supplementary Table 11.1). The flattened, syncarpous fruit, 6 mm long, consists of two carpels borne on a thin pedicel. It is possible that the fruits were schizocarpic as suggested by transverse section (Fig. 11.23m,n). An important question is whether the fruits developed from epigynous or from hypogynous flowers. If hypogynous, then some malvid families—for example, Brassicaceae, Sapindaceae—might come into consideration. There is no obvious swelling at the junction of the fruit and pedicel that would be interpreted as the position of hypogynous perianth. On the other hand, there is no obvious perianth bulge or scar at the apical side of the fruit. More detailed comparative work is needed to assess whether the resemblance to Apiaceae is more than superficial.

The relationships of Apiales within Campanulids are unclear. Most analyses suggest a sister relationship with Dipsacales + Paracryphiales with Bruniales sister to the entire clade (Tank and Donoghue 2010; Soltis et al. 2011; Fig. 11.20), although this topology is poorly supported. Tank and Donoghue (2010) found Escalloniaceae to be sister to the Bruniales, Apiales, Paracryphiales + Dipsacales clade, although this position is also not highly supported (see Soltis et al. 2011, and above, for alternate position of Escalloniaceae). Pittosporaceae, Araliaceae, and *Torricellia* share a base chromosome number of $x = 12$ with *Escallonia*. *Griselinia* has $x = 9$, a number found in some Apiaceae and common in campanulids (e.g., many Asterales and Adoxaceae). *Pennantia* (Pennantiaceae), previously placed in Icacinaceae, has been resolved as sister to the remainder of the order (Kårehed 2002; Winkworth et al. 2008a,b; Tank and Donoghue 2010; Soltis et al. 2011; Fig. 11.20). Torricelliaceae (including Melanophyllaceae and Aralidiaceae), Griseliniaceae, and Pittosporaceae are well supported as subsequent sisters to a well-supported clade of Araliaceae and Apiaceae + Myodocarpaceae (Tank and Donoghue 2010; Soltis et al. 2011), in line with earlier topologies (Bremer et al. 2002).

Araliaceae + Apiaceae + Myodocarpaceae is well supported: these plants have petroselenic acid in the seeds, umbelliferose, inflorescences that are variously umbellate (Fig. 11.22), flowers with an open calyx, a valvate corolla, and a stylopodium, and flattened, schizocarpic fruits, all likely synapomorphic. Winkworth et al. (2008a,b) resolved Myodocarpaceae as sister to Araliaceae, but this placement was poorly supported. Plunkett et al. (2004b) first clarified the positions of several "ancient araliads" and enigmatic hydrocotyloids of Apiaceae, resulting in the recognition of Myodocarpaceae. Myodocarpaceae contain *Myodocarpus*, *Delarbrea*, and *Pseudosciadium* (Plunkett and Lowry 2001; Plunkett et al. 2004a), and Lui et al. (2010) provided further morphological characters for recognizing the family. Mackinlayaceae have been included within Apiaceae in most recent treatments (APG III 2009; see also Stevens 2001 onwards; Nicolas and Plunkett 2009; Soltis et al. 2011). Additional phylogenetic and phylogenomic analyses of Apiales, Apiaceae, and Araliaceae include Plunkett et al. (1996a,b, 2004a,b), Downie et al. (1998, 2000a,b), Plunkett and Downie (1999), Henwood and Hart (2001), Chandler and Plunkett (2004), Chandler et al. (2007), Smith and Donoghue (2008), Winkworth et al. (2008a,b), Nicolas and Plunkett (2009), Downie et al. (2010), and R. Li et al. (2013). Several members of Apiales are illustrated in Figs. 11.22, 11.23.

BRUNIALES

Although resolving the position of Bruniaceae has been problematic (Olmstead et al. 1993; Savolainen et al. 2000b; D. Soltis et al. 2000; Albach et al. 2001c), multigene analyses have shown that Bruniaceae and Columelliaceae are sisters and are now placed in Bruniales (Bremer et al. 2002; Tank and Donoghue 2010; Soltis et al. 2011; Fig. 11.20). Bruniaceae are considered a well-defined family of 12 genera native to southern Africa. The family has at times been associated with Hamamelidaceae. However, most modern classifications have placed the family among what were considered the woody relatives of Saxifragaceae in Rosales (e.g., Greyiaceae, Escalloniaceae, Columelliaceae, Grossulariaceae of Cronquist 1981). However, Takhtajan (1997) placed the family with Grubbiaceae (now in Cornales) in Bruniales in Dilleniidae. Two genera, *Berzelia* (sometimes considered a separate family, Berzeliaceae) and *Brunia*, have standardly been sampled in molecular analyses, with one or the other genus used as a placeholder for Bruniaceae; both Tank and Donoghue (2010) and Soltis et al. (2011) found them to be sister taxa. *Brunia* is illustrated in Fig. 11.22.

Columelliaceae as previously circumscribed consist of the genus *Columellia* with four species native to the Andes. Cronquist (1981) considered the family taxonomically isolated, but placed them in Rosales with several other genera and families now recognized as part of the asterid clade. Takhtajan (1997) envisioned a relationship of *Columellia* to some of the same "escallonioid" taxa, but placed these families in his Hydrangeales (Cornidae). There was strong molecular support for a sister relationship of *Columellia* and *Desfontainia* (formerly Desfontainiaceae) (Bremer et al. 2002; Tank and Donoghue 2010; Soltis et al. 2011). *Desfontainia* is another enigmatic genus, usually included in Loganiaceae (Gentianales) (see Cronquist 1981). However, Takhtajan (1997) placed Desfontainiaceae in its own order near Hydrangeales and noted a possible close relationship between Desfontainiaceae and Columelliaceae, supported by similarities in wood anatomy (Mennega 1980). Given the close relationship between *Columellia* and *Desfontainia*, the two have been combined to form a single family, Columelliaceae (Bremer et al. 2002; APG II 2003; APG III 2009); that circumscription is followed here (see Chapter 12). The relationship of Bruniales to other asterids is still unclear. Most recent analyses have recovered Bruniales as sister to an Apiales, Paracryphiales, Dipsacales clade, but with low bootstrap support (62% in Tank and Donoghue 2010; <50% in Soltis et al. 2011).

PATTERNS OF CHARACTER EVOLUTION

The asterid clade affords an excellent opportunity to examine the evolution of several key morphological and chemical characters, including sympetaly, integument number, nucellar condition, floral symmetry, and iridoid production. Many of these non-DNA characters were examined in a phylogenetic context by Albach et al. (2001b) using the four-gene topology for this large clade (Albach et al. 2001c), and the evolution of floral symmetry was addressed by Donoghue et al. (1998), and Howarth and Donoghue (2005, 2009). However, because phylogenetic relationships within the asterid clade are now better understood (e.g., Olmstead et al. 2001; Bremer et al. 2002; Kårehed 2002; Xiang et al. 2002; Lundberg and Bremer 2003; Plunkett et al. 2004a; Jansen et al. 2007; Moore et al. 2007, 2010, 2011; Soltis et al. 2011; Ruhfel et al. 2014), we evaluate the evolution of several key features. We used a topology that closely reflects that in Soltis et al. (2011), but with specific subclades modified by increased taxon sampling to reflect recent phylogenetic analyses (e.g., Ericales, Kron et al. 2002, Schönenberger et al. 2005; Apiales, Plunkett and Lowry 2001, Kårehed 2002, and Plunkett et al. 2004a; Cornales, Xiang et al. 2002, 2011; Asterales, Lundberg and Bremer 2003, Lundberg 2009; Tank and Donoghue 2010; Lamiales, Olmstead et al. 2001, Refulio-Rodriguez and Olmstead 2014).

Sympetaly characterizes most asterids, and the core of this clade has been recognized as a natural group in classifications for more than 200 years on this basis (e.g., de Jussieu 1789; de Candolle 1813; Reichenbach 1827–1829). A suite of embryological and chemical characters also appears to be highly but not perfectly correlated with sympetaly: unitegmy (Warming 1879), tenuinucellate ovules (cf. Warming 1879; Philipson 1974), cellular endosperm formation (Sporne 1954; Dahlgren 1975), terminal endosperm haustoria (Dahlgren 1977), pollen grains released at the tricellular stage (Dahlgren 1975), and iridoids (Dahlgren et al. 1981).

Nearly all eudicots with sympetalous corollas are asterids. Notable exceptions are Plumbaginaceae (Caryophyllales), Caricaceae (Brassicales: rosids), and Cucurbitaceae (Cucurbitales: rosids), in which sympetalous corollas undoubtedly evolved independently. Even within asterids, homology of sympetaly is not certain because sympetalous corollas may arise through different ontogenetic pathways (Erbar 1991) and have been derived many times (at least four or more; Olmstead et al. 1992; Albach 1998). Consequently, additional information, particularly on the development of sympetalous corollas, is needed for many groups before the evolutionary history of sympetaly can be clarified.

Instead of examining the evolution of sympetaly itself, we therefore present hypotheses of the evolution of those characters typically associated with sympetaly in asterids: unitegmy, tenuinucellate ovules, cellular endosperm formation, trinucleate pollen, and iridoids. The evolution of these characters was studied by Albach et al. (2001b). Using Albach et al.'s (2001c) phylogenetic tree (see above), we provide an abbreviated discussion of these patterns in the following sections.

UNITEGMY

Unitegmy, although typically associated with sympetaly (cf. Warming 1879; Olmstead et al. 1993), also occurs frequently outside asterids (certain magnoliids, some members of rosids, Ceratophyllaceae; Philipson 1974; Corner 1976; Endress 2011a-c). Multiple transitions to unitegmic ovules are supported by the discovery of multiple ontogenetic pathways to unitegmy (Bouman and Calis 1977; Bouman 1984), although a general lack of information on ovule ontogeny hinders attempts to relate the distribution of unitegmy to specific mechanisms of unitegmic ovule formation. Our assignment of unitegmic versus bitegmic ovules is based on Philipson (1974), Corner (1976), Cronquist (1981), and Takhtajan (1997). The ancestral condition of asterids appears to be unitegmy based on the reconstruction presented here (Fig. 11.24). Most angiosperms outside the asterid clade have bitegmic ovules, except for Santalales, which are either unitegmic or totally lack integuments, along with drastically reduced ovules, in some members (Brown et al. 2010; see Endress 2011c). Both Caryophyl-

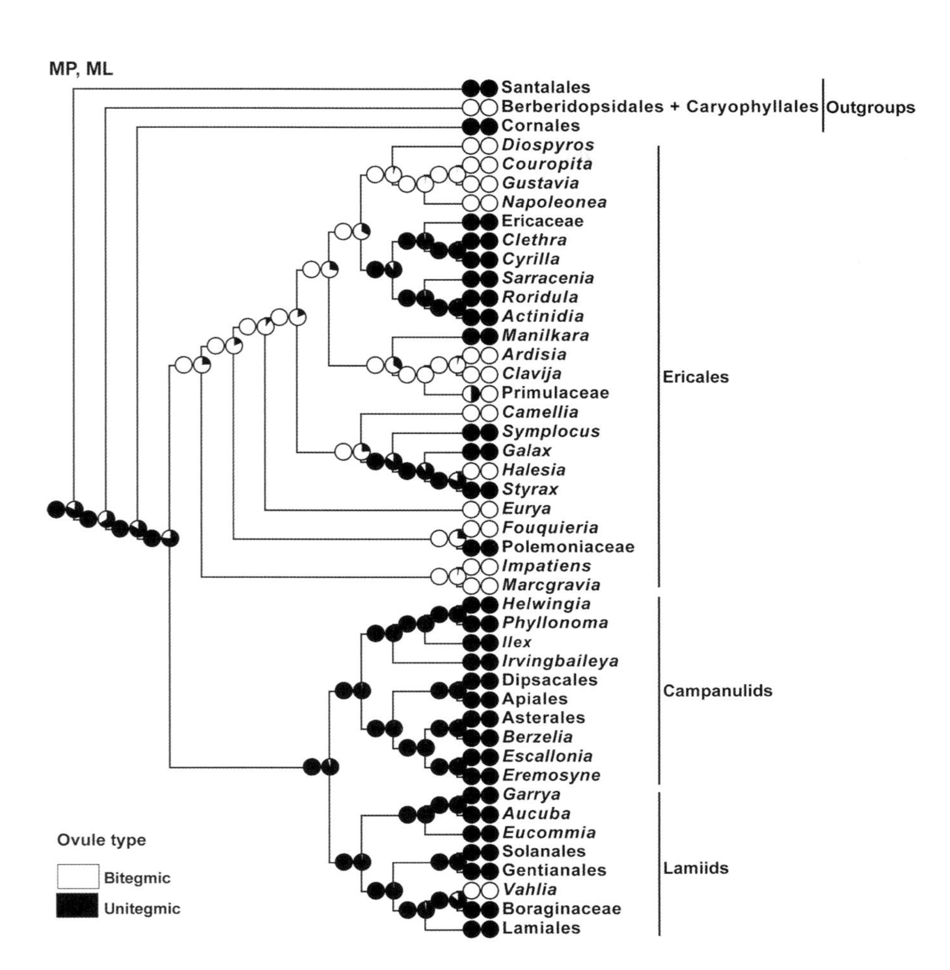

Figure 11.24. Reconstruction of ovule type in Asterids. Unitegmic ovules are most likely ancestral in the Asterids with a major reversal to bitegmic ovules in Ericales. Within Ericales, there have been further shifts to unitegmy; the entire Ericaceae + Actinidiaceae clade is ancestrally unitegmic. Vahliaceae is interesting, being the only member in our reconstruction of the Campanulid + Lamiid clade exhibiting bitegmic ovules.

lales and Berberidopsidales exhibit bitegmic ovules and are sister to asterids (Fig. 11.20).

Cornales are sister to all other asterids, and have unitegmic ovules. Likewise, the most recent common ancestor of campanulids and lamiids was clearly unitegmic, most likely a plesiomorphic state, although there were two shifts to bitegmy, one in Vahliaceae and one in Icacinaceae s.s. (*Emmotum*; Endress and Rapini 2014; not included here) (Fig. 11.24). However, the situation is more complex in Ericales, the immediate sister to the lamiid + campanulid clade. Bitegmy occurs frequently in Ericales (cf. Endress 2003b) and is reconstructed as the most probable ancestral state for this clade using both MP and ML (Fig. 11.24). Within Ericales, independent shifts to unitegmy have occurred on many occasions (perhaps six times using MP or ML) in (1) some Balsaminaceae (reported to have both bitegmy and unitegmy; Boesewinkel and Bouman 1991; Albach et al. 2001b; McAbee et al. 2005; Colombo et al. 2008; Kelley and Gasser 2009), (2) some Primulaceae (reported to have both bitegmy and unitegmy; Anderberg 1995; Albach et al. 2001b), (3) core Ericales (Ericaceae, Clethraceae, Cyrillaceae, Actinidiaceae, Roridulaceae, Sarraceniaceae), (4) Polemoniaceae, (5) Sapotaceae, and (6) Diapensiaceae–Styracaceae–Symplocaceae. The last clade is of interest because there may have been a reversal to bitegmy in *Halesia* (Styracaceae) (Fig. 11.24).

Four possible pathways from bitegmy to unitegmy have been proposed (Bouman and Calis 1977; Boesewinkel and Bouman 1991): (1) reduction of one integument, (2) fusion of two integument primordia, (3) shift of the inner integument along a subdermal outer integument, and (4) shift of a dermal outer integument along the inner integument. However, the developmental sequence for unitegmy has rarely been studied. Asterids, in general, appear to have lost one integument by integumentary shifting (pathway 3 or 4; Bouman and Schier 1979; Boesewinkel and Bouman 1991; see review by Albach et al. 2001b, for pathways in other angiosperms). Genes that cause loss or deformation of one integument have been identified in *Arabidopsis thaliana* (e.g., Robinson-Beers et al. 1992; Leon-Kloosterziel et al.

1994; Gaiser et al. 1995; Baker et al. 1997) and provide possible mechanisms for the transition from bitegmy to unitegmy. However, given the number of genes involved in ovule development, it seems likely that integument formation and transition to unitegmy may require changes in several genes (Leon-Kloosterziel et al. 1994; Baker et al. 1997). Hence, there may be several genetic mechanisms by which these transitions can be achieved.

Endress (2011c) suggested that pathway 2 may be more likely in asterids, where both integuments are present but not morphologically distinguishable. This is further corroborated by *Impatiens* (Balsaminaceae: Ericales) that shows both uni- and bitegmy (McAbee et al. 2005; Colombo et al. 2008; Kelley and Gasser 2009). *Coriaria* (Coriariaceae: Cucurbitales, rosids) also exhibits both uni- and bitegmy (Matthews and Endress 2006). It remains uncertain whether unitegmic ovules in asterids are developmentally and genetically homologous.

TENUINUCELLATE OVULES

Tenuinucellate ovules are defined by the absence of parietal tissue between the megaspore mother cell and the nucellar epidermis (Bouman 1984, and references therein). As with sympetaly and integument number, interpretation of tenuinucellate, or further reduced ovules versus crassinucellate ovules is not always clear because different processes may lead to similar nucellar structures (Endress 2003b); therefore some researchers have revised terminology to depict those differences in nucellar structure (see Endress 2011c). However, although information on nucellar development is critical for determining ontogenetic homology of nucellar structure, this information is lacking for most angiosperms, and our analysis relies on reports of tenuinucellate versus crassinucellate ovules taken from Philipson (1974), Corner (1976), Cronquist (1981), and Takhtajan (1997).

Tenuinucellate ovules occur throughout asterids, with only a few exceptions (i.e., some Cornales, many Apiales,

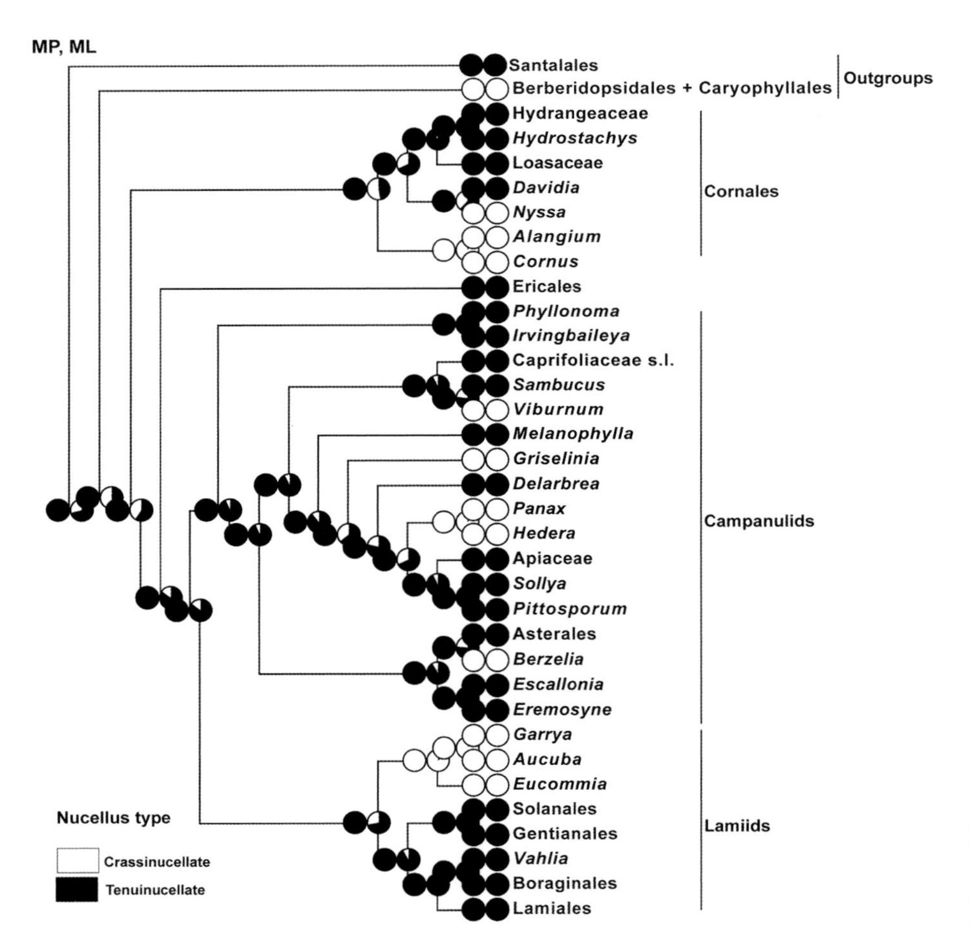

Figure 11.25. Reconstruction of nucellus type in Asterids. Possessing tenuinucellate ovules is plesiomorphic in the Asterids, although there have been at least seven reversals to crassinucellate ovules based on MP.

Viburnum, Garryales, and Icacinales [*Emmotum*, Endress and Rapini 2014]; Fig. 11.25); the ancestral state reconstructed here for asterids is tenuinucellate. Santalales, sister to the rest of the superasterids, also have mostly tenuinucellate ovules (Endress 2011c); our reconstruction suggests that the ancestral state for superasterids is also tenuinucellate (Fig. 11.22). However, angiosperms outside of asterids, including the immediate sister clade to the asterids, Caryophyllales + Berberidopsidales, have crassinucellate ovules (Endress 2011c; Fig. 11.25). Although tenuinucellate ovules characterize asterids in general and some clades are exclusively tenuinucellate (Lamiales, Solanales, Gentianales, Asterales, Escalloniales, and Ericales), crassinucellate ovules have evolved at least seven times considering our reconstruction (Fig. 11.25); they occur in at least some Cornales, Apiales, Dipsacales, Aquifoliales, Bruniales, and Garryales. Thus, multiple transitions between tenuinucellate and crassinucellate ovules are apparent (Fig. 11.25), and both types have been reported within single species of *Viburnum* (Philipson 1974). Transitions from tenuinucellate to crassinucellate ovules have occurred in *Berzelia* (Bruniaceae), *Griselinia* and Araliaceae (Apiales), *Viburnum* (Adoxaceae), and Garryales. *Nyssa* (Nyssaceae), *Alangium*, and *Cornus* (Cornaceae) in Cornales represent additional transitions to crassinucellate ovules. A single loss of parietal tissue leading to production of tenuinucellate ovules has been inferred for the Ericales-lamiids-campanulid clade (i.e., all asterids other than Cornales). A different ontogenetic pathway may, however, be responsible for tenuinucellate ovules in Cornales and some rosid families (see Philipson 1974). Our assessment of evolutionary patterns in ovule structure will be improved through additional information on ovule development (Albach et al. 2001b).

ENDOSPERM FORMATION

Patterns of endosperm formation may be related to nucellus development, because a nucellus without protection from parietal tissue may favor a cellular endosperm, whereas a well-developed nucellus may experience a delay in cell wall formation, resulting in free nuclear endosperm (e.g., Wunderlich 1959; Dahlgren 1975). Thus, cellular endosperm formation is expected in asterids in which tenuinucellate ovules predominate. However, cellular endosperm formation is also common in Magnoliales, Laurales, and Piperales and is proposed to be ancestral in the angiosperms (e.g., Coulter and Chamberlain 1903; Schürhoff 1926; Glisic 1928; Swamy and Ganapathy 1957; Wunderlich 1959; Friedman 1992, 1994). Therefore, cellular endosperm formation in the asterids may represent either the retention of the ancestral cellular endosperm or a reversal (e.g., Sporne 1954; Swamy and Ganapathy 1957; Dahlgren 1991). Recent studies of endosperm development reveal that traditional categories of endosperm formation—that is, cellular, nuclear, helobial—do not capture the diversity of developmental pathways, and patterns considered "nuclear," for example, in many species may not be ontogenetically homologous (Floyd et al. 1999; Floyd and Friedman 2000).

Because developmental data are available for relatively few species, here we report analyses of the asterids that are based on the traditional designations of endosperm formation as cellular versus nuclear, from Rao (1972), Corner (1976), Cronquist (1981), Kamelina (1984), Dahlgren (1991), and Takhtajan (1997), but we recognize that at least some such designations may require reevaluation. Cellular endosperm is reconstructed as ancestral for the asterids (see Albach et al. 2001b). Although rosids and Caryophyllales have nuclear endosperm, cellular endosperm is present in Santalales. Furthermore, cellular endosperm is widespread in Cornales and Ericales and is reconstructed as ancestral for both of these clades. The reconstructions of Albach et al. (2001b), as well as our reconstructions (not shown) based on more recent topologies, indicated many independent derivations of nuclear endosperm formation: once in Cornales (in *Alangium*), perhaps twice in Ericales, and multiple times in lamiids and campanulids. In lamiids, there were derivations of nuclear endosperm in Gentianales (see Wagenitz 1959), *Garrya*, Convolvulaceae, and Boraginaceae; in campanulids, nuclear endosperm is present in *Irvingbaileya* and most Apiales. Several of these derivations match reversals to crassinucellate ovules, supporting Wunderlich's (1959) hypothesis of a close association between nucellar structure and patterns of endosperm development. Multiple derivations of nuclear endosperm development from cellular have also been observed across the angiosperms as a whole (Floyd et al. 1999). Despite multiple transitions from cellular to nuclear endosperm in the asterids, no reversals were identified, indicating that perhaps transformations from nuclear to cellular are not possible in asterids. An evaluation of this hypothesis requires additional information on patterns and processes of endosperm development throughout the angiosperms.

TRICELLULAR POLLEN

In most angiosperms, pollen grains are shed when the microgametophyte reaches the bicellular stage. In some, however, including many asterids, pollen grains are shed at the tricellular stage. Release of pollen at the tricellular stage has been considered a selective advantage under certain circumstances (Stebbins 1974). However, this character is generally uniform within large clades recognized as orders or families. Data on bicellular versus tricellular pollen grains were obtained from Brewbaker (1967), Kamelina (1984), Eyde (1988), Takhtajan (1997), and Albach et al. (2001b). Pollen shed in a tricellular state occurs mostly in asterids, Caryophyllales, some rosid families (e.g., Brassicaceae), Alismatales, and Poaceae. The ancestral state for asterids appears to be bicellular pollen (Albach et al. 2001b); tricellular pollen has not been reported in Cornales, and it occurs only sporadically in Ericales. Additional derivations of tricellular pollen are found in campanulids, evolving independently in *Irvingbaileya* (Cardiopteridaceae), Asteraceae, *Corokia* (Argophyllaceae), and the clade of Apiales and Dipsacales (reconstruction not shown). The pattern of pollen evolution is more complex in lamiids, in which both bicellular and tricellular pollen are present in members of several large clades (i.e., Lamiales, Boraginaceae, and Gentianales) and occasionally even within a single genus (*Plantago*, *Ipomoea*; Brewbaker 1967). However, there are no

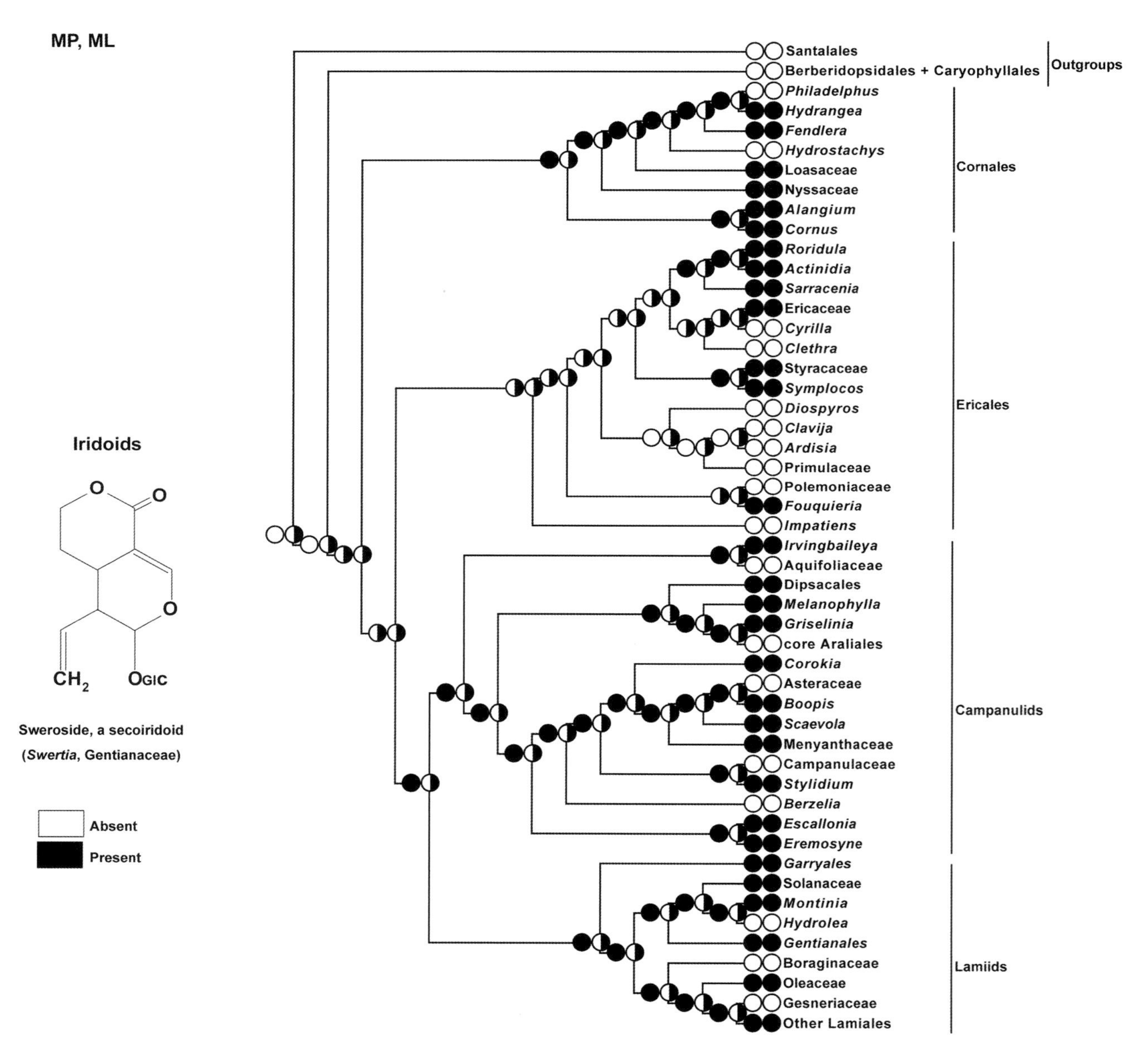

Figure 11.26. Reconstruction of the evolution of iridoids in Asterids. The evolution of iridoids in the asterids is mostly ambiguous in this reconstruction. However, the occurrence of iridoids in Cornales, as well as the Campanulid + Lamiid clade (Gentianidae), is ancestral in both of those clades according to our reconstructions. The situation in Ericales is less clear, where the ancestral state is equivocal for iridoid production. However, the entire Ebenaceae + Primulaceae clade is reconstructed as having lost iridoids.

apparent reversions from tricellular to bicellular pollen (see Albach et al. 2001b), supporting Brewbaker's (1967) suggestion of a unidirectional shift. Tricellular pollen appears to be associated with efficient cross-pollination (Stebbins 1974) and sporophytic self-incompatibility, shorter storage longevity, and an aquatic habit with submersed flowers (Brewbaker 1957, 1967).

IRIDOIDS

Iridoids are monoterpenes characterized by a cyclopentanopyran ring system with a double bond between C_3 and C_4 (Fig. 11.26). Iridoids may have several functions in plants, such as anti-herbivory (e.g., Bernays and De Luca 1981; Bowers and Puttick 1988) or inhibition of seed germination (Pardo et al. 1998). The occurrence of iridoids is highly correlated with sympetaly (Jensen et al. 1975), with only a few occurrences of iridoids outside asterids, in *Stigmaphyllon* (Malpighiaceae; Davioud et al. 1985), *Homalium* (Salicaceae; Ekabo et al. 1993), *Bhesa* (Centroplacaceae; Ohashi et al. 1993), *Ailanthus* (Simaroubaceae; Kosuge et al. 1994), *Liquidambar* (Altingiaceae; Plouvier 1964; Jiang and Zhou 1992), *Daphniphyllum* (Daphniphyllaceae; Inouye et al. 1966), and *Paeonia* (Paeoniaceae; Kolpalova and Popov 1994). Several clades within asterids do not produce iridoids, such as Solanales, some Apiales, Asteraceae (but see Wang and Yu 1998), and several families of Ericales (reviewed in Albach et al. 2001b).

Iridoids have been classified by their chemical properties, such as functional features or oxidation level (Kaplan and Gottlieb 1982), biosynthetic pathway (Jensen 1991), or a combination of the two (Jensen et al. 1975). However, the pathway followed in any given species is typically not known, and there may be several pathways to the same compound (Jensen 1991). Thus, an extensive classification based on biosynthetic pathway, although desirable, is not currently possible. However, two main, presumably mutually exclusive, pathways have been identified (Jensen 1991): one leading to secoiridoids and a second to iridoids decarboxylated at the C_4 position. We report analyses of the presence or absence of (1) iridoids generally and (2) secoiridoids; additional information on the distribution of iridoids can be found in Albach et al. (2001b). Data on the distribution of iridoids are from Kostecka-Madalska and Rymkiewick (1971), Jensen et al. (1975), Degut and Fursa (1980), El-Naggar and Beal (1980), Lammel and Rimpler (1981), Makboul (1986), Boros and Stermitz (1990, 1991), Jensen (1991, 1992), Nicoletti et al. (1991), Drewes et al. (1996), and von Poser et al. (1997).

Iridoids are widespread in asterids (Fig. 11.23); the evolution and diversification of iridoids and related alkaloids are complex. Nevertheless, some patterns are evident. The presence of iridoids is reconstructed as ancestral using MP in at least three of the four subclades of asterids (Cornales, campanulids, lamiids). However, whether iridoid production is the ancestral state for all asterids, as suggested by Olmstead et al. (1993), is equivocal using either MP or ML (Fig. 11.26). Most of the transitions in asterids involve the loss of iridoids (Fig. 11.26), as in *Hydrostachys* and *Philadelphus* of Cornales, at least two clades in lamiids, and five clades of campanulids (with perhaps a subsequent derivation in *Aster*; Wang and Yu 1998). In Ericales, the ancestral state is also equivocal using either MP or ML; however, iridoid evolution is clearly dynamic within this clade also, with numerous transitions between iridoid production and lack of iridoid production.

Secoiridoids, characterized by a ring fission between C_7 and C_8 and alkaloids derived from them, are mostly derived via loganin and secologanin (Cordell 1974) or their corresponding acids, but variants of this pathway have been recently reported (see Albach et al. 2001b). Secoiridoids and related alkaloids occur in Oleaceae, Gentianales, most Cornales, and many campanulids (Fig. 11.26) and have been reported for *Lamium album*, although these are formed via a different biosynthetic pathway (Damtoft et al. 1992). Aucubin and other iridoid compounds decarboxylated at C_4 seem also to have evolved independently in Lamiales, Garryales, and some Ericaceae (see Albach et al. 2001b). Thus, presence of aucubin alone is not characteristic of any particular clade.

FLORAL SYMMETRY

Most asterids have zygomorphic (bilaterally symmetric, monosymmetric) flowers. However, actinomorphic (radially symmetric, polysymmetric) flowers are found in many lineages, including some or most of Apocynaceae, Asteraceae, Boraginaceae s.l., Calyceraceae, Campanulaceae, Menyanthaceae, Oleaceae, Plocospermataceae, Rubiaceae, Tetrachondraceae, Apiales, Cornales, Dipsacales, Ericales, and Solanales. Given the wealth of floral developmental genetic data available for *Antirrhinum* (Plantaginaceae), asterids present ideal opportunities for investigations of evolutionary changes in floral symmetry (e.g., Endress 1992a, 1999; Donoghue et al. 1998; Reeves and Olmstead 1998). Donoghue et al. (1998) examined the evolution of floral symmetry in the asterids by grafting subclades of major subgroups of asterids onto a backbone tree based on *rbcL* sequences (Olmstead et al. 1993) and mapping character states using MacClade (Maddison and Maddison 1992). According to their analysis, the ancestral condition for asterids was an actinomorphic flower. Zygomorphic flowers originated independently at least eight times in asterids, in Asterales, Dipsacales, Solanales, Ericales, Lamiales, and a few Gentianales.

There are clear reversals from zygomorphy to actinomorphy within asterids. The greatest number of such shifts occurs in Lamiales, providing a model clade for more detailed examination of these reversals. Donoghue et al. (1998) therefore examined floral symmetry in more detail in Lamiales. Their most parsimonious reconstruction of ancestral states (with equal costs of going from zygomorphy to actinomorphy and vice-versa) indicated at least seven reversals from zygomorphy to actinomorphy.

We reanalyzed floral symmetry in Lamiales in light of more recent phylogenetic hypotheses (e.g., D. Soltis et al. 2000; Olmstead et al. 2001; Bremer et al. 2002; Soltis et al. 2011; Refulio-Rodriguez and Olmstead 2014), including additional actinomorphic taxa (*Ramonda*, Gesneriaceae). The plesiomorphic condition in Lamiales appears to be actinomorphic (basalmost taxa have actinomorphic corollas; Plocospermataceae, Oleaceae, Tetrachondraceae), as shown in Zhong and Kellogg (2015) with a shift to zygomorphy in the most recent common ancestor of the core Lamiales (i.e., the clade sister to *Polypremum* in our analysis) (Fig. 11.27), and Carlemanniaceae (not shown here; see Zhong and Kellogg 2015). Within the core Lamiales clade, there have been at least 14 reversals to actinomorphy (Fig. 11.27). Actinomorphic flowers occur in other Gesneriaceae (see Clark et al. 2011) as well, and more detailed character-state mappings of that family are needed to elucidate further transitions between zygomorphy and actinomorphy. An additional reversal to near actinomorphy may have occurred in *Byblis* (Fig. 11.27).

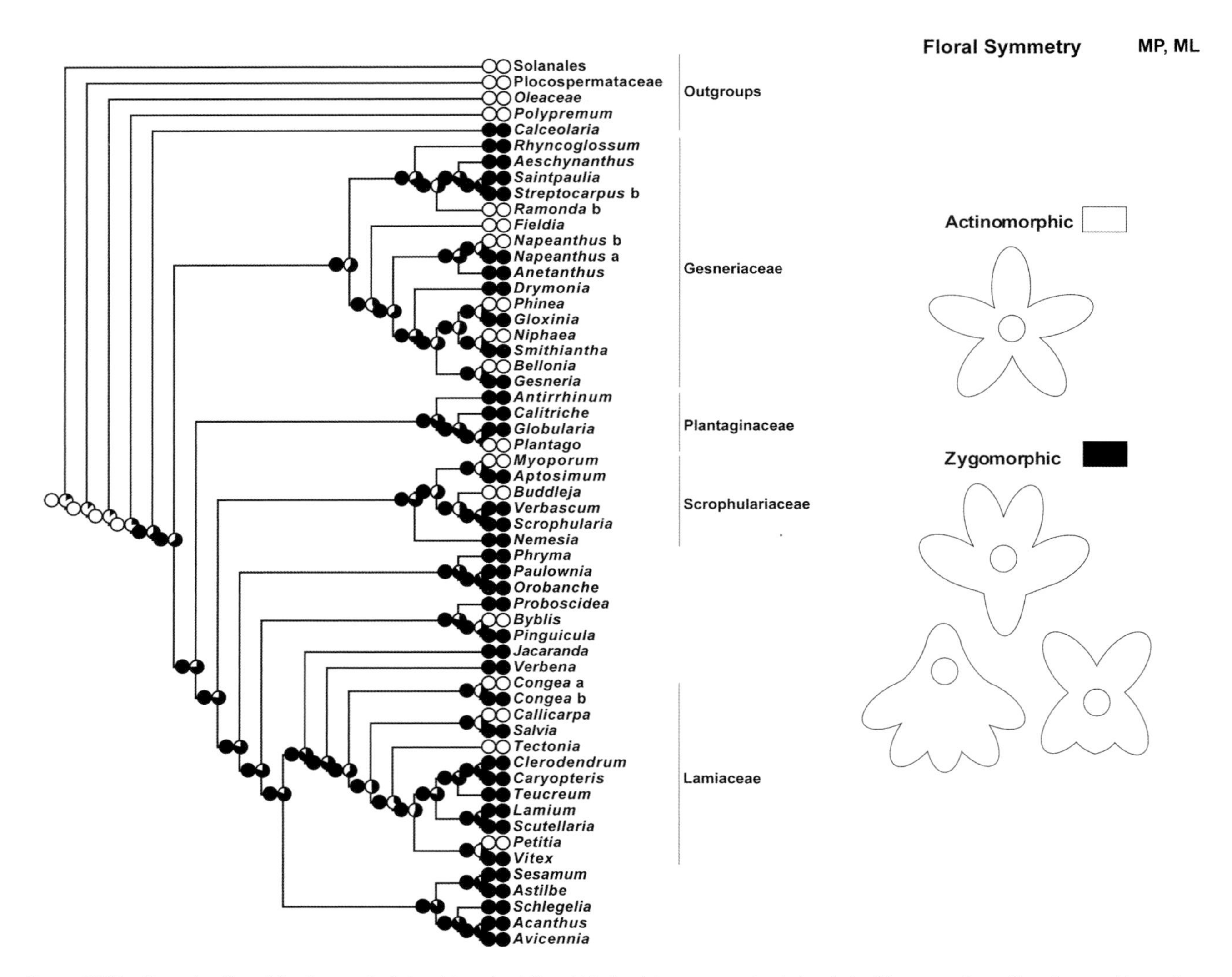

Figure 11.27. Reconstruction of floral symmetry in Lamiales using MP and ML. Lamiales are reconstructed as derived from ancestors with actinomorphic corollas. Bilaterally symmetric corollas first evolved with the common ancestor of Calceolariaceae + Acanthaceae according to MP. There have been at least 14 reversals to actinomorphic corollas in Lamiales according to the reconstruction shown here.

Donoghue et al. (1998) explored possible explanations for these reversals to actinomorphy. *Plantago* is an excellent organism in which to examine these shifts because it is well nested within a zygomorphic clade. This and other reversals to actinomorphy often co-occur with shifts from pentamery to tetramery and reduction in flower size. Other examples include reversals to actinomorphy in *Buddleja*, *Callicarpa*, and *Petitia*. A likely explanation is that actinomorphy arises in these taxa by union of the adaxial petals, accompanied by a shift in the orientation of the two lateral petals. Some derived actinomorphic flowers resemble mutants caused by loss-of-function mutations in genes such as *CYCLOIDEA* (*CYC*) (see Chapter 14), a gene that is needed for normal development of zygomorphic flowers (Carpenter and Coen 1990). Actinomorphic flowers in several genera of Gesneriaceae (e.g., *Ramonda*) and *Sibthorpia* (Plantaginaceae) may have arisen via this mechanism. More recent studies have suggested that the origin of zygomorphy from actinomorphy, for instance in Lamiales, occurs through the asymmetrical expression of CYC2-like genes (Zhong and Kellogg 2015) (see Chapter 14); multiple paralogous copies of CYC genes have arisen throughout Lamiales (Zhong and Kellogg 2015). Zhong and Kellogg (2015) showed that the origin of zygomorphy in Lamiales was strongly correlated with the evolution and subsequent diversification of bumblebees and euglossine bees.

12 Angiosperm Classification

INTRODUCTION

Until the late 1990s, most investigators, systematists and non-systematists alike, consulted one of a few classifications of angiosperms (e.g., Dahlgren 1980; Cronquist 1981; Takhtajan 1980, 1987, 1997; Thorne 1992a,b, 2001). Dahlgren et al.'s (1985) classification of monocots remained the system of choice for more than a decade. Although the systems are generally similar, they differed in several respects. For example, Takhtajan established many small families and orders not recognized by other authors (e.g., Cronquist 1981; Thorne 1992a,b, 2001; Heywood 1993). All of these classifications were based chiefly on morphological characters. Only Dahlgren (1980) used chemical data to a significant extent, and Thorne's (1992a,b, 2001) updates incorporated information from molecular phylogenetics. In addition, only Dahlgren tried to think about classification in a formal phylogenetic sense (Dahlgren and Rasmussen 1983; Dahlgren et al. 1985; see Chapter 3).

Relationships among major groups of taxa (recognized, for example, as orders, subclasses) were often depicted in terms of phylogenetic shrubs or "bubble" diagrams that represented an author's intuitive ideas of evolutionary relationships (see Chapter 3 and Fig. 3.7; Judd et al. 2002). In this sense, these classifications, and those tracing back to Bessey (1915), are semi-phylogenetic. However, the inferences of phylogeny were not based on topologies derived from the explicit phylogenetic analysis of either DNA or non-DNA characters. Irreconcilable differences among these classifications reflected subjective weights of characters imposed by the authors. Nonetheless, the classifications of the last 25 years of the twentieth century represented tremendous improvements over earlier classifications in many ways. They contained more information on obscure taxa, and more characters were described. For these reasons, the classifications of Cronquist (1981), Takhtajan (1997), Thorne (1992a,b, 2001), and Heywood (1993, 1998) continue to receive use.

The past two decades have seen major advances in reconstruction of angiosperm phylogenetic relationships at all taxonomic levels, from species to higher levels (reviewed in Chapter 3); in this chapter, we focus primarily on developments at the family level and above. As reviewed in Chapter 3, broad molecular phylogenetic analyses provided the general framework of angiosperm relationships, and major suprafamilial clades of angiosperm families were also identified.

THE ANGIOSPERM PHYLOGENY GROUP (APG) CLASSIFICATIONS

It became clear that no previous "traditional" classification system adequately reflected the relationships of angiosperms as reconstructed by molecular phylogenetics and that a dramatic change was needed. The molecular trees became the stimulus for a new classification proposed by a consortium of angiosperm systematists, the Angiosperm Phylogeny Group (APG). As a result of the APG's efforts, angiosperms became the first large clade of organisms to be reclassified on the basis of explicit phylogenetic analysis of DNA sequence data. Furthermore, the first APG classification developed by 29 systematists (APG 1998) represented a landmark in the history of angiosperm systematics because for the first time a classification was provided not by one or a few individuals, as had been the case for several hundred years, but by a broad representation of the systematics community.

The first APG system recognized 462 families placed in 40 putatively monophyletic orders and a few monophyletic higher groups. The general strategy of the APG authors, as discussed in APG (1998), was to focus on orders and, to a lesser extent, on families. In general, orders were quite widely circumscribed, especially in comparison with those of Takhtajan (1997). Several formal clades above the level of order were also recognized, using informal names rather than Linnaean ranks: monocots, eudicots, core eudicots, rosids, and asterids, as well as two major subclades of the rosid clade (eurosids I and eurosids II, or fabids and malvids in APG III [2009] and APG IV [2016]; see Chapter 10) and the asterid clade (euasterids I and euasterids II or lamiids and campanulids in APG III [2009] and APG IV [2016]; see Chapter 11). The APG authors conservatively recognized only those clades that received moderate to strong support in either phylogenetic analysis of *rbcL* gene sequences (Källersjö et al. 1998; Chase and Albert 1998) or an initial analysis of a three-gene dataset (D. Soltis et al. 1998).

In APG (1998), many families were not placed in an order because their positions were unknown or not strongly supported. When possible, these families of uncertain position were simply placed in the larger clade in which they appeared (e.g., asterids, rosids). For example, Adoxaceae, Bruniaceae, Icacinaceae, and several other families were placed within the euasterid II clade, but the positions of these families within euasterid II were unclear.

Some families could not be reliably placed in any supraordinal group and were therefore placed at the end of the classification in a list of eudicot families of "uncertain position." In APG (1998), 25 families could not be placed with confidence within any major clade. Other families could be placed within major clades, but not within any of the orders within that clade.

Many monophyletic groups were recognized at the family level in APG (1998). In many cases, family circumscriptions were unchanged or modified only slightly from longstanding concepts. In other cases, however, dramatic restructuring of familial boundaries was required. The APG (1998) authors also recognized the limitations of molecular analyses available at that time. Reclassification of some families required further molecular phylogenetic study that used additional taxon sampling; still other families awaited phylogenetic analysis.

APG (1998) provided names for previously unrecognized clades and much-needed nomenclatural stability. With the explosion of molecular systematic endeavors, different authors were using different names for the same clade. For example, the DNA-based asterid clade was variously referred to as Asteridae, Asteridae s.l., asterids, and euasterids.

The APG (1998) authors recognized that changes in familial circumscription and recognition of new orders would ultimately be needed, requiring a revised APG classification. What was surprising, however, was how quickly an updated APG classification was needed. After publication of APG (1998), rapid advances in angiosperm phylogenetics continued (see Chapter 3). Completion of multigene analyses for all major groups of angiosperms (e.g., Savolainen et al. 2000a; P. Soltis et al. 1999b; D. Soltis et al. 2000), as well as more focused studies of monocots (e.g., Chase et al. 2000a, 2004), asterids (Albach et al. 2001a; B. Bremer et al. 2002), and many orders and families (see other chapters of this book), necessitated an update of the APG (1998) classification, resulting in APG II (2003). APG II involved fewer authors (seven) than the initial APG (1998). With the changes provided in APG II, the number of angiosperm orders increased from 40 to 44, and the number of families decreased from 462 to 397, if all suggestions to combine families are followed. In many instances, APG II suggested that sister families be combined; for example, Platanaceae and Proteaceae (Proteaceae), Nymphaeaceae and Cabombaceae (Nymphaeaceae), and Illiciaceae and Schisandraceae (Schisandraceae). In these cases, investigators were given the choice in APG II to follow these broader circumscriptions or not follow them, providing flexibility of the system, and many investigators have chosen not to use the broader circumscriptions.

Note that in APG II, some families still remained unplaced. For example, Aphloiaceae, Ixerbaceae, Geissolomataceae, Picramniaceae, Strasburgeriaceae, and Vitaceae were unplaced to order in the rosids, and Bruniaceae, Columelliaceae, and Eremosynaceae were three of several families unplaced to order in euasterids II. In APG II, the placement of some taxa was considered uncertain across the angiosperms as a whole (e.g., Balanophoraceae, Rafflesiaceae, *Cytinus*). However, the molecular phylogenetic analysis of Barkman et al. (2004) placed Rafflesiaceae in Malpighiales in the rosid clade.

Further improvements in our understanding of evolutionary relationships across the angiosperms resulted in the revised APG III (2009) classification. In APG III (2009), the number of angiosperm orders was 57, and the number of families was 413. However, some families remained unplaced in APG III, and detailed investigations of major problematic clades (e.g., Malpighiales, Lamiales, Santalales) provided additional new insights into relationships, ultimately necessitating an updated classification. P. F. Stevens (Angiosperm Phylogeny Website; www.mobot.org/

MOBOT/research/APweb/) has continued to provide regular updates and changes to the overall angiosperm classification. Many of these proposed changes, as well as others, have been incorporated in APG IV (2016). APG IV differs only slightly from APG III, and only where new phylogenetic information or nomenclatural issues necessitated changes. In the classification we present here (APG IV 2016), there are 64 orders of angiosperms and 416 families.

RANKED VERSUS RANK-FREE CLASSIFICATION

For more than 250 years, the Linnaean hierarchy (e.g., kingdom, phylum, class, order, family, genus, species), published in 1753, has been the foundation of taxonomy. However, the Linnaean hierarchy was established a century before publication of Darwin's (1859) *Origin of Species*, which introduced the principle of descent with modification. During the past two decades, use of the Linnaean hierarchy has been challenged by investigators who seek to apply the principle of descent to the field of nomenclature. In fact, some have suggested that the Linnaean hierarchy be abandoned.

De Queiroz (1997) made a distinction between taxonomic and nomenclatural systems. A Linnaean system, whether of taxonomy or nomenclature, is based on the Linnaean hierarchy, whereas a phylogenetic system is based on the principle of descent (de Queiroz and Gauthier 1992; de Queiroz 1997). Phylogenetic classification is the representation of phylogenetic relationships, and its aim is to produce classifications that reflect phylogenetic relationships accurately and efficiently. A phylogenetic system of nomenclature is a body of principles and rules (nomenclatural conventions) governing taxonomic practice, "the components of which are unified by their relation to the central tenet of evolutionary descent" (de Queiroz and Gauthier 1992).

Hennig, well known for introducing principles of phylogeny reconstruction (Hennig 1950, 1965, 1966), also initiated the development of a phylogenetic system of classification (Hennig 1969, 1981, 1983), and many other authors subsequently contributed its development. In a series of papers, de Queiroz and Gauthier (1990, 1992, 1994) attempted to formulate nomenclatural conventions based on the principle of descent, and specific rules and recommendations of phylogenetic nomenclature are still being developed (e.g., de Queiroz and Gauthier 1990, 1992, 1994; Bryant 1994, 1996; Sundberg and Pleijel 1994; de Queiroz 1996; Lee 1996; Cantino et al. 1997; www.ohio.edu/phylocode/).

Until recently, most phylogenetic systems of classification have attempted to use the basic conventions of the Linnaean hierarchy. As several authors have noted, the familiar set of nested categories offered by the Linnaean system of classification (kingdom, phylum, class, order, etc.) can be used in phylogenetic classification to provide information on rank (that is, the relative nested positions of clades). However, Linnaean categories are not needed to provide this information to the reader and may, in fact, be a hindrance to faithful communication of phylogenetic relatedness. Linnaean categories do not contain any information about ancestry that is not already present in a branching diagram or in an indented list of names. Furthermore, assignment of taxa to categories is, alone, insufficient information to specify relationships. That is, knowing that one taxon is a family and another an order does not indicate that the family in question is nested within that given order.

Although Linnaean categories can be used in such a way that they are consistent with phylogenetic classification, several problems have been cited (reviewed in de Queiroz and Gauthier 1992). One problem is that of "mandatory categories." That is, certain categories (i.e., kingdom, phylum, class, order, family, and genus) are mandatory in the Linnaean system. Every named species must be assigned to a taxon at every one of these taxonomic levels, causing a proliferation of names in the case of monotypic taxa (see below). Furthermore, ancestors cannot be assigned to monophyletic taxa less inclusive than those originating with them. These ancestors are not part of clades that are less inclusive than the one stemming from that ancestral node or population. For example, the stem species of the angiosperms would be included in Magnoliophyta, but not in any of the subgroups of Magnoliophyta (Chapter 2).

The proliferation of categories that occurs as a direct result of the attempt to reflect fine levels of systematic resolution is easily illustrated. As an example, if we consider the current summary topology for angiosperms (Fig. 12.1), *Amborella* appears as sister to all other angiosperms. Hence, if angiosperms are treated as a phylum (Magnoliophyta), *Amborella* would belong to class Amborellopsida, and all remaining angiosperms, which are the sister to *Amborella*, would be part of a second class (here labeled Euangiopsida). Within the class Euangiopsida, Nymphaeaceae are sister to all other angiosperms. Nymphaeaceae would be placed in the order Nymphaeales, and their sister group (all remaining angiosperms) would be a separate order, here named Austrobaileyales. As illustrated in Figure 12.1, because the early-branching angiosperms form a graded series, the problem of proliferation of categories would continue. In a short time, we would exhaust the taxonomic categories afforded by the Linnaean system and would need to introduce many additional taxonomic

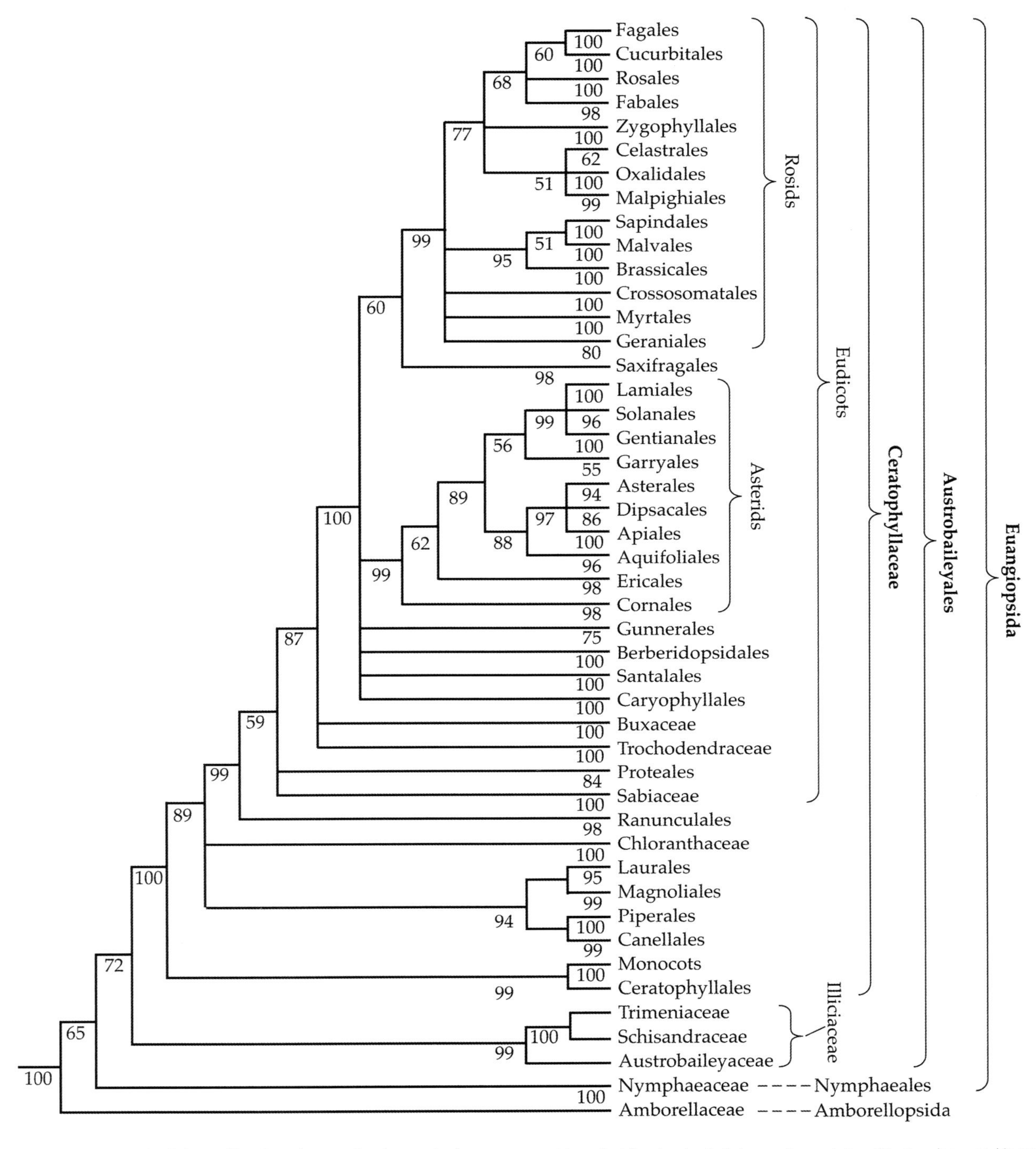

Figure 12.1. An example of the proliferation of categories that results from an attempt to reflect fine levels of phylogenetic resolution. The topology used is one possible tree based on a summary of the three-gene analysis of angiosperms (Soltis et al. 2000). Names in lower case letters (i.e., eudicots, rosids, asterids) are common names given to clades. Names in bold (e.g., Illiciaceae) represent an attempt to place the topology in a Linnaean-based hierarchical classification. We exhaust the taxonomic categories afforded by the Linnaean system very quickly and would need to introduce many more taxonomic categories (from Soltis et al. 2005).

categories. Crane and Kenrick (1997a) similarly noted this problem of proliferation of hierarchical levels in their attempt to translate recent topologies for green plants into a classification that follows Linnaean ranks.

Several modifications of the Linnaean system have been proposed to deal with this problem, but none is totally satisfactory. The addition of new taxonomic categories has been proposed (e.g., McKenna 1975; Farris 1976; Gaffney and Meylan 1988). Farris (1976), for example, proposed a method for producing new taxonomic categories that used rank-modifying prefixes. This method, although not widely used, could generate many taxonomic categories through the application of single-prefix as well as multi-prefix categories (reviewed in Kron 1997). Other investigators attempted to avoid the proliferation of names and categories by introducing systems that conveyed information regarding phylogenetic relationships without using Linnaean categories. Nelson (1972, 1973) noted that the sequence of taxon names in a list could be used to convey information about phylogenetic relationships. This method, which became known as the "sequencing convention," greatly reduced the number of taxa and categories, but it also left many clades unnamed (reviewed in de Queiroz 1997).

Because of these and other problems associated with the use of Linnaean ranks, and because categories themselves are not needed for conveying phylogenetic relationships, some authors have proposed that the best way to implement a phylogenetic classification is to abandon Linnaean categories. Phylogenetic nomenclature differs fundamentally from the Linnaean system (as governed by International Codes of Nomenclature) in that it does not require ranks above the level of species (hence, it is often referred to as "rank-free"). The idea of a phylogenetic system as an alternative to the traditional Linnaean system is not new. Hennig (1969, 1981, 1983), for example, used numerical prefixes rather than Linnaean categories in his taxonomies of insects and chordates. Other investigators subsequently contributed to development of the concept and application of a rank-free classification (Griffiths 1973, 1976; Farris 1976; Løvtrup 1977; Stevens 1984; Ax 1987; Ereshefsky 1994). More recently, de Queiroz and Gauthier (1990, 1992, 1994) further developed a phylogenetic system of nomenclature by discussing specific rules and principles. Their papers subsequently stimulated extensive discussion and debate (e.g., Bryant 1994, 1996; Wyss and Meng 1996; Cantino et al. 1997; de Queiroz 1997; Kron 1997; Cantino 1998, 2000; Hibbett and Donoghue 1998). There have also been strong counter-arguments against phylogenetic nomenclature (e.g., Liden and Oxelman 1996; Brummitt 1997; Stuessy 2000). Some authors stress, for example, that the problems noted with the Linnaean system can be addressed without abandoning the Linnaean hierarchy. Thus, many systematists who readily embrace phylogenetic principles continue to use the Linnaean hierarchy (see Wiley 1979, 1981; Wiley and Lieberman 2011; Eldredge and Cracraft 1980).

Phylogenetic nomenclature does not mandate that taxa be assigned to ranked categories such as genus, family, or order. Names such as Lamiaceae, Rosales, or Caryophyllales can be used, following phylogenetic nomenclature, without implying any taxonomic category; they are simply names of clades that do not have to correspond to ranks within a taxonomic hierarchy. Elimination of Linnaean ranks is not as dramatic a change as it might seem. For example, the APG classification of angiosperms is, in part, rank-free; above the category of order, the Linnaean hierarchy is not followed. The intent of the APG editors was not to use a rank-free system, but rather to name formally these higher levels only when their inter-relationships are clear. An advantage of phylogenetic nomenclature is that newly discovered clades can be named without changing the names of other taxa. A major problem with the Linnaean system is that ranks of taxa, and as a result names given to taxa, depend on their relative position to other groups of taxa. Thus, naming a new clade can require a cascading series of name changes (Kron 1997; Cantino 1998; Hibbett and Donoghue 1998; Cantino et al. 2007). A code of phylogenetic nomenclature, referred to as the PhyloCode, has been developed (www.ohiou.edu/phylocode/).

Other authors have proposed a compromise method of classification that includes elements of both Linnaean and phylogenetic nomenclature. For example, Sennblad and Bremer (2002) proposed a nomenclatural system for Apocynaceae that includes "standard" Linnaean tribal names, but, above that level, uses a variant of the definitions from the PhyloCode (Fig. 12.2) for an infrafamilial classification. Four rankless taxa are recognized, each ending in "ina" and each progressively more inclusive (Asclepiadoidina, Asclepiadacina, Euapocynoidina, Apocynoidina).

All APG classifications specifically assigned familial and ordinal names and hence are clearly Linnaean in intent. However, "families" and "orders" directly correspond to monophyletic groups and therefore could be incorporated into a rank-free system of classification (the PhyloCode Companion Volume, when published, will be a useful reference for these clades). Above the level of order, a rank-free system of classification has quietly slipped into wide usage in the angiosperms, even among ardent advocates of Linnaean classification.

Thus, plant systematists readily use terms such as monocots, eudicots, core eudicots, rosids, asterids, and others (fabids, malvids, lamiids, campanulids), all of which appear in the APG classifications and represent informal, yet highly meaningful, names for well-defined and well-

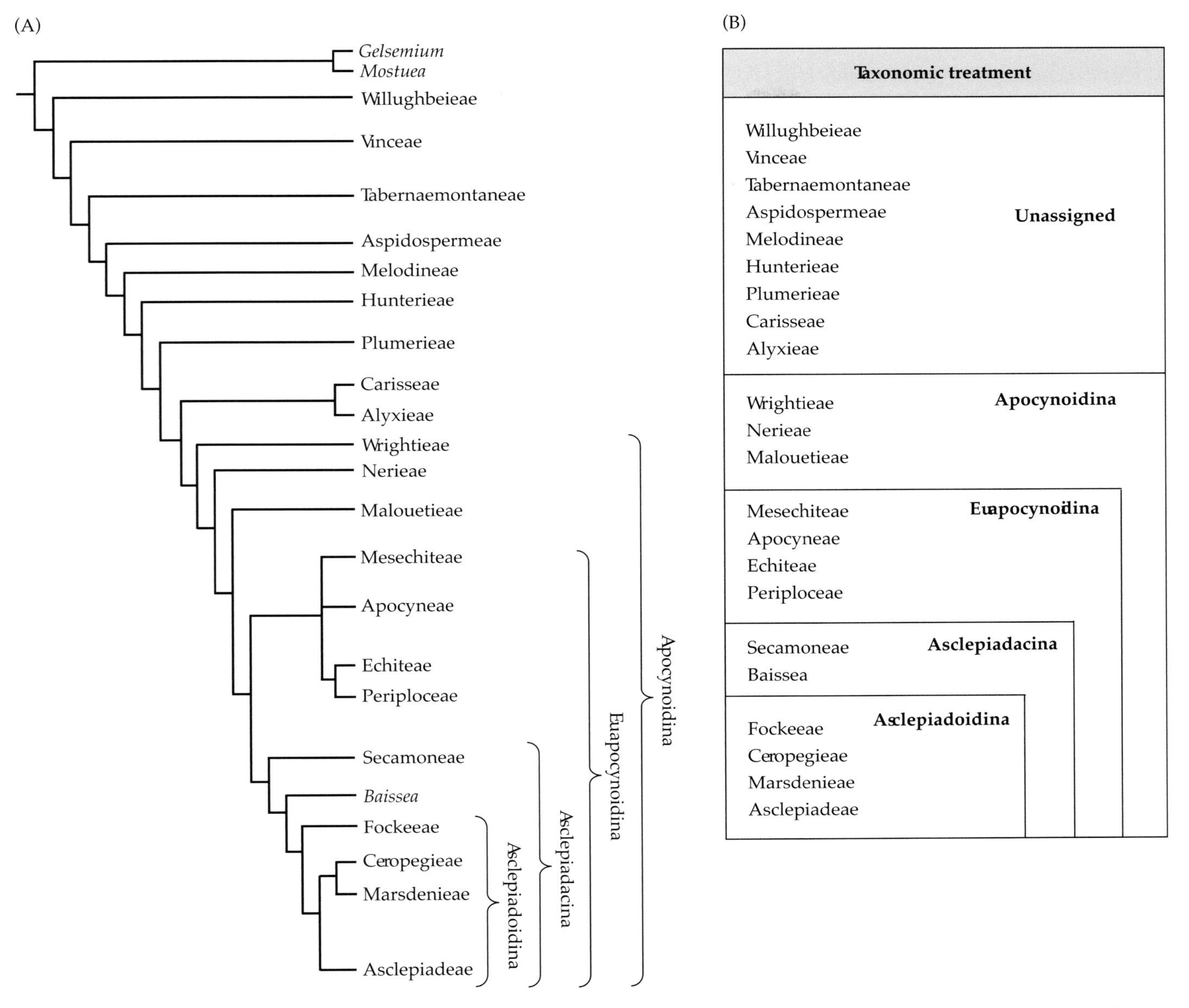

Figure 12.2. A "compromise" reclassification of Apocynaceae. Standard tribal names (Asclepiadeae, etc.) are given following the Linnaean system. However, four nonstandard taxa, Apocynoidina, Euapocynoidina, Asclepiadacina, and Asclepiadoidina, are also indicated (see text). (A) Nonstandard names superimposed on phylogenetic tree of Apocynaceae. (B) Phylogenetic treatment in table form. (Modified from Sennblad and Bremer 2002; from Soltis et al. 2005.)

supported clades. Forcing a Linnaean system of ranks and names onto the existing phylogenetic framework for angiosperms above the level of order would not help but hinder angiosperm systematics. That is, following the Linnaean system of classifying and naming the asterid and rosid clades as subclasses (for example) Asteridae and Rosidae, respectively, would force a cascade of name changes throughout the angiosperms, as well as establishment of new ranks to accommodate sister groups. Furthermore, many clade names, such as the term "eudicots," have become widely used, even among nonspecialists. What benefit would accrue, for example, from naming the eudicots as a class, and what implications would this change have for the nomenclature of the remaining angiosperms that form a grade sister to the eudicots? Thus, angiosperm systematics following the APG system (see also Stevens 2001 onwards, www.mobot.org/MOBOT/research/APweb/) currently has a hybrid system with orders, families, genera, and species following the Linnaean system of classification, but a rank-free system above the level of order.

Some investigators have been strong proponents of a rank-free system (e.g., Mishler 1999; Platnick 2009, 2012, 2013; see also de Queiroz and Donoghue 2011, 2013, and de Queiroz 2013; Cellinese et al. 2012), but acceptance of such a system will take time. The formal classification of tracheophytes (Cantino et al. 2007) treats large, inclusive clades of angiosperms, and ongoing efforts to provide phylogenetic definitions of clades corresponding to orders and families recognized by the APG classification are underway (de Queiroz et al. in prep.). However, it seems unlikely that a rank-free system will be widely used by angiosperm systematists at levels corresponding to that of genus and below in the near future. Researchers who make use of ranks and who currently recognize clades at the ranks of genus, subgenus, section, and series will continue to do so into the forseeable future. In contrast, clades that are currently recognized above the ordinal rank (or that are given only informal names) will likely soon be widely recognized using unranked formal names (either using the PhyloCode, or as yet unwritten new provisions to the existing ICN, which would allow such unranked formal names). Although the Linnaean system seems entrenched at the levels of order and below, rankless taxon names continue to be proposed and will likely continue to be used more and more frequently by plant systematists. Extension of the principles of rank-free phylogenetic classification to the species level, although logically consistent with the treatment of larger clades, is especially controversial and is unlikely to be resolved in the foreseeable future (see Cellinese et al. 2012).

ANGIOSPERMS/FLOWERING PLANTS (*ANGIOSPERMAE*)

Here we provide a summary of the APG IV (2016) classification, augmented with additional clade names, summaries of intraordinal phylogenies, and additional commentary. Summaries of interfamilial relationships within each order are portrayed using parentheses (similar to that in Stevens, Angiosperm Phylogeny Website; www.mobot.org/MOBOT/research/APweb/). Names in bold italics are from Cantino et al. (2007) and/or the upcoming PhyloCode Companion Volume (de Queiroz et al. in press).

SYMBOLS:

*new family placement or new family recognition since APG III (2009);
†newly recognized order for the APG system since APG III (2009);
§new circumscription of family or order since APG III (2009)

Amborellales Melikian, A. V. Bobrov & Zaytzeva (1999)
Amborellaceae Pichon (1948), *nom. cons.*
Nymphaeales Salisb. ex Bercht. & J.Presl (1820)
Cabombaceae Rich. ex A.Rich. (1822), *nom. cons.*
Hydatellaceae U.Hamann (1976)
Nymphaeaceae Salisb. (1805), *nom. cons.*
(Hydatellaceae (Cabombaceae, Nymphaeaceae))
Austrobaileyales Takht. ex Reveal (1992)
Austrobaileyaceae Croizat (1943), *nom. cons.*
Schisandraceae Blume (1830), *nom. cons.*
Trimeniaceae L.S.Gibbs (1917), *nom. cons.*
(Austrobaileyaceae (Trimeniaceae, Schisandraceae))

MESANGIOSPERMAE (INCLUDES ALL CLADES BELOW THIS NAME)

Magnoliids (*Magnoliidae*)

Laurales Juss. ex Bercht. & J.Presl (1820)
Atherospermataceae R.Br. (1814)
Calycanthaceae Lindl. (1819), *nom. cons.*
Gomortegaceae Reiche (1896), *nom. cons.*
Hernandiaceae Blume, *nom. cons.*
Lauraceae Juss. (1789), *nom. cons.*
Monimiaceae Juss. (1809), *nom. cons.*
Siparunaceae Schodde (1970)
(Calycanthaceae ((Siparunaceae (Gomortegaceae, Atherospermataceae)) (Monimiaceae (Hernandiaceae, Lauraceae))))
Magnoliales Juss. ex Bercht. & J.Presl (1820)
Annonaceae Juss. (1789), *nom. cons.*
Degeneriaceae I.W.Bailey & A.C.Sm. (1942), *nom. cons.*
Eupomatiaceae Orb., *nom. cons.*
Himantandraceae Diels (1917), *nom. cons.*
Magnoliaceae Juss. (1789), *nom. cons.*
Myristicaceae R.Br. (1810), *nom. cons.*
(Myristicaceae (Magnoliaceae ((Himantandraceae, Degeneriaceae) (Eupomatiaceae, Annonaceae))))
Canellales Cronquist (1957)
Canellaceae Mart. (1832), *nom. cons.*
Winteraceae R.Br. ex Lindl. (1830), *nom. cons.*
Piperales Bercht. & J.Presl (1820)
§Aristolochiaceae Juss. (1789), *nom. cons.* (includes Lactoridaceae Engl., *nom. cons.* and Hydnoraceae C.Agardh, *nom. cons.*)
Piperaceae Giseke (1792), *nom. cons.*
Saururaceae Rich. ex T.Lestib., *nom. cons.*
(Aristolochiaceae (Piperaceae, Saururaceae))

Independent Lineage: Unplaced to Larger Clade

Chloranthales Mart. (1835)
Chloranthaceae R.Br. ex Sims (1820), *nom. cons.*

Monocots (*Monocotyledoneae*)

Acorales Mart.
Acoraceae Martinov (1820)
Alismatales R.Br. ex Bercht. & J.Presl (1820)
Alismataceae Vent. (1799), *nom. cons.*
Aponogetonaceae Planch. (1856), *nom. cons.*
Araceae Juss. (1789), *nom. cons.*
Butomaceae Mirb. (1804), *nom. cons.*
§Cymodoceaceae Vines (1895), *nom. cons.*
Hydrocharitaceae Juss. (1789), *nom. cons.*
§Juncaginaceae Rich. (1808), *nom. cons.*
*Maundiaceae Nakai (1943) (now separated from Juncaginaceae)
Posidoniaceae Vines (1895), *nom. cons.*
Potamogetonaceae Bercht. & J.Presl (1823), *nom. cons.*
Ruppiaceae Horan. (1834), *nom. cons.*
Scheuchzeriaceae F.Rudolphi (1830), *nom. cons.*
Tofieldiaceae Takht. (1995)
Zosteraceae Dumort. (1829), *nom. cons.*
(Araceae (Tofieldiaceae ((Alismataceae (Hydrocharitaceae, Butomaceae)) (Scheuchzeriaceae (Aponogetonaceae (Juncaginaceae (Maundiaceae ((Posidoniaceae (Ruppiaceae, Cymodoceaceae)) (Zosteraceae, Potamogetonaceae)))))))))
Petrosaviales Takht. (1997)
Petrosaviaceae Hutch. (1934), *nom. cons.*
Dioscoreales R.Br. (1835) This clade remains problematic due to the uncertain relationships of Burmanniaceae, Taccaceae, and Thismiaceae. Some would include Burmanniaceae, Taccaceae, and Thismaceae in a broadly defined Dioscoreaceae. Until these relationships are resolved, APG IV follows APG III (as below).
Burmanniaceae Blume (1827), *nom. cons.*
Dioscoreaceae R.Br. (1810), *nom. cons.* (incl. Taccaceae Dumort., Thismiaceae J.Agardh.)
Nartheciaceae Fr. ex Bjurzon (1846)
(Nartheciaceae (Burmanniaceae, Dioscoreaceae))
Pandanales R.Br. ex Bercht. & J.Presl (1820)
Cyclanthaceae Poit. ex A.Rich. (1824), *nom. cons.*
Pandanaceae R.Br. (1810), *nom. cons.*
Stemonaceae Caruel (1878), *nom. cons.*
Triuridaceae Gardner (1843), *nom. cons.*
Velloziaceae J.Agardh (1858), *nom. cons.*
(Velloziaceae, Triuridaceae, Stemonaceae, (Pandanaceae, Cyclanthaceae))
Liliales Perleb (1826)
Alstroemeriaceae Dumort. (1829), *nom. cons.*
Campynemataceae Dumort. (1829)
Colchicaceae DC. (1804), *nom. cons.*
Corsiaceae Becc. (1878), *nom. cons.*
Liliaceae Juss. (1789), *nom. cons.*
Melanthiaceae Batsch ex Borkh. (1797), *nom. cons.*
Petermanniaceae Hutch. (1934), *nom. cons.*
Philesiaceae Dumort. (1829), *nom. cons.*
Ripogonaceae Conran & Clifford (1985), *nom. cons.*
Smilacaceae Vent. (1799), *nom. cons.*
(Corsiaceae (Campynemataceae (Petermanniaceae (Colchicaceae, Alstroemeriaceae)), Melanthiaceae, ((Philesiaceae, Rhipogonaceae) (Smilacaceae, Liliaceae))))
Asparagales Link (1829)
Amaryllidaceae J.St.-Hil. (1805), *nom. cons.*
Asparagaceae Juss. (1789), *nom. cons.*
Asphodelaceae Juss. (1789), *nom. cons., prop.* (=Xanthorrhoeaceae Dumort., *nom. cons.*)
Asteliaceae Dumort. (1829)
Blandfordiaceae R.Dahlgren & Clifford (1985)
Boryaceae M.W.Chase, Rudall & Conran (1997)
Doryanthaceae R.Dahlgren & Clifford (1985)
Hypoxidaceae R.Br. (1814), *nom. cons.*
Iridaceae Juss. (1789), *nom. cons.*
Ixioliriaceae Nakai (1943) (as 'Ixiolirionaceae'; spelling corrected)
Lanariaceae H.Huber ex R.Dahlgren
Orchidaceae Juss. (1789), *nom. cons.*
Tecophilaeaceae Leyb. (1862)
Xeronemataceae M.W.Chase, Rudall & M.F.Fay (2000)
(Orchidaceae ((Boryaceae (Blandfordiaceae (Lanariaceae (Asteliaceae, Hypoxidaceae)))) ((Ixioliriaceae, Tecophilaceae) (Doryanthaceae (Iridaceae (Xeronemataceae (Asphodelaceae (Amaryllidaceae, Asparagaceae))))))))

COMMELINIDS (*COMMELINIDAE*)

§**Arecales Bromhead (1840)** Now include Dasypogonaceae based on recent DNA sequence data that indicate Dasypogonaceae are well supported as sister to Arecaceae.
Arecaceae Bercht. & J.Presl (1820), *nom. cons.*
Dasypogonaceae Dumort. (1829)
Poales Small (1903)
Bromeliaceae Juss. (1789), *nom. cons.*
Cyperaceae Juss. (1789), *nom. cons.*
Ecdeiocoleaceae D.F.Cutler & Airy Shaw (1965)
Eriocaulaceae Martinov (1820), *nom. cons.*
Flagellariaceae Dumort. (1829), *nom. cons.*

Joinvilleaceae Toml. & A.C.Sm. (1970)
Juncaceae Juss. (1789), *nom. cons.*
Mayacaceae Kunth (1842), *nom. cons.*
Poaceae Barnhart (R.Br.) (1895), *nom. cons.*
Rapateaceae Dumort. (1829), *nom. cons.*
§Restionaceae R.Br. (1810), *nom. cons.* [including Centrolepidaceae Endl. (1836), *nom. cons.* and Anarthriaceae D.W.Cutler & Airy Shaw (1965)]
Thurniaceae Engl. (1907), *nom. cons.*
Typhaceae Juss. (1789), *nom. cons.*
Xyridaceae C.Agardh (1823), *nom. cons.*

((Typhaceae, Bromeliaceae) (Rapateaceae ((Mayacaceae (Eriocaulaceae, Xyridaceae)) (Thurniaceae (Juncaceae, Cyperaceae))) (Restionaceae)) (Ecdeiocoleaceae, Poaceae (Flagellariaceae, Joinvilleaceae)))))))

Commelinales Mirb. ex Bercht. & J.Presl (1820)
Commelinaceae Mirb. (1804), *nom. cons.*
Haemodoraceae R.Br. (1810), *nom. cons.*
Hanguanaceae Airy Shaw (1965)
Philydraceae Link (1821), *nom. cons.*
Pontederiaceae Kunth (1816), *nom. cons.*

((Commelinaceae, Hanguanaceae) (Philydraceae (Haemodoraceae, Pontederiaceae)))

Zingiberales Griseb. (1854)
Cannaceae Juss. (1789), *nom. cons.*
Costaceae Nakai (1941)
Heliconiaceae Vines (1895)
Lowiaceae Ridl. (1924), *nom. cons.*
Marantaceae R.Br. (1814), *nom. cons.*
Musaceae Juss. (1789), *nom. cons.*
Strelitziaceae Hutch. (1934), *nom. cons.*
Zingiberaceae Martynov (1820), *nom. cons.*

(Musaceae, (Strelitziaceae, Lowiaceae), Heliconiaceae, ((Cannaceae, Marantaceae) (Costaceae +Zingiberaceae)))

Independent Lineage: Unplaced to Major Clade (possibly sister to eudicots)

Ceratophyllales Link (1829)
Ceratophyllaceae Gray (1822), *nom. cons.*

Note: Ceratophyllales are not included in either monocots or eudicots.

Eudicots (*Eudicotyledoneae*)

Ranunculales Juss. ex Bercht. & J.Presl (1820)
Berberidaceae Juss. (1789), *nom. cons.*
Circaeasteraceae Hutch. (1926), *nom. cons.*
Eupteleaceae K.Wilh. (1910), *nom. cons.*
Lardizabalaceae R.Br. (1821), *nom. cons.*
Menispermaceae Juss. (1789), *nom. cons.*
Papaveraceae Juss. (1789), *nom. cons.*
Ranunculaceae Juss. (1789), *nom. cons.*

(Eupteleaceae (Papaveraceae ((Lardizabalaceae, Circaeasteraceae) (Menispermaceae (Berberidaceae, Ranunculaceae)))))

§Proteales Juss. ex Bercht. & J.Presl (1820) Proteales now include Sabiaceae based on recent complete plastid data analyes that show strong support for this relationship
Nelumbonaceae Bercht. & J.Presl (1820), *nom. cons.*
Platanaceae T.Lestib. (1826), *nom. cons.*
Proteaceae Juss. (1789), *nom. cons.*
Sabiaceae Blume (1851), nom. cons.

(Sabiaceae (Nelumbonaceae (Platanaceae, Proteaceae)))

Trochodendrales Takht. ex Cronquist (1981)
Trochodendraceae Eichl. (1865), *nom. cons.*

§Buxales Takht. ex Reveal (1996)
§Buxaceae Dumort. (1822), *nom. cons.* (including Haptanthaceae C. Nelson (2001) and Didymelaceae Leandri (1937))

CORE EUDICOTS (*GUNNERIDAE*)

Gunnerales Takht. ex Reveal (1992)
Gunneraceae Meisn. (1842), *nom. cons.*
Myrothamnaceae Nied. (1891), *nom. cons.*

PENTAPETALAE (DILLENIALES + SUPERROSIDS + SUPERASTERIDS)

Note: *Pentapetalae* is less inclusive than core eudicots but contains both superrosids and superasterids.

†Dilleniales DC. ex Bercht. & J.Presl (1820) (Placement unclear, see Chapter 9)
Dilleniaceae Salisb. (1807), *nom. cons.*

SUPERROSIDS (*SUPERROSIDAE*)

§Saxifragales Bercht. & J.Presl (1820) Based on the recent work of Bellot et al. (2016), we have included Cynomoriaceae in Saxifragales.
Altingiaceae Horan. (1841), *nom. cons.*
Aphanopetalaceae Doweld (2001)
Cercidiphyllaceae Engl. (1907), *nom. cons.*
Crassulaceae J.St.-Hil. (1805), *nom. cons.*
Cynomoriaceae Endl. ex Lindl. (1833), *nom. cons.*
Daphniphyllaceae Müll.-Arg. (1869), *nom. cons.*
Grossulariaceae DC. (1805), *nom. cons.*

Haloragaceae R.Br. (1814), *nom. cons.*
Hamamelidaceae R.Br. (1818), *nom. cons.*
Iteaceae J.Agardh (1858), *nom. cons.*
Paeoniaceae Raf. (1815), *nom. cons.*
Penthoraceae Rydb. ex Britton (1901), *nom. cons.*
Peridiscaceae Kuhlm. (1950), *nom. cons.*
Saxifragaceae Juss. (1789), *nom. cons.*
Tetracarpaeaceae Nakai (1943)

(Peridiscaceae ((Paeoniaceae (Altingiaceae (Hamamelidaceae (Cercidiphyllaceae, Daphniphyllaceae)))) ((Crassulaceae (Aphanopetalaceae (Tetracarpaeaceae (Penthoraceae, Haloragaceae)))) (Iteaceae (Grossulariaceae, Saxifragaceae))))); Cynomoriaceae unplaced within order.

Rosids (*Rosidae*)

Vitales Juss. ex Bercht. & J.Presl (1820)

Vitaceae Juss. (1789), *nom. cons.*

Note that not all recent analyses support a placement of Vitales as sister to eurosids.

Eurosids

Fabids (*Fabidae*)

Zygophyllales Link (1829)

Krameriaceae Dumort. (1829), *nom. cons.*
Zygophyllaceae R.Br. (1814), *nom. cons.*

COM clade

(Note that plastid data place the COM clade in fabids, whereas nuclear and mitochondrial genes predominantly favor a placement in the malvids, with some signal for fabids. This may repesent an ancient reticulation event, see Chapter 10, and if that hypothesis is confirmed, it would be appropriate to include the COM clade within both fabids and malvids.)

Celastrales Link

Celastraceae R.Br. (1814), *nom. cons.*
Lepidobotryaceae J.Léonard (1950), *nom. cons.*

Oxalidales Bercht. & J.Presl

Brunelliaceae Engl. (1897), *nom. cons.*
Cephalotaceae Dumort. (1829), *nom. cons.*
Connaraceae R.Br. (1818), *nom. cons.*
Cunoniaceae R.Br. (1814), *nom. cons.*
Elaeocarpaceae Juss. (1816), *nom. cons.*
Huaceae A.Chev. (1947)
Oxalidaceae R.Br. (1818), *nom. cons.*

(Huaceae ((Connaraceae, Oxalidaceae) (Cunoniaceae (Elaeocarpaceae (Brunelliaceae, Cephalotaceae)))))

§Malpighiales Juss. ex Bercht. & J.Presl

Achariaceae Harms (1897), *nom. cons.*
Balanopaceae Benth. & Hook.f. (1880), *nom. cons.*
Bonnetiaceae L.Beauvis. ex Nakai (1948)
Calophyllaceae J.Agardh (1858)
Caryocaraceae Voigt (1845), *nom. cons.*
Centroplacaceae Doweld & Reveal (2005)
Chrysobalanaceae R.Br. (1818), *nom. cons.*
Clusiaceae Lindl. (1836), *nom. cons.*
Ctenolophonaceae Exell & Mendonça (1951)
Dichapetalaceae Baill. (1886), *nom. cons.*
Elatinaceae Dumort. (1829), *nom. cons.*
Erythroxylaceae Kunth (1822), *nom. cons.*
§Euphorbiaceae Juss. (1789), *nom. cons.*
Euphroniaceae Marc.-Berti (1989)
Goupiaceae Miers (1862)
Humiriaceae A.Juss. (1829), *nom. cons.*
Hypericaceae Juss. (1789), *nom. cons.*
§Irvingiaceae Exell & Mendonça (1951), *nom. cons.* (incl. *Allantospermum* Forman)
§Ixonanthaceae Planch. ex Miq. (1858), *nom. cons.*
Lacistemataceae Mart. (1826), *nom. cons.*
Linaceae DC. ex Perleb (1818), *nom. cons.*
Lophopyxidaceae H.Pfeiff. (1951)
Malpighiaceae Juss. (1789), *nom. cons.*
Ochnaceae DC. (1811), *nom. cons.*
Pandaceae Engl. & Gilg (1912–1913), *nom. cons.*
Passifloraceae Juss. ex Roussel (1806), *nom. cons.*
*Peraceae Klotzsch (1859) (segregated from Euphorbiaceae)
Phyllanthaceae Martynov (1820), *nom. cons.*
Picrodendraceae Small (1917), *nom. cons.*
Podostemaceae Rich. ex Kunth (1816), *nom. cons.*
Putranjivaceae Meisn. (1842)
Rafflesiaceae Dumort. (1829), *nom. cons.*
Rhizophoraceae Pers. (1807), *nom. cons.*
Salicaceae Mirb. (1815), *nom. cons.*
Trigoniaceae A.Juss. (1849), *nom. cons.*
Violaceae Batsch (1802), *nom. cons.*

((Ctenolophonaceae (Erythroxylaceae, Rhizophoraceae)), Irvingiaceae, Pandaceae, (Ochnaceae ((Bonnetiaceae, Clusiaceae) (Calophyllaceae (Hypericaceae, Podostemaceae))))) (((Lophopyxidaceae, Putranjivaceae), Caryocaraceae, Centroplacaceae, (Elatinaceae, Malpighiaceae), (Balanopaceae ((Trigoniaceae, Dichapetalaceae) (Chrysobalanaceae, Euphroniaceae)))) ((Humiriaceae ((Achariaceae (Goupiaceae, Violaceae) (Passifloraceae (Lacistemataceae, Salicaceae))))) ((Peraceae (Rafflesiaceae, Euphorbiaceae)) ((Phyllanthaceae, Picrodendraceae) (Linaceae, Ixonanthaceae)))))

Nitrogen-Fixing Clade

Fabales Bromhead (1838)

Fabaceae Lindl. (1836), *nom. cons.*

Polygalaceae Hoffmanns. & Link (1809), *nom. cons.*

Quillajaceae D.Don (1831)

Surianaceae Arn. (1834), *nom. cons.*

(Quillajaceae (Fabaceae (Polygalaceae, Surianaceae)))

Rosales Perleb (1826)

Barbeyaceae Rendle (1916), nom. cons. *nom. cons.*

Cannabaceae Martynov (1820), *nom. cons.* (includes Celtidaceae)

Dirachmaceae Hutch. (1959)

Elaeagnaceae Juss. (1789), *nom. cons.*

Moraceae Gaudich. (1835), *nom. cons.*

Rhamnaceae Juss. (1789), *nom. cons.*

Rosaceae Juss. (1789), *nom. cons.*

Ulmaceae Mirb. (1815), *nom. cons.*

Urticaceae Juss. (1789), *nom. cons.*

(Rosaceae ((Rhamnaceae (Elaeagnaceae (Barbeyaceae, Dirachmaceae))) (Ulmaceae (Cannabaceae (Moraceae, Urticaceae)))))

§Cucurbitales Juss. ex Bercht. & J.Presl (1820)

Anisophylleaceae Ridl. (1922)

*Apodanthaceae Tiegh. ex Takht. (1987) (unplaced in APG III)

Begoniaceae C.Agardh, *nom. cons.*

Coriariaceae DC. (1824), *nom. cons.*

Corynocarpaceae Engl. (1897), *nom. cons.*

Cucurbitaceae Juss. (1789), *nom. cons.*

Datiscaceae Dumort., *nom. cons.*

Tetramelaceae Airy Shaw (1965)

(Anisophylleaceae ((Corynocarpaceae, Coriariaceae) (Cucurbitaceae (Tetramelaceae (Datiscaceae, Begoniaceae)))); Apodanthaceae unplaced

Fagales Engl.

Betulaceae Gray (1821), *nom. cons.*

Casuarinaceae R.Br. (1814), *nom. cons.*

Fagaceae Dumort. (1829), *nom. cons.*

Juglandaceae DC. ex Perleb (1818), *nom. cons.*

Myricaceae Rich. ex Kunth (1817), *nom. cons.*

Nothofagaceae Kuprian. (1962)

Ticodendraceae Gómez-Laur. & L.D.Gómez (1991)

(Nothofagaceae (Fagaceae ((Myricaceae (Rhoipteleaceae, Juglandaceae) (Casuarinaceae (Ticodendraceae, Betulaceae)))))

Malvids (*Malvidae*)

Geraniales Juss. ex Bercht. & J.Presl (1820)

Geraniaceae Juss. (1789), *nom. cons.*

§Francoaceae A.Juss. (1832), nom. cons. (incl. Bersamaceae Doweld, Greyiaceae Hutch., *nom. cons.*, Ledocarpaceae Meyen, Melianthaceae Horan., *nom. cons.*, Rhynchothecaceae Juss., and Vivianiaceae Klotzsch, *nom. cons.*)

Myrtales Juss. ex Bercht. & J.Presl (1820)

Alzateaceae S.A.Graham (1985)

Combretaceae R.Br. (1810), *nom. cons.*

Crypteroniaceae A.DC. (1868), *nom. cons.*

Lythraceae J.St.-Hil. (1805), *nom. cons.*

Melastomataceae Juss. (1789), *nom. cons.*

Myrtaceae Juss. (1789), *nom. cons.*

Onagraceae Juss. (1789), *nom. cons.*

Penaeaceae Sweet ex Guill. (1828), *nom. cons.*

Vochysiaceae A.St.-Hil. (1820), *nom. cons.*

(Combretaceae ((Onagraceae, Lythraceae) ((Vochysiaceae, Myrtaceae) (Melastomataceae (Crypteroniaceae (Alzateaceae, Penaeaceae))))))

Crossosomatales Takht. ex Reveal (1993)

Aphloiaceae Takht. (1985)

Crossosomataceae Engl. (1897), *nom. cons.*

Geissolomataceae A.DC , *nom. cons.*

Guamatelaceae S.H.Oh & D.Potter (2006)

Stachyuraceae J.Agardh (1858), *nom. cons.*

Staphyleaceae Martinov (1820), *nom. cons.*

Strasburgeriaceae Tiegh. (1908), *nom. cons.*

((Staphyleaceae (Guamatelaceae (Crossosomataceae, Stachyuraceae))) (Aphloiaceae (Geissolomataceae, Strasburgeriaceae)))

Picramniales Doweld (2001)

Picramniaceae Fernando & Quinn (1995)

Sapindales Juss. ex Bercht. & J. Presl

Anacardiaceae R.Br. (1818), *nom. cons.*

Biebersteiniaceae Schnizl.

Burseraceae Kunth (1824), *nom. cons.*

Kirkiaceae Takht. (1967)

Meliaceae Juss. (1789), *nom. cons.*

Nitrariaceae Lindl.

Rutaceae Juss. (1789), *nom. cons.*

Sapindaceae Juss. (1789), *nom. cons.* (includes Xanthocerataceae Buerki, Callm. & Lowry, as "Xanthoceraceae")

Simaroubaceae DC. (1811), *nom. cons.*

(Biebersteiniaceae, Nitrariaceae, ((Kirkiaceae (Anacardiaceae, Burseraceae)) (Sapindaceae (Simaroubaceae, Rutaceae, Meliaceae))))

§Huerteales Doweld (2001)

Dipentodontaceae Merr (1941), *nom. cons.*

Gerrardinaceae M.H.Alford (2006)

*Petenaeaceae Christenh., M.F.Fay & M.W.Chase (2010) (segregated from Malvaceae s. l.)

Tapisciaceae Takht. (1987)

((Gerrardinaceae, Petenaeaceae) (Tapisciaceae, Dipentodontaceae))

Malvales Juss. ex Bercht. & J.Presl (1820)

Bixaceae Kunth (1822), nom. cons.

§Cistaceae Juss. (1789), nom. cons. (including *Pakaraimaea* Maguire & P.S.Ashton, formerly of Dipterocarpaceae)

Cytinaceae A. Rich (1824)

§Dipterocarpaceae Blume (1825), nom. cons.

Malvaceae Juss. (1789), nom. cons.

Muntingiaceae C.Bayer, M.W.Chase & M.F.Fay (1998)

Neuradaceae Kostel (1835), nom. cons.

Sarcolaenaceae Caruel (1881), nom. cons.

Sphaerosepalaceae Tiegh. ex Bullock (1959)

Thymelaeaceae Juss. (1789), nom. cons.

(Neuradaceae (Thymelaeaceae (Sphaerosepalaceae, Bixaceae, (Cistaceae (Sarcolaenaceae, Dipterocarpaceae)), (Cytinaceae, Muntingiaceae), Malvaceae))

Brassicales Bromhead (1838)

Brassicaceae s.l. The clade Capparaceae + Cleomaceae + Brassicaceae s.s. should be named—it is well supported by molecular data and actually more easily diagnosed morphologically than the three segregate families. Those researchers using a rank-based system could treat it as Brassicariae (using the ending –ariae that has been proposed for superfamily).

Akaniaceae Stapf (1912), *nom. cons.*

Bataceae Mart. ex Perleb (1838)

Brassicaceae Burnett (1835), *nom. cons.*

§Capparaceae Juss. (1789), *nom. cons.*

Cleomaceae Berchtold & J. Presl (1825)

Caricaceae Dumort. (1829), *nom. cons.*

Emblingiaceae J. Agardh (1958)

Gyrostemonaceae Endl. (1841), *nom. cons.*

Koeberliniaceae Engl. (1895), *nom. cons.*

Limnanthaceae R.Br. (1833), *nom. cons.*

Moringaceae Martynov (1820), *nom. cons.*

Pentadiplandraceae Hutch. & Dalziel (1928)

§Resedaceae Martinov, *nom. cons.* (includes Borthwickiaceae J.X.Su, Wei Wang, Li Bing Zhang & Z.D.Chen (2012) and Stixidaceae Doweld, as "Stixaceae", and *Forchhammeria* Liebm.)

Salvadoraceae Lindl. (1836), *nom. cons.*

Setchellanthaceae Iltis (1999)

Tovariaceae Pax (1891), *nom. cons.*

Tropaeolaceae Bercht. & J.Presl (1820), *nom. cons.*

((Akaniaceae, Tropaeolaceae) ((Moringaceae, Caricaceae) Setchellanthaceae ((Limnanthaceae (Koeberliniaceae (Batidaceae, Salvadoraceae) (Emblingiaceae ((Pentadiplandraceae (Borthwickiaceae (Gyrostemonaceae, Resedaceae))) Tovariaceae (Capparaceae (Cleomaceae, Brassicaceae)))))))))

SUPERASTERIDS (*SUPERASTERIDAE*)

Berberidopsidales Doweld (2001)

Aextoxicaceae Engl. & Gilg (1920), *nom. cons.*

Berberidopsidaceae Takht. (1985)

Santalales R.Br. ex Bercht. & J.Presl (1820)

APG IV (2016) continues to follow the APG III (2009) treatment due to the continued lack of strong support for most relationships within the clade. That treatment follows.

Balanophoraceae Rich. (1822), *nom. cons.*

Loranthaceae Juss. (1808), *nom. cons.*

Misodendraceae J.Agardh (1858), *nom. cons.*

"Olacaceae" R.Br. (1818), *nom. cons.* (Paraphyletic as now recognized, and incl. Aptandraceae Miers, Coulaceae Tiegh., Erythropalaceae Planch. Ex Miq., Octoknemaceae Soler., Strombosiaceae Tiegh., and Ximeniaceae Horan.)

Opiliaceae Valeton (1886), *nom. cons.*

"Santalaceae" R.Br. (1810), *nom. cons.* (Family non-monophyletic due to placement of Balanophoraceae; including Amphorogynaceae Nickrent & Der, Cervantesiaceae Nickrent & Der, Comandraceae Nickrent & Der, Nanodeaceae Nickrent & Der)

Schoepfiaceae Blume (1850)

The treatment below follows Nickrent et al. (2010), with the addition of Balanophoraceae.

*Amphorogynaceae Nickrent & Der (2010) (now segregated from Santalaceae)

*Aptandraceae Miers (1853) (now segregated from Olacaceae)

Balanophoraceae Rich. (1822), *nom. cons.*

*Cervantesiaceae Nickrent & Der (2010) (now segregated from Santalaceae)

*Comandraceae Nickrent & Der (2010) (now segregated from Santalaceae)

*Coulaceae Tiegh. (1897) (now segregated from Olacaceae)

*Erythropalaceae Planch. ex Miq. (1856) (now segregated from Olacaceae)

§Loranthaceae Juss. (1808), *nom. cons.*

Misodendraceae J.Agardh (1858), *nom. cons.*

*Nanodeaceae Nickrent & Der (2010) (now segregated from Santalaceae)

*Octoknemaceae Tiegh. (1907) (now segregated from Olacaceae)

§Olacaceae Juss. ex R.Br. (1818), *nom. cons.*

Opiliaceae Valeton (1886), *nom. cons.*

§Santalaceae R.Br. (1810)

Schoepfiaceae Blume (1850)

*Strombosiaceae Tiegh. (1900) (now treated separately from Olacaceae and Erythropalaceae)
*Thesiaceae Vest (1818) (now treated separately from Santalaceae and also from recently proposed Cervantesiaceae)
*Viscaceae Batsch (1802) (now segregated from Loranthaceae and Santalaceae)
*Ximeniaceae Horan. (1834) (now segregated from Olacaceae)

(Strombosiaceae, Erythropalaceae, (Coulaceae (Olacaceae, Aptandraceae, Ximeniaceae (Octoknemaceae ((Loranthaceae (Misodendraceae, Schoepfiaceae)) (Opiliaceae (Comandraceae (Cervantesiaceae, Thesiaceae) (Nanodeaceae ((Viscaceae, Amphorogynaceae) (Balanophoraceae, Santalaceae)

Caryophyllales Juss. ex Bercht. & J.Presl (1820)

Achatocarpaceae Heimerl (1934), *nom. cons.*
Aizoaceae Martinov (1820), *nom. cons.*
Amaranthaceae Juss. (1789)
Anacampserotaceae Eggli & Nyffeler (2010)
Ancistrocladaceae Planch. ex Walp. (1851), *nom. cons.*
Asteropeiaceae Takht. ex Reveal & Hoogland (1990)
Barbeuiaceae Nakai (1942)
Basellaceae Raf. (1837), *nom. cons.*
Cactaceae Juss. (1789), *nom. cons.*
Caryophyllaceae Juss. (1789), *nom. cons.*
Didiereaceae Radlk. (1896), *nom. cons.*
Dioncophyllaceae Airy Shaw (1952), *nom. cons.*
Droseraceae Salisb. (1808), *nom. cons.*
Drosophyllaceae Chrtek, Slavíková & Studnicka (1989)
Frankeniaceae Desv. (1817), *nom. cons.*
Gisekiaceae Nakai (1942)
Halophytaceae S.Soriano (1984)
*Kewaceae Christenh.
§Limeaceae Shipunov ex Reveal (2005)
Lophiocarpaceae Doweld & Reveal (2008)
*Macarthuriaceae Christenh. (segregated from Molluginaceae)
*Microteaceae Schäferhoff & Borsch (2010) (segregated from Phytolaccaceae)
§Molluginaceae Bartl. (1825), *nom. cons.*
Montiaceae Raf. (1820)
Nepenthaceae Dumort. (1820), *nom. cons.*
Nyctaginaceae Juss. (1789), *nom. cons.*
*Petiveriaceae C. Agardh (1824) (including Rivinaceae; segregated from Phytolaccaceae)
Physenaceae Takht. (1985)
§Phytolaccaceae R.Br. (1818), *nom. cons.* (including Agdestidaceae Nakai)
Plumbaginaceae Juss. (1789), *nom. cons.*
Polygonaceae Juss. (1789), *nom. cons.*
Portulacaceae Juss. (1789), *nom. cons.*
Rhabdodendraceae Prance (1968)
Sarcobataceae Behnke (1997)
Simmondsiaceae Tiegh. (1899)
Stegnospermataceae Nakai (1942)
Talinaceae Doweld (2001)
Tamaricaceae Link. (1821), *nom. cons.*

(((Droseraceae (Nepenthaceae (Drosophyllaceae (Ancistrocladaceae, Dioncophylleaceae)))) ((Frankeniaceae, Tamaricaceae) (Polygonaceae, Plumbaginaceae))) (Rhabdodendraceae (Simmondsiaceae ((Asteropeiaceae, Physenaceae) (Macarthuriaceae (Microteaceae ((Caryophyllaceae (Achatocarpaceae, Amaranthaceae)) (Stegnospermataceae (Limeaceae ((Lophiocarpaceae (Kewaceae (Barbeuiaceae (Aizoaceae (Gisekiaceae (Sarcobataceae, Phytolaccaceae, Nyctaginaceae, Petiveriaceae)))))) (Molluginaceae (Montiaceae ((Halophytaceae (Didiereaceae, Basellaceae)) (Talinaceae (Anacampserotaceae (Portulacaceae, Cactaceae)))))))))))))))

Asterids (*Asteridae*)

§Cornales Link (1829)

§Cornaceae Bercht. & J.Presl., *nom. cons.*
Curtisiaceae Takht. (1987)
Grubbiaceae Endl. ex Meisn. (1839), *nom. cons.*
Hydrangeaceae Dumort. (1829), *nom. cons.*
Hydrostachyaceae Engl. (1894), *nom. cons.*
Loasaceae Juss. (1804), *nom. cons.*
*Nyssaceae Juss. ex Dumort. (1829), *nom. cons.* (segregated from Cornaceae)

((Cornaceae (Grubbiaceae, Curtisiaceae)) (Nyssaceae (Hydrostachyaceae (Hydrangeaceae, Loasaceae))))

Ericales Bercht. & J.Presl (1820)

Actinidiaceae Gilg & Werderm., *nom. cons.*
Balsaminaceae A.Rich. (1824), *nom. cons.*
Clethraceae Klotzsch (1851), *nom. cons.*
Cyrillaceae Lindl., *nom. cons.*
Diapensiaceae Lindl. (1836), *nom. cons.*
Ebenaceae Gürke (1891), *nom. cons.*
Ericaceae Juss. (1789), *nom. cons.*
Fouquieriaceae DC. (1828), *nom. cons.*
Lecythidaceae A.Rich. (1825), *nom. cons.*
Marcgraviaceae Bercht. & J.Presl, *nom. cons.*
Mitrastemonaceae Makino (1911)
Pentaphylacaceae Engl. (1897), *nom. cons.*
Polemoniaceae Juss. (1789), *nom. cons.*
Primulaceae Batsch ex Borkh. (1797), *nom. cons.*
Roridulaceae Martinov (1820), *nom. cons.*
Sapotaceae Juss. (1789), *nom. cons.*
Sarraceniaceae Dumort. (1829), *nom. cons.*

Sladeniaceae Airy Shaw (1964)
Styracaceae DC. & Spreng. (1821), *nom. cons.*
Symplocaceae Desf. (1820), *nom. cons.*
Tetrameristaceae Hutch. (1959)
Theaceae Mirb. ex Ker Gawl. (1816), *nom. cons.*

((Balsaminaceae (Marcgraviaceae, Tetrameristaceae)) ((Polemoniaceae, Fouquieriaceae), Lecythidaceae, ((Sladeniaceae, Pentaphylacaceae), (Sapotaceae (Ebenaceae, (Primulaceae)), (Mitrastemonaceae, Theaceae, (Symplocaceae (Styracaceae, Diapensiaceae)), ((Sarraceniaceae (Roridulaceae, Actinidiaceae)) (Clethraceae (Cyrillaceae, Ericaceae)))))))

Euasterids

Lamiids (*Lamiidae*)

†Boraginales Juss. ex Bercht. & J.Presl (1820)

Boraginaceae Juss. (1789), *nom. cons.* (incl. Codonaceae Weigend & Hilger)

Segregate families recognized by some workers (see Luebert et al. 2016; Stevens 2001 onwards):

Codonaceae Weigend & Hilger
Wellstediaceae Novák
Boraginaceae Jussieu
Hydrophyllaceae R. Brown
Namaceae Molinari
Heliotropiaceae Schrader
Cordiaceae R. Brown ex Dumortier
Hoplestigmataceae Gilg in Engler & Gilg
Coldeniaceae J. S. Mill. & Gottschling
Ehretiaceae Martius
Lennoaceae Solms

Garryales Mart. (1835)

Eucommiaceae Engl. (1907), *nom. cons.*
Garryaceae Lindl. (1834), *nom. cons.*

Gentianales Juss. ex Bercht. & J.Presl (1820)

Apocynaceae Juss. (1789), *nom. cons.*
§Gelsemiaceae Struwe & V.A.Albert (1995) (including *Pteleocarpa* Oliv. and Pteleocarpaceae Brummitt)
Gentianaceae Juss. (1789), *nom. cons.*
Loganiaceae R.Br. ex Mart. (1827), *nom. cons.*
Rubiaceae Juss. (1789), *nom. cons.*

(Rubiaceae (Gentianaceae (Loganiaceae (Gelsemiaceae, Apocynaceae))))

†Icacinales van Tieghem (1993)

§Icacinaceae Miers (1851), *nom. cons.* (10 genera moved to Metteniusaceae; see below)
Oncothecaceae Kobuski ex Airy Shaw (1965)

†Metteniusales Takht. (1997)

§Metteniusaceae H. Karst. ex Schnizl. (1860–1870) (now includes 10 genera formerly in Icacinaceae: *Apodytes, Calatola, Dendrobangia, Emmotum, Oecopetalum, Ottoschulzia, Pittosporopsis, Platea, Poraqueiba, Rhaphiostylis*; 11 genera total)

Solanales Juss. ex Bercht. & J.Presl (1820)

Convolvulaceae Juss. (1789), *nom. cons.*
Hydroleaceae R.Br.
Montiniaceae Nakai (1943), *nom. cons.*
Solanaceae Juss. (1789), *nom. cons.*
Sphenocleaceae T. Baskerv. (1839), *nom. cons.*

((Montiniaceae (Sphenocleaceae, Hydroleaceae)) (Convolvulaceae, Solanaceae))

Lamiales Bromhead (1838)

Acanthaceae Juss. (1789), *nom. cons.*
Bignoniaceae Juss. (1789), *nom. cons.*
Byblidaceae Domin (1922), *nom. cons.*
Calceolariaceae R.G.Olmstead (2001)
Carlemanniaceae Airy Shaw (1964)
§Gesneriaceae Rich. & Juss. (1816), *nom. cons.* (traditionally including *Peltanthera* Benth., although this genus now unplaced)
Lamiaceae Martinov (1820), *nom. cons.*
Lentibulariaceae Rich. (1808), *nom. cons.*
Linderniaceae Borsch, K.Müll. & Eb.Fisch. (2005)
Martyniaceae Horan. (1847), *nom. cons.*
*Mazaceae Reveal (2011) (segregated from Scrophulariaceae)
Oleaceae Hoffmanns. & Link (1809), *nom. cons.*
§Orobanchaceae Vent. (1799), *nom. cons.* (now includes Rehmanniaceae Reveal (2011))
Paulowniaceae Nakai (1949)
Pedaliaceae R.Br. (1810), *nom. cons.*
*Phrymaceae Schauer (1847), *nom. cons.*
§Plantaginaceae Juss. (1789), *nom. cons.*
Plocospermataceae Hutch. (1973)
Schlegeliaceae Reveal (1996)
§Scrophulariaceae Juss. (1789), *nom. cons.*
Stilbaceae Kunth (1831), *nom. cons.*
Tetrachondraceae Wettst. (1924)
Thomandersiaceae Sreem. (1977)
Verbenaceae J.St.-Hil. (1805), *nom. cons.*

(Plocospermataceae ((Carlemanniaceae, Oleaceae) (Tetrachondraceae ((Peltantheraceae Molinari [*Peltanthera* Roth] (Calceolariaceae, Gesneriaceae, # *Santago*)) (Plantaginaceae (Scrophulariaceae (Stilbaceae ((Byblidaceae, Linderniaceae) ((Lamiaceae (Mazaceae (Phrymaceae (Paulowniaceae, Orobanchaceae,)))) (Thomandersiaceae, Verbenaceae), Pedaliaceae, (Schlegeliaceae, Martyniaceae) Bignoniaceae, Acanthaceae, Lentibulariaceae))))))

†Vahliales Doweld [unplaced]

Vahliaceae Dandy (1959)

Campanulids (*Campanulidae*)

Aquifoliales Senft (1856)

Aquifoliaceae Bercht. & J.Presl (1835), *nom. cons.*

Cardiopteridaceae Blume (1847), *nom. cons.*

Helwingiaceae Decne. (1836)

Phyllonomaceae Small (1905)

Stemonuraceae Kårehed (2001)

((Cardiopteridaceae, Stemonuraceae) (Aquifoliaceae (Helwingiaceae, Phyllonomaceae)))

Asterales Link (1829)

Alseuosmiaceae Airy Shaw

Argophyllaceae Takht. (1987)

Asteraceae Bercht. & J.Presl (1820), *nom. cons.*

Calyceraceae R.Br. ex Rich. (1820), *nom. cons.*

Campanulaceae Juss. (1789), *nom. cons.*

Goodeniaceae R.Br. (1810), *nom. cons.*

Menyanthaceae Dumort., *nom. cons.*

Pentaphragmataceae J.Agardh (1858), *nom. cons.*

Phellinaceae Takht. (1967)

Rousseaceae DC. (1839)

Stylidiaceae R.Br. (1810), *nom. cons.*

((Rousseaceae, Campanulaceae) (Pentaphragmataceae ((Alseuosmiaceae (Phellinaceae, Argophyllaceae)) (Stylidiaceae (Menyanthaceae (Goodeniaceae (Calyceraceae, Asteraceae)))))))

Escalloniales R.Br. (1835)

Escalloniaceae R.Br. ex Dumort. (1829), *nom. cons.*

Bruniales Dumort. (1829)

Bruniaceae R.Br. ex DC, nom. cons.

Columelliaceae D.Don (1828), nom. cons.

Apiales Nakai (1930)

Apiaceae s.l. Note that the clade Apiaceae + Myodocarpaceae + Araliaceae is well supported by molecular data and is also easily diagnosed morphologically; some workers may want to recognize this clade.

§Apiaceae Lindl. (1836), *nom. cons.* (now includes Actinotaceae Konstantinova & Melikian (2005); family not noted in APG III)

Araliaceae Juss. (1789), *nom. cons.*

Griseliniaceae J.R.Forst. & G.Forst. ex A.Cunn. (1839)

Myodocarpaceae Doweld (2001)

Pennantiaceae J.Agardh (1858)

Pittosporaceae R.Br. (1814), *nom. cons.*

Torricelliaceae Hu (1934)

(Pennantiaceae (position?) (Torricelliaceae (Griseliniaceae (Pittosporaceae (Araliaceae (Myodocarpaceae, Apiaceae))))))

Paracryphiales Takht. ex Reveal (1992)

Paracryphiaceae Airy Shaw (1965)

Dipsacales Juss. ex Bercht. & J.Presl (1820)

Adoxaceae E.Mey. (1839), *nom. cons.*

Caprifoliaceae Juss. (1789), *nom. cons.*

TAXA OF UNCERTAIN POSITION

Atrichodendron Gagnep. (specimen poorly preserved, and thus difficult to know to which family it belongs, but it is definitely not Solanaceae where it was previously placed, S. Knapp, pers. com.)

Coptocheile Hoffmanns. (described in Gesneriaceae and may belong there but could be elsewhere in Lamiales)

Hirania Thulin (described in Sapindales and stated to be related to *Diplopeltis*, but may belong elsewhere; phylogenetic evidence is wanting)

Gumillea Ruiz & Pav. (originally placed in Cunoniaceae, where it does not belong; it may be close to Picramniales or Huerteales)

Keithia Spreng. (described as Capparaceae; may belong elsewhere in Brassicales)

Poilanedora Gagnep. (described as Capparaceae, but does not seem to belong there)

Rumpfia L. (only known from an illustration)

13 Parallel and Convergent Evolution

INTRODUCTION

Clarification of phylogenetic relationships at all levels across the angiosperm tree of life has had broad implications, facilitating, for example, a more accurate assessment of character evolution. These topologies have provided insights into the evolution of key characters. Several morphological and chemical features long thought to have evolved many times following traditional classifications are now established as having arisen once or only a few times. As reviewed earlier, glucosinolate production, a chemical defense long thought to have evolved recurrently (Cronquist 1981; but see Dahlgren 1975), was shown to have evolved only twice (in Brassicales and again in Putranjivaceae; Chapter 10). Similarly, symbiosis with nitrogen-fixing bacteria was thought to have evolved many times, but molecular analyses revealed instead that some set of predisposing traits for this character evolved only once, in the ancestor of the nitrogen-fixing clade (Chapter 10). However, within this one relatively small clade, nitrogen-fixing symbiosis may have evolved and/or been lost several times (see Chapter 10).

Here we overview three features that are highly prone to parallel or convergent evolution: parasitism, carnivory, and C_4 photosynthesis. Molecular phylogenetic investigations have helped to clarify the extent of parallelism or convergence of these traits and also have provided additional evolutionary insights. The possible parallel or convergent evolution of other features, including chemical characters (e.g., ellagic acid, iridoids), is examined in other chapters (Chapters 6 and 11).

PARASITISM IN ANGIOSPERMS

Although most angiosperms produce their own carbohydrates via photosynthesis (i.e., are autotrophic), some are heterotrophic. The latter comprise two types: mycoheterotrophs and haustorial parasites (see Nickrent et al. 1998; Selosse and Cameron 2010). However, the terms are confusing—the two types are sometimes treated as a single category, "parasites" (Nickrent et al. 1998), sometimes considered separately (Selosse and Cameron 2010), and sometimes viewed as part of a "mutualism–parasitism continuum" (Johnson et al. 1997; Merckx and Freudenstein 2010). We provide a summary of terms (Table 13.1; taken from Nickrent et al. 1998, and Selosse and Cameron 2010) and follow their strict definition in which a parasite is considered "a symbiotic association in which an organism obtains at least some of its nutrition directly from another organism." A restrictive definition in plants includes only parasites with haustorial connections to other plants, not mycoheterotrophs. The term parasite therefore includes hemiparasites, holoparasites, facultative parasites, and obligate parasites, but some would exclude mycoheterotrophs (see below and Parasitic Plant Connection, www.parasiticplants.siu.edu/).

Mycoheterotrophs derive nutrients indirectly from the host plant through a symbiotic relationship with mycorrhizae. In contrast, haustorial parasites directly penetrate host tissues via an haustorium (often a modified root). Categorizing mycoheterotrophs is problematic. Some do not consider these plants to be parasites. However, fully achlorophylous mycoheterotrophs would seem to fit the defini-

Table 13.1. Terminology associated with heterotrophs (modified from Nickrent 1998 and Selosse and Roy 2009)

Autotroph: an organism that is able to use atmospheric CO_2 as its sole carbon source, for example by way of photosynthesis.

Heterotroph: a plant that obtains carbon via a mechanism other than photosynthesis. Two major categories: mycoheterotrophs and haustorial parasites.

Haustorial parasite: plant that obtains minerals and carbon directly from the host plant using haustorium.

Hemiparasite: a plant that, although capable of performing photosynthesis, lives parasitically on other plants, from which it obtains mineral nutrients and water.

Holoparasite: a nonphotosynthetic plant that obtains all its water and nutrients from the host using a haustorium.

Mycoheterotroph (mycophyte, mycotroph, myco-heterotroph, saprophytic plant): a non-photosynthetic, non-chlorophyllous plant that obtains not only minerals but also carbon from its mycorrhizal fungus that is growing on the host plant.

Mixotroph (hemi-autotroph, partial mycoheterotroph): an autotrophic organism that combines its photosynthesis and a partial heterotrophy as carbon sources.

tion of parasite as well as holoparasites do (i.e., nutrients exclusively obtained from other organisms, no evidence of mutualism). Full mycoheterotrophs are completely dependent on fungi for carbon metabolites (Merckx and Freudenstein 2010; Selosse and Cameron 2010), but some apparently photosynthetic plants also obtain some carbohydrates from their fungal symbiont. This may depend upon the life stage of the plant (e.g., orchids where fungal association is critical early in the life cycle, but less so as the plant develops). Some Ericaceae (e.g., *Pyrola* and *Chimaphila*) are strongly mycorrhizal, but are green and photosynthetic, whereas others (e.g., *Monotropa*) lack chlorophyll and obtain all carbohydrates from a mycorrhizal fungus. The degree of nutritional dependence on the host varies among haustorial parasites (Nickrent et al. 1998). Hemiparasites photosynthesize during part of their life cycle and obtain mainly water and dissolved minerals from their hosts; holoparasites are non-photosynthetic. Both holoparasites and hemiparasites are briefly discussed in the chapters on rosids and asterids (Chapters 10 and 11).

Following these definitions, there are facultative hemiparasites—those that can survive without connection to a host during part of their life cycle (e.g., *Castilleja*, Orobanchaceae)—and obligate hemiparasites—those that cannot survive unless connected to their host (e.g., *Viscum*, Santalaceae of APG III [2009]; APG IV [2016]; Viscaceae of Nickrent et al. 2010). This use of hemiparasite is not completely satisfactory because it combines two different phenomena, but it is an improvement in that previous definitions were difficult to apply.

Following APG III (2009) and APG IV (2016), ~4,000 species in 265 genera and perhaps 17 families are considered to be haustorial parasites (see Parasitic Plant Connection, /www.parasiticplants.siu.edu/). Our estimate of the number of parasitic families (Table 13.2) differs from other recent estimates because of changes in classification in the angiosperms (APG III 2009; APG IV 2016; see Chapter 12), as well as recent changes to the classification of Santalales (Nickrent et al. 2010; below and Chapter 12). Some of these diverse parasites are illustrated in Fig. 13.1. Most are holoparasites; members of Santalales, Krameriaceae, and one genus of Lauraceae are hemiparasites, although single genera in Santalaceae and Loranthaceae appear to be holoparasites (below, Santalales parasites; Chapter 11). Orobanchaceae (sensu APG IV 2016; Chapter 12) include both hemi- and holoparasites.

Ascertaining the relationships of many parasitic plants has been problematic because morphological features associated with the parasitic habit are often extensively modified (e.g., Nickrent et al. 1998). Modifications include loss of diverse organs, including leaves, perianth parts, and integuments, as well as the loss of chlorophyll. Some morphological modifications are subject to parallelism and convergence; similar features have evolved in distantly related groups. Perhaps surprising, the parasitic habit was used by some (e.g., Cronquist 1981) to link Balanophoraceae, Hydnoraceae, and Rafflesiaceae with Santalales (only the former family is now known to be part of Santalales; Chapters 11, 12).

Initially, molecular systematics did not resolve the relationships of most holoparasites because the first wave of molecular analyses relied primarily on plastid genes (Chapter 3). One consequence of the evolution of holoparasitism is relaxation of the evolutionary constraints associated with photosynthetic function—these plants are non-photosynthetic and often have highly modified plastid genomes that lack many genes or have accelerated rates of molecular evolution, including, within genes, nucleotide deletions that are not in triplets (e.g., dePamphilis and Palmer 1990; Nickrent et al. 1998; Bungard 2004; Nickrent and Garcia 2009). Parasitic plants also exhibit extensive horizontal transfer of mtDNA (Won and Renner 2003; Bergthorsson et al. 2003, 2004; Sanchez-Puerta et al. 2008; Mower et al. 2010; Renner and Bellot 2012; Xi et al. 2013; Chapter 15).

As a result of modifications to the plastid genome, plastid genes that have been widely used to infer angiosperm phylogeny have been of limited value in placing holoparasites. There are exceptions; sequences of several plastid

Figure 13.1. Selected representatives of heterotrophic angiosperms. a. *Prosopanche americana* (R. Br.) Baill. (Aristolochiaceae), flower. b-c. *Cassytha filiformis* L. (Lauraceae) in fruit (b) and flower (c). d. *Corallorhiza striata* Lindl. (Orchidaceae), a mycoheterotroph. e. *Cynomorium coccineum* L. (Cynomoriaceae), habit. f. *Cytinus ruber* (Fourr.) Fritsch (Cytinaceae) growing with its host, *Cistus creticus* L. (Cistaceae). g. *Scybalium jamaicense* (Sw.) Schott & Endl. (Balanophoraceae), habit, a holoparasite. h. *Arceuthobium bicarinatum* Urb. (Viscaceae or Santalaceae s.l.), showing stem-parasitic habit. i. *Viscum minimum* Harv. (Viscaceae or Santalaceae s.l.), an internal holoparasite. j. *Santalum paniculatum* Hook. & Arn. (Santalaceae), branch with flowers, a hemiparasite with root haustoria. k. *Dendropemon constantiae* Krug & Urb. (Loranthaceae), showing stem-parasitic habit. l. *Chimaphila maculata* (L.) Pursh (Ericaceae), plant mycorrhizal. m. *Monotropa uniflora* L. (Ericaceae), a mycoheterotroph. n. *Cuscuta cuspidata* Engelm. & Gray (Convolvulaceae), habit, flowers, and fruits, a hemiparasite. o. *Castilleja indivisa* Engelm. (Orobanchaceae), a hemiparasite. p. *Orobanche canescens* C. Presl (Orobanchaceae), a holoparasite.

Table 13.2. Angiosperm families containing parasitic members, with degree of specialization, and general phylogenetic placement in the angiosperms

Family	Genera / Total Species	Specialization	Placement
Hydnoraceae	2 / 14–18	Holoparasite	Piperales
Lauraceae	1 / 20	Hemiparasite	Laurales
Cynomoriaceae	1 / 1–2	Holoparasite	Saxifragales
Balanophoraceae	11 / 32	Holoparasite	Santalales
Cytinaceae	1 / 7	Holoparasite	Rosid-Malvales
Rafflesiaceae	3 / 19	Holoparasite	Rosid-Malpighiales
Apodanthaceae	2 / 19	Holoparasite	Rosid-Cucurbitales
Krameriaceae	1 / 17	Hemiparasite	Rosid-Zygophyllales
Mitrastemonaceae	1 / 2	Holoparasite	Asterid-Ericales
Convolvulaceae	1 / 158	Hemiparasite	Asterid-Solanales
Boraginaceae	3 / 5	Holoparasite	Asterid-Boraginales
Orobanchaceae	78 / 1950	Hemi- and holoparasite	Asterid-Lamiales
Ximeniaceae	4 / 13	Hemiparasite	Santalales
Aptandraceae	8 / 34	Hemiparasite	Santalales
Olacaceae	3 / 57	Hemiparasite	Santalales
Octoknemaceae	1 / 14	Hemiparasite	Santalales
Schoepfiaceae	3 / 55	Hemiparasite	Santalales
Misodendraceae	1 / 8	Hemiparasite	Santalales
Loranthaceae	68 / 950	Hemiparasites	Santalales
Opiliaceae	11 / 29	Hemiparasite	Santalales
Santalaceae s.l.	44 / 990	Hemiparasite	Santalales
Comandraceae*	2	Hemiparasite	Santalales
Thesiaceae*	5	Hemiparasite	Santalales
Cervantesiaceae*	8	Hemiparasite	Santalales
Nanodeaceae*	2	Hemiparasite	Santalales
Amphorogynaceae*	9	Hemiparasite	Santalales
Viscaceae*	7	Hemiparasite	Santalales
Santalaceae s.s.*	11	Hemiparasite	Santalales

*Treated as distinct families by Nickrent et al. rather than as part of a broadly defined Santalales (see Chapters 11 and 12). Only the number of genera is given for the segregate families.

genes were used to resolve the relationships of Orobanchaceae, parasitic Lamiales (Olmstead et al. 2001), and the variably mycoparasitic monocot family Burmanniaceae (Caddick et al. 2002a). Holomycoparastic genera and species in otherwise photosynthetic families/genera have also typically been easily placed, presumably because their plastid genomes are largely intact due to the recent transition (e.g., *Geosiris*, Iridaceae and many orchid genera or species). Although parasitic taxa generally can be placed with nuclear (initially 18S and 26S rDNA were used) and mitochondrial DNA sequences, the placement of many parasitic plants has remained obscure until recently, despite the enormous success of molecular phylogenetics in clarifying relationships of other taxonomically troublesome plants.

We have plotted the occurrence of parasitism in the eudicots on the current estimate of angiosperm phylogeny. Parasitism has arisen independently many times in the angiosperms (Fig. 13.2a,b, with the following clades, or members thereof, each representing a putative separate origin: Convolvulaceae, Boraginaceae, Orobanchaceae, Cynomoriaceae, Cytinaceae, Balanophoraceae, Apodanthaceae, Mitrastemonaceae, Rafflesiaceae, Hydnoraceae (now considered part of Aristolochiaceae; APG IV 2016; Chapter 12), Lauraceae, Krameriaceae, and Santalales, a clade with parasitic members in several families (Nickrent et al. 1998, 2010; Smith et al. 2001; Barkman et al. 2004; see the Parasitic Plant Connection).

MYCOHETEROTROPHY IN ANGIOSPERMS

The terminology associated with mycoheterotrophs is problematic, as noted above (Table 13.1). Mycoheterotrophs should not be confused with mixotrophs (synonyms are hemi-autotrophs or partial mycoheterotrophs); mixotrophs are defined as autotrophic organisms that combine photosynthesis and partial heterotrophy as carbon sources (Table 13.1; Selosse and Roy 2009). Selosse and Roy (2009) further noted that a continuum from autotrophic to fully heterotrophic organisms exists. Among angiosperms, true mixotrophy is known only from some Ericaceae (pyrolids) and some Orchidaceae (Tedersoo et al. 2007; Selosse and Roy 2009). Mixotrophy is usually given as a clear example of a homoplastic trait (Tedersoo et al. 2007; Selosse and Roy 2009). Reversals from mixotrophy to autotrophy also occur (Selosse and Roy 2009).

A summary of mycoheterotrophs is given in Table 13.3. Morphological reduction and extensive secondary loss/rearrangements of multiple plastid loci have often hindered resolution of phylogenetic relationships of mycoheterotrophic plants (e.g., Merckx et al. 2009; Merckx and Freudenstein 2010; Benny et al. 2011; Braukmann and Stefanović 2012), but several recent publications have used complete plastomes with success (e.g., Logacheva et al. 2014; Mennes et al. 2015). Due to the strong rate heterogeneity of nuclear and mitochondrial genes of mycoheterotrophic plants (Merckx et al. 2009; Merckx and Freudenstein 2010; Benny et al. 2011; Lemarie et al. 2011a; Bromhan et al. 2013), this problem cannot be resolved via inclusion of additional types of markers (Merckx et al. 2009;

Table 13.3. A summary of mycoheterotrophs in angiosperms based on Merckx et al. (2013)

Clade	Number of origins of full mycoheterotrophy	Associated fungi
Petrosaviaceae	1	Glomeromycota
Triuridaceae	1	Glomeromycota
Burmanniaceae	>8	Glomeromycota
Thismiaceae	1–2	Glomeromycota
Corsiaceae	1	Glomeromycota
Iridaceae	1	Glomeromycota
Orchidaceae-Vanilloideae	2–3	Ascomycota and Basidiomycota
Orchidaceae-Orchidoideae	>9	Basidiomycota
Orchidaceae-Epidendroideae	>14	Ascomycota and Basidiomycota
Polygalaceae	1	Glomeromycota
Ericaceae	2–3	Ascomycota and Basidiomycota
Gentianaceae	4	Glomeromycota

Merckx and Freudenstein 2010). Hence, the phylogenetic relationships of some mycoheterotrophic plants remain poorly known. Many mycoheterotrophic genera have not yet been included in phylogenetic analyses (e.g., *Cheilotheca* [Ericaceae], *Corsiopsis* [Corsiaceae], *Epirixanthes* [Polygalaceae], *Miersiella* and *Marthella* [Burmanniaceae], *Kihansia*, *Peltophyllum*, *Seychellaria*, *Soridium*, and *Triuridopsis* [Triuridaceae], and several genera of Orchidaceae) (Merckx and Freudenstein 2010).

Following Merckx and Freudenstein (2010), outside of monocots there are at least seven independent origins of mycoheterotrophy: one in Polygalaceae, two or three in Ericaceae (Braukmann and Stefanović 2012), and four in Gentianaceae, for a total of 46 species. The remaining angiosperm mycoheterotrophs are all monocots (Merckx and Freudenstein 2010), in which mycoheterotrophy is present in all major clades except Acorales, Alismatales, and commelinids (Fig. 13.2b). The habit is found in Petrosaviales, Pandanales, Dioscoreales, Liliales, and Asparagales. Thus, mycoheterotrophy appeared early (there are fossil Triuridaceae at 90 mya) and often in monocot evolution. Within monocots, at least 411 species in 43 lineages evolved full mycoheterotrophy (Merckx and Freudenstein 2010). Orchidaceae contain the largest number of fully mycoheterotrophic species (~210), which represent over 30 independent occurrences (Merckx and Freudenstein 2010). Mycoheterotrophy may evolve from either autotrophic or mixotrophic ancestors, based on recent analyses (Selosse and Roy 2009) (Fig. 13.2).

PARASITES IN THE MAGNOLIIDS

Unlike many parasites, Hydnoraceae were easily placed with DNA data in early analyses (Nickrent et al. 2002; Fig. 13.2). Molecular analyses supported monophyly of Hydnoraceae and association of the family with Aristolochiaceae (Nickrent et al. 2002) as part of Piperales (Chapter 5). Phylogenetic analysis permitted the general placement of Hydnoraceae, but the precise relationship of Hydnoraceae within Piperales, until recently, remained unclear. Recent analyses (Naumann et al. 2013; Massoni et al. 2014) have continued to place haustorial Hydnoraceae within Piperales and nested within Aristolochiaceae, prompting a recircumscription of Aristolochiaceae to include Hydnoraceae (APG IV 2016; Chapter 12).

In contrast to morphology-based classifications, molecular analyses clearly do not suggest a close relationship between parasitic Aristolochiaceae (i.e., the former Hydnoraceae) and Rafflesiaceae (e.g., Cronquist 1981). Molecular analyses indicate, instead, that Rafflesiaceae in the traditional sense represent a polyphyletic assemblage with constituent members associated with various asterid and rosid families (see below). Thus, several floral morphological features shared by Rafflesiaceae and taxa formerly placed in Hydnoraceae must represent independent acquisitions. These features include flowers or inflorescences arising from the host, fleshy flowers, a perianth of a single series (three to five tepals), and a highly modified androecium.

Cassytha (Lauraceae) represents another independent origin of the parasitic habit among magnoliids (Fig. 13.2a). The genus of ~20 species of hemiparasites is distinctive because of its twining, herbaceous habit, whereas other Lauraceae are trees or shrubs. Although sometimes formerly placed in its own family, Cassythaceae, most treatments included *Cassytha* in Lauraceae (e.g., Cronquist 1981; Takhtajan 1987, 1997; see APG IV 2016; Chapter 12). Phylogenetic analyses (Rohwer and Rudolph 2005) indicated that *Cassytha* is nested among Lauraceae. *Cassytha* is typical of Lauraceae in features other than habit, including floral and fruit morphology, pollen structure, and alkaloid chemistry (Rohwer 1993). Thus, it has long been apparent that *Cassytha* is allied with Lauraceae. This genus therefore differs from many other parasitic lineages that are so highly modified morphologically that their relationships have been difficult, if not impossible, to ascertain from morphology alone.

ROSID PARASITES

Several lineages of parasitic plants that have been particularly difficult to place using morphology appear to be nested within rosids (Chapter 10; Fig. 13.2). As reviewed below, these rosid parasites include some of the best known of the parasitic plants, such as *Rafflesia* and other genera once placed in a broadly defined Rafflesiaceae (illustrated in Chapter 10).

The rosid parasite that has proved to be one of the easiest to place in molecular analyses is *Krameria* (Krameriaceae). *Krameria* comprises ~15 species of hemiparasitic shrubs and perennial herbs. Because of superficial floral similarities, some authors allied the genus with Polygalaceae and others with the caesalpinioid legumes (Cronquist 1981; Takhtajan 1997). In contrast to most parasites, plastid sequences were easily obtained for *Krameria*, and early molecular investigations indicated, with strong support, that *Krameria* (shown in Chapter 10) is sister to Zygophyllaceae (together comprising Zygophyllales). The relationship of Krameriaceae + Zygophyllaceae to other rosids was more difficult to disentangle and has only recently become clear with larger datasets (see Chapter 10)—Zygophyllales are sister to other fabids.

The affinities of *Rafflesia* have long been of special interest because *Rafflesia arnoldii* from Sumatra is renowned for having the largest flower in the world (Fig. 13.2a). As traditionally recognized, Rafflesiaceae were a large, morphologically diverse assemblage; historically there was considerable debate regarding the actual limits of the family. *Cytinus* has also been placed in Rafflesiaceae (e.g., Cronquist 1981). Cronquist (1981) also placed Mitrastemonaceae with Rafflesiaceae in his Rafflesiales, close to Santalales.

All these rosid parasites lack chlorophyll and are endoparasitic on the roots, or less often on the shoots, of other plants. Only the flowers or the short flowering shoot of the parasite emerge from the host plant (Fig. 13.1). The vegetative body of the plant is largely filamentous and resembles a fungal mycelium. For these reasons, these taxa have been extremely difficult to place. These parasites have modified plastid genomes, and it has been either impossible or problematic to place them with plastid genes. Molina et al. (2014) discussed the possible loss of the plastid genome in *Rafflesia lagascae*. However, recent studies using mtDNA and nuclear genes have now placed these taxa within the angiosperm framework tree; we review these results here.

The broadly circumscribed Rafflesiaceae formerly recognized based on morphology are clearly polyphyletic. Rafflesiaceae are now narrowly defined and limited to *Sapria*, *Rhizanthes*, and *Rafflesia* (Davis et al. 2007). Barkman et al. (2004), using mtDNA sequence data, placed *Rhizanthes* and *Rafflesia* within the rosid order Malpighiales, a finding reinforced by others (Davis and Wurdack 2004; Davis et al. 2007; Bendiksby et al. 2010) (Fig. 13.2a). Within this narrowly defined Rafflesiaceae, *Sapria* is sister to *Rhizanthes* + *Rafflesia* (Davis et al. 2007). This placement of Rafflesiaceae within Malpighiales was not suggested in previous classifications, although floral similarities between *Rafflesia* and Passifloraceae of Malpighiales had been noted (Barkman et al. 2004). *Rafflesia* flowers share morphological features with Passifloraceae, including a central column (termed an androgynophore in Passifloraceae) and a corona, characteristics that are unusual.

There is considerable interest in the evolutionary origin of the large flowers of *Rafflesia* (Fig. 13.1). There was an enormous (79-fold) and rapid increase in flower size in stem Rafflesiaceae that may reflect a switch to fly pollination (Davis et al. 2007, 2008). Barkman et al. (2008) suggested that there were considerable changes in flower size just within crown group *Rafflesia*, with both repeated increases and decreases.

Phylogenetic analysis of mitochondrial DNA sequence data has also placed *Cytinus* (Cytinaceae), another parasite once considered closely related to or part of Rafflesiaceae (Cronquist 1981; Takhtajan 1997). *Cytinus* usually is a parasite on *Cistus* and *Halimium* (Cistaceae) (Thorogood and Hiscock 2007). However, rather than a placement close to Rafflesiaceae or Malpighiales (see above), molecular results suggest, instead, that *Cytinus* is embedded within Malvales and sister to Muntingiaceae (Nickrent et al. 2004; Nickrent 2007).

The relationships of Apodanthaceae have been more difficult to ascertain than the parasites noted above. Apodanthaceae were considered part of either Rafflesiaceae (e.g., Cronquist 1981) or Rafflesiales (e.g., Takhtajan 1997). However, morphological studies indicated that Apodanthaceae (*Pilostyles* and *Apodanthes*) share with Malvales several noteworthy floral structures, including an androecial tube and a trend from normal stamens to synandria without thecal organization (Blarer et al. 2004). Using sequence data, Nickrent et al. (2004) suggested a relationship of Apodanthaceae with Malvales. A later DNA analysis (Barkman et al. 2007), in contrast, provided weak support for a position in Cucurbitales; more recent DNA studies provide additional and stronger support for this latter placement (see Chapter 10) (Filipowicz and Renner 2010). Some morphological features support this placement as well, including dioecy, extrose anthers, inferior ovary, and parietal placentation, all features common in Cucurbitales (Filipowicz and Renner 2010).

The syntheses of Barkman et al. (2004, 2007), Davis et al. (2007), and Nickrent et al. (2004), as well as other recent DNA-based analyses, represent major contributions to our understanding of the relationships of parasitic rosids. The placement of some of these families with various rosid lineages was totally unexpected; morphology-based taxonomic treatments did not suggest the placement of parasites with orders of rosids. One caveat, however, is that branch lengths to these parasitic taxa are often long and are typically based on a few mitochondrial genes. Additional nuclear-based analyses remain important for testing these hypotheses.

THE ENIGMATIC CYNOMORIACEAE

Cynomoriaceae comprise one genus (*Cynomorium*) with two species, both non-photosynthetic. The family was often included in, or considered closely related to, Balanophoraceae (e.g., Cronquist 1968; Takhtajan 1997), with the latter often associated with Santalales. Despite several DNA-based studies, the phylogenetic relationships of Cynomoriaceae remain uncertain. Initial analysis of 18S and

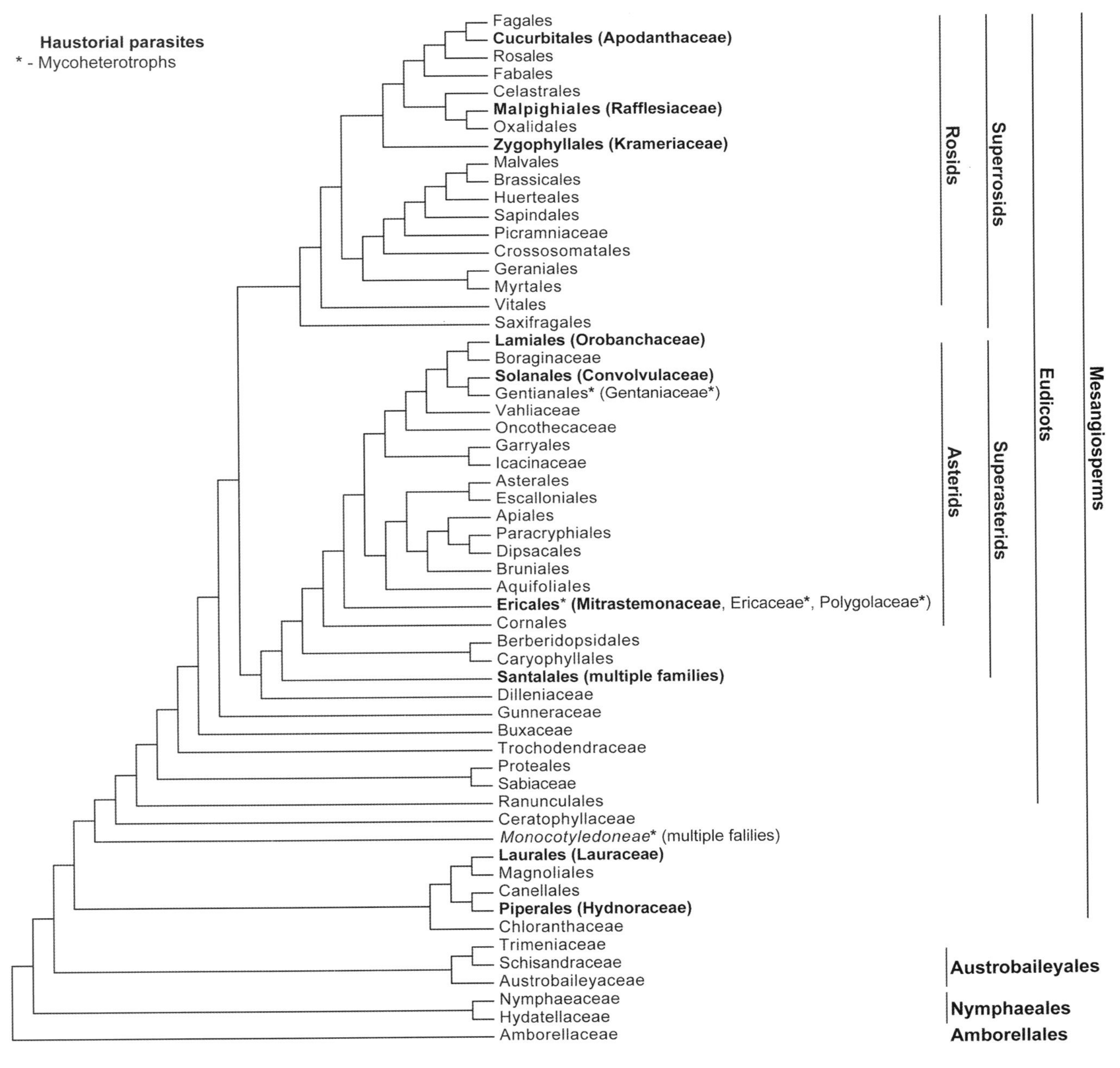

(*continued*)

Figure 13.2. Major occurrences of parasitism in angiosperm plant families (in bold); * = Mycoheterotrophs A. Across a broad summary tree for the angiosperms. B. In monocots. The summary trees are based on Soltis et al. (2011).

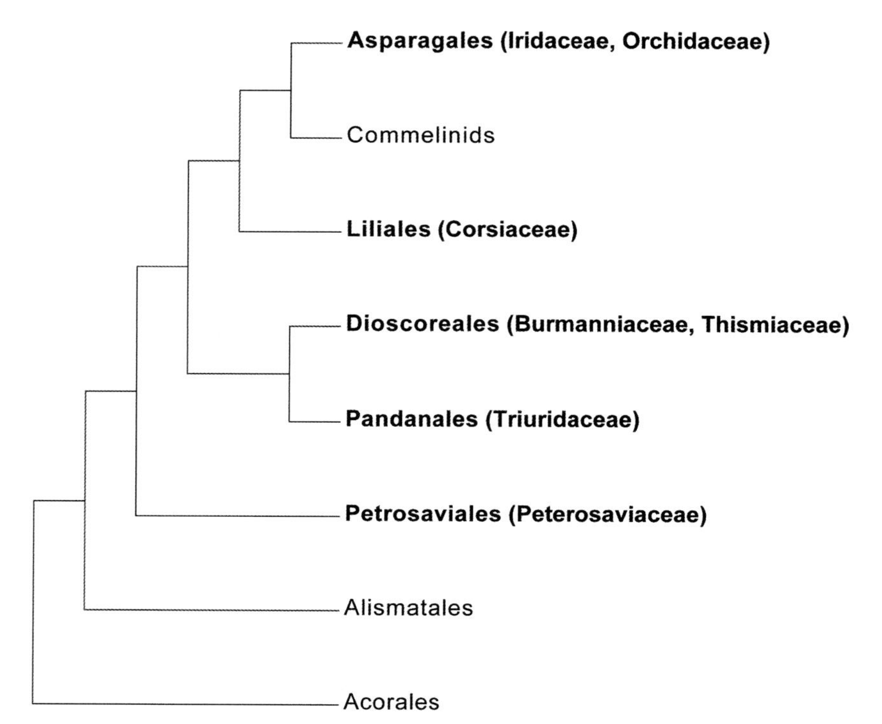

Figure 13.2. *(continued)*

26S rDNA and mitochondrial *matR* sequence data placed *Cynomorium coccineum* in Saxifragales (Nickrent et al. 2005). However, bootstrap support for this relationship was less than 50%, and some broad analyses placed Cynomoriaceae in Santalales (Jian et al. 2008), in agreement with traditional classifications (e.g., Cronquist 1981). However, other analyses using what remains of the inverted repeat of the plastid genome have suggested that the genus belongs in Rosales (Zhang et al. 2009; Moore et al. 2010); recent analyses of phytochrome genes indicate a placement somewhere with the rosids (Gitzendanner et al. unpubl.). Given the uncertain placement of *Cynomorium*, it was unplaced in earlier APG classifications, but recent evidence (Bellot et al. 2016) strengthens support for a placement in Saxifragales (Fig. 13.2; see APG IV 2016; see also Chapter 12).

SANTALALEAN PARASITES

Santalales (Chapter 11; parasitic plants illustrated in Fig. 13.1) provide an excellent model for studying the evolution of parasitism. The order contains non-parasites, hemiparasites, and holoparasites (Hürlimann and Stauffer 1957; Kraus et al. 1995; Nickrent and Malécot 2001). Much of Santalales was included in a paraphyletic Olacaceae (a basal grade) and a large Santalaceae. Nickrent et al. (2010) provide a revised classification of Santalales, dividing both Olacaceae and Santalaceae into a number of separate families, and we present that treatment here for comparison with APG III (2009) and APG IV (2016) (see Chapter 12). Balanophoraceae are also considered part of Santalales (see below).

Following the current summary tree for Santalales, the basal lineages are nonparasitic: Erythropalaceae, Strombosiaceae, and Coulaceae. Parasitism then arose once in the ancestor of the remaining Santalales (Fig. 13.3). Many of the parasitic taxa in Santalales are terrestrial (i.e., root parasites). Root parasitism appears to be the ancestral parasitic state in Santalales—it is found in Ximeniaceae and a narrowly defined Olacaceae, which are sisters to most remaining Santalales. Two other poorly understood early-diverging families are assumed to be root parasites (Octoknemaceae and Aptandraceae), but this remains unknown (Nickrent et al. 2010). Root parasites are also found in many other Santalales: Schoepfiaceae, Opiliaceae, Thesiaceae, Cervantesiaceae, and Nanodeaceae, as well as some Loranthaceae, Santalaceae, and Amphorogynaceae. Aerial parasitism (confined to trees) appears to have arisen five times in Santalales (e.g., Loranthaceae, Santalaceae, Misodendraceae, Amphorogynaceae, and Viscaceae; Fig. 13.3) (Nickrent et al. 2010). Based on dating analyses, the oldest example of this appears to be in Misodendraceae, ~80 mya (Vidal-Russell and Nickrent 2008).

Another parasitic member of Santalales is Balanophoraceae, previously placed in Rafflesiaceae in some classifications (e.g., Cronquist 1981). Several analyses have placed

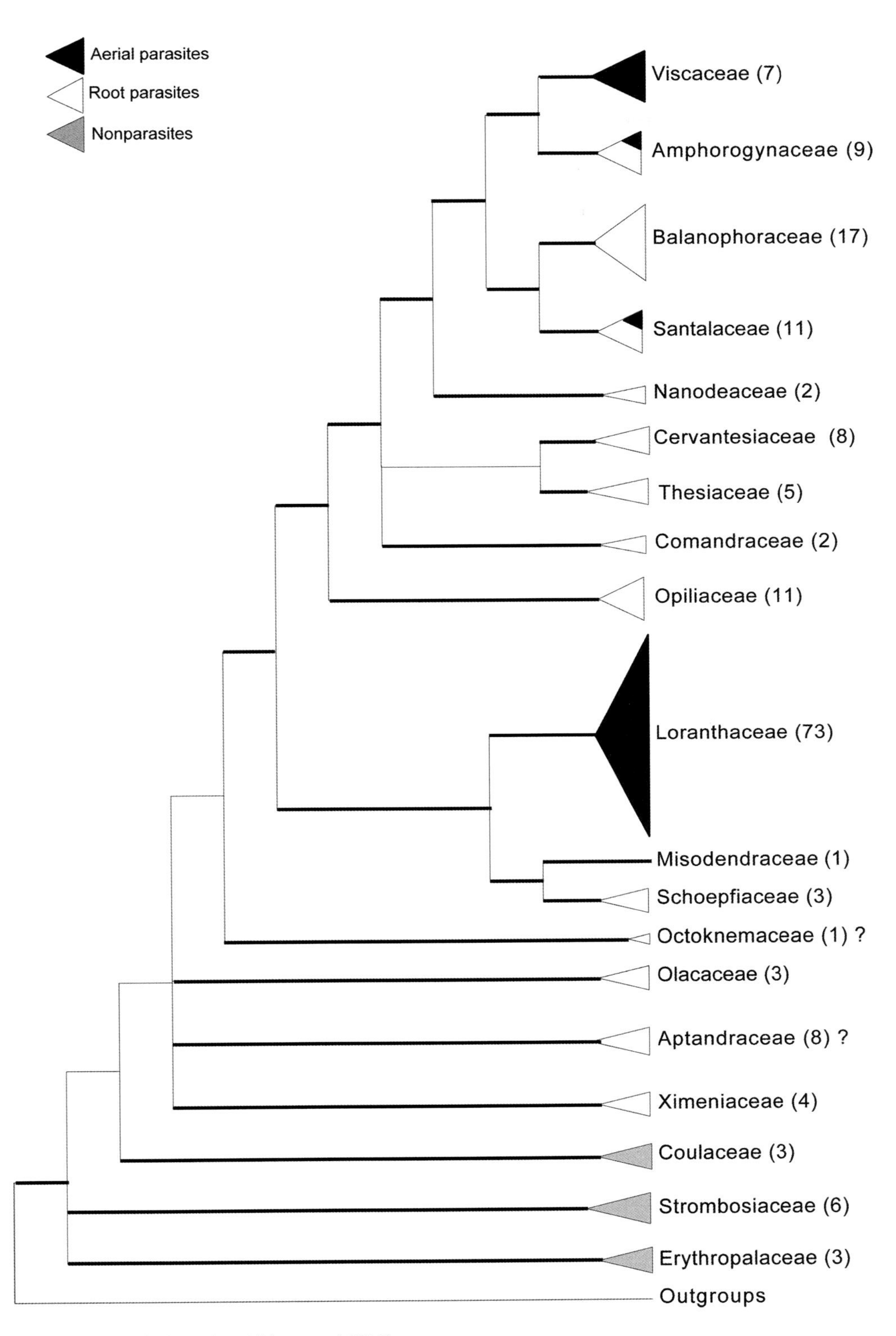

Figure 13.3. Parasitism in Santalales (redrawn from Nickrent et al. 2010).

Balanophoraceae in Santalales (e.g., H. Su et al. 2012; Su and Hu 2012), but the exact position of the family within the order has remained unclear. As a result, Nickrent et al. (2010) did not consider the family in their classification of Santalales. A recent analysis, however, suggests a relationship of the family to a narrowly defined Santalaceae (Nickrent et al. pers. comm.; Chapter 11).

ASTERID PARASITES

Several genera and families of parasites are clearly asterids (Fig. 13.2a)—*Cuscuta* (previously of Cuscutaceae, now placed in Convolvulaceae; Solanales), Orobanchaceae (Lamiales), and *Lennoa* (two species now included in Boraginaceae of APG III [2009] and APG IV [2016], Lennoaceae of Luebert et al. [2016], or Ehretiaceae of the Angiosperm Phylogeny Website, Stevens [2001 onwards]; see Chapter 12). These parasites have long been associated with asterids (Chapter 11) based on non-DNA evidence (i.e., obvious morphological synapomorphies derived from floral characters), but molecular results have provided additional important insights, below.

Cuscuta, comprising more than 100 species of hemiparasites and holoparasites, has long been considered related to, or derived from within, Convolvulaceae (Nickrent et al. 1998). Holoparasitic members lack thylakoids, chlorophyll, and RUBISCO, but retain the gene *rbcL*. Molecular results placed *Cuscuta* with Convolvulaceae (e.g., Soltis et al. 1997b, 2011; Nickrent et al. 1998; Stefanović et al. 2002, 2004), and the genus has now been included within Convolvulaceae (APG II 2003; APG III 2009; APG IV 2016; Chapter 12) as a separate subfamily, Cuscutoideae. Despite intensive research, the position of Cuscutoideae within the family has remained problematic (Stefanović et al. 2002; Stefanović and Olmstead 2004); Wright et al. (2011) indicated that, based on gynoecial characters, *Cuscuta* is closely related to the "bifid" clade (Dicranostyloideae). In *Cuscuta*, most changes in the plastid genome previously attributed to its parasitic mode of life can be explained as either a plesiomorphic condition within Convolvulaceae or as autapomorphies of particular species of *Cuscuta* (Stefanović et al. 2002).

Lennoa and *Pholisma* (former Lennoaceae) represent another lineage of parasites long considered asterids. Based on morphology, these genera have been associated either with Boraginaceae or Hydrophyllaceae (Cronquist 1981; Takhtajan 1987, 1997; Thorne 1992a,b; the latter included in the former in APG III 2009 and APG IV 2016). Cronquist (1981) placed Lennoaceae in Lamiales with Boraginaceae, Verbenaceae, and Lamiaceae because of similarities in the structure of the gynoecium. Phylogenetic analysis of mtDNA sequence data place *Lennoa* with Boraginaceae (Barkman et al. 2007): the genus is now included in that family by APG (APG II 2003; APG III 2009; APG IV 2016), but recent treatments have recognized eight or 11 segregate families from Boraginaceae. Luebert et al. (2016) again recognized Lennoaceae (see Chapters 11, 12). Stevens (2001 onwards, http://www.mobot.org/MOBOT/research/APweb/welcome.html) places *Lennoa* in Ehretiaceae (Boraginales), another segregate of a broadly defined Boraginaceae.

Many genera of parasitic plants have been associated with the former, broadly recognized Scrophulariaceae and Orobanchaceae. In fact, these parasites exhibit the greatest range in degree of specialization and were considered to represent an evolutionary series beginning with fully photosynthetic taxa and culminating with holoparasitism. *Lathraea*, *Harveya*, and *Hyobanche* were viewed as links or transitional taxa between parasites placed in Scrophulariaceae and the holoparasitic Orobanchaceae (Young et al. 1999).

Molecular phylogenetic analyses of the former broadly defined Scrophulariaceae have revealed a single well-supported clade that contains all of the parasites (Young et al. 1999; Olmstead et al. 2001; Refulio-Rodriguez and Olmstead 2014). This clade, containing taxa traditionally placed in either Scrophulariaceae or Orobanchaceae, is now considered a broadly defined Orobanchaceae (APG III 2009; APG IV 2016; see also Olmstead et al. 2001 and the current narrowly defined Scrophulariaceae; Chapters 11, 12).

All members of this broadly defined Orobanchaceae are parasitic except the genus *Lindenbergia*, which is nonparasitic and sister to the remainder of the family (Young et al. 1999; Olmstead et al. 2001). This is true unless *Rehmannia* is also included in Orobanchaceae—as some suggest (see Chapter 12). Molecular results indicate that hemiparasitism has evolved only once (McNeal et al. 2013); holoparasites have evolved independently from hemiparasites multiple times, with recent estimates indicating three times (dePamphilis et al. 1997; Nickrent et al. 1998; Young et al. 1999; Schneeweiss et al. 2004; Bennett and Mathews 2006; McNeal et al. 2013). However, multiple workers have stressed that the distinction between holoparasite and hemiparasite is not clear in this clade (see McNeal et al. 2013). There has long been interest in taxa considered transitional between hemi- and holoparasitic, and there may be different pathways by which holoparasitism evolves. Transcriptome evidence suggested that this may involve the retention and regulation of chlorophyll biosynthesis (Wickett et al. 2011) and that transitions to holoparasitism have occurred in at least three separate lineages of Orobanchaceae. Interestingly, host specificity has evolved in *Orobanche*, with races that are specific to particular hosts (Thorogood and Hiscock 2010).

The plastid genome of the holoparasite *Epifagus virginiana* (Orobanchaceae) has been investigated in detail; the species has become a model for the evolution of a reduced plastome in a non-photosynthetic angiosperm. Although the plant lacks chlorophyll, the cells retain plastids with DNA (dePamphilis and Palmer 1990). The plastid genome of *Epifagus virginiana* is much smaller than that of other angiosperms (71 kb vs. 156 kb for the asterid *Nicotiana* [tobacco; Solanaceae], which is typical of most angiosperms) and contains only 42 intact genes (dePamphilis and Palmer 1990; Wolfe et al. 1992). However, the greatly reduced plastid genome of *Epifagus* is nearly collinear with that of *Nicotiana* (dePamphilis and Palmer 1990). The plastid genome of another Orobanchaceae, *Cistanche deserticola*, is also greatly reduced in size (102,657 bp compared to a typical plastid genome of ~150 kb) and gene content; all genes required for photosynthesis except one have either been lost or become pseudogenized (X. Li et al. 2013). The plastid genome of *C. deserticola* differs from those of other holoparasitic plants in that it retains nearly a complete set of tRNA genes (X. Li et al. 2013). Several studies have provided evidence of horizontal gene transfer (HGT) involving plastid genes and the host plant of members of Orobanchaceae, including *Orobanche* and *Phelipanche* (Park et al. 2007) and *Cistanche* (X. Li et al. 2013). Wicke et al. (2013) provided complete plastomes for 10 photosynthetic and non-photosynthetic parasites and their non-parasitic sister from Orobanchaceae and reconstructed a history of convergent gene losses and genome reconfigurations.

Using mitochondrial DNA sequence data, Barkman et al. (2004) demonstrated that *Mitrastema* (Mitrastemonaceae), an enigmatic holoparasite sometimes placed in Rafflesiaceae (see "Rosid Parasites," above), is placed in the asterid order Ericales. A close relationship of *Mitrastema* to asterids had not been suggested previously. However, the genus exhibits a sympetalous corolla, which supports a placement with asterids (Chapter 11). In addition, a relationship of *Mitrastema* with Ericales is also supported by shared morphological features, including opposite and decussate leaves, parietal placentation, and circumscissile fruit dehiscence (Barkman et al. 2004). The exact placement of the family is unclear, but it is likely related to Theaceae, Symplocaceae, Styracaceae, and Diapensiaceae.

CARNIVORY

Carnivory in angiosperms is a highly integrated system that involves morphological modifications to attract, trap, kill, and ultimately digest (and absorb nutrients from) animal prey (Juniper et al. 1989; Barthlott et al. 2007). All carnivorous plants rely on photosynthesis; the main nutrients recovered from prey are not fixed carbon, but other elements (primarily nitrogen). Several highly divergent trap types have evolved, all relying on modified leaves. The most familiar traps are the pitcher, flypaper, and snap or steel trap (Fig. 13.4; Table 13.4). The pitcher trap is found in the well-known families Sarraceniaceae (Fig. 13.4) and Nepenthaceae, the New and Old World pitcher plants, respectively, as well as the less-well-known Australian pitcher plant family, Cephalotaceae (Figure 13.4).

Flypaper traps are present in several families, including Droseraceae (*Drosera*; Fig. 13.4), Drosophyllaceae (*Drosophyllum*), Dioncophyllaceae (*Triphyophyllum*; Fig. 13.4),

Table 13.4. Conventionally recognized carnivorous plants, including family and higher-level placement

Family	Genera	Number of Species	Trap Type
Roridulaceae (asterids—Ericales)	*Sarracenia*	8	Pitcher
	Darlingtonia	1	Pitcher
	Roridula	2	Flypaper
Byblidaceae (asterids—Lamiales)	*Byblis*	2	Flypaper
Lentibulariaceae (asterids—Lamiales)	*Pinguicula*	52	Flypaper
	Utricularia	~200	Bladder
	Genlisea	15	Lobster pot
Martyniaceae (asterids—Lamiales)	*Proboscidea*	9	Flypaper
Nepenthaceae (Caryophyllales)	*Nepenthes*	68	Pitcher
Dioncophyllaceae (Caryophyllales)	*Triphyophyllum*	1	Flypaper
Droseraceae (Caryophyllales)	*Aldrovanda*	1	Snap trap
	Drosera	110	Flypaper
	Dionaea	1	Snap trap
Drosophyllaceae (Caryophyllales)	*Drosophyllum*	1	Flypaper
Cephalotaceae (rosids—Oxalidales)	*Cephalotus*	1	Pitcher
Bromeliaceae* (monocots)	*Brocchinia*	2	Tank
	Catopsis	1	Tank

* Debated as to whether these are true carnivores.

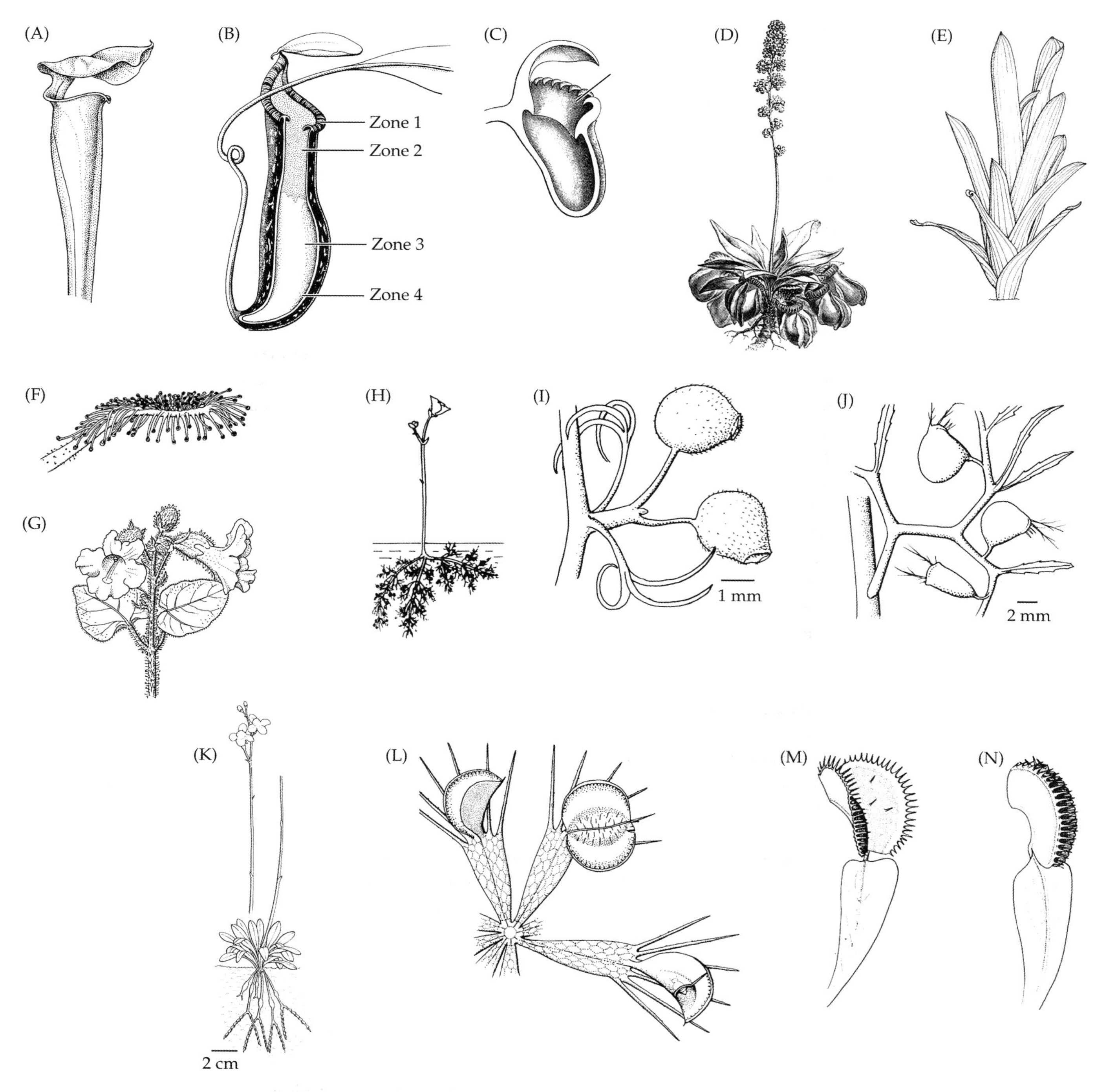

Figure 13.4. Examples of carnivorous plants in Caryophyllales (Droseraceae, Drosophyllaceae, Nepenthaceae), Ericales (Sarraceniaceae), Lamiales (Lentibulariaceae), Oxalidales (Cephalotaceae), and Poales (Bromeliaceae). a. *Cephalotus follicularis* Labill. (Cephalotaceae), leaves modified with an insect-catching pitcher. b. *Dionaea muscipula* J. Ellis ex L. (Droseraceae), leaf showing snap trap. c. *Aldrovanda vesiculosa* L. (Droseraceae), leaves with snap-traps. Genus is sister to *Dionaea* (b), and the snap trap is a synapomorphy. d. *Pinguicula casabitoana* J. Jiménez Alm. (Lentibulariaceae), an epiphyte with gland-headed hairs on the leaves. e. *Utricularia inflata* Walter (Lentibulariaceae): an aquatic carnivorous plant with bladder traps. f. *Drosera filiformis* Raf. (Droseraceae), leaf shows circinate vernation and is covered with gland-headed hairs that catch and digest insects. g. *Heliamphora heteroxdoxa* Steyerm. (Sarraceniaceae), traps and inflorescence. h. *Catopsis berteroniana* (Schult. & Schult.f.) Mez (Bromeliaceae), habit. i. *Nepenthes ventricosa* Blanco (Nepenthaceae), habit and pitcher (formed as an extension of the midrib). j. *Drosophyllum lusitanicum* (L.) Link (Drosophyllaceae), habit and reverse-circinate leaves (compared to those in *Drosera*). k. *Genlisea* (Lentibulariaceae), general habit of plant; the tall flowering stem has been cut so that the entire inflorescence can fit into the frame and remain in scale; traps project down beneath the basal leaves. l. *Aldrovanda vesiculosa* (Droseraceae), with several leaves. m. *Dionaea muscipula* (Droseraceae), leaf open, unstimulated. n. *Dionaea muscipula* (Droseraceae), leaf immediately after mechanical stimulation. From Soltis et al. (2005).

Lentibulariaceae (*Pinguicula*), Byblidaceae (*Byblis*), and Martyniaceae (*Proboscidea*; Fig. 13.4). Snap traps are found in *Dionaea* and the little-known *Aldrovanda* (Droseraceae; Fig. 13.4). *Philcoxia* (Plantaginaceae) from Brazil is also carnivorous (Fritsch et al. 2007; Pereira et al. 2012).

Less well known are the intricate bladder traps found in *Utricularia* (Lentibulariaceae) and the "lobster pot" traps of *Genlisea* (Fig. 13.4; also of Lentibulariaceae). The trap of *Utricularia* is highly sophisticated and involves negative pressure or suction within the "bladder," which is a thin-walled sac with an intricately constructed doorway that opens inwardly. The lobster pot traps of *Genlisea* consist of a swollen utricle or bulb (thought to serve as a digestive cavity). A tubular channel (5–20 mm long) leads to this bulb area, below which is a tubular channel with two helical arms (10–50 mm long). A slit, which extends into both arms, forms a long and narrow mouth through which protozoa enter the trap by chemotaxis. The trapping mechanism and habitat for *Philcoxia* (Plantaginaceae) are like those of *Genlisea*, and phosphatases digest nematodes (for overviews, see Juniper et al. 1989 and Barthlott et al. 1998).

Another type of carnivory (termed "protocarnivory") has also been observed in some groups (e.g., Stylidiaceae, Darnowski et al. 2006; and Roridulaceae). These plants trap prey but depend on assistance for digestion; they blur the distinction between carnivory and complete reliance on photosynthesis. *Roridula* (Roridulaceae, Ericales) is considered by many to be carnivorous, but it depends on a capsid bug, *Pomeridea*, to eat the insects it catches on its flypaper traps; the bug lives its entire life on the plant and defecates around its base, providing the plant with additional nitrogen. Many plants (e.g., various Solanaceae) have glandular hairs over their stems and leaves and could similarly be trapping insects that then decay once they fall to the ground around these plants, thereby providing additional nitrogen (Chase et al. 2009). Many plants may benefit from these "protocarnivorous" processes; they certainly warrant more study to determine whether animal proteins or their components are incorporated in plant tissues.

A mechanism approaching carnivory has also evolved in some monocots. Tank-forming Bromeliaceae (e.g., *Brocchinia* and *Catopsis*; Fig. 13.4) have absorptive trichomes that acquire nutrients from trapped animals or degrading plant debris (Givnish et al. 1984; Benzing et al. 1985). However, there is debate as to whether tank bromeliads are true carnivores; the plants lack special leaf zones for attracting prey and do not actively digest prey. Krol et al. (2012) used the term "passive traps" for *Brocchinia*, in much the same way that pitcher traps are considered passive. Krol et al. (2012) considered three bromeliads to be true carnivores, stating (p. 3) "Although among Bromeliaceae . . . only *Brocchinia reducta* has been shown to secrete phosphatase . . . and neither *B. hechtioides* nor *Catopsis berteroniana* produces proteases, these three species are established plant carnivores. Depending on a food-web to acquire nutrients, these plants provide habitats for frogs, insects (e.g., ants), other carnivorous plants (e.g., *Utricularia humboldtii*) and bacteria (including nitrogen-fixing bacteria), themselves exploiting whatever is left over: feces, animal or vegetable debris." In contrast, Renner and Specht (2013; p. 438) considered *Brocchinia* to be evolving towards carnivory, but not truly carnivorous, indicating that "although the tank trichomes of *Brocchinia* (Bromeliaceae) have been demonstrated to absorb water and nutrients, they do not produce enzymes. Such taxa may be examples of plants that are currently using certain morphologies derived for defense in a way that is evolving toward plant carnivory." Perhaps carnivory in such taxa would be better termed "passive" rather than "protocarnivorous," which implies a directionality to the phenomenon; passive carnivory could be a final evolutionary stage, a completely successful ecological strategy that is not on its way to another. Some pitcher plants (e.g., *Heliamphora*; Sarraceniaceae), also do not produce digestive enzymes and only absorb nutrients first broken down by bacteria in their watery traps. If production of digestive enzymes is essential to a species being defined as a carnivore, then species of *Heliamphora* and *Roridula* are not truly carnivorous. Like Chase et al. (2009), we view carnivory as a continuum in which every intermediate condition exists and strict, potentially misleading definitions (e.g., "protocarnivorous") should be avoided because they imply something that we do not know.

Carnivory in angiosperms has long been considered as having several independent origins (e.g., Darwin 1875). Modern taxonomic treatments also indicate multiple origins of carnivory, but both the actual extent to which multiple origins had occurred, as well as the relationships of carnivorous plants to other angiosperms, were much debated (e.g., Takhtajan 1980, 1987, 1997; Cronquist 1981; Thorne 1992a,b). Cronquist (1981) placed three morphologically divergent, carnivorous families—Nepenthaceae, Sarraceniaceae, and Droseraceae—in his Nepenthales. He based this treatment not only on the shared carnivorous habit, but also on the fact that some data suggested that the insect-catching leaves of all three families were homologous (Markgraf 1955). He also noted that Droseraceae and Nepenthaceae have similar pollen. Takhtajan (1997) also considered these three families of carnivorous plants to be closely related. Molecular phylogenetic results have confirmed a close relationship between Droseraceae and Nepenthaceae (both in Caryophyllales; Chapter 11), but Sarraceniaceae (Ericales, asterids; Chapter 9) are distantly related to these two families but close to Roridulaceae, which may be considered carnivorous.

Albert et al. (1992) first addressed the number of origins of carnivory using a tree based on *rbcL* sequences. The number was subsequently reanalyzed (Soltis et al. 2005b) using a better tree; we have reconsidered the number of origins again here using recent results. Multiple analyses (below) now suggest at least six origins of carnivory among eudicots (see Fig. 13.5): (1) *Cephalotus* (Cephalotaceae); (2) a single origin for the predisposition for carnivory in Caryophyllales (see below); (3) one origin in Ericales (*Roridula* of Roridulaceae and Sarraceniaceae); and three origins in Lamiales: (4) *Proboscidea* (Martyniaceae); (5) *Byblis* (Byblidaceae), and (6) Lentibulariaceae. As noted, several monocots may represent additional origins (Fig. 13.5).

Multigene topologies for asterids (Chapter 11) are crucial for determining the likely number of origins of carnivory. Also of importance are multigene analyses of carnivorous plants in Caryophyllales (Chapter 11). Although the results provide clear evidence for multiple origins of carnivory, there are also clades that are evolutionary hotspots for the carnivorous habit (Fig. 13.5). Caryophyllales (sensu APG IV 2016), for example, contain a large proportion of the carnivorous taxa (five genera in four families: Droseraceae, Drosophyllaceae, Nepenthaceae, and Dioncophyllaceae). The asterid order Lamiales represents another hotspot of carnivory with six genera in four families: Plantaginaceae, Martyniaceae, Lentibulariaceae, and Byblidaceae (Table 13.4). Two families of Ericales (Roridulaceae and Sarraceniaceae) are also potentially carnivorous; these appear to represent one origin (below). Only one species of the large rosid clade is carnivorous, *Cephalotus follicularis* (Cephalotaceae; Oxalidales). Carnivory is not present in any basal eudicot lineages or in basal angiosperm lineages, magnoliids, or Chloranthaceae—it is an innovation limited to core eudicots and perhaps monocots (which remains controversial, as noted).

Cameron et al. (2002) provided strong evidence for a single origin of carnivory within Caryophyllales (Fig. 13.6; see also Meimberg et al. 2000; Cuénoud et al. 2002). Within this carnivorous clade, two subclades are apparent. One subclade contains *Nepenthes* (Nepenthaceae) as sister to *Drosophyllum* (Drosophyllaceae) + (Dioncophyllaceae + Ancistrocladaceae). Dioncophyllaceae contain the carnivorous *Triphyophyllum* and two non-carnivorous genera; *Ancistrocladus* (Ancistrocladaceae) is not carnivorous (Fig. 13.6). The second subclade contains the carnivorous taxa *Dionaea* + *Aldrovanda* as sister to *Drosera* (Droseraceae). These results indicate a single origin of carnivory, followed by a loss of carnivory in *Ancistrocladus* and some members of Dioncophyllaceae. There are several different trap types within the carnivorous clade of Caryophyllales: pitcher (*Nepenthes*), flypaper (*Drosophyllum*, *Triphyophyllum*, *Drosera*), and snap trap (*Aldrovanda* and *Dionaea*).

Although *Roridula* and Sarraceniaceae are both in Ericales, it was initially thought likely that they represented separate origins of carnivory because the two lineages did not appear closely related in early molecular analyses (e.g., Albert et al. 1992; Chase et al. 1993; D. Soltis et al. 2000). Even with large datasets involving many taxa and multiple genes, relationships within Ericales have been difficult to resolve (see Chapter 11). However, analyses of asterids using four (Albach et al. 2001c), five (Anderberg et al. 2002), six (Bremer et al. 2002), or more genes (Soltis et al. 2011) all placed *Sarracenia* as sister to a clade of *Roridula* and the non-carnivorous *Actinidia* (Fig. 13.7). Support for this clade was weak in some studies (e.g., Albach et al. 2001c; Bremer et al. 2002). Anderberg et al. (2002) found strong support (BS = 100%) for *Actinidia* + *Roridula* and (BS = 89%) for *Sarracenia* + (*Actinidia* + *Roridula*). Soltis et al. (2011) further supported these relationships, indicating a single origin of carnivory in Ericales, followed by a loss in Actinidiaceae (Fig. 13.7). The trap types of *Roridula* and Sarraceniaceae differ, representing the flypaper and pitcher types, respectively. The presence of multiple trap types in one clade also occurs in Droseraceae and Lentibulariaceae.

Several carnivorous lineages are members of Lamiales: *Proboscidea* (Martyniaceae), *Byblis* (Byblidaceae), *Philcoxia* (Plantaginaceae), and genera of Lentibulariaceae. However, relationships within Lamiales have been difficult to resolve due to their low levels of divergence, which apparently is due to a recent and rapid radiation (Chapter 11). In early molecular analyses, these three families did not appear together as close relatives within Lamiales (e.g., Albert et al. 1992; Chase et al. 1993; Savolainen et al. 2000a,b; D. Soltis et al. 2000), reinforcing a view of separate origins. However, relationships among these three lineages were poorly supported, even with three- and four-gene analyses (D. Soltis et al. 2000; Albach et al. 2001c; Olmstead et al. 2001; Bremer et al. 2002). Based on more recent studies, however, Byblidaceae, Pedaliaceae, and Lentibulariaceae are each part of distinct subclades (e.g., Refulio-Rodriguez and Olmstead 2014; see Fig. 13.8), strongly suggesting that each represents a separate origin of carnivory. Again, different trap types are present in these carnivorous Lamiales. *Proboscidea* and *Byblis* have flypaper traps; *Philcoxia* has a lobster pot trap; and Lentibulariaceae have multiple trap types: flypaper (*Lentibularia*), bladder (*Utricularia*), and lobster pot (*Genlisea*) (Fig. 13.4).

Phylogeny reconstruction has provided important insights into the patterns of structural evolution associated with carnivory (Albert et al. 1992; Williams et al. 1994; Cameron et al. 2002). *Byblis* (Lamiales) and *Drosophyllum* (Caryophyllales), although distantly related (Fig. 13.5), have similar flypaper traps. The two genera share several characters, including woody habit (but only some species

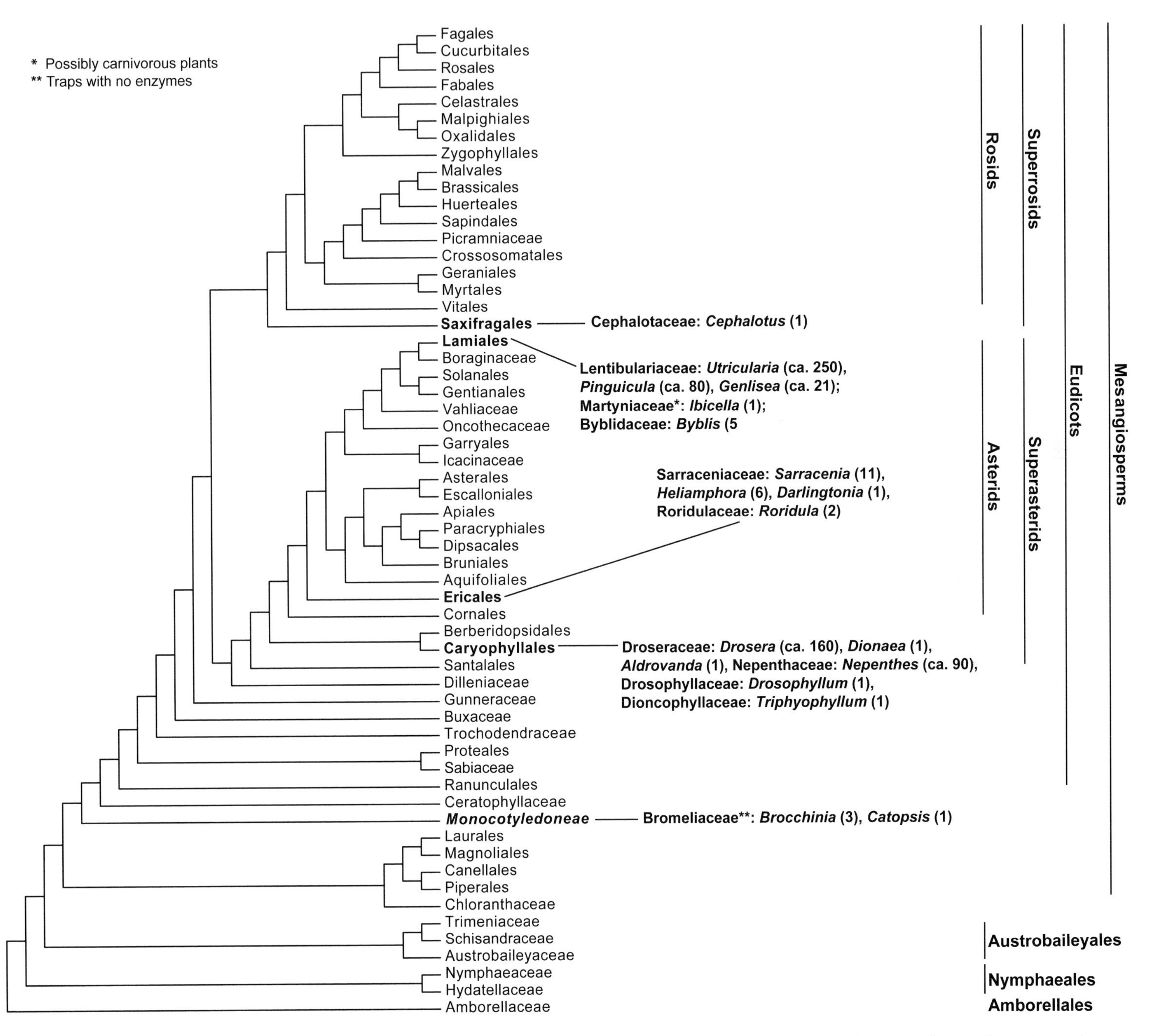

Figure 13.5. Occurrences of carnivory in the angiosperms. Clades containing carnivorous taxa are shown in bold. Monocots are indicated based on reports for Bromeliaceae (see text). The summary tree is based on Soltis et al. (2011).

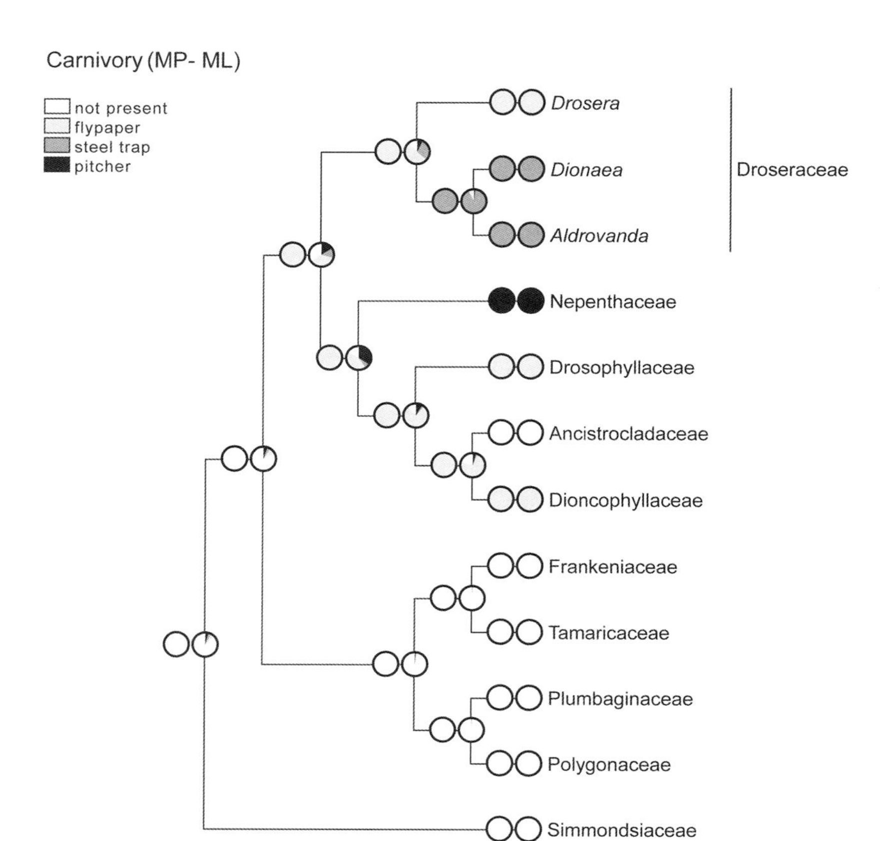

Figure 13.6. Reconstruction of the evolution of carnivory in Caryophyllales with different trap types considered as distinct character states. MP = ancestral state reconstruction using maximum parsimony; ML = ancestral state reconstruction using maximum likelihood.

Ericales: Carnivory (ML, MP)

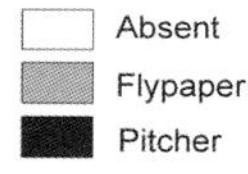

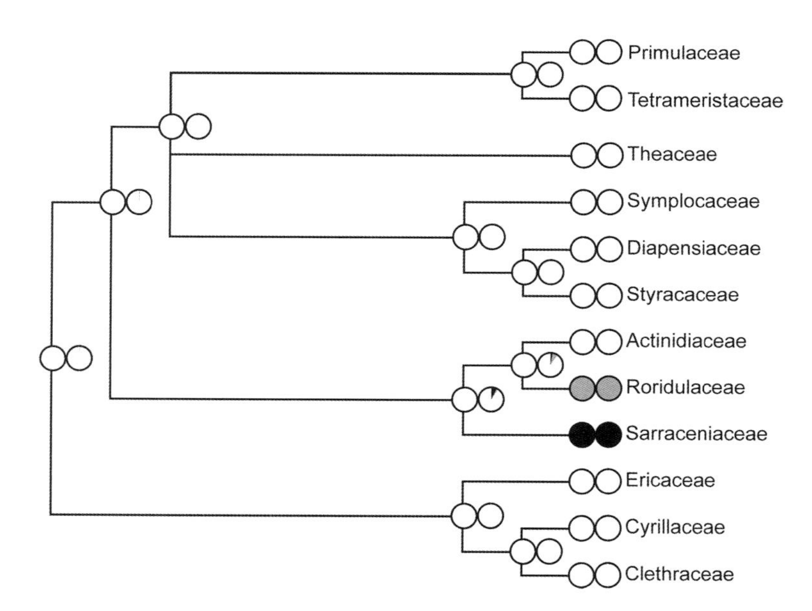

Figure 13.7. Reconstruction of the evolution of carnivory in Ericales with different trap types considered as distinct character states. MP = ancestral state reconstruction using maximum parsimony; ML = ancestral state reconstruction using maximum likelihood.

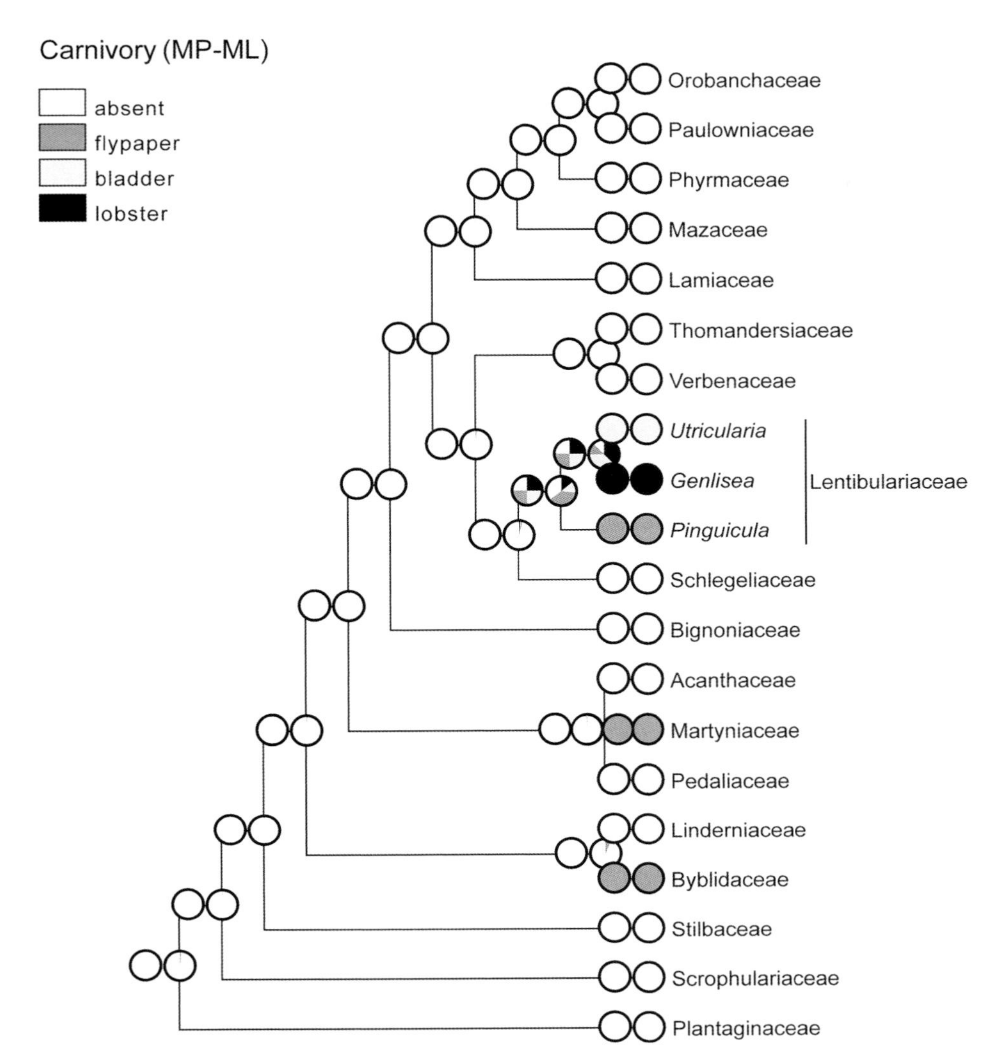

Figure 13.8. Reconstruction of the evolution of carnivory in Lamiales with different trap types considered as distinct character states. MP = ancestral state reconstruction using maximum parsimony; ML = ancestral state reconstruction using maximum likelihood.

of *Byblis*) and dimorphic stalked and sessile glands. In both genera, the stalked glands secrete mucilage that coats prey, whereas the sessile glands secrete digestive enzymes. However, *Byblis* differs from *Drosophyllum* in that its mucilage-secreting glands have unicellular stalks that protrude from reservoir cells of the epidermis, whereas the mucilage-secreting cells of *Drosophyllum* are multicellular and vascularized with tracheids and phloem (Fig. 13.9); glands of *Drosophyllum* are similar to those of *Triphyophyllum* (Dioncophyllaceae), to which it is closely related (Figs. 13.6, 13.9). The mucilage-secreting glands of both *Drosophyllum* and *Triphyophyllum* are also similar to those of *Drosera* (Droseraceae), also in Caryophyllales (Figs. 13.6, 13.9). There are other differences between the mucilage-secreting glands of these members of Caryophyllales and those of *Byblis*. The sessile digestive glands of *Byblis* are simple with four to eight head cells, whereas those of *Drosophyllum*, *Triphyophyllum*, and *Drosera* have multicellular stalks (Fig. 13.9). Glands of *Byblis* are more similar, structurally, to those of *Pinguicula* (Lentibulariaceae; Fig. 13.9). Therefore, the flypaper trapping mechanism differs across the angiosperms, reflecting its phylogenetically independent origins, despite remarkable functional similarities (Albert et al. 1992).

In the carnivore subclade of Caryophyllales (Figs. 13.4, 13.6), the flypaper trap may be the ancestral type (Fig. 13.6; see also Cameron et al. 2002). It is also noteworthy that *Ancistrocladus* has retained leaf glands similar to those of the related carnivorous families Droseraceae and Dioncophyllaceae, despite having lost the carnivorous condition. Cameron et al. (2002) provided strong support for a clade of *Aldrovanda* and *Dionaea*. *Aldrovanda* (Fig. 13.6), a little-known carnivore from several continents, is a rootless, submerged aquatic herb. Each leaf resembles a leaf of *Dionaea* (Fig. 13.4), having two lobes, with each lobe possessing roughly 20 trigger hairs. Darwin (1875), in fact,

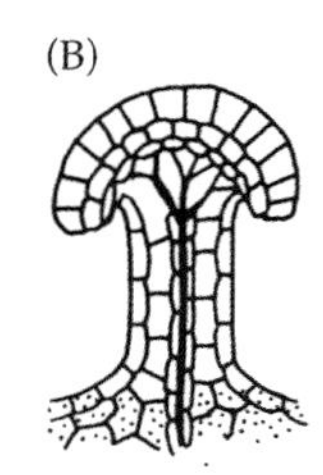

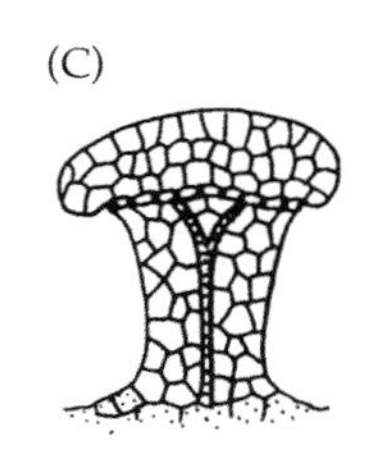

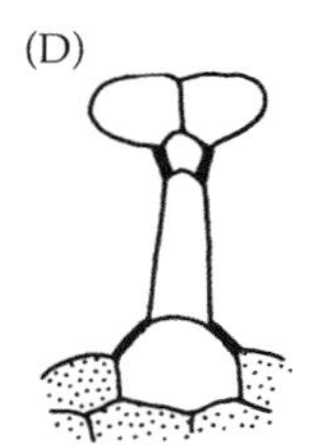

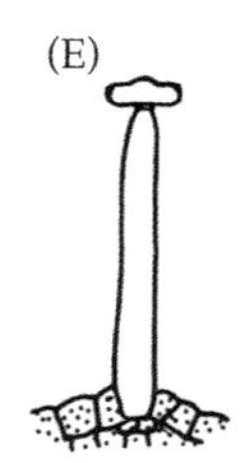

Figure 13.9. Glandular hairs in genera having flypaper traps (a–f) in members of Caryophyllales. a. *Drosera* (Droseraceae; Caryophyllales). b. *Drosophyllum* (Drosophyllaceae; Caryophyllales). c. *Triphyophyllum* (Dioncophyllaceae; Caryophyllales). d. *Pinguicula* (Lentibulariaceae; Lamiales). e. *Byblis* (Byblidaceae; Lamiales). f. *Proboscidea* (Martyniaceae; Lamiales). From Soltis et al. (2005).

first noted the similarity of the leaves of *Aldrovanda* and *Dionaea*. *Aldrovanda* was later shown to have a rapidly closing snap trap mechanism, somewhat similar to that of *Dionaea*. Cameron et al.'s (2002) results indicated a single origin of this snap trap mechanism.

Topologies for the carnivorous clade of Caryophyllales (Cuénoud et al. 2000; Meimberg et al. 2000; Cameron et al. 2002; Drysdale et al. 2007; Brockington et al. 2009) placed *Nepenthes* as sister to a clade of *Drosophyllum* + (Dioncophyllaceae + *Ancistrocladus*). Reconstructions (Fig. 13.6) indicate that the pitcher trap of *Nepenthes* was derived from the flypaper trap type in this clade. The pitcher traps of *Nepenthes* are produced at the tips of tendrils; this contrasts with the pitcher traps of other families (e.g., Sarraceniaceae, Cephalotaceae) in which the pitcher trap is an entire leaf (Fig. 13.4). Tendril formation is present throughout the subclade of carnivores to which *Nepenthes* belongs, with the possible exception of *Drosophyllum*, which has linear leaves. *Triphyophyllum*, which has a flypaper trap mechanism, has tendrils similar to those of *Nepenthes*, as does the non-carnivorous *Ancistrocladus* (Fig. 13.4). Hence, tendril formation may have been an ancestral trait for this subclade of carnivores.

Within Lentibulariaceae, three trap types have evolved. *Pinguicula* (butterworts; a flypaper trap) is sister to *Genlisea* (lobster pot) + *Utricularia* (bladder) (Jobson and Albert 2002; Jobson et al. 2003). Using the best current estimates of relationships and ML reconstruction, it is not clear that the flypaper trap is ancestral in Lentibulariaceae (Fig. 13.8); both the lobster pot and bladder traps are clearly derived, but their immediate ancestral state, as well as that of the family, is ambiguous. The fact that *Utricularia* contains several types of bladders adds to the structural and evolutionary complexity of Lentibulariaceae.

Jobson and Albert (2002) argued that the clade of *Utricularia* and *Genlisea* is substantially more species-rich and morphologically divergent than *Pinguicula*. They suggest that bladderworts have a flexible vegetative development that could be viewed as a key innovation (sensu Sanderson and Donoghue 1994; Givnish 1997; Hodges 1997a) that permits these plants to invade a broad range of nutrient-poor niches, ranging from water-saturated terrestrial to epiphytic and suspended aquatic. Jobson and Albert (2002) also found that bladderwort genomes evolve significantly faster across seven genes than do non-bladderwort genomes and proposed that increased cladogenesis in this instance is related to increased nucleotide substitution (see below and Ibarra-Laclette et al. 2013 for molecular evolution of the *Utricularia gibba* genome). Molecular investigations of Lentibulariaceae have also provided insights into character evolution, including habit (Jobson et al. 2004). The terrestrial habit is the most common for all three genera, whereas other species exhibit other habits, including affixed aquatic, suspended aquatic, and epiphytic. Jobson et al. (2004) found that the epiphytic habit evolved independently at least three times in the family (Fig. 13.4).

Some members of Lentibulariaceae have small genomes, and recently the complete genome of *Utricularia gibba* (only 82 megabases in size) was reported (Ibarra-Laclette et al. 2013). Despite its very small size, the genome of *U. gibba* has a gene number typical for an angiosperm, but differs from other plant genomes in having a drastic reduction in non-genic DNA. Although the genome is small, at least three rounds of whole-genome duplication (polyploidy; see Chapter 16) have occurred in *U. gibba* since its common ancestry with tomato (*Solanum*) and grape (*Vitis*). The architecture of the *U. gibba* genome indicates that only small amounts of intergenic DNA, and few or no active retrotransposons, are necessary in the genome of these plants for normal development and function.

C_4 PHOTOSYNTHESIS

Some chemical pathways seem to be restricted to only one or a few lineages and must therefore have originated only a few times. Betalain production, for example, is limited to the core Caryophyllales clade (Chapter 11). Similarly, glucosinolate production has evolved only twice, once in the Brassicales and again in Putranjivaceae (Malpighiales) (Chapter 10). Symbiosis with nodule-forming nitrogen-fixing bacteria appears to have evolved many times, but all within a single well-defined clade—the nitrogen-fixing clade of rosids—leading to the suggestion that the predisposition for nitrogen-fixing symbiosis evolved only once in the angiosperms (Soltis et al. 1995; see Chapter 10). In contrast, C_4 photosynthesis has evolved repeatedly in the angiosperms in distantly related groups, but often within a small number of species in a given group (Kellogg 1999; Sage et al. 1999, 2011). This is remarkable given that C_4 photosynthesis is a complex trait (see Hibberd and Covshoff 2010; Sage et al. 2011) that necessitates changes to several genes involved in multiple traits and processes, including stem and leaf anatomy as well as the transfer of metabolites and regulation by many enzymes.

The C_4 pathway is an adaptation to high light intensities, high temperatures, and a dry climate. The optimal temperature range for C_4 photosynthesis is much higher than that for C_3 photosynthesis. The C_4 photosynthetic pathway is well understood and developmentally and genetically complex (Edwards and Walker 1983; Sage et al. 2011). Typically, C_4 plants have a characteristic leaf anatomy (Kranz anatomy; Fig. 13.10) in which the leaves have an orderly arrangement of mesophyll cells around a layer of large bundle-sheath cells; together, these two cell types form two concentric layers or rings around the vascular bundle. Adding to the complexity are multiple types of Kranz anatomy (below).

Early attempts to explain the broad distribution of C_4 photosynthesis among the angiosperms involved hypotheses of repeated loss and lateral transfer. However, the most likely explanation is repeated evolution. Using the topologies for angiosperms available at that time, Kellogg (1999) made the first attempt to address the phylogenetic distribution of C_4 photosynthesis across the angiosperms and within particular clades, including asterids and Poaceae, in which many C_4 species occur. A reanalysis was then conducted by Soltis et al. (2005b) using better-supported topologies available at that time. More recently, Sage et al. (2011) have reconsidered the evolution of C_4, and we rely largely on those results here.

Whereas Kellogg's (1999) analyses indicated at least 31 origins of C_4 photosynthesis, including multiple origins within some large clades (monocots and asterids), Sage et al. (2011) reported 62 lineages of C_4 photosynthesis and stressed that with so many independent origins, C_4 should be considered "one of the most convergent of the complex evolutionary phenomena" (p. 1). Of these 62 origins, 36 (60%) are in eudicot lineages, and the other 26 (40%) are in monocots. However, there are no reports from basal angiosperm lineages, the magnoliids, or Chloranthales—as with carnivory (above), C_4 was clearly a later innovation in angiosperms, limited to monocots and eudicots.

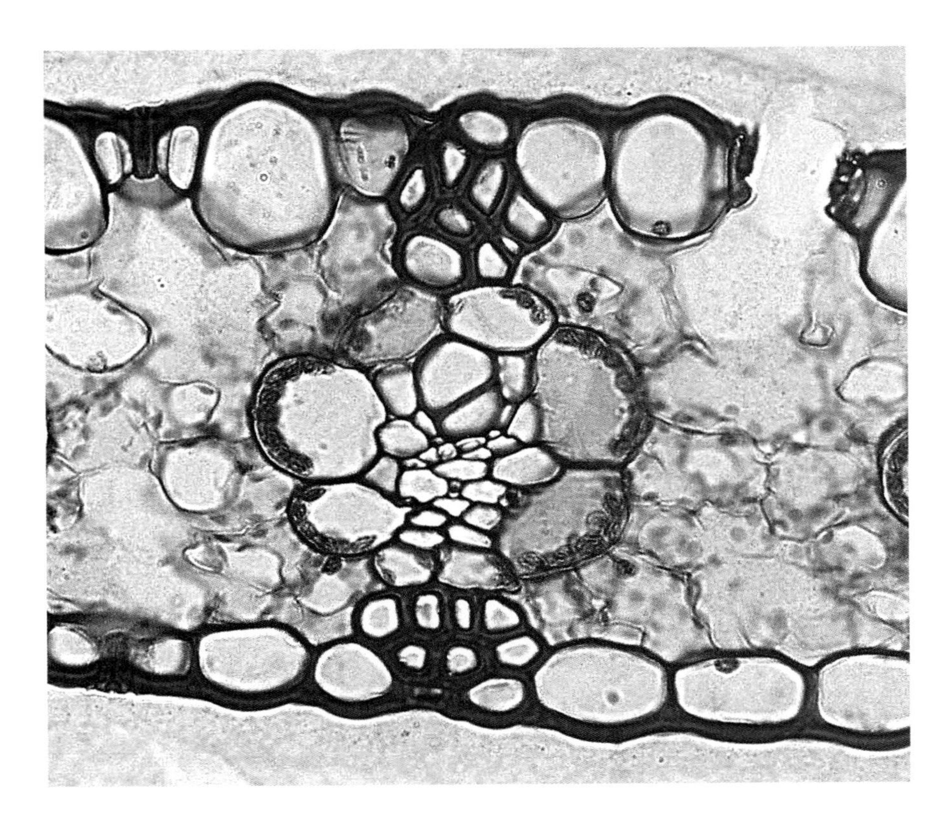

Figure 13.10. Stem cross section in *Zea mays* showing Kranz anatomy typical of C_4 plants.

C_4 has arisen in all major subclades of eudicots (Fig. 13.11). In the superrosid clade, it occurs in multiple rosid lineages (Brassicales-Cleomaceae; Malpighiales-Euphorbiaceae; Zygophyllales-Zygophyllaceae), but not within Saxifragales. Within superasterids, C_4 occurs in asterids (Boraginales-Boraginaceae; Lamiales-Acanthaceae, Scrophulariaceae; Asterales-Asteraceae), as well as Caryophyllales (Amaranthaceae including the formerly recognized Chenopodiaceae, Gisekiaceae, Molluginaceae, Nyctaginaceae, Polygonaceae, Portulacaceae; see below), but not in Santalales. In monocots, C_4 is limited to Alismatales (Hydrocharitaceae) and Poales (Cyperaceae, Poaceae). Thus, despite wide evolutionary occurrence, there are clear C_4 hotspots. Although C_4 is scattered across eudicots (Fig. 13.11), one clear eudicot evolutionary hot spot is Caryophyllales.

In addition to the many origins of C_4 photosynthesis on a broad scale, Kellogg (1999) and more recently Sage et al. (2011) also showed that multiple origins were frequent within several families, such as Asteraceae (four origins),

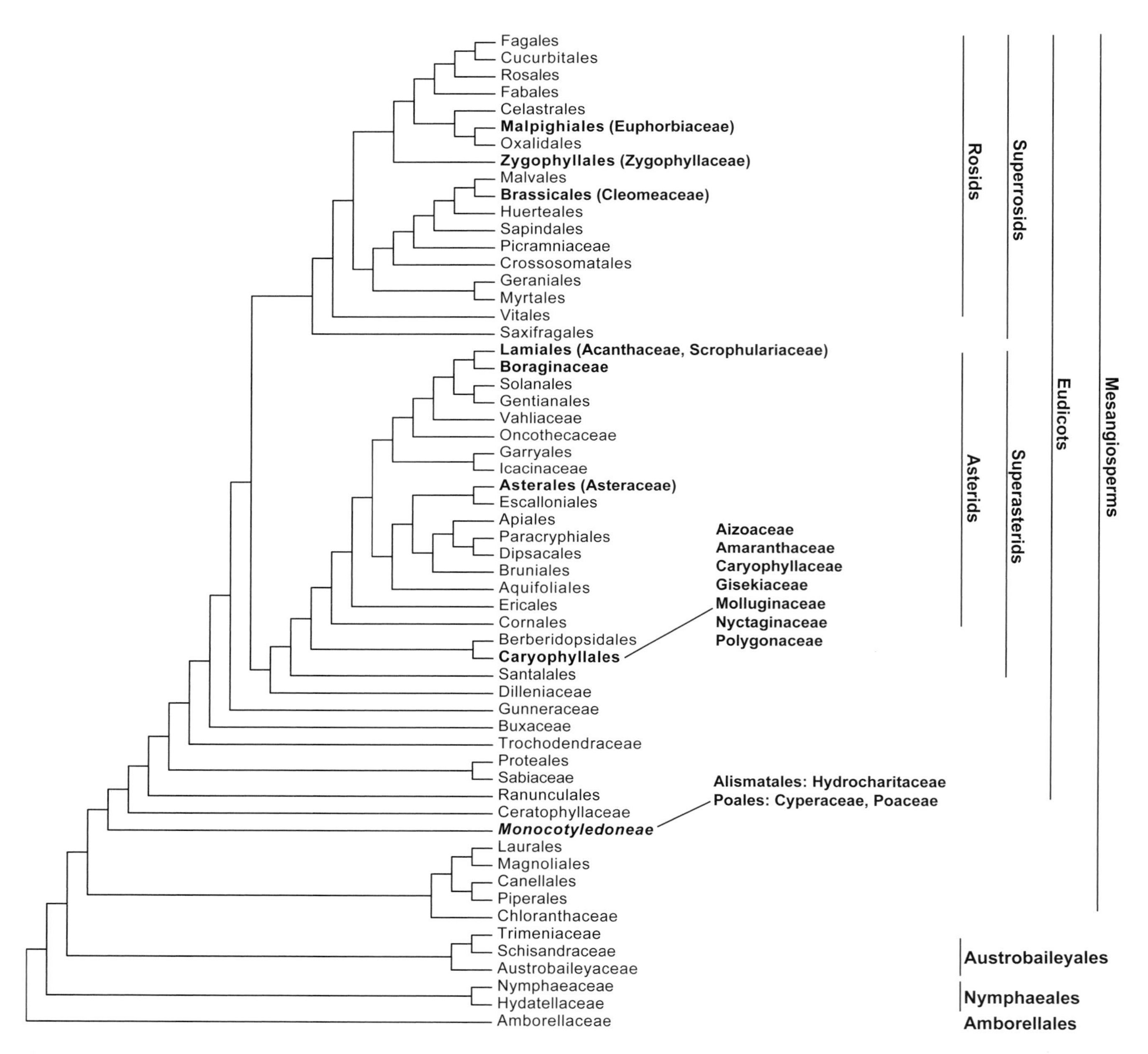

Figure 13.11. Major occurrences of C_4 photosynthesis in the angiosperms. Lineages with some members with C_4 photosynthesis are indicated in bold. The summary tree is based on Soltis et al. (2011).

Zygophyllaceae (two), Cleomaceae (three), Poaceae (18), and Cyperaceae (six). Hydrocharitaceae are noteworthy in that C_4 in both *Hydrilla* and *Egeria* clearly differs from that of all other C_4 lineages because in these two aquatic genera, rather than above-ground photosynthetic leaves and stems, leaves are submersed. These two genera also exhibit cellular differences from all other C_4 plants (Bowes 2011; Sage et al. 2011).

CARYOPHYLLALES

There is widespread variation in photosynthetic systems in Caryophyllales, and the pattern of origins and losses of C_4 photosynthesis in Caryophyllales has long been recognized as complex and highly labile (e.g., Pyankov et al. 2001). Sage et al. (2007, 2011) estimated 23 origins in this clade. Origins occur in multiple families: Aizoaceae, Caryophyllaceae, Amaranthaceae, Gisekiaceae, Molluginaceae, Nyctaginaceae, Polygonaceae, and Portulacaceae. In some of these families, there have been multiple origins. Amaranthaceae (including the former Chenopodiaceae) are particularly complex (represented here in Fig. 13.12).

POACEAE

Several possible origins of C_4 photosynthesis are indicated for Poaceae. The Grass Phylogeny Working Group II (GPWG II 2012) has provided not only a robust framework phylogenetic tree for Poaceae but also reassessed the number of origins of C_4 photosynthesis (see also Sage et al. 2011). Our review is a summary of the findings of GPWG II (2012) and Sage et al. (2011). The number of

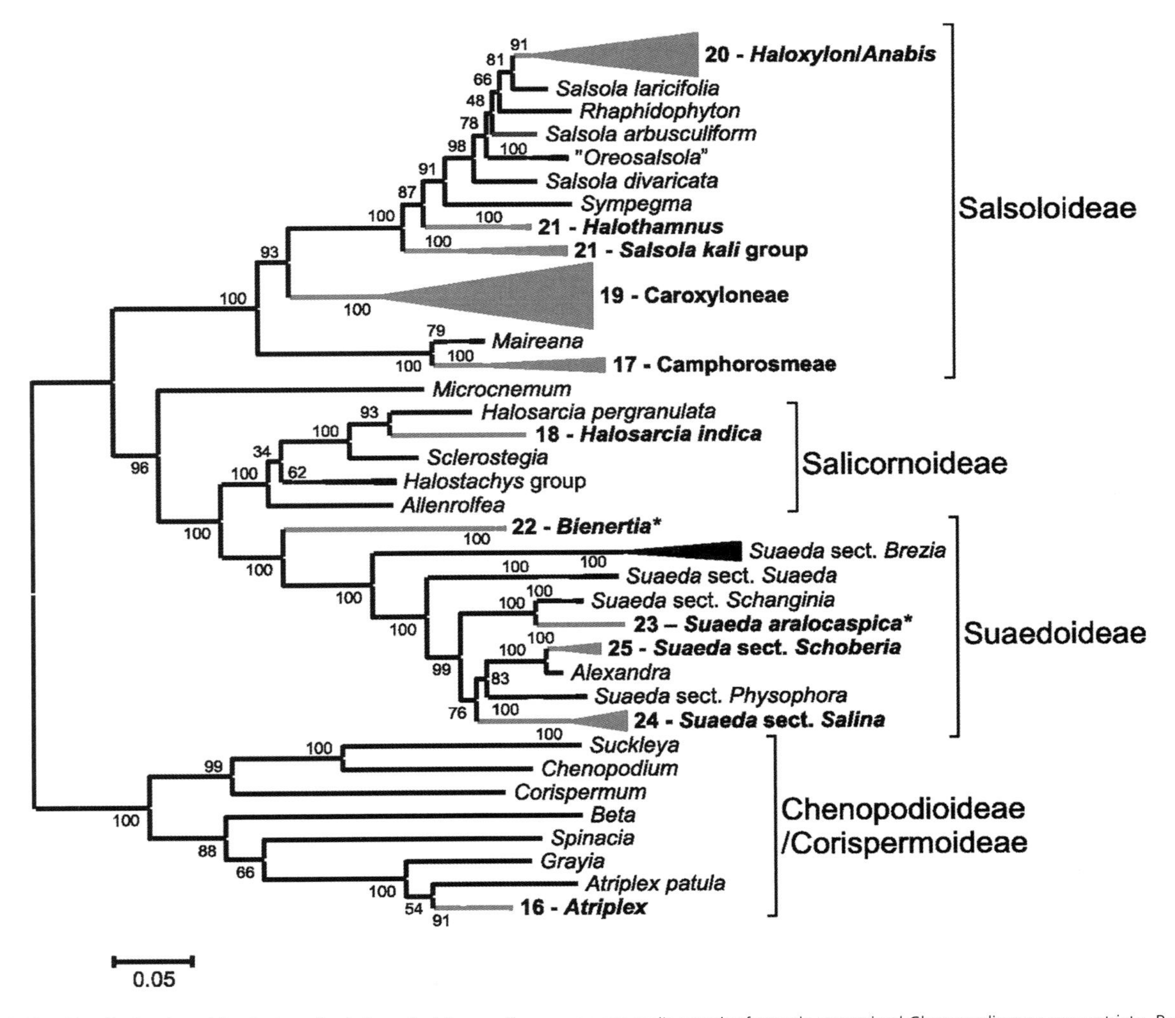

Figure 13.12. The distribution of C_4 photosynthesis in part of Amaranthaceae corresponding to the formerly recognized Chenopodiaceae sensu stricto. Bayesian support values are given near branches. Clades that have a single photosynthetic type are indicated as follows: gray = C_4; *Salsola arbusculiform* = C_3-C_4; black = C_3. Names of C_4 clades are in bold, and the numbers given beside C_4 clades refer to a lineage numbers given in Table 1 of Sage et al. (2011; table not shown here). Figure modified from Sage et al. (2011).

C_4 grasses is huge (~4600), but all examples occur in the PACMAD clade (see Chapter 7). GPWG II (2012) inferred 22–24 origins of the C_4 pathway and only one potential reversal—the different estimates were a result of the use of different underlying trees. The origins reported by the GPWG II (2012) conform to those identified in earlier analyses (e.g., Christin et al. 2009). As the GPWG II (2012) noted, these results indicate that the relationships of most C_4 grass clades are now well understood. These origins are shown in Fig. 13.13 and listed in Table 13.5.

Uncertainties in relationships and evolution of C_4 in grasses based on GPWG II (2012) are focused on two places. The first is in Boivinellinae (Panicoideae), which includes two C_4 genera, *Alloteropsis* and *Echinochloa*. These are not sisters, so they are likely distinct origins, but their closest C_3 relatives are unclear. A potential reversal from C_4 to C_3 occurs in this part of the tree, in *Alloteropsis*. The second area of uncertainty involves the Arthropogoninae clade (Panicoideae). These uncertainties result in the number of inferred origins in each group ranging from two to five (GPWG II 2012).

These groups also illustrate how it can be difficult to determine an unambiguous number of origins of C_4 photosynthesis in some lineages. In fact, attempts to do so may be misleading in that even a well-resolved phylogenetic tree does not overcome the problem of taking into account past photosynthetic transitions (GPWG II 2012). An important point stressed by GPWG II (2012) is that some of the inferred separate "origins" may not actually be independent. Certain key "precursor traits" likely evolved early in some grass lineage (or lineages), which in turn increased the possibility of evolving the C_4 phenotype. That is, various elements of the C_4 phenotype in some "independent" lineages were probably inherited from their common ancestor. This pattern of extended, parallel evolution of the C_4 pathway has been demonstrated in several eudicot groups (e.g., McKown et al. 2005; Christin et al. 2011) and is similar to the evolution of the capacity for symbiosis with nitrogen-fixing bacteria in rosids (Chapter 10).

Important insights into C_4 evolution have recently been obtained via the use of a densely sampled, large phylogenetic tree. Edwards and Smith (2010) used independent contrasts in a study of C_3/C_4 grasses and habitat features (also noted in Chapter 3; see Fig. 3.14). C_4 grasses dominate tropical and subtropical grasslands and savannas, whereas C_3 grasses typify cooler temperate grassland regions. This pattern has been attributed to C_4 physiology, and it has long been maintained that the evolution of the pathway enabled C_4 grasses to persist in warmer environments. Edwards and Smith (2010) found, however, that grasses are ancestrally a warm-adapted clade and that C_4 evolution actually was not correlated with shifts between temperate and tropical biomes (Fig. 3.14). Instead, based on an independent contrast analysis, 18 of 20 inferred C_4 origins were correlated with marked reductions in mean annual precipitation. Hence, precipitation may be the crucial underlying causal factor involved in the evolution of C_4.

Table 13.5. C_4 lineages identified by GPWG 2011, with their recommendations for a clear C_3 sister group that could be used for comparative studies of closely related C_3/C_4 species pairs

	C_4 lineage	**Clear C_3 sister for comparative work?**
1	Aristida	*Aristida longifolia*
2	***Stipagrostis***	***Sartidia***
3	**Chloridoideae**	**No, and not likely**
4	***Centropodia***	***Ellisochloa rangei***
5	***Eriachne***	***Isachne***
6	**Tristachyideae**	***Centotheceae/ Thysanolaeneae***
7	**Andropogoneae**	**No, and not likely**
8	***Reynaudia***	**No, and not likely**
9	*Axonopus*	*Streptostachys asperifolia*
10	*Paspalum*	No, and not likely
11	***Anthaenantia***	**Otachyriinae p.p.**
12	***Steinchisma***	***Steinchisma laxa***
13	*Arthropogon*	Not yet, but likely
14	*Mesosetum*	*Homolepis*
15	*Oncorachis*	Not yet, but likely
16	*Coleataenia 1*	Not yet, but likely
17	*Coleataenia 2*	*Triscenia*
18	*Digitaria*	No, and not likely
19	*Echinochloa*	Not yet, but likely
20	***Paraneurachne***	***Neurachne***
21	**MPC**	***Homopholis***
22–24	*Alloteropsis*	*Alloteropsis ecklonianа**

Bold indicates high confidence in that particular origin (both that it is an independent origin and that it is correctly placed; regular text indicates uncertainty in either or both). MPC = Melinidinae + Panicinae + Cenchrinae.
*Alternatively, this may represent a reversal to C_3 photosynthesis. From Grass Phylogeny Working Group (2012); see text for details.

DIFFERENT PATHWAYS

The C_4 pathway is not identical in all angiosperm clades. For example, C_4 plants differ in their decarboxylating enzymes

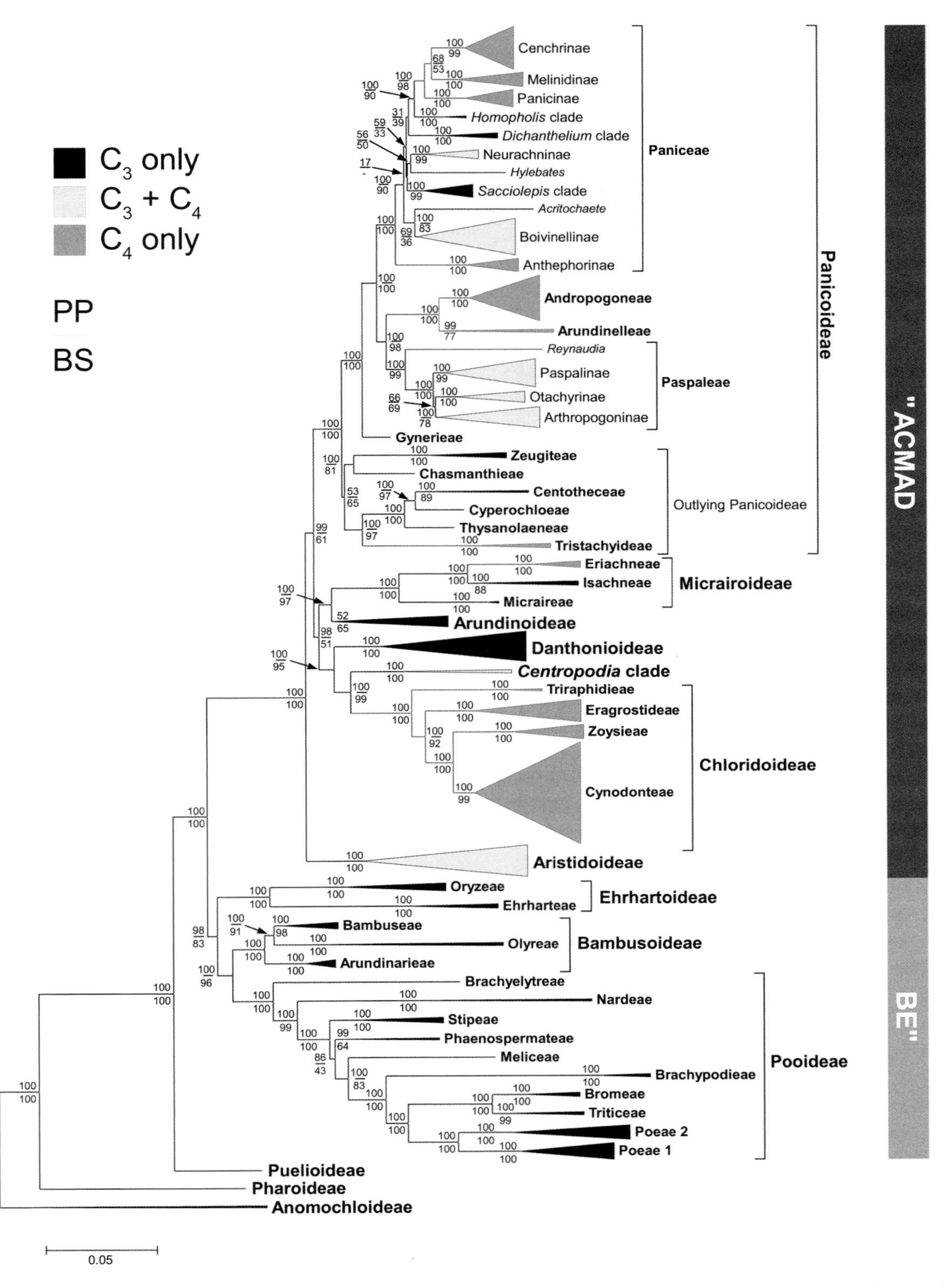

Figure 13.13. The distribution of C_4 photosynthesis in Poaceae on a summary tree of the family (see text). Numbers above branches are Bayesian posterior probability (PP) values; numbers below branches are ML bootstrap values. Shades designate clade composition in terms of photosynthetic type as shown in key. Supplementary Figure 2 from GPWG II (2012).

(Hatch et al. 1975; Sage et al. 2011), using NAD-malic enzyme (NAD-ME), NADP-malic enzyme (NADP-ME), or phosphoenolpyruvate carboxykinase (PCK) as the primary decarboxylase. NADP-ME is used by 43 of 62 clades, and NAD-ME is used by 20 (mostly eudicots); only grasses use PCK.

Kellogg (1999) and coworkers (Sinha and Kellogg 1996; Giussani et al. 2001) used Poaceae as an example of the variability in the C_4 pathway within a single clade. The NAD-malic enzyme (ME) type is present in *Centropodia* and Chloridoideae. *Eriachne* exhibits the NADP-ME subtype. Considering the Panicoideae, both the PCK and NAD-ME subtypes of C_4 photosynthesis have each evolved only once; the other origins of C_4 photosynthesis are NADP-ME (Giussani et al. 2001).

As another example of chemical variability, the enzymes involved in the evolution of C_4 photosynthesis within Caryophyllales are complex. Phylogenetic analysis of tribe Salsoleae (Amaranthaceae) revealed the complexities of this family. The molecular results provide strong support for the origin and evolution of two main groups, one having NAD-ME and the other NADP-ME C_4 photosynthesis (Pyankov et al. 2001). These NAD-ME and NADP-ME clades have not only different photosynthetic types but also different structural and photosynthetic characteristics in cotyledons. The topology also reveals two independent reversions to C_3 photosynthesis and loss of Kranz anatomy (Pyankov et al. 2001; see also Kadereit et al. 2003).

ANATOMY

The characteristic Kranz anatomy mostly associated with C_4 photosynthesis displays various patterns, as might be expected by its diverse phylogenetic occurrences. Sage et al. (2011) found extreme variability, including 22 Kranz anatomy cell types with additional variation in some of these cell types. Among eudicots, the most common type is termed atriplicoid, present in at least 20 of the 36 eudicot lineages that have C_4. There is also extensive variation in anatomy within monocots. In Cyperaceae and Poaceae, most clades have evolved a unique type of Kranz anatomy; the classic type that is often depicted represents seven lineages of grasses.

Thus, the C_4 pathway is not a single pathway, but several pathways that share phosphoenolpyruvate carboxylase (PEP carboxylase or PEPC) as a carbon acceptor. The use of PEPC as a carbon acceptor is common, and the activation of a PEPC pathway for C_4 photosynthesis may be relatively easy. The required anatomical changes may be more difficult to achieve, and this may be more of a limiting factor, and ultimately the causal agent, in the scattered and sporadic distribution of C_4 photosynthesis among flowering plants (Kellogg 1999). The absence of Kranz anatomy in some members of Caryophyllales (Pyankov et al. 2001; Voznesenskaya et al. 2001) demonstrates, however, that biochemical and anatomical components of C_4 photosynthesis can be decoupled (Kadereit et al. 2014).

INTERMEDIATES

Some angiosperms have features that are considered intermediate between the traits of C_3 and C_4 species, but these types may in fact be evolutionarily stable, final stages, not evolutionary intermediates between C_3 and C_4, as previously suggested (Monson et al. 1984; reviewed in Sage et al. 2011). At this point, however, the term C_3-C_4 intermediate is mainly used to refer to those plants possessing a photorespiratory CO_2-concentrating mechanism, in which expression of the photorespiratory enzyme glycine decarboxylase (GDC) is localized to bundle sheath cells (BSCs) (Monson 1999; Bauwe 2011; Sage et al. 2011). This localization of the enzyme GDC to the bundle sheath forces all glycine produced by photorespiration to move into the BSCs to complete the photorespiratory cycle, and the end result is enhanced photosynthetic efficiency (see Sage et al. 2011 for a review). In addition, C_3–C_4 intermediate species have evolved C_4-like traits, including close vein spacing and enlarged BSCs (Sage 2004). These evolutionary changes may indeed increase the chances of subsequent C_4 evolution (Bauwe 2011), but recent authors stress that they are also advantageous on their own. Many C_3–C_4 lineages are distinct from C_4 clades. In fact, numerous separate (21) origins of C_3–C_4 intermediate photosynthesis have now been reported (Sage et al. 2011). The first C_3–C_4 intermediate described was *Mollugo verticillata*, followed by the report of intermediate species in diverse eudicots and monocots (reviewed in Sage et al. 2011). Most intermediates are in genera that also include C_4 plants. However, roughly 25% of all intermediates occur in taxa that are not closely related to a C_4 lineage. For example, there are two occurrences of C_3–C_4 intermediates in Brassicaceae, in which no C_4 plants are known.

Flaveria is one of several genera that contains some species that are C_3 and others that are C_4, as well as several species that are considered C_3–C_4 intermediates. Based on phylogenetic results coupled with other evidence, Kopriva et al. (1996) favored a single origin of C_3–C_4 intermediate photosynthesis, with a subsequent origin of C_4, but subsequent work has changed this view. *Flaveria* has nine C_3–C_4 species. Using a phylogenetic approach, McKown et al.

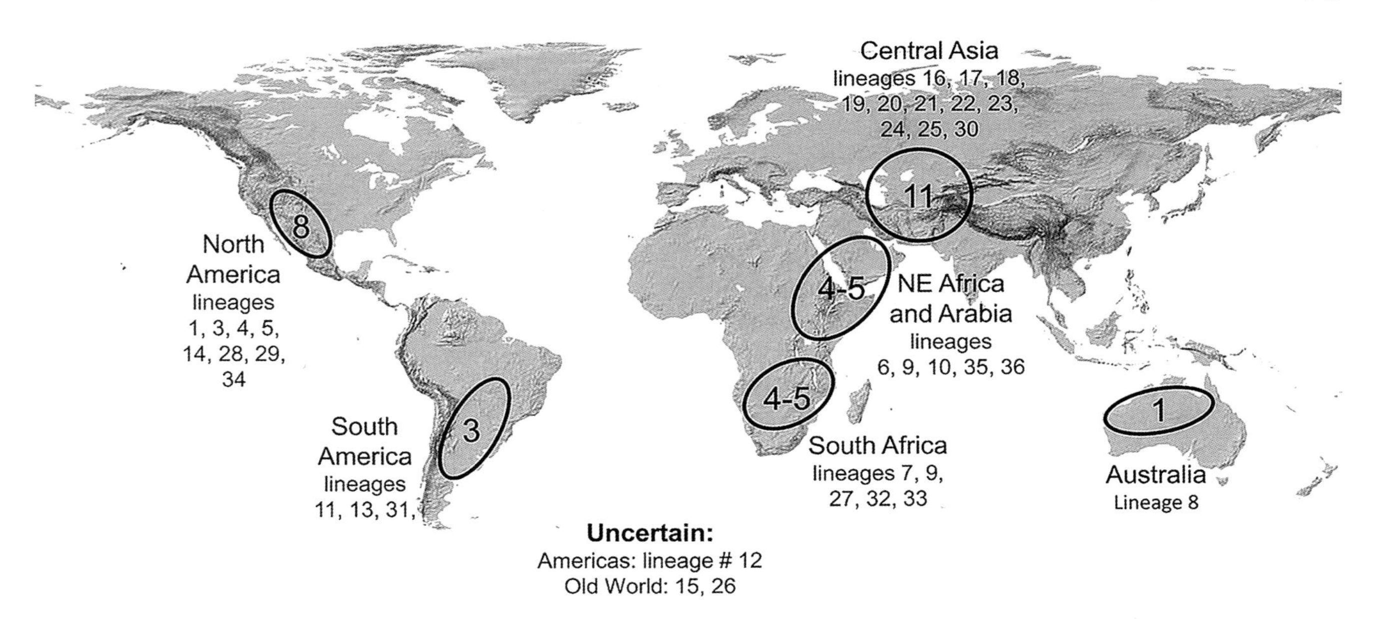

Figure 13.14. Geographic centers of origin of 35 of the 36 eudicot lineages examined by Sage et al. (2011). Numbers in circles are the numbers of lineages in a given region; lineage numbers are from Sage et al. (2011). Figure modified from Sage et al. (2011).

(2005) showed there are two subclades (A and B) and that C_3–C_4 intermediacy has evolved twice in the genus (once in each subclade) and is evolutionarily intermediate between C_3 and C_4 in clade A, but not in clade B. C_4-like photosynthesis is also derived once independently in each of these subclades from C_3 ancestors with high vein density (McKown and Dengler 2007). These examples show the difficultly in pinpointing the true "origin" of these pathways because several distinct lineages have independently built upon a shared ancestral set of what are considered key facilitating traits.

CENTERS OF ORIGIN

Sage et al. (2011) also estimated the geographic center of origin for 47 lineages of C_4 plants. The centers of origin for eudicot lineages were clustered in arid and semi-arid regions of southwestern North America, south-central South America, Central Asia, northeastern and southern Africa, and interior Australia (Fig. 13.14). However, the ancestral distributions of the monocot taxa with C_4 photosynthesis are more problematic to reconstruct. Nonetheless, the results suggested South America was a major center for origins of C_4 photosynthesis in monocots.

14 Floral Diversification

INTRODUCTION

In the past ten years, research on flower morphology and development, and the genes underlying floral structures, has progressed on a broad front. Comparative morphological studies of large clades have been carried out, in some cases in close parallel with phylogenetic studies that led to finer resolution of relationships, enabling new studies of flower evolution, a process of reciprocal illumination. Examples are analyses of Malpighiales (e.g., Matthews and Endress 2008, 2011, 2013; Matthews et al. 2012; Xi et al. 2012b; Endress et al. 2013), Ericales (Schönenberger et al. 2005, 2010; Schönenberger 2009; von Balthazar and Schönenberger 2013), and Fabales (Prenner 2004a; Bello et al. 2007, 2009, 2010). Prominent structural features, such as ovules (Endress 2011a), floral monosymmetry and asymmetry (Endress 2012), and multicarpellate gynoecia (Endress 2014), have been studied across all angiosperms, and insight has been gained into structural interrelationships of these traits. Moreover, evo-devo studies have been conducted on clades that are only distantly related to model species (e.g., Boyden et al. 2012; Yockteng et al. 2013). Fossil and extant flowers have been internally reconstructed with new techniques, especially new variants of tomography (Schönenberger 2005; Friis et al. 2009, 2011, 2013, 2014; Gamisch et al. 2013; Staedler et al. 2013; Nikolov et al. 2014), allowing greater insight into fossils of flowers that were previously known only from surface views (e.g., the Early Cretaceous water lily *Monetianthus*, Friis et al. 2009) and into large-flowered extant plants that cannot be studied in toto with microtome series (e.g., Rafflesiaceae, Nikolov et al. 2014). Evolutionary changes in flowers of closely related species and in the context of pollinator shifts have also received recent attention (e.g., Barrett 2008, 2013; Crane et al. 2010).

Flowers are exclusive to angiosperms. Although the flowers of all angiosperms are homologous in a general sense, floral organization exhibits different traits when the basal angiosperms (ANA grade and magnoliids; see Chapters 4 and 5) are compared with highly apomorphic groups, such as orchids or some asterids. These differences are not single characters, but entire sets of traits make up the evolutionary "behavior" of flowers. Such major evolutionary steps are often referred to as "key innovations" (e.g., Müller and Wagner 1991; Endress 2001b; Wagner 2001; Donoghue 2005; Ree 2005; Moore and Donoghue 2009; Vamosi and Vamosi 2010). Such innovations have been hypothesized to enable new radiations; consequently, the presence of an enhanced number of new forms is more likely to produce additional key innovations, which, in turn, can be seen as enhanced evolvability. Here we provide an overview of some of the "key innovations" in floral diversification.

In ANA grade and magnoliid angiosperms, flowers are flexible in organ number and organ phyllotaxis, and thus a wide range of organ numbers is possible, from one to thousands, in flowers with spiral or whorled or irregular organ position. In the course of angiosperm evolution, key innovations have involved the following changes: (1) synorganization (fusion or otherwise close association) of organs of the same kind and (2) synorganization between organs of different kinds (Endress 1990a, 1994a, 2001b). Preconditions for or consequences of this synorganization were restrictions to whorled phyllotaxis and to fixed and low numbers of organs. Examples of synorganization are syncarpy, sympetaly, and adnation of petals and stamens and of stamens and carpels. The most complex flowers, such as in Orchidaceae and Apocynaceae, are characterized by many and highly elaborated kinds of synorganization and by a uniform number of organs. Other key innovations include the differentiation of sepals and petals and the transition from crassinucellar (and bitegmic) to tenuinucellar (and unitegmic) ovules.

Tepals (or sepals and petals), stamens, and carpels are the structural elements, or modular structures, that compose a flower. Their number and arrangement can vary widely. However, if a flower is bisexual, a gynoecium is always in the center of the flower surrounded by an androecium. Thus, this sequence is part of the fixed floral organization (see next paragraph). Only one exception is known in which an androecium is regularly surrounded by a gynoecium in bisexual flowers: *Lacandonia* (Triuridaceae) (Márquez-Guzmán et al. 1989). However, the carpel whorl is so distorted in *Lacandonia* that it becomes difficult to determine what is morphologically (not simply topographically) inside versus outside (Endress 2014). Flowers with a whorl each of sepals, petals, stamens, and carpels have been used as the basis for the ABC model of flower development in molecular developmental genetics (Coen and Meyerowitz 1991; see below). In unisexual flowers, the stamens or the carpels are reduced to various degrees; they are either suppressed (initiated but not differentiated) or lost (not noticeably initiated) (cf. Tucker 1988; Diggle et al. 2011).

The crown group of angiosperms has existed for at least 135 million years (see Chapters 1–3). Flowers similar in structure to those of the basalmost extant clades have existed at least since that time. Since then, many evolutionary changes and radiations have taken place. Some floral traits have undergone many changes, whereas others have remained nearly static.

Thus, it is practical to distinguish different levels of structure, such as organization (bauplan), construction (architecture), and mode (style), if used in a general way (cf. Vogel 1954, 2012; Endress 1994a). Organization (bauplan) refers to features that are conservative (e.g., presence and pattern of disposition of floral organs: sepals, petals, stamens, carpels); construction (architecture) refers to shape of flowers (e.g., bowl-shaped flowers, tubular flowers, salverform flowers, lip flowers, revolver flowers); mode (style) refers to adaptation of flowers to specific pollinators (e.g., bird-pollinated flowers, moth-pollinated flowers, wind-pollinated flowers; colors and scents are also attributes of "mode"). The levels of organization, construction, and mode are in general increasingly unstable in evolutionary terms—that is, mode varies at shallower phylogenetic levels than organization.

That these concepts should be applied only in a loose way is shown by the fact that, for instance, a feature corresponding to organization may suddenly become labile in a group, as expected for an evolving group of organisms. An example is found in Brassicaceae, which consistently have flowers with four petals and six stamens. However, in *Lepidium* this combination has become unstable, and reductions to flowers with four or two stamens and without petals have occurred several times (Endress 1992a; Bowman et al. 1999; Karoly and Conner 2000; Mummenhoff et al. 2001).

PATHWAYS OF MORPHOLOGICAL EVOLUTIONARY INNOVATIONS

Innovations in floral evolution may be achieved through several pathways:

(1) Use of existing structures for new functions. An example is the odd staminode in flowers of Lamiales. Concomitant with the evolution of monosymmetry, the upper median stamen has been reduced to a staminode or has been lost. In a few genera, such as *Penstemon* (Plantaginaceae) and *Jacaranda* (Bignoniaceae), this staminode became secondarily enlarged and functions as a nectar guide for pollinating bees (Endress 1994a; Walker-Larsen and Harder 2000, 2001), and in *Jacaranda*, in addition, the staminode has secretory hairs that produce scent (Guimarães et al. 2008). This kind of evolutionary innovation was called "diverted development" by Crane and Kenrick (1997a).

(2) Synorganization of organs of the same kind and/or organs of different kinds. An example is the formation of pollinia with transport organs (translators) in Orchidaceae and Apocynaceae, a precondition of which was the synorganization of androecium and gynoecium by congenital fusion in orchids (gynostemium, or column) or postgenital fusion in Apocynaceae (gynostegium) (Vogel 1959, 1969; Endress 1990a, 1994a, 1997b, 2011c, 2016; Rudall and Bateman 2002).

(3) Ectopic gene expression. A putative example is the proposed evolutionary origin of carpels by ectopic development of ovules on microsporophylls in a Mesozoic group of gymnosperms, as proposed by Frohlich and Parker (2000) (see Chapter 1). Another example is the expression of floral genes in the bracts of *Cornus* (Zhang et al. 2008).

EVOLUTIONARY RADIATIONS AND PARALLEL EVOLUTION

Adaptive radiations are spectacular evolutionary events. Within angiosperms, many examples of adaptive floral radiations were triggered by different groups of pollinators, which have been studied in some detail (review in van der Niet and Johnson 2012). Two classic examples are reported by Vogel (1954, 2012) on various angiosperm genera in South Africa (e.g., *Bauhinia*, *Clerodendrum*, *Erica*, *Pelargonium*) and by Grant and Grant (1965) on Polemoniaceae, both on floral adaptation to different pollinators and accompanying structural, visual, and olfactorial diversification. Bees, butterflies, sphingids, flies, birds, and bats are involved in the rich diversification of many of these genera. More recently, it has become possible to base such studies on a phylogenetic framework, derived from molecular studies and allowing more rigorous reconstruction of evolutionary changes. A basic general work on this subject is the book by Givnish and Sytsma (1997), in which Givnish (1997) provided a general introduction to the field, and a number of authors discuss case studies, among them several on flowers.

The general lesson for flower evolution derived from such studies is that many structural (and other) changes are apparently simple whereas others are more complex. Such features may greatly differ in detail from group to group. However, some previously overlooked patterns may later emerge from a comparative survey of many groups.

In the past ten years, new studies combined with more detailed phylogenetic analyses and improved phylogenetic resolution have greatly refined this general insight. Iochrominae, a subtribe of Solanaceae with a high diversity of floral forms (Cocucci 1999) and pollination systems (encompassing hummingbirds, hymenopterans, lepidopterans and dipterans; Smith et al. 2008a,b), traditionally contained six genera (*Acnistus*, *Dunalia*, *Eriolarynx*, *Iochroma*, *Saracha*, and *Vassobia*) (Fig. 14.1). In a phylogenetic study using three nuclear loci, Smith and Baum (2006) found that only *Vassobia* is monophyletic. The other five genera were earlier mistaken as clades because of parallel evolution in floral forms. In the exceptionally diverse genus *Ruellia* (Acanthaceae), Tripp and Manos (2008), studying the degree of evolutionary transitions between different pollination systems (hummingbird-, bat-, bee-, and moth-adapted flowers) (Fig. 14.2), showed that evolutionary transitions occurred between all four pollination systems and in both directions, but in different frequencies. The directions hummingbird to bee, bee to moth, hummingbird to moth, and bee to bat are more common than all other directions. The transition from bat-pollinated flowers to any other flowers is the most unlikely direction in *Ruellia*. Bee flowers derived from hummingbird flowers are slightly different from plesiomorphous bee flowers in *Ruellia* (Tripp and Manos 2008). In contrast, in *Penstemon* and *Keckiella* (Plantaginaceae), which contain bee- and hummingbird-pollinated flowers, the common evolutionary direction is bee to hummingbird, and this transition may have occurred separately 21 times (Wilson et al. 2007).

A prominent insight from many studies in the past ten years is that the great flexibility of flower evolution is ever more clearly demonstrated by more detailed and better-resolved phylogenetic analyses (reviews in Endress 2011b,c). Multiple parallel evolution and "reversible" evolution have been demonstrated for floral features in almost all evolutionary studies of many clades (e.g., Onagraceae-*Fuchsia*, Berry et al. 2004; Solanaceae-Iochrominae, Smith and Baum 2006; Smith et al. 2008a,b; Acanthaceae-*Ruellia*, Tripp and Manos 2008; Euphorbiaceae-*Dalechampia*, Armbruster et al. 2009; Phyllanthaceae-Phyllantheae, Kawakita and Kato 2009; Bignoniaceae-Bignonieae, Alcantara and Lohmann 2010; oil flowers, Renner and Schaefer 2010 (Fig. 14.3); Malpighiaceae, Zhang et al. 2012; Orchidaceae-Orchideae, Inda et al. 2012; Orchidaceae-Oncidiinae, Neubig et al. 2012; Papadopoulos et al. 2013; Orchidaceae-*Disa*, Johnson et al. 2013; review of various angiosperms, van der Niet and Johnson 2012). The notion of "Brownian motion"-like evolution (Simpson 1944; Endress 1994a; Losos 2011) is thus more appropriate than ever, and a big open question is which evolutionary directions are less likely to occur than others (Barrett 2010a,b, 2013), which again differs among groups (e.g., Wilson et al. 2007; Tripp and Manos 2008).

A topic of special interest is the evolution of deceptive flowers that do not offer a reward to their pollinators, a phenomenon especially common in orchids, in which both sexual and food deception are widespread (Schiestl 2005; Gaskett 2011; Schiestl and Johnson 2013). Food deception appears to be ancestral in Orchideae (Cozzolino and Widmer 2005; Inda et al. 2012). The primarily olfactorial role of sexually deceptive orchids (Peakall et al. 2010; Schlüter et al. 2011; van der Niet et al. 2011), but also optical (Newman et al. 2012) and tactile aspects, have received recent attention (Francisco and Ascensão 2013). The role of scent in pollination was also analyzed in other families, such as in beetle-pollinated Araceae (Schiestl and Dötterl 2012).

VARIATION IN FLORAL FORM

FLORAL MODIFICATIONS

Floral form varies in many respects: shape, number, position, and histological differentiation (surface structure,

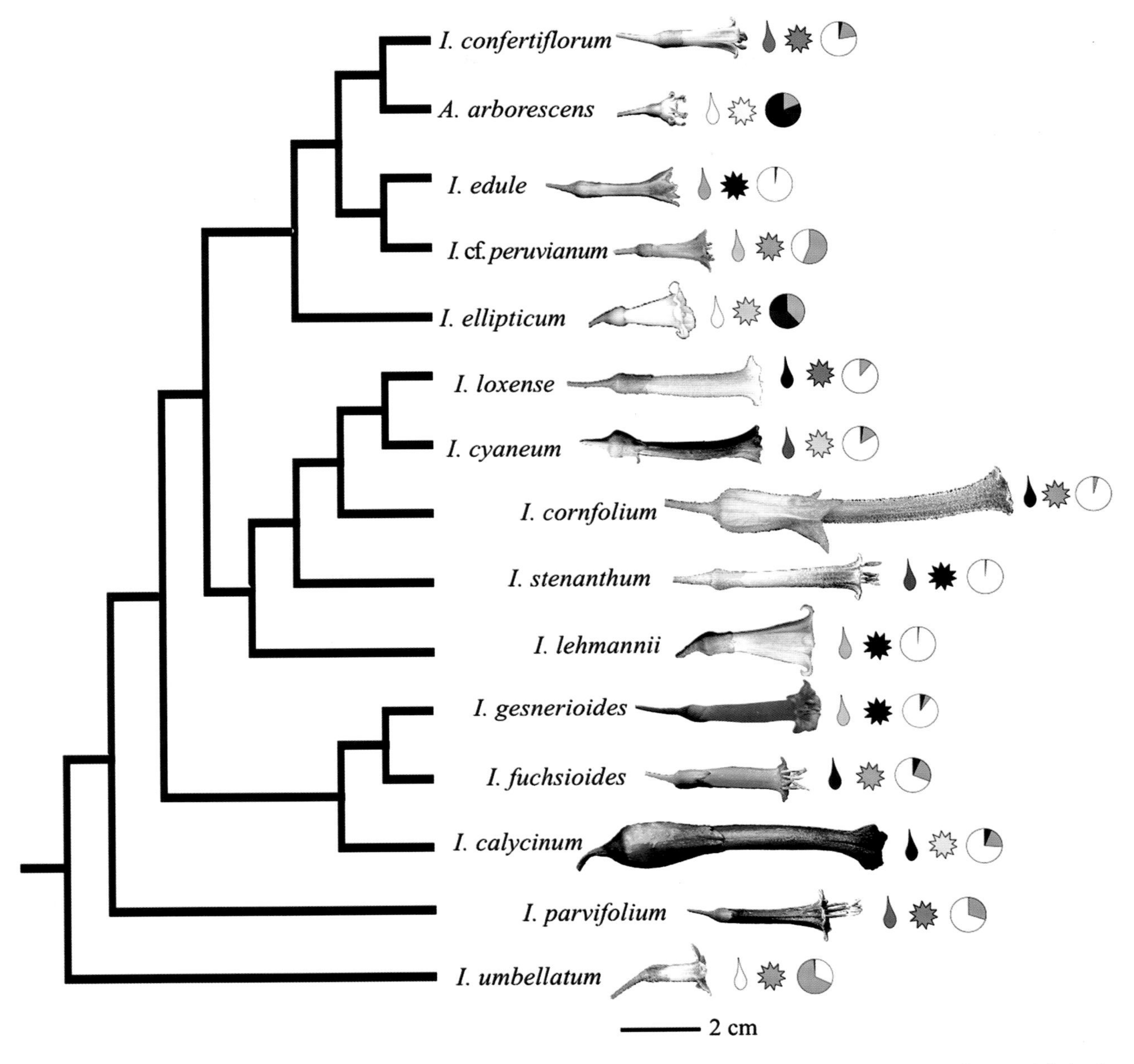

Figure. 14.1. Evolutionary radiation and floral diversification in *Iochroma* (Solanaceae). The droplet and star symbols indicate reward per flower and display size, respectively, with darker shading showing high values. The pie-graphs show proportion of pollinator importance from each of the four groups: hummingbirds (white), hymenopterans (light gray), lepidopterans (dark gray), and dipterans (black). (From Smith et al. 2008, p. 795.)

color, scent, secretions) of the floral organs themselves as well as the collective behavior of the organs, such as floral phyllotaxis, floral symmetry, and synorganization of floral organs. Here we review patterns of variation in floral form.

Floral Phyllotaxis

Floral phyllotaxis may be whorled (Fig. 14.4a-c), spiral (Fig. 14.4d-f), or unordered (Endress 1987a; Doust 2001). In spiral phyllotaxis, all subsequent organs are initiated with the same divergence angle (i.e., the angle of the two lines drawn from the floral center to the center of each of two subsequent organs). In whorled phyllotaxis, commonly a fixed number of organs is arranged in a whorl, and subsequent whorls alternate with each other. Thus, the divergence angle is constant within a whorl but different at the transition to the subsequent whorl. In complex whorls, the number of organs may be doubled (or multiplied) or halved from one whorl to the subsequent whorl (Staedler and Endress 2009; Endress 2011c). Unordered phyllotaxis often occurs when the number of organs is greatly increased and the ratio of the size of the floral apex to the size

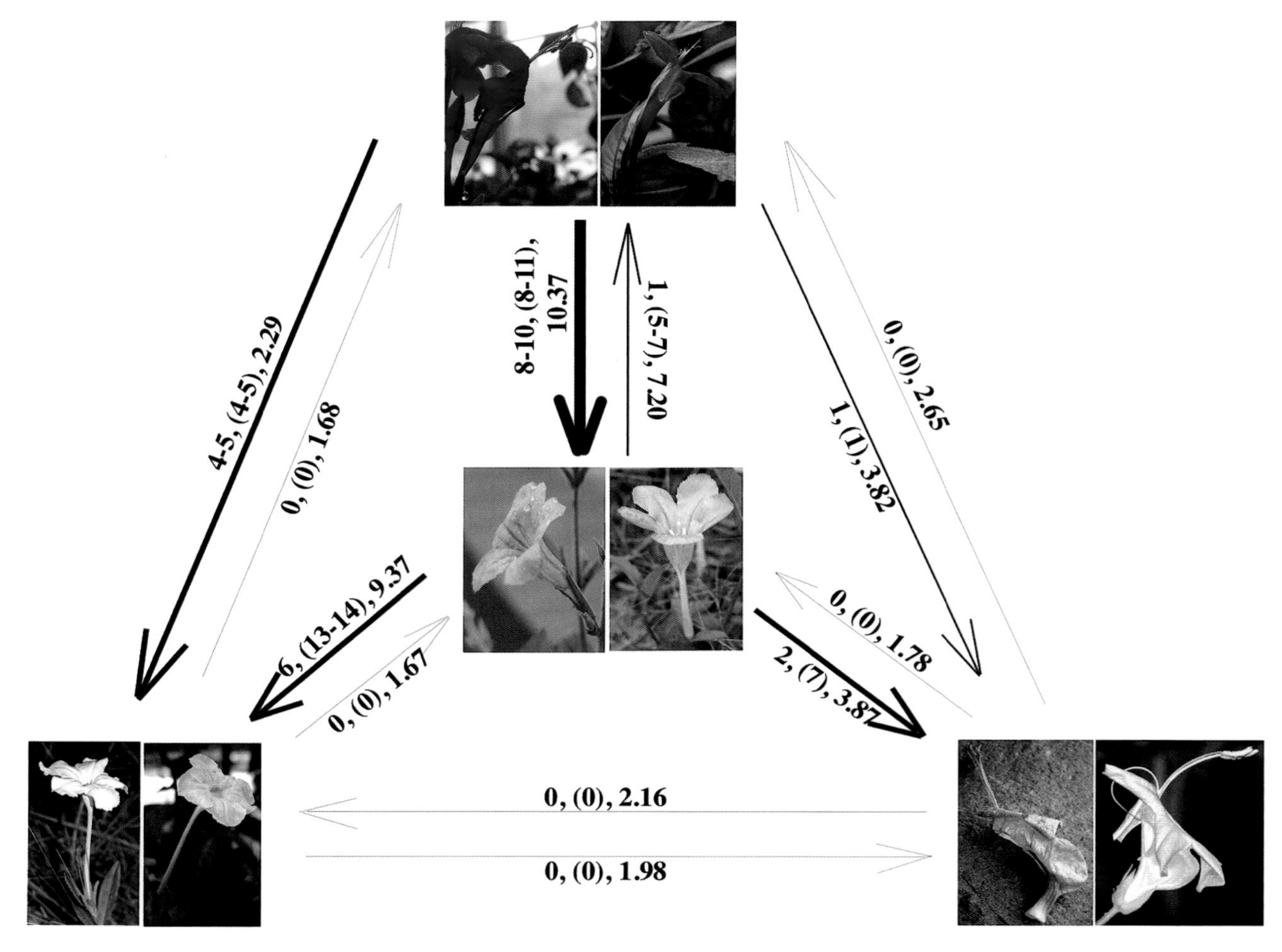

Figure 14.2. Transitions among pollination systems in *Ruellia* (Acanthaceae). Direction of arrow represents observed transitions and thickness of arrow represents relative number of transitions. Relative numbers of transitions are provided next to arrows with first set of numbers corresponding to 116-taxon analysis, second number (in parentheses) corresponding to 154-taxon analysis, and third numbers corresponding to Bayesian posterior expectations or transition rates based on stochastic character mapping. At the top and moving clockwise, the red-flowered *Ruellia humboldtiana* (left) and *R. mcvaughii* (right) are examples of hummingbird-adapted flowers, the yellow-flowered *R. bourgaei* (left) and *Ruellia sp.* nov. (right) (i.e., the pair in the center of the figure) represent bat-adapted flowers, the purple-flowered *R. nudiflora* (left) and *R. caroliniensis* (right) represent bee- or insect-adapted flowers, and the white-flowered *R. noctiflora* (left) and *R. megachlamys* (right) represent nocturnal moth-adapted flowers. (Modified frrom Tripp and Manos 2008, p. 1725).

of individual organs is increased or when the floral apex has an irregular shape.

A simple way to determine whether a pattern is whorled or spiral is to examine a young floral bud when all organs have been initiated but are still short. The organs are regularly disposed in sets of spiral lines (parastichies) going to the right and to the left. In whorled patterns, the steepness (inclination) of corresponding right and left parastichies is the same (Fig. 14.4a,b), but in spiral patterns the steepness is different (Fig. 14.4d,e). In addition, in whorled flowers the organs are at the same time disposed on straight radial lines (orthostichies) (Fig. 14.4c), but such orthostichies are absent in spiral flowers. In flowers with irregular phyllotaxis, there are neither orthostichies nor parastichies (Endress 2006; Endress and Doyle 2007). In the literature, the presence of parastichies has often been mistaken for spiral phyllotaxis. However, it should be clear that parastichies are characteristic of all regular phyllotaxis patterns, spiral and whorled. The spiral phyllotaxis pattern gets its name from the (onto)genetic spiral, which is a single line and encompasses all floral organs in the sequence of their initiation (sequence numbered in Fig. 14.4f).

These phyllotactic patterns may not be exclusive in an individual flower. Flowers of core eudicots are known to have whorled phyllotaxis; however, even in such flowers, phyllotaxis commonly begins spirally in the calyx and becomes whorled only in the corolla. Alternatively, flowers with an unordered phyllotaxis in the polymerous androecium or gynoecium often have an ordered phyllotaxis in the perianth. Another complication is that even in flowers with a completely whorled phyllotaxis, the organs may be initiated in a spiral sequence (Erbar and Leins 1985, 1997; Endress 1987a). However, the plastochrons (i.e., the time lapse between the initiation of two subsequent organs) seem then to be much shorter than in spiral phyllotaxis. If a perianth is lacking, the phyllotaxis of the remaining organs is often irregular (Endress 1989b, 1990a; see also Tucker 1991; Doust 2001). This observation suggests that the perianth sets the boundary conditions for phyllotaxic regularity of the organs of the inner floral region—that is, androecium and gynoecium.

In spiral phyllotaxis of flowers, the divergence angle—that is, the angle between two subsequently formed organs with the floral center, commonly approaches the value of 137.5 . . . ° (Fibonacci pattern). Developmentally, an activator-inhibitor mechanism exists in the floral apex, in which the inhibitor exponentially decays and produces this pattern (Meinhardt 1982). Auxin has been found to play a central role in phyllotaxis formation in vegetative shoots by acting as an activator of organ initiation (review in Traas 2013). Ecologically, the spiral pattern ensures that no two organs fall exactly in the same radius and is common when many organs are arranged on a short axis, e.g., the showy perianth organs in a flower (or the leaves in a vegetative rosette). In contrast, a whorled position is only favorable (e.g., for visual impact or absorption of radiation) for a small number of organs that can be displayed in one or two series in a rosette. Thus, it is not surprising that in core eudicots and monocots, in which the number of petals (or tepals) is generally low, the whorled pattern is almost exclusively present, whereas in the ANA grade, magnoliids, and some families of basal eudicots (e.g., Ranunculales), in which there is a large range of numbers of perianth mem-

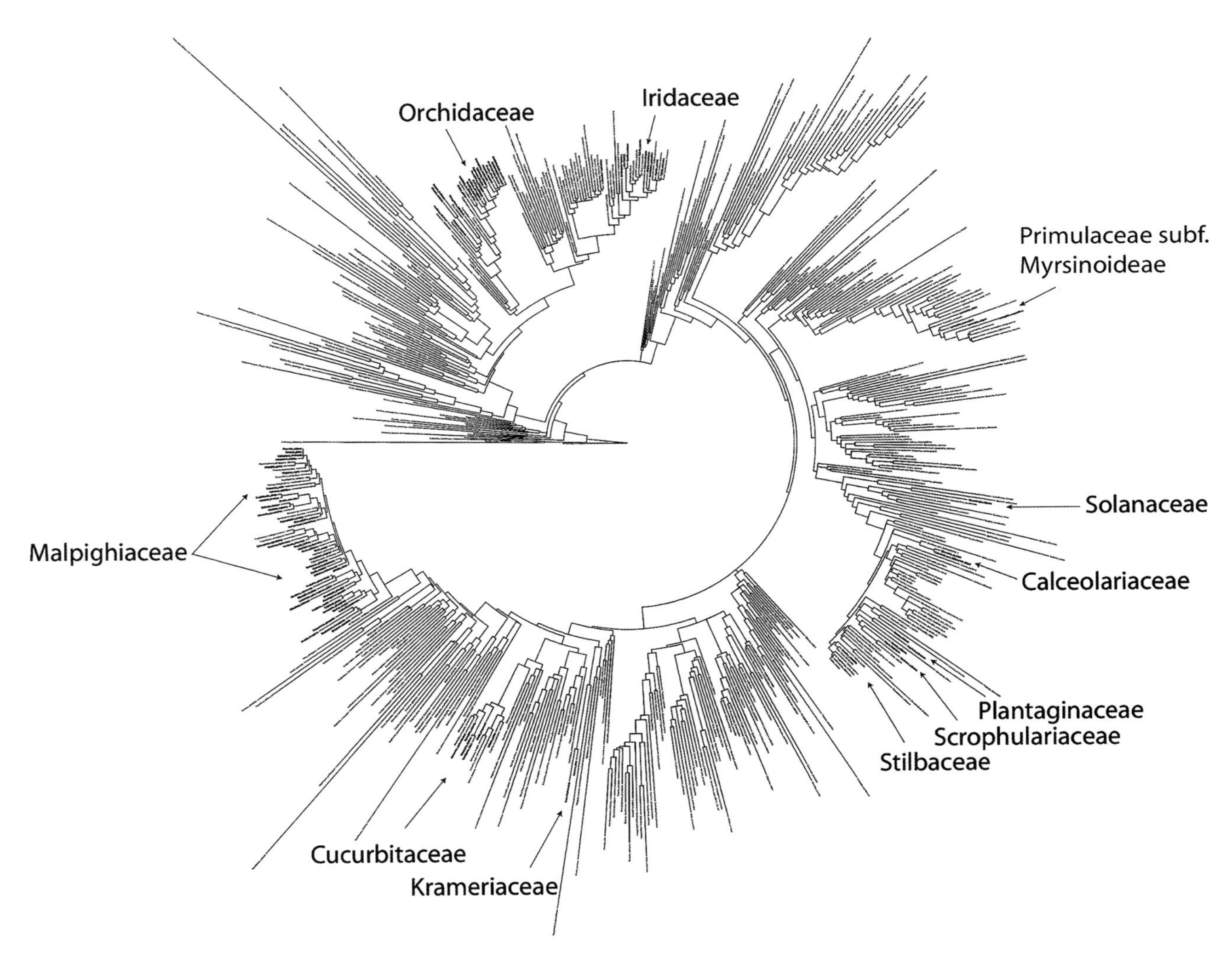

Figure 14.3. Overview of independent evolution of oil-offering flowers in angiosperms. Of known oil-offering taxa, the figure lacks the Plantaginaceae genera *Basistemon* and *Monopera*, the Scrophulariaceae *Colpias*, the Stilbaceae *Anastrabe* and *Bowkeria*, and numerous oil-offering orchids. (Modified from Renner and Schaefer 2010, p. 427.)

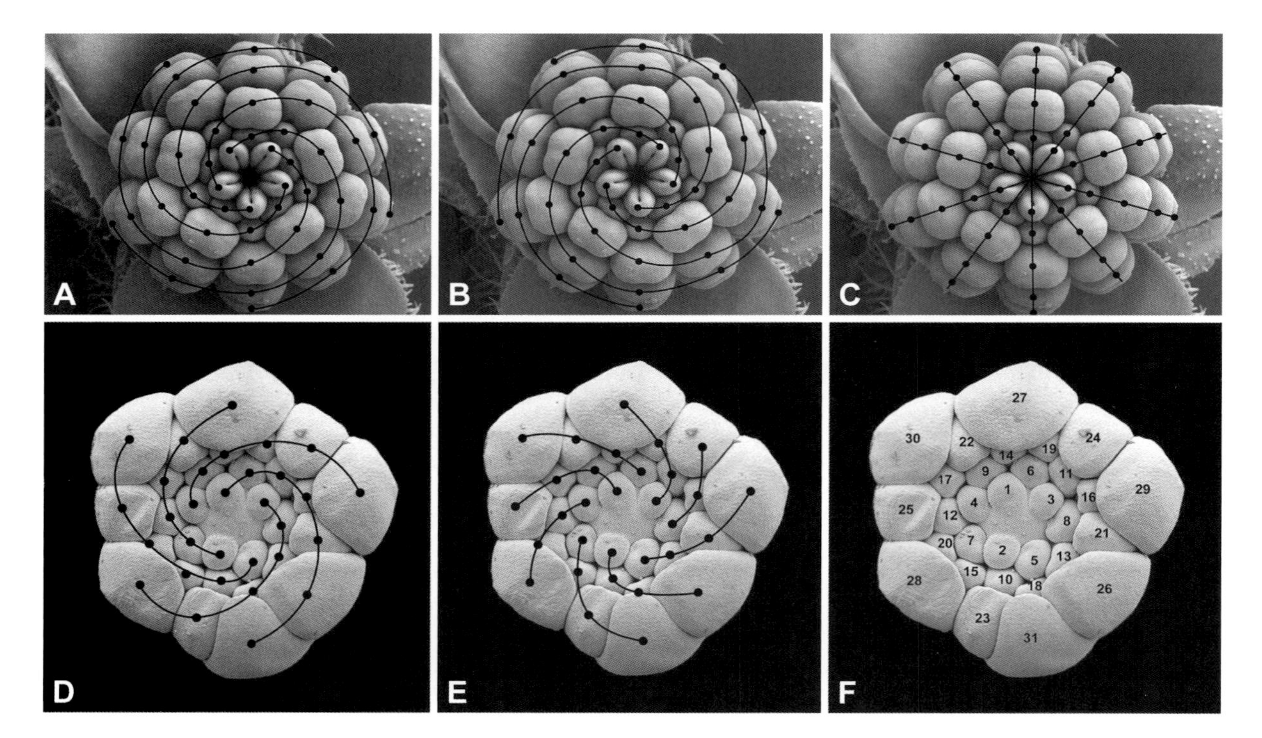

Figure 14.4. Floral phyllotaxis, as shown in floral buds with young organs. A-C. Whorled phyllotaxis in *Aquilegia vulgaris* (Ranunculaceae). D-F. Spiral phyllotaxis in *Austrobaileya scandens* (Austrobaileyaceae). In A the set of 5 counterclockwise-oriented parastichies is marked, in B the set of 5 clockwise-oriented parastichies is marked, in C the orthostichies are marked, in D the set of 5 counterclockwise-oriented parastichies is marked, in E the set of 8 clockwise-oriented parastichies is marked, in F all organs are numbered in the sequence of development (for practical reasons in the opposite direction of development, i.e. beginning in the center of the flower instead of the periphery), and the line (not drawn) connecting all subsequent organs would be the (onto)genetic spiral. A-F from Endress and Doyle (2007, p. 53).

bers, both phyllotaxis patterns occur (Endress 1987a). In some magnoliids, in extreme cases, both spiral and whorled patterns may even occur in the same species (e.g., *Drimys*, Winteraceae, Doust 2001).

The situation is more complicated in flowers than in vegetative shoots because, in contrast to vegetative shoots, in flowers there are repeated switches from one kind of organ to another. In whorled flowers these switches are abrupt. Because the width of different kinds of organs within a flower may differ considerably, there may also be a change in organ number from one whorl to the next. For instance, double positions may occur after the transition from the perianth to the androecium, such that instead of one organ in an expected position, there are two side by side (Endress 1994a; e.g., Nymphaeaceae, Aristolochiaceae, Butomaceae, Velloziaceae, Papaveraceae, Rosaceae, Zygophyllaceae, Phytolaccaceae). In most eudicots, toward the floral center the reverse takes place: the gynoecium generally has fewer organs than the inner whorl of the androecium (Endress 1987a).

Due to synorganization, the extension of "phyllotaxis theories" (Lacroix and Sattler 1988) to flowers is limited. For example, the association of different organs of two different whorls may be so intimate that the two types of organs always occur together, even if this arrangement is not expected in a normal phyllotactic situation. An example is the pentamerous terminal flowers of *Berberis vulgaris*, in which the five petals and five stamens are exactly in the same sector, although they are in spiral phyllotaxis (Endress 1987a). The petal-stamen complex forms a functional apparatus, in which the nectaries (on petal base) are close to the sensitive stamen base, where stamen movement is triggered by touch. Flower organization has changed during evolution. Flowers became more "closed" (in their genetic makeup and evolutionary behavior) and the organs more integrated (Endress 1987a, 1990a).

Floral Merism

Floral merism (mery, merosity) describes the number of organs in a whorl (in whorled flowers) or a series (in spiral flowers) (Eichler 1875–1878; Endress 1987a, 1994a; Ronse De Craene and Smets 1994; Endress and Doyle 2007; Staedler et al. 2007). This number is related to floral phyllotaxis and is dependent also on the ratio of organ primordium size (width) to the floral apex size (diameter) at the time the organs are formed. The smaller the ratio, the more organs are formed in a whorl or series. In ANA grade and magnoliid taxa and some early-diverging eudicots (especially Ranunculaceae, e.g., Schöffel 1932), floral merism may be highly variable within a species or variable even within an individual. In contrast, in nested clades of monocots and asterids, floral merism may be constant even (almost) for entire families (e.g., Apocynaceae) or orders (Lamiales); the genetic canalization of floral organ number (e.g., Huether 1969) allows increased synorganization of organs (Endress 1990a, 2006). Thus, in more nested clades of angiosperms, diversity is less in organ number than in organ plasticity and evolution of various precision mechanisms and specialization for a wide variety of pollinators (Stebbins 1970, 1974; Endress 1994a). However, merism may change within a flower, especially at the transition from perianth to androecium and from androecium to gynoecium, independent of phylogenetic position and other aspects of variability. (Diversity in floral merism is discussed in more detail in the chapters on basal angiosperms and early-diverging eudicots, Chapters 4–6, 8.)

Floral Symmetry

Most flowers are polysymmetric (actinomorphic, with several symmetry planes) (Fig. 14. 5a) or monosymmetric (zygomorphic, with one plane of symmetry) (Fig. 14.5b). In monosymmetric flowers, the plane of symmetry is usually vertical, and this position is commonly present from the beginning of development. Only in a few taxa is the plane of symmetry horizontal or oblique at the beginning and secondarily vertical only at anthesis by some bending or rotation of the flower. The flowers of ANA grade and magnoliid taxa are polysymmetric. Floral monosymmetry has evolved many times from polysymmetry (Endress 1999, 2012, Sargent 2004; Jabbour et al. 2009a,b; Citerne et al. 2010; Hileman 2014a,b) and has led to conspicuous evolutionary radiations (Sargent 2004) (Fig. 14.6). A general trend is that in more synorganized flowers, monosymmetry appears more as a key innovation than in groups with less synorganized flowers. Species-rich groups with exclusively or predominantly monosymmetric flowers are Fabaceae, Lamiales, Asteraceae, Orchidaceae, and Zingiberales (Endress 1999, 2001a, 2012; Ree and Donoghue 2000; Rudall and Bateman 2002); Asteraceae and Orchidaceae are the two most species-rich families of angiosperms.

Evolutionary reversals from monosymmetry to polysymmetry have occurred more often than previously believed. In the only slightly monosymmetric Malpighiaceae, such reversals occurred several times (W. Zhang et al. 2012, 2013), but even in Lamiales and Fabaceae with their pronouncedly monosymmetric flowers, a number of evolutionary reversals are known. These reversals have occurred primarily in phylogenetically relatively basal groups within these clades, such as Detarieae of Fabaceae (Cardoso et al. 2012a), Gesneriaceae of Lamiales (Endress 1998; Zhou et al. 2008; Clark et al. 2011), and Prostantheroideae of Lamiaceae (G. Godden et al. unpubl.). It may be assumed that in such basal clades monosymmetry is structurally and genetically less deeply rooted than in more derived clades so that reversals are more likely to occur. In some cases, approximate secondary polysymmetry evolved from monosymmetry by a transition from pentamerous to tetramerous flowers, such as in *Veronica* or *Plantago* (Plantaginaceae)

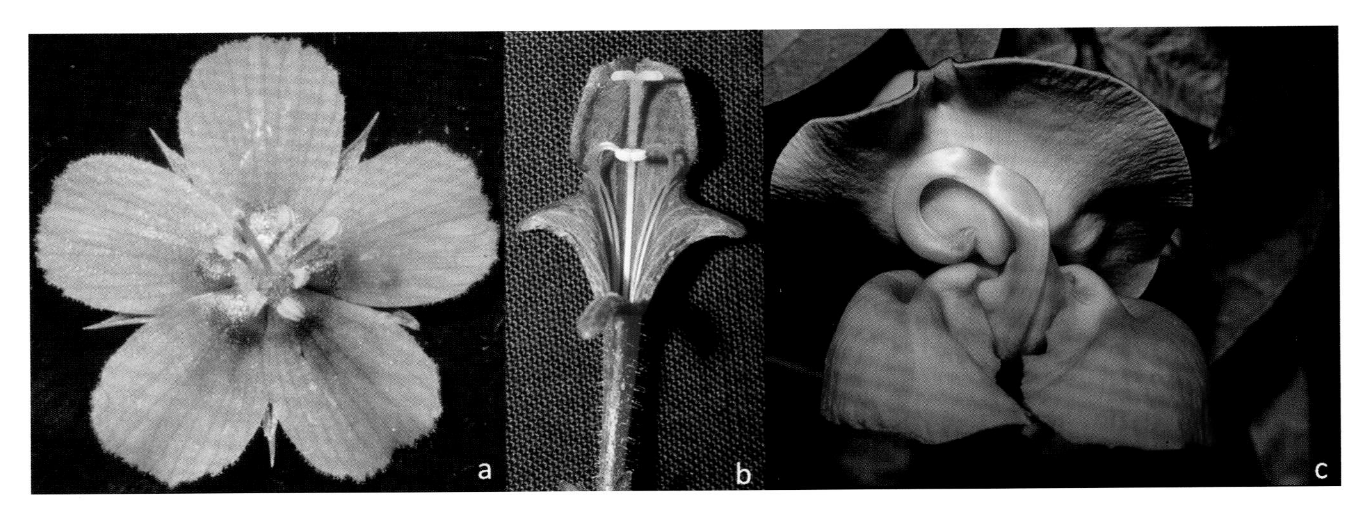

Figure 14.5. Floral symmetry. A. Polysymmetric flower. *Lysimachia arvensis* (Primulaceae). B. Monosymmetric flower. *Columnea* sp. (Gesneriaceae). The two stigmatic lobes and the four united anthers (by postgenital fusion) are exactly in the median plane. C. Asymmetric flower. *Sigmoidotropis speciosa* (Fabaceae).

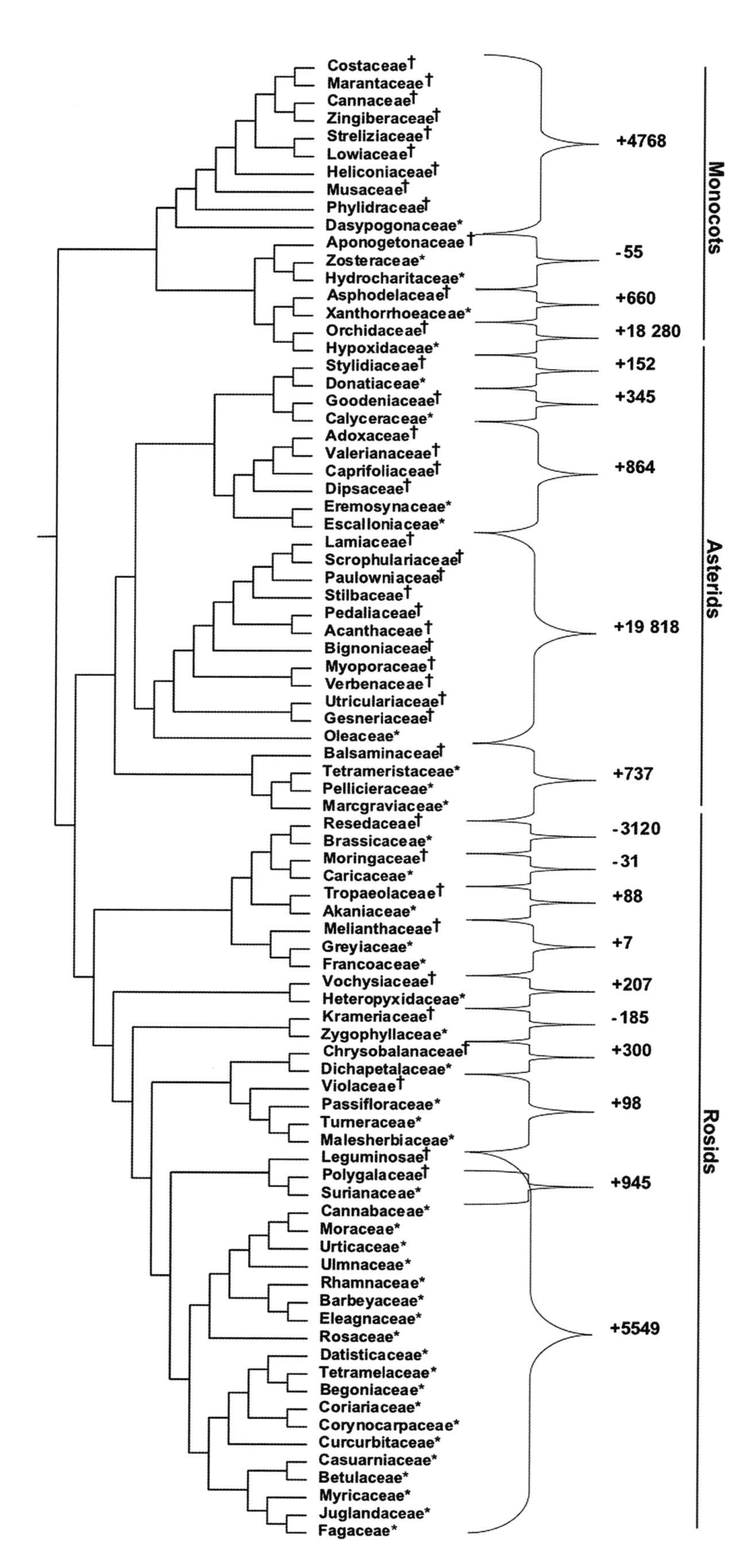

Figure 14.6. Floral symmetry. Increased species number in clades with monosymmetric flowers compared to their polysymmetric sisters (phylogeny adapted from Soltis et al. 2000). Brackets indicate the 19 sister group comparisons conducted by Sargent (2004). The number opposite each bracket indicates the difference in species number between the two sister groups (monosymmetric species minus polysymmetric species); a dagger indicates a monosymmetric family; an asterisk indicates a polysymmetric family. (Figure and legend modified from Sargent 2004, p. 605).

(Donoghue et al. 1998; Endress 1999; Preston et al. 2011) or Melianthaceae (Ronse De Craene et al. 2001).

Asymmetric flowers (without a symmetry plane) have evolved from monosymmetric flowers in a few clades, such as Fabaceae, Orobanchaceae, Orchidaceae, and Zingiberales (Endress 2012) (Fig. 14.5c), in some cases many times (Fabaceae, Prenner 2004b; Marazzi et al. 2006, 2007). Whereas the evolution of monosymmetry from polysymmetry has been studied in numerous molecular developmental studies, because one of the classical model organisms, *Antirrhinum*, is monosymmetric (e.g., Hileman et al. 2003; Preston and Hileman 2009), the transition from monosymmetry to asymmetry is almost untackled (Bartlett and Specht 2011; Hileman 2014a,b). Unfortunately, molecular developmental biologists often use the term "asymmetric" for monosymmetric and thus probably overlook the fact that there are truly asymmetric flowers.

The establishment of floral symmetry is to a great extent independent of floral phyllotaxis (Endress 1999). Monosymmetric flowers do not occur in basal angiosperms (except for monosymmetry by organ reduction, e.g., unistaminate flowers in Chloranthaceae or unilaterally bistaminate flowers in Piperaceae). Especially noteworthy and unusual are relatively highly elaborated monosymmetric flowers in basal eudicots (Ranunculaceae: *Aconitum*, *Delphinium*), which are based on spiral phyllotaxis (Schöffel 1932; Hiepko 1965b; Jabbour et al. 2009a). Probably all other elaborate monosymmetric flowers have whorled phyllotaxis. Unelaborated monosymmetric flowers occur in Proteaceae (Douglas and Tucker 1996a,b), Caryophyllales (some species of Cactaceae and Caryophyllaceae), rosids (some species of Capparaceae, Malvaceae, and Rutaceae), asterids (some species of Rubiaceae), and monocots (some species of Iridaceae and Amaryllidaceae) (Endress 1999).

Perianth

In most angiosperms, flowers have a perianth that commonly consists of one or two whorls or series of organs with protective and attractive functions. In eudicots they are commonly called sepals and petals, whereas in the ANA grade, magnoliids, and most monocots, all perianth organs are referred to as tepals (Hiepko 1965b; Endress 2008). Petals evolved many times in the eudicots (Irish 2009; Ronse De Craene and Brockington 2013). For Caryophyllales alone, Brockington et al. (2009) calculated at least nine separate origins of petals. Thus, petals across the eudicots are not homologous in a phylogenetic sense, but may be homologous in an organizational sense (for such so-called biological homology, see Wagner 1989, 2014).

Stamens and carpels are easy to define because of their specific functions. Stamens contain pollen sacs in which meiosis takes place and microspores are formed, and carpels contain ovules in which meiosis takes places and megaspores are formed. In contrast, tepals, or sepals and petals, are more difficult to define because their functions are less specific. Commonly, tepals or sepals have a mostly protective function for the young inner floral organs, and petals have a mostly attractive function for pollinators during anthesis. However, these functions can also be performed by other organs. Transfer of function has occurred between sepals and petals and between perianth and organs further inside the flowers and also outside the flowers (Baum and Donoghue 2002; Buzgo et al. 2005; Endress 2010a). In the majority of the eudicots, the perianth organs are present in two whorls or series. In these cases, sepals and petals are best defined by their position: sepals are the organs of the first whorl (or series), and petals are the organs of the second whorl (or series). Otherwise, there is no single characteristic that would always differentiate between sepals and petals. In cases with only one whorl (series) or more than two whorls (series), comparisons with the closest relatives that have two whorls (series) should help clarify homologies.

However, several traits tend to be more common in sepals or more common in petals, thus allowing a loose characterization. These traits can be grouped into four categories:

(1) Position: in a flower that contains sepals and petals, each organ type is commonly in one series, sepals are always in the outer series, and petals in the inner. However, if only one series or more than two series are present, distinguishing between them may be more problematic (see above).

(2) Structure: sepals tend to be robust, persistent, green, acute, smooth, with broad base and more narrow upper part, with three vascular strands in the floral base, and with only small intercellular spaces (Fig. 14.7a). In contrast, petals tend to be delicate, ephemeral, colored, obtuse or emarginate, papillate, with narrow base and broader upper part, with one vascular strand in the floral base, and with large intercellular spaces (Fig. 14.7b).

(3) Development: sepals tend to be initiated in spiral sequence (resulting in quincuncial aestivation) and differentiated early in flower development (Figs. 14.8a, 14.9a). In contrast, petals tend to be initiated almost simultaneously or in a rapid spiral sequence and differentiated late, often being delayed until shortly before anthesis (Fig. 14.9b).

(4) Function: sepals are commonly photosynthetic (green) protective organs for inner floral organs in bud and the young fruit; sometimes they are also involved in

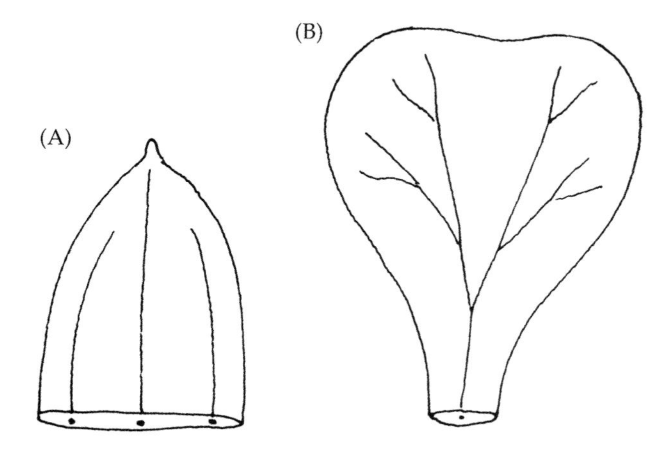

Figure 14.7. Perianth. Typical perianth elements in a choripetalous flower in some eudicots and some monocots, with vascular system (schematic diagrams) (from Endress 1994a; from Soltis et al. 2005). A. Sepal. B. Petal.

fruit dispersal. Petals are commonly attractive organs for pollinators at anthesis (shape, color, scent) and are only rarely green or persistent.

Petals that are characterized by the features assembled above are common in some core eudicots, especially in rosids (e.g., Geraniaceae; Endress 2010c) and Caryophyllales (e.g., Caryophyllaceae) (Rohweder 1967; Rohweder and Endress 1983; Endress 2011c). Among basal eudicots, only in some Ranunculales are similar petals present, especially in Ranunculaceae and Berberidaceae (Hiepko 1965b; Erbar et al. 1999); however, the petals of these families are mostly characterized by the presence of nectaries, whereas nectaries occur at other sites in the flowers of core eudicots. In Saxifragales, the perianth is diverse, from the type just described to absent.

In a number of other rosids and in many asterids, however, the petals are robust, early differentiated, with a broad base, and protective at least in later stages of bud development when the sepals stop growing (Endress 2011c) (Figs. 14.8b, 14.10). Olacaceae, of Santalales, have a double perianth, with the sepals small and the petals exhibiting both protective functions from early in bud and attractive functions at anthesis. It seems that in more derived Santalales the calyx is absent and only the petals remain (Endress 1994a).

In many basal eudicots the perianth is simple (uniseriate); the sepals are well differentiated (Proteaceae), weakly differentiated (Buxaceae, Myrothamnaceae, *Tetracentron*), rudimentary (*Trochodendron*), or lacking (Eupteleaceae) (Jäger-Zürn 1966; Endress 1986a, 2010a; Douglas and Tucker 1996a; von Balthazar and Endress 2002a,b; Soltis et al. 2003c; Chen et al. 2007; Ren et al. 2007b). The problem of sepal differentiation was discussed in a comparative developmental study by von Balthazar and Endress (2002a). However, the distinction between bracts and sepals in basal eudicots has not yet been comparatively examined critically.

In basal angiosperms, several groups show some features of petals in the inner perianth members. However, the distinctness of such features is much less clear than in eudicots. Therefore, use of the terms sepals and petals in this assemblage (Hiepko 1965b; Endress 2008) is often problematic. Among the ANA grade studied thus far (see Chapter 4), *Cabomba* (Nymphaeaceae) has inner perianth parts that come closest to the characterization of petals (Endress 2001c).

In monocots, outer and inner perianth members are commonly less different from each other than in eudicots, although the inner may have more petal-like features than the outer (Weber 1980). Therefore, use of these terms in monocots is problematic.

Molecular developmental genetic studies, in concert

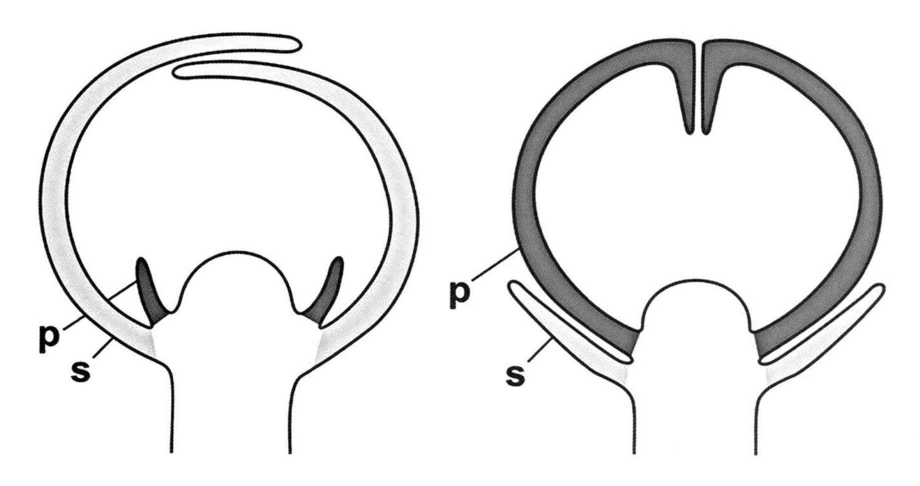

Figure 14.8. Perianth. Floral buds with different perianth differentiation. Longitudinal sections. A. Sepals (light gray) large, imbricate, function as protective organs, petals (dark gray) delayed after initiation. B. Sepals (s, light gray) remaining small, petals (p, dark gray) large, valvate, function as protective organs. (From Endress 2011c, p. 374.)

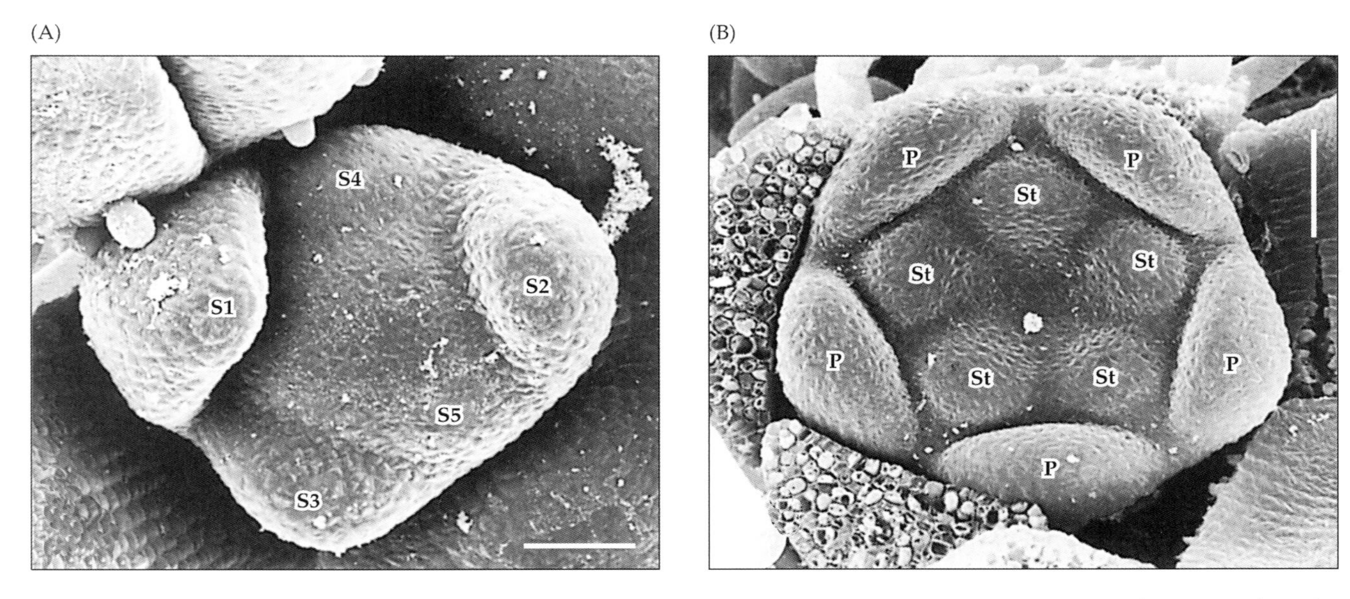

Figure 14.9. Perianth. Typical perianth development in core eudicots. *Asclepias curassavica* (Apocynaceae) (modified after Endress 1994a; from Soltis et al. 2005). Bars = 50 μm. A. Sepals (S1-S5) with successive initiation in spiral sequence. B. Petals (P) and stamens (St) with simultaneous initiation.

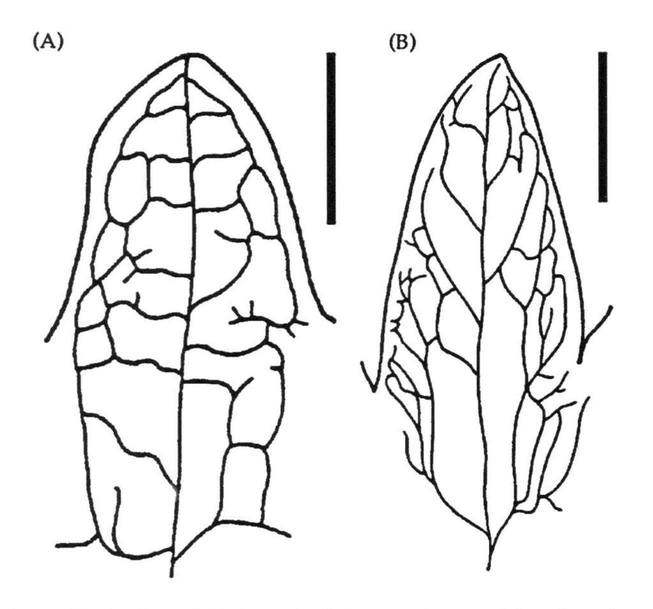

Figure 14.10. Perianth. Typical perianth elements in a sympetalous flower in core eudicots, with vascular system shown. *Solanum luteum* (Solanaceae) (modified from Rohweder and Endress 1983; from Soltis et al. 2005). Bars = 1 mm. A. Sepal. B. Petal.

with new comparative developmental studies, will hopefully provide improved insight into the evolution of petals and the homology of perianth organs (Albert et al. 1998; Irish and Kramer 1998; Kramer and Irish 2000; Zanis et al. 2002; Brockington et al. 2009; Irish 2009; Ronse De Craene and Brockington 2013; see case study on Aizoaceae, below).

Sympetaly and Floral Tubes

In asterids, petals are commonly congenitally united (sympetalous) (Fig. 14.5b). This has repercussions on petal structure, and some features are different from those of free petals. Such petals do not have narrow bases but are more often acute and have three main vascular bundles, and they are less delicate than those in flowers with free petals. In addition, their early development is less delayed than that of free petals, and they often take on a protective function of the inner organs in later bud stages by extending beyond the sepals. Thus, in asterids, petals have several features that are more characteristic of sepals (Endress 1994a) (Fig. 14.7a,b).

Sympetaly is a prominent key innovation in angiosperm evolution. It is a synapomorphy for euasterids (although it also occurs in some, but not all, members of Ericales and Cornales), one of the most species-rich major clades of angiosperms, in which some of the most complex flowers (asclepiads of Apocynaceae) and inflorescences (Asteraceae) have evolved (see Chapter 11). Sympetaly has also evolved in a number of other groups, with a lower number of genera and species than asterids, such as Caricaceae and some species of Crassulaceae, Malvaceae, and Rutaceae (El Ottra et al. 2013). In fact, sympetaly is present in rudimentary form—that is, as a very short basal zone, much more often than previously assumed (e.g., Trigoniaceae and Dichapetalaceae, Matthews and Endress 2008; Cucurbitaceae [if the so-called petals really correspond to petals], Matthews and Endress 2004; some Protieae of

Burseraceae, Bachelier and Endress 2009). Fusion of petals allows the formation of floral tubes, which, in turn, facilitates adaptive radiations correlated with different pollinators because the length and width of floral tubes can easily and rapidly be changed (Alexandersson and Johnson 2002). Petal union also allows for the construction of larger flowers, because sympetaly provides stability to the floral architecture. Thus, the largest flowers in the world are in groups with united petals (or tepals) (Davis et al. 2008). Examples are *Brugmansia* (Solanaceae) and *Fagraea* (Gentianaceae), both asterids, which attain flower lengths of more than 50 cm. *Aristolochia* (Piperales, magnoliids) and *Rafflesia* (Rafflesiaceae, Malpighiales, rosids, Nikolov et al. 2013), with flower sizes of a meter in length or diameter, respectively, have united tepals or sepals and petals. This architecture extends the upper end of a spectrum of diversification, which is narrower in plants with free perianth parts (see also Endress 1994a).

In asterids, two kinds of developmental patterns of petals have been distinguished, "late" and "early" sympetaly (Erbar 1991; Erbar and Leins 1996). In late sympetaly, the developing petals appear free at first and become united by a common base. In early sympetaly, at first a ring meristem appears, on which the individual petals then become visible. Late sympetaly is common in Gentianales and Lamiales, whereas early sympetaly is common in Dipsacales and Asterales. An interesting problem is whether these two developmental patterns could be (functionally or organizationally) dependent on the overall shape of the floral apex at the time of organ inception (Endress 1997b; Roehls 1998). Because in Gentianales and Lamiales the ovaries are predominantly superior and in Dipsacales and Asterales predominantly inferior, the early shape of the floral apex is more convex in the first group, but more concave in the latter. The concave shape could give the impression of a ring meristem at petal inception, whereas this is not the case in apices that are more convex. Erbar and Leins (2011) argue that this association may not be significant and mention some families with early sympetaly that have superior ovaries. However, the correlation still remains, and corolla development in asterids needs more detailed study.

In some asterids, in early floral development sympetaly is expressed but is no longer evident in mature flowers (e.g., some Pittosporaceae, Araliaceae) (Erbar and Leins 1988, 1995a). Developmental studies on seemingly choripetalous representatives of early-branching asterids (Cornales and Ericales) are needed. Perhaps more than one developmental step is necessary to form distinct sympetaly: (1) confluent meristems of the initiated petals, and (2) an intercalary elongation zone below the united petal meristems. It could be that in these seemingly choripetalous asterids, only step (1) has occurred but not step (2). This should also be studied in other asterids that seemingly have free petals (see also Erbar and Leins 2011).

In many groups with sympetalous flowers, the stamens are also fused with the corolla to a greater or lesser degree. This fusion enhances the firmness and stability of the floral tube even more and creates an intimate synorganization between androecium and corolla. For instance, in polysymmetric flowers, this fusion may create five separate nectar canals (if five stamens are present), and this results in the architecture of revolver flowers, in which pollinators have to move around the flower to exploit the entire nectar source and thus most likely have extensive contact with anthers and stigmas. The most complex revolver flowers are in Apocynaceae (Asclepiadoideae, Periplocoideae, Secamonoideae), in which five pollen units (groups of pollinia) can be attached to an insect by means of a complicated transport organ (translator with clip or glue). Other clades of Apocynaceae have much simpler flowers, and the stepwise evolution of complex flowers can be seen if the entire family is studied comparatively (Fallen 1986; Endress et al. 1996; Endress and Bruyns 2000; M. Endress 2001).

In some eudicots, sepals and petals are fused and form a synorganized perianth (e.g., Tropaeolaceae, Myrtaceae: especially *Eucalyptus*; Drinnan and Ladiges 1989). In monocots, outer and inner perianth whorls may also be fused (e.g., some Asparagaceae).

Floral Spurs

Floral spurs are hollow outgrowths of perianth organs and are present in corollas more often than in calyces. They do not characterize large clades, but they have evolved in many angiosperm clades, especially in eudicots (among monocots, mainly in some orchids). Examples of spurred flowers are *Aquilegia* (Ranunculaceae) (Hodges 1997a,b; Whittall and Hodges 2007), *Epimedium* (Berberidaceae), and *Halenia* (Gentianaceae) (von Hagen and Kadereit 2002), all with four or five spurs. Most spurred flowers, however, have only one spur, as in *Delphinium* (Ranunculaceae), some Antirrhineae (Plantaginaceae) (Sutton 1988), and *Heterotoma* (Campanulaceae) (Ayers 1990). These spurs mostly have a connection with nectar production and/or nectar presentation. In early-diverging eudicots with nectariferous petals (Ranunculaceae, Berberidaceae), nectar is produced at the end of the spur and stored in the spur, which is part of the petal. However, in Papaveraceae-Fumarioideae, the petals are not nectariferous, and nectar is produced at the base of the stamen filaments (e.g., *Corydalis*). In asterids, nectar is more often produced by a separate nectary disk and, if a spur is present, then secondarily stored in the spur. An exception is *Halenia* (Gentianaceae), in which nectar is produced directly in the spur, and the closest relatives have shallow petal nectaries. Another special case is *Diascia* (Scrophulariaceae), which has two oil-producing spurs per flower, from which specialized bees collect oil with their two front legs (Vogel 1974).

A conspicuous adaptive radiation often accompanies the origin of spurs, particularly in comparison to close relatives with spur-less flowers. The presence of spurs has therefore been proposed as a key innovation (Hodges 1997b). However, it has been argued that this is not the case for *Halenia* (Gentianaceae) (von Hagen and Kadereit 2002) because in this clade, radiation occurred long after spurs originated. As in other cases, a specific feature may apparently be a key innovation in one clade but not in another (Endress 2011c).

Androecium

The androecium comprises all stamens (the floral organs that contain microsporangia) of a flower. Stamens are not restricted to angiosperms, if they are defined in this general sense (Endress 2001b). However, angiosperm stamens have a thecal organization (Endress and Stumpf 1990) not present in other plant groups. Each stamen characteristically has an anther with two lateral thecae often situated on a more or less elongate sterile part, the filament. Each theca is differentiated into a collateral pair of pollen sacs (microsporangia). At maturity, each theca opens by a longitudinal slit between the two pollen sacs. All major groups of angiosperms are characterized by this type of stamen structure (Fig. 14.11a). In only rare cases have some small angiosperm clades diverged from this pattern (Endress and Hufford 1989; Hufford and Endress 1989; Endress and Stumpf 1990). Diversity of anther shapes based on variation of proportions of the basic pattern is also considerable (Bernhardt 1996; Endress 1996a) (Fig. 14.12).

Deviation from the basic anther pattern occurs in the opening mechanism and number of pollen sacs per stamen. In some groups (some Magnoliales, Laurales; basal eudicots [Trochodendraceae]; Saxifragales [Hamamelidaceae]; Endress and Hufford 1989; Hufford and Endress 1989), the longitudinal slit bifurcates at the ends, and each theca opens by two valves. There is a clear correlation between this form of dehiscence and the presence of a thick connective. Such anthers are prone to lose one pollen sac per theca (some Laurales, some Hamamelidaceae). Loss of one pollen sac per theca also occurs in other eudicots, but there is no obvious phylogenetic pattern. Rather, a functional correlation can be seen, as this reduction occurs, for example, in some plants with cleistogamous flowers. The genetics of tetra- vs. disporangiate anthers in *Microseris* (Asteraceae) was studied by Gailing and Bachmann (2000). Increase in number of microsporangia also occurs in some groups but without a distinct phylogenetic pattern (Figs. 14.11b,c). The most extreme forms of stamens are those with a complete loss of thecal organization; an irregular number of microsporangia is present, and each opens by an individual pore or all converge to one apical pore (Fig. 14.11d,e). Such forms are known from three parasitic (but unrelated) families, Apodanthaceae, Rafflesiaceae, and Santalaceae, and from *Polyporandra* of Icacinaceae (Endress and Stumpf 1990; Blarer et al. 2004; Nikolov et al. 2014), and perhaps *Boschia* and *Cullenia* of Malvaceae (Nyffeler and Baum 2000).

Stamen number is in general more diverse than numbers of the other three organs. In many groups with whorled floral phyllotaxis, stamens occur in two whorls (diplostemony), which results in a total of six stamens in many monocots and ten in many eudicots. A variant of diplostemony is obdiplostemony, in which the epipetalous (and

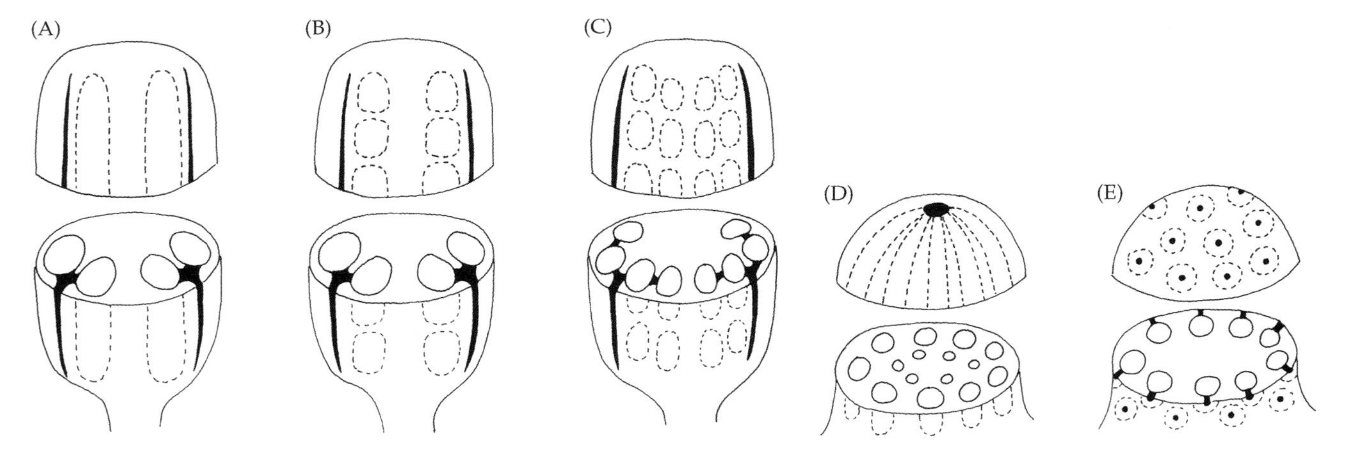

Figure 14.11. Androecium. Anther types in angiosperms (schematic diagrams) (from Endress and Stumpf 1990; from Soltis et al. 2005). A-C. With thecal organization. D, E. Without thecal organization. A. Basic type with two disporangiate thecae. B. Polysporangiate anther with transverse septa. C. Polysporangiate anther with transverse and longitudinal septa. D. Polysporangiate anther with a single central pore for all sporangia. E. Polysporangiate anther with numerous pores, one for each sporangium.

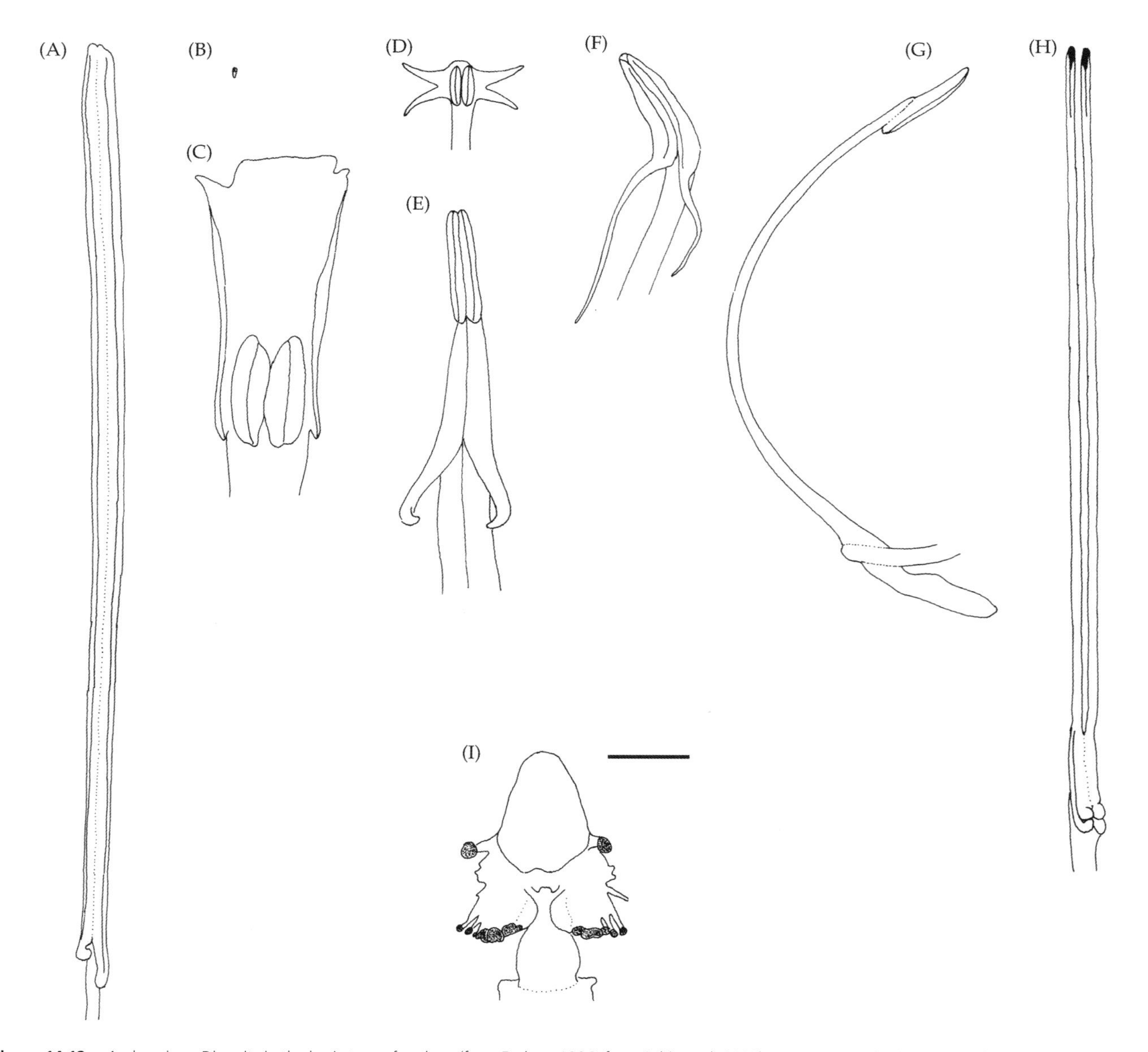

Figure 14.12. Androecium. Diversity in the basic type of anthers (from Endress 1996; from Soltis et al. 2005). A-B. Extremes in sizes. A. *Strelitzia reginae* (Strelitziaceae). B. *Piper peltatum* (Piperaceae). C-I. Diversity in proportions. C. *Costus igneus* (Costaceae). D. *Globba winitii* (Zingiberaceae). E. *Roscoea purpurea* (Zingiberaceae). F. *Thunbergia mysorensis* (Acanthaceae). G. *Salvia patens* (Lamiaceae). H. *Demosthenesia cordifolia* (Ericaceae). I. *Psychopsis papilio* (Orchidaceae). Bar = 4 mm.

not the episepalous) stamens appear to be the outer organs, with the result that the carpels—if they are isomerous with the stamens of a whorl—alternate with the epipetalous instead of the episepalous stamens. This is unexpected, as subsequent whorls of organs normally alternate with each other. The reason for this disturbance of normal alternation is apparently that epipetalous stamens tend to be reduced in size, so that the actual space for the carpels is larger between the epipetalous than between the episepalous stamens (Eckert 1965; Matthews and Endress 2002; Endress 2010b; Leins and Erbar 2010). Examples are many Geraniales and Oxalidales.

In many taxa, the number of stamens per flower has increased even more. In flowers with a spiral phyllotaxis, high numbers of stamens are attained by an extended period of stamen formation following the ontogenetic spiral. In contrast, in flowers with a whorled phyllotaxis, stamen number may increase through an increase in whorl number, often with double positions, for example, in Papaveraceae (Endress 1987a), or an increase in the number of floral sectors, for example, in some Crassulaceae or Araliaceae (Nuraliev et al. 2014). A different mode of development is by the formation of primary primordia, on which many secondary primordia, each giving rise to a complete stamen, are formed. Such primary primordia may be formed at the place where single stamens would be expected; they appear as several distinctive sectorial primordia or as a single ring primordium on which the individual stamen primordia are then not formed in whorls but in a less regular fashion. Notwithstanding the irregular stamen position, in androecia with primary and secondary primordia, centripetal and centrifugal patterns can be distinguished (Fig. 14.13, 14.14a,b), which are often characteristic at the family level (Corner 1946). Centripetal stamen initiation occurs, for example, in Mimosoideae-Fabaceae (Gemmeke 1982), but centrifugal in Capparaceae (Leins and Metzenauer 1979; Ronse De Craene and Smets 1997), Dilleniaceae (Corner 1946; Endress 1997a), and Lecythidaceae (Hirmer 1918; Endress 1994a). More rarely, the stamens appear to be initiated almost simultaneously, as in Achariaceae (Bernhard and Endress 1999).

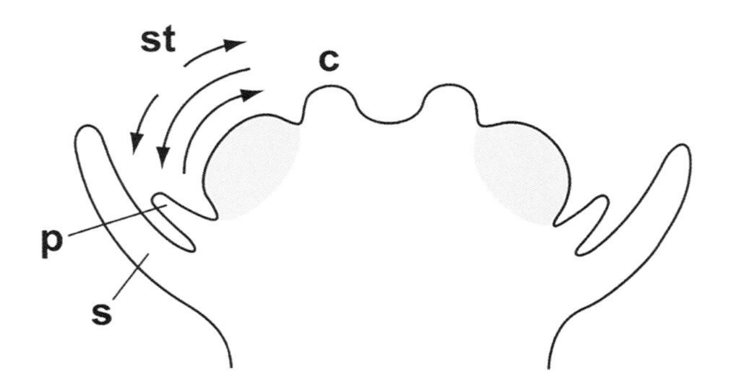

Figure 14.13. Androecium. Bud of polystemonous flower (schematic longitudinal section) with primary androecium primordium (st, gray) and possible directions of initiation of secondary primordia (stamens) indicated by arrows. (c: carpel, p: petal, s: sepal). (From Endress 2011c, p. 374.)

The centrifugal pattern is so unexpected that it has long been thought to characterize a major group of "dicots" (the dilleniids) (e.g., Cronquist 1957). However, phylogenetic studies have shown that centrifugal androecia evolved in very different clades of core eudicots. Thus, the pattern is of interest at a more shallow phylogenetic level than previously thought (see also Endress and Matthews 2012).

Stamens are typically free. However, synandry, the congenital union of stamens, occurs in many angiosperm groups. It is more prominent in unisexual than bisexual flowers because there is no gynoecium in the flower center, and thus the stamens may be united in a central structure that also involves the anthers (Fig. 14.15). Such extreme cases are present in Myristicaceae, Menispermaceae, Cucurbitaceae, and Euphorbiaceae (in the latter three families only in a few taxa). Cases in which only the filaments are involved in stamen union are more common in bisexual flowers, for example, diagnostic for some Amaranthaceae, Malvaceae, and Meliaceae (von Balthazar et al. 2006; Janka et al. 2008). In some families the stamens tend to be united groupwise as several bundles (e.g., Clusiaceae, Hypericaceae, Myrtaceae) (Fig. 14.16).

Gynoecium

Gynoecia are exclusive and omnipresent in angiosperms. Carpels, the structural elements of a gynoecium, are cup-shaped or scale-shaped in early development and later become tubular (ascidiate) or folded (plicate). Intermediate forms between ascidiate and plicate are most common (Endress and Igersheim 2000b; Endress and Doyle 2009). One to numerous ovules are formed near the margin of each carpel. During development the ovule(s) become(s) completely

(A) (B)

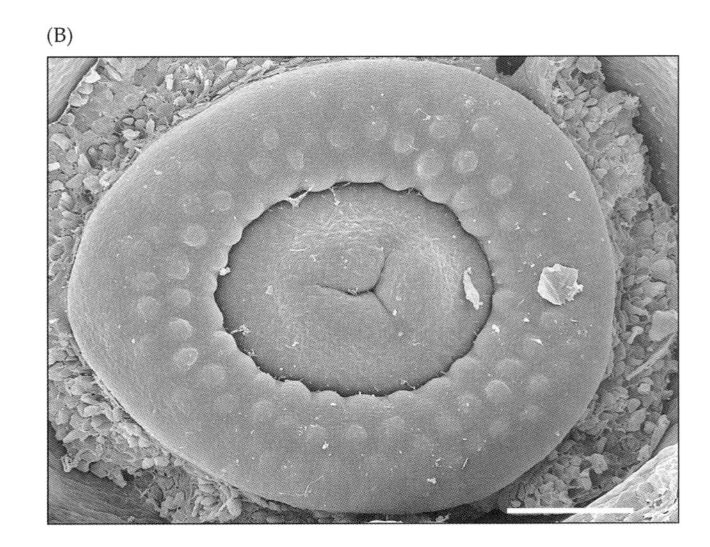

Figure 14.14. Androecium. Divergent development of polyandrous androecia. A. Centripetal. *Annona squamosa* (Annonaceae). Bar = 200 µm. B. Centrifugal. *Barringtonia cf. samoensis* (Lecythidaceae). Bar = 100 µm. P. Endress, from Soltis et al. (2005).

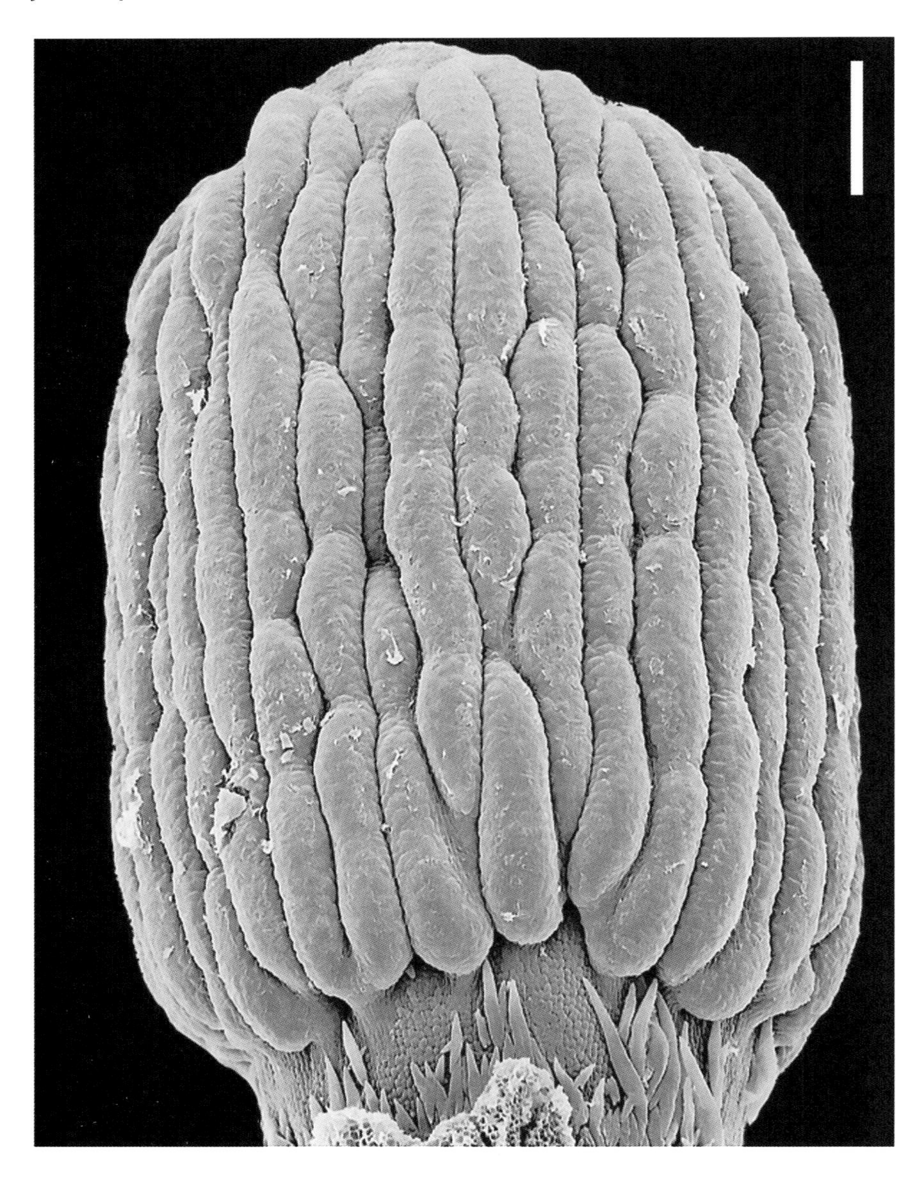

Figure 14.15. Androecium. Stamens completely united into a synandrium, and pollen sacs polysporangiate (portioned by transverse constrictions). *Myristica insipida* (Myristicaceae). Bar = 200 μm. P. Endress, from Soltis et al. (2005).

Figure 14.16. Androecium. *Montrouziera gabriellae* (Clusiaceae). Flower with stamens basally united into five bundles. P. Endress, from Soltis et al. (2005).

enclosed in the carpel in that the carpel flanks grow towards each other on top (in ascidiate carpels) or on the ventral side (in plicate carpels) and finally become sealed, so that the inner space with the ovule(s) is secluded from the outer world (angiospermy). Closure and "sealing" of the carpels is attained by secretion or by postgenital fusion. Endress and Igersheim (2000b) distinguished four types of angiospermy across the ANA grade, magnoliids, early-diverging eudicots, and early-diverging monocots: (1) closure by secretion, without postgenital fusion, (2) closure by postgenital fusion at the periphery, but with a continuous secretory canal without fusion, (3) closure by postgenital fusion at the entire periphery and a secretory canal that stops below the stigma, and (4) closure by complete postgenital fusion, without a secretory canal (see Fig. 4.6). Postgenital fusion (epidermal fusion) comes about by intimate cuticular connection or by interdentation of epidermal cells and sometimes also division of epidermal cells between two contiguous, coherent surfaces, with the result that the surfaces are no longer obvious (Sinha 2000). In the ANA grade, carpels are cup-shaped and sealed only by secretion (type 1); they have only one or a small number of carpels (Doyle and Endress 2000; Endress and Igersheim 2000b; Endress 2001a) (see also Chapter 4). In other angiosperm groups, carpel closure by partial or complete postgenital fusion is predominant (types 2–4) (Endress and Igersheim 1999, 2000b; Igersheim et al. 2001).

A major functional innovation of the advent of angiospermy is pollination on the surface of the carpel (the stigma), no longer on the surface of the ovule (via the micropyle), and, concomitantly, the formation of a pollen tube transmitting tract (PTTT). The PTTT almost always differentiates along the primary morphological surface of the carpel (which, however, is hidden from the outer world because of carpel closure; Endress 1994a). Pollen grains are often deposited in groups on the stigma, and then multiple pollen tubes grow synchronously along the PTTT. Consequently, the PTTT acts as an area where pollen tube (male gametophyte) competition and selection take place before fertilization (Mulcahy 1979; Mulcahy and Mulcahy 1987). In some cases, competition and selection of female gametophytes may also occur (Bachelier and Friedman 2011). Another innovation is that the stigma and PTTT act as sites where self-incompatibility reactions take place (Wheeler et al. 2001). Such prezygotic self-incompatibility is otherwise not well developed in non-angiosperm seed plants (Runions and Owens 1998). The speed of pollen tube growth increased in angiosperms compared to gymnosperms and increased again in monocots and eudicots compared to the ANA grade and magnoliids (Williams 2008, 2012).

Syncarpy and Compitum

Syncarpy is the congenital union of carpels. Two major aspects of syncarpy are (1) the extent of union, whether carpels are united only at the base or up to the top, with all possible transitions; and (2) the extent of union (confluence) of the inner morphological surface of the carpels; if the inner morphological surfaces are confluent, formation of a unified PTTT of all carpels, termed a compitum, is possible (Carr and Carr 1961; Endress 1982). A compitum is the precondition for centralized pollen tube selection, which is hypothesized to be more efficient than individual selection in each carpel (Endress 1982; Armbruster et al. 2002; X.-F. Wang et al. 2012).

The members of the ANA grade and magnoliids are largely apocarpous (Endress and Igersheim 2000b; Endress and Doyle 2009). In contrast, syncarpy is predominant in core eudicots and monocots. More than 80% of all angiosperm species are syncarpous, with the remainder either apocarpous or unicarpellate (Endress 1982). It has yet to be determined in more detailed studies how common the presence of a compitum in syncarpous gynoecia is. However, it seems to be the normal state in core eudicots and monocots (other than the early-diverging lineages, which are partly apocarpous, e.g., several families in Alismatales), although monocots appear to be originally syncarpous (Endress and Doyle 2009; Doyle and Endress 2011; Sokoloff et al. 2013b). In some apocarpous Alismatales, an unusual compitum has evolved below the free carpels or by postgenital carpel fusion (Igersheim et al. 2001; X.-F. Wang et al. 2002, 2012; Endress 2011c). Postgenital fusion of free carpels (or carpels free in the upper region), and thus compitum formation without syncarpy, is also known from a number of eudicot families (e.g., Walker 1975; Endress et al. 1983; Fallen 1986; Matthews and Endress 2005b; El Ottra et al. 2013).

Flowers with a syncarpous or unicarpellate gynoecium may exhibit an inferior ovary (Fig. 14.17), suggested to provide better protection against certain pollinators with potentially destructive mouth parts, such as beetles and birds (Grant 1950). Inferior ovaries have evolved many times. They are more common in derived angiosperms, although they are also present in some basal groups (e.g., Chloranthaceae, Gomortegaceae, Hernandiaceae) and are known since the Early Cretaceous (Friis et al. 1994b). In core eudicots and derived monocots, entire large clades may be characterized by inferior ovaries (e.g., Asterales, Dipsacales, Rubiaceae, Orchidaceae, Zingiberales). Developmentally they arise by carpel initiation on a slightly concave floral apex and differential growth of the floral base including the gynoecium base, which results in a reinforcement of the concavity (cf. Endress 1994a). Evolutionary reversals in exceptional cases are becoming increasingly known (see Endress 2011b). Single genera with superior ovaries in these groups suggest such reversals, and the change from inferior to superior may even be seen during ontogeny, as in *Gaertnera* (Rubiaceae; Igersheim et al. 1994). In Saxifragaceae, there is evolutionary oscillation between more superior and more inferior ovaries (Kuzoff et al. 1999; Soltis et al. 2001a).

Ovules

Ovules are constituent parts of the carpels, although they appeared before the carpels in the evolutionary history of seed plants. Ovules consist of a nucellus (megasporangium) and, in angiosperms, basically two integuments. In the nucellus, meiosis takes place, and the embryo sac (megagame-

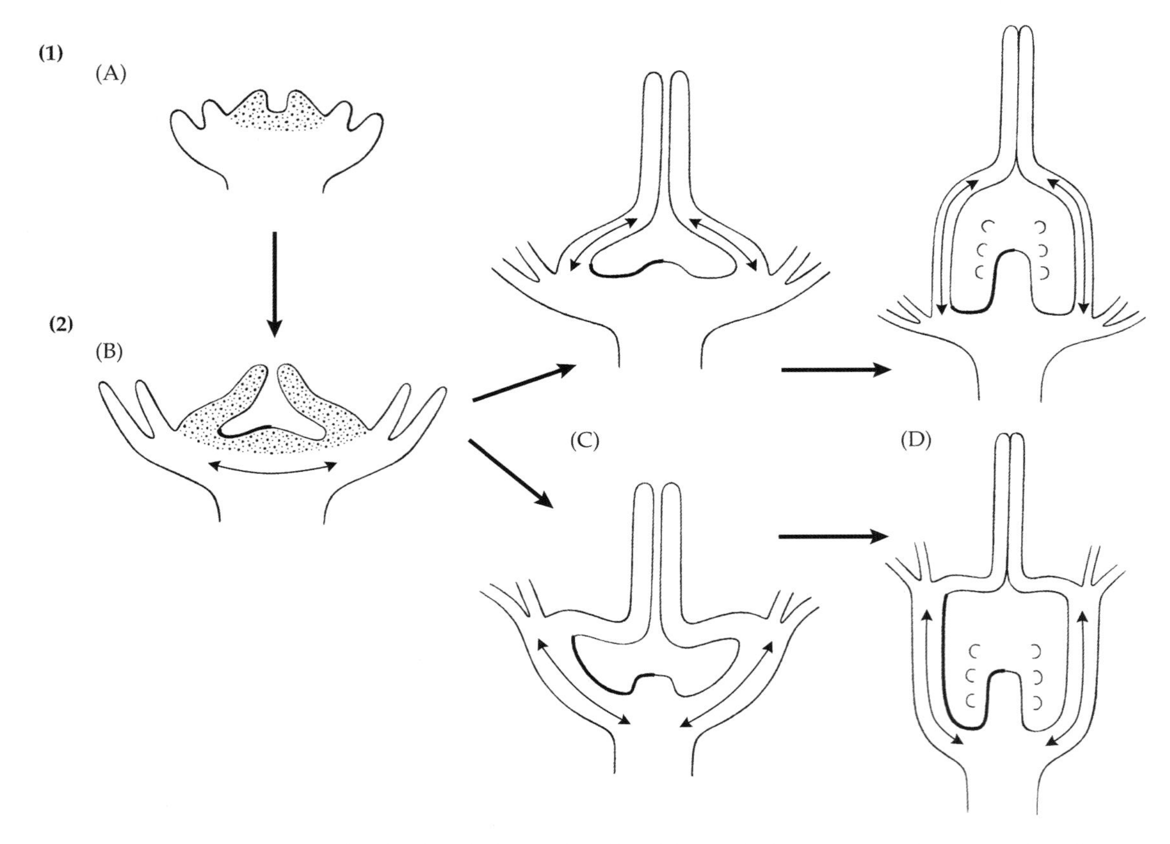

Figure 14.17. Gynoecium. Development of superior (1) and inferior (2) ovary by differential growth (C-D) from similar early stage (A-B) (modified from Endress 1994c; from Soltis et al. 2005).

tophyte) develops, in which fertilization of the egg cell takes place. One or both integuments form a micropyle, a narrow canal through which a pollen tube is chemotactically attracted (Herrero 2001). The pollen tube grows through the micropyle, reaches the nucellus apex, which it penetrates, grows into the embryo sac, and releases the male gametes. One male gamete commonly fuses with the egg cell to give rise to a zygote, and the other with the (diploid) nucleus of the central cell to give rise to a (triploid) endosperm.

Commonly, in angiosperms, crassinucellar (Fig. 14.18a) and tenuinucellar ovules (Fig. 14.18b) are distinguished. Tenuinucellar ovules are characterized by a single cell layer around the meiocyte in the nucellus, whereas in crassinucellar ovules there are two or more cell layers. The definition of the former is clear, whereas the latter is everything else, and everything else is really not the same. Crassinucellar ovules are predominant in angiosperms but they are not uniform in structure, so they should not be considered a single character. However, it has long been known that some larger groups of angiosperms are characterized by tenuinucellar ovules. In addition, tenuinucellar ovules commonly have only a single integument. The macrosystematic significance of this combination was emphasized by Philipson (1977). Among monocots, the phylogenetic distribution of these ovule types is less distinct (Rudall 1997).

Based on the results of molecular systematic studies since the 1990s, the tenuinucellar ovule is characteristic of some large clades, especially the asterids (cf. Endress et al. 2000a; Albach et al. 2001c; Endress 2002a; see Chapter 11). Several families characterized by tenuinucellar ovules that were once classified in Rosidae, Dilleniidae, or Hamamelidae (Cronquist 1981; Takhtajan 1997; see Chapter 3) have now been transferred to the asterids. However, tenuinucellar, unitegmic ovules, although predominant in asterids, cannot be considered a clear synapomorphy for asterids, because they do not occur in some early-branching clades (Endress et al. 2000a; Albach et al. 2001c; Endress 2002a, 2011a; Endress and Rapini 2014). In Cornales, ovules are crassinucellar or tenuinucellar and unitegmic (Satô 1976; Endress 2002a). In Ericales, ovules are tenuinucellar but fluctuate between bitegmic and unitegmic (Boesewinkel and Bouman 1991; Endress 2002a). In Garryales (with inclusion of part of Icacinaceae; Kårehed 2001), ovules are commonly crassinucellar but mostly unitegmic (e.g., Eckardt 1963; Kapil and Mohana Rao 1966; Satô 1976; Endress 2002a). There are also intermediate forms between crassinucellar and tenuinucellar, which may characterize families, for example, Convolvulaceae (see Endress 2002a). Endress (2011a) distinguished between tenuinucellar ovules (without any hypodermal tissue around the meiocyte in the nucellus) and incompletely tenuinucellar ovules (with hypodermal tissue lacking at the apex, but present at the flanks and/or below the meiocyte in the nucellus). Tenuinucellar ovules in this narrower sense are more or less restricted to groups of euasterids (campanulids and lamiids), whereas the "tenuinucellar" ovules in rosids and basal asterids are commonly incompletely tenuinucellar (Endress 2011a) (Fig. 14.19).

Tenuinucellar ovules are smaller than crassinucellar ovules, and thus the potential for flexibility in ovule number (between one and exceedingly many) per carpel is higher. The largest number of ovules per ovary occurs in some Orchidaceae, which have tenuinucellar ovules (over half a million in *Coryanthes*; see Nazarov and Gerlach 1997). In asterids with tenuinucellar ovules, some families, such as Solanaceae and Rubiaceae, also have clades with high ovule numbers. However, ovule number is a poorly explored character in these groups.

MORPHOLOGICAL ELABORATIONS IN FLOWERS POLLINATED BY ANIMALS

Many elaborations are based on synorganization of organs. An especially successful elaboration (innovation) was the evolution of sympetaly (see above), which allowed the easy building of different floral architectures (e.g., lipped flowers, tubular flowers of various lengths and widths, e.g., in Gentianales, Lamiales) and specialization for a variety of pollinators (e.g., bees, butterflies, sphingids, birds, bats) (e.g., Vogel 1990). Another example is the evolution of revolver flowers—that is, flowers that have several canals to reach the nectar, on which pollinators have to rotate around the center of the flower to gather all available nectar. This increases body contact with the anthers and stigma

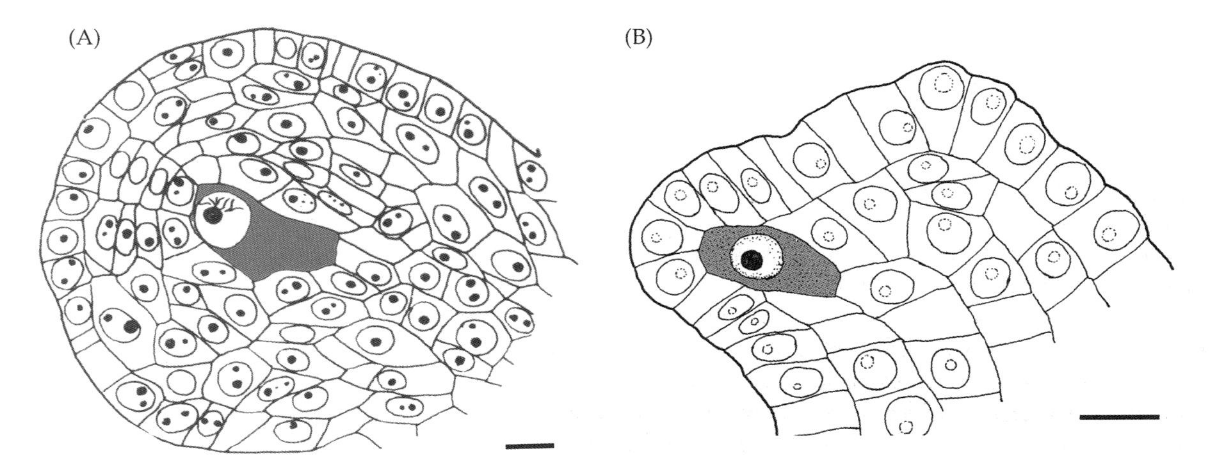

Figure 14.18. Ovules. Two major nucellus types, at the meiocyte stage, meiocyte highlighted dark grey. A. Crassinucellar. *Corylopsis willmottiae* (Hamamelidaceae) (modified from Endress 1977). B. Tenuinucellar. *Dermatobotrys saundersii* (Scrophulariaceae) (modified from Hakki 1977; from Soltis et al. 2005). Bars = 10 μm.

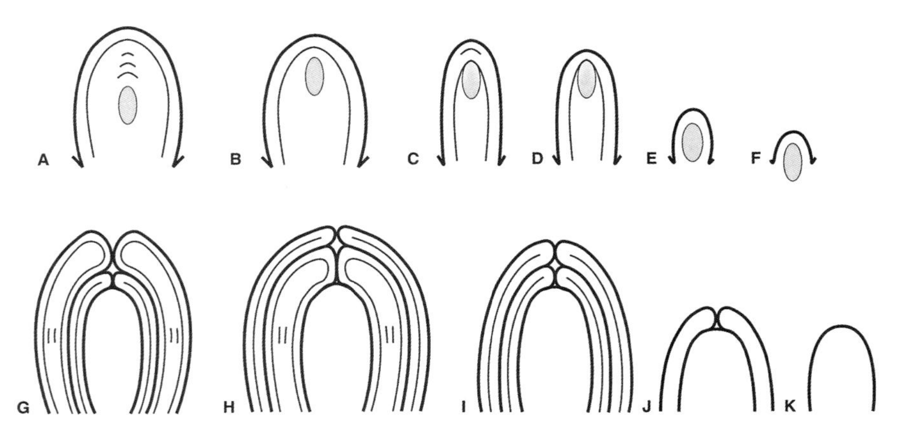

Figure 14.19. Ovules. Refined types of ovules of macrosystematic significance. Cell layers indicated by thin lines. A-F. Different nucellus shapes. A. Crassinucellar. B. Weakly crassinucellar. C. Pseudocrassinucellar. D. Incompletely tenuinucellar. E. Tenuinucellar. F. Reduced tenuinucellar. G-K. Different integument differentiation. G-I: Bitegmic, cell layers indiated by thin lines. G. Outer integument thicker than inner. H. Inner integument thicker than outer. I. Both integuments equally thick. J. Unitegmic. K. Ategmic. (Modified from Endress 2011a.)

and thus may enhance pollination success. Such flowers are common in Gentianales and Solanaceae (Endress 1994a).

Another principle is portioning of the pollen produced in a flower (or anther), resulting in staggered pollen presentation (Leins 2000). This is achieved by various means—for example, by elaborate secondary pollen presentation, as in Asterales (Leins and Erbar 1990; Erbar and Leins 1995b) or Fabales (Westerkamp and Weber 1999). In flowers with secondary pollen presentation, pollen transfer to the pollinators is not directly from the anthers, but rather pollen is first deposited on the style or in the tip of a keel before it is removed by pollinators. Secondary pollen presentation has evolved in many clades of angiosperms (Yeo 1992), some of which are species-rich, such as Asteraceae and Fabaceae. Pollen portioning is also found in flowers with poricidal anthers, which are buzz-pollinated by bees (see below). Pollination by pollinia has been successful in orchids and asclepiads (Pacini and Hesse 2002). Here, the entire pollen mass of a theca is transported as a solid parcel (pollinium), and thus no pollen is lost on the way from one flower to the next. In some subgroups of both orchids and asclepiads, pollen portioning is also attained by the formation of smaller units, such as tetrads or massulae instead of entire pollinia.

The evolution of monosymmetric flowers allowed enhanced precise mechanisms of pollen application to the body of a pollinator and subsequently to the stigma of a flower. This also led to a greater potential for diversification in pollination biology. Some large, successful clades have evolved monosymmetric flowers, such as Orchidaceae, Zingiberales, Fabales, and Lamiales (Endress 1999).

Various means of pollinator attraction, in addition to pollen, have evolved in flowers, such as nectar, oil, resin, and perfume, which are used for food, nest building, or attraction of mates. Whereas pollen and nectar are used by many pollinators, oil, resin, and perfume are used by some highly specialized bees (Vogel 1974, 1988; Simpson and Neff 1981; Steiner 1991). Some flowers pollinated by certain bees that collect pollen by "buzzing"—that is, by activity of flight muscles—have attained a "solanoid" overall shape, that is, they have a flat, expanded corolla, a cone-shaped androecium with large, showy, poricidal anthers and short filaments, and a punctiform stigma (Vogel 1978; Vaknin et al. 2001). Such flowers are known from many families of eudicots and monocots. In flowers that offer only pollen as a reward to pollinators, there is also an evolutionary trend to diversify stamen structure and function into showy "feeding stamens" and cryptic "pollinating stamens" (heteranthery), and in some cases even the pollen becomes different in the two stamen morphs (Vogel 1978; Endress 1997a; Barrett 2010a). Arrangement, size, shape, color, and scent of flowers are means of differential attraction of pollinators and a source for reproductive isolation, speciation, and thus diversification (Waser 1998; Chittka et al. 1999; Lunau 2000; Chittka and Thomson 2001; Prusinkiewicz et al. 2007).

A considerable repertoire of mating systems, apart from specialization to different pollinators, also plays a role in floral diversification, such as various kinds of gender distribution, self-incompatibility, self-compatibility, dichogamy, and herkogamy (e.g., Bertin and Newman 1993; Barrett 1995, 1998, 2002, 2003; Barrett et al. 1996, 2000; Holsinger 1996; Richards 1997; Harder and Barrett 2006). These specializations involve a number of ways to enhance outbreeding or keep a balance between outbreeding and inbreeding (Lloyd and Schoen 1992; Freeman et al. 1997; Schoen et al. 1997). Monoecy and dioecy are different ways of separating genders on one or on separate individuals (e.g., Renner and Won 2001). Dioecy seems to evolve commonly from monoecy (Renner and Ricklefs 1995; Sakai and Weller 1999; Weller and Sakai 1999; but see also Spigler and Ashman 2012). In dichogamy, both genders are produced by the same individual but not at the same time; both genders can be present in the same flowers (bisexual flowers) or in separate flowers (unisexual flowers); both protogyny (with the female function working first) and protandry (with the male function working first) occur (Lloyd and Webb 1986; Renner 2001). In herkogamy, both genders are produced by the same individual and at the same time but spatially separated on the same plant (Webb and Lloyd 1986). A special case of herkogamy is heterostyly, in which different floral morphs with long and short stamens or styles are produced on separate individuals (Barrett 1992). Commonly, these sexual system variants are flexible at relatively shallow levels of the phylogenetic hierarchy (Korpelainen 1998), but protogyny (in perfect flowers) is extensive in the ANA grade and magnoliids (Endress 1990b, 2001c, 2010b).

MORPHOLOGICAL REDUCTION IN FLOWERS THAT ARE NOT ANIMAL-POLLINATED

In some angiosperm groups, wind pollination is predominant, and flowers are concomitantly greatly altered in structure and behavior as compared to those of their animal-pollinated ancestors. The main trend is floral reductions of various sorts, but this is accompanied by an increase in pollen production (Wagenitz 1975), including increased pollen-to-ovule ratio (Friedman and Barrett 2009), which may lead to drastic changes in flower appearance. Major

(A)

(B)

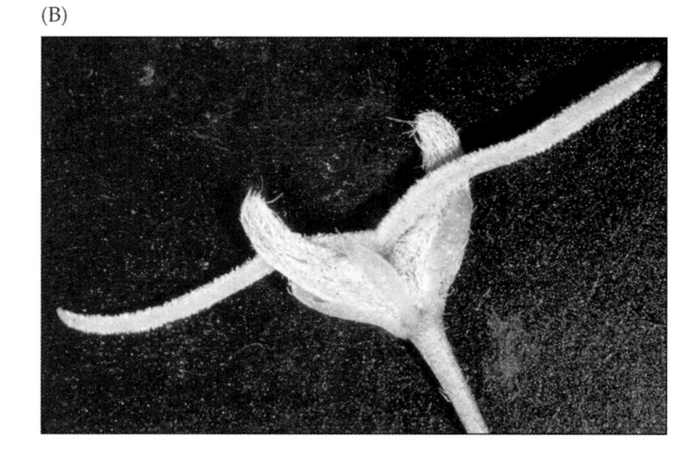

Figure 14.20. Drastically different appearance of flowers with two different pollination syndromes in two species of the same genus. A. *Acer platanoides* (Sapindaceae), insect-pollinated. B. *Acer negundo* (Sapindaceae), wind-pollinated. P. Endress, from Soltis et al. (2005).

wind-pollinated groups are Fagales and many Rosales (the urticoid families) among eudicots, and several Poales among monocots, grasses and sedges being the most prominent (Linder 1998). Wind pollination has also evolved in many smaller groups that are phylogenetically nested in larger animal-pollinated groups, such as single genera (e.g., *Dodonaea* in Sapindaceae, *Ambrosia* in Asteraceae, *Plantago* in Plantaginaceae, Preston et al. 2011) or single species. In *Acer negundo* (Fig. 14. 20b), in contrast to the insect-pollinated *Acer platanoides* (Fig. 14.20a), the flowers are unisexual, petals and nectary are lacking, and the stigma is much larger, thus providing a dramatic change in the superficial appearance. Water pollination is much less prominent in angiosperms, but it has evolved in several groups (Philbrick and Les 1996) and is also accompanied by floral reduction, in some groups extremely so (some genera of Hydrocharitaceae and other small families of Alismatales, *Callitriche* of Plantaginaceae) (Cook 1982).

In Fagales, which are predominantly wind-pollinated, a suite of traits, not only in floral morphology at anthesis but also in floral development before and after anthesis, is apparently functionally related to wind pollination. These traits include flowering early in the growing season, in the leafless state, and immaturity of the ovary and ovules at anthesis with the effect that pollination may precede fertilization by months. In extreme cases (e.g., *Corylus*), ovules are not even present at the time of pollination (Endress 1977; Thompson 1979), but this post-pollination delay in ovule formation can also occur in insect-pollinated groups, such as Orchidaceae. Whether wind pollination evolved within Fagales or their ancestors and how many reversals to insect pollination there were in Fagales is not clear (Manos et al. 2001).

EVOLUTIONARY INNOVATIONS AND DIVERSIFICATION IN FLOWERS OF THE MAJOR ANGIOSPERM GROUPS

Evolutionary innovations or idiosyncrasies in flowers often characterize large angiosperm clades (Fig. 14.21). In many basal angiosperms, including Magnoliales and Laurales, floral phyllotaxis is diverse and flexible, and the spiral pattern is not suppressed because synorganizations are largely lacking (Endress 1987a). Floral merism is also highly variable. The perianth is not clearly differentiated into sepals and petals, and sometimes it is even absent (see also Chapters 4–6). Stamens are often massive, with short filaments, thick connectives, and frequently valvate dehiscence. Uniaperturate pollen is predominant. The gynoecium is often apocarpous. Ovules are crassinucellar, bitegmic, and commonly anatropous.

In some species of the ANA grade, flowers are small (except for some more specialized taxa: Austrobaileyaceae, Nymphaeaceae); floral phyllotaxis is spiral (Amborellaceae, Austrobaileyales) or whorled (Nymphaeales). The perianth is not clearly differentiated into sepals and petals. Stamens open with longitudinal slits (not with valves; except for *Nuphar*, Nymphaeaceae; Hufford 1996a). Carpel closure (angiospermy) is not by postgenital fusion but by secretion (except Illiciaceae, Nymphaeaceae) (Doyle and Endress 2000, 2011; Endress and Igersheim 2000b; Endress 2001c; Endress and Doyle 2007, 2009).

In magnoliids, many floral features are similar to those in the ANA grade. However, stamens opening by valves, concomitant with a thick connective, is common (Magnoliales, Laurales), angiospermy is often by postgenital fusion, and floral trimery is widespread.

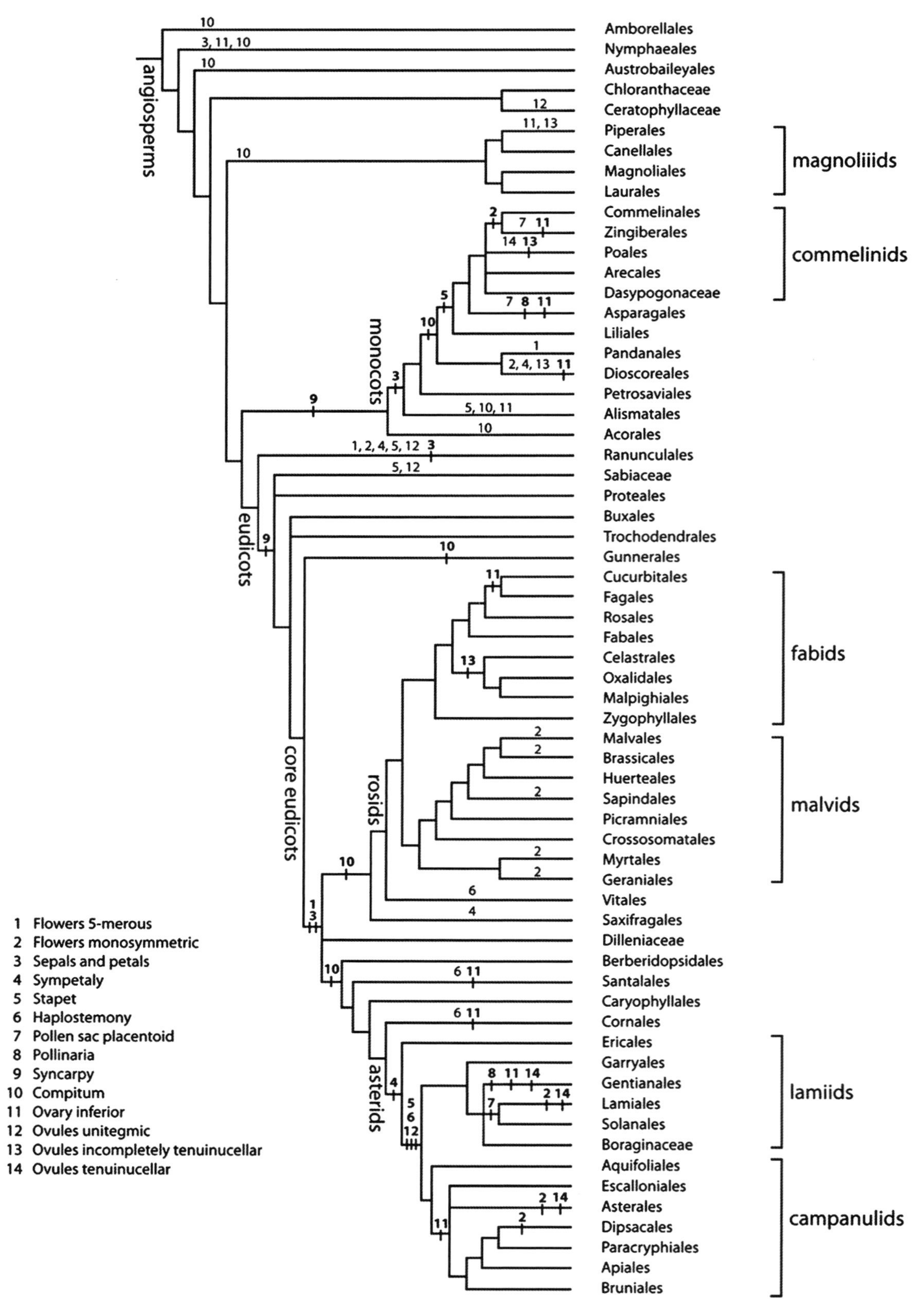

Figure 14.21. Innovations in flower evolution of angiosperms. Fourteen innovations are mentioned, most of them innovations in more than one clade. Probable key innovations are marked with a short line on the respective clades. Non-key innovations are marked without a short line. Cladogram of angiosperms from APG III (2009). (Modified from Endress 2011c, p. 371.)

In monocots, flowers are almost always trimerous, except in Pandanales in which dimery, tetramery, pentamery, and polymery also occur. There is a tendency for intimate connections of organs of different kinds within the same floral radius (Endress 1995a; Remizowa et al. 2010). The perianth is not always clearly differentiated into sepals and petals. One of the three outer perianth organs is often anterior in the median plane of the flower (Remizowa et al. 2013). Pollen is uniaperturate, as in the early-diverging angiosperms. Nectaries are present in the septa of carpels (septal nectaries) or on the tepal surface (tepal nectaries). Nectaries are lacking in some large groups (e.g., Poales). Often, floral bracts or inflorescence bracts are elaborated and play important roles as protective organs for flowers or entire inflorescences (e.g., Poaceae and Arecaceae) and/or as attractive organs for pollinators (e.g., Araceae, Bromeliaceae, Arecaceae, Pandanaceae, Zingiberales) (Endress 1995a).

In eudicots, floral phyllotaxis is commonly whorled, except for some Ranunculales and Nelumbonaceae (Hiepko 1965b; Endress and Doyle 2007; Endress 2010b,c). Flowers are often pentamerous, except for clades at the basal nodes. Perianths have sepals and petals in core eudicots and many Ranunculales. Stamens commonly have long filaments and anthers with thin connectives, each theca opening by a longitudinal slit. Eudicots are characterized by the presence of tricolpate or tricolpate-derived pollen (Doyle and Hotton 1991); an alternative name for the eudicots is tricolpates.

Early-diverging eudicots are labile in floral construction (see also Chapter 8). Spiral and whorled floral phyllotaxis occur side by side, sometimes in the same genus but more often among genera and families. Flowers with spirally arranged organs occur in many Ranunculaceae (Schöffel 1932; Hiepko 1965b), some Menispermaceae (Endress 1995b), Circaeasteraceae (Ren et al. 2004; Tian et al. 2006), *Nelumbo* (perianth) (Hayes et al. 2000), and sometimes Trochodendraceae (*Trochodendron* is spiral and whorled) (Endress 1990a). In taxa with spiral floral phyllotaxis, floral organ number is highly variable; in taxa with whorled flowers, di- and trimerous flowers predominate (Ranunculales, Proteaceae, Platanaceae, Buxaceae, *Tetracentron, Myrothamnus*) (Drinnan et al. 1994; Chen et al. 2007; von Balthazar and Schönenberger 2009; Endress 2010a). In some groups, thecae open by valves and not by longitudinal slits (a few Ranunculaceae, Eupteleaceae, Platanaceae, Trochodendraceae). Pollen is often tricolpate, or tricolporate in some (e.g., some Menispermaceae; Thanikaimoni 1984). Disk nectaries are lacking except for Buxaceae and Sabiaceae (von Balthazar and Endress 2002b); nectaries are present on petals or staminodes or are lacking. Ovules are mostly crassinucellar and bitegmic (but pseudocrassinucellar in Papaveraceae and Ranunculaceae, cf. Endress and Igersheim 1999).

In core eudicots, flowers are predominantly pentamerous or pentamerous-derived. Increase in stamen number, based on secondary stamen primordia superimposed upon the primary primordia, has occurred in several families. Pollen is mostly tricolporate or tricolporate-derived. Disk nectaries are common.

In rosids, flowers are almost always choripetalous. Ovules are mostly crassinucellar and bitegmic. In many groups, secondary increase in stamen number is common. Many floral details in rosid clades are poorly known, but comparative studies of several larger clades of rosids sensu APG have been carried out in the past 15 years (e.g., Ronse De Craene and Smets 1999; Matthews et al. 2001, 2012; Ronse De Craene et al. 2001; Schönenberger et al. 2001a; Matthews and Endress 2002, 2004, 2005a,b, 2008, 2011; Endress et al. 2013).

In asterids, flowers are almost always sympetalous, and, in many groups, stamens are also fused with the sympetalous corolla. At least in some derived asterids (Gentianales, Lamiales, Asterales, Dipsacales), stamens are in a single whorl. Carpels are commonly reduced to two (or three). Ovules are incompletely tenuinucellar or tenuinucellar and unitegmic. The great diversity in floral forms among asterids is mainly based on variation in the sympetalous corolla and the often monosymmetric flowers. More recent comparative structural studies on larger clades of asterids are by Schönenberger et al. (2010) and von Balthazar and Schönenberger (2013).

FLORAL DEVELOPMENTAL GENETICS

The vast diversity of floral forms reflects the interplay of underlying genetic programs that control elements of organization, construction, and mode. Here we review current understanding of the genetic underpinnings of these components of floral variation, emphasizing floral organ specification and number (organization), symmetry (construction), and color (mode). The most obvious shifts in floral morphology seem to be controlled by variation in expression of transcription factors, which may then trigger complex downstream cascades of genes that encode more specific attributes of floral features.

ORGANIZATION: PRESENCE AND DISPOSITION OF FLORAL ORGANS (FLORAL ORGAN IDENTITY)

Broad contributions from paleobotany, phylogenetics, genomics, developmental biology, and developmental genetics have yielded tremendous insight into Darwin's "abominable mystery" (see Friedman 2009)—the origin and rapid diversification of the angiosperms soon after their origin (see also reviews by P. Soltis et al. 2006, 2009; Soltis et al. 2007, 2008b). Much of floral diversity results from variation in the types and numbers of organs composing a flower: for example, tepals vs. sepals and petals, and indeterminate vs. fixed numbers of organs.

Whereas the central feature of the angiosperms is the flower, the origin of this complex structure and its subsequent diversification throughout angiosperm evolution still remain fundamental evolutionary questions. The past 25 years have seen tremendous new developments in our understanding of the genes involved in the making of a flower. Over that time, the well-known ABC model (now often referred to as the ABCE model, below; Fig. 14.22) of floral organ identity has been the unifying paradigm for floral developmental genetics (Coen and Meyerowitz 1991). This model proposed that floral organ identity is controlled by three major gene functions, A, B, and C, that act in combination to produce the floral organs; A-function alone specifies sepal identity, A- and B-functions together control petal identity; B- and C-functions together control stamen identity; C-function alone specifies carpel identity (Fig. 14.22).

Several genes act as key regulators of floral organ identity in model eudicots, such as *Arabidopsis thaliana* (Brassicaceae) and *Antirrhinum majus* (Plantaginaceae; snapdragon). In *Arabidopsis*, *APETALA1* (*AP1*) and *AP2* are the A-function genes, *AP3* and *PISTILLATA* (*PI*) are the B-function genes, and *AGAMOUS* (*AG*) is the C-function gene. In *Antirrhinum*, the homologous gene (or homolog) to *AP1* is *SQUAMOSA*. Details regarding A-function remain complex, however, with A-function not clearly documented except in *Arabidopsis*. The homologs of the A-function gene *AP2* in *Antirrhinum* are *LIPLESS1* and *LIPLESS2*, which may provide partial A-function in snapdragon (Keck et al. 2003). The B-function genes in *Antirrhinum* are *DEFICIENS* (*DEF*) and *GLOBOSA* (*GLO*), which are homologs of *AP3* and *PI*, respectively. The C-function gene in *Antirrhinum* is *PLENA* (*PLE*). All these genes, with the exception of *AP2* (and its homologs), are MADS-box genes (Theissen et al. 2000), a broad family of eukaryotic genes that encode transcription factors containing a highly conserved DNA-binding domain (MADS domain).

Extensions to the original ABC model accommodated subsequent discoveries, including the identification of additional MADS-box genes that control ovule identity (D-function; Colombo et al. 1995) and those that also contribute to sepal, petal, stamen, and carpel identity (E-function; Pelaz et al. 2000). We will not consider D-function further here, given that ovules are part of the carpels and not floral organs per se. However, E-function plays a major role in the formation of floral organs and is closely integrated with ABC functions. The E-function genes in *Arabidopsis* are *SEPALLATA1* (*SEP1*), *2*, *3*, and *4* (Pelaz et al. 2000). SEP proteins, together with the protein products of the ABC genes, are required to specify floral organ identity. The *SEP* genes are functionally redundant in their control of sepals, petals, stamens, and carpels (Fig. 14.22). Given the important role of E-function in organ specification, ABCE is a more appropriate name for the general model of floral organ identity than simply ABC.

Despite its joint derivation from phylogenetically divergent species (*Arabidopsis thaliana* and *Antirrhinum majus*; Bowman et al. 1989; Schwarz-Sommer et al. 1990; Coen and Meyerowitz 1991; Davies et al. 2006), the ABCE model is nonetheless based on phylogenetically derived eudicot model systems (D. Soltis et al. 2002a; P. Soltis et al. 2006). Data from additional eudicots, including *Petunia* (reviewed in Rijpkema et al. 2006) and *Gerbera* (reviewed in Teeri et al. 2006), complemented the data for *Arabidopsis* and *Antirrhinum*. Expression patterns of MADS-box genes in eudicots as well as grasses typically support the ABCE model. In particular, early studies revealed that strong expression of *AP3* and *PI* homologs in eudicots is typically limited to petals and stamens, where these genes are required for organ identity specification (Ma and dePamphilis 2000). However, over the past decade, it has become clear that the ABCE model developed in eudicot model plants does not apply to early-diverging angiosperm groups. Data produced in large part by the Floral Genome Project (D. Soltis et al. 2002a, 2007a; Albert et al. 2005; Kim et al. 2005; P. Soltis et al. 2006) provided some of the first crucial insights into the floral organ identity genes and their patterns of expression in several species of basal angiosperms (reviewed in P. Soltis et al. 2006; D. Soltis et al. 2007a).

Is the ABCE model applicable to basal angiosperms and thereby relevant to the origin and early evolution of the flower? As reviewed below (see also D. Soltis et al. 2002a, 2007a; Albert et al. 2005; P. Soltis et al. 2006, 2009), the answer is yes and no. The expression of MADS-box genes in the floral organs of basal angiosperms is generally consistent with the ABCE model. Importantly, however, the expression patterns in basal angiosperms are often broader than those in eudicot flowers. The floral morphology of many basal angiosperms provides a crucial hint to what may be a more appropriate model for these plants. In *Amborella* and some other basal angiosperms, floral organs are spirally arranged with a gradual transition from bracts to outer and inner tepals, from tepals to stamens, and finally to carpels (see Chapter 4). These gradual intergradations of floral organs cannot be easily explained by the classic

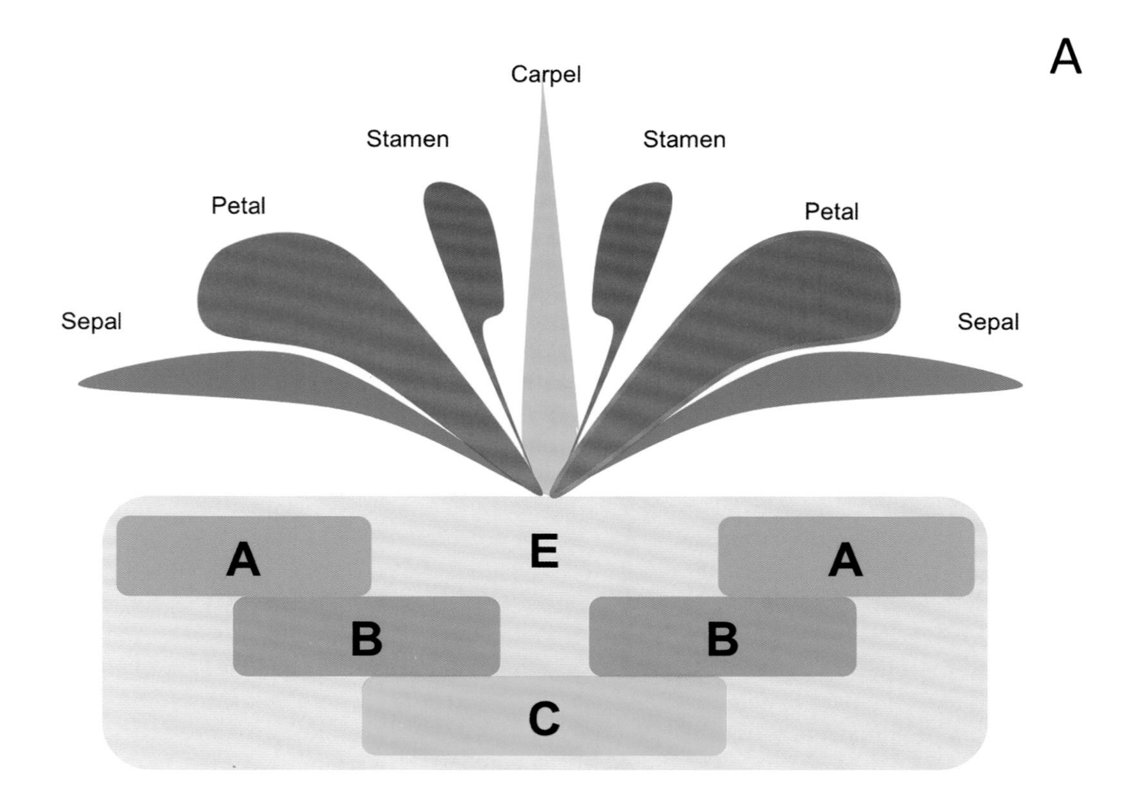

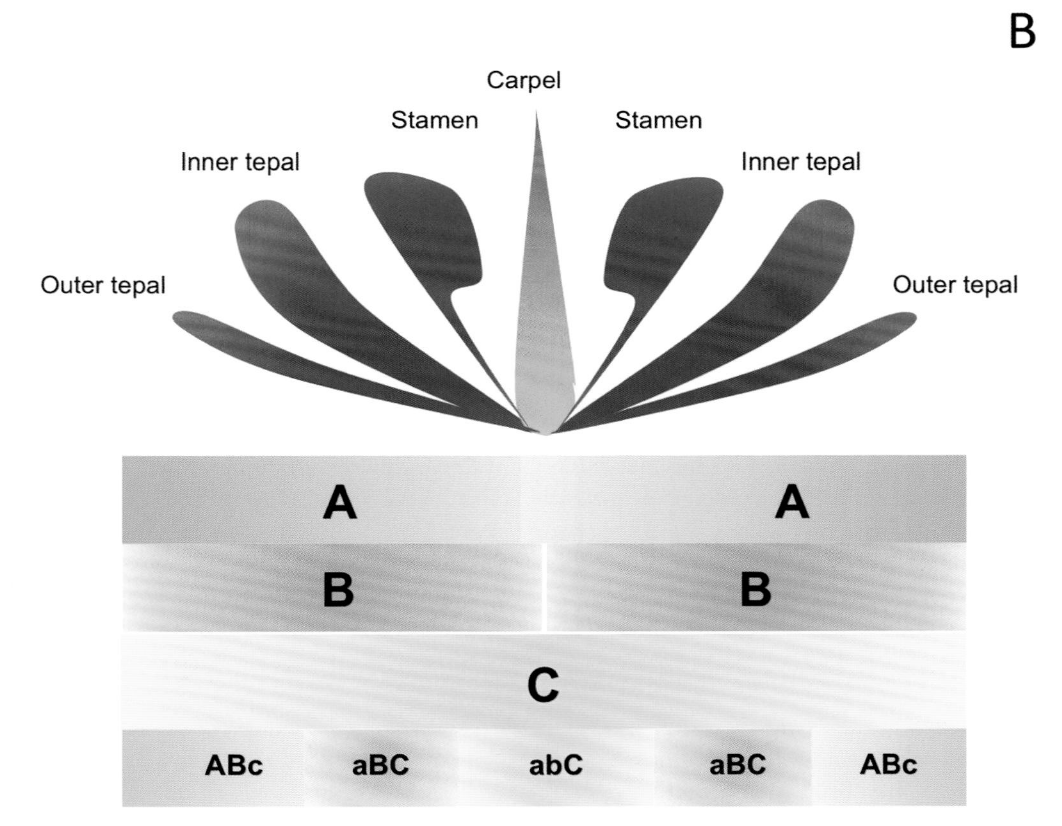

Figure 14.22. A. ABCE Model of floral organ identity. Sepals are produced where A function acts alone, petals where A and B functions overlap, stamens where B and C functions combine, and carpels where C function acts alone. In the eudicot genetic model plant *Arabidopsis thaliana*, *APETALA1* (*AP1*) and *APETALA2* (*AP2*) are the A-function genes, *APETALA3* (*AP3*) and *PISTILLATA* (*PI*) together specify B function, C function is specified by *AGAMOUS* (*AG*), and multiple *SEPALLATA* genes provide E function. (Modified from Chanderbali et al. 2010.) B. Fading borders model for floral organ identity. Unlike the classic ABCE model where there is a sharp boundary between gene expression corresponding to floral organs, in the fading borders model there is a gradual transition of gene function across floral organs. (Courtesy of A. Chanderbali.)

ABCE model as proposed for eudicots in which there are sharp boundaries between floral organs.

Developmental genetics studies conducted for basal angiosperms indicate a broader pattern of expression of B- (and to a lesser extent, C- and E-) function homologs in basal angiosperms than in eudicots (Kim et al. 2005). These results prompted the formulation of the fading borders model (Buzgo et al. 2004a, 2005) (Fig. 14.22b), which proposes that the gradual transitions in floral organ morphology result from a gradient in the level of expression of floral organ identity genes across the developing floral meristem. Weak expression at the margin of a gene's range of "activity" overlapping with the expression of another regulator in adjacent cells results in the formation of morphologically intermediate floral organs rather than organs that are clearly distinct.

Homologs of B-function genes, *AP3* and *PI*, are broadly expressed in tepals, stamens, and carpels in the ANA grade of angiosperms, including *Amborella* (Amborellales), water lilies (Nymphaeales), and *Illicium* (Austrobaileyales), as well as in magnoliids (e.g., *Magnolia*) (Kim et al. 2005; Fig. 14.23). Additional support for fading borders was provided at a transcriptome-wide level (Chanderbali et al. 2010; see below and Fig. 14.24).

The ABCE model as developed for eudicots is therefore not perfectly applicable to basal angiosperms and, by inference, the earliest angiosperms. The ABCE model of floral organ identity is typically considered the default program, with variants viewed as derivatives of this program. In fact, however, when gene expression profiles of floral-organ regulators are compared in a phylogenetic context, it is clear that the ABC model of *Arabidopsis* is evolutionarily derived. The ancestral flower had broad expression patterns of at least B-function regulators; broad and overlapping expression yielded morphologically intergrading floral organs, as seen in a number of extant basal angiosperms. Restriction of expression (and function) to specific regions of the floral meristem resulted in the discrete whorls of morphologically distinct floral organs that together characterize most of the eudicots and certainly all of the core eudicots. Further investigation of the evolution of the floral regulatory network should rely on the phylogenetic perspective that the ABCE model is derived.

But how was the genetic machinery necessary for specifying a flower assembled in the first place? Were the genes co-opted from other processes and integrated into a pathway gradually, or were they brought together more suddenly, perhaps through gene or whole-genome duplication? Parallel duplications of floral regulatory genes, consistent with evidence from broad-scale genomic analyses (Jiao et al. 2011, 2012), suggest whole-genome duplications in the common ancestor of extant angiosperms and the common ancestor of core eudicots. However, the mere duplication of a genome or set of genes was likely not coincident with the origin of morphological novelty. Certainly some time would have been needed for the assembly of a functional floral-organ specification program in the ancestral angiosperm. Furthermore, the duplication of the B-function homologs (*AP3* and *PI*) apparently occurred approximately 260 mya, 130 myr before the first fossil evidence of angiosperms. The process of assembling a new genetic program and its translation into morphological innovation merits further study.

Studies of floral developmental genetics have also provided additional insights into the evolution of the perianth. The traditional view of angiosperm flower evolution maintains that stamens and carpels evolved just once, whereas the sterile perianth organs may have arisen multiple times (e.g., Eames 1961; Takhtajan 1991). The rationale for this evolutionary reasoning is based on the longstanding view that angiosperms are derived from ancestors without petals, making the perianth an evolutionary novelty. The strong resemblance in some groups of sepals to foliar bracts and the similarity of petals to stamens has supported the view that sepals are derived from foliar bracts whereas petals are derived from stamens. Those petals that are stamen-derived (called andropetals by Takhtajan 1991) are primarily associated with what we now term eudicots. In contrast, the perianth of basal angiosperms often comprises morphologically similar organs, termed tepals (Endress 2001a)—these tepaloid organs could be assigned bracteal (bracteopetal) or staminal (andropetal) origins depending on whether sepal-like or petal-like features prevail.

Following these criteria, the tepals of Lauraceae have been considered bracteopetals (Albert et al. 1998; Ronse De Craene et al. 2003), but expression data for *Persea* (based on RT-PCR, in situ hybridization, and microarrays), coupled with developmental data, indicate that the "petals" of *Persea* and other Lauraceae are clearly of staminal origin (Chanderbali et al. 2006, 2009). Global patterns of gene expression in three phylogenetically pivotal angiosperms, a water lily (*Nuphar advena*), avocado (*Persea americana*, a magnoliid), and California poppy (*Eschscholzia californica*, a basal eudicot), along with the model core eudicot *Arabidopsis thaliana* and a cycad (*Zamia vazquezii*), support the fading borders model (Chanderbali et al. 2010). Transcriptional cascades exhibit broadly overlapping patterns of spatial gene expression deployed across the morphologically similar and intergrading floral organs typical of both water lily and avocado flowers (Fig. 14.24). In contrast to the patterns in these basal angiosperms, in both the basal eudicot California poppy and derived core eudicot *Arabidopsis*, spatially discrete transcriptional programs were observed in the morphologically distinct floral organs.

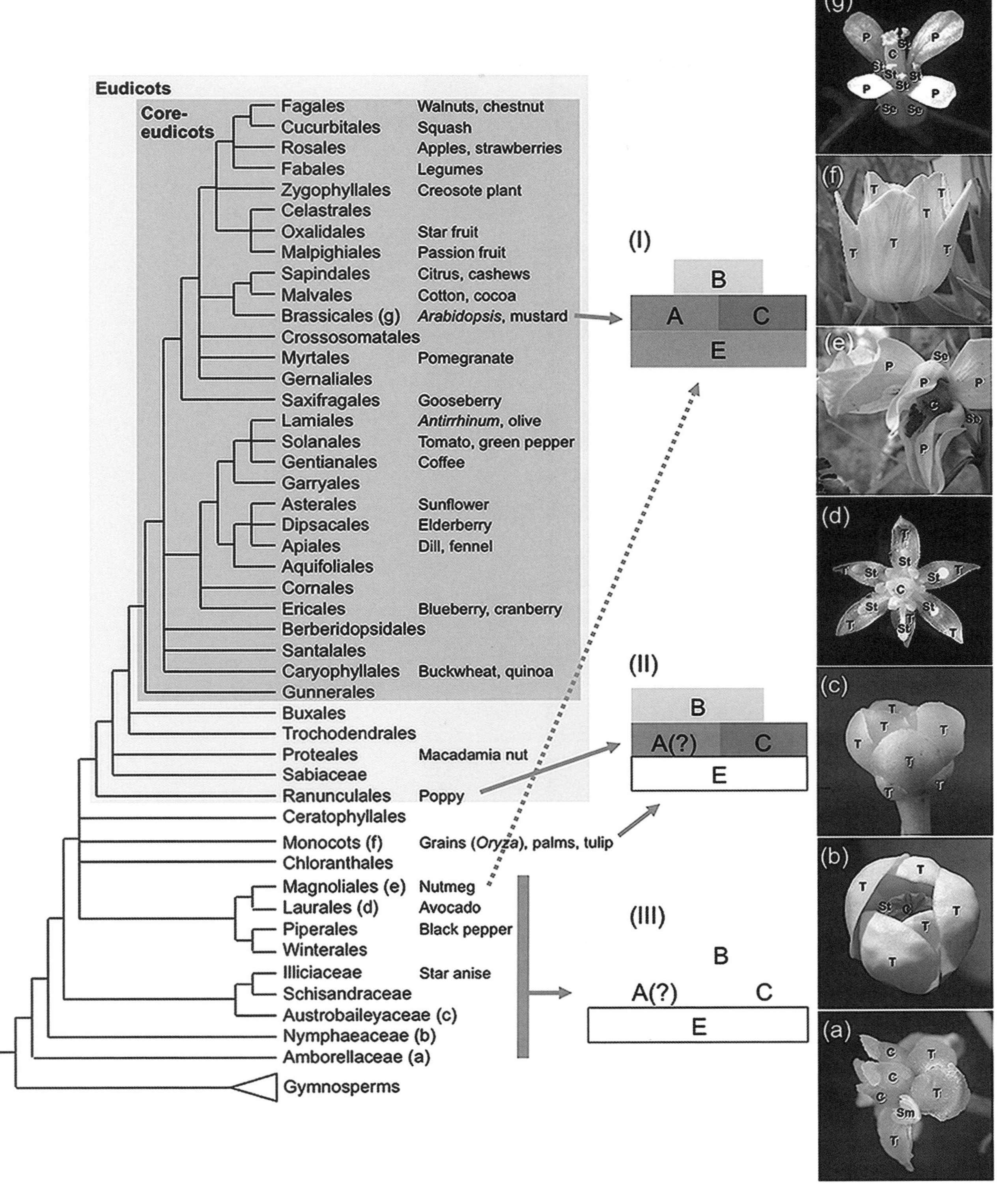

Figure 14.23. Phylogenetic summary of angiosperm floral developmental genetics. Known or postulated expression patterns are shown on the right for organ identity genes: (I) ABCE model developed for core eudicots—this model may apply to *Asimina* (Annonaceae), a member of the magnoliid clade. (II) Shifting boundary model which has been applied to some basal eudicots (see Kramer and Irish 2000). (III) Fading borders (see Fig. 14.22) proposed for basal angiosperms. (Modifed from Kim et al. 2005.)

Figure 14.24. Fading borders in action. Canalization of floral organ transcriptional programs during angiosperm diversification. (A) Flowers of *Nuphar* and *Persea* bear an undifferentiated perianth of petaloid organs (tepals), whereas in *Eschscholzia* and *Arabidopsis* flowers the perianth is differentiated into leaf-like outer sepals and colorful inner petals. (B) Log2 floral organ/leaf gene expression ratios ranked by organs of peak expression contrast the blurred boundaries between adjacent floral organs in *Nuphar* and *Persea* versus the sharp boundaries in *Arabidopsis* and *Eschscholzia*. Stamen-preferential genes generally are more broadly expressed, but more so in *Nuphar* and *Persea* than in the eudicots, whereas carpel-preferential genes generally are more spatially restricted. The scale of log2 ratios ranges from saturated yellow (1 and higher = at least two-fold up-regulated) to black (0 and lower = no change or down-regulated). (Modified from Chanderbali et al. 2010.)

In addition, deep evolutionary conservation in the genetic programs of putatively homologous floral organs appears to trace to those operating in gymnosperm reproductive cones. For example, angiosperm carpels and female cycad cones share conserved genetic features that are associated with the ovule developmental program common to both organs. In contrast, male cones of gymnosperms share transcriptomic features with both the angiosperm perianth and stamens. These results therefore support the evolutionary origin of the perianth of angiosperms from the male genetic program of other seed plants.

Floral Developmental Genetics in Aizoaceae: A Case Study of the Origins of Petaloidy

Caryophyllales exhibit exceptional levels of floral diversity and complex patterns of floral variation (Ronse De Craene 2013). Reconstructions of perianth differentiation across Caryophyllales reveal that the ancestral condition was most likely a pentamerous and apetalous flower with a uniseriate perianth in a quincuncial arrangement (Brockington et al. 2009). A differentiated perianth with sepals and petals has originated independently at least nine times, in Asteropeiaceae, Caryophyllaceae, Stegnospermataceae, Limeaceae, Lophiocarpaceae, Aizoaceae, Nyctaginaceae, Molluginaceae, and Portulacineae (Brockington et al. 2009). Perianth differentiation has been achieved by different mechanisms in these nine lineages, through either petaloid differentiation of the ancestral uniseriate perianth or the recruitment of other floral structures, either the androecium or the preceding bracts (Brockington et al. 2009). These separate origins of differentiated perianth, with the concomitant evolution of petaloidy from either androecial or bracteal organs, make Caryophyllales a valuable system to address the evolutionary developmental genetics of petaloidy, particularly the role of the canonical eudicot petal identity program in recurrent petal evolution (Brockington et al. 2011).

Perianth differentiation through sterilization and petaloid modification of the outer members of a centrifugally initiating androecium has arisen a minimum of four times in Caryophyllales, in Aizoaceae, Molluginaceae, Lophiocarpaceae, and Limeaceae. In these instances, androecial development proceeds centrifugally, and the basipetal members become progressively more sterile and petaloid (Brockington et al. 2009). Aizoaceae comprise four subfamilies, with Sesuvioideae and Aizooideae as successive sisters to Mesembryanthemoideae plus Ruschioideae (Klak et al. 2003). Sesuvioideae and Aizooideae exhibit an undifferentiated perianth comprising a single whorl of five tepals (Fig. 14.25), which are petaloid on their adaxial surfaces and sepaloid on their abaxial surfaces. In contrast, Mesembryanthemoideae and Ruschioideae display a differentiated perianth with an outer whorl of sepals (homologous to the petaloid tepals of Sesuvioideae/Aizooideae) and an inner whorl of putative andropetals. Therefore, a transfer of function has occurred from the adaxial tepal surface of Sesuvioideae and Aizooideae to the andropetals of Mesembryanthemoideae and Ruschioideae. Brockington et al. (2011) examined these differently derived petaloid organs for evidence of a shared core eudicot petal identity program, by isolating *AP3*, *PI*, and *AG* homologs from all four subfamilies and observing in situ expression patterns of these genes in two species representing the two distinct floral types, *Sesuvium portulacastrum* (Sesuvioideae: petaloid tepals) and *Delosperma napiforme* (Ruschioideae: petaloid staminodes).

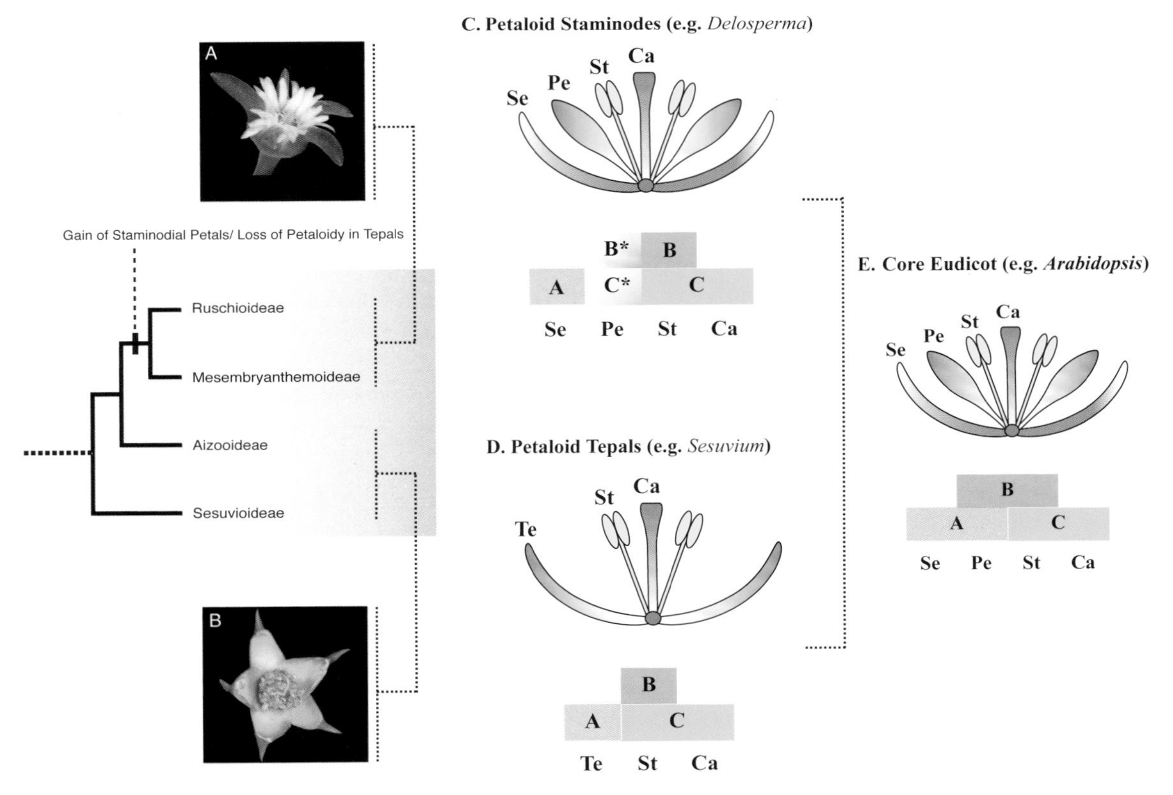

Figure 14.25. Phylogeny of Aizoaceae depicting the four major sub-families and the two different floral types: A = petaloid staminodes in *Delosperma napiforme* (Ruschoideae), B = petaloid tepals in *Sesuvium portulacastrum* (Sesuvioideae). C = summary of ABC expression patterns in floral type with petaloid staminodes, D = summary of ABC expression patterns in foral type with petaloid tepals, E = summary of ABC expression patterns in model eudicot species *Arabidopsis thaliana*. (Courtesy of S. Brockington.)

Given the concept of a conserved petal identity program within core eudicots, and a hypothesis of heterotopy in petal evolution, it might be expected that there would be similarities in MADS-box gene expression between petaloid tepals and petaloid stamens in members of Aizoaceae. Specifically, an evolutionary extension of the ABCE model predicts that *AP3* and *PI* homologs should be required for petal identity and be expressed throughout the development of the petaloid organ (Brockington et al. 2011). Contrary to this expectation, the homologs of *AP3* and *PI* are not expressed at any point in the development of the petaloid tepals in Aizoaceae, while the homologs *AP3* and *PI* are only transiently expressed in the petaloid staminodes in Aizoaceae. Similarly, the *AG* homolog is not expressed in petaloid tepals, but is transiently expressed in petaloid staminodes. Therefore, at the level of *AP3*, *PI*, and *AG* homolog gene expression, Brockington et al. (2011) found no evidence for homology between the petaloid tepals and petaloid staminodes of *S. portulacastrum* and *D. napiforme*. Significantly, neither petal class exhibits gene expression patterns consistent with the classic core eudicot petal identity program. Thus, Brockington et al. (2011) provide evidence for core eudicot petal development that is independent of the *AP3* and *PI* lineage of MADS-box genes and suggest that even within the core eudicots, different genetic control of petal identity has evolved within Caryophyllales in the context of its unusual floral evolutionary history (Fig. 14.25).

CONSTRUCTION: SHAPE AND FLORAL SYMMETRY

A major determinant of floral shape is symmetry, although additional factors, such as cell size and elongation, contribute to differences between actinomorphic bowl-shaped and tubular flowers. Because actinomorphy vs. zygomorphy represents a major dichotomy in floral morphology—and because relatively little is known about factors that influence other attributes of floral shape—we will confine our review here to symmetry.

The role of floral symmetry in angiosperm evolution has long been of interest to evolutionary biologists and systematists in large part due to its impact on plant-pollinator interactions (see Endress 2012). Here we consider the underlying developmental genetics of floral symmetry, addressing actinomorphy (radial symmetry, polysymmetry) and zygomorphy (bilateral symmetry, monosymmetry), with a focus on the underlying developmental genetics. Actinomorphy is considered the ancestral condition for flowering plants (e.g., Cronquist 1981). Not only is radial symmetry prevalent in extant basal angiosperm lineages (see Chapters 4 and 5), but this condition is also present in the earliest fossils (Dilcher 2000). In contrast, bilateral symmetry appears later in multiple fossils during the Paleocene and Eocene (Dilcher 2000) and is reconstructed as derived in phylogenetic reconstructions (e.g., Ronse De Craene et al. 2003; Soltis et al. 2005b).

Bilateral symmetry is widespread in the angiosperms, with reports from at least 38 families (Westerkamp et al. 2007) or more than 200 families, if bilateral symmetry by reduction is included (Endress 2012), and has evolved multiple times during angiosperm evolution, as long noted by taxonomists and as reconstructed on the backbone phylogeny of Soltis et al. (2011) (Hileman 2014a and redrawn as Fig. 14.26). For many lineages with bilateral symmetry, the same gene (*CYCLOIDEA*) has been implicated as the underlying genetic mechanism (see below). However, the actual number of transitions to bilateral symmetry is much higher than is apparent from a simple summary tree of angiosperms. For example, Citerne et al. (2010) found evidence for at least 70 transitions to bilateral symmetry (reviewed in Hileman 2014a): one event in basal angiosperms, 23 in monocots, and 46 in the eudicots. More detailed studies of the transitions to zygomorphy have also been conducted within individual angiosperm subclades. For example, Donoghue et al. (1998) found evidence for multiple transitions to zygomorphy as well as a reversal to actinomorphy in asterids. Floral symmetry across Lamiales (analyzed at the family level) involves one change from radial to bilateral symmetry, as well as one reversal back to radial symmetry (Schäferhoff et al. 2010).

Shifts to bilateral symmetry have resulted in increased diversification of angiosperms. In 15 of 19 comparisons of large sister groups (clades recognized as families) having radial and bilateral symmetry, the clades with bilateral symmetry are more species-rich than their sisters with radial symmetry (Sargent 2004). This difference is attributed to differences in pollination, with zygomorphy increasing pollinator specificity, thus ultimately promoting reproductive isolation. Sargent's (2004) results support longstanding views on the role of floral variation in angiosperm speciation and diversification.

Developmental Genetics of Floral Symmetry

The genetic control of floral symmetry has been investigated in a number of species, with the model asterid *Antirrhinum majus* (snapdragon) the earliest investigated system. In *Antirrhinum*, bilateral symmetry results from two genes, *CYCLOIDEA* (*CYC*) and *DICHOTOMA* (*DICH*) (Luo et al. 1996, 1999; Cubas et al. 1999). Since the initial identification of the role of *CYC* and *DICH*, much has been learned about floral symmetry genes in snapdragon—in fact, the

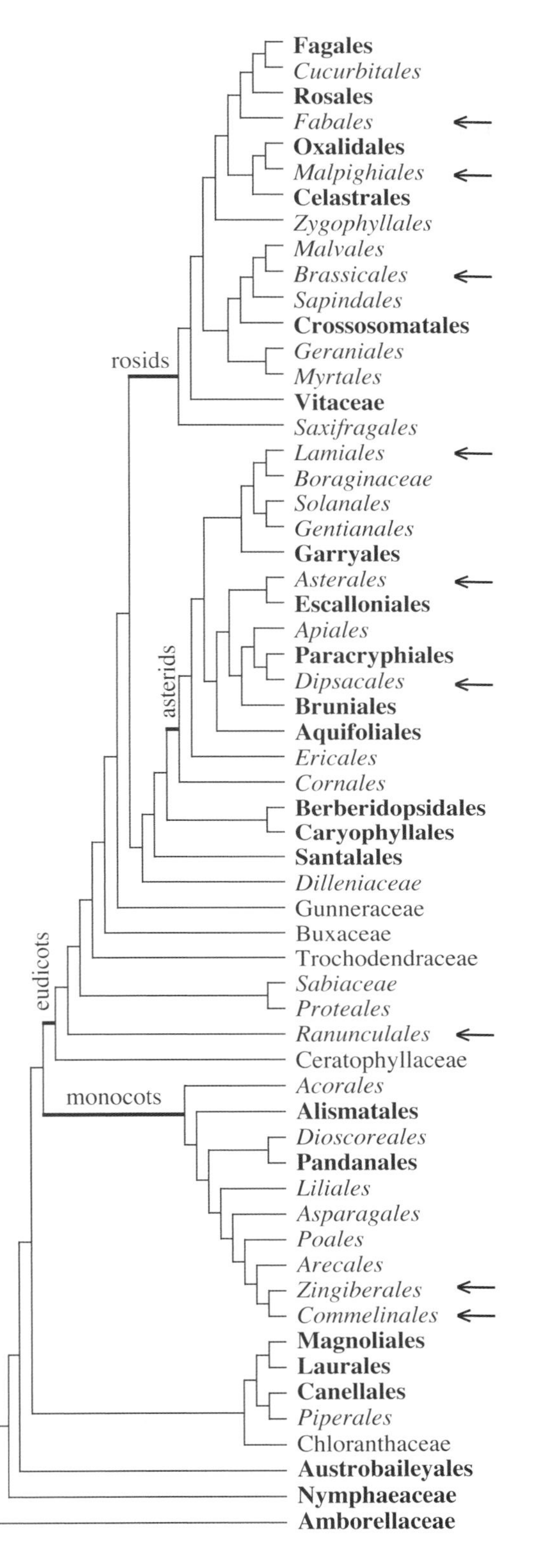

Figure 14.26. Summary phylogenetic tree for angiosperms (Soltis et al. 2011) showing evolutionary transitions in floral symmetry. Bold = lineages with radial floral symmetry. Plain text (including italics) = lineages with some bilateral symmetry. Italics = lineages with elaborate bilateral symmetry. Arrows = lineages with species for which CYCLOIDEA homologues have been implicated in transitions to bilateral symmetry. (Modified from Hileman 2014b with permission.)

genes and genetic interactions that control its bilateral symmetry are now generally well understood (reviewed in Hileman 2014a,b). A number of genes and gene interactions are needed for the development of bilateral floral symmetry. *CYC* and *DICH* are recent gene duplicates; both are transcription factors and function in a manner that is partially redundant, specifying adaxial floral identity in zygomorphic flowers (Fig. 14.27). Expression of *CYC* and *DICH* genes not only determines the shape of the dorsal petals in snapdragon (Fig. 14.27), but also controls the abortion of the dorsal stamen (which is sterile in these plants). The gene *DIVARICATA* (*DIV*), representing another family of transcription factors, specifies the distinctive ventral petal shape in the flowers of snapdragon (Fig. 14.27). Lastly, the role of all three genes (*CYC*, *DICH*, and *DIV*) depends on the interplay with another transcription factor, *RADIALIS* (*RAD*). *CYC* and *DICH* have a role in positively regulating *RAD*—as a result, the function of the RAD protein is largely in the dorsal portion of flowers, where *CYC* and *DICH* are expressed (Fig. 14.27). These data for *Antirrhinum* provide a crucial foundation for elucidating the evolution of floral symmetry and obtaining new insights across the angiosperms (Hileman 2014b).

As more lineages have been investigated using developmental genetic methods, it has been possible to garner new insights into the genetic underpinning of the evolution of zygomorphy across the multiple lineages that obtained this feature independently. These studies now indicate a "striking parallelism in the independent evolution of bilateral symmetry" (Hileman 2014b; p. 5). That is, a *CYC*-based genetic program has been employed repeatedly to specify dorsal floral identity across the angiosperms.

The developmental program of *Antirrhinum* may be conserved throughout Lamiales (Zhong and Kellogg 2015), and *CYC*-like genes may also be involved in the evolution of zygomorphy in Dipsacales and Asterales (Hileman 2014b). In Asteraceae, *CYC*-like genes are involved in the differentiation of ray flowers (with bilateral symmetry) from disc flowers (with radial symmetry) (Fig. 14.28). In fact, different paralogs of the *CYC* gene family confer zygomorphy in species of Asteraceae with independently derived zygomorphy (Chapman et al. 2012).

A *CYC* homolog also controls bilateral symmetry in the rosid clade based on studies of *Lotus japonicus* (Fabaceae; Feng et al. 2006). Thus, homologs of *CYC* have been independently recruited in the determination of floral symmetry in the two major lineages of eudicots, superasterids and superrosids. Based on phylogenetic analyses, it is clear that bilateral symmetry has evolved multiple times in rosids (see Fig. 14.26). Furthermore, developmental genetic data now indicate that *CYC* homologs have been recruited in at least three cases (Fabaceae, Brassicaceae, Malpighiaceae).

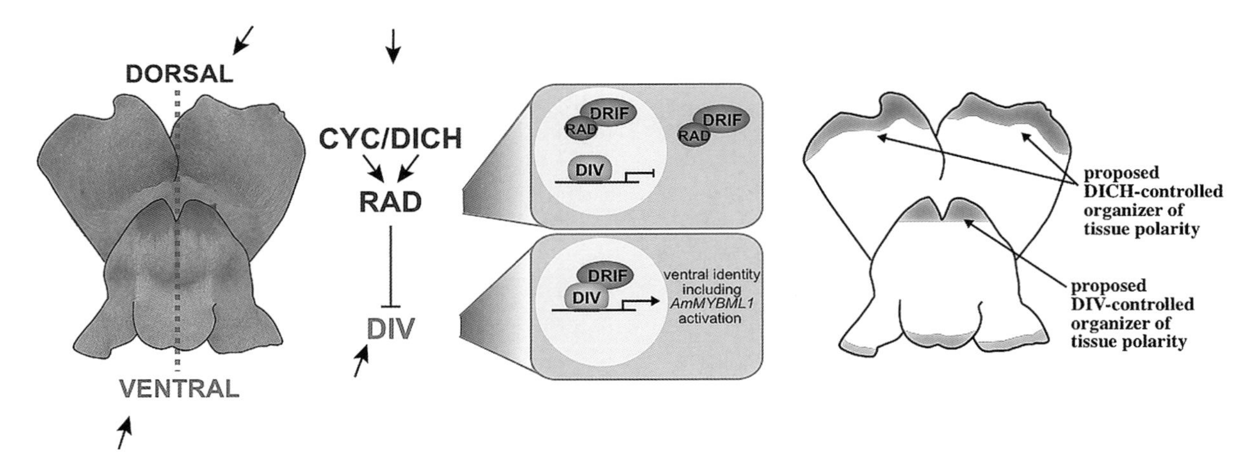

Figure 14.27. Developmental control of bilateral symmetry in *Antirrhinum*. (a) Image showing the dorsal and ventral sides of the flower and the line that divides the flower into left and right mirror images (bilateral symmetry). (b) Dorsal flower identity is specified by *CYCLOIDEA* (*CYC*) and *DICHOTOMA* (*DICH*), transcription factors that are partially redundant in function. *CYC* and *DICH* establish dorsal identity via activation of *RADIALIS* (*RAD*). Ventral floral identity is specified by *DIVARICATA* (*DIV*). Expression of *DIV* is excluded from the dorsal flower domain area via negative regulation by *RAD*. (c) In the ventral portion of the flower where RAD is absent (bottom), DIV-DRIF heterodimers form, bind DIV consensus binding sequences, and are able to regulate downstream genes necessary for ventral flower development. RAD, which is restricted to the dorsal part of the flower, antagonizes DIV function by competing with DIV for a DRIF transcription factor (top panel). RAD-DRIF heterodimers may form in the nucleus and/or cytoplasm, but RAD lacks the C-terminal DNA binding domain found in DIV. Therefore, RAD sequesters DRIF from interaction with DIV; DIV alone cannot regulate downstream target genes. (d) Modeling studies suggest that the dorsoventral identity genes, *CYC*, *DICH*, and *DIV*, not only control petal growth rates, but also influence the establishment of tissue polarity organizers that are critical for specific growth trajectories. Distal polarity organizers (gray—upper lobes of flower) are established early in floral development, and without DICH (plus CYC) contribution to the dorsal distal organizers of tissue polarity, models of floral development fail to produce asymmetric dorsal petal lobes. Central polarity organizers (gray—middle of flower) are established later in flower development at the ventral-lateral petal junction, and without DIV contribution to the central organizers, models of floral development do not produce a properly elongated corolla tube and ventral palate. (Figure and legend modified from Hileman 2014a, with permission.)

Although there have been fewer investigations of basal eudicots and monocots, the evidence also suggests that *CYC*-like genes are responsible for the transition to bilateral symmetry in those lineages as well (reviewed in Hileman 2014b). Developmental genetic data support a role for *CYC*-like genes in the evolution of bilateral symmetry in the basal eudicot *Capnoides* (Papaveraceae) (Kölsch and Gleissberg 2006; Damerval et al. 2007). Only a few monocot lineages have been investigated, but *CYC*-like genes are associated with a change to bilateral symmetry in Zingiberales and *Commelina* (Commelinaceae) (Bartlett and Specht 2011; Preston and Hileman 2012). Interestingly, in these monocots, differential expression of *CYC*-like genes occurs on the ventral side of the flower, whereas in eudicots *CYC*-like genes are involved in dorsal floral development (Hileman 2014b).

Many lineages contain multiple paralogs of *CYC*-like genes. For example, in some Fabaceae (e.g., *Lotus* and *Pisum* have been investigated in detail), there are three paralogs of *CYC*; two paralogs (*CYC1* and *CYC2*) function redundantly and are responsible for dorsal petal identity (e.g., Xu et al. 2013; Wang et al. 2008; reviewed in Hileman 2014b). Similarly, duplicate *CYC*-like genes play an important role in the bilateral symmetry of flowers of various asterid species, including *Lonicera* (Caprifoliaceae) (Howarth et al. 2011), members of Gesneriaceae (Gao et al. 2008; Zhou et al. 2008), and various Asteraceae (Broholm et al. 2008; Chapman et al. 2012).

The loss of bilateral symmetry and the reversion to radial symmetry, documented in many clades of angiosperms, may derive from multiple causes. A complete loss of *CYC*-like gene expression is one possible mechanism, but the typical pathway for this reversal in lineages studied so far involves a role for the multiple paralogs of *CYC*-like genes (see above: one *CYC*-like paralog is lost, and the other paralog has expanded expression across the floral center [e.g., as in *Plantago*; Preston et al. 2011]).

MODE: ADAPTATIONS TO POLLINATORS

Much of floral diversity has been shaped by interactions of angiosperms with their pollinators. Floral organization and construction reflect broad pollination syndromes (e.g., so-called "hummingbird flowers" vs. "bee flowers"; e.g., Fenster et al. 2004; Waser 2006), but additional features may also result from selection by pollinators and be crucial elements of floral diversity. Most prominent among these attributes—and the best studied—is color variation, particularly anthocyanins and their underlying genetic and biosynthetic pathway.

Figure 14.28. A flower head of *Galinsoga ciliata* (Asteraceae) consisting of 5 ligulate (monosymmetric) flowers and ca. 25 tubular (polysymmetric) flowers. The inner, tubular flowers have 5 small yellow petals. The peripheral, ligulate flowers have a monosymmetric corolla with 3 large white petals and two tiny, reduced petals.

The role of anthocyanins in controlling floral color has been known for over half a century (e.g., Hagen 1959), and the basic biosynthetic pathway yielding anthocyanins has likewise been understood for many years (see Koes et al. 2005; Rausher 2008). On a gross level, we have long known that a loss-of-function mutation at any step in the anthocyanin pathway will block pigment production, resulting in white-flowered mutants. Often, white-flowered plants represent the double recessive genotype of a Mendelian trait within a population, but white flowers on broader geographic and phylogenetic scales are likely not homologous due to the many ways in which anthocyanin production can be blocked. The control of different colors, however, is much more complex. Despite knowledge of the catalyzing enzymes in the pathway, the genes for many of the steps have, until recently, remained elusive. Moreover, the connection between altered gene expression (via either gene number or regulation), variation in floral color, and pollinator-mediated radiations is only now beginning to emerge (e.g., Whittall et al. 2006; Smith et al. 2008; Smith and Rausher 2011; Smith and Goldberg 2015). Here we review the key elements of anthocyanin genetics and how genetic variation in key enzymes leads to evolutionarily significant variation in flower color.

Anthocyanin production is controlled by complex interactions involving the biosynthetic pathway itself and a set of regulatory proteins (reviewed by Koes et al. 2005; Stommel et al. 2009). The biosynthetic pathway is controlled by six enzymes, beginning with chalcone synthase (CHS). The number of gene copies encoding these key enzymes varies among species, for example, from two CHS genes in *Zea mays* (Coe et al. 1981) to 12 in *Petunia* × *hybrida* (Koes et al. 1989). Regulatory genes control the tissues and/or developmental stage at which the enzyme-coding genes are expressed. A regulatory complex comprising MYB and bHLH MYC proteins plus WD40 repeat (WDR) proteins controls transcription of the anthocyanin structural genes (Ramsay and Glover 2005). Specific sets of *MYB*, *MYC*, and *WDR* genes have been identified in *Zea* and *Petunia* as coding the proteins that form the complex that activates the anthocyanin genes. Evolutionary analyses of flower color variation (e.g., Rauscher 2008; Smith and Rausher 2011) are extending understanding of this pathway and its interactions to natural systems.

15 The Evolution of Genome Size

Genome evolution in angiosperms is highly complex and multifaceted, and therefore our review cannot be comprehensive. A general review of genome evolution might include wide-ranging topics, such as rates and patterns of nucleotide substitution, origins of introns and novel genes, evolutionary forces that shape patterns of genomic diversity, origins and evolution of transposable elements (TEs), concerted evolution, origins and functions of small RNAs, interactions between nuclear and organellar genomes, and many more. However, many of these topics have not been addressed broadly within angiosperms, and information is often restricted to one or a few model species. Rather than attempting a comprehensive review of genome evolution in angiosperms, we instead emphasize the evolution of genome size, the dynamics and mechanisms of genome size evolution, and resulting variation in genome content and structure. Well over 50 genome sequences have been reported for plants, 47 of them angiosperms (Michael and Jackson 2013), and investigation of these sequences has provided enormous insights into the gene content and structure of plant genomes, with an emphasis on angiosperms. Despite enormous progress, there is still much to learn about the organization of plant genomes, gene function, and the noncoding space (Michael and Jackson 2013, p. 1).

Here we focus on the evolution of genome size across the angiosperms and land plants, as well as some of the factors that govern variation and evolution of genome size. We also mention briefly the diverse mechanisms of genome evolution, the role of repetitive elements in genome size dynamics, horizontal gene transfer (HGT) as a source of variation in gene content, and issues related to gene expression, such as methylation and epigenetics. It is impossible to consider genome evolution in the angiosperms without concomitantly considering polyploidy (e.g., Fedoroff 2000; Wendel 2000; Adams and Wendel 2005; Eckardt 2008; Diez et al. 2014; chapters in Soltis and Soltis 2012; chapters in Wendel et al. 2012; chapters in Chen and Birchler 2013). As a result, polyploidy (whole-genome duplication; WGD) is discussed here as a basic component of genome size variation and evolution, as is genome downsizing (polyploidy is covered in greater detail in Chapter 16).

Both chromosome number and genome size vary tremendously across the flowering plants (e.g., Fedorov 1969; Plant DNA C-Values Database http://www.rbgkew.org.uk/cval/homepage.html; Chromosome Counts Database http://ccdb.tau.ac.il/). This variation has stimulated a great deal of speculation about the original genome size and base chromosome number of the angiosperms, as well as about the patterns of genome and chromosomal evolution. Similar questions regarding patterns of genome evolution are of interest across all land plants.

Despite the frequency of polyploidy in plants, most estimates of the original base chromosome number for angiosperms are low, between $x = 6$ and 9 (e.g., Ehrendorfer et al. 1968; Stebbins 1971; Walker 1972; Raven 1975; Grant 1981), with a correspondingly small ancestral genome size (Leitch et al. 1998; Soltis et al. 2003b). However, analysis of the vast range of chromosome numbers and genome sizes encountered in angiosperms shows that genome size can vary independently of chromosome number (see Soltis et al. 2005b). Furthermore, reconstructing ancestral base chromosome numbers and genome sizes for angiosperms is complex (see Soltis et al. 2005b), due in large part to rampant polyploidy throughout angiosperm evolutionary history. Moreover, although the estimation of the ancestral base chromosome number for angiosperms was of paramount interest historically, with discussion by some of the leading botanists of the mid-1900s (e.g., Stebbins 1971; Raven 1975; Grant 1981), current efforts at inferring the ancestral angiosperm genome focus more on gene content and genome size and structure, rather than how the genome is packaged into chromosomes. We envision, however, a return to the evolution of "genome packaging," once greater understanding of the basic elements of genome evolution is achieved.

GENOME SIZE

A significant semantic problem concerns different uses of the term "genome" and "genome size" (Greilhuber et al. 2005). As originally defined (Winkler 1920), genome referred to a monoploid chromosome complement. Since a monoploid complement is defined as "having one chromosome set with the basic (x) number of chromosomes" (Rieger et al. 1991), it followed by definition that any polyploid taxon had three or more genomes.

However, an alternative meaning, now in common usage, uses genome as an interchangeable alternative for the 1C-value to refer to the DNA content of an unreplicated gametic nuclear complement, irrespective of ploidal level. Unless the meaning intended is clearly defined on each occasion, this can be confusing, especially when authors use both meanings for a polyploid taxon in the same paper.

To overcome this issue Greilhuber et al. (2005) tried to standardize the terminology; they distinguished between the "holoploid" genome size, which is the amount of DNA in the nucleus independent of ploidal level, and "monoploid" genome size in which the ploidal level is taken into account. Greilhuber et al. (2005) recognized that the term "genome size" was imprecise. Thus, it is always more accurate to state whether one is referring to a 1C, 2C, or 4C holoploid value or a monoploid C-value, for which Greilhuber et al. (2005) used the shorthand of 1Cx, 2Cx, or 4Cx-value. Note that for polyploids, monoploid genome sizes estimated in this way are always smaller than the holoploid C-values. For example, in the diploid *Triticum monococcum*, 2C = 12.45 pg, so the holoploid 1C-value = 12.45 ÷ 2 = 6.23 pg, which also equals the monoploid 1Cx-value. In the tetraploid *T. dicoccum*, in contrast, 2C = 24.05 pg, so holoploid 1C = 24.05 ÷ 2, which equals 12.03 pg, but the monoploid 1Cx-value is 24.05 ÷ 4, or 6.01 pg.

C-values (nearly 11,000) have been estimated for more than 7542 species of angiosperms (Bennett et al. 1997; Bennett and Leitch 2001, 2003, 2011; Hanson et al. 2001; Leitch and Hanson 2002; Garcia et al. 2014, http://data.kew.org/cvalues/), representing more than 2% of the approximately 350,000 species of flowering plants and approximately 60% of all angiosperm families (sensu APG 1998; APG II 2003; APG III 2009; APG IV 2016). The Plant DNA C-Values Database (http://data.kew.org/cvalues/) represents the largest collection of nuclear DNA amounts for any group of organism (reviewed in Leitch et al. 1998; Bennett and Leitch 2011; Leitch and Leitch 2013). We will focus our discussions and analyses on 1Cx-values.

C-values in angiosperms span a huge range (Greilhuber and Leitch 2013; see Plant DNA C-Values Database and Figs. 15.1, 15.2). These values have historically been reported in pg, but in this age of relatively facile and inexpensive genome sequencing, perhaps a better unit is megabases (Mb; 1 pg = 978 Mb). Some of the smallest reported values are for several species in the carnivorous genus *Genlisea* (0.06 pg, Fleischmann et al. 2014), *Cardamine amara* (Brassicaceae; 1C = 0.05 pg, Bennett and Smith 1991), and *Fragaria* (Rosaceae; 1C = 0.10 pg, Antonius and Ahokas 1996). *Arabidopsis thaliana*, a well-known model organism with a very small genome, has 1C = 0.16 pg (157 Mb; Bennett et al. 2003); the largest value is for *Paris japonica* (Melanthiaceae; 1C = 152.2 pg, Bennett and Smith 1976); other Melanthiaceae and Liliaceae also have high values (see below).

Despite this 3,000-fold range in DNA amount, the basic complement of genes required for normal growth and development appears to be essentially the same (i.e., even the 82-Mb genome of *Utricularia gibba* has an estimated 28,500 genes, similar to gene content estimated for *Arabidopsis*, papaya, grape, *Erythranthe*, and tomato [Ibarra-Laclette et al. 2013]), reflective of the decades-old "C-value paradox" (Thomas 1971). The apparent paradox high-

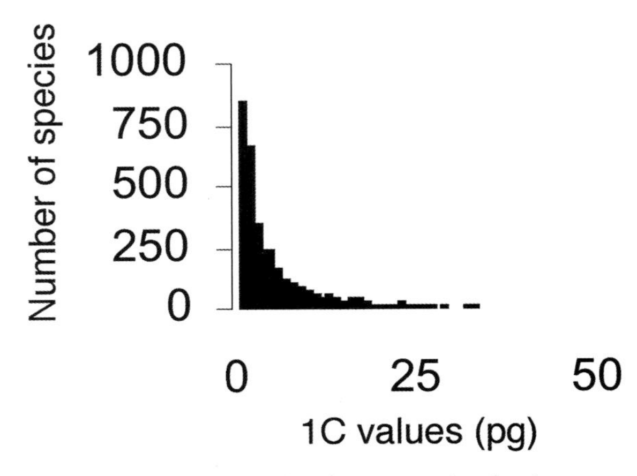

Figure 15.1. Distribution of 1C-values for 3,543 species of angiosperms (mean 1C = 6.25 pg; mode 1C = 0.6 pg; see Leitch et al. 1998); from D. Soltis et al. 2003b.

lights the lack of association between genome size, in both plants and animals, and organismal complexity. There is, however, much less variation in actual genic content among plant genomes; differences in amount of DNA largely reflect changes in the proportion of non-coding, repetitive DNA. This new understanding helps put to rest the primary observation that underlies the C-value paradox.

Several mechanisms have been proposed for the large variation in genome size in the angiosperms (reviewed in Kellogg and Bennetzen 2004; Grover and Wendel 2010; Leitch and Leitch 2013; Michael 2014). Repeated cycles of polyploidy may increase genome size (e.g., Leitch and Bennett 1997; D. Soltis et al. 2009; Otto and Whitton 2000; P. Soltis and Soltis 2000, 2009; Wendel 2000; see Chapter 16). In addition, transposable elements (TEs) also contribute to increases in genome size throughout eukaryotes (e.g., Flavell 1988; Bennetzen 2000, 2002, 2005, 2007; Fedoroff 2000; Sankoff 2001; Kidwell 2002; Kazazian 2004; Bennetzen et al. 2005; Michael and Jackson 2013; Michael 2014; see below for further discussion). Gregory (2001) suggested that the phrase "C-value paradox" be replaced by "C-value enigma" to indicate that the current challenge is to understand the mechanisms and forces that determine the amounts of repetitive DNA in a genome.

Leitch et al. (1998) calculated mean C-values for many angiosperm species and first considered the evolution of genome size in light of angiosperm phylogeny. Updated calculations (in Soltis et al. 2003d, 2005b) showed that despite the enormous range in values, most angiosperms have very small C-values, between 0.1 and 3.5 pg (Fig. 15.1). In fact, the modal C-value for angiosperms is actually quite low, 0.7 pg (~675 Mb). Large genomes are present in only a few diverse clades: various monocots (especially Asparagales and Liliales), some Santalales, and a few asterids. However, even in these clades, most members have small or intermediate-sized genomes; only two groups contain members with very large genomes (>35 pg, 50 times the modal value), Santalales and monocots. Considering only the six groups in which large to very large genome sizes have been reported, relatively few species in each group have large genomes. Furthermore, those species with large genomes tend to be restricted to the more derived families within each of these groups (Leitch et al. 1998). The most parsimonious explanation for these observations is that the ancestral angiosperms had small genomes and that the possession of large genomes is derived, with very large genomes derived independently multiple times (see Soltis et al. 2005b). However, given the recognition of frequent ancient WGDs in angiosperms, small genome sizes are, in many cases, not ancestral but rather have been derived via genome downsizing (see Chapter 16).

Within extant seed plants, a very small genome is unique to the angiosperms (Leitch et al. 1998, 2001). Extant gymnosperms are generally characterized by much larger C-values than is typical of angiosperms (see Soltis and Soltis 2013b), although some Gnetales do have small genomes (Burleigh et al. 2012; see below). The modal C-value for 152 gymnosperms is 15.8 pg, compared with a modal value of 0.6 for angiosperms (Leitch et al. 2001; this modal value for angiosperms is slightly lower than that reported by Leitch et al. 1998, 0.7 pg). Similarly, the modal C-value for 50 leptosporangiate ferns is 7.95 pg (Obermayer et al. 2002; Leitch and Leitch 2013), which is also higher than that for angiosperms.

C-values for phylogenetically pivotal taxa helped provide important insights into genome size evolution in angiosperms (e.g., *Amborella*, *Austrobaileya*, *Illicium*, *Ceratophyllum*; Hanson et al. 2001; Leitch and Hanson 2002). *Amborella* and Nymphaeales (the first two branches of angiosperm phylogeny; see Chapter 3) have low 1C values. Two species of *Nymphaea* have very small genomes (1C-values of 0.89, and 0.60 and 1.10, respectively; Leitch and Hanson 2002), and the water lily *Brasenia* has 1C = 1.25 (Diao et al. 2006). *Trithuria submersa* (Hydatellaceae; sister to other Nymphaeales) also has a small genome (1C = 1.37 pg) and yet is a clear polyploid with $2n = 56$, indicating that diploids in *Trithuria* likely have smaller C-values.

Soltis et al. (2003b, 2005b) reexamined genome size evolution in the angiosperms, as well as in seed plants in general. They used genome size estimates for diploids, avoided using means for large clades, and implemented the same categories of C-values established by Leitch et al. (1998): C-values less than or equal to 1.4 pg and 3.5 pg (2 and 5 times the modal C-value of 0.7 pg for angiosperms) were considered "very small" and "small," respectively; C-values of 3.51 to 13.99 pg were considered "intermediate"; C-values greater than or equal to 14.0 pg and 35 pg (20 and 50 times the modal C-value) were considered "large" and "very large" genomes, respectively. We will use these genome size categories throughout the remainder of this section of the chapter.

Average genome sizes for families can be misleading if values reported in the literature represent both diploids and polyploids, because often the highest values of genome size in a family are for polyploids. In *Magnolia*, for example, the reported C-values are 0.90 pg, 5.98 pg, and 7.1 pg. *Magnolia kobus* has a C-value of 0.90 pg and is diploid, with $2n = 38$, which is the lowest number for *Magnolia*. *Liriodendron*, the sister of *Magnolia*, also with $2n = 38$, has a C-value of 0.80, which is comparable to *M. kobus*. The highest C-values for *Magnolia* are from two separate reports for *M. soulangeana*, with $2n = 76$. The higher C-values for this species, 5.98 and 7.1, would therefore be attributed to polyploidy, although genome doubling alone cannot be responsible for this huge increase in C-value. In contrast, another paleopolyploid, the eudicot *Aesculus hippocastanum* (horse chestnut), has a very small genome (1C = 0.1 pg), but it, as well as the entire genus *Aesculus* (Sapindaceae), are ancient polyploids with $2n = 40$ (Stebbins 1950; Soltis and Soltis 1990). It is apparent, therefore, that not all presumed ancient polyploids necessarily have large genomes; ancient rounds of WGD are frequently accompanied by loss of much of the duplicated DNA, in some cases even reducing genome size over time to a lower value than that observed in related lineages that did not experience the WGD.

The results above illustrate the importance of considering multiple taxa when reconstructing ancestral genome sizes. For example, *Magnolia* and *Liriodendron* should both be represented in analyses of genome evolution, and *Magnolia* should be represented by the value for the diploid *M. kobus*. With this approach, the ancestral state for Magnoliaceae is reconstructed with a very small 1C-value (less than 1.5 pg). When a mean value is used for *Magnolia*, the family has a small C-value (1.5–3.6 pg). Mean values reported for large clades at higher taxonomic levels can also be misleading. The mean for Santalales is 1C = 12.7 pg (Leitch et al. 1998), but this value is strongly influenced by extremely high values for just one genus, *Viscum* (see below).

New analyses with an updated angiosperm topology and the addition of new C-values again found that the ancestral angiosperm genome size is very small (Fig. 15.2), regardless of the reconstruction method. Thus, our reconstructions of a revised and updated matrix reinforce the earlier conclusion (Leitch et al. 1998; Soltis et al. 2003b, 2005b) that the ancestral genome size of angiosperms was very small. This result is also recovered when alternative nuclear-based topologies for angiosperms are used (e.g., Zeng et al. 2014; results not shown here).

Not only do our new reconstructions indicate that the ancestral genome size of angiosperms is very small, they also indicate that ancestral genome sizes are very small for most major clades of the angiosperm tree of life (Fig. 15.2), including magnoliids, monocots, eudicots, and major subclades within the eudicots. Refinements of this conclusion may emerge from phylogenetic analyses that include many more terminals, capitalizing on the expanding C-values resources.

Broader reconstructions of genome size diversification across all embryophytes (land plants) similarly revealed that the ancestral genome size of angiosperms was very small (Leitch et al. 2005). Our new analyses here (Fig. 15.3a) reinforce this conclusion. Reports from the literature indicate that all three bryophyte lineages (liverworts, mosses, hornworts) are characterized by very small genomes (1C < 1.4 pg). Our reconstructions that include C-values for all three bryophyte groups indicate that the original genome size of land plants was likely very small (Fig. 15.3a). Although the accuracy of the reports for hornworts and liverworts has been questioned (Leitch et al. 2005), recently reported values continue to be very low (i.e., the average 1C-value recently reported for 24 hornworts was ~0.27 pg; Bainard and Villarreal 2013). Genome sizes are similarly small in liverworts (Temsch et al. 2010).

In contrast to bryophytes, diverse C-values are evident in the lycophytes. *Isoetes* has an intermediate C-value, and *Selaginella* has a very small C-value; the C-values of *Lycopodium* and *Huperzia* are small and large, respectively. The ancestor of the lycophytes is reconstructed as equivocal in our analyses—or with an intermediate genome size—but more analyses with more taxa are needed (Fig. 15.3a).

Following the lycophytes, our reconstructions favor an intermediate genome size (Fig. 15.3a) as ancestral for all remaining vascular plants (*Euphyllophytina* or euphyllophytes, Kenrick and Crane 1997). That is, an increase in genome size accompanied the origin of the euphyllophytes, and intermediate genome sizes are maintained throughout most lineages of euphyllophytes, with some increases and decreases. Again, the broader analyses highlight that the origin of the angiosperms involved a dramatic decrease in genome size (Fig. 15.3a; see Soltis and Soltis 2013b).

Within the monilophytes, several independent decreases and increases in genome size occurred. The evolution of Marsileaceae (represented by *Regnellidium* and *Pilularia*) + Salviniaceae (represented by *Salvinia*) within the ferns clearly involved a major decrease in genome size (Fig. 15.3a). In all reconstructions, ferns have an ancestral genome size that is intermediate; reductions to small genomes apparently occurred independently in *Ceratopteris* and *Asplenium* (Fig. 15.3a).

Considering gymnosperms, the mean C-values are generally high: cycads, 14.71 pg; *Ginkgo*, 9.95 pg; Gnetales, 7.23 pg; Pinaceae, 22.02 pg; and other conifers, 11.89 pg. The lowest C-values for gymnosperms are for Gnetales—*Gnetum* has a mean 1C = 3.38 pg, which represents a decrease in genome size just within the clade of extant gymnosperms (Soltis et al. 2003d; Leitch et al. 2005; Burleigh et al. 2012) (Fig. 15.3a).

There is also evidence for several independent increases in genome size across land plants. One occurred in *Huperzia* of the lycophytes; the largest increase may have occurred in the clade that contains Ophioglossaceae (*Ophioglossum*) + Psilotaceae (*Psilotum*; Fig. 15.3a). Although there are uncertainties in our analyses, our reconstructions indicate that gymnosperms as well as euphyllophytes had an ancestral genome size that was intermediate.

Genome size evolution across land plants as well as

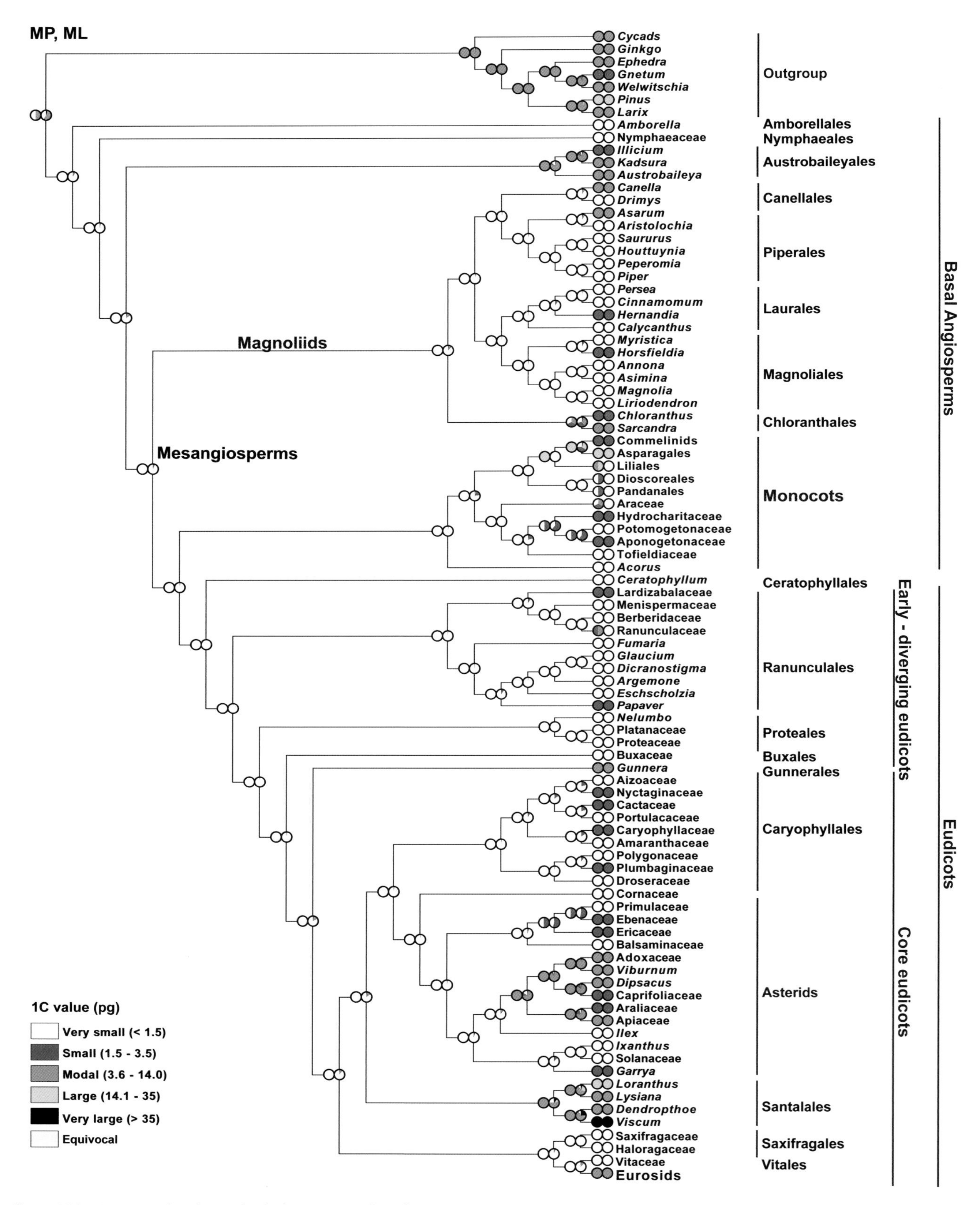

Figure 15.2. Reconstructions (MP and ML) of genome size diversification in the angiosperms. The general topology is from Soltis et al. (2011; see Chapter 3). 1C-values are from Bennett and Leitch (2001), Hanson et al. (2001a, 2001b), Leitch and Hanson (2002); see Soltis et al. (2003b). Ranges of values for Asparagales, commelinids, Dioscoreales, Liliales, and Pandanales are from Leitch et al. (1998); more detail for monocots is provided in Figure 15.3B.

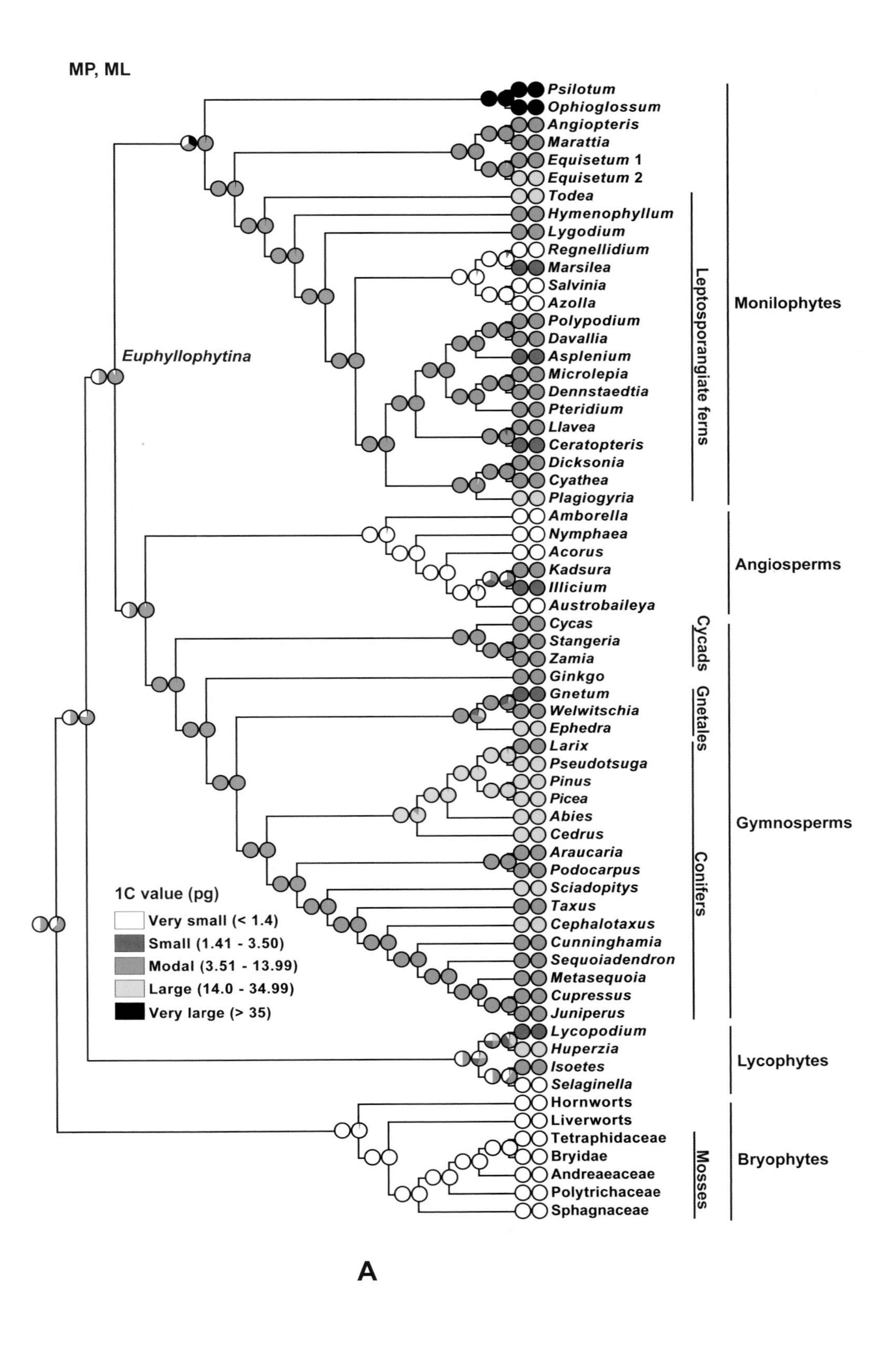

(continued)

Figure 15.3. A. Reconstructions (MP and ML) of genome size diversification in land plants (see Leitch et al. 2005 and Soltis et al. 2005). B. Reconstructions (MP and ML) of genome size diversification in the monocots. Families indicated in bold have some representatives with "very large" genomes (see text). The general topology is from Chapter 7 (see text). 1C-values for families are from Leitch et al. (1998) and Soltis et al. (2003b).

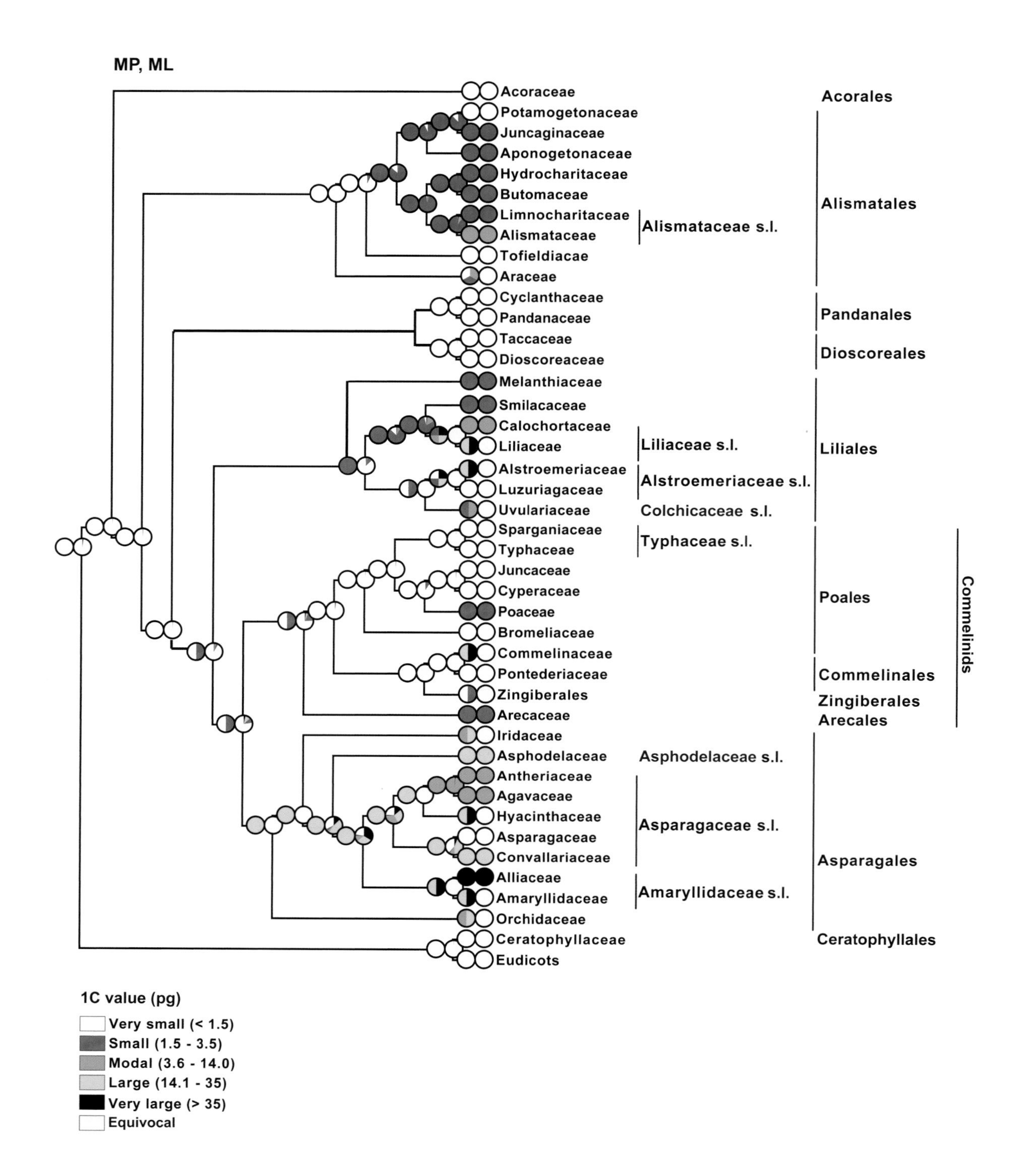

Figure 15.3. *(continued)*

within angiosperms was dynamic, with both increases and decreases; for example, dramatic changes can be seen just within the monocots (Leitch et al. 2005; Soltis et al. 2005b; Soltis and Soltis 2013b) (Fig. 15.3a,b). A consistent result is that the origin of the angiosperms coincided with a decrease in genome size (Figs 15.2, 15.3a). Our new reconstructions indicate that a very small genome size was ancestral throughout the basal angiosperms, the early-diverging eudicots, and also most of the major clades of core eudicots (e.g., Caryophyllales, rosids, asterids; Fig. 15.2). Our reconstructions indicate repeated increases in genome size across diverse lineages of flowering plants, but more detailed analyses of major subclades of angiosperms are needed to pinpoint the various increases and decreases.

GENETIC OBESITY HYPOTHESIS

Bennetzen and Kellogg (1997) first proposed that genome size evolution in plants would be largely unidirectional, with an overall pattern of increase due to the combined influence of polyploidy and the accumulation of retroelements. These authors suggested that plants have a "one way ticket to genetic obesity." Broad surveys of angiosperms appear to be in general agreement with the genetic obesity hypothesis, with very large genomes confined to taxa that occupy derived positions within larger clades (Leitch et al. 1998; Soltis et al. 2003b). However, these approaches are coarse-grained, and new mechanisms of DNA loss have been uncovered; thus, it has become evident that plant genomes often have a return ticket and can slim down (Bennetzen and Kellogg 1997; Cox et al. 1998; Wendel et al. 2002; Grover and Wendel 2010; Michael 2014).

The evolution of genome size at lower taxonomic levels has still only occasionally been examined in a phylogenetic context. In Pooideae (Poaceae), the ancestral grass genome size was inferred to be 2C = 3.5 pg (Kellogg 1998). Kellogg also provided evidence for a steady increase in genome size in Pooideae, ultimately leading to the much larger genome sizes (2C = 10.7 pg or more) in the Triticeae. However, genome size decreased in some genera, including *Oryza*, *Chloris*, and *Sorghum*. Furthermore, although increases in genome size in grasses are often due to polyploidy, proliferation of TEs also contributes, but is lineage-specific (e.g., Bennetzen 2007; Morgante et al. 2007; Charles et al. 2008). Moreover, rapid losses of TEs can, in fact, lead to reductions in genome size (Bennetzen 2005; Hawkins et al. 2009).

Proliferation of transposable elements has led to a threefold increase in genome size in *Gossypium* (cotton) in the 5–10 million years since the divergence of the major clades (Hawkins et al. 2006). This variation is due to lineage-specific expansions of different classes of TEs among species of *Gossypium* (Hawkins et al. 2006). However, in contrast to a model of unidirectional change in genome size, in a clade of *Gossypium* and allies (the cotton tribe, Gossypieae, Malvaceae), the frequency of decreases in genome size exceeded the number of increases (Fig. 15.4; Wendel et al. 2002). Independent decreases in genome size occurred in *Cienfuegosia*, *Gossypium raimondii*, the ancestor of *Gossypioides kirkii* + *Kokia drynarioides*, and the ancestor of *Thespesia*, with three separate increases, in *Gossypium herbaceum*, *Hampea appendiculata*, and *Thespesia populnea* (Fig. 15.4).

Lentibulariaceae, well known for their very small genomes (*Genlisea tuberosa* and *G. aurea* have 1C-values of 0.06 pg; Fleischmann et al. 2014), have experienced both expansions and contractions of genome size (Veleba et al. 2014). A new phylogenetic analysis of Lentibulariaceae (*Genlisea*, *Utricularia*, and *Pinguicula*) has served as the basis for examining the evolution of genome size, with 1C-values now available for 119 taxa. From an ancestral estimated genome size of 414 Mb, genome size increased in *Pinguicula*, in some cases due to polyploidy, with a few miniaturizations (Veleba et al. 2014). More dramatic miniaturizations, to genomes < 100 Mb, have occurred repeatedly in *Genlisea* and *Utricularia*. Despite these well-documented cases of genome miniaturization, the mechanisms and driving forces have not been identified. Carnivory, and an adaptation to low-nutrient habitats, has been proposed as a factor selecting for small genome sizes in these plants, yet no statistical association of genome size with any morphological or ecological variables was detected (Veleba et al. 2014). Likewise, increased mutation rates, suggested as a mechanism of genome contraction in *Utricularia* and *Genlisea*, have not been verified to drive genome shrinkage in *U. gibba* (Ibarra-Laclette et al. 2013).

Polyploids sometimes have genome sizes that are substantially smaller than those of diploid congeners. Using C-values for 3,021 species, Leitch and Bennett (2004) found that mean 1C-value does not increase in direct proportion with ploidy. This observation held true in comparisons conducted in several distinct clades and provides clear evidence for genome downsizing in many polyploids (Fig. 15.5). Furthermore, "aggressive purging" of long terminal repeat retrotransposons (LTRs) via multiple mechanisms provides evidence for genome downsizing, even in the wake of many mechanisms of genome expansion (Michael 2014). Thus, several lines of evidence indicate that genome size evolution in the angiosperms, as well as in embryophytes in general, is dynamic, with both increases and decreases having repeatedly occurred (e.g., Rabinowicz 2000; Wendel et al. 2002; Leitch et al. 2005; Bennetzen 2005; Grover and Wendel 2010; Leitch and Leitch 2013).

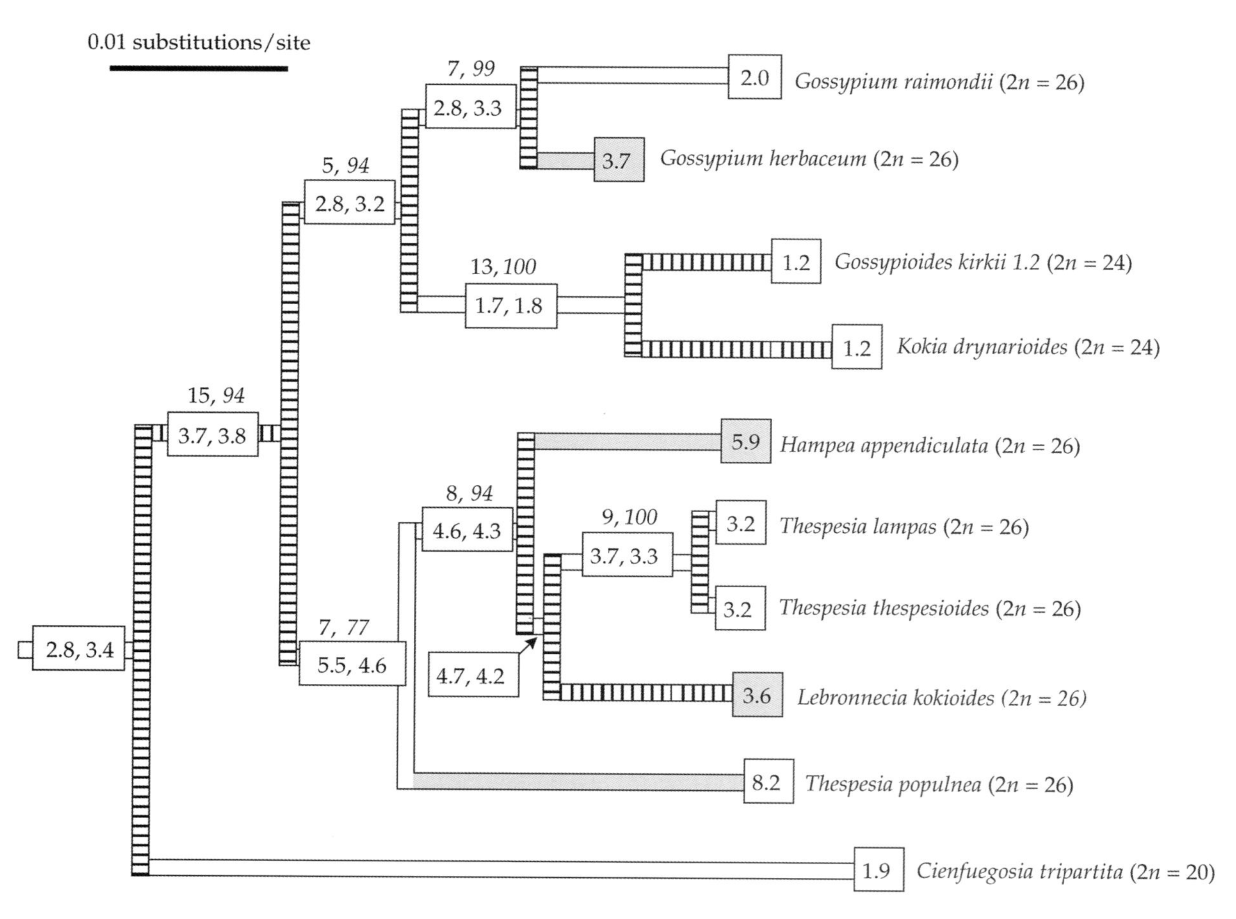

Figure 15.4. Reconstruction of the evolution of genome size in the cotton tribe (Gossypieae). The maximum likelihood and parsimony trees are identical and were inferred from *CesA1* sequence data, The number of character state changes and jackknife support (%; in italics) from maximum parsimony analysis are shown above each internal branch. The tree is rooted with *Malva sylvestris* (ingroup-outgroup branch length = 0.12). Genome sizes (in pg) are shown at branch tips before species names, which are followed by somatic chromosome numbers. Ancestral genome sizes were estimated using sum-of-squared-changes parsimony analysis (Maddison 1991) and a generalized least squared method (Martins and Hansen 1997). A Wagner parsimony (Swofford and Maddison 1987) reconstruction was also conducted, but these estimates are not shown. The former estimates are shown in boxes on the internal branches. Inferred genome size increases are shown by shaded branches, decreases are indicated by unfilled branches, and ambiguities or stasis are denoted by hatched branches. (Modified from Wendel et al. 2002.)

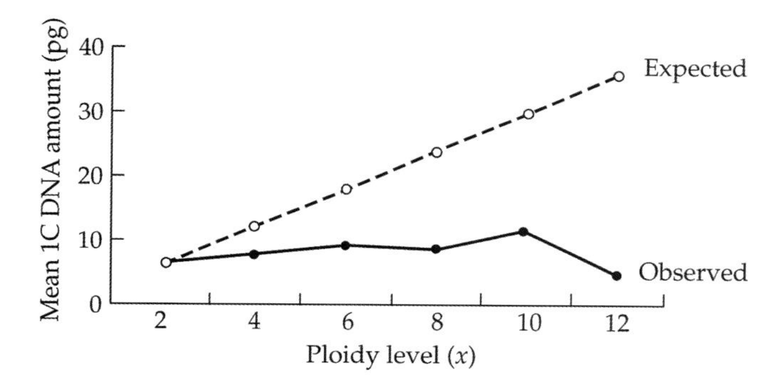

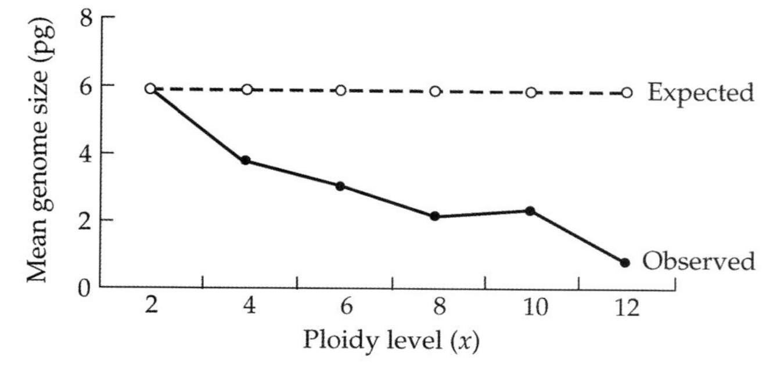

Figure 15.5. Genome downsizing. The relationship between genome size and ploidy for 3,021 angiosperm species. Open circles and dashed lines are expected values for 1C DNA content based on genome doubling with no loss of DNA; black dots and black lines are the actual observed values for polyploids, revealing loss of DNA as ploidy increases. (Original figure from Leitch and Bennett 2004; modified in Soltis et al. 2005.)

MECHANISMS OF GENOME SIZE EVOLUTION

There are several avenues by which changes in genome size can occur. It is difficult to consider genome evolution in angiosperms without considering the impact of polyploidy (e.g., Wendel 2000). Polyploidy may substantially increase genome size in a single step (see Chapter 16 for further discussion of polyploidy). Proliferation of repetitive elements also plays a major role in increasing genome size (Kidwell 2002), through the large-scale accumulation of retroelements, as in Poaceae (Bennetzen 1996, 2000, 2007; SanMiguel et al. 1996) and some non-angiosperm lineages, such as *Pinus* (Elsik and Williams 2000) and Norway spruce (Nystedt et al. 2013). Transposable elements contribute to increases in genome size throughout eukaryotes (e.g., Gaut 2002; Bennetzen 2000, 2005, 2007; Wendel 2000; Fedoroff 2000; Kidwell 2002; Kazazain 2004; Michael and Jackson 2013; Michael 2014). In fact, plant genomes are often packed with TEs.

Mechanisms of genome contraction are not as well known as mechanisms for increase, but our understanding of how decreases in genome size occur is improving (e.g., Vicient et al. 1999; Kirik et al. 2000; Petrov 2001, 2002; Bennetzen 2002, 2005; Frank et al. 2002; Hancock 2002; Hawkins et al. 2008; see Michael 2014 for recent review). Unequal recombination can slow the increase in genome size, and illegitimate recombination and other deletion processes may be the major mechanisms for decreases in genome size (e.g., Bennetzen 2002; Devos et al. 2002; Kellogg and Bennetzen 2004; Hawkins et al. 2008; Michael 2014). Whereas losses of large LTRs characterize some reductions in genome size (e.g., in *Oryza sativa*, Ma et al. 2004), decreases in genome size may also result from thousands of losses of small segments of DNA, as in *Arabidopsis thaliana* relative to *A. lyrata* (Hu et al. 2011).

Studies of microbial genomes suggest that downsizing of some genomes may be the result of homologous recombination at repeated genes, leading to the loss of large blocks of DNA as well as repeated sequences (Frank et al. 2002). Differences in double-stranded break repair may be responsible for some variation in genome size (Kirik et al. 2000). For example, differences in the processing of chromosomal ends have been reported in *Arabidopsis* and *Nicotiana* (tobacco) (Orel and Puchta 2003); the former has a genome size 20 times smaller than the latter. Free DNA ends are much more stable in tobacco, perhaps due to better protection of DNA break ends. Exonucleolytic degradation of DNA ends might be a driving force in the evolution of genome size (Orel and Puchta 2003). DNA resulting from WGD can also be "purged" through mechanisms of fractionation and neofunctionalization, as in *Utricularia gibba* (Ibarra-Laclette et al. 2013; Michael 2014). Extensive loss of duplicate loci in the recent (~80-year-old) allotetraploids *Tragopogon mirus* and *T. miscellus* (Asteraceae) supports this argument as a possible general mechanism of genome downsizing following WGD (e.g., Tate et al. 2006, 2009; Koh 2010; Buggs et al. 2012).

In some animals, insertion/deletion biases may lead to significant changes in genome size. A high rate of deletion has apparently occurred in *Drosophila* (Petrov 2002). Hancock (2002) has reconsidered the relationship between the level of repetitiveness in genomic sequences and genome size. A previously reported correlation between genome size and repetitiveness was generally confirmed, but with some deviations. Some of the variance in repetitive DNA observed may be the result of variation in the effectiveness of mechanisms for regulating slippage errors during replication. Recent analyses (Oliver et al. 2007) further demonstrate that the rate of genome size evolution is positively correlated with genome size, such that the fastest rates of genome size evolution occur in those species with the largest genomes. Further, genome size may have significant phenotypic effects (e.g., Greilhuber and Leitch 2013), with bursts of genome size due to TE proliferation suggested as an explanation for bursts of phenotypic evolution (e.g., Oliver and Greene 2009).

The driving forces behind changes in genome size remain unclear. It is tempting to invoke selection, and several hypotheses have been proposed for reductions in genome size (Leitch and Bennett 2004): (1) to reduce the nucleotypic effects of increased DNA amounts; (2) to reduce the biochemical costs associated with additional DNA amounts; and (3) to enhance polyploid genome stability. However, Oliver et al. (2007) argue that their analyses of correlated evolutionary rates and genome size explain the skewed distribution of genome sizes in eukaryotes without invoking selection against large genomes.

The interplay between mechanisms that increase and decrease genome size has ultimately shaped observed patterns of genome size evolution (Grover and Wendel 2010). Properties of genomes themselves may lead to expansions or contractions of genomes while population size, selection, and drift may augment or oppose genomic processes. For example, polyploidy, proliferation of TEs, decreased heterochromatin content, decreased recombination, decreased methylation, and epigenetic release of TEs, along with low effective population size and drift, will lead to increases in genome size. Counteracting these processes, however, are unequal homologous recombination, illegitimate recombination, increased heterochromatin, increased recombination, increased methylation, epigenetic suppression of TEs, high effective population size, and selection (Grover and

Wendel 2010). The magnitude and interaction of these synergistic or opposing forces have rarely been examined and offer exciting avenues for further research.

POACEAE AND BRASSICACEAE AS MODELS FOR STUDYING GENOME EVOLUTION

Both the grass family (Poaceae) and mustard family (Brassicaceae) have been well studied genomically because of the enormous economic importance of many of their members. These clades therefore provide useful models for evaluating the process of genome evolution. Here we use both systems and employ important reviews to summarize tendencies in genome evolution. Use of these and other models has revealed the complexities of plant gnome evolution, including broad synteny, as well as major structural changes and genome rearrangement at finer scales. Polyploidy coupled with chromosomal rearrangement results in considerable variation in chromosome number. Major changes in genome size have also resulted from expansion and contraction of repetitive elements.

POACEAE

At a grand scale, grass genomes are highly labile for genome size and chromosome number, yet retain considerable synteny, as reviewed initially nearly two decades ago (e.g., Devos and Gale 1997; Gale and Devos 1998; see Gaut and Ross-Ibarra 2008 and below). Here, we focus on fine-scale processes of genome evolution in Poaceae and discuss synteny below.

Despite a pattern of overall genome collinearity, at a finer scale, numerous rearrangements and deletions are evident. For example, following the origin of the rice genome via WGD (~70 mya), there have been a segmental duplication (5 mya), many large-scale chromosomal rearrangements, and deletions (Wang et al. 2005). Significantly, following the ancient WGD event, there appears to have been massive and rapid diploidization in which 30–65% of gene duplicates have been lost.

Gaut (2002) examined genome size evolution in grasses in a phylogenetic context. Genome size varies a great deal even when only diploids are considered, varying from 1C-values of 0.70 to 9.15 pg. Gaut (2002) noted that genomes of Poaceae vary a great deal in size, varying 36-fold between the smallest and largest diploid values. Even within subfamilies, there is eight-fold variation in size, indicating rapid change in genome size across the family. In addition, genome size varies across subclades. Chloridoid grasses have low DNA content (highest 1C-value = 1.68 pg) while Pooideae have larger genomes (smallest 1C-value = 1.22 pg). Panicoideae and Pooideae also appear to differ in genome size.

These results pose the question—what mechanisms contribute to these dramatic changes in genome size? Much of the increase in genome size in grasses appears to be due to the accumulation of repetitive elements (e.g., Meyers et al. 2001; Gaut 2002; Piegu et al. 2006; Bennetzen 2007; Charles et al. 2008; Brenchley et al. 2012; Han et al. 2013). Barley and rice have similar gene complements, but exhibit a 12-fold difference in DNA content, most of which is the result of variation in repetitive DNA. Even within *Oryza*, doubling of genome size relative to *O. sativa* has occurred in *O. australiensis* via expansion of LTRs (Piegu et al. 2006). Likewise, proliferation of TEs has contributed to genome size variation in wheat, with dynamic change occurring both prior to and since allopolyploidization (Charles et al. 2008). Comparing maize and sorghum, differences in the number and complement of retrotransposons explain much of the four-fold difference in DNA content. In maize, rapid accumulation of transposable elements has been a major aspect of genomic change (Wei et al. 2007).

Researchers in grass genomics have attempted to reconstruct the ancestral chromosomal complement of the grass genome. Wei et al. (2007) suggested that grass genomes evolved from an ancestor with $n = 12$ as the base chromosome number and that the structure of this genome was similar to that of the rice genome. Research by Salse et al. (2008; reviewed in Eckardt 2008) provides additional support for this basic model of grass genome evolution and suggests that the $n = 12$ ancestral chromosomes arose from WGD and rearrangement of a basic number of $x = 5$; several breakage/fusion events yielded the ancestral $n = 12$ chromosome complement (Fig. 15.6).

BRASSICACEAE

Brassicaceae, with over 4,100 species, provide another model for the study of genome evolution. As in many other large clades (e.g., Poaceae, with ~11,000 species, discussed above), genome size and chromosome number vary considerably. Examination of genome size in a phylogenetic context suggests a highly dynamic pattern of evolution, with numerous ups and downs in genome size (Johnston et al. 2005; Lysak et al. 2009). Lysak et al. (2009) reconstructed the ancestral genome size for Brassicaceae as very small (1C = 0.50 pg). Approximately 50% of the species

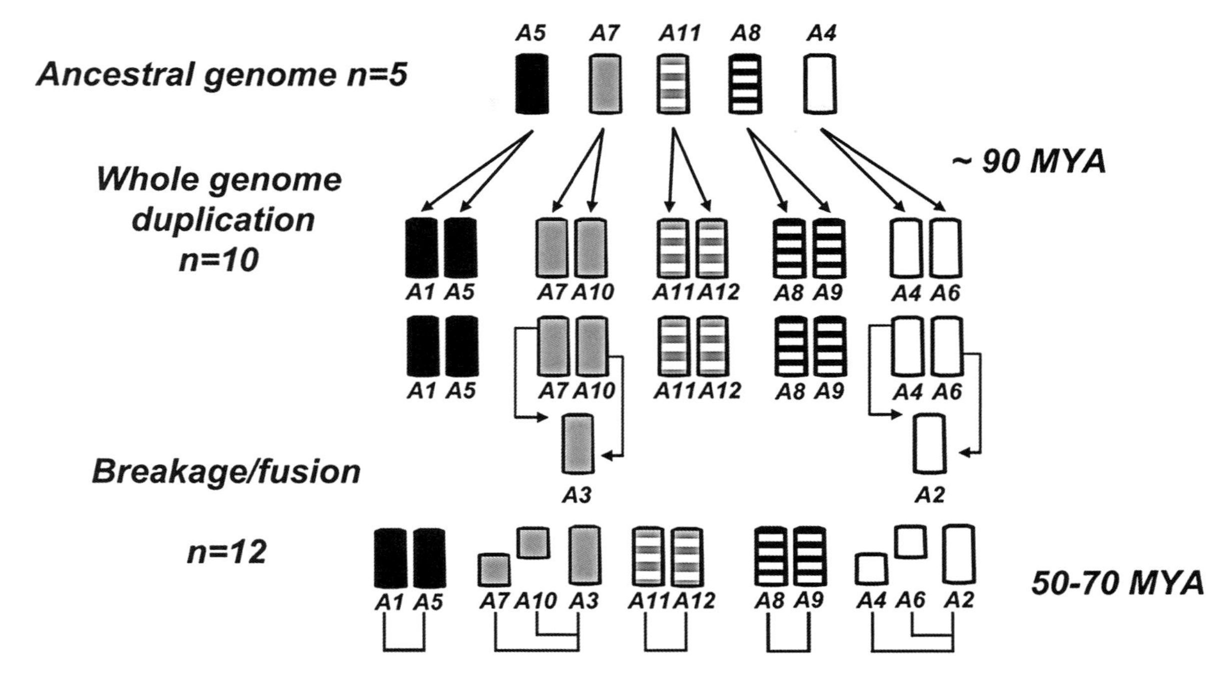

Figure 15.6. Model proposed for the ancestral grass genome and subsequent evolution. A common ancestor with $n = 5$ chromosomes underwent a whole-genome duplication resulting in an $n = 10$ intermediate. This event was then followed by two interchromosomal translocations and fusions that led to the construction of two new chromosomes, resulting in an $n = 12$ intermediate ancestor. The five ancestral chromosomes are named according to rice nomenclature. This model (from Salse et al. 2008) proposes how the genomes of rice, wheat, maize, and sorghum evolved from the $n = 12$ intermediate (Eckardt 2008; modified from Salse et al. 2008).

they analyzed showed a decrease in genome size from this ancestral state, whereas the remaining species showed an increase. This increase was generally moderate, but significant increases in C-value were found in tribes Anchonieae and Physarieae. Several studies have examined patterns of transposon evolution and changes in methylation in Brassicaceae. In a recent comparison, Seymour et al. (2014) examined DNA methylation and transposons from three species in Brassicaceae that vary in genome architecture, *Capsella rubella*, *Arabidopsis lyrata*, and *Arabidopsis thaliana*. Interestingly, they found evidence that lineage-specific expansion and contraction of transposon and repeat sequences are the main drivers of interspecific differences in DNA methylation in the family.

Brassica has emerged as a model for investigating the genomic changes that occur after polyploidization. Studies of synthetic lines of allopolyploid *Brassica napus* over multiple generations show evidence of rapid changes in gene expression, as well as genomic changes, including deletions, duplications, and translocations, many of which appear to be the consequence of homeologous recombination (Gaeta et al. 2007; Gaeta and Pires 2010). There is also evidence for changes in methylation (Lukens et al. 2006), rapid chromosomal change (Xiong et al. 2011), and major epigenetic changes (Lukens et al. 2006; reviewed in Diez et al. 2014).

Comparing the genome sequence of *Brassica oleracea* with that of its sister species, *B. rapa*, Liu et al. (2014) found evidence of numerous chromosome rearrangements, gene loss, and amplification of transposable elements. They also observed changes in gene expression, some of which trace to changes in patterns of alternative splicing.

Beyond genomic alterations, major changes have also been reported following polyploidy in *Arabidopsis*. Studies of both natural and synthetic *Arabidopsis* allotetraploids show that polyploidy can result in rapid gene silencing (Comai et al. 2000; Lee et al. 2001; Chang et al. 2010), global changes in patterns of gene expression (Chen 2007, 2010; Wang et al. 2006; Chi et al. 2012), epigenetic changes (Osborn et al. 2003; Madlung et al. 2005), changes in alternative splicing (Zhang et al. 2010), and consequent changes in the proteome (Ng et al. 2012).

SYNTENIC PATTERNS ACROSS ANGIOSPERMS

As the number of sequenced plant genomes continues to increase rapidly, new areas of comparative genomic inference have become feasible. Levels and patterns of ancient

WGD are being inferred from these genomes. Furthermore, assembled genomes (not just the gene space) are making it increasingly feasible to examine synteny (conservation of gene order on chromosome segments) both within clades, such as Brassicaceae and Poaceae, and in broader comparisons across the angiosperms. Extensive synteny across long periods of angiosperm evolutionary history has been one of the most important discoveries resulting from the comparison of plant genomes. Comparison of gene order in *Amborella* with genomes of peach (*Prunus*; Rosaceae), cacao (*Theobroma*; Malvaceae), and grape (*Vitis*; Vitaceae) showed synteny across diverse clades of *Superrosidae* (Fig. 15.7) and that synteny was retained from the common ancestor of rosids and *Amborella* (*Amborella* Genome Project 2013). Furthermore, this synteny has enabled additional studies of ancient polyploidy—comparisons among these superrosid genomes revealed that syntenic blocks had been preserved following the two WGDs that occurred early in eudicot evolution (Tang et al. 2008; Jiao et al. 2012; *Amborella* Genome Project 2013).

Similarly, on a broad phylogenetic scale, Guyot et al. (2012) compared the genomes of *Coffea* (coffee; Rubiaceae) and *Solanum* (tomato; Solanaceae), representing the asterid clade, with the genome of *Vitis*, a member of the rosid clade, and found considerable synteny at this scale (referred to as "macrosynteny") across the rosid and asterid clades. There was also evidence of genome restructuring. Unexpectedly, there was more conservation of synteny between coffee and grape (an asterid and a rosid) than between the two asterids (coffee and tomato) (Fig. 15.8).

It has long been recognized that gene order is highly conserved across the grass family (Poaceae) over ~60 myr of evolution (Gale and Devos 1998). Genetic mapping of nine genomes, including maize, rice, wheat, sorghum, sugar

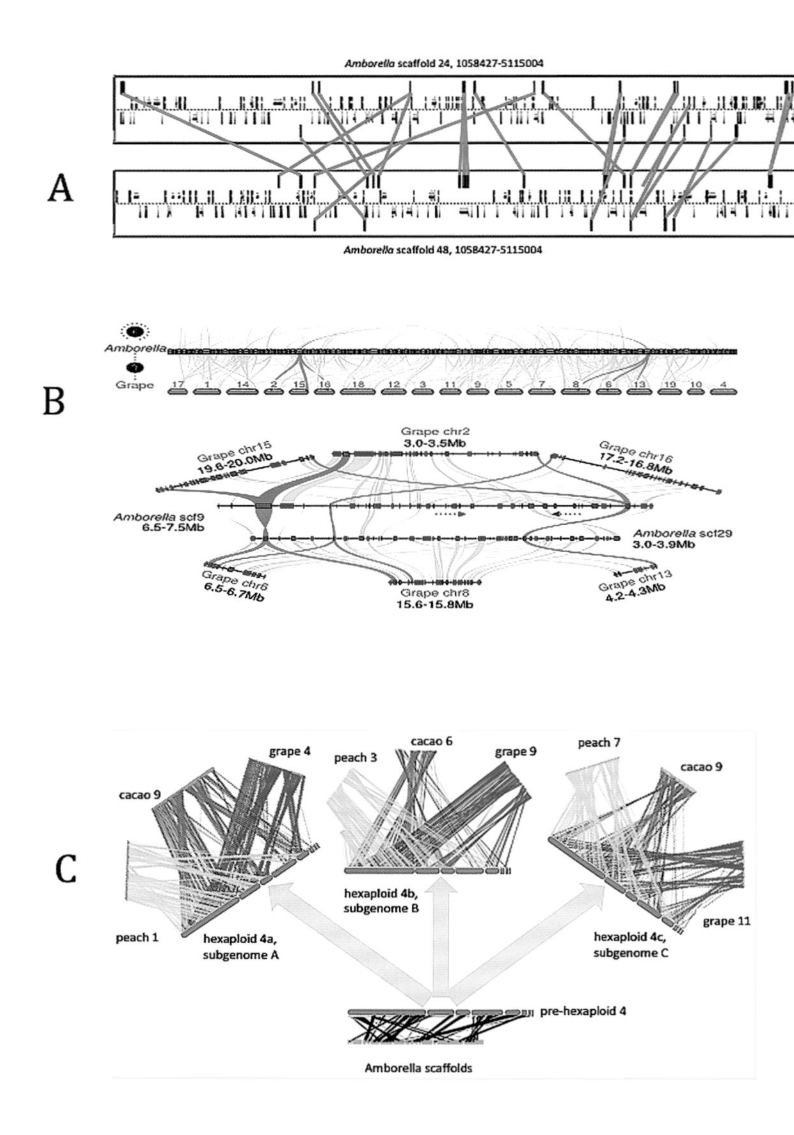

Figure 15.7. Synteny analysis of *Amborella*. (A) High-resolution analysis of intragenomic syntenic regions from *Amborella* putatively derived from the pre-angiosperm whole-genome duplication (WGD). Note the series of collinear genes between the two regions. (B) Macrosynteny and microsynteny between genomic regions in *Amborella* and *Vitis* (grape). (Top) Macrosynteny patterns between *Vitis* and *Amborella* and within *Amborella* scaffolds (only scaffolds 1–100 are shown). Each *Amborella* region aligns to up to three *Vitis* regions, which resulted from the "gamma" hexaploidization event in the eudicots (Jaillon et al. 2007). Syntenic regions within the *Amborella* genome were derived from the WGD prior to the origin of all extant angiosperms (Jiao et al. 2011). An exemplar set of blocks, showing two homeologous *Amborella* regions derived from this earlier WGD, aligns to three distinct grape regions (derived from "gamma"), with eight parallel regions in total. (Bottom) Microsynteny among the eight regions. Blocks represent genes with orientation of the same strand (blue) or reverse strand (green); shades represent matching gene pairs. (C) Gene order alignments between one of the seven ancestral core eudicot chromosomes and a subset of the *Amborella* scaffolds, as well as between the three post-hexaploidization copies of this chromosome and the peach, cacao, and grape chromosomes descending from it. Similar configurations were obtained for the other six ancestral chromosomes (Fig. 2 from the *Amborella* Genome Project 2013).

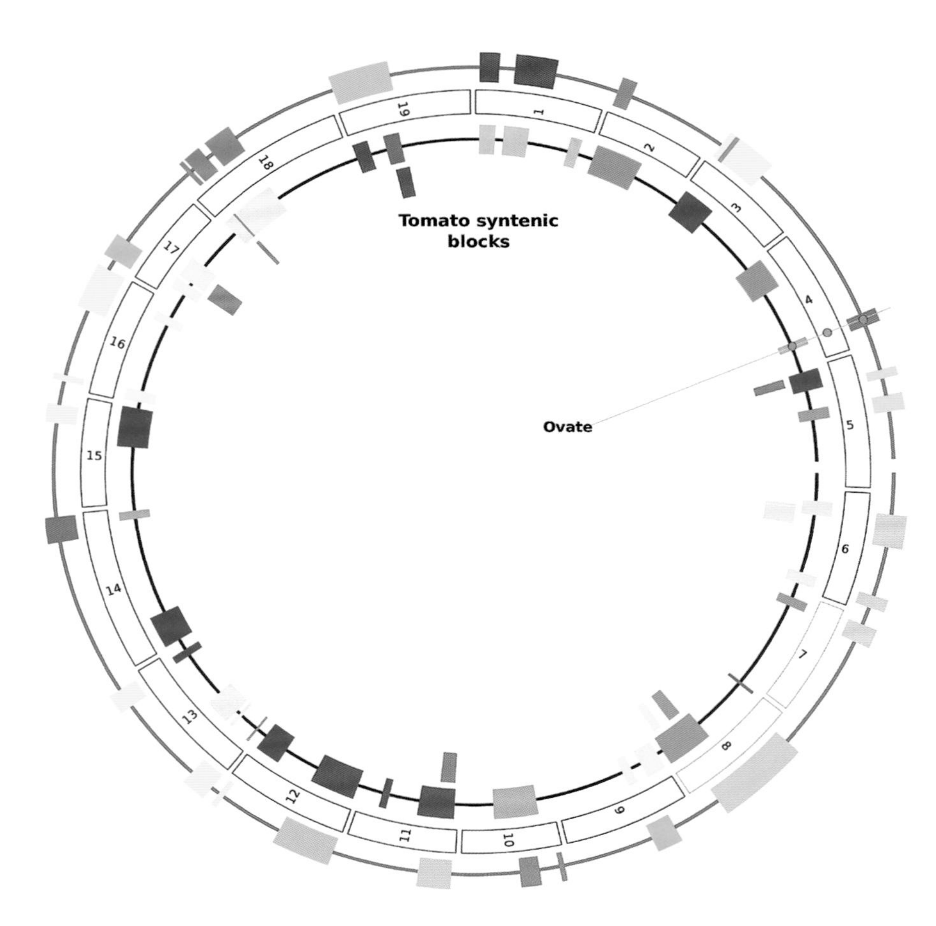

Figure 15.8. Synteny between the genomes of the asterids coffee (*Coffea*) and tomato (*Solanum*) compared to pseudochromosomes of the rosid grape (*Vitis*), shown as a circle diagram. The central circle represents the genome of *Vitis*; the outside and inside circles represent syntenic blocks in *Coffea* and *Solanum*, respectively. For syntenic blocks, each color represents a different linkage group in *Coffea* and *Solanum*. Red dots represent the relative location of the ovate genes on the physical map of the grape genome and on the genetic map of coffee and tomato. (From Guyot et al. 2012; Open Access; BioMed Central, original publisher.)

cane, foxtail millet, and others, yielded a conserved consensus grass genetic map (Fig. 15.9). Furthermore, these comparative genomic studies also revealed that some chromosomal rearrangements characterize certain groups of grasses (Devos and Gale 1997; Gale and Devos 1998). The high conservation of gene number and order in grasses prompted Bennetzen and Freeling (1997) to propose the idea of a "unified grass genome: synergism in synteny."

Mapping studies using genetic/genomic data as well as evidence from FISH (fluorescent in-situ hybridization) also revealed a high degree of synteny across genomes in Brassicaceae (e.g., Koch and Kiefer 2005; Cheng et al. 2012; Schmidt and Bancroft 2011). However, some genomes, such as *Arabidopsis thaliana*, exhibit considerable genomic reshuffling (Lysak et al. 2006). Using comparative genetic and cytogenetic analyses, researchers have attempted to reconstruct the "ancestral crucifer karyotype" of eight chromosomes with conserved genomic blocks (Lysak et al. 2006; Schranz et al. 2006). Although this sounds plausible, it must be recognized that this number of chromosomes does not itself represent the ancestral state for Brassicaceae; given the number of ancient WGDs known in this family (see above), there must have been ancestral species with higher numbers. The eight chromosome pairs hypothesized to have been present in the Brassicaceae ancestor are also the result of even older periods of chromosome reorganization/reduction—events that may have been similar to those postulated to have occurred in the process of going from the eight pairs in the Brassicaceae ancestor to the five pairs of modern *Arabidopsis thaliana*. Chromosomal evolution across tribes of Brassicaceae has also subsequently been reconstructed (Mandáková et al. 2010).

In their investigation of ancient polyploidy in monocots, Jiao et al. (2014) observed conservation of syntenic blocks between banana (*Musa acuminata*, Musaceae; Zingiberales) and the grasses rice (*Oryza sativa*) and sorghum (*Sorghum bicolor*). On a broader scale, synteny facilitated comparison of genomes and duplicated gene blocks among these genomes as well as with the oil palm (*Elaeis guineensis*) and two sequenced eudicot genomes included for comparison, grape (*Vitis vinifera*; Vitaceae) and sacred lotus (*Nelumbo nucifera*; Nelumbonaceae).

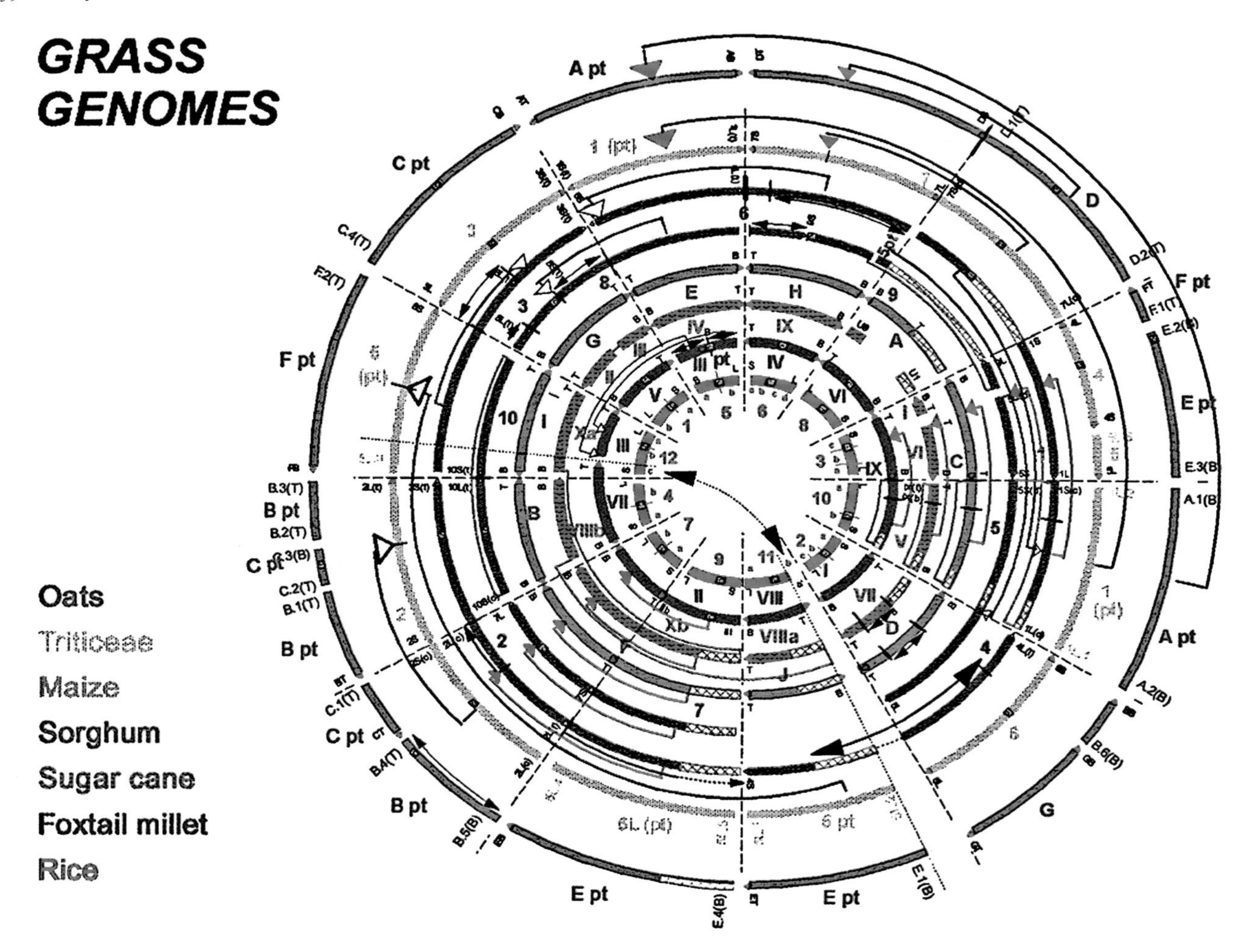

Figure 15.9. Synteny in grasses. A consensus comparative genetic map for grasses, depicted as a circle diagram. The seven genomes compared are color-coded. From Gale and Devos (1998).

THE ANCESTRAL ANGIOSPERM GENOME

The complete sequencing of the nuclear genome of *Amborella* (*Amborella* Genome Project 2013), sister to all other extant flowering plants (Chapters 3 and 4), has profound implications, permitting researchers to reconstruct the genome and gene content of the ancestral angiosperms (Adams 2013; *Amborella* Genome Project 2013). The ancestral angiosperm gene set comprised at least 14,000 genes. Compared with non-angiosperms, the *Amborella* genome has 1,179 gene lineages that first appeared in flowering plants, with about 4,000 gene lineages specific to seed plants (*Amborella* Genome Project 2013). The new gene lineages discovered in flowering plants may have had gene functions specific to angiosperms and therefore may have been crucial to their success and rapid diversification (Adams 2013; *Amborella* Genome Project 2013). It is also apparent that new functions were obtained by genes already present in plants—a good example being genes involved in floral development, which have homologs in other seed plants. Interestingly, orthologs of most floral genes were in existence long before a role was established in flowering—these genes were co-opted for floral function.

The *Amborella* genome contains at least 124 microRNA loci corresponding to 90 different families, with 27 of those likely present in the ancestral angiosperm, indicating that non-coding small RNAs were well represented in early angiosperms (Adams 2013; *Amborella* Genome Project 2013). Transposable elements in *Amborella* are "old and dead"—some TEs have apparently persisted for millions of years but are no longer active. These results contrast with other angiosperms in which TEs continue to proliferate, sometimes massively, leading to divergent families of TEs even within single species.

HORIZONTAL GENE TRANSFER: BACKGROUND

Horizontal gene transfer (HGT), also termed lateral gene transfer (LGT), refers to the transfer of DNA between different organisms in a manner other than by vertical descent from parent to offspring. It is best known as a major force in prokaryote evolution, but it has also played an important and perhaps underappreciated role in eukaryotic evolution (reviewed in Keeling and Palmer 2008); we mentioned HGT and its potential influence on phylogeny reconstruction briefly in Chapter 3 (see Fig. 3.4) and include this discussion here because of the role of HGT in altering gene content. HGT can occur between all three subcellular genomes of plants—nucleus, mitochondrion, and plastid. The process of HGT is complex in eukaryotes and not completely understood. The foreign DNA involved in the HGT event must pass through both the cell membrane and nuclear (organellar) membrane to reach the genome of the host cell. The number of well-supported cases of HGT in prokaryotes, as well as eukaryotes, including plants, has expanded rapidly in recent years, due in large part to increasingly large nuclear, as well as plastid and mitochondrial, datasets for many species.

OVERVIEW OF HGT IN PLANTS

Renner and Bellot (2012) provided an excellent recent review of HGT in eukaryotes with a thorough overview of plant-to-plant HGT events, which is our focus here. Much of our summary below is based on the extensive review of these authors. Increasing numbers of studies have indicated that all three plant genomes have experienced HGT and that this may be much more common than previously appreciated (Richardson and Palmer 2007; Bock 2010; Keeling and Palmer 2008; Renner and Bellot 2012). HGT may be particularly prevalent in the mitochondrial genome (Renner and Bellot 2012), and we review some of these cases below (e.g., Bergthorsson et al. 2003, 2004; Won and Renner 2003; Davis and Wurdack 2004; Barkman et al. 2007; Mower et al. 2010, 2012; Knoop et al. 2011; Sanchez-Puerta et al. 2011; Iorizzo et al. 2012; Radice 2012; Stegemann et al. 2012; W. Wang et al. 2012; T. Zhang et al. 2012; Chang et al. 2013; Cuenca et al. 2013; Greiner and Bock 2013; Guo and Mower 2013; Huang and Yue 2013; Rice et al. 2013; Richardson et al. 2013; Straub et al. 2013; Davis et al. 2014; Gao et al. 2014; Park et al. 2014; Smith 2014).

The causes of HGT in plants and other eukaryotes are not well understood, but parasitic plants appear to provide some of the strongest evidence of HGT in eukaryotes in general (e.g., Davis and Wurdack 2004; Mower et al. 2004; Nickrent et al. 2004; Park et al. 2007; Yoshida et al. 2010; see below). As many of these papers have stressed, HGT in these plants appears to be facilitated by the extremely close association between tissues of the parasites and their hosts. Kim et al. (2014) demonstrated that messenger RNAs representing thousands of genes moved in high frequency in both directions between *Cuscuta* and its host (*Arabidopsis* in this experiment), providing a potential mechanism for HGT.

THE EXTENT OF HGT

Recent genomic studies have provided the first insights into the extent of HGT in parasitic plants. A phylogenomic study suggested that in the holoparasite *Rafflesia cantleyi* (Rafflesiaceae), roughly 2% of the nuclear gene transcripts may have been acquired from its obligate host (Xi et al. 2012a). In a more recent and comprehensive analysis of mtDNA genes, Xi et al. (2013), using NGS to investigate *R. cantleyi* and additional parasites, including two of the closest relatives of *R. cantleyi*, conservatively estimated that 24–41% of the gene sequences show evidence of HGT in Rafflesiaceae, depending on the species examined. Significantly, the authors stressed that most of these transgenic sequences possess intact reading frames and appear to be actively transcribed—hence these transferred genes are potentially functional. Furthermore, some of these transgenes maintain synteny, suggesting that insertion is not random and that host genes have been displaced by foreign genes, perhaps via homologous recombination.

MECHANISMS OF HGT IN PLANTS

What are the possible means by which such events can occur? Renner and Bellot (2012) suggested the following possibilities: (1) vectors, such as bacteria, fungi, or phloem-attacking bugs; (2) transfer of entire mitochondria through plasmodesmata, facilitated by plant-to-plant contact; (3) illegitimate pollination followed by elimination of most foreign DNA except for a few mitochondria that might fuse with native mitochondria; and (4) natural transformation.

CASE STUDIES IN HGT: MTDNA

In their review of over 30 potential cases of mtDNA gene transfer in plants (see also Bergthorsson et al. 2003),

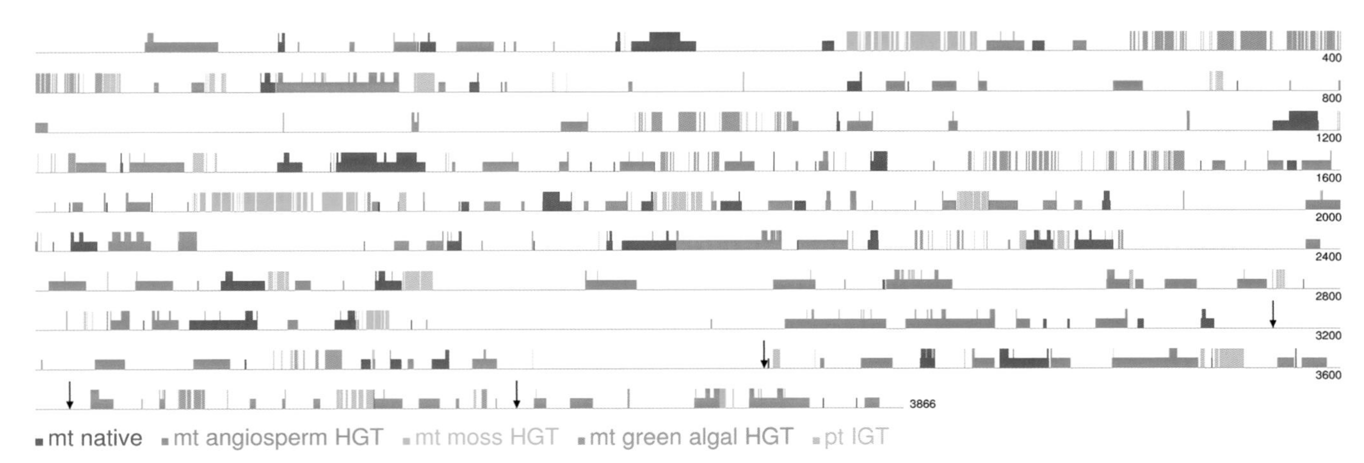

Figure 15.10. Foreign DNA in the *Amborella* mitochondrial genome as discussed by Rice et al. (2013). Map of the five mtDNA chromosomes of *Amborella trichopoda*, shown linearized and abutted (see vertical arrows). Shown here are regions of inferred organellar origin for which the ancestry was assignable (key: mt = mitochondrial; pt = plastid). Full-height boxes indicate genes. Half-height boxes indicate tracts of native and angiosperm-HGT DNA. Labeled black lines indicate mitochondrial genomes or partial genomes that have been horizontally transferred to *Amborella*. (Figure and legend modified from Rice et al. 2013; courtesy of D. Rice and J. Palmer.)

Renner and Bellot (2012) noted that the actual number of cases is weighted heavily by the large number of *Amborella* mtDNA genes putatively acquired from other angiosperms and mosses (e.g., Bergthorsson et al. 2004; Rice et al. 2013) (Fig. 15.10). Given the large number of events of HGT involving this one plant species, we begin with *Amborella*. The extent of proposed HGT in *Amborella* is incredible: the mtDNA of *Amborella* is huge (3.9 Mb), and the genome contains "six genome equivalents of foreign mitochondrial DNA, acquired from green algae, mosses, and other angiosperms. Many of these horizontal transfers were large, including acquisition of entire mitochondrial genomes from three green algae and one moss" (Rice et al. 2013; p. 1468) (Fig. 15.10). It remains unclear if *Amborella* is just a special case, or if there are other angiosperms with similar stories of massive HGT to the mtDNA genome. This is an important area of future study—only additional analyses of mtDNA genomes will address this question.

It is also noteworthy that at least seven of the examples reviewed by Renner and Bellot (2012) involve parasitic plants. The close connection of host and parasite must facilitate HGT events in plants. HGT has occurred in a number of phylogenetically distinct plant clades with parasites. Examples include an mtDNA intron in *Rafflesia* (Rafflesiaceae) (Davis and Wurdack 2004; Barkman et al. 2007) and mtDNA genes in *Pilostyles* (Apodanthaceae) (Barkman et al. 2007), *Cuscuta* (Convolvulaceae), *Mitrastemon* (Mitrastemonaceae) (Barkman et al. 2007), and several genera of Orobanchaceae (e.g., Mower et al. 2004, 2010).

There are also examples of mtDNA HGT that do not involve parasites, and some of these involve angiosperms as well as other plants. These include the transfer of mtDNA genes to the asterid *Actinidia* from an unspecified monocot (Bergthorsson et al. 2003) and to the gymnosperm *Gnetum* from an unknown asterid (Won and Renner 2003). Also reported is the transfer from an angiosperm to a fern: an mtDNA intron in *Botrychium* (grape fern; Ophioglossales in the fern clade) from an unknown member of Loranthaceae (Davis et al. 2005a).

Several mtDNA genes have been transferred: *rps2*, *rps11*, *matR*, *atp1*, and *atp6*. Introns of mtDNA genes have also resulted from HGT events, including the *nad1* second intron and the *cox1* intron, which may have experienced multiple HGT events (reviewed in Renner and Bellot 2012). In most of the examples of HGT involving mtDNA genes, the transferred genes are present in the mtDNA genome of the host and typically are nonfunctional. Several instances of chimeric mtDNA sequences involving the host mtDNA and foreign DNA copies have also been reported (Barkman et al. 2007; Hao et al. 2010; Mower et al. 2010; see Renner and Bellot 2012).

CASE STUDIES IN HGT: NUCLEAR AND PLASTID GENES

There are only a few examples of plant-to-plant HGT that involve nuclear genes. One of the better known involves *Striga hermonthica*, a parasitic plant (witchweed; Orobanchaceae) in which a nuclear gene was likely obtained via HGT from a monocot host plant (Yoshida et al. 2010). A more recent example involves the apparent HGT of a neochrome gene from hornworts to ferns; significantly, this gene transfer may have then played a significant role in the subsequent diversification of modern ferns (Li et al. 2014). As larger datasets of nuclear genes are acquired for more lineages via NGS, it will be possible to assess better the extent of the transfer of nuclear genes (as well as genes in general) via HGT.

The plastid genome seems to be rarely involved in HGT events, with the exception of frequent cases of plastid transfer between species via introgression. In a recent review, Huang and Yue (2013, p. 16) noted, "HGT in plastid genomes, though relatively rare, still exists" (see also Smith 2014 and Gao et al. 2014). An exciting finding indicated the possible conversion of functional mtDNA genes (*atp1*) by the homologous *atpA* gene from the plastid. Nine such cases have been reviewed (see Hao and Palmer 2009).

Angiosperm genomes are dynamic, with fluctuations in size mediated by a host of mechanisms. Here we have summarized patterns of genome size evolution across the angiosperms and addressed mechanisms of both genome expansion and contraction. Despite growing resources on genome size, such data are currently available for only approximately 2–3% of all angiosperm species, hampering the development of hypotheses to explain and understand patterns of genome size dynamics. Additional genome size measurements, perhaps via development of high-throughput and collaborative methods, are needed to aid our understanding of genome size evolution and to provide a foundation for prioritizing species for genome sequencing (e.g., Galbraith et al. 2011). An enhanced perspective on phylogenetic patterns of genome size, coupled with continued investigation of HGT, will allow further analysis and insight into factors governing the size, structure, and content of angiosperm genomes.

16 Polyploidy

INTRODUCTION

Biologists have long recognized that polyploidy (or whole-genome duplication, WGD) is a major evolutionary force in plants with seminal papers dating to the early and mid-1900s (e.g., Muntzing 1936; Darlington 1937; Clausen et al. 1945; Love and Love 1949; Stebbins 1950; Lewis 1980; Grant 1981; Levin 1983). Importantly, polyploidy is now recognized not just as a process restricted largely to plants, but also as a significant evolutionary force throughout eukaryotes (e.g., Mable 2003; Gregory and Mable 2005; Husband et al. 2013).

Genomic investigations have expanded appreciation of the association between WGD and evolutionary innovation. They have also played a major role in the dramatic modification of the longstanding polyploidy paradigm—that WGD is a process limited largely to plants. Genomic studies over the past 15 years have shown that all eukaryotes have genomes with considerable gene redundancy resulting in large part from ancient WGD. Diverse polyploid events ranging from ancient to recent have been documented in vertebrates (e.g., Van de Peer et al. 2010; Braasch and Postlethwait 2012; Cañestro 2012), fungi (Hudson and Conant 2012), and ciliates (Aury et al. 2006). Relying on the excellent review of Husband et al. (2013) that covers green plants, the incidence of polyploidy appears to be generally low in liverworts, hornworts, cycads, and conifers (see also Ahuja 2005), but is frequent in lycophytes, ferns, and angiosperms. Polyploidy is also prevalent in other photosynthetic groups, such as red algae (Husband et al. 2013).

Until recently it was difficult to determine the actual frequency of polyploidy in various plant lineages. For decades, workers used chromosome counts as a proxy to estimate diploidy versus polyploidy. The researcher would essentially provide a best guess for what the original base chromosome numbers might have been in a given lineage, and multiples of that were considered polyploids. Others tried to set a cut-off for which base numbers were themselves the result of polyploidy. For example, some set $x = 12$ (see below) as the lowest number indicating that a group of plants had polyploidy in their ancestry. Obviously, such arbitrary cut-offs could be problematic. Despite many obvious caveats, there were numerous attempts to estimate the frequency of polyploidy in various land plant clades using these methods. The angiosperms, in particular, were the focus of many attempts to estimate the occurrence of polyploidy with this general approach.

Stebbins (1950, 1971) estimated that ~30–35% of angiosperm species had formed via polyploidy. This estimate was one of the lowest suggested from that time period, but it differs from many others in that Stebbins was speaking about speciation that involved a change in ploidy, whereas others were concerned with the percentage of angiosperms that had polyploidy in their ancestry. This would result in different percentages because the former was concerned with formation and the latter with total numbers; a species may form via polyploidy (and so be included in Stebbins's estimate) but then undergo diversification, leading to a clade of perhaps hundreds of species with polyploidy in their background, but only one instance of involvement of polyploidy in speciation, which was Stebbins's focus. In what were among the earliest estimates for the percentage of angiosperms with polyploidy in their ancestry, Müntzing (1936) and Darlington (1937) speculated that roughly

50% of all angiosperm species were polyploid. To understand the variation in estimates among authors, it is important to understand the limitations of the methodology. As noted, estimates were based on chromosome counts and generally varied based on the cut-off chromosome number used by the author for scoring species as diploid or polyploid. Stebbins (1971, p. 124) suggested that "all genera or families with $x = 12$ or higher have been derived originally by polyploidy" and noted "that even the numbers $x = 10$ and $x = 11$ may often be of polyploid derivation." Grant (1981) suggested the cut-off number should be higher and hypothesized that angiosperms with chromosome numbers of $n = 14$ or higher were of polyploid origin. With this cut-off point, he determined that 47% of all angiosperms were of polyploid origin and further proposed that 58% of monocots and 43% of "dicots" (eudicots is a concept that came later; Chapters 3 and 8) were polyploid. In contrast, Goldblatt (1980) maintained that this estimate was too conservative and proposed that taxa with chromosome numbers above $n = 9$ or 10 are of polyploid origin. He calculated that at least 70–80% of monocots have polyploidy in their ancestry. Lewis (1980) applied an approach similar to Goldblatt's to "dicots" and estimated that 70–80% were polyploid. Masterson (1994) used the novel approach of comparing leaf guard cell size in fossil and extant taxa from a few angiosperm families (Platanaceae, Lauraceae, Magnoliaceae) to estimate occurrence of polyploidy through time. With this method, she estimated that 70% of all angiosperms had experienced one or more episodes of polyploidy in their evolutionary history.

The modern approaches to answering these questions rely on genomic data analyzed within a phylogenetic context. Operating on a similar basis to Stebbins's original estimate, Wood et al. (2009) estimated that 15% of speciation events for flowering plants and 31% for ferns directly involve a change in ploidy. Other investigators have similarly proposed that polyploid speciation is much more frequent than widely appreciated (e.g., Soltis et al. 2007d; Ramsey and Schemske 1998, 2002).

Genomic data have made it possible to avoid the use of chromosome number per se, which can be highly misleading. A major surprise was that the complete genome sequence of *Arabidopsis* revealed evidence for three polyploid events (Vision et al. 2000; Bowers et al. 2003). However, *Arabidopsis thaliana* with its small genome size and low chromosome number ($n = 5$) had been considered a typical diploid plant—which is in large part why it was chosen as the first plant for complete sequencing. Similarly, the complete nuclear genome of rice (*Oryza sativa*; Poaceae) contained evidence for polyploidy in a plant with a chromosome number ($n = 12$) considered by many to represent a model "diploid " plant (Paterson et al. 2004; Yu et al. 2005).

Genomic data also revealed a high level of gene redundancy that may result from multiple episodes of polyploidy. The crop *Brassica napus* was formed relatively recently (~7500 years ago) via allotetraploidy. When recent genome doubling events are considered together with a series of more ancient polyploidizations that occurred deeper in Brassicaceae, Brassicales, eudicots, and before the origin of the angiosperms, the number of genome doubling events involved is considerable. Chalhoub et al. (2014) estimated that the combined series of WGD events resulted in an aggregate 72X genome multiplication in *B. napus* since the origin of the flowering plants and, even considering loss of some duplicates, a very high overall duplicated gene content. Multiple rounds of genome doubling, however, have not resulted in substantial increases in the number of retained genes. Genomic data reveal considerable gene loss following multiple rounds of WGD, as in the genome of *Utricularia gibba* (Ibarra-Laclette et al. 2013), discussed in more detail below.

GENOMIC INSIGHTS INTO FREQUENCY OF ANCIENT POLYPLOIDY

Over the past decade, the availability of powerful genomic tools has made it possible to obtain not only novel insights into polyploidy as a process (e.g., Doyle et al. 2008; Soltis and Soltis 2009; D. Soltis et al. 2009; Grover et al. 2012; Madlung and Wendel 2013), but also much more precise estimates of the timing and frequency of ancient polyploidy in the ancestry of extant species. Continued progress in establishing the framework of angiosperm phylogeny has been critical for pinpointing where ancient polyploidy has occurred on the plant tree of life. This progress has accelerated in recent years, and undoubtedly we will see enormous continued progress in the years to come with the steady improvement of NGS technology, as well as rapidly decreasing costs.

Recent genomic investigations not only indicate that polyploidy is ubiquitous among angiosperms, but also suggest the presence of many major ancient WGD events (e.g., Van de Peer et al. 2009, 2010; D. Soltis et al. 2009). Indeed, the question is no longer what percent of angiosperms are polyploid, but how many WGD events occurred in any given lineage (D. Soltis et al. 2009). In this regard, at least 50 independent ancient WGDs are distributed across flowering plant phylogeny (Cui et al. 2006; D. Soltis et al. 2009; Van de Peer et al. 2009; M. S. Barker et al., University of Arizona, pers. comm.); we cover some of the major events below with a summary given in Fig. 16.1. However, additional events are being discovered on a regular basis, and

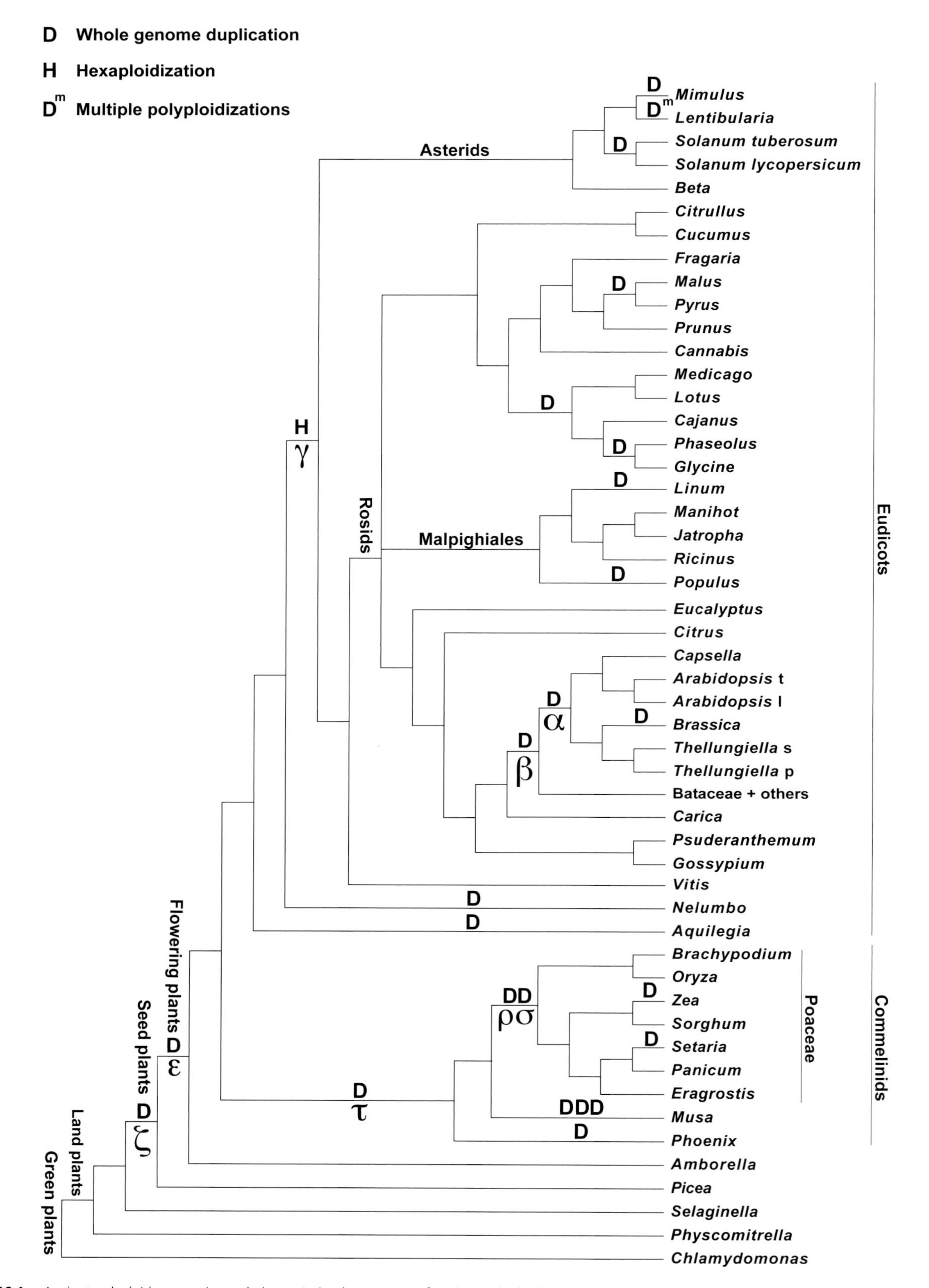

Figure 16.1. Ancient polyploidy events in seed plants. A simple summary of ancient polyploidy events with a focus on angiosperms. Nearly all taxa listed have completely sequenced nuclear genomes, although "Bataceae + others" are included as a placeholder to help illustrate the placement of β in Brassicales (see text and Fig. 16.2). The summary phylogeny is from Soltis et al. (2011). Some of the prominent ancient polyploidy events that have been designated with Greek letters are so indicated here (modified from CoGe website).

current understanding of paleopolyploidy events in angiosperm history can be found at the Comparative Genomics (CoGe) website (http://genomevolution.org/CoGe/).

Before the origin of the angiosperms as well as within the angiosperms, genomic data now indicate many ancient WGD events; many of these appear to be associated with the origins of major subclades. Some of these have been given Greek symbols (letters), which seem confusing to many researchers and hard to remember. We include those that have been used commonly in the literature for some of the most prominent WGD events as a point of reference (Fig. 16.1). It is perhaps a hopeless task to continue giving this sort of recognition to all published polyploid events; for example, in *Nicotiana* (Solanaceae), there are seven separate documented cases of polyploidy (Chase et al. 2003). Those involving large clades are worth noting, but it is clear that in angiosperms, polyploidy is ubiquitous.

Significantly, all angiosperm genomes sequenced to date exhibit evidence of ancient polyploidy. Genomic data also indicate that an ancient WGD event preceded the origin of all extant angiosperms, an event sometimes referred to as epsilon or ε (Jiao et al. 2011; *Amborella* Genome Project 2013), indicating that all living angiosperms are of ancient polyploid origin. This finding raises an important question: did WGD contribute to the success of the flowering plants? Genomic data also suggest an older WGD event that is associated with the origin of all extant seed plants, referred to as zeta or ζ (Jiao et al. 2011).

Two successive and perhaps closely associated WGDs apparently occurred during early eudicot evolution (Jiao et al. 2012; Chanderbali et al. 2017; gamma event, γ); these events are often referred to as a triplication, implying a single event, but as has been described from bread wheat, *Triticum aestivum* (e.g., Salamini et al. 2002; Marcussen et al. 2014; The International Wheat Genome Sequencing Consortium 2014), hexaploidy is a consequence of two rounds of genome doubling. Rather than simply referring to a triplication event as γ, perhaps the hexaploid ancestor of all core eudicots should be characterized as a product of γ1 and γ2, indicating two events close in time. These events are correlated with and perhaps in some way directly associated with the core eudicot radiation (70% of all angiosperm species; Chapter 8). However, the precise location of these events on the angiosperm tree of life is unknown, although Ming et al. (2013) raised the possibility of one WGD before divergence of the *Nelumbo* and core eudicot lineages and one WGD afterwards, but the predivergence WGD could be explained by either incomplete sorting of retained ancestral variation (e.g., Jones et al. 2013) or error in tree estimation.

Genomic studies have pinpointed two WGD events in Brassicales (Figs. 16.1, 16.2). One event (alpha or α) is associated with the origin of Brassicaceae (Schranz et al. 2011). A second event (beta or β) occurred deeper in Brassicales; following Schranz et al. (2011), this WGD appears to have occurred just after the branch to Limnanthaceae and prior to the origin of the subclade containing Koeberliniaceae, Bataceae, Salvadoraceae, Emblingiaceae, Pentadiplandraceae, Resedaceae, Gyrostemonaceae, Tovariaceae, Capparaceae, Cleomaceae, and Brassicaceae (see Chapter 10; Fig. 16.2). Note that the coverage of families in Schranz et al. (2011) does not reflect the current treatment presented in Chapters 10 and 12 (we have updated Fig. 16.2 accordingly).

There are also a number of major WGD events in the monocots. In fact, some authors have suggested that one ancient polyploid event (tau or τ; Jiao et al. 2014) may be associated with the origin of the monocots (Paterson et al. 2012). However, because of the limited sampling of monocots, it remains unclear as to where τ occurred. The event is known to be present in the commelinid clade sampled to date, and this is where we have placed the event (Jiao et al. 2014; Fig. 16.1). Analyses are needed for more monocot lineages to determine whether the event characterizes all monocots. Within the monocots, two WGD events are closely associated with the origin of the grass family (Poaceae); these are referred to as sigma or σ and rho or ρ (see Fig. 16.1) (Paterson et al. 2004, 2012; D'Hont et al. 2012; Jiao et al. 2014). Other ancient WGD events were detected in the genomes of *Musa* (Musaceae) and *Phoenix* (Arecaceae) (D'Hont et al. 2012; Jiao et al. 2014) (Fig. 16.1). More sampling and analyses are needed to determine where these WGD events occurred. Where does the earliest event identified in the *Musa* genome occur on the monocot tree? Does the event identified in *Phoenix* occur in a common ancestor of all palms?

The origins of several families are also associated with WGD events. Poaceae and Brassicaceae were noted above. Other examples include the origin of Asteraceae (Barker et al. 2008), Solanaceae (The Tomato Genome Consortium 2012), Cleomaceae (Schranz and Mitchell-Olds 2006), and papilionoid legumes (Papilionoideae or Faboideae; Cannon et al. 2015). Some of these examples, as well as others, could be discussed in much more detail to highlight the prevalence of polyploidy, but we are only briefly touching on these here. For example, the origin of Asteraceae appears to coincide with several ancient polyploidy events, with additional ancient events associated with some of the major subclades within this large family. Interestingly, Asteraceae also have two well-documented examples of young (less than 200 years old) polyploids (Hegarty et al. 2012; Soltis et al. 2012), making it an excellent test case

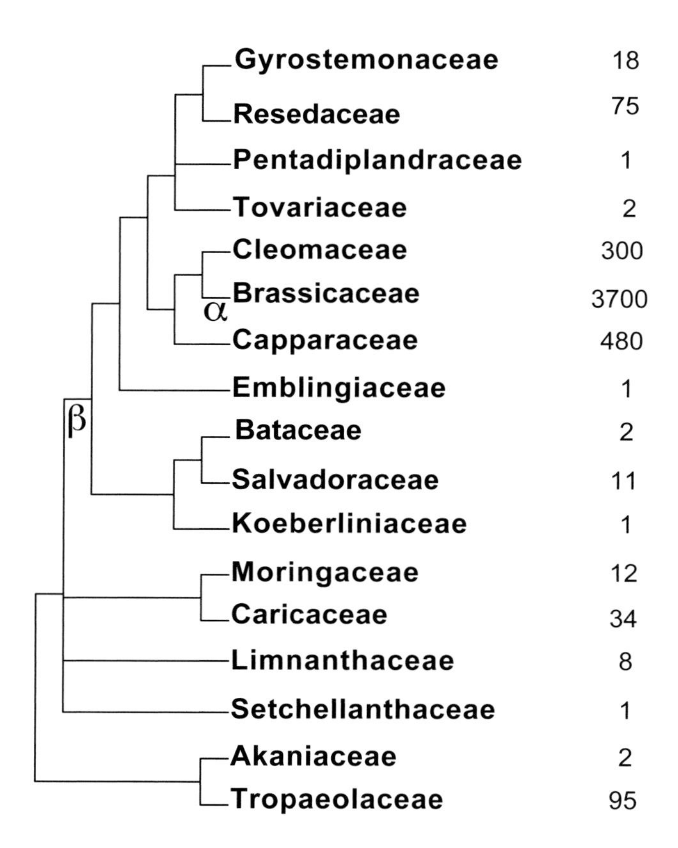

Figure 16.2. The placement of the α and β WGD events in Brassicales (modified from Schranz et al. 2011).

for investigating polyploidy over a continuum of ages from 40 million years (age of the family) to 40 generations (age of new *Tragopogon* polyploids). The complete genome sequence for *Arabidopsis thaliana* provided evidence for three WGD events, termed α, β, and γ (Vision et al. 2000; Bowers et al. 2003). The most recent of these events is α, which characterizes most Brassicaceae and is distinct from an independent event in Cleomaceae (Schranz and Mitchell Olds-2006; Schmidt and Bancroft 2011). Based on chromosome numbers, it was long thought that an ancient polyploid event characterized part of Solanaceae (reviewed in D. Soltis et al. 2009). Complete genome sequencing of tomato (*Solanum lycopersicum*) confirmed this, revealing that the tomato lineage has experienced two different genome "triplication" events. One of these is now known to be shared by most eudicots (see above; Jiao et al. 2012), but there is also a more recent one (The Tomato Genome Consortium 2012). As noted, triplication events are likely not single events but two polyploidy events in close succession.

The events reviewed above, as well as the ~50 that have been noted to date for angiosperms, certainly represent only the beginning in terms of ancient WGDs in flowering plants. With the explosion of genomic analyses, the next five to ten years will see more accurate placements of known events, as well as the discovery of many other WGD events. Complete genome sequences are not required to estimate WGD events. Lynch and Conery (2000) proposed a genomic-scale approach to estimate the occurrence and age of gene duplication events by assessing the frequency distribution of per-site synonymous divergence levels (*Ks*) for pairs of duplicate genes. Analyses of *Ks* plots have been useful in suggesting ancient WGD events (see Cui et al. 2006; Lynch and Conery 2000; Blanc and Wolfe 2004; Schlueter et al. 2004). Recent analyses of *Ks* plots generated using large genomic resources such as data from the 1000 plant transcriptome project (1KP) continue to reveal additional WGD events. These results suggest, for example, that multiple ancient polyploid events occurred within Lamiaceae (Godden et al. unpubl.); multiple events are also evident within Laurales and Magnoliales (Cui et al. 2006).

DOES POLYPLOIDY PROMOTE DIVERSIFICATION?

A question that has long been of interest is—do polyploid events promote a burst of speciation or diversification? There has been an interesting and stimulating recent debate on this subject. Mayrose et al. (2011) and Arrigo and Barker (2012) examined phylogenies at the genus level (usually) and concluded that polyploid lineages have higher extinction rates than diploid lineages. Based on these results, Arrigo and Barker (2012, p. 140) revived the old view of Stebbins (1950) and Wagner (1970), referring to "rarely successful" polyploid lineages and stating that "most (polyploidy lineages) are evolutionary dead-ends." Certainly new polyploids go extinct regularly at the population level, but a question that remains is, what is the fate of established polyploid species? Are they really more prone to extinction? Soltis et al. (2013, 2014a) pointed to numerous flaws in the logic and arguments of Mayrose et al. (2011, 2014) and Arrigo and Barker (2012). It is also noteworthy that using the Mayrose et al. (2011) methods, Zhan et al. (2014) found the opposite for fish—polyploids actually diversify more than diploids. One of the points of Soltis et al. (2014a) was that the sample sizes in Mayrose et al. (2011) were small and not broadly representative of angiosperms. A study of 63 of ~14,000 angiosperm genera (a clade of 350,000–400,000 species) as in Mayrose et al. (2011) is not enough to make strong generalizations regarding the fate of polyploid lineages. This poor sampling issue, coupled with the methodological problems of Mayrose et al. (2011), as well as the contrasting results of Zhan et al. (2014) indicate that broad generalizations regarding the fate of polyploids (as in Arrigo and Barker 2012) are simply premature. Much more work is needed.

Increased focus on the mechanisms that might increase (or decrease) diversification of polyploid lineages relative to diploid lineages should improve understanding of the role of polyploidy in diversification. For example, the Bateson-Dobzhansky-Muller model of the evolution of genetic incompatibility among allopolyploid populations (Orr 1996) may spur diversification immediately following genome doubling whereas sustained increase in diversification of a polyploid lineage may be due to the evolutionary innovations such as the flower (Jiao et al. 2011; *Amborella* Genome Project 2013) or nodulation and symbiotic nitrogen fixation (e.g., Cannon et al. 2015). Future research on the genes contributing to genetic incompatibilities and evolutionary innovations will help elucidate the role of polyploidy in diversification.

Another approach to examine the fate of polyploid lineages is to use big phylogenies and place ancient WGD events on these megatrees. Using genomic data, researchers have identified and pinpointed where many WGD events occurred on the angiosperm tree of life. These well-marked events provide the opportunity to investigate whether there are correspondences between WGD events and major bursts in species richness—the role of polyploidy in diversification. Soltis et al. (2014a,b) recently reviewed the topic and found that many ancient WGDs when placed on a phylogenetic tree are indeed associated with key diversification events in flowering plants. Many of these have already been noted and include the origins of angiosperms, eudicots, and Poaceae. Examination by eye—plotting polyploid events on phylogenetic trees for several families, including Poaceae and Solanaceae—suggested that ancient WGD was followed by a burst in speciation (D. Soltis et al. 2009). However, in many cases polyploidy was not immediately associated with a burst in species richness, which occurred several nodes later. Schranz et al. (2012) referred to this pattern as a "lag" and further noted that ancient polyploidy may be associated with an increase in plant diversity in other families (Asteraceae, Brassicaceae, Cleomaceae, Fabaceae). This lag can be seen associated with several major polyploid events, including the origin of the angiosperms (Fig. 16.3); Schranz et al. (2012) noted this lag as well for Brassicaceae and suggested (p. 147) that "ultimate success of the crown group does not only involve the WGD and novel key traits, but largely subsequent evolutionary phenomena including later migration events, changing environmental conditions and/or differential extinction rates." Another explanation for the putative lag is that before the benefits of polyploidy can take effect, the deleterious phenomena associated with recent polyploidy (genomic shock, gene copy number imbalances, etc.) have to be eliminated by diploidization, which may take several million years (De Smet et al. 2013). As discussed above, more research on the mechanisms contributing to increased diversification is necessary.

The methods of D. Soltis et al. (2009) and Schranz et al. (2012) were simple approaches, simply plotting WGD on trees and looking at the number of species that appeared afterwards. More recently, using a very large phylogenetic tree for angiosperms, Tank et al. (2015) used a novel approach and demonstrated with statistically significant support that several major ancient polyploid events are closely correlated with major bursts in diversification (Fig. 16.4). These include the origin of angiosperms, eudicots, mono-

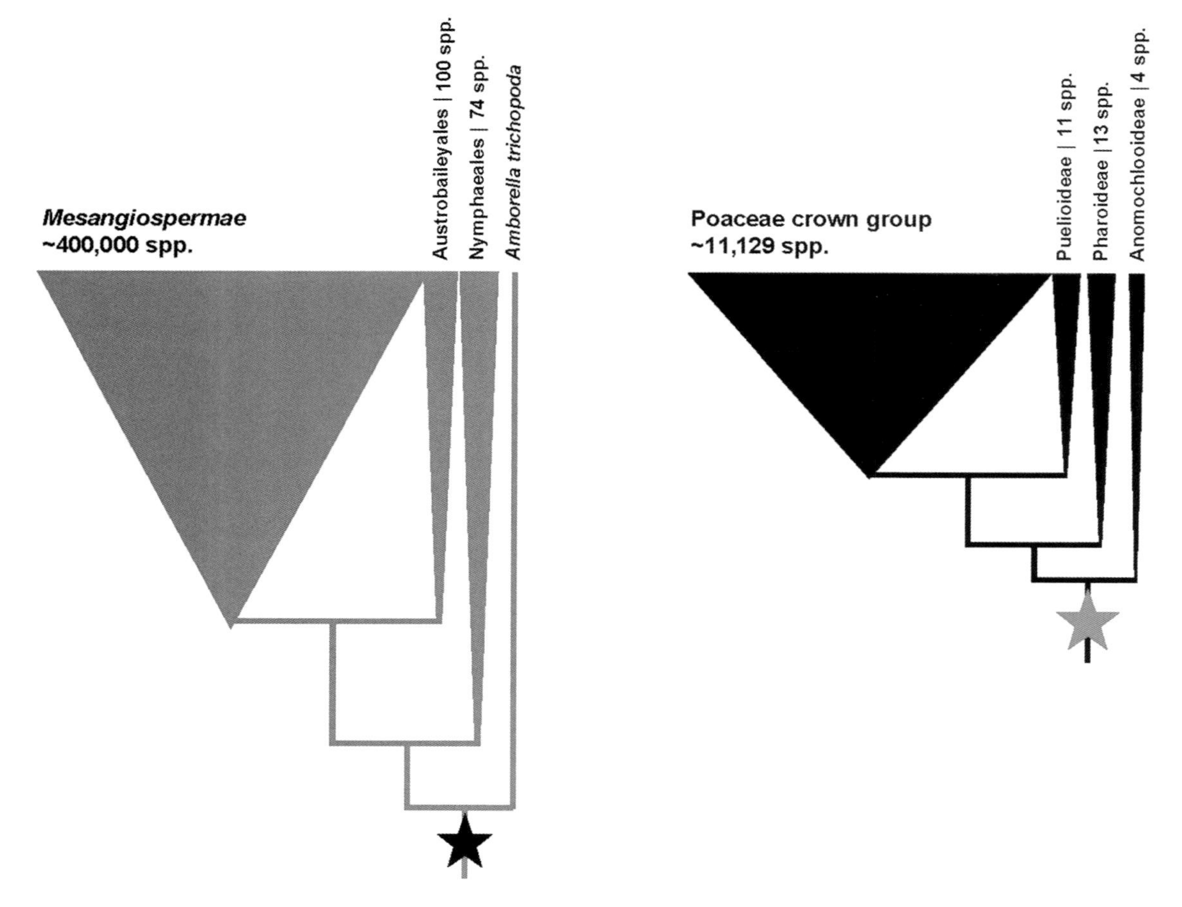

Figure 16.3. The diversification lag. Following WGD (stars) there may be a lag before diversification actually starts (modified from Schranz et al. 2012 and Soltis et al. 2014). This yields a phylogenetic pattern in which clades closest to the WGD event are small in size, ultimately followed by larger clades.

cots, and Asteraceae, and the β event within Brassicales. Some of these diversification bursts did not coincide exactly with the WGD event, but instead occurred a few nodes subsequent to that at which the WGD occurred, providing evidence for the "lag" effect. Improvements in the resolution of species phylogenies and associated estimation of divergence times and character evolution will help elucidate the length of these lags and the evolutionary changes occurring between genome duplication events and increased rates of diversification.

Many more phylogenomic analyses of evolutionary dynamics following WGD events are needed—there may be no single answer, but variation exists from group to group. In this regard, Estep et al. (2014) investigated polyploidy and diversification of the grass tribe Andropogoneae, which possesses numerous polyploids. At least 32% of the species sampled are allopolyploids. Using a large tree, Estep et al. (2014) found no evidence that polyploidy was followed by a net increase in diversification, although allopolyploidy is clearly a major mode of speciation in the clade.

ANCIENT POLYPLOIDY AND THE CRETACEOUS-TERTIARY BOUNDARY

Fawcett et al. (2009) dated WGD events using genomic data and noted that a number of ancient polyploid events occurred at the same time (~65 mya) in several diverse angiosperm lineages. This timing is close to the K-T boundary; they proposed that preexisting WGDs helped numerous plant lineages survive the K-T mass extinction (now known as the Cretaceous-Paleogene [K-Pg] extinction event). More examples of this association were provided by Vanneste et al. (2014), who built a stronger case for this association, noting that there was evidence for this relationship across 41 plant genomes from diverse lineages representing a number of families (Fig. 16.5).

It is often assumed that polyploidy must confer some type of advantage, and polyploidy has therefore been suggested to be the driver of diversification events (e.g., Levin 1983; Tank et al. 2015) or the reason behind survival of some lineages over others following the K-T (K-Pg) mass extinction event (Fawcett et al. 2009; Vanneste et al. 2014). However, Meyers and Levin (2006) provided an interesting twist on the idea that polyploidy confers "advantages." Rather than searching for "advantages" of polyploids to explain their high frequency as many other researchers have done, Meyers and Levin (2006, p. 1198) took a different approach and hypothesized that the abundance of polyploids may be the result of "a simple ratcheting process that does not require evolutionary advantages due to the biological properties of organisms." They show that the average ploidy within a lineage can continue to increase to the levels seen today, "even if there are ecological or physiological disadvantages to higher ploidy." However, this scenario does not take into account other aspects of what we now know about what happens subsequent to formation of polyploid plants, so we think that this ratchet effect, while

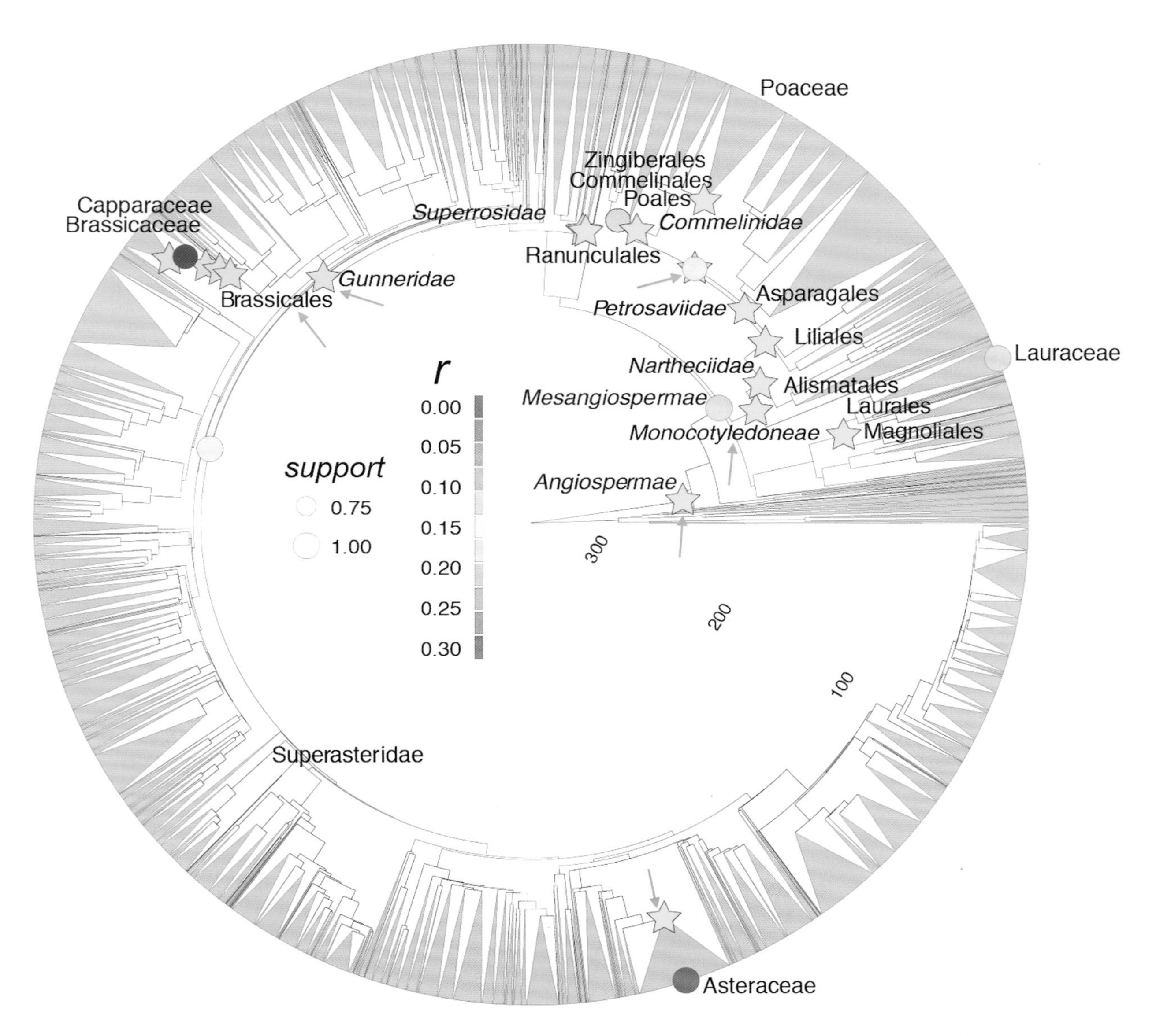

Figure 16.4. Use of a large phylogenetic tree for angiosperms to assess whether increased diversification accompanies WGD events. Collapsed clades are drawn as triangles proportional to family-level species richness. Stars = WGD events; circles show the location of increases in net diversification rate (r). Shades of color within circles reflect magnitudes of change in *r* from the immediate phylogenetic background to the shifted lineage, averaged across the distribution of bootstrap replicates; darker shades are associated with greater magnitudes of shifts. Size of circles indicates shift support as the relative frequency of trees in the distribution for which MEDUSA recovers a given shifted lineage. For some WGD events there is a significant association (arrows). Polyploidization is perfectly associated with an increase in diversification for Asteraceae and when *Commelinidae* + Asparagales was used as one of the putative placements of the τ event. Other polyploidization nodes show delayed diversification upticks: *Angiospermae*, *Gunneridae*, the monocot τ event (some of the placements), and all placements of the Brassicales β event. Modified from Tank et al. (2015).

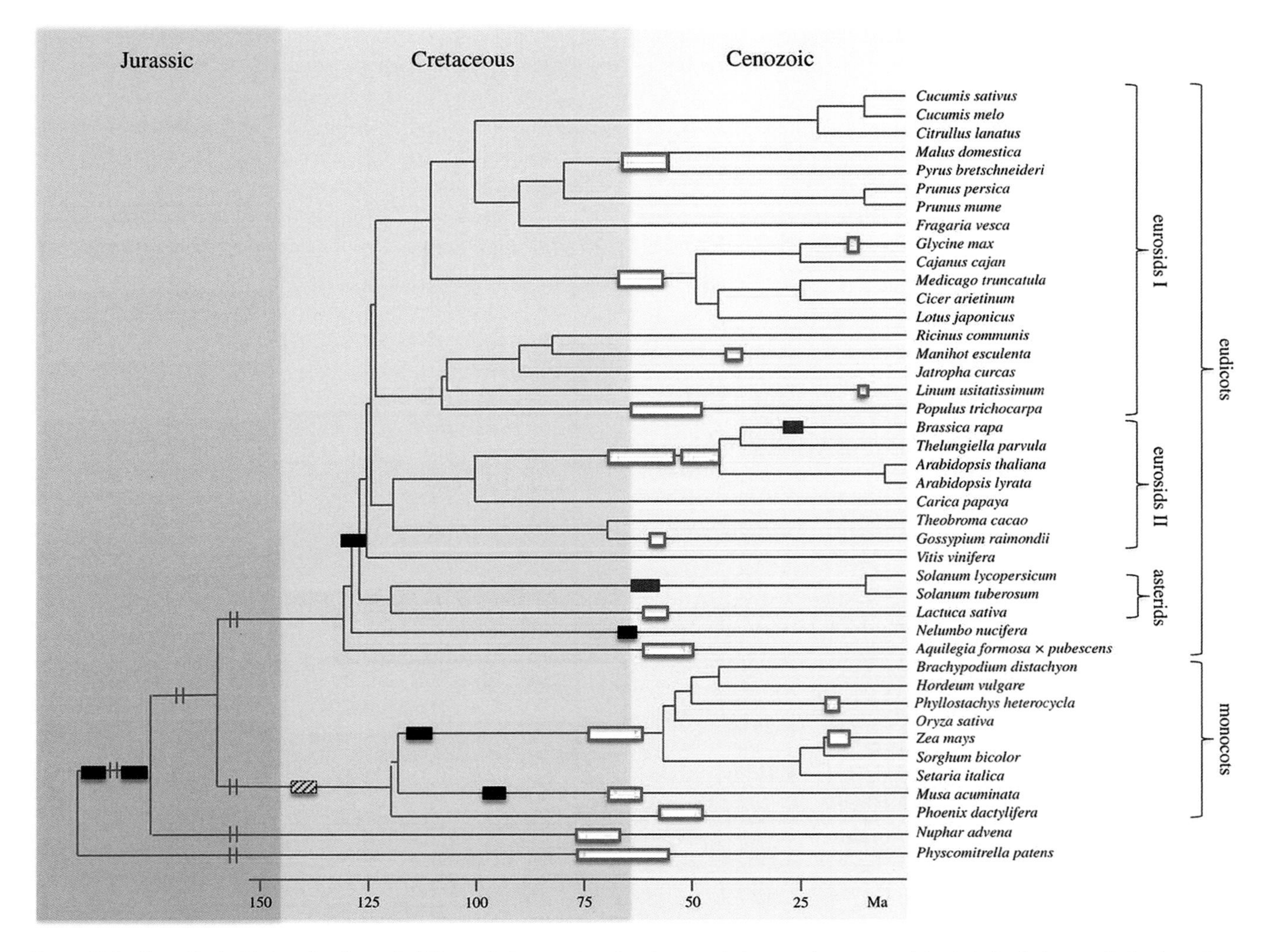

Figure 16.5. The estimated timing of ancient polyploidy events for a number of angiosperm clades seems to correspond closely to the Cretaceous-Tertiary boundary. White and gray bars represent 90% confidence intervals on dated tetraploid and hexaploid events, respectively. Black bars = WGD age estimates from literature. A possible WGD at the base of the monocot clade is indicated by a dashed bar (its exact phylogenetic placement remains unclear). (Modified from Vanneste et al. 2014.)

mathematically and theoretically sound, does not account for all the emerging facts.

GENOME EVOLUTION: FRACTIONATION FOLLOWING WGD

One of the more intriguing genome processes that occurs following polyploidy is genome fractionation—the loss of genes from one or the other parental genome following genome doubling. Lynch and Conery (2000) examined the evolutionary fate of duplicate genes and found that most gene duplicates are silenced within just a few million years—they further suggested that the stochastic silencing of duplicate genes may play a significant role in evolution.

The process of fractionation at a simple tetraploid level is shown diagrammatically in Fig. 16.6. A superb example of this process in action is seen in the carnivorous plant *Utricularia gibba* (bladderwort; Lentibulariaceae). The genome of this plant is very small (82 megabases). Despite its small size, the genome is the result of multiple WGD events, so this summary is also relevant to that discussion. Importantly, this finding is in line with many other studies that demonstrate that genome size and levels of polyploidy are often unrelated; soon after their formation, polyploids undergo episodes of reduction in genome size (Leitch and Bennett 2004). Genomic analyses of synteny indicate that *U. gibba* has experienced three sequential WGD events since its shared common ancestry with tomato (*Solanum*; Solanaceae) and grape (*Vitis*; Vitaceae) (Ibarra-Laclette et al. 2013). These analyses indicate that the genome of

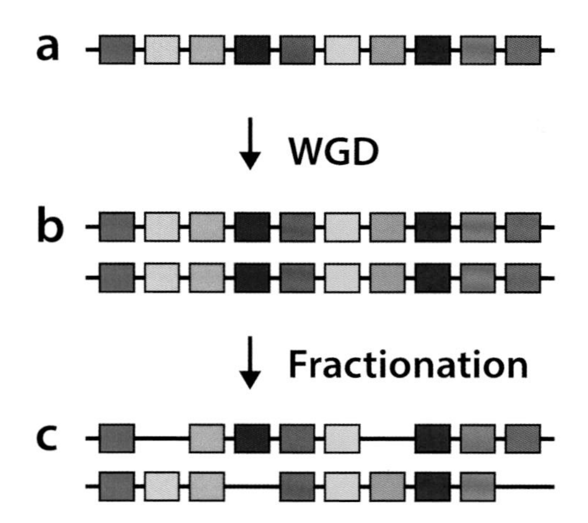

Figure 16.6. Fractionation following WGD. (a) In this simple example the initial diploid genome has 10 genes. Following one WGD (b) there are 20 genes in the tetraploid, which is then shown to fractionate into 16 genes (c). Figure and legend modified from Ibarra-Laclette et al. (2013).

U. gibba is 8x compared to the core eudicot ancestor, which is already considered palaeohexaploid (reviewed above; Jiao et al. 2012; see Ibarra-Laclette et al. 2013) (Fig. 16.6). When Ibarra-Laclette et al. (2013) compared the genome of *U. gibba* to that of tomato (*Solanum lycopersicum*), they found evidence for highly fractionated gene loss in *U. gibba* despite the numerous rounds of polyploidy. Almost two-thirds of the syntenic genes shared with *S. lycopersicum* appear to have been returned to single copy in *U. gibba*; however, these losses were spread across the different component genomes. This reduction in genes to something approaching the diploid condition is found in the 13 polyploid genomes studied by De Smet et al. (2013). This type of reduction in genes and chromosome number in many groups of herbaceous plants, particularly pronounced in annual taxa, is suggestive of selection being the driving force behind these processes, given its consistency across taxa after WGD.

REPEATED PATTERNS OF GENE RETENTION AND LOSS FOLLOWING POLYPLOIDY

Broad comparisons of angiosperm and even broader comparisons of eudicot genomes have provided the opportunity to address the fate of duplicated genes following WGD. Significantly, these broad genomic comparisons indicate a consistent pattern of gene loss/retention following polyploidization. Therefore, there may be some basic rules that govern the fate of duplicated genes following WGD (Paterson et al. 2006). Although most duplicated genes are lost following gene and genome duplications (Lynch and Conery 2000), maintenance of some genes in duplicate could occur as a consequence of copy number imbalances and neofunctionalization/subfunctionalization, and these could ultimately impact features of the organism itself (Freeling and Thomas 2006). Conversely, duplications of some genes may be deleterious. Studies in Asteraceae have compared gene retention after ancient events, 40–50 mya, with patterns of loss in young polyploids that are only 40 generations old (Buggs et al. 2011).

The De Smet et al. (2013) analysis described above documented consistent patterns of gene loss associated with structural and house-keeping genes, whereas those associated with plastid/mitochondrial functions and gene regulation (e.g., transcription factors) are maintained in multiple copies. Similar results have been observed by others in other genomic comparisons (e.g., Birchler 2012; Maere et al. 2005; Freeling et al. 2009; Chen et al. 2013). Retention of genes following duplication is hypothesized to be the result of both subfunctionalization and neofunctionalization (e.g., Force et al. 1999), but these processes may be difficult to distinguish for several reasons, including incomplete knowledge of ancestral function. In addition, although many instances of subfunctionalization have been documented (e.g., *Gossypium*), fewer cases of neofunctionalization have been reported. Perhaps what is being observed in recent studies is the effect of selection operating to eliminate extra gene copies that do not convey an advantage (including selectively neutral duplicated copies). Studies documenting that selection is active in polyploids resulting in diploidization and the loss and retention of particular categories of duplicated genes suggest that the ratchet process (above) may not completely govern the occurrence of polyploidy in angiosperms.

Descriptions of Major Clades

In the following descriptions of orders and major supraordinal clades, in parallel format, putative synapomorphies are indicated in **bold** and other useful diagnostic characters in *italics*. Characters found only rarely within the described taxa/clades are usually not mentioned. Descriptions are provided only for clades containing more than a single family, and style and format of the descriptions follow Judd et al. (2008). Note that under the ICN, the authority of a name stands, despite the circumscription of that named group adopted by the authors of this volume. Thus, for example, Magnoliidae Takhtajan is correct despite the fact his concept of Magnoliidae is very different from our own phylogenetic concept, which is defined formally as *Magnoliidae* W.S. Judd, P.S. Soltis & D.E. Soltis in Cantino et al. (2007) under the proposed PhyloCode. However, given that the PhyloCode names have not yet been validly published, here we follow the ICN for all names except those names that exist only under the proposed PhyloCode and were proposed as rankless; in those cases, we use the authorities from the treatments in Cantino et al. (2007) and the as-yet-unpublished Companion Volume to the PhyloCode.

CHAPTER 4

AMBORELLALES MELIKIAN, A.V. BOBROV & ZAYTZEVA

Composition: Amborellaceae.

NYMPHAEALES SALISB. EX BERCHT. & J. PRESL

Aquatic, usually rhizomatous, herbs; usually with branched sclereids; **stems with vascular bundles scattered** or ± in a single ring; *parenchymatous tissues aerenchymatous, usually with conspicuous air canals,* usually also with laticifers. Hairs simple, **usually mucilage-producing.** Leaves alternate and spiral, opposite and decussate, or occasionally whorled, simple, *often peltate or nearly so, entire to variously toothed or dissected*, with pinnate or palmate venation (and the latter with diverging major veins), or reduced and with a single vein, short- to more commonly long-petiolate, *with blade submerged, floating, or ± emergent*; stipules present or absent. Inflorescences of solitary, axillary flowers, or of several flowers and indeterminate, terminal. Flowers bisexual or unisexual (and plants then monoecious), *radial, usually with a long pedicel and floating or raised above the surface of the water. Perianth parts (tepals)* 4–12, distinct to connate, *imbricate*, often petal-like, in ± whorls, sometimes absent. Petal-like staminodia (often merely called petals) *8 to numerous*, inconspicuous to showy, often intergrading with stamens, or absent. Stamens 1 to numerous, whorled to irregular, distinct, often with staminodia; filaments free or adnate to petaloid staminodia, slender and well differentiated from anthers to laminar and poorly differentiated; anthers opening by longitudinal slits, the connective sometimes expanded; pollen grains usually monosulcate or lacking apertures. Carpels 2–3 to numerous, distinct or connate, ovary/ovaries superior to inferior (or nude), if connate then with several locules and placentation laminal (the ovules scattered on the partitions) or with a single locule (the ovule then solitary and apical); *stigmas often elongate and radiating on an expanded circular to marginally lobed and grooved disk, often surrounding the center of the receptacle, which is a knob or circular bump*, or stigmas forming a short or long crest or consisting of few to numerous, uniseriate hairs. Ovules anatropous to orthotropous, with 2 integuments, with a thick to thin nucellus, 1 to numerous per carpel. Nectaries lacking or rarely present on the staminodia, although a sweet fluid may also be secreted by the stigma. Fruit an aggregate of nuts or few-seeded indehiscent pods, a berry, variously dehiscent capsule, or achene; **seeds usually operculate (opening by a cap)**, often arillate; **endosperm ± lacking, but with abundant, starchy perisperm.**

AUSTROBAILEYALES TAKHT. EX REVEAL

Small trees, shrubs, or scrambling to twining lianas; parenchymatous tissues sometimes with crystal sand, *with spherical ethereal oil cells (pellucid dots),* usually also mucilage cells, and sometimes also branched sclereids. Hairs simple. Leaves alternate and spiral, or opposite and decussate, sometimes distally clustered on branches, simple, entire to variously toothed, with pinnate venation, *with pellucid dots (containing ethereal oils)*; stipules absent. Inflorescences determinate, axillary, often reduced to a solitary flower. Flowers bisexual to unisexual (and plants monoecious, dioecious, or polygamous), radial, the receptacle sometimes expanded or elongate. *Tepals few to numerous, opposite and decussate or spirally arranged, distinct, imbricate, the outer more bract or sepal-like, the inner petal-like*, sometimes intergrading with stamens. *Stamens few to numerous, spirally arranged*, distinct to connate (by filaments), sometimes with staminodia; *filaments short and thick to laminar, poorly or not at all differentiated with anthers*; anthers opening by longitudinal slits, the connective variously expanded or prolonged; pollen grains monosulcate, disulcate, tricolpate to hexacolpate, polyporate, or inaperturate. *Carpels* 1 to numerous, spirally arranged or seemingly in a single whorl, *distinct*, occasionally stipitate; *ovary or ovaries superior*, with lateral, apical, or ± basal placentation; *stigmas expanded, or elongate and decurrent on each style.* Ovules anatropous to campylotropous, with 2 integuments, with a thick nucellus, 1 to several in each carpel. Nectar absent or rarely produced at the base of the stamens. Fruit an aggregate of berries or follicles, or a drupe; seeds with dry/hard to fleshy coat; endosperm present, copious, **starchy**, homogeneous or ruminate.

CHAPTER 5

MESANGIOSPERMAE M. J. DONOGHUE, J. A. DOYLE & P. D. CANTINO

Trees, shrubs, lianas, vines, or herbs, sometimes parasitic (on other vascular plants or on fungi); parenchymatous tissues with or without pellucid dots (spherical ethereal oil cells, or larger, schizogenous or lysigenous cavities), with or without mucilage canals or cavities, resin or oil canals, or laticifers; stems with one to several rings of vascular bundles, or bundles scattered, with or without a vascular cambium. Hairs various. Leaves alternate and spiral or 2-ranked, rarely only 1 ranked and spiral, 3-ranked, or opposite and decussate, rarely opposite and 2-ranked, occasionally whorled, simple to variously compound, sometimes lobed or dissected, entire to variously toothed, usually with pinnate, palmate, or parallel venation, with the fine veins reticulate, but veins occasionally reduced or absent; stipules present or absent. Inflorescences determinate or indeterminate, usually terminal or axillary, sometimes reduced to a solitary flower. Flowers bisexual or unisexual (and plants then monoecious, dioecious, or polygamous), radial to bilateral, with receptacle small to variously expanded, with or without hypanthium, with perianth undifferentiated (tepals) or differentiated (calyx and corolla), or absent, the parts in whorls (of often 3, 4, or 5 parts) or occasionally spirally arranged. Tepals (if perianth undifferentiated) (1-) 2 to numerous, perhaps ancestrally **in whorls of 3**, distinct to connate, imbricate to valvate, sepaloid to petaloid. Sepals (if perianth differentiated) 2 to numerous, distinct to connate, imbricate to valvate. Petals (if perianth differentiated), 2 to numerous, distinct to connate, usually imbricate or valvate. Stamens 1 to numerous, spirally arranged or in whorls (these usually of 3, 4, or 5), distinct or connate (by filaments and/or anthers), laminar to well differentiated into filament and anther, sometimes with staminodia; filaments free and attached to receptacle, or adnate to gynoecium or perianth; anthers usually opening by longitudinal slits, flaps, or pores, the connective expanded or not; pollen grains usually in monads, but sometimes in tetrads or polyads, or even in pollinia, usually monosulcate, bisulcate, tricolpate, tricolporate, or variously porate, or inaperturate, the exine columellate to solid, sometimes reduced. Carpels 1 to numerous, **usually developmentally more or less plicate and sealed by postgenital fusion of the margins**, distinct to connate, and then with several styles, a branched style, or a single style, or style absent; ovary or ovaries superior or inferior (or nude), with marginal, parietal, intruded-parietal, axile, free-central, apical, or basal placentation; stigma various. Ovules anatropous to orthotropous, with 1 or 2 integuments, or integuments absent, with thin to thick nucellus, 1 to numerous per carpel or locule; **embryo sac 7-celled, 8-nucleate, i.e., the Polygonum type.** Nectaries various, or absent, but sometimes with pollen, oils, aromatic compounds, or resins as pollinator rewards. Fruit aggregate or multiple, or simple, extremely diverse, indehiscent to variously dehiscent or breaking into one (to several) -seeded segments, membranous or dry/hard to fleshy, and texture homogeneous to heterogeneous, sometimes associated with accessory structures; seeds with thin to thick, dry, hard to fleshy and colorful, or mucilaginous coat, sometimes flattened, winged, or with tuft of hairs, sometimes arillate; endosperm present to absent, homogeneous to ruminate, sometimes replaced by perisperm; embryo minute to large, with 1 or 2 (occasionally 3 or 4) cotyledons.

Composition: Magnoliidae, Chloranthaceae, *Monocotyledoneae*, *Eudicotyledoneae*.

Chloranthales Martius

Composition: Chloranthaceae

Magnoliidae Novák ex Takht.

Trees, shrubs, lianas, vines, or herbs, rarely parasitic (on other vascular plants); *parenchymatous tissues with pellucid dots (spherical ethereal oil cells)*, **with the phenylpropane compound asarone, the lignans galbacin and veraguesin, and the neolignan licarin**; stems with one to several rings of vascular bundles, occasionally with ± scattered bundles, vascular cambium present (developing in connection with outer ring of bundles, if several present) or absent. Hairs various. Leaves opposite and decussate, alternate and spiral or 2-ranked, occasionally whorled, *simple*, **entire** to serrate, rarely lobed, with pinnate or palmate venation (the latter usually with converging major veins), *with pellucid dots*; stipules present or absent. Inflorescences determinate or indeterminate, terminal or ± axillary, sometimes reduced to a solitary flower. Flowers bisexual or unisexual (and plants then monoecious, dioecious, or polygamous), *usually radial*, rarely bilateral, *with receptacle small to variously expanded*, sometimes forming a hypanthium, with *perianth undifferentiated (tepals) or differentiated (calyx and corolla), in whorls (often of 3 parts) or occasionally spirally arranged*, or absent. *Tepals* (if perianth undifferentiated) *3 to numerous*, *distinct* to less commonly connate, imbricate or valvate, *petaloid*. Sepals (if perianth differentiated) 2–4, distinct to connate, imbricate to valvate, sometimes petaloid or falling as a cap. *Petals* (if perianth differentiated) *6 to numerous*, *usually distinct*, imbricate or valvate. *Stamens* 1 to *numerous*, *spirally arranged or in whorls, these often 3-merous*, distinct or connate (by filaments), sometimes with staminodia, sometimes associated with a pair of glands; filaments free, attached to receptacle (which is sometimes modified, forming a hypanthium), ± adnate to gynoecium, or adnate to the connate sepals, *laminar to narrow, short and thick to elongate, usually poorly differentiated from anthers*, occasionally well differentiated; anthers usually opening by longitudinal slits or flaps, *the connective often expanded, thickened and/or appendaged*; pollen grains usually monads, occasionally tetrads or polyads, usually monosulcate, bisulcate, uniporate, or inaperturate, the exine sometimes reduced. Carpels 1 to numerous, *often with style poorly differentiated from ovary*, distinct to connate, and then with as many styles as carpels, a branched style, a single style, or reduced to absent style; ovary or ovaries superior to inferior (or to "nude"; see Cronquist 1981), with marginal, parietal, intruded-parietal, axile, apical, or basal placentation; *stigma often ± decurrent along style or elongate*, or less commonly small and peltate, globular, globular-lobate, capitate, 2-lobed, feathery-truncate, or brush-like. Ovules anatropous, hemitropous, campylotropous, or orthotropous, with 1 or 2 integuments, with thin to thick nucellus, 1 to numerous per carpel or locule. Nectaries on calyx tube, petals, staminodia, paired staminal glands, carpels, or nectar-producing structures absent. *Fruit an aggregate of follicles (opening ventrally or dorsally), samaras, achenes (these sometimes plumose), nuts, berries, or drupes, sometimes becoming ± fused at maturity, or enclosed by fleshy or dry/woody hypanthium, or a single nut, samara, berry, drupe, variously dehiscent capsule, or fleshy "follicle" (that also dehisces abaxially, exposing the single seed), sometimes associated with (or surrounded by) expanded, variously modified, fleshy or woody hypanthium* (or rarely, cupulate structure derived from connate bracts), the fruits occasionally appearing arillate (with fleshy structure derived from style); seeds with thin to thick, dry, hard to fleshy and colorful coat, sometimes flattened or winged, sometimes arillate; endosperm present, copious, homogeneous or ruminate, to absent, sometimes replaced by abundant perisperm; embryo often minute, less commonly large, with usually 2 (rarely 3 or 4) cotyledons.

Composition: Piperales, Canellales, Laurales, Magnoliales.

PIPERALES BERCHT. & J. PRESL

Shrubs, twining lianas/vines, **or herbs**, sometimes epiphytic or parasitic; *parenchymatous tissues usually with pellucid dots (spherical ethereal oil cells)*, sometimes with mucilage canals; crystals various; *stems often with nodes swollen or jointed*, **with widely spaced vascular bundles**, *in one, two, or several concentric rings, to ± scattered*, **or at least wood with broad rays**, but stems absent (and replaced by rhizome-like root, covered with tubercular outgrowths) in parasitic herbs. Hairs simple. Leaves alternate and 2-ranked or spiral, occasionally opposite or whorled, simple, entire, occasionally lobed, **often cordate, with palmate** to pinnate **venation** *(the former usually with converging major veins), with pellucid dots*, but leaves absent in parasitic herbs; stipules absent, or present and adnate to the petiole, and the petiole sometimes sheathing or ± basally expanded. Inflorescences determinate or indeterminate, axillary or terminal, sometimes displaced to a position opposite the leaf, sometimes reduced to a solitary flower, or arising endogenously from root in parasitic herbs. Flowers bisexual or unisexual (and plants then monoecious, dioecious, or polygamous), radial or bilateral (due to form of perianth), a hypanthium sometimes present, and perianth parts (when present) in whorls. *Sepals usually 3*, occasionally 4, distinct to connate, *valvate*, often petaloid, tubular and, if bilateral, S-shaped, or pipe-shaped, with spreading limb or thick and fleshy. Petals 3 but more commonly absent, when present distinct, imbricate, or very reduced. Stamens 1 to numerous, ancestrally in two 3-merous whorls, distinct to connate (and then monadelphous), poorly to well differentiated into filament and anther, occasionally with staminodia; filaments free, slightly to strongly adnate to the gynoecium, or strongly adnate to the connate sepals/hypanthium, very short to elongate, slender to stout; anthers opening by longitudinal slits; pollen grains in monads or tetrads, monosulcate, rarely bisulcate, sometimes inaperturate. Carpels 1 to 6, distinct to connate (then with a single style, several styles, or reduced to absent style); ovary superior or inferior (or nude), with lateral, parietal, intruded-parietal, axile, ± apical, or basal placentation; stigmas ± elongated along each style, globose, shallowly lobed, or brush-like. Ovules anatropous to orthotropous, with 1 or 2 integuments, with thin to ± thick nucellus, 1 to numerous per locule. Nectary absent or sometimes patches of glandular hairs on calyx tube. Fruit an aggregate of follicles or achenes, a drupe, berry, or variously dehiscent capsule; seed with dry coat, sometimes flattened or winged; endosperm present, copious, homogeneous, to scanty, and then with usually abundant perisperm.

CANELLALES CRONQUIST

Trees or shrubs; parenchymatous tissues with pellucid dots (spherical ethereal oil cells); crystals usually druses (when present); stems sometimes only with tracheids (i.e., vessels absent). **Hairs usually absent**, occasionally simple. *Leaves alternate and spiral or 2-ranked*, simple, entire, with pinnate venation, *with pellucid dots*, **with branched sclereids**; stipules absent. Inflorescences usually determinate, terminal or axillary, sometimes reduced to a solitary flower. Flowers bisexual or unisexual (and plants then dioecious or polygamo-dioecious), radial, the receptacle sometimes expanded, **the perianth differentiated into a calyx and corolla**, ± in whorls. Sepals usually 2–4, distinct to connate, sometimes falling as a cap, valvate or imbricate. Petals (2-) 5 to numerous, distinct to rarely slightly connate, imbricate. Stamens 6 to numerous, spirally arranged or whorled, distinct to connate (monadelphous); filaments free, arising from receptacle, thick to flattened, often poorly differentiated from anthers; anthers opening by longitudinal slits, the connective sometimes expanded; pollen grains usually monosulcate or uniporate, sometimes in tetrads. Carpels 1 to numerous, distinct or connate (and then with a single, short style); ovary or ovaries superior, with lateral or parietal placentation; stigma decurrent along each style or ± capitate (when carpels distinct) or globular-lobate (when carpels fused). Ovules anatropous or hemitropous, with 2 integuments, **the outer with only 2 to 4 cell layers**, a thick nucellus, 1 to numerous per locule. Nectary usually lacking. Fruit usually an aggregate of follicles or berries, sometimes becoming connate as they mature, or a berry; seeds with dry coat, **with a palisade exotesta**; endosperm present, copious, usually homogeneous.

LAURALES JUSS. EX BERCHT. & J. PRESL

Trees, shrubs, lianas, rarely twining parasitic vines; *parenchymatous tissues with pellucid dots (spherical ethereal oil cells)*; crystals various; **stems with unilacunar nodes** (with 1 to several traces). Hairs simple or ± stellate. **Leaves opposite (and decussate)** or alternate (and spirally arranged), rarely whorled, simple, rarely lobed, rarely palmately compound, entire to serrate, with pinnate or palmate venation (the latter usually with converging, but rarely diverging, major veins), *with pellucid dots*, but leaves very reduced in the parasitic *Cassytha*; stipules absent. Inflorescences ± determinate, terminal or axillary, occasionally reduced to a solitary flower. Flowers bisexual or unisexual (and plants then monoecious, dioecious, or polygamous), radial, **with a concave to cup-shaped or urn-shaped receptacle (hypanthium)** *that sometimes forms a floral roof. Perianth parts (tepals) 3 to numerous*, sometimes absent, *distinct* to very slightly connate, rarely strongly connate and calyptrate, imbricate, sometimes outer ones ± sepaloid, and innermost ± petaloid, in whorls, often of 3, or spirally arranged. Stamens 3 to numerous, spirally arranged or in whorls, these often 3-merous, distinct, often with staminodia (or staminodia restricted to carpellate flowers), *usually associated with a pair of glands* (likely of staminodial origin, often producing nectar or odor); filaments free, borne on inner surface to apex of hypanthium, short to elongate, poorly to well differentiated from anthers; *anthers usually opening by 1, 2, or 4 flaps*, but sometimes by longitudinal or transverse slits, the connective sometimes expanded and projecting; pollen grains in monads or occasionally tetrads, monosulcate or bisulcate, rarely syncolpate, or *more commonly inapertu-*

rate, often spinulose, with ± reduced exine. Carpels 1 to numerous, usually distinct, but occasionally connate and then with branched style; ovary or ovaries superior to inferior, *with apical and/or basal placentation*; stigma elongate, decurrent, globular, or capitate. Ovules anatropous, with 1 or 2 integuments, with thick nucellus, *solitary (or rarely 2, one functional and one abortive) in each carpel (or locule)*. Nectar produced by staminodia and/or paired staminal glands. Fruit an aggregate of achenes, these sometimes plumose, nuts or drupes, or a single nut, samara, drupe or berry, *often enclosed by fleshy or dry/woody hypanthium (= receptacle) that sometimes splits irregularly at maturity, exposing fruits, or associated with expanded, variously modified, fleshy or woody hypanthium (= receptacle, i.e., cupulate, with cupule often contrasting with fruit color)*, or rarely surrounded by cup-like fleshy structure derived from 2–3 connate bracteoles, the fruits occasionally appearing arillate (with aril-like structure developed from the style); seeds with thin to thick, ± dry coat; endosperm present, abundant, to absent, **embryo** small (**but elongate**) to large.

MAGNOLIALES JUSS. EX BERCHT. & J. PRESL

Trees, shrubs, or lianas; parenchymatous tissues with pellucid dots (spherical ethereal oil cells), sometimes with sclereids, often branched; crystals various; stems often with septate pith, *with nodes trilacunar to multilacunar*, **with stratified phloem**; bark sometimes with reddish sap. Hairs simple, T-shaped, variously branched, stellate, or peltate scales. **Leaves** *alternate*, **2-ranked** or less commonly spiral, simple, entire, rarely lobed, with pinnate venation, *with pellucid dots*; stipules present (and large, enclosing the terminal bud) or absent. Inflorescences determinate, in clusters of 2 or 3, *or frequently of solitary flowers*, terminal, axillary, or rarely supraxillary. Flowers bisexual or unisexual (and plants then dioecious), radial, *often fairly large, with often variously expanded (e.g., concave, flat, conic, globose, or elongated) receptacle*. Perianth parts undifferentiated, of tepals, or forming a calyx and corolla, spirally arranged or in whorls (usually of 3). Tepals (when perianth undifferentiated) 3 to numerous, distinct to connate, imbricate or valvate, petaloid. Sepals (if perianth differentiated) 2–3, distinct or connate (and then calyptrate, sometimes of indeterminate number), imbricate or valvate, sometimes interpreted as derived from bracts. *Petals* (if perianth differentiated) *6 to numerous, usually distinct*, imbricate or valvate, sometimes the outer differentiated from the inner, sometimes derived from staminodia. *Stamens* 2 *to numerous, spirally arranged* or whorled, *distinct or connate (monadelphous, with solid column)*, occasionally with staminodia; *filaments ± laminar to narrow, short and thick, not well differentiated from anther, free*; anthers opening by longitudinal slits, the connective expanded, thickened and/or appendaged; pollen grains usually monads, but occasionally tetrads or polyads, usually monosulcate to inaperturate. *Carpels 1 to numerous, distinct*, occasionally connate; ovary or ovaries superior, with marginal or ± basal placentation; stigma small, 2-lobed, ± decurrent along style, or elongate, extending along one side of carpel, or feathery-truncate. Ovules anatropous or occasionally campylotropous, with 2 integuments, with thick nucellus, 1 to numerous per carpel. Nectaries (or food bodies) on petals, nectar produced by carpels, or nectar-producing structures absent. *Fruit an aggregate of backwards-opening follicles, samaras, berries, or drupes, sometimes becoming ± fused at maturity, a berry, or fleshy "follicle" that also dehisces abaxially, exposing the single seed*, the fruits rarely sunken into fleshy receptacle; *seeds with dry, hard to fleshy and colorful coat, often arillate*; **endosperm** present, copious, **ruminate** or homogeneous.

CHAPTER 7

MONOCOTYLEDONEAE DE CANDOLLE

Herbs, often with sympodial growth, rhizomatous, sometimes with corms or bulbs, or less commonly shrubs or trees, and these with anomalous secondary growth (cambium producing ground-parenchyma with scattered vascular bundles), sometimes parasitic (on fungi) or epiphytic, **the prophyll single, adaxial**; parenchymatous tissues usually lacking pellucid dots (spherical ethereal oil cells, but these present in *Acorus*), occasionally with laticifers or resin canals; crystals various, but often with raphides, sometimes with silica bodies; **stems with scattered vascular bundles**, or occasionally with 2 or more rings, and **usually lacking vascular cambium; sieve cell plastids with cuneate protein crystals; radicle poorly developed, short-lived, the root system of adult plant adventitious.** Hairs various, but often simple, T-shaped, sometimes dendritic, stellate or peltate scales. *Leaves usually alternate*, spirally arranged or 2-ranked, 3-ranked, rarely opposite or whorled, sometimes in a basal rosette, *simple*, occasionally lobed, or rarely variously compound, **entire** or spinose-serrate, **with usually parallel venation**, sometimes pinnate-parallel, occasionally palmate (with major veins usually converging, and fine veins noticeably reticulate); stipules usually absent, but "stipular" tendrils occasionally present (*Smilax*), usually **sheathing at base, often lacking a petiole**, but sometimes with well-developed petiole, *and the leaf usually developing from a portion of the primordium just below the tip, the tip inactive or producing only a small terminal appendage*. Inflorescences determinate or indeterminate, terminal or axillary, sometimes reduced to a solitary flower, sometimes scapose. Flowers bisexual or unisexual, radial, bilateral, or without plane of symmetry (due to form of perianth and/or androecium), with receptacle usually small, occasionally expanded, with perianth undifferentiated (tepals) or differentiated (calyx and corolla), **in 2 whorls (often of 3 parts**, rarely of 2 or 4 parts), or absent. *Tepals* (if perianth undifferentiated) (1-) *2 to 6* (-10, or more numerous), distinct to connate, imbricate, valvate, or open, inconspicuous to petaloid, sometimes dimorphic, then often 1 differentiated from the others. *Sepals* (if perianth differentiated) *3*, distinct to slightly connate, usually imbricate, occasionally dimorphic. *Petals* (if perianth differentiated) *3*, distinct to connate, usually imbricate, occasionally dimorphic. Stamens 1 to numerous, but ancestrally 6, **in 2 whorls of 3**, distinct or connate (by filaments, rarely also anthers), sometimes with staminodia, filaments free, arising from receptacle, or adnate to gynoecium (or "grasping" gynoecium), or adnate to perianth, usually well differentiated into filament and anther, but variable in length; pollen grains usually monads, occasionally tetrads, *usually monosulcate or monoporate*, or derived conditions, sometimes inaperturate, the exine sometimes reduced. Carpels 1 to numerous, **but ancestrally 3 in a single whorl**, distinct to connate, then with as many styles as carpels, or a single branched or unbranched style; ovary or ovaries superior to inferior (or nude), with marginal, parietal, intruded-parietal, free-central, axile, apical, or basal placentation; stigma various. Ovules anatropous to orthotropous, with 1 or 2 integuments, with thin to thick nucellus, 1 to numerous per carpel or locule. *Nectaries in the septa of the ovary* (or glands between the distinct ovaries), atop ovary, on the bases of the perianth parts and/or stamens, on the staminodia, or absent. Fruit an aggregate of achenes, follicles, or drupes, or a variously dehiscent capsule, berry, drupe, samara, achene, grain, utricle, or drupaceous schizocarp; seeds various, sometimes with black crust, sometimes arillate, winged, or with tuft of hairs; endosperm present or absent, sometimes replaced by perisperm; embryo minute to large, **with a single cotyledon,** or undifferentiated.

Composition: Acorales, Alismatales, Petrosaviales, Dioscoreales, Pandanales, Liliales, Asparagales, Commelinidae.

ACORALES MARTIUS

Composition: Acoraceae.

ALISMATALES R. BR. EX BERCHT. & J. PRESL

Herbs or vines, often emergent, floating, or submerged aquatics (in fresh, brackish, or marine waters), usually rhizomatous, but sometimes from a tuber or corm; parenchymatous tissues sometimes with laticifers (milky sap), occasionally with mucilage or resin canals, *often aerenchymatous*; crystals various or absent; stems sometimes with vascular system much reduced, *often with air spaces, usually with 2 to numerous scales or hair-like structures at the nodes (usually within the leaf sheath)*. Hairs simple or absent. *Leaves alternate and spiral or 2-ranked*, occasionally opposite and decussate or whorled, borne along stem or in a basal rosette, *simple* (rarely deeply lobed or even compound), entire or occasionally spinose-serrate, with parallel, pinnate-parallel, pinnate, or palmate venation (then with major veins usually converging), the fine veins obscure to well developed, or with only a single vein, reticulate, the base ± sheathing; petiole absent (and leaf ± linear), or present, and then leaf ± well differentiated into a petiole and blade, or sheath and blade and then often ligulate; but leaves occasionally lost as a result of extreme vegetative reduction (*Lemna* and relatives); stipules absent, although sheath sometimes ± free and appearing stipulate. Inflorescences determinate or indeterminate, the axis sometimes thickened and associated with conspicuous bract (spathe), or sometimes with a whorl of bracts below flower, terminal or axillary, sometimes scapose, sometimes reduced to a pair of flowers or a solitary flower, rarely modified and pseudanthial. *Flowers* bisexual or unisexual (and plants then monoecious, dioecious, or polygamous), *radial* (or rarely bilateral, due to the perianth), *and perianth undifferentiated (of tepals) or differentiated into a calyx and corolla, parts in 1 or 2 whorls*, sometimes absent, a hypanthium occasionally present, the receptacle occasionally expanded. *Tepals 1–6*, distinct or connate, imbricate or valvate, ± inconspicuous, sepaloid or petaloid. *Sepals 3*, distinct, imbricate or valvate. *Petals 3*, distinct, imbricate, and sometimes crumpled. *Stamens 1 to numerous*, in 1 to several whorls or spirally arranged (when numerous, a secondary increase), distinct or connate (by filaments, and rarely also anthers), sometimes with staminodia; *filaments free or sometimes adnate to the perianth parts* (and then with each stamen adnate to claw of tepal, and sometimes the tepal interpreted as a connective outgrowth), elongate to very short or ± absent, well differentiated from anthers; anthers **extrorse**, opening by longitudinal slits or occasionally terminal pores, the connective sometimes expanded and/or forming an appendage (which in some taxa has been interpreted as a perianth part); pollen grains monosulcate, disulcate up to tetrasulcate, tetra- to polyporate, or inaperturate, sometimes lacking an exine, sometimes united into thread-like chains or pollen grains filamentous. *Carpels (1-) 3 to numerous* (due to secondary increase), occasionally pseudomonomerous, rarely each on a long stalk, distinct to connate, then with distinct styles, a single, well-branched style, or a single, unbranched style, or the styles sometimes reduced or divided; ovary or ovaries superior or inferior (or naked), with marginal, basal, apical, parietal, intruded-parietal, or axile placentation; stigmas punctate, capitate, globose to decurrent, ± feathery or tufted or funnel-form, or stigma 3-lobed. Ovules anatropous to campylotropous or orthotropous, with 1 or 2 integuments, with thick or thin nucellus, 1 to numerous per carpel/locule. *Nectaries in septa of ovary or basal and intercarpellary glands, or nectar produced by staminodia, or absent. Fruit an aggregate of follicles, achenes, or drupes, or a single ventricidal or septicidal capsule, fleshy and irregularly dehiscent capsule, berry, utricle, utricle-like or nut-like structure, or achene*; seeds occasionally appendaged; endosperm present or absent, the **embryo green, sometimes with large cotyledons.**

PETROSAVIALES TAKHT.

Composition: Petrosaviaceae.

DIOSCOREALES MARTIUS

Herbs or ± twining vines, sometimes mycoparasites, from rhizomes, often with tubers or tuber-like structures; crystals often raphides, sometimes druses; **stems with vascular bundles in 1 or 2 rings.** Hairs simple to stellate, sometimes with prickles. Leaves alternate and spiral or 2-ranked, rarely opposite, borne along stem to in a basal rosette, simple, sometimes palmately lobed (rarely even compound), entire, *with parallel to palmate venation* (then with major veins converging, rarely diverging), the fine veins obscure *to well developed, reticulate*, the base sheathing or not, the leaves sometimes reduced to scales; petiole absent or present; stipules absent, but sometimes with stipule-like flanges at base of petiole. Inflorescences determinate to indeterminate, terminal or axillary, sometimes scapose or appearing so, sometimes reduced to a solitary flower. Flowers bisexual or unisexual (and plants then monoecious or more commonly dioecious), usually radial, and perianth undifferentiated, parts in 1 or 2 whorls. *Tepals 3 or 6*, distinct to connate, imbricate or valvate, *usually petaloid*, sometimes inner whorl slightly differentiated from outer, occasionally corolla with 3 wings. *Stamens 3 or 6*, in 1 or 2 whorls, distinct or slightly connate (by their filaments) or strongly connate (by their anthers, forming a tube), sometimes inner whorl staminodial; **filaments** slightly to clearly **adnate to perianth,** elongate and slender to short or ± absent; anthers opening by longitudinal or transverse slits, the connective often expanded or extended beyond anther sacs and forming an appendage; pollen grains variable, usually monosulcate, disulcate, trichotomosulcate, uniporate (rarely 4- or 5-porate), or inaperturate. *Carpels 3, connate*, with 3 styles or a single branched to unbranched style; ovary superior to inferior, with axile, parietal, or intruded-parietal placentation, or axile below and parietal above (when ovary 3-locular below and 1-locular above), often 3-winged; stigma capitate, variously 3-lobed, or of 3 distinct stigmas, these variously shaped. Ovules anatropous to campylotropous, with 2 integuments, with thin or thick nucellus, 2 to numerous in each locule. *Nectaries in septa of ovary, on ovary apex*, on base of tepals, or absent. *Fruit a variably dehiscent capsule (often 3-winged), berry, or samara*, often associated with persistent perianth; seeds often winged, sometimes appendaged; endosperm present to absent.

PANDANALES R. BROWN EX BERCHT. & J. PRESL

Herbs (sometimes large), lianas, shrubs, or trees (without secondary growth), occasionally parasitic (on fungi), sometimes rhizomatous; crystals various, often with druses; parenchymatous tissues sometimes with mucilage or resin canals; stems sometimes appearing to be dichotomously branched, sometimes with prop-roots, with vascular bundles usually scattered, but sometimes only in 1 or 2 rings, occasionally with secondary growth. Hairs various, simple to branched, sometimes stellate. *Leaves alternate and spiral, often in 2 or 3 (rarely 4) ranks and these spiraled*, occasionally opposite or whorled, borne along stem, clustered at the stem-tip, reduced to scales in parasitic taxa, simple or sometimes bilobed or even palmately lobed, *entire to serrulate or spinose-serrate*, sometimes spiny also along midrib, with parallel, pinnate-parallel or palmate (with major veins converging) venation, the midvein (when present) single or forked, the fine veins obscure to well-developed and reticulate, the base usually sheathing; petiole absent (and leaves ± linear) or present (leaves differentiated into a sheath, petiole, and blade, sometimes 1-, 2- or 3-costate, palm-like), *sometimes plicate*; stipules absent. Inflorescences determinate to indeterminate, terminal or axillary, sometimes reduced to a solitary flower, sometimes associated with one to several large bracts (spathes), *sometimes with flowers densely packed on a thickened axis* (in spikes or heads). Flowers bisexual or unisexual (and plants then monoecious or dioecious), *usually radial, and perianth undifferentiated*, in 1 or 2 whorls, or absent, sometimes flowers sessile and adjacent flowers occasionally connate. *Tepals 3 or 4 to numerous*, distinct to connate, imbricate or valvate, petaloid, sepaloid or inconspicuous, sometimes reduced to a lobed cup. *Stamens 1 to numerous* (but ancestrally 6), in 1 or 2 whorls or a dense cluster, *distinct to connate* (by filaments), sometimes associated with staminodia, or staminodia restricted to carpellate flowers, these sometimes very elongate; filaments free or sometimes adnate to perianth, elongate to short or even absent; anthers opening by longitudinal or transverse slits, the connective sometimes forming an apical appendage; pollen grains in monads or tetrads, monosulcate to uniporate, or inaperturate. Carpels 1 to numerous, occasionally pseudomonomerous, distinct to connate, each carpel with its own style, or (when carpels connate) a single, short style, or style often absent; ovaries superior to inferior, with apical, basal, parietal, or axile placentation, or gynoecium of ± fused carpels and multiloculate, with a single ovule in each locule; stigma ± capitate, globular, truncate, brush-like, elongate, and sometimes strongly lobed. Ovules usually anatropous, with 2 integuments, with thick to thin nucellus, 1 to numerous per locule/carpel. *Nectaries usually absent*, occasionally with nectaries in the septa of the ovary. *Fruit an aggregate of follicles or achenes, or a single loculicidal capsule, drupe or berry, and these sometimes connate, forming multiple fruits, i.e., fleshy syncarps*; seeds rarely arillate; endosperm present, copious.

LILIALES PERLEB

Herbs or vines (often climbing by tendrils, twining, and/or prickles, sometimes merely scrambling), rarely subshrubs, rarely parasitic (on fungi), *from rhizomes, tubers, bulbs, or corms, with roots often contractile*, sometimes thickened and tuber-like; crystals various; stems with vascular bundles scattered or in 3 rings. Hairs usually simple; prickles sometimes present. *Leaves alternate and spiral or 2-ranked*, occasionally equitant, *occasionally whorled*, rarely opposite, borne along stem or in a basal rosette, simple, *entire*, spinose-serrate, serrulate, or rarely ± dentate, *with parallel to pinnate-parallel or palmate venation (then with major veins converging)*, the fine veins obscure to well developed, reticulate, the base sheathing or not, the leaves sometimes reduced to scales, sometimes resupinate (due to a basal twist, so abaxial surface is uppermost), rarely terminating in a tendril; petiole usually absent, but occasionally present; stipules absent, or present and represented by a pair of tendrils (fused to petiole). Inflorescences determinate or indeterminate, terminal or axillary (rarely terminal but deflexed to a lateral, and leaf-opposed, position as a result of sympodial growth), sometimes scapose or appearing so, sometimes reduced to a solitary flower, rarely modified into a tendril. Flowers bisexual or unisexual (and plants then dioecious or polygamous), radial or bilateral (due to form of perianth or perianth and androecium), *perianth undifferentiated (of tepals) or differentiated (with calyx and corolla), in 1 or 2 whorls. Tepals (3-) 6*, distinct to connate, imbricate, usually petaloid, rarely with one member differentiated from the others, occasionally the outer whorl somewhat differentiated from the inner, *often bicolored, variegated, or spotted. Sepals 3* (-10), distinct, imbricate or contorted, rarely petaloid. *Petals 3* (-8), distinct, imbricate or contorted, rarely reduced, rarely variegated or spotted. *Stamens (3-) 6* (to rarely numerous), in 2 (-6) whorls, distinct to connate (by filaments), rarely with a staminodium, and staminodia often present in carpellate flowers; filaments free or occasionally adnate to perianth, elongate to short or absent; anthers opening by longitudinal slits, or occasionally valves opening to form a peltate disk, or apical pores, the connective occasionally prolonged; pollen grains monosulcate, disulcate, rarely 1–4-porate, or sometimes inaperturate. *Carpels 3* (-10), *connate, with 3 (several) styles or a single branched or unbranched style;* ovary superior to inferior, with axile, parietal, intruded-parietal, or basal placentation, stigma ± decurrent on each style or style-branch, punctate, capitate, or ± 3-lobed. Ovules anatropous to campylotropous or orthotropous, with usually 2 integuments, with a thick to thin nucellus, 1 or 2 to numerous in each locule/carpel. **Nectaries on perianth parts (i.e., perigonal)**, at base of stamens, or absent. *Fruit a loculicidal, septicidal, ventricidal, or irregularly dehiscent capsule*, sometimes fleshy, *or berry*; *seeds with coat lacking phytomelan*, occasionally fleshy, sometimes flattened or winged, or with appendages, sometimes arillate; endosperm present, often copious.

ASPARAGALES LINK

Herbs (sometimes epiphytic), shrubs, or trees, occasionally scrambling or twining vines/lianas, occasionally parasites (on fungi), *from slender to tuberous rhizomes, corms, or bulbs, with roots often contractile*, sometimes thickened and tuber-like; crystals usually of raphides, pseudoraphides, or styloids; parenchymatous tissues occasionally with laticifers or resin canals, and occasionally with aloine cells (at phloem pole of vascular bundles of the leaf) producing colored sap; *stems sometimes with anomalous secondary growth* (cambium producing ground-parenchyma with scattered vascular bundles), *occasionally forming phylloclades*, rarely producing thorns, sometimes basally thickened and forming pseudobulbs, sometimes very reduced. Hairs various, but often simple, rarely with prickles. *Leaves alternate and spiral, or 2-* or occasionally *3-ranked*, sometimes equitant, rarely opposite or whorled, *borne along stem, distally clustered, or in a basal rosette,* simple, entire to spinose-serrate, *with usually parallel venation*, occasionally pinnate-parallel or palmate (then with major veins converging), the fine veins obscure to rarely well developed, reticulate, occasionally plicate, the base sheathing, the leaves sometimes reduced to scales, occasionally spinose, sometimes thickened and/or succulent; petiole usually absent, but rarely present; stipules absent. Inflorescences determinate or indeterminate, terminal or axillary, *sometimes scapose or appearing so*, sometimes reduced to a solitary flower. Flowers bisexual or unisexual (and plants then dioecious or polygamous), radial or bilateral (due to form of perianth and/or androecium), sometimes resupinate, *perianth usually undifferentiated, the parts in 2 whorls. Tepals* (4 or) 6 (or 8), *distinct to connate, imbricate, usually petaloid, sometimes with one member (labellum) differenti-*

ated from the others, occasionally bicolored, variegated, or spotted, and occasionally with a perigonal corona. *Stamens 1–6 (-8), in 1 or 2 whorls, but commonly and ancestrally 6, distinct to connate* (by filaments), sometimes with staminodia, and staminodia also present in carpellate flowers; filaments free or *sometimes adnate to perianth, or gynoecium (and then forming a column)*, elongate to short, occasionally with paired lobes or more elaborate structures that sometimes form a staminal corona; anthers opening by longitudinal slits, or rarely an apical pore; pollen grains in monads or tetrads, and the latter often fused (forming pollinia), monosulcate, trichotomosulcate, bisulcate, porate, or inaperturate. *Carpels usually 3, connate, with a single branched or unbranched style*, or style ± absent, the style branches (when present) occasionally petaloid; **ovary** superior **to inferior**, with axile, parietal, intruded-parietal, or basal placentation, stigmas elongate and decurrent to punctate, or expanded, complex and crest-like on each style branch, or stigma ± 3-lobed, capitate to punctate, or sometimes asymmetrical, highly modified, capitate, or forming a concave stigmatic cavity (and on lower side of column). Ovules anatropous, campylotropous, to nearly orthotropous, with 2 integuments, with a thick to thin nucellus, 1 to numerous in each locule/carpel. Nectaries usually in septa of ovary, occasionally at base of or on the filaments, occasionally on the tepals, or absent, but flowers occasionally producing oils, resins, or fragrances as rewards. *Fruit a loculicidal or septicidal capsule, berry, or nutlet* (then 3-angled or winged and samara-like), rarely rupturing early and exposing seeds; **seeds with coat often with phytomelan (or if lacking, then outer epidermis of seed coat usually obliterated)**, occasionally fleshy, sometimes flattened or winged, occasionally arillate, **the inner cell layers ± collapsing in development**; endosperm present or absent.

Commelinidae Takht.

Herbs or occasionally vines or lianas, to shrubs or trees (without secondary growth, but stem apex with large apical meristem), occasionally aquatic or epiphytic, from rhizomes, these sometimes tuberous, corms, or bulbs, or such structures absent and plants merely from fibrous root system, with roots occasionally tuber-like; crystals various, with or without raphides, sometimes with **silica bodies**, variously positioned, but often epidermal or **in bundle-sheath cells**; parenchymatous tissues **with cells with UV-fluorescent ferulic and coumaric acids**, sometimes with mucilage cells or canals, laticifers, or cells with aromatic ethereal oils; **epidermis often with strelitzia-type epicuticular waxes**; stems variously shaped, occasionally with swollen nodes, with vascular bundles scattered to reduced to 1 or 2 rings, occasionally sheathed by overlapping leaves. Hairs various, and plants occasionally spiny due to modified leaf segments, exposed fibers, sharp-pointed roots, or petiole outgrowths. Leaves alternate and spiral, 2-ranked (and sometimes also equitant), 3-ranked, or 1-ranked and spiral, borne along stem to in a basal rosette, or clustered at stem apices, simple but sometimes appearing lobed or compound due to tearing between segments, entire or spinose-serrate, with parallel, pinnate-parallel, or ± palmate venation (and then the major veins converging), or only a single vein, the fine veins usually obscure, sometimes plicate, the base sheathing, and leaf sometimes differentiated into sheath and blade, or sheath, petiole, and blade, the sheath open or closed, and sometimes with a ligule; stipules absent; the petiole (or leaf sheath) with or without air canals. Inflorescences determinate, indeterminate, or mixed, terminal or axillary, occasionally reduced to a solitary flower, sometimes scapose or appearing so, occasionally associated with expanded bracts, the individual flowers occasionally inconspicuous, arranged in spikelets. Flowers bisexual or unisexual (and the plants then monoecious or dioecious), radial, bilateral, or without plane of symmetry (due to form of perianth/androecium/gynoecium), perianth undifferentiated (of tepals) or differentiated (with calyx and corolla), in 1 or 2 whorls, or perianth absent. Tepals 1 to numerous (but ancestrally and commonly 6), distinct to variously connate, imbricate, valvate, or with open aestivation, petaloid, sepaloid, or bract-like, occasionally reduced to scales or bristles, occasionally the outer whorl somewhat differentiated from the inner. Sepals 2 or 3, distinct to connate, rarely slightly adnate to petals, imbricate, valvate, or with open aestivation, occasionally one member differentiated from the other two, occasionally petaloid. Petals 2 or 3, distinct to connate, imbricate, sometimes also crumpled, convolute, or valvate, occasionally divergent in color, shape, and size, sometimes one member differentiated from the other two, sometimes quickly self-digesting or wilting. Stamens 1 to numerous, or with only half a fertile stamen, in 1 or 2 whorls (and ancestrally 6), distinct to connate (by filaments), sometimes dimorphic or with staminodia, and staminodia sometimes present in carpellate flowers; filaments free or occasionally adnate to perianth, short to elongate; anthers opening by longitudinal slits or ± apical pores, occasionally sagittate or grasping the style; pollen grains in monads or tetrads, mono- to trisulcate, monoporate to pantoporate, or inaperturate, **with starch**, the exine sometimes very reduced. Carpels 2–3 (-10), but sometimes only 1 functional, rarely monomerous, connate, with as many styles as carpels, or a single, elongate to very short, branched to unbranched style; ovary superior to inferior, with axile, free-central, parietal, intruded-parietal, apical, or basal placentation, stigma punctate, capitate, funnel-form, sometimes fringed or 3-lobed, occasionally asymmetrical, modified with only one stigma functional, or 3 decurrent to very elongate, sometimes fringed or plumose, or capitate to punctate stigmas. Ovules anatropous to orthotropous, with 2 integuments, and a thick to thin nucellus, 1 to numerous in each locule or on each placenta, or only 1 per gynoecium. Nectaries in the septa of the ovary, 2 on ovary apex, or absent. Fruit a variously dehiscent capsule, follicle, berry, drupe, drupaceous schizocarp, nut, achene, or grain; seeds sometimes with an aril and/or operculum; *endosperm* present, copious, *starchy* (but non-starchy in palms), sometimes ± reduced, and then *with starchy perisperm*.

Composition: Arecales, Commelinales, Poales, Zingiberales.

ARECALES BROMHEAD

Composition: Arecaceae (= Palmae) and the recently added Dasypogonaceae. See descriptions of these two distinctive clades.

COMMELINALES MIRB. EX BERCHT. & J. PRESL

Herbs or rarely twining vines, sometimes aquatic, from rhizomes, these sometimes tuberous or purple to red in color, corms, or bulbs, or such structures absent and plants merely from fibrous root system, with roots occasionally tuber-like, sometimes purple to red; crystals various, sometimes with silica bodies; parenchymatous tissues sometimes with mucilage canals; stems sometimes with ± swollen nodes. Hairs simple to branched (variously dendritic to stellate), sometimes restricted to reproductive parts. *Leaves alternate and spiral or 2-ranked*, sometimes equitant, borne along stem, ± in a basal rosette, or clustered at stem apices, simple, entire or rarely spinose-serrate, *with parallel*, pinnate-parallel, *or ± palmate venation* (and then the major veins converging), the fine veins usually obscure, the base sheathing, and leaf sometimes differentiated into sheath, petiole, and blade, the sheath open or closed, occasionally with a ligule; petiole usually absent, but occasionally present; stipules absent. Inflorescences often determinate or less commonly indeterminate, terminal or axillary, occasionally reduced to a solitary flower, sometimes associated with a folded, leafy bract. Flowers bisexual or unisexual (and plants then dioecious or polygamous), sometimes tristylous, radial or bilateral (due to form of perianth, androecium, and gynoecium), *perianth undifferentiated (of tepals) or differentiated (with calyx and corolla), the parts in 2 whorls*. *Tepals 6*, distinct, sometimes with the 2 upper tepals of the inner whorl fused with the upper tepal of the outer whorl, or all 6 connate, and then sometimes with a slit along upper surface, imbricate or valvate, petaloid, sometimes multicolored or with spotted nectar guides, or with only the upper tepals of the inner whorl with a colored blotch, basal spots, or streaks (again, nectar guides), occasionally the outer whorl somewhat differentiated from the inner. *Sepals 3*, distinct, imbricate or with open aestivation. *Petals 3*, *distinct to connate*, imbricate, sometimes also crumpled, occasionally divergent in color, shape, and size, sometimes clawed, sometimes quickly self-digesting. Stamens 1–6, in 1 or 2 whorls, distinct to slightly connate (by filaments), sometimes dimorphic or with staminodia, and staminodia present in carpellate flowers; filaments free or occasionally adnate to perianth, elongate, sometimes with simple hairs; anthers opening by longitudinal slits, rarely apical pores, occasionally spirally twisted; pollen grains in monads or tetrads, mono- to trisulcate, di- to 8-porate, or inaperturate. *Carpels 3*, but sometimes only 1 functional, rarely pseudomonomerous, *connate*, *with a single, elongate to very short, unbranched style*; ovary superior or inferior, with axile or intruded-parietal placentation, stigma punctate, capitate, sometimes fringed or 3-lobed, or 3 adjacent stigmas. Ovules anatropous to orthotropous, with 2 integuments, and usually a ± thick nucellus, 1 to numerous in each locule/carpel. Nectaries in the septa of the ovary, or absent. Fruit a loculicidal capsule, berry, or nut (which is surrounded by persistent basal portion of perianth tube); seeds occasionally flattened and/or winged, sometimes with a conspicuous conical cap, sometimes with aril-like structure; endosperm present, copious.

POALES SMALL

Herbs or occasionally vines, *usually rhizomatous*, the rhizome sometimes corm-like or bulb-like, or such structures absent and plants merely from fibrous root system, sometimes epiphytic or wetland plants; crystals various or absent, **and raphides often absent**, also **with silica bodies associated with epidermis**, surrounding vascular bundles, or scattered in parenchyma, or such bodies absent; *stems round, polyhedral, or 3-angled, solid or hollow* in internodal regions, with vascular bundles scattered or reduced to merely 1 or 2 rings. Hairs simple, T-shaped, variously branched, stellate, or peltate scales. *Leaves alternate and spiral, 2-ranked (and occasionally also equitant) or 3-ranked, borne along stem or in a basal rosette*, simple, entire or spinose-serrate, *with parallel venation* or only a single vein, the fine veins obscure, the base sheathing, the leaves rarely terminating in a tendril, *often differentiated into sheath and blade*, sometimes reduced to the sheath, sometimes with a ligule present at junction of sheath and blade, rarely with petiole between blade and sheath, and sheath open or closed; stipules absent. Inflorescences usually indeterminate, occasionally determinate, *terminal*, *sometimes scapose or appearing so*, rarely appearing axillary, *the individual flowers often inconspicuous, arranged in spikelets*. Flowers bisexual or unisexual (and plants then monoecious or dioecious), radial or occasionally bilateral (due to form of perianth and androecium/gynoecium), *perianth undifferentiated (of tepals, these sometimes merely scales or bristles) or differentiated (with calyx and corolla)*, the parts usually in 1 or 2 whorls, or perianth absent. *Tepals 1 to numerous* (but ancestrally and commonly 6), distinct to slightly connate, imbricate or with open aestivation, petaloid, sepaloid, or bract-like, *sometimes reduced to scales or bristles*, these often persistent and associated with fruit. *Sepals 2–3*, distinct to connate, imbricate or valvate, sometimes with one member differentiated from the other two. *Petals 2–3*, distinct to connate, imbricate, sometimes quickly wilting, occasionally with 2 ± basal appendages. *Stamens 1–6* (-8), in 1 or 2 whorls, distinct to connate (by filaments), sometimes with staminodia; filaments free or sometimes adnate to perianth, short to elongate; anthers opening by longitudinal slits or ± apical pores, sometimes sagittate, the connective rarely expanded; pollen grains in monads or occasionally tetrads, usually monosulcate (the sulcus sometimes elongated and spiral), monoporate, or inaperturate, rarely disulcate or di- to several-porate. *Carpels 2–3*, but sometimes only 1 functional or pseudomonomerous, *connate, with a single elongate to very short, unbranched to ± branched style, or as many styles as carpels*; ovary superior to inferior, with axile, free-central, parietal, apical, or basal placentation, stigmas ± decurrent on, or elongated and covering, each style or style-branch, sometimes spirally twisted or plumose, or reduced and capitate or punctate. Ovules anatropous, hemianatropous, amphitropous, campylotropous, or orthotropous, with 2 integuments, with a thick to thin nucellus, 1 to numerous in each locule or placenta, or only 1 per gynoecium. *Nectaries* occasionally in septa of ovary, *more commonly absent*. Fruit a septicidal, loculicidal, or irregularly dehiscent capsule, berry, drupe, follicle, achene, grain, nut; seeds with coat often longitudinally ridged, sometimes with an operculum, sometimes winged or with a tuft of hairs; *endosperm* present, copious, *starchy*.

ZINGIBERALES GRISEB.

Large herbs or shrubs to banana-like trees (without secondary growth), from rhizomes, these occasionally tuber-like, or corms; crystals various, sometimes with raphides, and **silica bodies present in the sheath of vascular bundles;** parenchymatous tissues sometimes with secretory cells containing aromatic ethereal oils, terpenes, and phenylpropanoid compounds, occasionally with laticifers, mucilage cells, or canals; stems occasionally appearing thicker than they actually are due to overlapping sheathing leaves. Hairs simple or occasionally branched. *Leaves alternate and spiral or 2-ranked, or 1 ranked and spiraled* (i.e., spiromonistichous), usually borne along stem, **rolled into a tube in bud**, entire, **with ± pinnate-parallel venation, sometimes tearing between veins**, the fine veins obscure, **usually differentiated into a blade, petiole,** and sheath (but occasionally with only blade and sheath); stipules absent, but a ligule sometimes present at junction of sheath and petiole, occasionally with a pulvinus at junction of petiole and blade; **petiole (and leaf sheath) usually with air canals,** these separated into segments by diaphragms composed of stellate-shaped cells; sheaths open or closed. Inflorescences usually mixed (indeterminate but with determinate lateral units), terminal or axillary, occasionally on specialized reproductive shoots arising separately from rhizome, erect to pendulous, **often with large bracts. Flowers** bisexual or *unisexual* (and plants then monoecious), **bilateral** (due to form of perianth and androecium) *or without plane of symmetry* (i.e., asymmetrical), perianth undifferentiated (of tepals, although one member morphologically different from the other five) or differentiated (with calyx and corolla), the parts in 2 whorls. Tepals 6, with 5 connate and one (of either inner or outer whorl) distinct, imbricate, petaloid. *Sepals 3*, distinct or connate, occasionally slightly adnate to petals, imbricate, sometimes petaloid. *Petals 3*, distinct to connate, imbricate or convolute, often one member slightly to strongly differentiated from the other two. Stamens ancestrally 6, *but usually reduced to 5* (and then often with an inconspicuous staminodium), *reduced to 1 (and then with showy staminodia) or reduced to half a functional stamen (the other half of which is staminodial, and also with other showy staminodia)*, and inconspicuous staminodia also present in carpellate flowers, distinct, but staminodia sometimes connate; filaments free or occasionally adnate to perianth, slender; anthers opening by longitudinal slits, sometimes grasping the style; pollen grains monosulcate, disulcate, or pantoporate, often inaperturate, **the exine usually very reduced.** Carpels 3 (but sometimes only 1 or 2 functional), connate, with a single, branched or unbranched, occasionally petaloid style; **ovary inferior,** with axile or rarely parietal or free-central placentation, the stigmas 3, elongate or fringed, or stigma single, 3-lobed, capitate, or funnel-form, sometimes asymmetrical, modified with only one stigma functional. Ovules anatropous to campylotropous, with 2 integuments, with a thick nucellus, 1 to numerous in each locule. Nectaries in the septa of the ovary, 2 on ovary apex, or absent. Fruit a loculicidal or irregularly dehiscent capsule, berry, or schizocarp (of three drupes); seeds **with testa operculate, sometimes with an aril;** *endosperm* and **perisperm present** and copious to endosperm present to ± reduced, not as abundant as the perisperm, both *starchy*, the embryo usually with 1 cotyledon.

CHAPTER 8

EUDICOTYLEDONEAE M.J. DONOGHUE, J.A. DOYLE & P.D. CANTINO

Trees, shrubs, lianas, vines, or herbs, often rhizomatous, occasionally aquatic or epiphytic; crystals various; stems usually with a eustele, secondary growth (when present) by typical development of a vascular cambium, but occasionally anomalous, *and branches associated with 2 lateral prophylls*; parenchymatous tissues lacking spherical idioblasts with ethereal oils, but sometimes with larger, schizogenous to lysogenous cavities containing various aromatic compounds, occasionally with laticifers or resin, oils, or mucilage canals. Hairs various. Leaves alternate and spiral or 2-ranked, or opposite and decussate, rarely opposite and 2-ranked, or whorled, simple to variously compound, entire to variously toothed, sometimes lobed, with pinnate or palmate venation (the latter with converging or diverging major veins), occasionally with domatia or pellucid dots; stipules various, present or absent. Inflorescences determinate, indeterminate, or mixed, usually terminal or axillary, sometimes reduced to a solitary flower. Flowers bisexual or unisexual (and plants then monoecious, dioecious, or polygamous), with radial or bilateral symmetry (due to form of perianth and/or androecium), with or without a hypanthium, the perianth undifferentiated (of tepals) or differentiated (with calyx and corolla) and in 1 to several whorls, or spirally arranged, occasionally absent. Tepals usually 2 to numerous, distinct to connate, variously imbricate to valvate. Sepals 2 to numerous, distinct to connate, occasionally calyptrate, variously imbricate or valvate, or with open aestivation, occasionally absent. Petals 1 to numerous, sometimes absent, distinct to connate, variously imbricate or convolute, valvate, or with open aestivation, sometimes absent. Stamens 1 to numerous, distinct to connate (by filaments or filaments and anthers or by anthers only) and sometimes fasciculate, spirally arranged or in whorls, occasionally with staminodia; filaments free, usually arising from receptacle or hypanthium, or adnate to petals, variable in form, but usually well differentiated into anther and filament; anthers usually opening by longitudinal slits, occasionally by pores or flaps; pollen grains in monads or occasionally tetrads or polyads, **usually tricolpate, tricolporate, or modifications thereof**, e.g., triporate to pantoporate, etc. Carpels 1 to numerous, distinct to connate, with as many styles as carpels or a single style, either apically branched or unbranched; ovary superior or inferior, with variable placentation; stigma various. Ovules anatropous to orthotropous, with one or two integuments, with a thick to thin nucellus, 1 to numerous in each locule/placenta, or reduced to 1 per gynoecium. Nectaries variable or absent. Fruit extremely diverse, simple, aggregate, or multiple, often with accessory tissues; seeds diverse in form, sometimes arillate; endosperm present to absent, sometimes replaced by perisperm.

Composition: Ranunculales, Proteales, Trochodendrales, Buxales, *Gunneridae*.

RANUNCULALES JUSS. EX BERCHT. & J. PRESL

Trees, shrubs, lianas or vines (scrambling, twining), to herbs, sometimes rhizomatous; crystals various or absent; stems with vascular bundles in a ring, in several rings, or ± scattered, with vascular cambium normally developed or anomalous; tissues sometimes yellow-colored by berberine, sometimes with white to yellow, orange, or red (or clear) latex; **leaf epicuticular wax crystaloids of clustered tubules** (but absent in Eupteleaceae). Hairs simple, sometimes glandular. *Leaves alternate, usually spiral* but rarely 2-ranked, or occasionally opposite and decussate, *simple or variously compound, entire to variously toothed, sometimes lobed or dissected*, with usually pinnate or palmate venation (the latter with converging to diverging major veins) but rarely dichotomously veined, *lacking pellucid dots*, sometimes peltate, occasionally forming spines; stipules usually absent. Inflorescences determinate or indeterminate, terminal or axillary, occasionally reduced to a solitary flower. *Flowers* bisexual or unisexual (and plants then monoecious, dioecious, or rarely polygamous), *radial, occasionally merely biradial*, or sometimes bilateral (due to form of perianth), with perianth undifferentiated (of tepals) *or differentiated (with calyx and corolla)*, in 1 to several whorls or spirally arranged, *often 3-merous*, but occasionally absent, the receptacle short to elongated or enlarged. Tepals (2-) 3 to numerous, distinct, imbricate, usually petaloid. *Sepals 2 to numerous,* usually distinct, imbricate or valvate, sometimes petaloid, sometimes quickly deciduous. *Petals 3 to numerous*, sometimes absent, usually distinct, imbricate, sometimes wrinkled, *often producing nectar*, and occasionally with nectar spurs, sometimes the inner differentiated from the outer ones (and perhaps of staminodial origin). *Stamens (1-) 3 to numerous*, spirally arranged or in 1 to several whorls, distinct to occasionally connate (by filaments), rarely with staminodia, and staminodia sometimes present in carpellate flowers (when present); filaments free, short and thick to long and slender; anthers opening by longitudinal slits or 2 flaps, the connective not to sometimes expanded-thickened, usually not elongated beyond anther sacs; pollen grains tricolpate or tricolporate, occasionally with as many as 12 colpi, tri- to polyporate, occasionally spiral-aperturate. *Carpels 1 to numerous, distinct to connate*, with several styles, a single style, or style absent; ovary or ovaries superior, with lateral, parietal, intruded-parietal, basal, ± apical, or rarely axile placentation; stigmas elongated, decurrent along style, to capitate, punctate or lobed, or with 2 to many stigmas or stigma lobes. Ovules usually anatropous to campylotropous, occasionally orthotropous, with one or two integuments, with a usually thick nucellus, 1 to numerous in each carpel or on each placenta. *Nectar usually produced by the petals*, but sometimes one or more of the filaments with a basal nectar gland, or nectaries absent. *Fruit an aggregate of follicles (dry or fleshy), achenes, samaras, berries, or drupes, or a simple berry, nut, variously dehiscent capsule (often with valves alternating with replum-like structures), or schizocarp*; *seeds with abundant, occasionally ruminate endosperm*, sometimes arillate.

PROTEALES JUSS. EX BERCHT. & J. PRESL

Trees, shrubs, lianas, or aquatic, rhizomatous herbs; crystals various; parenchymatous tissues rarely with laticifers, air canals. Hairs simple to branched, sometimes also glandular. *Leaves alternate and spiral or 2-ranked*, occasionally opposite and decussate or whorled, but in *Nelumbo* the leaves and bracts in groups of 3 (1 bract on lower side of rhizome, and 1 bract on the upper side, immediately subtending the leaf), simple, pinnately or palmately lobed or dissected, or once to twice pinnately compound, entire to variously toothed, often lobed, with pinnate or palmate venation (the latter with diverging major veins), rarely peltate; stipules present or absent, sometimes conspicuous. Inflorescences determinate or indeterminate, terminal or axillary, occasionally reduced to a solitary flower. Flowers bisexual or unisexual (and plants then monoecious, dioecious, or polygamous), with radial or bilateral symmetry (due to form of perianth), the perianth parts undifferentiated (of tepals), and then the outer slightly differentiated from the inner ones, or differentiated (with calyx and corolla), spirally arranged or in 1 or 2 whorls, showy to inconspicuous, the receptacle small or expanded, flat-topped and obconic. *Tepals 4 or numerous, distinct to connate, imbricate or valvate, ± petaloid. Sepals 3–7, distinct to slightly connate, imbricate. Petals 3–7, distinct, imbricate or with open aestivation, sometimes unequal in size or very reduced*, sometimes lacking in carpellate flowers. Stamens 2 to numerous, distinct, spirally arranged or in a single whorl, sometimes with staminodia; filaments free or sometimes adnate to the perianth, very short to elongate; anthers opening by longitudinal slits, *the connective often thickened and projecting beyond the anther sacs or expanded into a flat-topped structure*; pollen grains usually tricolpate, tricolporate, or triporate. *Carpels 1 to numerous, distinct or connate*, each carpel with a single, ± absent to elongate style, or fused carpels with a single style; ovary or ovaries superior, but occasionally sunken into expanded receptacle, sometimes stipitate, with lateral, apical, or axile placentation; stigma elongate, extending along one side of style to globose or lobed. **Ovules** anatropous to orthotropous, with 1 or 2 integuments, with a thick nucellus, **1 or 2** to numerous in each carpel or locule. Nectary a disk, or absent. Fruit an aggregate of achenes, or a nut, follicle, achene, samara, or drupe (often flattened), sometimes associated with long bristles; seeds sometimes winged; **endosperm thin or absent.**

TROCHODENDRALES TAKHT. EX CRONQUIST

Composition: Trochodendraceae (incl. Tetracentraceae).

BUXALES TAKHT. EX REVEAL

Composition: Buxaceae (incl. Didymelaceae and Haptanthaceae).

CHAPTER 9

GUNNERIDAE D.E. SOLTIS, P.S. SOLTIS & W.S. JUDD

Composition: Gunnerales, *Pentapetalae*, Dilleniales (precise placement within *Gunneridae* uncertain).

GUNNERALES TAKHT. EX REVEAL

Shrubs (and then reviving after drought) or rhizomatous herbs, the latter sometimes very large; crystals sometimes of druses; **epidermis with mucilage cells,** *sometimes also resin-cells*; *stems and roots often with internal, N-fixing colonies of Nostoc*, and often with an anomalous vascular system (i.e., with separate, anastomosing vascular cylinders). Hairs simple or absent. Leaves alternate and spirally arranged, or opposite and decussate, *simple*, entire, *serrate, or dentate* (**and the teeth with hydathodes**), sometimes shallowly lobed, small to very large, *with palmate venation, often ± dichotomous*, clearly petiolate or not (and then the leaf base sheathing); **stipules present,** *as adaxial scales at or near petiole base, or as 2 projections, 1 on each side of leaf sheath*. **Inflorescences determinate, but appearing to be indeterminate,** i.e., **with raceme-like or spicate axes ending in a terminal flower,** axillary or terminal, **with inconspicuous, ± densely clustered flowers. Flowers** bisexual or **unisexual** *(and plants then monoecious, dioecious, or polygamous), radial*, with perianth differentiated (with calyx and corolla), in 1 or 2 whorls, or absent. *Sepals 2, distinct*, ± valvate, *often with marginal colleters*, **very small**, or absent. *Petals 2, distinct*, valvate or with open aestivation, **very small**, or absent. *Stamens 1–8*, distinct or connate (by filaments), in a single whorl; *filaments free, short*; anthers opening by longitudinal, **latrorse** slits, the connective sometimes prolonged; pollen grains in monads or tetrads, ± tricolpate or triporate. *Carpels 2–4, connate, with separate styles; ovary inferior or nude*, with apical or axile placentation; stigmas elongate. Ovules anatropous, with 2 integuments, with a thick nucellus, 1 in a single locule, or numerous, in each of 3 or 4 locules. *Nectaries absent. Fruit a drupe, nut or ventricidal capsule*; seeds with copious endosperm.

PENTAPETALAE D.E. SOLTIS, P.S. SOLTIS & W.S. JUDD

Trees, shrubs, lianas, vines, or herbs, occasionally aquatic or epiphytic; crystals various; stems usually with a eustele, secondary growth (when present) by typical development of a vascular cambium, but occasionally anomalous; parenchymatous tissues lacking spherical idioblasts with ethereal oils, but sometimes with larger, schizogenous to lysogenous cavities containing various aromatic compounds, occasionally with laticifers or resin, oils, or mucilage canals. Hairs various. Leaves alternate and spiral or 2-ranked, or opposite and decussate, rarely opposite and 2-ranked, or whorled, simple to variously compound, entire to variously toothed, sometimes lobed, with pinnate or palmate venation (the latter with converging or diverging major veins), occasionally with domatia or pellucid dots; stipules various, present or absent. Inflorescences determinate, indeterminate, or mixed, usually terminal or axillary, sometimes reduced to a solitary flower. Flowers bisexual or unisexual (and plants then monoecious, dioecious, or polygamous), radial or bilateral (due to form of perianth and/or androecium), with or without a hypanthium, **the perianth differentiated (with calyx and corolla) and usually in 2 whorls** (or this may be synapomorphic for *Gunneridae*), occasionally reduced to a single whorl (and then the parts frequently called tepals) or absent. **Sepals** *(often called tepals when petals lost) (3-) 4 or* **5** *(-6)*, distinct to connate, occasionally calyptrate, variously imbricate, valvate or with open aestivation. **Petals** *usually 4 or* **5**, but sometimes reduced to one, or with a secondary increase and numerous, sometimes absent, distinct to connate, variously imbricate or convolute, valvate or with open aestivation. Stamens 1 to numerous *(but ancestrally in 2 whorls, each usually of 5 members)*, distinct to connate (by filaments or filaments and/or anthers) and sometimes fasciculate, occasionally with staminodia; filaments free, usually arising from receptacle or hypanthium, or adnate to petals, variable in form; anthers usually opening by longitudinal slits, occasionally by pores or flaps; pollen grains in monads or occasionally tetrads or polyads, usually tricolpate, tricolporate, or modifications thereof, e.g., triporate, pantoporate. *Carpels* 1 to numerous, *connate* or less commonly distinct, with as many styles as carpels or a single style, either apically branched or unbranched; ovary superior or inferior, with variable placentation; stigma various. Ovules anatropous to orthotropous, with 1 or 2 integuments, with a thick to thin nucellus, 1 to numerous in each locule/placenta, or reduced to 1 per gynoecium. Nectaries variable or absent. Fruit extremely diverse; seeds sometimes arillate; endosperm present to absent, sometimes replaced by perisperm.

Composition: *Superrosidae*, *Superasteridae*, Dilleniales (precise placement within *Pentapetalae* uncertain).

DILLENIALES DE CANDOLLE EX BERCHT. & J. PRESL

Composition: Dilleniaceae.

Note: *Superrosidae* and *Superasteridae* will be treated in subsequent chapters (Chapters 10 and 11, respectively).

CHAPTER 10

SUPERROSIDAE W.S. JUDD, D.E. SOLTIS & P.S. SOLTIS

Trees, shrubs, lianas, vines, or herbs, often rhizomatous, occasionally aquatic, epiphytic or parasitic; crystals various; stems usually with a eustele, secondary growth (when present) by typical development of a vascular cambium, but occasionally anomalous; parenchymatous tissues lacking spherical idioblasts with ethereal oils, but sometimes with larger, schizogenous to lysigenous cavities containing various aromatic compounds, occasionally with laticifers or resin, oils, or mucilage canals or cavities. Hairs various. Leaves alternate and spiral or 2-ranked, or opposite and decussate, rarely opposite and 2-ranked, or whorled, simple to variously compound, entire to variously toothed, sometimes lobed, with pinnate or palmate venation (the latter with converging or diverging major veins), occasionally with domatia or pellucid dots; stipules various, present or absent. Inflorescences determinate, indeterminate, or mixed, usually terminal or axillary, sometimes reduced to a solitary flower. Flowers bisexual or unisexual (and plants then monoecious, dioecious, or polygamous), radial or bilateral (due to form of perianth and/or androecium), with or without a hypanthium, the perianth usually differentiated (with calyx and corolla) and in 2 whorls, occasionally reduced to a single whorl (and then the parts frequently called tepals) or absent. Sepals (often called tepals when petals lost) (3-) 4–5 (-10), distinct to connate, occasionally calyptrate, with various aestivation. *Petals usually 4 or 5, but sometimes reduced to one, or with a secondary increase and numerous, sometimes absent, distinct* to less commonly connate, with various aestivation. *Stamens 1 to numerous (but ancestrally in 2 whorls, each of 5 members),* distinct to connate (usually by filaments) and sometimes fasciculate, occasionally with staminodia; filaments free, usually arising from receptacle or hypanthium, or occasionally adnate to petals, variable in length and thickness; anthers usually opening by longitudinal slits, occasionally by pores or flaps; pollen grains in monads or occasionally tetrads or polyads, usually tricolpate, tricolporate, or modifications thereof, e.g., triporate, pantoporate, etc. Carpels 1 to numerous, connate or less commonly distinct, with as many styles as carpels or a single style, either apically branched or unbranched; ovary superior or inferior, with variable placentation; stigma various. *Ovules* anatropous to orthotropous, *with usually 2 integuments, with a usually thick nucellus,* 1 to numerous in each locule or on each placenta, or reduced to 1 per gynoecium. Nectaries variable or absent. Fruit various; seed sometimes arillate; endosperm present to absent.

Composition: Saxifragales, Rosidae.

SAXIFRAGALES BERCHT. & J. PRESL

Trees, shrubs, scrambling lianas, herbs, often rhizomatous, occasionally aquatic, or rarely root parasites; the branches sometimes differentiated into long and short shoots; crystals various, solitary or druses; stems and parenchymatous tissues occasionally with resin canals or sclerieds. Hairs simple, sometimes glandular, occasionally forming prickles, or branched or stellate. Leaves alternate and spirally arranged or less commonly 2-ranked, or opposite and decussate, occasionally whorled, rarely reduced, borne along the stems or clustered in basal rosettes, simple or less commonly variously palmately or pinnately compound, entire to variously toothed, occasionally lobed or dissected, with pinnate or palmate venation (the latter with diverging or less commonly converging major veins), sometimes succulent, occasionally aromatic (due to resin canals); stipules present or absent. Inflorescences determinate or indeterminate, terminal or axillary, occasionally reduced to a solitary flower. Flowers bisexual or unisexual (and plants then monoecious or dioecious), *usually radial*, but rarely bilateral (due to perianth, androecium, and gynoecium), *with or without variously developed hypanthium,* and perianth differentiated (with calyx and corolla, but petals sometimes lost, and then perianth parts often referred to as tepals), in usually 2 whorls (occasionally ± spirally arranged), showy or inconspicuous, and sometimes absent. Sepals (or tepals, if petals lost) (3-) 4–5 (-10), sometimes absent, distinct to connate, imbricate, valvate, or with open aestivation. Petals 4–5 (to rarely numerous), sometimes absent, distinct or less commonly connate, imbricate, valvate, or with open aestivation, occasionally clawed, rarely lobed or dissected. *Stamens* 4 to numerous (and *ancestrally of 2 whorls, each of 4 or 5 members*, but occasionally numerous and spirally arranged), distinct to rarely slightly connate (by filaments), staminodia occasionally present; filaments free, arising from receptacle or hypanthium, occasionally adnate to petals, slender to stout, very short to elongate; **anthers** opening by longitudinal slits (*and usually latrorse*) or flaps, usually **basifixed,** *often with a basal pit*, the connective sometimes shortly prolonged beyond anther sacs; pollen grains tri- to hexacolpate, di- or tricolporate or di- to polyporate. Carpels 1 to numerous (but often 2, 4 or 5), *distinct* to ± connate, *but usually at least distally distinct*, *with separate styles or less commonly a single style, with or without apical branches*, occasionally the styles postgenitally fused at stigma; ovary superior to inferior, with lateral, parietal, intruded-parietal, axile, apical-axile or apical placentation; stigmas various (i.e., elongate, decurrent along style, or reduced and capitate to punctate). Ovules anatropous to orthotropous, with usually 2 integuments, with a thick to thin nucellus, 1 to numerous per carpel, in each locule, or on each placenta. Nectary or nectaries a disk, gland associated with each carpel, inner surface of hypanthium, inner/basal surface of petals, staminodia, or absent. *Fruit an aggregate of follicles*, or a single variously dehiscent capsule, *follicle*, nut, achene, schizocarp, drupe or berry; seeds rarely flattened or winged; endosperm present, copious to scanty.

Rosidae Takht.

Trees, shrubs, lianas, vines, or herbs, often rhizomatous, occasionally aquatic, epiphytic, or parasitic; crystals various; stems usually with a eustele, secondary growth (when present) by typical development of a vascular cambium, but occasionally anomalous; parenchymatous tissues lacking spherical idioblasts with ethereal oils, but sometimes with larger, schizogenous to lysigenous cavities containing various aromatic compounds, occasionally with laticifers or resin, oils, or mucilage canals or cavities. Hairs various. Leaves alternate and spiral or 2-ranked, or opposite and decussate, rarely opposite and 2-ranked, or whorled, simple to variously compound, entire to variously toothed, sometimes lobed, with pinnate or palmate venation (the latter with converging or diverging major veins), occasionally with domatia or pellucid dots; stipules various, present or absent. Inflorescences determinate, indeterminate, or mixed, usually terminal or axillary, sometimes reduced to a solitary flower. Flowers bisexual or unisexual (and plants then monoecious, dioecious, or polygamous), radial or bilateral (due to form of perianth and/or androecium and gynoecium), with or without a hypanthium, the perianth differentiated (with calyx and corolla) and usually in 2 whorls, occasionally reduced to a single whorl (and then the parts frequently called tepals) or absent. Sepals (often called tepals when petals lost) (3-) 4–5 (-6), distinct to connate, occasionally calyptrate, variously imbricate or valvate, or with open aestivation. *Petals usually 4 or 5, but sometimes reduced to one, or with a secondary increase and numerous, sometimes absent, distinct* to less commonly connate, variously imbricate or convolute/contorted, valvate, or with open aestivation. *Stamens 1 to numerous (but ancestrally in 2 whorls, each usually of 4 or 5 members),* distinct to connate (usually by filaments) and sometimes fasciculate, occasionally with staminodia; filaments free, usually arising from receptacle or hypanthium, or occasionally adnate to petals, variable in form; anthers usually opening by longitudinal slits, occasionally by pores; pollen grains in monads or occasionally tetrads or polyads, usually tricolpate, tricolporate, or modifications thereof, e.g., triporate, pantoporate. Carpels 1 to numerous, connate or less commonly distinct, with as many styles as carpels or a single style, either apically branched or unbranched; ovary superior or inferior, with variable placentation; stigma various. *Ovules* anatropous to orthotropous, *with usually 2 integuments, with a usually thick nucellus,* 1 to numerous in each locule or on each placenta, or reduced to 1 per gynoecium. Nectaries variable or absent. Fruit variable; seeds sometimes arillate; endosperm present to absent.

Composition: Vitales, Malvidae, Fabidae.

VITALES JUSS. EX BERCHT. & J. PRESL

Composition: Vitaceae (incl. Leeaceae).

FABIDAE W.S. JUDD, D.E. SOLTIS & P.S. SOLTIS

Trees, shrubs, lianas, vines, or herbs, sometimes rhizomatous, rarely parasitic; crystals various; stems with or without internal phloem; parenchymatous tissues sometimes with schizogenous to lysigenous cavities containing various resinous compounds, laticifers, or secretory cavities or canals; stems rarely with anomalous secondary growth; vessel elements with or without vestured pits; *roots sometimes with N-fixing nodules* (with *Frankia,* or *Rhizobium* and relatives). Hairs various. Leaves alternate and spiral or 2-ranked, or opposite and decussate, or whorled, distributed along stem or in a basal rosette, simple to variously compound, entire to variously toothed, sometimes lobed, with pinnate or palmate venation (the latter with converging or diverging major veins), occasionally with domatia or pellucid/black dots; stipules various, present or absent. Inflorescences determinate, indeterminate, or mixed, usually terminal or axillary, sometimes reduced to a solitary flower. Flowers bisexual or unisexual (and plants then monoecious, dioecious, or polygamous), radial or bilateral (due to form of perianth and/or androecium and gynoecium), with or without a hypanthium, the perianth differentiated (with calyx and corolla) and usually in 2 whorls, occasionally reduced to a single whorl (and then the parts frequently called tepals) or absent, sometimes with a gynophore or androgynophore. Sepals (often called tepals when petals lost) 1–8 (to numerous, but 4 or 5 the most common), distinct to connate, rarely calyptrate, with various aestivation. Petals 2–8 (to numerous, but 4 or 5 the most common), often absent, distinct to less commonly connate, with various aestivation. Stamens 1 to numerous (but ancestrally in 2 whorls, each usually of 5 members), distinct to connate (by filaments or filaments and anthers) and sometimes fasciculate, occasionally with staminodia; filaments free, usually attached to receptacle or hypanthium, but occasionally adnate to petals; anthers usually opening by longitudinal slits, occasionally by transverse slits, or pores; pollen grains usually in monads, occasionally in tetrads or polyads, usually tricolpate, tricolporate, or modifications thereof, e.g., triporate to pantoporate. Carpels 1 to numerous, occasionally pseudomonomerous, distinct or connate, with as many styles as carpels or a single style, either apically branched or unbranched; ovary superior or inferior, with variable placentation, sometimes lobed; stigma various. Ovules anatropous to orthotropous, with 2 (or less commonly 1) integuments, with a thick (or less commonly thin) nucellus, 1 to numerous per carpel, or reduced to 1 per gynoecium. Nectaries variable or absent. Fruit a variously dehiscent capsule, legume, loment, follicle, schizocarp (of nutlets, samara-like segments, or of splitting segments, and thus somewhat capsule-like), achene, nut, samara, berry, indehiscent pod, drupe, or pome, or an aggregate of follicles, achenes, nutlets, drupelets, or multiple fruits (comprised of achenes or drupes), sometimes closely associated with accessory structures; seeds sometimes arillate; endosperm present to absent.

Composition: Zygophyllales, Celastrales, Malpighiales, Oxalidales, Fabales, Rosales, Cucurbitales, Fagales.

ZYGOPHYLLALES LINK

Trees, shrubs, or herbs, sometimes rhizomatous, autotrophic or hemiparasites; crystals various; parenchymatous tissues sometimes with sclereids or fibers; stems sometimes with swollen and/or jointed nodes, occasionally forming thorns. Hairs various. *Leaves alternate and spiral, or opposite and 2-ranked or occasionally decussate, simple, sometimes dissected, bifoliolate, trifoliolate, or pinnately compound,* the leaves or leaflets entire or serrate, with pinnate to palmate venation, occasionally reduced to a single vein, occasionally resinous; stipules absent or present, occasionally spinose. Inflorescences determinate or indeterminate, terminal or axillary, occasionally reduced to a solitary flower. Flowers usually bisexual, radial or bilateral (due to perianth, androecium and gynoecium), the perianth differentiated (with calyx and corolla), in 2 whorls. *Sepals usually 5, distinct or slightly connate, imbricate or valvate, green and all alike, or petaloid and dimorphic, with the 3 outer ones larger than the 2 inner. Petals usually 5, imbricate or convolute, rarely valvate, uniform in size, then distinct, sometimes clawed, or strongly dimorphic, then with the upper 3 clawed, usually connate by their claws, the 2 lower ones smaller, thickened, sessile, modified into lipid-secreting glands.* Stamens 4–15, in 1–3 whorls, distinct to slightly connate (by filaments); filaments free, sometimes adnate to petals, thick to slender, sometimes each with glandular appendage at base, occasionally with staminodia; anthers opening by longitudinal slits or pores; pollen grains usually tricolpate, tricolporate, triporate, or polyporate. *Carpels 2–5, but sometimes pseudomonomerous (and then with one fertile and one sterile carpel), connate, with a single style; ovary superior, often ridged or winged*, occasionally with gynophore, with axile or ± apical placentation; stigma usually punctate, capitate, or lobed. Ovules anatropous to orthotropous, with 2 integuments, with a thick nucellus, 1 to several in each locule. Nectary a disk at base of ovary, or absent and the pollinators rewarded by oil produced by modified petals. *Fruit a septicidal or loculicidal capsule, schizocarp, sometimes spiny or winged, or a nut with retrorsely barbed spines*, rarely a berry or drupe; seeds sometimes arillate; endosperm present or absent.

CELASTRALES LINK

Trees, shrubs, lianas (and climbing by means of twining stems, hook-like branches, or downward-pointing bud scales, or adventitious roots), or herbs, sometimes rhizomatous, rarely epiphytic; crystals rhomboidal or druses; parenchymatous tissues often with laticifers (associated with vascular bundles and/or in mesophyll), occasionally with tannin cells or mucilage cells in epidermis; stems often with laticifers, occasionally forming thorns. Hairs simple or branched to stellate. *Leaves alternate and spiral or 2-ranked, opposite and decussate, or rarely whorled, simple*, rarely unifoliolate (and then articulate, with a stipule-like structure and articulation at base of pulvinate petiolule), *entire to variously toothed, with usually pinnate venation*, but occasionally palmately veined (with converging major veins), rarely with domatia; stipules usually present, often minute, but absent in *Parnassia* and *Lepuropetalon*. *Inflorescences determinate* or rarely indeterminate, terminal or axillary, rarely epiphyllous, sometimes reduced to a solitary flower. Flowers bisexual or unisexual (and plants then monoecious, dioecious, or polygamous), *usually radial*, occasionally with a short hypanthium, with perianth differentiated (of calyx and corolla), in 2 whorls, occasionally with an androgynophore. *Sepals usually 4 or 5, distinct* to slightly connate, imbricate, valvate, or with open aestivation. *Petals usually 4 or 5, distinct* to occasionally ± connate, imbricate to occasionally valvate. *Stamens usually 3–5*, in a single whorl, distinct to connate (by filaments), occasionally alternating with staminodia (and these with stalked pseudonectaria in *Parnassia*), and staminodia also sometimes present in carpellate flowers; filaments free, borne on receptacle or hypanthium, ± slender, short to elongate; anthers opening by longitudinal to transverse slits; pollen grains usually tricolporate to triporate, occasionally in tetrads or polyads. *Carpels usually 2–5, connate*, with a single unbranched to strongly branched style, sometimes very short or even absent, or occasionally with separate styles (one per carpel); ovary superior or half-inferior, with usually axile (rarely intruded-parietal, parietal or basal) placentation; stigma or stigmas usually punctate, capitate, elongate or lobate. Ovules anatropous, with usually 2 integuments, with thick to thin nucellus, 1 to numerous per locule. **Nectary an** *intrastaminal to extrastaminal,* **variously developed disk,** occasionally lobate, sometimes absent, but nectar produced by staminodia in *Parnassia*. *Fruit a loculicidal and/or septicidal capsule*, schizocarp, *drupe*, *berry* (with flesh derived from mesocarp or mucilaginous arils), or samara, *often ± lobed;* seeds often arillate, sometimes winged; endosperm present or absent.

OXALIDALES BERCHT. & J. PRESL

Trees, shrubs, lianas (sometimes twining), or herbs, sometimes rhizomatous and rarely carnivorous; crystals solitary or druses; parenchymatous tissues occasionally with mucilage canals or cells, sometimes very sour because of high levels of soluble calcium oxalate. Hairs simple, sometimes glandular, 2-branched, occasionally stellate. Leaves alternate and spiral or 2-ranked, opposite and decussate, rarely whorled, distributed along stems or in a basal rosette, simple, *palmately or pinnately compound, or reduced and trifoliolate or unifoliolate* (and rarely the petiole succulent, rarely the leaflets lost and the petiole expanded and photosynthetic), entire to variously toothed, with pinnate or palmate venation (the latter with major veins diverging or converging), *often with prominent pulvini* and sometimes showing sleep movements, occasionally with domatia, in *Cephalotus* some leaves highly modified, pitcher-shaped, insectivorous, with a protective lid and strongly ribbed rim; *stipules present*, and then sometimes interpetiolar, or absent, occasionally also with stipule-like structures associated with the leaflets. *Inflorescences determinate* or occasionally indeterminate, axillary or terminal, occasionally cauliflorous, sometimes reduced to a solitary flower. *Flowers* bisexual or unisexual (and plants then dioecious or polygamous), *radial*, occasionally with a hypanthium, the perianth differentiated (of calyx and corolla, but petals occasionally lost, and sepals then often referred to as tepals), in 1 or 2 whorls, *sometimes heterostylous*. Sepals (3-) 4–6 (-10), *distinct to slightly connate*, rarely strongly connate, imbricate or valvate, occasionally petaloid (and then petals absent), rarely calyptrate. Petals (3-) 4–5 (-10), distinct to slightly connate, imbricate convolute or valvate, occasionally absent, occasionally accrescent or toothed to lobed. Stamens 4 to numerous (but ancestrally twice the number of petals), usually in 1 or 2 whorls, distinct to connate (by filaments), sometimes alternating with staminodia, or with staminodia in carpellate flowers; filaments free, from receptacle or hypanthium, short to elongate, *sometimes of unequal lengths*, sometimes incurved in bud; anthers opening by longitudinal slits or apical pores, the connective sometimes forming an appendage; pollen grains usually di- or tricolpate or di- or tricolporate. Carpels usually 2–6, connate to distinct, with a single unbranched to ± branched style, or with as many styles as carpels; ovary superior to ± inferior, with axile, basal or lateral placentation, sometimes lobed; stigma or stigmas capitate, punctate or elongate/decurrent. Ovules anatropous to hemitropous, with 2 integuments, with a thick or thin nucellus, usually 2 to numerous in each locule or carpel, or occasionally only 1 or a few at base of syncarpus ovary with single locule. Nectary a disk, sometimes lobed or toothed, a ring of glands, disk-like inner surface of hypanthium, or nectar produced by base of filaments. Fruit a variously dehiscent capsule, berry, drupe, indehiscent pod, samara, sometimes lobed or angled, or an aggregate of follicles (and these sometimes with hard endocarp separating from softer exocarp); seeds sometimes arillate, sometimes with fleshy coat, sometimes with the outer part of testa elastically turning inside out and ejecting the seed, occasionally winged or hairy; endosperm present or absent.

MALPIGHIALES JUSS. EX BERCHT. & J. PRESL

Trees, shrubs, lianas/vines (scrambling, twining, or with various kinds of tendrils), or herbs, sometimes succulent and cactus-like, occasionally parasitic (with mycelium-like vegetative body, branching within the host), or showing adaptations to aquatic (and then often ± thalloid) or mangrove (and then showing vivipary, root system specializations) habitats, rarely with glucosinolates (*Drypetes* and *Putranjiva*), often cyanogenic, sometimes with a cyclopentenoid ring system; crystals various, sometimes also with silica bodies; parenchymatous tissues sometimes with laticifers (white), secretory canals (branched or unbranched, or cavities (sap clear, brightly colored to nearly white, often pellucid or black dots containing resinous compounds, xanthones), mucilage canals, or various kinds of secretory cells, rarely with sclereids; stems occasionally with internal phloem or cortical and/or medullary vascular bundles, or showing anomalous secondary growth, sometimes dimorphic, rarely expanded and photosynthetic (phylloclades); vessel elements sometimes with vestured pits. Hairs various. *Leaves* alternate and spiral or 2-ranked, opposite and decussate or adjacent leaf pairs diverging from each other at an angle less than or greater than 90° (bijugate), or whorled, borne along the stems, distally clustered, or in basal rosettes, *simple*, occasionally lobed, rarely dissected, sometimes palmately compound or trifoliolate, rarely pinnately compound, entire to **variously toothed (and teeth, when present, often with vein entering tooth distally expanded, and apex thus congested, variously theoid, violoid, salicoid)**, with pinnate or palmate venation (the latter with converging or diverging major veins), sometimes with pellucid or black dots, rarely with pellucid lines and dots, sometimes with extrafloral nectar glands, occasionally reduced, with various ptyxis, occasionally folded longitudinally, and with 2 lines, one on each side of and parallel with midvein, evident when expanded; stipules present (and variously positioned, occasionally with colleters) or absent. Inflorescences determinate or indeterminate, terminal or axillary, occasionally reduced to a solitary flower, occasionally comprised of pseudanthial units (cyathia), rarely epiphyllous. Flowers bisexual or unisexual (and plants then monoecious, dioecious, or polygamous), *radial* or less commonly bilateral (due to form of the perianth or perianth, androecium and gynoecium), occasionally with a cup-shaped to tubular hypanthium, with perianth differentiated (of calyx and corolla), in usually 2 whorls, but occasionally ± spirally arranged (and then sepals and petals only poorly differentiated), or sometimes the petals absent (lost) and then sepals referred to as tepals, and in a single whorl, rarely perianth entirely lost, occasionally with a short to elongate gynophore or androgynophore and with an inconspicuous to showy, membranaceous, scaly, and/or filamentous corona, sometimes associated with conspicuous bracts. *Sepals (or tepals) 2–8* (rarely numerous), *but most commonly 4 or 5, distinct or slightly* (rarely strongly) *connate*, valvate, imbricate, or with open aestivation, occasionally with 2 conspicuous, abaxial, oil-producing glands per sepal, green or occasionally petaloid, occasionally ± vestigial or absent. *Petals 2–8* (rarely numerous), *but most commonly 4 or 5*, sometimes absent, *distinct to slightly connate* (rarely strongly connate), valvate, induplicate-valvate, imbricate or convolute, sometimes clawed or asymmetrical, sometimes hairy or with fringed or toothed margin, occasionally with a ± basal appendage or crumpled in the bud, rarely bilobed, all the same size, or 1 or 2 larger than the others, or the lowermost spurred, occasionally individually enclosing either a single stamen or a group of stamens in bud, occasionally ± papilionaceous, (with 2 lower petals forming a ± saccate keel, the upper petal a saccate/spurred standard, and 2 lateral, flat wings), rarely deliquescent. Stamens 1 to numerous, due to reduction (and then often opposite the petals) or secondary increase (and developmentally centripetal) from an ancestral 10, in 1 or 2 (rarely more numerous) whorls, distinct to connate (by filaments), sometimes fascicled, rarely inflexed in bud, occasionally with staminodia; filaments free, arising from receptacle, rim of the hypanthium, at the base of its inner surface, on an androgynophore, rarely slightly adnate to base of petals, also rarely adnate to gynoecium, slender to stout, very short to elongate; anthers opening by longitudinal slits, or rarely by terminal pores, rarely multiloculate and opening by a longitudinal flap, the connective occasionally forming a triangular, apical appendage, or quite broad (thus anther sacs well separated); pollen grains tri- to polycolpate, tri- to polycolporate, tri- to pantoporate, and the colpi rarely connate, or inaperturate, rarely released in diads. *Carpels 2–5* (occasionally to numerous), occasionally pseudomonomerous (so appearing to be a single carpel), *connate*, *terete to strongly lobed*, with a single, terminal to gynobasic, very short to elongate, unbranched to distally branched style, or with as many styles as carpels, and each style often bifid or several times divided, occasionally distally enlarged (and pollen collecting, i.e., with plunger pollination mechanism); *ovary superior* to occasionally inferior, *with axile, parietal, or intruded-parietal*, or occasionally ± basal or apical *placentation (and when axile, often with ± lobed ovary)*; stigma or stigmas extremely variable. Ovules anatropous to campylotropous or orthotropous, with 1 or 2 integuments, with a thick or thin nucellus, 1 to numerous per locule or on each placenta. Nectary a variously positioned disk or ring of glands, or a few glands or even a single gland, inner surface of hypanthium producing nectar or with annular nectary at apex of hypanthium, with glandular or spur-like structure on the stamens, with nectiferous staminodia, or absent, and then sometimes using pollen, oils, or terpenoid resins as floral rewards. Fruit a variously dehiscent capsule, schizocarp (with samara-like units, drupaceous units [and then contrasting with expanded and colorful receptacle], or the units elastically dehiscent from a persistent central column, and opening, so fruit intermediate between a schizocarp and capsule), drupe (sometimes with ridged pit, or rarely separate pits, occasionally the pit with resin cavities, and fruit rarely associated with a scaly cupule), berry, nut-like, and fruit often with a columella; seeds sometimes winged, arillate, or with a tuft of hairs, and the coat occasionally mucilaginous, rarely germinating while fruit still attached (some Rhizophoraceae); endosperm present (occasionally starchy) to absent.

FABALES BROMHEAD

Trees, shrubs, lianas/vines (climbing by tendrils or twining), or herbs, sometimes rhizomatous, rarely mycoparasitic; crystals prismatic, styloids, or druses; parenchymatous tissues occasionally with secretory canals or cavities; stems occasionally with anomalous secondary growth; roots often with N-fixing nodules (harboring *Rhizobium* and relatives). Hairs various. *Leaves alternate and spiral or 2-ranked*, rarely opposite and decussate, *simple, pinnately (or twice pinnately) compound, to palmately compound, trifoliolate, or unifoliolate*, occasionally the leaflets lost and the axis expanded and photosynthetic, rarely leaflets modified into tendrils, entire to serrate, *with pinnate* (rarely palmate) *venation*, rarely with pellucid dots, *often with well-developed pulvinus* (of leaf and individual leaflets), often showing sleep movements; extrafloral nectaries sometimes present; stipules present, inconspicuous to leaf-like, occasionally forming spines, or absent; sometimes with stipule-like structures also associated with the leaflets. Inflorescences indeterminate or occasionally determinate, terminal or axillary, occasionally reduced to a solitary flower. *Flowers usually bisexual, radial or bilateral* (due to form of perianth, androecium and gynoecium), *sometimes with a short, cup-shaped to flat and strongly lobed hypanthium*, with perianth differentiated (of calyx and corolla), in 2 whorls. Sepals usually 5, distinct to connate, imbricate or valvate, all alike or with the 2 lateral ones larger than the others and petaloid. **Petals** usually 3 or 5, distinct or connate, imbricate, contorted or valvate, **often clawed,** *when 5, all alike or the uppermost petal differentiated in size, shape, or coloration, the 2 lower petals often connate or sticking together and forming a keel, or widely flaring, and when only 3, with 2 upper and 1 lower, the latter keeled and often appendaged. Stamens* 1 to numerous, *but commonly 8 or 10*, and ancestrally 10 and in 2 whorls, distinct to connate (by their filaments, and then monadelphous or diadelphous), occasionally with staminodia; filaments free or sometimes adnate to the petals, ± slender; anthers opening by longitudinal slits or apical pores; pollen grains tricolporate, tricolpate, triporate, or polycolporate, occasionally in tetrads or polyads. **Carpels** usually *1–5*, **distinct or nearly so** to connate, occasionally pseudomonomerous, with a single, sometimes apically branched style, or each carpel with a separate, gynoterminal or gynobasic style, sometimes with a short gynophore; *ovary superior*, with lateral, basal or axile placentation; stigma punctate, capitate or truncate to elongate. Ovules anatropous to campylotropous, with usually two integuments, with a thick nucellus, 1 to numerous in each carpel or locule. Nectar produced by inner surface of hypanthium or a disk, or nectary absent. Fruit a loculicidal capsule, legume, samara, loment, follicle, indehiscent pod, achene, drupe, nut, berry, or an aggregate of nutlets; seeds sometimes arillate, sometimes with a U-shaped line, stiff hairs, or winged, *with ± large, green embryo*; endosperm present or absent.

ROSALES BERCHT. & J. PRESL

Trees, shrubs, lianas/vines (scrambling, with downward pointing branches, curved thorns, prickles, or with tendrils), or herbs, often rhizomatous; crystals various, *often also with cystoliths*; *parenchymatous tissues sometimes with laticifers (milky sap), or laticifers sometimes reduced (sap clear, mucilaginous) and/or restricted to bark*, sometimes with mucilage cavities; stems occasionally producing thorns; roots sometimes with N-fixing nodules (with *Frankia*), rarely contractile. Hairs simple, sometimes with mineralized cell walls, sometimes hooked, glandular, Y-shaped, stellate or of peltate scales, sometimes stinging or with prickles. Leaves alternate and spiral or 2-ranked, or opposite and decussate, simple, sometimes lobed, or palmately or pinnately compound, entire to variously toothed, *the teeth glandular or non-glandular (and then usually with single vein proceeding straight to apex)*, with pinnate or palmate venation (the latter with converging or diverging major veins), sometimes with domatia, *sometimes with cystoliths (and these globose to elongate)*, occasionally oblique or with extrafloral nectaries; *stipules present*, small to conspicuous, sometimes sheathing stem, distinct to variously fused, or absent. Inflorescences determinate or indeterminate, terminal or axillary, occasionally reduced to a solitary flower. Flowers bisexual or unisexual (and plants then monoecious, dioecious, or polygamous), *usually ± radial*, **often with a variously shaped,** *green to petaloid* **hypanthium,** *but lost in some* (i.e., some Ulmaceae, also Cannabaceae, Moraceae, Urticaceae and Barbeyaceae), with perianth differentiated (of calyx and corolla), in 2 whorls, *or also often with petals absent (an evolutionary loss; and then with tepals*, i.e., as in Ulmaceae, Cannabaceae, Moraceae and Urticaceae), sometimes closely associated (and eventually forming multiple fruits) and with various accessory tissues (fleshy inflorescence axis/axes, pedicels, perianth parts). **Sepals** (2-) *4–5* (-8), *distinct*, **valvate**, occasionally associated with an epicalyx. **Petals** *4–5* (-8), *distinct*, imbricate or contorted, *often ± concave or hooded*, **often ± clawed**, sometimes each petal "enclosing" a stamen, rarely with basal appendages, *often absent*. *Tepals* (0-) *4–5* (-10), distinct to connate, valvate or imbricate, sometimes becoming fleshy and associated with mature fruits. Stamens 1–10, in a single whorl, or occasionally 2 whorls, but sometimes ± numerous (with secondary increase; as in many Rosaceae), distinct; filaments free, arising from apex of hypanthium, and sometimes also slightly adnate to petal-bases, or arising from receptacle or slightly adnate to tepals (when hypanthium lost), *often opposite the petals or tepals*; slender, short to elongate, ± straight to sometimes incurved in bud (and then often elastically reflexed at anthesis); anthers opening by longitudinal slits; pollen grains tricolporate or di- to multiporate. Carpels 1 to numerous, sometimes pseudomonomerous, distinct to connate, with as many styles as carpels or with a single, variably branched (rarely unbranched) style; ovary superior to inferior, *with axile placentation and several locules, or reduced, with a unilocular gynoecium and apical or basal placentation* (when carpels connate), *or basal, lateral or apical placentation* (when ovaries distinct, or flower with only a single or seemingly single carpel), the gynoecium sometimes adnate to hypanthium; stigmas punctate, truncate, capitate or ± elongate. Ovules anatropous to campylotropous, with 2 integuments, with a thick or occasionally thin nucellus, 1 to few per carpel. *Nectary a disk, variably positioned on inner surface of hypanthium*, rarely nectar produced by petal appendages, *or often absent. Fruit a pome, drupe (and either dehiscent or indehiscent, with 1 to several pits)*, samaroid schizocarp, septicidal or septifragal *capsule*, samara, *or achene (and these sometimes with accrescent, dry to fleshy perianth or hypanthium), or sometimes an aggregate of follicles, achenes, drupelets* (which may be exposed, or enclosed by sometimes fleshy hypanthium), *or multiple fruits (with achenes or drupes associated with fleshy perianth and/or inflorescence axis);* **endosperm** present, and **scanty, or absent.**

CUCURBITALES JUSS. EX BERCHT. & J. PRESL

Trees, shrubs, herbs, or vines (climbing by tendrils), sometimes rhizomatous, occasionally tuberous; crystals solitary or in druses; parenchymatous tissues often with idioblasts or canals producing oxidized triterpenoid compounds (cucurbitacins); stems sometimes succulent, rarely dimorphic (orthotropic and plagiotropic), sometimes with bicollateral vascular bundles, ± 2 concentric rings, or with cortical and medullary bundles, **and wood (when present) with broad, multiseriate rays;** roots occasionally with N-fixing nodules (harboring *Frankia*). Hairs simple, glandular or eglandular, branched, stellate, or occasionally peltate scales, *often with calcified walls and a cystolith at the base. Leaves alternate, spiral* or 2-ranked, or opposite and decussate, rarely whorled, *simple, entire to variously toothed (and often with several veins entering tooth and ending at an expanded, ± translucent glandular apex), sometimes lobed or dissected*, rarely palmately or pinnately compound, **with palmate** or occasionally pinnate **venation** (the former with converging or diverging major veins), occasionally ± succulent, rarely strongly anisophyllous (i.e., with pairs of scale leaves alternating on the upper surface of the plagiotropic stem, and corresponding foliage leaves on the lower surface), symmetric to asymmetric; stipules present or absent. Inflorescences determinate or occasionally indeterminate, terminal or axillary, occasionally reduced to a solitary flower. Flowers bisexual or *more commonly unisexual* (and plants then monoecious, dioecious, or polygamous), *usually radial, sometimes with a short to elongate, variously shaped hypanthium*, with perianth differentiated (of a calyx and corolla) or occasionally undifferentiated (of tepals, a reversal), of 1 or 2 whorls. Tepals usually 2–5 (-10), usually distinct, imbricate or valvate, petaloid. Sepals usually 4 or 5, distinct to connate, imbricate, valvate or with open aestivation, rarely petaloid. Petals usually 4 or 5, distinct to connate, imbricate, contorted, valvate or with open aestivation, occasionally fringed, lobed or dissected, rarely absent. Stamens 3 to numerous, in usually 1 or 2 whorls, distinct to connate (merely by filaments or by filaments and anthers), occasionally inflexed in bud, rarely with staminodia, and staminodia sometimes present in carpellate flowers; filaments free, variously inserted on hypanthium or the receptacle, very short to elongate, occasionally unequal in length; anthers opening by longitudinal slits or apical pores, the connective occasionally extended, the locules sometimes bent, folded or variously convoluted; pollen grains usually tricolporate, triporate or with more numerous pores or furrows. *Carpels usually 2–5, connate*, gynoecium **with as many styles as carpels,** *or with a single, unbranched or variably branched style,* and style-branches sometimes more numerous than the number of carpels; *ovary half-inferior to inferior*, but rarely superior, *with parietal, intruded-parietal, axile* or apical *placentation*, sometimes with prominent wings or horns; stigma or stigmas capitate, ± lobate or expanded, curved to forming a twisted band, or extremely elongate. Ovules anatropous, with 1 or 2 integuments, with a thick nucellus, 1 or 2 per locule, *or more commonly numerous on each placenta or in each locule*. Nectary a ring of glands, variously developed on hypanthium, a disk, or absent. *Fruit a berry, the rind often leathery or hard, or fleshy to try, variously dehiscing, sometimes winged capsule*, drupe, samara, schizocarp (of nutlets, with fleshy, accrescent petals); seeds sometimes flattened, occasionally with fleshy coat or winged, sometimes with a ring of elongated testa cells that rupture, forming a cap (operculate); endosperm present (and scanty) or absent.

FAGALES ENGLER

Trees or shrubs, sometimes rhizomatous; crystals solitary or druses; stems occasionally grooved; **roots ectomycorrhizal**, sometimes with nodules (containing N-fixing *Frankia*). *Hairs simple, T-shaped, branched or stellate, often also gland-headed or peltate (often producing aromatic compounds)*. Leaves usually alternate and spiral or 2-ranked, or whorled, rarely opposite and decussate, simple or pinnately compound (rarely these reduced, trifoliolate or unifoliolate), entire to serrate or doubly serrate, sometimes lobed, *with pinnate venation* (and secondary veins forming loops or running into the serrations), occasionally with domatia, sometimes very reduced (and then scale-like, in ± connate whorls, forming a toothed sheath at each node); *stipules present, usually quickly deciduous, or absent.* **Inflorescences** *determinate (but appearing indeterminate) or indeterminate*, terminal or axillary, **of solitary or paniculate, erect or pendulous, spikes or spike-like axes, catkins or head-like clusters;** *the individual flowers in cymose clusters of 2 or 3 flowers, or only a single flower associated with each bract; inflorescence axes sometimes modified, forming a scaly to spiny cupule; inflorescence bracts and/or bracteoles sometimes conspicuous, variously modified, free or ± fused to each other and/or with flowers.* **Flowers** *usually unisexual* (and plants then monoecious, dioecious, or polygamous, *with staminate and carpellate flowers in the same inflorescence or in separate inflorescences)* or rarely bisexual, ± *radial*, **inconspicuous, with perianth reduced and undifferentiated (of tepals)**, *in 1 whorl, or absent.* **Tepals** 1–6, distinct to connate, ± imbricate, sometimes lobed, **very reduced** *or absent.* Stamens 1 to numerous, often in a single whorl, distinct to slightly connate (by filaments), occasionally divided, sometimes with staminodia in carpellate flowers; filaments free, short to elongate, ± slender; anthers opening by longitudinal slits, the connective occasionally extended; pollen grains tricolpate, tricolporate, triporoporate to polyporoporate (or polyporate). *Carpels 2 or 3* (-12), *connate, with a single, strongly branched style, or as many styles as carpels;* **ovary inferior** *(or nude)*, sometimes with intercalary meristematic activity around and/or beneath the gynoecium, forming a cuplike structure, *with axile placentation, but sometimes partitions incomplete and ovule or ovules borne at apex of incomplete septum*, and occasionally with false partitions, or with basal placentation; stigmas peltate to expanded or elongate/decurrent. **Ovules** anatropous to orthotropous, with 1 or 2 integuments, with a thick nucellus, **2 in each locule, but all but one aborting**, sometimes only one locule without ovules, *or sometimes only 1 ovule per gynoecium*, **poorly developed at pollination, fertilization delayed,** *and often with pollen tube entering ovule via chalazal end.* **Nectary absent. Fruit indehiscent, one-seeded,** *a drupe, and then with waxy or fleshy papillae, or with adnate bracts/bracteoles and these forming an outer husk, ± splitting to release the bony pit, or with no accessory structures, a round to 3-angled or occasionally ± winged nut, and then closely associated with a spiny to scaly, 2–8-valved or non-valved cupule, an achene, samara, or nut, and then associated with variously fused and developed bract-bracteole complex, a samara associated with (and initially surrounded by) 2 woody bracteoles* *or without accessory structures, or an achene associated with wing-like bracts and/or bracteoles*; seeds with embryo small to large, the cotyledons occasionally corrugated; endosperm present or absent.

MALVIDAE W.S. JUDD, D.E. SOLTIS & P.S. SOLTIS

Trees, shrubs, lianas, vines, or herbs, sometimes rhizomatous, rarely parasitic; crystals various; stems with or without internal phloem; parenchymatous tissues sometimes with schizogenous to lysigenous cavities containing various aromatic compounds, occasionally with laticifers, or resin or mucilage canals; stems rarely succulent or with anomalous secondary growth; vessel elements with or without vestured pits. Hairs various. Leaves alternate and spiral or 2-ranked, or opposite and decussate, or whorled, distributed along stem or in a basal rosette, simple to variously compound, entire to variously toothed, sometimes lobed, with pinnate or palmate venation (the latter with converging or diverging major veins), occasionally with domatia or pellucid dots; stipules various, present or absent. Inflorescences determinate, indeterminate, or mixed, usually terminal or axillary, sometimes reduced to a solitary flower. Flowers bisexual or unisexual (and plants then monoecious, dioecious, or polygamous), radial or bilateral (due to form of perianth and/or androecium), with or without a hypanthium, the perianth differentiated (with calyx and corolla) and usually in 2 whorls, occasionally reduced to a single whorl (and then the parts frequently called tepals) or absent, sometimes with a gynophore or androgynophore. Sepals (often called tepals when petals lost) 3–8 (but 4 or 5 the most common), distinct to connate, occasionally calyptrate, with various aestivation. Petals 1–8 (but 4 or 5 the most common), sometimes absent, distinct to less commonly connate, with various aestivation. Stamens 1 to numerous (but ancestrally in 2 whorls, each usually of 4 or 5 members), distinct to connate (by filaments) and sometimes fasciculate, occasionally with staminodia; filaments free, usually attached to receptacle or hypanthium, but occasionally adnate to petals; anthers usually opening by longitudinal slits, occasionally by pores; pollen grains usually in monads, usually tricolpate, tricolporate, or modifications thereof, e.g., triporate to pantoporate. Carpels 2 to numerous, occasionally pseudomonomerous, connate (but occasionally only by styles, and separating as they mature), with as many styles as carpels or a single style, either apically branched or unbranched; ovary superior or inferior, with variable placentation, sometimes lobed; stigma various. Ovules anatropous to orthotropous, with usually two integuments, with a usually thick nucellus, 1 to numerous per carpel, or reduced to 1 per gynoecium. Nectaries variable or absent. Fruit a variously dehiscent capsule (sometimes with a replum), silique, berry, drupe, dehiscent drupe, nut, indehiscent pod, samara, schizocarp (with dry segments that often open to release seeds, or of nutlets, achene-like, samara-like, or drupe-like segments), or an aggregate of follicles, drupes, or samaras (due to secondary separation of ovaries that are initially fused or at least fused by styles); seeds sometimes arillate; endosperm present to absent.

Composition: Myrtales, Geraniales, Crossosomatales, Picramniales, Sapindales, Huerteales, Brassicales, Malvales.

GERANIALES JUSS. EX BERCHT. & J. PRESL

Herbs, shrubs or small trees; crystals various; stems occasionally succulent. Hairs simple, often gland-headed, sometimes branched, occasionally stellate. *Leaves* alternate and spiral or opposite and decussate, distributed along the stem or in a basal rosette, *simple, often palmately or pinnately lobed or dissected, or trifoliolate, or palmately or pinnately compound*, entire to ± serrate, and teeth often glandular, with pinnate to ± palmate venation (with diverging major veins), the leaf bases sometimes sheathing; stipules present or absent. *Inflorescences determinate* or occasionally indeterminate, terminal or axillary, occasionally reduced to a solitary flower. *Flowers bisexual* or rarely unisexual (and plants then monoecious), radial or bilateral (due to form of perianth and often also androecium), *the perianth differentiated (with calyx and corolla), the parts in 2 whorls*, occasionally resupinate. *Sepals usually 4 or 5, distinct* to somewhat connate, ± imbricate, sometimes one saccate or prolonged backward into a nectar-producing spur. *Petals usually 4 or 5*, rarely absent, *distinct, imbricate and often contorted*, occasionally valvate. *Stamens 4–15*, in 1 or 2 whorls, *distinct* to slightly connate (by filaments), occasionally with staminodia; filaments free, very short to long and slender, sometimes of unequal lengths; anthers opening by longitudinal slits; pollen grains tricolpate or tricolporate, occasionally polyporate or inaperturate. **Carpels** *3–5, connate, with a single, short to elongated style*; *ovary superior* to half-inferior, ± **lobed** *and often with a prominent, elongate, persistent, terminal sterile column, with usually axile placentation*; stigmas distinct, ± elongate, or stigma truncate, punctate, or ± lobed. Ovules anatropous to campylotropous, with 2 integuments, with a thick nucellus, 1 to numerous in each locule. **Nectaries extrastaminal**, *often lobate disk, ring of glands*, a single sepal-associated gland, paired staminal appendages, or absent. *Fruit a schizocarp with 5 one-seeded segments that separate elastically from the persistent central column, and often opening to release the seed, or a variously dehiscent capsule*; seeds occasionally arillate or hairy; endosperm present to absent.

MYRTALES JUSS. EX BERCHT. & J. PRESL

Trees, shrubs, lianas, vines, or herbs, sometimes rhizomatous; crystals various, but raphides only in Onagraceae; **stems with internal phloem, with stratified secondary phloem; vessel elements with vestured pits.** Hairs various. **Leaves opposite and decussate**, alternate and spiral, or whorled, simple, entire to variously toothed, occasionally lobed, with pinnate or palmate venation (the latter with converging major veins), sometimes with domatia, sometimes with pellucid dots; *stipules small and often dissected into finger-like projections, or absent.* Inflorescences determinate or indeterminate, terminal or axillary, occasionally reduced to a solitary flower. Flowers bisexual or unisexual (and plants then monoecious, dioecious, or polygamous), radial or bilateral (due to form of perianth and/or androecium and gynoecium), **with variously developed hypanthium**, but sometimes lost (Vochysiaceae), with perianth differentiated (of calyx and corolla) and in 2 whorls. Sepals (2-) 3–8 (but 4 or 5 the most common numbers), distinct to connate, sometimes forming a circumscissile or irregularly rupturing cap, imbricate to valvate, sometimes one larger than the others, sometimes rather thick, or associated with external projections. Petals 1–8 (but 4–6 the most common), sometimes absent, distinct, less commonly connate, often ± clawed, sometimes forming a cap and adnate to the sepals, imbricate, valvate, or convolute, occasionally wrinkled. Stamens 1 to numerous, ancestrally in 2 whorls, but with secondary increase and decrease, distinct to less commonly connate (by filaments), and sometimes fasciculate, usually **inflexed in bud** (but occasionally straight), occasionally dimorphic, occasionally with staminodia; filaments usually free and arising from apex of hypanthium, but sometimes attached well to slightly below its summit, usually ± slender, sometimes geniculate, sometimes twisted at anthesis, bringing anthers to one side of flower, sometimes unequal in length; anthers opening by longitudinal slits or apical pores, the connective sometimes thickened or appendaged, sometimes with an apical secretory cavity, or a resin-secreting gland; pollen grains usually tricolporate, sometimes with the furrows fused, **often with 3 alternating pore-less furrows**, less commonly triporate (or biporate), sometimes associated with viscin threads. Carpels 2 to numerous, connate, *with a single style*; ovary superior to inferior, usually with axile placentation, less commonly apical or parietal with intruded placentae; *stigma* various, but *usually not or only obscurely lobed*, but less commonly clearly lobed to branched (Onagraceae). Ovules usually anatropous, with 2 integuments, usually with a thick nucellus, 1 to numerous in each locule (or 1 to a few, pendulous or lateral in the single locule). **Nectary** or nectaries at or near base of hypanthium, a disk atop ovary, **lining inner surface of hypanthium**, in a modified or spurred sepal, or absent (and then pollen or resin usually employed as rewards). Fruit a variously dehiscent capsule, berry, drupe, or nut; seed sometimes winged or with tuft of hairs, with seed coat sometimes mucilaginous; endosperm usually scanty or absent.

CROSSOSOMATALES TAKHT. EX REVEAL

Trees or shrubs; crystals solitary or druses. Hairs simple, sometimes glandular or T-shaped. Leaves alternate and spiral or 2-ranked, or opposite and decussate, simple, trifoliolate or pinnately compound, entire or variously toothed, sometimes lobed, *with pinnate* or rarely palmate *venation*; stipules present, minute to well developed, or absent, and stipule-like structures also associated with leaflets. Inflorescences determinate or indeterminate, terminal or axillary, sometimes reduced and fasciculate or a solitary flower. *Flowers* bisexual or occasionally unisexual (and plants then dioecious), *radial*, **often with a short to well developed and cup-shaped to tubular hypanthium**, with perianth differentiated (of calyx and corolla, but petals sometimes lost, and sepals then called tepals), in usually 1 or 2 whorls. *Sepals 4 or 5* (rarely to 10), *distinct* to slightly connate, imbricate, these sometimes differentiated, with the inner ones petaloid, or similar sized and all petaloid. *Petals 4 or 5 (-6), distinct*, occasionally clawed, imbricate, sometimes absent (lost). *Stamens 4–10, or numerous (by secondary increase), in 1–3 whorls, distinct*; filaments free, borne on receptacle or *more characteristically on hypanthium*, very short to elongate, thick to slender, and staminodia present in carpellate flowers; anthers opening by longitudinal slits, the connective occasionally appendaged; pollen grains ± tricolporate. *Carpels 1–5 (-9), distinct to connate, each carpel with its own style, or (when connate) with separate styles or a single, essentially absent or short to elongate style, and the styles (of free or connate carpels) sometimes apically fused and free proximally*; ovary superior to partly inferior, with lateral, intruded-parietal or axile placentation, sometimes stipitate; stigma capitate, punctate or lobate. Ovules anatropous or campylotropous, with 2 integuments, with a thick nucellus, 1 to numerous per carpel/locule/placenta. **Nectary** *an intrastaminal, often lobate disk or of separate, grooved glands,* **on inner surface of hypanthium**, or absent. Fruit an aggregate of follicles, or a berry, loculicidal capsule, membranous inflated-capsule (with carpels often apically separating) or indehiscent pod; **seeds with distinctive coat in which the cell walls of the many-layered testa are ± all lignified and thickened, often arillate**; endosperm present to absent.

PICRAMNIALES DOWELD

Composition: Picramniaceae.

SAPINDALES JUSS. EX BERCHT. & J. PRESL

Trees, shrubs, lianas (these occasionally with tendrils), or rarely herbs, sometimes rhizomatous, occasionally with thorns, spines or prickles, *often with bitter, triterpenoid substances;* crystals various, sometimes also with irregular silica bodies; bark, wood, and/or parenchymatous tissues sometimes with resin canals (resin clear to cream-colored, sometimes drying black) in association with vascular bundles (in stems, leaves, fruits, etc.) or with secretory cavities (pellucid dots) containing aromatic, ethereal oils, but sometimes merely with scattered secretory cells; stems rarely with anomalous secondary growth (multiple vascular cylinders); bark roughened to smooth, peeling. Hairs various. **Leaves** *alternate and spiral or opposite and decussate*, rarely whorled, **pinnately** *or occasionally palmately* **compound**, *or reduced and trifoliolate or unifoliolate* (then usually with an articulation at base of blade), rarely apparently simple (perhaps by loss of articulation), entire to variously toothed, occasionally lobed, with pinnate, twice-pinnate or palmate venation (the latter with diverging major veins), sometimes with domatia, *sometimes with pellucid dots or resin canals, and the rachis occasionally winged*; *stipules usually absent*. Inflorescences usually determinate, terminal or axillary, occasionally reduced to a solitary flower. *Flowers* bisexual or unisexual (and plants then monoecious, dioecious, or polygamous), *radial* or rarely bilateral (due to form of perianth, androecium, and gynoecium), occasionally with a short to cup-shaped hypanthium, with perianth differentiated (with calyx and corolla), in 2 whorls, but rarely petals absent and then in a single whorl. *Sepals usually 3–5, distinct* to slightly connate (rarely strongly connate), imbricate, valvate or with open aestivation. *Petals usually 3–5, distinct* or sometimes slightly connate (rarely strongly connate), imbricate, convolute or valvate, sometimes clawed or with an adaxial appendage, rarely absent. Stamens 4–10, occasionally numerous, rarely reduced to a single stamen, usually in 1 or 2 whorls, distinct to connate (by filaments, and then sometimes with apical appendages), with staminodia in carpellate flowers; filaments free, usually borne on receptacle, but occasionally from hypanthium, or rarely adnate to the corolla, short to elongate; anthers opening by longitudinal slits; pollen grains di- to hexacolporate or triporate, rarely the colpi fused. *Carpels usually (1-) 2–6*, occasionally numerous, sometimes only one functional, *connate* (at least in ovary region) *and with a single, short to elongate, unbranched to branched style, or as many styles as carpels, but sometimes with ovaries distinct and* **carpels only ± coherent or postgenitally connate by their styles**, *occasionally the ovary syncarpous but* **with separate styles that are distally postgenitally connate**, and the styles terminal or somewhat gynobasic; *ovary usually superior, usually with axile*, occasionally apical or basal *placentation*; stigma or stigmas minute, punctate, discoid to ± capitate, strongly lobed, or decurrent/elongate. Ovules anatropous to orthotropous, usually with 2 integuments, with a thick nucellus, 1 to numerous in each locule or 1 in the single fertile carpel, sometimes lacking a funiculus and broadly attached to a protruding portion of the placenta (obturator). **Nectary well-developed, an intrastaminal to extrastaminal disk or a ring of glands,** rarely absent. Fruit a drupe (with one to several pits), dehiscent drupe, variously dehiscent capsule, samara, variously developed berry (with mesocarp homogeneous or distinctly heterogeneous with a hard to leathery rind, with or without distinct partitions, and flesh or pulp derived from the ovary wall, multicellular, juice-filled hairs, or an aril), schizocarp that splits into nutlets (which are sometimes pendulous from a central column), samara-like or drupe-like segments, or an aggregate of follicles, drupes or samaras; seeds sometimes winged or with a fleshy coat, sometimes arillate, the embryo straight to folded and a radicular-pocket sometimes present in the coat; endosperm present or absent.

HUERTEALES DOWELD

Trees or shrubs; crystals various; epidermis with mucilage cells. Hairs simple or branched. *Leaves alternate and spiral* or occasionally 2-ranked, *simple*, or pinnately compound to bifoliolate, variously toothed, with pinnate or occasionally palmate venation; stipules present or absent, and sometimes also with stipule-like structures associated with leaflets. Inflorescences ± determinate, terminal or axillary. *Flowers* bisexual or unisexual (and plants then monoecious, dioecious, or polygamous), *radial*, *often with a short hypanthium*, with perianth differentiated (of calyx and corolla, but these sometimes only slightly differentiated, and corolla sometimes lost), in 1 or 2 whorls. *Sepals 4–8*, distinct to connate, imbricate or valvate. *Petals usually 4–7*, distinct, imbricate or valvate, rarely absent. *Stamens 4–12*, in 1 or 2 whorls, distinct; filaments free, slender; anthers opening by longitudinal slits; pollen grains tricolporate. *Carpels 2–5*, connate, with a single, branched or unbranched style, or styles as many as carpels; ovary superior to partly inferior, with axile, free-central, basal or apical placentation; stigma ± capitate or slightly lobed. Ovules anatropous, with 2 integuments, 1–2 per locule, 6 on free-central axis, or 4 at apex of the single-loculed gynoecium, or numerous. *Nectary a disk, ring of glands* or absent. Fruit a berry, drupe or 1-seeded capsule; seeds occasionally arillate; endosperm present.

MALVALES JUSS. EX BERCHT. & J. PRESL

Trees, shrubs, lianas, or herbs, sometimes rhizomatous, rarely root parasites; crystals solitary or in druses, sometimes also with cystoliths; **parenchymatous tissues with mucilage (slime) canals, cavities or cells**, occasionally with sclereids, *or with resin canals in pith, wood, and bark*; **stems usually with stratified phloem with fibrous and soft layers, or phloem with a network of fibers, and rays broad to wedge-shaped** (and these conditions evidently lost in Neuradaceae), occasionally with internal phloem; **vessel elements with vestured pits** (except Neuradaceae). Hairs simple, 2-branched, *fascicled, sometimes glandular, stellate or peltate scales*, rarely with prickles. *Leaves alternate and spiral or 2-ranked*, or occasionally opposite and decussate, *simple, occasionally palmately lobed or dissected, or palmately compound*, rarely unifoliolate, entire to variously toothed, with pinnate or palmate venation (the latter with converging or diverging major veins), sometimes with domatia and/or extrafloral nectaries, rarely reduced, scale-like and with a single vein; stipules present, and often conspicuous, or absent. Inflorescences determinate, indeterminate, or mixed, terminal or axillary (rarely supra-axillary), rarely cauliflorus, sometimes reduced to a single flower. *Flowers* bisexual or unisexual (and plants then monoecious, rarely dioecious), *usually radial*, occasionally with cup-shaped or ± tubular, often petaloid hypanthium (and then often with glandular-scales or petal-like appendages, in Thymelaeaceae), with perianth differentiated (of calyx and corolla), in 2 whorls, but sometimes the petals absent (and sepals then petaloid), occasionally with a short to elongate androgynophore, sometimes associated with conspicuous bracts that form an epicalyx. *Sepals* (2-) *3–5* (-6), distinct to connate, imbricate, convolute, valvate, or with open aestivation, sometimes the outer 2 distinctly narrower than the inner 3. *Petals* (3-) *5*, occasionally absent, *distinct*, imbricate, convolute, or valvate, occasionally crumpled. *Stamens 5 to numerous*, ancestrally in 2 whorls, *but often with a secondary increase (initiated in a centrifugal sequence)*, distinct to slightly connate (by filaments), and then sometimes fasciculate, to strongly connate and forming a tube around the gynoecium (monadelphous), sometimes with staminodia, these occasionally elongate and alternating with stamens or groups of stamens; filaments free, arising from receptacle, adnate to base of the petals, or attached to hypanthium, very short to elongate, ± slender; anthers opening by longitudinal slits or apical pores, and sometimes apparently only half anthers (and opening by a single slit), sometimes with connective extended and forming a terminal appendage; pollen grains usually tri- to pentacolporate, triporate to pantoporate, rarely in tetrads. *Carpels 2 to numerous*, sometimes pseudomonomerous, connate, *with a single*, gynoterminal or rarely ± gynobasic, *distally branched to unbranched style*, or rarely as many styles as carpels; *ovary superior* or rarely ± inferior, occasionally covered by projections, rarely strongly lobed, *placentation axile, intruded parietal, parietal*, or apical; stigma or stigmas punctate, capitate or lobed. Ovules anatropous to campylotropous or orthotropous, with 2 integuments, with a thick nucellus, 1 to numerous in each locule or on each placenta. *Nectary a disk, sometimes lobate, or of densely packed, multicellular, glandular hairs on sepals* or sometimes on petals or androgynophore. Fruit a loculicidal capsule, schizocarp (of nutlets), nut, indehiscent pod, drupe, or berry, or aggregate of follicles (resulting from the carpels of the syncarpus gynoecium separating as they mature into fruits), the calyx sometimes persistent and wing-like; seeds sometimes with hairs, arillate or with a fleshy coat, occasionally winged; endosperm present, **often with cyclopropenoid fatty acids**.

BRASSICALES BROMHEAD

Trees, shrubs, lianas/vines (scrambling, twining petioles or pedicels), or herbs, **usually with glucosinolates (mustard oil glucosides, which react with myrosinase to release mustard oils) and myrosin cells**, the endoplasmic reticulum often with dilated cisternae; crystals various, solitary, or druses; parenchymatous tissues rarely with laticifers (milky sap); pith rarely with mucilage canals; vessel elements often with vestured pits; trunk rarely massive and branches rarely forming thorns; roots often with root hairs in vertical files or vertical files absent. Hairs diverse, simple, T-shaped or variously branched, glandular or non-glandular, stellate, or peltate. *Leaves usually alternate and spiral*, or rarely opposite and decussate, along the stems or in basal rosettes, simple, often dissected or lobed, or trifoliolate, palmately, or pinnately compound, entire to serrate, with pinnate or palmate venation (the latter with diverging major veins), rarely succulent, spiny, or quite reduced; stipules present or absent. **Inflorescences indeterminate** or occasionally determinate, terminal or axillary, occasionally reduced to a solitary flower, sometimes the floral bracts absent. *Flowers bisexual* or occasionally unisexual (and plants then monoecious, dioecious, or polygamous), radial or bilateral (due to form of perianth and androecium), rarely resupinate, without or occasionally with a ± cup-shaped hypanthium, the perianth differentiated (of calyx and corolla), in 2 whorls, but rarely the perianth absent (e.g., carpellate flowers of *Batis*) or merely the corolla absent (Gyrostemonaceae), *the receptacle often prolonged, forming a short to elongated gynophore or androgynophore*. *Sepals* usually (3-) *4 or 5* (-8), distinct or less commonly connate, imbricate or valvate, occasionally calyptrate or forming a nectar-spur, rarely petaloid. *Petals* usually (2-) *4 or 5* (-8), *distinct* or rarely connate, valvate, imbricate, convolute, or with open aestivation, *often with a claw and limb*, occasionally the limb dissected, rarely each petal basally differentiated into a saccate pouch. Stamens (2-) 6 to numerous, usually in1 or 2 whorls, all ± the same length or the 2 outer ones shorter than the 4 inner (tetradynamous), or declinate, *distinct* to rarely connate (by filaments), rarely with staminodia and also sometimes staminodia in carpellate flowers; *filaments free*, emerging from receptacle, androgynophore, or hypanthium, or rarely adnate to corolla, elongate to very short, ± slender; anthers opening by longitudinal slits, occasionally coiled at dehiscence, rarely the connective slightly extending beyond anther sacs; usually tricolporate, tricolpate or dicolpate. *Carpels* 2–5 (to numerous), *connate*, rarely apically open, with a single, gynoterminal or occasionally gynobasic, very short to elongate, unbranched or distally branched style, or rarely as many styles as carpels; *ovary superior*, variously shaped and occasionally strongly lobed, *with axile, or parietal/intruded parietal placentation, and then often with the placentas forming a thick rim (replum) around the fruit and sometimes connected by a false septum (a thin partition lacking vascular tissue) that divides the ovary into 2 chambers*; stigma truncate, capitate, lobate, fimbriate, or elongate. Ovules anatropous to campylotropous, with 2 or occasionally only 1 integument, with a thick or thin nucellus, 1 to numerous per locule or placenta. Nectary a disk, gland, or appendage at base of filaments, or absent. Fruit a berry, achene, schizocarp (of achenes, nutlets, or drupelets), drupe (with 4 pits or a single pit), samara, variously dehiscent capsule, *often with 2 valves breaking away from a replum* and sometimes additionally with a persistent septum (the fruit then a silique), these short to elongate, globose to flattened, and gynoecia of adjacent flowers rarely fused (forming drupaceous syncarp); seeds with or without broad to narrow invagination, occasionally arillate, winged, or hairy, or with a fleshy or mucilaginous coat; embryo straight, curved or folded; endosperm present to absent.

CHAPTER 11

SUPERASTERIDAE W.S. JUDD, D.E. SOLTIS & P.S. SOLTIS

Trees, shrubs, lianas, vines, or herbs, occasionally succulent, aquatic, epiphytic, or parasitic; crystals various; stems usually with a eustele, secondary growth (when present) by typical development of a vascular cambium, but occasionally with internal phloem or anomalous; parenchymatous tissues lacking spherical idioblasts with ethereal oils, but sometimes with laticifers (usually milky latex), secretory canals or cavities (with oils, resins, etc.). Hairs various. Leaves alternate and spiral or 2-ranked, or opposite and decussate, or whorled, simple to variously compound, entire to variously toothed, sometimes lobed, with pinnate or palmate venation (the latter with converging or diverging major veins), occasionally with domatia, rarely with pellucid dots; stipules various, present or absent. Inflorescences determinate, indeterminate, or mixed, usually terminal or axillary, sometimes reduced to a solitary flower. Flowers bisexual or unisexual (and plants then monoecious, dioecious, or polygamous), radial or bilateral (due to form of perianth and/or androecium and gynoecium; see Endress 2012 for various possibilities in organ involvement in monosymmetry), with or without a hypanthium, the perianth differentiated (with calyx and corolla) and usually in 2 whorls, sometimes reduced to a single whorl (and then sometimes perianth-like structures recruited from staminodia or bracts), or absent. Sepals (or tepals) usually 4–5 (to numerous, or sometimes lost), distinct to connate, with various aestivation. Petals usually 4 or 5, sometimes absent, distinct to connate, with various aestivation. Stamens usually 2–5, less commonly numerous (a secondary increase, but ancestrally in 2 whorls, each 5 members), distinct to connate (by filaments or filaments and anthers) and occasionally fasciculate or with staminodia; filaments attached to receptacle, hypanthium, or adnate to petals; anthers usually opening by longitudinal slits, occasionally by pores; pollen grains in monads or occasionally tetrads, usually tricolpate, tricolporate, or modifications thereof, e.g., triporate, pantoporate. Carpels usually 2–5, sometimes numerous or reduced to a single carpel, or pseudomonomerous, usually connate, with as many styles as carpels or a single style, either apically branched or unbranched; ovary superior or inferior, with variable placentation; stigma various. Ovules anatropous to orthotropous, with usually 1 or 2 integuments, with a thick to thin nucellus, 1 to numerous per carpel, or reduced to 1 per gynoecium. Nectaries variable or absent. Fruit variable; seeds sometimes arillate; endosperm present to absent, sometimes replaced by perisperm.

Composition: Santalales, Caryophyllales, Berberidopsidales, Asteridae.

BERBERIDOPSIDALES DOWELD

Trees or scandent shrubs; crystals prismatic and/or of druses; parenchymatous tissues sometimes with sclereids; **stomata cyclocytic** (i.e., stoma surrounded by 4 or more subsidiary cells, in a narrow ring). *Hairs simple or ferruginous, stellate-peltate scales.* Leaves alternate and spiral or ± opposite and decussate, simple, entire or spinose-serrate or -dentate, with pinnate venation, but sometimes ± plinerved and thus slightly palmately veined; stipules absent.

Inflorescences determinate, but sometimes appearing indeterminate, terminal and/or axillary, sometimes reduced to a solitary flower. *Flowers bisexual or unisexual (and plants then dioecious), radial, the perianth differentiated (with calyx and corolla, in 2 whorls) or ± undifferentiated (with tepals, but outer ones sepal-like and inner ones petal-like, these intergrading and apparently arranged ± spirally*, this a reversal from a whorled ancestry), and sometimes (in bud) surrounded by a calyptrate bract. *Tepals ± numerous, distinct*, imbricate, the *outer sepaloid and inner petaloid. Sepals usually 5, distinct*, imbricate. *Petals usually 5, distinct*, imbricate. Stamens 5 to numerous, in a single whorl, *spirally arranged*, distinct, and staminodia also present in carpellate flowers; **filaments** free, **± short**; anthers opening by longitudinal slits, the connective sometimes extending beyond the pollen sacs; pollen grains tricolpate or tricolporate. Carpels 2–5, but sometimes only one functional, connate, with a single, often short style; ovary superior, with parietal or apical placentation; stigma variously lobed. Ovules anatropous, with 2 integuments, with a thick nucellus, 2 to numerous per carpel/placenta. Nectar produced by a ring of glands (alternating with stamens or staminodia) or a lobate disk. *Fruit a berry or drupe*; seed sometimes with a sarcotesta, sometimes arillate; endosperm present, sometimes ruminate.

SANTALALES R.BROWN EX BERCHT. & J.PRESL

Usually parasitic trees, shrubs, lianas (scrambling or with tendrils), *or herbs, with or without chlorophyll (and thus hemi- to holoparasites), thus usually with various haustorial attachment (or attachments) to the roots or branches of their hosts, and the plants terrestrial or epiphytic (then often parasitizing the tree on which they are growing)*, the vegetative body typical or much reduced, external to the host or internal and ramifying within the host, producing emergent inflorescences or flowers, occasionally forming a tuber that usually incorporates tissues of the host as well as the parasite (and in which the inflorescences form, and when emergent producing a collar-like structure), **with polyacetylenes**; crystals various, solitary or of druses, sometimes with cystoliths; parenchymatous tissues sometimes with mucilage cells, resin glands and/or laticifers, **often with sclereids** *and/or groups of silicified cells often present in mesophyll*; branches terete, quadrangular, angular, or flattened, with or without nodal constrictions, rarely forming thorns; *roots often with haustorial connections*, **lacking root hairs.** Hairs simple to dendritic, or stellate. *Leaves* opposite and decussate, whorled, subopposite, or alternate and spiral or 2-ranked, *simple, entire* (rarely toothed), with *pinnate* or rarely palmate *venation, sometimes merely scales with very reduced venation; stipules absent. Inflorescences determinate* or less commonly indeterminate, terminal, axillary, or cauliflorus, sometimes reduced to a solitary flower. Flowers bisexual or unisexual (and plants then monoecious or dioecious), radial or bilateral (due to form of perianth), showy to inconspicuous, sometimes with a hypanthium, with perianth differentiated (of calyx and corolla), in 2 whorls, *but the calyx often much reduced (a calyculus), or with only a single whorl of perianth parts (tepals, probably derived from petals, with the complete loss of the calyx)*, or perianth absent. **Sepals reduced to a cup-like rim** *(and 3–6 or often indeterminate in number)*, distinct to connate, **with open aestivation,** or absent. **Petals** *3–6* (-9), distinct or connate, usually **valvate**, showy to inconspicuous, *sometimes with adaxial hairs*, sometimes absent. *Tepals* (2-) *3–5* (-8), distinct to connate, *valvate*, rarely with open aestivation, ± *inconspicuous*, rarely adnate to the gynoecium. **Stamens** 1 to numerous, **when few then opposite the petals**, in a single whorl, distinct or connate (by filaments), sometimes unequal in length, occasionally with staminodia, or with staminodia in the carpellate flowers (these rarely elongate and feathery); filaments free, borne on receptacle or hypanthium, *or often adnate to the petals*, very short to elongate; anthers distinct or connate (forming synandrium), opening by longitudinal or transverse slits or apical pores (or slit-like pores), sometimes with only a single locule and then opening by a terminal slit or apical pore, or opening irregularly; pollen grains tricolpate to pentacolpate, tricolporate to 12-colporate, triporate to multiporate, or inaperturate, and spherical, triangular, or 3-lobed. *Carpels* (2-) *3–4* (-5), *connate*, rarely pseudomonomerous, *with a single, short to elongate style* or as many styles as carpels; *ovary* superior to *more commonly inferior*, with axile, *free-central (and then often with variably developed partitions, ovules pendulous from apex of axis) to basal placentation*; stigma punctate to capitate or lobate. **Ovules** anatropous to orthotropous, with 1 or 2 integuments, with a ± thin nucellus, *or ± undifferentiated*, **pendent**, *1 to few per carpel, or reduced to 1 or 2 per gynoecium*. Nectary a disk, sometimes lobate, a ring of glands, or absent. **Fruit a drupe**, *berry (with a single seed), often viscous*, or a nut, achene, samara, and sometimes with associated perianth; seeds with **small embryo** *and coat thin or absent*; endosperm present.

CARYOPHYLLALES JUSS. EX BERCHT. & J. PRESL

Trees, shrubs, lianas/vines (scrambling, twining, with hooked branches, or leaves with 2-hooked midrib or tendril), or ***herbs*** (synapomorphy of core Caryophyllineae), *often stem or leaf succulents, often with CAM or C4 metabolism, often halophytic* (sometimes with salt excretory glands), *sometimes carnivorous* (and then with sensitive or non-sensitive leaves, each forming a snap-trap, pitcher-trap, or covered with stalked and/or sessile, gland-headed hairs that produce mucilage and digestive enzymes), *often not mycorrhizal*; crystals various; parenchymatous tissues

often with yellow to red betalains (possible synapomorphy of core Caryophyllineae, but not in *Macarthuria*), but sometimes with anthocyanins or neither of these pigments, sometimes *with* ***acetogenic naphthaquinones*** (e.g., plumbagin, occasionally with sclereids; a synapomorphy of Polygonineae); ***stems often with successive cambia (producing concentric rings or bundles of xylem and phloem***; synapomorphy of Caryophyllineae), occasionally with cortical and/or medullary vascular bundles, sometimes with long and short shoots, and expanded and photosynthetic (as in Cactaceae), sometimes with a nodal line and/or swollen nodes, sometimes forming thorns; vessel elements sometimes with vestured pits; ***often with plastids of the sieve elements with a ± peripheral ring of proteinaceous filaments, and with or without a central protein crystalloid*** (which may be globular or angled; synapomorphy of core Caryophyllineae), but sometimes the plastids with starch grains. Hairs various, *but sometimes with pit-glands or sessile to stalked gland-headed hairs, these often vascularized*. Leaves alternate and spiral, or opposite and decussate, rarely whorled or pseudo-whorled, borne along stem or in a basal rosette, simple, rarely lobed or dissected, occasionally reduced and scale-like, **entire** to occasionally serrate or dentate, *with pinnate venation*, the system of veins well developed to obscure, reduced or even absent, occasionally parallel or appearing so, occasionally circinate or reverse-circinate, the lamina sometimes highly modified in connection with carnivory, succulence, or reduced and forming spines, the epidermis occasionally with large, bladder-like cells; stipules present (then sometimes forming a sheathing structure, ocrea, or occasionally scarious) or absent, and sometimes with stipule-like scales, bristles or hairs in leaf axils. Inflorescences determinate or indeterminate, terminal or axillary, sometimes reduced to a solitary flower. Flowers bisexual or unisexual (and plants then monoecious or dioecious), *usually radial*, sometimes with a short to elongate hypanthium, with perianth differentiated (of calyx and corolla), in 2 whorls, *but the petals early lost in the evolutionary history of this clade, so many have only tepals* (*characteristic of most Caryophyllineae*), in 1 or 2 whorls or spirally arranged, although a seemingly biseriate perianth has re-evolved by recruitment of other floral plant parts into the "perianth" (i.e., "petals" from staminodia, "sepals" from a pair of bracts) or by an increase in the number of tepals followed by differentiation of the inner ones from the outer ones, the flower sometimes sunken into the apex of a modified stem (most Cactaceae) or associated with often accrescent bracts or bracteoles and in some 2 bracteoles appearing to be a calyx, occasionally with an androgynophore. Sepals (or tepals) 3 to numerous, distinct to connate, imbricate, valvate, induplicate-valvate, plicate/contorted, or with open aestivation, green to colorful (± petaloid), sometimes fleshy, occasionally the proximal portion green (and accrescent) and the distal portion colorful, occasionally appearing to be petals (when associated with a calyx-like pair of bracteoles, as in Portulacaceae and relatives, or a calyx-like series of bracts (as in some Nyctaginaceae). Petals usually 5, distinct to connate, imbricate to contorted/convolute, *but commonly absent* (in Caryophyllineae). Petaloid staminodia (in most floras merely called "petals"), 4 or 5 to numerous, distinct to connate, ± imbricate, often bilobed, sometimes differentiated into a claw and limb separated by an appendaged joint. Stamens 3 to numerous, usually in 1 or 2 whorls, but sometimes spiral, often numerous by secondary increase (and centrifugal), distinct to connate (by filaments), often with staminodia; *filaments free*, usually attached to receptacle or hypanthium, and sometimes fused to staminodia, but sometimes adnate to perianth, ± slender, short to elongate; anthers opening by longitudinal slits, the wall **with outer secondary parietal cell layer developed directly into the endothecium**; pollen grains tricolpate to polycolpate, tricolporate, *hexaporate to polyporate/pantoporate*, sometimes in tetrads, the **exine often with fine spinules**. *Carpels* 2 to numerous, *connate* (or sometimes only slightly so), *with a single distally branched style* (**and style branches well developed), or with as many styles as carpels**, or gynoecium represented by only a single carpel, then style rarely gynobasic; ovary superior to inferior (and then sometimes with accessory tissue derived from the stem, in which the flower is sunken), with axile, parietal, *free-central, or basal placentation*; stigmas minute to capitate, dissected, or ***linear, often elongated, extending most of the length of the style or style-branch*** (a synapomorphy of Caryophyllineae). Ovules anatropous to *campylotropous* (synapomorphy of core Caryophyllineae) or orthotropous, with 2 integuments (rarely a single integument), with a thick to thin nucellus (but ovules always crassinucellate), 1 to numerous per carpel. *Nectary a disk, ring of glands, pits, ring on inner surface of hypanthium,* ***or nectar produced by glandular adaxial staminal bases*** (synapomorphy of core Caryophyllineae), or absent. Fruit a variably dehiscent capsule, berry (sometimes spiny), utricle, achene or nut, *often associated with fleshy or dry, sometimes papery or winged perianth parts or bracts*, or sometimes enclosed in persistent, leathery to fleshy, basal portion of perianth tube (which may have sticky gland-headed or hooked hairs); seeds sometimes with ornamented testa (individual cells clearly visible, bulging), sometimes arillate or hairy, rarely winged, the embryo straight *to curved*, ***but in core Caryophyllineae always curved***; endosperm present (homogeneous to rarely ruminate, *often starchy*) to ***absent, often replaced by starchy perisperm*** (synapomorphy of core Caryophyllineae).

Note: Ordinal clade composed of two major subclades, i.e., Polygonineae and Caryophyllineae, and the putative synapomorphies of these also indicated in the description (by ***bold/italics***).

Asteridae Takht.

Trees, shrubs, lianas, vines, and herbs, sometimes rhizomatous, occasionally aquatic, epiphytic, or parasitic; crystals various; stems usually with a eustele, secondary growth (when present) by typical development of a vascular cambium, but occasionally anomalous; parenchymatous tissues lacking spherical idioblasts with ethereal oils, but sometimes with laticifers or secretory cavities or canals (with resins, oils, etc.); **iridoids often present**. Hairs various. Leaves alternate and spiral or 2-ranked, or opposite and decussate, or whorled, simple to variously compound, entire to variously toothed, sometimes lobed, with pinnate or palmate venation (the latter with converging or diverging major veins), occasionally with domatia or pellucid dots; stipules various, present or absent. Inflorescences determinate, indeterminate, or mixed, usually terminal or axillary, sometimes reduced to a solitary flower. Flowers bisexual or unisexual (and plants then monoecious, dioecious, or polygamous), radial or bilateral (due to form of perianth and/or androeciumand gynoecium), usually without a hypanthium, the perianth differentiated (with calyx and corolla) and usually in 2 whorls, rarely reduced to a single whorl or absent. *Sepals* usually 4–5 *distinct to more frequently connate*, variously imbricate or valvate. *Petals usually 4 or 5*, rarely absent, *distinct to more frequently connate*, variously imbricate or convolute, or valvate. *Stamens usually 2–5, less commonly numerous* (a secondary increase, but ancestrally in 2 whorls, each usually of 4 or 5 members), distinct to connate (usually by filaments) and sometimes fasciculate, occasionally with staminodia; filaments free, arising from receptacle *or more commonly adnate to petals*, variable in form; anthers usually opening by longitudinal slits, occasionally by pores; pollen grains in monads or occasionally tetrads, usually tricolpate, tricolporate, or modifications thereof, e.g., triporate, pantoporate. *Carpels usually 2–5*, rarely numerous, *connate*, with as many styles as carpels or a single style, either apically branched or unbranched; ovary superior or inferior, with variable placentation; stigma various. **Ovules** anatropous to orthotropous, **with usually one integument, and with a thin nucellus**, 1 to numerous in each locule/placenta, or reduced to 1 per gynoecium. Nectaries various or absent. Fruit various; seeds sometimes arillate; endosperm present to absent.

Composition: Cornales, Ericales, *Lamiidae*, *Campanulidae*.

CORNALES LINK

Trees, shrubs, and occasionally lianas and vines (often climbing by adventitious roots), often rhizomatous, rarely submerged aquatics (i.e., *Hydrostachys*, and then with stem modified, disk-like, with adventitious roots, attaching plant to rocks); crystals druses or raphides, sometimes also with cystoliths (of calcium carbonate or silica); parenchymatous tissues with resin-canals or laticifers occasional (in association with vascular bundles or in the pith, flowers or fruits). *Hairs simple, Y- or T-shaped, or stellate, sometimes glandular, sometimes stinging (sharp-pointed with ± bulbous base, containing irritating fluid), sometimes tuberculate, scabrous or glochidate* (i.e., with rounded to pointed, short to elongate, often downward-hooked processes, often silicified or calcified). *Leaves* alternate and spiral or 2-ranked to opposite and decussate, rarely whorled, the petioles sometimes joined by a line or ridge, *simple*, entire to dentate or serrate, sometimes lobed or dissected, with pinnate or palmate venation (and the latter with major veins converging or diverging), but highly modified, simple to 2- to 3-times pinnatifid and covered with rounded, tuberculate, linear to scale-like or fringed outgrowths in *Hydrostachys*; *stipules absent*. *Inflorescences determinate*, terminal or axillary, occasionally reduced to a solitary flower. *Flowers* bisexual or unisexual (then plants monoecious or dioecious), *radial*, with perianth differentiated (of a calyx and corolla, but occasionally of tepals by reduction), in 1 or 2 whorls, or occasionally perianth completely absent, rarely the inflorescences with both reproductive and sterile, display flowers. *Sepals usually 4–5 (-12), distinct to slightly connate*, **often represented by small teeth or obsolete**, imbricate, valvate, *or with open aestivation*. *Petals usually 4–5 (-12), distinct to connate*, rarely calyptrate, imbricate, contorted, valvate, or with open aestivation, sometimes reduced or absent, occasionally with margins toothed, appendaged or with flaps. *Stamens (2-) 4 to numerous*, in 1 or 2 whorls, distinct to connate (by filaments), sometimes with staminodia (and these inconspicuous to petaloid, sometimes connate, highly modified); filaments free or sometimes adnate to (or closely associated with) corolla/perianth, ± slender or basally broadened, short to elongate; anthers opening by longitudinal slits, with connective rarely extending beyond pollen sacs; pollen grains tricolporate, **sometimes with H-shaped thin regions** (endoapertures), rarely inaperturate (in *Hydrostachys*). *Carpels usually 2–5 (-12), connate*, sometimes pseudomonomerous, with a single, short to elongate, branched or unbranched style, or as many styles as carpels; **ovary inferior to half-inferior** (to rarely only 1/5-inferior) but nude in *Hydrostachys*, with axile, parietal or intruded-parietal placentation; stigma (or stigmas) usually punctate, capitate, lobate, or elongate. Ovules usually anatropous to hemitropous, with 1 integument, with thick to thin nucellus, 1 in each locule, or 1 to numerous on each placenta. **Nectary a ± epigynous disk**, or nectar produced by fused and variously modified staminodia, or absent. **Fruit drupes**, *the pit often with a germination valve, 1- to few-seeded, sometimes ridged or winged, or a variably dehiscent, often ribbed capsule*, occasionally a berry, achene or nut; seeds occasionally winged; endosperm present or absent.

ERICALES BERCHT. & J. PRESL

Trees, shrubs, lianas (usually scrambling or weakly twining), and herbs, sometimes rhizomatous, rarely carnivorous (then with leaf modifications, i.e., pitcher-like traps or leaves with gland-headed hairs), rarely parasitic or mycoparasitic; crystals various solitary, styloids, or druses, sometimes also with silica bodies; parenchymatous tissues sometimes with sclereids, laticifers (milky sap), or secretory cavities containing yellow to red resinous materials, and then appearing as yellow-green, red, brown, or black dots to lines (on leaves, stems, and floral parts), occasionally with wandering fibers, rarely with mucilage canals; stems sometimes with cortical vascular bundles; black or dark-colored naphthaquinones occasionally present (in nearly all tissues). Hairs extremely diverse. *Leaves alternate and spiral or 2-ranked, less commonly opposite and decussate* or whorled, borne along the stem, clustered at shoot apices, or occasionally in a basal rosette, usually *simple*, but occasionally lobed to dissected or even pinnately compound, entire to variously toothed, **and the teeth** (when present) **often theoid (with a ± deciduous, opaque apical portion, or with an associated multicellular hair)**, *with pinnate* or less commonly palmate or ± parallel *venation*, sometimes with domatia, sometimes with nectar glands on abaxial surface or leaf base, occasionally with yellowish to red or black lines and dots, rarely with persistent, spinose petiole, reduced to scales in parasites; *stipules absent* or occasionally present, minute. Inflorescences determinate or indeterminate, terminal or axillary, occasionally cauliflorous, sometimes reduced to a solitary flower (and rarely scapose). Flowers bisexual or unisexual (and plants then dioecious or less commonly monoecious or polygamous), *radial* or less commonly bilateral (due to form of perianth and/or androecium and gynoecium), with perianth differentiated (of calyx and corolla), in 2 or occasionally 3 whorls, or rarely of tepals (and flowers reduced, wind pollinated), erect to pendulous, occasionally heterostylous or resupinate. *Sepals* (2-) 3–8 (-11), but *most commonly 4 or 5*, distinct or connate, imbricate, valvate or with open aestivation, occasionally dimorphic, occasionally enlarged in fruit, with a nectar spur, or intergrading with bracts, rarely petaloid. *Petals 3–8 (to numerous), but most commonly 4 or 5, connate* or less commonly distinct or nearly so, occasionally the 4 lower petals fused into 2 pairs, valvate or imbricate, sometimes contorted or convolute, occasionally with paired petaloid appendages (outgrowths from lower portion of corolla lobes), or wrinkled to lacerate along margins, rarely plicate, sometimes the flower pendulous and the corolla urn-shaped. *Stamens (2-) 4 or 5 to numerous, but ancestrally 10, in 2 whorls, but with secondary increase (± numerous) and decrease (loss of outer whorl, and thus stamens equaling the petals in number and opposite them, or loss of inner whorl, and thus stamens equaling the petals and alternating with them)*, distinct to connate (by filaments), sometimes fascicled, sometimes alternating with staminodia, or with androecium asymmetrical, prolonged on one side into a strap-like structure that arches over the ovary, formed from numerous staminodia (these producing nectar or "feeder" pollen), staminodia also sometimes in carpellate flowers; *filaments free* (and arising from receptacle) *or adnate to corolla*, short and broad to elongate, slender, occasionally with paired projections; anthers opening by longitudinal slits, ± apical pores, or valves, the anther often inverting in development, occasionally with terminal or dorsal, paired projections; pollen grains usually tri- to multicolporate, or 5- to 8-zonocolpate, or 4- to pantoporate, sometimes released in tetrads, occasionally with viscin threads. *Carpels 2 to numerous, connate*, with a single, sometimes hollow, short to elongate, unbranched to distally branched style, which is rarely distally expanded and shield- or umbrella-like, or rarely as many styles as carpels; ovary superior to inferior, sometimes with an apical depression, with axile, intruded-parietal or free-central placentation, the locules occasionally secondarily divided, **the placentae small to expanded and diffuse, extending into the locules**, *or if free-central*, **expanded**, ± *globose*, **filling or nearly filling the locule**; stigma or stigmas minute, punctate to capitate, slightly elongated or ± lobate. Ovules anatropous to campylotropous or rarely orthotropous, with 1 or 2 integuments, with a thin nucellus, 1 or 2 to numerous in each locule, or few to numerous on free-central placenta. Nectary at base of filaments or around base or apex of ovary, a well-developed, sometimes lobate disk, or nectar produced by staminodia, by nectariferous spur on one of the sepals, or nectary absent; rarely with nectar-producing bracts associated with flowers. Fruit a variously dehiscent, dry to fleshy capsule (this rarely associated with a fleshy calyx), berry, drupe (and then with a single or several pits, and the pits single to few-seeded, occasionally with a germination pore), indehiscent pod, samara or a schizocarp of nutlets; seeds occasionally with a large or depressed hilum, occasionally winged, occasionally arillate, the coat hard/shiny to fleshy, sometimes mucilaginous, thin, or even ± disappearing; embryo minute to large; endosperm present to absent, occasionally ruminate.

GENTIANIDAE R.G. OLMSTEAD, W.S. JUDD & P.D. CANTINO

Trees, shrubs, lianas, vines, and herbs, sometimes rhizomatous, rarely aquatic, epiphytic, and parasitic; crystals various; stems occasionally with internal phloem, anomalous secondary growth, or medullary bundles; parenchymatous tissues sometimes with laticifers or secretory canals or cavities (with aromatic compounds, resins, oils, etc.). Hairs various. Leaves alternate and spiral or 2-ranked, or opposite and decussate, or whorled, simple to variously compound, entire to variously toothed, sometimes lobed or dissected, with pinnate or palmate venation (the latter with converging or diverging major veins), rarely parallel or appearing so, occasionally with domatia, rarely with pellucid dots; stipules various, present or absent, occasionally with colleters. Inflorescences determinate, indeterminate, or mixed, usually terminal or axillary, sometimes reduced to a solitary flower. Flowers bisexual or unisexual (and plants then monoecious, dioecious, or polygamous), radial or bilateral (due to form of perianth and/or androecium and gynoecium, and then the corolla often bilabiate or unilabiate, often with an upper lip composed of 2 petals and a lower lip composed of 3 petals, with or without a dorsal slit), with or without a hypanthium, the perianth differentiated (with calyx and corolla) and usually in 2 whorls, rarely reduced to a single whorl or absent. Sepals usually 2–5, rarely numerous, in a single whorl, distinct to more frequently connate, with various aestivation, sometimes highly modified or reduced. Petals usually 4–5, rarely numerous or absent, **connate** or occasionally distinct (but then with developmental evidence of connate ancestry), with various aestivation. **Stamens usually 2–5, usually equaling the corolla lobes** *or fewer than the lobes*, distinct to connate (by filaments or filaments and anthers) occasionally with staminodia; **filaments usually adnate to corolla**; anthers usually opening by longitudinal slits, occasionally by pores, occasionally adnate to gynoecium (away from the stigma, or before the stigma is receptive), sometimes showing plunger-pollination (with pollen deposited onto gynoecium, and with secondary presentation to pollinators); pollen grains in monads or occasionally tetrads, usually tricolpate, tricolporate, or modifications thereof, e.g., triporate, polyporate. **Carpels usually 2,** but sometimes as many as 5 or rarely more (see Endress 2014), rarely pseudomonomerous, connate, with as many styles as carpels or a single style, either apically branched or unbranched, sometimes with modifications for pollen reception and presentation; ovary sometimes lobed, superior or inferior, with variable placentation; stigma various. Ovules anatropous to campylotropous, rarely orthotropous, **with one integument, with a usually thin nucellus** (or these characters may have evolved at the level of the Asteridae), 1 to numerous per carpel, or reduced to 1 per gynoecium. Nectaries variable or absent. Fruit a variously dehiscent capsule, berry, drupe, indehiscent pod, nut, achene, samara, schizocarp (of nutlets, achenes, or drupelets), or a pair of fleshy to dry follicles, berries or drupes, often associated with accessory structures; endosperm present to absent.

Composition: *Lamiidae* and *Campanulidae.*

LAMIIDAE TAKHT.

Trees, shrubs, lianas, vines, and herbs, sometimes rhizomatous, occasionally aquatic, epiphytic, and parasitic; crystals various; stems sometimes with internal phloem, occasionally with anomalous secondary growth; parenchymatous tissues sometimes with laticifers or secretory cavities (with resins, oils). Hairs various. Leaves alternate and spiral or 2-ranked, or opposite and decussate, or whorled, simple to variously compound, entire to variously toothed, sometimes lobed or dissected, with pinnate or palmate venation (the latter with converging or diverging major veins), rarely appearing ± parallel, occasionally with domatia, rarely with pellucid dots; stipules various, present or absent, sometimes with colleters. Inflorescences determinate, indeterminate, or mixed, usually terminal or axillary, sometimes reduced to a solitary flower. Flowers bisexual or unisexual (and plants then dioecious or polygamous), radial or bilateral (due to form of perianth and/or androecium and gynoecium, and then the corolla usually with an upper lip composed of 2 petals and a lower lip composed of 3 petals, with androecium usually of 4 stamens, 2 long and 2 short), without a hypanthium, the perianth differentiated (with calyx and corolla) and usually in 2 whorls, rarely reduced to a single whorl or absent. *Sepals usually 2–5,* rarely numerous, *in a single whorl,* distinct to more frequently connate, with various aestivation. Petals usually 4–5, rarely numerous or absent, distinct to more frequently connate, with various aestivation, *usually showing "late sympetaly," i.e., petals initiated as separate lobes, with later development of connation.* Stamens usually 2–5, usually equaling the corolla lobes or fewer than the lobes, distinct to connate (by filaments or filaments and anthers) occasionally with staminodia; *filaments usually adnate to petals*; anthers usually opening by longitudinal slits or a single, U- or V-shaped slit, occasionally by pores, occasionally adnate to gynoecium; pollen grains in monads or occasionally tetrads, usually tricolpate, tricolporate, or modifications thereof, e.g., triporate, polyporate. Carpels 2 (-5), but rarely pseudomonomerous, connate, with as many styles as carpels or a single style, either apically branched or unbranched; ovary sometimes lobed, superior or inferior, with variable placentation; stigma various. Ovules anatropous to hemitropous, rarely orthotropous, with one integument, with a usually thin nucellus, 1 to numerous per carpel, or reduced to 1 per gynoecium. Nectaries variable or absent. Fruit a variously dehiscent capsule, berry, drupe, indehiscent pod, samara, schizocarp (of nutlets, achenes, or drupelets), or a pair of fleshy to dry follicles, berries or drupes; endosperm present to absent.

Composition: Icacinales, Metteniusales, Boraginales, Vahliales, Garryales, Gentianales, Solanales, Lamiales.

ICACINALES TIEGH.

Trees, shrubs and lianas, sometimes bulbous succulents and thorny; often with druses or crystal sand; parenchymatous tissues sometimes with astrosclereids; **phloem stratified**; roots sometimes tuberous. Hairs simple to 2-branched or absent. Leaves alternate and spiral, or opposite and decussate, simple, but rarely lobed, entire, serrate, or glandular-toothed, with pinnate to palmate venation; stipules absent. Inflorescences indeterminate to less commonly determinate, terminal or axillary. Flowers bisexual or unisexual (and plants then monoecious or dioecious), *radial*, the perianth differentiated (of calyx and corolla), in 2 whorls. *Sepals usually 5, shortly connate, imbricate. Petals usually 5, distinct to connate*, imbricate, *with apices often inflexed*. Stamens usually 5, in a single whorl, distinct; filaments free or adnate to corolla, and alternating with the lobes/petals, short or elongate; anthers opening by longitudinal slits, the anther connective not well developed to greatly prolonged; pollen grains often tricolporate. Carpels 1 (possibly pseudomonomerous) or 5, connate, with a very short to elongate style or as many styles as carpels; ovary superior, with apical or apical-axile placentation; stigmas punctate to truncate or capitate. Ovules anatropous or campylotropous, with 1 or rarely 2 integuments, 1 or 2 per gynoecium (but only 1 maturing), or 2 in each locule. Nectary a lobed to unlobed disk or disk absent. *Fruit a drupe, sometimes slightly flattened, the pit ribbed or pocketed, with one to few seeds*; *endosperm present* or absent, *ruminate or homogeneous*.

METTENIUSALES TAKHT.

Composition: Metteniusaceae (*Metteniusa* H.Karst. and 10 genera formerly placed in Icacinaceae s.l.; *Apodytes* E. Mey ex Arn., *Calatola* Standl., *Dendrobangia* Rusby, *Emmotum* Ham., *Oecopetalum* Greenm. & C.H.Thomps., *Ottoschulzia* Urb., *Pittosporopsis* Craib, *Platea* Blume, *Poraqueiba* Aubl., *Rhaphiostylis* Planch. & Benth.).

BORAGINALES JUSS. EX BERCHT. & J. PRESL

Composition: Boraginaceae (incl. Coldeniaceae, Codonaceae, Cordiaceae, Ehretiaceae, Hoplestigmataceae, Lennoaceae, Heliotropiaceae, Hydrophyllaceae, Namaceae, Wellstediaceae).

VAHLIALES MARTIUS

Composition: Vahliaceae.

GARRYALES MARTIUS

Trees and shrubs; crystal sand; **parenchymatous tissues with laticifers (and latex ± elastic)**, often with sclereids. Iridoids (e.g., aucubin) present. Hairs simple. Leaves opposite and decussate or alternate and spiral, simple, entire to serrate, *with pinnate venation*, the petioles (of leaves at a single node, when opposite) basally connate; *stipules absent*. Inflorescences determinate (but often appearing indeterminate), terminal or axillary, rarely reduced to a solitary flower, in *Garrya* forming ± dangling catkins. **Flowers unisexual** (and **plants dioecious**), *radial*, the perianth differentiated (of calyx and corolla), the petals lost (and sepals then referred to as tepals), or perianth entirely absent (through loss), in 1 or 2 whorls (when present). *Sepals (or tepals, when petals absent) 2–4, sometimes minute or entirely absent, distinct or apically postgenitally connected*, **valvate**. *Petals 4* (in *Acuba*) *or absent* (in other genera), *distinct*, *valvate*. Stamens 4–12, in a single whorl, distinct; filaments free, arising from receptacle, short; anthers opening by longitudinal slits, the connective rarely forming an apical appendage; pollen grains usually tricolporate. *Carpels 2*, *connate*, **with a single short style**, or seemingly a single carpel (pseudomonomerous) and then with a single style; **ovary inferior** (or nude), **with ± apical placentation**; *stigmas ± elongate*. Ovules anatropous, with 1 integument, with thick nucellus, 1 or 2 **in the single locule**. Nectary a disk, in the carpellate flowers atop ovary, or absent (but vestigial disk in staminate flowers of *Garrya*). *Fruit a berry or* (in *Eucommia*) *a samara*; endosperm present.

GENTIANALES JUSS. EX BERCHT. & J. PRESL

Trees, shrubs, lianas/vines (twining or scrambling) and herbs, rarely mycoparasitic; crystals various; parenchymatous tissues sometimes with laticifers (latex mostly white); **stems with a nodal line**, *often with internal phloem*, occasionally winged or succulent (and plants then cactus-like); **vessel elements with vestured pits**. Hairs various. **Leaves opposite and decussate**, whorled, or occasionally alternate and spiral or 2-ranked, *simple*, rarely lobed, *usually entire*, *with usually pinnate venation*, occasionally with domatia, rarely reduced to scales (in mycoparasitic species); **stipules**

present, *sometimes interpetiolar (and connate)*, rarely leaf-like, **with colleters on adaxial surface**, *to very reduced or absent*, **but then colleters often present at base of petiole (or along nodal line)**. *Inflorescences determinate*, terminal or axillary, occasionally reduced to a solitary flower. *Flowers usually bisexual, sometimes heterostylous, usually radial*, with perianth differentiated (of calyx and corolla), in 2 whorls. *Sepals* usually 4 or 5, *usually connate*, valvate, imbricate or with open aestivation, *sometimes with colleters on adaxial surface or at base adaxially*. *Petals* usually 4 or 5, *connate*, valvate, imbricate or contorted, occasionally plicate, *the adaxial surface often pubescent*, the lobes occasionally fringed, and sometimes with coronal appendages or scales on inside or apex of tube. *Stamens usually 4 or 5, in a single whorl*, distinct to connate (by filaments and sometimes also anthers); *filaments adnate to corolla*, slender, short to elongate; anthers opening by longitudinal slits or occasionally terminal pores, distinct or sticking together and forming a ring around the stylar head, free, or sticking to the stylar head by intertwining trichomes and sticky secretions, or strongly adnate to stylar head by parenchymatous tissue, the connective occasionally with apical appendages, and staminal outgrowths sometimes present and petal-like, hood-like and horn-shaped; pollen grains usually tricolpate, tricolporate, di- or triporate, or polyporate, shed as monads, sometimes loosely sticking together by means of viscin, or as tetrads, also loosely sticking together, or strongly coherent and forming hardened masses (pollinia), then with translators (structures connecting pollinia from pollen sacs of adjacent anthers) present, often with a clip (corpusculum). *Carpels* 2 (-5), *connate*, fused in ovary region and with a single, sometimes apically branched (or twice-branched) style or separate styles, or with separate ovaries and only postgenitally fused in the style or stylar head; ovary or ovaries superior or inferior, with axile, parietal or intruded-parietal placentation (when connate) or with lateral placentation (when distinct); stigma or stigmas linear, capitate or lobed, the lobes occasionally spirally twisted, or sometimes highly modified, forming part of a stylar head, i.e., the apical portion of the style (and associated stigmas) expanded, forming a simple or complex, ± latitudinally differentiated head (and specialized for pollen deposition, viscin secretion, and pollen reception). Ovules anatropous to hemitropous, with 1 integument or rarely without integuments (see Endress 2014), with a thin (or secondarily thickened) nucellus, 1 to numerous in each locule, or 2 to numerous on each placenta. Nectar produced by a disk, surrounding ovary or atop ovary, or by glands, or by stigmatic chambers (and held in the hollows of the staminal tube or coronal appendages), or absent. Fruit a loculicidal or septicidal capsule, berry, drupe, schizocarp, indehiscent pod or pair of fleshy to dry follicles, berries or drupes. Seeds sometimes flattened, winged, or with a tuft of hairs; endosperm present or absent, **its development nuclear**.

SOLANALES JUSS. EX BERCHT. & J. PRESL

Trees, shrubs, vines (scrambling or twining) and herbs, sometimes rhizomatous, rarely ± hemiparasites; crystals solitary or druses; *stems often with internal phloem*, rarely with cortical air spaces (*Sphenoclea*); parenchymatous tissues sometimes with laticifers (milky sap). Hairs simple, Y-shaped, variously branched or stellate, sometimes glandular, occasionally with prickles. *Leaves alternate and spiral*, rarely opposite and decussate, *usually simple*, sometimes deeply lobed or even pinnately or palmately compound, entire to serrate, *with pinnate* or less commonly palmate *venation*, rarely reduced to scales; *stipules absent*. *Inflorescences usually determinate*, but rarely indeterminate (*Sphenoclea*), terminal or axillary, sometimes reduced to a solitary flower. *Flowers usually bisexual*, but rarely unisexual (and plants then dioecious, with carpellate flowers solitary and staminate flowers in inflorescences with cymose branches), *usually radial*, with perianth differentiated (of calyx and corolla), in 2 whorls. Sepals usually 5, distinct to connate, imbricate or valvate, rarely reduced and with open aestivation, sometimes expanding in fruit. *Petals usually 5, usually connate*, but rarely distinct (in Montiniaceae), **usually plicate (with longitudinal fold lines)** and with marginal portions of each lobe folded inwards and often convolute, the middle portion of lobes sometimes valvately arranged, sometimes merely imbricate or valvate. *Stamens usually 5*, in a single whorl, distinct, but sometimes the anthers sticking to each other; *filaments usually adnate to corolla* (but free in Montiniaceae), usually slender, elongate (but short and thick in Montiniaceae), occasionally abruptly glandular-swollen near base (*Hydrolea*), sometimes of unequal lengths; anthers opening by longitudinal slits or apical pores, **the pollen sacs with placentoids** (i.e., with a parenchymatic or tapetal protrusion into sporogenous tissue); pollen grains 3- to 5-colpate or -colporate, or polyporate. *Carpels* 2 (-5), *connate*, but sometimes deeply lobed, with a single, terminal to gynobasic, short to elongate, unbranched to apically branched style, or with as many styles as carpels; *ovary usually superior*, but rarely inferior or half-inferior, *with usually axile placentation*; stigma or stigmas ± capitate, discoid, lobed or linear. Ovules anatropous to hemitropous, with a single integument, with a thin nucellus, 1 or 2 to numerous per locule. Nectary an entire to lobate disk, or absent. Fruit a variously dehiscent capsule, berry, drupe, or schizocarp of nutlets; seeds sometimes flattened; endosperm present or absent.

LAMIALES BROMHEAD

Trees, shrubs, lianas (scrambling, twining, or with tendrils), *and herbs*, often rhizomatous, occasionally epiphytic and aquatic, occasionally hemi- to holoparasites, and then with haustorial attachments to the roots of their hosts, occasionally insectivorous (with sticky, mucilage-secreting and digestive hairs, and often with leaf margins enrolling in response to contact of glandular hairs with prey organism, or leaves tubular and spiraled, with downward pointing and digestive hairs and a basal chamber, or highly dissected, bearing prey-catching bladders, each with 2 sensitive valves forming a trapdoor entrance, and lined on the inside with branched digestive hairs); crystals various, occasionally also with cystoliths; parenchymatous tissues sometimes with sclereids or secretory cavities (containing resin or oils); stems when young either round or square in cross-section, occasionally with anomalous secondary growth (resulting in a 4- or many-lobed or furrowed xylem cylinder), their buds sometimes 2 to several and superposed; roots rarely forming pneumatophores; *plants often with oligosaccharides as carbohydrate storage molecules. Hairs simple, eglandular and gland-headed* (**cells in heads often with vertical walls only**, but absent from Plocospermataceae), variously forked, dendritic, stellate, or peltate scales, occasionally calcified or silicified and/or associated with a cystolith, rarely with prickles. Leaves alternate and spiral (rarely 2-ranked), opposite and decussate, or whorled, distributed along stems or in basal rosettes, simple, sometimes lobed or dissected, or pinnately, twice-pinnately or palmately compound, occasionally trifoliolate, rarely unifoliolate, entire to variously toothed, with pinnate or occasionally palmate venation (and the latter with converging or diverging major veins), rarely ± parallel-veined, occasionally with pellucid dots or with adaptations to catch and digest insects, rarely with salt-excretory glands (*Avicennia*), and leaves reduced to scales in some parasites; stipules absent, but sometimes with interpetiolar line or sheath connecting leaves; **mesophyll cells with protein inclusions in nucleus**. Inflorescences determinate, indeterminate, or mixed, terminal or axillary, occasionally reduced to a solitary flower. *Flowers bisexual* or rarely unisexual (and plants then dioecious to polygamous), radial or *more commonly bilateral* (due to form of perianth and/or androecium), with perianth differentiated (of calyx and corolla), in 2 whorls, rarely the corolla or both calyx and corolla absent. *Sepals 3–5* (rarely numerous), distinct to connate, imbricate, valvate, or with open aestivation, the lobes of equal size (and calyx radial) or unequal (and calyx variously bilateral, sometimes bilabiate), occasionally enlarging in fruit, rarely calyptrate, very rarely absent. *Petals 4 or 5* (rarely numerous), *connate*, imbricate, convolute, valvate or induplicate-valvate, the lobes all the same size (and flower radial, with 4 or less commonly 5 corolla lobes) *or unequal in size, with 2 petals forming an upper lobe, and 3 petals forming a lower lobe (i.e., flower held ± horizontally, ± bilabiate or at least with distinctive 2 + 3 pattern)*, occasionally with an upper lip formed by 2 petals and a larger, saccate lower lip formed by 2, strongly fused petals (i.e., bilateral and 2 + 2 pattern), sometimes with a basal nectar spur, rarely with a bulge obscuring the throat (personate), or rarely absent. *Stamens 4*, in a single whorl, distinct or sometimes adherent only by their anthers (in pairs or all 4 together), *usually with 2 long stamens and 2 shorter ones (didynamous), the 2 shorter occasionally staminodial*, and the fifth stamen occasionally represented by a minute to well-developed staminodium, *or merely 2 stamens* (without staminodia), rarely with 5 stamens, all functional, very rarely with only a single stamen (*Callitriche, Hippuris*) or ± numerous stamens (e.g., *Symphorema*); filaments *adnate to corolla*, slender, short to elongate; anthers opening by longitudinal slits (i.e., *with two slits, often converging apically and opening by an inverted ± V-shaped slit, or pollen sacs confluent and opening by a single distal slit oriented at right angles to the filament, ± curved to U-shaped*), *often ± sagittate*, occasionally asymmetrical, the connective occasionally expanded, the **locules with pollen sac placentoids** (i.e., placenta-like parenchymatous tissue extends from connective into the locule); pollen grains tricolpate to multicolpate, tricolporate to multicolporate, or inaperturate, rarely in tetrads or polyads. *Carpels 2, connate*, sometimes one sterile, rarely pseudomonomerous, *with a single, gynoterminal to gynobasic, unbranched to distally branched* (rarely twice branched) *style*; *ovary superior* to less commonly inferior, unlobed to ± 4-lobed, *with axile* (and then the placentas divided or not, and the locules sometimes secondarily divided, so number of locules is twice the number of stigmas), intruded-parietal, parietal, free-central, or basal *placentation*; stigma punctate to capitate, funnel-shaped, elongate/plumose, or 2-lobate (and the lobes sometimes sensitive) and one lobe sometimes reduced or lacking. Ovules anatropous to hemitropous, rarely orthotropous, with 1 integument, with a thin nucellus, 1 or 2 to numerous per carpel. Nectary a disk, a single gland, or several glands, nectar produced by a corolla spur or by hairs within corolla tube, or absent, and then occasionally rewarding pollinators by oil, produced by densely packed hairs on inner surface of saccate lower lip of corolla. Fruit a variably dehiscent, ± globose to elongated capsule, indehiscent pod, samara, berry, or drupe (with a single pit, or 2 or 4 pits), schizocarp of achenes, nutlets, or drupelets, the fruits occasionally with elongate hook-like appendages, spines or wings; seeds sometimes flattened and/or winged or with fringe or tuft of hairs, occasionally borne on a hook-shaped projection (and then with ballistic dispersal); endosperm present or absent.

CAMPANULIDAE M.J. DONOGHUE & P.D. CANTINO

Trees, shrubs, lianas, vines, or herbs, sometimes rhizomatous, occasionally aquatic; crystals various; stems occasionally with medullary bundles; parenchymatous tissues sometimes with laticifers or secretory canals (with ethereal oils, resins, etc.). Hairs various. Leaves alternate and spiral or 2-ranked, or opposite and decussate, or whorled, simple to variously compound, entire to variously toothed, sometimes lobed or dissected, with pinnate or palmate venation (the latter with converging or diverging major veins), rarely ± parallel, occasionally with domatia; stipules various, present or absent. Inflorescences determinate, indeterminate, or mixed, usually terminal or axillary, sometimes reduced to a solitary flower. Flowers bisexual or unisexual (and plants then monoecious, dioecious, or polygamous), radial or bilateral (due to form of perianth and/or androecium, and then often bilabiate or unilabiate, with or without a dorsal slit, a ligulate flower or a ray flower), with or without a hypanthium, the perianth differentiated (with calyx and corolla) and usually in 2 whorls, rarely reduced to a single whorl. Sepals most commonly 4 or 5, in a single whorl, distinct to more frequently connate, with various aestivation, sometimes highly modified or reduced. Petals most commonly 4 or 5, rarely absent, connate to distinct (*but then with developmental evidence of a ring primordium, so showing "early sympetaly," with various aestivation*). *Stamens usually 4 or 5, usually equaling the corolla lobes*, distinct to connate (by filaments or filaments and anthers) occasionally with staminodia; *filaments usually adnate to corolla*, rarely reduced to 2 stamens that are adnate to style; anthers usually opening by longitudinal slits, *often showing plunger-pollination* (with pollen deposited onto gynoecium, and with secondary presentation to pollinators); pollen grains in monads, usually tricolpate, tricolporate, or modifications thereof, e.g., triporate, polyporate. Carpels usually 2–5, but rarely pseudomonomerous, connate, with as many styles as carpels or a single style, either apically branched or unbranched, sometimes with modifications for pollen reception and presentation; ovary sometimes lobed, superior or inferior, with variable placentation; stigma various. Ovules anatropous to campylotropous, with one integument, with a usually thin nucellus, 1 to numerous per carpel, or reduced to 1 per gynoecium. Nectaries variable or absent. Fruit a variously dehiscent capsule, drupe, berry, nut, achene, or schizocarp (of 2 dry segments) often associated with accessory structures; endosperm present to absent.

Composition: Aquifoliales, Bruniales, Dipsacales, Paracryphiales, Apiales, Escalloniales, Asterales.

AQUIFOLIALES SENFT

Trees, shrubs, or vines (twining), sometimes rhizomatous; crystals solitary or druses; parenchymatous tissues rarely with laticifers (milky sap). Hairs simple, sometimes glandular. *Leaves alternate and spiral* or 2-ranked, *simple*, rarely dissected, entire to serrate, the teeth often theoid (i.e., with deciduous, glandular apex) or spinose, with pinnate or palmate venation; **stipules present**, often minute, occasionally branched or fimbriate, or absent. *Inflorescences determinate, axillary*, but sometimes reduced to a solitary flower, *occasionally borne on adaxial surface of leaves (on midrib). Flowers* bisexual or unisexual (plants dioecious or polygamous), *radial*, with perianth differentiated (of calyx and corolla), in 2 whorls, or petals lost (and then with tepals). Sepals usually 3–6, distinct to ± connate, imbricate, valvate, or with open aestivation. *Petals usually 3–6, usually ± connate*, rarely distinct, imbricate or valvate, rarely absent. *Stamens usually 3–6, in a single whorl*, distinct, often with staminodia in carpellate flowers; filaments free, borne on receptacle, or adnate to base of corolla, ± slender to short and stout; anthers opening by longitudinal slits; pollen grains 3–4-colporate or porate. *Carpels usually 2–6*, sometimes pseudomonomerous, *connate, usually with a single very short style, or lacking style*, or with 2, short to elongate styles; ovary superior or inferior, **with apical, apical-axile**, or intruded-parietal **placentation;** *stigma* elongate, *truncate, or capitate, often ± sitting atop ovary.* **Ovules** anatropous, with 1 integument, with a thin nucellus, **1 or 2 per locule**, or few on parietal placentae. Nectary a disk, nectar secreted by papillose swellings on adaxial petal surface, or absent. **Fruit a drupe,** *often with several pits, occasionally of different appearance on one side than the other, rarely samaroid*, or a berry; endosperm present.

ASTERALES LINK

Trees, shrubs, lianas/vines (twining, scrambling), or herbs, sometimes rhizomatous, rarely aquatics, storing carbohydrates as **inulin** (an oligosaccharide); crystals solitary or druses; *parenchymatous tissues sometimes with laticifers (milky or rarely colored sap) or resin canals*, occasionally with sclereids; stems occasionally with medullary vascular bundles. Hairs various, simple, glandular or eglandular, T-shaped, Y-shaped, variously branched, stellate or peltate scales, sometimes associated with cystoliths. *Leaves alter-*

nate and spiral (rarely 2-ranked) *or opposite and decussate*, or whorled, borne along stem, clustered distally on shoots, or in a basal rosette or the rosettes scattered along the stem, *simple*, but sometimes deeply lobed or dissected, rarely trifoliolate, entire to variously toothed, with pinnate or palmate venation (and the latter with diverging or converging major veins), sometimes with domatia; *stipules absent*. Inflorescences determinate or indeterminate, often aggregated into heads surrounded by an involucre of bracts, terminal or axillary, occasionally reduced to a solitary flower, occasionally scapose. Flowers bisexual or occasionally unisexual (and then monoecious, dioecious, or polygamous), sometimes sterile (and associated with fertile flowers, increasing the attractiveness of the inflorescence), radial or bilateral (due to form of the androecium and perianth or merely the perianth), sometimes with a hypanthium, with perianth differentiated (of calyx and corolla, but calyx often highly modified and sometimes reduced or lost), in 2 whorls, sometimes twisting 90 degrees (half resupinate) or 180 degrees in development (resupinate), rarely heterostylous. *Sepals* usually 5, distinct to connate, valvate, imbricate, or with open aestivation, *often highly modified*, forming a pappus, of 2 to numerous scales, awns, or bristles that are capillary, hairy, minutely barbed, or plumose, or sometimes reduced to a rim or absent. *Petals usually 5, connate* to rarely distinct, and then with tubular to bell-shaped corolla, or with a variously developed adaxial slit (and then 1 or 2-lipped), or with "ray flower" morphology, i.e., with upper lip ± lacking and lower lip elongated, ± 3-lobed, or with "ligulate flower" morphology, i.e., bilateral, elongated, tongue-like, ending in 5 small teeth, or bilabiate, the lobes (or petals, when distinct) **valvate, induplicate-valvate** or occasionally imbricate, one corolla lobe occasionally differentiated from the others, *the petal margins sometimes winged* (i.e., with marginal region differentiated from the central portion of the lobe), occasionally with tufts or fringes of hairs near the base, fringed adaxial appendages, or with the adaxial surface or the margins fringed. *Stamens usually* (2-) *5*, in a single whorl, *distinct to connate (by filaments, anthers, or both);* filaments free, attached to disk at apex of ovary, on rim of hypanthium, ± adnate to corolla tube, or rarely on receptacle, slender, short to elongate; *anthers distinct to connate*, opening by longitudinal slits, *often forming a tube into which the pollen is shed, and often with apical or basal appendages, and the style then growing through this tube, pushing out the pollen, picking up pollen with specialized hairs that later invaginate, with variously positioned and developed, non-invaginating hairs, or with a pollen-cup, and presenting it to floral visitors, after which the stigmas become receptive (i.e., plunger or brush pollination, a secondary pollen presentation mechanism)*, or occasionally reduced to 2 stamens that are adnate to the style (forming a sensitive column); pollen grains usually tricolporate, tricolpate, or triporate to pantoporate. Carpels usually 2–5, connate, with a single, unbranched or apically branched style, *often with pollen-collecting hairs or a pollen-collecting cup at or near apex*, rarely with as many styles as carpels; **ovary inferior to partly inferior**, rarely superior, with axile, parietal, intruded-parietal, free-central, apical or basal placentation; stigma or stigmas capitate, discoid, cylindrical, lobate to dissected, or elongated/decurrent, sometimes restricted to 2 lines, one on each side of each style branch. Ovules anatropous, with a single integument, with a thin nucellus, 1 to numerous in each locule, ± numerous on each placenta, or reduced to 1 per ovary. Nectary a disk or ring of glands above or around ovary, two glands, a ring of nectar-producing cavities between hypanthium and gynoecium, or absent. Fruit a variously dehiscent capsule, berry, drupe (with a single pit, or rarely several pits), indehiscent pod, nut, achene (then crowned by persistent calyx/pappus); endosperm present or absent.

ESCALLONIALES LINK

Composition: Escalloniaceae (incl. Polyosmaceae, Tribelaceae, Eremosynaceae).

BRUNIALES DUMORT.

Shrubs or trees; crystals prismatic or druses. Hairs simple. *Leaves alternate and spiral (and then reduced in size, ± ericoid), or opposite and decussate (and then the petioles joined at the base by a distinct nodal line or flange), simple*, entire to serrate (and then the teeth either glandular or spinose), *with pinnate venation, or reduced, with only a few ± parallel veins*; stipules absent or present and minute. Inflorescences determinate or indeterminate, terminal or axillary, sometimes reduced to a solitary flower. *Flowers bisexual, radial or only slightly bilateral* (due to form of androecium), with perianth differentiated (of calyx and corolla), in 2 whorls. Sepals (4-) 5, distinct or slightly connate, imbricate to valvate. Petals (4-) 5, distinct to connate, imbricate and sometimes contorted. Stamens 2–5, in a single whorl, distinct; filaments free, sometimes adnate to corolla, short to elongate; anthers opening by longitudinal slits, the connective sometimes expanded (and then the pollen sacs then twisted) or forming an apical appendage; pollen grains 3–10-colporate. *Carpels 2* (but sometimes pseudomonomerous), *connate*, with a single, unbranched or apically branched style; *ovary usually inferior or at least partly so*, with axile or intruded-parietal placentation; stigmas/stigma punctate, capitate or lobate. Ovules anatropous, with 1 integument, with thin to ± thick nucellus, 2 to numerous per locule. Nectary a disk or absent. Fruit a ± septicidal or septicidal and loculicidal capsule, berry, or indehiscent and nut-like; seeds occasionally arillate; endosperm present.

APIALES NAKAI

Trees, shrubs, lianas, or herbs, occasionally rhizomatous, and usually aromatic; crystals various, solitary or of druses; *parenchymatous tissues of roots, stems, bark, and leaves often with secretory canals containing ethereal oils and resins, triterpenoid saponins, coumarins, falcarinone* **polyacetylenes** (but condition in Pennantiaceae needs study), *monoterpenes and sesquiterpenes*; stems often proximally with cataphylls, often hollow in internodal regions, rarely forming thorns; plants often with umbelliferose (a trisaccharide) as carbohydrate storage product. Hairs various, sometimes also with prickles. *Leaves alternate and spiral*, rarely 2-ranked, rarely subopposite, simple, sometimes lobed or dissected, or pinnately or palmately compound, entire to serrate or dentate, with pinnate or palmate venation (and then with major veins diverging or less commonly converging), *the petioles often sheathing*; stipules present or absent. *Inflorescences determinate, often forming simple or compound umbels, or panicles, racemes or spikes of umbels*, occasionally condensed into a head, or variously cymose, terminal and/or axillary, rarely reduced to a solitary flower. **Flowers** bisexual or unisexual (and plants then monoecious, dioecious, or polygamous), *usually radial*, **small**, the perianth differentiated (of calyx and corolla), in 2 whorls, but rarely the sepals lost. **Sepals** usually 5, connate to distinct, imbricate, valvate, or with open aestivation, **often reduced or even ± obsolete. Petals** usually 5 (rarely up to 12), **connate (but with well-developed lobes) to distinct** (but with developmental evidence of a ring primordium), imbricate to valvate, sometimes inflexed. *Stamens usually 5*, in a single whorl, rarely numerous, *distinct* to slightly connate (by filaments), with staminodia often in carpellate flowers; filaments *free, arising from receptacle*, or ± adnate to corolla, very short to elongate, terete to flattened; anthers opening by longitudinal slits or terminal pores; pollen grains tricolporate. Carpels 2–5 (rarely numerous), connate, *with a single style or as many styles as carpels, often recurved, and the styles often swollen at base to form a nectar secreting structure (stylopodium) atop ovary;* ovary superior or inferior, with axile, intruded-parietal, or basal-parietal placentation; stigma or stigmas punctate, truncate, capitate, elongate or lobate. **Ovules** anatropous to campylotropous, with 1 integument, with a thin to less commonly thick nucellus, **2 in each locule (but only 1 fertile, or only 1 per locule**, or even only 1 locule of the gynoecium fertile), or 1 to numerous on parietal placentae. *Nectary a disk, stylopodium*, or absent. Fruit a loculicidal (rarely septicidal) capsule, berry, **drupe with 1–5 pits,** *or schizocarp with 2 dry (rarely drupaceous) segments often attached to an entire to deeply forked central stalk (carpophore) and usually with globular to elongated oil canals (vittae) present in the fruit wall*, which may be smooth, ribbed, covered with hairs, scales, bristles, or flattened or winged; seeds occasionally with viscid, resinous, usually red pulp, rarely winged; endosperm present, often with petroselinic acid.

PARACRYPHIALES TAKHT. EX REVEAL

Composition: Paracryphiaceae.

DIPSACALES JUSS. EX BERCHT. & J. PRESL

Trees, shrubs, lianas and herbs, sometimes rhizomatous; crystals various, solitary or druses. Hairs various. **Leaves opposite and decussate**, rarely whorled, *simple, sometimes lobed or divided, to pinnately compound or trifoliolate*, entire to variously toothed, with pinnate to palmate venation (and the latter with diverging major veins), sometimes with domatia; stipules absent or occasionally present, sometimes glandular. *Inflorescences determinate, terminal*, rarely with sterile flowers contributing to attractiveness. *Flowers usually bisexual*, radial or bilateral (due to form of perianth), with perianth differentiated (of calyx and corolla), in 2 whorls. *Sepals (2-) 5, connate*, imbricate or with open aestivation, sometimes reduced in size or modified and pappus-like, ± persistent. *Petals (4-) 5, connate*, imbricate or valvate, *all the same size, or often with 2 upper lobes and 3 lower lobes or with 4 upper lobes and 1 lower lobe. Stamens (1-) 4 or 5*, in a single whorl, but occasionally divided and appearing as 10, distinct; *filaments adnate to corolla*, slender, short to elongate; anthers opening by longitudinal slits; pollen grains tricolpate to tricolporate. *Carpels 2* **or** *3–5, connate*, with a single elongated or very short style, or several short styles; *ovary inferior to half-inferior*, with axile or ± apical placentation, sometimes only 1 locule fertile; stigma or stigmas capitate or strongly lobed. Ovules anatropous, with 1 integument, with a thin to thick nucellus, 1 to numerous in each locule, sometimes only one functional per gynoecium. *Nectar produced by closely packed, unicellular, glandular hairs on lower part of corolla tube, by glandular tissue atop ovary, by a cushion-like group of multicellular hairs*, or absent. Fruit a septicidal capsule, berry, drupe, or achene, rarely associated with persistent bracts or bracteoles; **seeds with vascularized testa**; endosperm present or absent.

Acknowledgments

In addition to the many people who provided photographs and figures, we also thank the many people who helped with various aspects of the manuscript. Ilia Leitch, Jeff Doyle, Jonathan Wendel, and Jim Leebens-Mack provided critical comments on several of the chapters. Many colleagues shared figures and information including Victor Albert, David Baum, Chuck Bell, Birgitta Bremer, Sam Brockington, Barbara Carlsward, Andre Chanderbali, Peter Crane, Chuck Davis, David Dilcher, Jeff Doyle, Jim Doyle, Erika Edwards, Lena Hileman, Scott Hodges, Dianella Howarth, Elizabeth Kellogg, Sangtae Kim, Ilia Leitch, Jim Leebens-Mack, Hong Ma, Susana Magallón, Mike Moore, David Morgan, Dan Nickrent, Paul Manos, Richard Olmstead, Jeff Palmer, Chris Pires, Susanne Renner, Danny Rice, Brad Ruhfel, Stephen Smith, Dennis Stevenson, Ken Sytsma, Miao Sun, Marshall Sunberg, David Tank, John Thompson, Erin Tripp, Yves Van de Peer, Kevin Vanneste, Jonathan Wendel, Joe Williams, Mi-Jeong Yoo. Terry Lott and Clare Carter helped assemble and proof the literature cited. Andy Sinauer provided permission to use all of the figures prepared by Sinauer and Associates from the Soltis et al. 2005 angiosperm volume. A special thanks to Carroll Wood for granting permission to use some of his line drawings of flowering plants. Amy McPherson helped obtain the permissions required from the *American Journal of Botany*. Finally, Kent Perkins and Norris Williams facilitated use of herbarium material in the preparation of the major clade descriptions.

Reference List

Aboy, H.E. 1936. A study of the anatomy and morphology of *Ceratophyllum demersum*. MSc. Thesis, Cornell University, Ithaca, NY.

Ackerman, J.D. 1983. On the evidence for a primitively epiphytic habit in orchids. *Systematic Botany* 8:474–476.

———. 2000. Abiotic pollen and pollination: Ecological, functional, and evolutionary perspectives. *Plant Systematics and Evolution* 222:167–185.

Adams, K.C. 2013. Genomic clues to the ancestral flowering plant. *Science* 342(6165):1456–1457.

Adams, K.C., Y.-L. Qiu, M. Stoutemyer, and J.D. Palmer. 2002. Punctuated evolution of mitochondrial gene content: High and variable rates of mitochondrial gene loss and transfer to the nucleus during angiosperm evolution. *Proceedings of the National Academy of Sciences USA* 99:9905–9912.

Adams, K.L., and J.F. Wendel. 2005. Polyploidy and genome evolution in plants. *Current Opinion in Plant Biology* 8(2):135–141.

Adams, S.P., T.P.V. Hartman, K.Y. Lim, M.W. Chase, M. D. Bennett, I.J. Leitch, and A.R. Leitch. 2001. Loss and recovery of *Arabidopsis*–type telomere repeat sequences 5'–(TTTAGGG)n–3' in the evolution of a major radiation of flowering plants. *Proceeding of the Royal Society of London, Series B* 268:1541–1546.

Agrawal, A.A., and M. Fishbein. 2006. Plant defense syndromes. *Ecology* 87:S132–S149.

———. 2008. Phylogenetic escalation and decline of plant defense strategies. *Proceedings of the National Academy of Sciences USA* 105:10057–10060.

Agrawal, A.A., M. Fishbein, R. Halitschke, A.P. Hastings, D. Rabosky, and S. Rasmann. 2009. Tempo of trait evolution in the milkweeds: Evidence for adaptive radiation. *Proceedings of the National Academy of Sciences USA* 106:18067–18072.

Agrawal, A.A., M. Fishbein, R. Jetter, J.-P. Salminen, J.B. Goldstein, A.E. Freitag, and J.P. Sparks. 2009. Phylogenetic ecology of leaf surface traits in the milkweeds (*Asclepias* spp.): Chemistry, ecophysiology, and insect behavior. *New Phytologist* 183:848–867.

Agrawal, A.A., M.J. Lajeunesse, and M. Fishbein. 2008. Evolution of latex and its constituent defensive chemistry in milkweeds (*Asclepias*). *Entomologia Experimentalis et Applicata* 128:126–138.

Agrawal, A.A., J.-P. Salminen, and M. Fishbein. 2009. Phylogenetic trends in phenolic metabolism of milkweeds (*Asclepias*). *Evolution* 63:663–673.

Ahuja, M.R. 2005. Polyploidy in gymnosperms: Revisited. *Silvae Genetica* 54(2):59–69.

Airy Shaw, H.K. 1964. Plagiopteraceae. *Kew Bulletin* 18:266.

———. 1965. On a new species of the genus *Silvianthus* Hook. f., and on the family Carlemanniaceae. *Kew Bulletin* 19:507–512.

Akkermans, A.D.L., and C. van Dijk. 1981. Non-leguminous root-nodule symbioses with Actinomycetes and *Rhizobium*. *In* W.J. Broughton [ed.], Nitrogen Fixation, vol. 1, Ecology, 57–103. Oxford University Press, Oxford, UK.

Albach, D.C. 1998. Phylogeny of the Asteridae s.l. Master's thesis, Washington State University, Pullman, WA.

Albach, D.C., D.E. Soltis, M.W. Chase, and P. S. Soltis. 2001c. Phylogenetic placement of the enigmatic *Hydrostachys*. *Taxon* 50:781–805.

Albach, D.C., P.S. Soltis, and D.E. Soltis. 2001b. Patterns of embryological and biochemical evolution in the asterids. *Systematic Biology* 26:242–262.

Albach, D.C., P.S. Soltis, D.E. Soltis, and R.G. Olmstead. 2001a. Phylogenetic analysis of the Asteridae s.l. based on sequences of four genes. *Annals of the Missouri Botanical Garden* 88:163–212.

Albert, V.A., A. Backlund, K. Bremer, M.W. Chase, J.R. Manhart, B.D. Mishler, and K.C. Nixon. 1994. Functional constraints and *rbcL* evidence for land plant phylogeny. *Annals of the Missouri Botanical Garden* 81:534–567.

Albert, V.A., M.H.G. Gustafsson, and L. Di Laurenzio. 1998. Ontogenetic systematics, molecular developmental genetics, and the angiosperm petal. *In* D.E. Soltis, P.S. Soltis, and J.J. Doyle [eds.], Molecular Systematics of Plants, II, DNA Sequencing, 349–374. Kluwer Academic Publishers, Boston, MA.

Albert, V.A., D.G. Oppenheimer, and C. Lindqvist. 2002. Pleiotropy, redundancy and the evolution of flowers. *Trends in Plant Science* 7:297–301.

Albert, V.A., D.E. Soltis, J.E. Carlson, W.G. Farmerie, P.K. Wall, D.C. Ilut, T.M. Solow, L.A. Mueller, L.L. Landherr, Y. Hu, M. Buzgo, S. Kim, M.-J. Yoo, M.W. Frohlich, R. Perl-Treves, S.E. Schlarbaum, B.J. Bliss, X. Zhang, S.D. Tanksley, D.G. Oppenheimer, P.S. Soltis, H. Ma, C.W. dePamphilis, and J.H. Leebens-Mack. 2005. Floral gene resources from basal angiosperms for comparative genomics research. *BMC Plant Biology* 5:5. doi:10.1186/1471-2229-5-5.

Albert, V.A., S.E. Williams, and M.W. Chase. 1992. Carnivorous plants: phylogeny and structural evolution. *Science* 257:1491–1495.

Albertson, R.C., J.A. Markert, P.D. Danley, and T.D. Kocher. 1999. Phylogeny of a rapidly evolving clade: The cichlid fishes of Lake Malawi, East Africa. *Proceedings of the National Academy of Sciences USA* 96:5107–5110.

Albesiano, S., and T. Terrazas. 2012. Cladistic analysis of *Trichocereus* (Cactaceae: Cactoideae: Trichocereeae) based on morphological data and chloroplast DNA sequences. *Haseltonia* 17:3–23.

Alcantara, S., and L.G. Lohmann. 2010. Evolution of floral morphology and pollination system in Bignonieae (Bignoniaceae). *American Journal of Botany* 97:782–796.

Alexandersson, R., and S.D. Johnson. 2002. Pollinator-mediated selection on flower-tube length in a hawkmoth-pollinated *Gladiolus* (Iridaceae). *Proceedings of the Royal Society of London B* 269:631–636.

Allard, M.W., M.M. Miyamoto, L. Jarecki, F. Kraus, and M.R. Tennant. 1992. DNA systematics and evolution of the artiodactyl family Bovidae. *Proceedings of the National Academy of Sciences USA* 89:3972–3976.

Allen, O.N., and E.K. Allen. 1976. Symbiotic Nitrogen Fixation in Plants. Cambridge University Press, Cambridge, UK.

Al-Shammary, K.I.A. 1991. Systematic studies of the Saxifragaceae, chiefly from the Southern Hemisphere. PhD thesis, University of Leicester, Leicester, UK.

Al-Shammary, K.I.A., and R.J. Gornall. 1994. Trichome anatomy of the Saxifragaceae s. l. from the southern hemisphere. *Botanical Journal of the Linnean Society* 114:99–131.

Al-Shehbaz, I.A. 1984. The tribes of Cruciferae (Brassicaceae) in the southeastern United States. *Journal of the Arnold Arboretum* 65:343–373.

———. 2012. A generic and tribal synopsis of the Brassicaceae (Cruciferae). *Taxon* 61(5):931–954.

Al-Shehbaz, I.A., M.A. Beilstein, and E.A. Kellogg. 2006. Systematics and phylogeny of the Brassicaceae (Cruciferae): An overview. *Plant Systematics and Evolution* 259(2–4):89–120.

Al-Shehbaz, I.A., and B.G. Schubert. 1989. The Dioscoreaceae of the southeastern United States. *Journal of the Arnold Arboretum* 70:57–95.

Alverson, A.J., X. Wei, D.W. Rice, D.B. Stern, K. Barry, and J.D. Palmer. 2010. Insights into the evolution of mitochondrial genome size from complete sequences of *Citrullus lanatus* and *Cucurbita pepo* (Cucurbitaceae). *Molecular Biology and Evolution* 27(6):1436–1448.

Alverson, W.S., K.G. Karol, D.A. Baum, M.W. Chase, S.M. Swensen, R. McCourt, and K.Y. Sytsma. 1998. Circumscription of the Malvales and relationships to other Rosidae: Evidence from *rbcL* sequence data. *American Journal of Botany* 85:876–887.

Alverson, W.S., B.A. Whitlock, R. Nyffeler, C. Bayer, and D.A. Baum. 1999. Phylogeny of the core Malvales: Evidence from *ndhF* sequence data. *American Journal of Botany* 86:1474–1486.

Amborella Genome Project [Albert, V.A., W.B. Barbazuk, C.W. dePamphilis, J.P. Der, J. Leebens-Mack, H. Ma, J.D. Palmer, S. Rounsley, D. Sankoff, S.C. Schuster, D.E. Soltis, P.S. Soltis, S.R. Wessler, R.A. Wing, J.S.S. Ammiraju, S. Chamala, A.S. Chanderbali, R. Determann, P. Ralph, J. Talag, L. Tomsho, B. Walts, S. Wanke, T.-H. Chang, T. Lan, S. Arikit, M.J. Axtell, S. Ayyampalayam, J.M. Burnette, III, E. De Paoli, J.C. Estill, N.P. Farrell, A. Harkess, Y. Jiao, K. Liu, W. Mei, B.C. Meyers, S. Shahid, E. Wafula, J. Zhai, X. Zhang, L. Carretero-Paulet, E. Lyons, H. Tang, C. Zheng, N.S. Altman, F. Chen, J.-Q. Chen, V. Chiang, B. Fogliani, C. Guo, J.Harholt, C. Job, D. Job, S. Kim, H. Kong, G. Li, L. Li, J. Liu, J. Park, X. Qi, L. Rajjou, V. Burtet–Sarramegna, R. Sederoff, Y.-H. Sun, P. Ulvskov, M. Villegente, J.-Y. Xue, T.-F. Yeh, X. Yu, J.J. Acosta, R.A. Bruenn, A. de Kochko, L.R. Herrera-Estrella, E. Ibarra-Laclette, M. Kirst, S.P. Pissis, V. Poncet, and Amborella Genome PJ. 2013. The *Amborella* genome and the evolution of flowering plants. *Science* 342:1467. doi:10.1126/science.1241089.

Ambrose, B.A., D.R. Lerner, P. Ciceri, C.M. Padilla, M.F. Yanofsky, and R.J. Schmidt. 1999. Genes specifying grass floral organ identity (abstract), XVI International Botanical Congress, St. Louis, Missouri.

———. 2000. Molecular and genetic analyses of the *Silky1* gene reveal conservation in floral organ specification between eudicots and monocots. *Molecular Cell* 5:569–579.

Anderberg, A.A. 1992. The circumscription of the Ericales, and their cladistic relationships to other families of "higher" dicotyledons. *Systematic Botany* 17:660–675.

———. 1993. Cladistic relationships among major clades of the Ericales. *Plant Systematics and Evolution* 184:207–231.

———. 1995. Phylogenetic interrelationships in the order Primulales, with special emphasis on the family circumscription. *Canadian Journal of Botany* 73:1699–1730.

Anderberg, A.A., C. Rydin, and M. Källersjö. 2002. Phylogenetic relationships in the order Ericales s.l.: Analyses of molecular data from five genes from the plastid and mitochondrial genomes. *American Journal of Botany* 89:677–687.

Anderberg, A.A., and B. Stahl. 1995. Phylogenetic interrelationships in the order Primulales, with special emphasis on the family circumscriptions. *Canadian Journal of Botany* 73:1699–1730.

Anderberg, A.A., B. Ståhl, and M. Källersjö. 1998. Phylogenetic relationships in the Primulales inferred from *rbcL* sequence data. *Plant Systematics and Evolution* 211:93–102.

———. 2000. Maesaceae, a new primuloid family in the order Ericales s.l. *Taxon* 49:183–187.

Anderson, C.L., K. Bremer, and E.M. Friis. 2005. Dating phylogenetically basal eudicots using *rbcL* sequences and multiple fossil reference points. *American Journal of Botany* 92(10):1737–1748.

Andersson, L., and A. Antonelli. 2005. Phylogeny of the tribe Cinchoneae (Rubiaceae), its position in Cinchinoideae, and description of a new genus, *Ciliosemina*. *Taxon* 54:17–28.

Andersson, L., and M.W. Chase. 2001. Phylogeny, relationships and classification of Marantaceae. *Botanical Journal of the Linnean Society* 135:275–287.

Andersson, L., and J.H. Rova. 1999. The *rps16* intron and the phylogeny of the Rubioideae (Rubiaceae). *Plant Systematics and Evolution* 214:161–186.

Andreasen, K., and B. Bremer. 2000. Combined phylogenetic analysis in the Rubiaceae–Ixoroideae: Morphology, nuclear and chloroplast DNA data. *American Journal of Botany* 87:1731–1748.

Angulo, D.F., R. Duno de Stefano, and G.W. Stull. Systematics of *Mappia* (Icacinaceae), an endemic genus of tropical America. *Phytotaxa* 116:1–18.

Antonelli, A. 2008. Higher level phylogeny and evolutionary trends in Campanulaceae subfam. Lobelioideae: Molecular signal overshadows morphology. *Molecular Phylogenetics and Evolution* 46:1–18.

———. 2009. Have giant lobelias evolved several times independently? Life form shifts and historical biogeography of the cosmopolitan and highly diverse subfamily Lobelioideae (Campanulaceae). *BMC Biology* 7:82. doi:10.1186/1741-7007-7-82.

Antonius, K., and H. Ahokas. 1996. Flow cytometric determination of polyploidy level in spontaneous clones of strawberries. *Hereditas* 124:285.

APG. 1998. An ordinal classification for the families of flowering plants. *Annals of the Missouri Botanical Garden* 85:531–553.

APG II. 2003. An update of the Angiosperm Phylogeny Group classification for the orders and families of flowering plants. *Botanical Journal of the Linnean Society* 141:399–436.

APG III. 2009. An update of the angiosperm phylogeny group classification for the orders and families of flowering plants: APG III. *Botanical Journal of the Linnean Society* 161:105–121.

APG IV. 2016. An update of the Angiosperm Phylogeny Group classification for the orders and families of flowering plants. APG IV. *Botanical Journal of the Linnean Society* 181:1–20.

Appelhans, M.S., P.J.A. Keßler, E. Smets, S.G. Razafimandimbison, and S.B. Janssens. 2012. Age and historical biogeography of the pantropically distributed Spathelioideae (Rutaceae, Sapindales). *Journal of Biogeography* 39(7):1235–1250.

Appelhans, M.S., E. Smets, S.G. Razafimandimbison, T. Haevermans, E.J. van Marle, A. Couloux, H. Rabarison, M. Randrianarivelojosia, and P.J.A. Kessler. 2011. Phylogeny, evolutionary trends and classification of the *Spathelia–Ptaeroxylon* clade: Morphological and molecular insights. *Annals of Botany* 107(8):1259–1277.

Applequist, W.L., and D.B. Pratt. 2005. The Malagasy endemic *Dendroportulaca* (Portulacaceae) is referable to *Deeringia* (Amaranthaceae): Molecular and morphological evidence. *Taxon* 54:681–687.

Applequist, W.L., W.L. Wagner, E.Z. Zimmer, and M. Nepokroeff. 2006. Molecular evidence resolving the systematic position of *Hectorella* (Portulacaceae). *Systematic Botany* 31:310–319.

Applequist, W.L., and R.S. Wallace. 2001. Phylogeny of the portulacaceous cohort based on *ndhF* sequence data. *Systematic Botany* 26:406–419.

Arabidopsis Genome Initiative (AGI). 2000. Analysis of the genome sequence of the flowering plant *Arabidopsis thaliana*. *Nature* 408:796–815.

Arakaki, M., P.-A. Christin, R. Nyffeler, A. Lendel, U. Eggli, R.M. Ogburn, E. Spriggs, M.J. Moore, and E.J. Edwards. 2011. Contemporaneous and recent radiations of the world's major succulent plant lineages. *Proceeding of the National Academy of Sciences USA* 108:8379–8384.

Arakaki, M., D.E. Soltis, and P. Speranza. 2007. New chromosome counts and evidence of polyploidy in *Haagerocereus* and related genera in tribe Trichocereeae and other tribes of Cactaceae. *Brittonia* 59:290–297.

Arber, A. 1920. Water Plants: A Study of Aquatic Angiosperms. Cambridge University Press, Cambridge, UK.

———. 1921. The leaf structure of Iridaceae, considered in relation to the phyllode theory. *Annals of Botany (London)* 35:301–336.

———. 1925. Monocotyledons: A Morphological Study. Cambridge University Press, Cambridge, UK.

Arber, E.A.N., and J. Parkin. 1907. On the origin of angiosperms. *Journal of the Linnean Society, Botany,* 38:29–80.

———. 1908. Studies on the evolution of the angiosperms: The relationship of the angiosperms to the Gnetales. *Annals of Botany* 22:489–515.

Argus, G.W. 1997. Infrageneric classification of *Salix* (Salicaceae) in the New World. *Systematic Botany Monographs* 52:1–121.

Arias, S., T. Terrazas, H. Arreola-Nava, M. Vazquéz-Sánchez, and K. Cameron. 2005. Phylogenetic relationships in *Peniocereus* (Cactaceae) inferred from plastid DNA sequence data. *Journal of Plant Research* 118:317–328.

Armbruster, W.S. 1992. Phylogeny and the evolution of plant-animal interactions. *BioScience* 42:12–20.

———. 1994. Evolution of plant pollination systems: Hypotheses and tests with the neotropical vine *Dalechampia*. *Evolution* 47:1480–1505.

———. 1996. Evolution of floral morphology and function: An integrative approach to adaptation, constraint, and compromise in *Dalechampia* (Euphorbiaceae). *In* D.G. Lloyd, and S.C.H. Barrett [eds.], Floral Biology. Studies on Floral Evolution in Animal-Pollinated Plants, 241–272. Chapman and Hall, New York, NY.

Armbruster, W.S., and B.G. Baldwin. 1998. Switch from specialized to generalized pollination. *Nature* 394:632. doi:10.1038/29210.

Armbruster, W.S., E.M. Debevec, and M.F. Willson. 2002. Evolution of syncarpy in angiosperms: Theoretical and phylogenetic analyses of the effects of carpel fusion on offspring quantity and quality. *Journal of Evolutionary Biology* 15:657–672.

Armbruster, W.S., J. Lee, and B.G. Baldwin. 2009. Macroevolutionary patterns of defense and pollination of *Dalechampia* vines: Adaptation, exaptation and evolutionary novelty. *Proceedings of the National Academy of Sciences USA* 106:18085–18090.

Ashworth, V.E.T.M. 2000. Phylogenetic relationships in Phoradendreae (Viscaceae) inferred from three regions of the nuclear ribosomal cistron. I. Major lineages and paraphyly of *Phoradendron*. *Systematic Botany* 25:349–370.

Asmussen, C.B., W.J. Baker, and J. Dransfield. 2000. Phylogeny of the palm family (Arecaceae) based on *rps*16 intron and *trnL–trnF* plastid DNA sequences. *In* K.L. Wilson, and D.A. Morrison [eds.], Monocots: Systematics and Evolution, 525–537. CSIRO, Melbourne, Australia.

Asmussen, C.B., and M.W. Chase. 2001. Coding and noncoding plastid DNA in palm systematics. *American Journal of Botany* 88:1103–1117.

Aulenback, K.R. 2009. Identification guide to the fossil plants of the Horseshoe Canyon Formation of Drumheller, Alberta. University of Calgary Press, Calgary, Canada.

Aury, J.-M., O. Jaillon, L. Duret, B. Noel, C. Jubin, B.M. Porcel, B. Segurens, V. Daubin, V. Anthouard, N. Aiach, O. Arnaiz, A. Billaut, J. Beisson, I. Blanc, K. Bouhouche, F. Camara, S. Duharcourt, R. Guigo, D. Gogendeau, M. Katinka, A.-M. Keller, R. Kissmehl, C. Klotz, F. Koll, A. Le Mouel, G. Lepere, S. Malinsky, M. Nowacki, J.K. Nowak, H. Plattner, J. Poulain, F. Ruiz, V. Serrano, M. Zagulski, P. Dessen, M. Betermier, J. Weissenbach, C. Scarpelli, V. Schaechter, L. Sperling, E. Meyer, J. Cohen, and P. Wincker. 2006. Global trends of whole-genome duplications revealed by the ciliate *Paramecium tetraurelia*. *Nature* 444(7116):171–178.

Avise, J.C. 1994. Molecular Markers, Natural History, and Evolution. Chapman and Hall, New York, NY.

Ax, P. 1987. The Phylogenetic System: The Systematization of Organisms on the Basis of their Phylogenesis. John Wiley and Sons, Chichester, UK.

Axelrod, D.I. 1952. A theory of angiosperm evolution. *Evolution* 6:29–60.

———. 1970. Mesozoic paleo-geography and early angiosperm history. *Botanical Review* 36:277–319.

———. 1972. Edaphic aridity as a factor in angiosperm evolution. *American Naturalist* 106:311–320.

Axsmith, B.J., E.L. Taylor, and T.N. Taylor. 1998. The limitations of molecular systematics: A palaeobotanical perspective. *Taxon* 47:105–108.

Ayers, T.J. 1990. Systematics of *Heterotoma* (Campanulaceae) and the evolution of nectar spurs in the New World Lobelioideae. *Systematic Botany* 15:296–327.

Azuma, T., T. Kajita, J. Yokoyama, and H. Ohashi. 2000. Phylogenetic relationships of *Salix* (Salicaceae) based on *rbcL* sequence data. *American Journal of Botany* 87:67–75.

Baas, P. 1984. Vegetative anatomy of *Berberidopsis* and *Streptothamnus* (Flacourtiaceae). *Blumea* 30:39–44.

Baas, P., R. Geesink, W.A. Van Heel, and H.J. Muller. 1979. The affinities of *Plagiopteron suaveolens* Griff. (Plagiopteraceae). *Grana* 18:69–89.

Baas, P., E. Wheeler, and M. Chase. 2000. Dicotyledonous wood anatomy and the APG system of angiosperm and classification. *Botanical Journal of the Linnean Society* 134:3–17.

Bachelier, J.B., and P.K. Endress. 2007. Development of inflorescences, cupules, and flowers in *Amphipterygium* and comparison with *Pistacia* (Anacardiaceae). *International Journal of Plant Sciences* 168(9):1237–1253.

———. 2008. Floral structure of *Kirkia* (Kirkiaceae) and its position in Sapindales. *Annals of Botany* 102(4):539–550.

———. 2009. Comparative floral morphology and anatomy of Anacardiaceae and Burseraceae (Sapindales), with a special focus on gynoecium structure and evolution. *Botanical Journal of the Linnean Society* 159(4):499–571.

Bachelier, J.B., P.K. Endress, and L.P. Ronse De Craene. 2011. Comparative floral structure and development of Nitrariaceae (Sapindales) and systematic implications. *In* L. Wanntorp, and L.P. Ronse De Craene [eds.], Flowers on the Tree of Life, 181–217. Cambridge University Press, Cambridge, UK.

Bachelier, J.B., and W.E. Friedman. 2011. Female gamete competition in an ancient angiosperm lineage. *Proceedings of the National Academy of Sciences USA* 108:12360–12365.

Bachmann, K. 1983. Evolutionary genetics and the genetic control of morphogenesis in flowering plants. *Evolutionary Biology* 15:157–208.

Backlund, A., and B. Bremer. 1997. Phylogeny of the Asteridae s.str. based on *rbcL* sequences, with particular reference to the Dipsacales. *Plant Systematics and Evolution* 207:225–254.

———. 1998. To be or not to be: Principles of classification and monotypic plant families. *Taxon* 47:391–401.

Backlund, A., and N. Pyck. 1998. Diervillaceae and Linnaeaceae, two new families of caprifolioids. *Taxon* 47:657–661.

Backlund, M., B. Oxelman, and B. Bremer. 2000. Phylogenetic relationships within the Gentianales based on *ndhF* and *rbcL* sequences with particular reference to the Loganiaceae. *American Journal of Botany* 87:1029–1043.

Bailey, C.D., M.A. Koch, M. Mayer. K. Mummenhoff, S.L. O'Kane, Jr., S.I. Warwick, M.D. Windham, and I.A. Al-Shehbaz. 2006. Toward a global phylogeny of the Brassicaceae. *Molecular Biology and Evolution* 23:2142–2160.

Bailey, I.W. 1944a. The development of vessels in angiosperms and its significance in morphological research. *American Journal of Botany* 31:421–428.

———. 1944b. The comparative morphology of the Winteraceae. *Journal of the Arnold Arboretum* 97–103.

———. 1953. Evolution of the tracheary tissue of land plants. *American Journal of Botany* 40:4–8.

———. 1956. Nodal anatomy in retrospect. *Journal of the Arnold Arboretum* 37:269–287.

Bailey, I.W., and B.G.L. Swamy. 1948. *Amborella trichopoda* Baill., a new morphological type of vesselless dicotyledon. *Journal of the Arnold Arboretum* 29(3):245–254.

———.1951. The conduplicate carpel and its initial trends of specialization. *American Journal of Botany* 38:373–379.

Baillon, H. 1866–1895. Histoire des Plantes. Hachette, Paris, France.

Bainard, J.D., and J.C. Villarreal. 2013. Genome size increases in recently diverged hornwort clades. *Genome* 56(8):431–435.

Baker, M.A., J.P. Rebman, B.D. Parfitt, D.J. Pinkava, and A.D. Zimmerman. 2009. Chromosome numbers in some cacti of western North America–VIII. *Haseltonia* 15:117–134.

Baker, S.C., K. Robinson-Beers, J.M. Villanueva, J.C. Gaiser, and C.S. Gasser. 1997. Interactions among genes regulating ovule development in *Arabidopsis thaliana*. *Genetics* 145:1109–1124.

Baker, W.J., C.B. Asmussen, S.C. Barrow, J. Dransfield, and T.A. Hedderson. 1999. A phylogenetic study of the palm family (Palmae) based on chloroplast DNA sequences from the *trnL–trnF* region. *Plant Systematics and Evolution* 219:111–126.

Bakker, F.T., F. Breman, and V. Merckx. 2006. DNA sequence evolution in fast evolving mitochondrial DNA *nad1* exons in Geraniaceae and Plantaginaceae. *Taxon* 55(4):887–896.

Bakker, F.T., D.D. Vassiliades, C.M. Morton, and V. Savolainen. 1988. Phylogenetic relationships of *Biebersteinia* Stephen (Geraniaceae) inferred from *rbcL* and *atpB* sequence comparisons. *Botanical Journal Linnean Society* 127:149–158.

Baldwin, B.G., and M.J. Sanderson. 1998. Age and rate of diversification of the Hawaiian silversword alliance (Compositae). *Proceedings of the National Academy of Sciences USA* 95:9402–9406.

Balogh, P. 1982. Generic redefinition in subtribe Spiranthinae (Orchidaceae). *American Journal of Botany* 69:1119–1132.

Barcenas, R.T., C. Yesson, and J.A. Hawkins. 2011. Molecular systematics of the Cactaceae. *Cladistics* 27:470–489.

Barker, M.S., N.C. Kane, M. Matvienko, A. Kozik, W. Michelmore, S.J. Knapp, and L.H. Rieseberg. 2008. Multiple paleopolyploidizations during the evolution of the Compositae reveal parallel patterns of duplicate gene retention after millions of years. *Molecular Biology and Evolution* 25(11):2445–2455.

Barker, N.P., L.G. Clark, J.I. Davis, M.R. Duvall, G.F. Guala, C. Hsiao, E.A. Kellogg, H.P. Linder, R.J. Mason-Gamer, S.Y. Mathews, M.P. Simmons, R.J. Soreng, R.E. Spangler, and Grass Phylogeny Working G. 2001. Phylogeny and subfamilial classification of the grasses (Poaceae). *Annals of the Missouri Botanical Garden* 88:373–457.

Barker, N.P., P.H. Weston, F. Rutschmann, and H. Sauquet. 2007. Molecular dating of the 'Gondwanan' plant family Proteaceae is only partially congruent with the timing of the break-up of Gondwana. *Journal of Biogeography* 34(12):2012–2027.

Barker, W.R., G.L. Nesom, P.M. Beardsley, and N.S. Fraga. 2012. A taxonomic conspectus of Phrymaceae: A narrowed circumscription for *Mimulus*, new and resurrected genera, and new names and combinations. *Phytoneuron* 39:1–60.

Barkman, T.J., M. Bendiksby, S.-H. Lim, K.M. Salleh, J. Nais, D. Madulid, and T. Schumacher. 2008. Accelerated rates of floral evolution at the upper size limit for flowers. *Current Biology* 18(19):1508–1513.

Barkman, T.J., G. Chenery, J.R. McNeal, J. Lyons-Weiler, and C.W. dePamphilis. 2000. Independent and combined analysis of sequences from all three genomic compartments converge to the root of flowering plant phylogeny. *Proceedings of the National Academy of Sciences USA* 97:13166–13171.

Barkman, T.J., S.-H. Lim, K.M. Salleh, and N.D.J. Nais. 2004. Mitochondrial DNA sequences reveal the photosynthetic relatives of *Rafflesia*, the world's largest flower. *Proceedings of the National Academy of Sciences USA* 101:787–792.

Barkman, T.J., J.R. McNeal, S.-H. Lim, G. Coat, H.B. Croom, N.D. Young, and C.W. dePamphilis. 2007. Mitochondrial DNA sequences suggest at least 11 origins of parasitism in Angiosperms and and reveals genomic chimerism in parasitic plants. *BMC Evolutionary Biology* 7:248. doi:10.1186/1471-2148-7-248.

Barraclough, T.G., and G. Reeves. 2005. The causes of speciation in flowering plant lineages: Species-level DNA trees in the African genus *Protea*. *Regnum Vegetabile* 143:31–46.

Barreda, V.D., L. Palazzesi, L. Katinas, J.V. Crisci, M.C. Tellería, K. Bremer, M.G. Passala, F. Bechis, and R. Corsolini. 2012. An extinct Eocene taxon of the daisy family (Asteraceae): Evolutionary, ecological and biogeographical implications. *Annals of Botany* 109:127–134.

Barreda, V.D., L. Palazzesi, M.C. Tellería, L. Katinas, and J.V. Crisci. 2010. Fossil pollen indicates an explosive radiation of basal Asteracean lineages and allied families during Oligocene and Miocene times in the Southern Hemisphere. *Review of Palaeobotany and Palynology* 160:102–110.

Barrett, C.F., J.I. Davis, J. Leebens-Mack, J.G. Conran, and D.W. Stevenson. 2013. Plastid genomes and deep relationships among the commelinid monocot angiosperms. *Cladistics* 29:65–87.

Barrett, C.F., C.D. Specht, J. Leebens-Mack, D.W. Stevenson, W.B. Zomlefer, and J.I. Davis. 2014. Resolving ancient radiations: Can complete plastid gene sets elucidate deep relationships among the tropical gingers (Zingiberales)? *Annals of Botany* 113:119–133.

Barrett, S.C.H. 1992. Evolution and Function of Heterostyly. Springer-Verlag, Berlin, Germany.

———. 1995. Mating-system evolution in flowering plants: Micro- and macroevolutionary approaches. *Acta Botanica Neerlandica* 44:385–402.

———. 1998. The evolution of mating strategies in flowering plants. *Trends in Plant Science* 3:335–341.

———. 2002. The evolution of plant sexual diversity. *Nature Reviews Genetics* 3:274–284.

———. 2003. Mating strategies in flowering plants: The outcrossing-selfing paradigm and beyond. *Philosophical Transactions of the Royal Society of London B* 358:991–1004.

——— [ed.]. 2008. Major evolutionary transitions in flowering plant reproduction. University of Chicago Press, Chicago, IL.

———. 2010a. Darwin's legacy: The forms, function and sexual diversity of flowers. *Philosophical Transactions of the Royal Society of London B* 365:351–368.

———. 2010b. Understanding plant reproductive diversity. *Philosophical Transactions of the Royal Society of London B* 365:99–109.

———. 2013. The evolution of plant reproductive systems: How often are transitions irreversible? *Proceedings of the Royal Society of London B* 280(1765):20130913. doi:10.1098/rspb.2013.0913.

Barrett, S.C.H., and S.W. Graham. 1997. Adaptive radiation in the aquatic plant family Pontederiaceae. Insights from phylogenetic analysis. *In* T.J. Givnish, and K.J. Sytsma [eds.], Molecular Evolution and Adaptive Radiation, 225–258. Cambridge University Press, Cambridge, UK.

Barrett, S.C.H., L.D. Harder, and A.C. Worley. 1996. The comparative biology of pollination and mating in flowering plants. *Philosophical Transactions of the Royal Society of London B* 351:1271–1280.

Barrett, S.C.H., L.K. Jesson, and A.M. Baker. 2000. The evolution and function of stylar polymorphisms in flowering plants. *Annals of Botany (Supplement A)* 85:253–265.

Barrier, M., B.G. Baldwin, R.H. Robichaux, and M.D. Purugganan. 1999. Interspecific hybrid ancestry of a plant adaptive radiation: Allopolyploidy of the Hawaiian silversword alliance (Asteraceae) inferred from floral homeotic gene duplications. *Molecular Biology and Evolution* 16:1105–1113.

Barrio, R.Á., A. Hernández-Machado, C. Varea, J.R. Romero-Arias, and E. Álvarez-Buylla. 2010. Flower development as an interplay between dynamical physical fields and generic networks. *PLoS ONE* 5(10):e13523. doi:10.1371/journal.pone.0013523.

Barthlott, W., S. Porembski, E. Fischer, and B. Gemmel. 1998. First protozoa-trapping plant found. *Nature* 392:447. doi:10.1038/33037.

Barthlott, W., S. Porembski, R. Seine, and I. Theisen. 2007. The Curious World of Carnivorous Plants: A Comprehensive Guide to Their Biology and Cultivation. Timber Press, Portland, OR.

Barthlott, W., and I. Theisen. 1995. Epicuticular wax ultra-structure and classification of Ranunculiflorae. *In* U. Jensen, and J.W. Kadereit [eds.], Plant Systematics and Evolution, 39–45. Springer, Vienna, Austria.

Bartlett, M.E., and C.D. Specht. 2011. Changes in expression pattern of the *TEOSINTE BRANCHED1*-like genes in the Zingiberales provide a mechanism for evolutionary shifts in symmetry accross the order. *American Journal of Botany* 98:1–17.

Basinger, J.F. 1976. *Paleorosa similkameenensis*, gen. et sp. nov., permineralized flowers (Rosaceae) from the Eocene of British Columbia. *Canadian Journal of Botany* 54:2293–2305.

Basinger, J.F., and D.C. Christophel. 1985. Fossil flowers and leaves of the Ebenaceae from the Eocene of southern Australia. *Canadian Journal of Botany* 63:1825–1843.

Basinger, J.F., and D.L. Dilcher. 1984. Ancient bisexual flowers. *Science* 224:511–513.

Bateman, R.M., and W.A. DiMichele. 1994. Saltational evolution of form in plants: A neoGoldschmidtian synthesis. *In* D.S. Ingram, and A. Hudson [eds.], Shape and Form in Plants and Fungi, 63–100. Academic Press, London, UK.

Bate-Smith, E.C. 1965. Recent progress in the chemical taxonomy of some phenolic constituents of plants. *Bulletin de la Société Botanique de France, Mémoire* 1965:16–28.

Bate-Smith, E.C., I.K. Ferguson, K. Hutson, S.R. Jensen, B.J. Nielsen, and T. Swain. 1975. Phytochemical interrelationships in the Cornaceae. *Biochemical Systematics and Ecology* 3:79–89.

Baum, B.R. 1992. Combining trees as a way of combining data sets for phylogenetic inference, and the desirability of combining gene trees. *Taxon* 41:3–10.

Baum, D.A. 1998. The evolution of plant development. *Current Opinions in Plant Biology* 1:79–86.

Baum, D.A., J. Doebley, V.F. Irish, and E.M. Kramer. 2002. Response: Missing links: The genetic architecture of flower and floral diversification. *Trends in Plant Science* 7:31–34.

Baum, D.A., and M.J. Donoghue. 2002. Transference of function, heterotopy and the evolution of plant development. *In* Q.C.B. Cronk [ed.], Developmental Genetics and Plant Evolution, 52–69. Taylor and Francis, London, UK.

Baum, D.A., R.L. Small, and J.F. Wendel. 1998. Biogeography and floral evolution of baobabs (*Adansonia*, Bombaceae) as inferred from multiple data sets. *Systematic Biology* 47:181–207.

Baum, D.A., and B.A. Whitlock. 1999. Plant development: Genetic clues to petal evolution. *Current Biology* 9:R525–R527.

Bauwe, H. 2011. Photorespiration: The bridge to C-4 photosynthesis. *C4 Photosynthesis and Related CO2 Concentrating Mechanisms* 32:81–108.

Bawa, S.B. 1970. Haloragaceae. *Bulletin of the Indian National Science Academy* 41:226–229.

Bayer, C. 1999. The bicolor unit—homology and transformation of an inflorescence structure unique to core Malvales. *Plant Systematics and Evolution* 214:187–198.

Bayer, C., M.W. Chase, and M.F. Fay. 1998. Muntingiaceae, a new family of dicotyledons with malvalean affinities. *Taxon* 47:37–42.

Bayer, C., M.F. Fay, A. De Bruijn, V. Savolainen, C.M. Morton, K. Kubitzki, and M.W. Chase. 1999. Support for an expanded family concept of Malvaceae within a recircumscribed order Malvales: A combined analysis of plastid *atpB* and *rbcL* sequences. *Botanical Journal of the Linnean Society* 129:267–303.

Bayer, C., and K. Kubitzki. 1996. Inflorescence morphology of some Australian Lasiopetaleae (Sterculiaceae). *Telopea* 6:721–728.

Beardsley, P.M., and R.G. Olmstead. 2002. Redefining Phrymaceae: The placement of *Mimulus*, tribe Mimuleae, and *Phryma*. *American Journal of Botany* 89:1093–1102.

Beardsley, P.M., S. Schoenig, and R.G. Olmstead. 2001. Radiation of *Mimulus* (Phrymaceae) in western North America: Evolution of polyploidy, woodiness, and pollination syndromes. *In* Botany 2001: Plants and People, 100, Abstracts. Albuquerque, NM.

Beardsley, P.M., S.F. Schoenig, J.B. Whittall, and R.G. Olmstead. 2004. Patterns of evolution in western North American *Mimulus* (Phrymaceae). *American Journal of Botany* 91:474–489.

Beaulieu, J.M., B.C. O'Meara, and M.J. Donoghue. 2013. Identifying hidden rate changes in the evolution of a binary morphological character: The evolution of plant habit in campanulid Angiosperms. *Systematic Biology* 62(5):725–737.

Beaulieu, J.M., R.H. Ree, J. Cavender-Bares, G.D. Weiblen, and M.J. Donoghue. 2012. Synthesizing phylogenetic knowledge for ecological research. *Ecology* 93:S4–S13.

Beaulieu, J.T., and M.J. Donoghue. 2013. Fruit evolution and diversification in campanulid Angiosperms. *Evolution* 67:3132–3144.

Bedell, H.G. 1980. A taxonomic and morphological reevaluation of the Stegnospermaceae (Caryophyllales). *Systematic Botany* 5:419–431.

Behnke, H.-D. 1967. Ultrastructure of sieve-element plastids in Caryophyllales (Centrospermae): Evidence for the delimitation and classification of the order. *Plant Systematics and Evolution* 126:31–54.

———. 1969. Die Siebröhren-Plastiden bei Monocotylen. *Naturwissenschaften* 55:140–141.

———. 1971. Zum Feinbau der Siebröhrenplastiden von *Aristolochia* und *Asarum* (Aristolochiaceae). *Planta* 97:62–69.

———. 1976a. A tabulated survey of some characters of systematic importance in centrospermous families. *Plant Systematics and Evolution* 125:95–98.

———. 1976b. Ultrastructure of sieve-element plastids in Caryophyllales (Centrospermae): Evidence for the delimitation and classification of the order. *Plant Systematics and Evolution* 126:31–54.

———. 1977. Zur skulptur der pollen-exine bei drei Centrospermen (*Gisekia*, *Limeum*, *Hectorella*), bei Gyrostemonaceen und Rhabdodendraceen. *Plant Systematics and Evolution* 128:227–235.

———. 1981. Sieve-element characters. *Nordic Journal of Botany* 1:381–400.

———. 1982. *Geocarpon minimum*: Sieve-element plastids as additional evidence for its inclusion in Caryophyllaceae. *Taxon* 31:45–47.

———. 1984. Ultrastructure of sieve-element plastids of Myrtales and allied groups. *Annals of the Missouri Botanical Garden* 71:824–831.

———. 1988a. Sieve-element plastids, phloem protein, and evolution of the flowering plants: III. Magnoliidae. *Taxon* 37:699–732.

———. 1988b. Sieve-element plastids and systematic relationships of Rhizophoraceae, Anisophylleaceae and allied groups. *Annals of the Missouri Botanical Garden* 75:1387–1409.

———. 1991. Sieve-element characters of *Ticodendron*. *Annals of the Missouri Botanical Garden* 78:131–134.

———. 1993. Further studies of the sieve-element plastids of the Caryophyllales including *Barbeuia*, *Corrigiola*, *Lyallia*, *Microtea*, *Sarcobatus*, and *Telephium*. *Plant Systematics and Evolution* 186:231–243.

———. 1995. Sieve-element plastids, phloem proteins, and evolution in the Ranunculanae. *Plant Systematics and Evolution, Supplement* 9:25–37.

———. 1997. Sarcobataceae – a new family of Caryophyllales. *Taxon* 46:495–507.

———. 1999. P-type sieve-element plastid present in members of the tribes Triplareae and Coccolobeae (Polygonaceae) renew the links between the Polygonales and the Caryophyllales. *Plant Systematics and Evolution* 214:15–27.

———. 2000. Forms and sizes of sieve-element plastids and evolution of the monocotyledons. *In* K.L. Wilson, and D.A. Morrison [eds.], Monocots: Systematics and Evolution, 163–188. CSIRO, Collingwood, Victoria, Australia.

Behnke, H.-D., and W. Barthlott. 1983. New evidence from ultrastructural and micro-morphological fields in angiosperm classification. *Nordic Journal of Botany* 3:343–66.

Behnke, H.-D., J. Treutlein, M. Wink, K. Kramer, C. Schneider, and P.C. Kao. 2000. Systematics and evolution of Velloziaceae, with special reference to sieve-element plastids and *rbcL* sequence data. *Botanical Journal of the Linnean Society* 134:93–129.

Beilstein, M.A., I.A. Al-Shehbaz, and E.A. Kellogg. 2006. Brassicaceae phylogeny and trichome evolution. *American Journal of Botany* 93(4):607–619.

Beilstein, M.A., I.A. Al-Shehbaz, S. Mathews, and E.A. Kellogg. 2008. Brassicaceae phylogeny inferred from *phytochrome A* and *ndhF* sequence data: Tribes and trichomes revisited. *American Journal of Botany* 95(10):1307–1327.

Beilstein, M.A., N.S. Nagalingum, M.D. Clements, S.R. Manchester, and S. Mathews. 2010. Dated molecular phylogenies indicate a Miocene origin for *Arabidopsis thaliana*. *Proceedings of the National Academy of Sciences USA* 107(43):18724–18728.

Bekaert, M., P.P. Edger, C.M. Hudson, J.C. Pires, and G.C. Conant. 2012. Metabolic and evolutionary costs of herbivory defense: Systems biology of glucosinolate synthesis. *New Phytologist* 196(2):596–605.

Bell, C.D., and M.J. Donoghue. 2005a. Dating the Dipsacales: Comparing models, genes, and evolutionary implications. *American Journal of Botany* 92(2):284–296.

———. 2005b. Phylogeny and biogeography of Valerianaceae (Dipsacales) with special reference to the South American valerians. *Organisms Diversity & Evolution* 5:147–159.

Bell, C.D., D.E. Soltis, and P.S. Soltis. 2005. The age of the angiosperms: A molecular timescale without a clock. *Evolution* 59(6):1245–1258.

———. 2010. The age and diversification of the Angiosperms re-revisited. *American Journal of Botany* 97:1296–1303.

Bello, A., M.W. Chase, R.G. Olmstead, N. Rønsted, and D. Albach. 2002. *Plantago* (Plantaginaceae) is the sister genus of the páramo endemic, *Aragoa* (Lamiales): Evidence from plastid *rbcL* and nuclear ITS sequence data. *Kew Bulletin* 57:585–597.

Bello, M.A., A. Bruneau, F. Forest, and J.A. Hawkins. 2009. Elusive relationships within order Fabales: Phylogenetic analyses using *matK* and *rbcL* sequence data. *Systematic Botany* 34:102–114.

Bello, M.A., J.A. Hawkins, and P.J. Rudall. 2007. Floral morphology and development in Quillajaceae and Surianaceae (Faales), the species-poor relatives of Leguminosae and Polygalaceae. *Annals of Botany* 101:1491–1505.

———. 2010. Floral ontogeny in Polygalaceae and ist bearing on the homologies of keeled flowers in Fabales. *International Journal of Plant Sciences* 171:482–498.

Bellot, S., N. Cusimano, S. Luo, G. Sun, S. Zarre, A. Gröger, E. Temsch, and S.S. Renner. 2016. Assembled plastid and mitochondrial genomes, as well as nuclear genes, place the parasite family Cynomoriaceae in the Saxifragales. *Genome Biology and Evolution* 8:2214–2230. doi:10.1093/gbe/evw147.

Bendiksby, M., T. Schumacher, G. Gussarova, J. Nais, K. Mat-Salleh, N. Sofiyanti, D. Madulid, S.A. Smith, and T. Barkman. 2010. Elucidating the evolutionary history of the Southeast Asian, holoparasitic, giant-flowered Rafflesiaceae: Pliocene vicariance, morphological convergence and character displacement. *Molecular Phylogenetics and Evolution* 57(2):620–633.

Bennett, J.R. 2008. Revision of *Solanum* section *Regmandra* (Solanaceae). *Edinburgh Journal of Botany* 65:69–112.

Bennett, J.R., and S. Mathews. 2006. Phylogeny of the parasitic plant family Orobanchaceae inferred from phytochrome A. *American Journal of Botany* 93(7):1039–1051.

Bennett, M.D. 1972. Nuclear DNA content and minimum generation time in herbaceous plants. *Proceedings of the Royal Society of London B* 181:109–135.

———. 1987. Variation in genomic form in plants and its ecological implications. *New Phytologist* 106:177–200.

———. 1998. Plant genome values: How much do we know? *Proceedings of the National Academy of Sciences USA* 95:2011–2016.

Bennett, M.D., A. Cox, and I.J. Leitch. 1997. Angiosperms DNA C-values database. http://www.rbgkew.org.uk/cval/database1.html.

Bennett, M.D., and I.J. Leitch. 1995. Nuclear DNA amounts in Angiosperms. *Annals of Botany* 80:169–196.

———. 2001. Angiosperm DNA C-values database (release 3.1 Sept. 2001) http://www.rbgkew.org.uk/cval/hompage.html.

———. 2003. Angiosperm DNA C-values database (release 3.1, Sept. 2001). http://www. rbgkew.org.uk/cval/homepage.html.

———. 2005. Nuclear DNA amounts in Angiosperms: Progress, problems and prospect. *Annals of Botany* 95:45–90.

Bennett, M.D., I.J. Leitch, and L. Hanson. 1998. DNA amounts in two samples of angiosperm weeds. *Annals of Botany* 82:121–134.

Bennett, M.D., I.J. Leitch, H.J. Price, and J.S. Johnston. 2003. Comparions with *Caenorhabditis* (~100 Mb) and *Drosophila* (~175 Mb) using flow cytometry show genome size in *Arabidopsis* to be ~157 MB and this 25% larger than the *Arabidopsis* genome initiative of ~125 Mb. *Annals of Botany* 91:547–557.

Bennett, M.D., and J.B. Smith. 1976. Nuclear DNA amounts in angiosperms. *Philosophical Transactions of the Royal Society of London B* 274:227–274.

———. 1991. Nuclear DNA amounts in angiosperms. *Philosophical Transactions of the Royal Society of London B* 334:309–345.

Bennett, M.D., J.B. Smith, and J.S. Heslop-Harrison. 1982. Nuclear DNA amounts in angiosperms. *Proceedings of the Royal Society of London B* 216:179–199.

Bennetzen, J.L. 1996. The contributions of retroelements to plant genome organization, function, and evolution. *Trends in Microbiology* 4:347–353.

———. 2000. Transposable element contributions to plant

gene and genome evolution. *Plant Molecular Biology* 42:251–269.

———. 2002. Mechanisms and rates of genome expansion and contraction in flowering plants. *Genetica* 115:29–36.

Bennetzen, J.L., and M. Freeling. 1997. The unified grass genome: Synergy in synteny. *Genome Research* 7(4):301–306.

Bennetzen, J.L., and E.A. Kellogg. 1997. Do plants have a one-way ticket to genomic obesity? *The Plant Cell* 9:1509–1514.

Bensel, C.R., and B.F. Palser. 1975. Floral anatomy in the Saxifragaceae sensu lato. II. Saxifragoideae and Iteoideae. *American Journal of Botany* 62:661–675.

Bentham, G. 1880. Notes on Euphorbiaceae. *Botanical Journal of the Linnean Society* 17:185–267.

Bentham, G., and J.D. Hooker. 1862–1883. Genera Plantarum. Reeve, London, UK.

Benzing, D.H. 1967. Developmental patterns in stem primary xylem of woody Ranales. I and II. *American Journal of Botany* 54:805–820.

Benzing, D.H., and J.T. Atwood. 1984. Orchidaceae: Ancestral habitats and current status in forest canopies. *Systematic Botany* 9:155–165.

Benzing, D.H., T. Givnish, and D. Bermudes. 1985. Absorptive trichomes in *Brocchinia reducta* (Bromeliaceae) and their evolutionary and systematic significance. *Systematic Botany* 10:81–91.

Berger, A. 1930. Crassulaceae. *In* A. Engler, and K. Prantl. [eds.], Die Natürlichen Pflanzenfamilien, ed. 2, 18a, 352–483. Engelmann, Leipzig, Germany.

Bergthorsson, U., K.L. Adams, B. Thomason, and J.D. Palmer. 2003. Widespread horizontal transfer of mitochondrial genes in flowering plants *Nature* 424(6945):197–201.

Bergthorsson, U., A.O. Richardson, G.J. Young, L.R. Goertzen, and J.D. Palmer. 2004. Massive horizontal transfer of mitochondrial genes from diverse land plant donors to the basal angiosperm *Amborella*. *Proceedings of the National Academy of Sciences USA* 101(51):17747–17752.

Bernays, E., and C. De Luca. 1981. Insect antifeedant properties of an iridoid glycoside: Ipolamiide. *Experientia* 37:1289–1290.

Bernhard, A. 1999. Flower structure, development, and systematics in Passifloraceae and in *Abatia* (Flacourtiaceae). *International Journal of Plant Sciences* 160:135–150.

Bernhard, A., and P.K. Endress. 1999. Androecial development and systematics in Flacourtiaceae s.l. *Plant Systematics and Evolution* 215:141–155.

Bernhardt, P. 1996. Anther adaptations in animal pollination. *In* W.G. D'Arcy, and R.C. Keating [eds.], The Anther: Form, Function and Phylogeny, 192–220. Cambridge University Press, Cambridge, UK.

———. 2000. Convergent evolution and adaptive radiation of beetle-pollinated angiosperms. *Plant Systematics and Evolution* 222:293–320.

Berry, P.E., W.J. Hahn, K.J. Sytsma, J.C. Hall, and A. Mast. 2004. Phylogenetic relationships and biogeography of *Fuchsia* (Onagraceae) based on noncoding nuclear and chloroplast DNA data. *American Journal of Botany* 91:601–614.

Berry, P.E., V. Savolainen, K. Sytsma, J.C. Hall, and M.W. Chase. 2001. *Lissocarpa* is sister to *Diospyros* (Ebenaceae). *Kew Bulletin* 56:725–729.

Berry, P.E., J.J. Skvarla, D. Partridge, and M.K. Macphail. 1990. *Fuchsia* pollen from the tertiary of Australia. *Australian Systematic Botany* 3(4):739–744.

Bertin, R., and C. Newman. 1993. Dichogamy in angiosperms. *Botanical Review* 59:112–152.

Besnard, G., R. Rubio de Casas, P.-A. Christin, and P. Vargas. 2009. Phylogenetics of *Olea* (Oleaceae) based on plastid and nuclear ribosomal DNA sequences: Tertiary climatic shifts and lineage differentiation times. *Annals of Botany* 104(1):143–160.

Bessey, C.E. 1897. The phylogeny and taxonomy of angiosperms. *Botanical Gazette* 24:145–178.

———. 1915. The phylogenetic taxonomy of flowering plants. *Annals of the Missouri Botanical Garden* 2:109–164.

Beusekom, C.F.V. 1971. Revision of *Meliosma* (Sabiaceae), section *Lorenzanea* excepted, living and fossil, geography and phylogeny. *Blumea* 19:355–529.

Bharathan, G. 1996. Reproductive development and nuclear DNA content in angiosperms. *American Journal of Botany* 83:440–451.

Bharathan, G., and E.A. Zimmer. 1995. Early branching events in monocotyledons – partial 18S ribosomal DNA sequence analysis. *In* P.J. Rudall, P.J. Cribb, D.F. Cutler, and C.J. Humphries [eds.], Monocotyledons: Systematics and Evolution, 81–107. Royal Botanic Gardens, Kew, UK.

Bhatnagar, A.K., and M. Garg. 1977. Affinities of *Daphniphyllum*: A palynological approach. *Phytomorphology* 27:92–97.

Bhatnagar, S.P., and A. Moitra. 1996. *Gymnosperms*. New Age International, New Delhi, India.

Biffin, E., E.J. Lucas, L.A. Craven, I. Ribeiro da Costa, M.G. Harrington, and M.D. Crisp. 2010. Evolution of exceptional species richness among lineages of fleshy-fruited Myrtaceae. *Annals of Botany* 106(1):79–93.

Bininda-Emonds, O.R.P., M. Cardillo, K.E. Jones, R.D.E. MacPhee, R.M.D. Beck, R. Grenyer, S.A. Price, R.A.

Vos, J.L. Gittleman, and A. Purvis. 2007. The delayed rise of present-day mammals. *Nature* 446:507–512.

Birchler, J.A. 2012. Genetic consequences of polyploidy in plants. *In* P.S. Soltis, and D.E. Soltis [eds.], Polyploidy and Genome Evolution, 21–23. Springer-Verlag, Berlin, Heidelberg, Germany.

Blanc, G., and K.H. Wolfe. 2004. Widespread paleopolyploidy in model plant species inferred from age distributions of duplicate genes. *Plant Cell* 16(7):1667–1678.

Blanc, P. 1986. Edification d'arbres par croissance d'etablissement de type monocotylédonien: l'exemple de Chloranthaceae. *In* Colloque international sur l'arbre, 101–123. Naturalia Monspeliensia, numéro hors série, Montpellier, France.

Blarer, A., D.L. Nickrent, and P.K. Endress. 2004. Comparative floral structure and systematics in Apodanthaceae (Rafflesiales). *Plant Systematics and Evolution* 245:119–142.

Bock, R. 2010. The give-and-take of DNA: Horizontal gene transfer in plants. *Trends in Plant Science* 15(1):11–22.

Boesewinkel, F.D., and F. Bouman. 1991. The development of bi- and unitegmic ovules and seeds in *Impatiens* (Balsaminaceae). *Botanische Jahrbücher für Systematik* 113:87–104.

Bogle, A.L. 1969. The genera of Portulacaceae and Basellaceae in the southeastern United States. *Journal of the Arnold Arboretum* 50:566–598.

———. 1970. Floral morphology and vascular anatomy of the Hamamelidaceae: The apetalous genera of Hamamelidoideae. *Journal of the Arnold Arboretum* 51:310–366.

———. 1986. The floral morphology and vascular anatomy of the Hamameliaceae subfamily Liquidambaroideae. *Annals of the Missouri Botanical Garden* 73:325–347.

———. 1989. The floral morphology, vascular anatomy, and ontogeny of the Rhodoleioideae (Hamamelidaceae) and their significance in relation to the 'lower' hamamelids. *In* P.R. Crane, and S. Blackmore [eds.], Evolution, Systematics and Fossil History of the Hamamelidae, 1, 201–226. Clarendon Press, Oxford, UK.

Bohs, L. 2001. Revision of *Solanum* section *Cyphomandropsis* (Solanaceae). *Systematic Botany Monographs* 61:1–85.

———. 2005. Major clade in *Solanum* based on *ndhF* sequence data. *Systematic Botany Monographs* 104:27–49.

———. 2007. Phylogeny of the *Cyphomandra* clade of the genus *Solanum* (Solanaceae) based on ITS sequence data. *Taxon* 56:1012–1026.

Bohs, L., and R.G. Olmstead. 1997. Phylogenetic relationships in *Solanum* (Solanaceae) based on *ndhF* sequences. *Systematic Botany* 22:5–17.

Boke, N.H. 1963. Anatomy and development of the flower and fruit of *Pereskia pititache*. *American Journal of Botany* 50:843–858.

———. 1964. The cactus gynoecium: A new interpretation. *American Journal of Botany* 51:598–610.

———. 1966. Ontogeny and structure of the flower and fruit of *Pereskia aculeata*. *American Journal of Botany* 53:534–542.

Boros, C.A., and F.R. Stermitz. 1990. Iridoids. An updated review, I. *Journal of Natural Products* 53:1055–1147.

———. 1991. Iridoids. An updated review, II. *Journal of Natural Products* 54:1172–1246.

Borsch, T. 1998. Pollen types in the Amaranthaceae: Morphology and evolutionary significance. *Grana* 37:129–142.

Borsch, T., K.W. Hilu, D. Quandt, V. Wilde, C. Neinhuis, and W. Barthlott. 2003. Non-coding plastid *trnT–trnF* sequences reveal a well resolved phylogeny of basal angiosperms. *Journal of Evolutionary Biology* 16(4):558–576.

Borsch, T., C. Loehne, and J. Wiersema. 2008. Phylogeny and evolutionary patterns in Nymphaeales: Integrating genes, genomes and morphology. *Taxon* 57:1052–1081.

Bortenschlager, S., A. Auimger, J. Blaha, and P. Simonsburger. 1972. Pollen morphology of Achatocarpaceae (Centrospermae). *Berichte des Naturwissenschaftlich–Medizinischen Vereins Innsbruck* 59:7–13.

Bose, M.N., P.K. Pal, and T.M. Harris. 1985. The *Pentoxylon* plant. *Philosophical Transactions of the Royal Society of London B* 310:77–108.

Bouchenak-Khelladi, Y., A.M. Muasya, and H.P. Linder. 2010. A revised evolutionary history of Poales: Origins and diversification. *Botanical Journal of the Linnean Society* 175:4–16.

Bouchenak-Khelladi, Y., G.A Verboom, V. Savolainen, and T.R. Hodkinson. 2010. Biogeography of the grasses (Poaceae): A phylogenetic approach to reveal evolutionary history in geographical space and geological time. *Botanical Journal of the Linnean Society* 162:543–557.

Boucher, L.D., S.R. Manchester, and W.S. Judd. 2003. An extinct genus of Salicaceae based on twigs with attached flowers, fruits, and foliage from the Eocene Green River Formation of Utah and Colorado, USA. *American Journal of Botany* 90(9):1389–1399.

Bouman, F. 1984. The ovule. *In* B.M. Johri [ed.], Embryology of Angiosperms, 123–157. Springer-Verlag, Berlin, Germany.

Bouman, F., and J.I.M. Calis. 1977. Integumentary shifting—a third way to unitegmy. *Berichte der Deutschen Botanischen Gesellschaft* 90(1):15–28.

Bouman, F., and S. Schier. 1979. Ovule ontogeny and seed coat development in *Gentiana* with a discussion on the evolutionary origin of the single integument. *Acta Botanica Neerlandica* 28:467–478.

Bousquet, J., S.H. Strauss, A.H. Doerksen, and R.A. Price. 1992. Extensive variation in evolutionary rate of *rbcL* gene-sequences among seed plants. *Proceedings of the National Academy of Sciences USA* 89:7844–7848.

Bowe, L., G. Coat, and C. dePamphilis. 2000. Phylogeny of seed plants based on all three genomic compartments: Extant gymnosperms are monophyletic and Gnetales' closest relatives are conifers. *Proceedings of the National Academy of Sciences USA* 97(8):4092–4097.

Bowers, J.E., B.A. Chapman, J. Rong, and A.H. Paterson. 2003. Unravelling angiosperm genome evolution by phylogenetic analysis of chromosomal duplication events. *Nature* 422(6930):433–438.

Bowers, M.D., and G.M. Puttick. 1988. Response of generalist and specialist insects to qualitative allelochemical variation. *Journal of Chemical Ecology* 14:319–334.

Bowes, G. 2011. Single-Cell C-4 photosynthesis in aquatic plants. *C4 Photosynthesis and Related CO2 Concentrating Mechanisms* 32:63–80.

Bowman, J.L. 1997. Evolutionary conservation of angiosperm flower development at the molecular and genetic levels. *Journal of Bioscience* 22:515–527.

———. 2000. Axial patterning in leaves and other lateral organs. *Current Opinion in Genetics and Development* 10:399–404.

Bowman, J.L., H. Brüggemann, J.-Y. Lee, and K. Mummenhoff. 1999. Evolutionary changes in floral structure within *Lepidium* L. (Brassicaceae). *International Journal of Plant Sciences* 160:917–929.

Bowman, J.L, S.K. Floyd, and K. Sakakibara. 2007. Green genes—Comparative genomics of the green branch of life. *Cell* 129:229–234.

Bowman, J.L., D.R. Smyth, and E.M. Meyerowitz. 1989. Genes directing flower development in *Arabidopsis*. *Plant Cell* 1(1):37–52.

———. 1991. Genetic interactions among floral homeotic genes of *Arabidopsis*. *Development* 112:1–20.

Boyden, G.S., M.J. Donoghue, and D.G. Howarth. 2012. Duplications and expression of *RADIALIS*-like genes in Dipsacales. *International Journal of Plant Sciences* 173:971–983.

Braasch, I., and J.H. Postlethwait. 2012. Polyploidy in fish and the teleost genome duplication. *In* P.S. Soltis, and D.E. Soltis [eds.], Polyploidy and Genome Evolution, 341–383. Springer-Verlag, Berlin, Germany.

Bradley, D., R. Carpenter, H. Sommer, N. Hartley, and E. Coen. 1993. Complementary floral homeotic phenotypes result from opposite orientations of a transposon at the *PLENA* locus of *Antirrhinum*. *Cell* 72:85–95.

Braem, G. 1993. Studies in the Oncidiinae. Discussion of some taxonomic problems with description of *Gudrania* Braem, gen. nov., and reinstatement of the genus *Lophiaris* Raffinesque. *Schlechteriana* 4:8–21.

Brandl, R., W. Mann, and M. Sprintzl. 1992. Estimation of the monocot-dicot age through tRNA sequences from the chloroplast. *Proceedings of the Royal Society of London B* 249:13–17.

Braukmann, T., and S. Stefanović. 2012. Plastid genome evolution in mycoheterotrophic Ericaceae. *Plant Molecular Biology* 79(1–2):5–20.

Bremer, B. 1987. The sister group of the paleo-tropical tribe Argostemmateae: A redefined neotropical tribe Hamelieae (Rubiaceae, Rubioideae). *Cladistics* 3:35–51.

———. 1996. Phylogenetic studies within Rubiaceae and relationships to other families based on molecular data. *Opera Botanica Belgica* 7:33–50.

Bremer, B., K. Andreason, and D. Olsson. 1995. Subfamilial and tribal relationships in the Rubiaceae based on *rbcL* sequence data. *Annals of the Missouri Botanical Garden* 82:383–397.

Bremer, B., K. Bremer, M.W. Chase, M.F. Fay, J.L. Reveal, D.E. Soltis, P.S. Soltis, P.F. Stevens, A.A. Anderberg, M.J. Moore, R.G. Olmstead, P.J. Rudall, K.J. Sytsma, D.C. Tank, K. Wurdack, J.Q.Y. Xiang, S. Zmarzty, and Angiosperm Phylogeny G. 2009. An update of the Angiosperm Phylogeny Group classification for the orders and families of flowering plants: APG III. *Botanical Journal of the Linnean Society* 161:105–121.

Bremer, B., K. Bremer, N. Heirdari, P. Erixon, R.G. Olmstead, M. Källersjö, A.A. Anderberg, and E. Barkhordarian. 2002. Phylogenetics of asterids based on 3 coding and 3 non-coding chloroplast DNA markers and the utility of non-coding DNA at higher taxonomic levels. *Molecular Phylogenetics and Evolution* 24:274–301.

Bremer, B., and T. Eriksson. 2009. Time tree of Rubiaceae: Phylogeny and dating the family, subfamilies, and tribes. *International Journal of Plant Sciences* 170:766–793.

Bremer, B., R.K. Jansen, B. Oxelman, M. Backlund, H. Lantz, and K.-J. Kim. 1999. More characters or more taxa for a robust phylogeny–case study from the coffee family (Rubiaceae). *Systematic Biology* 48:413–435.

Bremer, B., R.G. Olmstead, L. Struwe, and J.A. Sweere.

1994. *rbcL* sequences support exclusion of *Retzia*, *Desfontainia*, and *Nicodemia* from the Gentianales. *Plant Systematics and Evolution* 190:213–230.

Bremer, B., and L. Struwe. 1992. Phylogeny of the Rubiaceae and Loganiaceae: Congruence or conflict between morphological and molecular data. *American Journal of Botany* 79:1171–1194.

Bremer, K. 1987. Tribal interrelationships of the Asteraceae. *Cladistics* 3:210–253.

———. 1988. The limits of amino acid sequence data in angiosperm phylogenetic reconstruction. *Evolution* 42:795–803.

———. 1994. *Asteraceae: Cladistics and Classification*. Timber Press, Portland, OR.

———. 1996. Major clades and grades of the Asteraceae. *In* D.J.N. Hind, and H.J. Beentje [eds.], Compositae: Systematics, 1–7. Royal Botanic Gardens, Kew, UK.

———. 2000. Early Cretaceous lineages of monocot flowering plants. *Proceedings of the National Academy of Sciences USA* 97:4707–4711.

———. 2002. Gondwanan evolution of the grass alliance of families (Poales). *Evolution* 56:1374–1387.

Bremer, K., A. Backlund, B. Sennblad, U. Swenson, K. Andreasen, M. Hjertson, J. Lundberg, M. Backlund, and B.Bremer. 2001. A phylogenetic analysis of 100+ genera and 50+ families of euasterids based on morphological and molecular data with notes on possible higher level morphological synapomorphies. *Plant Systematics and Evolution* 229:137–169.

Bremer, K., E.M. Friis, and B. Bremer. 2004. Molecular phylogenetic dating of asterid flowering plants shows early Cretaceous diversification. *Systematic Biology* 53(3):496–505.

Bremer, K., and M.H. Gustafsson. 1997. East Gondwana ancestry of the sunflower alliance of families. *Proceedings of the National Academy of Sciences USA* 94:9188–9190.

Bremer, K., and R.K. Jansen. 1992. A new subfamily of the Asteraceae. *Annals of the Missouri Botanical Garden* 79:414–415.

Brenchley, R., M. Spannagl, M. Pfeifer, G.L.A. Barker, R. D'Amore, A.M. Allen, N. McKenzie, M. Kramer, A. Kerhornou, D. Bolser, S. Kay, D. Waite, M. Trick, I. Bancroft, Y. Gu, N. Huo, M.-C. Luo, S. Sehgal, B. Gill, S. Kianian, O. Anderson, P. Kersey, J. Dvorak, W.R. McCombie, A. Hall, K.F.X. Mayer, K.J. Edards, M.W. Bevan, and N. Hall. 2012. Analysis of the bread wheat genome using whole-genome shotgun sequencing. *Nature* 491:705–10.

Brenner, G.J. 1996. Evidence for the earliest stage of angiosperm pollen evolution: A paleoequatorial section from Israel. *In* D.W. Taylor, and L.J. Hickey [eds.], Flowering Plant Origin, Evolution, and Phylogeny, 91–115. Chapman and Hall, New York, NY.

Breuer, B., T. Stuhlfauth, H. Fock, and H. Huber. 1987. Fatty acids of some Cornaceae, Hydrangeaceae, Aquifoliaceae, Hamamelidaceae and Styracaceae. *Phytochemistry* 26:1441–1445.

Brewbaker, J.L. 1957. Pollen cytology and self-incompatibility systems in plants. *Journal of Heredity* 48:271–277.

———. 1967. The distribution and phylogenetic significance of binucleate and trinucleate pollen grains in the angiosperms. *American Journal of Botany* 54:1069–1083.

Briggs, B.G., and L. Johnson. 1979. Evolution of Myrtaceae–evidence from inflorescence structure. *Proceedings Linnean Society New South Wales* 102:157–256.

Briggs, B.G., A.D. Marchant, S. Gilmore, and C.L. Porter. 2000. A molecular phylogeny of Restionaceae and allies. *In* K.L. Wilson, and D.A. Morrison, [eds.], Monocots: Systematics and Evolution, 661–671. CSIRO, Melbourne, Australia.

Britton, T., B. Oxelman, A. Vinnersten, and K. Bremer. 2002. Phylogenetic dating with confidence intervals using mean pathlengths. *Molecular Phylogenetics and Evolution* 24:58–65.

Brizicky, G.K. 1962. The genera of Rutaceae in the southeastern United States. *Journal of the Arnold Arboretum* 43:1–22.

———. 1963. The genera of the Sapindales in the southeastern United States. *Journal of the Arnold Arboretum* 44:462–501.

Brockington, S.F., R. Alexandre, J. Ramdial, M.J. Moore, S. Crawley, A. Dhingra, K. Hilu, D.E. Soltis, and P.S. Soltis. 2009. Phylogeny of the Caryophyllales sensu lato: Revisiting hypotheses on pollination biology and perianth differentiation in the core Caryophyllales. *International Journal of Plant Sciences* 170(5):627–643.

Brockington, S.F., R. Alexandre, J. Ramdial, M.J. Moore, S. Crawley, A. Dhingra, K. Hilu, P.S. Soltis, and D.E. Soltis. 2010. Phylogeny of the Caryophyllales and the evolution of the perianth. *International Journal of Plant Sciences* 171:185–198.

Brockington, S.F., P. Dos Santos, B. Glover, and L.P. Ronse De Craene. 2013. Androecial evolution in Caryophyllales in light of a paraphyletic Molluginaceae. *American Journal of Botany* 100(9):1757–1778.

Brockington, S.F., P.J. Rudall, M.W. Frohlich, D.G. Oppenheimer, P.S. Soltis, and D.E. Soltis. 2012. 'Living stones' reveal alternative petal identity programs within the core eudicots. *Plant Journal* 69(2):193–203.

Brockington, S.F., R.H. Walker, B.J. Glover, P.S. Soltis, and

D.E. Soltis. 2011. Complex pigment evolution in the Caryophyllales. *New Phytologist* 190:854–864.

Bromham, L., P.F. Cowman, and R. Lanfear. 2013. Parasitic plants have increased rates of molecular evolution across all three genomes. *BMC Evolutionary Biology* 13:126. doi:10.1186/1471-2148-13-126.

Brown, R.H., D.L. Nickrent, and C.S. Gasser. 2010. Expression of ovule and integument-associated genes in reduced ovules of Santalales. *Evolution and Development* 12:231–240.

Brückner, C. 2000. Clarification of the carpel number in Papaverales, Capparales, and Berberidaceae. *Botanical Review* 66:155–307.

Brummitt, R.K. 1997. Taxonomy versus cladonomy, a fundamental controversy in biological systematics. *Taxon* 46:723–734.

Brummitt, R.K., H. Banks, M.A.T. Johnson, K.A. Doherty, K. Jones, M.W. Chase, and P.J. Rudall. 1998. Taxonomy of Cyanastroideae (Tecophilaeaceae): A multidisciplinary approach. *Kew Bulletin* 53:769–803.

Bruneau, A., J.J. Doyle, P. Herendeen, C. Hughes, G. Kenicer, G. Lewis, B. Mackinder, R.T. Pennington, M.J. Sanderson, M.F. Wojciechowski, S. Boatwright, G. Brown, D. Cardoso, M. Crisp, A. Egan, R.H. Fortunato, J. Hawkins, T. Kajita, B. Klitgaard, E. Koenen, M. Lavin, M. Luckow, B. Marazzi, M.M. McMahon, J.T. Miller, D.J. Murphy, H. Ohashi, L.P. de Queiroz, L. Rico, T. Saerkinen, B. Schrire, M.F. Simon, E.R. Souza, K. Steele, B.M. Torke, J.J. Wieringa, B.-E. van Wyk, and Legume Phylogeny Working G. 2013. Legume phylogeny and classification in the 21st century: Progress, prospects and lessons for other species-rich clades. *Taxon* 62(2):217–248.

Bruneau, A., F. Forest, P.S. Herendeen, B.B. Klitgaard, and G.P. Lewis. 2001. Phylogenetic relationships in the Caesalpinioideae (Leguminosae) as inferred from chloroplast *trnL* intron sequences. *Systematic Botany* 26:487–514.

Bruneau, A., M. Mercure, G.P. Lewis, and P.S. Herendeen. 2008. Phylogenetic patterns and diversification in the caesalpinioid legumes. *Botany-Botanique* 86(7):697–718.

Bruyns, P.V., M. Oliveira-Neto, G.F. Melo-de-Pinna, and C. Klak. 2014. Phylogenetic relationships in the Didiereaceae with special reference to subfamily Portulacarioideae. *Taxon* 63:1053–1064.

Bryant, H.N. 1994. Comments on the phylogenetic definition of taxon names and conventions regarding the naming of crown clades. *Systematic Biology* 43:124–130.

———. 1996. Explicitness, stability, and universality in the phylogenetic definition and usage of taxon names: A case study of the phylogenetic taxonomy of the Carnivora (Mammalia). *Systematic Biology* 45:174–189.

Buggs, R.J.A., L. Zhang, N. Miles, J.A. Tate, L. Gao, W. Wei, P.S. Schnable, W.B. Barbazuk, P.S. Soltis, and D.E. Soltis. 2011. Transcriptomic shock generates evolutionary novelty in a newly formed, natural allopolyploid plant. *Current Biology* 21(7):551–556.

Bungard, R.A. 2004. Photosynthetic evolution in parasitic plants: Insight from the chloroplast genome. *Bioessays* 26(3):235–247.

Burge, D.O., and S.R. Manchester. 2008. Fruit morphology, fossil history, and biogeography of *Paliurus* (Rhamnaceae). *International Journal of Plant Sciences* 169(8):1066–1085.

Burger, W. 1977. The Piperales and monocots-alternative hypotheses for the origin of the monocotyledonous flowers. *Botanical Review* 43:345–393.

Burke, J.M., A. Sanchez, K. Kron, and M. Luckow. 2010. Placing the woody tropical genera of Polygonaceae: A hypothesis of character evolution and phylogeny. *American Journal of Botany* 97:1377–1390.

Burleigh, J.G., M.S. Bansal, O. Eulenstein, S. Hartmann, A. Wehe, and T.J. Vision. 2011. Genome-scale phylogenetics: Inferring the plant tree of life from 18,896 gene trees. *Systematic Biology* 60:117–125.

Burleigh, J.G., W.B. Barbazuk, J.M. Davis, A.M. Morse, and P.S. Soltis. 2012. Exploring diversification and genome size evolution in gymnosperms through phylogenetic synthesis. *Journal of Botany* 2012. Article ID 292857, 6 pages, 2012. doi:10.1155/2012/292857.

Burleigh, J.G., K.W. Hilu, and D.E. Soltis. 2009. Inferring phylogenies with incomplete data sets: A 5-gene, 567-taxon analysis of Angiosperms. *BMC Evolutionary Biology* 9:61. doi:10.1186/1471-2148-9-61.

Burleigh, J.G., and S. Mathews. 2004. Phylogenetic signal in nucleotide data from seed plants: Implications for resolving the seed plant tree of life. *American Journal of Botany* 91:1599–1613.

———. 2007. Assessing systematic error in the inference of seed plant phylogeny. *International Journal of Plant Sciences* 168:125–135.

Burns-Balogh, P., and V. Funk. 1986. A phylogenetic analysis of the Orchidaceae. *Smithsonian Contributions to Botany* 61:1–79.

Burtt, B.L. 1974. Patterns of structural change in the flowering plants. *Transactions of the Botanical Society of Edinburgh* 42:133–142.

———. 1994. A commentary on some recurrent forms and changes of form in angiosperms. *In* D.S. Ingram, and A. Hudson [eds.], Shape and Form in Plants and Fungi, 143–152. Academic Press, London, UK.

Butterworth, C.A., and R.S. Wallace. 2004. Phylogenetic

studies of *Mammillaria* (Cactaceae)—insights from chloroplast sequence variation and hypothesis testing using the parametric bootstrap. *American Journal of Botany* 91:1086–1098.

Buzgo, M. 2001. Flower structure and development of Araceae compared with alismatids and Acoraceae. *Botanical Journal of the Linnean Society* 136:393–425.

Buzgo, M., and P.K. Endress. 2000. Floral structure and development of Acoraceae and its systematic relationships with basal angiosperms. *International Journal of Plant Sciences* 161:23–41.

Buzgo, M., D.E. Soltis, P.S. Soltis, and H. Ma. 2004. Towards a comprehensive integration of morphological and genetic studies of floral development. *Trends in Plant Science* 9(4):164–173.

Buzgo, M., P.S. Soltis, S. Kim, and D.E. Soltis. 2005. The making of a flower. *The Biologist* 52:149–154.

Buzgo, M., P.S. Soltis, and D.E. Soltis. 2004. Floral developmental morphology of *Amborella trichopoda* (Amborellaceae). *International Journal of Plant Sciences* 165:925–947.

Byng, J.W., B. Bernardini, J.A. Joseph, M.A. Chase, and T.M.A. Utteridge. 2014. Phylogenetic relationships of Icacinaceae focusing on the vining genera. *Biological Journal of the Linnean Society* 176:277–294.

Caddick, L.R., P.J. Rudall, P. Wilkin, T.A.J. Hedderson, and M.W. Chase. 2002a. Phylogenetics of Dioscoreales based on combined analyses of morphological and molecular data. *Botanical Journal of the Linnean Society* 138:123–144.

Caddick, L.R., P. Wilkin, P.J. Rudall, T.A.J. Hedderson, and M.W. Chase. 2002b. Yams reclassified: A recircumscription of Dioscoreaceae and Dioscoreales. *Taxon* 51:103–114.

Cai, Z., C. Penaflor, J.V. Kuehl, J. Leebens-Mack, J.E. Carlson, C.W. dePamphilis, J.L. Boore, and R.K. Jansen. 2006. Complete plastid genome sequences of *Drimys*, *Liriodendron*, and *Piper*: Implications for the phylogenetic relationships of magnoliids. *BMC Evolutionary Biology* 6:77. doi:10.1186/1471-2148-6-77.

Calderon-Urrea, A., Q.C.B. Cronk, R.M. Bateman, and J.A. Hawkins. 2002. Developmental Genetics and Plant Evolution. Taylor and Francis, London, UK.

Call, V.B. and D.L. Dilcher. 1992. Investigations of angiosperms from the Eocene of southeastern North America: Samaras of *Fraxinus wilcoxiana* Berry. *Review of Palaeobotany and Palynology* 74(3):249–266.

Calvente, A., D.C. Zappi, and L.G. Lohmann. 2011. Molecular phylogeny of tribe Rhipsalidae (Cactaceae) and taxonomic implications for *Schlumbergera* and *Hatiora*. *Molecular Phylogenetics and Evolution* 58:456–468.

Calvillo-Canadell, L., and S.R.S. Cevallos-Ferriz. 2007. Reproductive structures of Rhamnaceae from the Cerro del Pueblo (Late Cretaceous, Coahuila) and Coatzingo (Oligocene, Puebla) Formations, Mexico. *American Journal of Botany* 94:1658–1669.

Cameron, K.M., 2004. Utility of plastid *psaB* gene sequences for investigating intrafamilial relationships within Orchidaceae. *Molecular Phylogenetics and Evolution* 31:1157–1180.

———. 2006. A comparison of plastid *atpB* and *rbcL* gene sequences for inferring phylogenetic relationships within Orchidaceae. *In* J.T. Columbus, E.A. Friar, J.M. Porter, L.M. Prince, and M.G. Simpson [eds.], Monocots: Comparative Biology and Evolution, Excluding Poales, 447–464. Rancho Santa Ana Botanic Garden, Claremont, CA.

Cameron, K.M., M.W. Chase, W.R. Anderson, and H.G. Hills. 2001. Molecular systematics of Malpighiaceae: Evidence from plastid *rbcL* and *matK* sequences. *American Journal of Botany* 88:1847–1862.

Cameron, K.M., M.W. Chase, and P.J. Rudall. 2003. Recognition and recircumscription of Petrosaviales to include *Petrosavia* and *Japanolirion* (Petrosaviaceae s.l.). *Brittonia* 55:214–225.

Cameron, K.M., M.W. Chase, W.M. Whitten, P.J. Kores, D.C. Jarrell, V.A. Albert, T. Yukawa, H.G. Hills, and D.H. Goldman. 1999. A phylogenetic analysis of the Orchidaceae: Evidence from *rbcL* nucleotide sequences. *American Journal of Botany* 86:208–224.

Cameron, K.M., and W.C. Dickison. 1998. Foliar architecture of vanilloid orchids: Insights into the evolution of reticulate leaf venation in monocotyledons. *Botanical Journal of the Linnean Society* 128:45–70.

Cameron, K.M., K.J. Wurdack, and R.W. Jobson. 2002. Molecular evidence for the common origin of snap-trees among carnivorous plants. *American Journal of Botany* 89:1503–1509.

Campbell, C.S. 1985. The subfamilies and tribes of Gramineae (Poaceae) in the southeastern United States. *Journal of the Arnold Arboretum* 66:123–199.

Campbell, C.S., R.C. Evans, D.R. Morgan, T.A. Dickinson, and M.P. Arsenault. 2007. Phylogeny of subtribe Pyrinae (formerly the Maloideae, Rosaceae): Limited resolution of a complex evolutionary history. *Plant Systematics and Evolution* 266(1–2):119–145.

Canestro, C. 2012. Two rounds of whole genome duplication: Evidence and impact on the evolution of vertebrate innovations. *In* P.S. Soltis, and D.E. Soltis [eds.], Polyploidy and Genome Evolution, 309–340. Springer-Verlag, Berlin, Germany.

Cannon, S. 2008. Legume comparative genomics. *Genetics and Genomics of Soybean* 2:35–54.

Cannon, S.B., D. Ilut, A.D. Farmer, S.L. Maki, G.D. May, S.R. Singer, and J.J. Doyle. 2010. Polyploidy did not predate the evolution of nodulation in all legumes. *PLoS One* 5(7): e11630. doi:10.1371/journal.pone.0011630.

Cannon, S.B., G.D. May, and S.A. Jackson. 2009. Three sequenced legume genomes and many crop species: Rich opportunities for translational genomics. *Plant Physiology* 151(3):970–977.

Cannon, S.B., M.R. Mckain, A. Harkess, M.N. Nelson, S. Dash, M.K. Deyholos, Y. Peng, B. Joyce, C.N. Stewart Jr., M. Rolf, T. Kutchan, X. Tan, C. Chen, Y. Zhang, E. Carpenter, G.K.-S. Wong, J.J. Doyle, and J. Leebens-Mack. 2015. Multiple polyploidy events in the early radiation of nodulating and nonnodulating legumes. *Molecular Biology and Evolution* 32(1):193–210.

Cannon, S.B., L. Sterck, S. Rombauts, S. Sato, F. Cheung, J. Gouzy, X. Wang, J. Mudge, J. Vasdewani, T. Scheix, M. Spannagl, E. Monaghan, C. Nicholson, S.J. Humphray, H. Schoof, K.F.X. Mayer, J. Rogers, F. Quetier, G.E. Oldroyd, F. Debelle, D.R. Cook, E.F. Retzel, B.A. Roe, C.D. Town, S. Tabata, Y. Van de Peer, and N.D. Young. 2006. Legume genome evolution viewed through the *Medicago truncatula* and *Lotus japonicus* genomes. *Proceedings of the National Academy of Sciences USA* 103:14959–14964.

Canright, J.E. 1952. The comparative morphology and relationships of the Magnoliaceae. I. Trends of specialization in the stamens. *American Journal of Botany* 39:484–497.

———. 1955. The comparative morphology and relationships of the Magnoliaceae. IV. Wood and nodal anatomy. *Journal of the Arnold Arboretum* 36:119–140.

Cantino, P.D. 1992a. Toward a phylogenetic classification of the Labiatae. *In* R. M. Harley, and T. Teynolds [eds.], Advances in Labiate Science, 27–32. Royal Botanic Garden, Kew, UK.

———. 1992b. Evidence for a polyphyletic origin of the Labiatae. *Annals of the Missouri Botanical Garden* 79:361–379.

———. 1998. Binomials, hyphenated uninomials, and phylogenetic nomenclature. *Taxon* 47:425–429.

———. 2000. Phylogenetic nomenclature: Addressing some concerns. *Taxon* 49:85–93.

Cantino, P.D., J.A. Doyle, S.W. Graham, W.S. Judd, R.G. Olmstead, D.E. Soltis, P.S. Soltis, and M.J. Donoghue. 2007. Towards a phylogenetic nomenclature of Tracheophyta. *Taxon* 56:822–846.

Cantino, P.D., R.G. Olmstead, and S.J. Wagstaff. 1997. A comparison of phylogenetic nomenclature with the current system: A botanical case study. *Systematic Biology* 46:313–331.

Capasso, A., R. Urrnaga, L. Garofala, L. Sorrentino, and R. Aquino. 1996. Phytochemical and pharmacological studies on medicinal herb *Acicarpha tribuloides. International Journal of Pharmacognosy* 34:255–261.

Cardoso, D., H.C. de Lima, R.S. Rodrigues, L.P. de Queiroz, R.T. Pennington, and M. Lavin. 2012. The realignment of *Acosmium* sensu stricto with the dalbergioid clade (Leguminosae: Papilionoideae) reveals a proneness for independent evolution of radial floral symmetry among early-branching papilionoid legumes. *Taxon* 61(5):1057–1073.

Cardoso, D., L.P. de Queiroz, H.C. de Lima, E. Suganuma, C. van den Berg, and M. Lavin. 2013. A molecular phylogeny of the vataireoid legumes underscores floral evolvability that is general to many early-branching papilionoid lineages. *American Journal of Botany* 100(2):403–421.

Cardoso, D., L.P. de Queiroz, R.T. Pennington, H.C. de Lima, E. Fonty, M.F. Wojciechowski, and M. Lavin. 2012. Revisiting the phylogeny of papilionoid legumes: New insights from comprehensively sampled early-branching lineages. *American Journal of Botany* 99(12):1991–2013.

Cardoso, D., R.T. Pennington, L.P. de Queiroz, J.S. Boatwright, B.E. Van Wyk, M.F. Wojciechowski, and M. Lavin. 2013. Reconstructing the deep-branching relationships of the papilionoid legumes. *South African Journal of Botany* 89:58–75.

Carlquist, S. 1975. Ecological Strategies of Xylem Evolution. University of California Press, Berkeley, CA.

———. 1978. Vegetative anatomy and systematics of Grubbiaceae. *Botaniska Notiser* 131:117–126.

———. 1981. Wood anatomy of Nepenthaceae. *Bulletin of the Torrey Botanical Club* 108:324–330.

———. 1982. *Exospermum stipitatum* (Winteraceae): Observations on wood, leaves, flowers, pollen, and fruit. *Aliso* 10:257–277.

———. 1983. Wood anatomy of *Bubbia* (Winteraceae), with comments on origin of vessels in dicotyledons. *American Journal of Botany* 70:578–590.

———. 1988. Comparative Wood Anatomy. Springer-Verlag, Berlin, Germany.

———. 1990. Wood anatomy of *Ascarina* (Chloranthaceae) and the tracheid-vessel element transition. *Aliso* 13:447–462.

———. 1991. Wood and bark anatomy of *Ticodendron*: Comments on relationships. *Annals of the Missouri Botanical Garden* 78:97–104.

———. 1992. Wood anatomy and stem of *Chloranthus*: Summary of wood anatomy of Chloranthaceae, with comments on relationships, vessellessness, and the origin of monocotyledons. *IAWA Bulletin II* 13:3–16.

———. 1996. Wood anatomy of primitive angiosperms: New perspectives and syntheses. *In* W.D. Taylor, and R.J. Hickey [eds.], Flowering Plant Origin, Evolution and Phylogeny, 68–90. Chapman and Hall, New York, NY.

———. 1999a. Wood and stem anatomy of *Stegnosperma* (Caryophyllales): Phylogenetic relationships; nature of lateral meristems and successive cambial activity. *IAWA Journal* 20:149–163.

———. 1999b. Wood anatomy, stem anatomy, and cambial activity of *Barbeuia* (Caryophyllales). *IAWA Journal* 20:431–440.

———. 1999c. Wood, stem, and root anatomy of Basellaceae with relation to habit, systematics, and cambial variants. *Flora* 194:1–12.

———. 2000a. Wood and bark anatomy of *Takhtajania* (Winteraceae); phylogenetic and ecological implications. *Annals of the Missouri Botanical Garden* 87:317–322.

———. 2000b. Wood and stem anatomy of -phytolaccoid and rivinoid Phytolaccaceae (Caryophyllales): Ecology, systematics, nature of successive cambia. *Aliso* 19:13–29.

———. 2001a. Comparative Wood Anatomy. Springer-Verlag, Berlin, Germany.

———. 2001b. Wood and stem anatomy of Rhabdodendraceae is consistent with placement in Caryophyllales sensu lato. *IAWA Journal* 22:171–181.

———. 2003. Wood anatomy of Aextoxicaceae and Berberidopsidaceae is compatible with their inclusion in Berberidopsidales. *Systematic Botany* 28:317–325.

———. 2007. Successive cambia revisited: Ontogeny, histology, diversity, and functional significance. *Journal of the Torrey Botanical* Society 134:301–332.

———. 2010. Caryophyllales: A key group for understanding wood anatomy character states and their evolution. *Botanical Journal of the Linnean Society* 164:342–393.

———. 2012. How wood evolves: A new synthesis. *Botany-Botanique* 90:901–940.

Carlquist, S., and C.J. Boggs. 1996. Wood anatomy of Plumbaginaceae. *Bulletin of the Torrey Botanical Club* 123:135–147.

Carlquist, S., and E.L. Schneider. 2001. Vegetative anatomy of the New Caledonia endemic *Amborella trichopoda*: Relationships with the Illiciales and implications for vessel origin. *Pacific Science* 55:305–312.

———. 2002. The tracheid-vessel element transition in angiosperms involves multiple independent features: Cladistic consequences. *American Journal of Botany* 89:185–195.

Carlquist, S., and E.J. Wilson. 1995. Wood anatomy of *Drosophyllum* (Droseraceae): Ecological and phylogenetic considerations. *Bulletin of the Torrey Botanical Club* 122:185–189.

Carlson, S.E., D.G. Howarth, and M.J. Donoghue. 2011. Diversification of *CYCLOIDEA*-like genes in Dipsacaceae (Dipsacales): Implications for the evolution of capitulum inflorescences. *BMC Evolutionary Biology* 11:325. doi:10.1186/1471-2148-11-325.

Carlson, S.E., V. Mayer, and M.J. Donoghue. 2009. Phylogenetic relationships, taxonomy, and morphological evolution in Dipsacaceae (Dipscales) inferred by DNA sequence data. *Taxon* 58:1075–1091.

Carlsward, B.S., W.S. Judd, D.E. Soltis, S.R. Manchester, and P.S. Soltis. 2011. Putative morphological synapomorphies of Saxifragales and their major subclades. *Journal of the Botanical Research Institute of Texas* 5(1):179–196.

Carolan, J.C., I.L.I. Hook, M.W. Chase, J.W. Kadereit, and T.R. Hodkinson. 2006. Phylogenetics of *Papaver* and related genera based on DNA sequences from ITS nuclear ribosomal DNA and plastid *trnL* intron and *trnL-F* intergenic spacers. *Annals of Botany* 98(1):141–155.

Carpenter, R., and E. Coen. 1990. Floral and homeotic mutations produced by transposon mutagenesis in *Antirrhinum majus. Genes and Development* 4:1483–1493.

Carr, S.G.M., and D.J. Carr. 1961. The functional significance of syncarpy. *Phytomorphology* 11:249–256.

Cavalier-Smith, T. 1985a. The Evolution of Genome Size. John Wiley, Chichester, NY.

———. 1985b. Eukaryotic gene numbers, non-coding DNA and genome size. *In* T. Cavalier-Smith [ed.], The Evolution of Genome Size, 69–103. John Wiley, Chichester, NY.

Cellinese, N., D.A. Baum, and B.D. Mishler. 2012. Species and phylogenetic nomenclature. *Systematic Biology* 61:885–891.

Cellinese, N., S.A. Smith, E.J. Edwards, S.-T. Kim, R.C. Haberle, M. Avramakis, and M.J. Donoghue. 2009. Historical biogeography of the endemic Campanulaceae of Crete. *Journal of Biogeography* 36:1253–1269.

Cevallos-Ferriz, S.R.S., D.M. Erwin, and R.A. Stockey. 1993. Further observations on *Paleorosa similkameenensis* (Rosaceae) from the middle Eocene Princeton chert of British Columbia. *Review of Palaeobotany and Palynology* 78:277–292.

Cevallos-Ferriz, S.R.S., and R.A. Stockey. 1991. Fruits and seeds from the Princeton chert (Middle Eocene) of British-Columbia—Rosaceae (Prunoideae). *Botanical Gazette* 152(3):369–379.

Chacón J., N. Cusimano, and S.S. Renner. 2014. The evolution of Colchicaceae, with a focus on chromosome numbers. *Systematic Botany* 39:415–427.

Chadwell, T.B., S.J. Wagstaff, and P.D. Cantino. 1992. Pollen morphology of *Phryma* and some putative relatives. *Systematic Botany* 17:210–219.

Chalhoub, B., F. Denoeud, S. Liu, I.A.P. Parkin, H. Tang, X. Wang, J. Chiquet, H. Belcram, C. Tong, B. Samans, M. Correa, C. Da Silva, J. Just, C. Falentin, C.S. Koh, I. Le Clainche, M. Bernard, P. Bento, B. Noel, K. Labadie, A. Alberti, M. Charles, D. Arnaud, H. Guo, C. Daviaud, S. Alamery, K. Jabbari, M. Zhao, P.P. Edger, H. Chelaifa, D. Tack, G. Lassalle, I. Mestiri, N. Schnel, M.-C. Le Paslier, G. Fan, V. Renault, P.E. Bayer, A.A. Golicz, S. Manoli, T.-H. Lee, V.H.D. Thi, S. Chalabi, Q. Hu, C. Fan, R. Tollenaere, Y. Lu, C. Battail, J. Shen, C.H.D. Sidebottom, X. Wang, A. Canaguier, A. Chauveau, A. Berard, G. Deniot, M. Guan, Z. Liu, F. Sun, Y.P. Lim, E. Lyons, C.D. Town, I. Bancroft, X. Wang, J. Meng, J. Ma, J.C. Pires, G.J. King, D. Brunel, R. Delourme, M. Renard, J.-M. Aury, K.L. Adams, J. Batley, R.J. Snowdon, J. Tost, D. Edwards, Y. Zhou, W. Hua, A.G. Sharpe, A.H. Paterson, C. Guan, and P. Wincker. 2014. Early allopolyploid evolution in the post-Neolithic *Brassica napus* oilseed genome. *Science* 345(6199):950–953.

Chalk, L., and M.M. Chattaway. 1937. Identification of woods with included phloem. *Tropical Woods* 50:1–31.

Chanderbali, A.S., V.A. Albert, J. Leebens-Mack, N.S. Altman, D.E. Soltis, and P.S. Soltis. 2009. Transcriptional signatures of ancient floral developmental genetics in avocado (*Persea americana*; Lauraceae). *Proceedings of the National Academy of Sciences USA* 106:8929–8934.

Chanderbali, A.S., B.A. Berger, D.G. Howarth, D.E. Soltis, and P.S. Soltis. 2017. Evolution of eudicot floral diversity: genomics, genes, and gamma. *Philosophical Transactions of the Royal Society of London B* 372:20150509.

Chanderbali, A.S., H. van der Werff, and S.S. Renner. 2001. Phylogeny and historical biogeography of Lauraceae: Evidence from the chloroplast and nuclear genomes. *Annals of the Missouri Botanical Garden* 88:104–134.

Chanderbali, A.S., M.-J. Yoo, L.M. Zahn, S.F. Brockington, P.K. Wall, M.A. Gitzendanner, V.A. Albert, J. Leebens-Mack, N.S. Altman, H. Ma, C.W. dePamphilis, D.E. Soltis, and P.S. Soltis. 2010. Conservation and canalization of gene expression during angiosperm diversification accompany the origin and evolution of the flower. *Proceedings of the National Academy of Sciences USA* 107:22570–22575.

Chandler, G.T., and R.J. Bayer. 2000. Phylogenetic placement of the enigmatic western Australian genus *Emblingia* based on *rbcL* sequences. *Plant Species Biology* 15:67–72.

Chandler, G.T., and G.M. Plunkett. 2004. Evolution in Apiales: Nuclear and chloroplast markers together in (almost) perfect harmony. *Botanical Journal of the Linnean Society* 144:123–147.

Chandler, G.T., G.M. Plunkett, S.M. Pinney, L.W. Cayzer, and C.E.C. Gemmill. 2007. Molecular and morphological agreement in Pittosporaceae: Phylogenetic analysis with nuclear ITS and plastid *trnL–trnF* sequence data. *Australian Systematic Botany* 20:390–401.

Chang, P.L., B.P. Dilkes, M. McMahon, L. Comai, and S.V. Nuzhdin. 2010. Homoeolog-specific retention and use in allotetraploid *Arabidopsis suecica* depends on parent of origin and network partners. *Genome Biology* 11:R125. http://genomebiology.com/2010/11/12/R125.

Chang, S., Y. Wang, J. Lu, J. Gai, J. Li, P. Chu, R. Guan, and T. Zhao. 2013. The mitochondrial genome of soybean reveals complex genome structures and gene evolution at intercellular and phylogenetic levels. *PLoS One* 8(2): e56502. doi:10.1371/journal.pone.0056502.

Changzeng, W., and Y. Dequan. 1997. Diterpenoid, sesquiterpenoid and secoiridoid glucosides from *Aster auriculatus*. *Phytochemistry* 45:1483–1487.

Chapman, M.A., S. Tang, D. Draeger, S. Nambeesan, H. Shaffer, J.G. Barb, S.J. Knapp, and J.M. Burke. 2012. Genetic analysis of floral symmetry in Van Gogh's sunflowers reveals independent recruitment of *CYCLOIDEA* genes in the Asteraceae. *PLoS Genetics* 8(3): e1002628. doi:10.1371/journal.pgen.1002628.

Charlesworth, B., P. Sniegowski, and W. Stephan. 1995. The evolutionary dynamics of repetitive DNA in eukaryotes. *Nature* 371:215–220.

Chase, M.W. 1986. A reappraisal of the oncidioid orchids. *Systematic Botany* 11:477–491.

———. 1987. Systematic implications of pollinarium morphology in *Oncidium* Sw., *Odontoglossum* Kunth, and allied genera (Orchidaceae). *Lindleyana* 2:8–28.

———. 1988. Obligate twig epiphytes: A distinct subset of Neotropical orchidaceous epiphytes. *Selbyana*: 24–30.

———. 2001. The origin and biogeography of Orchidaceae. *In* A.M. Pridgeon, P.J. Cribb, M.W. Chase, and F. Rasmussen [eds.], Genera Orchidacearum, Vol. II: Orchidoideae (part I), 1–5. Oxford University Press, Oxford, UK.

———. 2004. Monocot relationships: An overview. *American Journal of Botany* 91(10):1645–1655.

Chase, M.W., and V.A. Albert. 1998. A perspective on the contribution of plastid *rbcL* DNA sequences to angiosperm phylogenetics. *In* D.E. Soltis, P.S. Soltis, and J.J. Doyle [eds.], Molecular Systematics of Plants, II: DNA Sequencing, 488–507. Kluwer, Boston, MA.

Chase, M.W., K.M. Cameron, R.L. Barrett, and J.V. Freudenstein. 2003. A phylogenetic classification of Orchidaceae. *In* K.M. Dixon, S.P. Kell, R.L. Barrett, and P.J. Cribb [eds.], Orchid Conservation, 69–89. Natural History Publications, Kota Kinabalu, Sabah, Malaysia.

Chase, M.W., K.M. Cameron, J.V. Freudenstein, A.M. Pridgeon, G. Salazar, C. van den Berg, and A. Schuiteman. 2015. An updated classification of Orchidaceae. *Botanical Journal of the Linnean Society*. 177:151–174.

Chase, M.W., K.M. Cameron, H.G. Hills, and D. Jarrell. 1994. Molecular systematics of the Orchidaceae and other lilioid monocots. *In* A. Pridgeon [ed.], Proceedings of the 14th World Orchid Conference, 61–73. HMSO, London, UK.

Chase, M.W., and A.V. Cox. 1998. Gene sequences, collaboration, and analysis of large data sets. *Australian Systematic Botany* 11:215–229.

Chase, M.W., A.Y. de Bruijn, G. Reeves, A.V. Cox, P.J. Rudall, M.A.T. Johnson, and L.E. Eguiarte. 2000a. Phylogenetics of Asphodelaceae (Asparagales): An analysis of plastid *rbcL* and *trnL-F* DNA sequences. *Annals of Botany* 86:935–956.

Chase, M.W., M.R. Duvall, H.G. Hills, J.G. Conran, A.V. Cox, L.E. Eguiarte, J. Hartwell, M.F. Fay, L.R. Caddick, K.M. Cameron, and S. Hoot. 1995a. Molecular phylogenetics of Lilianae. *In* P.J. Rudall, P.J. Cribb, D.F. Cutler, and C.J. Humphries [ed.], Monocotyledons: Systematics and Evolution, 109–137. Royal Botanic Gardens, Kew, UK.

Chase, M.W., M.F. Fay, D.S. Devey, O. Maurin, N. Rønsted, T.J. Davies, Y. Pillon, G. Petersen, O. Seberg, M.N. Tamura, C.B. Asmussen, K. Hilu, T. Borsch, J.I. Davis, D.W. Stevenson, J.C. Pires, T.J. Givnish, K.J. Sytsma, M.A. McPherson, S.W. Graham, and H.S. Rai. 2006. Multi-gene analyses of monocot relationships: A summary. *In* J. T. Columbus, E.A. Friar, J.M. Porter, L.M. Prince, and M.G. Simpson [eds.], Monocots: Comparative Biology and Evolution, Excluding Poales, 63–75. Rancho Santa Ana Botanic Garden, Claremont, CA.

Chase, M.W., M.F. Fay, and V. Savolainen. 2000b. Higher-level classification in the angiosperms: New insights from the perspective of DNA sequence data. *Taxon* 49:685–704.

Chase, M.W., and H.G. Hills. 1992. Orchid phylogeny, flower sexuality, and fragrance seeking. *BioScience* 42:43–49.

Chase, M.W., C.M. Morton, and J.A. Kallunki. 1999. Phylogenetic relationships of Rutaceae: A cladistic analysis of the subfamilies using evidence from *rbcL* and *atpB* sequence variation. *American Journal of Botany* 86:1191–1199.

Chase, M.W., and J.D. Palmer. 1988. Chloroplast DNA variation, geographical distribution and morphological parallelism in subtribe Oncidiinae (Orchidaceae). *American Journal of Botany* 75:163–164.

———. 1989. Chloroplast DNA systematics of lilioid of the lilioid monocots: Feasibility, resources, and an example from the Orchidaceae. *American Journal of Botany* 76:1720–1730.

———. 1992. Floral morphology and chromosome number in subtribe Oncidiinae (Orchidaceae): Evolutionary insights from a phylogenetic analysis of chloroplast DNA restriction site variation. *In* D.E. Soltis, P.S. Soltis, and J.J. Doyle [eds.], Molecular Systematics of Plants, 324–339. Chapman and Hall, New York, NY.

———. 1997. Leapfrog radiation in floral and vegetative traits among twig epiphytes in the orchid subtribe Oncidiinae. *In* T.J. Givnish, and K.J. Sytsma [eds.], Molecular Evolution and Adaptive Radiation, 331–352. Cambridge University Press, Cambridge, UK.

Chase, M.W., and J.S. Pippen. 1988. Seed morphology in the subtribe Oncidiinae (Orchidaceae). *Systematic Botany* 13:313–323.

Chase, M.W., P.J. Rudall, and J.G. Conran. 1996. New circumscriptions and a new family of asparagoid lilies: Genera formerly included in Anthericaceae. *Kew Bulletin* 51:667–680.

Chase, M.W., D.E. Soltis, R.G. Olmsted, D. Morgan, D.H. Les, B.D. Mishler, M.R. Duvall, R.A. Price, H.G. Hills, Y.-L. Qiu, K.A. Kron, J.H. Rettig, E. Conti, J.D. Palmer, J.R. Manhart, K.J. Sytsma, H.J. Michaels, W.J. Kress, K.G. Karol, W.D. Clark, M. Hedron, B.S. Gaut, R.K. Jansen, K.-J. Kim, C.F. Wimpee, J.F. Smith, G.R. Furnier, S.H. Strauss, Q.-Y. Xiang, G.M. Plunkett, P.S. Soltis, S.M. Swensen, S.E. Williams, P.A. Gadek, C.J. Quinn, L.E. Egguiarte, E. Golenberg, G.H. Learn, Jr., S.W. Graham, S.C.H. Barrett, S. Dayanandan, and V.A. Albert. 1993. Phylogenetics of seed plants: An analysis of nucleotide sequences from the plastid gene *rbcL*. *Annals of the Missouri Botanical Garden* 80:526–580.

Chase, M.W., D.E. Soltis, P.S. Soltis, P.J. Rudall, M.F. Fay, W.H. Hahn, S. Sullivan, J. Joseph, T. Givnish, K.J. Sytsma, and J.C. Pires. 2000a. Higher-level systematics of the monocotyledons: An assessment of current knowledge and a new classification. *In* K.L. Wilson, and D.A. Morrison [eds.], Monocots: Systematics and Evolution, 3–16. CSIRO, Melbourne, Australia.

Chase, M.W., D.W. Stevenson, P. Wilkin, and P.J. Rudall. 1995b. Monocot systematics: A combined analysis. *In* P.J. Rudall, P.J. Cribb, D.F. Cutler, and C.J. Humphries [eds.], Monocotyledons: Systematics and Evolution, 685–730. Royal Botanic Gardens, Kew, UK.

Chase, M.W., S. Zmarzty, M.D. Lledo, K.J. Wurdack, S.M. Swensen, and M.F. Fay. 2002. When in doubt, put it in Flacourtiaceae: A molecular phylogenetic analysis based on plastid *rbcL* DNA sequences. *Kew Bulletin* 57:141–181.

Chat, J., B. Jáuregui, R.J. Petit, and S. Nadot. 2004. Reticulate evolution in kiwi fruit (*Actinidia*, Actinidiaceae) identified by comparing their maternal and paternal phylogenies. *American Journal of Botany* 91:736–747.

Chatrou, L.W., M.D. Pirie, R.H.J. Erkens, T.L.P. Couvreur, K.M. Neubig, J.R. Abbott, J.B. Mols, J.W. Maas, R.M.K. Saunders, and M.W. Chase. 2012. A new subfamilial and tribal classification of the pantropical flowering plant family Annonaceae informed by molecular phylogenetics. *Botanical Journal of the Linnean Society* 169:5–40.

Chaw, S.M., C.L. Parkinson, Y. Cheng, T.M. Vincent, and J.D. Palmer. 2000. Seed plant phylogeny inferred from all three plant genomes: Monophyly of extant gymnosperms and origin of Gnetales from conifers. *Proceedings of the National Academy of Sciences USA* 97:4086–4091.

Chaw, S.M., A. Zharkikh, H.-M. Sung, T.-C. Lau, and W.-H. Li. 1997. Molecular phylogeny of extant gymnosperms and seed plant evolution: Analysis of nuclear 18S rRNA sequences. *Molecular Biology and Evolution* 14:56–58.

Chen, E.C.H., C.F.B.A. Najar, C. Zheng, A. Brandts, E. Lyons, H. Tang, L. Carretero-Paulet, V.A. Albert, and D. Sankoff. 2013. The dynamics of functional classes of plant genes in rediploidized ancient polyploids. *BMC Bioinformatics* 14(Suppl. 15):S19. http://www.biomedcentral.com/1471–2015/14/S15/S19.

Chen, L., Y. Ren, P.K. Endress, X.H. Tian, and X.H. Zhang. 2007. Floral development of *Tetracentron sinense* (Trochodendraceae) and its systematic significance. *Plant Systematics and Evolution* 264:183–193.

Chen, P., L. Chen, and J. Wen. 2011. The first phylogenetic analysis of *Tetrastigma* (Miq.) Planch., the host of Rafflesiaceae. *Taxon* 60(2):499–512.

Chen, S.-C., D.K. Kim, M.W. Chase, and J.-H. Kim. 2013. Networks in a large-scale phylogentic analysis: Reconstructing evolutionary history of Asparagales (Lilianae) based on four plastid genes. *PLoS One* 8(3): e59472. doi:10.1371/journal.pone.0059472.

Chen, Z.D., X.Q. Wang, H.Y. Sun, and Y. Han. 1998. Systematic position of the Rhoipteleaceae: Evidence from *rbcL* sequences. *Zhiwu Fenlei Xuebao* 36:1–5.

Chen, Z.J. 2007. Genetic and epigenetic mechanisms for gene expression and phenotypic variation in plant polyploids. *Annual Review of Plant Biology* 58:377–406.

———. 2010. Molecular mechanisms of polyploidy and hybrid vigor. *Trends in Plant Science* 15(2):57–71.

Chen, Z.J., and J.A. Birchler [eds.]. 2013. Polyploid and Hybrid Genomics. Wiley-Blackwell, Ames, IA.

Cheng, F., J. Wu, L. Fang, and X. Wang. 2012. Syntenic gene analysis between *Brassica rapa* and other Brassicaceae species. *Frontiers in Plant Science* 3:198. doi:10.3389/fpls.2012.00198. eCollection 2012.

Chesters, K.I.M. 1955. Some plant remains from the Upper Cretaceous and Tertiary of West Africa. *The Annals and Magazine of Natural History*, Ser. 12, 8:498–503.

Chittka, L., and J.D. Thomson [eds.]. 2001. Cognitive Ecology of Pollination. Animal Behavior and Evolution. Cambridge University Press, Cambridge, UK.

Chittka, L., J.D. Thomson, and N.M. Waser. 1999. Flower constancy, insect psychology, and plant evolution. *Naturwissenschaften* 86:361–377.

Citerne, H., F. Jabbour, S. Nadot, and C. Damerval. 2010. The evolution of floral symmetry. *Advances in Botanical Research* 54:85–137.

Cho, Y., J.P. Mower, Y.L. Qiu, and J.D. Palmer. 2004. Mitochondrial substitution rates are extraordinarily elevated and variable in a genus of flowering plants. *Proceedings of the National Academy of Sciences USA* 101:17741–17746.

Chorinsky, F. 1931. Vergleichende morphologishe Untersuchungen der Haargebilde bei Portulacaceae und Cactaceae. *Österreichische Botanische Zeitschrift* 80:308–327.

Christenhusz, M.J.M, S.F. Brockington, P.-A. Christin, and R.F. Sage. 2014. On the disintegration of Molluginaceae: A new genus and family (*Kewa*, Kewaceae) segregated from *Hypertelis*, and placement of *Macarthuria* in Macarthuriaceae. *Phytotaxa* doi:http://dx.doi.org/10.11646/phytotaxa.181.4.

Christenhusz, M.J.M., M.F. Fay, J.J. Clarkson, P. Gasson, J. Morales, J.B. Jimenez Barrios, and M.W. Chase. 2010. Petenaeaceae, a new angiosperm family in Huerteales with a distant relationship to *Gerrardina* (Gerrardinaceae). *Botanical Journal of the Linnean Society* 164(1):16–25.

Christin, P.-A., M. Arakaki, C.P. Osborne, A. Brautigam, R.F. Sage, J.M. Hibberd, S. Kelly, S. Covshoff, G.K. Wong, L. Hancock, and E.J. Edwards. 2014. Shared origins of a key enzyme during the evolution of C4 and CAM metabolism. *Journal of Experimental Botany* 65(13):3609–3621.

Christin, P.-A., and C.P. Osborne. 2013. The recurrent assembly of C-4 photosynthesis, an evolutionary tale. *Photosynthesis Research* 117(1–3):163–175.

Christin, P.-A, C.P. Osborne, D.S. Chatelet, J.T. Columbus, G. Besnard, T.R. Hodkinson, L.M. Garrison, M.S. Vorontsova, and E.J. Edwards. 2013. Anatomical enablers and the evolution of C-4 photosynthesis in grasses. *Proceedings of the National Academy of Sciences USA* 110(4):1381–1386.

Christin, P.-A., T.L. Sage, E.J. Edwards, R.M. Ogburn, R. Khoshravesh, and R.F. Sage. 2011. Complex evolutionary transitions and the significance of C3-C4 intermediate forms of photosynthesis in Molluginaceae. *Evolution* 65:646–660.

Christin, P.-A., N. Salamin, E.A. Kellogg, A. Vicentini, and G. Besnard. 2009. Integrating phylogeny into studies of C-4 variation in the grasses. *Plant Physiology* 149(1):82–87.

Christin, P.-A., N. Salamin, V. Savolainen, M.R. Duvall, and G. Besnard. 2007. C-4 photosynthesis evolved in grasses via parallel adaptive genetic changes. *Current Biology* 17(14):1241–1247.

Civeyrel, L., A. Lethomas, K. Ferguson, and M.W. Chase. 1998. Critical reexamination of palynological characters used to delimit Asclepiadaceae in comparison to the molecular phylogeny obtained from plastid *matK* sequences. *Molecular Phylogenetics and Evolution* 9:517–527.

Clark, J.L. 2005. A Monograph of *Alloplectus* (Gesneriaceae). *Selbyana* 25:182–209.

———. 2009. Systematics of *Glossoloma* (Gesneriaceae). *Systematic Botany Monographs* 89:1–126.

Clark, J.L., M.M. Funke, A.M. Duffy, and J.F. Smith. 2012. Phylogeny of a Neotropical clade in the Gesneriaceae: More tales of convergent evolution. *International Journal of Plant Sciences* 173:894–916.

Clark, J.L., P.S. Herendeen, L.E. Skog, and E.A. Zimmer. 2006. Phylogenetic relationships and generic boundaries in the *Episcieae* (Gesneriaceae) inferred from nuclear, chloroplast, and morphological data. *Taxon* 55:313–336.

Clark, J.L., E.H. Roalson, R.A. Pritchard, C.L. Coleman, V. Teoh, and J. Matos. 2011. Independent origin of radial floral symmetry in the Gloxinieae (Gesnerioideae: Gesneriaceae) is supported by the rediscovery of *Phinaea pulchella* in Cuba. *Systematic Botany* 36:757–767.

Clarkson, J.J., K.Y. Lim, A. Kovarik, M.W. Chase, S. Knapp, and A.R. Leitch. 2005. Long-term diploidization in allopolyploid *Nicotiana* section Repandae (Solanaceae). *New Phytologist* 168:241–252.

Clausen, J., D.D. Keck, and W.M. Hiesey. 1945. Experimental studies on the nature of species. II. Plant Evolution through amphidiploidy and autoploidy, with examples from the Madiinae. Carnegie Insitution, Washington, D.C.

Clausing, G., K. Meyer, and S. Renner. 2000. Correlations among fruit traits and evolution of different fruits within Melastomataceae. *Botanical Journal Linnaean Society* 133:303–326.

Clausing, G., and S. Renner. 2001. Molecular phylogenetics of Melastomataceae and Memecylaceae: Implications for character evolution. *American Journal of Botany* 88(3):486–498.

Clawson, M.L., A. Bourret, and D.R. Benson. 2004. Assessing the phylogeny of *Frankia-actinorhizal* plant nitrogen-fixing root nodule symbioses with *Frankia* 16S rRNA and glutamine synthetase gene sequences. *Molecular Phylogenetics and Evolution* 31(1):131–138.

Clayton, J.W., E.S. Fernando, P.S. Soltis, and D.E. Soltis. 2007. Molecular phylogeny of the tree-of-heaven family (Simaroubaceae) based on chloroplast and nuclear markers. *International Journal of Plant Sciences* 168(9):1325–1339.

Clegg, M. 1990. Dating the monocot-dicot divergence. *Trends Ecology and Evolution* 5:1–2.

Clement, J.S., and T.J. Mabry. 1996. Pigment evolution in the Caryophyllales: A systematic overview. *Botanica Acta* 109:360–367.

Clement, W.L., and M.J. Donoghue. 2007. Dissolution of *Viburnum* section *Megalotinus* (Adoxaceae) of Southeast Asia and its implications for morphological evolution and biogeography. *International Journal of Plant Sciences* 172:559–573.

Cocucci, A. 1999. Evolutionary radiation in Neotropical Solanaceae. *In* M. Nee. D.E. Symon, R.N. Lester, and J.P. Jessup [eds.], Solanaceae IV: Advances in Biology and Utilization, 9–22. Royal Botanic Gardens, Kew, UK.

Coe, E.H., S. McCormick, and S.A. Modena. 1981. White pollen in maize. *Journal of Heredity* 72:318–320.

Coen, E.S., and E.M. Meyerowitz. 1991. The war of the whorls: genetic interactions controlling flower development. *Nature* 353(6339):31–37.

Cohen, J.I. 2013. A phylogenetic analysis of morphological and molecular characters of Boraginaceae: Evolutionary relationships, taxonomy, and patterns of character evolution. *Cladistics* 30(2):139–169.

Coiffard, C., B. Gomez, and F. Thevenard. 2007. Early Cretaceous angiosperm invasion of western Europe and major environmental changes. *Annals of Botany* 100:545–553.

Collett, R. 1999. Arboretum collection makes headline news. *Bulletin of University of California, Santa Cruz Arboretum Association* 23:1–2.

Collinson, M.E., J.J. Hooker, and D.R. Grocke. 2003. Cobham lignite bed and penecontemporaneous macrofloras of Southern England: A record of vegetation and fire across Palaeocene-Eocene thermal maximum. *Geological Society of America Special Paper* 369:333–349.

Collinson, M.E., S.R. Manchester, and V. Wilde. 2012. Fossil fruits and seeds of the Middle Eocene Messel biota, Germany. *Abhandlungen der Senckenbergischen Naturforschenden Gesellschaft* 570:1–250.

Colombo, L., R. Battaglia, and M.M. Kater. 2008. *Arabidopsis* ovule development and its evolutionary conservation. *Trends in Plant Science* 13:444–450.

Colombo, L., J. Franken, E. Koetje, J. Vanwent, H.J.M. Dons, G.C. Angenent, and A.J. Vantunen. 1995. The *Petunia* MADS-box gene *FBP11* determines ovule identity. *Plant Cell* 7(11):1859–1868.

Comai, L., A.P. Tyagi, K. Winter, R. Holmes-Davis, S.H. Reynolds, Y. Stevens, and B. Byers. 2000. Phenotypic instability and rapid gene silencing in newly formed *Arabidopsis* allotetraploids. *Plant Cell* 12(9):1551–1567.

Conran, J.G., J.M. Bannister, D.E. Lee, R.J. Carpenter, E.M. Kennedy, T. Reichgelt, and R.E. Fordyce. In Press. An update of monocot macrofossil data, with an emphasis on Australia and New Zealand. *Botanical Journal of the Linnean Society*.

Conran, J.G., J.M. Bannister, D.C. Mildenhall, D.E. Lee, J. Chacon, and S.S. Renner. 2014. Leaf fossils of *Luzuriaga* and a monocot flower with in situ *Liliacidites contortus* Mildenh. sp. nov. pollen from the Early Miocene. *American Journal of Botany* 101:141–155.

Conran, J.G., M.W. Chase, and P.J. Rudall. 1997. Two new monocotyledon families Anemarrhenaceae and Behniaceae (Lilianae: Asparagales). *Kew Bulletin* 52:995–999.

Conti, E., T. Eriksson, J. Schönenberger, K.J. Sytsma, and D.A. Baum. 2002. Early Tertiary out-of-India dispersal of Crypteroniaceae: Evidence from phylogeny and molecular dating. *Evolution* 56(10):1931–1942.

Conti, E., A. Fischbach, and K.J. Sytsma. 1993. Tribal relationships in Onagraceae: Implications from *rbcL* sequence data. *Annals of the Missouri Botanical Garden* 80:672–685.

Conti, E., A. Litt, and K.J. Sytsma. 1996. Circumscription of Myrtales and their relationships to other rosids: Evidence from *rbcL* sequence data. *American Journal of Botany* 83:221–233.

Conti, E., A. Litt, P.G. Wilson, S.A. Graham, B.G. Briggs, L.A.S. Johnson, and K.J. Sytsma. 1997. Interfamilial relationships in Myrtales: Molecular phylogeny and patterns of morphological evolution. *Systematic Botany* 22(4):629–647.

Conti, E., F. Rutschmann, T. Eriksson, K.J. Sytsma, and D.A. Baum. 2004. Calibration of molecular clocks and the biogeographic history of Crypteroniaceae: A reply to Moyle. *Evolution* 58(8):1874–1876.

Conti, E., E. Suring, D. Boyd, J. Jorgensen, J. Grant, and S. Kelso. 2000. Phylogenetic relationships and character evolution in *Primula* L.: The usefulness of ITS sequence data. *Plant Biosystems* 134:385–392.

Contreras, V.R., R. Scogin, and C.T. Philbrick. 1993. A phytochemical study of selected Podostemaceae: Systematic implications. *Aliso* 13:513–520.

Cook, C.D.K. 1982. Pollination mechanisms in the Hydrocharitaceae. *In* J.J. Symoens, S.S. Hooper, and P. Compère [eds.], Studies on Aquatic Vascular Plants, 1–15. Royal Botanical Society of Belgium, Brussels, Belgium.

Corbett, S.R., and S.R. Manchester. 2004. Phytogeography and fossil history of *Ailanthus* (Simaroubaceae). *International Journal of Plant Sciences* 165:671–690.

Cordell, G.A. 1974. The biosynthesis of indole alkaloids. *Lloydia* 37:219–298.

Corner, E.J.H. 1946. Centrifugal stamens. *Journal of the Arnold Arboretum* 27:423–437.

———. 1976. The Seeds of Dicotyledons. Cambridge University Press, New York, NY.

Cornet, B., 1989. Late Triassic angiosperm like pollen from the Richmond Rift Basin of Virginia, U.S.A. *Palaeontographica Abteilung B* 213:37–87.

Correa, E, C. Jaramillo, S.R. Manchester, and M. Guteierrez. 2010. A fruit and leaves of rhamnaceous affinities from the Late Cretaceous (Maastrichtian) of Colombia. *American Journal of Botany* 97:71–79.

Cosner, M.E., R.K. Jansen, and T.G. Lammers. 1994. Phylogenetic relationships in the Campanulales based on *rbcL* sequences. *Plant Systematics and Evolution* 190:79–95.

Coulter, J.M., and C.J. Chamberlain. 1903. Morphology of Angiosperms. Appleton, New York, NY.

Couvreur, T.L.P., and W.J. Baker. 2013a. Global biogeography and diversification of palms sheds light on the evolution of tropical lineages. II Diversification history and origin of regional assemblages. *Journal of Biogeography* 40:286–298.

———. 2013b. Tropical rain forest evolution: Palms as a model group. *BMC Biology* 11:48. doi:10.1186/1741-7007-11-48.

Cox, A.V., G.J. Abdelnour, M.D. Bennett, and I.J. Leitch. 1998. Genome size and karyotype evolution in the slipper orchids (Cypripedioideae: Orchidaceae). *American Journal of Botany* 85:681–687.

Cox, A.V., A.M. Pridgeon, V.A. Albert, and M.W. Chase. 1997. Phylogenetics of the slipper orchids (Cypripedioideae: Orchidaceae): Nuclear rDNA ITS sequences. *Plant Systematics and Evolution* 208:197–223.

Cox, P.A. 1988. Hydrophilous pollination. *Annual Review of Ecology and Systematics* 19:261–280.

Cox, P.A., and C.J. Humphries. 1993. Hydrophilous pollination and breeding system evolution in seagrasses: A phylogenetic approach to the evolutionary ecology of the Cymodoceaceae. *Botanical Journal of the Linnean Society* 113:217–226.

Cozzolino, S., and A. Widmer. 2005. Orchid diversity: An evolutionary consequence of deception? *Trends in Ecology & Evolution* 20:487–493.

Crane, P.R. 1981. Betulaceous leaves and fruits from the British Upper Paleocene. *Botanical Journal of the Linnean Society* 83:103–136.

———. 1985. Phylogenetic analysis of seed plants and the origin of angiosperms. *Annals of the Missouri Botanical Garden* 72:716–793.

———. 1989. Patterns of evolution and extinction in vascular plants. *In* K.C. Allen, and D.E.G. Briggs [eds.], Evolution and the Fossil Record, 153–187. Belhaven Press, London, UK.

———. 2013. Ginkgo: The Tree That Time Forgot. Yale University Press, New Haven, CT.

Crane, P.R., and S. Blackmore [eds.]. 1989. Evolution, Systematics and Fossil History of the Hamamelidae. 2 vols. Clarendon Press, Oxford, UK.

Crane, P.R., E.M. Friis, and W.G. Chaloner [eds.]. 2010. Darwin and the evolution of flowers. *Philosophical Transactions of the Royal Society of London B* 365:345–543.

Crane, P.R., E.M. Friis, and K.R. Pedersen. 1989. Reproductive structure and function in Cretaceous Chloranthaceae. *Plant Systematics and Evolution* 165:211–226.

———. 1995. The origin and early diversification of angiosperms. *Nature* 374:27–33.

Crane, P.R., P. Herendeen, and E.M. Friis. 2004. Fossils and plant phylogeny. *American Journal of Botany* 91(10):1683–1699.

Crane, P.R., and P. Kenrick. 1997a. Problems in cladistic classification: Higher-level relationships in land plants. *Aliso* 15:87–104.

———. 1997b. Diverted development of reproductive organs: A source of morphological innovation in land plants. *Plant Systematics and Evolution* 206:161–174.

Crane, P.R., S.R. Manchester, and D.L. Dilcher. 1990. A preliminary survey of fossil leaves and well-preserved plant reproductive structures from the Sentinel Butte Formation (Paleocene) near Almont, North Dakota USA. *Fieldiana Geology* (20):1–64.

———. 1991. Reproductive and vegetative structure of *Nordenskioldia* (Trochodendraceae), a vessel-less dicotyledon from the Early Tertiary of the northern hemisphere. *American Journal of Botany* 78(10):1311–1334.

Crane, P.R., K. Raunsgaard, E.M. Friis, K.R. Pedersen, and A.N. Drinnan. 1993. Early Cretaceous (Early to Middle Albian) platanoid inflorescences associated with *Sapindopsis* leaves from the Potomac Group of eastern North America. *Systematic Botany* 18:328–344.

Crane, P.R., and R.A. Stockey. 1987. *Betula* leaves and reproductive structures from the middle Eocene of British Columbia, Canada. *Canadian Journal of Botany* 65:2490–2500.

Crane, P.R., and G.R. Upchurch. 1987. *Drewria Potomacensis* gen. et sp. nov. an Early Cretaceous member of Gnetales from the Potomac group of Virginia. *American Journal of Botany* 74:1722–1736.

Crepet, W.L. 1996. Timing in the evolution of derived floral characters: Upper Cretaceous (Turonian) taxa with tricolpate and tricolpate-derived pollen. *Review of Palaeobotany and Palynology* 90:339–359.

———. 2000. Progress in understanding angiosperm history, success and relationships: Darwin's abominably "perplexing phenomenon". *Proceedings of the National Academy of Sciences USA* 97:12939–12941.

Crepet, W.L., and K.C. Nixon. 1998. Fossil Clusiaceae from the Late Cretaceous (Turonian) of New Jersey and implications regarding the history of bee pollination. *American Journal of Botany* 85:1122–1133.

Crepet, W.L., K.C. Nixon, and C.P. Daghlian. 2013. Fossil Ericales from the Upper Cretaceous of New Jersey. *International Journal of Plant Sciences* 174(3):572–584.

Crepet, W.L., K.C. Nixon, and M.A. Gandolfo. 2004. Fossil evidence and phylogeny: The age of major angiosperm clades based on mesofossil and macrofossil evidence from Cretaceous deposits. *American Journal of Botany* 91(10):1666–1682.

———. 2005. An extinct calycanthoid taxon, *Jerseyanthus calycanthoides*, from the Late Cretaceous of New Jersey. *American Journal of Botany* 92:1475–1485.

Crisp, M.D., and L.G. Cook. 2007. A congruent molecular signature of vicariance across multiple plant lineages. *Molecular Phylogenetics and Evolution* 43(3):1106–1117.

Croizat, L. 1941. On the systematic position of *Daphniphyllum* and its allies. *Lingnan Science Journal* 20:79–103.

Cronk, Q.C.B, R.M. Bateman, J.A. Hawkins, and G.A. McCarthy [eds.]. 2002. Developmental Genetics and Plant Evolution. Taylor and Francis, London, UK.

Cronk, Q., I. Ojeda, and R.T. Pennington. 2006. Legume comparative genomics: Progress in phylogenetics and phylogenomics. *Current Opinion in Plant Biology* 9:99–103.

Cronquist, A. 1957. Outline of a new system of families and orders of dicotyledons. *Bulletin du Jardin Botanique National de Belgique* 27:13–40.

———. 1968. The Evolution and Classification of Flowering Plants. Houghton Mifflin, Boston, MA.

———. 1981. An Integrated System of Classification of Flowering Plants. Columbia University Press, New York, NY.

———. 1983. Some realignments in the dicotyledons. *Nordic Journal of Botany* 3:75–83.

———. 1984. A commentary on the definition of the order Myrtales. *Annals of the Missouri Botanical Garden* 71:780–782.

———. 1988. The Evolution and Classification of Flowering Plants. New York Botanical Garden, Bronx, NY.

Cronquist, A., and R.F. Thorne. 1994. Nomenclatural and taxonomic history. *In* H.-D. Behnke, and T.J. Mabry [eds.], Caryophyllales: Evolution and Systematics, 87–121. Springer-Verlag, Berlin, Germany.

Crowl, A.A., E. Mavrodiev, G. Mansion, R. Haberle, A. Pistarino, G. Kamari, D. Phitos, T. Borsch, and N. Cellinese. 2014. Phylogeny of Campanuloideae (Campanulaceae) with emphasis on the utility of nuclear pentatricopeptide repeat (ppr) genes. *PLoS One* 9(4): e94199. doi:10.1371/journal.pone.0094199.

Cubas, P. 2004. Floral zygomorphy, the recurring evolution of a successful trait. *Bioessays* 26(11):1175–1184.

Cubas, P., C. Vincent, and E. Coen. 1999. An epigenetic mutation responsible for natural variation in floral symmetry. *Nature* 401(6749):157–161.

Cuenca, A., G. Petersen, and O. Seberg. 2013. The complete sequence of the mitochondrial genome of *Butomus umbellatus*—a member of an early branching lineage of monocotyledons. *PLoS One* 8(4): e61552. doi:10.1371/journal.pone.0061552.

Cuénoud, P., M.A.D. Martinez, P.-A. Loizeau, R.E. Spichiger, and J.-F. Manen. 2000. Molecular phylogeny and biogeography of the genus *Ilex* L. (Aquifoliaceae). *Annals of Botany* 85:111–122.

Cuénoud, P., V. Savolainen, L.W. Chatrou, M. Powell, R.J. Grayer, and M.W. Chase. 2002. Molecular phylogenetics of Caryophyllales based on nuclear 18S rDNA and plastid *rbcL, atpB,* and *matK* DNA sequences. *American Journal of Botany* 89(1):132–144.

Cuerrier, A., L. Brouillet, and D. Barabé. 1997. Numerical analyses of the modern classifications of the flowering plants. *American Journal of Botany* 84:185. Abstract.

Culley, T.M., S.G. Weller, and A.K. Sakai. 2002. The evolution of wind pollination in angiosperms. *Trends in Ecology & Evolution* 17:361–369.

Dafni, A. 1983. Pollination of *Orchis caspia*–a nectarless plant which deceives the pollinators of nectariferous species from other plant families. *Journal of Ecology* 71:467–474.

———. 1987. Pollination in *Orchis* and related genera: Evolution from reward to deception. *In* J. Arditti [ed.], Orchid Biology, Reviews and Perspectives, 79–104. Cornell University Press, Utica, NY.

Dafni, A., and D.M. Calder. 1987. Pollination by deceit and floral mimesis in *Thelymitra antennifera* (Orchidaceae). *Plant Systematics and Evolution* 158:11–22.

Dafni, A., and Y. Ivri. 1981. Floral mimicry between *Orchis israelitica* Baumann and Dafni (Orchidaceae) and *Bellevalia flexuosa* Boiss. (Liliaceae). *Oecologia* 49:229–232.

Dahlgren, G. 1989. The last Dahlgrenogram: System of classification of the dicotyledons. *In* K. Tan, R.R. Mill, and T.S. Elias [eds.], Plant Taxonomy, Phylogeography and Related Subjects, 249–260. Edinburgh University Press, Edinburgh, UK.

———. 1991. Steps towards a natural system of the dicotyledons: Embryological characters. *Aliso* 13:107–165.

Dahlgren, R.M.T. 1975. The distribution of characters within an angiosperm system. I. Some embryological characters. *Botaniska Notiser* 128:181–197.

———. 1977. A Commentary on a diagrammatic presentation of the angiosperms in relation to the distribution of character states. *Plant Systematics and Evolution Supplement* 1:253–283.

———. 1980. A revised system of classification of the angiosperms. *Botanical Journal of the Linnean Society* 80:91–124.

———. 1983. General aspects of angiosperm evolution and macrosystematics. *Nordic Journal of Botany* 3:119–149.

———. 1988. Rhizophoraceae and Anisophylleaceae: Summary statement, relationships. *Annals of the Missouri Botanical Garden* 75:1259–1277.

Dahlgren, R.M.T., K. Bremer. 1985. Major clades of the Angiosperms. *Cladistics* 1:349–368.

Dahlgren, R.M.T., H.T. Clifford, and P.F. Yeo. 1985. The Families of the Monocotyledons: Structure, Evolution and Taxonomy. Springer-Verlag, Berlin, Germany.

Dahlgren, R.M.T., and F. Rasmussen. 1983. Monocotyledon evolution: Characters and phylogenetic estimation. *Evolutionary Biology* 16:255–395.

Dahlgren, R.M.T., S. Rosendal-Jensen, and B.J. Nielsen. 1981. A revised classification of the angiosperms with comments on correlation between chemical and other characters. *In* D.A. Young, and D.S. Seigler [eds.], Phytochemistry and Angiosperm Phylogeny, 149–204. Praeger Publishers, New York, NY.

Dahlgren, R.M.T., and R. Thorne. 1984. The order Myrtales: Circumscription, variation, and relationships. *Annals of the Missouri Botanical Garden* 71:633–699.

Damtoft, S., S.R. Jensen, and B.J. Nielsen. 1992. Biosynthesis of iridoid glucosides in *Lamium album. Phytochemistry* 31:135–137.

Dandy, J.E. 1927. The genera of Saxifragaceae. *Kew Bulletin of Miscellaneous Information* 1927:100–118.

Danilova, M.E. 1996. Anatomia seminum comparativa.

Tomus 5. Rosidae I. NAUKA, Leningrad, Russia. [In Russian.]
Darlington, C.D. 1937. Recent Advances in Cytology. P. Blakiston's Son & Co., Philadelphia. PA.
Darnowski, D.W., D.M. Carroll, B. Plachno, E. Kabanoff, and E. Cinnamon. 2006. Evidence of protocarnivory in triggerplants (*Stylidium* spp.; Stylidiaceae). *Plant Biology* 8(6):805–812.
Darwin, C. 1859. The Origin of Species. John Murray, London, UK.
———. 1875. Insectivorous Plants. John Murray, London, UK.
Davies, B., M. Cartolano, and Z. Schwarz-Sommer. 2006. Flower development: The *Antirrhinum* perspective. *Advances in Botanical Research: Developmental Genetics of the Flower* 44: 279–321.
Davies, T.J., T.G. Barraclough, M. Chase, P.S. Soltis, D.E. Soltis, and V. Savolainen. 2004. Darwin's abominable mystery: Insights from a supertree of the angiosperms. *Proceedings of the National Academy of Sciences USA* 101:1904–1909.
Davioud, E., F. Bailleul, P. Delaveau, and H. Jacquemin. 1985. Iridoids of Guyanese species of *Stigmaphyllon*. *Planta Medica* 51:78–79.
Davis, C.C., and W.R. Anderson. 2010. A complete generic phylogeny of Malpighiaceae inferred from nucleotide sequence data and morphology. *American Journal of Botany* 97(12):2031–2048.
Davis, C.C., W.R. Anderson, and M.J. Donoghue. 2001. Phylogeny of Malpighiaceae: Evidence form chloroplast *ndhF* and *trnL-F* nucleotide sequences. *American Journal of Botany* 88:1847–1846.
Davis, C.C., W.R. Anderson, and K.J. Wurdack. 2005. Gene transfer from a parasitic flowering plant to a fern. *Proceedings of the Royal Society of London B* 272(1578):2237–2242.
Davis, C.C., C.D. Bell, S. Mathews, and M.J. Donoghue. 2002. Laurasian migration explains Gondwanan disjunctions: Evidence from Malpighiaceae. *Proceedings of the National Academy of Sciences USA* 99(10):6833–6837.
Davis, C.C., and M.W. Chase. 2004. Elatinaceae are sister to Malpigiaceae; Peridiscaceae belong to Saxifragales. *American Journal of Botany* 91:262–273.
Davis, C.C., P.K. Endress, and D.A. Baum. 2008. The evolution of floral gigantism. *Current Opinion in Plant Biology* 11:49–57.
Davis, C.C., P.W. Fritsch, C.D. Bell, and S. Mathews. 2004. High-latitude tertiary migrations of an exclusively tropical clade: Evidence from Malpighiaceae. *International Journal of Plant Sciences* 165(4):S107–S121.
Davis, C.C., M. Latvis, D.L. Nickrent, K.J. Wurdack, and D.A. Baum. 2007. Floral gigantism in Rafflesiaceae. *Science* 315(5820):1812. doi:10.1126/science.1135260.
Davis, C.C., C.O. Webb, K.J. Wurdack, C.A. Jaramillo, and M.J. Donoghue. 2005. Explosive radiation of Malpighiales supports a mid-cretaceous origin of modern tropical rain forests. *American Naturalist* 165(3):36–65.
Davis, C.C., and K.J. Wurdack. 2004. Host-to-parasite gene transfer in flowering plants: Phylogenetic evidence from Malpighiales. *Science* 305(5684):676–678.
Davis, C.C., Z. Xi, and S. Mathews. 2014. Plastid phylogenomics and green plant phylogeny: Almost full circle but not quite there. *BMC Biology* 12:11. doi:10.1186/1741-7007-12-11.
Davis, G.L. 1967. Systematic Embryology of the Angiosperms. Wiley, New York, NY.
Davis, J.I., J.R. McNeal, C.F. Barrett, M.W. Chase, J.I. Cohen, M.R. Duvall, T.J. Givnish, S.W. Graham, G. Petersen, J.C. Pires, O. Seberg, D.W. Stevenson, and J. Leebens-Mack. 2013. Contrasting patterns of support among plastid genes and genomes for major clades of the monocotyledons. *In* P. Wilkin, and S. J. Mayo [eds.], Early Events in Monocot Evolution, 315–349. Cambridge University Press, New York, NY.
Davis, J.I., M.P. Simmons, D.W. Stevenson, and J.F. Wendel. 1998. Data decisiveness, data quality, and incongruence in phylogenetic analysis: An example from the monocotyledons using mitochondrial *atpA* sequences. *Systematic Biology* 47:282–310.
Davis, J.I., and R.J. Soreng. 1993. Phylogenetic structure in the grass family (Poaceae), as determined from chloroplast DNA restriction site variation. *American Journal of Botany* 80:1444–1454.
Davis, J.I., D.W. Stevenson, G. Petersen, O. Seberg, L.M. Campbell, J.V. Freudenstein, D.H. Goldman, C.R. Hardy, F.A. Michelangeli, M.P. Simmons, C.D. Specht, F. Vergara-Silva, and M.A. Gandolfo. 2004. A phylogeny of the monocots, as inferred from *rbcL* and *atpA* sequence variation. *Systematic Botany* 29(3):467–510.
Davis, P., and P. Kenrick. 2004. Fossil Plants. Smithsonian Books, Washington, D.C.
Dayanandan, S., P.S. Ashton, and R.B. Primack. 1999. Phylogeny of the tropical tree family Dipterocarpaceae based on nucleotide sequences of the chloroplast *rbcL* gene. *American Journal of Botany* 86:1182–1190.
Debry, R., and R.G. Olmstead. 2000. A simulation study of reduced tree-search effort in bootstrap resampling analysis. *Systematic Biology* 49:171–179.
De Candolle, A.P. 1813. Théorie Elémentaire de Botanique. Deterville, Paris, France.
———. 1824-1873. Prodromus Systematis Naturalis Regni Vegetabilis. Treuttel et Würtz, Paris, France.

Decombeix, A.-L., S.D. Klavins, E.L. Taylor, and T.N. Taylor. 2010. Seed plant diversity in the Triassic of Antarctica: A new anatomically preserved ovule from the Fremouw Formation. *Review of Palaeobotany and Palynology* 158:272–280.

Degut, A.V., and N.S. Fursa. 1980. Phenol compounds of *Digitalis ferruginea*. *Khimiya Prirodnykh Soedinenii* 3:417–418.

De la Torre-Barcena, J.E., S.-O. Kolokotronis, E.K. Lee, D.W. Stevenson, E.D. Brenner, M.S. Katari, G.M. Coruzzi, and R. DeSalle. 2009. The impact of outgroup choice and missing data on major seed plant phylogenetics using genome-wide EST data. *PLoS One* 4(6): e5764. doi:10.1371/journal.pone.0005764.

Del C. Zamaloa, M., M.A. Gandolfo, C.C. González, E.J. Romero, N.R. Cúneo, and P. Wilf. 2006. Casuarinaceae from the Eocene of Patagonia, Argentina. *International Journal of Plant Sciences* 167(6):1279–1289.

Dellaporta, S.L., and A. Calderon-Urrea. 1993. Sex determination in flowering plants. *Plant Cell* 5:1241–1251.

———. 1994. The sex determination process in maize. *Science* 266:1501–1505.

Delpino, F. 1890. Applicazione di nuovi criterii per la classificazione delle piante. Terza memoria. *Memorie della Accademia delle Scienze dell'Istituto di Bologna*, 565–599.

Denk, T., and G.W. Grimm. 2009. The biogeographic history of beech trees. *Review of Palaeobotany and Palynology* 158:83–100.

De Pamphilis, C.W., and J.D. Palmer. 1990. Loss of photosynthetic and chlororespiratory genes from the plastid genome of a parasitic flowering plant. *Nature* 348:337–339.

De Pamphilis, C.W., N.D. Young, and A.D. Wolfe. 1997. Evolution of plastid gene *rps2* in a lineage of hemiparasitic and holoparasitic plants: Many losses of photosynthesis and complex patterns of rate variation. *Proceedings of the National Academy of Sciences USA* 94:7367–7372.

De Queiroz, A., J. Gatesy. 2007. The supermatrix approach to systematics. *Trends in Ecology & Evolution* 22:34–41.

De Queiroz, K. 1997. The Linnaean hierarchy and the evolutionization of taxonomy, with emphasis on the problem of nomenclature. *Aliso* 15:125–144.

De Queiroz, K., and J. Gauthier. 1990. Phylogeny as central principle in taxonomy: Phylogenetic definitions of taxon names. *Systematic Zoology* 39:307–322.

———. 1992. Phylogenetic taxonomy. *Annual Review of Ecology and Systematics* 23:449–480.

———. 1994. Toward a phylogenetic system of biological nomenclature. *Trends in Ecology & Evolution* 9:27–31.

De Smet, R., K. Adams, K. Vandepoele, M. Van Montagu, S. Maere, and Y. Van de Peer. 2013. Convergent gene loss following gene and genome duplications create single-copy families in flowering plants. *Proceedings of the National Academy of Sciences USA* 110:2898–2903.

Der, J.P., and D.L. Nickrent. 2008. A molecular phylogeny of Santalaceae. *Systematic Botany* 33:107–116.

Deroin, T. 2000. Notes on the vascular anatomy of the fruit of *Takhtajania* (Winteraceae) and its interpretation. *Annals of the Missouri Botanical Garden* 87:398–406.

Dettmann, M.E. 1994. Cretaceous vegetation: The microfossil record. *In* R.S. Hill [ed.], History of the Australian Vegetation: Cretaceous to Recent, 143–170. Cambridge University Press, Cambridge, UK.

Dettmann, M.E., and D.M. Jarzen. 1996. Pollen of proteaceous-type from latest Cretaceous sediments, southeastern Australia. *Alcheringa* 20:103–160.

DeVore, M.L., and K.B. Pigg. 2007. A brief review of the fossil history of the family Rosaceae with a focus on the Eocene Okanogan Highlands of eastern Washington State, USA, and British Columbia, Canada. *Plant Systematics and Evolution* 266(1–2):45–57.

Devos, K.M., J.K.M. Brown and J.L. Bennetzen. 2002. Genome size reduction through illegitimate recombination counteracts genome expansion in *Arabidopsis*. *Genome Research* 12:1075–1079.

Devos, K.M., and M.D. Gale. 1997. Comparative genetics in the grasses. *Plant Molecular Biology* 35(1–2):3–15.

De Wilde, W. 1971. The systematic position of tribe Paropsieae, in particular the genus *Ancistrothyrsus*, and a key to the genera of Passifloraceae. *Blumea* 19:99–104.

Diao, Y., L. Chen, G.X. Yang, M.Q. Zhou, Y.C. Song, Z.L. Hu, and J.Y. Liu. 2006. Nuclear DNA C-values in 12 species in Nymphaeales. *Caryologia* 59(1):25–30.

Dickinson, T.A., and R. Sattler. 1974. Development of the epiphyllous inflorescence of *Phylloma integerrima* (Turez.) Loes.: Implications for comparative morphology. *Botanical Journal Linnaean Society* 69:1–13.

Dickison, W.C. 1981. Evolutionary relationships of the Leguminosae. *In* R.M.M. Polhill, and P.H. Raven [eds.], Advances in Legume Systematics, 1, 35–54. Royal Botanical Gardens, Kew, UK.

Dickison, W.C., and P. Baas. 1977. The morphology and relationships of *Paracryphia* (Paracryphiaceae). *Blumea* 23:417–438.

Dickison, W.C., and P.K. Endress. 1983. Ontogeny of the stem-node-leaf vascular continuum of *Austrobaileya*. *American Journal of Botany* 70:906–911.

Dickison, W.C., M.H. Hils, T.W. Lucansky, and W.L. Stern. 1994. Comparative anatomy and systematics of woody Saxifragaceae. Endl. *Botanical Journal of the Linnean Society* 114:167–182.

Diels, L.. 1930. Iridaceae. *In* A. Engler, and K. Prantl [eds.], Die Natürlichen Pflanzenfamilien, 2nd ed., Part 15a, 469–505. Engelmann, Leipzig, Germany.

Diez, C.M., K. Roessler, and B.S. Gaut. 2014. Epigenetics and plant genome evolution. *Current Opinion in Plant Biology* 18:1–8.

Diggle, P.K., V.S. Di Stilio, A.R. Gschwend, E.M. Golenberg, R.C. Moore, J.R.W. Russell, and J.P. Sinclair. 2011. Multiple developmental processes underlie sex differentiation in angiosperms. *Trends in Genetics* 27:368–376.

Dilcher, D.L. 1974. Approaches to identification of Angiosperm leaf remains. *Botanical Review* 40:1–157.

———. 1989. The occurrence of fruits with affinities to Ceratophyllaceae in lower and mid-Cretaceous sediments. *American Journal of Botany* 76:162.

———. 2000. Toward a new synthesis: Major evolutionary trends in the angiosperm fossil record. *Proceedings of the National Academy of Sciences USA* 97(13):7030–7036.

Dilcher, D.L., and P.R. Crane. 1984. *Archaeanthus*: An early angiosperm from the Cenomanian of the western interior of North America. *Annals of the Missouri Botanical Garden* 71:351–383.

Dilcher, D.L., and H. Wang. 2009. An Early Cretaceous fruit with affinities to Ceratophyllaceae. *American Journal of Botany* 96(12):2256–2269.

Di Stilio, V.S., E.M. Kramer, and D.A. Baum. 2005. Floral MADS-box genes and homeotic gender dimorphism in *Thalictrum dioicum* (Ranunculaceae)—a new model for the study of dioecy. *Plant Journal* 41(5):755–766.

D'Hont, A., F. Denoeud, J.-M. Aury, F.-C. Baurens, F. Carreel, O. Garsmeur, B. Noel, S. Bocs, G. Droc, M. Rouard, C. Da Silva, K. Jabbari, C. Cardi, J. Poulain, M. Souquet, K. Labadie, C. Jourda, J. Lengelle, M. Rodier-Goud, A. Alberti, M. Bernard, M. Correa, S. Ayyampalayam, M.R. McKain, J. Leebens-Mack, D. Burgess, M. Freeling, D. Mbéguié-A-Mbéguié, M. Chabannes, T. Wicker, O. Panaud, J. Barbosa, E. Hribova, P. Heslop-Harrison, R. Habas, R. Rivallan, P. Francois, C. Poiron, A. Kilian, D. Burthia, C. Jenny, F. Bakry, S. Brown, V. Guignon, G. Kema, M. Dita, C. Waalwijk, S. Joseph, A. Dievart, O. Jaillon, J. Leclercq, X. Argout, E. Lyons, A. Almeida, M. Jeridi, J. Dolezel, N. Roux, A.-M. Risterucci, J. Weissenbach, M. Ruiz, J.-C. Glaszmann, F. Quetier, N. Yahiaoui, and P. Wincker. 2012. The banana (*Musa acuminata*) genome and the evolution of monocotyledonous plants. *Nature* 488(7410):213–217.

Dobzhansky, T. 1973. Nothing in biology makes sense except in the light of evolution. *American Biology Teacher* 35:125–129.

Dodd, M.E., J. Silvertown, and M.W. Chase. 1999. Phylogenetic analysis of trait evolution and species diversity variation among angiosperm families. *Evolution* 53:732–744.

Donoghue, M.J. 1983. The phylogenetic relationships of *Viburnum*. *In* N. Platnick, and V.A. Funk [eds.], Advances in Cladistics, 143–166. Columbia University Press, New York, NY.

———. 1985. Pollen diversity and exine evolution in *Viburnum* and the Caprifoliaceae sensu lato. *Journal of the Arnold Arboretum* 66:421–469.

———. 1989. Phylogenies and the analysis of evolutionary sequences, with examples from seed plants. *Evolution* 43:1137–1156.

———. 2005. Key innovations, convergence, and success: Macroevolutionary lessons from plant phylogeny. *Paleobiology* 31:77–93.

Donoghue, M.J., and J.A. Doyle. 1989a. Phylogenetic studies of seed plants and angiosperms based on morphological characters. *In* K. Bremer, and H. Jörnvall [eds.], The Hierarchy of Life: Molecules and Morphology in Phylogenetic Studies, 181–193. Elsevier Science Publishers, Amsterdam, Netherlands.

———. 1989b. Phylogenetic analysis of angiosperms and the relationships of Hamamelidae. *In* P.R. Crane, and S. Blackmore [eds.], Evolution, Systematics, and Fossil History of Hamamelidae, 1, 17–45. Clarendon Press, Oxford, UK.

———. 2000. Seed plant phylogeny: Demise of the anthophyte hypothesis? *Current Biology* 10:R106–R109.

Donoghue, M.J., T. Eriksson, P.A. Reeves, and R.G. Olmstead. 2001. Phylogeny and phylogenetic taxonomy of Dipsacales, with special reference to *Sinadoxa* and *Tetradoxa* (Adoxaceae). *Harvard Papers in Botany* 6:459–479.

Donoghue, M.J., R.G. Olmstead, J.F. Smith, and J.D. Palmer. 1992. Phylogenetic relationships of Dipsacales based on *rbcL* sequences. *Annals of the Missouri Botanical Garden* 79:333–345.

Donoghue, M.J., R.H. Ree, and D.A. Baum. 1998. Phylogeny and evolution of flower symmetry in the Asteridae. *Trends in Plant Science* 3(8):311–317.

Douglas, A.W., and S.C. Tucker. 1996a. The developmental basis of diverse carpel orientations in Grevilleoideae (Proteaceae). *International Journal of Plant Sciences* 157:373–397.

———. 1996b. Comparative floral ontogenies among Persoonioideae including *Bellendena* (Proteaceae). *American Journal of Botany* 83:1528–1555.

Douglas, G.E. 1957. The inferior ovary. II. *Botanical Review* 23:1–46.

Douglas, N.A., and P.S. Manos. 2007. Molecular phylogeny of Nyctaginaceae: Taxonomy, biogeography, and characters associated with a radiation of xerophytic genera in North America. *American Journal of Botany* 94:856–872.

Doust, A.N. 2000. Comparative floral ontogeny in Winteraceae. *Annals of the Missouri Botanical Garden* 87:366–379.

———. 2001. The developmental basis of floral variation in *Drimys winteri* (Winteraceae). *International Journal of Plant Sciences* 162:697–717.

Doust, A.N., and P.E. Stevens. 2005. A reinterpretation of the staminate flowers of *Haptanthus*. *Systematic Botany* 30(4):779–785.

Douzery, E.J.P., A.M. Pridgeon, P.J. Kores, H.P. Linder, H. Kurzweil, and M.W. Chase. 1999. Molecular phylogenetics of Diseae (Orchidaceae): A contribution from nuclear ribosomal ITS sequences. *American Journal of Botany* 86:887–899.

Doweld, A.B. 1998. Carpology, seed anatomy and taxonomic relationships of *Tetracentron* (Tetracentraceae) and *Trochodendron* (Trochodendraceae). *Annals of Botany* 82:413–443.

Downie, S.R., D.S. Katz-Downie, and K. Cho. 1997. Relationships in the Caryophyllales as suggested by phylogenetic analyses of partial chloroplast DNA ORF2280 homolog sequences. *American Journal of Botany* 84:253–273.

Downie, S.R., D.S. Katz-Downie, and K. Spalik. 2000a. A phylogeny of Apiaceae tribe Scandiceae: Evidence from nuclear ribosomal DNA internal transcribed spacer sequences. *American Journal of Botany* 87:76–95.

Downie, S.R., D.S. Katz-Downie, and M.F. Watson. 2000b. A phylogeny of the flowering plant family Apiaceae based on chloroplast DNA *rpl16* and *rpoC1* intron sequences: Towards a suprageneric classification of subfamily Apioideae. *American Journal of Botany* 87:273–292.

Downie, S.R. and J.D. Palmer. 1992. Restriction site mapping of the chloroplast DNA inverted repeat: A molecular phylogeny of Asteridae. *Annals of the Missouri Botanical Garden* 79:266–238.

———. 1994. A chloroplast DNA phylogeny of the Caryophyllales based on structural and inverted repeat restriction sites variation. *Systematic Botany* 19:236–252.

Downie, S.R., S. Ramanath, D.S. Katz-Downie, and E. Llanas. 1998. Molecular systematics of Apiaceae subfamily Apioideae: Phylogenetic analyses of nuclear ribosomal DNA internal transcribed spacer and plastid *rpoC1* intron sequences. *American Journal of Botany* 85:563–591.

Downie, S.R., K. Spalik, D.S. Katz-Downie, and J.-P. Redurron. 2010. Major clades within Apiaceae subfamily Apioideae as inferred by phylogenetic analysis of nrDNA ITS sequences. *Plant Diversity and Evolution* 128:111–136.

Doyle, J.A. 1969. Cretaceous angiosperm pollen of Atlantic Coastal Plain and its evolutionary significance. *Journal of the Arnold Arboretum* 50:1–35.

———. 1978. Origin of angiosperms. *Annual Reviews in Ecology and Systematics* 9:365–392.

———. 1992. Revised palynological correlations of the lower Potomac Group (USA) and the Cocobeach sequence of Gabon (Barremian-Aptian). *Cretaceous Research* 13:337–349.

———. 1994. Origin of the angiosperm flower: A phylogenetic perspective. *Plant Systematics and Evolution, Supplement* 8:7–29.

———. 1996. Seed plant phylogeny and the relationships of Gnetales. *International Journal of Plant Sciences* 157:S3–S39.

———. 1998a. Phylogeny of vascular plants. *Annual Reviews in Ecology and Systematics* 29:567–599.

———. 1998b. Molecules, morphology, fossils, and the relationships of angiosperms and Gnetales. *Molecular Phylogenetics and Evolution* 9:448–462.

———. 2000. Paleobotany, relationships, and geographic history of Winteraceae. *Annals of the Missouri Botanical Garden* 87:303–316.

———. 2001. Significance of molecular phylogenetic analyses for paleobotanical investigations on the origin of angiosperms. *Palaeobotanist* 50:167–188.

———. 2005. Early evolution of angiosperm pollen as inferred from molecular and morphological phylogenetic analyses. *Grana* 44:227–251.

———. 2006. Seed ferns and the origin of Angiosperms. *Journal of the Torrey Botanical Society* 133:169–209.

———. 2008. Integrating molecular phylogenetic and paleobotanical evidence on origin of the flower. *International Journal of Plant Sciences* 169:816–843.

———. 2009. Evolutionary significance of granular exine structure in the light of phylogenetic analyses. *Review of Palaeobotany and Palynology* 156:198–210.

———. 2012. Molecular and fossil evidence on the origin of Angiosperms. *Annual Review of Earth and Planetary Sciences* 40:301–326.

Doyle, J.A., P. Bygrave, and A. Le Thomas. 2000. Implications of molecular data for pollen analysis in Annonaceae. *In* M.M. Harley, C.M. Morton, and S. Blackmore [eds.], Pollen and Spores: Morphology and Biology, 259–284. Royal Botanic Gardens, Kew, UK.

Doyle, J.A., and M.J. Donoghue. 1986. Seed plant phylogeny and the origin of the angiosperms: An experimental cladistic approach. *Botanical Review* 52:321–431.

———. 1987. The importance of fossils in elucidating seed plant phylogeny and macroevolution. *Review of Palaeobotany and Palynology* 50:63–95.

———. 1992. Fossils and seed plant phylogeny reanalyzed. *Brittonia* 44:89–106.

———. 1993. Phylogenies and angiosperm diversification. *Paleobiology* 19:141–167.

Doyle, J.A., M.J. Donoghue, and E. Zimmer. 1994. Integration of morphological and ribosomal RNA data on the origin of angiosperms. *Botanical Review* 52:321–431.

Doyle, J.A., H. Eklund, and P.S. Herendeen. 2003. Floral evolution in Chloranthaceae: Implications of a morphological phylogenetic analysis. *International Journal of Plant Sciences* 164:S365–S382.

Doyle, J.A., and P.K. Endress. 2000. Morphological phylogenetic analysis of basal angiosperms: Comparison and combination with molecular data. *International Journal of Plant Sciences* 161 (Supplement):S121–S153.

———. 2010. Integrating Early Cretaceous fossils into the phylogeny of living angiosperms: Magnoliidae and eudicots. *Journal of Systematics and Evolution* 48:1–35.

———. 2011. Tracing the evolutionary diversification of the flower in basal angiosperms. *In* L. Wanntorp, and L.P. Ronse De Craene [eds.], Flowers on the Tree of Life. Systematics Association Special Volume Series 80:88–119. Cambridge University Press, Cambridge, UK.

———. 2014. Integrating early Cretaceous fossils into the phylogeny of living Angiosperms: ANITA lines and relatives of Chloranthaceae. *International Journal of Plant Sciences* 175:555–600.

Doyle, J.A., and L.J. Hickey. 1976. Pollen and leaves from the mid-Cretaceous Potomac Group and their bearing on early angiosperm evolution. *In* C.B. Beck [ed.], Origin and Early Evolution of Angiosperms, 139–206. Columbia University Press, New York, NY.

Doyle, J.A., and C.L. Hotton. 1991. Diversification of early angiosperm pollen in a cladistic context. *In* S. Blackmore, and S.H. Barnes [eds.], Pollen and Spores: Patterns of Diversification, 169–195. Clarendon, Oxford, UK.

Doyle, J.A., and A. Le Thomas. 1997. Phylogeny and geographic history of Annonaceae. *Géographie Physique et Quaternaire* 51:353–361.

Doyle, J.A., S.R. Manchester, and H. Sauquet 2008. A seed related to Myristicaceae in the Early Eocene of southern England. *Systematic Botany* 33(4):636–646.

Doyle, J.A., H. Sauquet, T. Scharaschkin, and A. Le Thomas. 2004. Phylogeny, molecular and fossil dating, and biogeographic history of Annonaceae and Myristicaceae (Magnoliales). *International Journal of Plant Sciences* 165:S55–S67.

Doyle, J.J. 1994. Phylogeny of the legume family: An approach to understanding the origin of nodulation. *Annual Review of Ecology and Systematics* 25:325–349.

———. 2011. Phylogenetic perspectives on the origins of nodulation. *Molecular Plant-Microbe Interactions* 24(11):1289–1295.

Doyle, J.J., J.A. Chappill, C.C. Bailey, and T. Kajita. 2000. Towards a comprehensive phylogeny of legumes: Evidence from *rbcL* sequences and non-molecular data. *In* P.S. Herendeen, and A. Bruneau [eds.], Advances in Legume Systematics, Part 9, 1–20. Royal Botanic Gardens, Kew, UK.

Doyle, J.J., J.L. Doyle, J.A. Ballenger, E.E. Dickson, T. Kajita, and H. Ohashi. 1997. A phylogeny of the chloroplast gene *rbcL* in the Leguminosae: Taxonomic correlations and insights into the evolution of nodulation. *American Journal of Botany* 84:541–554.

Doyle, J.J., L.E. Flagel, A.H. Paterson, R.A. Rapp, D.E Soltis, P.S. Soltis, and J.F. Wendel. 2008. Evolutionary genetics of genome merger and doubling in plants. *Annual Review of Genetics* 42:443–461.

Doyle, J.J., and M.S. Luckow. 2003. The rest of the iceberg: Legume diversity and evolution in a phylogenetic context. *Plant Physiology* 131(3):900–910.

Dressler, R.L. 1981. The Orchids: Natural History and Classification. Harvard University Press, Cambridge, MA.

———. 1983. Classification of the Orchidaceae and their probable origin. *Telopea* 2:413–424.

———. 1993. Phylogeny and Classification of the Orchid Family. Cambridge University Press, New York, NY.

Dressler, R.L., and C.H. Dodson. 1960. Classification and phylogeny in the Orchidaceae. *Annals of the Missouri Botanical Garden* 47:25–68.

Drew, B.T., R. Gazis, P. Cabezas, K.S. Swithers, J. Deng, R. Rodriguez, L.A. Katz, K.A. Crandall, D.S. Hibbett, and D.E. Soltis. 2013. Lost branches on the tree of life. *PLoS Biology* 11: e1001636. doi:10.1371/journal.pbio.1001636.

Drew, B.T., B.R. Ruhfel, S.A. Smith, M.J. Moore, B.G. Briggs, M.A. Gitzendanner, P.S. Soltis, and D.E. Soltis. 2014. Another look at the root of the angiosperms reveals a familiar tale. *Systematic Biology* 63:368–382.

Drewes, S.E., L. Kayonga, T.E. Clark, T.D. Brackenbury, and C.C. Appleton. 1996. Iridoid molluscicidal compounds from *Apodytes dimidiata*. *Journal of Natural Products* 59:1169–1170.

Drinnan, A.N., P.R. Crane, E.M. Friis, and K.R. Pedersen. 1990. Lauraceous flowers from the Potomac Group (mid-Cretaceous) of eastern North America. *Botanical Gazette* 151:370–384.

Drinnan, A.N., P.R. Crane, and S.B. Hoot. 1994. Patterns of floral evolution in the early diversification of non-magnoliid dicotyledons (eudicots). *Plant Systematics and Evolution, Supplement* 8:93–122.

Drinnan, A.N., P.R. Crane, K.R. Pedersen, and E.M. Friis. 1991. Angiosperm flowers and tricolpate pollen of buxaceous affinity from the Potomac Group (mid-Cretaceous) of eastern North America. *American Journal of Botany* 78:153–176.

Drinnan, A.N., and P. Ladiges. 1989. Operculum development in *Eucalyptus cloeziana* and *Eucalyptus* informal subg. *Monocalyptus* (Myrtaceae). *Plant Systematics and Evolution* 166:183–196.

Drummond, A.J., S.Y.W. Ho, M.J. Phillips, and A. Rambaut. 2006. Relaxed phylogenetics and dating with confidence. *PLoS Biology* 4:699–710.

Drummond, A.J., and A. Rambaut. 2007. BEAST: Bayesian evolutionary analysis by sampling trees. *BMC Evolutionary Biology* 7:214. doi:10.1186/1471-2148-7-214.

Drummond, A.J., M.A. Suchard, D. Xie, and A. Rambaut. 2012. Bayesian phylogenetics with BEAUti and the BEAST 1.7. *Molecular Biology and Evolution* 29:1969–1973.

Drysdale, S.S., S. Newman, and K.W. Hilu. 2007. Phylogenetic utility of *trnK* intron in Caryophyllales. *In* Botany & Plant Biology. Joint Congress, 287. Abstract.

Du, L., J. Lykkesfeldt, C. Olsen, and B. Halkier. 1995. Involvement of cytochrome P450 in oxime production in glucosinolate biosynthesis as demonstrated by an *in vitro* microsomal enzyme system isolated from jasmonic acid-induced seedlings of *Sinapsis alba* L. *Proceedings of the National Academy of Sciences USA* 92:12505–12509.

Duangjai, S., R. Samuel, J. Munzinger, F. Forest, B. Wallnöfer, M.H.J. Barfuss, G. Fischer, and M.W. Chase. 2009. A multi-locus plastid phylogenetic analysis of the pantropical genus *Diospyros* (Ebenaceae), with an emphasis on the radiation and biogeographic origins of the New Caledonian species. *Molecular Phylogenetics and Evolution* 52:602–620.

Duangjai, S., B. Wallnofer, R. Samuel, J. Munzinger, and M.W. Chase. 2006. Generic delimitation and relationships in Ebenaceae sensu lato: Evidence from six plastid DNA regions. *American Journal of Botany* 93:1808–1827.

Duarte, J.M., P.K. Wall, P.P. Edger, L.L. Landherr, H. Ma, J.C. Pires, J. Leebens-Mack, and C.W. dePamphilis. 2010. Identification of shared single copy nuclear genes in *Arabidopsis*, *Populus*, *Vitis* and *Oryza* and their phylogenetic utility across various taxonomic levels. *BMC Evolutionary Biology* 10:61. doi:10.1186/1471-2148-10-61.

Duvall, M.R. 2000. Seeking the dicot sister group of the monocots. *In* K.L. Wilson, and D.A. Morrison [eds.], Monocots: Systematics and Evolution, 25–32. CSIRO Publishing, Collingwood, Victoria, Australia.

Duvall, M.R., M.T. Clegg, M.W. Chase, W.D. Lark, W.J. Kress, H.G. Hills, L.E. Eguiarte, J.F. Smith, B.S. Gaut, E.A. Zimmer, and G.H. Learn Jr. 1993a. Phylogenetic hypotheses for the monocotyledons constructed form *rbcL* sequences. *Annals of Missouri Botanical Garden* 80:607–619.

Duvall, M.R., G.H. Learn, L.E. Eguiarte, and M.T. Clegg. 1993b. Phylogenetic analysis of *rbcL* sequences identifies *Acorus calamus* as the primal extant monocotyledon. *Proceedings of the National Academy of Sciences USA* 90:4611–4644.

Duvall, M., S. Mathews, N. Mohammad, and T. Russell. 2006. Placing the monocots; conflicting signal from trigenomic analyses. *In* J.T. Columbus, E.A. Friar, J.M. Porter, L.M. Prince, and M.G. Simpson [eds.]. Monocots: Comparative Biology and Evolution, Excluding Poales, 77–88. Rancho Santa Ana Botanic Garden, Claremont, CA.

Eames, A.J. 1931. The vascular anatomy of the flower with refutation of the theory of carpel polymorphism. *American Journal of Botany* 47:147–188.

———. 1952. Relationships of the Ephedrales. *Phytomorphology* 2:79–100.

———. 1961. Morphology of the Angiosperms. McGraw Hill, New York, NY.

Eber, E. 1934. Karpellbau und Plazentationsverhältnisse in der Reihe der Helobiae. *Flora* 127:273–330.

Eckardt, T. 1963. Some observations on the morphology and embryology of *Eucommia ulmoides* Oliv. *Journal of the Indian Botanical Society* 42A:27–34.

———. 1964. Das Homologieproblem und Fälle strittiger Homologien. *Phytomorphology* 14:79–92.

———. 1965. Entwicklungsgeschichtliche und blütenanatomische Untersuchungen zum Problem der Obdiplostemonie. *Botanische Jahrbücher für Systematik* 85:523–604.

———. 1976. Classical morphological features of centrospermous families. *Plant Systematics and Evolution* 126:5–25.

Eckenwalder, J.E. 1996. Systematics and evoulution of *Populus*. *In* R.F. Stettler, H.D. Bradsha, Jr., P.E. Heilman, and T.M. Hinckley [eds.], Biology of *Populus* and its Implications for Management and Conservation,

7–32. National Research Council of Canada, Ottawa, Ontario, Canada.

Eckert, G. 1965. Entwicklungsgeschichtliche und blütenanatomische Untersuchungen zum Problem der Obdiplostemonie. *Botanische Jahrbücher für Systematik* 85:523–604.

Edwards, E.J., P. Nyffeler, and M.J. Donoghue. 2005. Basal cactus phylogeny: Implications of *Pereskia* (Cactaceae) paraphyly for the transtition to the cactus life form. *American Journal of Botany* 92:1177–1188.

Edwards, E.J., C.P. Osborne, C.A.E. Stroemberg, S.A. Smith, W.J. Bond, P.-A. Christin, A.B. Cousins, M.R. Duvall, D.L. Fox, R.P. Freckleton, O. Ghannoum, J. Hartwell, Y. Huang, C.M. Janis, J.E. Keeley, E.A. Kellogg, A.K. Knapp, A.D.B. Leakey, D.M. Nelson, J.M. Saarela, R.F. Sage, O.E. Sala, N. Salamin, C.J. Still, B. Tipple, and C.G. Consortium. 2010. The origins of C-4 grasslands: Integrating evolutionary and ecosystem science. *Science* 328(5978):587–591.

Edwards, E.J., and S.A. Smith. 2010. Phylogenetic analyses reveal the shady history of C-4 grasses. *Proceedings of the National Academy of Sciences USA* 107:2532–2537.

Edwards, E.J., C.J. Still, and M.J. Donoghue. 2007. The relevance of phylogeny to studies of global change. *Trends in Ecology & Evolution* 22:243–249.

Edwards, G.E., and D.A. Walker. 1983. C_3 and C_4: Mechanism, and Cellular and Environmental Regulation, of Photosynthesis. Blackwell Scientific Publications, Oxford, UK.

Ehrendorfer, F., F. Krendl, E. Habeler, and W. Sauer. 1968. Chromosome numbers and evolution in primitive angiosperms. *Taxon* 17:337–468.

Eichler, A. 1875–1878. Blüthendiagramme, I/II. Engelmann, Leipzig, Germnay.

Ekabo, O.A., N.R. Farnsworth, T. Santisuk, and V. Reutrakul. 1993. Phenolic, iridoid and ionyl glycosides from *Homalium ceylanicum*. *Phytochemistry* 32:747–754.

Eklund, H. 1999. Big survivors with small flowers: Fossil history and evolution of Laurales and Chloranthaceae. *Acta Universitatis Upsaliensis Dissertation*, Uppsala, Sweden.

Eklund, H., J.A. Doyle, and P.S. Herendeen. 2004. Morphological phylogenetic analysis of living and fossil Chloranthaceae. *International Journal of Plant Science* 165:107–151.

Eklund, H., E.M. Friis, and K.R. Pedersen. 1997. Chloranthaceous floral structures from the Late Cretaceous of Sweden. *Plant Systematics and Evolution* 207:13–42.

Eklund, H., and J. Kvaček. 1998. Lauraceous inflorescences and flowers from the Cenomanian of Bohemia (Czech Republic, Central Europe). *International Journal of Plant Sciences* 159:668–86.

Eldredge, N., and J. Cracraft. 1980. Phylogenetic Patterns and the Evolutionary Process. Columbia University Press, New York, NY.

Eldredge, N., and S.J. Gould. 1972. Punctuated equilibria: An alternative to phyletic gradualism. *In* T.J.M. Schopf [ed.], Models in Paleobiology, 82–115. Freeman, San Francisco, CA.

Elias, T.S. 1970. The genera of Ulmaceae in the southeastern United States. *Journal of the Arnold Arboretum* 51:18–40.

———. 1971. The genera of Fagaceae in the southeastern United States. *Journal of the Arnold Arboretum* 52:159–195.

———. 1972. The genera of Juglandaceae in the southeastern United States. *Journal of the Arnold Arboretum* 53:26–51.

Eliasson, U.H. 1988. Floral morphology and taxonomic relations amount the genera of Amaranthaceae in the New World and the Hawaiian Islands. *Botanical Journal of the Linnean Society* 96:235–283.

Ellis, B., D.C. Daly, L.J. Hickey, J.D. Mitchell, K.R. Johnson, P. Wilf, and S.L. Wing. 2009. Manual of Leaf Architecture. Cornell University Press, Ithaca, NY.

Ellison, A.M., E.D. Butler, E.J. Hicks, R.F. Naczi, P.J. Calie, C.D. Bell, and C.C. Davis. 2012. Phylogeny and biogeography of the carnivorous plant family Sarraceniaceae. *PLoS One* 7(6): e39291. doi:10.1371/journal.pone.0039291.

Ellison, A.M., and N.J. Gotelli. 2009. Energetics and the evolution of carnivorous plants–Darwin's 'most wonderful plants in the world'. *Journal of Experimental Botany* 60:19–42.

El-Naggar, L.J., and J.L. Beal. 1980. Iridoids. A review. *Journal of Natural Products* 43:649–707.

El Ottra, J.H.L., J.R. Pirani, and P.K. Endress. 2013. Fusion within and between whorls of floral organs in *Galipeinae* (Rutaceae): Structural features and evolutionary implications. *Annals of Botany* 111(5):821–837.

Elsik, C.G., and C.G. Williams. 2000. Retroelements contribute to the excess low-copy-number DNA in pine. *Molecular General Genetics* 264:47–55.

Elvander, P.E. 1984. The taxonomy of *Saxifraga* (Saxifragaceae) section *Boraphila* subsection *Integrifoliae* in Western North America. *Systematic Botany Monograph* 3:1–44.

Endress, M.E. 2001. Apocynaceae and Asclepiadaceae: United they stand. *Haseltonia* 8:2–9.

———. 2003. Morphology and angiosperm systematics in the molecular era. *Botanical Review* 68:545–570.

Endress, M.E., and P. Bruyns. 2000. A revised classification of the Apocynaceae s.l. *Botanical Review* 66:1–56.

Endress, M.E., B. Sennblad, S. Nilsson, L. Civeyarel, M.W. Chase, S. Huysmans, E. Grafström, and B. Bremer. 1996. A phylogenetic analysis of Apocynaceae s. str. and some related taxa in Gentianales: A multidisciplinary approach. *Opera Botanica Belgica* 7:59–102.

Endress, M.E., and W.D. Stevens. 2001. The renaissance of the Apocynaceae s. l.: Recent advances in systematics, phylogeny and evolution: introduction. *Annals of the Missouri Botanical Garden* 88:517–522.

Endress, P.K. 1967. Systematische Studie über die verwandtschaftlichen Beziehungen zwischen den Hamamelidaceen und Betulaceen. *Botanische Jahrbücher für Systematik* 87:431–525.

———. 1976. Die Androeciumanlage bei polyandrischen Hamamelidaceen und ihre systematische Bedeutung. *Botanische Jahrbücher für Systematik* 97:436–457.

———. 1977. Evolutionary trends in the Hamamelidales-Fagales-group. *Plant Systematics and Evolution, Supplement* 1:321–347.

———. 1978. Blütenontogenese, Blütenabgrenzung und systematische Stellung der perianthlosen Hamamelidoideae. *Botanische Jahrbücher für Systematik* 100:249–317.

———. 1980a. Ontogeny, function and evolution of extreme floral construction in Monimiaceae. *Plant Systematics and Evolution* 134:79–120.

———. 1980b. The reproductive structures and systematic position of the *Austrobaileyaceae*. *Botanische Jahrbücher für Systematik* 101:393–433.

———. 1982. Syncarpy and alternative modes of escaping disadvantages of apocarpy in primitive angiosperms. *Taxon* 31:48–52.

———. 1984. The flowering process in the Eupomatiaceae (Magnoliales). *Botanische Jahrbücher für Systematik* 104:297–319.

———. 1986. An entomophily syndrome in Juglandaceae: *Platycarya strobilacea*. *Veröffentlichungen des Geobotanischen Institutes der Eidgenoessische Technische Hochschule Stiftung Rübel in Zuerich* (87):100–111.

———. 1986a. Reproductive structures and phylogenetic significance of extant primitive Angiosperms. *Plant Systematics and Evolution* 152(1–2):1–28.

———. 1986b. Floral structure, systematics and phylogeny in Trochodendrales. *Annals of the Missouri Botanical Garden* 73:297–324.

———. 1987a. Floral phyllotaxis and floral evolution. *Botanische Jahrbücher für Systematik* 108:417–438.

———. 1987b. The early evolution of the angiosperm flower. *Trends in Ecology & Evolution* 2:300–304.

———. 1987c. The Chloranthaceae: Reproductive structures and phylogenetic position. *Botanische Jahrbücher für Systematik* 109:153–226.

———. 1989a. Aspects of evolutionary differentiation of the Hamamelidaceae and lower Hamamelididae. *Plant Systematics and Evolution* 162(1–4):193–211.

———. 1989b. Chaotic floral phyllotaxis and reduced perianth in *Achlys* (Berberidaceae). *Botanica Acta* 102:159–163.

———. 1990a. Patterns of floral construction in ontogeny and phylogeny. *Biological Journal of the Linnean Society* 39:153–175.

———. 1990b. Evolution of reproductive structures and functions in primitive angiosperms (Magnoliidae). *Memoirs of the New York Botanical Garden* 55:5–34.

———. 1992a. Evolution and floral diversity: The phylogenetic surroundings of *Arabidopsis* and *Antirrhinum*. *International Journal of Plant Sciences* 153:S106–S122.

———. 1992b. Primitive Blüten: Sind Magnolien noch zeitgemäss? *Stapfia* 28:1–10.

———. 1993. Cercidiphyllaceae. *In* K. Kubitzki, J.G. Rohwer, and V. Bittrich. [eds.], The Families and Genera of Vascular Plants, II, 250–252. Springer-Verlag, Berlin, Germany.

———. 1994a. Floral structure and evolution of primitive angiosperms: Recent advances. *Plant Systematics and Evolution* 192:79–97.

———. 1994b. Evolutionary aspects of the floral structure in *Ceratophyllum*. *Plant Systematics and Evolution, Supplement* 8:175–183.

———. 1994c. Diversity and Evolutionary Biology of Tropical Flowers. Cambridge University Press, New York, NY.

———. 1994d. Shapes, sizes and evolutionary trends in stamens of Magnoliidae. *Botanische Jahrbücher für Systematik* 115:429–460.

———. 1995a. Major evolutionary trends of monocot flowers. *In* P.J. Rudall, P.J. Cribb, D.F. Cutler, and C.J. Humphries [eds.], Monocotyledons: Systematics and Evolution, 43–79. Royal Botanic Gardens, Kew, UK.

———. 1995b. Floral structure and evolution in Ranunculanae. *Plant Systematics and Evolution, Supplement* 9:47–61.

———. 1996a. Diversity and evolutionary trends in angiosperm anthers. *In* W.G. D'Arcy, and R.C. Keating [eds.], The Anther: Form, Function and Phylogeny, 92–110. Cambridge University Press, New York, NY.

———. 1996b. Homoplasy in angiosperm flowers. *In* M.J. Sanderson, and L. Hufford [eds.], Homoplasy and the Evolutionary Process, 301–323. Academic Press, Orlando, FL.

———. 1997a. Relationships between floral organization, architecture, and pollination mode in *Dillenia* (Dilleniaceae). *Plant Systematics and Evolution* 206:99–118.

———. 1997b. Evolutionary biology of flowers: Prospects for the next century. *In* K. Iwatsuki, and P.H. Raven [eds.], Evolution and Diversification of Land Plants, 99–119. Springer, Tokyo, Japan.

———. 1998. *Antirrhinum* and Asteridae: Evolutionary changes of floral symmetry. *Symposium Series Society of Experimental Biology* 53:133–140.

———. 1999. Symmetry in flowers: Diversity and evolution. *International Journal of Plant Sciences* 160:S3–S23.

———. 2001a. The flowers in extant basal angiosperms and inferences on ancestral flowers. *International Journal of Plant Sciences* 162(5):1111–1140.

———. 2001b. Origins of flower morphology. *Journal of Experimental Zoology (Molecular Development and Evolution)* 291:105–115.

———. 2001c. Evolution of floral symmetry. *Current Opinion in Plant Biology* 4(1):86–91.

———. 2002a. Morphology and angiosperm systematics in the molecular era. *Botanical Review* 68:545–570.

———. 2002b. What should a "complete" morphological phylogenetic analysis entail? *In* T.F. Stuessy, E. Hörandl, and V. Mayer [eds.], Deep Morphology: Toward a Renaissance in Plant Morphology, 131–164. Koeltz, Königstein, Germany.

———. 2003a. Morphology and angiosperm systematics in the molecular era. *Botanical Review* 68:545–570.

———. 2003b. What should a "complete" morphological phylogenetic analysis entail? Problems and promises. *In* T.F. Stuessy, E. Hörandl, and V. Mayer [eds.], Deep Morphology, Toward a Renaissance of Morphology in Plant Systematics, 131–164. Gantner Verlag, Ruggell, Liechtenstein.

———. 2003c. Early floral development and the nature of calyptra in Eupomatiaceae. *International Journal of Plant Sciences* 164:489–503.

———. 2006. Angiosperm floral evolution: Morphological developmental framework. *Advances in Botanical Research* 44:1–61.

———. 2008. Perianth biology in the basal grade of extant Angiosperms. *International Journal of Plant Sciences* 169(7):844–862.

———. 2010a. Flower structure and trends of evolution in eudicots and their major subclades. *Annals of the Missouri Botanical Garden* 97(4):541–583.

———. 2010b. The evolution of floral biology in basal angiosperms. *Philosophical Transactions of the Royal Society of London B* 365:411–421.

———. 2010c. Synorganisation without organ fusion in the flowers of *Geranium robertianum* (Geraniaceae) and its not so trivial obdiplostemony. *Annals of Botany* 106:687–695.

———. 2011c. Angiosperm ovules: Diversity, development, evolution. *Annals of Botany* 107(9):1465–1489.

———. 2011a. Evolutionary diversification of the flowers in angiosperms. *American Journal of Botany* 98(3):370–96.

———. 2011b. Changing views of flower evolution and new questions. *In* L. Wanntorp, and L.P.R. DeCraene [eds.], Flowers on the Tree of Life, 120–141. Cambridge University Press, Cambridge, UK.

———. 2012. The immense diversity of floral monosymmetry and asymmetry across Angiosperms. *Botanical Review* 78(4):345–397.

———. 2014. Multicarpellate gynoecia in angiosperms: Occurrence, development, organization and architectural constraints. *Botanical Journal of the Linnean Society* 174(1):1–43.

———. 2016. Development and evolution of extreme synorganization in angiosperm flowers and diversity: a comparison of Apocynaceae and Orchidaceae. *Annals of Botany* 117:749–767.

Endress, P.K., P. Baas, and M. Gregory. 2000a. Systematic plant morphology and anatomy—50 years of progress. *Taxon* 49:401–434.

Endress, P.K., C.C. Davis, and M.L. Matthews. 2013. Advances in the floral structural characterization of the major subclades of Malpighiales, one of the largest orders of flowering plants. *Annals of Botany* 111(5):969–985.

Endress, P.K., and J.A. Doyle. 2007. Floral phyllotaxis in basal angiosperms—development and evolution. *Current Opinion in Plant Biology* 10:52–57.

———. 2009. Reconstructing the ancestral Angiosperm flower and its initial specializations. *American Journal of Botany* 96:22–66.

———. 2015. Ancestral traits and specializations in the flowers of the basal grade of living angiosperms. *Taxon* 64:1093–1116.

Endress, P.K., and L. Hufford. 1989. The diversity of stamen structures and dehiscence patterns among Magnoliidae. *Botanical Journal of the Linnean Society* 100:45–85.

Endress, P.K., and A. Igersheim. 1997. Gynoecium diversity and systematics of the Laurales. *Botanical Journal of the Linnean Society* 125:93–168.

———. 1999. Gynoecium diversity and systematics of the basal eudicots. *Botanical Journal of the Linnean Society* 130:305–393.

———. 2000a. The reproductive structures of the basal angiosperm *Amborella trichopoda* (Amborellaceae). *International Journal of Plant Sciences* 161:S237–S248.

———. 2000b. Gynoecium structure and evolution in basal angiosperms. *International Journal of Plant Sciences* 161:S211–S223.

Endress, P.K., A. Igersheim, F.B. Sampson, and G.E. Schatz. 2000b. Floral structure of *Takhtajania* and its systematic position in Winteraceae. *Annals of the Missouri Botanical Garden* 87:347–365.

Endress, P.K., M. Jenny, and M.E. Fallen. 1983. Convergent elaboration of apocarpous gynoecia in higher advanced dicotyledons (Sapindales, Malvales, Gentianales). *Nordic Journal of Botany* 3:293–300.

Endress, P.K., and M.L. Matthews. 2006. First steps towards a floral structural characterization of the major rosid subclades. *Plant Systematics and Evolution* 260(2–4):223–251.

———. 2012. Progress and problems in the assessment of flower morphology in higher-level systematics. *Plant Systematics and Evolution* 298(2):257–276.

Endress, P.K., and A. Rapini. 2014. Floral structure of *Emmotum* (Icacinaceae sensu stricto or Emmotaceae), a phylogenetically isolated genus of lamiids with a unique pseudotrimerous gynoecium, bitegmic ovules and monosporangiate thecae. *Annals of Botany* 114:945–959.

Endress, P.K., and F.B. Sampson. 1983. Floral structure and relationships of the Trimeniaceae (Laurales). *Journal of the Arnold Arboretum* 64:447–473.

Endress, P.K., and S. Stumpf. 1990. Non-tetrasporangiate stamens in the angiosperms: Structure, systematic distribution and evolutionary aspects. *Botanische Jahrbücher für Systematik* 112:193–240.

———. 1991. The diversity of stamen structures in lower Rosidae (Rosales, Fabales, Proteales, Sapindales). *Botanical Journal of the Linnean Society* 107(3):217–293.

Engler, A. 1930. Saxifragaceae. *In* A. Engler, and K. Prantl [eds.], Die Natürlichen Pflanzenfamilien, 2nd ed., 18a, 74–226. Engelmann, Leipzig, Germany.

Erbar, C. 1991. Sympetaly: A systematic character. *Botanische Jahrbücher für Systematik* 112:417–451.

Erbar, C., S. Kusma, and P. Leins. 1999. Development and interpretation of nectary organs in Ranunculaceae. *Flora* 194:317–332.

Erbar, C., and P. Leins. 1981. Zur Spirale in Magnolienblüten. *Beiträge zur Biologie der Pflanzen* 56:225–241.

———. 1983. Zur Sequenz von Blütenorganen bei einigen Magnoliiden. *Botanische Jahrbücher für Systematik* 103:433–449.

———. 1985. Studien zur Organsequenz in Apiaceen–Blüten. *Botanische Jahrbücher für Systematik* 105: 379–400.

———. 1988. Blütenentwicklungsgeschichtliche Studien an *Aralia* und *Hedera* (Araliaceae). *Flora* 180:391–406.

———. 1995a. An analysis of the early floral development of *Pittosporum tobira* (Thunb.) Aiton and some remarks on the systematic position of the family Pittosporaceae. *Feddes Repertorium* 106:463–473.

———. 1995b. Portioned pollen release and the syndromes of secondary pollen presentation in the Campanulales-Asterales-complex. *Flora* 190:323–338.

———. 1996. Distribution of the character states "early" and "late" sympetaly within the "Sympetalae Tetracyclicae" and presumably related groups. *Botanica Acta* 109:427–440.

———. 1997. Different patterns of floral development in whorled flowers, exemplified by Apiaceae and Brassicaceae. *International Journal of Plant Sciences* 158:S49–S64.

———. 2011. Synopsis of some important, non-DNA character states in the asterids with special reference to sympetly. *Plant Diversity and Evolution* 129:93–123.

Erdtman, G. 1952. Pollen Morphology and Plant Taxonomy. Almqvist and Wiksell, Stockholm, Sweden.

———. 1966. Pollen Morphology and Plant Taxonomy. Hafner, New York, NY.

Erdtman, G., P. Leins, R. Melville, and C. Metcalfe. 1969. On the relationships of *Emblingia*. *Botanical Journal of the Linnean Society* 62:169–186.

Ereshefsky, M. 1994. Some problems with the Linnean hierarchy. *Philosophy of Science* 61:186–205.

Eriksson, O., E.M. Friis, K.R. Pederson, and P.R. Crane. 2000. Seed size and dispersal systems of Early Cretaceous angiosperms from Famalicão, Portugal. *International Journal of Plant Sciences* 161:319–329.

Ernst, W.R. 1963. The genera of Hamamelidaceae and Platanaceae in the southeastern United States. *Journal of the Arnold Arboretum* 44:193–210.

———. 1964. The genera of Berberidaceae, Laridzabalaceae, and Menispermaceae in the southeastern United States. *Journal of the Arnold Arboretum* 45:1–35.

Ettlinger, M., and A. Kjaer. 1968. Sulfur compounds in plants. *Recent Advances in Phytochemistry* 1:59–144.

Evans, R.C., and T.A. Dickinson. 1999a. Floral ontology and morphology in subfamily Amygdaloideae T. and G. (Rosaceae). *International Journal of Plant Sciences* 160:955–979.

———. 1999b. Floral ontology and morphology in subfamily Spiraeoideae Endl. (Rosaceae). *International Journal of Plant Sciences* 160:981–1012.

Evans, R., C. Evans, and C.S. Campbell. 2002. The origin of the apple subfamily (Maloideae; Rosaceae) is clarified by DNA sequence data from duplicated GBSSI genes. *American Journal of Botany* 89:1478–1484.

Eyde, R.H. 1964. Inferior ovary and generic affinities of *Garrya*. *American Journal of Botany* 51:1083–1092.

———. 1988. Comprehending *Cornus*: Puzzles and progress in the systematics of the dogwoods. *Botanical Review* 54:233–351.

Eyde, R.H., and C.C. Tseng. 1971. What is the primitive floral structure of Araliaceae? *Journal of the Arnold Arboretum* 52:205–239.

Eyde, R.H., and Q.-X. Xiang. 1990. Fossil mastixioid (Cornaceae) alive in eastern Asia. *American Journal of Botany* 52:205–239.

Fallen, E.M. 1986. Floral structure in the Apocynaceae: Morphological, functional, and evolutionary aspects. *Botanische Jahrbücher für Systematik* 106:245–286.

Fan, C., and Q.-Y. (J.) Xiang. 2003. Phylogenetic analyses of Cornales based on 26S rRNA and combined 26S *rDNA-matK-rbcL* sequence data. *American Journal of Botany* 90(9):1357–1372.

Farrell, B.D. 1998. "Inordinate Fondness" explained: Why are there so many beetles? *Science* 281(5376):555–559.

Farris, J.S. 1976. Phylogenetic classification of fossils with recent species. *Systematic Zoology* 25:271–282.

———. 1988. Computer program and documentation Hennig 86. Port Jefferson, NY.

———. 1996. *Jac*. Volume 4.4. Swedish Museum of Natural History, Stockholm, Sweden.

———. 2002. RASA attributes highly significant structure to randomized data. *Cladistics* 18:334–353.

Farris, J.S., V.A. Albert, M. Källersjö, D. Lipscomb, and A.G. Kluge. 1996. Parsimony jackknifing outperforms neighbor-joining. *Cladistics* 12:99–124.

Fawcett, J.A., S. Maere, and Y. Van de Peer. 2009. Plants with double genomes might have had a better chance to survive the Cretaceous-Tertiary extinction event. *Proceedings of the National Academy of Sciences USA* 106(14):5737–5742.

Fay, M.F., C. Bayer, W. Alverson, A. De Bruijn, S. Swensen, and M. Chase. 1998a. Plastid *rbcL* sequences indicate a close affinity between *Diegodendron* and *Bixa*. *Taxon* 47:43–50.

Fay, M.F., B. Bremer, G.T. Prance, M. Van Der Bank, D. Bridson, and M.W. Chase. 2000b. Plastid *rbcL* sequence data show *Dialypetalanthus* to be a member of Rubiaceae. *Kew Bulletin* 55:853–864.

Fay, M.F., K.M. Cameron, G.T. Prance, M.D. Lledo, and M.W. Chase. 1997a. Familial relationships of *Rhabdodendron* (Rhabdodendraceae): Plastid *rbcL* sequences indicate a caryophyllid placement. *Kew Bulletin* 52(4):923–932.

Fay, M.F., and M.J.M. Christenhusz. 2010. Brassicales—an order of plants characterised by shared chemistry. *Curtis's Botanical Magazine* 27 (Part 3): 165–196.

———. 2012. Ranunculales—buttercups, poppies and their relatives. *Curtis's Botanical Magazine* 29(Part 3): 222–234.

Fay, M.F., R.G. Olmstead, J.E. Richardson, E. Santiago, G.T. Prance, and M.W. Chase. 1998b. Molecular data support the inclusion of *Duckeodendron celastroides* in Solanaceae. *Kew Bulletin* 53:203–212.

Fay, M.F., P.J. Rudall, S. Sullivan, K.L. Stobart, A.Y. De Bruijn, G. Reeves, F. Qamaruz-Zaman, W.-P. Hong, J. Joseph, W.J. Hahn, J.G. Conran, and M.W. Chase. 2000a. Phylogenetic studies of Asparagales based on four plastid DNA loci. *In* K.L. Wilson, and D.A. Morrison [eds.], Monocots: Systematics and Evolution, 360–371. CSIRO Publishing, Collingwood, Victoria, Australia.

Fay, M.F., S.M. Swenson, and M.W. Chase. 1997b. Taxonomic affinities of *Medusagyne oppositifolia* (Medusagynaceae). *Kew Bulletin* 52(1):111–120.

Fedoroff, N. 2000. Transposons and genome evolution in plants. *Proceedings of the National Academy of Sciences USA* 97(13):7002–7007.

Fedorov, A. 1969. Chromosome numbers of flowering plants. Academy of Sciences, Leningrad, USSR.

Feduccia, A. 2003. 'Big bang' for tertiary birds? *Trends in Ecology & Evolution* 18:172–176.

Feild, T.S., N.C. Arens, and T.E. Dawson. 2003b. The ancestral ecology of angiosperms: Emerging perspectives from extant basal lineages. *International Journal of Plant Sciences* 164:S129–S142.

Feild, T.S., N.C. Arens, J.A. Doyle, T.E. Dawson, and M.J. Donoghue. 2004. Dark and disturbed: A new image of early angiosperm ecology. *Paleobiology* 30:82–107.

Feild, T.S., P.J. Franks, and T.L. Sage. 2003a. Ecophysiological shade adaptation in the basal angiosperm, *Austrobaileya scandens* (Austrobaileyaceae). *International Journal of Plant Sciences* 164:313–324.

Feild, T.S., M.A. Zwieniecki, T. Brodribb, T. Jaffre, M.J. Donoghue, and N.M. Holbrook. 2000a. Structure and function of tracheary elements in *Amborella trichopoda*. *International Journal of Plant Sciences* 161:705–712.

Feild, T.S., M.A. Zwieniecki, and N.M. Holbrook. 2000b. Winteraceae evolution: An ecophysiological perspective. *Annals of the Missouri Botanical Garden* 87:323–334.

Felsenstein, J. 1978a. The number of evolutionary trees. *Systematic Zoology* 27:27–33.

———. 1978b. Cases in which parsimony or compatibility methods will be positively misleading. *Systematic Zoology* 27:401–410.

———. 1985. Confidence limits on phylogenies: An approach using the bootstrap. *Evolution* 39:783–791.

———. 1988. Phylogenies from molecular sequences: Inferences and reliability. *Annual Review of Genetics* 22:521–565.

Feng, X., B. Tang, T.M. Kodrul, and J. Jin. 2013. Winged fruits and associated leaves of *Shorea* (Dipterocarpaceae) from the Late Eocene of South China and their phytogeographic and paleoclimatic implications. *American Journal of Botany* 100:574–581.

Feng, X., Z. Zhao, Z.X. Tian, S.L. Xu, Y.H. Luo, Z.G. Cai, Y.M. Wang, J. Yang, Z. Wang, L. Weng, J.H. Chen, L.Y. Zheng, X.Z. Guo, J.H. Luo, S.S. Sato, S. Tabata, W. Ma, X.L. Cao, X.H. Hu, C.R. Sun, and D. Luo. 2006. Control of petal shape and floral zygomorphy in *Lotus japonicus*. *Proceedings of the National Academy of Sciences USA* 103(13):4970–4975.

Fenster, C.B., W.S. Armbruster, P. Wilson, M.R. Dudash, and J.D. Thomson. 2004. Pollination syndromes and floral specialization. *Annual Review of Ecology and Evolutionary Systematics* 35:375–403.

Fenster, C.B., S. Martén-Roriguez, and D.W. Schemske. 2009. Pollination syndromes and the evolution of floral diversity in *Iochroma* (Solanaceae). *Evolution* 63:2758–2762.

Ferguson, D.K. 1989. A survey of the Liquidambaroideae (Hamamelidaceae) with a view to elucidating its fossil record. *In* P.R. Crane, and S. Blackmore. [eds.], Evolution, Systematics and Fossil History of the Hamamelidae, 1, 249–272. Clarendon Press, Oxford, UK.

Ferguson, D.K., R. Zetter, and K.N. Paudayal. 2007. The need for the SEM in palaeopalynology. *Comptes Rendus Palevol* 6:423–430.

Ferguson, I.K. 1966. The Cornaceae in the southeastern United States. *Journal of the Arnold Arboretum* 47:106–116.

———. 1977. Cornaceae. *World Pollen and Spore Flora* 6:1–34.

Feuer, S. 1991. Pollen morphology and the systematic relationships of *Ticodendron incognitum*. *Annals of the Missouri Botanical Garden* 78:143–151.

Feuer, S.M, and J. Kuijt. 1980. Fine Structure of Mistletoe Pollen. III. Large-flowered Neotropical Loranthaceae and their Australian relatives. *American Journal of Botany* 67:34–50.

———. 1985. Fine structure of mistletoe pollen VI. Small-flowered Neotropical Loranthaceae. *Annals of the Missouri Botanical Garden* 72(2):187–212.

Filipowicz, N., and S.S. Renner. 2010. The worldwide holoparasitic Apodanthaceae confidently placed in the Cucurbitales by nuclear and mitochondrial gene trees. *BMC Evolutionary Biology* 10:219. doi:10.1186/1471-2148-10-219.

Finet, C., R.E. Timme, C.F. Delwiche, and F. Marleta. 2010. Multigene phylogeny of the Green lineage reveals the origin and diversification of Land Plants. *Current Biology* 20(24):2217–2222.

Fior, S., P.O. Karis, G. Casazza, L. Minuto, and F. Sala. 2006. Molecular phylogeny of the Caryophyllaceae (Caryophyllales) inferred from chloroplast matK and nuclear rDNA ITS sequences. *American Journal of Botany* 93:399–411.

Fishbein, M. 2001. Evolutionary innovation and diversification in the flowers of Asclepiadaceae. *Annals of the Missouri Botanical Garden* 88:603–623.

Fishbein, M., D. Chuba, C. Ellison, R. Mason-Gamer, and S.P. Lynch. 2011. Phylogenetic relationships of *Asclepias* (Apocynaceae) inferred from non-coding chloroplast DNA sequences. *Systematic Botany* 36(4):1008–1023.

Fishbein, M., C. Hibsch-Jetter, D.E. Soltis, and L. Hufford. 2001. Phylogeny of Saxifragales (angiosperms, eudicots): Analysis of a rapid, ancient radiation. *Systematic Biology* 50(6):817–847.

Fishbein, M., and D.E. Soltis. 2004. Further resolution of the rapid radiation of Saxifragales (angiosperms, eudicots) supported by mixed-model Bayesian analysis. *Systematic Botany* 29:883–891.

Flavell, R. 1988. Repetitive DNA and chromosome evolution in plants. *Philosophical Transactions of the Royal Society of London B* 312:227–242.

Fleischmann, A., T.P. Michael, F. Rivadavia, A. Sousa, W. Wang, E.M. Temsch, J. Greilhuber, K.F. Müller, and G. Heubl. 2014. Evolution of genome size and chromosome numbers in the carnivorous plant genus *Genlisea* (Lentibulariaceae). *Annals of Botany* 114(8):1651–1663.

Floyd, S.K., and W.E. Friedman. 2000. Evolution of endosperm developmental patterns among basal flowering plants. *International Journal of Plant Sciences* 161:S57–S81.

———. 2001. Developmental evolution of endosperm in basal angiosperms: Evidence from *Amborella* (Amborellaceae), *Nuphar* (Nymphaeaceae), and *Illicium* (Illiciaceae). *Plant Systematics and Evolution* 228:153–169.

Floyd, S.K., V.T. Lerner, and W.E. Friedman. 1999. A developmental and evolutionary analysis of embryology in *Platanus* (Platanaceae), a basal eudicot. *American Journal of Botany* 86:1523–1537.

Force, A., M. Lynch, F.B. Pickett, A. Amores, Y.L. Yan, and J. Postlethwait. 1999. Preservation of duplicate genes by complementary, degenerative mutations. *Genetics* 151:1531–1545.

Ford, V.A., J. Lee, B.G. Baldwin, and L.D. Gottlieb. 2006. Species divergence and relationships in *Stephanomeria* (Compositae): PgiC phylogeny compared to prior biosystematics studies. *American Journal of Botany* 93:480–490.

Fragnière, Y., S. Bétrisey, L. Cardinaux, M. Stoffel, and G. Kozlowski. 2015. Fighting their last stand? A global analysis of the distribution and concervation status of gymnosperms. *Journal of Biogeography* 42:809–820.

Francisco, A., and L. Ascensão. 2013. Structure of the osmophore and labellum micromorphology in the sexually deceptive orchids *Ophrys bombyliflora* and *Ophrys tenthredinifera* (Orchidaceae). *International Journal of Plant Sciences* 174:619–636.

Frank, A.C., H. Amiri, and S.G.E. Andersson. 2002. Genome deterioration: Loss of repeated sequences and accumulation of junk DNA. *Genetica* 115:1–12.

Frank, A.R., B.J. Cochrane, and J.R. Garey. 2013. Phylogeny, biogeography, and infrageneric classification of *Harrisia* (Cactaceae). *Systematic Botany* 38:210–223.

Franz, E. 1908. Beiträge zur Kenntnis der Portulacaceen und Basellaceen. *Botanische Jahrbücher für Systematik* 97:1–46.

Franzke, A., D. German, I.A. Al-Shehbaz, and K. Mummenhoff. 2009. *Arabidopsis* family ties: Molecular phylogeny and age estimates in Brassicaceae. *Taxon* 58(2):425–437.

Freeling, M. 2009. Bias in plant gene content following different sorts of duplication: Tandem, whole-genome, segmental, or by transposition. *Annual Review of Plant Biology* 60:433–453.

Freeling, M., and B.C. Thomas. 2006. Gene-balanced duplications, like tetraploidy, provide predictable drive to increase morphological complexity. *Genome Research* 16(7):805–814.

Freeman, D., J. Doust, A. El-Keblawy, K. Migla, and E. McArthur. 1997. Sexual specialization and inbreeding avoidance in the evolution of dioecy. *Botanical Review* 63:65–92.

Freitag, H., and G. Kadereit. 2014. C-3 and C-4 leaf anatomy types in Camphorosmeae (Camphorosmoideae, Chenopodiaceae). *Plant Systematics and Evolution* 300(4):665–687.

Freshwater, D.W., S. Fredericq, B.S. Butler, M.H. Hommersand, and M.W. Chase. 1994. A gene phylogeny of the red algae (Rhodophyta) based on plastic *rbcL*. *Proceedings of the National Academy of Sciences USA* 91:7281–7285.

Freudenstein, J.V., and M.W. Chase. 2015. Phylogenetic relationships within Epidendroideae (Orchidaceae), one of the great flowering plant radiations; progressive specialization and diversification. *Annals of Botany* 115:665–681.

Freudenstein, J.V., D.M. Senyo, and M.W. Chase. 2000. Mitochondrial DNA and relationships in the Orchidaceae. *In* K.L. Wilson, and D.A. Morrison [eds.], Monocots: Systematics and Evolution, 421–429. CSIRO Publishing, Collingwood, Victoria, Australia.

Friedman, J., and S.C.H. Barrett. 2009. Wind of change: New insights on the ecology and evolution of pollination and mating in wind-pollinated plants. *Annals of Botany* 103:1515–1527.

Friedman, W.E. 1990. Sexual reproduction in *Ephedra nevadensis* (Ephedraceae): Further evidence of double fertilization in a nonflowering seed plant. *American Journal of Botany* 77:1582–1598.

———. 1992. Evidence of a pre-angiosperm origin of endosperm: Implications for the evolution of flowering plants. *Science* 255:336–339.

———. 1994. The evolution of embryogeny in seed plants and the developmental origin and early history of endosperm. *American Journal of Botany* 81:1468–1486.

———. 1996. Biology and evolution of the Gnetales. *International Journal of Plant Sciences* 157:S1–S125.

———. 2006. Embryological evidence for developmental lability during early angiosperm evolution. *Nature* 441:337–340.

———. 2008. Hydatellaceae are water lilies with gymnospermous tendencies. *Nature* 453:94–97.

———. 2009. The meaning of Darwin's "Abominable Mystery". *American Journal of Botany* 96:5–21.

Friedman, W.E., and J.B. Bachelier. 2013. Seed development in *Trimenia* (Trimeniaceae) and its bearing on the evolution of embryo-nourishing strategies in early flowering plant lineages. *American Journal of Botany* 100:906–915.

Friedman, W.E, J.B. Bachelier, and J.I. Hormaza. 2012. Embryology in *Trithuria submersa* (Hydatellaceae) and relationships between embryo, endosperm, and perisperm in early-diverging flowering plants. *American Journal of Botany* 99:1083–1095.

Friedman, W.E., S.C.H. Barrett, P.K. Diggle, V.F. Irish, and L. Hufford. 2008. Whither plant evo-devo? *New Phytologist* 178:468–472.

Friedman, W.E., and S.K. Floyd. 2001. Perspective: The origin of flowering plants and their reproductive biology–a tale of two phylogenies. *Evolution* 55:217–231.

Friedman, W.F., C. Moore, and M.D. Purugganan. 2004. The evolution of plant development. *American Journal of Botany* 91(10):1726–1741.

Friedman, W.E, and K.C. Ryerson. 2009. Reconstructing the ancestral female gametophyte of angiosperms: Insights from *Amborella* and other ancient lineages of flowering plants. *American Journal of Botany* 96:129–143.

Friis, E.M. 1983. Upper Cretaceous (Senonian) floral structures of juglandalean affinity containing *Normapolles* pollen. *Review of Palaeobotany and Palynology* 39:161–188.

———. 1988. *Spirematospermum chandlerae* sp. nov., an extinct species of Zingiberaceae in the North American Cretaceous. *Tertiary Research* 9:7–12.

Friis, E.M., and P.R. Crane. 1989. Reproductive structures of Cretaceous Hamamelidae. *In* P. R. Crane, and S. Blackmore [eds.], Evolution, Systematics and Fossil History of the Hamamelidae, 1, 155–174. Clarendon Press, Oxford, UK.

Friis, E.M., P.R. Crane, and K.R. Pedersen. 1986. Floral evidence for Cretaceous chloranthoid angiosperms. *Nature* 320:163–164.

———. 1988. Reproductive structures of Cretaceous Platanaceae. *Biologiske Meddelelser Kongelige Danske Videnskabernes Selskab* 31:1–56.

———. 1997. Fossil history of magnoliid angiosperm. *In* K. Iwatsuki, and P.H. Raven [eds.], Evolution and Diversification of Land Plants, 121–156. Springer-Verlag, New York, NY.

———. 2011. Early Flowers and Angiosperm Evolution. Cambridge University Press, New York, NY.

Friis, E.M., P.R. Crane, K.R. Pedersen, S. Bengtson, P.C.J. Donoghue, G.W. Grimm, and M. Stampanoni. 2007. Phase-contrast x-ray microtomography links Cretaceous seeds with Gnetales and Bennettitales. *Nature* 450:549–552.

Friis, E.M., J.A. Doyle, P.K. Endress, and Q. Leng. 2003. *Archaefructus*–angiosperm precursor or specialized early angiosperm? *Trends in Plant Science* 8:369–373.

Friis, E.M., H. Eklund, K.R. Pedersen, and P.R. Crane. 1994a. *Virginianthus calycanthoides* gen. et sp. nov.—a calycanthaceous flower from the Potomac Group (Early Cretaceous) of eastern North America. *International Journal of Plant Sciences* 155:772–85.

Friis, E.M., and P.K. Endress. 1990. Origin and evolution of angiosperm flowers. *Advances in Botanical Research* 17:99–162.

Friis, E.M., F. Marone, K.R. Pedersen, P.R. Crane, and M. Stampanoni. 2014. Three-dimenional visualisation of fossil flowers, fruits, seeds and other plant remains using synchrotron radiation x-ray tomographic microscopy (SRXTM): New insights into Cretaceous plant diversity. *Journal of Palaeontology* 88:684–701.

Friis, E.M., K.R. Pedersen, and P.R. Crane. 1994b. Angiosperm floral structures from the Early Cretaceous of Portugal. *Plant Systematics and Evolution, Supplement* 8:31–49.

———. 1999. Early angiosperm diversification: The diversity of pollen associated with angiosperm reproductive structures in Early Cretaceous floras from Portugal. *Annals of the Missouri Botanical Garden* 86(2):259–296.

———. 2000. Reproductive structure and organization of basal angiosperms from the early Cretaceous (Barremian or Aptian) of western Portugal. *International Journal of Plant Sciences* 161:5169–5182.

———. 2001. Fossil evidence of waterlilies (Nymphaeales) in the Early Cretaceous. *Nature* 410:357–360.

———. 2004. Araceae from the Early Cretaceous of Portugal: Evidence on the emergence of monocotyledons. *Proceedings of the National Academy of Sciences USA*, 101(16):565–570.

———. 2010. Diversity in obscurity: Fossil flowers and the early history of angiosperms. *Philosophical Transactions of the Royal Society of London B* 365:369–382.

Friis, E.M., K.R. Pedersen, and P.K. Endress. 2013. Floral structure of extant *Quintinia* (Paracryphiales, campanulids) compared with the Late Cretaceous *Silvianthemum* and *Bertilanthus*. *International Journal of Plant Sciences* 174:647–664.

Friis, E.M., K.R. Pedersen, and J. Schonenberger. 2003. *Endressianthus*, a new normapolles-producing plant genus of Fagalean affinity from the Late Cretaceous of Portugal. *International Journal of Plant Sciences* 164(5 Supplement):S201–S223.

———. 2006. *Normapolles* plants: A complex of extinct fagalean lineages. *Plant Systematics and Evolution* 260:107–140.

Friis, E.M., K.R. Pedersen, M. von Balthazar, G.W. Grimm, and P.R. Crane. 2009. *Monetianthus mirus* gen. et sp. nov, a nymphaealean flower from the early Cretaceous of Portugal. *International Journal of Plant Sciences* 170:1086–1101.

Fritsch, P.W. 2001. Phylogeny and biogeography of the flowering plant genus *Stryax* (Styracaceae) based on chloroplast DNA restriction sites and DNA sequences of the internal transcribed spacer region. *Molecular Phylogenetics and Evolution* 19:387–408.

Fritsch, P.W., S.R. Manchester, R.D. Stone, B.C. Cruz and F. Almeda. 2014. Northern Hemisphere origins of the amphi-Pacific tropical plant family Symplocaceae. *Journal of Biogeography*. doi:10.1111/jbi.12442.

Frohlich, M.W. 1999. MADS about Gnetales. *Proceedings of the National Academy of Sciences USA* 96:8811–8813.

Frohlich, M.W., and M.W. Chase. 2007. After a dozen

years of progress the origin of angiosperms is still a great mystery. *Nature* 450:1184–1189.

Frohlich, M.W., and E.M. Meyerowitz. 1997. The search for flower homeotic gene homologs in basal angiosperms and Gnetales: A potential new source of data on the evolutionary origin of flowers. *International Journal of Plant Science* 158(6 Supplement):S131–S142.

Frohlich, M.W., and D.S. Parker. 2000. The mostly male theory of flower evolutionary origins: From genes to fossils. *Systematic Botany* 25:155–170.

Frumin, S., H. Eklund, and E.M. Friis. 2004. *Mauldinia hirsuta* sp. nov., a new member of the extinct genus *Mauldinia* (Lauraceae) from the Late Cretaceous (Cenomanian-Turonian) of Kazakhstan. *International Journal of Plant Sciences* 165:883–95.

Fukuhara, T., and S.-I. Tokumaru. 2014. Inflorescence dimorphism, heterodichogamy and thrips pollination in *Platycarya strobilacea* (Juglandaceae). *Annals of Botany* 113(3):467–476.

Fulton, M., and S.A. Hodges. 1999. Floral isolation between *Aquilegia formosa* and *Aquilegia pubescens*. *Proceedings of the Royal Society of London B* 266:2247–2252.

Funk, V.A., A.A. Anderberg, B.G. Baldwin, R.J. Bayer, J.M. Bonifacino, I. Breitwieser, L. Brouillet, R. Carbajal, R. Chan, A.X.P. Coutinho, and D.J. Crawford. 2009. Compositae Metatrees: The Next Generation. *In* V.A. Funk, A. Susana, T.F. Stuessy, and R.J. Bayer [eds.], Systematics, Evolution, and Biogeography of Compositae, 747–777. International Association for Plant Taxonomy (IAPT), Vienna, Austria.

Funk, V.A, G. Sancho, N. Roque, C.L. Kelloff, I. Ventosa-Rodriguez, M. Diazgranados, J.M. Bonifacino, and R. Chan. 2014. A phylogeny of the Gochnatieae: Understanding a critically placed tribe in the Compositae. *Taxon* 63:859–882.

Funk, V.A., A. Susanna, T.F. Steussy, and R.J. Bayer [eds.]. 2009a. Systematics, Evolution, and Biogeography of Compositae. International Association for Plant Taxonomy (IAPT), Vienna, Austria.

Funk, V.A., A. Susanna, T.F. Stuessy, and H. Robinson. 2009b. Classification of Compositae. *In* V.A. Funk, A. Susanna, T.F. Stuessy, and R.J. Bayer, [eds.], Systematics, Evolution and Biogeography of the Compositae, 171–189. International Association for Plant Taxonomy (IAPT), Vienna, Austria.

Funk, V.A., A. Susanna, T.F. Stuessy and H. Robinson. 2012. Classification of Compositae. Adalwyd 18 Tachwedd 2012.

Furness, C.A., and P.J. Rudall. 2004. Pollen aperture evolution—a crucial factor for eudicot success? *Trends in Plant Science* 9(3):154–158.

Furness, C.A., P.J. Rudall, and F.B. Sampson. 2002. Evolution of microsporogenesis in angiosperms. *International Journal of Plant Sciences* 163:235–260.

Fuse, S., and M.N. Tamura. 2000. A phylogenetic analysis of the plastid *matK* gene with an emphasis on Melanthiaceae sensu lato. *Plant Biology* 2:415–427.

Futey, M.K., M.A. Gandolfo, M.C. Zamaloa, and R. Cúneo. 2012. Arecaceae fossil fruits from the Paleocene of Patagonia, Argentina. *Botanical Review* 78:205–234.

Gadek, P.A., E.S. Fernando, C.J. Quinn, S.B. Hoot, T. Terrazas, M.C. Sheahan, and M.W. Chase. 1996. Sapindales: Molecular delimitation and infraordinal groups. *American Journal of Botany* 83:802–811.

Gadek, P.A, and C.J. Quinn. 1993. An analysis of relationships within the Cupressaceae sensu stricto based on *rbcL* sequences. *Annals of the Missouri Botanical Garden* 80:581–586.

Gadek, P.A., C.J. Quinn, J.E. Rodman, K.G. Karol, E. Conti, R.A. Price, and E.S. Fernando. 1992. Affinities of the Australian endemic Akaniaceae: New evidence from *rbcL* sequences. *Australian Systematic Botany* 5:717–734.

Gaeta, R.T., and J.C. Pires. 2010. Homoeologous recombination in allopolyploids: The polyploid ratchet. *New Phytologist* 186(1):18–28.

Gaeta, R.T., S.-Y. Yoo, J.C. Pires, R.W. Doerge, Z.J. Chen, and T.C. Osborn. 2009. Analysis of gene expression in resynthesized *Brassica napus* allopolyploids using arabidopsis 70-mer oligo microarrays. *PLoS ONE* 4(3): e4760. doi:10.1371/journal.pone.0004760.

Gaffney, E.S., and P.A. Meylan. 1988. A phylogeny of turtles. *In* M.J. Benton [ed.], The Phylogeny and Classification of Tetrapods, I: Amphibians, Reptiles, and Birds, 157–219. Clarendon Press, Oxford, UK.

Gailing, O., and K. Bachmann. 2000. The evolutionary reduction of microsporangia in *Microseris* (Asteraceae): Transition genotypes and phenotypes. *Plant Biology* 2:455–461.

Gaiser, J.C., K. Robinson-Beers, and C.S. Gasser. 1995. The *Arabidopsis SUPERMAN* gene mediates asymmetric growth of the outer integument of ovules. *Plant Cell* 7:333–345.

Galasso, G., E. Banfi, F. de Mattia, F. Grassi, S. Sgorbati, and M. Labra. 2009. Molecular phylogeny of *Polygonum* L. s.l. (Polygonoideae, Polygonaceae), focusing on European taxa: Preliminary results and systematic considerations based on *rbcL* plastidial sequence data. *ATTI della Societa Italiana di Scienze Naturali e del Museo Civico di Storia Naturale di Milano* 150:113–148.

Galbraith, D.W., J.L. Bennetzen, E.A. Kellogg, J.C. Pires, and P.S. Soltis. 2011. The genomes of all angiosperms: A call for a coordinated global census. *Journal of Botany* 2011: Article ID 646198, 10 pages. doi:10.1155/2011/646198.

Gale, M.D., and K.M. Devos. 1998. Comparative genetics in the grasses. *Proceedings of the National Academy of Sciences USA* 95(5):1971–1974.

Galen, C. 1999. Why do flowers vary? *BioScience* 49:631–640.

Galimba, K.D., T.R. Tolkin, A.M. Sullivan, R. Melzer, G. Theissen, and V.S. Di Stilio. 2012. Loss of deeply conserved C-class floral homeotic gene function and C- and E-class protein interaction in a double-flowered ranunculid mutant. *Proceedings of the National Academy of Sciences USA* 109(34):E2267–E2275.

Gamisch, A., Y.M. Staedler, J. Schönenberger, G.A. Fischer, and H.P. Comes. 2013. Histological and micro-CT evidence of stigmatic rostellum receptivity promoting auto-pollination in the Madagascan orchid *Bulbophyllum bicoloratum*. *PLoS ONE* 8(8): e72688. doi:10.1371/journal.pone.0072688.

Gandolfo, M.A. 1998. *Tylerianthus crossmanensis* gen. et sp. nov. (aff. Hydrangeaceae) from the Upper Cretaceous of New Jersey. *American Journal of Botany* 85: 376–386.

Gandolfo, M.A., and R.N. Cúneo. 2005. Fossil Nelumbonaceae from the La Colonia Formation (Campanian-Maastrichtian, Upper Cretaceous), Chubut, Patagonia, Argentina. *Review of Palaeobotany and Palynology* 133:169–178.

Gandolfo, M.A., E.J. Hermsen, M.C. Zamaloa, K.C. Nixon, C.C. Gonzalez, P. Wilf, N.R. Cuneo, and K.R. Johnson. 2011. Oldest known *Eucalyptus* macrofossils are from South America. *PLoS ONE* 6(6): e21084. doi:10.1371/journal.pone.0021084.

Gandolfo, M.A., K.C. Nixon, and W.L. Crepet. 1998a. A new fossil flower from the Turonian of New Jersey: *Dressiantha bicarpellata* gen. et sp. nov. (Capparales). *American Journal of Botany* 85(7):964–974.

Gandolfo, M.A., K. Nixon, and W. Crepet. 1998b. *Tylerianthus crossmanensis* gen. et sp. nov. (aff. Hydrangeaceae) from the Upper Cretaceous of New Jersey. *American Journal of Botany* 85:376–86.

———. 2000. Monocotyledons: A review of their Early Cretaceous record. *In* K.I. Wilson, and D.A. Morrison [eds]., Monocots: Systematics and Evolution, 44–51. CSIRO, Melbourne, Australia.

———. 2002. Triuridaceae fossil flowers from the Upper Cretaceous of New Jersey. *American Journal of Botany* 89:1940–1957.

———. 2004. The oldest complete fossil flowers of Nymphaeaceae and implications for the complex insect entrapment pollination mechanisms in Early Angiosperms. *Proceedings of the National Academy of Sciences USA* 101:8056–8060.

Gandolfo, M.A., K. Nixon, W.L. Crepet, D.W. Stevenson, and E.M. Friis. 1998c. Oldest known fossils of monocotyledons. *Nature* 394:532–533.

Gao, C., X. Ren, A.S. Mason, H. Liu, M. Xiao, J. Li, and D. Fu. 2014. Horizontal gene transfer in plants. *Functional & Integrative Genomics* 14(1):23–29.

Gaskett, A.C. 2011. Orchid pollination by sexual deception: Pollinator perspectives. *Biological Reviews* 86:33–75.

Gastony, G.J., and D.E. Soltis. 1977. Chromosome studies of *Parnassia* and *Lepuropetalon* from the eastern United States. A new base number for *Parnassia*. *Rhodora* 79:573–578.

Gaussen, H. 1946. Les Gymnospermes, actuelles et fossiles. *Travaux du Laboratoire Forestier Toulouse* 1:1–26.

Gaut, B.S., S.V. Muse, W.D. Clark, and M.T. Clegg. 1992. Relative rates of nucleotide substitution at the *rbcL* locus of monocotyledonous plants. *Journal of Molecular Evolution* 35:292–303.

Gauthier R. 1950. The nature of the inferior ovary in the genus *Begonia*. *Contributions de l'Institute de Botanique, Université de Montréal* 66:1–93.

Gavin, C.C., and H.W. Kenneth. 2008. Turning a hobby into a job: How duplicated genes find new functions. *Nature Reviews Genetics* 9:938–950.

Gehrke, B., and H.P. Linder. 2009. The scramble for Africa: Pan-temperate elements on the African high mountains. *Proceedings of the Royal Society of London B* 276(1667):2657–2665.

Gemmeke, V. 1982. Entwicklungsgeschichtliche Untersuchungen an Mimosaceen–Blüten. *Botanische Jahrbücher für Systematik* 103:185–210.

Gerwick, C., and G.J. Williams III. 1978. Temperature and water regulation of gas exchange of *Opuntia polyacantha*. *Oecologia* 35:149–159.

Giannasi, D.E., G. Zurawski, G. Learn, and M.T. Clegg. 1992. Evolutionary relationships of the Caryophyllidae based on comparative *rbcL* sequences. *Systematic Botany* 17:1–15.

Gibbs, R.D. 1957. The Mäule reaction, lignin, and the relationships between woody plants. *In* K.V. Thimann [ed.], The Physiology of Forest Trees, 269–312. Ronald, New York, NY.

Gibson, A.C. 1979. Anatomy of *Koeberlinia* and *Canotia* revisited. *Madroño* 26:1–12.

Gibson, A.C., and P.S. Nobel. 1986. The Cactus Primer. Harvard University Press, Cambridge, MA.

Gilg, E. 1925. Flacourtiaceae. *In* A. Engler, and K. Prantl [eds.], Die Natürlichen Pflanzenfamilien, 2nd ed., 21, 377–457. Engelmann, Leipzig, Germany.

Gilg, E., and R. Pilger. 1905. Rutaceae. *In* R. Pilger [ed.], Beiträge zur Flora der Hylea nach den Sammlungen von E. Ule. *Verhandlungen des Botanischen Vereins der Provinz Brandenburg* 47:152–154.

Giussani, L.M., J.H. Cota-Sanchez, F.O. Zuloaga, and E.A. Kellogg. 2001. A molecular phylogeny of the grass subfamily Panicoideae (Poaceae) shows multiple origins of C_4 photosynthesis. *American Journal of Botany* 88:1993–2012.

Givnish, T.J. 1997. Adaptive radiation and molecular systematics: Issues and approaches. *In* T.J. Givnish, and K.J. Sytsma [eds.], Molecular Evolution and Adaptive Radiation, 1–54. Cambridge University Press, Cambridge, UK.

———. 2014. New evidence on the origin of carnivorous plants. *Proceedings of the National Academy of Sciences USA* 112(1). doi/10.1073/pnas.1422278112.

Givnish, T.J., M. Ames, J.R. McNeal, M.R. McKain, P.R. Steele, C.W. dePamphilis, S.W. Graham, J.C. Pires, D.W. Stevenson, W.B. Zomlefer, B.G. Briggs, M.R. Duvall, M.J. Moore, J.M. Heaney, D.E. Soltis, P.S. Soltis, K. Thiele, and J.H. Leebens-Mack. 2010. Assembling the tree of the Monocotyledons: Plastome sequence phylogeny and evolution of Poales. *Annals of the Missouri Botanical Garden* 97:584–616.

Givnish, T.J., E.L. Burkhardt, R. Happel, and J. Weintraub. 1984. Carnivory in the bromeliad *Brocchinia reducta,* with a cost/benefit model for the general restriction of carnivorous plants to sunny, moist, nutrient-poor habitats. *American Naturalist* 124:479–497.

Givnish, T.J., T.M. Evans, J.C. Pires, and K.J. Sytsma. 1999. Polyphyly and convergent morphological evolution in Commelinales and Commelinidae: Evidence from *rbcL* sequence data. *Molecular Phylogenetics and Evolution* 12:360–385.

Givnish, T.J., K.C. Millam, A.R. Mast, T.B. Paterson, T.J. Theim, A.L. Hipp, J.M. Henss, J.F. Smith, K.R. Wood, and K.J. Sytsma. 2009. Origin, adaptive radiation and diversification of Hawaiian lobeliads (Asterales: Campanulaceae). *Proceedings of the Royal Society of London B* 276:407–416.

Givnish, T.J., R.A. Montgomery, and G. Goldstein. 2004. Adaptive radiation of photosynthetic physiology in the Hawaiian lobeliads: Light regimes, static light responses, and whole-plant compensation points. *American Journal of Botany* 91:228–246.

Givnish, T.J., J.C. Pires, S.W. Graham, M.A. McPherson, L.M. Prince, T.B. Paterson, H.S. Rai, E.H. Roalson, T.M. Evans, W.J. Hahn, K.C. Millam, A.W. Meerow, M. Molvray, P.J. Kores, H.E. O'Brien, J.C. Hall, W.J. Kress, and K.J. Sytsma. 2006. Phylogeny of the monocots based on the highly informative plastid gene *ndhF*: Evidence for widespread concerted convergence. *In* J.T. Columbus, E.A. Friar, J.M. Porter, L.M. Prince, and M.G. Simpson [eds.], Monocots: Comparative Biology and Evolution. Excluding Poales, 28–51. Rancho Santa Ana Botanical Garden, Claremont, CA.

Givnish, T.J., and K.J.E. Sytsma [eds.]. 1997. Molecular Evolution and Adaptive Radiation. Cambridge University Press, Cambridge, UK.

Givnish, T.J., K.J.E. Sytsma, J.F. Smith, W.J. Hahn, D.H. Benzing, and E.M. Burkhardt. 1997. Molecular evolution and adaptive radiation in *Brocchinia* (Bromeliaceae: Pitcairnioideae) atop tepuis of the Guyana Shield. *In* T.J. Givnish, and K.J. Sytsma [eds.], Molecular Evolution and Adaptive Radiation, 259–312. Cambridge University Press, Cambridge, UK.

Glisic, L.M. 1928. Development of the female gametophyte and endosperm in *Haberlea rhodopensis* (Friv.). *Bulletin de l'Institute et Jardin Botanique, Université Belgrade* 1:1–13.

Goldberg, E.E., J.R. Kohn, R. Lande, K.A. Robertson, S.A. Smith, and B. Igic. 2010. Species selection maintains self-incompatibility. *Science* 330:493–495.

Goldblatt, P. 1976. Cytotaxonomic studies in the tribe Quillajeae (Rosaceae). *Annals of the Missouri Botanical Garden* 63:200–206.

———. 1980. Polyploidy in angiosperms: Monocotyledons. *In* W.H. Lewis [ed.], Polyploidy, Basic Life Sciences, Vol. 13, 219–239. Plenum Press, New York, NY.

———. 1990. Phylogeny and classification of the Iridaceae. *Annals of the Missouri Botanical Garden* 77:607–627.

———. 1991. An overview of the systematics, phylogeny and biology of the southern African Iridaceae. *Contributions from the Bolus Herbarium* 13:1–74.

Goldblatt, P., J.E. Henrich, and P.J. Rudall. 1984. Occurrence of crystals in Iridaceae and allied families and their phylogenetic significance. *Annals of the Missouri Botanical Garden* 71:1013–1020.

Goldblatt, P., J.C. Manning, and P. Bernhardt. 1998. Adaptive radiation of bee-pollinated *Gladiolus* species (Iridaceae) in Southern Africa. *Annals of the Missouri Botanical Garden* 85:492–517.

———. 2001. Radiation of pollination systems in *Gladiolus* (Iridaceae: Crocoideae) in Southern Africa. *Annals of the Missouri Botanical Garden* 88:713–734.

Goldblatt, P., V. Savolainen, O. Porteous, I. Sostaric, M. Powell, G. Reeves, J.C. Manning, and T.G. Barraclough. 2002. Radiation in the Cape flora and the phylogeny of peacock irises *Moraea* (Iridaceae) based on four plastid DNA regions. *Molecular Phylogenetics and Evolution* 25:341–360.

Goldenberg, R., D.S. Penneys, F. Almeda, W.S. Judd, and F.A. Michelangeli. 2008. Phylogeny of *Miconia* (Melastomataceae): Patterns of stamen diversification in a megadiverse Neotropical genus. *International Journal of Plant Sciences* 169(7):963–979.

———. 2010. Phylogeny of *Miconia* (Melastomataceae): Patterns of stamen diversification in a megadiverse Neotropical genus. *Cladistics* 26(2):211–212.

Goloboff, P. 1993. NONA. Computer programs and documentation. Tucumán, Argentina.

Goloboff, P.A, S.A. Catalano, J.M. Mirande, C.A. Szumik, J.S. Arias, M. Kallersjo, and J.S. Farris. 2009. Phylogenetic analysis of 73 060 taxa corroborates major eukaryotic groups. *Cladistics* 25:211–230.

Goloboff, P.A., J.S. Farris, and K.C. Nixon. 2008. TNT, a free program for phylogenetic analysis. *Cladistics* 24:774–786.

Gómez-Laurito, J., and P. Gómez. 1989. *Ticodendron*: A new tree from Central America. *Annals of the Missouri Botanical Garden* 76:1148–1151.

González, F.A., J. Betancur, O. Maurin, J.V. Freudenstein, and M.W. Chase. 2007. Metteniusaceae, an early diverging family in the lamiid clade. *Taxon* 56:795–800.

González, F.A., and P.J. Rudall. 2001. The questionable affinities of *Lactoris*: Evidence from branching pattern, inflorescence morphology, and stipule development. *American Journal of Botany* 88:2143–2150.

———. 2003. Structure and development of the ovule and seed in Aristolochiaceae, with particular reference to *Saruma*. *Plant Systematics and Evolution* 241:223–244.

———. 2010. Flower and fruit characaters in the early-divergent lamiid family Metterniusaceae, with particular reference to the evolution of pseudomonomery. *American Journal of Botany* 97:191–206.

González, F.A., and D.W. Stevenson. 2002. A phylogenetic analysis of the subfamily Aristolochioideae (Aristolochiaceae). *Botanical Review* 26:25–58.

Goodman, M., J. Czelusniak, G.W. Moore, A.E. Romero-herrera, and G. Matsuda. 1979. Fitting the gene lineage into its species lineage, a parsimony strategy illustrated by cladograms constructed from globin sequences. *Systematic Zoology* 28:132–163.

Goremykin, V., V. Bobrova, J. Pahnke, A. Troitsky, A. Antonov, and W. Martin. 1996. Noncoding sequences from the slowly evolving chloroplast inverted repeat in addition to *rbcL* data do not support Gnetalean affinities of angiosperms. *Molecular Biology and Evolution* 13:383–396.

Goremykin, V., S. Hansmann, and W. Martin. 1997. Evolutionary analysis of 58 proteins encoded in six completely sequenced chloroplast genomes: Revised molecular estimates of two seed plant divergence times. *Plant Systematics and Evolution* 206:337–351.

Goremykin, V., K.I. Hirsch-Ernst, S. Wölfl, and F.H. Hellwig. 2003. Analysis of the *Amborella trichopoda* chloroplast genome sequence suggests that *Amborella* is not a basal angiosperm. *Molecular Biology and Evolution* 20:1499–1505.

———. 2004. The chloroplast genome of *Nymphaea alba*: Whole-genome analyses and the problem of identifying the most basal angiosperm. *Molecular Biology and Evolution* 21:1445–1454.

Goremykin, V.V., S.V. Nikiforova, P.J. Biggs, B. Zhong, P. Delange, W. Martin, S. Woetzel, R.A. Atherton, P.A. McLenachan, and P.J. Lockhart. 2013. The evolutionary root of flowering plants. *Systematic Biology* 62:50–61.

Goremykin, V.V., S.V. Nikiforova, and O.R.P. Bininda-Emonds. 2010. Automated removal of noisy data in phylogenomic analyses. *Journal of Molecular Evolution* 71:319–331.

Górniak, M., O. Paun, and M.W. Chase. 2010. Phylogenetic relationships within Orchidaceae based on a low-copy nuclear coding gene, *Xdh*: Congruence with plastid DNA results. *Molecular Phylogenetics and Evolution* 56:784–795.

Gosner, M.E., R.K. Jansen, and T.G. Lammers. 1994. Phylogenetic relationships in the Campanulales based on *rbcL* sequences. *Plant Systematics and Evolution* 190:79–95.

Gottsberger, G. 1977. Some aspects of beetle pollination in the evolution of flowering plants. *Plant Systematics and Evolution* 1:211–226.

———. 1988. The reproductive biology of primitive angiosperms. *Taxon* 37:630–643.

Gottschling, M., and H.H. Hilger. 2001. Phylogenetic analysis and character evolution of *Ehretia* and *Bourreria* (Ehretiaceae, Boraginales) and their allies based on ITS1 sequences. *Botanische Jahrbücher für Systematik* 123:249–268.

Gottschling, M., H.H. Hilger, M. Wolf, and N. Diane. 2001. Secondary structure of the ITS1 transcript and its application in a reconstruction of the phylogeny of Boraginales. *Plant Biology* 3:629–636.

Gottschling, M., F. Luebert, H.H. Hilger, and J.S. Miller. 2014. Molecular delimitations in the Ehretiaceae (Boraginales). *Molecular Phylogenetics and Evolution* 72:1–6.

Gottschling, M., J.S. Miller, M. Weigend, and H.H. Hilger. 2005. Congruence of a phylogeny of Cordiaceae

(Boraginales) inferred from ITS1 sequence data with morphology, ecology, and biogeography. *Annals of the Missouri Botanical Garden* 92:425–437.

Gottwald, H., and N. Parameswaran. 1968. Das sekundäre Xylem und die systematische Stellung der Ancistrocladaceae und Dioncophyllaceae. *Botanische Jahrbücher für Systematik* 88: 49–69.

Govaerts, R. 2001. How many species of seed plants are there? *Taxon* 50:1085–1090.

GPWG (Grass Phylogeny Working Group). 2000. A phylogeny of the grass family (Poaceae), as inferred from eight character sets. *In* S.W.L. Jacobs, and J. Everett [eds.], Grasses: Systematics and Evolution, 3–7. CSIRO Publishing, Collingwood, Victoria, Australia.

Graham, A. 1996. A contribution to the geologic history of the Compositae. *In* D.J.N. Hind, and H.J. Beentje [eds.], Compositae: Systematics, Proceedings of the International Compositae Conference, Vol. 1., 123–140. Royal Botanic Gardens, Kew, UK.

———. 2008. Fossil record of the Rubiaceae. *Annals of the Missouri Botanical Garden*. 96:90–108.

Graham, J., S. Sagitov, and B. Oxelman. 2012. Statistical inference of allopolyploid species networks in the presence of incomplete lineage sorting. *Systematic Biology* 62(3):467–478.

Graham, L.E., C.F. Delwiche, and B.D. Mishler. 1991. Phylogenetic connections between the "green algae" and the "bryophytes". *Advances in Bryology* 4:213–214.

Graham, S.A. 1964. The genera of Rhizophoraceae and Combretaceae in the southeastern United States. *Journal of the Arnold Arboretum* 45:285–301.

———. 1984. Alzateaceae, a new family of Myrtales in the American tropics. *Annals of the Missouri Botanical Garden* 71:757–779.

———. 2013. Fossil records in the Lythraceae. *Botanical Review* 79(1):48–145.

Graham, S.A., J.V. Crisci, and P.C. Hoch. 1993b. Cladistic analysis of the Lythraceae sensu lato based on morphological characters. *Botanical Journal of the Linnean Society* 113:1–33.

Graham, S.W., and W.J.D Iles. 2009. Different Gymnosperm outgroups have (mostly) congruent signal regarding the root of flowering plant phylogeny. *American Journal of Botany* 96:216–227.

Graham, S.W., and R.G. Olmstead. 2000. Utility of 17 chloroplast genes for inferring the phylogeny of the basal angiosperms. *American Journal of Botany* 87:1712–1730.

Graham, S.W., P.A. Reeves, A.C.E. Burns, and R.G. Olmstead. 2000. Microstructural changes in noncoding chloroplast DNA: Interpretation, evolution, and utility of indels and inversions in basal angiosperm phylogenetic inference. *International Journal of Plant Sciences* 161 (6 Supplement):S83–S96.

Graham, S.W., J.M. Zgurski, M.A. McPherson, D.M. Cherniawsky, J.M. Saarela, E.F.C. Horne, S.Y. Smith, W.A. Wong, H.E. O'Brien, V.L. Biron, J.C. Pires, R.G. Olmstead, M.W. Chase, and H.S. Rai. 2006. Robust inference of monocot deep phylogeny using an expanded multigene plastid data set. *Aliso* 22:3–21.

Grande, L. 1984. Paleontology of the Green River Formation, with a review of the fish fauna. 2nd ed. *Geological Survey of Wyoming, Bulletin* 63:1–333.

Grant, D.P., P. Cregan, and R.C. Shoemaker. 2000. Genome organization in dicots: Genome duplication in *Arabidopsis* and synteny between soybean and *Arabidopsis*. *Proceedings of the National Academy of Sciences USA* 97:4168–4173.

Grant, V. 1950. The protection of ovules in flowering plants. *Evolution* 4:179–201.

———. 1981. Plant Speciation. Columbia University Press, New York, NY.

Grant, V., and K.A. Grant. 1965. Flower Pollination in the *Phlox* Family. Columbia University Press, New York, NY.

Grass Phylogeny Working G, II. 2012. New grass phylogeny resolves deep evolutionary relationships and discovers C4 origins. *The New Phytologist* 193(2):304–312.

Graur, D., L. Duret, and M. Gouy. 1996. Phylogenetic position of the order Lagomorpha (rabbits, hares and allies). *Nature* 379:333–335.

Gray, A. 1875. A conspectus of the North American Hydrophyllaceae. *Proceedings of the American Academy* 10:312–332.

Graybeal, A. 1998. Is it better to add taxa or characters to a difficult phylogenetic problem? *Systematic Biology* 47:9–17.

Grayer, R.J., M.W. Chase, and M.S.J. Simmonds. 1999. A comparison between chemical and molecular characters for the determination of phylogenetic relationships among plant families: An appreciation of Hegnauer's "Chemotaxonomie der Pflanzen". *Biochemical Systematics and Evolution* 27:369–393.

Grayum, M.H. 1987. A summary of evidence and arguments supporting the removal of *Acorus* from the Araceae. *Taxon* 36:723–729.

Green, P.B. 1999. Expression of pattern in plants: Combining molecular and calculus-based biophysical paradigms. *American Journal of Botany* 86:1059–1076.

Greenberg, A.K., and M.J. Donoghue. 2011. Molecular systematics and character evolution in Caryophyllaceae. *Taxon* 60:1637–1652.

Gregory, T.R. 2001. Coincidence, coevolution, or causation? DNA content, cell size, and the C-value enigma. *Biological Reviews* 76:65–101.

Greilhubr, J., J. Doležel, M.A. Lysák, and M.D. Bennett. 2005. The orgin, evolution and proposed stabilization of the terms 'Genome Size' and 'C-value' to describe nuclear NDA contents. *Annals of Botany* 95:255–260.

Greiner, S., and R. Bock. 2013. Tuning a menage a trois: co-evolution and co-adaptation of nuclear and organellar genomes in plants. *Bioessays* 35(4):354–365.

Griffith, M.P. 2004. The origins of an important cactus crop, *Opuntia ficus-indica* (Cactaceae): New molecular evidence. *American Journal of Botany* 91:1915–1921.

———. 2008. *Pereskia*, Portulaceae, photosynthesis, and phylogenies: Implications for early Cactaceae. *Haseltonia* 14:37–45.

Griffith, M.P., and J.M. Porter. 2009. Phylogeny of Opuntioideae. *International Journal of Plant Sciences* 170:107–116.

Griffiths, G.C.D. 1973. Some fundamental problems in biological classification. *Systematic Zoology* 22(4): 338–343.

———. 1974. On the foundations of biological systematics. *Acta Biotheoretica* 23:85–131.

———. 1976. The future of Linnean nomenclature. *Systematic Zoology* 25:168–173.

Grimaldi, D. 1999. The co-radiations of pollinating insects and angiosperms in the Cretaceous. *Annals of the Missouri Botanical Garden* 86:373–406.

Grime, J., and M. Mowforth. 1982. Variation in genome size–an ecological interpretation. *Nature* 299:151–153.

Grimm, G.W., and T. Denk. 2008. ITS evolution in *Platanus* (Platanaceae): Homoeologues, pseudogenes and ancient hybridization. *Annals of Botany* 101(3):403–419.

Grímsson, F., T. Denk, and R. Zetter. 2008. Pollen, fruits, and leaves of *Tetracentron* (Trochodendraceae) from the Cainozoic of Iceland and western North America and their palaeobiogeographic implications. *Grana* 47(1):1–14.

Grímsson, F., R. Zetter, H. Halbritter, and G.W. Grimm. 2014. *Aponogeton* pollen from the Cretaceous and Paleogene of North America and West Greenland: Implications for the origin and palaeobiogeography of the genus. *Review of Palaeobotany and Palynology* 200:161–187.

Grimsson, F., R. Zetter, and C.-C. Hofmann. 2011. *Lythrum* and *Peplis* from the Late Cretaceous and Cenozoic of North America and Eurasia: New evidence suggesting early diversification within the Lythraceae. *American Journal of Botany* 98(11):1801–1815.

Groeninckx, I., S. Dessein, H. Ochoterena, C. Persson, T.J. Motley, J. Karehed, G. Bermer, S. Huysmans, and E. Smets. 2009. Phylogeny of the herbaceous tribe Spermacoceae (Rubiaceae) based on plastid DNA data. *Annals of the Missouri Botanical Garden* 96:109–132.

Groppo, M., J.R. Pirani, M.L.F. Salatino, S.R. Blanco, and J.A. Kallunki. 2008. Phylogeny of Rutaceae based on two noncoding regions from cpDNA. *American Journal of Botany* 95(8):985–1005.

Grover, C.E., J.P. Gallagher, E.P. Szadkowski, M.J. Yoo, L.E. Flagel, and J.F. Wendel. 2012. Homoeolog expression bias and expression level dominance in allopolyploids. *New Phytologist* 196(4):966–971.

Grover, C.E., and J.F. Wendel. 2010. Recent insights into mechanisms of genome size change in plants. *Journal of Botany* 2010: Article ID 382732, 8 pages. doi:10.1155/2010/382732.

Grudzinskaja, I.A. 1967. Ulmaceae and reasons for distinguishing Celtidoideae as a separate family Celtidaceae Link. *Botanicheskii Zhurnal* 52:1723–1749.

Gruenstaeudl, M., E. Urtubey, R.K. Jansen, R. Samuel, M.H.J. Barfuss, and T.F. Stuessy. 2009. Phylogeny of Barnadesioideae (Asteraceae) inferred from DNA sequence data and morphology. *Molecular Phylogenetics and Evolution* 51:572–587.

Guimarães, E., L.C. di Stasi, and R.C.S. Maimoni-Rodella. 2008. Pollination biology of *Jacaranda oxyphylla* with an emphasis on staminode function. *Annals of Botany* 102:699–711.

Guisinger, M.M., J.V. Kuehl, J.L. Boore, and R.K. Jansen. 2008. Genome-wide analyses of Geraniaceae plastid DNA reveal unprecedented patterns of increased nucleotide substitutions. *Proceedings of the National Academy of Sciences USA* 105(47):18424–18429.

———. 2011. Extreme reconfiguration of plastid genomes in the angiosperm family Geraniaceae: Rearrangements, repeats, and codon usage. *Molecular Biology and Evolution* 28(1):583–600.

Guo, S.-Q., M. Xiong, C.-F. Ji, Z.-R. Zhang, D.-Z. Li, and Z.-Y. Zhang. 2011. Molecular phylogenetic reconstruction of *Osmanthus* Lour. (Oleaceae) and related genera based on three chloroplast intergenic spacers. *Plant Systematics and Evolution* 294:57–64.

Guo, W., and J.P. Mower. 2013. Evolution of plant mitochondrial intron-encoded maturases: Frequent lineage-specific loss and recurrent intracellular transfer to the nucleus. *Journal of Molecular Evolution* 77(1–2):43–54.

Guralnick, L.J., A. Cline, M. Smith, and R.F. Sage. 2008. Evolutionary physiology: The extent of C4 and CAM photosynthesis in the genera *Anacampseros* and *Grahamia* of the Portulacaceae. *Journal of Experimental Botany* 59:1735–1742.

Gustafsson, M.H.G. 1995. Petal venation in the Asterales and related orders. *Botanical Journal Linnean Society* 118:1–18.

Gustafsson, M.H.G., and V.A. Albert. 1999. Inferior ovaries and angiosperm diversification. *In* P.M. Hollingsworth, R.M. Bateman, and R.J. Gornall [eds.], Molecular Systematics and Plant Evolution, 403–431. Taylor and Francis, London, UK.

Gustafsson, M.H.G., A. Backlund, and B. Bremer. 1996. Phylogeny of the Asterales sensu lato based on *rbcL* sequences with particular reference to the Goodeniaceae. *Plant Systematics and Evolution* 199:217–242.

Gustafsson, M.H.G., V. Bittrich, and P.F. Stevens. 2002. Phylogeny of Clusiaceae based on *rbcL* sequences. *International Journal of Plant Sciences* 163:1045–1054.

Gustafsson, M.H.G., and K. Bremer. 1995. Morphology and phylogenetic interrelationships of the Asteraceae, Calyceraceae, Campanulaceae, Goodeniaceae and related families (Asterales). *American Journal of Botany* 82:250–265.

———. 1997. The circumscription and systematic position of Carpodetaceae. *Australian Systematic Botany* 10:855–872.

Guyot, R., F. Lefebvre-Pautigny, C. Tranchant-Dubreuil, M. Rigoreau, P. Hamon, T. Leroy, S. Hamon, V. Poncet, D. Crouzillat, and A. de Kochko. 2012. Ancestral synteny shared between distantly-related plant species from the asterid (*Coffea canephora* and *Solanum* sp.) and rosid (*Vitis vinifera*) clades. *BMC Genomics* 13:103. doi:10.1186/1471-2164-13-103.

Haber, J.M. 1966. The comparative anatomy and morphology of the flowers and inflorescences of the Proteaceae. III. Some African taxa. *Phytomorphology* 16:490–527.

Haberle, R.C., A. Dang, T. Lee, C. Peñaflor, H. Cortes-Burns, A. Oestreich, L. Raubeson, N. Cellinese, E.J. Edwards, S.-T. Kim, W.M.M. Eddie, and R.K. Jansen. 2009. Taxonomic and biogeographic implications of a phylogenetic analysis of the Campanulaceae based on three chloroplast genes. *Taxon* 58:715–734.

Hackett, S.J., R.T. Kimball, S. Reddy, R.C.K. Bowie, E.L. Braun, M.J. Braun, J.L. Chojnowski, W.A. Cox, K.-L. Han, J. Harshman, C.J. Huddleston, B.D. Marks, K.J. Miglia, W.S. Moore, F.H. Sheldon, D.W. Steadman, C.C. Witt, and T. Yuri. 2008. A phylogenomic study of birds reveals their evolutionary history. *Science* 320:1763–1768.

Hagen, C.W. 1959. Influence of genes controlling flower color on relative quantities of anthocyanins and flavonols in petals of *Impatiens balsamina*. *Genetics* 44:787–793.

Haggard, C., and B.H. Tiffney. 1997. The flora of the early Miocene Brandon Lignite, Vermont, USA, VIII. *Caldesia* (Alismataceae). *American Journal of Botany* 84:239–252.

Hahn, W.J. 2002. A molecular phylogenetic study of the Palmae (Arecaceae) based on *atpB*, *rbcL* and 18S nrDNA sequences. *Systematic Biology* 51:91–112.

Hajibabaei, M., J. Xia, and G. Drouin. 2006. Seed plant phylogeny: Gnetophytes are derived conifers and a sister group to Pinaceae. *Molecular Phylogenetics and Evolution* 40:208–217.

Hakki, M.I. 1977. Über die Embryologie, Morphologie und systematische Zugehörigkeit von *Dermatobotrys saundersii* Bolus. *Botanische Jahrbücher für Systematik* 98:93–119.

Halkier, B.A., and J. Gershenzon. 2006. Biology and biochemistry of glucosinolates. *Annual Review of Plant Biology* 57:303–333.

Hall, A.E., A. Fiebig, and D. Preuss. 2002. Beyond the *Arabidopsis* genome: Opportunities for comparative genomics. *Plant Physiology* 129:1439–1447.

Hall, J.C. 2008. Systematics of Capparaceae and Cleomaceae: An evaluation of the generic delimitations of *Capparis* and *Cleome* using plastid DNA sequence data. *Botany-Botanique* 86(7):682–696.

Hall, J.C., K.J. Sytsma, and H.H. Iltis. 2002. Phylogeny of Capparaceae and Brassicaceae based on chloroplast sequence data. *American Journal of Botany* 89:1826–1842.

Hallier, H. 1905. Provisional scheme for the natural (phylogenetic) system of the flowering plants. *New Phytologist* 4:151–162.

Ham, van der, R.W.J.M. 1989. New observations on the pollen of *Ctenolophon* Oliver (Ctenolophonaceae), with remarks on the evolutionary history of the genus. *Review of Palaeobotany and Palynology* 59:153–160.

Hamann, U. 1975. Neue Untersuchungen zur Embryologie und Systematik der Centrolepidaceae. *Botanische Jahrbücher für Systematik* 96:154–191.

Hamby, R.K., and E.A. Zimmer. 1988. Ribosomal RNA sequences for inferring phylogeny within the grass family (Poaceae). *Plant Systematics and Evolution* 160:29–37.

———. 1992. Ribosomal RNA as a phylogenetic tool in plant systematics. *In* P.S. Soltis, D.E. Soltis, and J.J. Doyle [eds.], Molecular Systematics of Plants, 50–91. Chapman and Hall, New York, NY.

Han, Y., S. Qin, and S.R. Wessler. 2013. Comparison of class 2 transposable elements at superfamily resolution reveals conserved and distinct features in cereal grass genomes. BMC Genomics 14:71. doi:10.1186/1471-2164-14-71.

Hancock, J.M. 2002. Genome size and the accumulation of simple sequence repeats: Implications of new data from genome sequencing projects. *Genetica* 115:93–103.

Hansen, A., S. Hansmann, T. Samigullin, A. Antonov, and W. Martin. 1999. *Gnetum* and the angiosperms: Molecular evidence that their shared morphological characters are convergent, rather than homologous. *Molecular Biology and Evolution* 16:1006–1009.

Hansen, C.H., L. Du, P. Naur, C.E. Olsen, K.B. Axelsen, A.J. Hick, J.A. Pickett, and B.A. Halkier. 2001a. CYP83B1 is the oxime-metabolizing enzyme in the glucosinolate pathway. *Journal of Biological Chemistry* 276:24790–24796.

Hansen, C.H., U. Wittstock, C.E. Olsen, A.J. Hick, J.A. Pickett, and B.A. Halkier. 2001b. Cytochrome P450 CYP79F1 from *Arabidopsis* catalyzes the conversion of dihomomethionine and trihomomethionine to the corresponding aldoximes in the biosynthesis of aliphatic glucosinolates J. *Journal of Biological Chemistry* 276:11078–11085.

Hansen, H.V. 1992. Studies in the Calyceraceae with a discussion of its relationships to Compositae. *Nordic Journal of Botany* 12:63–75.

Hansen, T.F., W.S. Armbruster, and L. Antonsen. 2000. Comparative analysis of character displacement and spatial adaptations as illustrated by the evolution of *Dalechampia* blossoms. *American Naturalist* 156 (Supplement):S17–S34.

Hanson, L., R.L. Brown, A. Boyd, M.A.T. Johnson, and M.D. Bennett. 2003. First nuclear DNA c-values for 28 angiosperm genera. *Annals of Botany* 91(1):31–38.

Hanson, L., A. McMahon, M.A.T. Johnson, and M.D. Bennett. 2001. First nuclear DNA C-values for another 25 angiosperm families. *Annals of Botany* 88:851–858.

Hao, G., R.M.K. Saunders, and M.-L. Chye. 2000. A phylogenetic analysis of the Illiciaceae based on sequences of internal transcribed spacers (ITS) of nuclear ribosomal DNA. *Plant Systematics and Evolution* 223:81–90.

Hao, W.L., and J.D. Palmer. 2009. Fine-scale mergers of chloroplast and mitochondrial genes create functional, transcompartmentally chimeric mitochondrial genes. *Proceedings of the National Academy of Sciences USA* 106(39):16728–16733.

Hao, W.L., A.O. Richardson, Y. Zheng, and J.D. Palmer. 2010. Gorgeous mosaic of mitochondrial genes created by horizontal transfer and gene conversion. *Proceedings of the National Academy of Sciences USA* 107(50):21576–21581.

Hapeman, J.R., and K. Inoue. 1997. Plant-pollinator interactions and floral radiation in *Platanthera* (Orchidaceae). *In* T.J. Givnish, and K.J. Sytsma [eds.], Molecular Evolution and Adaptive Radiation, 433–454. Cambridge University Press, Cambridge, UK.

Harbaugh, D.T., and B.G. Baldwin. 2007. Phylogeny and biogeography of the sandalwoods (*Santalum*, Santalaceae): Repeated dispersals throughout the Pacific. *American Journal of Botany* 94:1028–1040.

Harborne, J. 1982. Introduction to Ecological Chemistry, 2nd Ed. Academic Press, Inc., New York, NY.

Harder, L.D., and S.C.H. Barrett [eds.]. 2006. Ecology and Evolution of Flowers. Oxford University Press, Oxford, UK.

Hardig, T.M., C.K. Anttila, and S.J. Brunsfeld. 2010. A phylogenetic analysis of *Salix* (Salicaceae) based on *matK* and ribosomal DNA sequence data. *Journal of Botany* 2010: Article ID 197696. doi:10.1155/2010/197696.

Harland, W.B., R.L. Armstrong, A.V. Cox, L.E. Craig, A.G. Smith, and D.G. Smith. 1989. A Geologic Timescale. Cambridge University Press, Cambridge, UK.

Harms, H. 1930. Hamamelidaceae. *In* A. Engler, and K. Prantl [eds.], Die Natürlichen Pflanzenfamilien, 2nd ed., 18a, 330–343. Engelmann, Leipzig, Germany.

———. 1934. Reihe Centrospermae. *In* A. Engler and K. Prantl [eds.], Die Natürlichen Pflanzenfamilien, 2nd ed., 16c, 1–6. Engelmann, Leipzig, Germany.

Harpke, D., and A. Peterson. 2006. Non-concerted evolution in *Mammillaria* (Cactaceae). *Molecular Phylogenetics and Evolution* 41:579–593.

Harrington, M.G., K.J. Edwards, S.A. Johnson, M.W. Chase, and P.A. Gadek. 2005. Phylogenetic inference in Sapindaceae sensu lato using plastid *matK* and *rbcL* DNA sequences. *Systematic Botany* 30(2):366–382.

Harrington, M.G., and P.A. Gadek. 2009. A species well travelled—the *Dodonaea viscosa* (Sapindaceae) complex based on phylogenetic analyses of nuclear ribosomal ITS and ETSf sequences. *Journal of Biogeography* 36(12):2313–2323.

———. 2010. Phylogenetics of hopbushes and pepperflowers (*Dodonaea*, *Diplopeltis* – Sapindaceae), based on nuclear ribosomal ITS and partial ETS sequences incorporating secondary-structure models. *Australian Systematic Botany* 23(6):431–442.

Harris, S.A., J.P. Robinson, and B.E. Juniper. 2002. Genetic clues to the origin of the apple. *Trends in Genetics* 18:426–430.

Harris, T.M. 1941. *Caytonanthus*, the microsporophyll of *Caytonia*. *Annals of Botany* 5:47–58.

———. 1951. The relationships of the Caytoniales. *Phytomorphology [Delhi]* 1:29–39.

———. 1964. The Yorkshire Jurassic Flora. II. Caytoniales, Cycadales, and Pteridosperms. British Museum (Natural History), London, UK.

Hasebe, M., R. Kofuji, M. Ito, M. Kato, K. Iwatsuki, and K. Ueda. 1992. Phylogeny of gymnosperms inferred from *rbcL* gene sequences. *Botanical Magazine* 105:673–679.

Hasegawa, M., H. Kishino, and T. Yano. 1985. Dating of the human-ape split by a molecular clock of mitochondrial DNA. *Journal of Molecular Evolution* 21: 160–174.

Hassan, N.S., J. Thiede, and S. Liede-Schumann. 2005. Phylogenetic analysis of Sesuvioideae (Aizoaceae) inferred from nrDNA internal transcribed spacer (ITS) sequences and morphological data. *Plant Systematics and Evolution* 255:121–143.

Hatch, M.D., T. Kagawa, and S. Craig. 1975. Subdivision of C_4-pathway species based on differing C_4 acid decarboxylating systems and ultrastructural features. *Australian Journal of Plant Physiology* 2:111–118.

Haudry, A., A.E. Platts, E. Vello, D.R. Hoen, M. Leclercq, R.J. Williamson, E. Forczek, Z. Joly-Lopez, J.G. Steffen, K.M. Hazzouri, K. Dewar, J.R. Stinchcomb, D.J. Schoen, X. Wang, J. Schmutz, C.D. Town, P.P. Edger, J.C. Pires, K.S. Schumaker, D.E. Jarvis, T. Mandakova, M.A. Lysak, E. van den Bergh, M.E. Schranz, P.M. Harrison, A.M. Moses, T.E. Bureau, S.I. Wright, and M. Blanchette. 2013. An atlas of over 90,000 conserved noncoding sequences provides insight into crucifer regulatory regions. *Nature Genetics* 45(8):891–898.

Hayes, V., E.L. Schneider, and S. Carlquist. 2000. Floral development of *Nelumbo nucifera* (Nelumbonaceae). *International Journal of Plant Sciences* 161:S183–S191.

Heard, J., and K. Dunn. 1995. Symbiotic induction of a MADS-box gene during development of alfalfa root nodules. *Proceedings of the National Academy of Sciences USA* 92:5273–5277.

Heckman, D.S., D.M. Geiser, B.R. Eidell, R.L. Stauffer, N.L. Kardos, and S.B. Hedges. 2001. Molecular evidence for the early colonization of land by fungi and plants. *Science* 293:1129–1133.

Hegarty, M.J., R.J. Abbott, and S.J. Hiscock. 2012. Allopolyploid speciation in action: The origins and evolution of *Senecio cambrensis*. *In* P.S. Soltis, and D.E. Soltis [eds.], Polyploidy and Genome Evolution, 245–270. Springer-Verlag, Berlin, Germany.

Hegnauer, R. 1962–1994. Chemotaxonomie der Pflanzen, 1–11a. Birkhäuser, Basel, Switzerland.

Heimsch, C., JR. 1942. Comparative anatomy of the secondary xylem in the Gruinales and Terebinthales of Wettstein with reference to taxonomic grouping. *Lilloa* 8:83–198.

Helsen, P., R.A. Browne, D.J. Anderson, P. Verdyck, and S. Van Dongen. 2009a. Galapagos *Opuntia* (prickly pear) cacti: Extensive morphological diversity, low genetic variability. *Biological Journal of the Linnean Society* 96:451–461.

Helsen, P., P. Verdyck, A. Tye, K. Desender, N. Van Houtte, and S. Van Dongen. 2009b. Isolation and characterization of polymorphic microsatellite markes in Galapagos prickly pear (*Opuntia*) cactus species. *Molecular Ecology Notes* 7:454–456.

Hempel, A.L., P.A. Reeves, R.G. Olmstead, and R.K. Jansen. 1995. Implications of *rbcL* sequence data for higher order relationships of the Loasaceae and the anomalous aquatic plant *Hydrostachys* (Hydrostachyaceae). *Plant Systematics and Evolution* 194:25–37.

Hennig, W. 1950. Grundzüge einer Theorie der phylogenetischen Systematik. Deutscher Zentralverlag, Berlin, Germany.

———. 1965. Phylogenetic systematics. *Annual Review of Entomology* 10:97–116.

———. 1966. Phylogenetic Systematics. University of Illinois Press, Urbana, IL.

———. 1969. Die Stammesgeschichte der Insekten. Kramer, Frankfurt, Germany.

———. 1981. Insect Phylogeny. John Wiley and Sons, New York, NY.

———. 1983. Stammesgeschichte der Chordaten. *Fortschritte der Zoologischen Systematik und Evolutionsforschung* 2:1–208.

Henrickson, J. 1967. Pollen morphology of the Fouquieriaceae. *Aliso* 6:137–160.

Henslow, G. 1893. A theoretical origin of the endogens from the exogens through self-adaptation to an aquatic habitat. *Journal of the Linnean Society, Botany* 29:485–528.

Henwood, M.J., and J.M. Hart. 2001. Towards an understanding of the phylogenetic relationships of Australian Hydrocotyloideae (Apiaceae). *Edinburgh Journal of Botany* 58:269–289.

Herbert, J., M.W. Chase, M. Moller, and R.J. Abbott. 2006. Nuclear and plastid DNA sequences confirm the placement of the enigmatic *Canacomyrica monticola* in Myricaceae. *Taxon* 55(2):349–357.

Herendeen, P.S., and P.R. Crane. 1995. The fossil history of the monocotyledons. *In* P.J. Rudall, P.J. Cribb, D.F. Cutler, and C.J. Humphries [eds.], Monocotyledons: Systematics and Evolution, 1–21. Royal Botanic Gardens, Kew, UK.

Herendeen, P., P.R. Crane, and A. Drinnan. 1995. Fagaceous flowers fruits and cupules from the Campanian (Late Cretaceous) of central Georgia, USA. *International Journal of Plant Sciences* 156:93–116.

Herendeen, P.S., W.L. Crepet, and K.C. Nixon. 1993.

Chloranthus-like stamens from the Upper Cretaceous of New Jersey. *American Journal of Botany* 80:865–871.

———. 1994. Fossil flowers and pollen of Lauraceae from the Upper Cretaceous of New Jersey. *Plant Systematics and Evolution* 189:29–40.

Hermsen, E.J., K.C. Nixon, and W.L. Crepet. 2006. The impact of extinct taxa on understanding the early evolution of Angiosperm clades: An example incorporating fossil reproductive structures of Saxifragales. *Plant Systematics and Evolution* 260(2–4):141–169.

Hernández-González, O., and O.B. Villareal. 2007. Crassulacean acid metabolism photosynthesis in columnar cactus seedlings during ontogeny: The effect of light on nocturnal acidity accumulation and chlorophyll fluorescence. *American Journal of Botany* 94:1344–1351.

Hernández-Hernández, T., J.W. Brown, B.O. Schlumpberger, L.E. Eguiarte, and S. Magallón. 2014. Beyond aridification: Multiple explanations for the elevated diversification of cacti in the New World Succulent Biome. *New Phytologist* 202:1382–1397.

Hernández-Hernández, T., H.M. Hernández, J.A. De-Nova, R. Puente, L.E. Eguiarte, S. Magallón. 2011. Phylogenetic relationships and evolution of growth form in Cactaceae (Caryophyllales, Eudicotyledoneae). *American Journal of Botany* 98:44–61.

Herrera, F.A., C.A. Jaramillo, D.L. Dilcher, S.L. Wing, and C. Gomez-N. 2008. Fossil Araceae from a Paleocene Neotropical rainforest in Colombia. *American Journal of Botany* 95(12):1569–1583.

Herrera, F.A., S.R. Manchester, S B. Hoot, K M. Wefferling, M.R. Carvalho, and C. Jaramillo. 2011. Phytogeographic implications of fossil endocarps of Menispermaceae from the Paleocene of Colombia. *American Journal of Botany* 98:2004–2017.

Herrera, F., S.R. Manchester, J. Velez-Juarbe, and C. Jaramillo. 2014. Phytogeographic history of the Humiriaceae (part 2). *International Journal of Plant Sciences* 175(7):828–840.

Herrero, M. 2001. Ovary signals for directional pollen tube growth. *Sexual Plant Reproduction* 14:3–17.

Hershkovitz, M.A. 1993. Revised circumscriptions and subgeneric taxonomies of *Calandrinia* and *Montiopsis* (Portulacaceae) with notes on phylogeny of the portulacaceous alliance. *Annals of the Missouri Botanical Garden* 80:333–365.

Hershkovitz, M.A., and E.A. Zimmer. 1997. On the evolutionary origins of the cacti. *Taxon* 46:217–232.

Hertwick, K., M. Kinney, S. Stuart, O. Maurin, S. Mathews, M.W. Chase, M. Gandolfo, and J.C. Pires. 2015. Phylogenetics, divergence times, and diversification from three genomics partitions in monocots. *Botanical Journal of the Linnean Society* 178:375–393.

Hesse, M., and R. Zetter. 2007. The fossil pollen record of Araceae. *Plant Systematics and Evolution* 263:93–115.

Heubl, G., G. Bringmann, and H. Meimberg. 2006. Molecular phylogeny and character evolution of carnivorous plant families in Caryophyllales—Revisited. *Plant Biology* 8(6):821–830.

Heuertz, M., S. Carnevale, S. Fineschi, F. Sebastiani, J.F. Hausman, L. Paule, and G.G. Vendramin. 2006. Chloroplast DNA phylogeography of European ashes, *Fraxinus* sp. (Oleaceae): Roles of hybridization and life history traits. *Molecular Ecology* 15:2131–2140.

Heywood, V.A. 1993. Flowering Plants of the World. B.T. Batsford, London, UK.

———. 1998. Flowering Plants of the World., 2nd edition. B.T. Batsford Ltd., London, UK.

Hibberd, J.M., and S. Covshoff. 2010. The regulation of gene expression required for C-4 photosynthesis. *Annual Review of Plant Biology* 61:181–207.

Hibbett, D.S., and M.J. Donoghue. 1998. Integrating phylogenetic analysis and classification in fungi. *Mycologia* 90:347–356.

Hibsch-Jetter, C., D.E. Soltis, and T.D. Macfarlane. 1997. Phylogenetic analysis of *Eremosyne pectinata* (Saxifragaceae s.l.) based on *rbcL* sequence data. *Plant Systematics and Evolution* 204:225–232.

Hickey, L.J., and J.A. Doyle. 1977. Early Cretaceous fossil evidence for angiosperm evolution. *Botanical Review* 43:3–104.

Hickey, L.J., and D.W. Taylor. 1991. The leaf architecture of *Ticodendron* and the application of foliar characters in discerning its relationships. *Annals of the Missouri Botanical Garden* 78:105–130.

———. 1996. Origin of the Angiosperm Flower. *In* D.W. Taylor, and L.J. Hickey [eds.], Flowering Plant Origin, Evolution, and Phylogeny, 176–231. Chapman and Hall, New York, NY.

Hickey, L.J., and A.D. Wolfe. 1975. The bases of angiosperm phylogeny: Vegetative morphology. *Annals of the Missouri Botanical Garden* 62:538–589.

Hiepko, P. 1965a. Vergleichend–morphologische und entwicklungsgeschichtliche Untersuchungen über das Perianth bei den Polycarpicae. *Botanische Jahrbücher für Systematik* 84:359–508.

———. 1965b. Das zentrifugale Androeceum der Paeoniaceae. *Berichte der Deutschen Botanischen Gesellschaft* 77:427–435.

Hileman, L.C. 2014a. Trends in flower symmetry evolution revealed through phylogenetic and developmental genetic advances. *Philosophical Transactions of the*

Royal Society of London B 370(1662). doi:10.1098/rstb.2013.0348.

———. 2014b. Bilateral flower symmetry—how, when and why? *Current Opinion in Plant Biology* 17:146–152.

Hileman, L.C., E.M. Kramer, and D.A. Baum. 2003. Differential regulation of symmetry genes and the evolution of floral morphologies. *Proceedings of the National Academy of Sciences USA* 100:12814–12819.

Hilger, H.H., and N. Diane. 2003. A systematic analysis of Heliotropiaceae (Boraginales) based on *trnL* and ITS1 sequence data. *Botanische Jahrbücher für Systematik* 125:19–51.

Hill, C.R., and P.R. Crane. 1982. Evolutionary cladistics and the origin of angiosperms. *In* K.A. Joysey, and A.E. Friday [eds.], Problems of Phylogenetic Reconstruction, 269–361. Academic Press, London, UK.

Hillis, D.M. 1995. Approaches for assessing phylogenetic accuracy. *Systematic Biology* 44:3–16.

———. 1996. Inferring complex phylogenies. *Nature* 383:130–131.

Hillis, D.M., J. Huelsenbeck, and D. Swofford. 1994. Hobgoblin of phylogenetics? *Nature* 369:363–364.

Hillis, D.M., D.D. Pollock, J.A. McGuire, and D.J. Zwickl. 2003. Is sparse taxon sampling a problem for phylogenetic inference? *Systematic Biology* 52:124–126.

Hilton. J., and R.M. Bateman. 2006. Pteridosperms are the backbone of seed-plant phylogeny. *Journal of the Torrey Botanical Society* 133:119–168.

Hilu, K.W., T. Borsch, K. Muller, D.E. Soltis, P.S. Soltis, V. Savolainen, M. Chase, M. Powell, L.A. Alice, R. Evans, H. Sauquet, C. Neinhuis, T. Slotta, J.G. Rohwer, C.S. Campbell, and L.W. Catrou. 2003. Inference of angiosperm phylogeny based on *matK* sequence information. *American Journal of Botany* 90(12):1758–1776.

Hirmer, M. 1918. Beiträge zur Morphologie der polyandrischen Blüten. *Flora* 110:140–192.

Hitchcock, C.L., A. Cronquist, M. Owenbey, and J.W. Thompson. 1961. Vascular Plants of the Pacific Northwest. University of Washington Press. Seattle, WA.

Hodges, S.A. 1997a. Floral nectar spurs and diversification. *International Journal of Plant Sciences* 158 (6 Supplement):S81–S88.

———. 1997b. Rapid radiation due to a key innovation in columbines (Ranunculaceae: *Aquilegia*). *In* T.J. Givnish, and K.J. Sytsma [eds.], Molecular Evolution and Adaptive Radiation, 391–405. Cambridge University Press, Cambridge, UK.

Hodges, S.A., and M.L. Arnold. 1995. Spurring plant diversification: Are floral nectar spurs a key innovation? *Proceedings of the Royal Society of London B* 262(1365):343–348.

Hodges, S.A., and N.J. Derieg. 2009. Adaptive radiations: From field to genomic studies. *Proceedings of the National Academy of Sciences USA* 106: 9947–9954.

Hodges, S.A., J.B. Whittall, M. Fulton, and J.Y. Yang. 2002. Genetics of floral traits influencing reproductive isolation between *Aquilegia formosa* and *Aquilegia pubescens*. *American Naturalist* 159 (Supplement):S51–S60.

Hörandl, E., and K. Emadzade. 2011. The evolution and biogeography of alpine species in *Ranunculus* (Ranunculaceae): A global comparison. *Taxon* 60(2):415–426.

———. 2012. Evolutionary classification: A case study on the diverse plant genus *Ranunculus* L. (Ranunculaceae). *Perspectives in Plant Ecology Evolution and Systematics* 14(4):310–324.

Hörandl, E., O. Paun, J.T. Johansson, C. Lehnebach, T. Armstrong, L.X. Chen, and P. Lockhart. 2005. Phylogenetic relationships and evolutionary traits in *Ranunculus* s.l. (Ranunculaceae) inferred from ITS sequence analysis. *Molecular Phylogenetics and Evolution* 36(2):305–327.

Hofberger, J.A., E. Lyons, P.P. Edger, J.C. Pires, and M.E. Schranz. 2013. Whole genome and tandem duplicate retention facilitated glucosinolate pathway diversification in the mustard family. *Genome Biology and Evolution* 5(11):2155–2173.

Hoffmann, M.H., K.B. von Hagen, E. Hörandl, M. Roser, and N.V. Tkach. 2010. Sources of the Arctic flora: Origins of Arctic species in *Ranunculus* and related genera. *International Journal of Plant Sciences* 171(1):90–106.

Hoffmann, P., H. Kathriarachchi, and K.J. Wurdack. 2006. A phylogenetic classification of Phyllanthaceae (Malpighiales; Euphorbiaceae sensu lato). *Kew Bulletin* 61(1):37–53.

Hofmann, U. 1977. Die Stellung von *Stegnosperma* innerhalb der Centrospermen. *Berichte der Deutschen Botanischen Gesellschaft* 90:39–52

Holsinger, K.E. 1996. Pollination biology and the evolution of mating systems in flowering plants. *Evolutionary Biology* 29:107–149.

Honma, T., and K. Goto. 2001. Complexes of MADS-box proteins are sufficient to convert leaves into floral organs. *Nature* 409:525–529.

Hooker, J.D. 1862–67. Ampelideae. *In* G. Bentham, and J.D. Hooker [eds.], Genera Plantarum, 386–388. Reeve & Co., London, UK.

Hoot, S.B. 1991. Phylogeny of the Ranunculaceae based on epidermal microcharacters and micromorphology. *Systematic Botany* 16:741–755.

———. 1995. Phylogeny of the Ranunculaceae based on *atpB*, *rbcL*, and 18S nuclear ribosomal DNA sequence data. *Plant Systematics and Evolution, Supplement* 9:241–251.

Hoot, S.B., and P.R. Crane. 1995. Interfamilial relationships in the Ranunculidae based on molecular systematics. *Plant Systematics and Evolution, Supplement* 9:119–131.

Hoot, S.B., A. Culham, and P.R. Crane. 1995a. The utility of *atpB* gene sequences in resolving phylogenetic relationships: Comparison within *rbcL* and 18S ribosomal DNA sequences in the Lardizabalaceae. *Annals of the Missouri Botanical Garden* 82:194–207.

———. 1995b. Phylogenetic relationships of the Lardizabalaceae and Sargentodoxaceae: Chloroplast and nuclear DNA sequence evidence. *Plant Systematics and Evolution, Supplement* 9:195–199.

Hoot, S.B., and A.W. Douglas. 1998. Phylogeny of the Proteaceae based on *atpB* and *atpB–rbcL* intergenic spacer region sequences. *Australian Systematic Botany* 11:301–320.

Hoot, S.B., J.W. Kadereit, F.R. Blattner, K.B. Jork, A.E. Schwarzbach, and P.R. Crane. 1997. Data congruence and phylogeny of the Papaveraceae s. l. based on four data sets: *atpB* and *rbcL* sequences, *trnK* restriction sites and morphological characters. *Systematic Botany* 22:575–590.

Hoot, S.B., J. Kramer, and M.T.K. Arroyo. 2008. Phylogenetic position of the South American dioecious genus *Hamadryas* and related Ranunculeae (Ranunculaceae). *International Journal of Plant Sciences* 169(3):433–443.

Hoot, S.B., S. Magallón, and P.R. Crane. 1999. Phylogeny of basal eudicots based on three molecular data sets: *atpB*, *rbcL*, and 18S nuclear ribosomal DNA sequences. *Annals of the Missouri Botanical Garden* 86:1–32.

Hoot, S.B., A.A. Reznicek, and J.D. Palmer. 1994. Phylogenetic relationships in *Anemone* (Ranunculaceae) based on morphology and chloroplast DNA. *Systematic Botany* 19:169–200.

Hoot, S.B., H. Zautke, D.J. Harris, P.R. Crane, and S.S. Neves. 2009. Phylogenetic patterns in Menispermaceae based on multiple chloroplast sequence data. *Systematic Botany* 34(1):44–56.

Horak, K.E. 1981. Anomalous secondary thickening in *Stegnosperma* (Phytolaccaceae). *Bulletin of the Torrey Botanical Club* 108:189–197.

Horn, J.W. 2007. Dilleniaceae. *In* K. Kubitzki, C. Bayer, and P.F. Stevens [eds.], Flowering Plants. Eudicots, 132–154. Springer-Verlag, Berlin–Heidelberg, Germany.

———. 2009. Phylogenetics of Dilleniaceae using sequence data from four plastid loci (*rbcl*, *infa*, *rps4*, *rpl16* intron). *International Journal of Plant Sciences* 170(6):794–813.

Horn, J.W., B.W. van Ee, J.J. Morawetz, R. Riina, V.W. Steinmann, P.E. Berry, and K.J. Wurdack. 2012. Phylogenetics and the evolution of major structural characters in the giant genus *Euphorbia* L. (Euphorbiaceae). *Molecular Phylogenetics and Evolution* 63(2):305–326.

Howarth, D.G., and M.J. Donoghue. 2005. Duplications in CYC-like genes from Dipsacales correlate with floral form. *International Journal of Plant Sciences* 166(3):357–370.

———. 2006. Phylogenetic analysis of the "'ECE" (CYC/TB1) clade reveals duplications predating the core eudicots. *Proceedings of the National Academy of Sciences USA* 103(24):9101–9106.

———. 2009. Duplications and expression of *DIVARICATA*-like genes in Dipsacales. *Molecular Biology and Evolution* 26:1245–1258.

Howarth, D.G., T. Martins, E. Chimney, and M.J. Donoghue. 2011. Diversification of *CYCLOIDEA* expression in the evolution of bilateral flower symmetry in Caprifoliaceae and *Lonicera* (Dipsacales). *Annals of Botany* 107(9):1521–1532.

Hu, J.-M., M. Lavin, M.F. Wojciechowski, and M.J. Sanderson. 2000. Phylogenetic systematics of the tribe Millettieae (Leguminosae) based on chloroplast *trnK*/*matK* sequences and its implications for evolutionary patterns in Papilionoideae. *American Journal of Botany* 87:418–430.

———. 2002. Phylogenetic analysis of nuclear ribosomal ITS/5.8S sequences in the tribe Millettieae (Fabaceae): *Poecilanthe–Cyclolobium*, the core Millettieae, and the *Callerya* group. *Systematic Botany* 27:722–733.

Huang, J., and J. Yue. 2013. Horizontal gene transfer in the evolution of photosynthetic eukaryotes. *Journal of Systematics and Evolution* 51(1):13–29.

Huber, H. 1969. Die Samenmerkmale und Verwandtschaftsverhältnisse der Liliifloren. *Mitteilungen der Botanischen Staatssammlung München* 8:219–538.

———. 1977. The treatment of monocotyledons in an evolutionary system of classification. *Plant Systematics and Evolution, Supplement* 1:285–298.

———. 1993. *Neurada*, eine Gattung der Malvales. *Sendtnera* 1:7–10.

Huber, K.A. 1980. Morphologische und entwicklungsgeschichtliche Untersuchungen an Blüten und Blütenständen von Solanaceen und von *Nolana paradoxa* Lindl. (Nolanaceae). Dissertationes Botanicae 55. J. Cramer, Vaduz, Liechtenstein.

Hudson, C.M., and G.C. Conant. 2012. Yeast as a window into changes in genome complexity due to polyploidization. *In* P.S. Soltis, and D.E. Soltis [eds.], Polyploidy and Genome Evolution, 293–308. Springer-Verlag, Berlin, Germany.

Huelsenbeck, J.P., B. Larget, R.E. Miller, and F. Ronquist. 2002. Potential applications and pitfalls of Bayesian inference of phylogeny. *Systematic Biology* 51:673–688.

Huelsenbeck, J.P., F. Ronquist, R. Nielson, and J.P. Bolback. 2001. Bayesian inference of phylogeny and its impact on evolutionary biology. *Science* 294:2310–2314.

Huether, C.A., Jr. 1968. Exposure of natural genetic variability underlying the pentamerous corolla constancy in *Linanthus androsaceus* ssp. *androsaceus*. *Genetics* 60:123–146.

———. 1969. Constancy of the pentamerous corolla phenotype in natural populations of *Linanthus*. *Evolution* 23:572–588.

Hufford, L. 1992. Rosidae and their relationships to other nonmagnoliid dicotyledons: A phylogenetic analysis using morphological and chemical data. *Annals of the Missouri Botanical Garden* 79:218–248.

———. 1995. Patterns of ontogenetic evolution in perianth diversification of *Besseya* (Scrophulariaceae). *American Journal of Botany* 82:655–680.

———. 1996a. The origin and early evolution of angiosperm stamens. *In* W.G. D'Arcy, and R.C. Keating [eds.], The Anther: Form, Function, and Phylogeny, 58–91. Cambridge University Press, Cambridge, UK.

———. 1996b. Ontogenetic evolution, clade diversification, and homoplasy. *In* M.J. Sanderson, and L. Hufford [eds.], Homoplasy: The Recurrence of Similarity in Evolution, 271–301. Academic Press, San Diego, CA.

———. 2001. Ontogeny and morphology of the fertile flowers of Hydrangeaceae and allied genera of tribe Hydrangeae (Hydrangeaceae). *Botanical Journal of the Linnean Society* 137:139–187.

Hufford, L., and P.R. Crane. 1989. A preliminary phylogenetic analysis of the "lower" Hamamelidae. *In* P.R. Crane, and S. Blackmore [eds.], Evolution, Systematics and Fossil History of the Hamamelidae, 1, 175–192. Clarendon Press, Oxford, UK.

Hufford, L., and W.C. Dickison. 1992. A phylogenetic analysis of Cunoniaceae. *Systematic Botany* 17: 181–200.

Hufford, L., and P.K. Endress. 1989. The diversity of anther structures and dehiscence patterns among Hamamelididae. *Botanical Journal of the Linnean Society* 99:301–346.

Hufford, L., M. L. Moody, and D.E. Soltis. 2001. A phylogenetic analysis of Hydrangeaceae based on sequences of the plastid gene *matK* and their combination with *rbcL* and morphological data. *International Journal of Plant Sciences* 162:835–846.

Hughes, N.F. 1994. The Enigma of Angiosperm Origins. Cambridge University Press, Cambridge, UK.

Hürlimann, H. and H.U. Stauffer. 1957. *Daenikera*, eine neue Santalaceen–Gattung. *Vierteljahrsschrift der Naturforschenden Gesellschaft Zürich* 102:332–336.

Husband, B.C., S.J. Baldwin, and J. Suda. 2013. The incidence of polyploidy in natural plant populations: Major patterns and evolutionary processes. *In* J.F. Wendel, J. Greilhuber, J. Doležel, and I.J. Leitch [eds.], Plant genome diversity, 255–276. Springer-Verlag Wien, New York, NY.

Hutchinson, J. 1934. The Families of Flowering Plants. Oxford University Press, Oxford, UK.

———. 1959. The Families of Flowering Plants, 2nd ed. Oxford University Press, Oxford, UK.

———. 1967. The Genera of Flowering Plant. Clarendon Press, Oxford, UK.

———. 1973. The Families of Flowering Plants, 3rd ed. Clarendon Press, Oxford, UK.

Huynh, K-L. 1976. L'arrangement du pollen du genre *Schisandra* (Schisandraceae) et sa signification phylogénique chez les Angiospermes. *Beiträge zur Biologie der Pflanzen* 52:227–253.

Ibarra-Laclette, E., E. Lyons, G. Hernandez-Guzman, C. Anahi Perez-Torres, L. Carretero-Paulet, T.-H. Chang, T. Lan, A.J. Welch, M.J. Abraham Juarez, J. Simpson, A. Fernandez-Cortes, M. Arteaga-Vazquez, E. Gongora-Castillo, G. Acevedo-Hernandez, S.C. Schuster, H. Himmelbauer, A.E. Minoche, S. Xu, M. Lynch, A. Oropeza-Aburto, S. Alan Cervantes-Perez, M. de Jesus Ortega-Estrada, J. Israel Cervantes-Luevano, T.P. Michael, T. Mockler, D. Bryant, A. Herrera-Estrella, V.A. Albert, and L. Herrera-Estrella. 2013. Architecture and evolution of a minute plant genome. *Nature* 498(7452):94–98.

Ickert-Bond, S.M., and J. Wen. 2013. A taxonomic synopsis of Altingiaceae with nine new combinations. *PhytoKeys* 31(2013):21–61.

Igersheim, A., M. Buzgo, and P.K. Endress. 2001. Gynoecium diversity and systematics in basal monocots. *Botanical Journal of the Linnean Society* 136:1–65.

Igersheim, A., and P.K. Endress. 1997. Gynoecium diversity and systematics of the Magnoliales and winteroids. *Botanical Journal of the Linnean Society* 124:213–271.

Igersheim, A., C. Puff, P. Leins, and C. Erbar. 1994. Gynoecial development of *Gaertnera* Lam. and of presumably allied taxa of the Psychotrieae (Rubiaceae): Secondarily "superior" vs. inferior ovaries. *Botanische Jahrbücher für Systematik* 116:401–414.

Iles, W.J.D., C. Lee, D.D. Sokoloff, M.V. Remizowa, S.R. Yadav, M.D. Barrett, R.L. Barret, T.D. Macfarlane, P.J. Rudall, and S.W. Graham. 2014. Reconstructing the age and historical biogeography of the ancient flowering-plant family Hydatellaceae (Nymphaeales). *BMC Evolutionary Biology* 14:102. doi:10.1186/1471-2148-14-102.

Iles, W.J.D., S.Y. Smith, M.A. Gandolfo, and S.W. Graham. 2015. A review of monocot fossils suitable for molecular dating analyses. *Botanical Journal of the Linnean Society* 178:346–374.

Iles, W.J.D., S.Y. Smith, and S.W. Graham. 2013. A well-supported phylogenetic framework for the monocot order Alismatales reveals multiples losses of the plastid NADH dehydrogenase complex and a strong long-branch effect. *In* P. Wilkin, and S.J. Mayo [eds.], Early Events in Monocot Evolution, 1–28. Cambridge University Press, Cambridge, UK.

Inda, L.A., M. Pimentel, and M.W. Chase. 2012. Phylogenetics of tribe Orchideae (Orchidaceae: Orchidoideae) based on combined DNA matrices: Inferences regarding timing of diversification and evolution of pollination systems. *Annals of Botany* 110(1):71–90.

Inda, L.A., P. Torrecilla, P. Catalan, and T. Ruiz-Zapata. 2008. Phylogeny of *Cleome* L. and its close relatives *Podandrogyne* Ducke and *Polanisia* Raf. (Cleomoideae, Cleomaceae) based on analysis of nuclear ITS sequences and morphology. *Plant Systematics and Evolution* 274(1–2):111–126.

Ingrouille, M.J., M.W. Chase, M.F. Fay, D. Bowman, M. Van Der Bank, and A. De Bruijn. 2002. Systematics of Vitaceae from the viewpoint of plastid *rbcL* DNA sequence data. *Botanical Journal of the Linnean Society* 138:421–432.

Inouye, H., S. Ueda, M. Hirabayashi, and N. Shimoka-Wa. 1966. Studies on the monoterpene glucosides of *Daphniphyllum macropodum*. *Yakugaku Zasshi* 86:943–947.

International Peach Genome Initiative. 2013. The high-quality draft genome of peach (*Prunus persica*) identifies unique patterns of genetic diversity, domestication and genome evolution. *Nature Genetics* 45:487–494.

Iorizzo, M., D. Senalik, M. Szklarczyk, D. Grzebelus, D. Spooner, and P. Simon. 2012. De novo assembly of the carrot mitochondrial genome using next generation sequencing of whole genomic DNA provides first evidence of DNA transfer into an Angiosperm plastid genome. *BMC Plant Biology* 12:61. doi:10.1186/1471-2229-12-61.

Irish, V.F. 2009. Evolution of petal identity. *Journal of Experimental Botany* 60:2517–2527.

Irish, V.F., and E.M. Kramer. 1998. Genetic and molecular analysis of angiosperm flower development. *Advances in Botanical Research* 28:199–230.

Jabbour, F., C. Damerval, and S. Nadot. 2008. Evolutionary trends in the flowers of Asteridae: Is polyandry an alternative to zygomorphy? *Annals of Botany* 102(2):153–165.

Jabbour, F., L.P. Ronse De Craene, S. Nadot, and C. Damerval. 2009. Establishment of zygomorphy on an ontogenic spiral and evolution of perianth in the tribe Delphinieae (Ranunculaceae). *Annals of Botany* 104(5):809–822.

Jabbour, F., S. Nadot, and C. Damerval. 2009. Evolution of floral symmetry: A state of the art. *Comptes Rendues Biologies* 332:219–231.

Jabbour, F., and S.S. Renner. 2011. *Consolida* and *Aconitella* are an annual clade of *Delphinium* (Ranunculaceae) that diversified in the Mediterranean basin and the Irano-Turanian region. *Taxon* 60(4):1029–1040.

———. 2012a. A phylogeny of Delphinieae (Ranunculaceae) shows that *Aconitum* is nested within *Delphinium* and that Late Miocene transitions to long life cycles in the Himalayas and Southwest China coincide with bursts in diversification. *Molecular Phylogenetics and Evolution* 62(3):928–942.

———. 2012b. Spurs in a spur: Perianth evolution in the Delphinieae (Ranunculaceae). *International Journal of Plant Sciences* 173(9):1036–1054.

Jack, T. 2001. Plant development going MADS. *Plant Molecular Biology* 46:515–520.

Jacobs, B., F. Lens, and E. Smets. 2009. Evolution of fruit and seed characters in the *Diervilla* and *Lonicera* clades (Caprifoliaceae, Dipsacales). *Annals of Botany* 104:253–276.

Jacques, F.M.B. 2009a. Fossil history of the Menispermaceae (Ranunculales). *Annales de Paléontologie* 95:53–69.

Jäger-Zürn, I. 1967. Infloreszenz- und blütenmorphologische, sowie embryologische Untersuchungen an *Myrothamnus* Welw. *Beiträge zur Biologie der Pflanzen* 42:241–271.

———. 1997. Embryological and floral studies in *Weddellina squamulosa* Tul. (Podostemaceae, Tristichoideae). *Aquatic Botany* 57:151–182.

Jaillon, O., J.-M. Aury, B. Noel, A. Policriti, C. Clepet, A. Casagrande, N. Choisne, S. Aubourg, N. Vitulo, C. Jubin, A. Vezzi, F. Legeai, P. Hugueney, C. Dasilva, D. Horner, E. Mica, D. Jublot, J. Poulain, C. Bruyere, A. Billault, B. Segurens, M. Gouyvenoux, E. Ugarte, F. Cattonaro, V. Anthouard, V. Vico, C. Del Fabbro, M. Alaux, G. Di Gaspero, V. Dumas, N. Felice, S. Paillard, I. Juman, M. Moroldo, S. Scalabrin, A. Canaguier, I. Le Clainche, G. Malacrida, E. Durand, G. Pesole,

V. Laucou, P. Chatelet, D. Merdinoglu, M. Delledonne, M. Pezzotti, A. Lecharny, C. Scarpelli, F. Artiguenave, M.E. Pe, G. Valle, M. Morgante, M. Caboche, A.-F. Adam-Blondon, J. Weissenbach, F. Quetier, and P. Wincker. 2007. The grapevine genome sequence suggests ancestral hexaploidization in major angiosperm phyla. *Nature* 449(7161):463–467.

Jaing, Z., and R. Zhou. 1992. Distribution of the iridoid compounds in the Hamamelidae. *Zhongguo Yaoke Daxue Xuebao* 23:140–143.

Janka, H., M. von Balthazar, W.S. Alverson, D.A. Baum, J. Semir, and C. Bayer. 2008. Structure, development and evolution of the androecium in Adansonieae (core Bombacoideae, Malvaceae s.l.). *Plant Systematics and Evolution* 275:69–91.

Jansen, R.K., Z. Cai, L.A. Raubeson, H. Daniell, C.W. dePamphilis, J. Leebens-Mack, K.F. Mueller, M. Guisinger-Bellian, R.C. Haberle, A.K. Hansen, T.W. Chumley, S.-B. Lee, R. Peery, J.R. McNeal, J.V. Kuehl, and J.L. Boore. 2007. Analysis of 81 genes from 64 plastid genomes resolves relationships in angiosperms and identifies genome-scale evolutionary patterns. *Proceedings of the National Academy of Sciences USA* 104(49):19369–19374.

Jansen, R.K., and K.-J. Kim. 1996. Implications of chloroplast DNA data for the classification and phylogeny of the Asteraceae. *In* D. Hind, and H. Beentje [eds.], Compositae: Systematics, 317–339. Proceedings of the International Compositae Conference, Royal Botanic Gardens, Kew, UK.

Jansen, R.K., H.J. Michaels, and J.D. Palmer. 1991. Phylogeny and character evolution in the Asteraceae based on chloroplast DNA restriction site mapping. *Systematic Botany* 16:98–115.

Jansen, R.K., H.J. Michaels, R.S. Wallace, K.J. Kim, S.C. Keeley, L.E. Watson, and J.D. Palmer. 1992. Chloroplast DNA Variation in the Asteraceae: Phylogenetic and Evolutionary Implications. Chapman and Hall, New York, NY.

Jansen, R.K., and J.D. Palmer. 1987. A chloroplast DNA inversion marks an ancient evolutionary split in the sunflower family, Asteraceae. *Proceedings of the National Academy of Sciences USA* 84: 5818–5822.

———. 1988. Phylogenetic implications of chloroplast DNA restriction site variation in the Mutisieae (Asteraceae). *American Journal of Botany* 75:751–764.

Jansen, S., T. Watanabe, P. Caris, K. Geuten, F. Lens, N. Pyck, and E. Smets. 2004. The distribution and phylogeny of aluminum accumulating plants in the Ericales. *Plant Biology* 6:498–505.

Janssen, T., and K. Bremer. 2004. The age of major monocot groups inferred from 800+*rbcL* sequences. *Botanical Journal of the Linnean Society* 146:385–398.

Janssens, S.B., E.B. Knox, S. Huysmans, E.F. Smets, and V.S.F.T. Merckx. 2009. Rapid radiation of *Impatiens* (Balsaminaceae) during Pliocene and Pleistocene: Results of a global climate change. *Molecular Phylogenetics and Evolution* 52:806–824.

Jaramillo, M.A., P.S. Manos, and E.A. Zimmer. 2004. Phylogenetic relationships of the perianthless Piperales: Reconstructing the evolution of floral development. *International Journal of Plant Sciences* 165:403–416.

Jarvis, E.D., S. Mirarab, A.J. Aberer, B. Li, P. Houde, C. Li, S.Y. Ho, B.C. Faircloth, B. Nabholz, J.T. Howard, A. Suh, C.C. Weber, R.R. da Fonseca, J. Li, F. Zhang, H. Li, L. Zhou, N. Narula, L. Liu, G. Ganapathy, B. Boussau, M.S. Bayzid, V. Zavidovych, S. Subramanian, T. Gabaldón, S. Capella-Gutiérrez, J. Huerta-Cepas, B. Rekepalli, K. Munch, M. Schierup, B. Lindow, W.C. Warren, D. Ray, R.E. Green, M.W. Bruford, X. Zhan, A. Dixon, S. Li, N. Li, Y. Huang, E.P. Derryberry, M.F. Bertelsen, F.H. Sheldon, R.T. Brumfield, C.V. Mello, P.V. Lovell, M. Wirthlin. M.P.C. Schneider, F. Prosdocimi, J.A. Samaniego, A.M.V. Velazquez, A. Alfaro-Núnez, P.F. Campos, B. Petersen, T. Sicheritz-Ponten, A. Pas, T. Bailey, P. Scofield, M. Bunce, D.M. Lambert, Q. Zhou, P. Perelman, A.C. Driskell, B. Shapiro, Z. Xiong, Y. Zeng, S. Liu, Z. Li, B. Liu, K. Wu, J. Xiao, X. Yinqi, Q. Zheng, Y. Zhang, H. Yang, J. Wang, L. Smeds, F.E. Rheindt, M. Braun, J. Fjeldsa, L. Orlando, F.K. Barker, K.A. Jønsson, W. Johnson, K.-P. Koepfli, S. O'Brien, D. Haussler, O.A. Ryder, C. Rahbek, E. Willerslev, G.R. Graves, T.C. Glenn, M. McCormack, D. Burt, H. Ellegren, P. Alström, S.V. Edwards, A. Stamatakis, D.P. Mindell, J. Cracraft, E.L. Braun, T. Warno, W. Jun, M.T.P. Gilbert, and G. Zhang. 2014. Whole-genome analyses resolve early branches in the tree of life of modern birds. *Science* 346:1320–1331.

Jensen, S.R. 1991. Plant iridoids, their biosynthesis and distribution in angiosperms. *In* J.B. Harborne, and F.A. Tomas-Barberan [eds.], Ecological Chemistry and Biochemistry of Plant Terpenoids, 133–158. Clarendon Press, Oxford, UK.

———. 1992. Systematic implications of the distribution of iridoids and other chemical compounds in the Loganiaceae and other families of the Asteridae. *Annals of the Missouri Botanical Garden* 79:284–302.

Jensen, S.R., S.E. Lyse-Peterson, and B.J. Nielsen. 1979. Novel bis-iridoid glucosides from *Dipsacus sylvestris. Phytochemistry* 18:273–277.

Jensen, S.R., B.J. Nielsen, and R. Dahlgren. 1975. Iridoid compounds, their occurrence and systematic importance in the angiosperms. *Botaniska Notiser* 128:148–180.

Jensen, S.R., L. Ravnkilde, and J. Schripsema. 1998. Unedoside derivatives in *Nuxia* and their biosynthesis. *Phytochemistry* 47:1007–1011.

Jensen, U. 1995. Secondary compounds of the Ranunculiflorae. *Plant Systematics and Evolution, Supplement* 9:85–97.

Jensen, U., S.B. Hoot, J.T. Johansson, and K. Kosuge. 1995. Systematics and phylogeny of the Ranunculaceae–a revised family concept on the basis of molecular data. *Plant Systematics and Evolution, Supplement* 9:273–280.

Jetz, W., G.H. Thomas, J.B. Joy, K. Hartmann, and A.O. Mooers. 2012. The global diversity of birds in space and time. *Nature* 491:444–448.

Jian, S., P.S. Soltis, M.A. Gitzendanner, M.J. Moore, R. Li, T.A. Hendry, Y.-L. Qiu, A. Dhingra, C.D. Bell, and D.E. Soltis. 2008. Resolving an ancient, rapid radiation in Saxifragales. *Systematic Biology* 57:38–57.

Jiang, Z., and R. Zhou. 1992. Distribution of the iridoid compounds in the Hamamelidae. *Zhongguo Yaoke Daxue Xuebao* 23:140–143.

Jiao, Y., J. Leebens-Mack, S. Ayyampalayam, J.E. Bowers, M.R. McKain, J. McNeal, M. Rolf, D.R. Ruzicka, E. Wafula, N.J. Wickett, X. Wu, Y. Zhang, J. Wang, Y. Zhang, E.J. Carpenter, M.K. Deyholos, T.M. Kutchan, A.S. Chanderbali, P.S. Soltis, D.W. Stevenson, R. McCombie, J.C. Pires, G.K.-S. Wong, D.E. Soltis, and C.W. dePamphilis 2012. A genome triplication associated with early diversification of the core eudicots. *Genome Biology* 13:R3. doi:10.1186/gb-2012-13-1-r3.

Jiao, Y., J. Li, H. Tang, and A.H. Paterson. 2014. Integrated syntenic and phylogenomic analyses reveal an ancient genome duplication in monocots. *The Plant Cell* 26(7):2792–2802.

Jiao, Y., N.J. Wickett, S. Ayyampalayam, A.S. Chanderbali, L. Landherr, P.E. Ralph, L.P. Tomsho, Y. Hu, H. Liang, P.S. Soltis, D.E. Soltis, S.W. Clifton, S.E. Schlarbaum, S.C. Schuster, H. Ma, J. Leebens-Mack, and C.W. dePamphilis. 2011. Ancestral polyploidy in seed plants and angiosperms. *Nature* 473(7345):97–100.

Jobson, R.W., and V.A. Albert. 2002. Molecular rates parallel diversification contrasts between carnivorous plant sister lineages. *Cladistics* 18:127–136.

Jobson, R.W., J. Playford, K.M. Cameron, and V.A. Albert. 2003. Molecular phylogenetics of Lentibulariaceae inferred from plastid *rps*16 intron and *trnL–F* DNA sequences: Implications for character evolution and biogeography. *Systematic Botany* 28:157–171.

Johansen, B., L.B. Pedersen, M. Skipper, and S. Frederiksen. 2002. MADS-box gene evolution–structure and transcription patterns. *Molecular Phylogenetics and Evolution* 23:458–480.

Johanssen, L.B. 2005. Phylogeny of *Orchidantha* (Lowiaceae) and the Zingiberales based on six DNA regions. *Systematic Botany* 30:106–117.

Johansson, J.T. 1995. A revised chloroplast DNA phylogeny of the Ranunculaceae. *Plant Systematics and Evolution, Supplement* 9:253–261.

Johansson, J.T., and R.K. Jansen. 1993. Chloroplast DNA variation and phylogeny of the Ranunculiflorae. *Plant Systematics and Evolution* 187:29–49.

Johnson, C.A.S., and B.G. Briggs. 1984. Myrtales and Myrtaceae—a phylogenetic analysis. *Annals of the Missouri Botanical Garden* 71:700–756.

Johnson, K.R. 1996. Description of seven common plant megafossils from the Hell Creek Formation (Late Cretaceous: late Maastrichtian), North Dakota, South Dakota, and Montana. *Proceedings of the Denver Museum of Nature & Science* 3:1–47.

Johnson, L.A., L.M. Chan, T.L. Weese, L.D. Busby, and S. McMurry. 2008. Nuclear and cpDNA sequences combined provide strong inference of higher phylogenetic relationships in the phlox family (Polemoniaceae). *Molecular Phylogenetics and Evolution* 48:997–1012.

Johnson, L.A., J.L. Schultz, D.E. Soltis, and P.S. Soltis. 1996. Monophyly and generic relationships of Polemoniaceae based on *matK* sequences. *American Journal of Botany* 83:1207–1224.

Johnson, L.A., and D.E. Soltis. 1995. Phylogenetic inference using *matK* sequences. *Annals of the Missouri Botanical Garden* 82:149–175.

Johnson, L.A., D.E. Soltis, and P.S. Soltis. 1999. Phylogenetic relationships of Polemoniaceae inferred from 18S ribosomal DNA sequences. *Plant Systematics and Evolution* 214:65–89.

Johnson, N.C., J.H. Graham, and F.A. Smith. 1997. Functioning of mycorrhizal associations along the mutualism-parasitism continuum. *New Phytologist* 135(4):575–586.

Johnson, S.D., N. Hobbhahn, and B. Bytebier. 2013. Ancestral deceit and labile evolution of nectar production in the African orchid genus *Disa*. *Biology Letters* 9. doi:10.1098/rsbl.2013.0500.

Johnson, S.D., H.P. Linder, and K.E. Steiner. 1998. Phylogeny and radiation of pollination systems in *Disa* (Orchidaceae). *American Journal of Botany* 85:402–411.

Johnson, S.D., and K. Steiner. 2000. Generalization versus specialization in plant pollination systems. *Trends Ecology and Evolution* 15:140–143.

Johnston, J.S., A.E. Pepper, A.E. Hall, Z.J. Chen, G. Hodnett, J. Drabek, R. Lopez, and H.J. Price. 2005. Evolution of genome size in Brassicaceae. *Annals of Botany* 95(1):229–235.

Johri, B.M., K.B. Ambegaokar, and S. Srivastava. 1992. Comparative embryology of Angiosperms, 2 vols. Springer-Verlag, Berlin, Germany.

Jones, E., D.A. Simpson, T.R. Hodkinson, M.W. Chase, and J.A.N. Parnell. 2007. The Juncaceae-Cyperaceae interface: A combined plastid sequence analysis. *In* J.T. Columbus, E.A. Friar, J.M. Porter, L.M. Prince, and M G. Simpson [eds.], Monocots: Comparative Biology and Evolution-Poales, 55–61. Rancho Santa Ana Botanic Garden, Claremont, CA.

Jones, J.H. 1986. Evolution of the Fagaceae: The implications of foliar features. *Annals of the Missouri Botanical Garden* 73:228–275.

Jones, S.S., S.V. Burke, and M.R. Duvall. 2014. Phylogenomics, molecular evolution, and estimated ages of lineages from the deep phylogeny of Poaceae. *Plant Systematics and Evolution* 300:1421–1436.

Jordan, G.J. and M.K. Macphail. 2003. A Middle-Late Eocene inflorescence of Caryophyllaceae from Tasmania, Australia. *American Journal of Botany* 90: 761–768.

Jud, N.A., and L.J. Hickey. 2013. *Potomacapnos apeleutheron* gen. et sp nov., a new Early Cretaceous Angiosperm from the Potomac Group and its implications for the evolution of eudicot leaf architecture. *American Journal of Botany* 100(12):2437–2449.

Judd, W.S. 1998. The Smilacaceae in the southeastern United States. *Harvard Papers in Botany* 3:147–169.

Judd, W.S., C.S. Campbell, E.A. Kellogg, and P.F. Stevens. 1999. Plant Systematics: A Phylogenetic Approach. Sinauer, Sunderland, MA.

Judd, W.S., C.S. Campbell, E.A. Kellogg, P.F. Stevens, and M.J. Donoghue. 2002. Plant Systematics: A Phylogenetic Approach, 2nd ed. Sinauer, Sunderland, MA.

———. 2008. Plant Systematics: A Phylogenetic Approach, 3rd ed. Sinauer, Sunderland, MA.

Judd, W.S., and K.A. Kron. 1993. Circumscription of Ericaceae (Ericales) as determined by preliminary cladistic analyses based on morphological, anatomical and embryological features. *Brittonia* 45:99–114.

Judd, W.S., R.W. Kron, and M.J. Donoghue. 1994. Angiosperm family pairs: Preliminary cladistic analyses. *Harvard Papers in Botany* 5:1–51.

Judd, W.S., and S.R. Manchester. 1997. Circumscription of Malvaceae (Malvales) as determined by a preliminary cladistic analysis of morphological, anatomical, palynological, and chemical characters. *Brittonia* 49:384–405.

Judd, W. S., and R.G. Olmstead. 2004. a survey of tricolpate (eudicot) phylogenetic relationships. *American Journal of Botany* 91(10):1627–1644.

Judd, W.S., D.E. Soltis, and P.S. Soltis. 2013a. Malvidae. *In* K. de Queiroz, P.D. Cantino, and J. Gauthier [eds.], Phylonyms: A Companion to the PhyloCode. University of California Press, Berkeley, CA.

———. 2013b. Rosidae. *In* K. de Queiroz, P.D. Cantino, and J. Gauthier [eds.], Phylonyms: A Companion to the PhyloCode. University of California Press, Berkeley, CA.

———. 2013c. Fabidae. *In* K. de Queiroz, P.D. Cantino, and J. Gauthier [eds.], Phylonyms: A Companion to the PhyloCode. University of California Press, Berkeley, CA.

Juniper, B.E., R.J. Robins, and D.M. Joel. 1989. The Carnivorous Plants. Academic Press, London, UK.

Jussieu, A.L. de. 1789. Genera plantarum secundum ordines naturales disposita. Herissant and Barrios, Paris, France.

Kadereit, G., T. Borsch, K. Weising, and H. Freitag. 2003. Phylogeny of Amaranthaceae and Chenopodiaceae and the evolution of C_4 photosynthesis. *International Journal of Plant Sciences* 164:959–986.

Kadereit, G., M. Lauterbach, M.D. Pirie, R. Arafeh, and H. Freitag. 2014. When do different C4 leaf anatomies indicate independent C4 origins? Parallel evolution of C4 leaf types in Camphorosmeae (Chenopodiaceae). *Journal of Experimental Botany* 65(13):3499–3511.

Kadereit, J.W. 1993. Papaveraceae. *In* K. Kubitzki, J. Rohwer, and V. Bittrich [eds.], The Families and Genera of Vascular Plants, 2, 494–506. Springer-Verlag, Berlin. Germany.

Kadereit, J.W., F.R. Blattner, K.B. Jork, and A.E. Schwarzbach. 1994. Phylogenetic analysis of the Papaveraceae s.1. (incl. Fumariaceae, Hypecoaceae and *Pteridophyllum*) based on morphological characters. *Botanische Jahrbücher für Systematik* 116:361–390.

———. 1995. The phylogeny of the Papaveraceae sensu lato: Morphological, geographical and ecological implications. *Plant Systematics and Evolution, Supplement* 9:133–145.

Kadereit, J.W., and C. Erbar. 2011. Evolution of gynoecium morphology in Old World Papaveroideae: A combined phylogenetic/ontogenetic approach. *American Journal of Botany* 98:1243–1251.

Kadereit, J.W., C.D. Preston, and F.J. Valtuena. 2011. Is welsh poppy, *Meconopsis cambrica* (L.) Vig. (Papaveraceae), truly a *Meconopsis*? *New Journal of Botany* 1(2):80–87.

Kadereit, J.W., A.E. Schwarzbach, and K.B. Jork. 1997. The phylogeny of *Papaver* s.l. (Papaveraceae): Polyphyly or monophyly? *Plant Systematics and Evolution* 204(1–2):75–98.

Kadereit, J.W., and K.J. Sytsma. 1992. Disassembling

Papaver: A restriction site analysis of chloroplast DNA. *Nordic Journal of Botany* 12:205–217.

Kajita, T., H. Ohashi, Y. Tateishi, C.D. Bailey, and J.J. Doyle. 2001. *RbcL* and legume phylogeny, with particular reference to Phaseoleae, Millettieae, and allies. *Systematic Botany* 26:515–536.

Källersjö, M., V.A. Albert, and J.S. Farris. 1999. Homoplasy increases phylogenetic structure. *Cladistics* 15: 91–93.

Källersjö, M., G. Bergqvist, and A.A. Anderberg. 2000. Generic realignment in primuloid families of the Ericales s.1.: A phylogenetic analysis based on DNA sequences from three chloroplast genes and morphology. *American Journal of Botany* 87:1325–1341.

Källersjö, M., J.S. Farris, M. Chase, B. Bremer, M.F. Fay, C.J. Humphries, G. Peterson, O. Seberg, and K. Bremer. 1998. Simultaneous parsimony jackknife analysis of 2538 *rbcL* DNA sequences reveals support for major clades of green plants, land plants, seed plants, and flowering plants. *Plant Systematics and Evolution* 213:259–287.

Källersjö, M., J.S. Farris, A.G. Kluge, and C. Bult. 1992. Skewness and permutation. *Cladistics* 8:275–287.

Kamelina, O.P. 1984. On the embryology of the genus *Escallonia* (Escalloniaceae). *Botanicheskii Zhurnal* 69:1304–1316.

Kane, N.C., N. Gill, M.G. King, J.E. Bowers, H. Berges, J. Gouzy, E. Bachlava, N.B. Langlade, Z. Lai, M. Stewart, J.M. Burke, P. Vincourt, S.J. Knapp, and L.H. Rieseberg. 2011. Progress towards a reference genome for sunflower. *Botany-Botanique* 89:429–437.

Kanno, A., H. Saeki, T. Kameya, H. Saedler, and G. Theissen. 2003. Heterotopic expression of class B floral homeotic genes supports a modified ABC model for tulip (*Tulipa gesneriana*). *Plant Molecular Biology* 52:831–841.

Kapil, R.N., and P.R. Mohana Rao. 1966. Studies on the Garryaceae. II. Embryology and systematic position of *Garrya* Douglas ex Lindley. *Phytomorphology* 16: 564–578.

Kaplan, D.R. 1967. Floral morphology, organogenesis and interpretation of the inferior ovary in *Downingia bacigalupii*. *American Journal of Botany* 54:1274–1290.

Kaplan, M.A.C., and O.R. Gottlieb. 1982. Iridoids as systematic markers in dicotyledons. *Biochemical Systematics and Ecology* 10:239–347.

Kaplan, M.A.C., J. Ribeiro, and O.R. Gottlieb. 1991. Chemogeographical evolution of terpenoids in Icacinaceae. *Phytochemistry* 30:2671–2676.

Kårehed, J. 2001. Multiple origins of the tropical forest tree family Icacinaceae. *American Journal of Botany* 88:2259–2274.

———. 2002. Evolutionary studies in asterids emphasising euasterids. Acta Universitatis Upsaliensis, Uppsala, Sweden.

Kårehed, J., J. Lundberg, B. Bremer, and K. Bremer. 1999. Evolution of the Australasian families Alseuosmiaceae, Argophyllaceae and Phellinaceae. *Systematic Botany* 24:660–682.

Karol, K.G., Y. Suh, G.E. Schatz, and E. Zimmer. 2000. Molecular evidence for the phylogenetic position of *Takhtajania* in the Winteraceae: Inference from nuclear ribosomal and chloroplast gene spacer sequences. *Annals of the Missouri Botanical Garden* 87:414–432.

Karoly, K., and J.K. Conner. 2000. Heritable variation in a family-diagnostic trait. *Evolution* 54:1433–1438.

Kathriarachchi, H., P. Hoffmann, R. Samuel, K.J. Wurdack, and M.W. Chase. 2005. Molecular phylogenetics of Phyllanthaceae inferred from five genes (plastid *atpB*, *matK*, 3 '*ndhF*, *rbcL*, and nuclear *PHYC*). *Molecular Phylogenetics and Evolution* 36(1):112–134.

Kato, M. 1990. Ophioglossaceae: A hypothetical archetype for the angiosperm carpel. *Botanical Journal of the Linnean Society* 102:303–311.

Kawakita, A., and M. Kato. 2009. Repeated independent evolution of obligate pollination mutualism in the Phyllantheae-*Epicephala* association. *Proceedings of the Royal Society of London B* 276:417–426.

Kawakita, A., A. Takimura, T. Terachi, T. Sota, and M. Kato. 2004. Cospeciation analysis of an obligate pollination mutualism: Have *Glochidion* trees (Euphorbiaceae) and pollinating *Epicephala* moths (Gracillariidae) diversified in parallel? *Evolution* 58(10): 2201–2214.

Kazazian, H.H., Jr. 2004. Mobile elements: Drivers of genome evolution. *Science* 303:1626–1632.

Keating, R. 1973. Pollen morphology and relationships of Flacourtiaceae. *Annals of the Missouri Botanical Garden* 60:273–305.

———. 2000. Anatomy of the young vegetative shoot of *Takhtajania perrieri* (Winteraceae). *Annals of the Missouri Botanical Garden* 87:335–346.

Keck, E., P. McSteen, R. Carpenter, and E. Coen. 2003. Separation of genetic functions controlling organ identity in flowers. *Embo Journal* 22(5):1058–1066.

Keefe, J.M., and J.M.F. Moseley. 1978. Wood anatomy and phylogeny of *Paeonia* section *Moutan*. *Journal of the Arnold Arboretum* 59:274–297.

Keeley, S.C., Z.H. Forsman, and R. Chan. 2007. A phylogeny of the "evil tribe" (Vernonieae: Compositae) reveals Old/New World long distance dispersal: Support from separate and combined congruent datasets (*trnL–F*, *ndhF*, ITS). *Molecular Phylogenetics and Evolution* 44:89–103.

Keeling, P.J., and J.D. Palmer. 2008. Horizontal gene transfer in eukaryotic evolution. *Nature Reviews Genetics* 9(8):605–618.

Keller, J.A., P.S. Herendeen, and P.R. Crane. 1996. Fossil flowers of the Actinidiaceae from the Campanian (Late Cretaceous) of Georgia. *American Journal of Botany* 83:528–541.

Kelley, D.R., and C.S. Gasser. 2009. Ovule development: Genetic trends and evolutionary considerations. *Sexual Plant Reproduction* 22:229–234.

Kellogg, E.A. 1998. Relationships of cereal crops and other grasses. *Proceedings of the National Academy of Sciences USA* 95:2005–2010.

———. 1999. Phylogenetic aspects of the evolution of C_4 photosynthesis. *In* R.F. Sage, and R.K. Monson [eds.], C_4 Plant Biology, 411–444. Academic Press, New York, NY.

———. 2000. The grasses: A case study in macroevolution. *Annual Reviews in Ecology and Systematics* 31:217–238.

———. 2001. Evolutionary history of the grasses. *Plant Physiology* 125:1198–1205.

———. 2002. Are macroevolution and microevolution qualitatively different? *In* Q.C.B. Cronk, R.M. Bateman, and J.A. Hawkins [eds.], Developmental Genetics and Plant Evolution, 70–84. Taylor and Francis, London, UK.

———. 2003. It's all relative. *Nature* 422:383–384.

Kellogg, E.A., and J.L. Bennetzen. 2004. The evolution of nuclear genome structure in plants. *American Journal of Botany* 91(10):1709–1725.

Kellogg, E.A., and N.D. Juliano. 1997. The structure and function of RUBISCO and their implications for systematic studies. *American Journal of Botany* 84:413–428.

Kellogg, E.A., and H.P. Linder. 1995. Phylogeny of Poales. *In* P.J. Rudall, P.J. Cribb, D.F. Cutler, and C.J. Humphries [eds.], Monocotyledons: Systematics and Evolution, 511–542. Royal Botanic Gardens, Kew, UK.

Kelly, L.M., and F. Gonzalez. 2003. Phylogenetic relationships in Aristolochiaceae. *Systematic Botany* 28: 236–249.

Kenrick, P., and P.R. Crane. 1997. The Origin and Early Diversification of Land Plants. Smithsonian Institution Press, Washington, D.C.

Kidwell, M.G. 2002. Transposable elements and the evolution of genome size in eukaryotes. *Genetica* 115:49–63.

Kierzkowski, D., N. Nakayama, A.-L. Routier-Kierzkowska, A. Weber, E. Bayer, M. Schorderet, D. Reinhardt, C. Kuhlemeier, and R.S. Smith. 2012. Elastic domains regulate growth and organogenesis in the plant shoot apical meristem. *Science* 335:1096–1099.

Kim, D.K., J.H. Kim. 2011. Molecular phylogeny of tribe *Forsythieae* (Oleaceae) based on nuclear ribosomal DNA internal transcribed spacers and plastid DNA *trnl-F* and *matK* gene sequences. *Journal of Plant Resources* 124:339–347.

Kim, J.-H., D.-G. Kim, F. Forest, M.F. Fay, and M.W. Chase. 2010. Molecular phylogenetics of Ruscaceae sensu lato and related families (Asparagales) based on plastid and nuclear DNA sequences. *Annals of Botany* 106:775–790.

Kim, J.S., J.-K. Hong, M.W. Chase, M.F. Fay, and J.-H. Kim. 2013. Familial relationships of the monocot order Liliales based on a molecular phylogenetic analysis using four plastid loci: *matK*, *rbcL*, *atpB* and *atpF-H*. *Botanical Journal of the Linnean Society* 172:5–21.

Kim, K.-J., and R.K. Jansen. 1995. *ndhF* sequence evolution and the major clades in the sunflower family. *Proceedings of the National Academy of Sciences USA* 92:10379–10383.

———. 1998. Chloroplast DNA restriction site variation and phylogeny of the Berberidaceae. *American Journal of Botany* 85:1766–1778.

Kim, K.-J., R.K. Jansen, and R.G. Olmstead. 1994. Multiple origins of sympetaly in dicots. *American Journal of Botany* 81(6, Supplement):165. Abstract.

Kim, K.-J., R.K. Jansen, R.S. Wallace, H.J. Michaels, and J.D. Palmer. 1992. Phylogenetic implications of *rbcL* sequence variation in the Asteraceae. *Annals of the Missouri Botanical Garden* 79:428–445.

Kim, S., V.A. Albert, M.-J. Yoo, J.S. Farris, M. Zanis, P.S. Soltis, and D.E. Soltis. 2004. Pre-angiosperm duplication of floral genes and regulatory tinkering at the base of flowering plants. *American Journal of Botany* 91:2102–2118.

Kim, S., J. Koh, M.J. Yoo, H.Z. Kong, Y. Hu, H. Ma, P.S. Soltis, and D.E. Soltis. 2005. Expression of floral MADS-box genes in basal angiosperms: Implications for the evolution of floral regulators. *Plant Journal* 43(5):724–744.

Kim, S., C.-W. Park, Y.-D. Kim, and Y. Suh. 2001. Phylogenetic relationships in family Magnoliaceae inferred from *ndhF* sequences. *American Journal of Botany* 88:717–728.

Kim, S., D.E. Soltis, P.S. Soltis, M.J. Zanis, and Y. Suh. 2004. Phylogenetic relationships among early-diverging eudicots based on four genes: Were the eudicots ancestrally woody? *Molecular Phylogenetics and Evolution* 31:16–30.

Kim, S., and Y. Suh. 2013. Phylogeny of Magnoliaceae based on ten chloroplast DNA regions. *Journal of Plant Biology* 56:290–305.

Kim, S.-C, J.S. Kim, M.W. Chase, M.F. Fay, and J.-H.

Kim. In Press. Molecular phylogentic relationships and circumscription of Melanthiaceae (Liliales). *Botanical Journal of the Linnean Society.*

Kim, S.-T., and M.J. Donoghue. 2008. Molecular Phylogeny of *Persicaria* (Polygonaceae). *Systematic Botany* 33:77–86.

Kim, Y.-D., and R.K. Jansen. 1995. Phylogenetic implications of chloroplast DNA variation in the Berberidaceae. *Plant Systematics and Evolution, Supplement* 9:341–349.

Kirchheimer, F. 1957. Die Laubgewächse der Braunkohlenzeit. Veb Wilhelm Knapp Verlag, Halle (Salle). 783 pp.

Kirik, A., S. Salomon, and H. Puchta. 2000. Species-specific double-strand break repair and genome evolution in plants. *EMBO Journal* 19:5562–5566.

Kishino, H., and M. Hasegawa. 1989. Evaluation of the maximum likelihood estimate of the evolutionary tree topologies from DNA sequence data, and the branching order in hominoidea. *Journal of Molecular Evolution* 29:170–179.

Kishino, H., J.L. Thorne, and W.J. Bruno. 2001. Performance of a divergence time estimation method under a probabilistic model of rate evolution. *Molecular Biology and Evolution* 18:352–361.

Kjaer, A. 1973. The natural distribution of glucosinolates: A uniform group of sulfur containing glucosides. *In* G. Bendz, and J. Santesson [ed.], Chemistry in Botanical Classification, 229–234. Academic Press, New York, NY.

Klak, C., P.V. Bruyns, and T.A.J. Hedderson. 2007. A phylogeny and new classification for Mesembryanthemoideae. *Taxon* 56:737–756.

Klak, C., A. Khunou, G. Reeves, and T. Hedderson. 2003. A phylogenetic hypothesis for the Aizoaceae (Caryophyllales) based on four plastid DNA regions. *American Journal of Botany* 90(10):1433–1445.

Klavins, S.D., T.N. Taylor, and E.L. Taylor. 2002. Anatomy of *Umkomasia* (Corystospermales) from the Triassic of Antarctica. *American Journal of Botany* 89:664–676.

Klopfer, K. 1973. Florale Morphogenese und Taxonomie der Saxifragaceae sensu lato. *Feddes Repertorium* 84:475–516.

Knapp, M., K. Stockler, D. Havell, F. Delsuc, F. Sebastiani, and P.J. Lockhart. 2005. Relaxed molecular clock provides evidence for long-distance dispersal of *Nothofagus* (southern beech). *PLoS Biology* 3(1):38–34. e14. doi:10.1371/journal.pbio.0030014.

Knapp, S. 2002. *Solanum* section *Geminata* (G. Don) Walpers (Solanaceae). *Flora Neotropica* 84:1–405.

———. 2010. On 'various contrivances': Pollination, phylogeny and flower form in the Solanaceae. *Philosophical Transactions of the Royal Society of London B* 1539:449–460.

———. 2013. A revision of the Dulcamaroid clade of *Solanum* L. (Solanaceae). *Phytokeys* 22:1–428.

Kneip, C., P. Lockhart, C. Voss, and U.-G. Maier. 2007. Nitrogen fixation in eukaryotes—new models for symbiosis. *BMC Evolutionary Biology* 7:55. doi:10.1186/1471-2148-7-55.

Knobloch, E., and D.-H. Mai. 1986. Monographie der Früchte und Samen in der Kreide von Mitteleuropa. Academia, Praha, Czech Republic.

Knoop, V., U. Volkmar, J. Hecht, and F. Grewe. 2011. Mitochondrial genome evolution in the plant lineage. *Plant Mitochondria* 1:3–29.

Koch, M. 2003. Molecular phylogenetics, evolution and population biology in the Brassicaceae. *In* V.K. Sharma, and A. Sharma [eds.], Plant Genome: Biodiversity and Evolution. Science Publishers, Inc., Enfield, NH.

Koch, M., I.A. Al-Shehbaz, and K. Mummenhoff. 2003. Molecular systematics, evolution, and population biology in the mustard family (Brassicaceae). *Annals of the Missouri Botanical Garden* 90(2):151–171.

Koch, M., J. Bishop, and T. Mitchell-Olds. 1999. Molecular systematics and evolution of *Arabidopsis* and *Arabis*. *Plant Biology* 1:529–537.

Koch, M., B. Haubold, and T. Mitchell-Olds. 2001. Molecular systematics of the Brassicaceae: Evidence from coding plastid *matK* and nuclear *Chs* sequences. *American Journal of Botany* 88(3):534–544.

Koch, M., and M. Kiefer. 2005. Genome evolution among cruciferous plants: A lecture from the comparison of the genetic maps of three diploid species—*Capsella rubella*, *Arabidopsis lyrata* subsp. *petraea*, and *A. thaliana*. *American Journal of Botany* 92(4):761–767.

Koch, M., K. Mummenhoff, and I.A. Al–Shehbaz. 2003. Molecular systematics, evolution, and population biology in the mustard family: A review of a decade of studies. *Annals of the Missouri Botanical Garden* 90:151–171.

Koes, R.E., C.E. Spelt, and J.N. Mol. 1989. The chalcone synthase multigene family of *Petunia hybrid* (V30): Differential, light-regulated expression during flower development and UV light induction. *Plant Molecular Biology* 12:213–225.

Koes, R.E., W. Verweij, and F. Quattrocchio. 2005. Flavonoids: A colorful model for the regulation and evolution of biochemical pathways. *Trends in Plant Sciences* 10:236–242.

Kofuji, R., K. Ueda, K. Yamaguchi, and T. Shimizu. 1994. Molecular phylogeny in the Lardizabalaceae. *Journal of Plant Research* 107(1087):339–348.

Koi, S., Y. Kita, Y. Hirayama, R. Rutishauser, K.A. Huber, and M. Kato. 2012. Molecular phylogenetic analysis of Podostemaceae: Implications for taxonomy of major groups. *Botanical Journal of the Linnean Society* 169(3):461–492.

Kolpalova, M.V., and D.M. Popov. 1994. Study of the amounts of iridoids in *Paeonia anomala* L. (Paeoniaceae). *Khimiko-Farmatsevticheskii Zhurnal* 28:24–26.

Kong, H.-Z., and Z. Chen. 2000. Phylogeny of *Chloranthus* (Chloranthaceae) inferred from sequence analysis of nrDNA ITS region. *Acta Botanica Sinica* 42:762–764.

Kong, H.-Z., Z. Chen, and A.-M. Lu. 2002a. Phylogeny of *Chloranthus* (Chloranthaceae) based on nuclear ribosomal ITS and plastid *trnL-F* sequence data. *American Journal of Botany* 89:940–946.

Kong, H.-Z., A.-M. Lu, and P.K. Endress. 2002b. Floral organogenesis of *Chloranthus sessilifolius*, with special emphasis on the morphological nature of the androecium of *Chloranthus* (Chloranthaceae). *Plant Systematics and Evolution* 232:181–188.

Koontz, J.A., and D.E. Soltis. 1999. DNA sequence data reveal polyphyly of Brexioideae (Brexiaceae; Saxifragaceae sensu lato). *Plant Systematics and Evolution* 219:199–208.

Kopriva, S., C.-C. Chu, and H. Bauwe. 1996. Molecular phylogeny of *Flaveria* as deduced from the analysis of nucleotide sequences encoding the H-protein of the glycine cleavage system. *Plant Cell and Environment* 19:1028–1036.

Kores, P.J., P.H. Weston, M. Molvray, and M.W. Chase. 2000. Phylogenetics relationships within the Diurideae (Orchidaceae): Inferences from plastid *matK* DNA sequences. *In* K.L. Wilson, and D.A. Morrison [eds.], Monocots: Systematics and Evolution, 449–456. CSIRO Publishing, Collingwood, Victoria, Australia.

Korotkova, N., J.V. Schneider, D. Quandt, A. Worberg, G. Zizka, and T. Borsch. 2009. Phylogeny of the eudicot order Malpighiales: Analysis of a recalcitrant clade with sequences of the *petD* group II intron. *Plant Systematics and Evolution* 282(3–4):201–228.

Korpelainen, H. 1998. Labile sex expression in plants. *Biological Review* 73:157–180.

Kostecka-Madalska, O., and A. Rymkiewick. 1971. Further research for plants containing aucubin. *Farmacia Polonica* 27:899–903.

Kosuge, K. 1994. Petal evolution in Ranunculaceae. *Plant Systematics and Evolution, Supplement* 8:185–191.

Kosuge, K., K. Mistunaga, K. Loike, and T. Ohmoto. 1994. Studies on the constituents of *Ailanthus integrifolia*. *Chemical and Pharmaceutical Bulletin* 42: 1669–1671.

Kosuge, K., F.-D. Pu, and M. Tamura. 1989. Floral morphology and relationships of *Kingdonia*. *Acta Phytotaxonomica Geobotanica* 40:61–67.

Kosuge, K., and M. Tamura. 1989. Ontogenetic studies on petals of the Ranunculaceae. *Journal of Japanese Botany* 64:65–67.

Kotilainen, M., P. Elomaa, A. Uimari, V. Albert, D. Yu, and T.H. Teeri. 2000. *GRCD1*, an *AGL2*-like MADS box gene, participates in the C function during stamen development in *Gerbera hybrida*. *Plant Cell* 12:1893–1902.

Kral, R.B. 1983. The Xyridaceae in the southeastern United States. *Journal of the Arnold Arboretum* 64:421–429.

Kramer, E.M. 2009. *Aquilegia*: A new model for plant development, ecology, and evolution. *Annual Review of Plant Biology* 60:261–277.

Kramer, E.M., V.S. Di Stilio, and P.M. Schlüter. 2003. Complex patterns of gene duplication in the *APETALA3* and *PISTILLATA* lineages of the Ranunculaceae. *International Journal of Plant Sciences* 164:1–11.

Kramer, E.M., R.L. Dorit, and V.F. Irish. 1999. Molecular evolution of genes controlling petal and stamen development: Duplication and divergence within the *APETALA3* and *PISTILLATA* MADS-box gene lineages. *Genetics* 149:765–783.

Kramer, E.M., and S.A. Hodges. 2010. *Aquilegia* as a model system for the evolution and ecology of petals. *Philosophical Transactions of the Royal Society of London B* 365(1539):477–490.

Kramer, E.M., L. Holappa, B. Gould, M.A. Jaramillo, D. Setnikov, and P.M. Santiago. 2007. Elaboration of B gene function to include the identity of novel floral organs in the lower eudicot *Aquilegia*. *Plant Cell* 19(3):750–766.

Kramer, E.M., and V.F. Irish. 1999. Evolution of genetic mechanisms controlling petal development. *Nature* 399(6732):144–148.

———. 2000. Evolution of the petal and stamen developmental programs: Evidence from comparative studies of the lower eudicots and basal angiosperms. *International Journal of Plant Sciences* 161:S29–S40.

Kramer, E.M., M.A. Jaramillo, and V. Di Stilio. 2004. Patterns of gene duplication and functional evolution during the diversification of the *AGAMOUS* subfamily of MADS box genes in angiosperms. *Genetics* 166:1011–1023.

Kramer, E.M., and E.A. Zimmer. 2006. Gene duplication and floral developmental genetics of basal eudicots. Advances in Botanical Research: Incorporating Advances in Plant Pathology. *Developmental Genetics of the Flower* 44:353–384.

Kraus, F., and M.M. Miyamoto. 1991. Rapid cladogenesis among the pecoran ruminants—evidence from mitochondrial DNA sequences. *Systematic Zoology* 40:117–130.

Kraus R., P. Trimborn, and H. Ziegler. 1995. *Tristerix aphyllus*, a holoparasitic Loranthacea. *Naturwissenschaften* 82:150–151l.

Kress, W.J. 1990. The phylogeny and classification of the Zingiberales. *Annals of the Missouri Botanical Garden* 77:698–721.

Kress, W.J., L.M. Prince, W.J. Hahn, and E.A. Zimmer. 2001. Unraveling the evolutionary radiation of the families of the Zingiberales using morphological and molecular evidence. *Systematic Biology* 50:926–944.

Krol, E., B.J. Plachno, L. Adamec, M. Stolarz, H. Dziubinska, and K. Trebacz. 2012. Quite a few reasons for calling carnivores 'the most wonderful plants in the world'. *Annals of Botany* 109(1):47–64.

Kron, K.A. 1996. Phylogenetic relationships of Empetraceae, Epacridaceae, Ericaceae, Monotropaceae, and Pyrolaceae: Evidence from nucleotide ribosomal 18S sequence data. *Annals of Botany* 77:293–303.

———. 1997. Exploring alternative systems of classification. *Aliso* 15:105–112.

Kron, K.A., and M.W. Chase. 1993. Systematics of the Ericaceae, Empetraceae, Epacridaceae and related taxa based on *rbcL* sequence data. *Annals of the Missouri Botanical Garden* 80:735–741.

Kron, K.A., and W.S. Judd. 1997. Systematics of the *Lyonia* group (Andromedeae, Ericacae) and the use of species of terminals in higher-level cladistic analyses. *Systematic Botany* 22:479–492.

Kron, K.A., W.S. Judd, and D.M. Crayn. 1999. Phylogenetic relationships of Andromedeae (Ericaceae subfam. Vaccinioideae). *American Journal of Botany* 86: 1290–1300.

Kron, K.A., W.S. Judd, P.F. Stevens, D.M. Crayn, A.A. Anderberg, P.A. Gadek, C.J. Quinn, and J.L. Luteyn. 2002. Phylogenetic classification of Ericaceae: Molecular and morphological evidence. *Botanical Review* 68:335–423.

Kron, K.A., and J.L. Luteyn. 2005. Origins and biogeographic patterns in Ericaceae: New insights from recent phylogenetic analyses. *Biologiske Skrifter* 55:479–500.

Kubitzki, K. 1987. Origin and significance of trimerous flowers. *Taxon* 36:21–28.

———. 1993. Myrothamnaceae. *In* K. Kubitzki, O. Rohwer, and V. Bittrich [eds.], The Families and Genera of Vascular Plants, II, 468–469. Springer-Verlag, Berlin, Germany.

Kubitzki, K., J. Rohwer, and V. Bittrich [eds.]. 1993. The Families and Genera of Vascular Plants, II. Springer-Verlag, Berlin, Germany.

Kubitzki, K., P. Von Sengbusch, and H.-H. Poppendiek. 1991. Parallelism, its evolutionary origin and systematic significance. *Aliso* 13:191–206.

Kuijt, J. 1969. The Biology of Parasitic Flowering Plants. University of California Press, Berkeley, CA.

———. 1982. The Viscaceae in the southeastern United States. *Journal of the Arnold Arboretum* 63:401–410.

Kuzoff, R.K., and C.S. Gasser. 2000. Recent progress in reconstructing angiosperm phylogeny. *Trends in Plant Science* 5:330–336.

Kuzoff, R.K., L. Hufford, and D.E. Soltis. 2001. Structural homology and developmental transformations associated with ovary diversification in *Lithophragma* (Saxifragaceae). *American Journal of Botany* 88:196–205.

Kuzoff, R.K., D.E. Soltis, L. Hufford, and P.S. Soltis. 1999. Phylogenetic relationships within *Lithophragma* (Saxifragaceae): Hybridization, allopolyploidy, and ovary diversification. *Systematic Botany* 24:598–615.

Kuzoff, R.K., J. Sweere, D. Soltis, P. Soltis, and E. Zimmer. 1998. The phylogenetic potential of entire 26S rDNA sequences in plants. *Molecular Biology and Evolution* 15:251–263.

Kvaček, Z., S.R. Manchester, and M. Akhmetiev. 2005. Review of the fossil history of *Craigia* (Malvaceae s.l.) in the Northern Hemisphere based on fruits and co-occurring foliage. *In* M.A. Akhmetiev, and A.B. Herman [eds.], Modern Problems of Palaeofloristics, Palaeophytogeography and Phytostratigraphy, 114–140. GEOS, Moscow, Russia.

Kvaček, Z., S.R. Manchester, R. Zetter, and M. Pingen. 2002. Fruits and seeds of *Craigia bronnii* (Malvaceae-Tilioideae) and associated flower buds from the late Miocene Inden Formation, Lower Rhine Basin, Germany. *Review of Palaeobotany and Palynology* 119:311–324.

Kynast, R.G., J.A. Joseph, J. Pellicer, M.M. Ramsay, and P.J. Rudall. 2014. Chromosome behavior at the base of the angiosperm radiation: Karyology of *Trithuria submersa* (Hydatellaceae, Nymphaeales). *American Journal of Botany* 101(9):1447–1455.

Lacroix, C., and R. Sattler. 1988. Phyllotaxis theories and tepal-stamen superposition in *Basella rubra*. *American Journal of Botany* 75:906–917.

Lagercrantz, U. 1998. Comparative mapping between *Arabidopsis thaliana* and *Brassica nigra* indicates that *Brassica* genomes have evolved through extensive genome replication accompanied by chromosome fusions and frequent rearrangements. *Genetics* 150:1217–1228.

Lagercrantz, U., and D.J. Lydiate. 1996. Comparative genome mapping in *Brassica*. *Genetics* 144:1903–1910.

Lagomarsino, L.P., A. Antonelli, N. Muchhala, A. Timmermann, S. Mathews, and C.C. Davis. 2014. Phylogeny, classification, and fruit evolution of the

species-rich Neotropical bellflowers (Campanulaceae: Lobelioideae). *American Journal of Botany* 101:2097–2112.

Lammel, G., and H. Rimpler. 1981. Iridoids in *Clerodendrum thomsonae*, Verbenaceae. *Zeitschrift fur Naturforschung C: Bioscience* 36:708–713.

Lammers, T.G. 1992. Circumscription and phylogeny of the Campanulales. *Annals of the Missouri Botanical Garden* 81:388–413.

Lammers, T.G., T.F. Stuessy, and M. Silva. 1986. Systematic relationships of the Lactoridaceae, an endemic family of the Juan Fernandez Islands, Chile. *American Journal of Botany* 152:243–266.

Lancaster, L.T. 2010. Molecular evolutionary rates predict both extinction and speciation in temperate angiosperm lineages. *BMC Evolutionary Biology* 10:162. doi:10.1186/1471-2148-10-162.

Landsmann, J., E.S. Dennis, T.J.V. Higgins, C.A. Appleby, A.A. Kortt, and W.J. Peacock. 1986. Common evolutionary origin of legume and non-legume plant haemoglobins. *Nature* 324:166–168.

Langley, C.H., and W. Fitch. 1974. An estimation of the constancy of the rate of molecular evolution. *Journal of Molecular Evolution* 3:161–177.

Laroche, J., P. Li, L. Maggia, and J. Bousquet. 1997. Molecular evolution of Angiosperm mitochondrial introns and exons. *Proceedings of the National Academy of Sciences USA* 94:5722–5727.

Las Peñas, M.L., J.D. Urdampilleta, G. Bernardello, and E.R. Forni-Martins. 2009. Karyotypes, heterochromatin, and physical mapping of 18S-26S rDNA in Cactaceae. *Cytogenetic Genome Resources* 124:72–80.

Las Peñas, M.L., J.D. Urdampilleta, B. López-Castro, F. Santiñaque, R. Kiesling, and G. Bernardello. 2014. Classical and molecular cytogenetics and DNA content in *Maihuenia* and *Pereskia* (Cactaceae). *Plant Systematics and Evolution* 300:549–558.

Lavin, M., P.S. Herendeen, and M.F. Wojciechowski. 2005. Evolutionary rates analysis of Leguminosae implicates a rapid diversification of lineages during the tertiary. *Systematic Biology* 54(4):575–594.

Lavin, M., B.P. Schrire, G. Lewis, R.T. Pennington, A. Delgado-Salinas, M. Thulin, C.E. Hughes, A.B. Matos, and M.F. Wojciechowski. 2004. Metacommunity process rather than continental tectonic history better explains geographically structured phylogenies in legumes. *Philosophical Transactions of the Royal Society of London B* 359(1450):1509–1522.

Lee, E.K., A. Cibrian-Jaramillo, S.-O. Kolokotronis, M.S. Katari, A. Stamatakis, M. Ott, J.C. Chiu, D.P. Little, D.W. Stevenson, W.R. McCombie, R.A. Martienssen, G. Coruzzi, and R. DeSalle. 2011. A functional phylogenomic view of the Seed Plants. *PLoS Genetics* 7 (12): e1002411. doi:10.1371/journal.pgen.1002411.

Lee, H.S., and Z.J. Chen. 2001. Protein-coding genes are epigenetically regulated in *Arabidopsis* polyploids. *Proceedings of the National Academy of Sciences USA* 98(12):6753–6758.

Lee, I., D.S. Wolfe, S. Nilsson, and D. Weigel. 1997. A LEAFY co-regulator encoded by unusual floral organs. *Current Biology* 7:95–104.

Lee, M.S.Y. 1996. The phylogenetic approach to biological taxonomy: Practical aspects. *Zoologica Scripta* 25:187–190.

Lee, S.L., and J. Wen. 2001. A phylogenetic analysis of *Prunus* and the Amygdaloideae (Rosaceae) using ITS sequences of nuclear ribosomal DNA. *American Journal of Botany* 88:150–160.

Leebens-Mack, J., L.A. Raubeson, L.Y. Cui, J.V. Kuehl, M.H. Fourcade, T.W. Chumley, J.L. Boore, R.K. Jansen, and C.W. dePamphilis. 2005. Identifying the Basal Angiosperm node in chloroplast genome phylogenies: Sampling one's way out of the Felsenstein zone. *Molecular Biology and Evolution* 22:1948–1963.

Legume Phylogeny Working Group [A. Bruneau, J.J. Doyle, P.S. Herendeen, C. Hughes, G. Kenicer, G. Lewis, B. Mackinder, R.T. Pennington, M.J. Sanderson, and M.F. Wojciechowski]. 2013. Legume phylogeny and classification in the 21st century: Progress, prospects and lessons for other species-rich clades. *Taxon* 62: 217–248.

———. 2013. Towards a new classification system for legumes: Progress report from the 6th International Legume Conference. *South African Journal of Botany* 89:3–9.

Lehman, N.L. and R. Sattler. 1993. Homeosis in floral development of *Sanguinaria canadensis* and *S. canadensis* "multiplex" (Papaveraceae). *American Journal of Botany* 80:1323–1335.

Leinfellner, W. 1969. Über die Karpelle verschiedener Magnoliales. VIII. Überblick über alle Familien der Ordnung. *Österreichische Botanische Zeitschrift* 117:107–127.

Leins, P. 1972. Das Karpell im ober–und unterständigen Gynoeceum. *Berichte der Deutschen Botanischen Gesellschaft* 85:291–294.

———. 2000. Blüte und Frucht. Schweizerbart, Stuttgart, Germany.

Leins, P., and C. Erbar. 1985. Ein Beitrag zur Blütenentwicklung der Aristolochiaceen, einer Vermittlergruppe zu den Monokotylen. *Botanische Jahrbücher für Systematik* 107:343–368.

———. 1990. On the mechanisms of secondary pollen presentation in the Campanulales-Asterales-complex. *Botanica Acta* 103:87–92.

———. 1991. Fascicled androecia in Dilleniidae and some remarks on the *Garcinia* androecium. *Botanica Acta* 104:336–344.

———. 1994. Flowers in Magnoliidae and the origin of flowers in other subclasses of the angiosperms. II. The relationships between the flowers of Magnoliidae, Dilleniidae, and Caryophyllidae. *Plant Systematics and Evolution, Supplement* 8:209–218.

———. 2010. Flower and Fruit. Schweizerbart, Stuttgart, Germany.

Leins, P., and G. Metzenauer. 1979. Entwicklungsgeschichtliche Untersuchungen an *Capparis*–Blüten. *Botanische Jahrbücher für Systematik* 100:542–554.

Leitch, I.J., and M.D. Bennett. 1997. Polyploidy in angiosperms. *Trends in Plant Science* 2:470–476.

———. 2004. Genomic downsizing in polyploid plants. *Botanical Journal of the Linnean Society* 82:651–663.

Leitch, I.J., M.W. Chase, and M.D. Bennett. 1998. Phylogenetic analysis of DNA C-values provides evidence for a small ancestral genome size in flowering plants. *Annals of Botany* 82:85–94.

Leitch, I.J., and L. Hanson. 2002. DNA C-values in seven families fill phylogenetic gaps in the basal angiosperms. *Botanical Journal of the Linnean Society* 140:175–179.

Leitch, I.J., L. Hanson, K.Y. Lim, A. Kovarik, M.W. Chase, J.J. Clarkson, and A.R. Leitch. 2008. The ups and downs of genome size evolution in polyploid species of *Nicotiana* (Solanaceae). *Annals of Botany* 101:805–814.

Leitch, I.J., L. Hanson, M. Winfield, J. Parker, and M.D. Bennett. 2001. Nuclear DNA C-values complete familial representation in gymnosperms. *Annals of Botany* 88:843–849.

Leitch, I.J., and A.R. Leitch. 2013. Genome size diversity and evolution in land plants. *In* I.J. Leitch, J. Greilhuber, J. Doležel, and J.F. Wendel [eds.], Plant Genome Diversity, Volume 2, Physical Structure, Behaviour and Evolution of Plant Genomes, 307–322. Springer–Verlag, Wien, Germany.

Leitch, I.J., D. E. Soltis, P. S. Soltis, and M.D. Bennett. 2005. Evolution of DNA amounts across land plants (Embryophyta). *Annals of Botany* 95:207–217.

Lemaire, B., S. Huysmans, E. Smets, and V. Merckx. 2011a. Rate accelerations in nuclear 18S rDNA of mycoheterotrophic and parasitic angiosperms. *Journal of Plant Research* 124:561–576.

Lemaire, B., P. Vandamme, V. Merckx, E. Smets, and S. Dessein. 2011b. Bacterial leaf symbiosis in angiosperms: Host specificity without co-speciation. *PLoS One* 6(9): e24430. doi:10.1371/journal.pone.0024430.

Lemke, D. 1988. A synopsis of Flacourtiaceae. *Aliso* 12:29–43.

Lens, F., J. Kårehed., P. Baas, S. Jansen, D. Rabaey, S. Huysmans, T. Hamann, and E. Smets. 2008. The wood anatomy of polyphyletic Icacinaceae s.l., and their relationships within asterids. *Taxon* 57:525–552.

Lens, F., J. Schönenberger, P. Baas, S. Jansen, and E. Smets. 2007. The role of wood anatomy in phylogeny reconstruction of Ericales. *Cladistics* 23:229–254.

Leon-Kloosterziel, K.M., C.L. Keijzer, and M. Koorneef. 1994. A seed shape mutant of *Arabidopsis* that is affected in integument development. *The Plant Cell* 6:385–392.

Les, D.H. 1988. The origin and affinities of the Ceratophyllaceae. *Taxon* 37:326–435.

———. 1993. Ceratophyllaceae. *In* K. Kubitzki, J.G. Rohwer, and V. Bittrich [eds.], The Families and Genera of Vascular Plants, II, 246–250. Springer-Verlag, Berlin, Germany.

Les, D.H., M.A. Cleland, and M. Waycott. 1997a. Phylogenetic studies in Alismatidae, II: Evolution of marine angiosperms (seagrasses) and hydrophily. *Systematic Botany* 22:443–463.

Les, D.H., D.K. Garvin, and C.F. Wimpee. 1991. Molecular evolutionary history of ancient aquatic angiosperms. *Proceedings of the National Academy of Sciences USA* 88:10119–10123.

Les, D.H., and R.R. Haynes. 1995. Systematics of the subclass Alismatidae: A synthesis of approaches. *In* P.J. Rudall, P.J. Cribb, D.F. Cutler, and C.J. Humphries [eds.], Monocotyledons: Systematics and Evolution, 353–377. Royal Botanic Gardens, Kew, UK.

Les, D.H., E. Landolt, and D.J. Crawford. 1997b. Systematics of the Lemnaceae (duckweeds): Inferences from micromolecular and morphological data. *Plant Systematics and Evolution* 204:161–177.

Les, D.H., M.L. Moody, S.W.L. Jacobs, and R.J. Bayer. 2002. Systematics of seagrasses (Zosteraceae) in Australia and New Zealand. *Systematic Botany* 27:468–484.

Les, D.H., C.T. Philbrick, and R.A. Novelo. 1997c. The phylogenetic position of riverweeds (Podostemaceae): Insights from *rbcL* sequence data. *Aquatic Botany* 57:5–27.

Les, D.H., and E.L. Schneider. 1995. The Nymphaeales, Alismatidae, and the theory of an aquatic monocotyledon origin. *In* P.J. Rudall, P.J. Cribb, D.F. Cutler, and C.J. Humphries [eds.], Monocotyledons: Systematics and Evolution, 23–42. Royal Botanic Gardens, Kew, UK.

Les, D.H., E. L. Schneider, D. J. Padgett, P.S. Soltis, D.E. Soltis, and M. Zanis. 1999. Phylogeny, classification and floral evolution of water lilies (Nymphaeaceae; Nymphaeales): A synthesis of non-molecular, *rbcL*, *matK*, and 18S rDNA data. *Systematic Botany* 24:28–46.

Les, D.H., and N.P. Tippery. 2013. In time and with water . . . The systematics of alismatid monocotyledons. *In* P. Wilkin, and S.J. Mayo [eds.], Early Events in Monocot Evolution, 118–164. Cambridge University Press, Cambridge, UK.

Luebert, F., L. Cecchi, M.F. Frohlich, M. Gottschling, C. Matt Guilliams, K.E. Hasenstab-Lehman, H.H. Hilger, J.S. Miller, M. Mittelbach, M. Nazaire, M. Nepi, D. Nocentini, D. Ober, R.G. Olmstead, F. Selvi, M.G. Simpson, K. Sutory, B. Valdéz, G.K. Walden, and M. Weigend. 2016. Familial classification of the Boraginales. *Taxon* 65:502–522.

Levin, D.A. 1983. Polyploidy and novelty in flowering plants. *American Naturalist* 122:1–25.

Levin, R.A., N.R. Myers, and L. Bohs. 2006. Phylogenetic relationships among the "spiny Solanums" (*Solanum* subgenus *Leptostemonum*, Solanaceae). *American Journal of Botany* 93:157–169.

Levin, R.A., W.L. Wagner, P.C. Hoch, M. Nepokroeff, J.C. Pires, E.A. Zimmer, and K.J. Sytsma. 2003. Family-level relationships of Onagraceae based on chloroplast *rbcL* and *ndhF* data. *American Journal of Botany* 90:107–115.

Lewis, G.P. 1998. *Caesalpinia*, a revision of the *Poincianella–Erythrostemon* group. Royal Botanic Gardens, Kew, UK.

Lewis, G.P., B.B. Simpson, and J.L. Neff. 2000. Progress in understanding the reproductive biology of the Caesalpinioideae (Leguminosae). *In* P.S. Herendeen, and A. Bruneau [eds.], Advances in Legume Systematics, part 9, 65–78. Royal Botanic Gardens, Kew, UK.

Lewis, W.H. 1980. Polyploidy in angiosperms: Dicotyledons. *In* W.H. Lewis [ed.], Polyploidy: Basic Life Sciences, Vol. 13, 241–268. Plenum Press, New York, NY.

Li, F.-W., J.C. Villarreal, S. Kelly, C.J. Rothfels, M. Melkonian, E. Frangedakis, M. Ruhsam, E.M. Sigel, J.P. Der, J. Pittermann, D.O. Burge, L. Pokornyk, A. Larsson, T. Chen, S. Weststrand, P. Thomas, E. Carpenter, Y. Zhang, Z. Tian, L. Chen, Z. Yan, Y. Zhu, X. Sun, J. Wang, D.W. Stevenson, B.J. Crandall-Stotler, A.J. Shaw, M.K. Deyholos, D.E. Soltis, S.W. Graham, M.D. Windham, J.A. Langdale, G.K.-S. Wong, S. Mathews, and K.M. Pryer. 2014. Horizontal transfer of an adaptive chimeric photoreceptor from bryophytes to ferns. *Proceedings of the National Academy of Sciences USA* 111(18):6672–6677.

Li, H.-F., S.-M. Chaw, C.-M. Du, and Y. Ren. 2011. Vessel elements present in the secondary xylem of *Trochodendron* and *Tetracentron* (Trochodendraceae). *Flora* 206(6):595–600.

Li, J. 2008. Molecular phylogenetics of Hamamelidaceae: Evidence from DNA sequences of nuclear and chloroplast genomes. *In* A.K. Sharma, and A. Sharma [eds.], Plant Genome Biodiversity and Evolution. Volume 1, Part E, Phanerogams-Angiosperm, 227–250. Science Publishers, Delhi, India.

Li, J., and A.L. Bogle. 2001. A new suprageneric classification system of the Hamamelidoideae based on morphology and sequences of nuclear and chloroplast DNA. *Harvard Papers in Botany* 5(2):499–515.

Li, J., A.L. Bogle, and A.S. Klein. 1999a. Phylogenetic relationships in the Hamamelidaceae: Evidence from the nucleotide sequences of the plastid gene *matK*. *Plant Systematics and Evolution* 218:205–219.

———. 1999b. Phylogenetic relationships of the Hamamelidaceae inferred from sequences of internal transcribed spacers (ITS) of nuclear ribosomal DNA. *American Journal of Botany* 86:1027–1037.

Li, J., H. Huang, and T. Sang. 2002. Molecular phylogeny and infrageneric classification of *Actinidia* (Actinidiaceae). *Systematic Botany* 27:408–415.

Li, R., P.-F. Ma, J. Wen, and T.-S. Yi. 2013. Complete sequencing of five Araliaceae chloroplast genomes and the phylogenetic implications. *PLoS One* 8(10): e78568. doi:10.1371/journal.pone.0078568.

Li, R., and J. Wen. 2013. Phylogeny and biogeography of *Dendropanax* (Araliaceae), an amphi-Pacific disjunct genus between tropical/subtropical Asia and the Neotropics. *Systematic Botany* 38:536–551.

Li, R.-Q., Z.-D. Chen, A.-M. Lu, D.E. Soltis, P.S. Soltis, and P.S. Manos. 2004. Phylogenetic relationships in Fagales based on DNA sequences from three genomes. *International Journal of Plant Sciences* 165(2):311–324.

Li, X., T.-C. Zhang, Q. Qiao, Z. Ren, J. Zhao, T. Yonezawa, M. Hasegawa, M.J.C. Crabbe, J. Li, and Y. Zhong. 2013. Complete chloroplast genome sequence of holoparasite *Cistanche deserticola* (Orobanchaceae) reveals gene loss and horizontal gene transfer from its host *Haloxylon ammodendron* (Chenopodiaceae). *PLoS One* 8(3): e58747. doi:10.1371/journal.pone.0058747.

Liden, M., and B. Oxelman. 1996. Do we need "phylogenetic taxonomy"? *Zoologica Scripta* 25:183–185.

Linder, H.P. 1998. Morphology and the evolution of wind pollination. *In* S. J. Owens, and P. J. Rudall [eds.], Reproductive Biology in Systematics, Conservation and Economic Botany, 123–135. Royal Botanic Gardens, Kew, UK.

———. 2000. Vicariance, climate change, anatomy and phylogeny of Restionaceae. *Botanical Journal of the Linnean Society* 134:159–177.

Linder, H.P., B.G. Briggs, and L.A.S. Johnson. 2000. Restionaceae: A morphological phylogeny. *In* K.L. Wilson, and D.A. Morrison [eds], Monocots: Systematics and Evolution, 653–660. CSIRO, Collingwood, Victoria, Australia.

Lippok, B., A.A. Gardine, P.S. Williamson, and S.S. Renner. 2000. Pollination by flies, bees, and beetles of *Nuphar ozarkana* and *N. advena* (Nymphaeaceae). *American Journal of Botany* 87:898–902.

Litt, A.J. 1999. Floral Morphology and Phylogeny of Vochysiaceae. Ph. D. Dissertation, University of New York. New York, NY.

Litt, A., and V. F. Irish. 2003. Duplication and diversification in the *APETALA1/FRUITFULL* floral homeotic gene lineage: Implications for the evolution of floral development. *Genetics* 165:821–833.

Litt, A., and D.W. Stevenson. 2003. Floral development and morphology of Vochysiaceae. I. The structure of the gynoecium. *American Journal of Botany* 90:1533–1547.

Liu, B., and J.F. Wendel. 2003. Epigenetic phenomena and the evolution of plant allopolyploids. *Molecular Phylogenetics and Evolution* 29(3):365–379.

Liu, F.-G., R. Liu, M. Miyamota, N. Freire, P. Ong, M. Tennant, T. Young, and K. Gugel. 2001. Molecular and morphological supertrees for eutherian (placental) mammals. *Science* 291:1786–1789.

Liu, L., L. Kubatko, D.K. Pearl, and S.V. Edwards. 2009. Coalescent methods for estimating phylogenetic trees. *Molecular Phylogenetics and Evolution* 53:320–328.

Liu, L., L. Yu, and S.V. Edwards. 2010. A maximum pseudo-likelihood approach for estimating species trees under the coalescent model. *BMC Evolutionary Biology* 10:302. doi:10.1186/1471-2148-10-302.

Liu, S., Y. Liu, X. Yang, C. Tong, D. Edwards, I.A.P. Parkin, M. Zhao, J. Ma, J. Yu, S. Huang, X. Wang, J. Wang, K. Lu, Z. Fang, I. Bancroft, T.-J. Yang, Q. Hu, X. Wang, Z. Yue, H. Li, L. Yang, J. Wu, Q. Zhou, W. Wang, G.J. King, J.C. Pires, C. Lu, Z. Wu, P. Sampath, Z. Wang, H. Guo, S. Pan, L. Yang, J. Min, D. Zhang, D. Jin, W. Li, H. Belcram, J. Tu, M. Guan, C. Qi, D. Du, J. Li, L. Jiang, J. Batley, A.G. Sharpe, B.-S. Park, P. Ruperao, F. Cheng, N.E. Waminal, Y. Huang, C. Dong, L. Wang, J. Li, Z. Hu, M. Zhuang, Y. Huang, J. Huang, J. Shi, D. Mei, J. Liu, T.-H. Lee, J. Wang, H. Jin, Z. Li, X. Li, J. Zhang, L. Xiao, Y. Zhou, Z. Liu, X. Liu, R. Qin, X. Tang, W. Liu, Y. Wang, Y. Zhang, J. Lee, H.H. Kim, F. Denoeud, X. Xu, X. Liang, W. Hua, X. Wang, J. Wang, B. Chalhoub, and A.H. Paterson. 2014. The *Brassica oleracea* genome reveals the asymmetrical evolution of polyploid genomes. *Nature Communications* 5. doi:10.1038/ncomms4930.

Liu, X., S.R. Manchester, and J. Jianhua. 2014. *Alnus* subgenus *Alnus* in the Eocene of western North America based on leaves, associated catkins, pollen, and fruits. *American Journal of Botany* 101(11):1925–1943.

Liu, X.-Q., S.M. Ickert-Bond, L.-Q. Chen, and J. Wen. 2013. Molecular phylogeny of *Cissus* L. of Vitaceae (the grape family) and evolution of its pantropical intercontinental disjunctions. *Molecular Phylogenetics and Evolution* 66(1):43–53.

Lledó, M.D., M.B. Crespo, K.M. Cameron, M.F. Fay, and M.W. Chase. 1998. Systematics of Plumbaginaceae based upon cladistic analysis of *rbcL* sequence data. *Systematic Botany* 23:21–29.

Lledó, M.D., P.O. Karis, M.B. Crespo, M.F. Fay, and M.W. Chase. 2001. Phylogenetic position and taxonomic status of the genus *Aegialitis* and subfamilies Staticoideae and Plumbaginoideae (Plumbaginaceae): Evidence from plastid DNA sequences and morphology. *Plant Systematics and Evolution* 229:107–124.

Lloyd, D.G. 1982. Selection of combined versus separate sexes in seed plants. *American Naturalist* 120:571–585.

Lloyd, D.G., and S.C.H. Barrett. 1996. Floral Biology. Studies on Floral Evolution in Animal-Pollinated Plants. Chapman and Hall, New York, NY.

Lloyd, D.G., and D. Schoen. 1992. Self- and cross-fertilization in plants. I. Functional dimensions. *International Journal of Plant Science* 153:358–369.

Lloyd, D.G., and C.J. Webb. 1986. The avoidance of interference between the presentation of pollen and stigmas in angiosperms. I. Dichogamy. *New Zealand Journal of Botany* 24:135–162.

Loconte, H. 1996. Comparison of alternative hypotheses for the origin of the Angiosperms. *In* D.W. Taylor, and L.J. Hickey [eds.], Flowering Plant Origin, Evolution, and Phylogeny. Chapman and Hall, New York, NY.

Loconte, H., M. Campbell, and D.W. Stevenson. 1995. Ordinal and familial relationships of ranunculid genera. *Plant Systematics and Evolution, Supplement* 9:99–118.

Loconte, H., and J.R. Estes. 1989. Phylogenetic systematics of Berberidaceae and Ranunculales (Magnoliidae). *Systematic Botany* 14:565–579.

Loconte, H., and D.W. Stevenson. 1990. Cladistics of the Spermatophyta. *Brittonia* 42:197–211.

———. 1991. Cladistics of the Magnoliidae. *Cladistics* 7:267–296.

Loehne, C., T. Borsch, and J.H. Wiersema. 2007. Phylogenetic analysis of Nymphaeales using fast-evolving and noncoding chloroplast markers. *Botanical Journal of the Linnean Society* 154:141–163.

Loehne, C., J.H. Wiersema, and T. Borsch. 2009. The unusual *Ondinea*, actually just another Australian water-lily of *Nymphaea* subg. *Anecphya* (Nymphaeaceae). *Willdenowia* 39:55–58.

Loehne, C., M.-J. Yoo, T. Borsch, J. Wiersema, V. Wilde, C.D. Bell, W. Barthlott, D.E. Soltis, and P.S. Soltis. 2008. Biogeography of Nymphaeales: Extant patterns and historical events. *Taxon* 57:1123–1146.

Logacheva, M.D., M.I. Schelkunov, M.S. Nuraliev, T.H. Samigullin, and A.A. Penin. 2014. The plastid genome of mycoheterotrophic monocot *Petrosavia stellaris* exhibits both gene losses and multiple rearrangements. *Genome Biology and Evolution* 6:238–246.

Long, A.G. 1977. Lower carboniferous pteridosperm cupules and the origin of angiosperms. *Royal Society of Edinburgh Transactions* 70:37–61.

Long, R.W. 1970. The genera of Acanthaceae in the southeastern United States. *Journal of the Arnold Arboretum* 51:257–309.

Losos, J.B. 2011. Seeing the forest for the trees: The limitation of phylogenies in comparative biology. *American Naturalist* 177:709–727.

Lott, T.A., S.R. Manchester, and D.L. Dilcher. 1998. A unique and complete polemoniaceous plant from the middle Eocene of Utah, USA. *Review of Palaeobotany and Palynology* 104:39–49.

Love, A., and D. Love. 1949. The geobotanical significance of polyploidy. I. Polyploidy and latitude. *Portugaliae Acta Biologica, Serie A. Morfologia, Fisiologia, Genetica e Biologia Geral* 273–352.

Løvtrup, S. 1977. The Phylogeny of Vertebrata. Wiley, London, UK.

Lowry, P.P., G.M. Plunkett, and A.A. Oskolski. 2001. Early lineages in *Apiales*: Insights from morphology, wood anatomy and molecular data. *Edinburgh Journal of Botany* 58:207–220.

Lu, Y., J.-H. Ran, D.-M. Guo, Z.-Y. Yang, and X.-Q. Wang. 2014. Phylogeny and divergence times of Gymnosperms inferred from single-copy nuclear genes. *PLoS One* 9(9). doi:10.1371/journal.pone.0107679.

Lui, M., G.M. Plunkett, and P.P. Lowry. 2010. Fruit anatomy provides structural synapomorphies to help define Myodocarpaceae (Apiales). *Systematic Botany* 35:675–681.

Lukens, L.N., J.C. Pires, E. Leon, R. Vogelzang, L. Oslach, and T. Osborn. 2006. Patterns of sequence loss and cytosine methylation within a population of newly resynthesized *Brassica napus* allopolyploids. *Plant Physiology* 140(1):336–348.

Lunau, K. 2000. The ecology and evolution of visual pollen signals. *Plant Systematics and Evolution* 222:89–111.

Lundberg, J. 2001. The asteralean affinity of the Mauritian *Roussea* (Rousseaceae). *Botanical Journal of the Linnean Society* 137:267–276.

———. 2009. Asteraceae and relationships within Asterales. *In* V.A. Funk, A. Susanna, T.F. Stuessy, and R.J. Bayer [eds.], Systematics, Evolution, and Biogeography of Compositae, 157–169. IAPT, Vienna, Austria.

Lundberg, J., and K. Bremer. 2002. A phylogenetic study of the order Asterales using one large morphological and three molecular data sets. *International Journal of Plant Sciences* 164:553–578.

Luo, D., R. Carpenter, L. Copsey, C. Vincent, J. Clark, and E. Coen. 1999. Control of organ asymmetry in flowers of *Antirrhinum*. *Cell* 99(4):367–376.

Luo, D., R. Carpenter, C. Vincent, L. Copsey, and E. Coen. 1996. Origin of floral asymmetry in *Antirrhinum*. *Nature* 383(6603):794–799.

Lüthy, B., and P. Matile. 1984. The mustard oil bomb: Rectified analysis of the subcellular organisation of the myrosinase system. *Biochemie und Physiologie der Pflanzen* 179:5–12.

Lynch, M., and J.S. Conery. 2000. The evolutionary fate and consequences of duplicate genes. *Science* 290(5494):1151–1155.

Lyons-Weiler, J., G.A. Hoelzer, and R.J. Tausch. 1996. Relative apparent synapomorphy analysis (RASA). I. The statistical measure of phylogenetic signal. *Molecular Biology and Evolution* 13:749–757.

Lysak, M.A., P.F. Franz, H.B.M. Ali, and I. Schubert. 2001. Chromosome painting in *Arabidopsis thaliana*. *Plant Journal* 28:689–697.

Lysak, M.A., M.A. Koch, J.M. Beaulieu, A. Meister, and I.J. Leitch. 2009. The dynamic ups and downs of genome size evolution in Brassicaceae. *Molecular Biology and Evolution* 26(1):85–98.

Lysak, M.A., and C. Lexer. 2006. Towards the era of comparative evolutionary genomics in Brassicaceae. *Plant Systematics and Evolution* 259(2–4):175–198.

Lysak, M.A., A. Pecinka, and I. Schubert. 2003. Recent progress in chromosome painting of *Arabidopsis* and related species. *Chromosome Research* 11: 195–204.

Ma, H. 1998. To be, or not to be, a flower-control of floral meristem identity. *Trends in Genetics* 14:26–32.

Ma, H., and C.W. dePamphilis. 2000. The ABCs of floral evolution. *Cell* 101(1):5–8.

Mabberley, D.J. 1993. The Plant Book: A Portable Dictionary of the Vascular Plants, 2nd ed. Cambridge University Press, Cambridge, UK.

MacGinitie, H.D. 1941. Middle Eocene flora from the central Sierra Nevada. *Carnegie Institute Washington Publication* 534. Washington, DC.

———. 1953. Fossil Plants of the Florissant Beds, Colorado. *Carnegie Institution of Washington Publication* 599. Washington, D.C.

Macphail, M.K., and R.S. Hill. 1994. K-Ar dated palynofloras in Tasmania; 1, Early Oligocene, *Proteacidites tuberculatus* Zone sediments, Wilmot Dam, northwestern Tasmania. *Papers Proceeding Royal Society Tasmania* 128:1–15.

Macphail, M.K., G. Jordan, F. Hopf, and E. Colhoun. 2012. When did the mistletoe family Loranthaceae become extinct in Tasmania? Review and conjecture. *Terra Australis* 34: 255–269.

Maddison, D.R. 1991. The discovery and importance of multiple islands of most-parsimonious trees. *Systematic Zoology* 40:315–328.

Maddison, W.P. 1990. A method for testing the correlated evolution of two binary characters: Are gains or losses concentrated on certain branches of a phylogenetic tree? *Evolution* 44:539–557.

———. 1997. Gene trees in species trees. *Systematic Biology* 46:523–536.

Maddison, W.P., and D.R. Maddison. 1992. MacClade: Analysis of phylogeny and character evolution. Sinauer, Sunderland, MA.

———. 2011. Mesquite: A modular system for evolutionary analysis. Version 2.75 http://mesquiteproject.org.

———. 2015. Mesquite: A modular system for evolutionary analysis. Version 303 http://mesquiteproject.org.

Madlung, A., A.P. Tyagi, B. Watson, H.M. Jiang, T. Kagochi, R.W. Doerge, R. Martienssen, and L. Comai. 2005. Genomic changes in synthetic *Arabidopsis* polyploids. *Plant Journal* 41(2):221–230.

Madlung, A., and J.F. Wendel. 2013. Genetic and epigenetic aspects of polyploid evolution in plants. *Cytogenetic and Genome Research* 140(2–4):270–285.

Maere, S., S. DeBoldt, J. Raes, T. Casneuf, M. Van Montagu, M. Kuiper, and Y. Van de Peer. 2005. Modeling gene and genome duplications in eukaryotes. *Proceedings of the National Academy of Sciences USA* 102:5454–5459.

Magallón, S. 2010. Using fossils to break long branches in molecular dating: A comparison of relaxed clocks applied to the origin of Angiosperms. *Systematic Biology* 59:384–399.

Magallón, S., and A. Castillo. 2009. Angiosperm diversification through time. *American Journal of Botany* 96(1):349–365.

Magallón, S., P.R. Crane, and P.S. Herendeen. 1999. Phylogenetic pattern, diversity, and diversification of eudicots. *Annals of the Missouri Botanical Garden* 86:297–372.

Magallón, S., S. Gómez-Acevedo, L.L. Sánchez-Reyes, and T. Hernández-Hernández. 2015. A metacalibrated timetree documents the early rise of flowering plant phylogenetic diversity. *New Phytologist* 2015. doi:10.1111/nph.13264.

Magallón, S., K. Hilu, and D. Quandt. 2013. Land plant evolutionary timeline: Gene effects are secondary to fossil constraints in relaxed clock estimation of age and substitution rates. *American Journal of Botany* 100:566–573.

Magallón, S., and M.J. Sanderson. 2002. Relationships among seed plants inferred from highly conserved genes: Sorting conflicting phylogenetic signals among ancient lineages. *American Journal of Botany* 89:1991–2006.

Magallón-Puebla, S., P.S. Herendeen, and P.R. Crane. 1997. *Quadriplatanus georgianus* gen. et sp. nov.: Staminate and pistillate platanaceous flowers from the late Cretaceous (Coniacian–Santonian) of Georgia, USA. *International Journal of Plant Sciences* 158:373–394.

Magallón-Puebla, S., P. S. Herendeen, and P. K. Endress. 1996. *Allonia decandra*: Floral remains of the tribe Hamamelideae (Hamamelidaceae) from Campanian strata of Southeastern USA. *Plant Systematics and Evolution* 202:177–198.

Magin, N. 1977. Das Gynoeceum der Apiaceae: Modell und Ontogenie. *Berichte der Deutschen Botanischen Gesellschaft* 90:S53–S66.

Mai, D.H. 1968. Zwei ausgestorbene Gattungen im Tertiär Europas und ihre Florengeschichtliche Bedeutung. *Palaeontographica Abteilung* B 123:184–199.

———. 1970. Subtropische Elemente im europäischen Tertiär I. Die *Gattungen Gironniera, Sarcococca, Illicium, Evodia, Ilex, Mastixia, Alangium, Symplocos* und *Rehderodendron. Paläontologische Abhandlungen Abt.* B 3:441–503, pls. 58–69.

———. 1993. On the extinct Mastixiaceae (Cornales) in Europe. *Geophytology* 23:53–63.

———. 1995. *Tertiäre Vegetationsgeschichte Europas*. Gustav Fischer Verlag, Jena, Germany.

Mai, D.H., and E. Martinetto. 2006. A reconsideration of the diversity of *Symplocos* in the European Neogene on the basis of fruit morphology. *Review of Palaeobotany and Palynology* 140:1–26.

Maia, V.H., M.A. Gitzendanner, P.S. Soltis, G.K.-S. Wong, and D.E. Soltis. 2014. Angiosperm phylogeny based on 18S/26S rDNA sequence data: Constructing a large data set using next-generation sequence data. *International Journal of Plant Sciences* 175:613–650.

Majure, L.C., W.S. Judd, P.S. Soltis, and D.E. Soltis. 2012c. Cytogeography of the *Humifusa* clade of *Opuntia* s.s. Mill. 1754 (Cactaceae: Opuntioideae): Correlations with geographic distributions and morphological differentiation of a polyploid complex. *Comparative Cytogenetics* 6:53–77.

Majure, L.C., and R. Puente. 2014. Phylogenetic relationships and morphological evolution in *Opuntia* s. str. and closely related members of tribe Opuntieae. *Succulent Plant Research* 8:9–30.

Majure, L.C., R. Puente, M.P. Griffith, W.S. Judd, P.S. Soltis, and D.S. Soltis. 2012a. Phylogeny of *Opuntia* s.s. (Cactaceae): Clade delineation, geographic origins, and reticulate evolution. *American Journal of Botany* 99:847–864.

Majure, L.C., R. Puente, M.P. Griffith, D.E. Soltis, and W.S. Judd. 2013b. *Opuntia lilae*, another *Tacinga* hidden in *Opuntia* s.l. *Systematic Botany* 38:444–450.

Majure, L.C., R. Puente, and D.J. Pinkava. 2012b. Miscellaneous chromosome counts in Opuntieae DC. (Cactaceae) with a compilation of counts for the group. *Haseltonia* 18:67–78.

Majure, L.C., and E. Ribbens. 2012. Chromosome counts of *Opuntia* (Cactaceae), prickly pear cacti, in the Midwestern United States and environmental factors restricting the distribution of *Opuntia fragilis*. *Haseltonia* 17:58–65.

Majure, L.C., D.E. Soltis, P.S. Soltis, and W.S. Judd. 2013a. A case of mistaken identity, *Opuntia abjecta*, long-lost in synonymy under the Caribbean species, *O. triacantha*, and a reassessment of the enigmatic *O. cubensis*. *Brittonia* 66(2):118–130.

Makboul, A.M. 1986. Chemical constituents of *Verbena officinalis*. *Fitoterapia* 57:50–51.

Malek, O., K. Lattig, R. Hiesel, A. Brennicke, and V. Knoop. 1996. RNA editing in bryophytes and a molecular phylogeny of land plants. *EMBO Journal* 14:1403–1411.

Manchester, S.R. 1986. Vegetative and reproductive morphology of an extinct plane tree (Platanaceae) from the Eocene of western North America. *Botanical Gazette* 147:200–226.

———. 1987. The fossil history of the Juglandaceae. *Missouri Botanical Garden Monographs in Systematic Botany* 21:1–137.

———. 1988. Fruits and seeds of *Tapiscia* (Staphyleaceae) from the middle Eocene of Oregon, USA. *Tertiary Research* 9:59–66.

———. 1989. Systematics and fossil history of the Ulmaceae. *In* P.R. Crane, and S. Blackmore [eds.], Evolution, Systematics, and Fossil History of the Hamamelidae, Volume 2: 'Higher' Hamamelida. Systematics Association Special Volume no. 40B, 221–252. Clarendon Press, Oxford, UK.

———. 1991. *Cruciptera*, a new Juglandaceous winged fruit from the Eocene and Oligocene of Western North America. *Systematic Botany* 16(4):715–725.

———. 1992. Flowers, fruits and pollen of *Florissantia*, an extinct malvalean genus from the Eocene and Oligocene of western North America. *American Journal of Botany* 79:996–1008.

———. 1994a. Inflorescence bracts of fossil and extant *Tilia* in North America, Europe, and Asia—patterns of morphologic divergence and biogeographic history. *American Journal of Botany* 81(9):1176–1185.

———. 1994b. Fruits and seeds of the Middle Eocene Nut Beds flora, Clarno Formation, Oregon. *Palaeontographica Americana* 58:1–205.

———. 1999. Biogeographical relationships of North American Tertiary floras. *Annals of the Missouri Botanical Garden* 86:472–522.

———. 2002. Leaves and fruits of *Davidia* (Cornales) from the Paleocene of North America. *Systematic Botany* 27(2):368–382.

———. 2014. Revisions to Roland Brown's North American Paleocene Flora. *Acta Musei Nationalis Pragae Series B—Historia Naturalis* 70(3–4):153–210.

Manchester, S.R., M.A. Akhmetiev, and T.M. Kodrul. 2002. Leaves and fruits of *Celtis aspera* (Newberry) comb. nov (Celtidaceae) from the Paleocene of North America and eastern Asia. *International Journal of Plant Sciences* 163(5):725–736.

Manchester, S.R., and I.J. Chen. 2006. *Tetracentron* fruits from the Miocene of western North America. *International Journal of Plant Sciences* 167(3):601–605.

Manchester, S.R., I. Chen, and T.A. Lott. 2012. Seeds of *Ampelocissus*, *Cissus*, and *Leea* (Vitales) from the Paleogene of western Peru and their biogeographic significance. *International Journal of Plant Sciences* 173(8):933–943.

Manchester, S.R., Z.-D. Chen, A.-M. Lu, and K. Uemura. 2009. Eastern Asian endemic seed plant genera and their paleogeographic history throughout the Northern Hemisphere. *Journal of Systematics and Evolution* 47(1):1–42.

Manchester, S.R., Z. Chen, and Z. Zhou. 2006. Wood anatomy of *Craigia* (Malvales) from southeastern Yunnan, China. *International Association of Wood Anatomists Journal* 27:129–136.

Manchester, S.R., M.E. Collinson, and K. Goth. 1994. Fruits of the Juglandaceae from the Eocene of Messel, Germany and implications for early Tertiary phytogeographic exchange between Europe and western North America. *International Journal of Plant Sciences* 155: 388–394.

Manchester, S.R., and P.R. Crane. 1987. A new genus of Betulaceae from the Oligocene of western North America. *Botanical Gazette* 148:263–273.

Manchester, S.R., P.R. Crane, and L. Golovneva. 1999. An extinct genus with affinities to extant *Davidia* and

Camptotheca (Cornales) from the Paleocene of North America and Eastern Asia. *International Journal of Plant Sciences* 160:188–207.

Manchester, S.R., and D.L. Dilcher. 1982. Pterocaryoid fruits (Juglandaceae) in the Paleogene of North America and their evolutionary and biogeographic significance. *American Journal of Botany* 69:275–286.

———. 1997. Reproductive and vegetative morphology of *Polyptera* (Juglandaceae) from the Paleocene of Wyoming and Montana. *American Journal of Botany* 84:649–663.

Mancheser, S.R., and R.M. Dillhoff. 2004. *Fagus* (Fagaceae) fruits, foliage, and pollen from the middle Eocene of Pacific northwestern North America. *Canadian Journal of Botany* 82:1509–1517.

Manchester, S.R. and M.J. Donoghue. 1995. Winged fruits of Linnaeeae (Caprifoliaceae) in the Tertiary of western North America: *Diplodipelta* gen. nov. *International Journal of Plant Sciences* 156:709–722.

Manchester, S.R., F. Grímsson, and R. Zetter. 2015. Assessing the fossil record of asterids in the context of our current phylogenetic framework. *Annals of the Missouri Botanical Garden* 100(4):329–363. doi:10.3417/2014033.

Manchester, S.R. and L.J. Hickey. 2007. Reproductive and vegetative organs of *Browniea* gen. n. (Nyssaceae) from the Paleocene of North America. *International Journal of Plant Sciences* 167(4):897–908.

Manchester, S.R., W.S. Judd, and B. Handley. 2006. Foliage and fruits of early poplars (Salicaceae: *Populus*) from the Eocene of Utah, Colorado, and Wyoming. *International Journal of Plant Sciences* 167(4):897–908.

Manchester, S.R., D.K. Kapgate, and J. Wen. 2013. Oldest fruits of the grape family (Vitaceae) from the Late Cretaceous deccan cherts of India. *American Journal of Botany* 100(9):1849–1859.

Manchester, S.R., and W.J. Kress. 1993. Bananas (Musaceae): *Ensete oregonense* sp. nov. from Eocene of Western North America, and its phytogeographic significance. *American Journal of Botany* 80(11):1264–1272.

Manchester, S.R., T.M. Lehman, and E.A. Wheeler. 2010. Fossil palms (Arecaceae, Coryphoideae) associated with juvenile herbivorous dinosaurs in the upper Cretaceous Aguja Formation, Big Bend National Park, Texas. *International Journal of Plant Sciences* 171(6):679–689.

Manchester, S.R., and R.B. Miller. 1978. Tile cells and their occurrence in malvalean fossil woods. *IAWA Bulletin II* 2–3:23–28.

Manchester, S.R. and E. O'Leary. 2010. Distribution and identification of fin-winged fruits. *Botanical Review* 76:1–82.

Manchester, S.R., K.B. Pigg, and P.R. Crane. 2004. *Palaeocarpinus dakotensis* sp. n. (Betulaceae: Coryloideae) and associated staminate catkins, pollen and leaves from the Paleocene of North Dakota. *International Journal of Plant Sciences* 165:1135–1148.

Manchester, S.R., V. Wilde, and M.E. Collinson. 2007. Fossil cashew nuts from the Eocene of Europe. Biogeographic links between Africa and South America. *International Journal of Plant Sciences* 168:1199–1206.

Manchester, S.R., Q.-Y. (J) Xiang, and Q.-P. Xiang. 2007. *Curtisia* (Cornales) from the Eocene of Europe and its phytogeographical significance. *Botanical Journal of the Linnean Society* 155:127–134.

Manchester, S.R., X.-P. Xiang, and Q.-Y. (J.) Xiang. 2010. Fruits of Cornelian Cherries (Cornaceae: *Cornus* Subg. *Cornus*) in the Paleocene and Eocene of the Northern Hemisphere. *International Journal of Plant Sciences* 171(8):882–891.

Mandakova, T., and M.A. Lysak. 2008. Chromosomal phylogeny and karyotype evolution in x=7 *Crucifer* species (Brassicaceae). *Plant Cell* 20(10):2559–2570.

Manen, J.F., G. Barriera, P.-A. Loizeau, and Y. Naciri. 2010. The history of extant *Ilex* species (Aquifoliaceae): Evidence of hybridization within a Miocene radiation. *Molecular Phylogenetics and Evolution* 57:961–977.

Manen, J.F., A. Natali, and F. Ehrendorfer. 1994. Phylogeny of Rubiaceae-Rubieae inferred from the sequence of a cp-DNA intergene region. *Plant Systematics and Evolution* 190:195–211.

Manhart, J.R., and J.H. Rettig. 1994. Gene sequence data. *In* H.-D. Behnke, and T.J. Mabry [eds.], Caryophyllales: Evolution and Systematics, 235–246. Springer-Verlag, Berlin, Germany.

Manning, W.E. 1938. The morphology of the flowers of the Juglandaceae. I. The inflorescence. *American Journal of Botany* 25(6):407–419.

———. 1940. The morphology of the flowers of the Juglandaceae. II. The pistillate flowers and fruit. *American Journal of Botany* 27(10):839–852.

Manos, P.S. 1997. Systematics of *Nothofagus* (Nothofagaceae) based on rDNA spacer sequences (ITS): Taxonomic congruence with morphology and plastid sequences. *American Journal of Botany* 84:1137–1155.

Manos, P.S., P.S. Soltis, D.E. Soltis, S.R. Manchester, S.-H. Oh, C.D. Bell, D.L. Dilcher, and D.E. Stone. 2007. Phylogeny of extant and fossil Juglandaceae inferred from the integration of molecular and morphological data sets. *Systematic Biology* 56(3):412–430.

Manos, P.S., and A.M. Stanford. 2001. The biogeography of Fagaceae: Tracking the Tertiary history of temperate and subtropical forests of the Northern Hemisphere. *International Journal of Plant Sciences* 162:S77–S93.

Manos, P.S., and K.P. Steele. 1997. Phylogenetic analyses of "higher" Hamamelidae based on plastid sequence data. *American Journal of Botany* 81:1407–1419.

Manos, P.S., and D.E. Stone. 2001. Evolution, phylogeny, and systematics of the Juglandaceae. *Annals of the Missouri Botanical Garden* 88:231–269.

Manos, P.S., Z.-K. Zhou, and C.H. Cannon. 2001. Systematics of Fagaceae: Phylogenetic tests of reproductive trait evolution. *International Journal of Plant Sciences* 162:1361–1379.

Marazzi, B., E. Conti, and P.K. Endress. 2007. Diversity in anthers and stigmas in the buzz-pollinated genus *Senna* (Leguminosae, Cassiinae). *International Journal of Plant Sciences* 168:371–391.

Marazzi, B., P.K. Endress, L. Paganucci de Queiroz, and E. Conti. 2006. Phylogenetic relationships within *Senna* (Leguminosae, Cassiinae) based on three chloroplast DNA regions: Patterns in the evolution of floral asymmetry and extrafloral nectaries. *American Journal of Botany* 93:288–303.

Marcussen, T., S.R. Sandve, L. Heier, M. Spannagl, M. Pfeifer, The International Wheat Genome Sequencing Consortium, K.S. Jakobsen, B.B.H. Wulff, B. Steuernagel, K.F.X. Mayer, and O.-A. Olsen. 2014. Ancient hybridizations among the ancestral genomes of bread wheat. *Science* 345(6194). doi:10.1126/science.1250092.

Markgraf, F. 1955. Über Laubblatt-Homologien und verwandtschaftliche Zusammenhänge bei Sarraceniales. *Planta* 46:414–446.

Márquez-Guzmán, J., M. Engleman, A. Martínez-Mena, E. Martínez, and C. Ramos. 1989. Anatomia reproductiva de *Lacandonia schismatica* (Lacandoniaceae). *Annals of the Missouri Botanical Garden* 76:124–127.

Marquinez, X., L.G. Lohmann, M.L.F. Salatino, A. Salatino, and F. Gonzalez. 2009. Generic relationships and dating of lineages in Winteraceae based on nuclear (ITS) and plastid (*rpS16* and *psbA–trnH*) sequence data. *Molecular Phylogenetics and Evolution* 53:435–449.

Martin, H.A. 1977. The History of *Ilex* (Aquifoliaceae) with special reference to Australia: Evidence from pollen. *Australian Journal of Botany* 25:655–673.

———. 2003. The history of the family Onagraceae in Australia and its relevance to biogeography. *Australian Journal of Botany* 51(5):585–598.

Martin, H.A., M.K. Macphail, and A.D. Partridge. 1996. Tertiary *Alangium* (Alangiaceae) in eastern Australia: Evidence from pollen. *Review of Palaeobotany and Palynology* 94:111–122.

Martin, W., A. Gierl, and H. Saedler. 1989. Molecular evidence for pre-Cretaceous angiosperm origins. *Nature* 339:46–48.

Martin, W., D.J. Lydiate, H. Brinkmann, G. Forkmann, H. Saedler, and R. Cerff. 1993. Molecular phylogenies in angiosperm evolution. *Molecular Biology and Evolution* 10:140–162.

Martínez-Millán, M. 2010. Fossil record and age of the Asteridae. *Botanical Review* 76:83–135.

Martínez-Millán, M., W.L. Crepet, and K.C. Nixon. 2009. *Pentapetalum trifasciculandricus* gen. et sp. nov., a thealean fossil flower from the Raritan Formation, New Jersey, USA (Turonian, Late Cretaceous). *American Journal of Botany* 96:933–949.

Martins, E.P., and T.F. Hansen. 1997. Phylogenies and the comparative method: A general approach to incorporating phylogenetic information into the analysis of interspecific data. *American Naturalist* 149:646–667.

Martins, L., C. Oberprieler, and F.H. Hellwig. 2003. A phylogenetic analysis of Primulaceae s.l. based on internal transcribed spacer (ITS) sequence data. *Plant Systematics and Evolution* 237:75–85.

Martins, T.R., and T.J. Barkman. 2005. Reconstruction of Solanaceae phylogeny using the nuclear gene SMAT. *Systematic Botany* 30:435–447.

Marx, H.E., N. O'Leary, Y.-W. Yuan, P. Lu-Irving, D.C. Tank, M.E. Mulgura, and R.G. Olmstead. 2010. A molecular phylogeny and classification of Verbenaceae. *American Journal of Botany* 97:1647–1663.

Massoni, J., F. Forest, and H. Sauquet. 2014. Increased sampling of both genes and taxa improves resolution of phylogenetic relationships within Magnoliidae, a large and early-diverging clade of angiosperms. *Molecular Phylogenetics and Evolution* 70:84–93.

Mast, A.R., S. Kelso, A.J. Richards, D. Lang, D.M.S. Feller, and E. Conti. 2001. Phylogenetic relationships in *Primula* L. and related genera (Primulaceae) based on noncoding chloroplast DNA. *International Journal of Plant Sciences* 162:1381–1400.

Mast, A.R., E.F. Milton, E.H. Jones, R.M. Barker, W.R. Barker, and P.H. Weston. 2012. Time-calibrated phylogeny of the woody Australian genus *Hakea* (Proteaceae) supports multiple origins of insect-pollination among bird-pollinated ancestors. *American Journal of Botany* 99(3):472–487.

Masterson, J. 1994. Stomatal size in fossil plants: Evidence for polyploidy in majority of angiosperms. *Science* 264(5157):421–424.

Mathews, S. 2009. Phylogenetic relationships among seed plants: Persistent questions and the limits of molecular data. *American Journal of Botany* 96:228–236.

Mathews, S., M.D. Clements, and M.A. Beilstein. 2010. A duplicate gene rooting of seed plants and the phylogenetic position of flowering plants. *Philosophi-*

cal Transactions of the Royal Society of London B 365:383–395.

Mathews, S., and M.J. Donoghue. 1999. The root of angiosperm phylogeny inferred from duplicate phytochrome genes. *Science* 286:947–950.

———. 2000. Basal angiosperm phylogeny inferred from duplicate phytochromes A and C. *International Journal of Plant Sciences* 161(6 Supplement):S41–S55.

Matthews, M.L., M.D.C.E. Amaral, and P.K. Endress. 2012. Comparative floral structure and systematics in Ochnaceae s.l. (Ochnaceae, Quiinaceae and Medusagynaceae; Malpighiales). *Botanical Journal of the Linnean Society* 170(3):299–392.

Matthews, M.L., and P.K. Endress. 2002. Comparative floral structure and systematics in Oxalidales (Oxalidaceae, Connaraceae, Brunelliaceae, Cephalotaceae, Cunoniaceae, Elaeocarpaceae, Tremandraceae). *Botanical Journal of the Linnean Society* 140:321–381.

———. 2004. Comparative floral structure and systematics in Cucurbitales (Corynocarpaceae, Coriariaceae, Tetramelaceae, Datiscaceae, Begoniaceae, Cucurbitaceae, Anisophylleaceae). *Botanical Journal of the Linnean Society* 145:129–185.

———. 2005a. Comparative floral structure and systematics in Celastrales (Celastraceae, Parnassiaceae, Lepidobotryaceae). *Botanical Journal of the Linnean Society* 149(2):129–194.

———. 2005b. Comparative floral structure and systematics in Crossosomatales (Crossosomataceae, Stachyuraceae, Staphyleaceae, Aphloiaceae, Geissolomataceae, Ixerbaceae, Strasburgeriaceae). *Botanical Journal of the Linnean Society* 147(1):1–46.

———. 2006. Floral structure and systematics in four orders of rosids, including a broad survey of floral mucilage cells. *Plant Systematics and Evolution* 260:223–251.

———. 2008. Comparative floral structure and systematics in Chrysobalanaceae s.l. (Chrysobalanaceae, Dichapetalaceae, Euphroniaceae, Trigoniaceae; Malpighiales). *Botanical Journal of the Linnean Society* 157(2):249–309.

———. 2011. Comparative floral structure and systematics in Rhizophoraceae, Erythroxylaceae and the potentially related Ctenolophonaceae, Linaceae, Irvingiaceae and Caryocaraceae (Malpighiales). *Botanical Journal of the Linnean Society* 166(4):331–416.

———. 2013. Comparative floral structure and systematics of the clade of Lophopyxidaceae and Putranjivaceae (Malpighiales). *Botanical Journal of the Linnean Society* 172(4):404–448.

Matthews, M.L., P.K. Endress, J. Schönenberger, and E.M. Friis. 2001. A comparison of floral structures of Anisophylleaceae and Cunoniaceae and the problem of their systematic position. *Annals of Botany* 88:439–455.

Maurin, O., A.P. Davis, M. Chester, E.F. Mvungi, Y. Jaufeerally-Fakim, and M.F. Fay. 2007. Towards a phylogeny for *Coffea* (Rubiaceae): Identifying well-supported lineages based on nuclear and plastid DNA sequences. *Annals of Botany* 100:1565–1583.

Mauseth, J.D. 2004. Wide-band tracheids are present in almost all species of Cactaceae. *Journal of Plant Research* 117:69–76.

———. 2006. Structure-function relationships in highly modified shoots of Cactaceae. *Annals of Botany* 98:901–926.

Mavrodiev, E.V., M. Gitzendanner, A.K. Calaminus, R.M. Baldini, P.S. Soltis, and D.E. Soltis. 2012. Molecular phylogeny of *Tragopogon* L. (Asteraceae) based on seven nuclear loci (Adh, GapC, LFY, AP3, PI, ITS, and ETS). *Journal of Plant Taxonomy and Geography* 67(2):111–137.

Mayer, V., M. Moller, M. Perret, and A. Weber. 2003. Phylogenetic position and generic differentiation of Epithemateae (Gesneriaceae) inferred from plastid DNA sequence data. *American Journal of Botany* 90:321–329.

Mayr, E.M., and A. Weber. 2006. Calceolariaceae: Floral development and systematic implications. *American Journal of Botany* 93:327–343.

Mayr, G. 2013. The age of the crown group of passerine birds and its evolutionary significance—molecular calibrations versus the fossil record. *Systematics and Biodiversity* 11:7–13.

Mayrose, I., S.H. Zhan, C.J. Rothfels, N. Arrigo, M.S. Barker, L.H. Rieseberg, and S.P. Otto. 2014. Methods for studying polyploid diversification and the dead end hypothesis: A reply to Soltis et al. (2014). *New Phytologist*. doi:10.1111/nph.13192.

Mayrose, I., S.H. Zhan, C.J. Rothfels, K. Magnuson-Ford, M.S. Barker, L.H. Rieseberg, and S.P. Otto. 2011. Recently formed polyploid plants diversify at lower rates. *Science* 333(6047):1257–1257.

McAbee, J.M., T.A. Hill, D.J. Skinner, A. Izhaki, B.A. Hauser, R.J. Meister, G. Venugopala Reddy, E.M. Meyerowitz, J.L. Bowman, and C.S. Gasser. 2006. ABERRANT TESTA SHAPE encodes a KANADI family member, linking polarity determination to separation and growth of Arabidopsis ovule integuments. *Plant Journal* 46:522–531.

McAbee, J.M., R.K. Kuzoff, and C.S. Gasser. 2005. Mechanisms of derived unitegmy among *Impatiens* species. *The Plant Cell* 17:1674–1684.

McClain, A.M., and S.R. Manchester. 2001. *Dipteronia* (Sapindaceae) from the Tertiary of North America and implications for the phytogeographic history of the Aceroideae. *American Journal of Botany* 88(7):1316–1325.

McDade, L. 1992. Pollinator relationships, biogeography, and phylogenetics. *BioScience* 42:21–26.

McDade, L., S.E. Masta, M.L. Moody, and E. Waters. 2000. Phylogenetic relationships among Acanthaceae: Evidence from two genomes. *Systematic Botany* 25:106–121.

McKain, M.R., N. Wickett, Y. Zhang, S. Ayyampalayam, W.R. McCombie, M.W. Chase, J.C. Pires, C.W. dePamphilis, and J. Leebens-Mack. 2012. Phylogenomic analysis of transcriptome data elucidates co-occurrence of a paleopolyploid event and the origin of bimodal karyotypes in Agavoideae (Asparagaceae). *American Journal of Botany* 99:397–406.

McKenna, M.C. 1975. Toward a phylogenetic classification of the Mammalia. *In* W.P. Luckett, and F.S. Szalay [eds.], Phylogeny of the Primates: A Multidisciplinary Approach, 21–46. Plenum, New York, NY.

McLean, R., and M. Evans. 1934. The Mäule Reaction and the systematic position of the Gnetales. *Nature* 134:936–937.

McMahon, M.M., and M.J. Sanderson. 2006. Phylogenetic supermatrix analysis of GenBank sequences from 2228 papilionoid Legumes. *Systematic Biology* 55:818–836.

McNeal, J.R., J.R. Bennett, A.D. Wolfe, and S. Matthews. 2013. Phylogeny and origins of holoparasitism in Orobanchaceae. *American Journal of Botany* 100:971–983.

Meerow, A.W., M.F. Fay, C.L. Guy, Q.-B. Li, F.Q. Qamaruz-Zaman, and M.W. Chase. 1999. Systematics of Amaryllidaceae based on cladistic analyses of plastid *rbcL* and *trnL–F* sequence data. *American Journal of Botany* 86:1325–1345.

Meeuse, A.D.J. 1975. Floral evolution as a key to angiosperms descent. *Acta Botanica Indica* 3:1–18.

Meimberg, H., P. Dittrich, G. Bringmann, J. Schlauer, and G. Heubl. 1999. Molecular phylogeny of Caryophyllidae s.i. based on *matK* sequences with special emphasis on carnivorous taxa. *Plant Biology* 2:218–228.

Meimberg, H., and G. Heubl. 2006. Introduction of a nuclear marker for phylogenetic analysis of Nepenthaceae. *Plant Biology* 8:831–840.

Meimberg, H., A. Wistuba, P. Dittrich, and G. Heubl. 2001. Molecular phylogeny of Nepenthaceae based on cladistic analysis of plastid *trnK* intron sequence data. *Plant Biology* 3:164–175.

Meinhardt, H. 1982. Models of Biological Pattern Formation. Academic Press, London, UK.

Melchior, H. 1964. Guttiferales. *In* H. Melchior [ed.], A. Engler's Syllabus der Pflanzenfamilien, 12th ed., 156–175. Gebrüder Borntraeger, Berlin-Nikolassee, Germany.

Meller, B. 2006. Comparative investigation of modern and fossil *Toricellia* fruits—a disjunctive element in the Miocene and Eocene of Central Europe and the USA. *Beitraege Paläontologie* 30:315–327.

Mello-Silva, R., D.Y.A.C. Santos, M.L.F. Salatino, L.B. Motta, M.B. Cattai, D. Sasaki, J. Lovo, P.B. Pita, C. Rocini, C.D.N. Rodrigues, M. Zarrei, and M.W. Chase. 2011. Five vicariant genera from Gondwana: The Velloziaceae as shown by molecules and morphology. *Annals of Botany* 108:87–102.

Melville, R. 1962. A new theory of the angiosperm flower. I. The gynoecium. *Kew Bulletin* 17:1–50.

———. 1963. A new theory of the angiosperm flower. II. The androecium. *Kew Bulletin* 17:51–63.

———. 1969. Leaf venation patterns and the origin of angiosperms. *Nature* 224:121–125.

———. 1971. Red Data Book. 5. Angiospermae. International Union of Conservation Nature Natural Resources, Survival Service Commission. Arts Graphiques, Heliographia, Lausanne, Switzerland.

Mennega, A.M.W. 1980. Anatomy of the secondary xylem. *In* A.J.M. Leeuwenberg [ed.], Die Natürlichen Pflanzenfamilien: Fam. Loganiaceae, 112–161. Duncker and Humblot, Berlin, Germany.

Mennes, C.B., V.K.Y. Lam, P.J. Rudall, S.P. Lyon, S.W. Graham, E.F. Smets, and V.S.F.T. Merckx. 2015. Ancient Gondwana break-up explains the distribution of the mycoheterotrophic family Corsiaceae (Liliales). *Journal of Biogeography* 42(6):1123–1136. doi:10.1111/jbi.12486.

Mennes, C.B., E.F. Smets, S.N. Moses, and V.S.F.T. Merchx. 2013. New insights in the long-debated evolutionary history of Triuridaceae (Pandanales). *Molecular Phylogenetic Evolution* 69:994–1004.

Merckx, V., F.T. Bakker, S. Huysmans, and E. Smets. 2009. Bias and conflict in phylogenetic inference of mycoheterotrophic plants: A case study in Thismiaceae. *Cladistics* 25(1):64–77.

Merckx, V., and J.V. Freudenstein. 2010. Evolution of mycoheterotrophy in plants: A phylogenetic perspective. *New Phytologist* 185(3):605–609.

Merckx, V., C.B. Mennes, K.G. Peary, and J. Geml. 2013. Evolution and diversication. *In* V. Merckx [ed.], Mycoheterotrophy: The Biology of Plants Living on Fungi, 215–244. Springer, New York, NY, USA.

Merckx V., P. Schols, H. Maas-van den Kamer, P. Maas, S. Huysmans, and E. Smets. 2006. Phylogeny and evolution of Burmanniaceae (Dioscoreales) based on nuclear and mitochondrial data. *American Journal of Botany* 93:1684–1698.

Meredith, R.W., J.E. Janecka, J. Gatesy, O.A. Ryder, C.A. Fisher, E.C. Teeling, A. Goodbla, E. Eizirik, T.L.L. Simao, T. Stadler, D.L. Rabosky, R.L. Honeycutt, J.J. Flynn, C.M. Ingram, C. Steiner, T.L. Williams, T.J. Robinson, A. Burk-Herrick, M. Westerman, N.A. Ayoub, M.S. Springer, and W.J. Murphy. 2011. Impacts of the Cretaceous terrestrial revolution and KPg extinction on mammal diversification. *Science* 334:521–524.

Merino Sutter, D., and P.E. Endress. 2003. Female flower and cupule structure in Balanopaceae, an enigmatic rosid family. *Annals of Botany* 92:459–469.

Metcalfe, C.R., and L. Chalk. 1950. Anatomy of the Dicotyledons. Leaves, Stem, and Wood in Relation to Taxonomy with Notes on Economic Uses. Clarendon Press, Oxford, UK.

———. 1988/1989. Anatomy of the Dicotyledons, 2nd ed. Oxford University Press, Oxford, UK.

Meyen, S.V. 1988. Origin of the Angiosperm gynoecium by gamoheterotopy. *Botanical Journal of the Linnean Society* 97:171–178.

Meyer, H.W., and S.R. Manchester. 1997. The Oligocene Bridge Creek flora of the John Day Formation, Oregon. *University of California Publications in Geological Sciences* 141:1–195.

Meyer, K.M., S.B. Hoot, and M.T.K. Arroyo. 2010. Phylogenetic affinities of South American anemone (Ranunculaceae), including the endemic segregate genera, *Barneoudia* and *Oreithales*. *International Journal of Plant Sciences* 171(3):323–331.

Meyerowitz, E.M. 1997. The search for homeotic gene homologs in basal angiosperms and Gnetales: A potential new source of data on the evolutionary origin of flowers. *International Journal of Plant Sciences* 158:S131–S142.

Meyers, B.C., S.V. Tingey, and M. Morgante. 2001. Abundance, distribution and transcriptional activity of repetitive elements in the maize genome. *Genome Research* 11:1660–1676.

Michael, T.P. 2014. Plant genome size variation: Bloating and purging DNA. *Briefings in Functional Genomics* 13(4):308–317. doi:10.1093/bfgp/elu005.

Michael, T.P., and S. Jackson. 2013. The First 50 Plant Genomes. *Plant Genome* 6(2):1–7.

Michelangeli, F.A., D.S. Penneys, J. Giza, D.E. Soltis, M.H. Hils, and J.D. Skean, Jr. 2004. A preliminary phylogeny of the tribe Miconieae (Melastomataceae) based on nrITS sequence data and its implication on inflorescence position. *Taxon* 53:279–290.

Miikeda, O., K. Kita, T. Handa, and T. Yukawa. 2006. Phylogenetic relationships of *Clematis* (Ranunculaceae) based on chloroplast and nuclear DNA sequences. *Botanical Journal of the Linnean Society* 152(2):153–168.

Millan, M., and W. Crepet. 2014. The fossil record of the Solanaceae revisited and revised–the fossil record of Rhamnaceae enhanced. *Botanical Review* 80(2): 73–106.

Miller, N.G. 1971. The genera of Polygalaceae in the southeastern United States. *Journal of the Arnold Arboretum* 52:267–284.

Miller, R.B. 1975. Systematic anatomy of the xylem and comments on the relationships of the Flacourtiaceae. *Journal of the Arnold Arboretum* 56:20–102.

Miller, R.E., T.R. Buckley, and P.S. Manos. 2002. An examination of the monophyly of morning glory taxa using Bayesian phylogenetic inference. *Systematic Biology* 51:740–753.

Ming, R., R. VanBuren, Y. Liu, M. Yang, Y. Han, L.-T. Li, Q. Zhang, M.-J. Kim, M.C. Schatz, M. Campbell, J. Li, J.E. Bowers, H. Tang, E. Lyons, A.A. Ferguson, G. Narzisi, D.R. Nelson, C.E. Blaby-Haas, A.R. Gschwend, Y. Jiao, J.P. Der, F. Zeng, J. Han, X.J. Min, K.A. Hudson, R. Singh, A.K. Grennan, S.J. Karpowicz, J.R. Watling, K. Ito, S.A. Robinson, M.E. Hudson, Q. Yu, T.C. Mockler, A. Carroll, Y. Zheng, R. Sunkar, R. Jia, N. Chen, J. Arro, C.M. Wai, E. Wafula, A. Spence, Y. Han, L. Xu, J. Zhang, R. Peery, M.J. Haus, W. Xiong, J.A. Walsh, J. Wu, M.-L. Wang, Y.J. Zhu, R.E. Paull, A.B. Britt, C. Du, S.R. Downie, M.A. Schuler, T.P. Michael, S.P. Long, D.R. Ort, J.W. Schopf, D.R. Gang, N. Jiang, M. Yandell, C.W. dePamphilis, S.S. Merchant, A.H. Paterson, B.B. Buchanan, S. Li, and J. Shen-Miller. 2013. Genome of the long-living sacred lotus (*Nelumbo nucifera* Gaertn.). *Genome Biology* 14:R41. doi:10.1186/gb-2013-14-5-r41.

Mirarab, S., R. Reaz, M.S. Bayzid, T. Zimmermann, M.S. Swenson, and T. Warnow. 2014. ASTRAL: Genome-scale coalescent-based species tree estimation. *Bioinformatics* 30(17):i541–i548. doi:10.1093/bioinformatics/btu462.

Misa Ward, N., and R. Price. 2002. Phylogenetic relationships of Marcgraviaceae: Insights from three chloroplast genes. *Systematic Botany* 27:149–160.

Mishler, B.D. 1994. Cladistic analysis of molecular and morphological data. *American Journal of Physical Anthropology* 94:143–156.

———. 1999. Getting rid of species? *In* R.A. Wilson [ed.], Species. New Interdisciplinary Essays, 307–315. MIT Press, London, UK.

Mishler, B.D., M.J. Donoghue, and V.A. Albert. 1991. The decay index as a measure of relative robustness within a cladogram. Hennig X (Annual meeting of the Willi Hennig Society), Toronto, Canada. Abstract.

Misof, B., S. Liu, K. Meusemann, R.S. Peters, A. Donath, C. Mayer, P.B. Frandsen, J. Ware, T. Flouri, R.G. Beutel, O. Niehuis, M. Petersen, F. Izquierdo-Carrasco, T. Wappler, J. Rust, A.J. Aberer, U. Aspock, H. Aspock, D. Bartel, A. Blanke, S. Berger, A. Bohm, T.R. Buckley, B. Calcott, J. Chen, F. Friedrich, M. Fukui, M. Fujita, C. Greve, P. Grobe, S. Gu, Y. Huang, L.S. Jermiin, A.Y. Kawahara, L. Krogmann, M. Kubiak, R. Lanfear, H. Letsch, Y. Li, Z. Li, J. Li, H. Lu, R. Machida, Y. Mashimo, P. Kapli, D.D. McKenna, G. Meng, Y. Nakagaki, J.L. Navarrete-Heredia, M. Ott, Y. Ou, G. Pass, L. Podsiadlowski, H. Pohl, B.M. von Reumont, K. Schutte, K. Sekiya, S. Shimizu, A. Slipinski, A. Stamatakis, W. Song, X. Su, N.U. Szucsich, M. Tan, X. Tan, M. Tang, J. Tang, G. Timelthaler, S. Tomizuka, M. Trautwein, X. Tong, T. Uchifune, M.G. Walzl, B.M. Wiegmann, J. Wilbrandt, B. Wipfler, T.K.F. Wong, Q. Wu, G. Wu, Y. Xie, S. Yang, Q. Yang, D.K. Yeates, K. Yoshizawa, Q. Zhang, R. Zhang, W. Zhang, Y. Zhang, J. Zhao, C. Zhou, L. Zhou, T. Ziesmann, S. Zou, Y. Li, X. Xu, Y. Zhang, H. Yang, J. Wang, J. Wang, K.M. Kjer, and X. Zhou. 2014. Phylogenomics resolves the timing and pattern of insect evolution. *Science* 346:763–767.

Mitchell-Olds, T., and M.J. Clauss. 2002. Plant evolutionary genomics. *Current Opinions in Plant Biology* 5:74–79.

Mithen, R., R. Bennett, and J. Marquez. 2010. Glucosinolate biochemical diversity and innovation in the Brassicales. *Phytochemistry* 71(17–18):2074–2086.

Mølgaard, P. 1985. Caffeic acid as a taxonomic marker in dicotyledons. *Nordic Journal of Botany* 5:203–213.

Molina, J., K.M. Hazzouri, D. Nickrent, M. Geisler, R.S. Meyer, M.M. Pentony, J.M. Flowers, P. Pelser, J. Barcelona, S.A. Inovejas, I. Uy, W. Yuan, O. Wilkins, C.-I. Michel, S. LockLear, G.P. Concepcion, and M.D. Purugganan. 2014. Possible loss of the chloroplast genome in the parasitic flowering plant *Rafflesia lagascae* (Rafflesiaceae). *Molecular Biology and Evolution* 31(4):793–803.

Mols, J.B., B. Gravendeel, L.W. Chatrou, M.D. Pirie, P.C. Bygrave, M.W. Chase, and P.A. Kessler. 2004. Identifying clades in Asian Annonaceae: Monophyletic genera in the polyphyletic Miliusieae. *American Journal of Botany* 91:590–600.

Money, L.L., I.W. Bailey, and B.G.L. Swamy. 1950. The morphology and relationships of the Monimiaceae. *Journal of the Arnold Arboretum* 31:372–404.

Monson, R.K., G.E. Edwards, and M.S.B. Ku. 1984. C-3-C-4 intermediate photosynthesis in plants. *BioScience* 34(9):563–566, 571–574.

Montgomery, R.A., and T.J. Givnish. 2008. Adaptive radiation of photosynthetic physiology in the Hawaiian lobeliads: Dynamic photosynthetic responses. *Oecologia* 155:455–467.

Moody, M., L. Hufford, D.E. Soltis, and P.S. Soltis. 2001. Phylogenetic relationships of Loasaceae subfamily Gronovioideae inferred from *matK* and ITS sequence data. *American Journal of Botany* 88:236–336.

Moore, B.R., and M.J. Donoghue. 2009. A Bayesian approach for evaluating the impact of historical events on rates of diversification. *Proceedings of the National Academy of Sciences USA* 106:4307–4312.

Moore, M.J, C.D. Bell, P.S. Soltis, and D.E. Soltis. 2007. Using plastid genome-scale data to resolve enigmatic relationships among Basal Angiosperms. *Proceedings of the National Academy of Sciences USA* 104:19363–19368.

Moore, M.J., N. Hassan, M.A. Gitzendanner, R.A. Bruenn, M. Croley, A. Vandeventer, J.W. Horn, A. Dhingra, S.F. Brockington, M. Latvis, J. Ramdial, R. Alexandre, A. Piedrahita, Z. Xi, C.C. Davis, P.S. Soltis, and D.S. Soltis. 2011. Phylogenetic analysis of the plastid inverted repeat for 244 species: Insights into deeper-level angiosperm relationships from a long, slowly evolving sequence region. *International Journal of Plant Sciences* 172(4):541–558.

Moore, M.J., P.S. Soltis, C.D. Bell, J.G. Burleigh, and D.E. Soltis. 2010. Phylogenetic analysis of 83 plastid genes further resolves the early diversification of eudicots. *Proceedings of the National Academy of Sciences USA* 107:4623–4628.

Moreau, C.S., C.D. Bell, R. Vila, S.B. Archibald, and N.E. Pierce. 2006. Phylogeny of the ants: Diversification in the age of Angiosperms. *Science* 312(5770):101–104.

Morgan, D.R., and D.E. Soltis. 1993. Phylogenetic relationships among Saxifragaceae sensu lato based on *rbcL* sequence data. *Annals of the Missouri Botanical Garden* 80:631–660.

Morgan, D.R., D.E. Soltis, and K. Robertson. 1994. Systematic and evolutionary implications of *rbcL* sequence variation in Rosaceae. *American Journal of Botany* 81:890–903.

Morgan, M. 2000. Evolution of interactions between plants and their pollinators. *Plant Species Biology* 15:249–259.

Morley, R J. 1982. Fossil pollen attributable to *Alangium* Lamarck (Alangiaceae) from the Tertiary of Malaysia. *Review of Palaeobotany and Palynology* 36(1/2):65–94.

Mort, M.E., and D.E. Soltis. 1999. Phylogenetic relationships and the evolution of ovary position in *Saxifraga* Section *Micranthes*. *Systematic Botany* 24:139–147.

Mort, M.E., D.E. Soltis, P.S. Soltis, J. Francisco-Ortega, and A. Santos-Guerra. 2001. Phylogenetic relationships and evolution of Crassulaceae inferred from *matK* sequence data. *American Journal of Botany* 88:76–91.

Mort, M.E., P.S. Soltis, D.E. Soltis, and M.L. Mabry. 2000. Comparison of three methods for estimating internal support on phylogenetic trees. *Systematic Biology* 49:160–171.

Morton, C.M. 2011. Newly sequenced nuclear gene (*Xdh*) for inferring angiosperm phylogeny. *Annals of the Missouri Botanical Garden* 98(1):63–89.

Morton, C.M., M.W. Chase, and K.G. Karol. 1997. Phylogenetic relationships of two anomalous dicot genera, *Physena* and *Asteropeia*: evidence from *rbcL* plastid DNA sequences. *Botanical Review* 63:231–239.

Morton, C.M., M.W. Chase, K.A. Kron, and S.M. Swensen. 1996. A molecular evaluation of the monophyly of the order Ebenales based upon *rbcL* sequence data. *Systematic Botany* 21:567–586.

Moulia, B. 2013. Plant biomechanics and mechanobiology are convergent paths to flourishing interdisciplinary research. *Journal of Experimental Botany* 64:4617–4633.

Mouradov, A.T., T. Glassick, L. Murphy, B. Fowler, S. Majla, and R.D. Teasdale. 1998. NEEDLY, a *Pinus radiata* ortholog of *FLORICAULA/LEAFY* genes, expressed in both reproductive and vegetative meristems. *Proceedings of the National Academy of Sciences USA* 95:6537–6542.

Mower, J.P., K. Jain, and N.J. Hepburn. 2012. The role of horizontal transfer in shaping the plant mitochondrial genome. *In* L. Maréchal-Drouard [ed.], *Advances in Botanical Research*, Volume 63: Mitochondrial Genome Evolution, 41–69. Elsevier, Netherlands.

Mower, J.P., S. Stefanović, W. Hao, J.S. Gummow, K. Jain, D. Ahmed, and J.D. Palmer. 2010. Horizontal acquisition of multiple mitochondrial genes from a parasitic plant followed by gene conversion with host mitochondrial genes. *BMC Biology* 8:150. doi:10.1186/1741-7007-8-150.

Mower, J.P., S. Stefanović, G.J. Young, and J.D. Palmer. 2004. Plant genetics—Gene transfer from parasitic to host plants. *Nature* 432(7014):165–166.

Muasya, A.M., D.A. Simpson, M.W. Chase, and A. Culham. 2001. A phylogenetic analysis of *Isolepis* and allied genera (Cyperaceae) based on plastid *rbcL* and *trnL–F* DNA sequence data. *Systematic Botany* 26:342–353.

Muasya, A.M., D.A. Simpson, A. Culham, and M.W. Chase. 1998. An assessment of suprageneric phylogeny in Cyperaceae using *rbcL* DNA sequences. *Plant Systematics and Evolution* 211:257–271.

Mueller, K.F., T. Borsch, and K.W. Hilu. 2006b. Phylogenetic utility of rapidly evolving DNA at high taxonomical levels: Contrasting *matK*, *trnT–F*, and *rbcL* in basal angiosperms. *Molecular Phylogenetics and Evolution* 41:99–117.

Mueller, K.F., T. Borsch, L. Legendre, S. Porembski, and W. Barthlott. 2006a. Recent progress in understanding the evolution of carnivorous Lentibulariaceae (Lamiales). *Plant Biology* 8:748–757.

Muellner, A.N., R. Samuel, S.A. Johnson, M. Cheek, T.D. Pennington, and M.W. Chase. 2003. Molecular phylogenetics of Meliaceae (Sapindales) based on nuclear and plastid DNA sequences. *American Journal of Botany* 90(3):471–480.

Muellner, A.N., V. Savolainen, R. Samuel, and M.W. Chase. 2006. The mahogany family "out-of-Africa": Divergence time estimation, global biogeographic patterns inferred from plastid *rbcL* DNA sequences, extant, and fossil distribution of diversity. *Molecular Phylogenetics and Evolution* 40(1):236–250.

Mulcahy, D. 1979. The rise of the angiosperms: A genecological factor. *Science* 206:20–23.

Mulcahy, D., and G. Mulcahy. 1987. The effects of pollen competition. *American Science* 75:44–50.

Mulholland, D.A., P. Cheplogoi, and N.R. Crouch. 2003. Secondary metabolites from *Kirkia acuminata* and *Kirkia wilmsii* (Kirkiaceae). *Biochemical Systematics and Ecology* 31(7):793–797.

Müller, G.B., and G.P. Wagner. 1991. Novelty in evolution. *Annual Reviews in Ecology and Systematics* 22:229–256.

Muller, J. 1981. Fossil pollen records of extant Angiosperms. *Botanical Review* 47(1):1–142.

Muller, K., and T. Borsch. 2005. Phylogenetics of Amaranthaceae based on matK/trnK sequence data: Evidence from parsimony, likelihood, and Bayesian analyses. *Annals of the Missouri Botanical Garden* 92:66–102.

Mummenhoff, K., H. Brüggemann, and J.L. Bowman. 2001. Chloroplast DNA phylogeny and biogeography of *Lepidium* (Brassicaceae). *American Journal of Botany* 88:2051–2063.

Munro, S.L., and H.P. Linder. 1997. The embryology and systematic relationships of *Prionium serratum* (Juncaceae: Juncales). *American Journal of Botany* 84:850–860.

Müntzing, A. 1930. Outlines to a genetic monograph of the genus *Galeopsis*. With special regard to the nature and inheritance of partial sterility. *Hereditas* 13(2/3):185–341.

———. 1936. The evolutonary significance of autopolyploidy. *Hereditas* 21(2/3):263–378.

Murata, J., T. Ohi, S.G. Wu, D. Darnaedi, T. Sugawara, T. Nakanishi, and H. Murata. 2001. Molecular phylogenetics of *Aristolochia* (Aristolochiaceae) inferred from *matK* sequences. *Acta Phytotaxonomica Geobotanica* 52:75–83.

Murbeck, S. 1912. Untersuchungen über den Blütenbau der Papaveraceen. *Kungl Svenska Vetenskapsakademiens Handlingar* 50:1–168.

Mustoe, G.E. 2002. *Hydrangea* fossils from the early Tertiary Chuckanut Formation. *Washington Geology* 30(3/4):17–20.

Nagasawa, N., M. Miyoshi, Y. Sano, H. Satoh, H. Hirano, H. Sakai, and Y. Nagato. 2003. *SUPERWOMAN1* and *DROOPING LEAF* genes control floral organ identity in rice. *Development* 130:705–718.

Nakai, T. 1942. Notulae ad plantas Asiae orientalis. XVIII. *Journal of Japanese Botany* 18:91–120.

Nandi, O.I. 1998a. Ovule and seed anatomy of Cistaceae and related Malvanae. *Plant Systematics and Evolution* 209:239–264.

———. 1998b. Floral development and systematics of Cistaceae. *Plant Systematics and Evolution* 212:107–134.

Nandi, O.I., M.W. Chase, and P.K. Endress. 1998. A combined cladistic analysis of angiosperms using *rbcL* and nonmolecular data sets. *Annals of the Missouri Botanical Garden* 85:137–212.

Nash, G. 1903. A revision of the family Fouquieriaceae. *Bulletin of the Torrey Botanical Club* 30:449–459.

Naumann, J., K. Salomo, J.P. Der, E.K. Wafula, J.F. Bolin, E. Maass, L. Frenzke, M.-S. Samain, C. Neinhuis, C.W. dePamphilis, and S. Wanke. 2013. Single-copy nuclear genes place haustorial Hydnoraceae within Piperales and reveal a Cretaceous origin of multiple parasitic Angiosperm lineages. *PLoS One* 8(11): e79204. doi:10.1371/journal.pone.0079204.

Naylor, G.J.P., and W.M. Brown. 1997. Structural biology and phylogenetic estimation. *Nature* 388:527–528.

Nazarov, V.V., and G. Gerlach. 1997. The potential seed productivity of orchid flowers and peculiarities of their pollination systems. *Lindleyana* 12:188–204.

Negrón-Ortiz, V. 2007. Chromosome numbers, nuclear DNA content, and polyploidy in *Consolea* (Cactaceae), an endemic cactus of the Caribbean Islands. *American Journal of Botany* 94:1360–1370.

Nei, M., S. Kumar, and K. Takahashi. 1998. The optimization principle in phylogenetic analysis tends to give incorrect topologies when the number of nucleotides or amino acids used is small. *Proceedings of the National Academy of Sciences USA* 95:12390–12397.

Neinhuis, C., S. Wanke, K.W. Hilu, K. Mueller, and T. Borsch. 2005. Phylogeny of Aristolochiaceae based on parsimony, likelihood, and Bayesian analyses of *trnL–trnF* sequences. *Plant Systematics and Evolution* 250:7–26.

Nelson, G. 1972. Phylogenetic relationship and classification. *Systematic Zoology* 21:227–231.

———. 1973. Classification as an expression of phylogenetic relationships. *Systematic Zoology* 22:344–359.

Neubig, K.M., W.M. Whitten, N.H. Williams, M.A. Blanco, L. Endara, J.G. Burleigh, K. Silvera, J.C. Cushman, and M.W. Chase. 2012. Generic recircumscriptions of Oncidiinae (Orchidaceae: Cymbidieae) based on maximum likelihood analysis of combined DNA datasets. *Botanical Journal of the Linnean Society* 168:117–146.

Newman, E., B. Anderson, and S.D. Johnson. 2012. Flower colour adaptation in a mimetic orchid. *Proceedings of the Royal Society of London B* 279:2309–2313.

Ng, D.W.K., C. Zhang, M. Miller, Z. Shen, S.P. Briggs, and Z.J. Chen. 2012. Proteomic divergence in *Arabidopsis* autopolyploids and allopolyploids and their progenitors. *Heredity* 108(4):419–430.

Nichols, D.J. 2002. Palynology and palynostratigraphy of the Hell Creek Formation in North Dakota: A microfossil record of plants at the end of Cretaceous time. *Geological Society of America Special Paper* 361:393–456.

Nickerson, J., and G. Drouin. 2004. The sequence of the largest subunit of RNA polymerase II is a useful marker for inferring seed plant phylogeny. *Molecular Phylogenetics and Evolution* 31:403–415.

Nickrent, D.L. 1997. The Parasitic Plant Connection. Last updated, December 20, 2014. http://www.parasiticplants.siu.edu/.

———. 2002. Orígenes filogenéticos de las plantas parásites. *In* J.A. López–Sáez, P. Catalán, and L. Sáez [eds]. Plantas parasites de la Peninsula Ibérica e Islas Baleares, 29–56. Mundi-Prensa Libros., S. A., Madrid, Spain.

———. 2007. Cytinaceae are sister to Muntingiaceae (Malvales). *Taxon* 56(4):1129–1135.

Nickrent, D.L., A. Blarer, Y.-L. Qiu, D.E. Soltis, P.S. Soltis, and M. Zanis. 2002. Molecular data and the relationships of the enigmatic angiosperm Hydnoraceae. *American Journal of Botany* 89:1809–1817.

Nickrent, D.L., A. Blarer, Y.-L. Qiu, R. Vidal-Russell, and F.E. Anderson. 2004. Phylogenetic inference in Rafflesiales: The influence of rate heterogeneity and horizontal gene transfer. *BMC Evolutionary Biology* 4:40. doi:10.1186/1471-2148-4-40.

Nickrent, D.L., J.P. Der, and F.E. Anderson. 2005. Discovery of the photosynthetic relatives of the "Maltese mushroom" *Cynomorium*. *BMC Evolutionary Biology* 5:38. doi:10.1186/1471-2148-5-38.

Nickrent, D.L., and R.J. Duff. 1996. Molecular studies of parasitic plants using ribosomal RNA. *In* M.T.

Moreno, J.I. Cubero, D. Berner, D. Joel, L.J. Musselman, and C. Parker [eds.], Advances in Parasitic Plant Research, 28–52. Junta de Andalucia, Direccion General de Investigacion Agraria, Cordoba, Spain.

Nickrent, D.L., R.J. Duff, A.F. Colwell, A.D. Wolfe, N.D. Young, K.E. Steiner, and C.W. dePamphilis. 1998. Molecular phylogenetics and evolutionary studies of parasitic plants. *In* P.S. Soltis, D.E. Soltis, and J.J. Doyle [eds.], Molecular Systematics of Plants, II. DNA Sequencing, 211–241. Kluwer, Boston, MA.

Nickrent, D.L., and M.A. Garcia. 2009. On the brink of holoparasitism: Plastome evolution in dwarf mistletoes (*Arceuthobium*, Viscaceae). *Journal of Molecular Evolution* 68(6):603–615.

Nickrent, D.L., and V. Malécot. 2001. A molecular phylogeny of Santalales. *In* A. Fer, P. Thalouarn, D. Joel, L.J. Musselman, C. Parker, and J.A.C. Verklejj [eds.], Proceedings of the 7th International Parasitic Weed Symposium, 69–74. Faculté des Sciences, Université de Nantes, Nantes, France.

Nickrent, D.L., V. Malécot, R. Vidal-Russell, and J.P. Der. 2010. A revised classification of Santalales. *Taxon* 59(2):538–558.

Nickrent, D.L., C.L. Parkinson, J.D. Palmer, and R.J. Duff. 2000. Multigene phylogeny of land plants with special reference to bryophytes and the earliest land plants. *Molecular Biology and Evolution* 17:1885–1895.

Nickrent, D.L., and D.E. Soltis. 1995. A comparison of angiosperm phylogenies from nuclear 18S rDNA and *rbcL* sequences. *Annals of the Missouri Botanical Garden* 82:208–234.

Nickrent, D.L. and E.M. Starr. 1994. High rates of nucleotide substitution in nuclear small-subunit (18S) rDNA from holoparasitic flowering plants. *Journal of Molecular Evolution* 39:62–70.

Nicolas, A.N., and G.M. Plunkett. 2009. The demise of subfamily Hydrocotyloideae (Apiaceae) and the realignment of its genera across the entire order Apiales. *Molecular Phylogenetics and Evolution* 53:134–151.

Nicoletti, M., A. Di Fabio, M. Seralini, J.A. Garbarino, and M.C. Chamy. 1991. Iridoids from *Loasa tricolor*. *Biochemical Systematics and Evolution* 19:167–170.

Nikolov, L.A., P.K. Endress, M. Sugumaran, S. Sasirat, S. Vessabutr, E.M. Kramer, and C.C. Davis. 2013. Developmental origins of the world's largest flowers, Rafflesiaceae. *Proceedings of the National Academy of Sciences USA* 110(46):18578–18583.

Nikolov, L.A., Y.M. Staedler, S. Manickam, J. Schönenberger, P.K. Endress, E.M. Kramer, and C.C. Davis. 2014. Floral structure and development in Rafflesiaceae with emphasis on their exceptional gynoecia. *American Journal of Botany* 101(2):225–243.

Nishida, H., and M. Nishida. 1988. *Protomonimia kasainakajhongii* gen. et sp. nov.: A permineralized magnolialean fructification from the mid-Cretaceous of Japan. *Botanical Magazine* 101:397–437.

Nishida, H., K.B. Pigg, and J.F. Rigby. 2003. Palaeobotany: Swimming sperm in an extinct Gondwanan plant. *Nature* 422:396–397.

Nishino, E. 1978. Corolla tube formation in four species of Solanaceae. *Botanical Magazine, Tokyo* 91:263–277.

———. 1983. Corolla tube formation in the Primulaceae and Ericales. *Botanical Magazine, Tokyo* 96:319–342.

Nixon, K.C. 1999. The parsimony ratchet: A rapid means for analyzing large data sets. *Cladistics* 15:407–414.

———. 2000. Winclada. Program and documentation. K.C. Nixon, Ithaca, NY.

Nixon, K.C., and W.L. Crepet. 1993. Late Cretaceous fossil flowers of ericalean affinity. *American Journal of Botany* 80:616–623.

Nixon, K.C., W.L. Crepet, D.W. Stevenson, and E.M. Friis. 1994. A reevaluation of seed plant phylogeny. *Annals of the Missouri Botanical Garden* 81:484–533.

Nobel, P.S. 1988. Environmental biology of agaves and cacti. Cambridge University Press, Cambridge, UK.

Nowicke, J.W. 1968. Palyotaxonomic study of the Phytolaccaceae. *Annals of the Missouri Botanical Garden* 55:294–364.

———. 1975. Pollen morphology in the order Centrospermae. *Grana* 15:51–77.

———. 1994. Pollen morphology and exine ultrastructure. *In* H.-D. Behnke, and T.J. Mabry [eds.], Caryophyllales. Evolution and Systematics, 167–222. Springer-Verlag, Berlin, Germany.

———. 1996. Pollen morphology, exine structure and the relationships of Basellaceae and Didiereaceae to Portulacaceae. *Systematic Biology* 21:187–208.

Nowicke, J.W., and J.J. Skvarla. 1977. Pollen morphology and the relationship of the Plumbaginaceae, Polygonaceae and Primulaceae to the order Centrospermae. *Smithsonian Contributions to Botany* 37:1–64.

———. 1979. Pollen morphology: The potential to influence in higher order systematics. *Annals of the Missouri Botanical Garden* 66:633–700.

———. 1984. Pollen morphology and the relationships of *Simmondsia chinensis* to the order Euphorbiales. *American Journal of Botany* 71:210–215.

Nuraliev, M.S., G.V. Degtjareva, D.D. Sokoloff, A.A. Oskolski, T.H. Samigullin, and C.M. Valiejo-Roman. 2014. Flower morphology and relationships of *Schefflera subintegra* (Araliaceae, Apiales): An evolutionary step towards extreme floral polymery. *Botanical Journal of the Linnean Society* 175:553–597.

Nyffeler, R. 2007. The closest relatives of cacti: Insights

from phylogenetic analyses of chloroplast and mitochondrial sequences with special emphasis on relationships in the tribe Anacampseroteae. *American Journal of Botany* 94:89–101.

Nyffeler, R., and D.L. Baum. 2000. Phylogenetic relationships of the durians (Bombacaceae–Durioneae or /Malvaceae/Helicteroideae/Durioneae) based on chloroplast and nuclear ribosomal DNA sequences. *Plant Systematics and Evolution* 224:55–82.

Nyffeler, R., and U. Eggli. 2010. Disintegrating Portulacaceae: A new familial classification of the suborder Portulacinae (Caryophyllales) based on molecular and morphological data. *Taxon* 59:227–240.

Obermayer, R., I.J. Leitch, L. Hanson, and M.D. Bennett. 2002. Nuclear DNA C-values in 30 species double the familial representation in pteridophytes. *Annals of Botany* 90(2):209–217.

Ocampo, G., and J.T. Columbus. 2010. Molecular phylogenetics of suborder Cactineae (Caryophyllales), including insights into photosynthetic diversification and historical biogeography. *American Journal of Botany* 97:1827–1847.

———. 2012. Molecular phylogenetics, historical biogeography, and chromosome number evolution in *Portulaca* (Portulacaceae). *Molecular Phylogenetics and Evolution* 63:97–112.

Ocampo, G., N.K. Koteyeva, E.V. Voznesenskaya, G.E. Edwards, T.L. Sage, R.F. Sage, and J.T. Columbus. 2013. Evolution of leaf anatomy and photosynthetic pathways in Portulacaceae. *American Journal of Botany* 100:2388–2402.

Oginuma, K., Z. Giu, and Z.-S. Yue. 1995. Karyomorphology of *Rhoiptelea* (Rhoipteleaceae). *Acta Phytotaxonomica et Geobotanica* 46:147–151.

Ogundipe, O.T., and M.W. Chase. 2009. Phylogenetic analyses of Amaranthaceae based on *matK* DNA sequence data with emphasis on west African species. *Turkish Journal of Botany* 33:153–161.

Oh, S.-H. 2010. Phylogeny and systematics of Crossosomatales as inferred from chloroplast *atpB*, *matK*, and *rbcL* sequences. *Korean Journal of Plant Taxonomy* 40(4):208–217.

Oh, S.-H, and P.S. Manos. 2008. Molecular phylogenetics and cupule evolution in Fagaceae as inferred from nuclear *CRABS CLAW* sequences. *Taxon* 57(2):434–451.

Oh, S.-H, and D. Potter. 2006. Description and phylogenetic position of a new Angiosperm family, Guamatelaceae, inferred from chloroplast *rbcL*, *atpB*, and *matK* sequences. *Systematic Botany* 31(4):730–738.

Ohashi, K., T. Tanikawa, Y. Okumura, K. Kawazoe, N. Tatara, M. Minato, H. Shibuya, I. Kitagawa, A. Shimoyama, M. Yamadaki, Y. Nakazawa, S. Yahara, and T. Nohara. 1993. Indonesian medicinal plants: X. Chemical structures of four new triterpene–glycosides, gongganosides D, E, F, and G, and two secoiridoid–glucosides from the bark of *Bhesa paniculata* (Celastraceae). *Shoyakugaku Zasshi* 47:56–59.

O'Kane, S.L., and I.A. Al-Shehbaz. 2003. Phylogenetic position and generic limits of *Arabidopsis* (Brassicaceae) based on sequences of nuclear ribosomal DNA. *Annals of the Missouri Botanical Garden* 90:603–612.

Oleson, J.M., M. Alarcón, B.K. Ehlers, J.J. Aldasoro, and C. Roquet. 2012. Pollination, biogeography and phylogeny of oceanic island bellflowers (Campanulaceae). *Perspectives in Plant Ecology Evolution and Systematics* 14:169–182.

Olmstead, R.G. 2013. Phylogeny and biogeography in Solanaceae, Verbenaceae and Bignoniaceae: A comparison of continental and intercontinental diversification patterns. *Botanical Journal of the Linnean Society* 171:80–102.

Olmstead, R.G., and L. Bohs. 2006. Summary of molecular systematic research in Solanaceae: 1982–2006. *Acta Horticulturae* 745:255–268.

Olmstead, R.G., L. Bohs, H.A. Migid, E. Santiago-Valentin, V.F. Garcia, and S.M. Collier. 2008. A molecular phylogeny of the Solanaceae. *Taxon* 57:1159–1181.

Olmstead, R.G., B. Bremer, K.M. Scott, and J.D. Palmer. 1993. A parsimony analysis of the Asteridae sensu lato based on *rbcL* sequences. *Annals of the Missouri Botanical Garden* 80:700–722.

Olmstead, R.G., C.W. dePamphilis, A.D. Wolfe, N.D. Young, W.J. Elisons, and A. Reeves. 2001. Disintegration of the Scrophulariaceae. *American Journal of Botany* 88:348–361.

Olmstead, R.G., K.-J. Kim, R.K. Jansen, and S.J. Wagstaff. 2000. The phylogeny of the Asteridae sensu lato based on chloroplast *ndhF* gene sequence. *Molecular Phylogenetics and Evolution* 16:96–112.

Olmstead, R.G., H.J. Michaels, K.M. Scott, and J.D. Palmer. 1992. Monophyly of the Asteridae and identification of their major lineages inferred from DNA sequences of *rbcL*. *Annals of the Missouri Botanical Garden* 79:249–265.

Olmstead, R.G., and P.A. Reeves. 1995. Evidence for the polyphyly of the Scrophulariaceae based on chloroplast *rbcL* and *ndhF* sequences. *Annals of the Missouri Botanical Garden* 82:176–193.

Olmstead, R.G., P.A. Reeves, and A.C. Yen. 1998. Patterns of sequence evolution and implications for parsimony analysis of chloroplast DNA. *In* D.E. Soltis, P.S. Soltis, and J.J. Doyle [eds.], Molecular Systematics of Plants II: DNA Sequencing, 164–187. Kluwer Academic Publishers, Boston, MA.

Olmstead, R.G., and J.A. Sweere. 1994. Combining data in phylogenetic systematics: An empirical approach using three molecular data sets in the Solanaceae. *Systematic Biology* 43:467–481.

Olmstead, R.G., J.A. Sweere, R.E. Spangler, L. Bohs, and J.D. Palmer. 1999. Phylogeny and provisional classification of the Solanaceae based on chloroplast DNA. *In* M. Nee, D.E. Symon, R.N. Lester, and J.P. Jessop [eds.], Solanaceae IV, 111–137. Royal Botanic Gardens, Kew, UK.

Olson, M.E. 2002a. Combining data from DNA sequences and morphology for a phylogeny of Moringaceae (Brassicales). *Systematic Botany* 27:55–73.

———. 2002b. Intergeneric relationships within the Caricaceae-Moringaceae clade (Brassicales) and potential morphological synapomorphies of the clade and its families. *International Journal of Plant Sciences* 163:51–65.

O'Quinn, R., and L. Hufford. 2005. Molecular systematics of Montieae (Portulacaceae): Implications for taxonomy, biogeography and ecology. *Systematic Botany* 30:314–331.

Orel, N., and H. Puchta. 2003. Differences in the processing of DNA ends in *Arabidopsis thaliana* and tobacco: Possible implications for genome evolution. *Plant Molecular Biology* 51:523–531.

Orgel, F. 1980. Selfish DNA: The ultimate parasite. *Nature* 284:604–607.

Orr, H.A. 1996. Dobzhansky, Bateson and the genetics of speciation. *Genetics* 144:1331–1335.

Ortiz, R.C., E.A Kellogg, and H.V. Werff. 2007. Molecular phylogeny of the moonseed family (Menispermaceae): Implications for morphological diversification. *American Journal of Botany* 94(8):1425–1438.

Osborn, J.M. 2000. Pollen morphology and ultrastructure of gymnospermous anthophytes. *In* M.M. Harley, C.M. Morton, and S. Blackmore [eds.], Pollen and Spores: Morphology and Biology, 163–185. Royal Botanic Gardens, Kew, UK.

Osborn, T.C., J.C. Pires, J.A. Birchler, D.L. Auger, Z.J. Chen, H.S. Lee, L. Comai, A. Madlung, R.W. Doerge, V. Colot, and R.A. Martienssen. 2003. Understanding mechanisms of novel gene expression in polyploids. *Trends in Genetics* 19(3):141–147.

Osborne, R., M.A. Calonje, K.D. Hill, L. Stanberg, and D.W. Stevenson. 2012. The world list of Cycads. *Memoirs of the New York Botanical Garden* 106:480–510.

Osmond, B., T. Neales, and G. Stange. 2008. Curiosity and context revisited: Crassulacean acid metabolism in the Anthropocene. *Journal of Experimental Botany* 59:1489–1502.

Osmond, C.B., D.L. Nott, and P.M. Firth. 1979. Carbon assimilation patterns and growth of the introduced CAM plant *Opuntia inermis* in eastern Australia. *Oecologia* 40:331–350.

Otto, S.P., and J. Whitton. 2000. Polyploidy incidence and evolution. *Annual Review of Genetics* 34:401–437.

Oxelman, B., M. Backlund, and B. Bremer. 1999. Relationships of the Buddlejaceae s.l. investigated using parsimony Jackknife and Branch Support Analysis of chloroplast *ndhF* and *rbcL* sequence data. *Systematic Botany* 24:164–182.

Oxelman, B., M. Liden, and D. Berglund. 1997. Chloroplast *rps16* intron phylogeny of the tribe Sileneae (Caryophyllaceae). *Plant Systematics and Evolution* 206:393–410.

Pacini, E., and M. Hesse. 2002. Types of pollen dispersal units in orchids, and their consequences for germination and fertilization. *Annals of Botany* 89:653–664.

Page, M., and R. Johnstone. 1992. Variation across species in the size of the nuclear genome supports the junk-DNA explanation for the C-value paradox. *Proceedings of the Royal Society of London B* 249:119–124.

Pagel, M.D. 1998. Inferring evolutionary processes from phylogenies. *Zoologica Scripta* 26:331–348.

———. 1999. The maximum likelihood approach to reconstructing ancestral character states of discrete characters on phylogenies. *Systematic Biology* 48:612–622.

Palmer, J.D. 1985. Evolution of chloroplast and mitochondrial DNA in plants and algae. *In* R.J. MacIntyr [ed.], Molecular Evolutionary Genetics, 131–238. Plenum Publishing, New York, NY.

———. 1987. Chloroplast DNA evolution and biosystematic uses of chloroplast DNA variation. *American Naturalist* 130(Supplement):S6–S29.

————. 1992. Mitochondrial DNA in plant systematics: Applications and limitations. *In* P.S. Soltis, D.E. Soltis, and J.J. Doyle [eds.], Molecular Systematics of Plants, 36–49. Chapman and Hall, New York, NY.

Palmer, J.D., J.M. Nugent, and L.A. Herbon. 1987. Unusual structure of *Geranium* chloroplast DNA: A triple-sized inverted repeat, extensive gene duplications, multiple inversions, and two repeat families. *Proceedings of the National Academy of Sciences USA* 84(3):769–773.

Palmer J.D., D.E. Soltis, and M.W. Chase. 2004. The plant tree of life: An overview and some points of view. *American Journal of Botany* 91(10):1437–1445.

Pan, A.D., B.F. Jacobs, and E.D. Currano. 2014. Dioscoreaceae fossils form the late Oligocene and early Miocene of Ethiopia. *Botanical Journal of the Linnean Society* 175(1):17–28.

Panero, J.L., and V.A. Funk. 2002. Toward a phylogenetic subfamilial classification for the Compositae (Asteraceae). *Proceedings of the Biological Society of Washington* 115:909–922.

———. 2008. The value of sampling anomalous taxa in phylogenetic studies: Major clades of the Asteraceae revealed. *Molecular Phylogenetics and Evolution* 47:757–782.

Pant, D.D. 1977. The plant of *Glossopteris*. *Journal of the Indian Botanical Society* 56:1–23.

Papadopoulos, A.S.T., M.P. Powell, F. Pupulin, J. Warner, J.A. Hawkins, N. Salamin, L. Chittka, N.H. Williams, W.M. Whitten, D. Loader, L.M. Valente, M.W. Chase, and V. Savolainen. 2013. Convergent evolution of floral signals underlies the success of Neotropical orchids. *Proceedings of the Royal Society of London B* 280. doi:10.1098/rspb.2013.0960.

Pardo, F., F. Perich, R. Torres, and F.D. Monache. 1998. Phytotoxic iridoid glucosides from the roots of *Verbascum thapsus*. *Journal of Chemical Ecology* 24:645–653.

Parenti, L.R. 1980. A phylogenetic analysis of the land plants. *Biological Journal of the Linnean Society* 13:225–242.

Park, J.-M., J.-F. Manen, and G.M. Schneeweiss. 2007. Horizontal gene transfer of a plastid gene in the non-photosynthetic flowering plants *Orobanche* and *Phelipanche* (Orobanchaceae). *Molecular Phylogenetics and Evolution* 43(3):974–985.

Park, S., T.A. Ruhlman, J.S.M. Sabir, M.H.Z. Mutwakil, M.N. Baeshen, M.J. Sabir, N.A. Baeshen, and R.K. Jansen. 2014. Complete sequences of organelle genomes from the medicinal plant *Rhazya stricta* (Apocynaceae) and contrasting patterns of mitochondrial genome evolution across asterids. *BMC Genomics* 15:405. doi:10.1186/1471-2164-15-405.

Parkinson, C.L., K.L. Adams, and J.D. Palmer. 1999. Multigene analyses identify the three earliest lineages of extant flowering plants. *Current Biology* 9:1485–1488.

Parkinson, C.L., J.P. Mower, Y.L. Qiu, A.J. Shirk, K.M. Song, N.D. Young, C.W. dePamphilis, and J.D. Palmer. 2005. Multiple major increases and decreases in mitochondrial substitution rates in the plant family Geraniaceae. *BMC Evolutionary Biology* 5:73. doi:10.1186/1471-2148-5-73.

Patel, V.C., J.J. Skvarla, and P.H. Raven. 1984. Pollen characters in relation to the delimitation of Myrtales. *Annals of the Missouri Botanical Garden* 71:858–969.

Paterson, A.H., J.E. Bowers, and B.A. Chapman. 2004. Ancient polyploidization predating divergence of the cereals, and its consequences for comparative genomics. *Proceedings of the National Academy of Sciences USA* 101(26):9903–9908.

Paterson, A.H., B.A. Chapman, J.C. Kissinger, J.E. Bowers, F.A. Feltus, and J.C. Estill. 2006. Many gene and domain families have convergent fates following independent whole-genome duplication events in *Arabidopsis*, *Oryza*, *Saccharomyces* and *Tetraodon*. *Trends in Genetics* 22(11):597–602.

Paterson, A.H., Z. Wang, J. Li, and H. Tang. 2012. Ancient and recent polyploidy in the monocots. *In* P.S. Soltis, and D.E. Soltis [eds.], Polyploidy and Genome Evolution, 93–108. Springer-Verlag, Berlin, Germany.

Patterson, C.D., D.M. Williams, and C.J. Humphries. 1993. Congruence between molecular and morphological phylogenies. *Annual Review of Ecology and Systematics* 24:153–188.

Paun, O., C. Lehnebach, J.T. Johansson, P. Lockhart, and E. Hörandl. 2005. Phylogenetic relationships and biogeography of *Ranunculus* and allied genera (Ranunculaceae) in the Mediterranean region and in the European Alpine system. *Taxon* 54(4):911–930.

Peakall, R., D. Eberst, J. Poldy, R.A. Barrow, W. Francke, C.C. Bower, and F.P. Schiestl. 2010. Pollinator specificity, floral odour chemistry and the phylogeny of Australian sexually deceptive *Chiloglottis* orchids: Implications for pollinator-driven speciation. *New Phytologist* 188:437–450.

Pedersen, K.R., E.M. Friis, P.R. Crane, and A.N. Drinnan. 1994. Reproductive structures of an extinct platanoid from the Early Cretaceous (latest Albian) of eastern North America. *Review of Palaeobotany and Palynology* 80:291–303.

Pelaz, S., G.S. Ditta, E. Baumann, E. Wisman, and M.F. Yanofsky. 2000. B and C floral organ identity functions require *SEPALLATA* MADS-box genes. *Nature* 405(6783):200–203.

Pellicer, J., M.F. Fay, and I.J. Leitch. 2010. The largest eukaryoptic genome of them all? *Botanical Journal of the Linnean Society* 164:10–15.

Pellicer, J., L.J. Kelly, I.J. Leitch, W.B. Zomlefer, and M.F. Fay. 2013. A universe of dwarfs and giants: Genome size and chromosome evolution in the monocot family Melanthiaceae. *New Phytologist* 201:1484–1497.

Pennington, R.T., M. Lavin, H. Ireland, B.B. Klitgaard, J. Preston, and J.-M. Hu. 2001. Phylogenetic relationships of basal papilionoid legumes based upon sequences of the chloroplast intron *trnL*. *Systematic Botany* 26:537–556.

Peralta, I.E., D.M. Spooner, and S. Knapp. 2008. Taxonomy of wild tomatoes and their relatives (*Solanum* sect. *Lycopersicoides*, sect. *Juglandifolia*, set. *Lycopersicon*; Solanaceae). *Systematic Botany Monographs* 84:1–186.

Perkins, J. 1925. Übersicht über die Gattungen der Monimiaceae. Engelmann, Leipzig, Germany.

Perret, M., A. Chautems, A. Onofre de Araujo, and N. Salamin. 2013. Temporal and spatial origin of Gesneriaceae in the New World inferred from plastid DNA sequences. *Botanical Journal of the Linnaean Society* 171:61–79.

Perrier de La Bâthie, H. 1933. Les Brexiées de Madagascar. *Bulletin de la Société Botanique de France* 80:198–204.

———. 1942. Au sujet des affinités des *Brexia*, des Celastracées, et de deux *Brexia* nouveaux de Madagascar. *Bulletin de la Société Botanique de France* 89:219–221.

Persson, C. 2001. Phylogenetic relationships in Polygalaceae based on plastid DNA sequences from the *trnL–F* region. *Taxon* 50(3):763–779.

Petrov, D.A. 2001. Evolution of genome size: New approaches to an old problem. *Trends in Genetics* 17:23–28.

———. 2002. DNA loss and evolution of genome size in *Drosophila*. *Genetica* 115:81–91.

Philbrick, C.T., and D.H. Les. 1996. Evolution of aquatic angiosperm reproductive systems. *BioScience* 46:813–826.

Philippe, H., F. Delsuc, H. Brinkmann, and N. Lartillot. 2005. Phylogenomics. *Annual Review of Ecology Evolution and Systematics* 36:541–562.

Philipson, W.R. 1970. Constant and variable features of the Araliaceae. *In* N.K.B. Robson, D.F. Cutler, and M. Gregory [eds.], New Research in Plant Anatomy, 87–100. Academic Press, London, UK.

———. 1974. Ovular morphology and the major classification of the dicotyledons. *Botanical Journal of the Linnean Society* 68:89–108.

———. 1977. Ovular morphology and the classification of dicotyledons. *Plant Systematics and Evolution, Supplement* 1:123–140.

Pigg, K.B., R.M. Dillhoff, M.L. DeVore, and W.C. Wehr. 2007. New diversity among the Trochodendraceae from the Early/Middle Eocene Okanogan Highlands of British Columbia, Canada, and northeastern Washington State, United States. *International Journal of Plant Sciences* 168(4):521–532.

Pigg, K.B., S.M. Ickert-Bond, and J. Wen. 2004. Anatomically preserved *Liquidambar* (Altingiaceae) from the middle Miocene of Yakima Canyon, Washington state, USA, and its biogeographic implications. *American Journal of Botany* 91(3):499–509.

Pigg, K.B., R.A. Stockey, and S.L. Maxwell. 1993. *Paleomyrtinaea princetonensis* gen. et sp. nov., permineralized myrtaceous fruits and seeds from the Princeton chert and related Myrtaceae from Almont, North Dakota. *Canadian Journal of Botany* 71(1):1–9.

Pingen, M., Z. Kvaček, and S.R. Manchester. 2001. Früchte und Samen von *Craigia bronnii* aus dem Obermiozän von Hambach (Niederrheinische Bucht—Deutschland) Vorläufige Mitteilung. *Documenta Naturae* 138:1–7.

Pires, J.C., M.F. Fay, W.S. Davis, L. Hufford, J. Rova, M.W. Chase, and K.J. Sytsma. 2001. Molecular and morphological phylogenetic analyses of Themidaceae (Asparagales). *Kew Bulletin* 56:601–626.

Pires, J.C., I.J. Maureira, T.J. Givnish, K.J. Sytsma, O. Seberg, G. Petersen, J.I. Davis, D.W. Stevenson, P.J. Rudall, M.F. Fay, and M.W. Chase. 2006. Phylogeny, genome size, and chromosome evolution of Asparagales. *In* J.T. Columbus, E.A. Friar, J.M. Porter, L.M. Prince, and M.G. Simpson [eds.], Monocots: Comparative Biology and Evolution, Excluding Poales, 267–304. Rancho Santa Ana Botanic Garden, Claremont, CA.

Pires, J.C., and K. Sytsma. 2002. A phylogenetic evaluation of a biosystematic framework: *Brodiaea* and related petaloid monocots (Themidaceae). *American Journal of Botany* 89:1342–1359.

Pirie, M.D., and J.A. Doyle. 2012. Dating clades with fossils and molecules: The case of Annonaceae. *Botanical Journal of the Linnean Society* 169:84–116.

Planchon, J.E. 1887. Ampelideae. *In* A. de Candolle, and C. de Candolle [eds.], *Monographiae Phanerogamarum* 5(2):305–654. Masson, Paris, France.

Plouvier, V. 1964. Recherche de l'Arbutoside et de l'Asperuloside chez quelques Rubiacées. Présence du Monotropeoside chez le *Liquidambar* (Hamamelidacées). *Comptes Rendus de l'Académie des Sciences Paris* 258:735–737.

———. 1992. Chimiotaxinomie des Caprifoliaceae et relations avec quelques familles voisines. *Bulletin du Muséum d'Histoire Naturelle, sect. B, Adansonia* 14:461–472.

Plouvier, V., and J. Favre-Bonvin. 1971. Les iridoides et sécoiridoides: Repartition, structure, propriétés, biosynthèse. *Phytochemistry* 10:1697–1722.

Plumstead, E.P. 1956. Bisexual fructifications borne on *Glossopteris* leaves from South Africa. *Palaeontographica Abteilung B* 100:1–25.

Plunkett, G.M. 2001. Relationship of the order Apiales to subclass Asteridae: A re-evaluation of morphological characters based on insights from molecular data. *Edinburgh Journal of Botany* 58:183–200.

Plunkett, G.M., G.T. Chandler, P.P. Lowry II, S.M. Pinney, and T.S. Sprenkle. 2004a. Recent advances in understanding Apiales and a revised classification. *South Africa Journal of Botany* 70:371–381.

Plunkett, G.M., and S.R. Downie. 1999. Major lineages within Apiaceae subfamily Apioideae: A comparison of chloroplast restriction site and DNA sequence data. *American Journal of Botany* 86:1014–1026.

Plunkett, G.M., and P.P. Lowry, Jr. 2001. Relationships among "ancient araliads" and their significance for the systematics of Apiales. *Molecular Phylogenetics and Evolution* 19:259–276.

Plunkett, G.M., D.E. Soltis, and P.S. Soltis. 1997b. Evolutionary patterns in Apiaceae: Inferences based on *matK* sequence data. *Systematic Botany* 21:477–495.

Plunkett, G.M., D.E. Soltis, P.S. Soltis, and R.E. Brooks. 1995. Phylogenetic relationships between Juncaceae and Cyperaceae: Insights from *rbcL* sequence data. *American Journal of Botany* 82:520–525.

———. 1996a. Evolutionary patterns in Apiaceae: Inferences based on *matK* sequence data. *Systematic Botany* 21:477–495.

———. 1996b. Higher level relationships of Apiales (Apiaceae and Araliaceae) based on phylogenetic analysis of *rbcL* sequences. *American Journal of Botany* 83:499–515.

———. 1997a. Clarification of the relationship between Apiaceae and Araliaceae based on *matK* and *rbcL* sequence data. *American Journal of Botany* 84: 567–580.

Plunkett, G.M, J. Wen, and P.P. Lowry II. 2004b. Infrafamilial relationships in Araliaceae: Insights from plastid (*trnL–trnF*) and nuclear (ITS) sequence data. *Plant Systematics and Evolution* 245:1–39.

Pollock, D.D., D.J. Zwickl, J.A. McGuire, and D.M. Hillis. 2002. Increased taxon sampling is advantageous for phylogenetic inference. *Systematic Biology* 51:664–671.

Poncet, V., M. Couderc, C. Tranchant-Dubreuil, C. Gomez, P. Hamon, S. Hamon, Y. Pillon, J. Munzinger, and A. de Kochko. 2012. Microsatellite markers for *Amborella* (Amborellaceae), a monotypic genus endemic to New Caledonia. *American Journal of Botany* 99:E411–E414.

Poncet, V., F. Munoz, J. Munzinger, Y. Pillon, C. Gomez, M. Couderc, C. Tranchant-Dubreuil, S. Hamon, and A. De Kochko. 2013. Phylogeography and niche modelling of the relict plant *Amborella trichopoda* (Amborellaceae) reveal multiple Pleistocene refugia in New Caledonia. *Molecular Ecology* 22:6163–6178.

Posluszny, U., and P.B. Tomlinson. 2003. Aspects of inflorescence and floral development in the putative basal angiosperm *Amborella trichopoda* (Amborellaceae). *Canadian Journal of Botany* 81:28–39.

Potgieter, K., and V.A. Albert. 2001. Phylogenetic relationships within Apocynaceae s.l. based on *trnL* intron and *trnL-F* spacer sequences and propagule characters. *Annals of the Missouri Botanical Garden* 88:523–549.

Potter, D., T. Eriksson, R.C. Evans, S.H. Oh, J.E.E. Smedmark, D.R. Morgan, M. Kerr, K.R. Robertson, M. Arsenault, T.A. Dickinson, and C.S. Campbell. 2007. Phylogeny and classification of Rosaceae. *Plant Systematics and Evolution* 266(1–2):5–43.

Potter, D., F. Gao, P.E. Bortiri, S.H. Oh, and S. Baggett. 2002. Phylogenetic relationships in Rosaceae inferred from chloroplast *matK* and *trnL–trnF* nucleotide sequence data. *Plant Systematics and Evolution* 231(1–4):77–89.

Prance, G.T. 1968. The systematic position of *Rhabdodendron* Gilg and Pilg. *Bulletin du Jardin Botanique National de Belgique* 38:127–146.

Prasad, V., C.A.E. Strömberg, H. Alimohammadian, and A. Sahni. 2005. Dinosaur coprolites and the early evolution of grasses and grazers. *Science* 310:1177–1180.

Prasad, V., C.A E. Strömberg, A.D. Leaché, B. Samant, R. Patnaik, L. Tang, D.M. Mohabey, S. Ge, and A. Sahni. 2011. Late Cretaceous origin of the rice tribe provides evidence for early diversification in Poaceae. *Nature Communications* 2:480. doi:10.1038/ncomms1482.

Prebble, J.M., C.N. Cupido, H.M. Meudt, and P.J. Garnock-Jones. 2011. First phylogenetic and biogeographical study of the southern bluebells (*Wahlenbergia*, Campanulaceae). *Molecular Phylogenetics and Evolution* 59:636–648.

Prenner, G. 2004a. Floral development of *Polygala myrtifolia* (Polygalaceae) and its similarities with Leguminosae. *Plant Systematics and Evolution* 249:67–76.

———. 2004b. The asymmetric androecium in Papilionoideae (Leguminosae): Definition, occurrence, and possible systematic value. *International Journal of Plant Sciences* 165: 499–510.

Preston, J.C., and L.C. Hileman. 2009. Developmental genetics of floral symmetry evolution. *Trends in Plant Science* 14(3):147–154.

Preston, J.C., C.C. Martinez, and L.C. Hileman. 2011. Gradual disintegration of the floral symmetry gene network is implicated in the evolution of a wind-pollination syndrome. *Proceedings of the National Academy of Sciences USA* 108:2343–2348.

Price, H.J. 1988. Nuclear DNA content variation within angiosperm species. *Evolutionary Trends in Plants* 2:53–60.

Price, R., and J. Palmer. 1993. Phylogenetic relationships of the Geraniaceae and Geraniales from *rbcL* sequence comparisons. *Annals of the Missouri Botanical Garden* 80:661–671.

Prusinkiewicz, P., Y. Eramus, B. Lane, L.D. Harder, and E. Coen. 2007. Evolution and development of inflorescence architectures. *Science* 316:1452–1456.

Pryer, K.M., H. Schneider, A.R. Smith, R. Cranfill, P.G. Wolf, J.S. Hunt, and S.D. Sipes. 2001. Horsetails and ferns are a monophyletic group and the closest living relatives to seed plants. *Nature* 409:618–622.

Puff, C., and A. Weber. 1976. Contributions to the morphology, anatomy, and karyology of *Rhabdodendron* and a reconsideration of the systematic position of Rhabdodendraceae. *Plant Systematics and Evolution* 125:195–222.

Purvis, A. 1995. A modification to Baum and Ragan's method for combining phylogenetic trees. *Systematic Biology* 44:251–255.

Pyankov, V.I., E.G. Artyusheva, G.E. Edwards, C.C.J. Black, and P.S. Soltis. 2001. Phylogenetic analysis of tribe Salsoleae (Chenopodiaceae) based on ribosomal ITS sequences: Implications for the evolution of photosynthesis types. *American Journal of Botany* 88:1189–1198.

Qiu, H., H.S. Yoon, and D. Bhattacharya. 2013. Algal endosymbionts as vectors of horizontal gene transfer in photosynthetic eukaryotes. *Frontiers in Plant Science* 4:366. doi:10.3389/fpls.2013.00366.

Qiu, Y.-L., M.W. Chase, S.B. Hoot, E. Conti, P.R. Crane, K.J. Sytsma, and C.R. Parks. 1998. Phylogenetics of the Hamamelidae and their allies: Parsimony analyses of nucleotide sequences of the plastid gene *rbcL*. *International Journal of Plant Science* 159:891–905.

Qiu, Y.-L., M.W. Chase, D.H. Les, and C.R. Parks. 1993. Molecular phylogenetics of the Magnoliidae: Cladistic analyses of nucleotide sequences of the plastid gene *rbcL*. *Annals of the Missouri Botanical Garden* 80:587–606.

Qiu, Y.-L., O. Dombrovska, J. Lee, L.B. Li, B.A. Whitlock, F. Bernasconi-Quadroni, J.S. Rest, C.C. Davis, T. Borsch, K.W. Hilu, S.S. Renner, D.E. Soltis, P.S. Soltis, M.J. Zanis, J.J. Cannone, R.R. Gutell, M. Powell, V. Savolainen, L.W. Chatrou, and M.W. Chase. 2005. Phylogenetic analyses of basal angiosperms based on nine plastid, mitochondrial, and nuclear genes. *International Journal of Plant Sciences* 166:815–842.

Qiu, Y.-L., J. Lee, F. Bernasconi-Quadroni, D.E. Soltis, P.S. Soltis, M. Zanis, Z. Chen, V. Savolainen, and M.W. Chase. 1999. The earliest angiosperms: Evidence from mitochondrial, plastid and nuclear genomes. *Nature* 402:404–407.

Qiu, Y.-L., J.-Y. Lee, F. Bernasconi-Quadroni, D.E. Soltis, P.S. Soltis, M. Zanis, E. Zimmer, Z. Chen, V. Savolainen, and M. Chase. 2000. Phylogeny of basal angiosperms: Analyses of five genes from three genomes. *International Journal of Plant Sciences* 161:S3–S27.

Qiu, Y.-L., J. Lee, B.A. Whitlock, F. Bernasconi-Quadroni, and O. Dombrovska. 2001. Was the ANITA rooting of the angiosperm phylogeny affected by long branch attraction? *Molecular Biology and Evolution* 18:1745–1753.

Qiu, Y.-L., L. Li, B. Wang, Z. Chen, O. Dombrovska, J. Lee, L. Kent, R. Li, R.W. Jobson, T.A. Hendry, D.W. Taylor, C.M. Testa, and M. Ambros. 2007. A nonflowering land plant phylogeny inferred from nucleotide sequences of seven chloroplast, mitochondrial, and nuclear genes. *International Journal of Plant Sciences* 168:691–708.

Qiu, Y.-L., L. Li, B. Wang, Z. Chen, V. Knoop, M. Groth-Malonek, O. Dombrovska, J. Lee, L. Kent, J. Rest, G.F. Estabrook, T.A. Hendry, D.W. Taylor, C.M. Testa, M. Ambros, B. Crandall-Stotler, R.J. Duff, M. Stech, W. Frey, D. Quandt, and C.C. Davis. 2006. The deepest divergences in land plants inferred from phylogenomic evidence. *Proceedings of the National Academy of Sciences USA* 103:15511–15516.

Qiu, Y.-L., L. Li, B. Wang, J.-Y. Xue, T.A. Hendry, R.-Q. Li, J.W. Brown, Y. Liu, G.T. Hudson, and Z.-D. Chen. 2010. Angiosperm phylogeny inferred from sequences of four mitochondrial genes. *Journal of Systematics and Evolution* 48:391–425.

Quibell, C.F. 1972. Comparative and Systematic Anatomy of Carpenterieae (Philadelphiaceae). Ph.D. Dissertation, University of California, Berkeley, CA.

Rabinowicz, P.D. 2000. Are obese plant genomes on a diet? *Genome Research* 10:893–894.

Radice, R. 2012. A Bayesian approach to modelling reticulation events with application to the ribosomal protein gene rps11 of flowering plants. *Australian & New Zealand Journal of Statistics* 54(4):401–426.

Ragan, M.A. 1992. Phylogenetic inference based on matrix representation of trees. *Molecular Phylogenetics and Evolution* 1:53–58.

Ragan, M.A., C.J. Bird, E.L. Rice, R.R. Gutell, C.A. Murphy, and R.K. Singh. 1994. A molecular phylogeny of the marine red algae (Rhodophyta) based on the nuclear small-subunit ribosomal-RNA gene. *Proceedings of the National Academy of Sciences USA* 91:7276–7280.

Rahn, K. 1996. A phylogenetic study of the Plantaginaceae. *Botanical Journal Linnaean Society* 120:145–198.

Ramsay, N.A., and B.J. Glover. 2005. MYB-bHLH-WD40 protein complex and the evolution of cellular diversity. *Trends Plant Sciences* 10:63–70.

Ramsey, J., and D.W. Schemske. 1998. Pathways, mechanisms, and rates of polyploid formation in flowering plants. *Annual Review of Ecology and Systematics* 29:467–501.

———. 2002. Neopolyploidy in flowering plants. *Annual Review of Ecology and Systematics* 33:589–639.

Ramshaw, J.A.M., D.L. Richardson, B.T. Meatyard, R. Brown, H.M. Richardson, E.W. Thompson, and D. Boulter. 1972. The time of origin of the flowering plants determined by using amino acid sequence data of cytochrome C. *New Phytologist* 71:773–779.

Ran, J.-H., H. Gao, and X.-Q. Wang. 2010. Fast evolution of the retroprocessed mitochondrial *rps3* gene in Conifer II and further evidence for the phylogeny of gymnosperms. *Molecular Phylogenetics and Evolution* 54:136–149.

Rao, P.R.M. 1972. Embryology of *Nyssa sylvatica*, and systematic consideration of the family Nyssaceae. *Phytomorphology* 22:8–21.

Rasmussen, D.A., E.M. Kramer, and E.A. Zimmer. 2009. One size fits all? Molecular evidence for a commonly inherited petal identity program in Ranunculales. *American Journal of Botany* 96:96–109.

Raubeson, L.A., and R.K. Jansen. 1992. A rare chloroplast-DNA structural mutation is shared by all conifers. *Biochemical Systematics and Ecology* 20: 17–24.

Rauscher, M.D. 2008. Evolutionary transitions in flower color. *International Journal of Plant Sciences* 169:7–21.

Raven, P.H. 1975. The bases of angiosperm phylogeny: Cytology. *Annals of the Missouri Botanical Garden* 62:724–764.

———. 1979. Onagraceae as a model of plant evolution. *In* L. Gottlieb, and S. Jain [eds.], Plant Evolutionary Biology, 85–107. Chapman and Hall, London, UK.

Ray, J. 1703. Methodus plantarum, emendata et aucta. Smith and Walford, London, UK.

Record, S.J. 1933. The woods of *Rhabdodendron* and *Duckeodendron*. *Tropical Woods* 33:6–10.

Ree, R.H. 2005. Detecting the historical signature of key innovations using stochastic models of character evolution and cladogenesis. *Evolution* 59:257–265.

Ree, R.H., and M.J. Donoghue. 2000. Inferring rates of change in flower symmetry in asterid angiosperms. *Systematic Biology* 48:633–641.

Reeves, G., M.W. Chase, P. Goldblatt, P.J. Rudall, M.F. Fay, A.V. Cox, B. Lejeune, and T. Souza-Chies. 2001. Molecular systematics of Iridaceae: Evidence from four plastid DNA regions. *American Journal of Botany* 88:2074–2087.

Reeves, P.A., and R.G. Olmstead. 1998. Evolution of novel morphological, ecological, and reproductive traits in a clade containing *Antirrhinum*. *American Journal of Botany* 86:1301–1315.

Refulio-Rodriguez, N.F., and R.G. Olmstead. 2014. Phylogeny of Lamiidae. *American Journal of Botany* 101(2):287–299.

Reichenbach, H.G.L. 1827–1829. *In* J.C. Moessler's Handbuch der Gewächskunde, 2nd ed., 3 vols. Johann Friedrich Hammerich, Altona, Germany.

Reid, E.M., and M.E.J. Chandler. 1926. Catalogue of Cainozoic plants in the Department of Geology vol. 1, The Bembridge Flora. British Museum (Natural History), London.

———. 1933. The London Clay Flora. British Museum (Natural History), London.

Remane, A. 1956. Die Grundlagen des natürlichen Systems, der vergleichenden Anatomie und der Phylogenetik. Akademische Verlagsbuchhandlung Geest & Portig, Leipzig, Germany.

Remizowa, M.V., P.J. Rudall, V.V. Choob, and D.D. Sokoloff. 2013. Racemose inflorescences of monocots: Structural and morphogenetic interaction at the flower/inflorescence level. *Annals of Botany* 112:1553–1566.

Remizowa, M.V., D.D. Sokoloff, and P.J. Rudall. 2010. Evolutionary history of the monocot flower. *Annals of the Missouri Botanical Garden* 97:617–645.

Ren, Y., H.-L. Chang, and P.K. Endress. 2010. Floral development in Anemoneae (Ranunculaceae). *Botanical Journal of the Linnean Society* 162(1):77–100.

Ren, Y., L. Chen, X.H. Tian, X.H. Zhang, and A.M. Lu. 2007. Discovery of vessels in *Tetracentron* (Trochodendraceae) and its systematic significance. *Plant Systematics and Evolution* 267(1–4):155–161.

Ren, Y., T.-Q. Gu, and H.-I. Chang. 2011. Floral development of *Dichocarpum*, *Thalictrum*, and *Aquilegia* (Thalictroideae, Ranunculaceae). *Plant Systematics and Evolution* 292(3–4):203–213.

Ren, Y., H.-F. Li, L. Zhao, and P.K. Endress. 2007. Floral morphogenesis in *Euptelea* (Eupteleaceae, Ranunculales). *Annals of Botany* 95:22–40.

Ren, Y., Z.-J. Li, H.-L. Chang, Y.-J. Lei, and A.-M. Lu. 2004. Floral development of *Kingdonia* (Ranunculaceae s. l., Ranunculales). *Plant Systematics and Evolution* 247(3–4):145–153.

Renner, S.S. 1999. Circumscription and phylogeny of the Laurales: Evidence from molecular and morphological data. *American Journal of Botany* 86:1301–1315.

———. 2001. How common is heterodichogamy? *Trends in Ecology & Evolution* 16:595–597.

Renner, S.S., and S. Bellot 2012. Horizontal gene transfer in eukaryotes: Fungi-to-plant and plant-to-plant transfers of organellar DNA. *In* R. Bock, and V. Knoop [eds.], Genomics of Chloroplasts and Mitochondria, Advances in Photosynthesis and Respiration 35, 223–235. Springer, Netherlands.

Renner, S.S., and A.S. Chanderbali. 2000. What is the

relationship among Hernandiaceae, Lauraceae, and Monimiaceae, and why is this question so difficult to answer? *International Journal of Plant Sciences* 161(6 Supplement):S109–S119.

Renner, S.S., G. Clausing, and K. Meyer. 2001. Historical biogeography of Melastomataceae: The roles of Tertiary migration and long-distance dispersal. *American Journal of Botany* 88:1290–1300.

Renner, S.S., and K. Meyer. 2001. Melastomataceae come full circle: Biogeographic reconstruction and molecular clock dating. *Evolution* 55:1315–1324.

Renner, S.S., and R.E. Ricklefs. 1995. Dioecy and its correlates in the flowering plants. *American Journal of Botany* 82:596–606.

Renner, S.S., and H. Schaefer. 2010. The evolution and loss of oil-offering flowers: New insights from dated phylogenies for angiosperms and bees. *Philosophical Transactions of the Royal Society of London B* 365:423–435.

Renner, S.S., and H.S. Won. 2001. Repeated evolution of dioecy from monoecy in Siparunaceae (Laurales). *Systematic Biology* 50:700–712.

Renner, T., and C.D. Specht. 2011. A sticky situation: Assessing adaptations for plant carnivory in the Caryophyllales by means of stochastic character mapping. *International Journal of Plant Sciences* 172(7):889–901.

Retallack, G., and D.L. Dilcher. 1981. Arguments for a glossopterid ancestry of angiosperms. *Palaeobiology* 7:54–67.

Rettig, J.H., H.D. Wilson, and J.M. Manhart. 1992. Phylogeny of the Caryophyllales—Gene sequence data. *Taxon* 41:201–209.

Reveal, J.L., and M. Chase. 2011. APGIII: Bibliographical information and synonymy of Magnoliidae. *Phytotaxa* 19:71–134.

Reymanówna, M. 1968. On seeds containing *Eucommiidites troedssonii* pollen from the Jurassic of Grojec, Poland. *Botanical Journal of the Linnean Society* 61:147–152.

———. 1973. The Jurassic flora from Grojec near Krakow in Poland. Part II. Caytoniales and anatomy of *Caytonia*. *Acta Palaeobotanica* 14:45–87.

Rice, D.W., A.J. Alverson, A.O. Richardson, G.J. Young, M. Virginia Sanchez-Puerta, J. Munzinger, K. Barry, J.L. Boore, Y. Zhang, C.W. dePamphilis, E.B. Knox, and J.D. Palmer. 2013. Horizontal transfer of entire genomes via mitochondrial fusion in the Angiosperm *Amborella*. *Science* 342(6165):1468–1473.

Rice, K.A., M.J. Donoghue, and R.G. Olmstead. 1997. Analyzing large data sets: *rbcL* 500 revisited. *Systematic Biology* 46:157–178.

Richards, A.J. 1997. Plant Breeding Systems, 2nd ed. Chapman & Hall, London, UK.

Richardson, A.O., and J.D. Palmer. 2007. Horizontal gene transfer in plants. *Journal of Experimental Botany* 58(1):1–9.

Richardson, A.O., D.W. Rice, G.J. Young, A.J. Alverson, and J.D. Palmer. 2013. The "fossilized" mitochondrial genome of *Liriodendron tulipifera*: Ancestral gene content and order, ancestral editing sites, and extraordinarily low mutation rate. *BMC Biology* 11:29. doi:10.1186/1741-7007-11-29.

Richardson, J.E., L.W. Chatrou, J.B. Mols, R.H.J. Erkens, and M.D. Pirie. 2004. Historical biogeography of two cosmopolitan families of flowering plants: Annonaceae and Rhamnaceae. *Philosophical Transactions of the Royal Society of London B* 359:1495–1508.

Richardson, J.E., M.F. Fay, Q.C.B. Cronk, D. Bowman, and M.W. Chase. 2000. A molecular phylogenetic analysis of Rhamnaceae using *rbcL* and *trnL–F* plastid DNA sequences. *American Journal of Botany* 87:1309–1324.

Riechmann, J.L., and E.M. Meyerowitz. 1997. MADS domain proteins in plant development. *Biological Chemistry* 378:1079–1109.

Rieger, R., A. Michaelis, and M.M. Green. 1991. Glossary of genetics, 5th ed. Springer-Verlag, Berlin, Germany.

Rieseberg, L.H., and D.E. Soltis. 1991. Phylogenetic consequences of cytoplasmic gene flow in plants. *Evolutionary Trends in Plants* 5:65–84.

Rijpkema, A., T. Gerats, and M. Vandenbussche. 2006. Genetics of floral development in *Petunia*. *Advances in Botanical Research, Developmental Genetics of the Flower* 44:237–278.

Riser, J.P., II, W.M. Cardinal-Mcteague, J.C. Hall, W.J. Hahn, K.J. Sytsma, and E.H. Roalson. 2013. Phylogenetic relationships among the North American cleomoids (Cleomaceae): A test of *Iltis*'s reduction series. *American Journal of Botany* 100(10):2102–2111.

Ritland, K., and M.T. Clegg. 1987. Evolutionary analysis of plant DNA sequences. *American Naturalist* 130:S74–S100.

Ritz, C.M., J. Reiker, G. Charles, P. Hoxey, D. Hunt, M. Lowry, W. Stuppy, and N. Taylor. 2013. Molecular phylogeny and character evolution in terete-stemmed Andean opuntias (Cactaceae–Opuntioideae). *Molecular Phylogenetics and Evolution* 65:668–681.

Rivadavia, F., K. Kondo, M. Kato, and M. Hasebe. 2003. Phylogeny of the sundews, *Drosera* (Droseraceae), based on chloroplast *rbcL* and nuclear 18S ribosomal DNA sequences. *American Journal of Botany* 90:123–130.

Robertson, K.R. 1972a. The genera of Geraniaceae in

the southeastern United States. *Journal of the Arnold Arboretum* 53:182–201.

———. 1972b. The Malpighiaceae in the southeastern United States. *Journal of the Arnold Arboretum* 53: 101–112.

———. 1974. The genera of Rosaceae in the southeastern United States. *Journal of the Arnold Arboretum* 55:303–332, 344–401, 600–662.

———. 1975. The genera of Oxalidaceae in the southeastern United States. *Journal of the Arnold Arboretum* 57:205–216.

Robinson, H. 1985. Observations on fusion and evolutionary variability in the angiosperm flower. *Systematic Botany* 10:105–109.

Robinson, H., and P. Burns-Balogh. 1982. Evidence for a primitively epiphytic habit in Orchidaceae. *Systematic Botany* 7:353–358.

Robinson, S., A. Burian, E. Couturier, B. Landrein. M. Louveaux, E.D. Neumann, A. Peaucelle, and N. Nakayama. 2013. Mechanical control of morphogenesis at the shoot apex. *Journal of Experimental Botany* 64:4729–4744.

Robinson-Beers, K., R.E. Pruitt, and C.S. Gasser. 1992. Ovule development in wild-type *Arabidopsis* and two female-sterile mutants. *Plant Cell* 4:1237–1249.

Rodenburg, W.F. 1971. A revision of the genus *Trimenia* (Trimeniaceae). *Blumea* 19:3–15.

Rodman, J.E. 1990. Centrospermae revisited, part 1. *Taxon* 39:383–393.

———. 1991a. A taxonomic analysis of glucosinolate-producing plants. I. Phenetics. *Systematic Botany* 16:598–618.

———. 1991b. A taxonomic analysis of glucosinolate-producing plants, II. Cladistics. *Systematic Botany* 16:619–699.

———. 1994. Cladistic and phenetic studies. *In* H.-D. Behnke, and T. J. Mabry [eds.], Caryophyllales: Evolution and Systematics, 279–301. Springer, Berlin, Germany.

Rodman, J.E., K.G. Karol, R.A. Price, E. Conti, and K.J. Sytsma. 1994. Nucleotide sequences of *rbcL* confirm the capparalean affinity of the Australian endemic Gyrostemonaceae. *Australian Systematic Botany* 7:57–69.

Rodman, J.E., K.G. Karol, R.A. Price, and K.J. Sytsma. 1996. Molecules, morphology, and Dahlgren's expanded order Capparales. *Systematic Botany* 21:289–307.

Rodman, J.E., M.K. Oliver, R.R. Nakamura, J.U.J. McClammer, and A.H. Bledsoe. 1984. A taxonomic analysis and revised classification of Centrospermae. *Systematic Botany* 9:297–323.

Rodman, J.E., R.A. Price, K. Karol, E. Conti, K.J. Sytsma, and J.D. Palmer. 1993. Nucleotide sequences of the *rbcL* gene indicate monophyly of mustard oil plants. *Annals of the Missouri Botanical Garden* 80:686–699.

Rodman, J.E., P.S. Soltis, D.E. Soltis, K.J. Sytsma, and K.G. Karol. 1998. Parallel evolution of glucosinolate biosynthesis inferred from congruent nuclear and plastid gene phylogenies. *American Journal of Botany* 85:997–1006.

Roels, P. 1998. Phylogenetic Position and Delimitation of the order Dipsacales. A Multidisciplinary Approach. Doctoral dissertation, University of Leuven, Belgium.

Roels, P., L.P. Ronse De Craene, and E.F. Smets. 1997. A floral ontogenetic investigation of the Hydrangeaceae. *Nordic Journal of Botany* 17:235–254.

Roels, P., and E. Smets. 1996. A floral ontogenetic study in Dipsacales. *International Journal of Plant Sciences* 157:203–218.

Rogers, G.K. 1983. The genera of Alismataceae in the southeastern United States. *Journal of the Arnold Arboretum* 64:383–420.

———. 1984. The Zingiberales (Cannaceae, Marantaceae, and Zingiberaceae) in the southeastern United States. *Journal of the Arnold Arboretum* 65:5–55.

———. 1985. The genera of Phytolaccaceae in the southeastern United States. *Journal of the Arnold Arboretum* 66:1–37.

Rohweder, O. 1967. Centrospermen-Studien 3. Blütenentwicklung und Blütenbau bei Silenoideen (Caryophyllaceen). *Botanische Jahrbücher für Systematik* 86:130–185.

Rohweder, O., and P.K. Endress. 1983. Samenpflanzen. Thieme, Stuttgart, Germany.

Rohwer, J.G. 1993. Lauraceae. *In* K. Kubitzki, J. Rohwer, and V. Bittrich [eds.], The Families and Genera of Vascular Plants, II, 366–391, Springer-Verlag, Berlin, Germany.

Rohwer, J.G., and B. Rudolph. 2005. Jumping genera: The phylogenetic positions of *Cassytha*, *Hypodaphnis*, and *Neocinnamomum* (Lauraceae) based on different analyses of *trnK* intron sequences. *Annals of the Missouri Botanical Garden* 92(2):153–178.

Rokas, A., B.L. Williams, N. King, and S.B. Carroll. 2003 Genome-scale approaches to resolving incongruence in molecular phylogenies. *Nature* 425:798–804.

Romeike, A. 1978. Tropane alkaloids: Occurrence and systematic importance in angiosperms. *Botaniska Notiser* 131:85–96.

Rönblom, K., and A.A. Anderberg. 2002. Phylogeny of Diapensiaceae based on molecular data and morphology. *Systematic Botany* 27:383–395.

Ronquist, F. 1996. Matrix representation of trees, redundancy, and weighting. *Systematic Biology* 45:247–253.

Ronse De Craene, L.P. 1992. The androecium of the

Magnoliophytina: Characterization and systematic importance. Doctoral dissertation, University of Leuven, Belgium.

———. 2004. Floral development of *Berberidopsis corallina*: A crucial link in the evolution of flowers in the core eudicots. *Annals of Botany* 94:1–11.

———. 2007. Are petals sterile stamens or bracts? The origin and evolution of petals in the core eudicots. *Annals of Botany* 100:621–630.

———. 2008. Homology and evolution of petals in the core eudicots. *Systematic Botany* 33:301–325.

———. 2013. Reevaluation of the perianth and androecium in Caryophyllales: Implications for flower evolution. *Plant Systematics and Evolution* 299: 1599–1636.

Ronse De Craene, L.P., and S.F. Brockington. 2013. Origin and evolution of petals in angiosperms. *Plant Ecology and Evolution* 146:5–25.

Ronse De Craene, L.P., and E. Haston. 2006. The systematic relationships of glucosinolate-producing plants and related families: A cladistic investigation based on morphological and molecular characters. *Botantical Journal of the Linnean Society* 151:453–494.

Ronse De Craene, L.P., H.P. Linder, T. Dlamini, and E.F. Smets. 2001. Evolution and development of floral diversity of Melianthaceae, an enigmatic Southern African family. *International Journal of Plant Sciences* 162:59–82.

Ronse De Craene, L.P., P. Louis, P.S. Soltis, and D.E. Soltis. 2003. Evolution of floral structure in basal angiosperms. *International Journal of Plant Sciences* 164: S329–S363.

Ronse De Craene, L.P., and E.F. Smets. 1992. Complex polyandry in the Magnoliatae: Definition, distribution and systematic value. *Nordic Journal of Botany* 12:621–649.

———. 1993. The distribution and systematic relevance of the androecial character polymery. *Botanical Journal of the Linnean Society* 113:285–350.

———. 1994. Merosity in flowers: Definition, origin, and taxonomic significance. *Plant Systematics and Evolution* 191:83–104.

———. 1995. The distribution and systematic relevance of the androecial character oligomery. *Botanical Journal Linnean Society* 118:193–247.

———. 1997. A floral ontogenetic study of some species of *Capparis* and *Boscia*, with special emphasis on the androecium. *Botanische Jahrbücher für Systematik* 119:231–255.

———. 1998a. The distribution and systematic relevance of the androecial character oligomery. *Botanical Journal of the Linnean Society* 118:193–247.

———. 1998b. Meristic changes in gynoecium morphology, exemplified by floral ontogeny and anatomy. *In* S.J. Owens, and P.J. Rudall [eds.], Reproductive Biology in Systematics, Conservation and Economic Botany, 85–112. Royal Botanic Gardens, Kew, UK.

———. 1998c. Notes on the evolution of androecial organisation in the Magnoliophytina (angiosperms). *Botanica Acta* 111:77–86.

———. 1999. Similarities in floral ontogeny and anatomy between the genera *Francoa* (Francoaceae) and *Greyia* (Greyiaceae). *International Journal of Plant Sciences* 160:377–393.

Ronse De Craene, L.P., E.F. Smets, and P. Vanvinckenroye. 1998. Pseudodiplostemony, and its implications for the evolution of the androecium in the Caryophyllaceae. *Journal of Plant Research* 111:25–43.

Ronse De Craene, L.P., D.E. Soltis, and P.S. Soltis. 2003. Evolution of floral structures in basal angiosperms. *International Journal of Plant Sciences* 164:S329–S363.

Ronse De Craene, L.P., and W. Stuppy. 2010. Floral development and anatomy of *Aextoxicon punctatum* (Aextoxicaceae-Berberidopsidales)—an enigmatic tree at the base of core eudicots. *International Journal of Plant Sciences* 171:244–257.

Ronse De Craene, L.P., P. Van Vinckenroye, and E.F. Smets. 1997. A study of floral morphological diversity in *Phytolacca* (Phytolaccaceae) based on early floral ontogeny. *International Journal of Plant Sciences* 158: 57–72.

Roquet, C., L. Sáez, J.J. Aldasoro, A. Susanna, M.L. Alarcón, and N. Garcia-Jacas. 2008. Natural delineation, molecular phylogeny and floral evolution in *Campanula*. *Systematic Botany* 33:203–217.

Roquet, C., I. Sanmartín, N. Garcia-Jacas, L. Sáez, A. Susanna, N. Wikstrom, and J.J. Aldasoro. 2009. Reconstructing the history of Campanulaceae with a Bayesian approach to molecular dating and dispersal-vicariance analyses. *Molecular Phylogenetics and Evolution* 52: 575–587.

Rosatti, T.J. 1984. The Plantaginaceae in the southeastern United States. *Journal of the Arnold Arboretum* 65:533–562.

———. 1986. The genera of Sphenocleaceae and Campanulaceae in the southeastern United States. *Journal of the Arnold Arboretum* 67:1–64.

———. 1987. The genera of Pontederiaceae in the southeastern United States. *Journal of the Arnold Arboretum* 68:35–71.

———. 1989. The genera of suborder Apocynineae (Apocynaceae and Asclepiadaceae) in the southeastern United States. *Journal of the Arnold Arboretum* 70:307–401, 443–514.

Rosenberg, M.S., and S. Kumar. 2001. Incomplete taxon sampling is not a problem for phylogenetic inference. *Proceedings of the National Academy of Sciences USA* 98:10751–10756.

Rothwell, G.W., W.L. Crepet, and R.A. Stockey. 2009. Is the Anthophyte hypothesis alive and well? New evidence from the reproductive structures of Bennettitales. *American Journal of Botany* 96:296–322.

Rothwell, G.W., and R. Serbet. 1994. Lignophyte phylogeny and the evolution of spermatophytes: A numerical cladistic analysis. *Systematic Botany* 19:443–482.

Rothwell, G.W., and R.A. Stockey. 2002. Anatomically preserved *Cycadeoidea* (Cycadeoidaceae), with a reevaluation of systematic characters for the seed cones of Bennettitales. *American Journal of Botany* 89:1447–1458.

Rothwell, G.W., M.R. Van Atta, H.E. Ballard JR., and R.A. Stockey. 2003. Molecular phylogenetic relationships among Lemnaceae and Araceae using the chloroplast *trnL–trnF* intergenic spacer. *Molecular Phylogenetics and Evolution* 30:378–385.

Rudall, P.J. 1994. Anatomy and systematics of Iridaceae. *Botanical Journal of the Linnean Society* 114:1–21.

———. 1997. The nucellus and chalaza in monocotyledons: Structure and systematics. *Botanical Review* 63:140–181.

———. 2000. 'Cryptic' characters in monocotyledons: Homology and coding. *In* R. Scotland, and R.T. Pennington [eds.], Homology and Systematics: Coding Characters for Phylogenetic Analysis. Taylor and Francis, London, UK.

———. 2003. Monocot pseudanthia revisited: Floral structure of the mycoheterotrophic family Triuridaceae. *International Journal of Plant Sciences* 164(4 Supplement):S307–320.

———. 2006. How many nuclei make an embryo sac in flowering plants? *Bioessays* 28:1067–1071.

Rudall, P.J., and R.M. Bateman. 2002. Roles of synorganisation, zygomorphy and heterotopy in floral evolution: The gynostemium and labellum of orchids and other lilioid monocots. *Biological Review* 77:403–441.

———. 2006. Morphological phylogenetic analysis of Pandanales: Testing contrasting hypotheses of floral evolution. *Systematic Botany* 31:223–238.

Rudall, P.J., and M.W. Chase. 1996. Systematics of Xanthorrhoeaceae sensu lato: Evidence for polyphyly. *Telopea* 6:629–647.

Rudall, P.J., P.J. Cribb, D.F. Cutler, and C.J. Humphries [eds.]. 1995. Monocotyledons: Systematics and Evolution. Royal Botanical Gardens, Kew, UK.

Rudall, P.J., J. Cunniff, P. Wilkin, and L.R Caddick. 2005. Evolution of dimery, pentamery and the monocarpellary condition in the monocot family Stemonaceae (Pandanales). *Taxon* 54:701–711.

Rudall, P.J., C.A. Furness, M.W. Chase, and M.F. Fay. 1997. Microsporogenesis and pollen sulcus type in Asparagales (Lilianae). *Canadian Journal of Botany* 75:408–430.

Rudall, P.J., M.V. Remizowa, A.S. Beer, E. Bradshaw, D.W. Stevenson, T.D. Macfarlane, R.E. Tuckett, S.R. Yadav, and D.D. Sokoloff. 2008. Comparative ovule and megagametophyte development in Hydatellaceae and water lilies reveal a mosaic of features among the earliest Angiosperms. *Annals of Botany* 101:941–956.

Rudall, P.J., M.V. Remizowa, G. Prenner, C.J. Prychid, R.E. Tuckett, and D.D. Sokoloff. 2009. Nonflowers near the base of extant Angiosperms? Spatiotemporal arrangement of organs in reproductive units of Hydatellaceae and its bearing on the origin of the flower. *American Journal of Botany* 96:67–82.

Rudall, P.J., D.D. Sokoloff, M.V. Remizowa, J.G. Conran, J.I. Davis, T.D. Macfarlane, and D.W. Stevenson. 2007. Morphology of Hydatellaceae, an anomalous aquatic family recently recognized as an early-divergent angiosperm lineage. *American Journal of Botany* 94:1073–1092.

Rudall, P.J., D.W. Stevenson, and H.P. Linder. 1999. Structure and systematics of *Hanguana*, a monocotyledon of uncertain affinity. *Australian Systematic Botany* 12:311–330.

Rudall, P.J., K.L. Stobart, W.-P. Hong, J.G. Conran, C.A. Furness, G.C. Kite, and M.W. Chase. 2000. Consider the lilies: Systematics of Liliales. *In* K.L. Wilson, and D.A. Morrison [eds.], Monocots: Systematics and Evolution, 347–359. CSIRO, Collingwood, Victoria, Australia.

Ruhfel, B.R., V. Bittrich, C.P. Bove, M.H.G. Gustafsson, C.T. Philbrick, R. Rutishauser, Z. Xi, and C.C. Davis. 2011. Phylogeny of The Clusioid clade (Malpighiales): Evidence from the plastid and mitochondrial genomes. *American Journal of Botany* 98(2):306–325.

Ruhfel, B.R., M.A. Gitzendanner, P.S. Soltis, D.E. Soltis, and J.G. Burleigh. 2014. From algae to angiosperms-inferring the phylogeny of green plants (Viridiplantae) from 360 plastid genomes. *BMC Evolutionary Biology* 14:23. doi:10.1186/1471-2148-14-23.

Ruhfel, B.R., P.F. Stevens, and C.C. Davis. 2013. Combined morphological and molecular phylogeny of the Clusioid clade (Malpighiales) and the placement of the ancient rosid macrofossil *Paleoclusia*. *International Journal of Plant Sciences* 174(6):910–936.

Runions, C.J., and J.N. Owens. 1998. Evidence of prezygotic self-incompatibility in a conifer. *In* S.J. Owens, and P.J. Rudall [eds.], Reproductive Biology in Systematics, Conservation and Economic Botany, 255–264. Royal Botanic Gardens, Kew, UK.

Rydin, C., and M. Källersjö. 2002. Taxon sampling and seed plant phylogeny. *Cladistics* 18:485–513.

Rydin, C., M. Källersjö, and E.M. Friis. 2002. Seed plant relationships and the systematic position of Gnetales based on nuclear and chloroplast DNA: Conflicting data, rooting problems and the monophyly of conifers *International Journal of Plant Sciences* 163: 197–214.

Rydin, C., and P. Korall. 2009. Evolutionary relationships in *Ephedra* (Gnetales), with implications for Seed Plant phylogeny. *International Journal of Plant Sciences* 170:1031–1043.

Saarela, J.M., P.J. Prentis, H.S. Rai, and S.W. Graham. 2008. Phylogenetic relationships in the monocot order Commelinales, with a focus on Philydraceae. *Botany* 86:719–731.

Saarela, J.M., H.S. Rai, J.A. Doyle, P.K. Endress, S. Mathews, A.D. Marchant, B.G. Briggs, and S.W. Graham. 2007. Hydatellaceae identified as a new branch near the base of the angiosperm phylogenetic tree. *Nature* 446:312–315.

Sadowski, E.-M., L.J. Seyfullah, F. Sadowski, A. Fleischmann, H. Behling, and A.R. Schmidt. 2014. Carnivorous leaves from Baltic amber. *Proceedings of the National Society of Science of the United States of America USA* 112(1):190–195.

Sage, R.F. 2004. The evolution of C-4 photosynthesis. *New Phytologist* 161(2):341–370.

Sage, R.F., P.-A. Christin, and E.J. Edwards. 2011. The C-4 plant lineages of planet Earth. *Journal of Experimental Botany* 62(9):3155–3169.

Sage, R.F., M. Li, and R.K. Monson. 1999. The taxonomic distribution of C_4 photosynthesis. *In* R.F. Sage, and R.K. Monson [eds.], C_4 Plant Biology, 551–584. Academic Press, San Diego, CA.

Sage, R.F., T.L. Sage, R.W. Pearcy, and T. Borsch. 2007. The taxonomic distribution of C4 photosynthesis in Amaranthaceae sensu stricto. *American Journal of Botany* 94(12):1992–2003.

Sajo, M.G., R. Mello-Silva, and P.J. Rudall. 2013. Anther, ovule and embryological characters in Velloziaceae in relation to the systematics of Pandanales. *In* P. Wilkin, and S.J. Mayo [eds.], Early Events in Monocot Evolution, 304–314. Cambridge University Press, Cambridge, UK.

Sakai, A.K., and S.G. Weller. 1999. Gender and sexual dimorphism in flowering plants: A review of terminology, biogeographic patterns, ecological correlates, and phylogenetic approaches. *In* M.A. Geber, T.E. Dawson, and L.F. Delph [eds.], Sexual and Gender Dimorphism in Flowering Plants, 1–31. Springer-Verlag, Heidelberg, Germany.

Salamin, N., M.W. Chase, T.R. Hodkinson, and V. Savolanien. 2003. Assessing internal support with large phylogenetic DNA matrices. *Molecular Phylogenetics and Evolution* 27:528–539.

Salamin, N., T.R. Hodkinson, and V. Savolainen. 2002. Building supertrees: An empirical assessment using the grass family (Poaceae). *Systematic Biology* 51:136–150.

Salamini, F., H. Özkan, A. Brandolini, R. Schäfer-Pregl, and W. Martin. 2002. Genetics and geography of wild cereal domestication in the near east. *Nature Reviews Genetics* 3:429–441.

Salazar, G.A., M.W. Chase, M.A. Soto Arenas, and M.J. Ingrouille. 2003. Phylogenetics of Cranichideae with an emphasis on Spiranthinae (Orchidaceae: Orchidoideae): Evidence from plastid and nuclear DNA sequences. *American Journal of Botany* 90:777–795.

Salazar, J., and K. Nixon. 2008. New discoveries in the Canellaceae in the Antilles: How phylogeny can support taxonomy. *Botanical Review* 74:103–111.

Salisbury, E.J. 1919. Variation in *Eranthis hyemalis, Ficaria verna,* and other members of the Ranunculaceae, with special reference to trimery and the origin of the perianth. *Annals of Botany* 33:47–79.

Sampson, F.B. 2000. Pollen diversity in some modern magnoliids. *International Journal of Plant Sciences* 161(6 Supplement):S193–S210.

Samuel, R., H. Kathriarachchi, P. Hoffmann, M. Barfuss, K.J. Wurdack, and M.W. Chase. 2005. Molecular phylogenetics of Phyllanthaceae: Evidence from plastid *matK* and nuclear *PHYC* sequences. *American Journal of Botany* 92(1):132–141.

Samylina, V.A. 1960. Angiosperms from Lower Cretaceous deposits of Kolyma River. *Bot. Zhurnal* 45(3):335–352. [In Russian.]

Sánchez, A., and K. Kron. 2008. Phylogenetics of Polygonaceae with an emphasis on the evolution of Eriogonoideae. *Systematic Botany* 33:87–96.

Sanchez, A., T.M. Schuster, and K.A. Kron. 2009. A large-scale phylogeny of Polygonaceae based on molecular data. *International Journal of Plant Sciences* 170:1044–1055.

Sánchez, D., S. Arias, and T. Terrazas. 2014. Phylogenetic relationships in *Echinocereus* (Cactaceae, Cactoideae). *Systematic Botany* 39:1183–1196.

Sanchez-Puerta, M.V., C.C. Abbona, S. Zhuo, E.J. Tepe, L. Bohs, R.G. Olmstead, and J.D. Palmer. 2011. Multiple recent horizontal transfers of the cox1 intron in Solanaceae and extended co-conversion of flanking exons. *BMC Evolutionary Biology* 11:277. doi:10.1186/1471-2148-11-77.

Sanderson, M.J. 1991. In search of homoplastic tendencies: Statistical inference of topological patterns in homoplasy. *Evolution* 45:351–358.

———. 1993. Reversibility in evolution: A maximum likelihood approach to character gain/loss bias in phylogenies. *Evolution* 47:236–252.

———. 1997. A nonparametric approach to estimating divergence times in the absence of rate constancy. *Molecular Biology and Evolution* 14:1218–1231.

———. 1998. Estimating rate and time in molecular phylogenies: Beyond the molecular clock? *In* D.E. Soltis, P.S. Soltis, and J.J. Doyle [eds.], Molecular Systematics of Plants, II, 242–264. Kluwer, Boston, MA.

———. 2002. Estimating absolute rates of molecular evolution and divergence times: A penalized likelihood approach. *Molecular Biology and Evolution* 19:101–109.

Sanderson, M.J., and M.J. Donoghue. 1992. The suitability of molecular and morphological evidence in reconstructing plant phylogeny. *In* P.S. Soltis, D.E. Soltis, and J.J. Doyle [eds.], Molecular Systematics of Plants, 340–368. Chapman and Hall, New York, NY.

———. 1994. Shifts in diversification rate with the origin of angiosperms. *Science* 264:1590–1593.

Sanderson, M.J., and J.A. Doyle. 2001. Sources of error and confidence intervals in estimating the age of angiosperms from *rbcL* and 18S rDNA data. *American Journal of Botany* 88:1499–1516.

Sanderson, M.J., A. Purvis, and C. Henze. 1998. Phylogenetic supertrees: Assembling the trees of life. *Trends Ecology and Evolution* 13:105–109.

Sanderson, M. J., J. L. Thorne, N. Wikström, and K. Bremer. 2004. Molecular evidence on plant divergence times. *American Journal of Botany* 91(10): 1656–1665.

Sanderson, M.J., M.F. Wojciechowski, J.-M. Hu, T. Sher Khan, and S.G. Brady. 2000. Error, bias, and long-branch attraction in data for two chloroplast photosystem genes in seed plants. *Molecular Biology and Evolution* 17:782–797.

Sang, T., D.J. Crawford, and T.F. Stuessy. 1995. Documentation of reticulate evolution in peonies (*Paeonia*) using internal transcribed spacer sequences of nuclear ribosomal DNA: Implications for biogeography and concerted evolution. *Proceedings of the National Academy of Sciences USA* 92:6813–6817.

———. 1997. Chloroplast DNA phylogeny, reticulate evolution, and biogeography of *Paeonia* (Paeoniaceae). *American Journal of Botany* 84:1120–1136.

Sankoff, D. 2001. Gene and genome duplication. *Current Opinions in Genetics and Development* 11:681–684.

SanMiguel, P.A., A. Tikhonov, Y.K. Jin, N. Motochoulskaia, D. Zakharov, A. Melake-Berhan, P.S. Springer, K.J. Edwards, M. Lee, Z. Avramova, and J. Bennetzen. 1996. Nested retrotransposons in the intergenic regions of the maize genome. *Science* 274:765–768.

Santi, C., D. Bogusz, and C. Franche. 2013. Biological nitrogen fixation in non-legume plants. *Annals of Botany* 111(5):743–767.

Sargent, R.D. 2004. Floral symmetry affects speciation rates in angiosperms. *Proceedings of the Royal Society of London B* 271(1539):603–608.

Sarich, V., and A.C. Wilson. 1967. Rates of albumin evolution in primates. *Proceedings of the National Academy of Sciences* USA 58:142–148.

Sarkinen, T., L. Bohs, R.G. Olmstead, and S. Knapp. 2013. A phylogenetic framework for evolutionary study of the nightshades (Solanaceae): A dated 1000-tip tree. *BMC Evolutionary Biology* 13:214. doi:10.1186/1471-2148-3-24.

Sassi, M., and T. Vernoux. 2013. Auxin and self-organization of the shoot apical meristem. *Journal of Experimental Botany* 64:2579–2592.

Sato, S., S. Tabata, H. Hirakawa, E. Asamizu, K. Shirasawa, S. Isobe, T. Kaneko, Y. Nakamura, D. Shibata, K. Aoki, M. Egholm, J. Knight, R. Bogden, C. Li, Y. Shuang, X. Xu, S.Pan, S. Cheng, X. Liu, Y. Ren, J. Wang, A. Albiero, F. Dal Pero, S. Todesco, J. Van Eck, R.M. Buels, A. Bombarely, J.R. Gosselin, M. Huang, J.A. Leto, N. Menda, S. Strickler, L. Mao, S. Gao, I.Y. Tecle, T. York, Y. Zheng, J.T. Vrebalov, J. Lee, S. Zhong, L.A. Mueller, W.J. Stiekema, P. Ribeca, T. Alioto, W. Yang, S. Huang, Y. Du, Z. Zhang, J. Gao, Y. Guo, X. Wang, Y. Li, J. He, C. Li, Z. Cheng, J. Zuo, J. Ren, J. Zhao, L. Yan, H. Jiang, B. Wang, H. Li, Z. Li, F. Fu, B. Chen, B. Han, Q. Feng, D. Fan, Y. Wang, H. Ling, Y. Xue, D. Ware, W.R. McCombie, Z.B. Lippman, J.-M. Chia, K. Jiang, S. Pasternak, L. Gelley, M. Kramer, L.K. Anderson, S.-B. Chang, S.M. Royer, L.A. Shearer, S.M. Stack, J.K.C. Rose, Y. Xu, N. Eannetta, A.J. Matas, R. McQuinn, S.D. Tanksley, F. Camara, R. Guigo, S. Rombauts, J. Fawcett, Y. Van de Peer, D. Zamir, C. Liang, M. Spannagl, H. Gundlach, R. Bruggmann, K. Mayer, Z. Jia, J. Zhang, Z. Ye, G.J. Bishop, S. Butcher, R. Lopez-Cobollo, D. Buchan, I. Filippis, J. Abbott, R. Dixit, M. Singh, A. Singh, J.K. Pal, A. Pandit, P.K. Singh, A.K. Mahato, V. Dogra, K. Gaikwad, T.R. Sharma, T. Mohapatra, N.K. Singh, M. Causse, C. Rothan, T. Schiex, C. Noirot, A. Bellec, C. Klopp, C. Delalande, H. Berges, J. Mariette, P. Frasse, S. Vautrin, M. Zouine, A. Latche, C. Rousseau, F. Regad, J.-C. Pech, M. Philippot, M. Bouzayen, P. Pericard, S. Osorio, A. Fernandez del Carmen, A. Monforte, A. Granell, R. Fernandez-Munoz, M. Conte, G. Lichtenstein,

F. Carrari, G. De Bellis, F. Fuligni, C. Peano, S. Grandillo, P. Termolino, M. Pietrella, E. Fantini, G. Falcone, A. Fiore, G. Giuliano, L. Lopez, P. Facella, G. Perrotta, L. Daddiego, M. Bryan, G. Orozco, X. Pastor, D. Torrents, K.N.V.M.G.M. van Schriek, R.M.C. Feron, J. van Oeveren, P. de Heer, L. daPonte, S. Jacobs-Oomen, M. Cariaso, M. Prins, M.J.T. van Eijk, A. Janssen, M.J.J. van Haaren, S.-H. Jo, J. Kim, S.-Y. Kwon, S. Kim, D.-H. Koo, S. Lee, C.-G. Hur, C. Clouser, A. Rico, A. Hallab, C. Gebhardt, K. Klee, A. Joecker, J. Warfsmann, U. Goebel, S. Kawamura, K. Yano, J.D. Sherman, H. Fukuoka, S. Negoro, S. Bhutty, P. Chowdhury, D. Chattopadhyay, E. Datema, S. Smit, E.W.M. Schijlen, J. van de Belt, J.C. van Haarst, S.A. Peters, M.J. van Staveren, M.H.C. Henkens, P.J.W. Mooyman, T. Hesselink, R.C.H.J. van Ham, G. Jiang, M. Droege, D. Choi, B.-C. Kang, B.D. Kim, M. Park, S. Kim, S.-I. Yeom, Y.-H. Lee, Y.-D. Choi, G. Li, J. Gao, Y. Liu, S. Huang, V. Fernandez-Pedrosa, C. Collado, S. Zuniga, G. Wang, R. Cade, R.A. Dietrich, J. Rogers, S. Knapp, Z. Fei, R.A. White, T.W. Thannhauser, J.J. Giovannoni, M. Angel Botella, L. Gilbert, R. Gonzalez, J.L. Goicoechea, Y. Yu, D. Kudrna, K. Collura, M. Wissotski, R. Wing, H. Schoof, B.C. Meyers, A.B. Gurazada, P.J. Green, S. Mathur, S. Vyas, A.U. Solanke, R. Kumar, V. Gupta, A.K. Sharma, P., Khurana, J.P. Khurana, A.K. Tyagi, T. Dalmay, I. Mohorianu, B. Walts, S. Chamala, W.B. Barbazuk, J. Li, H. Guo, T.-H. Lee, Y. Wang, D. Zhang, A.H. Paterson, X. Wang, H. Tang, A. Barone, M.L. Chiusano, M.R. Ercolano, N. D'Agostino, M. Di Filippo, A. Traini, W. Sanseverino, L. Frusciante, G.B. Seymour, M. Elharam, Y. Fu, A. Hua, S. Kenton, J. Lewis, S. Lin, F. Najar, H. Lai, B. Qin, C. Qu, R. Shi, D. White, J. White, Y. Xing, K. Yang, J. Yi, Z. Yao, L. Zhou, B.A. Roe, A. Vezzi, M. D'Angelo, R. Zimbello, R. Schiavon, E. Caniato, C. Rigobello, D. Campagna, N. Vitulo, G. Valle, D.R. Nelson, E. De Paoli, D. Szinay, H.H. de Jong, Y. Bai, R.G.F. Visser, R.M.K. Lankhorst, H. Beasley, K. McLaren, C. Nicholson, C. Riddle, G. Gianese, and C. Tomato Genome. 2012. The tomato genome sequence provides insights into fleshy fruit evolution. *Nature* 485(7400):635–641.

Satô, Y. 1976. Embryological studies of some cornaceous plants. *Science Reports of the Tôhoku University, Series IV* 37:117–130.

Sattler, R. 1973. Organogenesis of Flowers: A Photographic Text-Atlas. University of Toronto Press, Toronto, Canada.

Sauquet, H., J.A. Doyle, T. Scharaschkin, T. Borsch, K.W. Hilu, L.W. Chatrou, and A. Le Thomas. 2003. Phylogenetic analysis of Magnoliales and Myristicaceae based on multiple data sets: Implications for character evolution. *Botanical Journal of the Linnean Society* 142:125–186.

Sauquet, H., S.Y.W. Ho, M.A. Gandolfo, G.J. Jordan, P. Wilf, D.J. Cantrill, M.J. Bayly, L. Bromham, G.K. Brown, R.J. Carpenter, D.M. Lee, D.J. Murphy, J.M.K. Sniderman, and F. Udovicic. 2012. Testing the impact of calibration on molecular divergence times using a fossil-rich group: The case of *Nothofagus* (Fagales). *Systematic Biology* 61(2):289–313.

Savolainen, V., M.W. Chase, S.B. Hoot, C.M. Morton, D.E. Soltis, C. Bayer, M.F. Fay, A.Y. De Bruijn, S. Sullivan, and Y.L. Qiu. 2000c. Phylogenetics of flowering plants based on combined analysis of plastid *atpB* and *rbcL* gene sequences. *Systematic Biology* 49:306–362.

Savolainen, V., M.W. Chase, C.M. Morton, D.E. Soltis, C. Bayer, M.F. Fay, A. De Bruijn, S. Sullivan, and Y.-L. Qiu. 2000a. Phylogenetics of flowering plants based upon a combined analysis of plastid *atpB* and *rbcL* gene sequences. *Systematic Biology* 49:306–362.

Savolainen, V., M.W. Chase, N. Salamin, D.E. Soltis, P.S. Soltis, A. Lopez, O. Fedrigo, and G.J.P. Naylor. 2002. Phylogeny reconstruction and functional constraints in organellar genomes: Plastid versus animal mitochondrion. *Systematic Biology* 51:638–647.

Savolainen, V., M.F. Fay, D.C. Albach, A. Backlund, M. Van der Bank, K.M. Cameron, S.A. Johnson, M.D. Lledó, J.-C. Pintaud, M. Powell, M.C. Sheahan, D.E. Soltis, P.S. Soltis, P. Weston, W.M. Whitten, J. Wurdack, and M.W. Chase. 2000b. Phylogeny of the eudicots: A nearly complete familial analysis based on *rbcL* gene sequences. *Kew Bulletin* 55:257–309.

Savolainen, V., J.V. Manen, E. Douzery, and R. Spichiger. 1994. Molecular phylogeny of families related to Celastrales based *rbcL* 5' flanking regions. *Molecular Phylogenetics and Evolution* 3:27–37.

Savolainen, V., R. Spichiger, and J.F. Manen. 1997. Polyphyletism of Celastrales deduced from a non-coding chloroplast DNA region. *Molecular Phylogenetics and Evolution* 7:145–157.

Sawada, M. 1971. Floral vascularization of *Paeonia japonica* with some consideration on systematic position of the Paeoniaceae. *Botanical Magazine (Tokyo)* 84:51–60.

Schaal, B.A., and W.J. Leverich. 2001. Plant population biology and systematics. *Taxon* 50:679–696.

Schaefer, H., and S.S. Renner. 2011. Phylogenetic relationships in the order Cucurbitales and a new classification of the gourd family (Cucurbitaceae). *Taxon* 60(1):122–138.

Schäferhoff, B., A. Fleischmann, E. Fischer, D.C. Albach, T. Borsch, G. Heubl, and K.F. Muller. 2010. Towards resolving Lamiales relationships: Insights from rapidly evolving chloroplast sequences. *BMC Evolutionary Biology* 10:352. doi:10.1186/1471-2148-10-52.

Schäferhoff, B, K.F. Mueller, and T. Borsch. 2009. Caryophyllales phylogenetics: Disentangling Phytolaccaceae and Molluginaceae and description of Microteaceae as a new isolated family. *Willdenowia* 39:209–228.

Schatz, G.E. 2000. The rediscovery of a Malagasy endemic: *Takhtajania perrieri* (Winteraceae). *Annals of the Missouri Botanical Garden* 87:297–302.

Schick, B. 1980. Untersuchungen über die Biotechnik der Apocynaceenblüte. I. Morphologie und Funktion des Narbenkopfes. *Flora* 170:394–432.

———. 1982. Untersuchungen über die Biotechnik der Apocynaceenblüte. II. Bau und Funktion des Bestäubungsapparates. *Flora* 172:347–371.

Schiestl, F.P. 2005. On the success of a swindle: Pollination by deception in orchids. *Naturwissenschaften* 92:255–264.

Schiestl, F.P., and S. Dötterl. 2012. The evolution of floral scent and olfactory preferences in pollinators: Coevolution or pre-existing bias? *Evolution* 66:2042–2055.

Schiestl, F.P., and S.D. Johnson. 2013. Pollinator-mediated evolution of floral signals. *Trends in Ecology & Evolution* 28:307–315.

Schinz, H. 1893. Amaranthaceae. *In* A. Engler, and K. Prantl [eds.], Die Natürlichen Pflanzenfamilien, 1st ed., III, 1a, 91–118. Engelmann, Leipzig, Germany.

Schlueter, J.A., P. Dixon, C. Granger, D. Grant, L. Clark, J.J. Doyle, and R.C. Shoemaker. 2004. Mining EST databases to resolve evolutionary events in major crop species. *Genome* 47: 868–876.

Schlumpberger, B.O., and S.S. Renner. 2012. Molecular phylogenetics of *Echinopsis* (Cactaceae): Polyphyly at all levels and convergent evolution of pollination modes and growth forms. *American Journal of Botany* 99:1335–1349.

Schlüter, P.M., S. Xu, V. Gagliardini, E. Whittle, J. Shanklin, U. Grossniklaus, and F.P. Schiestl. 2011. Stearoyl-acyl carrier protein desaturases are associated with floral isolation in sexually deceptive orchids. *Proceedings of the National Academy of Sciences of USA* 108:5696–5701.

Schmid, R. 1964. Die systematische Stellung der Dioncophyllaceae. *Botanische Jahrbücher für Systematik* 83:1–56.

———. 1978. Actinidiaceae, Davidiaceae, and Paracryphiaceae: Systematic considerations. *Botanische Jahrbücher für Systematik* 100:196–204.

Schmidt, R., and I. Bancroft. 2011. Perspectives on genetics and genomics of the Brassicaceae. *Genetics and Genomics of the Brassicaceae* 9:617–632.

Schneeweiss, G.M., A. Colwell, J.M. Park, C.G. Jang, and T.F. Stuessy. 2004. Phylogeny of holoparasitic *Orobanche* (Orobanchaceae) inferred from nuclear ITS sequences. *Molecular Phylogenetics and Evolution* 30(2):465–478.

Schneider, E.L. 1979. Pollination biology of the Nymphaeaceae. *In* D.M. Caron [ed.], Proceedings of the Fourth International Symposium on Pollination, 419–430. Maryland Agricultural Experimental Station Special Miscellaneous Publication 1, College Park, MD.

Schneider, E.L., and S. Carlquist. 1996. Vessels in *Brasenia* (Cabombaceae): New perspectives on vessel origin in primary xylem of angiosperms. *American Journal of Botany* 83:1236–1240.

Schneider, E.L., S. Carlquist, and A. Kohn. 1995. Vessels in Nymphaeaceae: *Nuphar, Nymphaea,* and *Ondinea*. *International Journal of Plant Sciences* 156:857–862.

Schneider, E.L., S.C. Tucker, and P.S. Williamson. 2003. Floral development in the Nymphaeales. *International Journal of Plant Sciences* 164:S279–S292.

Schneider, H., E. Schuettpelz, K.M. Pryer, R. Cranfill, S. Magallon, and R. Lupia. 2004. Ferns diversified in the shadow of Angiosperms. *Nature* 428(6982):553–557.

Schneider, J.V., P. Bissiengou, M.C.E. Amaral, A. Tahir, M.F. Fay, M. Thines, M.S.M. Sosef, G. Zizka, and L.W. Chatrou. 2014. Phylogenetics, ancestral state reconstruction, and a new infrafamilial classification of the pantropical Ochnaceae (Medusagynaceae, Ochnaceae s.str., Quiinaceae) based on five DNA regions. *Molecular Phylogenetics and Evolution* 78:199–214.

Schnitzler, J., T.G. Barraclough, J.S. Boatwright, P. Goldblatt, J.C. Manning, M.P. Powell, T. Rebelo, and V. Savolainen. 2011. Causes of plant diversification in the cape biodiversity hotspot of South Africa. *Systematic Biology* 60(3):343–357.

Schodde, R. 1970. Two new suprageneric taxa in the Monimiaceae alliance (Laurales). *Taxon* 19:324–332.

Schoen, D.J., M.T. Morgan, and T. Bataillon. 1997. How does self-pollination evolve? Inferences from floral ecology and molecular genetic variation. *In* J. Silvertown, M. Franco, and J.L. Harper [eds.], Plant Life Histories, 77–101. Cambridge University Press, Cambridge, UK.

Schöffel, K. 1932. Untersuchungen über den Blütenbau der Ranunculaceen. *Planta* 17:315–371.

Scholz, H. 1964. Sapindales. *In* H. Melchior [ed.], A. Engler's Syllabus der Pflanzenfamilien, ed.12, 2, 267–268. Borntraeger, Berlin, Germany.

Schönenberger, J. 2005. Rise from the ashes—the reconstruction of charcoal fossil flowers. *Trends in Plant Science* 10:436–443.

———. 2009. Comparative floral structure and systematics of Fouquieriaceae and Polemoniaceae

(Ericales). *International Journal of Plant Sciences* 170:1132–1167.

Schönenberger, J., A.A. Anderberg, and K.J. Sytsma. 2005. Molecular phylogenetics and patterns of floral evolution in the Ericales. *International Journal of Plant Sciences* 166:265–288.

Schönenberger, J., and E. Conti. 2003. Molecular phylogeny and floral evolution of Penaeaceae, Oliniaceae, Rhynchocalycaceae, and Alzateaceae (Myrtales). *American Journal of Botany* 90(2):293–309.

Schönenberger, J., E.M. Friis, M.L. Matthews, and P.K. Endress. 2001. Cunoniaceae in the Cretaceous of Europe: Evidence from fossil flowers. *Annals of Botany* 88:423–437.

Schönenberger, J., K.R. Pedersen, and E.M. Friis. 2001. Normapolles flowers of fagalean affinity from the Late Cretaceous of Portugal. *Plant Systematics and Evolution* 226(3–4):205–230.

Schönenberger, J., M. von Balthazar, and K.J. Sytsma. 2010. Diversity and evolution of floral structure among early diverging lineages in the Ericales. *Philosophical Transactions of the Royal Society of London B* 365:437–448.

Schönenberger, J., M. von Balthazar, M. Takahashi, X.-H. Xiao, P.R. Crane, and P.S. Herendeen. 2012. *Glandulocalyx upatoiensis*, a fossil flower of Ericales (Actinidiaceae/Clethraceae) from the Late Cretaceous (Santonian) of Georgia, USA. *Annals of Botany* 109:921–936.

Schopf, J.M. 1976. Morphologic interpretations of fertile structures in glossopterid gymnosperms. *Review of Palaeobotany and Palynology* 21:25–64.

Schranz, M.E., P.R. Edger, J.C. Pires, D.M.V. Dam, C.W. Wheat, and N.M. van Dam. 2011. Comparative genomics in the Brassicales: Ancient genome duplications, glucosinolate diversification and Pierinae herbivore radiation. *In* D. Edwards, J. Batley, I. Parkin, and C. Kole [eds.], Genetics, genomics and breeding of oilseed brassicas, 206–218. Science Publishers, Lebanon, USA.

Schranz, M.E., M.A. Lysak, and T. Mitchell-Olds. 2006. The ABC's of comparative genomics in the Brassicaceae: Building blocks of crucifer genomes. *Trends in Plant Science* 11(11):535–542.

Schranz, M.E., and T. Mitchell-Olds. 2006. Independent ancient polyploidy events in the sister families Brassicaceae and Cleomaceae. *Plant Cell* 18(5):1152–1165.

Schranz, M.E., S. Mohammadin, and P.P. Edger. 2012. Ancient whole genome duplications, novelty and diversification: The WGD Radiation Lag-Time Model. *Current Opinion in Plant Biology* 15(2):147–153.

Schranz, M.E., B.-H. Song, A.J. Windsor, and T. Mitchell-Olds. 2007. Comparative genomics in the Brassicaceae: A family-wide perspective. *Current Opinion in Plant Biology* 10(2):168–175.

Schuettpelz, E., S.B. Hoot, R. Samuel, and F. Ehrendorfer. 2002. Multiple origins of Southern Hemisphere *Anemone* (Ranunculaceae) based on plastid and nuclear sequence data. *Plant Systematics and Evolution* 231: 143–151.

Schürhoff, P.N. 1926. Die Zytologie der Blütenpflanzen. Enke, Stuttgart, Germany.

Schwarzbach, A.E., and L.A. McDade. 2002. Phylogenetic relationships of the mangrove family Avicenniaceae based on chloroplast and nuclear ribosomal DNA sequences. *Systematic Botany* 27:84–98.

Schwarzbach, A.E., and R.E. Ricklefs. 2000. Systematic affinities of Rhizophoraceae and Anisophylleaceae, and intergeneric relationships within Rhizophoraceae, based on chloroplast DNA, nuclear ribosomal DNA, and morphology. *American Journal of Botany* 87:547–564.

Schwarz-Sommer, Z., P. Huijser, W. Nacken, H. Saedler, and H. Sommer. 1990. Genetic control of flower development in *Antirrhinum majus*. *Science* 250(4983): 931–936.

Scogin, R. 1977. Anthocyanins of the Fouquieriaceae. *Biochemical Systematics and Ecology* 5(4):265–267.

———. 1978. Leaf phenolics of the Fouquieriaceae. *Biochemical Systematics and Ecology* 6(4):297–298.

Scotland, R.W., R.G. Olmstead, and J.R. Bennett. 2003. Phylogeny reconstruction: The role of morphology. *Systematic Biology* 52:539–548.

Scotland, R.W., J.A. Sweere, P.A. Reeves, and R.G. Olmstead. 1995. Higher level systematics of Acanthaceae determined by chloroplast DNA sequences. *American Journal of Botany* 82:266–275.

Seberg, O., G. Petersen, A.S. Barfod, and J.I. Davis. 2010. Diversity, Phylogeny and Evolution in the Monocotyledons. Aarhus University Press, Denmark.

Seberg, O., G. Petersen, J.I. Davis, J.C. Pires, D.W. Stevenson, M.W. Chase, M.F. Fay, D.S. Devey, T. Jørgensen, K.J. Sytsma, and Y. Pillon. 2012. Phylogeny of the Asparagales, based on three plastid and two mitochondrial genes. *American Journal of Botany* 99:875–889.

Sede, S.M., S.I. Durnhofer, S. Morello, and F. Zapata. 2013. Phylogenetics of *Escallonia* (Escalloniaceae) based on plastid DNA sequence data. *Botanical Journal of the Linnaean Society* 173:442–451.

Seelanan, T., A. Schnabel, and J. Wendel. 1997. Congruence and consensus in the cotton tribe (Malvaceae). *Systematic Botany* 22:259–290.

Segraves, K.A. and J.N. Thompson. 1999. Plant polyploidy and pollination: Floral traits and insect visits to diploid and tetraploid *Heuchera grossulariifolia*. *Evolution* 53:1114–1121.

Segura, S., L. Scheinvar, G. Olalde, O. Leblanc, S. Filardo, A. Muratalla, C. Gallegos, and C. Flores. 2007. Genome sizes and ploidy levels in Mexican cactus pear species *Opuntia* (Tourn.) Mill. series Streptacanthae Britton et Rose, Leucotrichae DC., Heliabravoanae Scheinvar and Robustae Britton et Rose. *Genetic Resources of Crop Evolution* 54:1033–1041.

Selosse, M.-A., and D.D. Cameron. 2010. Introduction to a virtual special issue on mycoheterotrophy: New Phytologist sheds light on non-green plants. *New Phytologist* 185(3):591–593.

Selosse, M.-A., and M. Roy. 2009. Green plants that feed on fungi: Facts and questions about mixotrophy. *Trends in Plant Science* 14(2):64–70.

Sennblad, B., and B. Bremer. 1996. The familial and subfamilial relationships of Apocynaceae and Asclepiadaceae evaluated with *rbcL* data. *Plant Systematics and Evolution* 202:153–175.

———. 2002. Classification of Apocynaceae s.l. according to a new approach combining Linnaean and phylogenetic taxonomy. *Systematic Biology* 51:389–409.

Senters, A.E., and D.E. Soltis. 2003. Phylogenetic relationships in *Ribes* (Grossulariaceae) inferred from ITS sequence data. *Taxon* 52:51–66.

Seymour, D.K., D. Koenig, J. Hagmann, C. Becker, and D. Weigel. 2014. Evolution of DNA methylation patterns in the Brassicaceae is driven by differences in genome organization. *PLoS Genetics* 10(11): e1004785–e1004785.

Sharma, V.K. 1968. Floral morphology, anatomy, and embryology of *Coriaria nepalensis* Wall. with a discussion of the interrelationships of the family Coriariaceae. *Phytomorphology* 18:143–153.

Sheahan, M.C., and M.W. Chase. 1996. A phylogenetic analysis of Zygophyllaceae R. Br. based on morphological, anatomical, and *rbcL* DNA sequence data. *Botanical Journal of the Linnean Society* 122:279–300.

———. 2000. Phylogenetic relationships within Zygophyllaceae based on DNA sequences of three plastid regions, with special emphasis on Zygophylloideae. *Systematic Botany* 25:371–384.

Shi, T., H. Huang, and M.S. Barker. 2010. Ancient genome duplications during the evolution of kiwifruit (*Actinidia*) and related Ericales. *Annals of Botany* 106:497–504.

Shi, X., D.W.K. Ng, C. Zhang, L. Comai, W. Ye, and Z.J. Chen. 2012. Cis- and trans-regulatory divergence between progenitor species determines gene-expression novelty in *Arabidopsis* allopolyploids. *Nature Communications* 3:950. doi:10.1038/ncomms1954.

Shimada, S., T.Y. Inoue, and M. Sakuta. 2005. Anthocyanidin synthase in non-anthocyanin-producing Caryophyllales species. *The Plant Journal* 44:950–959.

Shimada, S., H. Otsuki, and M. Sakuta. 2007. Transcriptional control of anthocyanin biosynthetic genes in the Caryophyllales. *Journal of Experimental Botany* 58:957–967.

Shimada, S., K. Takahashi, Y. Sato, and M. Sakuta. 2004. Dihydroflavonol 4-reductase cDNA from non-anthocyanin-producing species in the Caryophyllales. *Plant and Cell Physiology* 45:1290–1298.

Shindo, S., K. Sakakibara, R. Sano, K. Ueda, and M. Hasebe. 2001. Characterization of a *FLORICAULA/LEAFY* homologue of *Gnetum parviflorum* and its implications for the evolution of reproductive organs in seed plants. *International Journal of Plant Sciences* 162:1199–1209.

Shipunov, A., and E. Shipunova. 2011. *Haptanthus* story: Rediscovery of enigmatic flowering plant from Honduras. *American Journal of Botany* 98(4):761–763.

Shore, J.S., K.L. McQueen, and S.L. Little. 1994. Inheritance of plastid DNA in the *Turnera ulmifolia* complex. *American Journal of Botany* 81:1636–1639.

Shulaev, V., S.S. Korban, B. Sosinski, A.G. Abbott, H.S. Aldwinckle, K.M. Folta, A. Iezzoni, D. Main, P. Arus, A.M. Dandekar, K. Lewers, S.K. Brown, T.M. Davis, S.E. Gardiner, D. Potter, and R.E. Veilleux. 2008. Multiple models for Rosaceae genomics. *Plant Physiology* 147:985–1003.

Shulaev, V., D.J. Sargent, R.N. Crowhurst, T.C. Mockler, O. Folkerts, A.L. Delcher, P. Jaiswal, K. Mockaitis, A. Liston, S.P. Mane, P. Burns, T.M. Davis, J.P. Slovin, N. Bassil, R.P. Hellens, C. Evans, T. Harkins, C. Kodira, B. Desany, O.R. Crasta, R.V. Jensen, A.C. Allan, T.P. Michael, J.C. Setubal, J.-M. Celton, D.J.G. Rees, K.P. Williams, S.H. Holt, J.J.R. Rojas, M. Chatterjee, B. Liu, H. Silva, L. Meisel, A. Adato, S.A. Filichkin, M. Troggio, R. Viola, T.-L. Ashman, H. Wang, P. Dharmawardhana, J. Elser, R. Raja, H.D. Priest, D.W. Bryant, Jr., S.E. Fox, S.A. Givan, L.J. Wilhelm, S. Naithani, A. Christoffels, D.Y. Salama, J. Carter, E.L. Girona, A. Zdepski, W. Wang, R.A. Kerstetter, W. Schwab, S.S. Korban, J. Davik, A. Monfort, B. Denoyes-Rothan, P. Arus, R. Mittler, B. Flinn, A. Aharoni, J.L. Bennetzen, S.L. Salzberg, A.W. Dickerman, R. Velasco, M. Borodovsky, R.E. Veilleux, and K.M. Folta. 2011. The genome of woodland strawberry (*Fragaria vesca*). *Nature Genetics* 43(2):109–116.

Simmons, M.P., and J. Gatesy. 2015. Coalescence vs. concatenation: Sophisticated analyses vs. first principles applied to rooting the angiosperms. *Molecular Phylogenetics and Evolution* 91:98–122.

Simmons, M.P., V. Savolainen, C.C. Clevinger, R.H. Archer, and J.I. Davis. 2001. Phylogeny of the Celastraceae inferred from 26S nuclear ribosomal DNA, phytochrome B, *rbcL*, *atpB*, and morphology. *Molecular Phylogenetics and Evolution* 19:353–366.

Simpson, B.B., and J.L. Neff. 1981. Floral rewards: Alternatives to pollen and nectar. *Annals of the Missouri Botanical Garden* 68:301–322.

Simpson, D. 1995. Relationships within Cyperales. *In* P.J. Rudall, P.J. Cribb, D.F. Cutler, and C.J. Humphries [eds.], Monocotyledons: Systematics and Evolution, 750. Royal Botanic Gardens, Kew, UK.

Sims, H.J., P.S. Herendeen, and P.R. Crane. 1998. New genus of fossil Fagaceae from the Santonian (Late Cretaceous) of central Georgia USA. *International Journal of Plant Sciences* 159(2):391–404.

Sims, H.J., P.S Herendeen, R. Lupia, R.A. Christopher, and P.R. Crane. 1999. Fossil flowers with *Normapolles* pollen from the Upper Cretaceous of southeastern North America. *Review of Palaeobotany and Palynology* 106:131–151.

Sinha, N. 2000. The response of epidermal cells to contact. *Trends in Plant Science* 5:233–234.

Sinha, N., and E. Kellogg. 1996. Parallelism and diversity in multiple origins of C_4 photosynthesis in the grass family. *American Journal of Botany* 83:1458–1470.

Skipper, M. 2002. Genes from the *APETALA3* and *PISTILLATA* lineages are expressed in developing vascular bundles of the tuberous rhizome, flowering stem and flower primordia of *Eranthis hyemalis*. *Annals of Botany* 89:83–88.

Skvarla, J.J., and J.W. Nowicke. 1976. Ultrastructure of pollen exine in centrospermous families. *Plant Systematics and Evolution* 126:55–78.

———. 1982. Pollen fine structure and relationships of *Achatocarpus* Triana and *Phaulothamnus* A. Gray. *Taxon* 31:244–249.

Skvarla, J.J., B.L. Turner, V.C. Patel, and A.S. Tomb. 1977. Pollen morphology in the Compositae and in morphologically related families. *In* V.H. Heywood, J.B. Harborne, and B.L. Turner [eds.], The Biology and Chemistry of the Compositae, 141–201. Academic Press, London, UK.

Sleumer, H. 1954. Flacourtiaceae. *In* C.G.G.J. van Steenis [ed.], Flora Malesiana, series 1, 2–106. Noordhoff-Kolff N.V., Djakarta, Indonesia.

———. 1980. Flacourtiaceae. *Flora Neotropica* 22:1–499.

Sloan, D.B., D.A. Triant, N.J. Forrester, L.M. Bergner, M. Wu, and D.R. Taylor. 2014. A recurring syndrome of accelerated plastid genome evolution in the angiosperm tribe Sileneae (Caryophyllaceae). *Molecular Phylogenetics and Evolution* 72:82–89.

Smissen, R.D., J.C. Clement, P.J. Garnock-Jones, and G.K. Chambers. 2002. Subfamilial relationships within Caryophyllaceae as inferred from 5' *ndhF* sequences. *American Journal of Botany* 89:1336–1341.

Smith, A.R., K.M. Pryer, E. Schuettpelz, P. Korall, H. Schneider, and P.G. Wolf. 2006. A classification for extant ferns. *Taxon* 55:705–731.

Smith, A.R., K.M. Pryer, P.G. Wolf, R. Cranfill, and H. Schneider. Submitted. An ordinal and familial classification for extant lycophytes and moniliophytes. *Taxon*.

Smith, D.L., T.J. Barkman, and C.W. dePamphilis. 2001. Hemiparasitism. *In* S.A. Levin [ed.], Encylcopedia of Biodiversity, vol. 3, 317–328. Academic Press, San Diego, CA.

Smith, D.R. 2014. Mitochondrion-to-plastid DNA transfer: It happens. *New Phytologist* 202(3):736–738.

Smith, G.H. 1928. Vascular anatomy of Ranalian flowers. II. Ranunculaceae (continued), Menispermaceae, Calycanthaceae, Annonaceae. *Botanical Gazette* 85:152–177.

Smith, J.F. 1996. Tribal relationships with the Gesneriaceae: A cladistic analysis of morphological data. *Systematic Botany* 21:497–513.

———. 2000a. A phylogenetic analysis of tribes Beslerieae and Napeantheae (Gesneriaceae) and evolution of fruit types: Parsimony and maximum likelihood analyses of *ndhF* sequences. *Systematic Botany* 25:72–81.

———. 2000b. Phylogenetic resolution within the tribe Episcieae (Gesneriaceae): Congruence of ITS and *ndhF* sequences from parsimony and maximum-likelihood analyses. *American Journal of Botany* 87:883–897.

Smith, J.F., and S. Atkinson. 1998. Phylogenetic analysis of the tribes Gloxinieae and Gesnerieae (Gesneriaceae): Data from *ndhF* sequences. *Selbyana* 19:122–131.

Smith, J.F., and C.L. Carroll. 1997. Phylogenetic relationships of the Episcieae (Gesneriaceae) based on *ndhF* sequences. *Systematic Botany* 22:713–724.

Smith, J.F., and J.L. Clark. 2013. Molecular phylogenetic analyses reveal undiscovered monospecific Genera in the tribe Episcieae (Gesneriaceae). *Systematic Botany* 38:451–463.

Smith, J.F., W.J. Kress, and E.A. Zimmer. 1993. Phylogenetic analysis of the Zingiberales based on *rbcL* sequences. *Annals of the Missouri Botanical Garden* 80:620–630.

Smith, L.B., and C.E. Wood. 1975. The genera of Bromeliaceae in the southeastern United States. *Journal of the Arnold Arboretum* 56:375–397.

Smith, S.A., and J.M. Beaulieu. 2009. Life history influences rates of climatic niche evolution in flowering plants. *Proceedings of the Royal Society of London B* 276:4345–4352.

Smith, S.A., J.M. Beaulieu, and M.J. Donoghue. 2009. Mega-phylogeny approach for comparative biology: An alternative to supertree and supermatrix approaches. *BMC Evolutionary Biology* 9. doi:10.1186/1471-2148-9-37.

———. 2010. An uncorrelated relaxed-clock analysis suggests an earlier origin for flowering plants. *Proceedings of the National Academy of Sciences USA* 107:5897–5902.

Smith, S.A., J.M. Beaulieu, A. Stamatakis, and M.J. Donoghue. 2011. Understanding Angiosperm diversification using small and large phylogenetic trees. *American Journal of Botany* 98:404–414.

Smith, S.A., and M.J. Donoghue. 2008. Rates of molecular evolution are linked to life history in flowering plants. *Science* 322:86–89.

Smith, S.A., and B.C. O'Meara. 2012. TreePL: Divergence time estimation using penalized likelihood for large phylogenies. *Bioinformatics* 28:2689–2690.

Smith, S.D. 2010. Using phylogenetics to detect pollinator-mediated floral evolution. *New Phytologist* 188:354–363.

Smith, S.D., C. Ané, and D.A. Baum. 2008. The role of pollinator shifts in the floral diversification of *Iochroma* (Solanaceae). *Evolution* 62:793–806.

Smith, S. D., and D.A. Baum. 2006. Phylogenetics of the florally diverse Andean clade Iochrominae (Solanaceae). *American Journal of Botany* 93:1140–1153.

Smith, S.D., and E.E. Golberg. 2015. Tempo and mode of flower color evolution. *American Journal of Botany* 102:1014–1025.

Smith, S.D., S.J. Hall, P.R.S. Izquierdo, and D.A. Baum. 2008. Comparative pollination biology of sympatric and allopatric Andean *Iochroma* (Solanaceae). *Annals of the Missouri Botanicall Garden* 95:600–617.

Smith, S.D., and M.D. Rausher. 2011. Gene loss and parallel evolution contribute to species difference in flower color. *Molecular Biology and Evolution* 28:2799–2810.

Smith, S.Y., M.E. Collinson, and P.J. Rudall. 2008. Fossil *Cyclanthus* (Cyclanthaceae, Pandanales) from the Eocene of Germany and England. *American Journal of Botany* 95:688–699.

Smith, S.Y., M.E. Collinson, P.J. Rudall, and D.A. Simpson. 2010. Cretaceous and Paleogene fossil record of Poales: Review and current research. *In* O. Seberg, G. Petersen, A.S. Barfod, and J.I. Davis [eds.], Diversity, Phylogeny, and Evolution in the Monocotyledons, 333–356. Aarhus University Press, Århus, Denmark.

Smith, S.Y., M.E. Collinson, D.A. Simpson, P.J. Rudall, F. Marone, and M. Stampanoni. 2009. Elucidating the affinities and habitat of ancient, widespread Cyperaceae: *Volkeria messelensis* gen. et sp. nov., a fossil mapanioid sedge from the Eocene of Europe. *American Journal of Botany* 96(8):1506–1518.

Smith, S.Y., and R.A. Stockey. 2003. Aroid seeds from the Middle Eocene Princeton Chert (*Keratosperma allenbyense*, Araceae): Comparisons with extant Lasioideae. *International Journal of Plant Sciences* 164:239–250.

———. 2007. Establishing a fossil record for the perianthless Piperales: *Saururus tuckerae* sp. nov. (Saururaceae) from the Middle Eocene Princeton Chert. *American Journal of Botany* 94:1642–1657.

Sogo, A., and H. Tobe. 2008. Mode of pollen tube growth in pistils of *Ticodendron incognitum* (Ticodendraceae, Fagales) and the evolution of chalazogamy. *Botanical Journal of the Linnean Society* 157(4):621–631.

Sokoloff, D.D., M.V. Remizowa, A.S. Beer, S.R. Yadav, T.D. Macfarlane, M.M. Ramsay, and P.J. Rudall. 2013. Impact of spatial constraints during seed germination on the evolution of Angiosperm cotyledons: A case study from tropical Hydatellaceae (Nymphaeales). *American Journal of Botany* 100:824–843.

Sokoloff, D.D., M.V. Remizowa, T.D. Macfarlane, and P.J. Rudall. 2008. Classification of the early-divergent angiosperm family Hydatellaceae: One genus instead of two, four new species and sexual dimorphism in dioecious taxa. *Taxon* 57:179–200.

Sokoloff, D.D., M.V. Remizowa, and P.J. Rudall. 2013. Is syncarpy an ancestral condition in monocots and core eudicots? *In* P. Wilkin, and S.J. Mayo [eds.], Early Events in Monocot Evolution. Systematics Association Special Volume 83: 60–81. Cambridge University Press, Cambridge, UK.

Soltis, D.E., V.A. Albert, S. Kim, M.-J. Yoo, P.S. Soltis, M.W. Frohlich, J. Leebens-Mack, H. Kong, K. Wall, C. dePamphilis, and H. Ma. 2005. Evolution of the Flower. *In* R. Henry [ed.], Diversity and Evolution of Plants, 165–200. CABI Publishing, Wallingford, UK.

Soltis, D.E., V.A. Albert, J. Leebens-Mack, C.D. Bell, A.H. Paterson, C. Zheng, D. Sankoff, C.W. dePamphilis, P.K. Wall, and P.S. Soltis. 2009. Polyploidy and angiosperm diversification. *American Journal of Botany* 96(1):336–348.

Soltis, D.E., V.A. Albert, J. Leebens-Mack, J.D. Palmer, R.A. Wing, C.W. dePamphilis, H. Ma, J.E. Carlson, N. Altman, S. Kim, P.K. Wall, A. Zuccolo, and P.S. Soltis. 2008. The *Amborella* genome: An evolutionary reference for plant biology. *Genome Biology* 9:402. doi:10.1186/gb-2008-9-3-402.

Soltis, D.E., V.A. Albert, V. Savolainen, K. Hilu, Y.-L. Qiu, M.W. Chase, J.S. Farris, S. Stefanović, D.W. Rice, J.D. Palmer, and P.S. Soltis. 2004. Genome-scale data, angiosperm relationships, and "ending incongruence": A cautionary tale in phylogenetics. *Trends in Plant Science*, 9(10):477–483.

Soltis, D.E., C.D. Bell, S. Kim, and P.S. Soltis. 2008b. Origin and early evolution of Angiosperms. *Annals of the New York Academy of Sciences* 1133:3–25.

Soltis, D.E., R.J.A. Buggs, B. Barbazuk, S. Chamala, M. Chester, J.P. Gallagher, P.S. Schnable, and P.S. Soltis. 2012. The early stages of polyploidy: Rapid and repeated evolution in *Tragopogon*. *In* P.S. Soltis, and D.E. Soltis [eds.], Polyploidy and Genome Evolution, 271–292. Springer-Verlag, Berlin, Germany.

Soltis, D.E., A.S. Chanderbali, S. Kim, M. Buzgo, and P.S. Soltis. 2007. The ABC model and its applicability to basal angiosperms. *Annals of Botany* 100(2):155–163.

Soltis, D.E., J.W. Clayton, C.C. Davis, M.A. Gitzendanner, M. Cheek, V. Savolainen, A.M. Amorim, and P.S. Soltis. 2007. Monophyly and relationships of the enigmatic family Peridiscaceae. *Taxon* 56(1):65–73.

Soltis, D.E., M. Fishbein, and R.K. Kuzoff. 2003c. Reevaluating the evolution of epigyny: Data from phylogenetics and floral ontogeny. *International Journal of Plant Sciences* 164(5 Supplement):S251–S264.

Soltis, D.E., M.A. Gitzendanner, and P.S. Soltis. 2007. A 567–taxon data set for angiosperms: The challenges posed by bayesian analyses of large data sets. *International Journal of Plant Sciences* 168:137–157.

Soltis, D.E., A. Grable, D. Morgan, P.S. Soltis, and R. Kuzoff. 1993. Molecular systematics of Saxifragaceae *sensu stricto*. *American Journal of Botany* 80:1056–1081.

Soltis, D.E., C. Hibsch-Jetter, P.S. Soltis, M. Chase, and J.S. Farris. 1997b. Molecular phylogenetic relationships among angiosperms: An overview based on *rbcL* and 18S rDNA sequences. *In* K. Iwatsuki, and P.H. Raven [eds.], Evolution and Diversification of Land Plants, 157–178. Springer, Tokyo, Japan.

Soltis, D.E., and L. Hufford. 2002. Ovary development and diversification in Saxifragaceae. *International Journal of Plant Sciences.* 163:277–293.

Soltis, D.E., I. Leitch, P.S. Soltis, and M. Bennett. 2003. Genome size evolution in angiosperms. *American Journal of Botany* 90:1596–1606.

Soltis, D.E., M.E. Mort, R. Kuzoff, M. Zanis, M. Fishbein, and L. Hufford. 2001b. Elucidating deep-level phylogenetic relationships in Saxifragaceae using sequences for six chloroplastic and nuclear DNA regions. *Annals of the Missouri Botanical Garden* 88:669–693.

Soltis, D.E., M.E. Mort, M. Latvis, E.V. Mavrodiev, B.C. O'Meara, P.S. Soltis, J.G. Burleigh, and R.R. de Casas. 2013. Phylogenetic relationships and character evolution analysis of Saxifragales using a supermatrix approach. *American Journal of Botany* 100(5):916–929.

Soltis, D.E., M.E. Mort, P.S. Soltis, C. Hibsch-Jetter, E.A. Zimmer, and D. Morgan. 1999. Phylogenetic relationships of the enigmatic angiosperm family Podostemaceae inferred from 18S rDNA and *rbcL* sequence data. *Molecular Phylogenetics and Evolution* 11:261–272.

Soltis, D.E., M.C. Segovia-Salcedo, I. Jordon-Thaden, L. Majure, N.M. Miles, E.V. Mavrodiev, W. Mei, M.B. Cortez, P.S. Soltis, and M.A. Gitzendanner. 2014a. Are polyploids really evolutionary dead-ends (again)? A critical reappraisal of Mayrose et al. (2011). *New Phytologist* 202(4):1105–1117.

Soltis, D.E., A.E. Senters, M.J. Zanis, S. Kim, J.D. Thompson, P.S. Soltis, L.P.R. Decraene, P.K. Endress, and J.S. Farris. 2003. Gunnerales are sister to other core eudicots: Implications for the evolution of pentamery. *American Journal of Botany* 90(3):461–470.

Soltis, D.E., S.A. Smith, N. Cellinese, K.J. Wurdack, D.C. Tank, S.F. Brockington, N.F. Refulio-Rodriguez, J.B. Walker, M.J. Moore, B.S. Carlsward, C.D. Bell, M. Latvis, S. Crawley, C. Black, D. Diouf, Z. Xi, C.A. Rushworth, M.A. Gitzendanner, K.J. Sytsma, Y.-L. Qiu, K.W. Hilu, C.C. Davis, M.J. Sanderson, R.S. Beaman, R.G. Olmstead, W.S. Judd, M.J. Donoghue, and P.S. Soltis. 2011. Angiosperm phylogeny: 17 genes, 640 taxa. *American Journal of Botany* 98:704–730.

Soltis, D.E., and P.S. Soltis. 1990. Isozyme evidence for ancient polyploidy in primitive angiosperms. *Systematic Botany* 15:328–337.

———. 1997. Phylogenetic relationships in Saxifragaceae sensu lato: A comparison of topologies based on 18S rDNA and *rbcL* sequences. *American Journal of Botany* 84:504–522.

———. 1998. Choosing an approach and an appropriate gene for phylogenetic analysis. *In* D.E. Soltis, P.S. Soltis, and J.J. Doyle [eds.], Molecular Systematics of Plants, II: DNA Sequencing, 1–42. Kluwer, Boston, MA.

———. 1999. Phylogenetic analyses of large data sets: Approaches using the angiosperms. *In* M. Kato [ed.], The Biology of Diversity, 91–103. Springer, Tokyo, Japan.

———. 2000. Contributions of molecular systematics to studies of molecular evolution. *Plant Molecular Biology* 42:45–75.

———. 2003. The role of phylogenetics in comparative genetics. *Plant Physiology* 132:1790–1800.

———. 2004. *Amborella* not a "basal angiosperm"? Not so fast. *American Journal of Botany* 91:997–1001.

———. 2004. The origin and diversification of angiosperms. *American Journal of Botany* 91(10):1614–1626.

Soltis, D.E., P.S. Soltis, V. Albert, C. dePamphilis, M. Frohlich, H. Ma, D. Oppenheimer, and G. Theissen. 2002b. Missing links: The genetic architecture of the flower and floral diversification. *Trends in Plant Science* 7(1): 22–31.

Soltis, D.E., P.S. Soltis, M.D. Bennett, and I.J. Leitch. 2003b. Evolution of genome size in the Angiosperms. *American Journal of Botany* 90(11):1596–1603.

Soltis, D.E., P.S. Soltis, M.W. Chase, M. Mort, D. Albach, M. Zanis, V. Savolainen, W. Hahn, S. Hoot, M. Fay, M. Axtell, S. Swensen, K. Nixon, and J. Farris. 2000. Angiosperm phylogeny inferred from a combined data set of 18S rDNA, *rbcL* and *atpB* sequences. *Botanical Journal of the Linnean Society* 133:381–461.

Soltis, D.E., P.S. Soltis, P.K. Endress, and M.W. Chase. 2005. Phylogeny and Evolution of Angiosperms. Sinauer Associates, Inc., Sunderland, MA.

Soltis, D.E., P.S. Soltis, and J.H. Leebens-Mack [eds.]. 2006. Developmental Genetics of the Flower. Advances in Botanical Research Series. Volume 44. Elsevier Limited, London, UK.

Soltis, D.E., P.S. Soltis, D.R. Morgan, S.M. Swensen, B.C. Mullin, J.M. Dowd, and P.G. Martin. 1995. Chloroplast gene sequence data suggest a single origin of the predisposition for symbiotic nitrogen fixation in angiosperms. *Proceedings of the National Academy of Sciences USA* 92:2647–2651.

Soltis, D.E., P.S. Soltis, M.E. Mort, M.W. Chase, V. Savolainen, S.B. Hoot, and C.M. Morton. 1998. Inferring complex phylogenies using parsimony: An empirical approach using three large DNA data sets for angiosperms. *Systematic Biology* 47:32–42.

Soltis, D.E., P.S. Soltis, D.L. Nickrent, L.A. Johnson, W.J. Hahn, S.B. Hoot, J. A. Sweere, R.K. Kuzoff, K.A. Kron, M. Chase, S.M. Swenson, E. Zimmer, S.M. Chaw, L.J. Gillespie, W.J. Kress, and K. Sytsma. 1997a. Angiosperm phylogeny inferred from 18S ribosomal DNA sequences. *Annals of the Missouri Botanical Garden* 84:1–49.

Soltis, D.E., P.S. Soltis, D.W. Schemske, J.F. Hancock, J.N. Thompson, B.C. Husband, and W.S. Judd. 2007. Autopolyploidy in angiosperms: Have we grossly underestimated the number of species? *Taxon* 56(1):13–30.

Soltis, D.E., P.S. Soltis, and M. Zanis. 2002a. Phylogeny of seed plants based on evidence from eight genes. *American Journal of Botany* 89:1670–1681.

Soltis, D.E., M. Tago-Nakazawa, Q.-X. Xiang, S. Kawano, J. Murata, M. Wakabayashi, and C. Hibsch-Jetter. 2001a. Phylogenetic relationships and evolution in *Chrysosplenium* (Saxifragaceae) based on *matK* sequence data. *American Journal of Botany* 88:883–893.

Soltis, D.E., C.J. Visger, and P.S. Soltis. 2014b. The polyploidy revolution then . . . and now: Stebbins revisited. *American Journal of Botany* 101(7):1057–1078.

Soltis, P.S., S.F. Brockington, M.-J. Yoo, A. Piedrahita, M. Latvis, M.J. Moore, A.S. Chanderbali, and D.E. Soltis. 2009. Floral variation and floral genetics in Basal Angiosperms. *American Journal of Botany* 96(1):110–128.

Soltis, P.S., and D.E. Soltis. 1998. Molecular evolution of 18S rDNA in angiosperms: Implications for character weighting in phylogenetic analysis. *In* D.E. Soltis, P.S. Soltis, and J.J. Doyle [eds.], Molecular Systematics of Plants II: DNA Sequencing, 188–210. Kluwer, Boston, MA.

———. 2000. The role of genetic and genomic attributes in the success of polyploids. *Proceedings of the National Academy of Sciences USA* 97:7051–7057.

———. 2001. Molecular systematics: Assembling and using the tree of life. *Taxon* 50:663–678.

———. 2009. The role of hybridization in plant speciation. *Annual Review of Plant Biology* 60:561–588.

———. 2012. [eds.] Polyploidy and genome evolution. Springer-Verlag, Berlin, Germany.

———. 2013. A conifer genome spruces up plant phylogenomics. *Genome Biology* 14(6):122. http://genomebiology.com/2013/14/6/122.

———. 2013. Angiosperm phylogeny: A framework for studies of genome evolution. *In* I. Leitch, J. Dolezel, and J. Greilhuber [eds.], Diversity of Plant Genomes. Springer, Vienna, Austria.

Soltis, P.S., D.E. Soltis, and M.W. Chase. 1999b. Angiosperm phylogeny inferred from multiple genes as a research tool for comparative biology. *Nature* 402: 402–404.

Soltis, P.S., D.E. Soltis, M.W. Chase, P. K. Endress, and P.R. Crane. 2004. The diversification of flowering plants. *In* J. Cracraft, and M.J. Donoghue [eds.] Assembling the Tree of life, 154–167. Oxford University Press, Oxford, UK.

Soltis, P.S., D.E. Soltis, S. Kim, A. Chanderbali, and M. Buzgo. 2006. Expression of floral regulators in basal angiosperms and the origin and evolution of ABC-function. *Advances in Botanical Research: Incorporating Advances in Plant Pathology* 44:483–506.

Soltis, P.S., D.E. Soltis, V. Savolainen, P.R. Crane, and T.G. Barraclough. 2002. Rate heterogeneity among lineages of tracheophytes: Integration of molecular and fossil data and evidence for molecular living fossils. *Proceedings of the National Academy of Sciences USA* 99:4430–4435.

Soltis, P.S., D.E. Soltis, P. Wolf, D. Nickrent, S.-M. Chaw, and R.L. Chapman. 1999a. The phylogeny of land plants inferred from 18S rDNA sequences: Pushing the limits of rDNA signal? *Molecular Biology and Evolution* 16:1774–1784.

Soltis, P.S., D.E. Soltis, M.J. Zanis, and S. Kim. 2000. Basal lineages of angiosperms: Relationships and implications for floral evolution. *International Journal of Plant Sciences* 161(Supplement):S97–S107.

Soreng, R.J., and J.I. Davis. 1998. Phylogenetics and character evolution in the grass family (Poaceae): Simultaneous analysis of morphological and chloroplast DNA restriction site character sets. *Botanical Review* 64:1–85.

Soros, C.L., and D.H. Les. 2002. Phylogenetic relationships in the Alismataceae. *In* Botany 2002: Botany in the curriculum, 152. Abstracts [Madison, Wisconsin].

Sosa, V., and M.W. Chase. 2003. Phylogenetics of Crossosomataceae based on *rbcL* sequence data. *Systematic Botany* 28:96–105.

Soza, V.L., J. Brunet, A. Liston, P.S. Smith, and V.S. Di Stilio. 2012. Phylogenetic insights into the correlates of dioecy in meadow-rues (*Thalictrum*, Ranunculaceae). *Molecular Phylogenetics and Evolution* 63(1):180–192.

Specht, C.D. 2006. Gondwanan vicariance or dispersal in the tropics? The biogeographic history of the tropical monocot family Costaceae (Zingiberales). *In* J.T. Columbus, E.A. Friar, J.M. Porter, L.M. Prince., and M.G. Simpson [eds.], Monocots: Comparative Biology and Evolution, Excluding Poales, 633–644. Rancho Santa Ana Botanic Garden, Claremont, CA.

Spencer, K.C., and D.S. Seigler. 1985a. Cyanogenic glycosides of *Malesherbia*. *Biochemical Systematics and Ecology* 13:421–431.

———. 1985b. Cynaogenic glycosides and systematics of the Flacourtiaceae. *Biochemical Systematics and Ecology* 13:433–435.

Sperling, C.R., and V. Bittrich. 1993. Basellaceae. *In* K. Kubitzki, J.G. Rohwer, and V. Bittrich [eds.], The Families and Genera of Vascular Plants, II, 143–146. Springer-Verlag, Berlin, Germnay.

Sperry, J.S. 1983. Observations on the structure and function of hydathodes in *Blechnum lehmanii*. *American Fern Journal* 73:65–72.

Sperry, J.S., K.L. Nicols, J.E.M. Sullivan, and S.E. Eastlack. 1994. Xylem embolism in ring-porous, diffuse-porous, and coniferous trees of northern Utah and interior Alaska. *Ecology* 75:1736–1752.

Spigler, P.B., and T.-L. Ashman. 2012. Gynodioecy to dioecy: Are we there yet? *Annals of Botany* 109:531–543.

Spongberg, S.A. 1971. The Staphyleaceae in the southeastern United States. *Journal of the Arnold Arboretum* 52:196–203.

———. 1972. The genera of Saxifragaceae in the southeastern United States. *Journal of the Arnold Arboretum* 53:409–498.

———. 1978. The genera of Crassulaceae in the southeastern United States. *Journal of the Arnold Arboretum* 59:197–248.

Spooner, D.M., G.J. Anderson, and R.K. Jansen. 1993. Chloroplast DNA evidence for the interrelationships of tomatoes, potatoes, and pepinos (Solanaceae). *American Journal of Botany* 80:676–688.

Spooner, D.M., K. McLean, G. Ramsay, R. Waugh, and G.J. Bryan. 2005. A single domestication for potato based on multilocus amplified fragment length polymorphism genotyping. *Proceedings of the National Academy of Sciences USA* 102:14694–14699.

Spooner, D.M., F. Rodriguez, Z. Polgar, H.E. Ballard, and S.H. Jansky. 2008. Genomic origins of potato polyploids: GBSSI gene sequencing data. *Crop Science* 48:S27–S36.

Spooner, D.M., R.G. van den Berg, A. Rodríguez, J. Bamberg, R.J. Hijmans, and S.I. Lara-Cabrera. 2004. Wild potatoes (*Solanum* section *Petota*) of North and Central America. *Systematic Botany Monographs* 68:1–209.

Sporne, K.R. 1954. A note on nuclear endosperm as a primitive character among dicotyledons. *Phytomorphology* 4:275–278.

Sprague, E.F. 1925. The classification of dicotyledons. II. Evolutionary progressions. *Journal of Botany* 63:105–113.

Sprent, J.I. 1994. Evolution and diversity in the legume-rhizobium symbiosis: Chaos theory? *Plant and Soil* 161:1–10.

———. 2005. Biological nitrogen fixation associated with angiosperms in terrestrial ecosystems. *In* H. Bassiri-Rad [ed.], Nutrient Acquisition by Plants, an Ecological Perspective, 89–116. Ecological Studies, Vol. 181. Springer-Verlag, Berlin, Germany.

Sprent, J.I., J.K. Ardley, and E.K. James. 2013. From North to South: A latitudinal look at legume nodulation processes. *South African Journal of Botany* 89:31–41.

Springer, M.S., E.C. Teeling, O. Madsen, M.J. Stanhope, and W.W. De Jong. 2001. Integrated fossil and molecular data reconstruct bat echolocation. *Proceedings of the National Academy of Sciences USA* 98:6241–6246.

Srivastava, S.K. 1972. Pollen genus *Erdtmanipollis* Krutzsch 1962. *Pollen et Spores* 14:309–322.

Stacey, D.S., and M.D. Rausher. 2011. Gene loss and parallel evolution contribute to species difference in flower color. *Molecular Biology and Evolution* 28(10): 2799–2810.

Staedler, Y.M., and P.K. Endress. 2009. Diversity and lability of floral phyllotaxis in the pluricarpellate families of core Laurales (Gomortegaceae, Atherospermataceae, Siparunaceae, Monimiaceae). *International Journal of Plant Sciences* 170:522–550.

Staedler, Y.M., D. Masson, and J. Schönenberger. 2013. Plant tissues in 3D via X-ray tomography: Simple contrasting methods allow high resolution imaging. *PLoS One* 8(9): e75295. doi:10.1371/journal.pone.0075295.

Staedler, Y.M., P.H. Weston, P.K. Endress. 2007. Floral phyllotaxis and floral architecture in Calycanthaceae (Laurales). *International Journal of Plant Sciences* 168:285–306.

Stafford, H.A. 1994. Anthocyanins and betalains: Evolution of the mutually exclusive pathways. *Plant Science* 101:91–98.

Stamatakis, A. 2006. RAxML-VI-HPC: Maximum likelihood-based phylogenetic analyses with thousands of taxa and mixed models. *Bioinformatics* 22:2688–2690.

———. 2014. RAxML Version 8: A tool for Phylogenetic Analysis and Post-Analysis of Large Phylogenies. Bioinformatics 2014. doi:10.1093/bioinformatics/btu033.

Stamatakis, A., P. Hoover, and J. Rougemont. 1979. Macroevolution: Pattern and Process, Freeman, San Francisco, CA.

———. 2008. A Rapid Bootstrap Algorithm for the RAxML Web Servers. *Systematic Biology* 57:758–771.

Stanley, S.M. 1975. A theory of evolution above the species level. *Proceedings of the National Academy of Sciences USA* 72:646–650.

Stebbins, G.L, Jr. 1950. Variation and Evolution in Plants. Columbia University Press, New York, NY.

———. 1967. Adaptive radiation and trends of evolution in higher plants. *Evolution Biology* 1:101–142.

———. 1970. Adaptive radiation in angiosperms. I. Pollination mechanisms. *Annual Review of Ecology and Systematics* 1:307–326.

———. 1971. Chromosomal Variation in Higher Plants. Arnold, London, UK.

———. 1974. Flowering Plants: Evolution Above the Species Level. Belknap Press, Cambridge, MA.

———. 1976a. Chromosome, DNA and plant evolution. *Evolutionary Biology* 9:1–34.

———. 1976b. Chromosomal Evolution in Higher Plants. Addison-Wesley, London, UK.

Stefanović, S., M. Jager, J. Deutsch, J. Broutin, and M. Masselot. 1998. Phylogenetic relationships of conifers inferred from partial 28S rRNA gene sequences. *American Journal of Botany* 85:688–697.

Stefanović, S., L. Krueger, and R.G. Olmstead. 2002. Monophyly of the Convolvulaceae and circumscription of their major lineages based on DNA sequences of multiple chloroplast loci. *American Journal of Botany* 89:1510–1522.

Stefanović, S., and R.G. Olmstead. 2004. Testing the phylogenetic position of a parasitic plant (*Cuscuta*, Convolvulaceae, Asteridae): Bayesian inference and the parametric bootstrap on data drawn from three genomes. *Systematic Biology* 53(3):384–399.

Stefanović, S., D.W. Rice, and J.D. Palmer. 2004. Long branch attraction, taxon sampling, and the earliest angiosperms: *Amborella* or monocots? *BMC Evolutionary Biology* 4(35). doi:10.1186/1471-2148-4-5.

Stegemann, S., M. Keuthe, S. Greiner, and R. Bock. 2012. Horizontal transfer of chloroplast genomes between plant species. *Proceedings of the National Academy of Sciences USA* 109(7):2434–2438.

Steiner, K.E. 1991. Oil flowers and oil bees: Further evidence for pollinator adaptation. *Evolution* 45: 1493–1501.

Stellari, G.M., A. Jaramillo, and E.M. Kramer. 2004. Evolution of the *APETALA3* and *PISTILLATA* lineages of MADS-box containing genes in the basal angiosperms. *Molecular Biology and Evolution* 21:506–519.

Stern, W.L. 1974. Comparative anatomy and systematics of woody Saxifragaceae: *Escallonia*. *Botanical Journal of the Linnaean Society* 68:1–20.

Stern, W.L., G.K. Brizicky, and R.H. Eyde. 1969. Comparative anatomy and relationships of Columelliaceae. *Journal of the Arnold Arboretum* 50:36–75.

Stevens, P.F. 1984. Metaphors and topology in the development of botanical systematics 1690–1960, or the art of putting new wine in old bottles. *Taxon* 33:169–211.

———. 2001. Angiosperm Phylogeny Website. Version 12, July 2012 [and more or less continuously updated since]. http://www.mobot.org/MOBOT/research/APweb/.

———. 2004. Angiosperm Phylogeny Website. Version 5, May 2004. http://www.mobot.org/MOBOT/research/APweb/.

———. 2007. Clusiaceae–Guttiferae. *In* K. Kubitzki [ed.], The Families and Genera of Vascular Plants. IX., 48–66. Springer-Verlag, Berlin, Germany.

———. 2012. Angiosperm Phylogeny Website. Version 12. Missouri Botancial Garden, St. Louis, MO.

Stevenson, D.W., and H. Loconte. 1995. Cladistic analysis of monocot families. *In* P.J. Rudall, P.J. Cribb, D.F. Cutler, and C.J. Humphries [eds.], Monocotyledons: Systematics and Evolution, 543–578. Royal Botanic Gardens, Kew, UK.

Stewart, W.N., and G.W. Rothwell. 1993. Paleobotany

and the Evolution of Plants, 2nd edition. Cambridge University Press, Cambridge, UK.

Stockey, R.A. 2006. The fossil record of basal monocots. *Aliso* 22:91–106.

Stockey, R.A., G.L. Hoffman, and G.W. Rothwell. 1997. The fossil monocot *Limnobiophyllum scutatum*: Resolving the phylogeny of Lemnaceae. *American Journal of Botany* 84:355–368.

Stockey, R.A., B.A. LePage, and K.B. Pigg. 1998. Permineralized fruits of *Diplopanax* (Cornaceae, Mastixioideae) from the middle Eocene Princeton chert of British Columbia. *Review of Palaeobotany and Palynology* 103:223–234.

Stommel, J.R., G.J. Lightbourn, B.S. Winkel, and R J. Griesbach. 2009. Transcription factor families regulate the anthocyanin biosynthetic pathway in *Capsicum annuum*. *Journal of the American Society of Horticultural Science* 134:244–251.

Straub, S.C.K., R.C. Cronn, C. Edwards, M. Fishbein, and A. Liston. 2013. Horizontal transfer of DNA from the mitochondrial to the plastid genome and its subsequent evolution in Milkweeds (Apocynaceae). *Genome Biology and Evolution* 5(10):1872–1885.

Straub, S.C.K., M.J. Moore, P.S. Soltis, D.E. Soltis, A. Liston, and T. Livshultz. 2014. Phylogenetic signal detection from an ancient rapid radiation: Effects of noise reduction, long branch attraction, and model selection in crown clade Apocynaceae. *Molecular Phylogenetics and Evolution* 80:169–185.

Struck, M. 1997. Floral divergence and convergence in the genus *Pelargonium* (Geraniaceae) in southern Africa: Ecological and evolutionary considerations. *Plant Systematics and Evolution* 208:71–97.

Struwe, L., V.A. Albert, and B. Bremer. 1994. Cladistics and family level classification of the Gentianales. *Cladistics* 10:175–206.

Stuessy, T.F. 2000. Taxon names are *not* defined. *Taxon* 49:231–233.

Stuhlfauth, T., H. Fock, H. Huber, and K. Klug. 1985. The distribution of fatty acids including petroselinic and tariric acids in the fruit and seed oils of the Pittosporaceae, Araliaceae, Umbelliferae, Simarubaceae and Rutaceae. *Phytochemistry* 13:447–453.

Stull, G.W., F. Herrera, S.R. Manchester, C. Jaramillo, and B.H. Tiffney. 2012. Fruits of an "Old World" tribe (Phytocreneae; Icacinaceae) from the Paleogene of North and South America. *Systematic Botany* 37(3):784–794.

Stull, G.W., M.J. Moore, V.S. Mandala, N.A. Douglas, H.-R. Kates, X. Qi, S.F. Brockington, P.S. Soltis, D.E. Soltis, and M.A. Gitzendanner. 2013. A targeted enrichment strategy for massively parallel sequencing of angiosperm plastid genomes. *Applications in Plant Sciences* 1(2). doi:http://dx.doi.org/10.3732/apps.1200497.

Stull, G.W., B. Roger Moore, and S.R. Manchester. 2011. Fruits of Icacinaceae from the Eocene of southeastern North America and their biogeographic implications. *International Journal of Plant Sciences* 172:935–947.

Su, H.-J., and J.-M. Hu. 2012. Rate heterogeneity in six protein-coding genes from the holoparasite *Balanophora* (Balanophoraceae) and other taxa of Santalales. *Annals of Botany* 110:1137–1147.

Su, H.-J., J. Murata, and J.-M. Hu. 2012. Morphology and phylogenetics of two holoparasitic plants, *Balanophora japonica* and *Balanophora yakushimensis* (Balanophoraceae), and their hosts in Taiwan and Japan. *Journal of Plant Research* 125(3):317–326.

Su, J.-X., W. Wang, L.-B. Zhang, and Z.-D. Chen. 2012. Phylogenetic placement of two enigmatic genera, *Borthwickia* and *Stixis*, based on molecular and pollen data, and the description of a new family of Brassicales, Borthwickiaceae. *Taxon* 61(3):601–611.

Suh, Y., L.B. Thien, H.E. Reeve, E.A. Zimmer. 1993. Molecular evolution and phylogenetic implications of internal transcribed spacer sequences of ribosomal DNA in Winteraceae. *American Journal of Botany* 80:1042–1055.

Sun, G., D.L. Dilcher, H. Wang, Z. Chen. 2011. A eudicot from the Early Cretaceous of China. *Nature* 471(7340):625–628.

Sun, G., D.L. Dilcher, S.L. Zheng, and Z.K. Zhou. 1998. In search of the first flower: A Jurassic angiosperm, *Archaefructus*, from northeast China. I. *Science* 282:1692–1695.

Sun, G., Q. Li, D.L. Dilcher, S. Zheng, K.C. Nixon, and X. Wang. 2002. Archaefructaceae, a new basal angiosperm family. *Science* 296:899–904.

Sun, Y.-X., M.J. Moore, A.-P. Meng, P.S. Soltis, D.E. Soltis, J.-Q. Li, and H.-C. Wang. 2013. Complete plastid genome sequencing of Trochodendraceae reveals a significant expansion of the inverted repeat and suggests a paleogene divergence between the two extant species. *PLoS One* 8(4):e60429. Epub 2013 Apr 5.

Sunberg, P., and G. Pleijel. 1994. Phylogenetic classification and the definition of taxon names. *Zoologica Scripta* 23:19–25.

Sundstrom, J., and P. Engstrom. 2002. Conifer reproductive development involves B-type MADS-box genes with distinct and different activities in male organ primorida. *Plant Journal* 31:161–169.

Surange, K.R., and S. Chandra. 1975. Morphology of the gymnospermous fructifications of the *Glossopteris* flora and their relationships. *Palaeontographica Abteilung B* 149:153–180.

Sutton, D.A. 1988. A Revision of the Tribe Antirrhineae. British Museum (Natural History), London/Oxford University Press, Oxford, UK.

———. 1989. The Daphniphyllales: A systematic review. *In* P.R. Crane, and S. Blackmore [eds.], Evolution, Systematics and Fossil History of the Hamamelidae, 1, 285–291. Clarendon Press, Oxford, UK.

Suzuki, Y., G.V. Glazko, and M. Nei. 2002. Overcredibility of molecular phylogenies obtained by Bayesian phylogenetics. *Proceedings of the National Academy of Sciences USA* 99:16138–16143.

Swamy, B.G.L. 1964. Macrogametic ontogeny in *Schisandra chinensis. Journal of the Indian Botanical Society* 43:391–396.

Swamy, B.G.L., and I.W. Bailey. 1949. The morphology and relationships of *Cercidiphyllum. Journal of the Arnold Arboretum* 30:187–210.

Swamy, B.G.L., and P.M. Ganapathy. 1957. On endosperm in dicotyledons. *Botanical Gazette* 119:47–50.

Swensen, S.M. 1996. The evolution of actinorhizal symbioses: Evidence for multiple origins of the symbiotic association. *American Journal of Botany* 83:1503–1512.

Swensen, S.M., and D.R. Benson. 2008. Evolution of actinorhizal host plants and *Frankia* endosymbionts. *In* K. Pawlowski, and W.E. Newton [eds.], Nitrogen-fixing Actinorhizal Symbioses, Nitrogen Fixation: Origins, Applications, and Research Progress, Volume 6, 73–104. Springer, Netherlands.

Swensen, S.M., J.N. Luthi, and L.H. Rieseberg. 1998. Datiscaceae revisited: Monophyly and the sequence of breeding system evolution. *Systematic Botany* 23:157–169.

Swensen, S.M., and B.C. Mullin. 1997. The impact of molecular systematics on hypotheses for the evolution of root nodule symbioses and implications for expanding symbioses to new host plant genera. *Plant and Soil* 194:185–192.

Swensen, S.M., B.C. Mullin, and M.W. Chase. 1994. Phylogenetic affinities of the Datiscaceae based on an analysis of nucleotide sequences from a plastid *rbcL* gene. *Systematic Botany* 19:157–168.

Swisher, C.C., Y.-Q. Wang, X.-L. Wang, X.X. Wang, and Y. Wang. 1999. Cretaceous age for the feathered dinosaurs of Liaoning, China. *Nature* 400:58–61.

Swofford, D.L. 1991. PAUP: Phylogenetic Analysis Using Parsimony. Version 3.1. Illinois Natural History Survey, Champaign, IL.

———. 1993. Phylogenetic Analysis Using Parsimony. Version 3.1.1. Illinois Natural History Survey, Champaign, IL.

———. 1999. PAUP* Phylogenetic Analysis Using Parsimony, version 4.0. Sinauer, Sunderland, MA.

Swofford, D.L., and W.P. Maddison. 1987. Reconstructing ancestral character states under Wagner parsimony. *Mathematical Biosciences* 87:199–229.

Swofford, D.L., G.J. Olsen, P.J. Waddell, and D.M. Hillis. 1996. Phylogenetic inference. *In* D.M. Hillis, C. Moritz, and B.K. Mable [eds.], Molecular Systematics, 407–514. Sinauer, Sunderland, MA.

Sykorova, E., K.Y. Lim, Z. Kunicka, M.W. Chase, M.D. Bennett, J. Fajkus, and A.R. Leitch. 2003. Telomere variability in the monocotyledonous plant order Asparagales. *Proceedings of the Royal Society of London B* 270:1893–1904.

Sytsma, K.J., A. Litt, M.L. Zjhra, J.C. Pires, M. Nepokroeff, E. Conti, J. Walker, and P.G. Wilson. 2004. Clades, clocks, and continents: Historical and biogeographical analysis of Myrtaceae, Vochysiaceae, and relatives in the Southern Hemisphere. *International Journal of Plant Sciences* 165(4 Supplement):S85–S105.

Sytsma, K.J., J. Morawetz, J.C. Pires, M. Nepokroeff, E. Conti, M. Zjhra, J.C. Hall, and M.W. Chase. 2002. Urticalean rosids: Circumscription, rosid ancestry, and phylogenetics based on *rbcL*, *trnL–F*, and *ndhF* sequences. *American Journal of Botany* 89:1531–1546.

Sytsma, K.J., and J.C. Pires. 2001. Plant systematics in the next 50 years—re-mapping the new frontier. *Taxon* 50:713–732.

Sytsma, K.J., and J.F. Smith. 1988. Molecular systematics of Onagraceae: Examples from *Clarkia* and *Fuchsia*. *In* P.S. Soltis, D.E. Soltis, and J.J. Doyle [eds.], Molecular Systematics of Plants. Chapman and Hall, New York, NY.

Sytsma, K.J., J. Smith, and L. Gottlieb. 1990. Phylogenetics in *Clarkia* (Onagraceae): Restriction site mapping of chloroplast DNA. *Systematic Botany* 15:280–295.

Sytsma, K.J., J.B. Walker, J. Schönenberger, and A.A. Anderberg. 2006. Phylogenetics, biogeography, and radiation of Ericales. *In* Botany 2006—Looking to the Future—Conserving the Past, 71. Abstract.

Takahashi, K., and M. Nei. 2000. Efficiencies of fast algorithms of phylogenetic inference under the criteria of maximum parsimony, minimum evolution and maximum likelihood when a large number of sequences are used. *Molecular Biology and Evolution* 17:1251–1258.

Takahashi, M., P.R. Crane, and S.R. Manchester. 2002. *Hironoia fusiformis* gen. et sp. nov.; A cornalean fruit from Kamikitaba locality (Upper Cretaceous, Lower Coniacian) in northeastern Japan. *Journal of Plant Research* 115(6):463–473.

Takahashi, M., E.M. Friis, K. Uesugi, Y. Suzuki, and P.R. Crane. 2008. Floral evidence of Annonaceae from the Late Cretaceous of Japan. *International Journal of Plant Sciences* 169:890–898.

Takahashi, M., P.S. Herendeen, and P.R. Crane. 2001. Lauraceous flowers from the Kamikitaba locality (Lower Coniacian; Upper Cretaceous) of Northeast Japan. *Journal of Plant Research* 114:429–34.

Takhtajan, A. 1969. Flowering Plants: Origin and Dispersal. Smithsonian Institution Press, Washington, D.C.

———. 1980. Outline of the classification of flowering plants (Magnoliophyta). *Botanical Review* 46:225–359.

———. [ed.]. 1985. Anatomia seminum comparativa. Tomus 1. Liliopsida seu Monocotyledones. NAUKA, Leningrad, USSR. [In Russian.]

———. 1987. System of Magnoliophyta. Academy of Sciences USSR, Leningrad, USSR.

———. 1991. Evolutionary Trends in Flowering Plants. Columbia University Press, New York, NY.

———. 1997. Diversity and Classification of Flowering Plants. Columbia University Press, New York, NY.

———. 2009. Flowering Plants, 2nd ed. Springer, New York, NY.

Tamura, M.N. 1965. Morphology, ecology, and phylogeny of the Ranunculaceae. IV. Reproductive organs. *Science Reporter Osaka University* 14:53–71.

———. 1993. Ranunculaceae. *In* K. Kubitzki, J. Rohwer, and V. Bittrich [eds.], Families and Genera of Vascular Plants, II, 563–583. Springer-Verlag, Berlin, Germany.

———. 1998. Calochortaceae, Liliaceae, Melanthiaceae, Nartheciaceae, and Trilliaceae. *In* K. Kubitzki [ed.], Families and Genera of Vascular Plants III, 164–172, 343–353, 369–391, 444–451. Springer-Verlag, Berlin, Germany.

Tamura, M.N., J. Yamashita, S. Fuse, and M. Haraguchi. 2004. Molecular phylogeny of monocotyledons inferred from combined analysis of *matK* and *rbcL* gene sequences. *Journal of Plant Research* 117:109–120.

Tanaka, N., K. Uehara, and J. Murata. 2013. Evolution of floral traits in relation to pollination mechanisms in Hydrocharitaceae. *In* O. Seberg, G. Petersen, A.S. Barfod, and J.I. Davis [eds.], Diversity, Phylogeny, and Evolution in the Monocotyledons, 165–184. Aarhus University Press, Århus, Denmark.

Tang, H., J.E. Bowers, X. Wang, R. Ming, M. Alam, and A.H. Paterson. 2008. Perspective—Synteny and collinearity in plant genomes. *Science* 320(5875):486–488.

Tang, Y. 1994. Embryology of *Plagiopteron suaveolens* Griffith (Plagiopteraceae) and its systematic implications. *Botanical Journal Linnaean Society* 116:145–157.

Tank, D.C., P.M. Beardsley, S.A. Kelchner, and R.G. Olmstead. 2006. Review of the systematics of Scrophulariaceae s.l. and their current disposition. *Australian Systematic Botany* 19:289–307.

Tank, D.C., and M.J. Donoghue. 2010. Phylogeny and phylogenetic nomenclature of the Campanulidae based on expanded sample of genes and taxa. *Systematic Botany* 35:425–441.

Tank, D.C., J.M. Egger, and R.G. Olmstead. 2009. Phylogenetic classification of subtribe Castillejinae (Orobanchaceae). *Systematic Botany* 34:182–197.

Tank, D.C., and R.G. Olmstead. 2008. From annuals to perennials: Phylogeny of subtribe Castillejinae (Orobanchaceae). *American Journal of Botany* 95:608–625.

———. 2009. The evolutionary origin of a second radiation of annual *Castilleja* species in South America: The role of long-distance dispersal and allopolyploidy. *American Journal of Botany* 96:1907–1921.

Taylor, D.W. 1990. Paleobiogeographic relationships of angiosperms from the Cretaceous and early Tertiary of the North American area. *Botanical Review* 56:249–417.

Taylor, D.W., and W.L. Crepet. 1987. Fossil floral evidence of Malpighiaceae and an early plant-pollinator relationship. *American Journal of Botany* 74(2):274–286.

Taylor, D.W., and L.J. Hickey. 1992. Phylogenetic evidence for the herbaceous origin of angiosperms. *Plant Systematics and Evolution* 180:137–156.

———. 1996a. Evidence for and implications of an herbaceous origin for angiosperms. *In* D.W. Taylor, and L.J. Hickey [eds.], Flowering Plant Origin, Evolution and Phylogeny, 232–266. Chapman and Hall, New York, NY.

———. [eds.]. 1996b. Flowering Plant Origin, Evolution and Phylogeny. Chapman and Hall, New York, NY.

Taylor, E.L., and T.N. Taylor. 1992. Reproductive biology of the Permian Glossopteridales and their suggested relationship to flowering plants. *Proceedings of the National Academy of Sciences USA* 89:11495–11497.

———. 2009. Seed ferns from the late Paleozoic and Mesozoic: Any angiosperm ancestors lurking there? *American Journal of Botany* 96:237–251.

Taylor, M.L., T.D. Macfarlane, and J.H. Williams. 2010. Reproductive ecology of the basal angiosperm *Trithuria submersa* (Hydatellaceae). *Annals of Botany* 106:909–920.

Taylor, R.L. 1965. The genus *Lithophragma* (Saxifragaceae). *University of California Publications in Botany* 37:1–122.

Taylor, T.N., and S. Archangelsky. 1985. The Cretaceous pteridosperms *Ruflorinia* and *Ktalenia* and implications of cupule and carpel evolution. *American Journal of Botany* 72:1842–1853.

Taylor, T.N., G.M. Del Fueyo, and E.L. Taylor. 1994. Permineralized seed fern cupules from the Triassic of

Antarctica: Implications for cupule and carpel evolution. *American Journal of Botany* 81:666–677.
Taylor, T.N., E.L. Taylor, M. Krings. 2009. Paleobotany: the Biology and Evolution of Fossil Plants. Academic Press, San Diego, CA.
Tedersoo, L., P. Pellet, U. Koljalg, and M.-A. Selosse. 2007. Parallel evolutionary paths to mycoheterotrophy in understorey Ericaceae and Orchidaceae: Ecological evidence for mixotrophy in Pyroleae. *Oecologia* 151(2):206–217.
Teeri, T.H., M. Kotilainen, A. Uimari, S. Ruokolainen, Y.P. Ng, U. Malm, E. Pollanen, S. Broholm, R. Laitinen, P. Elomaa, and V.A. Albert. 2006. Floral developmental genetics of *Gerbera* (Asteraceae). *Advances in Botanical Research: Developmental Genetics of the Flower* 44:323–351.
Templeton, A.R. 1983. Phylogenetic inference from restriction endonuclease cleavage site maps with particular reference to the evolution of human and apes. *Evolution* 37:221–244.
Temsch, E.M., J. Greilhuber, and R. Krisai. 2010. Genome size in liverworts. *Preslia* 82(1):63–80.
Ter Welle, B.J.H. 1976. Silica grains in woody plants of the Neotropics, Especially Surinam. *In* P. Baas, A.J. Bolton, and. M. Catling [eds.], Wood Structure in Biological and Technological Research, 107–142. Leiden Botanical Series Number 3. Leiden University Press, Leiden, Netherlands.
Thanikaimoni, G. 1984. Ménispermacées: palynologie et systématique. *Travaux de l'Institut Français de Pondichéry, Section Scientifique et Technique* 18:1–135.
The International Wheat Genome Sequencing Consortium (IWGSC). 2014. A chromosome-based draft sequence of the hexaploid bread wheat (*Triticum aestivum*) genome. *Science* 345(6194). doi:10.1126/science.1251788.
The Plant List. 2010. Version 1. http://www.theplantlist.org/. Accessed 1 January 2016.
Theis, N., M.J. Donoghue, and J. Li. 2008. Phylogenetics of the Capriolieae and *Lonicera* (Dipsacales) based on chlorplast DNA sequences. *Systematic Botany* 33:776–783.
Theissen, G., A. Becker, A. Di Rosa, A. Kanno, J.T. Kim, T. Munster, K.U. Winter, and H. Saedler. 2000. A short history of MADS-box genes in plants. *Plant Molecular Biology* 42(1):115–149.
Theissen, G., A. Becker, K.U. Winter, T. Muenster, C. Kirchner, and H. Saedler. 2002. How the land plants learned their floral ABCs: The role of MADS-box genes in the evolutionary origin of flowers. *In* Q.C. Cronk, R.M. Bateman, and J.M. Hawkins [eds.], Developmental Genetics and Plant Evolution, 173–206. Taylor and Francis, London, UK.
Theissen, G., and R. Melzer. 2007. Molecular mechanisms underlying origin and diversification of the angiosperm flower. *Annals of Botany* 100:603–619.
Theissen, G., and H. Saedler. 1999. The golden decade of molecular floral development (1990–1999): A cheerful obituary. *Developmental Genetics* 25:181–193.
———. 2001. Floral quartets. *Science* 409:469–471.
Thiede, J., S.A. Schmidt, and B. Rudolph. 2007. Phylogenetic implication of the chloroplast *rpoC1* intron loss in the Aizoaceae (Caryophyllales). *Biochemical Systematics and Ecology* 35:372–380.
Thien, L.B., H. Azuma, and S. Kawano. 2000. New perspectives on the pollination biology of Basal Angiosperms. *International Journal of Plant Sciences* 161:S225–S235.
Thien, L.B., P. Bernhardt, M.S. Devall, Z.-D. Chen, Y.-B. Luo, J.-H. Fan, L.-C. Yuan, and J.H. Williams. 2009. Pollination biology of basal angiosperms (ANITA grade). *American Journal of Botany* 96:166–182.
Thien, L.B., T.L. Sage, T. Jaffre, P. Bernhardt, V. Pontieri, P.H. Weston, D. Malloch, H. Azuma, S.W. Graham, M.A. McPherson, H.S. Rai, R.F. Sage, and J.L. Dupre. 2003. The population structure and floral biology of *Amborella trichopoda* (Amborellaceae). *Annals of the Missouri Botanical Garden* 90:466–490.
Thieret, J.W., and J.O. Luken. 1996. The Typhaceae in the southeastern United States. *Harvard Papers in Botany* 8:27–56.
Thomas, C.A. 1971. The genetic organization of chromosomes. *Annual Review of Genetics* 5:237–256.
Thomas, H.H. 1925. The Caytoniales, a new group of Angiospermous plants from the Jurassic rocks of Yorkshire. *Philosophical Transactions of the Royal Society of London B–Containing Papers of a Biological Character* 213:299–363.
———. 1934. The nature and origin of the stigma. *New Phytologist* 33:173–198.
———. 1936. Paleobotany and the origin of the angiosperms. *Botanical Review* 2:397–418.
———. 1955. Mesozoic pteridosperms. *Phytomorphology* 5:177–185.
———. 1958. *Lidgettonia*, a new type of fertile *Glossopteris*. *Bulletin of the British Museum (Natural History), Geology* 3:179–189.
Thomas, N., S. Archangelsky, and T. Archangelsky. 1985. The Cretaceous pteridosperms *Ruflorinia* and *Ktalenia* and implications on cupule and carpel evolution. *American Journal of Botany* 72:1842–1853.
Thomasson, J.R. 1987. Fossil Grasses: 1820–1986 and

Beyond. *In* T. R. Soderstrom, K. W. Hilu, C. S. Campbell, and M. E. Barkworth [eds.], Grass Systematics and Evolution, 159–171. Smithsonian Press, Washington, D.C.

Thompson, J.D., T.J. Gibson, F. Plewniak, F. Jeanmougin, and D.G. Higgins. 1997. The ClustalX windows interface: Flexible strategies for multiple sequence alignment aided by quality analysis tools. *Nucleic Acids Research* 24:4876–4882.

Thompson, J.N. 1994. The Co-evolutionary Process. University of Chicago Press, Chicago, IL.

———. 2013. Relentless Evolution. University of Chicago Press, Chicago, IL.

Thompson, J.N., and O. Pellmyr. 1992. Mutualism with pollinating seed parasites amid co-pollinators: Constraints on specialization. *Ecology* 73:1780–1791.

Thompson, J.N., C. Schwind, P.R. Guimaraes, Jr., and M. Friberg. 2013. Diversification through multitrait evolution in a coevolving interaction. *Proceedings of the National Academy of Sciences USA* 110(28): 11487–11492.

Thompson, M.M. 1979. Growth and development of the pistillate flower and nut in "Barcelona" filbert. *Journal of the American Horticultural Society* 104:427–432.

Thorne, J.L., and H. Kishino. 2002. Divergence time estimation and rate evolution with multilocus data sets. *Systematic Biology* 51:689–702.

Thorne, R.F. 1968. Synopsis of a putatively phylogenetic classification of the flowering plants. *Aliso* 6:57–56.

———. 1974. A phylogenetic classification of the Annoniflorae. *Aliso* 8:147–209.

———. 1976. A phylogenetic classification of the Angiospermae. *Evolutionary Biology* 9:35–106.

———. 1983. Proposed new realignments in the angiosperms. *Nordic Journal of Botany* 3:85–117.

———. 1992a. Classification and geography of the flowering plants. *Botanical Review* 58:225–348.

———. 1992b. An updated phylogenetic classification of the flowering plants. *Aliso* 13:365–389.

———. 2000. The classification and geography of the monocotyledon subclasses Alismatidae, Liliidae, and Commelinidae. *In* B. Nordenstam, G. El Ghazaly, and M. Kassas [eds.], Plant Systematics for the 21st Century, 75–124. Portland Press, Portland, OR.

———. 2001. The classification and geography of the flowering plants: Dicotyledons of the class Angiospermae (subclasses Magnoliidae, Ranunculidae, Caryophyllidae, Dilleniidae, Rosidae, Asteridae, and Lamiidae). *Botanical Review* 66:441–647.

Thornhill, A.H., L.W. Popple, R.J. Carter, S.Y.W. Ho, and M.D. Crisp. 2012. Are pollen fossils useful for calibrating relaxed molecular clock dating of phylogenies? A comparative study using Myrtaceae. *Molecular Phylogenetics and Evolution* 63(1):15–27.

Thorogood, C.J., and S.J. Hiscock. 2007. Host specificity in the parasitic plant *Cytinus hypocistis*. *Research Letters in Ecology* 84234. doi:10.1155/2007/84234.

———. 2010. Compatibility interactions at the cellular level provide the basis for host specificity in the parasitic plant *Orobanche*. *The New Phytologist* 186(3):571–575.

Thuiller, W., S. Lavergne, C. Roquet, I. Boulangeat, B. Lafourcade, and M.B. Araujo. 2011. Consequences of climate change on the tree of life in Europe. *Nature* 470:531–534.

Thulin, M., B. Bremer, J. Richardson, J. Niklasson, M.F. Fay, and M.W. Chase. 1998. Family relationships of the enigmatic rosid genera *Barbeya* and *Dirachma* from the Horn of Africa Region. *Plant Systematics and Evolution* 213:103–119.

Tian, X., L. Zhang, and Y. Ren. 2005. Development of flowers and inflorescences of *Circaeaster* (Circaeasteraceae, Ranunculales). *Plant Systematics and Evolution* 256(1–4):89–96.

Tiffney, B.H. 1985. The Eocene North-Atlantic Land-Bridge—its importance in Tertiary and modern phytogeography of The Northern Hemisphere. *Journal of the Arnold Arboretum* 66:243–273.

Tiffney, B.H., and S.R. Manchester. 2001. The use of geological and paleontological evidence in evaluating plant phylogeographic hypotheses in the Northern Hemisphere Tertiary. *International Journal of Plant Sciences* 162(6 Supplement):S3–S17.

Tillich, H.-J. 1985. Keimlingsbau und verwandtschaftliche Beziehungen der Araceae. *Gleditschia* 13:63–73.

———. 1995. Seedling and systematics in monocotyledons. *In* P.J. Rudall, P.J. Cribb, D.F. Cutler, and C.J. Humphries [eds.], Monocotyledons: Systematics and Evolution, 303–352. Royal Botanic Gardens, Kew, UK.

Tillich, H.-J., R. Tuckett, and E. Facher. 2007. Do Hydatellaceae belong to the monocotyledons or basal angiosperms? Evidence from seedling morphology. *Willdenowia* 37:399–406.

Tobe, H. 1991. Reproduction morphology, anatomy, and relationship of *Ticodendron*. *Annals of the Missouri Botanical Garden* 78:135–142.

———. 2013. Floral morphology and structure of *Phyllonoma* (Phyllonomaceae): Systematic and evolutionary implications. *Journal of Plant Resources* 126:709–718.

Tobe, H., T. Jaffré, and P.H. Raven. 2000. Embryology of *Amborella* (Amborellaceae): Descriptions and polarity of character states. *Journal of Plant Research* 113:271–280.

Tobe, H., Y. Kimoto, and N. Prakash. 2007. Development and structure of the female gametophyte in *Austrobaileya scandens* (Austrobaileyaceae). *Journal of Plant Research* 120:431–436.

Tobe, H., and N.R. Morin. 1996. Embryology and circumscription of Campanulaceae and Campanulales: A review of literature. *Journal of Plant Research* 109: 425–435.

Tobe, H., and P.H. Raven. 1983. An embryological analysis of the Myrtales: Its definition and characteristics. *Annals of the Missouri Botanical Garden* 70:71–94.

———. 1984. The embryology and relationships of *Rhynchocalyx* Olv. (Rhynchocalycaceae). *Annals of the Missouri Botanical Garden* 71:836–843.

———. 1988. Seed morphology and anatomy of Rhizophoraceae, inter- and infrafamilial relationships. *Annals of the Missouri Botanical Garden* 75:1319–1342.

———. 1989. The embryology and systematic position of *Rhabdodendron* (Rhabdodendraceae). *In* K. Tan, R.R. Mill, and T.S. Elias [eds.], Plant Taxonomy, Phytogeography, and Related Subjects, 233–248. Edinburgh University Press, Edinburgh, UK.

Tobe, H., and B. Sampson. 2000. Embryology of *Takhtajania* (Winteraceae) and a summary statement of embryological features for the family. *Annals of the Missouri Botanical Garden* 87:389–397.

Tokuoka, T. 2012. Molecular phylogenetic analysis of Passifloraceae sensu lato (Malpighiales) based on plastid and nuclear DNA sequences. *Journal of Plant Research* 125(4):489–497.

Tokuoka, T., and H. Tobe. 2001. Ovules and seeds in subfamily Phyllanthoideae (Euphorbiaceae): Structure and systematic implications. *Journal of Plant Research* 114:75–92.

Tomlinson, P.B. 1962. Phylogeny of the Scitamineae–morphological and anatomical considerations. *Evolution* 10:192–213.

———. 1969. Commelinales-Zingiberales. *In* C.R. Metcalfe [ed.], Anatomy of Monocotyledons, 295–421. Clarendon Press, Oxford, UK.

———. 1974. Development of the stomatal complex as a taxonomic character in the monocotyledons. *Taxon* 23:109–128.

———. 1995. Non-homology of vascular organisation in monocotyledons and dicotyledons. *In* P.J. Rudall, P.J. Cribb, D.F. Cutler, and C.J. Humphries [eds.], Monocotyledons: Systematics and Evolution, 589–622. Royal Botanic Gardens, Kew, UK.

Torices, R. 2010. Adding time-calibrated branch lengths to the Asteraceae supertree. *Journal of Systematics and Evolution* 48:271–278.

Torrey, J.G., and R.H. Berg. 1988. Some morphological features for generic characterization among the Casuarinaceae. *American Journal of Botany* 75:864–874.

Townrow, J.A. 1962. On *Pteruchus* a microsporophyll of the Corystospermaceae. *Bulletin of the British Museum (Natural History), Geology* 6:287–320.

Traas, J. 2013. Phyllotaxis. *Development* 140:249–253.

Traverse, A. 1988. Paleopalynology. Unwin Hyman, Boston, MA.

Trevisan, L. 1988. Angiosperm pollen (monosulcate-trichotomosulcate phase) from the very early Lower Cretaceous of Southern Tuscany (Italy): Some aspects. *In* Seventh International Palynological Congress (Brisbane), 165. Abstracts.

Trias-Blasi, A., J.A.N. Parnell, and T.R. Hodkinson. 2012. Multi-gene region phylogenetic analysis of the Grape family (Vitaceae). *Systematic Botany* 37(4):941–950.

Trift, I., M. Källersjö, and A.A. Anderberg. 2002. The monophyly of *Primula* (Primulaceae) evaluated by analysis of sequences from the chloroplast gene *rbcL*. *Systematic Botany* 27:396–407.

Tripp, E.A., and P.S. Manos. 2008. Is floral specialization an evolutionary dead end? Pollination system transitions in *Ruellia* (Acanthaceae). *Evolution* 62: 1712–1737.

Tschan, G.F., T. Denk, and M. von Balthazar. 2008. *Credneria* and *Platanus* (Platanaceae) from the Late Cretaceous (Santonian) of Quedlinburg, Germany. *Review of Palaeobotany and Palynology* 152:211–236.

Tsitrone, A., M. Kirkpatrick, and D.A. Levin. 2003. A model for chloroplast capture. *Evolution* 57:1776–1782.

Tsou, C.-H. 1995. Embryology of Theaceae—Anther and ovule development of *Adinandra*, *Cleyera*, and *Eurya*. *Journal of Plant Research* 108:77–86.

Tu, T., M.O. Dillon, H. Sun, and J. Wen. 2008. Phylogeny of *Nolana* (Solanaceae) of the Atacama and Peruvian deserts inferred from plastid markers and the nuclear *LEAFY* second intron. *Molecular Phylogenetics and Evolution* 49:561–573.

Tucker, G.C. 1987. The genera of Cyperaceae in the southeastern United States. *Journal of the Arnold Arboretum* 68:361–445.

———. 1989. The genera of Commelinaceae in the southeastern United States. *Journal of the Arnold Arboretum* 70:97–130.

Tucker, S.C. 1960. Ontogeny of the floral apex of *Michelia fuscata*. *American Journal of Botany* 47:266–277.

———. 1988. Loss versus suppression of floral organs. *In* P. Leins, S.C. Tucker, and P.K. Endress [eds.], Aspects of Floral Development, 69–82. Cramer, Berlin, Germany.

———. 1989. Overlapping organ initiation and common primordia in flowers of *Pisum sativum* (Leguminosae: Papilionoideae). *American Journal of Botany* 76: 714–729.

———. 1991. Helical floral organogenesis in *Gleditsia*, a primitive caesalpinioid legume. *American Journal of Botany* 78:1130–1149.

———. 1997. Floral evolution, development, and convergence: The hierarchical-significance hypothesis. *International Journal of Plant Sciences* 158(6 Supplement): S143–S161.

———. 1999. Evolutionary lability of symmetry in early floral development. *International Journal of Plant Sciences* 160(6 Supplement):S25–S39.

———. 2000. Organ loss in detarioid and other leguminous flowers, and the possibility of saltatory evolution. *In* P.S. Herendeen, and A. Bruneau [eds.], Advances in Legume Systematics, 9, 107–120. Royal Botanic Gardens, Kew, UK.

Tucker, S.C., and A.W. Douglas. 1996. Floral structure, development, and relationships of paleoherbs: *Saruma, Cabomba, Lactoris,* and selected Piperales. *In* D.W. Taylor, and L.J. Hickey [eds.], Flowering Plant Origin, Evolution, and Phylogeny, 141–175. Chapman and Hall, New York, NY.

Tucker, S.C., O.L. Stein, and K.S. Derstine. 1985. Floral development in *Caesalpinia* (Leguminosae). *American Journal of Botany* 72:1424–1434.

Turner, B., J. Munzinger, S. Duangjai, E.M. Temsch, R. Stockenhuber, M.H.J. Barfuss, M.W. Chase, and R. Samuel. 2013. Molecular phylogenetics of New Caledonian *Diospyros* (Ebenaceae) using plastid and nuclear markers. *Molecular Phylogenetics and Evolution* 69:740–763.

Tuteja, R., R.K. Saxena, J. Davila, T. Shah, W. Chen, Y.-L. Xiao, G. Fan, K.B. Saxena, A.J. Alverson, C. Spillane, C. Town, and R.K. Varshney. 2013. Cytoplasmic male sterility–associated chimeric open reading frames identified by mitochondrial genome sequencing of four *Cajanus* genotypes. *DNA Research* 20(5):485–495.

Tyree, M.T., and F. Ewers. 1991. The hydraulic architecture of trees and other woody plants. *New Phytologist* 119:345–360.

———. 1996. Hydraulic architecture of wood tropical plants. *In* S.S. Mulkey, R.L. Chazdon, and A.P. Smith [eds], Tropical Forest Ecophysiology, 217–243. Chapman and Hall, New York, NY.

Tzeng, T.Y., and C.H. Yang. 2001. A MADS box gene from lily (*Lilium longiflorum*) is sufficient to generate dominant negative mutation by interacting with *PISTILLATA (PA)* in *Arabidopsis thaliana. Plant and Cell Physiology* 42:1156–1168.

Ueda, K., A. Hanyun, T. Shiuchi, A. Seo, H. Okubo, and M. Hotta. 1997. Origin of Podostemaceae, a marvelous aquatic flowering plant family. *Plant Species Biology* 110(1097):87–92.

Uhl, N.W., J. Dransfield, J.I. Davis, M.A. Luckow, K.S. Hansen, and J.J. Doyle. 1995. Phylogenetic relationships among palms: Cladistic analyses of morphological and chloroplast DNA restriction site variation. *In* P.J. Rudall, P.J. Cribb, D.F. Cutler, and C.J. Humphries [eds.], Monocotyledons: Systematics and Evolution, 623–661. Royal Botanic Gardens, Kew, UK.

Upchurch, G.R., Jr. 1984. Cuticle evolution in Early Cretaceous angiosperms from the Potomac Group of Virginia and Maryland. *Annals of the Missouri Botanical Garden* 71:522–550.

Upchurch, G.R., Jr., P.R. Crane, and A.N. Drinnan. 1994. The megaflora from the Quantico locality (Upper Albian), Lower Cretaceous Potomac Group of Virginia. *Virginia Museum of Natural History Memoir* 4:1–57.

Vaknin, Y., S. Gan-Mor, A. Bechar, B. Ronen, and D. Eisikowitch. 2001. Are flowers morphologically adapted to take advantage of electrostatic forces in pollination? *New Phytologist* 152:301–306.

Vamosi, J.C., and S.M. Vamosi. 2010. Key innovations within a geographical context in flowering plants: Towards resolving Darwin's abominable mystery. *Ecology Letters* 13:1270–1279.

Van de Peer, Y., J.A Fawcett, S. Proost, L. Sterck, and K. Vandepoele. 2009. The flowering world: A tale of duplications. *Trends in Plant Science* 14(12):680–688.

Van de Peer, Y., S. Maere, and A. Meyer. 2010. 2R or not 2R is not the question anymore. *Nature Reviews Genetics* 11(2):166. doi:10.1038/nrg2600-c2.

van der Bank, M., M.F. Fay, and M.W. Chase. 2002. Molecular phylogenetics of Thymelaeaceae with particular reference to African and Australian genera. *Taxon* 51(2):329–339.

———. 2002. Molecular phylogenetics of Thymelaeaceae with particular reference to African and Australian genera. *Taxon* 51:329–339.

Van der Ham, R. 1989. New observations on the pollen of *Ctenolophon* Oliver (Ctenolophonaceae), with remarks on the evolutionary history of the genus. *Review of Palaeobotany and Palynology* 59(1–4):153–160.

van der Niet, T., D.M. Hansen, and S.D. Johnson. 2011. Carrion mimicry in a South African orchid: Flowers attract a narrow subset of the fly assemblage on animal carcasses. *Annals of Botany* 107:981–992.

van der Niet, T., and S.D. Johnson. 2012. Phylogenetic evidence for pollinator-driven diversification of angiosperms. *Trends in Ecology & Evolution* 27:353–361.

van der Pijl, L., and C.H. Dodson. 1966. Orchid Flowers, Their Pollination and Evolution. University of Miami Press, Coral Gables, FL.

van Heel, W.A. 1981. A SEM-investigation on the development of free carpels. *Blumea* 27:499–522.

———. 1983. The ascidiform early development of free carpels, a S.E.M.-investigation. *Blumea* 28:231–270.

———. 1984. Flowers and fruits in Flacourtiaceae. V. The seed anatomy and pollen morphology of *Berberidopsis* and *Streptothamnus*. *Blumea* 30:31–37.

Vanvinckenroye, P., and E. Smets. 1996. Floral ontogeny of five species of *Talinum* and of related taxa (Portulacaceae). *Journal of Plant Research* 109:387–402.

van Vliet, G.J.C.M., and P. Baas. 1984. Wood anatomy and classification of the Myrtales. *Annals of the Missouri Botanical Garden* 71:783–800.

Vanneste, K., S. Maere, and Y. Van de Peer. 2014. Tangled up in two: A burst of genome duplications at the end of the Cretaceous and the consequences for plant evolution. *Philosophical Transactions of the Royal Society of London B* 369(1648). pii: 20130353. doi:10.1098/rstb.2013.0353.

Varassin, I.G., D.S. Penneys, and F.A. Michelangeli. 2008. Comparative anatomy and morphology of nectar-producing Melastomataceae. *Annals of Botany* 102(6): 899–909.

Varshney, R.K., W. Chen, Y. Li, A.K. Bharti, R.K. Saxena, J.A. Schlueter, M.T.A. Donoghue, S. Azam, G. Fan, A.M. Whaley, A.D. Farmer, J. Sheridan, A. Iwata, R. Tuteja, R.V. Penmetsa, W. Wu, H.D. Upadhyaya, S.-P. Yang, T. Shah, K.B. Saxena, T. Michael, W.R. McCombie, B. Yang, G. Zhang, H. Yang, J. Wang, C. Spillane, D.R. Cook, G.D. May, X. Xu, and S.A. Jackson. 2012. Draft genome sequence of pigeonpea (*Cajanus cajan*), an orphan legume crop of resource-poor farmers. *Nature Biotechnology* 30(1):83–89.

Varshney, R.K., C. Song, R.K. Saxena, S. Azam, S. Yu, A.G. Sharpe, S. Cannon, J. Baek, B.D. Rosen, B. Tar'an, T. Millan, X. Zhang, L.D. Ramsay, A. Iwata, Y. Wang, W. Nelson, A.D. Farmer, P.M. Gaur, C. Soderlund, R.V. Penmetsa, C. Xu, A.K. Bharti, W. He, P. Winter, S. Zhao, J.K. Hane, N. Carrasquilla-Garcia, J.A. Condie, H.D. Upadhyaya, M.-C. Luo, M. Thudi, C.L.L. Gowda, N.P. Singh, J. Lichtenzveig, K.K. Gali, J. Rubio, N. Nadarajan, J. Dolezel, K.C. Bansal, X. Xu, D. Edwards, G. Zhang, G. Kahl, J. Gil, K.B. Singh, S.K. Datta, S.A. Jackson, J. Wang, and D.R. Cook. 2013. Draft genome sequence of chickpea (*Cicer arietinum*) provides a resource for trait improvement. *Nature Biotechnology* 31(3):240–246.

Vaudois-Miéja, N. 1983. Extension paléogéographique en Europe de l'actuel genre asiatique *Rehderodendron* Hu (Styracacées). *Comptes rendus des séances de l'Academie des Sciences. Serie II* 296(1):125–130.

Velasco, R., A. Zharkikh, J. Affourtit, A. Dhingra, A. Cestaro, A. Kalyanaraman, P. Fontana, S.K. Bhatnagar, M. Troggio, D. Pruss, S. Salvi, M. Pindo, P. Baldi, S. Castelletti, M. Cavaiuolo, G. Coppola, F. Costa, V. Cova, A. Dal Ri, V. Goremykin, M. Komjanc, S. Longhi, P. Magnago, G. Malacarne, M. Malnoy, D. Micheletti, M. Moretto, M. Perazzolli, A. Si-Ammour, S. Vezzulli, E. Zini, G. Eldredge, L.M. Fitzgerald, N. Gutin, J. Lanchbury, T. Macalma, J.T. Mitchell, J. Reid, B. Wardell, C. Kodira, Z. Chen, B. Desany, F. Niazi, M. Palmer, T. Koepke, D. Jiwan, S. Schaeffer, V. Krishnan, C. Wu, V.T. Chu, S.T. King, J. Vick, Q. Tao, A. Mraz, A. Stormo, K. Stormo, R. Bogden, D. Ederle, A. Stella, A. Vecchietti, M.M. Kater, S. Masiero, P. Lasserre, Y. Lespinasse, A.C. Allan, V. Bus, D. Chagne, R.N. Crowhurst, A.P. Gleave, E. Lavezzo, J.A. Fawcett, S. Proost, P. Rouze, L. Sterck, S. Toppo, B. Lazzari, R.P. Hellens, C.-E. Durel, A. Gutin, R.E. Bumgarner, S.E. Gardiner, M. Skolnick, M. Egholm, Y. Van de Peer, F. Salamini, and R. Viola. 2010. The genome of the domesticated apple (*Malus* x *domestica* Borkh.). *Nature Genetics* 42(10):833–839.

Veleba, A., P. Bures, L. Adamec, P. Smarda, I. Lipneroval, and L. Horoval. 2014. Genome size and genomic GC content evolution in the miniature genome-sized family Lentibulariaceae. *New Phytologist* 203(1):22–28.

Verde, I., A.G. Abbott, S. Scalabrin, S. Jung, S. Shu, F. Marroni, T. Zhebentyayeva, M.T. Dettori, J. Grimwood, F. Cattonaro, A. Zuccolo, L. Rossini, J. Jenkins, E. Vendramin, L.A. Meisel, V. Decroocq, B. Sosinski, S. Prochnik, T. Mitros, A. Policriti, G. Cipriani, L. Dondini, S. Ficklin, D.M. Goodstein, P. Xuan, C. Del Fabbro, V. Aramini, D. Copetti, S. Gonzalez, D.S. Horner, R. Falchi, S. Lucas, E. Mica, J. Maldonado, B. Lazzari, D. Bielenberg, R. Pirona, M. Miculan, A. Barakat, R. Testolin, A. Stella, S. Tartarini, P. Tonutti, P. Arus, A. Orellana, C. Wells, D. Main, G. Vizzotto, H. Silva, F. Salamini, J. Schmutz, M. Morgante, D.S. Rokhsar, and Int Peach Genome I. 2013. The high-quality draft genome of peach (*Prunus persica*) identifies unique patterns of genetic diversity, domestication and genome evolution. *Nature Genetics* 45(5):487–494.

Vergara-Silva, F., S. Espinosa-Matias, B.A. Ambrose, S. Vazquez-Santana, A. Martinez-Mena, J. Marquez-Guzman, E. Martinez, E.M. Meyerowitz, and E.R. Alvarez-Buylla. 2003. Inside-out flowers characteristic of *Lacandonia schismatica* evolved at least before its divergence from a closely related taxon, *Triuris brevistylis*. *International Journal of Plant Sciences* 164:345–357.

Vicentini, A., J.C. Barber, S.S. Aliscioni, L.M. Giussani, and E.A. Kellogg. 2008. The age of the grasses and clusters of origins of C(4) photosynthesis. *Global Change Biology* 14(12):2963–2977.

Vicient, C.M., A. Suoniemi, K. Anamtamat-Jonsson, J. Tanskanen, A. Beharav, E. Nevo, and A.H. Schulman. 1999. Retrotransposon BARE–1 and its role in genome evolution in the genus *Hordeum*. *Plant Cell* 11:1769–1784.

Vidal-Russell, R., and D.L. Nickrent. 2008. The first mistletoes: Origins of aerial parasitism in Santalales. *Molecular Phylogenetics and Evolution* 47(2):523–537.

Vink, W. 1988. Taxonomy in Winteraceae. *Taxon* 37:691–698.

———. 1993. Winteraceae. *In* K. Kubitzki, J. Rohwer, and V. Bittrich [eds.], The Genera and Families of Vascular Plants, II, 630–638. Springer-Verlag, Berlin, Germany.

Vinnersten, A., and K. Bremer. 2001. Age and biogeography of major clades in Liliales. *American Journal of Botany* 88:1695–1703.

Vishnu-Mittre, P. 1953. A male flower of the Pentoxyleae with remarks on the structure of the female cones of the group. *Palaeobotanist* 2:75–84.

Vishnu-Mittre, P. 1963. Pollen morphology of Indian Amaranthaceae. *Journal of the Indian Botanical Society* 42:86–101.

Vision, T.J., D.G. Brown, and S.D. Tanksley. 2000. The origins of genomic duplications in *Arabidopsis*. *Science* 290(5499):2114–2117.

Vöchting, H. 1886. Über Zygomorphie und deren Ursachen. *Jahrbuch der Wissenschaftlichen Botanik* 17: 297–346.

Vogel, S. 1954. Blütenbiologische Typen als Elemente der Sippengliederung, dargestellt anhand der Flora Südafrikas. *Botanische Studien* 1:1–338.

———. 1959. Organographie der Blüten kapländischer Ophrydeen mit Bemerkungen zum Koaptations–Problem I/II. *Abhandlungen der Akademie der Wissenschaften und der Literatur, Mainz, Jahrbuch* 1960: 81–94.

———. 1969. Über synorganisierte Blütensporne bei einigen Orchideen. *Österreichische Botanische Zeitschrift* 116:244–262.

———. 1974. Ölblumen und ölsammelnde Bienen. *Tropische und Subtropische Pflanzenwelt* 7:1–267.

———. 1978. Evolutionary shifts from reward to deception in pollen flowers. *In* A.J. Richards [ed.], The Pollination of Flowers by Insects, 89–96. Academic Press, London, UK.

———. 1988. Die Ölblumensymbiosen: Parallelismus und andere Aspekte ihrer Entwicklung in Raum und Zeit. *Zeitschrift fur Zoologische Systematik und Evolutions–Forschung* 26:341–362.

———. 1990. Radiación adaptiva del sindrome floral en las familias neotropicales. *Boletín de la Academia Nacional de Ciencias (Córdoba, Argentina)* 59:5–30.

———. 2000. The floral nectaries of Malvaceae s.l.: A conspectus. *Kurtziana* 28:155–171.

———. 2012. Floral-biological syndromes as elements of diversity within tribes in the Flora of South Africa. Shaker Verlag GmbH, Germany. [English translation of Vogel 1954].

von Balthazar, M., W.S. Alverson, J. Schönenberger, and D.A. Baum. 2004. Comparative floral development and androecium structure in Malvoideae (Malvaceae s.l.). *International Journal of Plant Sciences* 165(4):445–473.

von Balthazar, M., and P.K. Endress. 2002a. Development of inflorescences and flowers in Buxaceae and the problem of perianth interpretation. *International Journal of Plant Sciences* 163:847–876.

———. 2002b. Reproductive structures and systematics of Buxaceae. *Botanical Journal of the Linnean Society* 140:193–228.

von Balthazar, M., P.K. Endress, and Y.-L. Qiu. 2000. Phylogenetic relationships in Buxaceae based on nuclear internal transcribed spacers and plastid *ndhF* sequences. *International Journal of Plant Sciences* 161:785–792.

von Balthazar, M., G.E. Schatz, and P.K. Endress. 2003. Female flowers and inflorescences of Didymelaceae. *Plant Systematics and Evolution* 237:199–208.

von Balthazar, M., and J. Schönenberger. 2009. Floral structure and organization in Platanaceae. *International Journal of Plant Sciences* 170(2):210–225.

———. 2013. Comparative floral structure and systematics in the balsaminoid clade including Balsaminaceae, Marcgraviaceae and Tetrameristaceae (Ericales). *Botanical Journal of the Linnean Society* 173:325–386.

von Balthazar, M., J. Schönenberger, W.S. Alverson, H. Janka, C. Bayer, and D.A. Baum. 2006. Structure and evolution of the androecium in the Malvatheca clade (Malvaceae s.l.) and implications for Malvaceae and Malvales. *Plant Systematics and Evolution* 260:171–197.

von Hagen, B., and J.W. Kadereit. 2002. Phylogeny and flower evolution of the Swertiinae (Gentianaceae–Gentianeae): Homoplasy and the principle of variable proportions. *Systematic Botany* 27:548–572.

Von Mering, S., and J.W. Kadereit. 2010. Phylogeny,

systematics, and recircumscription of Juncaginaceae—a cosmopolitan wetland family. *In* O. Seberg, G. Petersen, A.S. Barfod, and J.I. Davis [eds.], Diversity, Phylogeny, and Evolution in the Monocotyledons, 55–79. Aarhus University Press, Århus, Denmark.

von Poser, G.L., M.E. Toffoli, M. Sobral, and A.T. Henriques. 1997. Iridoid glucosides substitution patterns in Verbenaceae and their taxonomic implication. *Plant Systematics and Evolution* 205:265–287.

Voznesenskaya, E.V., V.R. Franceschi, O. Kiirats, H. Freitag, and G.E. Edwards. 2001. Kranz anatomy is not essential for terrestrial C_4 plant photosynthesis. *Nature* 414:543–546.

Wagenitz, G. 1959. Die systematische Stellung der Rubiaceae. *Botanische Jahrbücher für Systematik* 79:17–35.

———. 1975. Blütenreduktion als ein zentrales Problem der Angiospermen–Systematik. *Botanische Jahrbücher für Systemtik* 96:448–470.

Wagenitz, G., and B. Laing. 1984. Die Nektarien der Dipsacales und ihre systematische Bedeutung. *Botanische Jahrbücher für Systematik* 104:483–507.

Wagner, G.P. 1989. The biological homology concept. *Annual Reviews in Ecology and Systematics* 20:51–69.

———. 1996. Homologues, natural kinds and the evolution of modularity. *American Zoology* 36:36–43.

———. 2001. The Character Concept in Evolutionary Biology. Academic Press, San Diego, CA.

———. 2014. Homology, Genes, and Evolutionary Innovation. Princeton University Press, Princeton, NJ.

Wagner, W.H.J. 1970. Biosystematics and evolutionary noise. *Taxon* 19(2):146–151.

Wagner, W.L., P.C. Hoch, and P.H. Raven. 2007. Revised classification of the Onagraceae. *Systematic Botany Monographs* 83:1–240.

Wagstaff, S.J., L. Hickerson, R. Spangler, P.A. Reeves, and R.G. Olmstead. 1998. Phylogeny of Labiatae s.l., inferred from cpDNA sequences. *Plant Systematics and Evolution* 209:265–274.

Wagstaff, S.J., and R.G. Olmstead. 1997. Phylogeny of Labiatae and Verbenaceae inferred from *rbcL* sequences. *Systematic Botany* 22:165–179.

Walbot, V. 2000. A green chapter in the book of life. *Nature* 408:794–795.

Walker, D.B. 1975. Postgenital carpel fusion in *Catharanthus roseus* (Apocynaceae). I. Light and scanning electron microscope study of gynoecial ontogeny. *American Journal of Botany* 62:457–467.

Walker, J.W. 1972. Chromosome numbers, phylogeny, phytogeography of the Annonaceae and their bearing on the (original) basic chromosome number of angiosperms. *Taxon* 21:57–65.

Walker, J.W., and J.A. Doyle. 1975. The bases of angiosperm phylogeny: Palynology. *Annals of the Missouri Botanical Garden* 62:664–723.

Walker, J.W., and A.G. Walker. 1984. Ultrastructure of Lower Cretaceous angiosperm pollen and the origin and early evolution of flowering plants. *Annals of the Missouri Botanical Garden* 71:464–521.

Walker-Larsen, J., and L.D. Harder. 2000. The evolution of staminodes in angiosperms: Patterns of stamen reduction, loss, and functional re-invention. *American Journal of Botany* 87:1367–1384.

———. 2001. Vestigial organs as opportunities for functional innovation: The example of the *Penstemon* staminode. *Evolution* 55:477–487.

Wallander, E. 2008. Systematics of *Fraxinus* (Oleaceae) and evolution of dioecy. *Plant Systematics and Evolution* 273:25–49.

Wallander, E., and V.A. Albert. 2000. Phylogeny and classification of Oleaceae based on *rps16* and *trnL-F* sequence data. *American Journal of Botany* 87:1827–1841.

Wang, H.-C., M.J. Moore, P.S. Soltis, C.D. Bell, S.F. Brockington, R. Alexandre, C.C. Davis, M. Latvis, S.R. Manchester, and D.E. Soltis. 2009. Rosid radiation and the rapid rise of Angiosperm-dominated forests. *Proceedings of the National Academy of Sciences USA* 106:3853–3858.

Wang, J.L., L. Tian, H.S. Lee, N.E. Wei, H.M. Jiang, B. Watson, A. Madlung, T.C. Osborn, R.W. Doerge, L. Comai, and Z.J. Chen. 2006. Genomewide nonadditive gene regulation in *Arabidopsis* allotetraploids. *Genetics* 172(1):507–517.

Wang, W., Z.-D. Chen, Y. Liu, R.-Q. Li, and J.-H. Li. 2007a. Phylogenetic and biogeographic diversification of Berberidaceae in the northern hemisphere. *Systematic Botany* 32(4):731–742.

Wang, W., A.-M. Lu, Y. Ren, M.E. Endress, and Z.-D. Chen. 2009. Phylogeny and classification of Ranunculales: Evidence from four molecular loci and morphological data. *Perspectives in Plant Ecology Evolution and Systematics* 11(2):81–110.

Wang, W., R.D.C. Ortiz, F.M.B. Jacques, X.-G. Xiang, H.-L. Li, L. Lin, R.-Q. Li, Y. Liu, P.S. Soltis, D.E. Soltis, and Z.-D. Chen. 2012. Menispermaceae and the diversification of tropical rainforests near the Cretaceous-Paleogene boundary. *New Phytologist* 195(2):470–478.

Wang, W., H.-C. Wang, and Z.-D. Chen. 2007b. Phylogeny and morphological evolution of tribe Menispermeae (Menispermaceae) inferred from chloroplast and nuclear sequences. *Perspectives in Plant Ecology Evolution and Systematics* 8(3):141–154.

Wang, C., and D. Yu. 1997. Diterpenoid, sesquiterpenoid

and secoiridoids glucosides from *Aster auriculatus*. *Phytochemistry* 45:1483–1497.

Wang, W., Y. Wu, and J. Messing. 2012. The mitochondrial genome of an aquatic plant, *Spirodela polyrhiza*. *PLoS One* 7(10):e46747. doi:10.1371/journal.pone.0046747.

Wang, X., 2010. *Schmeissneria*: An angiosperm from the Early Jurassic. *Journal of Systematics and Evolution* 48:326–335.

Wang, X.-F., W.S. Armbruster, and S.-Q. Huang. 2012. Extra-gynoecial pollen-tube growth in apocarpous angiosperms is phylogenetically wide-spread and probably adaptive. *New Phytologist* 193:253–260.

Wang, X.-F., Y.-B. Tao, and Y.-T. Lu. 2002. Pollen tubes enter neighbouring ovules by way of receptacle tissue, resulting in increased fruit set in *Sagittaria potamogetiflora* Merr. *Annals of Botany* 89:791–796.

Wang, X.-Q., and J.-H. Ran. 2014. Evolution and biogeography of gymnosperms. *Molecular Phylogenetics and Evolution* 75:24–40.

Wang, Y.-H., D.K. Ferguson, G.-P. Feng, Y.-F. Wang, S.G. Zhilin, C.-S. Li, P.-T. Svetlana, J. Yang, and A.G. Ablaev. 2009. The phytogeography of the extinct angiosperm *Nordenskioeldia* (Trochodendraceae) and its response to climate changes. *Palaeogeography Palaeoclimatology Palaeoecology* 280(1–2):183–192.

Wanke, S., M.A. Jaramillo, T. Borsch, M.-S. Samain, D. Quandt, and C. Neinhuis. 2007. Evolution of Piperales—*matK* gene and *trnK* intron sequence data reveal lineage specific resolution contrast. *Molecular Phylogenetics and Evolution* 42:477–497.

Wannan, B.S., J.T. Waterhouse, P.A. Gadek, and C.J. Quinn. 1985. Biflavonyls and the affinities of *Blepharocarya*. *Biochemical Systematics and Ecology* 13: 105–108.

Wanntorp, L. 2006. Molecular systematics and evolution of the genus *Gunnera*. *In* A.K. Sharma., and A. Sharma [eds.], Plant Genome Biodiversity and Evolution. Volume 1, Part C. Phanerogams (Angiosperm-Dicotyledons), 419–435. Science Publishers, Enfield, NH.

Wanntorp, L., H.-E. Wanntorp, and M. Källersjö. 2002. Phylogenetic relationships of *Gunnera* based on nuclear ribosomal DNA ITS region, *rbcL* and *rps16* intron sequences. *Systematic Botany* 27:512–521.

Ward, N.M., and R.A. Price. 2002. Phylogenetic relationships of Marcgraviaceae: Insights from three chloroplast genes. *Systematic Botany* 27:149–160.

Warming, E. 1879. Haandbog I den systematiske botanik. P. G. Philipsens Forlag, Copenhagen, Denmark.

Waser, N.M. 1998. Pollination, angiosperm speciation, and the nature of species boundaries. *Oikos* 81:198–201.

———. 2006. Specialization and generalization in plant-pollinator interactions: A historical perspective. *In* N.M. Waser, and J. Ollerton [eds.]. Plant-pollinator Interactions from Specialization to Generalization, 3–17. University of Chicago Press, Chicago, IL.

Waser, N.M., L. Chittka, M.V. Price, N.M. Williams, and J. Ollerton. 1996. Generalization in pollination systems, and why it matters. *Ecology* 77:1043–1060.

Watson, L.E., P.L. Bates, T.M. Evans, M.M. Unwin, and J.R. Estes. 2002. Molecular phylogeny of subtribe Artemisiinae (Asteraceae), including *Artemisia* and its allied and segregate genera. *BMC Evolutionary Biology* 2:17. doi:10.1186/1471-2148-2-7.

Webb, C.J., and D.G. Lloyd. 1986. The avoidance of interference between the presentation of pollen and stigmas in angiosperms. II. Herkogamy. *New Zealand Journal of Botany* 24:163–178.

Weber, A. 1980. Die Homologie des Perigons der Zingiberaceen. Ein Beitrag zur Morphologie und Phylogenie des Monokotylen-Perigons. *Plant Systematics and Evolution* 133:149–179.

Weberling, F. 1970. Weitere Untersuchungen zur Morphologie des Unterblattes bei den Dikotylen: V. Piperales. *Beiträge zur Biologie der Pflanzen* 46:403–434.

Webster, G.L. 1967. The genera of Euphorbiaceae in the southeastern United States. *Journal of the Arnold Arboretum* 48:303–430.

Weese, T., and L. Bohs. 2007. A three-gene phylogeny of the genus *Solanum* (Solanaceae). *Systematic Botany* 32:445–463.

Wefferling, K.M., S.B. Hoot, and S.S. Neves. 2013. Phylogeny and fuit evolution in Menispermaceae. *American Journal of Botany* 100(5):883–905.

Weigend, M., and H.H. Hilger. 2010. Codonaceae—a newly required family name in Boraginales. *Phytotaxa* 10:26–30.

Weigend, M., F. Luebert, M. Gottschling, L.P. Couvreur, H.H. Hilger, and J.S. Miller. 2014. From capsules to nutlets—phylogenetic relationships in the Boraginales. *Cladistics* 30:508–518.

Weigend, M., O. Mohr, and T.J. Motley. 2002. Phylogeny and classification of the genus *Ribes* (Grossulariaceae) based on 5S-NTS sequences and morphological and anatomical data. *Botanische Jahrbücher für Systematik* 124:163–182.

Weller, S.G., and A.K. Sakai. 1999. Using phylogenetic approaches for the analysis of plant breeding system evolution. *Annual Reviews of Ecology and Systematics* 30:167–199.

Weller, S.G., A. Sakai, A. Rankin, A. Golonka, B. Kutcher, and K. Ashby. 1998. Dioecy and the evolution of pollination systems in *Schiedea* and *Alsinidendron* (Caryophyllaceae: Alsinoideae) in the Hawaiian Islands. *American Journal of Botany* 85:1377–1388.

Wen, J., Z.-L. Nie, A. Soejima, and Y. Meng. 2007. Phylogeny of Vitaceae based on the nuclear *GAI1* gene sequences. *Canadian Journal of Botany-Revue Canadienne De Botanique* 85(8):731–745.

Wendel, J.F. 2000. Genome evolution in polyploids. *Plant Molecular Biology* 42:225–249.

Wendel, J.F., R.C. Cronn, J.S. Johnston, and H.J. Price. 2002. Feast and famine in plant genomes. *Genetica* 115:37–47.

Wendel, J.F., J. Greilhuber, J. Dolezel, and I.J. Leitch [eds.]. 2012. Plant Genomes, Their Residents, and Their Evolutionary Dynamics. Springer-Verlag, Wien, Austria.

Wendling, B.M., K.E. Galbreath, and E.G. DeChaine. 2011. Resolving the evolutionary history of *Campanula* (Campanulaceae) in western North America. *PLoS One*. doi:10.1371/journal.pone.0023559.

Werner, G.D.A., W.K. Cornwell, J.I. Sprent, J. Kattge, and E.T. Kiers. 2014. A single evolutionary innovation drives the deep evolution of symbiotic N-2-fixation in Angiosperms. *Nature Communications* 5:4087. doi:10.1038/ncomms5087.

Westerkamp, C., and R. Classen-Bockhoff. 2007. Bilabiate flowers: The ultimate response to bees? *Annals of Botany* 100(2):361–374.

Westerkamp, C., and A. Weber. 1999. Keel flowers of the Polygalaceae and Fabaceae: A functional comparison. *Botanical Journal of the Linnean Society* 129:207–221.

Weston, P.H. 1995. Proteaceae. *In* Flora of Australia 16. CSIRO, Melbourne, Australia.

Wettstein, R.V. 1907. Handbuch der systematischen Botanik, II. Franz Deuticke, Vienna, Austria.

Wheeler, E.A., T.M. Lehman, and P.E. Gasson. 1994. *Javelinoxylon*, an Upper Cretaceous dicotyledonous tree from Big-Bend National Park, Texas, with presumed malvalean affinities. *American Journal of Botany* 81(6):703–710.

Wheeler, M.J., V.E. Franklin-Tong, and F.C.H. Franklin. 2001. The molecular and genetic basis of pollen–pistil interactions. *New Phytologist* 151:565–584.

Whibley, A.C., N.B. Langlade, C. Andalo, A.I. Hanna, A. Bangham, C. Thebaud, and E. Coen. 2006. Evolutionary paths underlying flower color variation in *Antirrhinum*. *Science* 313:963–966.

Whitehead, D.R. 1969. Wind pollination in the angiosperms: Evolutionary and environmental considerations. *Evolution* 23:28–35.

———. 1983. Wind pollination: Some ecological and evolutionary perspectives. *In* L. Real [ed.], Pollination Biology, 97–108. Academic Press, Orlando, FL.

Whitson, M., and P.S. Manos. 2005. Untangling *Physalis* (Solanaceae) from the physaloids: A two-gene phylogeny of Physalinae. *Systematic Botany* 30: 216–230.

Whittall, J.B., and S.A. Hodges. 2007. Pollinator shifts drive increasingly long nectar spurs in columbine flowers. *Nature* 447:706–709.

Whittall, J.B., C. Voelckel, D.J. Kliebenstein, and S.A. Hodges. 2006. Convergence, constraint and the role of gene expression during adaptive radiation: Floral anthocyanins in *Aquilegia*. *Molecular Ecology* 15:4645–4657.

Whitten, W.M., N.H. Williams, and M.W. Chase. 2000. Subtribal and generic relationships of Maxillarieae (Orchidaceae) with emphasis on Stanhopeinae: Combined molecular evidence. *American Journal of Botany* 87:1842–1856.

Wible, J.R., G.W. Rougier, M.J. Novacek, and R.J. Asher. 2007. Cretaceous eutherians and Laurasian origin for placental mammals near the K/T boundary. *Nature* 447:1003–1006.

Wicke, S., K.F. Mueller, C.W. de Pamphilis, D. Quandt, N.J. Wickett, Y. Zhang, S.S. Renner, and G.M. Schneeweiss. 2013. Mechanisms of functional and physical genome reduction in photosynthetic and nonphotosynthetic parasitic plants of the Broomrape family. *Plant Cell* 25(10):3711–3725.

Wickett, N.J., L.A. Honaas, E.K. Wafula, M. Das, K. Huang, B. Wu, L. Landherr, M.P. Timko, J. Yoder, J.H. Westwood, and C.W. dePamphilis. 2011. Transcriptomes of the parasitic plant family Orobanchaceae reveal surprising conservation of chlorophyll synthesis. *Current Biology* 21(24):2098–2104.

Wickett, N.J., S. Mirarab, N. Nguyen, T. Warnow, E. Carpenter, N. Matasci, S. Ayyampalayam, M. Barker, G.J. Burleigh, M.A. Gitzendanner, B. Ruhfel, E. Wafula, J. Der, S.W. Graham, S. Mathews, M. Melkonian, D.E. Soltis, P.S. Soltis, C. Rothfels, L. Pokorny, J. Shaw, L. DeGironimo, D. Stevenson, B. Surek, J.C. Villarreal, B. Roure, H. Philippe, C. W. dePamphilis, T. Chen, M. Deyholos, J. Wang, Y. Zhang, Z. Tian, Z. Yan, X. Wu, X. Sun, G. Ka-Shu Wong, and J. Leebens-Mack. 2014. A phylotranscriptomics analysis of the origin and early diversification of land plants. *Proceedings of the National Academy of Sciences USA* 111(45):E4859–E4868.

Wiegrefe, S.J., K.J. Sytsma, and R.P. Guries. 1998. The Ulmaceae, one family or two? Evidence from chloroplast DNA restriction site mapping. *Plant Systematics and Evolution* 210:249–270.

Wieland, G.R. 1918. The origin of dicotyls. *Science* 48:18–21.

Wiens, D., and B.A. Barlow. 1971. The cytogeography and

relationships of the viscaceous and eremolepidaceous mistletoes. *Taxon* 20:313–332.

Wiens, J.J. 1998. Combining data sets with different phylogenetic histories. *Systematic Biology* 47:568–581.

Wiens, J.J., and M.C. Morrill. 2011. Missing data in phylogenetic analysis: Reconciling results from simulations and empirical data. *Systematic Biology* 60:719–731.

Wikström, N., and P. Kenrick. 2001. Evolution of Lycopodiaceae (Lycopsida): Estimating divergence times from *rbcL* gene sequences by use of nonparametric rate smoothing. *Molecular Phylogenetics and Evolution* 19:177–186.

Wikström, N., V. Savolainen, and M.W. Chase. 2001. Evolution of the angiosperms: Calibrating the family tree. *Proceedings of the Royal Society of London B* 268:2211–2220.

———. 2003. Angiosperm divergence times: Congruence and incongruence between fossils and sequence divergence estimates. *In* P.C.J. Donoghue, and M.P. Smith [eds.], Telling the Evolutionary Time: Molecular Clocks and the Fossil Record, 142–165. Taylor and Francis, London, UK.

Wilcox, T.P., D.J. Zwickl, T.A. Heath, and D.M. Hillis. 2002. Phylogenetic relationships of the dwarf boas and a comparison of Bayesian and bootstrap measures of phylogenetic support. *Molecular Phylogenetics and Evolution* 25:361–371.

Wilde V., Z. Kvaček, and J. Bogner. 2005: Fossil leaves of the Araceae from the European Eocene and notes to other aroid fossils. *International Journal of Plant Sciences* 166:157–183.

Wiley, E.O. 1979. An annotated Linnean hierarchy, with comments on natural taxa and competing systems. *Systematic Zoology* 28:308–337.

———. 1981. Phylogenetics: The Theory and Practice of Phylogenetic Systematics. John Wiley and Sons, New York, NY.

Wilf, P., C.C. Labandeira, W.J. Kress, C.L. Staines, D.M. Windsor, A.L. Allen, and K.R. Johnson. 2000. Timing the radiations of leaf beetles: Hispines on gingers from latest Cretaceous to recent. *Science* 289(5477): 291–294.

Williams, J.H. 2008. Novelties of the flowering plant pollen tube underlie diversification of a key life history stage. *Proceedings of the National Academy of Sciences USA* 105:11259–11263.

———. 2009. *Amborella trichopoda* (Amborellaceae) and the evolutionary developmental origins of the Angiosperm progamic phase. *American Journal of Botany* 96:144–165.

———. 2012. Pollen tube growth rates and the diversification of flowering plant reproductive cycles. *International Journal of Plant Sciences* 173:649–661.

Williams, J.H., and W.E. Friedman. 2002. Identification of diploid endosperm in an early angiosperm lineage. *Nature* 415:522–526.

———. 2004. The four-celled female gametophyte of *Illicium* (Illiciaceae; Austrobaileyales): Implications for understanding the origin and early evolution of monocots, eumagnoliids, and eudicots. *American Journal of Botany* 91:332–351.

Williams, N.H., M.W. Chase, T. Fulcher, and W.M. Whitten. 2001. Molecular systematics of the Oncidiinae based on evidence from four DNA regions: Expanded circumscriptions of *Cyrtochilum*, *Erycina*, *Otoglossum* and *Trichocentrum* and a new genus (Orchidaceae). *Lindleyana* 16:113–139.

Williams, S.E. 1976. Comparative sensory physiology of the Droseraceae: Evolution of a plant sensory system. *Proceedings of the American Philosophical Society* 120:187–204.

Williams, S.E., V.A. Albert, and M.W. Chase. 1994. Relationships of Droseraceae: A cladistic analysis of *rbcL* sequence and morphological data. *American Journal of Botany* 81:1027–1037.

Williamson, P.S., and E.L. Schneider. 1993a. Cabombaceae. *In* K. Kubitzki, J. G. Rohwer, V. Bittrich [eds.], The Families and Genera of Vascular Plants, II, 157–161. Springer-Verlag, Berlin, Germany.

———. 1993b. Nelumbonaceae. *In* K. Kubitzki, J. Rohwer, and V. Bittrich [eds.], The Families and Genera of Vascular Plants, II, 470–473. Springer-Verlag, Berlin, Germany.

Willis, C.G., B. Ruhfel, R.B. Primack, A.J. Miller-Rushing, and C.C. Davis. 2008. Phylogenetic patterns of species loss in Thoreau's woods are driven by climate change. *Proceedings of the National Academy of Sciences USA* 105:17029–17033.

Willis, J.C. 1966. A Dictionary of the Flowering Plants and Ferns. 7th ed., revised H. K. Airy Shaw. Cambridge University Press, Cambridge, UK.

Wilson, K.A. 1960a. The genera of Arales in the southeastern United States. *Journal of the Arnold Arboretum* 41:47–72.

———. 1960b. The genera of Hydrophyllaceae and Polemoniaceae in the southeastern United States. *Journal of the Arnold Arboretum* 41:197–212.

Wilson, K.L., and D.A. Morrison [eds.]. 2000. Monocot Systematics and Evolution. CSIRO, Melbourne, Australia.

Wilson, M.A., B. Gaut, and M.T. Clegg. 1990. Chloroplast DNA evolves slowly in the palm family (Arecaceae). *Molecular Biology and Evolution* 7:303–314.

Wilson, P., and J.D. Thomson. 1996. How do flowers diverge? *In* D.G. Lloyd, and S.C.H. Barrett [eds.], Floral Biology. Studies on Floral Evolution in Animal–Pollinated Plants, 88–111. Chapman and Hall, New York, NY.

Wilson, P., A.D. Wolfe, W.S. Armbruster, and J.D. Thomson. 2007. Constrained lability in floral evolution: Counting convergent origins of hummingbird pollination in *Penstemon* and *Keckiella*. *New Phytologist* 176:883–890.

Wilson, P.G. 2011. Myrtaceae. *In* K. Kubitzki [ed.], The Families and Genera of Flowering Plants. X. Flowering Plants: Eudicots, Sapindales, Cucurbitales, Myrtaceae, 212–271. Springer-Verlag, Berlin, Germany.

Wilson, P.G., M.M. O'Brien, P.A. Gadek, and C.J. Quinn. 2001. Myrtaceae revisited: A reassessment of infrafamilial groups. *American Journal of Botany* 88:2013–2025.

Wilson, P.G., M.M. O'Brien, M.M. Heslewood, and C.J. Quinn. 2005. Relationships within Myrtaceae sensu lato based on a *matK* phylogeny. *Plant Systematics and Evolution* 251(1):3–19.

Wilson, T.K. 1966. The comparative anatomy of the Canellaceae: IV. Floral morphology and conclusions. *American Journal of Botany* 53:336–343.

Wing, S.L., F.A. Herrera, C.A. Jaramillo, C. Gómez-Navarro, P. Wilf, and C.C. Labandeira. 2009. Late Paleocene fossils from the Cerrejón Formation, Colombia, are the earliest record of Neotropical rainforest. *Proceedings of the National Academy of Sciences USA* 106(44):18627–18632.

Wing, S.L., and L.J. Hickey. 1984. The *Platycarya* perplex and the evolution of the Juglandaceae. *American Journal of Botany* 71(3):388–411.

Winkler, H. 1920. Verbreitung und Ursache der Parthenogenesis im Pflanzen- und Tierreiche. Gustav Fischer Verlag, Jena, Germany.

Winkworth, R.C., C.D. Bell, and M.J. Donoghue. 2008b. Mitochondrial sequence data and Dipsacales phylogeny mixed models, partioned Bayesian analyses, and model selection. *Molecular Phylogenetics and Evolution* 46:830–843.

Winkworth, R.C., J. Lundberg, and M.J. Donoghue. 2008a. Toward a resolution of Campanulid phylogeny, with special reference to the placement of Dipsacales. *Taxon* 57:53–65.

Winter, A.N., and I.I. Shamrov. 1991. The development of the ovule and embryo sac in *Nuphar lutea* (Nymphaeaceae). *Botaniceskij Zhurnal* 76:378–390.

Winter, K., M. Garcia, and J.A.M. Holtum. 2011. Drought-stress-induced up-regulation of CAM in seedlings of a tropical cactus, *Opuntia elatior*, operating predominantly in the C3 mode. *Journal of Experimental Botany* 62:4037–4042.

Winter, K.U., A. Becker, T. Münster, J.T. Kim, H. Saedler, and G. Theissen. 1999. MADS–box genes reveal that gnetophytes are more closely related to conifers than to flowering plants. *Proceedings of the National Academy of Sciences USA* 96:7342–7347.

Withner, C.L. 1941. Stem anatomy and phylogeny of the Rhoipteleaceae. *American Journal of Botany* 28:872–878.

Wojciechowski, M.F., M. Lavin, and M.J. Sanderson. 2004. A phylogeny of legumes (Leguminosae) based on analyses of the plastid *matK* gene resolves many well-supported subclades within the family. *American Journal of Botany* 91(11):1846–1862.

Wolfe, J.A. 1975. Some aspects of plant geography of Northern Hemisphere during Late Cretaceous and Tertiary. *Annals of the Missouri Botanical Garden* 62(2):264–279.

Wolfe, K.H., M. Gouy, Y.-W. Yang, P.M. Sharp, and W.-H. Li. 1989. Date of the monocot–dicot divergence estimated from chloroplast DNA sequence data. *Proceedings of the National Academy of Sciences USA* 86: 6201–6205.

Wolfe, K.H., C.W. Morden, and J.D. Palmer. 1992. Function and evolution of a minimal plastid genome from a nonphotosynthetic parasitic plant. *Proceedings of the National Academy of Sciences USA* 89:10648–10652.

Wolter-Filho, W., A.I. Da Rocha, M. Yoshida, and O.R. Gottlieb. 1985. Ellagic-acid derivatives from *Rhabdodendron macrophyllum*. *Phytochemistry* 24:1991–1993.

———. 1989. Chemosystematics of *Rhabdodendron*. *Phytochemistry* 28:2355–2357.

Won, H., and S.S. Renner. 2003. Horizontal gene transfer from flowering plants to *Gnetum*. *Proceedings of the National Academy of Sciences USA* 100(19): 10824–10829.

Wood, C.E. 1961. The genera of Ericaceae in the southeastern United States. *Journal of the Arnold Arboretum* 42:10–80.

———. 1974. A Student's Atlas of Flowering Plants: Some Dicotyledons of Eastern North America, Generic Flora of the Southeastern U.S. Project. Harper and Row, New York, NY.

Wood, C.E., and R.B. Channell. 1960. The genera of Ebenales in the southeastern United States. *Journal of the Arnold Arboretum* 41:1–35.

Wood, T.E., N. Takebayashi, M.S. Barker, I. Mayrose, P.B. Greenspoon, and L.H. Rieseberg. 2009. The frequency of polyploid speciation in vascular plants.

Proceedings of the National Academy of Sciences USA 106(33):13875–13879.

Worberg, A., M.H. Alford, D. Quandt, and T. Borsch. 2009. Huerteales sister to Brassicales plus Malvales, and newly circumscribed to include *Dipentodon*, *Gerrardina*, *Huertea*, *Perrottetia*, and *Tapiscia*. *Taxon* 58(2):468–478.

Worberg, A., D. Quandt, A.-M. Barniske, C. Lohne, K.W. Hilu, and T. Borsch. 2007. Phylogeny of basal eudicots: Insights from non-coding and rapidly evolving DNA. *Organisms Diversity & Evolution* 7(1):55–77.

Worsdell, W.C. 1908. The affinities of *Paeonia*. *Journal of Botany* 46:114–116.

Wright, M.A.R., M. Welsh, and M. Costea. 2011. Diversity and evolution of the gynoecium in *Cuscuta* (dodders, Convolvulaceae) in relation to their reproductive biology: Two styles are better than one. *Plant Systematics and Evolution* 296(1–2):51–76.

Wu, C.I., and W.H. Li. 1985. Evidence for higher rates of nucleotide substitution in rodents than in man. *Proceedings of the National Academy of Sciences USA* 82:1741–1745.

Wu, C.-S., and S.-M. Chaw. 2014. Highly rearranged and size-variable chloroplast genomes in conifers II clade (cupressophytes): Evolution towards shorter intergenic spacers. *Plant Biotechnology Journal* 12(3):344–353.

Wu, C.-S., Y.-N. Wang, S.-M. Liu, and S.-M. Chaw. 2007. Chloroplast genome (cpDNA) of *Cycas taitungensis* and 56 cp protein-coding genes of *Gnetum parvifolium*: Insights into cpDNA evolution and phylogeny of extant seed plants. *Molecular Biology and Evolution* 24:1366–1379.

Wu, F., and S.D. Tanksley. 2010. Chromosomal evolution in the plant family Solanaceae. *BMC Genomics* 11:182. doi:10.1186/1471-2164-11-82.

Wunderlich, R. 1959. Zur Frage der Phylogenie der Endospermtypen bei den Angiospermen. *Österreichische Botanische Zeitschrift* 106:203–483.

Wurdack, J.J., and R.B. Kral. 1982. The genera of Melastomataceae in the southeastern United States. *Journal of the Arnold Arboretum* 63:429–439.

Wurdack, K.J, and C.C. Davis. 2009. Malpighiales phylogenetics: Gaining ground on one of the most recalcitrant clades in the angiosperm tree of life. *American Journal of Botany* 96(8):1551–1570.

Wurdack, K.J., P. Hoffmann, R. Samuel, A. Debruijn, M. Van Der Bank, and M.W. Chase. 2004. Molecular phylogenetic analysis of Phyllanthaceae (Phyllanthoideae pro parte, Euphorbiaceae sensu lato) using plastid *rbcL* DNA sequences. *American Journal of Botany* 91(11):1882–1900.

Wurdack, K.J., and J.W. Horn. 2001. A reevaluation of the affinities of the Tepuianthaceae: Molecular and morphological evidence for placement in the Malvales. *In* Botany 2001: Plants and People, 151. Abstracts.

Wyss, A.R., and J. Meng. 1996. Application of phylogenetic taxonomy to poorly resolved crown clades: A stem-modified node-based definition of *Rodentia*. *Systematic Biology* 45:559–568.

Xi, Z., L. Liu, J.S. Rest, and C.C. Davis. 2014. Coalescent versus concatenation methods and the placement of *Amborella* as sister to water lilies. *Systematic Biology* 63(6):919–932.

Xi, Z., Y. Wang, R.K. Bradley, M. Sugumaran, C.J. Marx, J.S. Rest, and C.C. Davis. 2013. Massive mitochondrial gene transfer in a parasitic flowering plant clade. *PLoS Genetics* 9(2):e1003265. doi:10.1371/journal.pgen.1003265.

Xi, Z.-X., R.K. Bradley, K.J. Wurdack, K.M. Wong, M. Sugumaran, K. Bomblies, J.S. Rest, and C.C. Davis. 2012. Horizontal transfer of expressed genes in a parasitic flowering plant. *BMC Genomics* 13:227. doi:10.1186/1471-2164-13-27.

Xi, Z.-X., B.R. Ruhfel, H. Schaefer, A.M. Amorim, M. Sugumaran, K.J. Wurdack, P.K. Endress, M.L. Matthews, P.F. Stevens, S. Mathews, and C.C. Davis. 2012. Phylogenomics and a posteriori data partitioning resolve the Cretaceous Angiosperm radiation Malpighiales. *Proceedings of the National Academy of Sciences USA* 109(43):17519–17524.

Xiang, Q.-X., M.L. Moody, D.E. Soltis, C.-Z. Fan, and P.S. Soltis. 2002. Relationships within Cornales and circumscription of Cornaceae–*matK* and *rbcL* sequence data and effects of outgroups and long branches. *Molecular Phylogenetics and Evolution* 24:35–57.

Xiang, Q.-Y., D.T. Thomas, and Q.P. Xiang. 2011. Resolving and dating the phylogeny of Cornales—effects of taxon sampling data partitions, and fossil calibrations. *Molecular Phylogenetics and Evolution* 59:123–138.

Xiang, Z.Y., D.E. Soltis, D.R. Moran, and P.S. Soltis. 1993. Phylogenetic relationships of *Cornus* L. sensu lato and putative relatives inferred from *rbcL* sequence data. *Annals of the Missouri Botanical Garden* 80: 723–734.

Xiong, Z., R.T. Gaeta, and J.C. Pires. 2011. Homoeologous shuffling and chromosome compensation maintain genome balance in resynthesized allopolyploid *Brassica napus*. *Proceedings of the National Academy of Sciences USA* 108(19):7908–7913.

Xu, S., Y. Luo, Z. Cai, X. Cao, X. Hu, J. Yang, and D. Luo. 2013. Functional diversity of CYCLOIDEA-like TCP genes in the control of zygomorphic flower development in *Lotus japonicas*. *Jour-*

nal of Integrated Plant Biology 55:221–231. doi:10.1111/j.1744-7909.2012.01169x.

Xue, J.-H., W.-P. Dong, T. Cheng, and S.-L. Zhou. 2012. Nelumbonaceae: Systematic position and species diversification revealed by the complete chloroplast genome. *Journal of Systematics and Evolution* 50(6):477–487.

Yakovlev, M.S., and M.D. Yoffe. 1957. On some peculiar features in the embryogeny of *Paeonia* L. *Phytomorphology* 7:74–82.

Yamada, T., R. Imaichi, and M. Kato. 2001b. Developmental morphology of ovules and seeds of Nymphaeales. *American Journal of Botany* 88:963–974.

Yamada, T., H. Tobe, R. Imaichi, and M. Kato. 2001a. Developmental morphology of the ovules of *Amborella trichopoda* (Amborellaceae) and *Chloranthus serratus* (Chloranthaceae). *Botanical Journal of the Linnaean Society* 137:277–290.

Yamada, T., K. Uehara, M. Ito, and M. Kato. 2003. Expression pattern of an *INNER NO OUTER* homologue in Genus *Nymphaea* (Nymphaeaceae) and its implication for the evolution of outer integument. *Development, Genes and Evolution* 213:510–513.

Yamashita, J., and M.N. Tamura. 2000. Molecular phylogeny of the Convallariaceae (Asparagales). *In* K.L. Wilson, and D.A. Morrison [eds.], Monocots: Systematics and Evolution, 387–400. CSIRO, Collingwood, Victoria, Australia.

Yamazaki, T. 1974. A system of Gamopetalae based on the embryology. *Journal of the Faculty of Science, University of Tokyo, III* 11:263–281.

Yang, Z., and B. Rannala. 2012. Molecular phylogenetics: Principles and practice. *Nature Reviews Genetics* 13:303–314.

Yang, Z.S. 1997. PAML: A program for package for phylogenetic analysis by maximum likelihood. *CABIOS* 15:555–556.

Yang, Z.S., S. Kumar, and M. Nei. 1995. A new method of inference of ancestral nucleotide and amino acid sequences. *Genetics* 141:1641–1650.

Yashina, S., S. Gubin, S. Maksimovich, A. Yashina, E. Gakhova, and D. Gilichinsky. 2012. Regeneration of whole fertile plants from 30,000-y-old fruit tissue buried in Siberian permafrost. *Proceedings of the National Academy of Sciences USA* 109:4008–4013.

Yeo, P.F. 1992. Secondary pollen presentation: Form, function and evolution. *Plant Systematics and Evolution, Supplement* 6:1–269.

Yesson, C., R.T. Bárcenas, H.M. Hernández, L. Ruiz-Maqueda Mde, V.M. Prado, A. Rodríguez, and J.A. Hawkins. 2011. DNA barcodes for Mexican Cactaceae, plants under pressue from wild collecting. *Molecular Ecology Resources* 11:775–783.

Yockteng, R., A.M.R. Almeida, K. Morioka, E.R. Alvarez-Buylla, and C.D. Specht. 2013. Molecular evolution and patterns of duplication in the *SEP/AGL6* lineage of the Zingiberales: A proposed mechanism for floral diversification. *Molecular Biology and Evolution* 30:2401–2422.

Yoo, M.-J., C.D. Bell, P.S. Soltis, and D.E. Soltis. 2005. Divergence times and historical biogeography of Nymphaeales. *Systematic Botany* 30:693–704.

Yoo, M.-J., P.S. Soltis, and D.E. Soltis. 2010. Expression of floral MADS–box genes in two divergent water lilies: Nymphaeales and *Nelumbo*. *International Journal of Plant Sciences* 171(2):121–146.

Yoshida, O. 1962. Embryologische Studien über *Schisandra chinensis* Baillon. *Journal of the College of Arts and Sciences, Chiba University, Natural Sciences Series* 3:459–462.

Yoshida, S., S. Maruyama, H. Nozaki, and K. Shirasu. 2010. Horizontal gene transfer by the parasitic plant *Striga hermonthica*. *Science* 328(5982):1128. doi:10.1126/science.1187145.

Young, D.A. 1981. Are the angiosperms primitively vesselless? *Systematic Botany* 6:313–330.

Young, N.D., and C.W. de Pamphilis. 2005. Rate variation in parasitic plants: Correlated and uncorrelated patterns among plastid genes of different function. *BMC Evolutionary Biology* 5:16. doi:10.1186/1471-2148-5-6.

Young, N.D., K.E. Steiner, and C.W. dePamphilis. 1999. The evolution of parasitism in the Scrophulariaceae/Orobanchaceae: Plastid gene sequences refute an evolutionary transition series. *Annals of the Missouri Botanical Garden* 86:876–893.

Yu, D., M. Kotilainen, E. Pollanen, M. Mehto, P. Elomaa, Y. Helariutta, V.A. Albert, and T.H. Teeri. 1999. Organ identity genes and modified patterns of flower development in *Gerbera hybrida*. *Plant Journal* 17:51–62.

Yu, J., J. Wang, W. Lin, S.G. Li, H. Li, J. Zhou, P.X. Ni, W. Dong, S.N. Hu, C.Q. Zeng, J.G. Zhang, Y. Zhang, R.Q. Li, Z.Y. Xu, S.T. Li, X.R. Li, H.K. Zheng, L.J. Cong, L. Lin, J.N. Yin, J.N. Geng, G.Y. Li, J.P. Shi, J. Liu, H. Lv, J. Li, Y.J. Deng, L.H. Ran, X.L. Shi, X.Y. Wang, Q.F. Wu, C.F. Li, X.Y. Ren, J.Q. Wang, X.L. Wang, D.W. Li, D.Y. Liu, X.W. Zhang, Z.D. Ji, W.M. Zhao, Y.Q. Sun, Z.P. Zhang, J.Y. Bao, Y.J. Han, L.L. Dong, J. Ji, P. Chen, S.M. Wu, J.S. Liu, Y. Xiao, D.B. Bu, J.L. Tan, L. Yang, C. Ye, J.F. Zhang, J.Y. Xu, Y. Zhou, Y.P. Yu, B. Zhang, S.L. Zhuang, W.B. Wei, B. Liu, M. Lei, H. Yu, Y.Z. Li, H. Xu, S.L. Wei, X.M. He, L.J. Fang, Z.J. Zhang, Y.Z. Zhang, X.G. Huang, Z.X. Su, W. Tong, J.H. Li, Z.Z. Tong, S.L. Li, J. Ye, L.S. Wang, L. Fang, T.T. Lei, C. Chen, H. Chen, Z. Xu,

H.H. Li, H.Y. Huang, F. Zhang, H.Y. Xu, N. Li, C.F. Zhao, L.J. Dong, Y.Q. Huang, L. Li, Y. Xi, Q.H. Qi, W.J. Li, W. Hu, Y.L. Zhang, X.J. Tian, Y.Z. Jiao, X.H. Liang, J.A. Jin, L. Gao, W.M. Zheng, B.L. Hao, S.Q. Liu, W. Wang, L.P. Yuan, M.L. Cao, J. McDermott, R. Samudrala, G.K.S. Wong, and H.M. Yang. 2005. The genomes of *Oryza sativa*: A history of duplications. *PLoS Biology* 3(2):266–281.

Yuan, L.-C., Y.-B. Luo, L.B. Thien, J.-H. Fan, H.-L. Xu, and D. Chen. 2007. Pollination of *Schisandra henryi* (Schisandraceae) by female, pollen-earing *Megommata* species (Cecidomyiidae, Diptera) in South-central China. *Annals of Botany* 99:451–460.

Zahn, L.M., H. Kong, J.H. Leebens-Mack, S. Kim, P.S. Soltis, L.L. Landherr, D.E. Soltis, and C.W. dePamphilis. 2005. The evolution of the *SEPALLATA* subfamily of MADS-box genes: A pre-angiosperm origin with multiple duplications throughout angiosperm history. *Genetics* 169:2209–2223.

Zanis, M.J., D.E. Soltis, P.S. Soltis, S. Matthews, and M.J. Donoghue. 2002. The root of the angiosperms revisited. *Proceedings of the National Academy of Sciences USA* 99:6848–6853.

Zanis, M.J., P.S. Soltis, Y.-L. Qiu, E. Zimmer, and D.E. Soltis. 2003. Phylogenetic analyses and perianth evolution in basal angiosperms. *Annals of the Missouri Botanical Garden* 90:129–150.

Zanne, A.E., D.C. Tank, W.K. Cornwell, J.M. Eastman, S.A. Smith, R.G. FitzJohn, D.J. McGlinn, B.C. O'Meara, A.T. Moles, P.B. Reich, D.L. Royer, D.E. Soltis, P.F. Stevens, M. Westoby, I.J. Wright, L. Aarssen, R.I. Bertin, A. Calaminus, R. Govaerts, F. Hemmings, M.R. Leishman, J. Oleksyn, P.S. Soltis, N.G. Swenson, L. Warman, and J.M. Beaulieu. 2014. Three keys to the radiation of angiosperms into freezing environments. *Nature* 506(7486):89–92.

Zapata, F. 2013. A multilocus phylogenetic analysis of *Escallonia* (Escalloniaceae): Diversification in montane South America. *American Journal of Botany* 100:526–545.

Zavada, M.S., and S.E. de Villiers. 2000. Pollen of the Asteraceae from the Paleocene-Eocene of South Africa. *Grana* 39(1):39–45.

Zavada, M.S., and D.L. Dilcher. 1986. Comparative pollen morphology and its relationship to phylogeny of pollen in the Hamamelidae. *Annals of the Missouri Botanical Garden* 73:348–381.

Zeng, L., Q. Zhang, R. Sun, H. Kong, N. Zhang, and H. Ma. 2014. Resolution of deep angiosperm phylogeny using conserved nuclear genes and estimates of early divergence times. *Nature Communications* 5(4956). doi:10.1038/ncomms5956.

Zetter, R., F. Grímsson, C. Hofmann, and G. Grimm. 2014. Paleogene Loranthaceae from West Greenland and Eurasia. P. 318. Abstract Book, European Paleobotany and Palynology Conference, Padoa, Italy.

Zhan, S.H., L. Glick, C.S. Tsigenopoulos, S.P. Otto, and I. Mayrose. 2014. Comparative analysis reveals that polyploidy does not decelerate diversification in fish. *Journal of Evolutionary Biology* 27(2):391–403.

Zhang, G., C. Li, Q. Li, B. Li, D.M. Larkin, C. Lee, J.F. Storz, A. Antunes, M.J. Greenold, R.W. Meredith, A. Ödeen, J. Cui, Q. Zhou, L. Xu, H. Pan, Z. Wang, L. Jin, P. Zhang, H. Hu, W. Yang, J. Hu, J. Xiao, Z. Yang, Y. Liu, Q. Xie, H. Yu, J. Lian, P. Wen, F. Zhang, H. Li, Y. Zeng, Z. Xiong, S. Liu, L. Zhou, Z. Huang, N. An, J. Wang, Q. Zheng, Y. Xiong, G. Wang, B. Wang, J. Wang, Y. Fan, R.R. da Fonseca, A. Alfaro-Núnez, M. Schubert, L. Orlando, T. Mourier. J.T Howard, G. Ganapathy, A. Pfenning, O. Whitney, M.V. Rivas, E. Hara, J. Smith, M. Farré, J. Narayan, G. Slavov, M.N. Romanov, R. Borger, J.P. Machado, I. Khan, M.S. Springer, J. Gatesy, F.G. Hoffmann, J.C. Opazo, O. Håstad, R.H. Saywer, H. Kim, K.-W. Kim, H.J. Kim, S. Cho, N. Li, Y. Huang, M.W. Bruford, X. Zhan, A. Dixon, M.F. Bertelsen, E. Derryberry, W. Warren, R.K. Wilson, S. Li, D.A. Ray, R.E. Green, S.J. O'Brien, D. Griffin, W.E. Johnson, D. Haussler, O.A. Ryder, E. Willerslev, G.R. Graves, P. Alström, J. Fjeldså, D.P. Mindell, S.V. Edwards, E.L. Braun, C. Rahbek, D.W. Burt, P. Houde, Y. Zhang, H. Yang, J. Wang, Avian Genome Consortium, E.D. Jarvis, M.T.P. Gilbert, and J. Wang. 2014. Comparative genomics reveals insights into avian genome evolution and adaptation. *Science* 346(6215):1311–320.

Zhang, L.B. and S. Renner. 2003. The deepest splits in Chloranthaceae as resolved by chloroplast sequences. *International Journal of Plant Science* 164: S383–S392.

Zhang, P.G., S.Z. Huang, A.-L. Pin, and K.L. Adams. 2010. Extensive divergence in alternative splicing patterns after gene and genome duplication during the evolutionary history of *Arabidopsis*. *Molecular Biology and Evolution* 27(7):1686–1697.

Zhang, Q., A. Antonelli, T.S. Feild, and H.-Z. Kong. 2011. Revisiting taxonomy, morphological evolution, and fossil calibration strategies in Chloranthaceae. *Journal of Systematics and Evolution* 49:315–329.

Zhang, S.-D., D.E. Soltis, Y. Yang, D.-Z. Li, and T.-S.Yi. 2011. Multi-gene analysis provides a well-supported phylogeny of Rosales. *Molecular Phylogenetics and Evolution* 60(1):21–28.

Zhang, T., Y. Fang, X. Wang, X. Deng, X. Zhang, S. Hu, and J. Yu. 2012. The complete chloroplast and mitochondrial genome sequences of *Boea hygrometrica*: Insights into the evolution of plant organellar genomes. *PLoS One* 7(1): e30531. doi:10.1371/journal.pone.0030531.

Zhang, W., E.M. Kramer, and C.C. Davis. 2012. Similar genetic mechanisms underlie the parallel evolution of floral phenotypes. *PLoS ONE* 7 (4). doi:10.1371/journal.pone.0036033.

Zhang, W., V.W. Steinmann, L. Nikolov, E.M. Kramer, and C.C. Davis. 2013. Divergent genetic mechanisms underlie reversals to radial floral symmetry from diverse zygomorphic flowered ancestors. *Frontiers in Plant Science* 4:302. doi:10.3389/fpls.2013.00302.

Zhang, W., Q.-Y. (J.) Xiang, D.T. Thomas, B.M. Wiegmann, M.W. Frohlich, and D.E. Soltis. 2008. Molecular evolution of *PISTILLATA*-like genes in the dogwood genus *Cornus* (Cornaceae). *Molecular Phylogenetics and Evolution* 47:175–195.

Zhang, Z.-H., C.-Q. Li, and J. Li. 2009. Phylogenetic placement of *Cynomorium* in Rosales inferred from sequences of the inverted repeat region of the chloroplast genome. *Journal of Systematics and Evolution* 47(4):297–304.

Zhao, B., L. Liu, D. Tan, and J. Wang. 2010. Analysis of phylogenetic relationships of Brassicaceae species based on *Chs* sequences. *Biochemical Systematics and Ecology* 38(4):731–739.

Zhao, D., Q. Yu, C. Chen, and H. Ma. 2001. Genetic control of reproductive meristems. *In* M.T. McManus, and B. Veit. [eds.] Meristematic Tissues in Plant Growth and Development, 89–142. Sheffield Academic Press, Sheffield, UK.

Zhong, B., L. Liu, and D. Penny. 2013. Origin of land plants using the multispecies coalescent model. *Trends in Plant Science* 18:492–496.

Zhong, B., L. Liu, Z. Yan, and D. Penny. 2014. The multispecies coalescent model and land plant origins: A reply to Springer and Gatesy. *Trends in Plant Science* 19:270–272.

Zhong, B., T. Yonezawa, Y. Zhong, and M. Hasegawa. 2010. The position of Gnetales among Seed Plants: Overcoming pitfalls of chloroplast phylogenomics. *Molecular Biology and Evolution* 27:2855–2863.

Zhou, X.-R., Y.-Z. Wang, J.F. Smith, and R. Chen. 2008. Altered expression of *TCP* and *MYB* genes relating to the floral developmental transition from initial zygomorphy to actinomorphy in *Bournea* (Gesneriaceae). *New Phytologist* 178:532–543.

Zhou, Z., D. Hong, Y. Niu, G. Li, Z. Nie, J. Wen, and H. Sun. 2013. Phylogenetic signal and biogeographic analyses of the Sino-Himalayan endemic genus *Cyananthus* (Campanulaceae) and implications for the evolution of its sexual system. *Molecular Phylogenetics and Evolution* 68:482–497.

Zhu, H.Y., H.K. Choi, D.R. Cook, and R.C. Shoemaker. 2005. Bridging model and crop legumes through comparative genomics. *Plant Physiology* 137:1189–1196.

Zhu, X.Y., M.W. Chase, Y.L. Qiu, H.Z. Kong, D.L. Dilcher, J.H. Li, and Z.D. Chen. 2007. Mitochondrial *matR* sequences help to resolve deep phylogenetic relationships in rosids. *BMC Evolutionary Biology* 7:217. doi:10.1186/1471-2148-17.

Zimmer, E.A., E.H. Roalson, L.E. Skog, J.K. Boggan, and A. Idnurm. 2002. Phylogenetic relationships in the Gesnerioideae (Gesneriaceae) based on nrDNA ITS and cpDNA *trnL-F* and *trnE-T* spacer region sequences. *American Journal of Botany* 89:296–311.

Zomlefer, W.B., N.H. Williams, W.M. Whitten, and W.S. Judd. 2001. Generic circumscription and relationships in the tribe Melanthieae (Lilliales, Melanthiaceae), with emphasis on *Zigadenus*: Evidence from ITS and *trnLF* sequence data. *American Journal of Botany* 88:1657–1669.

Zona, S. 1997. The genera of Palmae (Arecaceae) in the southeastern United States. *Harvard Papers in Botany* 11:71–107.

Zuckerkandl, E. 2002. Why so many noncoding nucleotides? The eukaryote genome as an epigenetic machine. *Genetica* 115:105–129.

Zuckerkandl, E., and L. Pauling. 1962. Molecular disease, evolution, and genetic heterogeneity. *In* M. Kasha, and B. Pullman [eds.], Horizons in Biochemistry, 198–225. Academic Press, New York, NY.

Zwickl, D.J., and D.M. Hillis. 2002. Increased taxon sampling greatly reduces phylogenetic error. *Systematic Biology* 51:588–598.

Photo Credits

If credits do not appear in a figure legend, the credit is below. All credits for the color plates showing seed plant diversity are found below. The photographs remain under copyright control of the photographers. Their reproduction and/or distribution without the expressed permission of the photographer(s) is in violation of copyright laws.

CHAPTER 1

Figure 1.1. a, b. *Cycas circinalis* (W. Judd); c. *Zamia furfuracea* (W. Judd); d. *Nageia nagi* (W. Judd); e. *Araucaria subulata* (W. Judd); f, g. *Gnetum gnemon* (W. Judd); h. *Ginkgo biloba* (S. Darwin).

CHAPTER 4

Figure 4.7. a. *Amborella trichopoda*, shoot and lateral inflorescences with immature fruits (S. Zona); b. *A. trichopoda*, staminate inflorescences, leaves with chloranthoid teeth (P. Endress); c. *A. trichopoda*, staminate flower (S. Kim); d. *A. trichopoda*, close-up of carpellate flower (S. Kim); e. *A. trichopoda*, flower bud showing bracts (S. Kim); f. *A. trichopoda* (S. Kim), top view of carpellate flower (S. Kim); g. *A. trichopoda*, fruits (drupes) (P. Endress).
Figure 4.10. a. *Brasenia schreberi* J.F.Gmel. (W. Judd); b. *Trithuria submersa* (D. Stevenson, from the web); c. *Victoria amazonica* (W. Judd); d. *Nymphaea odorata* (W. Judd); e. *Nuphar advena* (W. Judd); f. *Nymphaea* sp. (hybrid) (W. Judd).
Figure 4.12. *Nymphaea tuberosa* (K. Robertson & D. Nickrent).
Figure 4.15. a. *Austrobaileya scandens* (P. Endress); b. *Trimenia moorei* (P. Endress); c. *Illicium floridanum* (K. Neubig); d. *Kadsura japonica* (K. Neubig); f. *Schisandra rubriflora* (S. Zona).

CHAPTER 5

Figure 5.2. a. *Liriodendron tulipifera* (W. Judd); b. *Asimina parviflora* (K. Neubig); c. *Magnolia virginiana* (W. Judd); d. *Myristica fragrans* (W. Judd); e. *Annona squamosa* (W. Judd); f. *Myristica insipida* (W. Judd); g. *Cananga odorata* (W. Judd); h. *Monoon longifolium* (W. Judd); i. *Magnolia champaca* (W. Judd).
Figure 5.3. a. *Tasmannia insipida* (J. Dark); b. *Canella winterana* (W. Judd); c. *Drimys winteri* (S. Zona); d. *Aristolochia maxima* (W. Judd); e. *Persea americana* (S. Zona); f. *Hernandia nymphaeifolia* (W. Judd).
Figure 5.4. *Chloranthus serratus* (J.R. Abbott); b. *Sarcandra glabra* (J.R. Abbott); c. *Hedyosmum nutans* (W. Judd).

CHAPTER 6

Figure 6.3. *Amborella trichopoda* (S. Kim).

CHAPTER 7

Figure 7.2. a. *Acorus calamus* (K. Robertson & D. Nickrent); b. *Japonolirion osense* (Richard Wilford, sent by M. Chase); c. *Sagittaria lancifolia*, incl. insert (W. Judd); d. *Hydrocleys nymphoides* (W. Judd); e. *Butomus umbellatus* (M. Chase); f. *Triglochin maritima* (M. Chase); g. *Ruppia megacarpa* (M. Chase); h. *Potamogeton nodosus* (K. Robertson & D. Nickrent); i. *Potamogeton americanus* (W. Judd); j. *imnobium spongia* (W. Judd); k. *Syringodium filiforme* (W. Judd); l. *Anthurium superbum* (W. Judd).
Figure 7.5. a. *Burmannia coelestis* (S. Zona); b. *Dioscorea bulbifera* (W. Judd); c. *Dioscorea bulbifera* (S. Zona); d. *Tacca integrifolia* (W. Judd); e. *Thismia* sp. (S. Zona); f. *Aletris lutea* (W.M. Whitten); g. *Vellozia bahiana* (S. Zona); h. *Pandanus tectorius* (W. Judd); i. *Stemona tuberosa* (M. Chase); j. *Carludovica palmata* (M. Chase).
Figure 7.6. a. *Gloriosa superba* (W. Judd); b. *Uvularia grandiflora* (W. Judd); c. *Tulipa* sp. (W. Judd); d. *Lilium catesbaei* (W. Judd); e. *Tricyrtis hirta* (W. Judd); f. *Schoenocaulon dubium* (W. Judd); g. *Trillium maculatum* (W. Judd); h. *Lapageria rosea* (M. Chase); i. *Smilax smallii* (W. Judd); j. *Smilax auriculata* (W. Judd); k. *Bomarea dulcis* (M. Chase).
Figure 7.7 a. *Asparagus aethiopicus* (W. Judd); b. *Dianella tasmanica* (W. Judd); c. *Lachenalia aloides* (W. Judd); d. *Hypoxis juncea* (W. Judd); e. *Cordyline terminalis* (W. Judd); f. *Iris lutescence* (M. Martinez-Azorin) and *Ixia* sp. (M. Martinez-Azorin); g. *Cyrtopodium aureum* (W. Judd); h. *Dichelostemma ida-maia* (W. Judd); i. *Xanthorrhoea resinosa* (W. Judd).
Figure 7.8. a. *Beaucarnea recurvata* (W. Judd, including insert); b. *Convallaria majalis* (W. Judd); *Convallaria majalis*, unlabeled insert with "b." (D. Nickrent); c. *Dracaena draco* (W. Judd); d. *Astelia neocaledonica* (M. Chase); e. *Xeronema callistemon* (M. Chase); f. *Doryanthes palmeri* (M. Chase); g. *Odontostomum hartwegii* (M. Chase); h. *Borya sphaerocephala* (M. Chase).
Figure 7.10. a. *Agapanthus africanus* (W. Judd); b. *Agave macroantha* (W. Judd); c. *Hesperoaloe campanulata* (W. Judd); d. *Agave lechuguilla* (W. Judd); e. *Allium tricoccum* (W. Judd); f. *Allium canadense* (D. Nickrent); g. *Crinium americanum* (W. Judd); h. *Aloe saponaria* (W. Judd); i. *Crinum* 'Ellen Bosanquet' (W. Judd); j. *Aloe dyeri* (W. Judd).
Figure 7.13. a. *Roystonea borinquena* (W. Judd); b. *Veitchia arecina* (W. Judd); c. *Sabal palmetto* (W. Judd); d. *Serenoa repens* (S. Zona); e. *Metroxylon vitiense* (W. Judd); f. *Nypa fruticans* (W. Judd); g. *Pseudophoenix sargentii* (W. Judd); h. *Kingia australis* (M. Chase); i. *Callisia graminea* (W. Judd); j. *Tradescantia fluminensis* (K. Neubig); k. *Philydrum lanuginosum* (M. Chase); l, m. *Lachnanthes caroliniana* (W. Judd).
Figure 7.14. a. *Heliconia lingulata* (W. Judd); b. *Musa ornata* (W. Judd); c. *Musa balbisiana* (W. Judd); d. *M. balbisiana* (W. Judd); e. *Strelitzia nicolai* (W. Judd); f. *Strelitzia reginae* (W. Judd); g. *Ravenala madagascariensis* (W. Judd); h. *Canna x indica* (W. Judd); i. *Costus lucanusianus* (W. Judd); j. *Costus malortieanus* (W. Judd).
Figure 7.15. a. *Typha angustifolia* and *T. latifolia* (W. Judd); b. *Vriesea sintenisii* (W. Judd); c. *Eriocaulon decangulare* (W. Judd); d. *Eriocaulon decangulare* (S. Zona); e. *Xyris platylepis* (W. Judd); f. *Mayaca fluviatilis* (Santiaga Madrinan—sent by M. Chase); g. *Juncus dichotomus* (W. Judd); h. *Carex raynoldsii* (L.C. Majure); i. *Carex hamata* (L.C. Majure); j. *Anarthria scabra* (M. Chase); k. *Anarthria scabra* (M. Chase); l. *Chondropetalum tectorum* (W. Judd); m. *Centrolepis aristata* (M. Chase); n. *Flagellaria indica* (M. Chase); o. *Ecdeiocolea monostachya* (M. Chase); p. *Joinvillea gaudichaudiana* (M. Chase); q. *Coix lacryma-jobi* (L.C. Majure); r. *Lolium perenne* (K. Neubig); s. *Uniola paniculata* (W. Judd).

CHAPTER 8

Figure 8.2. *Rumex* (M. Sundberg).
Figure 8.3. a. *Grevillea robusta* (W. Judd); b. *Nelumbo nucifera* (W. Judd); c. *Platanus occidentalis* (W. Judd); d. *Euptelea pleiosperma* (J.R. Abbott); e. *Ficaria verna* (W. Judd); f, g. *Argemone albiflora* (W. Judd); h. *Cocculus carolinus* (K. Neubig); i. *Fumaria officinalis* (W. Judd); j. *Cocculus carolinus* (K. Neubig).
Figure 8.5. a. *Buxus microphylla* (W. Judd); b. *Meliosma impressa* (W. Judd); c. *Trochodendron aralioides* (J. R. Abbott); d. *Berberis julianae* (B. Carlsward); e. *Ceratophyllum demersum* (W. Judd); f. *Ceratophyllum echinatum* (J. R. Abbott); g. *Ceratophyllum* sp. (D. McMurran, Ulrich Collection); h. *Ceratophyllum* sp. (S. Manchester, UF Collection).

CHAPTER 9

Figure 9.2. a. *Dillenia indica* (W. Judd); b. *Hibbertia scandens* (J. R. Abbott); c, d. *Gunnera manicata* (W. Judd).

CHAPTER 10

Figure 10.3. a. *Cercidiphyllum japonicum* (W. Judd); b. *Liquidambar styraciflua* (K. Robertson and D. Nickrent); c. *Ribes hirtellum* (W. Judd); d. *Saxifraga stolonifera* (K. Neubig); e. *Sedum spurium* (W. Judd); f. *Fothergilla gardenia* (W. Judd); g. *Itea virginica* (K. Neubig); h. *Paeonia lactiflora* (B. Carlsward); i. *Leea coccinea* (W. Judd).

Figure 10.5. a-c. (S. Manchester); d. *Hamamwilsonia* (K. Pigg); e-n. (S. Manchester); o. *Ulmus okanaganensis* (T. Dillhoff); p-u. (S. Manchester).
Figure 10.13. a. *Parnassia grandifolia* (W. Judd); b. *Euonymus americanus* (W. Judd); c. *Oxalis debilis* (K. Neubig); d. *Averrhoa carambola* (W. Judd); e. *Cephalotus follicularia* (J.R. Abbott); f. *Elaeocarpus hygrophyllus* (W. Judd); g. *Weinmannia pinnata* (W. Judd); h. *Krameria lanceolata* (K. Neubig); i. *Bulnesia arborea* (W. Judd).
Figure 10.15. a. *Prunus domestica* (W. Judd); b. *Rosa bracteata* (W. Judd); c. *Sorbus americana* (W. Judd); d. *Colubrina arborescens* (W. Judd); e. *Artocarpus heterophyllus* (W. Judd); f. *Humulus lupulus* (W. Judd); g. *Ulmus parvifolia* (K. Neubig); h. *Pilea formonensis* (J.D. Skean); i. *Cecropia peltata* (W. Judd); j. *Elaeagnus pungens* (W. Judd); k. *Suriana maritima* (W. Judd); l. *Polygala myrtifolia* (W. Judd); m. *Crotalaria pallida* (K. Neubig); n. *Leucaena leucocephala* (W. Judd); o. *Calliandra haematocephala* (W. Judd).
Figure 10.17. a. *Quercus laevis*, including insert (W. Judd); b. *Ostrya virginana* (W. Judd); c. *Alnus incana* subsp. *rugosa* (W. Judd); d. *Gymnostoma nobile* (W. Judd); e. *Juglans nigra* (K. Neubig); f. *Carya glabra* (W. Judd); g. *Myrica cerifera* (W. Judd); h. *Nothofagus cunninghamii* (T. Schuster); i. *Cucurbita pepo* (W. Judd); j. *Begonia* sp. (W. Judd).
Figure 10.20. a. *Mammea americana* (W. Judd); b. *Clusia rosea* (W. Judd); c. *Erythroxylum coca* (K. Neubig); d. *Rafflesia priceri* (R. Beaman); e. *Cnidoscolus chayamansa* (W. Judd); f. *Euphorbia cyathophora* (W. Judd); g. *Geobalanus oblongifolius* (S. Zona); h. *Hypericum tetrapetalus* (W. Judd); i. *Malpighia emarginata* (W. Judd); j. *Ochna thomasiana* (W. Judd); k. *Passiflora alata* (W. Judd); l. *Phyllanthus juglandifolium* (W. Judd); m. *Linum kingii* (B. Carlsward); n. *Drypetes lateriflora* (W. Judd); o. *Flacourtia indica* (W. Judd); p. *Salix caroliniana* (K. Neubig); q. *Rhizophora mangle* (N. Garcia); r. *Turnera subulata* (W. Judd); s. *Viola sororia* (W. Judd); t. *Marathrum schiedeanum* (L.C. Majure).
Figure 10.21. a-d (S. Manchester); e. *Acer* (K. Johnson); f-l (S. Manchester).
Figure 10.25. a. *Geranium maculatum* (K. Robertson and D. Nickrent); b. *Melianthus comosus* (K. Neubig); c. *Picramnia pentandra* (W. Judd); d. *Staphylea trifolia* (W. Judd); e. *Miconia angustilamina* (W. Judd); f. *Chamerion angustifolium* (W. Judd); g. *Decodon verticillatus* (W. Judd); h. *Acca sellowiana* (K. Neubig); i. *Combretum indicum* (W. Judd); j. *Ailanthus altissima* (W. Judd); k. *Bursera simaruba* (W. Judd); l. *Rhus typhina* (W. Judd); m. *Chukrasia tabularis* (W. Judd); n. *Citrus x aurantium* (W. Judd); o. *Harpullia pendula* (W. Judd).
Figure 10.26. a. *Batis maritima* (W. Judd); b. *Berteroa incana*, incl. insert (W. Judd); c. *Brassica oleracea* (W. Judd, but the unlabeled insert showing flowers of this species is by K. Neubig); d. *Capparis flexuosa* (W. Judd); e, f. *Carica papaya* (W. Judd); g. *Moringa olifera*, incl. insert (W. Judd); h. *Tropaeolum majus* (W. Judd); i. *Cleome domingensis* (W. Judd); j. *Phaleria octandra* (W. Judd); k. *Shorea faguetioides* (J.R. Abbott); l. *Bixa orellana* (W. Judd); m. *Crocanthemum carolinianum* (W. Judd); n. *Lavatera assurgentiflora* (W. Judd); o. *Abutilon theophrasti* (W. Judd).

CHAPTER 11

Figure 11.3. a. *Nestronia umbellula* (L.C. Majure); b. *Phoradendron leucarpum* (K. Neubig); c. *Agonandra macrocarpa* (D. Nickrent); d. *Misodendron angulatum* (D. Nickrent, for Glatzel); e. *Tristerix penduliflorus* (L.C. Majure); f. *Corynaea crassa* (L.C. Majure); g. *Ximenia americana* (P. Corogin); h. *Aptandra zenkeri* (D. Nickrent).
Figure 11.5. a. *Aexitoxicon punctatum* (N. García); b. *Berberidopsis corallina* (N. García); c. *Mirabilis jalapa* (W. Judd); d. *Mesambryanthemum cordifolium* (W. Judd); e. *Simmondsia chinensis* (S. Zona); f. *Stegnosperma cubense* (J.R. Abbott); g. *Gomphrena serrata* (K. Neubig); h. *Silene scouleri* (M. Pajuelo).
Figure 11.6. a. *Nepenthes truncata* (B. Carlsward); b. *Drosera capillaris* (K. Neubig); c. *Drosera brevifolia* (K. Neubig); d. *Plumbago capensis* (W. Judd); e. *Tamarix chinensis* (L.C. Majure); f. *Coccoloba uvifera* (W. Judd); g. *Antigonon leptopus* (K. Neubig); h. *Phytolacca heterotepala* (W. Judd); i. *Rivina humilis* (W. Judd).
Figure 11.7. a. *Pereskia grandifolia* (L.C. Majure); b. *Hylocereus undatus* (W. Judd); c. *Opuntia chlorotica* (L.C. Majure); d. *Didierea trollii* (L.C. Majure); e. *Alluaudia procera* (L.C. Majure); f. *Portulaca grandiflora* (W. Judd); g. *Portulaca amilis* (K. Neubig); h. *Basella alba* (W. Judd); i. *Claytonia virginica* (K. Robertson & Dan Nickrent); j. *Talinum fruticosum* (W. Judd).
Figure 11.11. a. *Cornus alternifolia* (W. Judd); b. *Cornus foemina* (K. Neubig); c. *Philadelphus inodorus* (W. Judd); d. *Mentzelia floridana* (W. Judd); e. *Davidia involucrata* (W. Judd); f. *Nyssa ogeche* (W. Judd).
Figure 11.13. a. *Impatiens pallida* (W. Judd); b. *Marcgravia rubra* (W. Judd); c. *Gustavia superba* (W. Judd); d. *Phlox drummondii* (W. Judd); e. *Fouquieria splendens* (K. Neubig); f. *Ternstroemia gymnanthera* (W. Judd); g. *Manilkara zapota* (W. Judd); h. *Diospyros digyna* (W. Judd); i. *Diospyros ebenum* (W. Judd); j. *Bonellia macrocarpa* (W. Judd).
Figure 11.14. a. *Gordonia lasianthus* (W. Judd); b. *Halesia carolina* (D. Nickrent); c. *Symplocos hartwegii* (W. Judd); d. *Saurauia* sp. (W. Judd); e. *Roridula gorgonias* (B. Rice); f. *Clethra mexicana* (W. Judd); g. *Sarracenia minor* (W. Judd); h. *Cyrilla racemiflora* (W. Judd); i. *Eubotrys racemosus* (W. Judd); j. *Sarcodes sanguinea* (S. Zona).

Figure 11.16. a. *Eucommia ulmoides* (W. Judd); b. *Garrya elliptica* (W. Judd); c. *Garrya fadyenii* (W. Judd); d. *Pyrenacantha malvifolia* (T. Choo & G. Stull); e. *Calotropis procera* (W. Judd); f. *Nerium oleander* (W. Judd); g. *Gelsemium sempervirens* (W. Judd); h. *Sabatia grandiflora* (K. Neubig); i. *Spigelia loganioides* (W. Judd); j. *Strychnos spinosa* (W. Judd); k. *Cubanola domingensis* (W. Judd); l. *Portlandia proctorii* (W. Judd); m. *Rondeletia odorata* (W. Judd).

Figure 11.17. a, b. (S. Manchester); c-e. Ericaceae pollen tetrad (F. Grimsson); f-h. (S. Manchester); i. *Pyrenacantha grandiflora* (G. Stull); j-l. *Palaeophytocrene foveolata* (M. Collinson); m-p. (S. Manchester); q. (J. Horiuchi).

Figure 11.18. a. *Hydrolea corymbosa* (W. Judd); b. *Ipomoea pes-caprae* (W. Judd); c. *Iochroma cynea* (W. Judd); d. *Physalis walteri* (W. Judd); e. *Solanum tuberosum* (W. Judd); f. *Wigandia urens* (K. Neubig); g. *Tournefourtia staminea* (W. Judd).

Figure 11.19. a. *Cartrema americana* (W. Judd); b. *Polypremum procumbens* (W. Judd); c. *Calceolaria chelidonioides* (W. Judd); d. *Columnea schiedeana* (K. Neubig); e. *Nuttalanthus canadensis* (W. Judd); f. *Plantago lanceolata* (W. Judd); g. *Leucophyllum frutescens* (W. Judd); h. *Verbascum virgatum* (B. Carlsward); i. *Torenia fournieri* (W. Judd); j. *Uncarina grandidieri* (W. Judd); k. *Thunbergia grandiflora* (W. Judd); l. *Ruellia tweedieana* (W. Judd); m. *Kigelia africana* (W. Judd); n. *Spathodea campanulata* (W. Judd); o. *Schlegelia parasitica* (W. Judd); p. *Utricularia cornuta* (W. Judd); q. *Pinguicula primuliflora* (K. Neubig); r. *Lamium amplexicaule* (K. Neubig); s. *Callicarpa americana* (W. Judd); t. *Lantana montividensis* (W. Judd); u. *Agalinus fasciculata* (W. Judd); v. *Erythranthe guttata* (S. Zona).

Figure 11.21. a. *Ilex glabra* (K. Neubig); b. *Helwingia chinensis* (D. Mosquin); c. *Wahlenbergia marginata* (K. Neubig); d. *Lobelia martagon* (W. Judd); e. *Stylidium torticarpum* (M. Chase); f. *Nymphoides cristata* (W. Judd); g. *Scaevola plumieri* (W. Judd); h. *Calycera crassifolia* (M. Manske); i. *Vernonia gigantea* (W. Judd); j. *Chrysogonum virginianum* (W. Judd); k. *Helenium pinnatifidum* (W. Judd); l. *Hieracium aurantiacum* (W. Judd).

Figure 11.22. a, b. *Escallonia rubra* (B. Carlsward); c. *Brunia* sp. (W. Judd); d. *Pittosporum pentandrum* (W. Judd); e. *Pittosporum tenuifolium* (W. Judd); f. *Aralia nudicaulis* (W. Judd); g. *Myodocarpus fraxinifolius* (S. Zona); h. *Myodocarpus fraxinifolius* (W. Judd); i. *Heracleum lanatum* (K. Neubig); j. *Daucus carota* (K. Neubig); k. *Viburnum obovatum* (W. Judd); l. *Diervilla lonicera* (W. Judd); m. *Valeriana scandens* (L.C. Majure); n. *Dipsacus fullonum* (W. Judd).

Figure 11.23. a-c. *Ilex* pollen (F. Grimsson); d-e. *Diervilla* pollen (F. Grimsson); f-j. (S. Manchester); k-l. *Araliaecarpum* (L. Golovneva).

CHAPTER 13

Figure 13.1. a. *Prosopanche americana* (D. Nickrent); b. *Cassytha filiformis* (W. Judd); c. *Cassytha filiformis* (D.Nickrent); d. *Corallorhiza stricta* (K. Neubig); e. *Cynomorium coccineum* (Jose Quiles); f. *Cytinus ruber* (W.M. Whitten); g. *Scybalium jamaicense* (W. Judd); h. *Arceuthobium bicarinatum* (W. Judd); i. *Viscium minimum* (S. Zona); j. *Santalum paniculatum* (D. Nickrent); k. *Dendropemon constantiae* (W. Judd); l. *Chimaphila maculata* (W. Judd); m. *Monotropa uniflora* (W.M. Whitten); n. *Cuscuta cuspidata* (J.R. Abbott); o. *Castilleja indivisa* (W. Judd); p. *Orobanche canescens* (W.M. Whitten).

Figure 13.10. Kranz anatomy (P.J. Schulte, University of Nevada).

CHAPTER 14

Figure 14.5A. a. *Lysimachia arvensis* (P. Endress); b. *Columnea* sp. (P. Endress); c. *Vigna speciosa* (P. Endress).

Figure 14.16. *Montrouziera gabriellae* (P. Endress).

Figure 14.28. *Galinsoga ciliata* (P. Endress).

Index

Page numbers in italics refer to figures and tables.